LIST OF MEMBERS
12th edition
2001
LISTE DES MEMBRES
12ème édition

T0186897

INTERNATIONAL SOCIETY FOR SOIL MECHANICS
AND GEOTECHNICAL ENGINEERING
SOCIÉTÉ INTERNATIONALE DE MÉCANIQUE DES SOLS
ET DE LA GÉOTECHNIQUE

Address of the Secretariat/Adresse du Secrétariat
International Society for Soil Mechanics and Geotechnical Engineering
City University
Northampton Square
London EC1V 0HB, UK
Telephone: +44 20 7040 8154
Fax: +44 20 7040 8832
E-mail: secretariat@issmge.org

INTERNATIONAL SOCIETY FOR SOIL MECHANICS AND GEOTECHNICAL ENGINEERING
SOCIÉTÉ INTERNATIONALE DE MÉCANIQUE DES SOLS ET DE LA GÉOTECHNIQUE

ISBN 90 5809 500 2
Published by / Publié par
A.A. Balkema Publishers, a member of Swets & Zeitlinger Publishers
The Netherlands/Pays-Bas

ISSMGE CORPORATE MEMBERS

DISCLAIMER

ISSMGE MEMBERSHIP TABLE DECEMBER 2001

Member Societies	Members	Africa	Asia	Australasia	Europe	N. America	S. America
Albania	22				22		
Argentina	44						44
Australia	658			658			
Austria	98				98		
Azerbaijan	13				13		
Bangladesh	33		33				
Belgium	70				70		
Bolivia	13						13
Brazil	835						835
Bulgaria	52				52		
Canada	889					889	
Chile	53						53
China	106		106				
Colombia	30						30
Costa Rica	58						58
Croatia	126				126		
CTGA	28	28					
Czech & Slovak Republics	43				43		
Denmark	308				308		
Ecuador	29						29
Egypt	20	20					
Estonia	30				30		
Finland	177				177		
France	746				746		
Germany	929				929		
Ghana	33	33					
Greece	173				173		
Hungary	84				84		
Iceland	10				10		
India	225		225				
Indonesia	20		20				
Iran	58		58				
Ireland	28				28		
Israel	35		35				
Italy	223				223		
Japan	1390		1390				
Kazakhstan	18		18				
Kenya	21	21					
Korean Republic	146		146				
Latvia	31				31		
Lithuania	40				40		
Macedonia	72				72		
Mexico	200					200	
Morocco	12	12					
Nepal	15		15				
Netherlands	325				325		
New Zealand	260			260			
Nigeria	25	25					
Norway	339				339		
Pakistan	80		80				
Paraguay	17						17
Peru	25						25
Poland	298				298		
Portugal	181				181		
Romania	53				53		
Russia	235				235		
Slovenia	127				127		
South Africa	390	390					
South East Asia	230		230				
Spain	230				230		
Sri Lanka	45		45				
Sudan	25	25					
Sweden	766				766		
Switzerland	250				250		
Syria	17		17				
Tunisia	10	10					
Turkey	181				181		
United Kingdom	870				870		
USA	2322					2322	
Venezuela	17						17
Vietnam	18		18				
Yugoslavia	29				29		
Zimbabwe	17	17					
TOTAL (73)	15626	581	2436	918	7159	3411	1121

ALBANIA

Secretary: Mr. Shkodrani Neritan, Albanian Geotechnical Society, Naim, Frasheri, Nr. 36 Tirana, Albania.
E: Neritans@yahoo.com

Total number of members: 22

Bozo, L., Professor, Dr. Chief of Geotechnical Department, Civil Engineering Faculty of Polytechnic University, Muhamet Gjollesha St. 54, Tirana. T: +355 4244970, +355 4229045, +355 382136140

Shkodrani, N., Geotechnical Department, Civil Engineering Faculty of Polytechnic University. T: +355 382041855

Allkja, S., Geologist, President of the Geostudio 2000 Firm, St. Muhamet Gjollesha, P.11/1. T/F: +355 244981

Arapi, E., Civil Engineer, Member of the ALTEA Laboratory Staff

Bida, A., Civil Engineer, Assistant to the Civil Engineering Faculty of Polytechnic University

Canja, A., Civil Engineer, Member of the ALTEA Laboratory Staff

Gela, K., Civil Engineer, President of the Iliriadia Firm. T: +0382151353

File, D., Technician at the Transport Infrastructure Department, Civil Engineering Faculty of Polytechnic University

Goga, K., Hydraulic Engineer, Dr. Pedagogue at Civil Engineering Faculty of Polytechnic University, Private Consulting Office. T: +241379. Chief Editorial Office of Albania Geotechnic Review

Hajdari, V., Assistant at the Geotechnical Department

Hajdini, D., Civil Engineer, Member of the MAK ALBANIA Staff, Tirana

Kellezi, L., Doctor of Science, Technical University of Denmark

Kokona, H., Assistant at the Geotechnical Department

Kokthi, A., Assistant at the Geotechnical Department

Kuka, H., Civil Engineer, Member of the ALTEA Laboratory Staff

Lila, B., Geologist, Director of Mineral Service, Tirana. T: +23116

Marko, L., Civil Engineer, Designer, Private Office. T: +367944

Mishaxhiu, M., Civil Engineer, Part-Time Pedagogue at Geotechnical Department. T: +221365

Muceku, U., Geologist, M.Sc., Director of Engineering, Geological Directory for Albanian Geological Service, Sami Frashkri St.Tirana. T: +226530

Naska, N., Civil Engineer, Director of Investment, Directory at Public Service Ministry, Part-Time Pedagogue at Civil Engineering Faculty of Polytechnic University. T: +226948

Spahiu, E., Technician Laboratory at Geotechnical Department of Civil Engineering Faculty of Polytechnic University. T: +229045

Vorfi, J., Part-Time Pedagogue at Geotechnical Department, Member of the ALTEA Laboratory Staff, Frosina Plaku St. 18, Tirana. T: +226493

2

ARGENTINA/ARGENTINE

Secretary: E.R. Redolfi, Pablo Mirizzi 471 B° Parque Vélez Sarsfield, 5016 Córdoba, Provincia de Córdoba República, Argentina. T: (0351)485073, E: eredolfi@si.cordoba.com.ar, eredolfi@com.uncor.edu

Total number of members: 67

Abbona, P., Arrt Ingenieros Consultores Pablo Mirizzi 471 B°Parque Vélez Sarsfield 5016 Córdoba Provincia de Córdoba. T: (351)4685073, E: arrt@arnet.com.ar

Andrés, F.O., Holdrich 124 8000 Bahía Blanca Provincia de Buenos Aires

Armagnac, P.B.A., Fernandez 2081 Hurlinghan Provincia de Buenos Aires. T: (11)46627227

Barchiesi, A.M., Río Negro 246 5500 Mendoza Provincia de Mendoza. T: (261)4240414

Bessone, C., Cochabamba 1774 2000 Rosario Provincia de Santa Fe

Bolognesi, A., Valentín Virasoro 878 1405 Buenos Aires. T: (11)49821849, E: ajlbic@infovia.com.ar

Bonifazi, J., Bonifazi-Garcia Martin Av.J.B.Justo 7120 1407 Buenos Aires. T: (11)46362877, E: bgings@infovia.com.ar

Bustinza, J.A., Red Ingenieria SRL Antártida Argentina 240 8300 Neuquén Provincia de Neuquen. T: (299)4432156, E: reding@infovia.com.ar

Calo, J., Dardo Rocha 140 1870 Avellaneda Provincia de Buenos Aires

Capdevilla, E., IATASA SA Tacuarí 32 7°piso 1071 Buenos Aires. T: (11)43433417, E: capdevilarod@ciudad.com.ar

Compagnucci, D.J., Ingenieria Del Suelo SRL Neuquén 480 8000 Bahía Blanca Provincia de Buenos Aires. T: (291) 4512919, E: ingsuelo@bblanca.com.ar

Compagnucci, J.P., Compagnucci-Diaz SRL Lanín 255 6300 Neuquén Provincia de Neuquén. T: (299)4431411, E: compagnu@surlan.com.ar

Cordó, O.V., Av., Libertador 1109 – Oeste 5400 San Juan Provincia de San Juan. T: (264)4228666, E: ocordo@ eicam.unsj.edu.ar

Delprato, H.F., Cochabamba 1774 2000 Santa Fe Provincia de Santa Fe. T: (341)4812578, E: bdsuelos@sinectis.com.ar

Elena, A.M., Lazzari 4049 B° Tablada Park 5000 Córdoba Provincia de Córdoba

Fabbri, A.J., Junta 2507 AP of 11 3000 Santa Fe Provincia de Santa Fe. T: (342)4520872, E: ajfabbri@arnet.com.ar

Ferrero, V.H., GeoS Pampayasta 2191 5014 Córdoba Provincia de Córdoba. T: (351)4657108, E: geos@cordoba.com.ar

Filloy, J.E., Río Neuquen 926 8324 Cipolletti Provincia de Río Negro

Flores, R.M., Calle 14 N° 576 1900 La Plata Provincia de Buenos Aires

Giuliani, F., 9 de julio 192 8224 Cipolletti Provincia de Río Negro

Goldemberg, H., Paso 292 1° G 1640 Martinez Provincia de Buenos Aires. T: (11)47985136, E: geotecnica@datamarkets.com.ar

Goldemberg, J.J., Geotecnica – Cientec SA Lavalleja 847 1414 Buenos Aires. T: (11)48620547, E: geotecnica@ datamarkets.com.ar

Gómez, O.I., Ingenieria Del Suelo SRL Neuquén 480 8000 Bahía Blanca Provincia de Buenos Aires. T: (291)4525589, E: ingsuelo@bblanca.com.ar

Guerrero, J.L., Gelly y Obes 415 1706 Haedo Provincia de Buenos Aires

Gulian, J.R., Av., Santa Fe 2534 12°piso Dpto I 1425 Buenos Aires

Guindin, M.R., Cullen 5690 11°piso of.F 1431 Buenos Aires

Gutiérrez, G., Corrientes 135 3500 Resistencia Provincia de Chaco

Hazan, R.R., Holmberg 2119 1430 Buenos Aires. T: (11) 45437401, E: info@inghazan.com.ar

Heredia, S., Antártida Argentina 422 8300 Neuquén Provincia de Neuquén

Karakachoff, C.A., Calle 1 N°716 1900 La Plata Provincia de Buenos Aires

Keim, A., Calle 8 N°3775 (497 y 498) 1897 M.B.Gonnet Provincia de Buenos Aires. T: (221)4712408, E: azuk@ infovia.com.ar

Laborde, J.H., San Martín 996 15°piso Dpto D 1004 Buenos Aires

Leoni, A., Calle 9 N°1805 1900 La Plata Provincia de Buenos Aires. E: ebit@ba.net

López, A., Cimentaciones Argentinas SA Viamonte 740 3°piso 1053 Buenos Aires

López, O.A., Pasco 125 1834 Temperley Provincia de Buenos Aires

Lozada, R., Calle 26 N°1724 1900 La Plata Provincia de Buenos Aires

Marinelli, N., Calle 9 N°1805 1900 La Plata Provincia de Buenos Aires

Mendiguren, E., Almirante Brown 1198 1708 Morón Provincia de Buenos Aires. T: (11)46294059, E: mendi@movi.com.ar

Micucci, C.A., Paraguay 2499 1121 Buenos Aires. T: (11) 49617498, E: smicucci@sinectis.com.ar

Moll, L.L., Av.Vélez Sarsfield 23 4°piso B 5000 Córdoba Provincia de Córdoba

Morbidoni, N.P., Córdoba 2953 3000 Santa Fe Provincia de Santa Fe. T: (342)4554484, E: nmorbiodoni@megasfe.com.ar

Moreno, O.F., Quiroga 418 9103 Rawson Provincia de Chubut

Nadeo, J.R., Calle 120 N°202 esq., 36 1900 La Plata Provincia de Buenos Aires. T: (221)4823155, E: nadeosa@netverk.com.ar

Nuñez, E., Maipú 42.of.224 1084 Buenos Aires. T: (11) 43428360

Orlandi, S.G., Teniente Castillo 419B., Km 8 B° Centenario Comodoro Rivadavia Provincia de Chubut. T: (297) 4535800, E: sorlandi@infovia.com.ar

Pardina, M.E., 25 de Mayo 2051 5° piso 3000 Santa Fe Provincia de Santa Fe. T: (342)4597694, E: hectorcozo@ arnet.com.ar

Pérez, A.L., Pueyrredón 1040 1748 General Rodríguez Provincia de Buenos Aires. T: (237)4840656

Prieto, J.R., Pellegrini 846 5519 Guaymallén Provincia de Mendoza. T: (261)4455001, E: flia_prieto@cpsarg.com

Pujol, A., Antártida Argentina 240 8300 Neuquén Provincia de Neuquén

Pons, M.E., Corrientes 319 1°piso 4000 San Miguel de Tucumán Provincia de Tucumán. T: (381)4218140, E: marupons@arnet.com.ar

Redolfi, E.R., Brasil 10 7°piso Dpto A 5000 Córdoba Provincia de Córdoba. T: (0351)4690201, E: eredolfi@si.cordoba.com.ar

Reyes, S.A., Moreno 387 8000 Bahía Blanca Provincia de Buenos Aires. T: (291)4524723, E: ryainenieria@infovia.com.ar

Rocca, R.J., Ilolay 3174 B°Bajo Palermo 5009 Córdoba Provincia de Córdoba. T: (351)4840113, E: rjrocca@ gtwing.efn.uncor.edu

Sayapín, P., A.J., Carranza 1410 4°piso Dpto C 1414 Buenos Aires

Serrano, C.H., GeoS Pampayasta 2191 5014 Córdoba Provincia de Córdoba. T: (351)4657108, E: geos@cordoba.com.ar

Sfriso, A., TRIAXIS SRL Maipú 42 of.234 1084 Buenos Aires. T: (11)43425447, E: asfriso@triaxis.com.ar

Silva, E.G., Azopardo 549 1661 Bella Vista Provincia de Buenos Aires. T: (11)46681144, E: esilva@unimoron.edu.ar

Speziale, J.F., Tacuarí 32 8°piso 1071 Buenos Aires. E: speziale@iatasa.com.ar

Suárez, J.A., Calle 7 N°1076 9°piso 1900 La Plata Provincia de Buenos Aires. T: (221)4297039, E: jorsuarez@gualtieri. com.ar

Terzariol, R.E., Langer 830 B°Residencial Velez Sarsfield 5016 Córdoba Provincia de Córdoba. T: (351)4617477, E: arrt@arnet.com.ar

Trevisan, S., Consulting Engineering Diagonal 74 N°2831 1900 La Plata Provincia de Buenos Aires. T: (221)4510014

Torres, P.L., Maipú 548 2000 Rosario Provincia de Santa Fe. T: (341)4405979, E: toryver@citynet.net.ar

Vardé, O., Varde & Asociados SA Av., Quintana 585 4°piso 1129 Buenos Aires. T: (11)48040192, E: varde@sinectis. com.ar

Vercelli, H.J., Torres Y Vercelli SRL Maipu 548 2000 Rosario Provincia de Santa Fe. T: (341)4405979, E: toryver@ citynet.net.ar

Yema, C.A., 9 de julio 192 8324 Cipolletti Provincia de Río Negro

Zarazaga, C.H., Luis I., Vélez 42 B°Rogelio Martínez 5000 Córdoba Provincia de Córdoba

Zeballos, M.E., Anacreonte 1938., B° Los Paraisos 5008 Córdoba 5000 Córdoba. T: (351)4682808, E: mzeballos@ com.uncor.edu

AUSTRALIA

Secretariat: Australian Geomechanics Society, PO Box E303, KINGSTON, ACT 2604, Australia.
T: +61 2 6270 6558, F: +61 2 6273 2358, E: valerie_lee@ieaust.org.au, www.australiangeomechanics.org

Due to Australian Privacy Legislation, the list of Australian members of ISSMGE cannot be published. Further information and assistance can be obtained from the Australian Geomechanics Society Secretariat.

5

AUSTRIA/AUTRICHE

Secretary: M. Fross, Ass.Prof.Dipl.-Ing.Dr.techn., Institut für Grundbau und Bodenmechanik, TU Wien, Karlsplatz 13, 1040 Wien. T: +43(0)1-58801-22110, F: +43(0)1-58801-22198, E: m.fross@tuwien.ac.at

Total number of members: 98

Adam, D., Dipl.-Ing.Dr.techn., Univ.Ass., Institut für Grundbau und Bodenmechanik, TU Wien, Karlsplatz 13, 1040 Wien. T: 01-58801-22111, F: 01-58801-22198, E: dietmar.adam@tuwien.ac.at

Ausserer, G., Dipl.-Ing., STRABAG AG, Herbststraße 6-10, 1160 Wien. T: 01-49112-1331, F: 01-49112-1312, E: gert.ausserer@bauholding.at

Balasubramaniam, A.S., Prof., AIT, P.O. Box 4 Klong Luang, Pathumthani 12120, THAILAND

Behon, G., OBR Dipl.-Ing., Amt der Nö Landesregierung, 3910 Zwettl. T: 02822-544060, F: 02822-544060-13

Biberschik, P., BR h.c. Dipl.-Ing., Ziv.Ing.f.Bauwesen, Büro Pauser, Münichreiterstr. 31, 1130 Wien. T: 01-877-32-04-0

Binder, B., Dipl.-Ing., Ziv.-Ing.f.Bauwesen, Fasangasse 25, 1030 Wien. T: 01-783328-15, Telex 112009

Blovsky, S., Dipl.-Ing., Univ.Ass., Institut für Grundbau und Bodenmechanik, TU Wien, Karlsplatz 13, 1040 Wien. T: 01-58801-22112, F: 01-58801-22198, E: blovsky@tuwien.ac.at

Brandecker, H., Dr.Phil., Ber.f.Angew.Geologie, A.Breitnerstraße 1, 5020 Salzburg. T: 0662-20125

Brandl, H., o.Univ.Prof.Dipl.-Ing.Dr.techn.(*Chairman*), Institut für Grundbau und Bodenmechanik, TU Wien, Karlsplatz 13, 1040 Wien. T: 01-58801-22100, F: 01-58801-22198, E: heinz.brandl@tuwien.ac.at

Breit, K., Dipl.-Ing., Flurgasse 10, 3434 Katzeldorf/Tulbing. T: 0664-3380922

Breymann, H., BR h.c. Dipl.-Ing.Dr.techn., Ziv.-Ing.f. Bauwesen, bvfs, Alpenstraße 157, 5020 Salzburg. T: 0662-621758-300

Brezinschek, H., Dipl.-Ing., STBR, Lainzer Straße 113, 1130 Wien. T/F: 01-877-30-85

Dalmatiner, J., Dipl.-Ing.Dr.techn., Ziv.-Ing.f.Bauwesen, Leechgasse 37, 8010 Graz. T: 0316-381915-0, F: 0316-381915-22

Deix, F., SR Dipl.-Ing.Dr.techn., Rathausstraße 13, 1081 Wien. T: 01-4000-82761, F: 01-4000-9982761

Döllerl, A., SR i.R. Dipl.-Ing., Anton Kriegergasse 191, 1230 Wien. T: 01-8893167

Falk, E., Dipl.-Ing.Dr.techn., Bauleiter Spezialtiefbau, Peinlichgasse 7, 8010 Graz. T: 03137-3767, F: 03137-3788

Fleischmann, A., Dipl.-Ing., Mariagrüner Straße 48, 8043 Graz. T/F: 0316-388181

Flögl, W., Dipl.-Ing.,Dr.techn., Haus der Technik, Stockhofstraße 32, 4020 Linz. T: 0732-664832-14

Fock, G., Dipl.-Ing., SGS Geotechnik Ges.m.b.H., Lärchauerstraße 1, 4020 Linz. T: 0732-770214, F: 0732-7702144

Fuchsberger, M., em.o.Univ.Prof.Dipl.-Ing.M.Sc., Inst.f.-Bodenmechanik und Grundbau, TU Graz, Rechbauerstr. 12, 8010 Graz. T: 0316-8736239, F: 0316-8736427

Garber, G., Dipl.-Ing.Dr.techn., Ziv.-Ing.f.Bauwesen, Leechgasse 37, 8010 Graz. T: 0316-381915-0, F: 0316-381915-22

Geutebrück, E., Dr.mont.Mag.rer.nat., Kagraner Platz 39, 1220 Wien. T: 01-202-4717, F: 01-2024718

Goriupp, H., Dipl.-Ing., Rebenweg 43F, 8054 Seiersberg. T: 0316-283016, F: 0316-8772131

Grund, K., Dipl.-Ing.Dr.techn., Auf dem Wingertsberg 21, D-65817 Eppstein 1

Hatz, D., Dipl.-Ing.Dr.techn., Bahnstraße 14/2, 1140 Wien. T: 01-972235

Hazivar, W.K., Dipl.-Ing.Dr.techn., Ziv.-Ing.f.Bauwesen, 7000 Eisenstadt, Carlonegasse 1. T: 02682-62841

Heidinger, Ch., Dipl.-Ing., Braunhubergasse 4a/6, 1110 Wien. T: 01-7495430

Henögl, O., Dipl.-Ing. OR, Inst. f. Bodenmechanik u. Grundbau, TU Graz, Rechbauerstraße 12, 8010 Graz. T: 0316-873-6238, F: 0316-873-6239, E: Henoegl@ibg.TU-Graz.ac.at

Hillisch, Ch., Dipl.-Ing., Geotechnical Engineer, Leopold Figlstraße 9/5/3, 2361 Laxenburg. T: 02236-73268

Hofmann, R., Dipl.-Ing.Dr.techn., Ingenieurkonsulent für Bauwesen, Rudolf Hochmayergasse 28/40, 2380 Perchtoldsdorf. T/F: 01-8658943, E: rhofmann@everyday.com

Hohla, M., Dipl.-Ing. Prok., PORR AG, Söllheimer Str. 4, 5028 Salzburg. T: 0662-46955411, F: 0662-46955420

Hondl, A., SR i.R., Dipl.-Ing.Dr.techn., Sinagasse 1-19/6, 1220 Wien. T: 01-2359912

Hörhan, R., Min.-Rat Dipl.-Ing., Kahlenbergerstraße 34, 1190 Wien. T: 01-3718614

Janda, J., Dir.Dipl.-Ing., A.PORR AG, Absberggasse 47, 1103 Wien. T: 01-79525-0

Janku, H., Ob.Ing.Dir.Dipl.-Ing., Scheffergasse 35c/1, 2340 Mödling. T: 02236-269262

Kautz, H., Dipl.-Ing., Mühlsangergasse 26, 1110 Wien. T: 01-7687287

Kienberger, H., HTL-Dir.i.R.Dipl.-Ing.Dr.techn., Ziv.-Ing.f.-Bauwesen, Schloßstraße 17, 5760 Saalfelden. T: 06582-2066

Knittel, A., Dipl.-Ing.,Ziv.-Ing.f.Bauwesen, Schöndorferplatz 6, 5600 Hallein. T: 06245-88284-0, F: 06245-88284-20, E: office@ppp.co.at

Kolymbas, D., o.Univ.Prof.Dipl.-Ing.Dr.techn., Inst. f. Geotechnik und Tunnelbau, TU Innsbruck, Technikerstraße 13/III, 6020 Innsbruck. T: 0512-507-6671, F: 0512-507-2996

Kopf, F., Dipl.-Ing.Dr.techn.Univ.Ass., Institut für Grundbau und Bodenmechanik, TU Wien, Karlsplatz 13, 1040 Wien. T: 01-58801-22113, F: 01-58801-22198, E: f.kopf@tuwien.ac.at

Kraml, G., HTL-Prof.Dipl.-Ing.Dr.techn., Ziv.-Ing.f. Bauwesen, Castellezgasse 15, 1020 Wien. T: 01-216-39-17, F: 01-219-73-01

Krejci, E., HTL-Prof.Dipl.-Ing., Ziv.-Ing.f.Bauwesen, Buchenweg 1/10, 2362 Biedermannsdorf. T: 02236-719912

Lackner, K., Dipl.-Ing.Dr.techn., Consulting Engineer, Alberstraße 8, 8010 Graz. T: 0316-382434, F: 0316-382834

Leberbauer, P., Prok.Dipl.-Ing., Siemens AG, Wolfgang-Pauli-Str.2, 4021 Linz. T: 0732-1707-501, F: 0732-1707-505, E: peter.leberbauer@siemens.at

Led, M., Dipl.-Ing., Pod parekm Str. 10, CZ-400 11 Usti nad Labem

Leitner, K., Dipl.-Ing., Gentzgasse 10/5/17, 1180 Wien. T: 01-3109359

Liebl, W., Dipl.-Ing., Bergstraße 32, 4310 Mauthausen. T: 01-79525-486, F: 01-79525-499

Macho, H., Dipl.-Ing., Schultestraße 19, 4020 Linz. T: 0732-602212

Macho, T., Dipl.-Ing., Schultestraße 19, 4020 Linz. T: 0664-5164125, F: 0732-7720-2918, E: thomas.macho@bps.at

Makovec, F., em.o.Univ.Prof.Dr., Geologe, Schinaweisgasse 17, 1140 Wien. T: 01-9444782, F: 01-5044235

Mannsbart, G., Dipl.-Ing., Schachermayerstr. 1B, P.O.B. 675, 4021 Linz. T: 0732-6983-5380, F: 0732-6983-5353

Martak, L., OStBR Hon.Prof.Dipl.-Ing.Dr.techn. (*Vice-Chairman*), MA 29, Wilhelminenstraße 93a, 1160 Wien. T: 01-488-29-96-900, F: 01-488297291

Monarth, W., Dipl.-Ing., Ziv.-Ing.f.Bauwesen, Ingenieurbüro Schickl & Partner, Blindengasse 26, 1080 Wien. T: 01-4054260-6, F: 01-4074712

Müller, W., HTL-Prof.Dipl.-Ing., Mühlhofstraße 2, 3503 Krems. T: 02732-71525

Nußbaumer, H., Baurat h.c. Dir. Dipl.-Ing.Dr.techn., Siemens AG,Wolfgang-Pauli-Str.2, 4021 Linz. T: 0732-1707-500, F: 0732-1707-505, E: hans.nußbaumer@siemens.at

Oplusstil, G., Dr., Österreichische Draukraft, Kohldorferstraße 98, 9020 Klagenfurt. T: 0463-202-451

Pelzl, M., Dipl.-Ing., Ziv.-Ing.f.Bauwesen, Eichenstraße 20, A-1120 Wien. E: pelzl@ppp.co.at

Pfaffenwimmer, F., wHR i.R.Dipl.-Ing., Ghegastraße 19, 4020 Linz. T: 0732-576232

Plankel, A., Dipl.-Ing., Ziv.-Ing.f.Bauwesen, Baumweg 3, 6923 Lauterach. T: 05574-39811

Pototschnik, M., Dipl.-Ing., Drillgasse 13, 1230 Wien. T: 01-8899590

Pötscher, R., Dipl.-Ing., Ziv.-Ing.f.Bauwesen, Tegethoffplatz 3/III, 8010 Graz. T: 0316-382488-10, F: 0316-382488-9

Pregl, O., o.Univ.Prof.Dipl.-Ing.Dr.nat.techn.,Universität f. Bodenkultur, Feistmantelstraße 4, 1180 Wien. T: 01-47654-5551, F: 01-4789242, E: mail@pregl.boku.ac.at

Preindl, P., Dipl.-Ing.Dr.nat.techn., Alpine Bau Ges.m.b.H., Oberlaaer Straße 276, 1239 Wien. T: 01-61079-0, F: 01-61070-44

Prinzl, F., Dipl.-Ing.Dr.techn., ZT-Büro Geoconsult, Sterneckstraße 55, 5020 Salzburg. T: 0662-881621-0

Prodinger, W., Dipl.-Ing.Dr.techn., Ziv.-Ing.f.Bauwesen, Tegethoffplatz 3/III, 8010 Graz. T: 0316-382488

Proksch, E., OSR i.R.Dipl.-Ing., Steudelgasse 37-39/1/13, 1100 Wien. T: 01-6060480

Püchl, G., HTL-Prof.i.R.Dipl.-Ing., Ziv.-Ing.f.Bauwesen, Margaretenstraße 21/1/6, 1040 Wien. T: 01-5873281

Raschendorfer, H., Dipl.-Ing., Herzogbergstraße 37, 2345 Brunn a. Geb. T: 02236/33662

Regler, R.-M., Mag.jur.Dipl.-Ing., Fehlingergasse 4, Haus A, 1130 Wien. T: 01:8045201. T: (Büro) 01-50105-4000, F: 01-50206-233

Renezeder, W., Dr.techn., Herbeckstraße 75/5/3, 1180 Wien

Schaller, C., Dipl.-Ing., Franz Josef Straße 60. T: 02236-21743

Schimmerl, J., ao.Univ.Prof.Dipl.-Ing.Dr.techn.Dr.Ing., Ziv.-Ing.f.Bauwesen, Rudolfplatz 6, 1010 Wien. T: 01-6315780*

Schimpf, H., Dipl.-Ing.,Ziv.-Ing.f.Bauwesen, VEG, Parkring 12, 1010 Wien. T: 01-51538-2111, F: 01-53113-50612

Schippinger, K., Dipl.-Ing.Dr.techn.,Ziv.-Ing.f.Bauwesen, Wilhelm Raabe Gasse 14, 8010 Graz. T: 0316-682144, F: 0316-682144-9

Schneider, H., Dipl.-Chem.Dir., Obere Au 7, 4491 Niederneukirchen. T: 07224-7116

Schober, W., o.Univ.Prof.Dipl.-Ing.Dr.techn., Technikerstraße 13, 6020 Innsbruck. T: 0512-507-6684, F: 0512-507-2996

Scholz, P., Dipl.-Ing., Sandgasse 3, 4055, Pucking Coö. T: 07229-79489

Schönlaub, W., Dipl.-Ing., Pacassistr. 17, 1130 Wien. T: 01-8041081, F: 01-51538-2199

Schwab, E., Dipl.-Ing.Dr.techn., Schlägergasse 4, 1130 Wien. T: 01-887949, Arsenalresearch, Pf. 8, 1031 Wien. T: 01-79747-544, F: 01-79747-592

Schweiger, H.F., Ao.Univ.-Prof.Dipl.-Ing.Dr.techn.M.Sc., Inst. f. Bodenmechanik u. Grundbau, TU Graz, Rechbauerstr. 12, 8010 Graz. T: 0316-873-6234

Seltenhammer, U., Dir.i.R. Dipl.-Ing., Bechadeweg 6, 2371 Hinterbrühl

Semprich, St., o.Univ.Prof.Dipl.-Ing.Dr.techn., Inst.f. Bodenmechanik und Grundbau, TU Graz, Rechbauerstraße 12, 8010 Graz. T: 0316-873-6230, F: 0316-873-6232, E: semprich@ibg.TU-Graz.ac.at

Sochatzy, G., SR Dipl.-Ing., MA 29, Wilhelminenstraße 93a, 1160 Wien. T: 01-48829-96900

Stadler, G., o.Univ.Prof.Dipl.-Ing.Dr.mont., Inst. f. Baubetrieb und Bauwirtschaft, TU Graz, Lessingstraße 25/II, 8010 Graz. T: 0316-873-6250

Steigenberger, J., Dipl.-Ing.Dr.techn., Forschungsinstitut der V.Ö.Z., Reisner Str. 53, 1030 Wien. T: 01-7146681-46, F: 01-7146681-66

Stempkowski, H., Dipl.-Ing., ziv.ing. f.Bauwesen Haiteingergasse 29, 1180 Wien. T: 01-4705614,0699 -m 154721

Stockhammer, P., BRh.c.Dir.Dipl.-Ing., Keller Grundbau Ges.m.b.H., Mariahilferstraße 129, 1150 Wien. T: 01-892 35-26, F: 01-892-37-11

Stundner, H., HTL-Prof.Dipl.-Ing., Zipsgasse 11, 2344 Maria Enzersdorf. T: 02236-49675, E: herbert.stundner@htl.minic.ac.at

Swoboda, G., Dipl.-Ing., Tongasse 11/14, 1030 Wien. T: 01-7518945

Tiefenbacher, H., Min.Rat Dipl.-Ing.Dr.techn., BMfwA, Stubenring 1, 1010 Wien. T: 01-7500-5853

Tropper, W., Dipl.-Ing.Dr.techn., Ingenieurkonsulent f. Bauwesen, Fischnalerstraße 19, 6020 Innsbruck. T: 05222-84192

Unterberger, W., Dipl.-Ing.Dr.techn., iC Consulenten Ziviltechniker GmbH, Zollhausweg 1, A-5101 Bergheim. T: 0662-450773, F: 0662-450773-5, E: office@ic-salzburg.at

Waibel, P., HTL-Prof.Dipl.-Ing.Dr.techn.,Zivilingenieur für Bauwesen, Mariahilferstraße 20, 1070 Wien. T: 01-5242980-0, F: 01-5242980-4

Werner, G., Dipl.-Ing., Doka Espana Encofrados S.A., Julio Palacios, 18 Pol. Ind. Butarque, 28914 Leganés – Madrid. T: (91)6857802, F: (91)6857501

Wimmer, W., wHRi.R. Dipl.-Ing., Kärnter Str. 12, 4020 Linz. T: 0732-6584-2177, F: 0732-6584-2918

Würger, E., HTL-Dir.i.R.Dipl.-Ing.Dr.techn., Ziv.-Ing.f. Bauwesen, Fimbingergasse 3, 1230 Wien. T: 01-8694992, F: 01 8652740

Zanier, L., Dipl.-Ing., Salurnerstraße 56 A, 6330 Kufstein. T: 05372-62196, F: 05372-65090

BANGLADESH

Secretary: Dr. M. Hossain Ali, Bangladesh Society for Geotechnical Engineering, Room 307, Civil Engineering Building, BUET, Dhaka-1000, BANGLADESH. T/F: +880 2 966 5639, F: +880 2 863 026, E: hali@ce.buet.edu

Total number of members: 27

Abedin, Dr. M.Z., Professor, Department of Civil Engineering, Bangladesh University of Engineering & Technology (BUET), Dhaka-1000, Bangladesh. T: 880-2864640-44, ext. 7215 (off), 880-2-864640-44/ext. 7144

Ahmed, M., Managing Director, M. Ahmed & Associates Ltd., 519 Baitul Aman Housing Society, Road 12, Adabor, Shaymoli, Dhaka, Bangladesh. T: 880-29117589 (off), 880-2-9114328 (res)

Al-Hussaini, Dr. T.M., Assistant Professor, Department of Civil Engineering, Bangladesh University of Engineering & Technology (BUET), Dhaka-1000, Bangladesh. T: 880-2-500223 (res), 880-2-86464-4/ext. 6103 (off), E: tahmeed@ce.buet.edu

Ali, Dr. M.H., Professor, Department of Civil Engineering, Bangladesh University of Engineering & Technology (BUET), Dhaka-1000, Bangladesh. T: 880-2861070 (res), 880-2-866833/ext. 7284 (off), E: hali@ce.buet.edu

Ameen, Dr. S.F., Professor, Department of Civil Engineering, Bangladesh University of Engineering & Technology (BUET), Dhaka-1000, Bangladesh. T: 880-2-9669284 (res), 880-2-866833/ext. 74969 (off)

Amin, M.N., Managing Director, Foundation Consultants Ltd., GPO Box 609, Dhaka-1000, Bangladesh. T: 880-2-9558600 (off), 880-2-9880889 (res)

Ansary, Dr. M.A., Assistant Professor, Department of Civil Engineering, Bangladesh University of Engineering & Technology (BUET), Dhaka-1000, Bangladesh. T: 880-2-9116836 (res), 880-2-864640-4/ext. 6216, E: ansary@ce.buet.edu

Bashar, M.A., Associate Professor, Department of Civil Engineering, Bangladesh Institute of Technology (BIT), Khulna-9203, Bangladesh. T: 880-41-774780, E: cebitk@bdonline.com

Choudhury, Dr. J.R., Professor, Department of Civil Engineering, Bangladesh University of Engineering and Technology (BUET), Dhaka-1000, Bangladesh. T: 880-2-66833/ext. 7288 (off), 880-2-865258 (res), E: jrc@bangla.net

Bosunia, Dr. S.Z., Professor, Department of Civil Engineering, Bangladesh University of Engineering & Technology (BUET), Dhaka-l000, Bangladesh. T: 880-2-8616833

Dastidar, A.G., Consulting Geotechnical Engineer, Eastern Housing Ltd.

Daulah, I.U., Managing Director, DIRD Private Ltd., House 9, Road 51, Gulshan, Dhaka-1217, Bangladesh. T: 880-2-9886747 (off), F: 880-2-883621, E: dird@citechco.net

Hoque, Dr. E., Assistant Professor, Department of Civil Engineering, Bangladesh University of Engineering & Technology BUET, Dhaka-1000, Bangladesh, T/F: 880-2-9662352 (res), 880-2-864640-44/ext. 6154 (off), E: ehoq@ce.buet.edu

Islam, M.S., Lecturer, Department of Civil Engineering, Bangladesh University of Engineering & Technology (BUET), Dhaka-1000, Bangladesh

Karim, M.F., Geological Survey of Bangladesh, Pioneer Road, Segun Bagicha, Dhaka, Bangladesh. T: 880-2-418676 (off), 880-2-418533 (off)

Khan, Dr. A.J., Construction and Development Co., 4 Dhanmondi Road 4A, Dhaka, Bangladesh. T: 880-2-506870 (off), 880-2-506588 (res)

Majumder, P.U., Managing Director, Multitech Ltd., 52 New Eskaton Road, TMC Building (1st Floor), Dhaka, Bangladesh. T: 880-2-417197 (off), 880-2-500638 (res)

Manzoor, A.S.M., 145/1/D Azimpur Road, 4th floor, Dhaka-1205

Muqtadir, Dr. A., Professor, Department of Civil Engineering, Bangladesh University of Engineering & Technology (BUET), Dhaka-1000, Bangladesh. T: 880-2866833/ext. 7528 (off), 880-2-866833/ext. 7501 (res), 880-2-9665607 (res), E: mugtadir@ce.buet.edu

Sadeque, A., SDE, PWD Design Division-2, Room 512 (3rd Floor), Purta Bhavan, Segun Bagicha, Dhaka-1000

Safiullah, Dr. A.M.M., Professor, Department of Civil Engineering, Bangladesh University of Engineering & Technology (BUET), Dhaka-1000, Bangladesh. T: 880-2860333 (res), 880-2-866833/ext. 7213 (off), E: safi@ce.buet.edu

Seraj, Dr. S.M., Associate Professor, Department of Civil Engineering, Bangladesh University of Engineering & Technology (BUET), Dhaka-1000, Bangladesh. T: 880-2866833/ext. 7489 (off), 880-2-885961 (res), E: smsera-i@bangla.net, sms@bangla.net

Serajuddin, M., Geotechnical & Material Specialist, House No. 25, Block A, Road No. 18, Banani Model Town, Dhaka-1213, Bangladesh. T: 880-2-601318 (res), 880-2-9334486 (off), E: smsera-i@bangla.net

Siddique, Dr. A., Professor, Department of Civil Engineering, Bangladesh University of Engineering & Technology (BUET), Dhaka-1000, Bangladesh. T: 880-2-864640/ext. 7488 (off), 880-2-864640/ext. 7263 (res), E: abusid@ce.buet.edu

Siddiquee, Dr. M.S.A., Associate Professor, Department of Civil Engineering, Bangladesh University of Engineering & Technology (BUET), Dhaka-1000, Bangladesh. T: 880-2864640-4/ext. 7575 (off), 880-2-864640-4/ext. 7375 (res), E: sid@ce.buet.edu, sid@bangla.net

Uddin, Dr. M.K., House No. 296, Jahanara Lodge, Eidgah Road 15 (Old), 8/A (New), Dhanmondi (West), Dhaka-1209, Bangladesh. T: 9116349 (res), 880-2864640/ext. 6195 (off)

Yasin, Dr. S.J.M., Assistant Professor, Department of Civil Engineering, Bangladesh University of Engineering & Technology (BUET), Dhaka-1000, Bangladesh. T: 880-29008208 (res), 880-2-864640-4/ext. 6330 (off), E: yasin@ce.buet.edu

8

Secretary: Prof. Dr. Ir Patrick Mengé, Laboratory of Soil Mechanics, Ghent University, Technologiepark 9, 9052 Zwijnaarde, BELGIUM. T: +32/9/264.57.23, F: +32/9/264.58.49, E: Patrick.Menge@rug.ac.be

Total number of members: 87

Areias, L., Hertogstraat 141-701, 3001 Heverlee. T: +32/16/40.65.92, F: +32/16/40.65.92, E: lou.areias@rug.ac.be

Aruengt, C., Geologica S.A., Rue Phocas Lejeune 4, 5032 Gembloux. T: +32/81/56.63.56, F: +32/81/56.78.72, E: be_gembloux_geologica_geotechnique@sgsgroup.com

Bauduin, C., Besix, Gemeenschappenlaan 100, 1200 Brussel. T: +32/2/402.62.46, F: +32/2/402.62.30, E: cbauduin@besix.com

Bauwen, I., Fundex N.V., Kustlaan 118, 8380 Zeebrugge. T: +32/50/54.41.64, F: +32/50/54.79.02, E: fundex.be@fundexgroup.com

Bernard, A., Dredging International N.V., Bosstraat 28, 9111 Belsele. T: +32/3/779.68.37, F: +32/3/252.68.31

Boileau, R., Avenue Rene Stevens 15, 1160 Bruxelles. E: reneboileau@tiscalinet.be

Bolle, A., ULg. – Inst. Genie Civil – Bat. B52/3, Chemin Des Chevreuils 1, 4000 Liege (Sart Tilman). T: +32/4/366.93.32, F: +32/4/366.93.26, E: Albert.Bolle@ulg.ac.be

Bottiau, M., Socofonda N.V., Rue De Verdun 750, 1130 Bruxelles. T: +32/2/215.02.16, F: +32/2/215.11.93, E: m.bottiau@socofonda.be

Cathie, D., Sage Engineering S.A./N.V., Avenue Vandendriessche 18, 1150 Bruxelles. T: +32/2/776.03.10, F: +32/2/776.03.19, E: DCCATHIE@sage-becom

Charlier, R., ULg. – Inst. Genie Civil – Bat. B52/3, Chemin Des Chevreuils 1, 4000 Liege (Sart Tilman). T: +32/4/366.93.34, F: +32/4/366.93.26, E: Robert.Charlier@ulg.ac.be

Coelus, G., De Waal Palen, N.V., Voshol 6a, 9160 Lokeren. T: +32/9/340.55.00, F: +32/9/340.55.18, E: kris.hoffelinck@dewaalpalen.be

Cortvrindt, G., Franki Geotechnics B Nv-Sa, Av. Edgard Frankginoul 2, 1480 Saintes. T: +32/2/391.46.46, F: +32/2/391.46.47, E: GC@franki-geotechnics.be

Cremer, J.-M., Bureau D'Etudes Greisch S.A., Allee Des Noisetiers 25, 4031 Liege (Angleur)

D'Alvise, L., Fac. Polytechn., Mecanique Des Roches, Rue Du Joncquois 53, 7000 Mons. T: +32/65/37.45.17, F: +32/65/37.46.00, E: laurentdalvise@fpms.ac.be

De Cock, F., Hunselveldweg 33, 1750 Lennik. T: +32/54/32.80.98, F: +32/54/34.41.08, E: flordecock.geo.be@attglobal.net

De Cock, S., UCL – Unite Genie Civil, Place Du Levant (Bât.Vinci) 1, 1348 Louvain-La-Neuve. T: +32/10/47.21.12, F: +32/10/47.21.79, E: decock@gc.ucl.ac.be

De Jaeger, J., UCL – Unite Genie Civil, Place Du Levant (Bât.Vinci) 1, 1348 Louvain-La-Neuve. T: +32/10/47.21.22, F: +32/10/47.21.79, E: dejaeger@ecam.be

De Lathauwer, W., Abtus/Bvtos, Wetstraat (Res. Palace) 155 B.1, 1040 Brussel. T: +32/2/287.31.40, F: +32/2/287.31.44, E: willy.delathauwer@vici.fgov.be

De Rouck, J., Weg- & Waterbouw – Rug, Mollekens 8, 9860 Oosterzele. T: +32/9/264.54.89, F: +32/9/264.58.37, E: Julien.DeRouck@rug.ac.be

De Schrijver, J., Arbed Damwand Belgie N.V., Industrielaan 2, 3900 Overpelt. T: +32/11/80.08.90, F: +32/11/80.08.95

De Smedt, F., Vub–Vakgroep Hydrologie En Waterbouwkunde, Pleinlaan 2, 1050 Brussel. T: +32/2/629.35.47, F: +32/2/629.30.22, E: fdesmedt@vub.ac.be

Debacker, P., Rdb/Service De Stabilite, Avenue de la Toison d'or, 1060 Bruxelles. T: +32/2/541.65.11, F: +32/2/541.65.10, E: RDBStab@regie.fed.be

Debauche, P., Tractebel Development, Avenue Ariane 1, 1200 Bruxelles. T: +32/2/773.70.71, F: +32/2/773.79.90, E: Pierre.Debauche@tractebel.com

Dellicour, L., S.A. Bagon, Rue G. Dekelver 61, 1160 Bruxelles. T: +32/2/660.00.77, F: +32/2/673.94.45

Delmote, S., Thales Geosolutions (Belgium) Sa/Nv, R. Van Den Driesschelaan 18, 1150 Brussel. T: +32/2/776.03.10, F: +32/2/776.03.19, E: SDE@sage-becom

Detry, J., Tractebel Development, Avenue Ariane 5, 1200 Bruxelles. T: +32/2/773.70.71, F: +32/2/773.79.90, E: Jacques.Detry@tractebel.com

D'Haene, P., Verdeyen & Moenaert, Avenue Des Cerisiers 15, 1030 Bruxelles. T: +32/2/743.12.80, F: +32/2/743.12.89

D'Hemricourt, J., Ecole Royale Militaire, Av. De La Renaissance 30, 1040 Bruxelles. T: +32/2/735.51.52

D'Hoore, S., Adinco B.V.B.A., Woluwelaan 13-15, 1800 Vilvoorde. T: +32/2/252.02.55, F: +32/2/252.40.30

Dochy, C., Stib, Rue Colonel Bourg 162/BT2, 1140 Bruxelles. T: +32/2/515.23.08, F: +32/2/515.32.54, E: dochy@stib.irisnet.be

Duchene, P., Fac. Sciences Agronomiques, Passage Des Deportes 2, 5030 Gembloux. T: +32/81/62.21.92, F: +32/81/31.45.44

Dupont, A., Fundex N.V., Kustlaan 118, 8380 Zeebrugge. T: +32/50/54.41.64, F: +32/50/54.79.02, E: fundex.be@fundexgroup.com

Duvieusart, J.-C., Gevalco Sprl., Rue Du Printemps 83, 1380 Lasne. T: +32/2/654.02.42, F: +32/2/652.39.30, E: gevalco@skynet.be

Everaerts, W., De Neef Eco-Invest, Industriepark 8, 2220 Heist Op Den Berg. T: +32/15/24.93.60, F: +32/15/24.80.72

Gille, H., Soletanche Bachy France Nv, Oudergemlaan 68/14, 1040 Brussel. T: +32/2/733.20.12, F: +32/2/734.76.81, E: hgille.bachy@attglobal.net

Goossens, D., Bureau Seco, Aarlenstraat 53, 1040 Brussel. T: +32/2/238.22.11, F: +32/2/238.22.61, E: d.goossens@Seco.be

Goris, T., Ecole Royale Militaire, Av. De La Renaissance 30, 1000 Bruxelles. T: +32/2/737.64.15, F: +32/2/737.64.12, E: goris@cobo.rma.ac.be

Haegeman, W., RUG – Labo Grondmechanica, Technologiepark 9, 9052 Zwijnaarde. T: +32/9/264.57.24, F: +32/9/264.58.49, E: Wim.Haegeman@rug.ac.be

Holeyman, A., UCL – Unite Genie Civil, Place Du Levant (Bât.Vinci) 1, 1348 Louvain-La-Neuve. T: +32/10/47.21.18, F: +32/10/47.21.79, E: holeyman@gc.ucl.ac.be

Huergo, P.J., U.L.B. – Laboratoire J. Verdeyen, Av. Adolphe Buyl 87, 1050 Bruxelles. T: +32/2/650.27.37, F: +32/2/650.27.43, E: phuergo@ulb.ac.be

Janssens, R., Hogeschool Antwerpen, Dept. Iwt (2C), Paardenmarkt 92, 2000 Antwerpen. T: +32/3/231.50.36, F: +32/3/231.86.70, E: rob.janssens@vt4.be

Jewell, R.A., Geosyntec Consultants, Rue Tenbosch 14, 1000 Bruxelles. T: +32/2/646.89.69, F: +32/2/646.12.38, E: rjewell@gcsa.be

Joye, P., Ulb – Mecanique Des Sols, Av. F. Roosevelt (Cp 194/2) 50, 1050 Bruxelles. T: +32/2/650.27.35, F: +32/2/650.27.43, E: pjoye@ulb.ac.be

Labiouse, V., Labo De Mecanique Des Roches – Epfl, Ecole Polytechn. Federale Lausanne, 1015 Lausanne (Swiss). T: +41/021/693.23.23, F: +41/021/693.41.53, E: Vincent.Labiouse@epfl.ch

Legrand, C., CSTC-WTCB-BBRI, Lozenberg 7, 1932 Sint-Stevens-Woluwe. T: +32/2/655.77.11, F: +32/2/653.07.29, E: christian.legrand@bbri.be

Maekelberg, W., Tuc Rail N.V. – Dienst Geotechniek, Frankrijkstraat 91, 1070 Brussel. T: +32/2/529.71.21, F: +32/2/529.79.43, E: wim.maekelberg@tucrail.be

Maertens, J., Jan Maertens BVBA, H. Consciencestraat 4, 2340 Beerse. T: +32/14/61.39.01, F: +32/14/61.90.71, E: jan.maertens.bvba@skynet.be

Maertens, L., Besix, Gemeenschappenlaan 100, 1200 Brussel. T: +32/2/402.65.63, F: +32/2/402.65.30, E: lmaertens@besix.com

Mengé, P., Dredging International N.V. Scheldedyk 30, 2070 Zwyndrecht. T: 09 264 57 23, F: 03 779 68 37, E: Menge.Patrick@dredging.com

Mertens De Wilmars, A., UCL – Unite Genie Civil, Place Du Levant (Bât.Vinci) 1, 1348 Louvain-La-Neuve. T: +32/10/47.21.12, F: +32/10/47.21.79, E: mertensdewilmars@gc.ucl.ac.be

Miller, J.-P., Tractebel Development-Geotechnologie, Avenue Ariane 7, 1200 Bruxelles. T: +32/2/773.70.74, F: +32/2/733.79.90, E: jean-pierre.miller@tractebel.com

Nomérange, J., M.E.T. – Direction Geotechnique, Rue Cote D'Or 253, 4000 Liege. T: +32/4/254.58.11, F: +32/4/253.04.05, E: jnomerange@met.wallonie.be

Nuyens, J., S.C. Orex, Rue Du Page 14, 1050 Bruxelles. T: +32/2/538.28.55, F: +32/2/537.39.40, E: jean.nuyens@orex.be

Ourth, A.-S., Fac. Agronorique, Resist. Materiaux, Passage Des Deportes 2, 5030 Gembloux. T: +32/81/62.21.65, F: +32/81/61.45.44, E: ourth.as@fsagx.ac.be

Peiffer, H., Alpha Studieburo, Taxanderlei 48, 2900 Schoten. T: +32/3/658.87.14, F: +32/3/680.14.27, E:Alpha.Buro@skynet.be

Pohl, E., Smet Boring N.V., Bd. De L'Humanite 104, 1190 Forest. T: +32/2/344.02.13, F: +32/2/344.64.25, E: etienne.pohl@smetf-c.be

Poorteman, F., De Waal Palen N.V., Voshol 6A, 8160 Lokeren. T: 09 340 55 00, F: 09 340 55 18, E: kris.hoffelinck@dewaalpalen.be

Roisin, V., 2Ic – Scrl, Avenue Des Ericas 8, 1640 Rhode-St-Genese. T: +32/2/358.42.13, F: +32/2/358.63.28

Schiffmann, J., Setesco, Avenue Maurice 50/BT 7, 1050 Bruxelles. T: +32/2/644.01.60, F: +32/2/640.05.73, E: setesco@pophost.eunet.be

Schreiner, C., Ispc Profilarbed, R. De Luxembourg 66, 4009 Esch/Alzette. T: +352-5313.3105, F: +352-5313.3290, E: claude.schreinder@profilarbed.lu

Schroeder, C., Universite De Liege – L.G.I.H., Sart Tilman B 19, 4000 Liege. T: +32/4/366.22.16, F: +32/4/366.22.02, E: Christian.Schroeder@ulg.ac.be

Silence, F., Psi, Kersenbomenlaan 47, 3090 Overijse. T: +32/2/657.24.23, F: +32/2/657.24.23

Simon, G., M.E.T. – Direct. Geotechnique, Rue Cote D'Or 253, 4000 Liege. T: +32/4/254.58.11, F: +32/4/253.04.05, E: gsimon@met.wallonie.be

Simons, R., Sbe N.V., Slachthuisstraat 71, 9100 Sint-Niklaas. T: +32/3/777.95.19, F: +32/3/777.98.79, E: sbe@glo.be

Sterckse, L., ADVISON bvba, Edgard Sohiestraat 55, B-1560 Hoeilaart. T: +32/2/657.46.40, F: +32/2/657.46.41, E: info@advison.be

Théwissen, F., M.E.T. – D.421 Direction Geotechnique, Rue Cote D'Or 253, 4000 Liege. T: +32/4/254.58.11, F: +32/4/253.04.05

Thimus, J.-F., UCL – Unite Genie Civil, Place Du Levant (Bât.Vinci) 1, 1348 Louvain-La-Neuve. T: +32/10/47.21.12, F: +32/10/47.21.79, E: thimus@gc.ucl.ac.be

Thomas, R., Sbe N.V., Slachthuisstraat 71, 9100 Sint-Niklaas. T: +32/3/777.95.19, F: +32/3/777.98.79, E: sbe@glo.be

Thooft, K., De Nayer Instituut, Jan De Nayerlaan 5, 2860 St-KaTijne-Waver. T: +32/15/31.69.44, F: +32/15/ 31.74.53, E: kth@denayer.wenk.be

Trève, C., Cfe N.V., Herrmann-Debrouxlaan 40-42, 1160 Brussel. T: +32/2/661.12.25, F: +32/2/661.14.03, E: Ctreve@cfe.be

Van Alboom, G., M.V.G. – Afdeling Geotechniek, Tramstraat 52, 9052 Zwijnaarde T: +32/9/240.75.11, F: +32/9/240.75.00, E: Gauthier.VanAlboom@lin.vlaanderen.be

Van Calster, P., Geologica N.V., Tervuursesteenweg 200, 3060 Bertem. T: +32/16/49.00.39, F: +32/16/48.14.19, E: paulvc@compaqnet.be

Van Den Broeck, M., Dredging International, Scheldedijk 30, 2070 Zwijndrecht. T: 09 264 5723???, E: van.den.broeck.marc@dredging.com

Van Hoecke, H., Hogeschool Gent, Rijvisschestraat 29, 9052 Zwijnaarde. T: +32/9/243.87.87, F: +32/9/ 243.87.77

Van Impe, P., RUG – Labo Grondmechanica, Technologiepark 9, 9052 Zwijnaarde, Ghent. T: +32/9/264.57.32, F: +32/9/264.58.49, E: Peter.VanImpe@rug.ac.be

Van Impe, W., RUG – Labo Grondmechanica, Technologiepark 9, 9052 Zwijnaarde, Ghent. T: +32/9/264.57.23, F: +32/9/264.58.49, E: William.VanImpe@rug.ac.be

Van Rooten, C., Centre De Recherche Routieres, Boulevard De La Woluwe 42, 1200 Bruxelles. T: +32/2/775.82.20, F: +32/2/772.33.74, E: BRRC@BRRC.be

Van Tricht, K., De Nayer Instituut, Jan De Nayerlaan 5, 2860 Sint-KaTijne-Waver. T: +32/15/31.69.44, F: +32/15/31. 74.53, E: kth@denayer.wenk.be

Van Vooren, B., Dsi Nv, Industrieweg 25, 3190 Boortmeerbeek. T: +32/16/60.77.60, F: +32/16/60.77.66

Vanden Berghe, J.-F., Thales. GeoSolutions (Belgium), Av. R. Vandemdriessche 18, 1150 Bruscelles. T: 02 776 06 23, F: 02 776 03 19, E: JFVandenBerghe@sage-be.com

Vander Linden, J., Eurodril Belgium S.A., Chaussee D'Alsemberg 21, 1630 Linkebeek. T: +32/2/331.03.17, F: +32/2/347.14.62

Verbrugge, J.-C., Laboratoire De Mecanique Des Sols – ULB, Av. Adolphe Buyl 87 CPI 194/2, 1050 Bruxelles. T: +32/2/650.27.35, F: +32/2/650.27.43, E: jvebrug@ulb.ac.be

Verhelst, F., Lhoist, Parc Des Collines 50, 1300 Waver. T: +32/10/233.808, F: +32/10/233.850, E: frederik.verhelst@lhoist.com

Verheulpen, P., Aed – Ditp – Ccn, Rue De Progres 80 bte 1, 1030 Bruxelles. T: +32/2/204.29.25, F: +32/2/204.15.02, E: gverheulpen@mrbc.irisnet.be

Welter, P., M.E.T. – Direction Geotechnique, Rue Cote D'Or 253, 4000 Liege. T: +32/4/254.58.11, F: +32/4/253.04.05, E: pwelter@met.wallonie.be

Winand, A., S.N.C.B./Dept. Infrastructure, Rue De France 85, 1070 Bruxelles. T: +32/2/525.43.93, F: +32/2/525.23.00

Zaczek, Y., Tractebel Development-Geotechnologie, Avenue Ariane 7, 1200 Bruxelles. T: +32/2/773.70.69, F: +32/2/733.79.90, E: Yannick.Zaczek@tractebel.be

BRAZIL

Secretary: Alberto Sayão, Associação Brasileira de Mecânica dos Solos e Engenharia Geotécnica (ABMS),
Av. Prof. Almeida Prado, 532 – IPT, Prédio 54 – DEC, 05508-901 – São Paulo, SP.
T: +55 11 3768 7325, F: +55 11 3768 7325, E: abms@ipt.br

Total number of members: 883

Abitante, E., Av. Cristovao Colombo, 1761, 90560-004 – Porto Alegre, RS. E: etel-edgar@ca.conex.com.br

Aboim, R.D. de, Rua Voluntarios da Patria, 429-Ap.601, 22270-000 – Rio de Janeiro, RJ

Abrahao, R.A. Rua Domingos Antonio Ciccone, 51, 04710-220 – São Paulo, SP

Abramento, M., Av. Francisco de Assis Diniz, 55, 06030-380 – Osasco, SP. T: 11 9910 7630, 11 3681 7538, E: mabramen@ipt.br

Abrao, P.C., Rua Caraibas, 544-Ap.92 B., 05020-000 – São Paulo, SP. T: 11 3872 2076, 11 3872 2076, E: geo@originet.com.br

Abreu, A.E.S., de Av. Pompeia, 368-Ap.21, 05022-000 – São Paulo, SP

Abreu, F.L.R. de, Av. Parada Pinto, 3696-Ap.72-A, 02611-001 – São Paulo, SP

Abreu, F.R.S. de, Rua Marques de São Vicente, 194 Ap.404, 22451-040 – Rio de Janeiro, RJ

Abreu, J.R.S., Av. Joana Angelica, 1040-1º Andar, 40050-000 – Salvador, BA

Aguiar, P.R., Alameda Jauaperi, 1096 – Conjunto 182, 04523-014 – São Paulo, SP

Alarcon, C.A., Av. Angelica 1814 Cjs 902/903, 01228-200 – São Paulo, SP

Albiero, J.H., Rua São Bento, 700 – Conj. 91, 14801-300 – Araraquara, SP. T: 16 222 1040, F: 16 222 1768, E: albiero@sunrise.com.br

Albuquerque, P.J.R. de, Rua Maria Terci, 248-Apto.51, 18085-610 – Sorocaba, SP. T: 15 228 6629, E: pjra@uol.com.br

Albuquerque J., F.S., Av. Alberto Lamego, 2000-Cct, 28015-620 – Rio de Janeiro, RJ. T: 24 726 3738, F: 24 726 3730, E: saboya@uenf.br

Alcantara, P.B., Rua Visconde de Cairo, S/N-Ap.201/Bl.09, 48900-000 – Juazeiro, BA

Alencar J., J.A. de, Travessa Rui Barbosa, 656-Ap.1201, 66053-260 – Belem, PA

Almeida, C.G.R. de, Rua Barcelona, 240 Apto.32-Santa Lucia, 30350-260 – Belo Horizonte, MG. T: 31 9971 1279, 31 3344 3224, 31 3541 4009, E: cargus@terra.com.br

Almeida, D.A. de, Alameda Franca, 1433-Ap.81/Cerq. Ces., 01422-001 – São Paulo, SP

Almeida, G.C. de, Rua Amaro Ferreira, 194, 38065-170 – Uberaba, MG. T: 34 3313 9400

Almeida, M. de Souza S. de, R.Engº Carlos Euler, 77/905-B. Tijuca, 22793-260 – Rio de Janeiro, RJ. T: 21 562 7222, 21 431 1019, E: marcioal@openlink.com.br

Almeida Neto, J.A. de, Av. São Remo, 463-Apto.22B, 05360-150 – São Paulo, SP. T: 11 3735 5060, E: albuquerquenet@bol.com.br

Alonso, U.R., Rua Flavio Queiroz de Morais, 217, 01249-030 – São Paulo, SP

Alphageos Tecnologia Aplicada S/A, Rua Joao Ferreira de Camargo, 703, 06460-060 – Barueri, SP

Altrichter, G., Rua Prof. Oswaldo Teixeira, 84-Apto.51, 05617-020 – São Paulo, SP

Alvarenga, M.M., Av. Pref Dulcidio Cardoso 800/606-B 02, 22793-010 – Rio de Janeiro, RJ. T: 21 562 7962, 21 9684 4226, E: mmalvare@civil.ee.ufrj.br

Alves, A.M. de Lima, Rua Jose do Patrocinio, 369-Apto.201, 96010-500 – Pelotas, RS. T: 21 703 5139

Alves, M.C.M., Rua Alvaro Chaves, 38/704-Laranjeiras, 22231-220 – Rio de Janeiro, RJ. E: tina@civil.ee.ufrj.br

Amann, K.A.P., Rua dos Autonomistas,3-Jd. Itacolomy, 09402-520 – Ribeirao Pires, SP

Amaral, A.B.T. do, Rua Saldanha Marinho 1884, 80410-150 – Curitiba, PR

Amaral, C.P. do, Rua Das Laranjeiras, 457-Bl. B/Ap.804, 20920-020 – Rio de Janeiro, RJ

Amorim J., W.M. de, R.Br.Souza Leao 221/1503-Bl.B, 51030-300 – Recife, PE

Andrade, C.M.M. de, Rua Sen.Cezar Lacerda Vergueiro 531/153, 05435-060 – São Paulo, SP

Andrade, M.H. de, Noronha Rua Prof. Gastao Bahiana 575/102 Bl 2, 22071-030 – Rio de Janeiro, RJ

Andrade, P.I. de, Rua Soares Cabral, 80/1301, 22240-070 – Rio de Janeiro, RJ

Andrade, R.M. de, Rua Cupertino Durao, 135-Apto. 101-Leblon, 22441-030 – Rio de Janeiro, RJ

Andreo, C.S., Rua Benedito Wenceslau Ferreira, 108, 13219-801 – Jundiai, SP

Andres, M.A., Rua Cel.Xavier de Toledo,123-8º Andar, 01048-100 – São Paulo, SP. T: 11 255 6600, F: 11 255 6663, E: maroandres@uol.com.br

Angelino Neto, C., Rua Luis Gois, 1675, 04043-350 – São Paulo, SP. T: 11 276 9353, F: 11 5078 9695, E: constantinoan@osite.com.br

Anjos, A.P.V. dos, Rua T 62, 225-Ap.704/Setor Bela Vista, 74823-330 – Goiania, GO

Antoniutti Neto, L., Av. Arthur da Silva Bernardes, 1785-Portao, 80320-300 – Curitiba, PR. E: insitu@insitu.com.br

Aoki, N., Rua Haddock Lobo, 1141-Ap.51, 01414-003 – São Paulo, SP. E: nelson.aoki@originet.com.br

Apoio Asses.Proj.Fundacoes S.C. Ltda., Av. Brig Faria Lima 1794 And 5 Cj 56, 01452-001 – São Paulo, SP. T: 11 3814 1755, F: 11 3031 9490, E: apoioapf@osite.com.br

Arabe, L.C.G., Rua Rafael de Lourenco, 201-Sta. Maria, 38400-140 – Uberlandia, MG

Aragao, C.J.G. de, Av. Boa Viagem 3312 Ap.62 Boa Viagem, 51020-000 – Recife, PE

Araruna J., J.T., Rua Marques de São Vicente, 225/S-301L, 22453-900 – Rio de Janeiro, RJ

Araujo, E.S. de, Rua Sa Ferreira, 83-Ap.702, 22071-100 – Rio de Janeiro, RJ

Araujo, J.M.C. de, Praca N Senhora de Belem 70, 91712-240 – Porto Alegre, RS

Araujo, L.F.F. de, Av. Alvares Cabral, 374-Sala 609-Centro, 30170-000 – Belo Horizonte, MG

Araujo, L.G. de, Rua Dr. Claudio Lima, 149-Rosario, 35400-000 – Ouro Preto, MG. T: 31 3551 2951, E: luiz@em.ufop.br

Araujo, R.N.A., Av. Nego, 585-Apto.401-Tambai, 58039-101 – Joao Pessoa, PB

Araujo, S.P.M. de, Av. do Contorno, 6437-Cj.305-306 S.Pedro, 30110-110 – Belo Horizonte, MG. T: 31 32811492, 31 3223 1460, E: spma@zipmail.com.br

Araujo, S.M.D., Sqn 309-Bloco P-Apto.607-Asa Norte, 70755-160 – Brasilia, DF

Arduino, E.G. do Amaral, Av. Sernambetiba, 3600-Ap.201-Bl. 5, 22630-010 – Rio de Janeiro, RJ. E: edgard@geotec.coppe.ufrj.br

Armelin, J.L., Rua 227-A 281 Ap.301 Setor Universit, 74610-060 – Goiania, GO. T: 62 283 6122, 62 261 7651, E: armelin@furnas.com.br

Arnold, G.P., Rua Washington Luis, 675, 90010-460 – Poa, RS. T: 51 3287 2155

Arnouldi, D., Rua Lima E Silva, 19/704, 90050-101 – Porto Alegre, RS. T: 51 2224 4672

Ashcar, C., Rua Manoel da Nóbrega, 211, Cj.11, 04001-081 – São Paulo, SP. T: 11 288 1765, F: 11 288 1765, E: geocel@ig.com.br

Ashcar, C., Rua Bueno Brandao,134-Apto.71, 04509-020 – São Paulo, SP

Ashcar, R., Rua Hans Nobiling, 161/Ap.81/ Jd. Europa, 01455-060 – São Paulo, SP

Assis, A.P. de, Univers de Brasilia-Dept. Eng. Civil-Ft., 70910-900 – Brasilia, DF. T: 61 273 7313, F: 61 272 0732, E: aassis@unb.br

Assoc Bras Geologia de Engenharia ABGE, Av. Prof. Almeida Prado, 532-Ipt/Digeo, 05508-901 – São Paulo, SP

Assuiti, C.M.P., Br 116-Km 439-Bairro Arapongal, 11900-000 – Registro, SP

Augustini, P.S., Rua Tupi, 15-Vila Valparaizo, 09060-140 – Santo Andre, SP. T: 4426 7588, E: engos@zaz.com.br

Aun, G.V., Rua Conselheiro Zacarias, 86, 01429-020 – São Paulo, SP. T: 11 3887 9172, F: 11 3887 1849

Avila, I.A. de, Av. Raulino Cotta Pacheco, 70-Apto.902, 38401-090 – Uberlandia, MG. T: 34 3214 4533. T: 34 3235 9949, E: igoravila@yahoo.com

Avila, J.P. de, Caixa Postal 3002, 30112-970 – Belo Horizonte, MG

Azambuja, M.A.E. de, Rua Ferreira Viana 55 Ap.1, 90670-100 – Porto Alegre, RS

Azevedo, C.M.D. de, Rua Amavel Costa, 10-Jaragua, 31270-470 – Belo Horizonte, MG. T: 31 3334 8356, E: dazevedo.bhz@zaz.com.br

Azevedo Jr, N., Rua Arthur de Azevedo, 2103-Cj. 31, 05404-015 – São Paulo, SP. T: 11 3031 3571, E: nbazevedo@uol.com.br

Balbino, J.L., Rua Hermano Ribeiro da Silva, 611, 13063-510 – Campinas, SP. T: 19 3241 2699, E: balbino@mpc.com.br

Baptista, H.M., Rua Irma Dulce, 72-Apto.402-Brotas, 40290-195 – Salvador, BA. T: 71 356 9870, 71 240 1048, E: heliomb@e-net.com.br

Baras, V., Rua 15 de Novembro, 2051-Ap.21, 80050-000 – Curitiba, PR

Barata, F.E., Rua Prudente de Morais, 331-Ap.101, 22420-041 – Rio de Janeiro, RJ

Baratta, R., Rua Marcelino da Silva, 140 Fundesp, 06600-000 – Jandira, SP. T: 11 4789 3400, 11 3742 4808, F: 11 4789 2647, E: fundesp@fundesp.com.br

Barbosa, M.C. Rua, Guimaraes Rosa, 203/1203-B. Tijuca, 22793-090 – Rio de Janeiro, RJ

Barletta, S. de Castro, Av. dos Bancarios 49 Ap.22, 11030-301 – Santos, SP

Barreto, G.W., Rua Jose Carroci, 685 Distr. Industrial, 13602-120 – Araras, SP. T: 19 541 9444, F: 19 542 8800, E: egfundacoes@linkway.com.br

Barros, J.M. de Camargo, Av. Jorge Joao Saad 547-Bl. 1-Ap.23, 05618-001 – São Paulo, SP. T: 11 3767 4638, E: jmbarros@ipt.br

Barros, W.T. de, Rua Conselheiro Olegario, 19-Ap.503, 20271-090 – Rio de Janeiro, RJ

Barroso, E.V., Rua Dr. Celso Dias Gomes, 134, 24358-200 – Niteroi, RJ

Barth, R.E., Av. Presidente Vargas, 482-Sala 1113, 20071-000 – Rio de Janeiro, RJ

Barton, N.R., Rua Miguel de Almeida Prado, 14, 05578-040 – São Paulo, SP

Bastos, C.A.B., Rua Barao de Santa Tecla, 834-Apto.303, 96010-140 – Pelotas, RS. T: 53 233 6620, E: bastos@dmc.furg.br

Bauchspiess, A.H., Qnd 38 Casa 25, 72120-380 – Taguatinga, DF

Becker, L. de Bona, Rua Pedro Weingartner, 140-Apto. 804, 90430-140 – Porto Alegre, RS. T: 51 331 5406, E: leobecker@spovo.net

Beheregaray, L.C., Av. Duque de Caxias, 1917-Sala 11-Cx. P.32, 97500-000 – Uruguaiana, RS

Beim, J.W., Av. Lucio Costa, 4420-Bloco 8-Apto.102, 22630-011 – Rio de Janeiro, RJ. T: 21 434 1692, F: 21 434 2939, E: pdirj@pdi.com.br

Belincanta, A., Rua Francisco Glicerio, 896-Apto.61, 87030-050 – Maringa, PR

Benegas, E.Q., Estrada dos Tres Rios, 1416/204-Bl.6, 22745-003 – Rio de Janeiro, RJ. T: 21 293 5898, 21 3392 5386, E: ebenegas@pcrj.rj.gov.br

Benites, M.B., Rua Luiz de Camoes, 903, 90620-150 – Porto Alegre, RS. T: 51 223 4877, E: benites@solotec.com.br

Bento, R.C., Rua Rio Grande do Sul, 393-Apto.601, 37701-001 – Pocos de Caldas, MG. T: 35 3721 1243, 35 3722 3186, F: bento@pocos-net.com.br

Benvenuto, C., Rua Joao da Cruz Melao, 131-Jd. Leonor, 05621-020 – São Paulo, SP. T: 11 3742 0804, E: benvenuto@uol.com.br

Benvenutti, M., Av. Imp.Tereza Cristina, 946-Apto.32, 13095-160 – Campinas, SP

Berberian, D., Shis Qi 05 Cj 03 Casa 22, 71615-030 – Brasilia, DF

Bernardes, G. de Paula, Av. Ariberto Pereira Cunha, 333/Unesp-Dec, 12500-000 – Guaratingueta, SP

Bernardi, E., Rua Barao de São Gabriel, 184, 05085-060 – São Paulo, SP

Bernardo, G., Rua Fernandes Sampaio, 80, 02041-010 – São Paulo, SP

Bernini, E.C., Rua Henrique Monteiro, 47-Pinheiros, 05423-020 – São Paulo, SP

Bernucci, L.L.B., Rua Butirapoa 66 Alto da Lapa, 05059-030 – São Paulo, SP. T: 11 3818 5171, 11 3818 5485, F: 11 3818 5716, E: liedi@usp.br

Bezerra, R.L., Rua Joao Juliao Martins, 179, 58109-090 – Campina Grande, PB. T: 83 310 1305, 83 333 2995, E: leidimar@dec.ufpb.br

Biagi, R.A. de Rua Arnaldo Vitaliano, 1450-Apto.153, 14091-220 – Ribeirao Preto, SP

Bica, A.V.D., Rua Cariri, 120, 91900-560 – Porto Alegre, RS

Bilfinger, W., Rua Tuim, 465-Ap.111-Moema, 04514-101 – São Paulo, SP

Bock, E.I., Av. Atlantica, 3102-Ap.1201, 22070-000 – Rio de Janeiro, RJ. T: 21-262 8886, 21-255 1007, E: basebock@ig.com.br

Bogossian, F., Rua Bela, 1128-São Cristovao, 20930-380 – Rio de Janeiro, RJ. T: 21 580 1273, F: 21 580 1026, E: francisgeo@openlink.com.br

Bongiovanni, S., Av. Dom Antonio, 2100/D. C. Biol.-Unesp, 19800-000 – Assis, SP

Bonuccelli, T. de Jesus, Rua Domingos Faro, 408-Jd. Alvorada, 13562-320 – São Carlos, SP. T: 16 271 7424, E: stebone@terra.com.br

Borba, L.M., Rua Altinopolis, 373-Ap.182/Agua Fria, 02334-001 – São Paulo, SP

Borga, P.C., R:Marques de São Vicente,96-Ap.402-Bl.C, 22451-040 – Rio de Janeiro, RJ

Borges, A.F., Rua Paissandu,370-Ap.102-Laranjeiras, 22210-080 – Rio de Janeiro, RJ

Borma, L. de Simone, Estrada da Taquara, 666/Condom. Repouso, 25645-570 – Petropolis, RJ

Boscov, M.E.G., Rua Girassol, 584-Ap.401 A, 05433-001 – São Paulo, SP

Boszczowski, R.B., Rua Rubens Carlos Assuncao, 231-Jd. Social, 82530-080 – Curitiba, PR. E: roberta.bomfim@bol.com.br

Bottin Filho, I.A., Rua Teixeira de Carvalho, 311, 90880-300 – Porto Alegre, RS. E: sultepa@zaz.com.br

Braga, A.S.A., Rua Cel. Melo de Oliveira, 566-Pompeia, 05011-040 – São Paulo, SP

Branco, C.J.M. da Costa, Rua Para, 1110, 86010-400 – Londrina, PR. E: mecsolos@sercomtel.com.br

Branco, J. de Jesus, Av. Alexandre Ribeiro Guimaraes, 706, 38408-050 – Uberlandia, MG. T: 34-3236 6124, 34-3235 7973, E: engex@triang.com.br

Branco, J.S. de Barros, Av. Fernando Simões Barbosa, 22 Sala 611, 51020-390 – Recife, PE. T: 81 3466 8254, F: 81 3466 5343, E: sagepa@hotlink.com.br

Bressani, L.A., Rua Silva Jardim, 588, 90450-070 – Porto Alegre, RS

Brito, S.N.A. de, Al.da Serra, 420-Sala 207, 34000-000 – Nova Lima, MG

Brito Filho, H.G. de, Rua Olegarinha 47 Bloco 2 Ap.808, 20560-200 – Rio de Janeiro, RJ

Brochsztain, I., Estrada Municipal, 2900/Pr. Vermelha Sul, 11680-000 – Ubatuba, SP

Brugger, P.J., Estrada da Taquara, 666-Casa 23, 25645-570 – Petropolis, RJ

Brunelli, N.A., Rua General Marques, 654-Sala 01, 97670-000 – São Borja, RS. T: 55 431 1434, F: 55 431 2721, E: brunenge@gpsnet.com.br

Bueno, B. de Souza, Rua Romano Clapis, 188, 13670-000 – S.Rita Passa Quatro, SP. T: 12-273 9505

Burgos, P.C., Rua Aristides Novis, 2/Ufba/Dctm-Geotec., 40210-630 – Salvador, BA. T: 71 247 1860, F: 71 247 7860, E: burgos@geotec.eng.ufba.br

Busch, R.G., Rua Bartira, 1099-Apto.62-Perdizes, 05009-000 – São Paulo, SP

Busch, R.F., Rua Rodolpho Hatschbach, 1581/Amoco, 81450-070 – Curitiba, PR. E: rfbusch@amoco.com.br

Buzzetti, D.R.C., Rua Alvaro Rodrigues, 163-Sala 3, 04582-000 – São Paulo, SP

Calapodopulos, Y., Av. D.Maria Mesquita Mota E Silva 331, 05657-000 – São Paulo, SP

Calijuri, M.L., Universidade Federal de Vicosa/Dec, 36571-000 – Vicosa, MG

Camacho, M.E.A., Rua Pedro Alves, 15-Santo Cristo, 20220-280 – Rio de Janeiro, RJ. T: 21 223 4277, 21 253 3540-3646, E: tecnosolo@ax.apc.erg

Campelo, N. de Souza, Rua São Sebastiao,14-Pres.Vargas, 69025-370 – Manaus, AM. E: ncampelo@internext.com.br

Campos, G.C. de, Rua Francisco Ferreira, 40-V. Dionisia, 02670-060 – São Paulo, SP. T: 11-3767 4436, 11-3859 5406, E: gisleine@ipt.br

Campos, J. de Oliveira, Rua 3 (Tres), Nº 10-Jardim Portugal, 13504-018 – Rio Claro, SP

Campos, L.E.P. de, Rua Vicente Batalha 382/501-Costa Azul, 41760-030 – Salvador, BA. E: ledmundo@geotec.eng.ufba.br

Campos, R.A.B., Rua Alcobaca 1210 B São Francisco, 31255-210 – Belo Horizonte, MG. T: 31 3490 4355, F: 31 3490 4399, E: diretoria@engesolo.com.br

Campos, T.M.P. de, R. Marques de São Vicente, 225/Puc E.C., 22453-900 – Rio de Janeiro, RJ

Campuzano, L.F.A., Grupo Habitacional Aeropuerto, 245, - - Asuncion – Paraguay. E: lfa@uninet.com.py

Capellao, M.R.T., Rua Reseda, 10-Apto.202-Lagoa, 22471-230 – Rio de Janeiro, RJ

Caproni Jr, N., Furnas-Caixa Postal 457, 74001-970 – Goiania, GO. T: 62 259 4845. E: caproni@furnas.com.br

Caputo, A.N., Rua Jacques Felix, 223-V. N. Conceicao, 04509-000 – São Paulo, SP

Cardoso, A.P.P., Rua Riachuelo, 785-Apto.53-Centro, 90010-270 – Porto Alegre, RS

Cardoso, C., Rua Rogerio Fajardo, 235-Anchieta, 30310-450 – Belo Horizonte, MG

Cardoso, J.V., Rua Prof. Magalhaes Drumont 254 Ap.901, 30350-000 – Belo Horizonte, MG

Cardoso, R.R., Rua Vicente Machado, 81-Ap.82/Centro, 80420-010 – Curitiba, PR

Cardoso, R.R., Rua Vicente Machado 81 Ap.82, 40000-000 – Curitiba, PR

Carletto, M.F.W., Rua Simao Alvares, 555- Apto.63-B, 05417-030 – São Paulo, SP

Carlo Jr, V.D., Rua General Vitorino Monteiro, 112, 05053-060 – São Paulo, SP

Carneiro, B.J.I., Av. Jaura, 386-Centro, 69025-000 – Manaus, AM

Carneiro, C.O., Rua Estevao Bayao, 53-Ap.101, 80240-260 – Curitiba, PR

Carrion, C.W.U., Rua Afonso Celso, 694-Ap.151-Bl. B, 04119-060 – São Paulo, SP

Carvalho, A.C.M.G., Rua Fala Amendoeira, 454-Ap.1603, 22793-090 – Rio de Janeiro, RJ

Carvalho, C.S., Rua Rosa Broseguini, 17-Pq. Principes, 06030-350 – Osasco, SP

Carvalho, D. de, Caixa Postal 6011, 13081-970 – Campinas, SP. T: 19-3788 1032, E: david@agr.unicamp.br

Carvalho, L.F.M., Av. Brig. Faria Lima, 1478-Conj. 907, 01451-913 – São Paulo, SP. T: 11 3031 0621, F: 11 3031 8672, E: lfcarvalho@sci.com.br

Carvalho, M.F. de, Campus do Ininga-Ct, 64049-550 – Teresina, PI. T: 86 215 5712, 86 222 3930, E: furtado@ufpi.br

Carvalho, M. de Fatima, Rua Teixeira Barros, 800/208-Bl.A/ Brotas, 40275-410 – Salvador, BA

Carvalho, N.F. de, Rua Napoleao Laureano, 4-Icarai, 24220-350 – Niteroi, RJ

Carvalho, R.L.G. de, Rua Bororos, 804-Conj. 07/Vila Izabel, 80320-260 – Curitiba, PR

Carvalho, S.R.L., Rua Pereira da Silva, 93- Ap.604, 22221-140 – Rio de Janeiro, RJ

Casagrande, M.D.T., Rua da Republica, 193-Apto.302-C. Baixa, 90050-321 – Porto Alegre, RS. T: 51 228 6983, E: michec@zaz.com.br

Castello, R.R., Rua Romulo Leao Castelo, 106, 29052-740 – Vitoria, ES. T: 27 3225 0622, F: 27 3315 4514, E: renocstl@tropical.com.br

Castier, G., Av. Atlantica 1602-Ap.601-Copacabana, 22021-000 – Rio de Janeiro, RJ

Castro, C.F.C. de, Rua Silvana, 198, 04513-000 – São Paulo, SP. T: 11-3848 9941, E: cfbmc@uol.com.br

Castro, D.C. de, Rua Prof. Julio Lohman, 110, 22611-170 – Rio de Janeiro, RJ

Castro, G.R. de, Rua Serra Das Vertentes, 67-Jd. Amalia, 05890-250 – São Paulo, SP. T: 11 5821 0694, 11 5821 0693, F: 11 5821 0694, E: grcastro@usp.br

Cavalcanti, M. do Carmo Reis, Rua Sousa Cruz, 51-Ap. 403-Tijuca, 20510-280 – Rio de Janeiro, RJ

Cavalcanti J., Demostenes de A., Rua Frei Paulo, 455-Ap. 1002, 49050-330 – Aracaju, SE

Cavicchia, L.R., Rua Conego Bento, 276, 13487-079 – Limeira, SP

Cechella J., A., Rua Tuiuti, 940, 97015-660 – Santa Maria, RS. T: 55 223 1303, F: 55 225 1104, E: geo2000@zaz.com.br

Celestino, T.B., Rua Bela Cintra, 986-14º Andar, 01415-000 – São Paulo, SP. E: tbcelest@usp.br

Cella, P.R.C., Av. Prof. Almeida Prado, 532-Ipt/Digeo, 05508-901 – São Paulo, SP

Cepollina, M., Rua Jose Yazigi 383, 05658-020 – São Paulo, SP. T: 11 543 1044, F: 11 532 1726, E: cepollina@sti.com.br

Ceratti, J.A.P., Rua José Gertum, 135-Ap.301, 91330-450 – Porto Alegre, RS. E: lavap@cpgec.ufrgs.br

Cerqueira, C. de Azevedo Gusmao, Rua Itacuruca, 38-Ap. 502 – Tijuca, 20510-150 – Rio de Janeiro, RJ

Cesar, E.P.V., Rua Alexandre Gran Bell, 370/Jd. Bandeir, 86063-250 – Londrina, PR. T: 43 328 4179, F: 43 348 7533, E: politecnica@onda.com.br

Cesar, R.R., Rua Alexandre Grahan Bell, 370 B, 86063-250 – Londrina, PR. T: 43 328 4179, F: 43 348 7533, E: politecnica@onda.com.br

Chamecki, P.R., Rua Desemb. Vieira Cavalcante, 585/302, 80510-090 – Curitiba, PR. T: 41-267 4839, 41-336 5117, E: chamecki@lactec.org.br

Chammas, R., Av. Prudente de Morais, 135-Cj. 201-202, 30380-000 – Belo Horizonte, MG

Chmielewski, J.R., Rua Padre Primo Maria Vieira, 337, 11500-200 – Cubatao, SP. T: 13-3361 4151

Chwist, A.L., Rua Fernandes de Barros, 1220, 80040-200 – Curitiba, PR

Cia. do Metropolitano de São Paulo/Metro, Rua Augusta, 1626/Coord. Geot.-Pci/Cdg, 01304-902 – São Paulo, SP

Cia. Energetica de Minas Gerais, Av. Barbacena 1200 Caixa Postal 992, 30190-131 – Belo Horizonte, MG. T: 31 3299 3001, 31 3225 2393, E: mrocha@cemig.com.br

Cia. Hidroeletrica S. Francisco-Chesf, Rua Dr. Elphego Jorge de Sousa 333, 50761-260 – Recife, PE. T: 81 3229 2480, F: 81 3228 3011, E: aurelio@chesf.gov.br

Cintra, J.C.A., Escola de Engenharia de S.Carlos-Usp, 13560-250 – São Carlos, SP

Clemente, L.G., Av. Cesar Lacerda de Vergueiro, 11, 11030-220 – Santos, SP

Coelho, L.C., Rua Dr. Lauro Wolff Valente, 126, 81070-010 – Curitiba, PR. T: 41 345 1424, 41 345 1273, F: 41 345 2550, E: lccoelho@netpar.com.br
Coelho Jr, M.A.S., Av. Portugal, 474-Apto.51-Brooklin, 04559-001 – São Paulo, SP
Coelho Netto, A.L., Rua Itacuruça, 116-Casa 8- Tijuca, 20510-150 – Rio de Janeiro, RJ. T: 21 9689 9254, 21 512 8050, F: 21 598 9004, E: alcoenet@igeo.com.ufrj
Collet, H.B., Rua General Osorio, 206-Ap.503, 28625-630 – Nova Friburgo, RJ. E: hbcollet@uol.com.br
Colmanetti, J.P., Sqn 610-Bl. K-Apto.302-Colina-Asa Norte, 70910-900 – Brasilia, DF
Conciani, W., Av. Cel. Escolastico, 515-Ap.803, 78010-200 – Cuiaba, MT. E: conciani@dec.ufpb.br
Consoli, N.C., Rua Marechal Floriano, 4-Apto.302, 95520-000 – Osorio, RS. T: 51 316 3552, E: consoli@vortex.ufrgs.br
Construcoes e Com. Camargo Correa S/A, Rua Funchal 160-Vila Olimpia, 04551-060 – São Paulo, SP
Contemat Engenharia e Geotecnia S/A, Rua Fonseca Teles, 40-São Cristovao, 20940-200 – Rio de Janeiro, RJ. T: 21 3890 4000, E: mario.sergio@concremat.com.br
Contiero, L.C. da Cunha, Rua Vereador Terezio Meireles, 276, 91240-390 – Porto Alegre, RS
Cordeiro, D.D.R., Dr. Cyro Lopes Pereira, 1010-Apto.501, 29060-020 – Vitoria, ES. E: ddcordeiro@ig.com.br
Corradi Neto, P., Rua Parana, 2171-Quadra 88, 19293-000 – Primavera, SP
Correa, C.N., Rua Plinio Colas, 388-Apto.71, 02435-030 – São Paulo, SP. T: 11 3873 2500, 11 3496 7573, E: cncorrea@zaclisfalconi.com.br
Correa, L.S., Av. Felicissimo Cardoso, Bl.2 835-Apto.608, 22631-360 – Rio de Janeiro, RJ
Correa, R.S., Rua Morgado Mateus 76 Ap.84-V.Mariana, 04015-050 – São Paulo, SP
Cortes, H.V. de Menezes, Rua Des. Alfredo Russel, 173-Ap. 507, 22431-030 – Rio de Janeiro, RJ
Cortopassi, R.S., Sqn 108-Bloco H-Ap.105, 70744-080 – Brasilia, DF
Costa, A.A. da, Rua Dr. Pedro de Souza Ponde, 25-Apto.201, 40155-270 – Salvador, BA. T: 71-345 0035, 71-247 4354, F: 71-248 3642, E: approj@svn.com.br
Costa, A. de Albuquerque, Av. Presidente Wilson 76 Ap.174, 11065-200 – Santos, SP. T: 13 3237 1054
Costa, A.F. da Rua, 20 Sul-Lote9-Bl.B-Ap.503-Aguas Clara, 72030-100 – Brasilia, DF
Costa, C.M.C. da, Rua Cristina, 256-Ap.101-Carmo, 30330-130 – Belo Horizonte, MG
Costa, E.B.M., Caixa Postal 121, 69908-970 – Rio Branco, AC
Costa, E.J.P. da, Av. Angelica 1814 Cjs 902/903, 01228-000 – São Paulo, SP
Costa, F.L.C. Mac Dowell da, Rua Eng. Cesar Grillo, 190/B. da Tijuca, 22640-150 – Rio de Janeiro, RJ
Costa, J.C. da, Rua Marechal Floriano, 106-Apto.1302, 40110-010 – Salvador, BA. T: 71-328 0097, 71-247 7292, E: tecssa@svn.com.br
Costa, L.M. da, Rua Dr. Jose Maria, 517-Ap.1102-Rosarinho, 52041-000 – Recife, PE
Costa, O., Rua Atlantica, 96 – Jardim America, 01440-000 – São Paulo, SP. T: 11 3825 5800, 11 3280 5547
Costa, R.V.F. da, Rua Fiandeiras, 545-Apto.41-Bloco A, 04545-003 – São Paulo, SP
Costa, R.V. da, Rua Joaquim Nabuco, 201-Terreo, 22080-030 – Rio de Janeiro, RJ. T: 21 523 4466, E: ricardovital@bol.com.br
Costa, T.C.D. da, Rua Snapp, 03-Prox.Apass. S.Raimundo, 67010-420 – Ananindeua, PA
Costa, W.L. da, Rua 1130 Qd. 235 Lote 14-Setor Marista, 74180-090 – Goiania, GO
Costa Filho, L. de Moura, Rua Adolfo Lutz, 103-Apto.501-Gavea, 22451-120 – Rio de Janeiro, RJ. T: 21 533 1260, F: 21 240 7305, E: lmcosta@ism.com.br
Costalonga, M.A. dos Reis, Rua Tabapua, 240-Apt. 74-Itaim Bibi, 04553-000 – São Paulo, SP
Couso Jr, E., Rua Baronesa de Itu, 858, 01231-000 – São Paulo, SP

Coutinho, B.R., Rua Capitao Cesar de Andrade, 168Apt.707, 22431-010 – Rio de Janeiro, RJ
Coutinho, R.Q., Rua Muniz Tavares, 81-Ap.501, 52050-170 – Recife, PE. E: roberto@proacad.ufpe.br
Couto Jr, P.P.R., Rua Mario Portela, 161-Bloco 3-Apto. 1703, 22241-000 – Rio de Janeiro, RJ. T: 21-225 0126, E: pparga@zaz.com.br
Cruz, P.T. da, R Hadock Lobo 1663 Ap.182, 01414-003 – São Paulo, SP
Cunha, E.P.V. da, Rua Cardoso de Almeida, 1593, 05013-001 – São Paulo, SP. T: 11 3673 2400, E: pabst@uol.com.br
Cunha, J.E.V., Rua Acu, 501-Ap.901-Tirol, 59020-110 – Natal, RN
Cunha, R.P. da, Unb-Colina-Bl. G-Ap.105/Asa Norte, 70919-970 – Brasilia, DF. T: 61 273 7313, E: rpcunha@unb.br
Dalmoro, R.V., Rua Agostino Fochesato, 75, 95099-010 – Caxias do Sul, RS. T: 54 226 2268
Dantas, B.T., Rua da Bandeira, 489-Apto.61-Cabral, 80035-270 – Curitiba, PR. T: 41-263 3424, 41-339 2064, E: bruno.dantas@utp.br
Danziger, B.R.R., Vice Gov. Rubens Berardo, 65/304/Bl.2, 22451-070 – Rio de Janeiro, RJ. E: ragoni@civil.uff.br
Danziger, F.A.B.R., Vice Gov. Rubens Berardo, 65/304/Bl.2, 22451-070 – Rio de Janeiro, RJ. E: danziger@pec.coppe.ufrj.br
Dau, F.N.F., R Paulo Derly Streel 212 B Esp Santo, 91770-640 – Porto Alegre, RS
David, D. de, Rua Olavo Bilac, 110/405, 90040-310 – Porto Alegre, RS. T: 51 3267 1619
Debas, L.F., Rua Antonio Freitas Barbosa, 56, 81110-110 – Curitiba, PR
Decourt, L., Av. Brigadeiro Faria Lima, 1620-Andar 2, 01451-001 – São Paulo, SP. E: decourt@decourt.com.br
Degow, J., Av. do Contorno 3513 B Santa Efigenia, 30110-090 – Belo Horizonte, MG. T: 31 3481 1611, F: 31 3481 9992, E: seeblang.bh@terra.com.br
Dell'Avanzi, E., Rua Cristiano Strobel, 2479-Sobrado 24, 81750-000 – Curitiba, PR. E: avanzi@cesec.ufpr.br
Dellis, M.T., Br 324-Km 0-Retiro-Tecnosolos S/A, 40330-730 – Salvador, BA. T: 71-244 4844, F: 71 233 7737
Desideri, F.A.P., Rua Dias da Rocha, 76-Apto.301-Copacabana, 22051-020 – Rio de Janeiro, RJ
Dias, C.R.R., Rua Joao Manoel, 38-Ap.403, 96211-060 – Rio Grande, RS
Dias, M.R., Rua Erico Verissimo, 240/408, 90160-180 – Porto Alegre, RS. T: 51 227 3182
Dias, P.H.V., Rua 5 de Julho, 26-Ap.302/Copacabana, 22051-030 – Rio de Janeiro, RJ. T: 21 242 9696, 21 256 3302, F: 21 242 9696, E: phdias@connection.com.br
Dib, P.S., Sqs-303 Bloco F-Ap.106, 70336-060 – Brasilia, DF
Doro-Altan, V., Rua Banibas, 142-Alto de Pinheiros, 05460-010 – São Paulo, SP
Dragone, M.R.R.B., Lgo.Senador Raul Cardoso, 250-Apto. 61B, 04021-070 – São Paulo, SP. T: 11 3849 2166, 11 5539 1230, E: rdragone@engesolos.com.br
Dreyer, W.C., Rua Chaves Faria, 282, 25955-090 – Teresopolis, RJ
Drosemeyer, A., Rua Adalberto Aranha, 47-Apto.603, 20540-140 – Rio de Janeiro, RJ
Dupont do Brasil S/A, Alameda Itapecuru, 506-Alphaville, 06454-080 – Barueri, SP
Dutra, J.B. da Silva, Rua Riachuelo, 783, 96470-000 – Pinheiro Machado, RS
Dyminski, A.S., Rua Leonidas do Amaral Ferreira, 71, 80050-410 – Curitiba, PR. E: andrea@cesec.ufpr.br
Ehrlich, M., Rua Jose Higino, 214-Ap.702-Tijuca, 20520-200 – Rio de Janeiro, RJ. T: 21 562 7193, F: 21 290 1730, E: me@geotec.coppe.ufrj.br
Eigenheer, L.P.Q.T., Av. Pres Juscelino Kubitschek 1830 11A, 04543-000 – São Paulo Torre Iv, SP
Elias Filho, A., Rua Butanta, 461-6 Andar-Conj.61, 05424-140 – São Paulo, SP
Esposito, T. de Jesus, Rua Henrique Passini, 738 Ap.401-Serra, 30220-380 – Belo Horizonte, MG. T: 31 3238 1794, F: 31 3238 1742, E: esposito@etg.ufmg.br

Esquivel, E.R., Rua Thomaz Nogueira Gaia, 1796, 14020-290 – Ribeirao Preto, SP

Estacas Franki Ltda., Av. Rio Branco, 311-Andar 10, 20040-009 – Rio de Janeiro, RJ. T: 21 297 7130, E: frankibr@franki.com.br

Etesco – Construçoes e Comercio Ltd., Rodovia Raposo Tavares-Km 14.5, 05577-100 – São Paulo, SP

Faccin J., E., Rua Henrique Monteiro, 47-Pinheiros, 05423-020 – São Paulo, SP. T: 11 3814 2055, 11 3022 7343, E: berfac@terra.com.br

Fahel, A.R.S., Lote Santos Dumont, Rua H, Casa 1027, 88000-000 – Florianopolis, SC

Falcao, J.M., Rua Barao de Lucena, 16-Botafogo, 22260-020 – Rio de Janeiro, RJ. T: 21 266 3382, 21 295 3120, F: 21 537 2246, E: etg@osite.com.br

Falconi, F.F., Rua Embaixador Leao Veloso, 102, 05003-030 – São Paulo, SP. T: 11 9932 1885, 11 3873 2500, E: ffalconi@dialdata.com.br

Fanaya, A.P.A., 2715-Apto.195, 80710-000 – Curitiba, PR

Fanton, J. Vicente, Rua Henrique Monteiro 47 Pinheiros, 05423-020 – São Paulo, SP

Faria, L.A.G. de, Rua Humberto de Campos, 745, 90060-660 – Porto Alegre, RS

Farias, K.T.C. de, Rua Pedrina Mª Silva Valente, 44/Ap.21, 05782-450 – São Paulo, SP. T: 3744 0780

Farias, M.M. de, Sqn 115, Bloco J, Apto.305, 70772-100 – Brasilia, DF. T: 61 273 7313, F: 61 272 0732, E: muniz@unb.br

Farias, W.B. de, Rua Joao Fernandes Vieira 600 Ap.605B, 50050-200 – Recife, PE

Favaro Neto, P., Rua Manoel dos Santos Barreto, 75-Ap.44, 80530-250 – Curitiba, PR

Feijo, R.L., Av. Glaucio Gil, 505-Apto.201, 22795-170 – Rio de Janeiro, RJ. T: 21 589 5757, 21 487 3140, E: rlfeijo@uerj.br

Fernandes, G., Av. Yolanda Guimarães, 85-Passagem, 35421-000 – Mariana, MG. T: 31 3557 5059

Ferrari, I., Rua Maranhao 598 7A Cj73, 01240-000 – São Paulo, SP

Ferraz, N., Rua Sen Cesar Lacerda Vergueiro 494, 05435-010 – São Paulo, SP. T: 11 3813 8524

Ferreira, A.C.G., Rua Armando Coelho de Freitas,16-Ap.201, 22620-090 – Rio de Janeiro, RJ

Ferreira, A.A., Rua Garson Tinoco, 76-Apto. 32-Santana, 02408-020 – São Paulo, SP. E: argimiroferreira@hotmail.com.br

Ferreira, C. de Castro, Av. Arthur da Silva Bernardes, 1785-Portao, 80320-300 – Curitiba, PR

Ferreira, C.S.P., Rua Gouveia de Barros, 227-Sto. Amaro, 50180-030 – Recife, PE

Ferreira, C.V., Rua Capitao Joao Antonio, 12-Ap.06, 17013-470 – Bauru, SP

Ferreira, E.S., Sqs 109-Bloco "E" Apto.116-Asa Sul, 70372-050 – Brasilia, DF. T: 61 443 6589, E: engsoltec@hotmail.com.br

Ferreira, R.R., Rua Visconde de Piraja, 12-Ap.402, 22410-000 – Rio de Janeiro, RJ

Ferreira, R.P., Trav. Corneta Lopes, 2-Ap.802-Campo Grande, 40080-350 – Salvador, BA

Feuerharmel, M.R., Rua São Benedito, 925/104, 91410-410 – Porto Alegre, RS. T: 51 3338 4891

Figliola, M., Rua Dr. Jose Higino 71 Alto da Mooca, 03189-040 – São Paulo, SP

Figueiredo, C.M.M. de, Rua Corinto, 431-Apto.64-B, 05586-060 – São Paulo, SP

Flecha, H., Sia Trecho 04-Lote 2000-Sala 207, 71200- 040 – Brasilia, DF

Flechtman, J.S., Rua Pedroso Alvarenga, 1221-9°And.I.Bibi, 04531-012 – São Paulo, SP. T: 11 3168 8080, E: jaime@koema.com.br

Fleming, P.M., Rua Manoel Antonio Pinto 1200 Ap.41B, 05663-020 – São Paulo, SP. E: pmf@urano.cdtn.br

Fleury, S.V., Rua T-47,355-Apto.1702-St.Oeste, 74140-120 – Goiania, GO. T: 62 283 6122, 62 242 9218, E: sergiovf@furnas.com.br

Foa, A., Estrada do Coco, Km 10, 5-Vila Diabrandis, 42800-992 – Camacari, BA. E: jse@e-net.com.br

Foa, R.J., Av. Prof. Joaquim Barreto, 841/887, 06700-000 – Cotia, SP

Foa, S.B., Rua Leonor Calmon, 74-Apto.1402, 41620-280 – Salvador, BA

Folle, D., Rua Lava Pes, 998-Apto.02-Centro, 99010-170 – Passo Fundo, RS

Fonseca, A.P., Rua dos Limadores, 465-Bangu, 21830-000 – Rio de Janeiro, RJ

Fonseca, A.L., Rua 227, Nr.280-Apto.904-St. Universitario, 74605-080 – Goiania, GO. T: 62 202 0600, E: sete@cultura.com.br

Fonseca, E.C., Rua Aristides Novis, 2-Federaçao/Ufba, 40210-630 – Salvador, BA. E: evan@geotec.eng.ufba.br

Fontolan Filho, F., Rua Hipolito da Costa 411 Boqueirao, 81650-280 – Curitiba, PR

Fontoura, S.A.B. da, Rua Marques de São Vicente, 225-Puc-Rio, 22453-900 – Rio de Janeiro, RJ. T: 21 540 7749, 21-274 2281, F: 21 540 7424, E: fonoura@civ.puc-rio.br

Fornazier, A.Y., Av. Boa Viagem, 4398-Apto.801, 51021-000 – Recife, PE

Foti, A., Rua Belem, 181-Apto.171-V.Assuncao, 09030-120 – Santo Andre, SP

Fragelli, C.L.R., Rua Melanio Garcia Barbosa, 280-Apto. 601, 79150-000 – Maracaju, MS

Fraiha Neto, S.H., Av. Gov. Jose Malcher, 1649-Ap.602, 66060-230 – Belem, PA

Franca, F. de Oliveira, Praca Otavio Mangabeira 94 Sala 508, 45600-000 – Itabuna, BA

Franca, P.C.T., Rua Castro Delgado, 211, 05465-010 – São Paulo, SP

Franciss, F.O., Rua Sacopa, 274-Ap.201-Lagoa, 22471-180 – Rio de Janeiro, RJ. T: 21 266 4177, E: progeo@attglobal.net

Franco, J.A. de Mello, Rua Lopes Quintas, 255-Ap.202 B, 22460-010 – Rio de Janeiro, RJ

Franco Filho, J.M.M., Rua Dr. Nelson de Souza Pinto, 723, 82200-060 – Curitiba, PR

Francoso, N.C.T., Rua Martim Francisco, 204-Ap.162, 01226-000 – São Paulo, SP

Frederico, V.J., Rua Amauri Teixeira Leite, 20 Cpc 270, 11620-992 – São Sebastião, SP. T: 12 3865 3146

Freire, E.P., Campus Universitario-Rod.Mg 179-Km, 0, 37130-000 – Alfenas, MG

Freire, J.A.S., Rua Fortunato Mori, 280-B. Vianelo, 13207-150 – Jundiai, SP. T: 11 4521 3500, E: civilsolo@terra.com.br

Freitas, A.A. de, Rua Rep.do Libano, 66-Apto.500, 60160-140 – Fortaleza, CE

Freitas Jr, M.S., Rua Japao 85 Ap.72, 04530-070 – São Paulo, SP. T: 11 3167 2426, E: mfreitas65@hotmail.com

Frigerio, G.P., Rua Flauzino Marques, 67-Apto.23 A, 13562-310 – São Carlos, SP

Frota, C.A. da, Rua Kl Nº 10 Morada do Sol – Aleixo, 69080-780 – Manaus, AM

Frydman, M., Av. Afonso Tauany, 98-Apto.202-B-Tijuca, 22621-310 – Rio de Janeiro, RJ

Fujii, J., Av. Leonardo da Vinci, 211-Ap.132, 04313-000 – São Paulo, SP

Fumio, B.L.C., Av. Nicolau Zarvos, 1925, 16400-000 – Lins, SP. T: 14 520 3200, 14 522 3763, E: bernardo@fpte.br

Furtado, G.M. de Carvalho, Rua Casemiro de Abreu, 307-Barequecaba, 11600-000 – São Sebastiao, SP

Futai, M.M., Rua General Urquiza, 117-Apto.503-Leblon, 22351-040 – Rio de Janeiro, RJ. E: massão@geotec.coppe.ufrj.br

Galvao, T.C. de Brito, Rua Cardeal Stepinac, 761, 31170-220 – Belo Horizonte, MG. E: cassia@etg.ufmg.br

Gama, E.M. da, Rua Espírito Santo, 35-7 Andar, 30160-030 – Belo Horizonte, MG

Garcia, A.E.G., Facultad de Ing.-Iet-Jh Reissig 565, 11300-300 – Montevideo – Uruguay

Gazda, R.J., Rua Jose de Alencar, 399, 83321-230 – Pinhais, PR. T: 41 667 3455, 41 352 7005, E: robertojosegazda@uol.com.br

Gehling, W.Y.Y., Av. Lageado, 1370-Apto.201, 90480-110 – Porto Alegre, RS

Genevois, B.B.P., Praca do Carmo 5, 53120-000 – Olinda, PE. E: bbpg@npd.ufpe.br

Geo Geotecnia, Engenharia e Obras Ltda., Rua Bugaerti, 326-Vila Vermelha, 04298-020 – São Paulo, SP
Gerscovich, D.M.S., Av. Monsenhor Ascaneo, 176/101-B. Tijuca, 22621-060 – Rio de Janeiro, RJ. T: 21 493 6213, F: 21 495 8476, E: deniseg@uerj.br
Giacheti, H.L., Rua Almeida Brandao, 16-27, 17044-060 – Bauru, SP
Gilberto, T.J., Rua Acacio Lima, 501, 14404-265 – Franca, SP. T: 16 3722 6342, 16 3701 8200, F: 16 3701 8200, E: thales@solocon.com.br
Gimenes, E. de Avila, Rua Costa Rica, 295-Apto.301A, 30320-030 – B. Horizonte, MG. E: evandro_gimenes@yahoo.com
Gitelman, M., Rua Conde de Iraja 118-V.Mariana, 04119-906 – São Paulo, SP. T: 11 5085 5295, E: ff_são@uol.com.br
Gobara, W., Av. Prof. Almeida Prado, 532-Ipt/Dec, 05508-901 – São Paulo, SP. T: 11 3767 4811, F: 11 3767 4090, E: wgobara@ipt.br
Godoy, N.S. de, Rua Jeronimo da Veiga, 228-Itaim Bibi, 04536-001 – São Paulo, SP. T: 11 3079 3566, E: maiagodoy@sti.com.br
Goldbach, S., Rua Aperana, 10-Apto.101-Leblon, 22450-190 – Rio de Janeiro, RJ. E: sergio@uol.com.br
Golombek, M., Rua Baronesa de Itu 858, 01231-000 – São Paulo, SP
Golombek, S., Rua Baronesa de Itu, 858, 01231-000 – São Paulo, SP. T: 11 3662 1005, 11 3672 5453, E: conultrix@zaz.com.br
Gomes, C.C., Rua Silva Jatahy, 400-Apto.1101-A, 60165-070 – Fortaleza, CE. E: carisia@ufc.br
Gomes, M. do Carmo Vorcaro, Rua Bambina, 164-Ap.201, 22251-050 – Rio de Janeiro, RJ
Gomes, P.F.A.P., Rua J.2, N. 21-Nova Republica/Dist.Ind, 69075-640 – Manaus, AM
Gomes, R.C., Rua do Bispo, 316/605-Tijuca, 20521-062 – Rio de Janeiro, RJ
Gomes, R.C., Rua Benedito Valadares, 245-Rosario, 35400-000 – Ouro Preto, MG. T: 31 3559 1546, F: 31 3559 1548, E: romero@em.ufop.br
Goncalves, H.H.S., Alameda Equador, 197-Residencial Ii, 06470-060 – Barueri, SP. T: 11 3818 5207, 11 4191 4904, F: 11 5531 9907, E: helesilv@usp.br
Gonçalves, F.L., Rua Luis Goes, 1902 Apto.14, 04043-200 – Mirandópolis, SP. T: 11 289 3605, 11 5581 4306, E: fgleyser@usp.br
Gonçalves, V.F., Rua Major Barbosa, 182-Apto.301, 30240-870 – Belo Horizonte, MG. T: 31 3482 7313
Goncalves Jr, A., Av. Jonia, 326-Ap.21-Jardim Brasil, 04634-010 – São Paulo, SP. E: goncalves_jr@uol.com.br
Gonçalves Jr, E.M., Av. Jacutinga, 242-Apto.111, 04515-030 – São Paulo, SP
Gotlieb, I.D., Rua Natingui, 1053-Vila Madalena, 05443-002 – São Paulo, SP
Gotlieb, M., Rua Joao Batista Cardoso, 258, 05449-040 – São Paulo, SP. T: 11 3-31 5539, 11 3021 1820, F: 11 3819 8577, E: mgasolo@uol.com.br
Gouvea, M.A. de Souza, Rua Marechal Deodoro, 1105Ap. 1003, 36015-460 – Juiz de Fora, MG
Grandis, I., Av. Padre Pereira de Andrade, 545-Bl.D, 05480-900 – São Paulo, SP
Gripp, M. de Fatima Andrade, Caixa Postal 1406, 30123-970 – Belo Horizonte, MG
Guarana, C.A., Av. Gal.Mac Arthur, 1146-Piso Superior, 05338-001 – São Paulo, SP. T: 11 3719 1881, E: rual@picture.com.br
Guatteri, G., Rua Banibas, 142-Alto de Pinheiros, 05460-010 – São Paulo, SP
Guazzelli, M.C., Av. dos Carinas, 666-Moema, 04086-011 – São Paulo, SP
Guerra, L.C., Av. Rondon Pacheco, 1215-B. Altamira, 38400-242 – Uberlandia, MG. T: 34 3214 4533, F: 34 3214 4702, E: sondotec@tring.com.br
Guidicini, G. Rua Joao Lira, 84-Ap.901-Leblon, 22430-210 – Rio de Janeiro, RJ
Guimaraes, J.P., Rua Mauricio F. Klabin, 357-Ap.112, 04120-020 – São Paulo, SP

Guimaraes, M.L.V., Rua Dr. Zuquim, 1321-Santana, 02035-012 – São Paulo, SP. T: 11 6979 5075, F: 11 6973 9873, E: magproj@attglobal.net
Guimaraes, M.C.A.B., Rua David Gebara, 60-Apto.151, 05642-040 – São Paulo, SP. T: 11 3815 0355, 11 3773 9409, E: mcabg@uol.com.br
Guimaraes, R.C., Furnas Centrais Eletricas-Br 153-Km 1290, 74001-970 – Goiania, DF. T: 62 283 6122, E: renatocg@furnas.com.br
Guimaraes, R.B., Rua Marques de São Vicente, 194/Ap.404, 22451-040 – Rio de Janeiro, RJ
Guimarães, A.M. de Vasconcellos, Rua Oscar Freire, 1799-Apto.804, 05409-011 – São Paulo, SP. T: 11 3067 0788, E: andre.vasconcellos@poli.usp.br
Guimaraes Filho, J.D., Rua Martiniano Lemos Leite, 680, 06700-000 – Cotia, SP
Guimaraes Neto, J.D., Rua Aibi, 89-Ap.43, 05054-010 – São Paulo, SP
Gusmao, A.D., Av. Agamenon Magalhaes, 2901-Espinheiro, 52021-170 – Recife, PE. E: gusmao.alexandre@torricelli.com.br
Gusmao Filho, J. de Azevedo, Av. Agamenon Magalhaes, 2901 Espinheiro, 52021-170 – Recife, PE. T: 81-3241 5964, F: 81 3241 7576, E: gusmaojaime@torricelli.com.br
Gusso, L.C., Rua Jose de Alcantara, 444-V.do Encontro, 04324-000 – São Paulo, SP. T: 11 5588 3150, E: tecno@wm.com.br
Hachich, W.C., Rua Antonio de Gouveia Giudice, 1233, 05460-001 – São Paulo, SP. T: 11 3023 4024, 11 3022 3730, F: 11 3023 4024, E: whachich@usp.br
Haik, R., Av. John F Kennedy, 203-Jd. Das Nacoes, 12030-200 – Taubate, SP
Hebmuller, A.L., Av. Alegrete, 488-Apto.302, 90460-100 – Porto Alegre, RS
Heine, U.H. de Rocha Nery, Rua Pres Carlos de Campos 115/1004, 22231-080 – Rio de Janeiro, RJ
Heineck, K.S., Rua Garibaldi, 902-Apto.206, 90035-052 – Porto Alegre, RS
Hernando Filho, C.F., Rua São Carlos do Pinhal, 318-Apto. 91, 01333-900 – São Paulo, SP
Hessing, J.M., Av. Epitacio Pessoa, 683-Ap.52, 11030-603 – Santos, SP
Holanda, J.P.S. de, Rua Carlos Pereira Falcao, 811-Ap.101, 51021-350 – Recife, PE. T: 81 3326 5415, 81 3465-3464
Hori, K., Rua Afonso de Freitas, 523-Ap.71, 04006-052 – São Paulo, SP
Hosken, J.E.M., Av. Jose Cabalero, 245-And. 5/Sala 52, 09040-210 – Santo Andre, SP. T: 11 4992 3022, E: hosken@mhosken.com.br
Humes, C., Alameda Casa Branca 1177 Ap.2 B, 01408-001 – São Paulo, SP. T: 11 3841 4902, 11 3062 7072, E: chumes@cnec.com.br
Ignatius, S.G., Rua Lisboa, 1194-Ap.61-Pinheiros, 05413-001 – São Paulo, SP. T: 11 3083 0371, 11 3767 4811, E: ignatius@ipt.br
Incorpe Engenharia e Comercio S/A, Rua Libano, 66-3°Andar-Itapoa, 31710-030 – Belo Horizonte, MG
Inforcato, Joao Batista Rua Sebastiao Santos de Oliveira, 155, 13420-687 – Piracicaba, SP
Instituto de Pesquisas Tecnologicas, IPT-Cid. Univers./Caixa Postal 0141, 01064-970 – São Paulo, SP
Iorio, W.R., Rua Edward Joseph, 122-Cj.74-Morumbi, 05709-020 – São Paulo, SP
Iwamoto, R.K., Rua Rio Jatai, Quadra 35-Casa 10, 69053-020 – Manaus, AM. E: shimizu@zipmail.com.br
Iyomasa, W.S., Av. Prof. Almeida Prado 532-Digeo-Cx.29, 05508-901-SO Paulo-Sp, SP. T: 11 3767 4372, 11 3726 7863, F: 11 3767 4767, E: wsi@ipt.br
Jacob, J.H., Rua Bruno Filgueira, 2045-Ap.604, 80730-380 – Curitiba, PR. E: datageo@mps.com.br
Jobim Filho, H., Rua Francisco Manuel, 328, 97015-260 – Santa Maria, RS
Josetti Neto, F.A., Rua 05 Casa 05-Setor Oeste, 78000-000 – Cuiaba, MT
Juca, J.F.T., R. Jose Nunes da Cunha, 678/ 1101 Piedade, 54410-280 – Jaboatao, PE. E: jucah@npd.ufpe.br

Junqueira, L.A., Rua J29-Q.59-Lotes 27/28 Setor Jao, 74673-450 – Goiania, GO. T: 62 204 2971, E: solotec@h.com.br

Kaimoto, L.S.A., Rua Dr. Carlos A. Campos, 295, 04750-060 – São Paulo, SP

Kaminski, S., Av. Das Industrias 400 B Anchieta, 90200-290 – Porto Alegre, RS. T: 51 371 1022, E: serki@serki.com.br

Kange, M.C., Alameda Lirios 84 Residencial 5, 06539-230 – Santana de Parnaiba, SP. T: 11 6163 5050, 11 4153 1221, E: mckange@yahoo.com

Kanji, M.A., Alameda Santos, 2527-Ap.55, 01419-001 – São Paulo, SP. E: makanji@usp.br

Karez, M.B., Rua Joaquim Lustosa, 113-Ap.701, 30310 410 – Belo Horizonte, MG

Katsutani, L.T., Rua Argentina, 180, 19160-000 – Alvares Machado, SP

Kawamura, N., P.O. Box 362040-Geoconsult-G.Engineers, 09362-040 – San Juan/Puerto Rico. E: kawamura@geoserv.isgs.uiuc.edu

Kenan, A.A.N., Rua Afonso Bras, 408-Conj. 203, 04511-001 – São Paulo, SP. T: 11 3842 1565, E: contag@uol.com.br

Kochen, R., Rua Amalia de Noronha, 161, 05410-020 – São Paulo, SP. T: 11 3083 5563, 11 3083 5561, F: 11 3082 6517, E: geocompany-rk@uol.com.br

Komesu, I., Rua Conde de Assumar, 282, 02255-020 – São Paulo, SP

Kormann, A,C.M., Av. Joao Gualberto, 1034/Ap.601/A.Gloria, 80030-000 – Curitiba, PR. E: aless@cesec.ufpr.br

Koshima, A., Rua Salvador Nascimento 222, 05363-110 – São Paulo, SP. T: 11 3813 8700, F: 11 3031 7967, E: novatecna@novatecna.com.br

Kreische, C.E., Rua Brasilia, 709, 85504-400 – Pato Branco, PR. T: 46 224 3965, F: 46 225 3230, E: carlos@whiteduck.com.br

Kutner, M., Rua Armando Pinto, 115, 05442-060 – São Paulo, SP. T: 11 3873 4420, T: 11 9172 5606, F: 11 3865 6999, E: sarub@uol.com.br

Labaki Neto, D.C., Rua Veriano Pereira, 7-Andar 11/Saude, 04144-030 – São Paulo, SP. T: 11 5585 9984, E: dlabaki@attglobal.net

Lacerda, W.A., Av. Rui Barbosa, 170-Apto.1405, 22250-020 – Rio de Janeiro, RJ. T: 21 562 7195, F: 21 290 1730, E: willyl@globo.com

Lana, M.S., Escola de Minas-Demin-Campus Universitar, 35400-000 – Ouro Preto, MG

Laprovitera, H., Av. Boa Viagem 3020 Ap. 601, 51020-000 – Recife, PE

Lazaro, A.A., Rua Joao Gualberto de Almeida Pires, 48/5, 02343-020 – São Paulo, SP

Leme, C.R. de Moraes, Rua Antonio Alves Magan 110, 01251-150 – São Paulo, SP

Lemes, F.M., Rua Majestic, 225-Cumbica, 07221-060 – Guarulhos, SP

Leoni, G.L.M., Rua Cupertino Durão, 132-Apto.404, 22441-030 – Rio de Janeiro, RJ. E: gleoni@geotec.coppe.ufrj.br

Lima, A.P., Rua Vilhena de Moraes, 240-Apto.106-Ii, 22793-140 – Rio de Janeiro, RJ. T: 21 3325 6087, E: andrepl@globo.com

Lima, D.S. de, Rua Dr. Augusto de Miranda, 1092/86, 05026-001 – São Paulo, SP. T: 11 3872 1448, F: 11 3872 1448, E: deniseli@terra.com.br

Lima, L.S. de Andrade, Rua Soares da Costa, 300/1003-Tijuca, 20520-100 – Rio de Janeiro, RJ

Lima, M.J.C.P. de, Rua Silva Guimaraes, 73-Ap.604, 20521-200 – Rio de Janeiro, RJ

Lima, M.C., Unb-Colina-Bloco K-Apto.302-Asa Norte, 70910-900 – Brasilia, DF. E: geopsqsa@terra.com.br

Lima Filho, S.C., P R. Vilhena de Moraes, 240-Bl.02-Apt.106, 22793-140 – Rio de Janeiro, RJ. T: 21 533 1260, 21 3325 6087, E: splima@ism.com.br

Lima Jr, N.R., Rua 118, N° 42, Setor Sul, 74085-400 – Goiania, GO

Lima Sobrinho, O. de Oliveira, R.Coronel Manoel Cecilio 425, 79004-610 – Campo Grande, MS. E: lima_jr@yahoo.com

Linheiro Jr, A.C.C., Rua Miguel Gustavo, 270-Brotas, 40285-010 – Salvador, BA

Lins, A.S.T., Av. Deputado Silvio Teixeira, 1120, 49025-100 – Aracaju, SE. T: 79-217 3398

Lionço, A., Rua Bela Cintra, 1450-Apto.91, 01415-001 – São Paulo, SP

Lira, E.N. de Souza, R.S-4-Esquina C/T-65-Ap.401 Ed.Meridiani, 74823-450 – Goiania, GO. T: 62 283 6122, 62 255 6036, E: enslira@furnas.com.br

Lobo, A. da Silva, Rua Jose Ferreira Marques, 12–16, 17044-570 – Bauru, SP. T: 14 234 8260, F: 14 221 6112, E: lobo@bauru.unesp.br

Lobo, A.C. da Costa, Rua Samin, 211-Iraja, 21235-210 – Rio de Janeiro, RJ

Locali, N.J., Rua Quintino Bocaiuva, 553/Sta. Catarina, 13466-300 – Americana, SP. T: 19 461 6401, E: estatec@acia.com.br

Loiola, F.L.P., Rua Marques de Olinda, 80-Apto.308-Bl.1, 22251-040 – Rio de Janeiro, RJ

Lopes, F. de Rezende, Rua Barao de Jaguaribe, 19 Ap.301, 22421-000 – Rio de Janeiro, RJ. E: flopes@pec.coppe.ufrj.br

Lopes, G.S., Rua Cristiano Teixeira Sales, 43-Apto.403, 30450-630 – Belo Horizonte, MG

Lopes, L. da Silva, Rua Bento Gonçalves, 297-Sala 903, 99010-010 – Passo Fundo, RS

Lopes, P.C.C., Rua Bela, 1128 – São Cristovao, 20930-380 – Rio de Janeiro, RJ

Lopes, S., Rua Antonio Gentil, 547, 83412-030 – Colombo, PR

Lopes Jr, D., Rua Piracuama, 404-Apto.22-Perdizes, 05017-040 – São Paulo, SP

Lopez, O.N., Av. Mascotes, 1160-Apto.101, 04363-010 – São Paulo, SP

Lorenzi, G.A., Av. Carlos Gomes, 711, 85806-230 – Cascavel, PR

Louro, R.R., Av. Rui Barbosa, 741 # 506, 28015-520 – Campos de Goitacazes, RJ. T: 24 723 4115, F: 24 723 4115, E: louro@censanet.com.br

Loyola, J.M.T., Rua Alexandre de Gusmao,764-B. Taruma, 82530-050 – Curitiba, PR

Lozano, M.H., Rua Visconde de Ourem, 282, 04632-020 – São Paulo, SP. T: 11 5034 3848, E: mauro@dynamis.com.br

Lucena, F.B. de, Rua Capitao Joao Alves de Lira, 1112, 58101-281 – Campina Grande, PB

Lucena, S.M. de, Rua Sebastiao Alves, 225-Ap.1701, 52060-100 – Recife, PE

Ludemann, S.M., R.Min.Alvaro Souza Lima, 253-Apto.704-B14, 04664-020 – São Paulo, SP. T: 11 535 4745, 11-5546 0456, E: sergiludemann@uol.com.br

Luz, P.A. de Cerqueira, Rua Harmonia, 445-Apt.52, 05435-000 – São Paulo, SP

Maccarini, M., C.P.476-Campus Universitario/Trindade, 88010-970 – Florianopolis, SC

Macedo, I.A.T. de, Rua Euzebio da Mota, 961-Apto.22, 80530-260 – Curitiba, PR. T: 41 254 2914, F: 41 254 2914, E: bidimsul@terra.com.br

Macedo, M. de Canossa, Rua Eca de Queiroz, 1001-Apto.304, 90670-020 – Porto Alegre, RS

Macedo, R.S.V., dos Santos, Rua Adalberto Aranha, 47-Ap.603 Tijuca, 20540-140 – Rio de Janeiro, RJ

Machado, C.R., Rua 501, 110, 88330-000 – Balneario Camboriu, SC. T: 47 367 3700, E: machaco@nts.com.br

Machado, S.L., Rua Teixeira Barros, 800-Bloco A, Apto.208, 40275-410 – Salvador, BA

Maffei, C.E.M., Alameda Equador, 197-Residencial Ii, 06470-060 – Barueri, SP. T: 11 5531 7416, 11 4191 4904, F: 11 5531 9907, E: maffei@maffeiengenharia.com.br

Mafra, J.M.Q., Rua Eng. Zoroastro Torres, 237-Ap.702, 30350-260 – Belo Horizonte, MG. T: 31 3335 3888, 31 3296 1529, E: mafra@gold.com.br

Maia, C.M.M., Rua Jeronimo da Veiga, 228–Itaim Bibi, 04536-001 – São Paulo, SP

Maia, E.C.D., Trav D.Pedro I 463 Bairro do Umarizal, 66050-100 – Belem, PA

Maia, G. de Brito, Rua Esmeraldino Bandeira, 132-Ap. 901, 52011-090 – Recife, PE. T: 81 3241 5964, E: gusmao.gilmar@torricelli.com.br

Maia, L. de Souza C., Rua Cupertino Durao, 109/901-Leblon, 22441-030 – Rio de Janeiro, RJ. T: 21 533 7158, 21 294 5884, E: kcgeo@openlink.com.br

Maia, P.C. de Almeida, Rua Tubira, 8-Ap.209- Leblon, 22441-070 – Rio de Janeiro, RJ

Manrubia, H., R.Francisco de Paula Alvarenga, 57, 06283-070 – Osasco, SP

Mantilla, J.N.R. de, Rua Dom Modesto Augusto, 102-Ap.200, 30535 430 – Belo Horizonte, MG

Marangon, M., Rua Antonio Pinto Pereira, 295/101A, 36016-570 – Juiz de Fora, MG. T: 32 3232 4620, E: marangon@civil.ufjf.br

Marco, L.A.A., do, Rua Jubiaba 150 Alto de Pinheiros, 05444-030 – São Paulo, SP

Maria, F.C.M. de Santa, Av. Gal. Guedes da Fontoura, 1091-Apto.102, 22621-241 – Rio de Janeiro, RJ. T: 21 491 2273, F: 21 491 2273, E: pesantamaria@openlink.com.br

Maria, L.B. de Santa, Rua Amaral, 131, 20510-080 – Rio de Janeiro, RJ

Maria, P.E.L. de Santa, Av. Gal.Guedes da Fontoura, 1091-Apto.102, 22621-241 – Rio de Janeiro, RJ. T: 21 2491 2273, F: 21 2491 2273, E: pesantamaria@openlink.com.br

Mariano, G.T., Rua Jerico, 275, 05435-040 – São Paulo, SP

Mariano, M. de, Rua Jerico, 271-Vila Madalena, 05435-040 – São Paulo, SP

Marinho, F.A.M., R.Coronel Oscar Porto, 167-Ap.63 Paraiso, 04003-000 – São Paulo, SP. E: fmarinho@usp.br

Marino, R.C.M., Rua Angelina Maffei Vita, 647-Ap.61, 01455-070 – São Paulo, SP. T: 11 303571, 11 3812 6387, E: rgmarino@ig.com.br

Maroni, L.G., Rua Iguatemi, 192-And. 17/Cj. 173/174, 01451-010 – São Paulo, SP

Marques, A.G., Av. Alvaro Otacilio, 2865 Ed. Escuna, 57035-180 – Maceio, AL. T: 82 221 6290, 82 231 1012, F: 82 326 6324, E: agmgeot@uol.com.br

Marques, A.C.M., Av. Brigadeiro Faria Lima, 2413 And 1, 01452-000 – São Paulo, SP

Marques, J.P., Rua Constantino de Souza 1145 Ap.32, 04605-003 – São Paulo, SP

Marques, J.A.F., Av. Macuco, 654-Apto.81-Moema, 04523-001 – São Paulo, SP

Martinati, L.R., Rua dos Campineiros, 173-Ap.33/Mooca, 03167-020 – São Paulo, SP

Martins, I.S.M., Praia Icarai, 281-Apto.501, 24230-004 – Niteroi, RJ. E: ian@geotec.coppe.ufrj.br

Martins, C.C., Rua dos Inconfidentes, 280-Ap.201, 30140-120 – Belo Horizonte, MG

Martins, C.R. de Souza, Rua Croata, 427-Ap.93-Lapa, 05056-020 – São Paulo, SP

Martins, F.B., Rua Carlos Julio Becken, 66, 90670-020 – Porto Alegre, RS. T: 51 3267 1587, E: flaviaabm@pucrs.br

Martins, J.A.T., Rua Viscondessa de Campinas, 15, 13092-350 – Campinas, SP

Martins, M. de Almeida, Av. Bartolomeu Mitre, 390/ 101-Leblon, 22431-000 – Rio de Janeiro, RJ. T: 21 249 1994, E: mamfam@uol.com.br

Martins Filho, G., Rua Ceara, 8-Permanente, 68464-000 – Tucurui, PA

Marzionna, J.D., Rua Adalivia de Toledo, 310-Apto.33, 05683-000 – São Paulo, SP. E: engeos@usway.com.br

Massad, E., R.Hamilton Ary Exel Jr., 08-Jd.Guadalupe, 06026-150 – São Paulo, SP

Massad, F., Rua Moncorvo Filho 293 City Butanta, 05507-060 – São Paulo, SP

Matos, M.M. de, Rua Cardeal Cagliori, 252, 05454-030 – São Paulo, SP. T: 11 3021 2441, E: Miltondematos@uol.com.br

Matos, W.D. de, Shin-Lago Norte-Qi4 Conj.6 Casa 13, 71510-260 – Brasilia, DF

Mattar Jr, D., Rua Iperoig, 749-Ap.11-Perdizes, 05016-000 – São Paulo, SP

Mattos, E.F.O. de, Rua Rio São Pedro, 24-Apto.702-Graca, 40150-350 – Salvador, BA

Maza Jr, J.L., Aos 8-Bloco A-Ap.304, 70660-081 – Brasilia, DF

Mbt Brasil-Master Builders Technologies Av. Firestone, 581-Bairro Industrial, 09290-055 – Santo Andre, SP

Medeiros, L.V. de, Rua Marques de São Vicente, 225/Puc-Rio, 22453-900 – Rio de Janeiro, RJ. T: 21 9761 4378, F: 21 671 4248, E: lmedeiros@globo.com.br

Medero, G.M., Rua Esperediao Medeiros, 133, 91330-020 – Porto Alegre, RS

Medina, J. de, R. Constante Ramos, 167-Apto.601, 22051-010 – Rio de Janeiro, RJ

Meireles, E. de Barros, Rua Dona Maria, 96 Apto.401-Tijuca, 20541-030 – Rio de Janeiro, RJ

Meller, F.C., Rua Engenheiro Figueiredo 52, 04012-150 – São Paulo, SP. T: 11 5571 0880

Mello, F.M. de, Estrada da Gavea Pequena, 952/Ap.109 A, 20531-420 – Rio de Janeiro, RJ. T: 21 562 7301, F: 21 562 7718, E: fmiguez@civil.ee.ufrj.br

Mello, J.R.C. de, Rua Miraima, 134-Bairro Itanhanga, 22641-580 – Rio de Janeiro, RJ. T: 21 3876 5527, 21 492 7741, E: jrcmello@vento.com.br

Mello, K.S. de, Rua Americo Vespucio, 101/208, 90550-031 – Porto Alegre, RS. T: 51 98359365

Mello, L.G.F.S. de, Rua Cardeal Arcoverde, 1749-Cj.103-Bl.A, 05407-002 – São Paulo, SP. T: 11 3034 2333, 11 3883 1640, F: 11 3034 2334, E: lgmello@vecttor.com.br

Mello, V.F.B. de, Rua Madressilva, 43-Brooklin, 04704-070 – São Paulo, SP. T: 11 533 5939, F: 11 533 9601, E: vmello7@ibm.net

Mello Jr, L.P. de, Rua dos Artistas, 134-Morro da Gloria, 36035-130 – Juiz de Fora, MG

Melo, A.C. de, Rua Sebastiao Malta Arcoverde, 115-A, 52060-070 – Recife, PE

Mendes, J.B. de Carvalho, Rua Pirapetinga, 239-Ap.1001-Serra, 30220-150 – Belo Horizonte, MG

Mendes, M.C., Estrada Das Ubaias, 311-Apto.602-Bloco A, 52061-080 – Recife, PE

Mendonca, J.C., Av. Castelo Branco, 572, 77600-000 – Paraiso do Tocantins, TO

Mendonça, A.A. de, Rua Dom Aristides Porto, 95/Apto.201, 30535-450 – Belo Horizonte, MG. T: 31 3375 7314, 31 3376 6053, E: geoestruturar@uol.com.br

Mendoza, J.F.P., Sicilia 941 Casi Colon, PARAGUAY – Asuncion

Menescal, R. de Abreu, Rua Fausto Cabral, 920-Ap.604, 60155-410 – Fortaleza, CE. E: rogeriom@roadnet.com.br

Menezes, G., Rua Batatais 333 Ap.81 Jd Paulista, 01423-010 – São Paulo, SP

Menezes, M.S. de Souza, Rua Djalma Ramos 362 Ap.602, 40150-380 – Salvador, BA

Menezes, R.S., Rua Antenor Tupinamba, 214/402-Pituba, 41820-220 – Salvador, BA

Menezes, S.M., Cx.Postal 37-Departamento de Engenharia, 37200-000 – Lavras, MG. E: stelio@dec.feis.unesp.br

Merighi, V.A., Av. Deputado Emilio Carlos, 966-Apt.43, 02720-100 – São Paulo, SP

Miguel, M.G., Rua Belo Horizonte, 433-Apto.503, 86020-060 – Londrina, PR. T: 43 371 4455, 43 324 9649, E: miriam@uel.br

Miguel, R.A., Rua Alfredo Guedes, 1949-Sala 504, 13416-901 – Piracicaba, SP. T: 19-433 1804, F: 19-433 1804

Milititsky, J., Rua Santo Inacio, 56-Apto.902, 90570-150 – Porto Alegre, RS. T: 51 332 4189, F: 51 332 4189, E: jarbasmil@terra.com.br

Minette, E., Univer. Federal Vicosa – Dec/C. Univers., 36570-000 – Vicosa, MG

Mio, G. de, Rua Noemia Sampaio de Souza, 55-Casa 7, 13562-360 – São Carlos, SP. E: demio@terra.com.br

Mioto, Jose Augusto Rua Prof. Arthur Ramos, 222-Ap.121, 01454-010 – São Paulo, SP

Miranda, A.N. de, Rua Eduardo Bezerra, 1276, 60130-271 – Fortaleza, CE. T: 85 227 8174, E: miranda@cagece.com.br

Miranda Jr, G., Rua Eng. Candido Gomide, 557, 13070-200 – Campinas, SP. T: 19 3243 2164, E: gentil@operamail.com

Mitsuse, C.T., Rua Onofre Silveira, 50, 04334-100 – São Paulo, SP

Monteiro, D. de Almeida, Rua Jose Coelho, 41-Tatuape, 03347-030 – São Paulo, SP
Monteiro, H.J.A., Rua Pedroso de Alvarenga, 725-Apto.121, 04531-011 – São Paulo, SP. T: 11 3079 2781, E: heliomon@uol.com.br
Monteiro, P.F. de F, Rua Visconde de Piraja, 164-Ap.1001, 22410-000 – Rio de Janeiro, RJ. T: 21 2297 7130, 21 2227 2413, F: 21 2220 2247, E: pfmoneiro@franki.com.br
Monteiro, V.E.D., Rua Francisco de Cunha, 1846-Apto.1101, 51020-041 – Recife, PE
Montez, F.T., Rua Serimbura, 320-Sala 15, 12243-360 – S. Jose dos Campos, SP. T: 12 3942 4316, F: 12 3922 6878, E: huesker@hueker.com.br
Mora, M.R., Rua Dr. Homem de Melo, 852-Casa 6, 05007-002 – São Paulo, SP
Moraes, J.M. de, Av. Dom Joaquim, 779, 96020-260 – Pelotas, RS
Moraes, J.T.L. de, Rua Santa Zoe, 255-Alto da Boa Vista, 04742-040 – São Paulo, SP
Moraes, M. da Cunha, Sqs-302 Bl.B Ap.606, 70338-020 – Brasilia, DF. T: 61 226 0606, E: mcmengenharia@zaz.com.br
Moraes, R. de Sousa, Av. Gov. Carlos de Lima Cavalcanti, 9, 50070-110 – Recife, PE
Moraes, S.Z., Sqs 302/B-606, 70338-020 – Brasilia, DF
Moraes Jr, L.J. de, Rua Barão de Mesquita, 314-Ss 102-Tijuca, 20540-003 – Rio de Janeiro, RJ. T: 21 567 0822, 21 225 1831, F: 21 568 3115, E: luciano@geotechnia.com.br
Morais, L.A.P. de, Rua Maestro Egidio J.L. Pinto, 165/S. 16, 12245-470 – São Jose dos Campos, SP
Moreira, J.E., Av. Pref. Dulcidio Cardoso, 800/108 Bl.3, 22620-311 – Rio de Janeiro, RJ
Moreira Filho, C., Qnd-40 Lote 2-Loja 2-Taguatinga Norte, 72120-390 – Taguatinga, DF
Moretti, A.O., Rua Fidalga, 563-Cj.22-A-Vila Madalena, 05432-070 – São Paulo, SP. T: 11 3811 9534, 11 3819 1386, F: 11 3815 8999, E: moretti.ec@uol.com.br
Moretti, M.R., Rua Otávio Nébias, 182-Apto.121, 04002-011 – São Paulo, SP. T: 11 3889 8672, F: 11 3889 8672, E: moretti@uninet.com.br
Mori, M., Rua Nicolau Zarvos, 118, 04356-080 – São Paulo, SP
Mori, R.T., Rua Caviana, 140-Fundos-Vila Guarani, 04307-010 – São Paulo, SP. T: 11 5017 2138, F: 11 5017 5037, E: ruimori@attglobal.net
Morita, L., Rua dos Democratas, 461-Ap.114, 04305-000 – São Paulo, SP
Mota, J.L.C.P. da, Rua Humaita, 334-Apto.1009- Humaita, 22261-001 – Rio de Janeiro, RJ. T: 21 539 1303, 21 537 6545, E: campinho@sondotecnica.com.br
Mota, N.M.B., Sqs 402-Bloco P-Apto.204-Asa Sul, 70236-160 – Brasilia, DF
Motidome, M.J., Rua Antonio Vieira de Medeiros, 101, 05425-060 – São Paulo, SP
Motta, L.M.G. da, Rua Vitor Meireles, 518, 20950-230 – Rio de Janeiro, RJ
Moura, A.C. de Figueiredo, Av. Yervant Kissajikian, 260, 04657-000 – São Paulo, SP
Moura, C.L.A., Av. Yervant Kissajikian, 260, 04657-000 – São Paulo, SP
Moura, P.R.M., Rua Elisa Silveira, 511, 04152-000 – São Paulo, SP. T: 11 5071 2986, 11 5583 0327, E: prmoura@usp.br
Muhlen, R. Von, Rua Gen. Portinho, 1067-Ap.02, 98005-050 – Cruz Alta, RS. T: 55 324 2214, E: ricardo@laguna.com.br
Murakami, C.A., Alameda Barros, 150/Ap.142-B/S. Cecilia, 01232-000 – São Paulo, SP
Nacci, D.C., Rua Fernandes Vieira, 634/1002, 90035-090 – Poa, RS. T: 51 311 5763, E: diegoydati@terra.com.br
Nakao, H., Rua Laiana, 717, 05470-000 – São Paulo, SP. T: 11 3022 3946, F: 11 3022 4967, E: hiromiti_nakao@hotmail.com
Namba, M., Rua Dr. Jose Candido de Souza, 356, 04518-050 – São Paulo, SP. E: makotonamba@originet.com.br

Napoles Neto, A.D.F., Rua Mateus Grou, 314/Ap.41-Pinheiros, 05415-040 – São Paulo, SP
Nascimento, A.S.A., do, Rua Roberto Dias Lopes, 225-Apto.1002, 22010-110 – Rio de Janeiro, RJ. T: 21 541 2531, E: asergio@civ.puc-rio.br
Nascimento, N.A., Rua Augusto Stelfeld 873/202, 80730-140 – Curitiba, PR
Nascimento, O.C., Rua Manoel Romao da Silva, 41-Ponto Novo, 49095-000 – Aracaju, SE
Natorf, C.L., Rua Miguel Couto, 207-Cj. 3, 90850-050 – Porto Alegre, RS
Navajas, S., Ipt/Dec-Cid.Universitaria – Cx.P. 0141, 01064-970 – São Paulo, SP
Navarro, R., Rua Campos do Jordao, 179, 05516-040 – São Paulo, SP
Negro, A.A., Rua Paraiba, 170, 09521-070 – São Caetano do Sul, SP
Negro Jr, A., Rua Girassol, 1033, 05433-002 – São Paulo, SP. T: 11 3819 0099, E: bureau@bureauprojetos.com.br
Neme, P.A., Rua Demostenes, 1195-Campo Belo, 04614-014 – São Paulo, SP. E: latina@osite.com.br
Nery Filho, J.V., Praca Xv de Novembro, 40-Ap.81, 13024-180 – Campinas, SP
Neves, M. das, Rua Licinia Teixeira de Souza, 533/Ap. 03, 13033-660 – Campinas, SP
Nieble, C.M., Rua Maria da Gra, 333-Casa 10, 05465-040 – São Paulo, SP. T: 11 3815 5416, E: geral@matraeng.com.br
Nilsson, T.U., Av. Marechal Floriano Peixoto, 7971-S/2, 81650-000 – Curitiba, PR. T: 41 284 1746, E: tnf@techemail.com
Niyama, S., Rua Conde de Sousel, 185, 05436-130 – São Paulo, SP. T: 11 3767 4421, E: sniyama@ipt.br
Nobre, M. de Melo M., Av. Aristeu de Andrade, 40-Ap.702, 57021-090 – Maceio, AL
Nogami, J.S., Epusp/D. Transporte/C.P.61548/C. Univers., 05424-970 – São Paulo, SP
Nogueira, C. de Lyra, Rua Itacolomy, 450, 35400-000 – Ouro Preto, MG
Nogueira, J.B., Escola de Engenharia de S.Carlos-Usp, 13560-250 – São Carlos, SP. T: 16 273 9506, 16 271 4695, E: jbnogueira@linkway.com.br
Nogueira, J.R., Rua Americo Brasiliense, 88-Ap.121, 13025-230 – Campinas, SP
Nogueira, M.C.N.G., Rua Santa Cruz, 1063, 04121-001 – São Paulo, SP
Nogueira, R., Rua Santa Cruz, 1063-Vila Mariana, 04121-001 – São Paulo, SP
Nouh, J.R., Rua Joaquim Tavora, 1020-Ap.51 B, 04015-012 – São Paulo, SP
Novais, F.L. de, Rua Ouro Preto 1707 Sto Agostinho, 30170-041 – Belo Horizonte, MG
Noya, J.M.C., Rua General Canabarro, 78-Ap.401, 20271-200 – Rio de Janeiro, RJ
Nunes, C.M.F., Av. Peixoto de Castro, 199, 12600-000 – Lorena, SP
Nunes, M.S., R.Carmelino Pinto Coelho 105 Liberdad, 31270-510 – Belo Horizonte, MG. E: msnunes@pucminas.br
Odebrecht, E., Rua Machado de Assis, 277/502-America, 89204-390 – Joinville, SC
Olavo, J.M., Rua Jose de Alencar, 399, 83321-230 – Pinhais, PR
Olavo, L.H.F., Rua Pe.Germano Mayer, 131-Apto.1102, 80050-270 – Curitiba, PR
Oliveira, A.T.J. de, Rua Hermano Barros Silva, 5664-Candeias, 54440-100 – Recife, PE. T: 81 9139 5664, 81 3469 3103
Oliveira, A. de, Rua Guacu, 186, 01256-040 – São Paulo, SP
Oliveira, H.G. de, Rua Murtinho Nobre, 261, 05502-050 – São Paulo, SP
Oliveira, J.L. de, Av. São Remo, 463-Apto.22B, 05360-150 – São Paulo, SP. T: 11 3735 5060, E: jeloliveira@uol.com.br
Oliveira, J.T.R. de, Av. Beira Mar, 1651-Bairro Novo, 53130-000 – Olinda, PE. T: 81 3271 8224, 81 3429 5429, E: jtro@npd.ufpe.br

Oliveira, L.A.K. de, Rua Mostardeiro, 856-Apto.303, 90430-000 – Porto Alegre, RS

Oliveira, M.A. de, Rua do Mármore, 89, 08420-470 – São Paulo, SP. T: 11 4446 2089, 11 6961 8363, F: 11 3873 2500, E: marquinhosz@bol.com.br

Oliveira, R.E.I. de, Av. Francisco Deslandis, 699-Ap.302, 30310-530 – Belo Horizonte, MG

Oliveira, W. de, Rua Prof. Andrade Nogueira 194, 17400-000 – Garca, SP

Omomo, A.M., Vale Das Palmeiras, Lote 21, 93900-000 – Ivoti, RS. T: 51 371 1022, 51 564 1126, E: al_m_o@yahoo.com.br

Ono, S., Higs 712-Bloco P-Casa 04, 70361-766 – Brasilia, DF. T: 61 426 3160, 61 245 7476, F: 61 345 7784, E: ono@abordo.com.br

Orlando, C., Al.Casa Branca, 1060-Apto.62-Jd.Paulista, 01408-000 – São Paulo, SP. T: 11 289 2234, T: 11 3061 9962, E: celsorlan@uol.com.br

Ortigao, A., Av. Niemeyer, 925-Ap.1203 C, 22450-221 – Rio de Janeiro, RJ. E: ortigao@openlink.com.br

Oyarzun, G.W.R., Rua Arandu, 205-Cj.505-Brooklin Novo, 04562-030 – São Paulo, SP. T: 11 5506 9496, E: reccius@sti.com.br

Pacheco, E.B., Rua Senambetiba, 3360-Bl.2-Apto.604, 20630-010 – Rio de Janeiro, RJ

Pacheco, L.C.D., Rua Antonio Pinto Pereira, 18, 36016-570 – Juiz de Fora, MG

Pacheco, M.P., Rua Baronesa de Pocone 141/604 Bl.2, 22471-270 – Rio de Janeiro, RJ. T: 21 9143 4073, 21 2539 5639, F: 21 535 3180, E: mpacheco@netyet.com.br

Paiva, P.R. de, Rua Mario Arruda, 34, 12900-000 – Braganca Paulista, SP

Paladino, L., Rua Pasteur, 515-Jardim São Caetano, 09580-450 – São Caetano do Sul, SP

Palmeira, E.M., Univ. Brasilia/Dec/Ft- Predio Sg-12 Geot, 70910-900 – Brasilia, DF

Palocci, A., Furnas Centr Eletr Caixa Postal 457, 74001-970 – Goiania, GO

Paraiso, S.C., Av. Brasil, 691 11° Andar Sta.Efigenia, 30140-000 – Belo Horizonte, MG. T: 31 3222 1970, F: 31 3213 7204, E: geomec.bhz@zaz.com.br

Parreira, A.B., Rua Juan Lopes 786 Ap.31, 13567-020 – São Carlos, SP

Paula, C.G. de, Rua Ilha Grande 120 Pam-pulha, 31555-030 – Belo Horizonte, MG. E: cgp@rural.com.br

Paula, L.A.L. de, Rua Raja Gabaglia, 1001-Apto.404-C.Jardim, 30380-090 – Belo Horizonte, MG. T: 31 275-2479

Paula, M.C. de, Rua Henrique Passini, 50-Ap.02/Serra, 30220-380 – Belo Horizonte, MG. T: 31 3221 7409, E: marcocp@bis.com.br

Peixoto, A.S.P., R.Aviador Ribeiro de Barros,4-20 Ap.31 B, 17045-490 – Bauru, SP, E: annapeixoto@yahoo.com

Peixoto, F.J.L., Rua Clarimundo de Melo, 308-C 1/202, 20740-321 – Rio de Janeiro, RJ

Penna, A.S.P.D., Alameda Rio Negro, 1105-Sala 11, 06454-000 – Barueri, SP. T: 11 4195 8385, E: penna@damascopenna.com.br

Pereira, A. da Costa, Rua Gen. Oliveira Galvao, 1047-Tirol, 59015-120 – Natal, RN

Pereira, A.G., Henriques Rua Bejamin Batista, 197-Apto.503, 22461-120 – Rio de Janeiro, RJ. E: ghislane@openlink.com.br

Pereira, D.F. da Costa, Estrada do Arraial, 2823-Ap.1402, 52051-380 – Recife, PE

Pereira, J.H.F., Colina, Bloco H, 70910-900 – Brasilia, DF

Pereira, L.R., Rua Montesquieu,105-Apto.52-Vl.Mariana, 04116-190 – São Paulo, SP

Pereira, M. dos Santos G., R.Jose Janarelli 210-V.Progredior, 05615-000 – São Paulo, SP. T: 11 3662 1005, F: 11 3667 3530, E: consultrix@zaz.com.br

Perfurasolo Empreit. de Construcoes Ltda., Rua Hugo Vitor da Silva, 32-American., 04340-040 – São Paulo, SP

Pessoa, R. de Pinho, Rua Gal. Gustavo Cordeiro de Farias, 380, 59010-180 – Natal, RN. T: 84 201 5991, E: rogepinho@sol.com.br

Petrobras-Petroleo Brasileiro S/A Rua Gen Canabarro 500 And 8 Maracana, 20271-201 – Rio de Janeiro, RJ

Pimenta, C., Alameda Casa Branca, 755-Ap.131, 01408-001 – São Paulo, SP

Pimenta, J.A. de Mattos, Rua Joao Moura, 870-Apto.151A, 05412-002 – São Paulo, SP

Pinezi, L.C., Rua Joel Jorge de Melo, 192-Apt° 94, 04128-080 – São Paulo, SP

Pinheiro, R.J.B., Travessa Cassel, 290 Apto.304, 97050-110 – Santa Maria, RS. T: 55 225 1905, 55 220 8432, E: rinaldo@ct.ufsm.br

Pinto, C. de Sousa, Av. Afranio Peixoto, 137, 05507-000 – São Paulo, SP. E: cspinto@usp.br

Pinto, C.P., Rua Santa Clara 248 Ap.401 Copacabana, 22041-010 – Rio de Janeiro, RJ

Pires Filho, C.J., Rua Barata Ribeiro, 615-Ap.202, 22051-000 – Rio de Janeiro, RJ

Pitta, H.F., Av. Bernardo Vieira de Melo, 5206/303, 54450-020 – Jaboatao Guararapes, PE

Polido, U.F., Av. Saturnino Rangel Mauro, 340, 29060-770 – Vitoria, ES. T: 27 3345 6482, 27 3227 3792, E: geoconsult@geoconsult.com.br

Politano, C.F., Av. Rui Barbosa, 310-Apto.203, 27521-190 – Resende, RJ. T: 24 3354 6663, E: politano@resenet.com.br

Pontes Filho, I.D. da Silva, Rua Antonio Falcao, 504/1004 - Boa Viagem, 51020-240 – Recife, PE. E: ivaldo@etseccpb.upc.es

Poroca, J. dos Santos, Rua Antonio de Sa Leitao, 108-Casa 7, 51020-090 – Recife, PE

Porto, R.S.M., Rua Arapore, 670, 05608-001 – São Paulo, SP. T: 11 3032 2530

Postiglione, P., Rua dos Correeiros, 92-1Ra. Esq., 1100-100 – Lisboa-Portugal

Prado, C.M. de Almeida, Rua Dom Armando Lombardi, 80-Apto.53 D, 05616-100 – São Paulo, SP

Prates, C., Rua Padre Agostinho 2055 Ap.801, 80710-000 – Curitiba, PR

Prates Jr, C., Rua Padre Agostinho, 2055-Ap.801, 80710-000 – Curitiba, PR

Presa, E.P., Rua Barao de Loreto, 18, 40150-270 – Salvador, BA

Prietto, P.D.M., Rua Anchieta, 784-Ap.101, 96015-420 – Pelotas, RS

Priszkulnik, S., Rua Madre Maria Angelica Rezende, 60, 05448-040 – São Paulo, SP

Progeo Engenharia Ltda., Rua Maria Beatriz, 894-Nova Barroca, 30555-140 – Belo Horizonte, MG. T: 31 3312 1348, E: miranda@progeo.com.br

Protendit Construcoes e Comercio Ltda., Rua Ana do Sacramento Andrade 530, 02289-000 – São Paulo, SP

Pugliese, P.B., Rua Aldino Del Nero, 75-Pq. Arnold Sch, 13566-588 – São Carlos, SP

Puppi, R.F.K., Rua Marechal Deodoro 882, 83601-020 – Campo Largo, PR

Quadros, E.F. de, Av. Prof. Almeida Prado, 532-Ipt/Digeo, 05508-901 – São Paulo, SP. E: equadros@ipt.br

Quaresma, A.R., Al Raja Gabaglia, 271, 04551-090 – São Paulo, SP. T: 11 3849 2166, F: 11 3849 7875

Quaresma Filho, A.R., Alameda Raja Gabaglia 271, 04551-090 – São Paulo, SP. T: 11 3849 2166, 11 3032 7544, F: 11 3849 7875, E: aquaresma@engesolos.com.br

Quebaud, S., Rua Jornalista Orlando Dantas, 59, 22231-010 – Rio de Janeiro, RJ. T: 21 552 5422, E: squebaud@usu.br

Queiroz, L. de Araujo, Alameda Sarutaia 381 16An Ap 162, 01403-010 – São Paulo, SP

Queiroz, P.I.B. de, Rua Girassol, 964-Apto.74, 05433-002 – São Paulo, SP

Ramalho, G.G.C., Rua Benedito Valadares, 63, 35400-000 – Ouro Preto, MG

Ramires, M.C.P., Rua Costa, 361, 90110-270 – Porto Alegre, RS. T: 51 231 9235, E: mirtes@via-rs.net

Ramos, M. de Oliveira, Av. Henrique Dodswarth, 83/206-Lagoa, 23061-030 – Rio de Janeiro, RJ

Ramos, O.G., Praia do Flamengo, 88-Ap.304, 22210-030 – Rio de Janeiro, RJ. T: 21 558 0261

Ranzini, S.M.T., Rua Peixoto Gomide 996 3An Sl330, 01409-000 – São Paulo, SP

Ratton, R.B., Rua Alexandre Farah, 542-Amambai, 79008-020 – Campo Grande, MS. T: 67 761 5532, 67 725 7176

Ravaglia, A., Rua Banibas, 142-Alto Pinheiros, 05460-010 – São Paulo, SP

Re, G., Al.Min Rocha Azevedo 976 Ap.61, 01410-002 – São Paulo, SP

Reffatti, M.E., Rua Osvaldo Aranha, 522-Apto.34, 90035-190 – Porto Alegre, RS. T: 51 3312 5005, E: gingome@ig.com.br

Rego, A.A., Av. Sete de Setembro 5485 Ap.801-Batel, 80240-001 – Curitiba, PR

Reis, J.M. dos, Shigs, 707-Bl.F-Casa 4, 70351-706 – Brasilia, DF

Reis, J.H.C., dos, Rua Nicoleta Stella Germano, 51-Apto.404, 13561-090 – São Carlos, SP. E: jeselay@hotmail.com

Reis, M. dos, Rua da Ternura, 33, 06700-000 – Cotia, SP

Reis Filho, P., Av. Prudente de Morais 44 And 2, 30380-000 – Belo Horizonte, MG

Remy, J.P.P., Rua Moises Amelio, 23-Andar 4°, 28613-210 – Nova Friburgo, RJ. T: 24 522 0766, F: 24 523 1363, E: jpremy@mecasolo.com.br

Rezende, J.A. de, Rua Rio Grande do Sul, 1576-Ap.302, 30170-111 – Belo Horizonte, MG

Rezende, L.R. de, Rua 9, Nr. 1496-Setor Marista, 74150-130 – Goiania, GO. T: 62 281 3693, E: ribeiro@unb.br

Rezende, M.E.B., Rua Roosevelt de Oliveira, 767-Apto.403, 38400-610 – Uberlandia, MG. E: melisa@ufu.br

Ribas, J.B.M., Rua Republica do Peru, 53-Ap.301, 22021-040 – Rio de Janeiro, RJ

Ribeiro, A.T.F., Rua 152, N. 175- Bairro Laranjal, 27255-020 – Volta Redonda, RJ

Ribeiro, O. de Sousa, Rua Ceara, 1431-Cj. 1201, 30150-311 – Belo Horizonte, MG. T: 31 3273 3255, F: 31 3273 5553, E: tenge@uol.com.br

Ribeiro Neto, F., Rua Dr. Paulo Vieira, 383-Apto.2203, 01257-000 – São Paulo, SP. T: 11 283 7348, F: 11 283 7464, E: franribeironeto@uol.com.br

Rios, A.S., Av. Brasil, 1438-Cj. 1302-Funcionarios, 30140-003 – Belo Horizonte, MG

Riscado, J.N. de Souza, Rua Carlos Lacerda, 622-C.dos Goitacazes, 28025-660 – Rio de Janeiro, RJ. T: 24 733 7357, 24 733 0982, E: riscado@rol.com.br

Ritter, E., Rua Gal. Urquiza, 190-Casa, 22431-040 – Rio de Janeiro, RJ. T: 21-587 7849, F: 21-512 6316, E: ritter@uerj.br

Rocha, G.E. da, Rua 227-A N° 107-Ap.505 – Qd. 67A, 74610-060 – Goiania S. Universit, GO

Rocha, J.A.B. da, Rua Lourival Goncalves Oliveira, 105, 30570-565 – Belo Horizonte, MG. T: 31 3374 2833, T: 31 3262 0278, E: consolos@uai.com.br

Rocha, M.G., Rua do Ouro, 1138-Ap.1601/B. Serra, 30220-000 – Belo Horizonte, MG. T: 31 3299 3001, 31 3225 2393, E: mrocha@cemig.com.br

Rocha, R., Rua Lisboa, 1194-Apto.154, 05413-001 – São Paulo, SP. T: 11 3767 4437, 11 3088 0046, E: rrocha@ipt.br

Rocha, W.M. da, R Felipe dos Santos, 825-Cj. 704, 30180-160 – Belo Horizonte, MG

Rocha Filho, P., Rua Fernando Nogueira de Souza, 110/Cob., 22620-380 – Rio de Janeiro, RJ

Rodrigues, E.V., Rua Joao Moura, 945-Apto.73, 05412-002 – São Paulo, SP

Rodrigues, J.M. de Azevedo, Rua General Garzon 28/503-Lagoa, 22470-010 – Rio de Janeiro, RJ

Rodrigues, L.H.B., Av. Imperatriz Leopoldina, 1845 Ap.41, 05305-007 – São Paulo, SP. T: 11 9172 7655, E: homerobonini@uol.coml.br

Rodrigues, M.A., Av. Agua Fria, 1816-Cj. 4, 02332-000 – São Paulo, SP

Rodrigues, W.B. de Souza, R.Paulo Firmeza, 330 Ap.201 S.Joao Tauap, 60130-421 – Fortaleza, CE. T: 85 227 2134, 85 227 1714, F: 85 257 7912, E: waldirrodrigues@secrel.com.br

Rohde, L., Rua Erico Verissímo, 441 Ap.413, 90160-181 – Porto Alegre, RS. T: 51 231 1863, E: lurohde@bol.com.br

Romanel, C., R. Marques São Vicente, 225/Puc-Rio/Dec, 22453-900 – Rio de Janeiro, RJ. E: romanel@civ.puc.rio.br

Rosa, R.L. de, Av. Benedito Castilho de Andrade, 1007, 13212-070 – Jundiai, SP. T: 4582 7588, E: derosareynaldo@uol.com.br

Rosas, M.M. de Castro, Rua Cosme Velho, 9-Ap.201, 22241-090 – Rio de Janeiro, RJ

Rossi, G.M., Av. Beira Mar, 1400-Ap.802-B.Piedade, 54410-000 – Jaboatao Guararapes, PE. T: 81 3339 0322, 81 3361 1504, E: frossi@elogica.com.br

Rotta, G.V., Rua Tiradentes, 690/202, 98800-000 – Santo Angelo, RS. T: 51 9973 0188

Rozenbaum, D., Rua Desembargador do Valle, 258, 05010-040 – São Paulo, SP. T: 11 3872 1718, E: fundacta@fundacta.com.br

Ruffier, A.P., Rua Castro Barbosa, 65-Bloco 2-Apto.901, 20540-230 – Rio de Janeiro, RJ. T: 21 598 6149, 21 571 6183, E: aureo@cepel.br

Ruiz, A.S., C/Torroella de Montgri 3, 13-A, 08027-027 – Barcelona – Espanha

Ruiz, M.D., Rua Bennett, 439, 05464-010 – São Paulo, SP. T: 11-3815 7133, 11-3022 5839, E: murillo@engecorps.com

Rusilo, L.C., Rua Mandiuba, 342-Agua Rasa, 03158-070 – São Paulo, SP

Russo Neto, L., Rua Santa Catarina, 427-Ap.14, 80620-100 – Curitiba, PR. E: russo@rla01.pucpr.br

Ruzzante Neto, F., Av. Afonso Vergueiro, 1810-Apto.111-B, 18040-000 – Sorocaba, SP

Sa, C.T., Sqn 216 Bloco C-Apto.102-Asa Norte, 75835-030 – Brasilia, DF

Sa, T.J. de Abreu e Lima e, Av. Canal de Setubal, 30-Piedade, 54310-320 – Jaboatao Guararapes, PE. T: 81-3241 1299

Saccab, C., Rua Natividade, 124-V.Uberabinha, 04513-020 – São Paulo, SP. T: 11 3842 0328, F: 11 3045 6243, E: csacab@sagesse.com.br

Sadowski, G.R., Rua Prof. Guilherme Milward, 246, 05506-000 – São Paulo, SP

Saes, J.L., Rodovia Sp-274, N° 2353, 06683-000 – Itapevi, SP

Sales, L.F.P., Rua Poços de Caldas, 301-Centro, 88103-030 – São Jose, SC

Sales, M.M., Rua 25 A-Quadra 75A-Lote 17/18-Apto.1002, 74070-150 – Goiania, GO. E: msales@eec.ufg.br

Salgado, R., West Lafayette In 47907-1284, USA – USA – West Lafayette,

Salioni, C., Rua Iramaia, 181, 01450-020 – São Paulo, SP. T: 11 3097 0910, E: geosonda@geosonda.com.br

Salioni Jr, C., Rua Iramaia, 181, 01450-020 – São Paulo, SP. T: 11 3097 0910, E: clovis-salioni@uol.com.br

Samara, V., Rua Duarte da Costa, 208, 05080-000 – São Paulo, SP

Sana, L.F., Av. Bage, 1313-Apto.202-Petropolis, 90460-080 – Porto Alegre, RS

Sandroni, S.S., Rua Santa Clara, 431-Bloco B1-1002, 22041-010 – Rio de Janeiro, RJ. T: 21 2518 0202, E: geoprojetos@geoprojetos.com.br

Santoro, E., Av. Prof. Almeida Prado, 532-Ipt/Digeo, 05508-901 – São Paulo, SP

Santos, A.C.B. dos, Rua Abelia,226-Cobertura 2-I.Governador, 21941-010 – Rio de Janeiro, RJ

Santos, C.T., Rua Vereador Constante Pinto, 281, 82510-240 – Curitiba, PR. T: 41 256 8174, E: cats@cwb.com.br

Santos, E.H. dos, Rua Eurico de Souza Leao, 26-Apto.202, 50721-100 – Recife, PE. T: 81 3423 7355, E: qualimax@hotlink.com.br

Santos, L.A.C.B. dos, Rua Senador Dantas,19-S/907-Cinelandia, 20031-200 – Rio de Janeiro, RJ. T: 21 544 5724, 21 285 4328, F: 21 544 5724, E: batista@envirogeo.com.br

Santos, M.C.V. dos, Sqn 209 Bloco G Ap.305, 70854-070 – Brasilia, DF. E: mt@cr-df.rnp.br

Santos, W.M., Rod.Raposo Tavares, 3175, Km 13,5 Ap.138-D, 05577-100 – São Paulo, SP. T: 11 3818 5459, 11 3735 7120, E: wmourasantos@uol.com.br

Santos Filho, M.G. dos, Rua Alves Guimaraes, 682-Ap.163, 05410-001 – São Paulo, SP. T: 11 3815 7133, 11 3088 9255, E: mauro@engecorps.com

Santos Filho, W.M. dos, Av. Magalhaes Barata, 223-B. São Braz, 66040-170 – Belém, PA

Santos Jr, O.F. dos, Ufrgn/Dec/Centro de Tecnologia, 59072-970 – Natal, RN

Santos Neto, P.M., Sqn 205-Bloco G-Ap.102, 70843-070 – Brasilia, DF

Saraiva, R.M., Rua Barros Cassal, 693/602-Bl. Delta, 90035-030 – Porto Alegre, RS

Saramago, R.P., Rua Prof. Lara Vllela, 185/Ap.904-Inga, 24210-590 – Niteroi, RJ

Sare, A.R., Av. Conselheiro Furtado, 348, 66025-160 – Belem, PA

Sayao, A.S.F.J., Rua Marques de São Vicente 225 S/301L, 22453-900 – Rio de Janeiro, RJ. E: sayao@civ.puc-rio.br

Scanbrasil – L&S Europa Importadora Ltda., Rua Principal, 05-Praia de Carapebus, 29164-434 – M-Serra/Vitoria, ES

Schilling, G.H., Rua Coelho Neto, 44-Ap.502-Bloco 2, 22231-110 – Rio de Janeiro, RJ

Schimidt, L.A., Rua Silva Jardim, 939-Ap.01, 90450-071 – Porto Alegre, RS

Schmidt, C.A.B., Rua Haddock Lobo, 39-Apto.903, 20260-130 – Rio de Janeiro, RJ

Schmidt, C.F., Rua Dona Marinha, 444, 02460-080 – São Paulo, SP. T: 11 3034 2333, 11 6973 4134, E: cristina@vecttor.com.br

Schmidt Filho, A.C., Rua 7 de Setembro, 58E-Centro, 89801-140 – Chapeco, SC

Schnaid, F., Av. Osvaldo Aranha, 99/3° Andar/Ufrs-Dec, 90046-900 – Porto Alegre, RS. E: fernando@vortex.ufrgs.br

Schultz, C.C., Rua Divino Salvador, 166-Apto.92, 04078-000 – São Paulo, SP. T: 11 4358 7500, E: cintiaschultz@hotmail.com

Schultze, J.P.S., Rua Marcilio Dias, 2596, 96400-000 – Bage, RS

Scott, H.E.P., Rua Augusta, 257-7 Andar, 01305-000 – São Paulo, SP

Seitenfus, R.A., Rua Passo Fundo, 59, 90035-040 – São Leopoldo, RS. T: 51 9978 6725, 51 592 7881, E: renato@indus.unisinos.br

Senna Jr, R.S. de, Rua Byron Ortiz Araujo, 145 V. Sacobucci, 13567-230 – São Carlos, SP. E: savoi@uol.com.br

Seraphim, L.A., Rua Dr Alves de Banho, 666-Apto.103-A, 13030-580 – Campinas, SP. E: seraphim@fec.unicamp.br

Serman, C., Rua Conde de Itaguai, 13-Ap.503, 20511-200 – Rio de Janeiro, RJ

Sieira, A.C.C.F., Av. Ministro Afranio Costa, 255-Ap.202, 22621-220 – Rio de Janeiro, RJ

Signer, S., R Peixoto Gomide 493 Ap.221 Cer Cesar, 01409-001 – São Paulo, SP. E: shoshana@uol.com.br

Silva, A.M., Veloso e, Rua Paqueta, 1945, 15025-180 – São Jose Rio Preto, SP

Silva, C.E. da, Estrada da Gavea 681 Bl.3 Ap.1001, 22610-000 – Rio de Janeiro, RJ

Silva, C.M., Sqs 316-Bloco G-Apto.501, 70000-000 – Brasilia, DF

Silva, E.F.N. da, Rua Dona Francisca, 3479-Apto.3, 89221-001 – Joinville, SC

Silva, E.A. da, Av. Caiapo, 924-Setor Sta. Genoveva, 74672-400 – Goiania, GO

Silva, H.H.R. da, Rua Caiowaa, 1194-Ap.97, 05018-001 – São Paulo, SP

Silva, J.C.B.J. da, Rua Praia de Aratuba/Q. 17-Lote 16, 42700-000 – Lauro de Freitas, BA

Silva, J.M.J. da, R.Olivia Menelau 388-Imbiribeira, 51170-110 – Recife, PE

Silva, J.V.O. da, Rua Das Creoulas, 101-Ap.502/Graças, 52011-270 – Recife, PE

Silva, L.T.P. da, Rua Antonio José Menezes, 615, 91750-000 – Cachoeira do Sul, RS. T: 51 722 5140

Silva, L.A.A. da, Rua Bernardino de Campos, 982-Brooklin, 04620-003 – São Paulo, SP. T: 11 3818 5187, 11 241 7846, E: layres@usp.br

Silva, L.C.R. da, Rua Muriae, 104-Ap.32, 04269-020 – São Paulo, SP

Silva, M.F., Rua Cel. Moreira Cesar, 165-Apto. 1103, 24230-051 – Niteroi, RJ. T: 21 7102689, E: furtadosilva@mailbox.urbi.com.br

Silva, M.T. da, Av. Caiapo, 924-Setor Santa Genoveva, 74672-400 – Goiania, GO

Silva, M.A.M. Sa Santos, Rua Itambe 96 Ap.163, 01239-000 – São Paulo, SP

Silva, P.A.B.A., Trav.Juca Barreto, 93-Apto.306, 49015-200 – Aracaju, SE

Silva, P.R. da, Cond.Imperio dos Nobres, Q.2-Cj.B-Casa 17, 73017-009 – Sobradinho, DF. T: 61 302 2011, E: parosil@bol.com.br

Silva, R.W.D. da, Scln 105-Bloco A-Nr. 44-Salas 4/6, 70734-510 – Brasilia, DF

Silva, S.G. da, Rua Franca Pinto 1255 Ap.162 B, 04016-035 – São Paulo, SP

Silva, W. da, Av. Maringa, 1433-V.Emiliano Perneta, 83325-360 – Pinhais, PR

Silva Filho, J.F. da, Rua Bernardo Guimaraes, 455-Apt.401, 30140-080 – Belo Horizonte, MG

Silva Jr, J.M. da, Rua Conselheiro Lafaiete, 1952, 31035-560 – Belo Horizonte, MG

Silva Neto, A.J. da, Rua Paulo Barreto, 34 Ap.302/Botafogo, 22280-010 – Rio de Janeiro, RJ. T: 21 2211 8903, 21 2266 3529, E: armando.neto@lightrio.com.br

Silveira, A., Rua São Jose, 409-Alto da Boa Vista, 04739-001 – São Paulo, SP. T: 11 5546 0851, F: 11 247 9143, E: silveiracongeo@uol.com.br

Silveira, A.A. da, Av. Juruce, 766-Ap.32, 04080-012 – São Paulo, SP. E: arsilvei@amcham.com.br

Silveira, G.C., Av. Osvaldo Aranha, 350-Ap.801/Bom Fim, 90035-190 – Porto Alegre, RS

Silveira, H.S. da, Rua Cel. Jose Justino, 425, 37470-000 – São Lourenço, MG

Silveira, J. Rua Flores, 207, 93800-000 – Sapiranga, RS. T: 51 9984 7860

Silveira, J.E.S., Av. Prudente de Moraes 44 Cj.201 E 202, 30380-000 – Belo Horizonte, MG. T: 31 3344 3511, 31 3223 2410, E: jernani@net.em.com.br

Silveira, J. da, Rua Cristovao Barcelos, 281-Cob. 01, 22245-110 – Rio de Janeiro, RJ

Simoes, G.F., Rua Boaventura, 771-Ap.402-Bl. L, 31270-020 – Belo Horizonte, MG. T: 31 3238 1792, E: gfsimoes@etg.ufmg.br

Simoes, P.R.M., Av. Otavio Mangabeira, 11881-M1/Casa 32, 41650-000 – Salvador, BA. T: 71 240 1048, 71 367 2740, E: geotechnique@e-net.com.br

Soares, E., Rua Visconde da Luz, 88-Ap.61, 04537-070 – São Paulo, SP

Soares, J.M.D., Rua Honorio Magno, 441, 97070-450 – Santa Maria, RS

Soares, M.M., Av. A N.159 Ap.1001 Ed Joao Paulo Ii, 78058-090 – Cuiaba, MT. T: 65 621 6742, 65 642 4235, F: 65 621 3179, E: nacon@terra.com.br

Soares, V.B., Rua Prof. Batista Leite, 229-B. Roger, 58020-600 – Joao Pessoa, PB

Sobral, A.C. dos Santos, R. Frei Duarte Jorge de Mendonça, 100/81, 05725-060 – São Paulo, SP. T: 11 3746 6897, E: acsobral@uol.com.br

Sobral, H.S., R.Eng° Ademar Fontes, 282 Ap.701 Pituba, 41820-240 – Salvador, BA

Sodre, D.J.R., Rua Esperança, 196 Ed. Itapeva-Ap.63, 12243-700 – São José dos Campos, SP. T: 12 39476803, 12 3947 6802, F: 91 211 1608, E: dsore@ufpa.br

Sousa, L.M.R. e, Lnec-Av. do Brasil, 1700-066, - - Lisboa–Portugal

Sousa, L.A. de, Av. Ministro Victor Konder, 56, 88301-280 – Itajai, SC

Souza, A.A.C., Rua Joaquim Nabuco, 201-Terreo, 22080-030 – Rio de Janeiro, RJ

Souza, G.B. de, Rua Joao Martins da Silva, 831-Caseb, 44038-430 – Feira de Santana, BA

Souza, H.G. de, Rua Das Ostras, Lote 4-Quadra 48, 24358-360 – Rio de Janeiro, RJ

Souza, J.L.C. de, Rua 5 de Julho, 336-Cob. 01/Copacabana, 22051-030 – Rio de Janeiro, RJ

Souza, M. de, Al.Joaquim Eugenio de Lima, 310-Apto.66, 01403-000 – São Paulo, SP

Souza, N.M. de, Colina Unb-Bl. H-Ap.608, 70910-900 – Brasilia, DF. E: newton@fpk.tu-berlin.de

Souza, R.N., Rua Baronesa de Itu, 858, 01231-000 – São Paulo, SP. T: 11 3662 1005, 11 3726 3795, E: rubenei@usp.br

Souza Filho, J.M. de, Sia Trecho 2 Lotes 850/60, 71200-020 – Brasilia, DF

Sozio, L.E., Rua Cel. Conrado Siqueira Campos, 133/23, 04704-140 – São Paulo, SP

Spada, J.L.G., Rua Fruto do Mato, 109-Jacarepagua, 22750-480 – Rio de Janeiro, RJ

Speceng Eng.E Fundacoes Especiais Ltda., Rua Maestro Carlos Cruz, 176-V.Indiana, 05585-020 – São Paulo, SP

Spotti, A.P., Rua Rui Barbosa, 210-Apto.302, 25963-090 – Teresopolis, RJ

Stancati, G., Av. Dr. Carlos Botelho, 1465-Eesc/Usp, 13560-250 – São Carlos, SP

Stefenoni, S.A., Rua Gil Martins de Oliveira, 185, 29047-600 – Vitoria, ES

Stein, J.H., Rua Raul de Leoni, 66-Apto.205, 25610-330 – Petropolis, RJ. T: 21 544 5724, 24 245 6360, E: stein@env.rogeo.com.br

Steinmeyer, C.M., Rua Dr. Zuquim, 1321-Santana, 02035-012 – São Paulo, SP. T: 11 6979 5075, F: 11 6973 9873, E: magproj@uol.com.br

Strauss, M., Rua Monte Cristo, 100/48, 91750-000 – Porto Alegre, RS. T: 51 9815 2385

Stucchi, F.R., Rua Fabia, 442-Andar 4, 05051-030 – São Paulo, SP. T: 11 3862 1236, E: egt@egteng.com.br

Taga, E., Rua Humberto de Campos, 67-Ap.131, 04311-080 – São Paulo, SP

Taioli, F., Rua Francisco Perrote, 406, 05531-000 – São Paulo, SP

Tanaka, A., Av. Dr. Ariberto Pereira da Cunha 333, 12500-000 – Guaratingueta, SP

Targas, D.N., Rua Urano, 9-Ap.11-Aclimaçao, 01529-010 – São Paulo, SP. T: 11 232 2622, 11 278 0900, E: deboratargas@uol.com.br

Tavares, A.X., Rua Nunes Machado 2040, 80220-070 – Curitiba, PR

Tayar, F. de Barros, Rua Dr. Mario Ferraz 95 Ap.83, 01453-010 – São Paulo, SP

Tecnosolo Eng e Tecnol Solos e Mat S/A, Rua Sara, 18-Santo Cristo, 20220-090 – Rio de Janeiro, RJ

Tecnotest – Tecnologia de Estacas Ltda., Rua Aquidabam, 394-B. Padre Eustaquio, 30720-420 – Belo Horizonte, MG

Tecper Engenh. de Solos e Fundacoes Ltda., Rua Juari, 75, 04446-160 – São Paulo, SP. T: 11 5611 6747, F: 11 5611 2351, E: tecper@uol.com.br

Teixeira, A.A.H., Rua Managua, 82-Cidade Jardim, 05601-050 – São Paulo, SP

Teixeira, A.H., Rua Managua, 82-Cidade Jardim, 05601-050 – São Paulo, SP. T: 11 3817 5877, F: 11 3813 3984, E: aht@originet.com.br

Teixeira, C.Z., Rua Hermilio Alves, 235-Apto.101-Bl.1, 30010-070 – Belo Horizonte, MG. E: zampier@estrelar.com.br

Teixeira, D.C.L., Av. Flor de Santana 104 Parnamirim, 52060-290 – Recife, PE

Teixeira, P.F., Rua Clovis Bevilacqua, 289, 13075-040 – Campinas, SP

Teixeira, R.S., Rua Goiás, 1777-Apto.304, 86020-410 – Londrina, PR. T: 43 371 4455, 43 344 2984, E: raquel@uel.br

Teixeira, W.R., Av. Cidade Jardim, 3141-Quinta 96, 12233-900 – São Jose dos Campos, SP

Telles, I. de Araujo, Travessa Timbo, 3107-Marco, 66095-750 – Belem, PA

Themag Engenharia Ltda., Rua Bela Cintra 986 And 8, 01415-000 – São Paulo, SP

Tibana, S., Rua Pereira da Silva, 121-Ap.1002, 22221-140 – Rio de Janeiro, RJ

Tibrand, L., Rua Principal, 05-Praia de Carapebus, 29164-434 – M-Serra/Vitoria, ES

Torres, R.R., Av. Venancio Aires, 449/512, 90040-193 – Porto Alegre, RS. T: 51 3221 2281

Tozatto, J.H.F., Rua Amaral, 40-Apto.403, 20510-080 – Rio de Janeiro, RJ

Tsukahara, C.N., Rua dos Curupias, 100-Apto.61, 04344-050 – São Paulo, SP. T: 11 3034 2333, 11 5011 1048, E: claratsukahara@hotmail.com

Tuzzolo, F.P., Av. Cons. Nebias, 340-Conjunto 47, 11015-002 – Santos, SP

Val, E.C. do, Rua Beatriz Galvão, 111 Apto.52-Sumaré, 01257-100 – São Paulo, SP. T: 11-3875 6086, T: 11 3862 6238, E: edoval@dvs.com.br

Vallejos, D.V.P., Rua Bernardo Monteiro, 1000-Hibisco, 32017-170 – Contagem, MG

Vargas, M., Rua Bela Cintra, 986-Andar 15, 01415-000 – São Paulo, SP

Vargas Jr, E. do A., Praca Santos Dumont 138-Apto.707 A, 22470-060 – Rio de Janeiro, RJ. E: vargas@nciv.puc-rio.br

Vargens, J.R. da Costa, Av. Magalhaes Neto, 91-Ap.402-Pituba, 41820-021 – Salvador, BA. T: 71 248 7409, E: rvargens@sec.ba.gov.br

Vasconcellos, C.A. de, Rua Tupi, 343-Ap.151, 01233-001 – São Paulo, SP. T: 11 3371 7339, 11 3662 0670, F: 11 3371 7464, E: cbt@metrosp.com.br

Vasconcelos, M.A.H. de Barros, Rua Jose Antonio Coelho, 435-Ap.102 B, 04011-061 – São Paulo, SP

Vasconcelos, M. de Fatima Costa de, Rua General Belegarde, 245, 20710-000 – Rio de Janeiro, RJ

Vaz, L.F., Rua Bela Cintra, 986-Andar 7, 01415-000 – São Paulo, SP

Vazquez, R.H.R., Manuel Garcia 22-P.4º/Dtq. 15, 1284-284 – Cap.Feder./Argentina

Velloso, D. de A., Rua Visconde de Piraja, 357-Ap.801, 22410-003 – Rio de Janeiro, RJ. T: 21 562 8041, 21 267 4945, E: davelloso@openlink.com.br

Velloso, M.H., Av. Adolplho Vasconcelos, 497-Ap.208, 22793-380 – Barra da Tijuca, RJ. T: 21 537 8334, 21 498 0630, F: 21 579 1691, E: mhammes@br.odebrecht.com

Velloso, R.Q., Rua Miranda Rosa, 14, 22641-330 – Rio de Janeiro, RJ

Velloso Filho, S.M.P., Av. Cremona, 436-Bairro Badeirantes, 31340-520 – Belo Horizonte, MG. T: 31 3213 5453, E: pvelloso.dhe@teres.com.br

Vertamatti, E., Cta-Ita-Iei, 12228-900 – São Jose dos Campos, SP

Vertematti, J.C., Rua Jacarei,166-Pq.D. Henrique Iii, 06700-000 – Cotia, SP

Viana, P.M.F., Rua Jose Riga, 235-Samambaia, 13566-340 – São Carlos, SP

Vianna, A. de Camargo, Av. Djalma Dutra, 254-Centro, 14800-000 – Araraquara, SP. T: 16 3334 2141, E: ac_vianna@bol.com.br

Vianna, I.L., Av. Brasil, 1438-Cj. 1302, 30140-003 – Belo Horizonte, MG. T: 31 3213 1333, 31 3225 9375, E: ivianna@intersolo.com.br

Vicente, F.A., Av. Duque de Caxias, 3404, 86010-200 – Londrina, PR. T: 43 322 6862, 43 3391172, E: basestac@sercomtel.com.br

Vidal, D. de Mattos, Cta-Ita-Ieig, 12228-900 – São Jose dos Campos, SP. E: delma@infra.ita.cta.br

Vidal, I.G., Al. Amazonas, 363-Ap.104 Ii Alphaville, 06454-070 – Barueri, SP. T: 11 4166 3040, E: indiara.vidal@ig.com.br

Vidal, M.L., Rua Italia 1910, 14801-350 – Araraquara, SP

Viebrantz, V., Rua 7 de Setembro, 2570-Apto.302-Centro, 89102-400 – Blumenau, SC

Vieira, A., Rua Uruguai, 527 Ap.102-Tijuca, 20510-060 – Rio de Janeiro, RJ

Vieira, E., Sqs 207-Bloco H-Apto.504, 70253-080 – Brasilia, DF. E: erico.vieira@apls.com.br

Vieira, L.O.M., Rua Gustavo Corçao, 845-Apto.103-Recreio, 22790-150 – Rio de Janeiro, RJ

Vieira, M.C.R., Av. Marcos Konder, 1313-Sala 206, 88301-120 – Itajai, SC. T: 47 224 1537, 47 348 1036, E: ceciliafpolis@terra.com.br

Vieira, M.V.C.M., Rua São Tome, 144-Vila Progresso, 79080-390 – Campo Grande, MS

Vieira, M.A.G., Rua Caetés, 880-Apto.51, 05016-081 – São Paulo, SP
Vilar, O.M., Av. Dr. Carlos Botelho 1465, 13560-250 – São Carlos, SP
Villar, L.F. de Souza Av. Carandai, 467-Apto.202, 30130-060 – Belo Horizonte, MG
Villefort, L.F.C.B., Rua Teixeira de Freitas, 490-Apto.403, 30350-180 – Belo Horizonte, MG. T: 31 3275 2417, 31 3293 3201, E: lvillefort@golder.com.br
Vincenzo Jr, M. de, Rua Dom Constantino Barradas, 120-Apto.62, 04134-110 – São Paulo, SP
Viotti, C.B., Av. Bandeirantes, 665-Ap.1001, 30315-000 – Belo Horizonte, MG. T: 31 3225 9673, F: 31 3225 9673, E: cbviotti@rbeep.com.br
Viviani, E., Rua Floriano Peixoto, 216-Vila Prado, 13574-420 – São Carlos, SP
Volpe, J.C., Rua Barao do Triunfo, 576-Sala 404, 90130-100 – Porto Alegre, RS
Volpini, D.O., Rua Almir Tamandare, 859-B. Gutierrez, 30430-150 – Belo Horizonte, MG. T: 31 3335 8091, E: dvolpini@terra.com.br
Watanabe, M., Rua Catao, 523-Lapa, 05049-000 – São Paulo, SP. T: 11-3873 3399, E: ept01@ept.com.br
Weege, M.E.B., Av. Nacoes Unidas, 13797/14 And./Pred. 2, 04794-000 – São Paulo, SP
Winz, H. de Castro, Rua Dr. Nogueira Martins 634 Ap.93, 04143-020 – São Paulo, SP
Wolle, C.M., Rua Darwin 368-Apto.121-Bloco A, 04741-010 – São Paulo, SP
Wunder, E., Rua Afonso de Freitas, 320/Apto.161, 04006-051 – São Paulo, SP

Ximenes Filho, N. da Cunha, Av. Recife 5305 Estancia, 50781-000 – Recife, PE
Yassuda, A.J., Rua Girassol, 1033, 05433-002 – São Paulo, SP. T: 11 3819 0099, E: bureau@bureauprojetos.com.br
Yoshida, E.M., Rua Andre Saraiva, 710-Apto.22, 05626-001 – São Paulo, SP. T: 11-3501 0763, E: edimassa@ig.com.br
Yoshikawa, N.K., Ipt/Digeo/C.P. 0141/Cidade Universitaria, 01064-970 – São Paulo, SP
Zacarin, P.D., Rua Garibaldi, 1108-Apt.162, 14010-170 – Ribeirao Preto, SP
Zaclis, E., Rua Jose Maria Lisboa, 331-Ap.81, 01423-000 – São Paulo, SP. T: 11 3873 2500, 11 3884 5258, E: efzaclis@terra.com.br
Zavaleta, K.M.S., Rua Esteves Junior, 496-Apto.215-Centro, 88015-530 – Florianopolis, SC. T: 58 222 9810, E: kzavalet@matrix.com.br
Zendron Neto, A., Rua Germano Schaefer, 46, 88350-170 – Brusque, SC
Zillmann, A., Rua Vitor Buono, 49-Campos Maia, 12400-000 – Pindamonhangaba, SP. T: 11 9249 5231, 12 242 5993, E: azillmann@uol.com.br
Zingano, A.C., Av. Osvaldo Aranha, 99-Apto.507, 90035-190 – Porto Alegre, RS. T: 51 316 3594, 51 351 7654, F: 51 316 3394, E: andrezin@ufrgs.br
Zirlis, A.C., Rua Capiberibe, 647-Jardim Aeroporto, 04631-000 – São Paulo, SP. E: zirlis@solotrat.com.br
Zorzi, L., Rua Riveira, 202-Ap.402, 90670-160 – Porto Alegre, RS
Zorzi, O.R., Rua Barao de Itapagipe, 385/606-Bl. 1, 20261-000 – Rio de Janeiro, RJ

24

BULGARIA/BULGARIE

Secretary: Prof. Dr. Trifon Germanov, Secretary, Bulgarian Society for SMGE, University of Architecture, Civil Engineering & Geodesy Department of Geotechnics, 1 Christo Smirnenski Boul. BG-1164 Sofia. T: +359 2 63321/434/245, F: +359 2 656863, E: germanov_fte@uacg.acad.bg

Total number of members: 55

Alexiew, A., Professor, University of Architecture and Civil Engineering & Geodesy, Boul.Vitosha 43 BG-1000 Sofia. T: +359 2 9810692

Alexiew, D., Dr.-Ing., Huesker Synthetic & Co., Stationsweg 23, D-48712 Gescher GERMANY. T: +49 2542 2352, E: d.alexiew@gmx.net

Angelova, R., Research Associate Professor, Bulgarian Academy of Sciences, Geotechnical Division. Acad.G. Bonchev Str., Blok 24 BG-1113 Sofia. T: +359 2 9792262, F: 724638, E: angelova@geology.bas.bg

Baloushev, E., Associate Professor, University of Architecture and Civil Engineering & Geodesy, 14 Greben Planina Str., BG-1421 Sofia. T: +359 2 650429

Bejkoff, M., Dipl-Eng Professor, 79 Exarch Yosif Str., BG-1000 Sofia. T: +359 2 9834160

Borissov, Ju., Research Assistant, Road Research Institute, Complex Mladost-IA Bl.513/3 ap.64 BG-1729 Sofia. T: +359 2 761455

Boshinova, A., Chief Assistant Professor, University of Architecture, Civil Engineering & Geodesy, Complex "Slavia" Blok 25 entr.A., BG-1618 Sofia. T/F: +359 2 567968, E: ASSIA_FTE@uacg.acad.bg

Boshinov, B., Professor, University of Mining and Geology, Complex "Slavia" Blok 25 entr.A, BG-1618 Sofia. T/F: +359 2 567968

Boshinova-Popova, I., Assistant Professor, University of Mining and Geology, Complex "Slavia" Blok 25 entr.A, BG-1618 Sofia. T: +359 2 9672364

Christow, Ch., Professor Dr.-Ing., D-76275 Ettlingen, Neuwiesenrebenstr 13 Germany. T: 07243 14009, F: 07243 12076

Denev, D., Associate Professor, University of Architecture, Civil Engineering & Geodesy, Complex Mladost 2 Blok 235-B ap.114 BG 1799 Sofia. T: +359 2 752448, E: DOBRIN_DENEV@hotmail.com

Dontchev, P., Associate Professor, Geotechnical and Geoecological Laboratory Ltd., P.O. Box 409, BG-7000 Russe. T: +359 82 823436, E: north-east@starcointer.net

Etimov, T., Professor, University of Architecture, Civil Engineering & Geodesy, Ul.Rakovsky 130 BG-1000 Sofia. T: +359 2 880971

Evlogiev, I., Research Associate Professor, Bulgarian Academy of Sciences, 28 Tzar Osvoboditel str., P.O. Box 501, BG-7000 Russe. T: +359 82 223255, E: evlogiev@insoft. inetg.bg

Evstatiev, D., Research Professor, Bulgarian Academy of Sciences, Geotechnical Division. Acad.G.Bonchev Str., Blok 24 BG-1113 Sofia. T: +359 2 9792262, F: +359 2 724638, E: geodimo@geology.bas.bg

Filipov, K., Chief Assistant Professor, University of Architecture, Civil Engineering & Geodesy, Complex Mladost Blok 102-VIII., ap.121 BG-1156 Sofia. T: +359 2 719841

Frangov, G., Research Associate Professor, Bulgarian Academy of Sciences, Geological Institute. Acad.G.Bonchev Str., Blok 24 BG-1113 Sofia. T: +359 2 9572244, F: +359 2 718148, E: frangov@geology.bas.bg

Galaboff, N., Chief Assistant Professor, University of Architecture, Civil Engineering & Geodesy, Ul. "Banat" 12 BG-1407 Sofia. T: +359 2 624328

Germanov, T., Professor, University of Architecture, Civil Engineering & Geodesy, Complex Lyulin, Blok 224, ap.34 BG-1336 Sofia. T: +359 2 243232, E: germanov_fte@ uacg.acad.bg

Hamova, M., Associate Professor, High School of Construction Engineering, Vishneva str., 18 BG-1164 Sofia. T: +359 2 328084, E: hamova@yahoo.com

Ivanov, I., Chief Assistant Professor, University of Architecture and Civil Engineering& Geodesy, Bojanski vodopad str., Bl.241 Vh.9 ap.120 BG-1404 Sofia. T: +359 2 588685, E: IJI_fte@uacg.acad.bg

Ilov, G., Associate Professor, University of Architecture and Civil Engineering, Complex Mladost-2 Blok 235-A ap.20 BG-1799 Sofia. T: +359 2 752408, E: gilov@ techno-link.com

Ivanova, V., Dr. Research Assistant, University of Mining and Geology, Boul. Samokov 16 ap.33 BG-1113 Sofia. T: +359 2 721116

Jellev, J., Associate Professor, University of Architecture, Civil Engineering & Geodesy, Complex Sv.Troitsa Bl.145-B ap.53 BG-1309 Sofia. T: +359 2 2928729

Kalchev, I., Associate Professor, University of Architecture, Civil Engineering & Geodesy, 30 J.Boucher Blv. BG-1164 Sofia. T: +359 2 662104, E:kaltchev@mail. orbitel.bg

Karatchorov, P., Research Associate Professor, Bulgarian Academy of Sciences Geotechnical Research Station, Ul.Volov 30 P.O. Box 433, BG-7000 Russe. T: 234179, F: 230367

Karastanev, D., Research Associate Professor, Bulgarian Academy of Sciences, Geotechnical Division, Acad.G. Bonchev Str., Blok 24 BG-1113 Sofia. T: +359 2 7132262, F: +359 2 724638, E: doncho@geology.bas.bg

Kirov, B., Associate Professor, University of Architecture and Civil Engineering, Ul.Krivolak 48-II BG-1421 Sofia. T: +359 2 667350, E: kirov_fte@uacg.acad.bg

Kissiov, P., Associate Professor, Higher Transport School, Kv.Slatina Ul.Geo Milev, Blok 69-B BG-1577 Sofia. T/F: +359 2 709338

Kolev, M., Eng. Specialist, REA-Direction "Central Road and Bridges Laboratory", Complex Mladost 1A Bl.502 ap.104 BG-1729 Sofia. T: +359 2 760959

Kolev, Tch., MS Civil Engineer, Director Vodno Stopanstvo Ltd., Han Asparuh 58 BG-1000 Sofia. T: +359 2 986498, F: +359 2 9813745, E: vodno@ttm.bg

Kostov, V., Chief Assistant Professor, University of Architecture, Civil Engineering & Geodesy, Boul. "Eng. Ivan Ivanov" 70 ap.15 BG-1303 Sofia. T: +359 2 313001, E: kostov_fte@uacg.acad.bg

Markov, G., Dr. Eng., Energoproject, Ul.Platchkovski Manastir 21-A BG-1505 Sofia. T: +359 2 445755, E: markov_gt@ yahoo.com

Mihaylov, A., Assistant Professor, High School of Construction Engineering, Compex Lagera Bl.39 ap.14, BG-1612 Sofia. T: +359 2 9520266, E: vvisu@applet-bg.com

Mihova, L., Dr. Chief Assistant Professor, University of Architecture, Civil Engineering & Geodesy, Blvd. Sv. Naum 43-B, BG-1421 Sofia. T: +359 2 669378, E: MIHOVA_ FTE@uacg.acad.bg

Petkova, V., Chief Assistant Professor, University of Architecture, Civil Engineering & Geodesy, "Drushba 1" Bl.159-B ap.418 BG-1592 Sofia. T: +359 2 9783395

Petrov, P., Civil Engineer, Ministry of regional development & urbanization, Ul Kiril I Metodi 17-19 BG-1202 Sofia. T: +359 2 9877726, F: +359 2 9872517

Popov, N., Civil Engineer, University of Architecture, Civil Engineering & Geodesy, Ul.Stara Stena 1, BG-1421 Sofia. T: +359 2 666029

Sadgorski, W., Dr.-Ing., Rotbuchenstr. 73 D-8000 Munchen 90 Germany. T: 089 690797

Slavov, P., Research Assistant, Bulgarian Academy of Sciences, 28 Tzar Osvoboditel str., P.O. Box 501 BG-7000 Russe. T: +359 82 223255, E: petko@insoft.inetg.bg

Sofev, R., Chief Assistant Professor, University of Architecture, Civil Engineering, Boul. Tsar Boris III 257-B BG-1619 Sofia. T: +359 2 570958

Stakev, M., Research Associate Professor, Building Research Institute, Boul. Cherni Vrah 19A, BG-1421 Sofia. T: +359 2 9633911, E: e_stakev@techno-link.com

Stefanoff, G. (Chairman) Professor, University of Architecture, Civil Engineering & Geodesy, Ul.Latinka 35 BG-1113 Sofia. T: +359 2 658972

Stefanoff, St., Chief Assistant Professor, University of Mining and Geology, Ul.Latinka 35 BG-1113 Sofia. T: +359 2 664317, E: teddy_st@hotmail.com

Stoeva, P., Professor, University of Mining and Geology, Complex Lyulin, Bl.425 ap.76 BG-1336 Sofia. T: +359 2 248057

Todorov, M., Assistant Professor, University of Architecture Civil Engineering & Geodesy, Complex Borovo Bl.5 ap.4 BG-1618 Sofia. T: +359 2 589321, E: miros_fte@uacg.acad.bg

Toshev, D., Associate Professor, University of Architecture, Civil Engineering & Geodesy, Complex Mladost Blok 235 A BG-1799 Sofia. T: +359 2 744172, F: +359 2 656683, E: d_toshev_fhe@uacg.acad.bg

Toshkov, E., (Vice-Chairman) Professor, Building Research Institute, Ul.Elin Pelin 26, BG-1164 Sofia. T: +359 2 666091

Tzekov, G., Engineer, Baugrund Institut Knierium Gmbx., Complex Liulin Bl.409 B BG-1336 Sofia. T: +359 2 253480

Wenkov, W., Professor, University of Architecture Civil Engineering & Geodesy, Ul.Dimitar Dimov 18 BG-1164 Sofia. T: +359 2 663374

Collective members

University of Architecture, Civil Engineering & Geodesy, 1 Chr. Smirnenski Boul. BG-1421 Sofia. T: +359 2 63321, F: +359 2 655863

"ENERGOPROEKT" Plc – Hydropower & Geotechnics Unit, 51 J.Bourchier Blvd. BG-1407 Sofia. T: +359 2 9 600 600, F: +359 2 963 41 72, E: hydro@enpro.bg

"OVERGAS Engineering" Ltd., 5 Philip Kutev Str., BG-1407 Sofia. T: +359 2 9603 444, F: +359 2 9622193, E: overgaso@aster.net

TERRA CONSULT Ltd., Rodina 3 Blok1/G, BG-7006 Russe, T/F: 1359 82 270413

Geotechnical and Geoecological Laboratory Ltd., P.O.Box 409, BG-7000 Russe. T: +359 82 823436, E: north-east@starcointer.net

CANADA

Director General: Dr James Graham, The Canadian Geotechnical Society, P.O. Box 937, Alliston, Ontario, Canada L9R 1W1. T: 705 434 0916, F: 705 434 0917, E: cgs@cgs.ca, Website: www.cgs.ca

Total number of members: 589

Agar, J.G., O'Connor Assoc Environmental Inc., 1000-639-5th Avenue SW, Calgary, AB, T2P 0M9. T: 403294-4200, F: 403-294-4240

Agensky, N., Inspec-Sol Inc., 200-4600 Cote Vertu, Montreal, QC, H4S 1C7. T: 514-333-5151, F: 514-333-4674

Aggarwal, S.K., 1175 Potter Drive, Manotick, ON, K4M 1E2. T: 613-723-2411, F: 613-727-9580

Ahlfield, K., 10621 Hollymount Dr., Richmond, BC, V7E 4Z3. T: 604-273-0311, F: 604-279-4300

Ahmad, K.S.Q., 43 Major Oak Terrace, Scarborough, ON, M1V 3E4. F: 416-235-5240

Ahmad, S.A., R.R. #3, Schomberg, ON, L0G 1T0. T: 416-213-1255

Allen, D.C., B.C. Rail, P.O. Box 8770, Vancouver, BC, V6B 4X6. T: 604-984-5102, F: 604-984-5352

Alston, C., 102 Senator Reesor's Drive, Markham, ON, L3P 3E5. T: 905-415-1995, F: 905-415-1996

Aly, A., 1007-1176 Ouellette Avenue, Windsor, ON, N9A 6S9. T: 519-258-7599, F: 519-734-6888

Antunes, P.J., Saskatchewan Hwys & Transportation, 5th Floor, 350-3rd Avenue North, Saskatoon, SK, S7K 2H6

Arabshahi, H., 307-3033 Ospika Blvd. South, Prince George, BC, V2N 4L5. T: 250-564-3243, F: 250-562-7045

Armstrong, R., 1411 Seventh Avenue NW, Unit 216, Calgary, AB, T2N 0Z3. F: 403-247-4811

Arteau, J., IRSST, 505 de Maisonneuve Ouest, Montreal, QC, H3A 3C2. T: 514-288-1551, F: 514-288-9399

Arvidson, W.D., 140 Oakfield Place S.W., Calgary, AB, T2V 0J2. T: 403-251-5654, F: 403-251-9196

Atkinson, C.J.W., 149 Iroquois Avenue, London, ON, N6C 2K9. T: 519-685-6400, F: 519-685-0943

Au, V.C.S., Block 2, 11/F, FLAT C, Cotton Tree Mansions, Site 7, Whampoa Gardens, Kowloon, Hong Kong. T: 852-2330-1480, F: 852-2330-1564

Authier, J., B-Sol Ltee, 50 William-Dobel, Bai-Comeau, QC, G4Z 1T7. T: 418-296-6788, F: 418-296-4247

Ayres, B.K., 207-280 Heritage Way, Saskatoon, SK, S7H 5R2

Azam, S., 133 Michener Park, Edmonton, AB, T6H 4M4

Babuin, C.G., 1-2235 West 40th Avenue, Vancouver, BC, V6M 1W7. T: 604-685-0275, F: 604-684-6241

Baikie, L.D., Dalhousie University, Dept. of Civil Engineering, P.O. Box 1000, Halifax, NS, B3J 2X4. T: 902-494-3958, F: 902-494-3108

Balanko, L.A., HWA GeoSciences Inc., 19730-64th Avenue West, Suite 200, Lynnwood, WA, 980365957 USA. T: 604-684-4384, F: 604-684-5124

Balins, J.K., Golder Associates Ltd., 500-4260 Still Creek Drive, Burnaby, BC, V5C 6C6. T: 604-298-6623, F: 604-298-5253

Barbour, S.L., University of Saskatchewan, Dept. of Civil Engineering, Saskatoon, SK, S7N 0W0. T: 306-966-5369, F: 306-966-5427

Barker, J.F., University of Waterloo, Department of Earth Sciences, Waterloo, ON, N2L 3G1. T: 519-885-1211, Ext. 2103, F: 519-746-7484

Barlow, J.P., 5315-143 Street, Edmonton, AB, T6H 4E3. T: 403-436-2152, F: 403-435-8425

Barnes, D.J., 242-52152 Rge. Road 210, Sher-wood Park, AB, T8G 1A5. T: 403-436-2152, F: 403-435-8425

Barone, F., 600 Main Street, Glen Williams, ON, L7G 3T6. F: 905-567-6561

Bartle, H., Ministry of Forests, 2100 Labieux Road, Nanaimo, BC, V9T 6E9. T: 250-751-7073

Bathurst, R., Royal Military College, Department of Civil Engineering, PO Box 17000 STN Forces, Kingston, ON K7K 7B4. T: 613-541-6000, Ext. 6479, F: 613-545-8336, E: bathurst-r@rmc.ca

Beadle, M.E., 59 Notre Dame Crescent, London, ON, N6J 2G2

Beaton, N.F., 9595 Pinewell Cres., Richmond, BC, V6A 2C7. T: 604-525-2656

Beaudoin, P., 414 Des Prés ouest, Ste-Odile-sur-Rimouski, QC, G5L 7B5. T: 418-722-3770

Becker, D.E., Golder Associates, 2180 Meadowvale Blvd., Mississauga, ON, L5N 5S3. T: 905-567-4444, F: 905-567-6561

Bedard, C., 904 Genereux, Repentigny, QC, J5Y 1T7. T: 514-636-4102, F: 514-636-8447

Bedell, P.R., Golder Associates, 500 Nottinghill Road, London, ON, N6K 3P1. T: 519-471-9600, F: 519-471-4707

Been, K., 11 Easthorpe View, Bottesford, Nottinghamshire, NG13 0DL UK

Belliveau, D.A., 78 Murphy Crescent, Regina, SK, S4X 1B7. T: 306-789-4463, F: 306-347-8595

Belyea, P.S., AGRA Earth & Environmental, 80 Driscoll Crescent, Moncton, NB, E1E 3R8. T: 506-859-1940, F: 506-857-9974

Benko, B., Golder Associates Ltd., 202-2790 Gladwin Road, Abbotsford, BC, V2T 4S8. T: 604-820-8786, F: 604-850-8756

Bertrand, D.G., 17975, rue Foster, Pierrefonds, QC, H9K 1L2

Bews, B.E., 202 Zeman Court, Saskatoon, SK, S7K 7W5. T: 306-966-5410, F: 306-966-5427

Biggar, K.W., University of Alberta, Department of Civil Engineering, Edmonton, AB, T6G 2G7. T: 403-492-2534, F: 403-492-8198

Birn, L.J., 4691 Penhallow Road, Mississauga, ON, L5V 1G2

Blais, J.-C., 82 Beausoleil, Gatineau, QC, J8T 7G8. F: 819-243-4629

Blanchet, R., 591 Le Breton, Longueuil, QC, J4G 1R9. T: 514-674-4901, F: 514-674-3370

Blanchette, G., 1364 Adelard-Plourde, Chicoutimi, QC, G7H 6J4. T: 418-696-4500, Ext. 6893, F: 418-696-6896

Bleakney, M.R., NB Department of Transportation, P.O. Box 6000, Fredericton, NB, E3B 5H1. T: 506-453-2674, F: 506-457-6714

Block, J., Block & Associates Ltd., 18 Sweetwood Bay, Winnipeg, MB, R2V 2S2. T: 204-334-5356, F: 204-339-7976

Bobey, L.W.M., 623 Wotherspoon Close, Edmonton, AB, T6M 2K2. T: 403-438-1460, F: 403-437-7124

Boncompain, B., 3-521 Rue Prince Arthur West, Montreal, QC, H2X 1T6. T: 514-849-2671, F: 514-849-2671

Boone, L.C., Box 3091, Manuels, NF, A1W 1B6. T: 709-576-1428, F: 709-576-2126

Boone, S., Golder Associates Ltd., 2180 Meadowvale Blvd., Mississauga, ON, L5N 5S3. T: 905-567-4444, F: 905-567-6561

Borowy, B.J., 19 Cityview Circle, Barrie, ON, L4N 7V2, F: 705-730-1555

Bosdet, B.W., Golder Associates, 100-388 First Avenue, Kamloops, BC, V2C 6W3. T: 250-828-6116, F: 250-828-1215

Botham, L.C., Golder Associates Ltd., 209-2121 Airport Drive, Saskatoon, SK, S7L 6W5. T: 306-665-7989, F: 306-665-3342

Bouchard, R., 6332, Lapointe, Laterriere, QC, G7N 1S9. T: 418-547-5716, F: 418-547-0374

Bouclin, G., 32 Leroux, Coteau-du-lac, QC, J0P 1B0. T: 514-763-2751, F: 514-371-5227

Boulanger, R.W., 4237 Dogwood Place, Davis, CA, 95616 USA. T: 916-752-2947, F: 916-752-8924

Bousquet, R., 903-2525-B, ave du Havre des Iles, Laval, QC, H7W 5C5. T: 514-393-1000, F: 514-393-9540

Bouthot, M., SAGEOS, 3000, rue Boulle, St-Hyacinthe, QC, J2S 1H9. T: 450-771-4608, F: 450-778-3901

Bozzo, I., Hydro-Quebec, Barrages et Ouvrages civils, 855, rue Ste-Catherine est 18 etage, Montreal, QC, H2L 4P5. T: 514-840-3000, F: 514-840-5112

Brachman, R.W.I., University of Alberta, Dept. of Civil Environmental Eng., 220 Civil/Electrical, Eng. Bldg, Edmonton, AB, T6G 2G7. T: 780-492-5112, F: 780-492-8198

Branco, P., Thurber Engineering Ltd., 101-170 Evans Avenue, Etobicoke, UN, M8Z 5Y6. T: 416-503-3600, F: 416-503-3010

Brinovec, W.V., 11 Camwood Crescent, Don Mills, ON, M3A 3L3. T: 416-444-0001

Brockbank, W.J., Reinforced Earth Company Ltd., The Enterprise Centre, 229-1550 Enterprise Road, Mississauga, ON, L4W 4P4. T: 905-564-0896, Ext. 307, F: 905-564-2609

Brosseau, H.N., 4198 Hingston Ave., Montreal, QC, H4A 2J7. T: 514-333-5151

Brotherton, M., 29 Watson St., Red Deer, AB, T4N 5X7

Brown, B.A., 17 Potato Road, Enfield, NH, 03748 USA. T: 603-448-1562, F: 603-448-3216

Brown, J.D., P.O. Box 213, 66 Main Street, Wolfville, NS, B0P 1X0. T: 902-468-7777, F: 902-468-9009

Bruch, Phil, 266 Waterloo Cres., Saskatoon, SK, S7H 4G5. T: 306-975-0444, F: 306-955-2446

Bruno, B., UQAT-VRSTM, 445 bowl De l'Universite, Rouyn-Noranda, QC, J9X 5E4. T: 819-762-0971, Ext. 2531, F: 819-797-6672

Buck, G.F., Thurber Engineering Ltd., 210-4475 Viewmont Avenue, Victoria, BC, V8Z 6L8. T: 604-727-2201, F: 604-727-3710

Budkowska, B.B., University of Windsor, Civil & Environmental Eng. Dept., 401 Sunset, Windsor, ON, N9B 3P4. T: 519-253-4232, Ext. 250, F: 519-971-3686

Burnotte, F., Université de Sherbrooke, Département de Génie Civil, 2500 boul. Universite, Sherbrooke, QC, J1K 2R1. T: 819-821-8000, Ext. 2992, F: 819-821-7974

Burwash, W.J., 2648 Signal Hill Drive SW, Calgary, AB, T3H 2T8. T: 403-246-1500, F: 403-252-4884

Cabral, A.R., 2410, Prospect, Sherbrooke, QC, J1J 4G2. T: 819-821-7906

Cameron, R., 128 Cote Bay, Fort McMurray, AB, T9H 4R9. T: 403-790-5920, F: 403-790-5657

Campbell, C., Morrow Environmental Consultants, 9531-42nd Avenue, Edmonton, AB, T6E 5R2

Capozio, N.U., 147 Rue Mozart, Dollard des Ormeaux, QC, H9G 2Z7. T: 514-737-9139, F: 514-737-2526

Carey, E., Carey Geotechnical Engineering, 503 Chebucto Street, P.O. Box 129, Baddeck, NS, B0E 1B0. T: 902-295-2251, F: 902-295-2256

Carnaffan, P., 100 Larkin Drive, Nepean, ON, K2J 1C1. T: 613-738-0708, F: 613-738-0721

Carrier, L., 2530 Degas, St-Emile, QC, G3E 1M4

Cascante, G., University of Waterloo, 200 University Avenue, Dept. of Civil Engineering, Waterloo, ON, N2L 3G1. T: 519-888-4567, Ext. 2098, F: 519-888-6197

Cassie, J.W., BGC Engineering Inc., #1170, 840-7th Avenue SW, Calgary, AB, T2P 3G2. T: 403-250-5185, F: 403-250-5330

Chai, Y.J., 923 Burley Drive, Edmonton, AB, T6R 1X5. T: 403-496-8931, F: 403-496-4671

Chan, D.H.-K., University of Alberta, Box 60102, U.A. Postal Outlet, Edmonton, AB, T6G 2S4. T: 780-492-4725, F: 403-492-8198

Chan, P., 111 Wintermute Blvd., Scarborough, ON, M1W 3M8

Chan, W.K.D., 8532-10th Avenue NW, Edmonton, AB, T6K 1X1. T: 403-435-8425, F: 403-435-8425

Chatillon, M., 784 de Namur, St. Lambert, QC, J4S 1Z5. T: 514-255-6375, F: 514-875-2666

Chatterji, P.K., Thurber Engineering Ltd., 101-170 Evans Avenue, Etobicoke, ON, M8Z 5Y6. T: 416-503-3600, F: 416-503-3010

Chenaf, D., Royal Military College of, Civil Engineering Department, P.O. Box 17000, Stn Forces, Kingston, ON, K7K 7B4. T: 613-541-6000 x6603, F: 613-545-8336

Cheng, E., 439 Weldrick Road East, Richmond Hill, ON, L4B 2M5

Cherniawski, M.J., 4021-113 Avenue, Edmonton, AB, T5W 0R2. T: 403-436-2152, F: 403-435-8425

Chiu, B., Trow Consulting Engineers Ltd., 15 Cuddy Blvd., London, ON, N5V 3Y3. T: 519-453-1480, F: 519-453-1551

Choquet, P., Roctest Ltee., 665 Pine, St. Lambert, QC, J4P 2P4. T: 514-465-1113, F: 514-465-1938

Chung, E.Y.-F., 128 Shaughnessy Pl, Waterloo, ON, N2T 1C8. T: 519-885-6220, F: 519-885-3490

Clark, J.I., 6 Dover Place, St. John's, NF, A1B 2P5. T: 709-739-6251, F: 709-737-4706

Collins, G., 1294 Huntingdale Court, Gloucester, ON, K1J 1B1. T: 613-227-5864

Collins, R.T., AGRA Earth & Environmental Ltd., 5406-52nd Avenue, P.O. Box 1518, Lloydminster, AB, S9V 1K5. T: 708-875-8975, F: 780-875-1970

Comeau, S., Brunswick Eng Group 1996 Ltd., 334 Dover Street, P.O. Box 760, Campbellton, NB, E3N 3H2. T: 506-789-1224, F: 506-789-1222

Cooke, B.H., 96 Orchard View Blvd., Toronto, ON, M4R 1C2. T: 905-678-7820, F: 905-678-7131

Corbett, I., 13014 Dixie Road, R.R. #4, Brampton, ON, L6T 3S1. T: 905-796-2650

Corkum, A.G., 205 Newton Place, 8515-112th Street, Edmonton, AB, T6G 1K7. T: 403-250-8850, F: 403-291-0186

Cormier, Y., 12 Patrick Street, Moncton, NB, E1C 0A8. F: 506-634-8104

Costa, J.M.A., 2562 Pollard Drive, Mississauga, ON, L5C 3H1. T: 416-567-4444, F: 416-567-6561

Coyne, L.C., Golder Associates Ltd., 2180 Meadowvale Blvd., Mississauga, ON, L5N 5S3. T: 905-567-4444, F: 905-567-6561

Crans, C.G., 408 Traviss Drive, Newmarket, ON, L3Y 7J7. T: 416-881-2882, F: 905-881-2564

Crawford, H.S., Thurber Engineering Ltd., 190-550 71st Avenue SE, Calgary, AB, T2H 0S6. T: 403-253-9217, F: 403-252-8159

Crooks, J.H.A., Golder Associates, 1000-940 6th Avenue SW, 10th Floor, Calgary, AB, T2P 3T1. T: 403-299-5646, F: 403-299-5606

Crowe, R.E., ECI Technologies, 910-100 Millside Drive, Milton, ON, L9T 5E2. T: 905-878-6231, F: 905-878-0235

Cruden, D.M., University of Alberta, Dept. of Civil Engineering, Edmonton, AB, T6G 2G7. T: 780-492-5923, F: 780-492-8198

Csanyi, L.S., 202-830 East 6th Avenue, Vancouver, BC, V5T 1M1. T: 604-874-1245, F: 604-874-2358

Cunningham, M.I., 47 Locheland Crescent, Nepean, ON, K2G 6H2. T: 613-224-5864, F: 613-224-9928

Curran, J.H., 31 Balsam Avenue, Toronto, ON, M4E 3B5. T: 416-698-5402, F: 416-698-0908

Dahlman, A.E., 36090 Regal Parkway, Abbotsford, BC, V3G 1L1. T: 604-294-3811, F: 604-294-4664

Dascal, O., 303-4444 ouest, rue Sherbrooke, Montreal, QC, H3Z 1E4. T: 514-289-7461, F: 514-289-7342

Davies, M.P., 7755 Teakwood Place, Vancouver, BC, V5S 4A5. T: 604-473-5304, F: 604-273-4387

Davis, J.B., Golder Associates, 2180 Meadowvale Blvd., Mississauga, ON, L5N 5S3. T: 905-567-4444, F: 905-567-6561

DeGroot, D.J., University of Massachusetts, 38 Marston Hall, Amherst, MA, USA 01003. T: 413-545-0088, F: 413-545-2840

Denby, N.P., Cook Pickering & Doyle Ltd., 141 East 7th Avenue, Vancouver, BC, V5T 1M5. T: 604-879-0494, F: 604-879-6522

Deschenes, J.-H., 820 Buchanan, St. Laurent, QC, H4L 2V1. T: 1-514-747-0209

Dielemans, H.J., 785 D'Andrea Court, Windsor, ON, N9G 2N3. T: 519-945-5500, F: 519-945-4481

Dietrich, J.B., Peto MacCallum Ltd., 25 Sixth Avenue, Kitchener, ON, N2C 1P9. T: 519-893-7500, F: 519-893-0654

Dimitriu, D.V.S., 618-11873 Tecumseh Road, Tecumseh, ON, N8N 2K1, F: 519-969-0160

Dixon, D.A., 26 Shields Street, Winnipeg, MB, R2C 4E1. T: 204-753-2311

Diyaljee, V.A., 2408-113A Street, Edmonton, AB, T6J 4Y2
D'Onofrio, B., 8 West Palm Court, North York, ON, M9M 1R7. T: 416-745-5168, F: 416-745-5158
Donolo, Jr., L., Petrifond Foundation Co. Ltd., 8320 Blvd. St. Laurent, Montreal, QC, H2P 2M3. T: 514-387-7838, F: 514-387-9684
Downs, D.P., Levelton Engineering Ltd., 1935 Bollinger Road, Nanaimo, BC, V9S 5W9. T: 604-475-1077, F: 604-753-1203
Dowse, B.E.W., 34 Evergreen Bay S.W., Calgary, AB, T2Y 3E9. T: 403-248-4331, F: 403-278-5848
Drouin, R., Fondex Outaouais, 123 Jean-Proulx, Hull, QC, J8Z 1T4. T: 819-778-1770, F: 819-778-6302
Dundas, D.H., 102 Coles Avenue, Woodbridge, ON, L4L 1L9. T: 416-235-3482, F: 416-235-5240
Dussault, M., 1080 Rue Vanier, St. Laurent, QC, H4L 1S9. T: 514-624-4901, F: 514-624-3370
Dussault, R.G., 3120 Rue Savard, St. Laurent, QC, H4K 1T8, F: 514-336-3198
Dyck, R.A., 838 Oakerwald Avenue, Winnipeg, MB, R3T 1N1. T: 204-475-4133, F: 204-477-9194
Dyregrov, A.O., Dyregrov Consultants, 1666 Dublin Avenue, Winnipeg, MB, R3H 0H1. T: 204-632-7252, F: 204-632-1442
Eason, A., 124 Larchdale Crescent, Winnipeg, MB, R2K 0C1. T: 204-475-4133, F: 204-477-9194
Edmond, P.L., 16 Filion Crescent, Kanata, ON, K2M 1V6
Egyir, P.K., 4567 Vellencher Road, Prince George, BC, V2L 2V8. T: 250-565-6152, F: 250-565-6671
Eigenbrod, K.D., Lakehead University, Dept. of Civil Engineering, Thunder Bay, ON, P7B 5E1. T: 807-343-8363, F: 807-343-8928
Eivemark, M.M., Jacques Whitford & Associates, 1-3771 North Fraserway, Burnaby, BC, V5J 5G5. T: 604-436-3014, F: 604-436-3074
El Naggar, M.H., University of Western Ontario, Dept. of Civil Engineering, Geotechnical Research Centre, London, ON, N6A 5B9. T: 519-861-3344
Emery, J.J., John Emery Geotechnical Eng. Ltd., 1-109 Woodbine Downs Blvd., Etobicoke, ON, M9W 6Y1. T: 416-213-1060, Ext. 226, F: 416-213-1070
Ennis, S., Norwest Mine Services, 1022-475 Howe Street, Vancouver, BC, V6C 2B3. T: 604-602-8992, F: 604-602-8951
Evans, R.J.R., Paine and Associates, 17505-106th Avenue, Edmonton, AB, T6N 1C8. T: 403-462-1288, F: 403-450-1994
Fabius, M., DST Consulting Engineers Inc., 605 Hewitson Street, Thunder Bay, ON, P7B 5V5. T: 807-623-2929, F: 807-623-1792
Faghihi, M., 3571 Moresby Drive, Richmond, BC, V7C 4G7, F: 604-241-1145
Fellenius, B.H., 735 Ludgate Court, Ottawa, ON, K1J 8K8. T: 613-748-3232, F: 613-748-7402
Ferris, G.W., 642 Sandringham Place NW, Calgary, AB, T3K 3V7. T: 403-263-2556
Filson, H.L., 1011-11 Innovation Blvd., Saskatoon, SK, S7N 3H5. T: 306-975-6036, F: 306-975-4594
Finn, W.D., Liam, University of British Columbia, Department of Civil Engineering, 2324 Main Mall, Vancouver, BC, V6T 1Z4. T: 604-822-4938, F: 604-822-6901
Fisher, D.G., Haddad Geotechnical Inc., 7321 Victoria Park Ave., Unit 21, Markham, ON, L3R 2Z8. T: 416-475-0951, F: 416-475-8338
Fitzell, T.P., Golder Associates Ltd., 500-4260 Still Creek Dr., Burnaby, BC, V5C 6C6. T: 604-298-6623, F: 604-298-5253
Fleming, I.R., 548 Main Street, Glen Williams, ON, L7G 3T2. T: 905-469-3400, F: 905-469-3404
Fortier, R., Assistant Professor, Department de, geologie et de genie geologique, Pavillon Pouliot, Universite Laval, Sainte-Foy, QC, G1K 7P4. T: 418-656-2746, F: 418-656-7339
Forward, R., 9 Crystal Court, Grand Bay, Westfield, NB, E5K 2C1. T: 506-457-3200, F: 506-452-7652
Fredlund, D.G., University of Saskatchewan, Dept. of Civil Engineering, 57 Campus Drive, Saskatoon, SK, S7N 5A9. T: 306-966-5374, F: 306-966-5427
Froc, G.D.L., AGRA Earth & Environmental Ltd., 608 McLeod Street, Regina, SK, S4N 4Y1. T: 306-721-7100

Gadsby, J.W., Gadsby Consultants Ltd., 14-2425 Edgemount Blvd., North Vancouver, BC, V6P 2L2. T: 604-669-7780, F: 604-669-1779
Gaffran, P.C., 1101-945 Jervis Street, Vancouver, BC, V6E 2B8. T: 604-528-2706, F: 604-528-1940
Gagné, J., Jacques Gagné Experts-Conseils Inc., 1143, Lamy, Cap-de-la-madeleine, QC, G8V 1R7. T: 819-370-8158, F: 819-370-6016
Gan, J.K., M.D.Haug & Associates Ltd., 1-320 Jessop Avenue, Saskatoon, SK, S7N 1Y6. T: 306-966-2735, F: 306-651-3676
Gareau, L., Golder Associates Ltd., 209-2121 Airport Drive, Saskatoon, SK, S7L 6W5. T: 306-665-7989, F: 306-665-3342
Garga, V.K., 1204-211 Wurtemburg Street, Ottawa, ON, K1N 8R4. T: 613-562-5800, Ext. 6143, F: 613-562-5173
Gaskin, P.N., Queens University, Dept. of Civil Engineering, Kingston, ON, K7L 3N6. T: 613-533-2134, F: 613-545-2128
Georgiou, D.N., Trow Consulting Engineers Ltd., 807 Harold Cres., Thunder Bay, ON, P7C 5H8. T: 807-623-9495, F: 807-623-8070
Gerry, B.S., 66 Regent Street, Lindsay, ON, K9V 3V1. T: 519-324-9144
Gervais, R., 501 de la Rabastaliere est, St-Bruno, QC, J3V 2A9. T: 514-840-3000, Ext. 5430, F: 514-840-5112
Ghaly, A., Union College, Civil Engineering Dept., Schenectady, NY, 123082311 USA. T: 508-388-6515, F: 518-388-6789
Goel, H., 115 Romfield Circuit, Thornhill, ON, L3T 3H7. T: 905-940-8509, F: 905-940-8192
Gohl, W.B., Pacific Geodynamics Inc., 1168 Skana Drive, Delta, BC, V4M 2L4. T: 604-943-0350, F: 604-943-6190
Gomez, J., 241 Second Avenue North 335, Sudbury, ON, P3B 4A7
Gonsalves, L., 51 Fulbert Crescent, Scarborough, ON, M1S 1C5
Gonsalves, S.E.M., Trow Consulting Engineers Ltd., 1595 Clark Boulevard, Brampton, ON, L6T 4V1. T: 905-793-9800, F: 905-793-0641
Good, A.B., 3450 Emerald Drive, N. Vancouver, BC, V7R 3B7. T: 604-528-8182, F: 604-528-1883
Gorman, A.E., 34 Rose Way, Markham, ON, L3P 3V1. T: 416-785-5110, F: 416-785-5120
Gossen, K.A., 1202-311 Sixth Avenue North, Saskatoon, SK, S7K 7A9
Gourdeau, M., 58, Rue Isaie, Beauport, QC, G1C 2S5. T: 418-644-7036, F: 418-644-9662
Grabinsky, M.W., University of Toronto, 35 St. George Street, Department of Civil Engineering, Toronto, ON, M5S 1A4. T: 416-978-7130, F: 416-978-6813
Graham, C., 5359 Dalhurst Crescent, NW, Calgary, AB, T3A 1P6
Graham, J., 74 Montclair Bay, Winnipeg, MB, R3T 4B3. T: 204-474-9682, F: 204-474-7513
Gray, B.R., 2846 Termini Terrace, Mississauga, ON, L5M 5S3. T: 416-785-5110, F: 416-785-5120
Gray, B.S., 10 Cove Court, P.O. Box 247, Munster, ON, K0A 3P0. T: 819-775-4602, F: 819-775-4912
Gray, M.N., P.O. Box 187, 10226 Old Pine Crest Road, Norval, ON, L0P 1K0. T: 905-823-9060, F: 905-855-8173
Green, H., Paul, Kacqies Wjotfprd, 500-703 Sixth Avenue SW, Calgary, AB, T2P 0T9. T: 403-263-7113
Gregoire, M., 22-62 Scurfield Blvd., Winnipeg, MB, R3Y 1M5. T: 204-489-2964, Ext. 23, F: 204-489-3014
Griffiths, F.J., Jacques, Whitford Limited, 200-2781 Lacaster Road, Ottawa, ON, K1B 1A7. T: 613-738-0708, F: 613-738-0721
Guillemette, R., Cogemat Inc., 201 Blainville Ouest, Ste-Therese, QC, J7E 1Y4. T: 514-435-6159, F: 514-435-2407
Gutwein, M.D., Levelton Engineering Ltd., 103- 34609 Delair Rd., Abbotsford, BC, V2S 2E1. T: 604-855-0206, F: 604-853-1186
Haché, J.G.A.R., Jacques, Whitford & Associates Ltd., 2781 Lancaster Rd., Ste. 200, Ottawa, ON, K1B 1A7. T: 613-738-0708, F: 613-738-0721
Hall, B.E., 2871 Lyndene Road, N. Vancouver, BC, V7R 1E2. T: 604-436-3014, F: 604-436-3752
Hammamji, Y., 11255 Suzor Cote, Montreal, QC, H3M 2H4. T: 514-840-3000, Ext. 4596, F: 514-840-3199

Hammoud, A.M., 672 Longworth Road, London, ON, N6K 4W2. T: 519-471-9600, F: 519-471-4707

Hanna, A.M., Concordia University, Dept. Bldg., Civil & Envir. Eng., 1455 de Maisonneuve W., Montreal, QC, H3G 1M8. T: 514-848-7808

Harrington, E.J., 1295 Haywood Avenue, West Vancouver, BC, V7T 1V2. T: 604-439-0922, F: 604-439-9189

Harris, M.C., 157-52559 Hwy. 21, Sherwood Park, AB, T8A 4S6. T: 403-438-1460, F: 403-437-7125

Hartford, D.N.D., B.C. Hydro, 6911 Southpoint Drive, (A02) Burnaby, BC, V3N 4X8. T: 604-528-2423, F: 604-528-2444

Haug, M.D., 232-111 Research Drive, Saskatoon, SK, S7N 3R2. T: 306-934-7527, F: 306-934-7528

Hawson, H., Golder Associates, 500-4260 Still Creek Drive, Burnaby, BC, V5C 6C6. T: 604-298-6623, F: 604-294-3811

Hawton, K.E., Knight Piesold, 34 Commerce Crescent, P.O. Box 10, North Bay, ON, P1B 8G8. T: 705-476-2165, F: 705-474-8095

Hayes, J.A., Site Investigation Services Ltd., 785 The Kingsway, Peterborough, ON, K9J 6W7. T: 705-743-6850, F: 705-743-6854

Hefny, A.M., Nanyang Technological University, School of Civil & Structural Eng., Singapore, Singapore, 639798. T: 657906309, F: 657910676

Heinz, H.K., 76 Deermoss Crescent SE, Calgary, AB, T2J 6P4. T: 403-253-9217, F: 403-252-8159

Heystee, R.J., Ontario Power Generation, 700 University Avenue, Toronto, ON, M5G 1X6. T: 416-592-5078, F: 416-592-4485

Hill, D.W., 310 MacBeth Crescent, West Vancouver, BC, V7T 1V7. T: 604-684-4384, F: 604-684-5124

Ho, D.Y.F., An-Geo Environmental Consultants, 204-8708 48th Avenue, Edmonton, AB, T6E 5L1. T: 403-450-3377, F: 403-450-3232

Ho, K.S., 28 Brimwood Crescent, Richmond Hill, ON, L4B 4B6. T: 1-416-751-6565, F: 416-751-7592

Ho, R.H.C., 11-A Caroline Garden, 101 Caroline Hill Road, Hong Kong. T: 852-2828-5757, F: 852-5827-18236

Holubec, I., 1117 Notley Crescent, Oakville, ON, L6M 1H4. T: 905-825-8793

Horvath, R.G., P.O. Box 43, Waterdown, ON, L0R 2H0. T: 416-525-9140 x24914, F: 416-529-9688

Houle, A.C., R.R. 1, 7935 Jock Trail, Richmond, ON, K0A 2Z0. T: 613-692-8686, F: 613-692-6411

Howie, J.A., 952 West 22nd Avenue, Vancouver, BC, V5Z 2A1. T: 604-736-6210, F: 604-254-4664

Hryhoruk, C.D., 104 Niagara Street, Winnipeg, MB, R3N 0T9

Huang, C.S., Chih S. Huang & Associates, Inc., 11-2750 Fourteenth Avenue, Markham, ON, L3R 0B6. T: 416-475-0784, F: 416-475-5127

Huard, N., 368, du Bosquet, Rimouski, QC, G5L 8V8. T: 418-723-1144, F: 418-722-4691

Huber, F., Greater Vancouver Regional Dist., 4330 Kingsway, Burnaby, BC, V5H 4G8. T: 604-436-6945, F: 604-432-6297

Hughes, J.M., 804-938 Howe Street, Vancouver, BC, V6Z 1N9, F: 604-331-4452

Hui, K.-S., 2739 East 6th Avenue, Vancouver, BC, V5M 1R6

Huma, I., 76 Shendale Drive, Rexdale, ON, M9W 2B5. T: 416-793-9800, F: 416-740-8723

Inglis, D.J., 4765 Spurraway Road, Kamloops, BC, V2H 1M6. T: 604-672-6721

Jagani, M.M., 63 Connery Crescent, Markham, ON, L3S 4E6. T: 905-294-3079, F: 905-474-0601

Jagdat, R., 81 Rakewood Crescent, Scarborough, ON, M1V 1V6. T: 416-492-4000, F: 416-750-3926

James, M., 3721 St-Ambroise, Montreal, QC, H2C 2C4

Jansons, K.J., 680 Telstar Avenue, Sudbury, ON, P3E 5N2. T: 705-674-9681, F: 705-674-8271

Janzen, P., 634 Oxford Street, Winnipeg, MB, R3M 3K1. T: 204-488-1614

Jardine, K.V., 5 Sunny Lane, Quispamsis, NB, E2G 1N5

Jean, P., 3260 Chemin de la Riviere, Jonquiere, QC, G7X 7V6. T: 418-547-5716, F: 418-547-0374

Jenkins, A.K., 46 Edgeview Drive NW, Calgary, AB, T3A 4V3. T: 406-267-6100

Jenkins, R.H., 2125 County Road 7, R.R. #4, Picton, ON, K0K 2T0

Jinks, A.R., AGRA Earth & Environmental Ltd., 219 18th Street S.E., Calgary, AB, T2E 6J5. T: 403-248-4331, F: 403-248-2188

Jo, K.-Y., 3030, rue Lechasseur, #7, St-Foy, QC, G1W 1L1. T: 418-656-2206

Johnson, K.R., 398 Hickling Trail, Barrie, ON, L4M 6B2. T: 705-739-8355, F: 705-739-8369

Johnston, K.L., 406-280 Cypress Street, Nanaimo, BC, V9S 1R9. T: 250-756-2256, F: 250-756-2686

Jones, K.W., EBA Engineering Consultants, 14535-118 Avenue, Edmonton, AB, T5L 2M7. T: 403-451-2121, F: 403-454-5688

Jorden, E., Geotechnology Ltd., 23 Roslyn Drive, Dartmouth, NS, B2W 2M2. T: 902-435-4939, F: 902-435-5840

Juneau, R., 406 Ch. du Roy, St. Augustin, QC, G3A 1W8. T: 418-845-0858, F: 418-845-0300

Jurgens, E.I., 2586 Robin Drive, Mississauga, ON, L5K 2H9. T: 416-252-5315, F: 416-252-4599

Justason, M., Berminghammer Foundation Equip., Wellington St. Marine Ter., Hamilton, ON, L8L 4Z9. T: 905-528-0425, F: 905-528-6187

Kack, G.J., MHPM Project Managers Inc., 101-3027 Harvester Road, Burlington, ON, L7N 3G7. T: 905-639-2425, F: 905-639-2810

Keegan, T., Senior Geotechnical Engineer, CN West, Fl. 16 10004-104th Avenue, Edmonton, AB, T5J 0K2. T: 403-421-6935, F: 403-421-6120

Keil, L.D., Palmas del Mar, Candelero Drive, Fairway Courts #788, Humacao, 00791 Puerto Rico. T: 403-932-6288, F: 403-932-6597

Keith, S., 73 Earl Grey Road, Toronto, ON, M4J 3L6. T: 604-685-0147, F: 604-685-0543

Kellestine, W.M., 43 Kirk Drive, London, ON, N6P 1E2. T: 519-471-9600, F: 519-471-4707

Kelley, B.L., #728 67th Avenue SE, Apt. 102, Calgary, AB, T2V 0M1. T: 403-259-6683, F: 403-233-8700

Kelly, D.J., R.R. 4, Stratford, ON, N5A 6S5. T: 519-893-7500, F: 519-893-0654

Kelly, S., GeoNorth Engineering Ltd., 302-1777 Third Avenue, Prince George, BC, V2L 3G7. T: 250-564-4304, F: 250-564-9323

Kenyon, R.M., KGS Group, 865 Waverley Street, 3rd Floor, Winnipeg, MB, R3T 5P4. T: 204-896-1209, F: 204-896-0754

Kerr, T.F., Knight Piesold, 34 Commerce Crescent, P.O. Box 10, North Bay, ON, P1B 8G8. T: 705-476-2165, F: 705-474-8095

Kim, T.C., 4257 Camaro Court, Mississauga, ON, L4W 3R1. T: 416-235-3506, F: 416-235-5240

King, A.D., P.O. Box 55, Brigus, NF, A0A 1K0

King, G.L., R.R. #3, Prince Albert, SK, S6V 5R1. T: 306-953-3509, F: 306-953-3533

King, L.C., 2834-6715 Hunterview Drive NW, Calgary, AB, T2K 5C8. T: 403-270-0399, F: 403-270-9200

King, R.D., M.J. O'Connor & Assoc. Ltd., 1000-639 5th Avenue SW, Calgary, AB, T2P 0M9. T: 403-294-4200, F: 403-294-4240

Kingerski, D.E., The City of Winnipeg, 310-10 Fort Street, Winnipeg, MB, R3C 4X5. T: 204-986-5159

Kirkham, T.L., Manitoba Hydro, Transmission & Civil, 1100 Waverley Street, Winnipeg, MB, R3C 2P4. T: 204-474-4570

Klohn, E.J., 3412 Canterbury Drive, Surrey, BC, V4P 2N5

Knapik, D., 5 Guenette Crescent, Spruce Grove, AB, T7X 3G8

Knight, M.A., Department of Civil Engineering, University of Waterloo, Waterloo, ON, N2L 3G1. T: 519-888-4678, Ext. 3919, F: 519-888-6137

Kochan, D.H., Director of Environmental Services, 600-9th Street, Canmore, AB, T1W 2T2. T: 403-687-1504, F: 403-678-1524

Kofoed, W., J.R. Paine & Associates Ltd., 14 Burns Road, Whitehorse, YT, Y1A 4Y9. T: 867-668-4648

Kolakowski, E., 2430 Rosemary Drive, Mississauga, ON, L5C 1X2. T: 905-896-8203, F: 905-896-3154

Konrad, J.-M., Universite Laval, Departement de Genie Civil, Ste Foy, QC, G1K 7P4. T: 418-656-3878, F: 418-656-2928

Korynkiewicz, W., 5 Howarth Avenue, Scarborough, ON, M1R 1H3. T: 905-415-1995, F: 905-415-1996

Kosar, Keith M., 14621-84A Avenue, Surrey, BC, V3S 7V5. T: 604-591-2555

Kostaschuk, R.N., 148 West 16th Avenue, Vancouver, BC, V5Y 1Y7. T: 604-685-0543, F: 604-685-0147

Kouicem, A., Genilab BSLG inc., 192 Industrielle, Rimouski, QC, G5M 1A5. T: 418-724-7030, F: 418-724-7057

Koutsoftas, D.C., Dames & Moore, 221 Main Street, Suite 600, San Francisco, CA, 94105 USA. T: 415-243-3840, F: 415-882-9261

Kowalewich, E.M.L., 26 Silvercreek Manor NW, Calgary, AB, T3B 5L3. T: 713-931-8674, F: 713-931-3246

Kozicki, P., Ground Engineering Ltd., 415-7th Avenue, Regina, SK, S4N 4P1. T: 306-569-9075, F: 306-565-3677

Krahn, J., Geo-Slope International Ltd., 633-6th Avenue SW, Suite 1400, Calgary, AB, T2P 2Y5. T: 403-213-5333, F: 403-266-4851

Kramer, G.J.E., 2249 Lakeside Drive, Sarnia, ON, N7T 7H4. T: 519-869-6263

Kreycir, P.J., Anchor Shoring & Caissons Ltd., 3445 Kennedy Road, Scarborough, ON, M1V 4Y3. T: 416-292-1401, F: 416-292-1124

Krzywicki, H., Fondex Ontario Limited, 350-179 Colonnade road, Nepean, ON, K2E 7T4. T: 613-727-0895, F: 613-727-0581

Kumala, V., Dywidag Systems Ltd., 5-65 Bowes Road, Concord, ON, L4K 1H5. T: 905-669-4952, F: 905-669-2148

Kupper, A., 4723-139th Street, Edmonton, AB, T6H 3Z3

Kwan, W.W.W., GeoKwan Engineering Ltd., 103A Scurfield Blvd., Winnipeg, MB, R3Y 1M5. T: 204-488-8103, F: 204-488-8102

Kwok, C.C.K., Jacques Whitford, 703-6th Avenue SW, Suite 500, Calgary, AB, T3G 3S1. T: 403-781-4135, F: 403-263 7116

Lacasse, S., Director, Norwegian Geotechnical Institute, PO Box 3930 Ullevaal Hageby, N-0806 Oslo, Norway. T: 011-47-22-02-30-00, F: 011-47-22-23-04-48

Lach, P.R., 6484-178 St., Edmonton, AB, T5T 2J4. T: 780-436-2152, F: 403-435-8425

Lahti, L.R., 10 Queens Quay West, Suite SPH. 1-09, Toronto, ON, M5J 2R9. T: 416-567-4444, F: 416-567-6561

Laidlaw, J., 3038 West 19th Avenue, Vancouver, BC, V6L 1E7. T: 604-879-9266, F: 604-879-5014

Landine, P., Cameco Corporation, 2121-11th Street West, Saskatoon, SK, S7M 1J3. T: 306-956-6413, F: 306-956-6201

Lardner, W.E., Deep Foundations Contractors Inc., 29 Ruggles Avenue, Thornhill, ON, L3T 3S4. T: 905-881-2882, F: 905-881-2564

LaRochelle, P., 2528 Hospitalieres, Sillery, QC, G1T 1V7. T: 418-651-3610, F: 418-651-2617

Lau, R.K.M., Ronel Engineering Ltd., 8716-51 Avenue 2nd Floor, Edmonton, AB, T6E 5E8. T: 403-466-6888, F: 403-466-7117

Laurin, P., Tototnto Transit Commission, 1138 Bathurst Street, Inglis Building, Toronto, ON, M5R 3H2. T: 905-712-4771, F: 905-712-0515

Lauzon, M., 3980 Beauchemin, Brossard, QC, J4Z 2N4. T: 514-670-8043, F: 514-670-3390

Law, D.J., Thurber Consultants Ltd., 200-9636-51st Avenue, Edmonton, AB, T6E 6A5. T: 403-438-1460, F: 403-437-7125

Laxdal, J., AMEX, 2227 Douglas Road, Burnaby, BC, V5C 5A9. T: 604-473-5322, F: 604-294-4664

Leahy, D., 1208 Rousseau, Sillery, QC, G1S 4H1. T: 418-621-9700, F: 418-621-9090

LeBoeuf, D., Universite Laval, Dép de génie civil, Pavillon Pouliot, 2908-B, Faculté des sciences et de génie, Quebec, QC, G1K 7P4. T: 418-656-2131, F: 418-656-2928

Lee, R. (K-C), 39 Adel Drive, St. Catharines, ON, L2M 3W9. T: 905-641-1932

Lefebvre, G., Université de Sherbrooke, Département de Génie Civil, 2500 boul. Universite, Sherbrooke, QC, J1K 2R1. T: 819-821-7107, F: 819-821-7974

Leitch, M., 402-1510 Bathurst Street, Toronto, ON, M5P 3H3. T: 416-652-5696

Leroueil, S., 1590 Cote Ross, Ste-Foy, QC, G1W 3L5. T: 418-656-2601, F: 418-656-2928

Leroux, S., LVM-Fondatec Inc., 1200 boul. St-Martin, bureau 300, Laval, QC, H7S 2E4. T: 514-281-51513, F: 514-668-5532

Lessard, G., 4205 Northcliffe, Montreal, QC, H4A 3L2. T: 514-285-5511, F: 514-285-5521

Letavay, M., 231-5820 Vedder Road, Chilliwak, BC, V2R 1C4. T: 604-793-8401, F: 604-793-8402

Lewkowich, G.F., 1407 Roseann Drive, Nanaimo, BC, V9T 4L3. T: 250-754-7552, F: 250-754-3876

Lewycky, D.M., City of Edmonton, Engineering Services, 11404-60th Avenue, Edmonton, AB, T6H 1J5. T: 780-496-6773, F: 780-944-7653

Li, L., University of British Columbia, Dept. of Civil Engineering, 2324 Main Mall, Vancouver, BC, V6T 1Z4. T: 604-822-1820, F: 604-822-6901

Lieszkowszky, I.P., Geo-Ltd., 90 Nolan Crt., Units 17 & 18, Markham, ON, L3R 4L9. T: 416-474-9255, F: 416-474-9267

Lighthall, P.C., 4093 West 14th Ave., Vancouver, BC, V6R 2X3. T: 604-273-0311, F: 604-279-4300

Lingnau, B.E., 4651 East Michigan Avenue, Phoenix, AZ, 85032 USA. T: 602-997-6391, F: 602-943-5508

List, B.R., 129 Burns Place, Ft. McMurray, AB, T9K 2G3. T: 780-791-9723, F: 403-790-4847

Livingstone, G., Syncrude Ltd., P.O. Bag 4009 M.D. 0062, Fort McMurray, AB, T9H 3L1. T: 403-790-8110, F: 403-790-4847

Lo, G.W.-B., 43 Queenscourt Drive, Scarborough, ON, M1T 2J4. T: 416-497-9812, F: 416-497-0163

Lo, K.Y., University of Western Ontario, Department of Civil Engineering, Faculty of Eng. Science, London, ON, N6A 5B9. T: 519-661-2125, F: 519-661-3942

Lord, E.R.F., 315 Greystone Place, Sherwood Park, AB, T8A 3E9. T: 403-449-6907, F: 403-449-6805

Lou, J.K., 3586 West 17th Avenue, Vancouver, BC, V6S 1A1. T: 604-528-2405

Lovbakke, D.E., 127 Wolverine Drive, Fort McMurray, AB, T9H 4L5. T: 403-790-7807, F: 403-790-7550

Lowry, D.K., Armtec, 15 Campbell Road, Guelph, ON, N1H 6P2. T: 519-822-0210 x265, F: 519-822-1160

Lui, L.W.H., 3244 Waneta Place, Vancouver, BC, V5M 3H7. T: 604-255-0828, F: 604-255-0817

Lukajic, B., P.O. Box 186, Station A, Toronto, ON, M5W 1B2. T: 416-445-4844, F: 416-445-9475

Maber, C.T., Thurber Consultants, 210-4475 Viewmont Avenue, Victoria, BC, V8Z 6L8. T: 604-727-2201, F: 604-727-3710

MacDonald, D.H., R.R. #1, 15114 Niagara Pkwy., Niagara-on-the-Lake, ON, L0S 1J0. T: 905-262-5295, F: 905-262-5374

Machin, D., 49 Harding Blvd. East, Richmond Hill, ON, L4C 1S7. T: 705-495-1383, F: 705-474-8055

MacLellan, M.E., Klohn-Crippen Consultants Ltd., 114-6815-8th Street N.E., Calgary, AB, T2E 7H7. T: 403-274-3424, F: 403-274-5349

MacLeod, D.R., R.R.1, Box 50, Dunvegan, ON, K0C 1J0. T: 613-225-3664

MacNeil, B., Jacques Whitford Associates, 3 Spectacle Lake Drive, Dartmouth, NS, B3B 1W8

Madjar, H., Terratech, 455 Rene Levesque Ouest, 10e etage, Montréal, QC, H2Z 1Z3. T: 514-393-1000, Ext. 7710, F: 514-393-9540

Malott, M.S., McNally and Sons Ltd., 1855 Barton Street East, Box 3338 Station C, Hamilton, ON, L8H 7L8. T: 905-549-6561, F: 905-549-3548

Mann, P.T., 25 Highgate Drive, Box 962, St. George, ON, N0E 1N0. T: 519-893-7500, F: 519-893-0654

Marcotte, M., Solmers International, 2160, Chemin du Tremblay #205, Longueuil, QC, J4N 1A8. T: 450-448-0870, F: 450-448-1070

Marmulak, J.P., 29 Topshee Drive, Sydney, NS, B1S 2L1. T: 902-564-1855, F: 902-564-8756

Marron, J.-C., 1182 Rue du Perche, Boucherville, QC, J4B 6V4. T: 514-874-0272, F: 514-874-2642

Marttila, R., Site Investigation Services Lt, 785 The Kingsway, Peterborough, ON, K9J 6W7. T: 705-743-6850, F: 705-743-6854

Marvin, T.K., 16 Regency Court, Fall River, NS, B2T 1E7. T: 902-468-6486, F: 902-468-4919

Massiera, M., Universite de Moncton, Ecole de Genie, Moncton, NB, E1A 3E9. T: 506-858-4141, F: 506-858-4082

Matich, M.A.J. (Fred), 79 Bywood Drive, Islington, ON, M9A 1M2. T: 416-239-0821, F: 416-244-5375

McBeath, G.A., 6057 Fraser Street, Halifax, NS, B3J 3N5. T: 902-468-6486, F: 902-468-4919

McCammon, N.R., Golder Associates, 500-4260 Still Creek Dr., Burnaby, BC, V5C 6C6. T: 604-298-6623, F: 604-298-5253

McCormick, M., 17331-107th Avenue, Edmonton, AB, T5S 1E5. T: 780-484-3377, F: 403-435-8425

McDonald, P.D., 483 Samuel Harper Court, P.O. Box 494, Mt. Albert, ON, L0G 1M0. T: 416-881-2882, F: 416-881-2564

McGaghran, S.M., 371 Bronte Street Unit 11, Milton, ON, L9T 3K5. T: 905-567-9994, F: 905-567-9541

McGlone, P.M.A., AGRA Earth & Enviromental, 5-3300 Merrittville Hwy., S.S. #1, Thorold, ON, L2V 4Y6. T: 905-687-6616, F: 905-687-6620

McIntyre, J., DST Consulting Engineers Inc., 605 Hewitson Street, Thunder Bay, ON, P7B 5V5. T: 807-623-2929, F: 604-294-4664

McKee, R.C.E., O'Connor Associates, 2150 Winston Park Drive, Ste. 200W, Oakville, ON, L6H 5V1. T: 905-829-3330, F: 905-829-3404

McKenna, G., Syncrude Ltd., P.O. Bag 4009 MD M203, Fort McMurray, AB, T9H 3L1. T: 780-790-8324, F: 780-790-5657

McKeown, S., 2735 Chalice Road NW, Calgary, AB, T2L 1C7. T: 403-299-5600, F: 493-299-4631

McLachlin, D.M., 508-2175 Marine Drive, Oakville, ON, L6L 5L5. T: 905-567-4444, F: 905-567-6561

McLaughlin, T.A., J.R. Paine & Associates Ltd., 7710-102nd Avenue, Peace River, AB, T8S 1M5. T: 402-624-4966, F: 403-624-3430

McLeod, B.C., Levelton Associates, 520 Dupplin Road, Victoria, BC, V8Z 1C1. T: 250-475-1000, F: 250-475-2211

McQuinn, D.R., Jacques Whitford & Associates Ltd., 3 Spectacle Lake Dr., Dartmouth, NS, B3B 1W8. T: 902-468-0425, F: 902-468-9009

McRoberts, E.C., 226 Wolf Willow Crescent, Edmonton, AB, T5T 1T2

McRostie, G.C., McRostie Genest St-Louis & Assoc., 201-1755 Woodward Drive, Ottawa, ON, K2C 0P9. T: 613-228-7088, F: 613-228-0986

Mejia, C.A., 1116 Premier Street, North Vancouver, BC, V7J 2H3. T: 604-684-4384, F: 604-684-5124

Merkosky, R.R., UMA Engineering Ltd., 17007-107th Avenue, Edmonton, AB, T5S 1G3. T: 780-486-7657, F: 780-486-7070

Merleau, M.A., Merlex Engineering Ltd., 120 Progress Court, P.O. Box 885, North Bay, ON, P1B 8K1. T: 705-476-2550, F: 705-476-8882

Militano, G., 1224 Fleet Avenue, Winnipeg, MB, R3M 1M1. T: 204-488-2997, F: 204-489-8261

Milligan, V., Golder Associates, 2180 Meadowvale Blvd., Mississauga, ON, L5N 5S3. T: 905-567-4444, F: 905-567-6561

Milne-Epp, J.D., 404-10732-86th Avenue, Edmonton, AB, T6E 2M9

Mimura, D.W., 60 Freestone Way, Saprae Creek, Fort McMurray, AB, T9H 5B4. T: 780-790-4753, F: 780-790-5657

Mirza, C., 20 Trudy Road, Willowdale, ON, M2J 2Y9. T: 416-441-2560, F: 416-441-4161

Mitchell, A.J., BC Gas Utility Ltd., 2-1111 West Georgia Street, Vancouver, BC, V6E 4M4. T: 604-443-6851, F: 604-443-6850

Mitchell, D.W., 34, 5760 Hampton Place, Vancouver, BC, V6T 2G1

Mitchell, D.W., 34-5760 Hampton Place, Vancouver, BC, V6T 2Ga. T: 604-222-8153

Mitchell, R.J., Queens University, Ellis Hall, Kingston, ON, K7L 3N6. T: 613-545-2133, F: 613-545-2128

Mlynarek, J., SAGEOS, 3000, rue Boullé St., St-Hyacinthe, QC, J2S 1H9. T: 514-778-1870, F: 514-778-3901

Mollard, J.D., J.D. Mollard and Associates Ltd., 810 Avord Tower, 2002 Victoria Avenue, Regina, SK, S4P 0R7. T: 306-352-8811, F: 306-352-8820

Monkman, M.J.L., 49 Harding Blvd. East, Richmond Hill, ON, L4C 1S7

Moore, I.D., University of Western Ontario, Dept. of Civil Engineering, Ellis Hall, Queen's University, Kingston, ON, K7L 3N6. T: 613-533-3160, F: 613-533-2128, E: moore@civil@queensu.ca

Mordhorst, C., GeoSystems Engineering Inc., 7802 Barton, Lenexa, Kansas, 66214 USA. T: 913-962-0909, F: 913-962-0909

Morgenstern, N.R., University of Alberta, Dept. of Civil Engineering, Edmonton, AB, T6G 2G7. T: 780-492-5127, F: 780-492-8198

Morissette, L., 41 Wychwood Drive, Aylmer, QC, J9H 4E2. T: 819-776-4446, F: 819-776-9920

Morrison, J.A., 3008-111 B St., Edmonton, AB, T6J 3Y2. T: 403-427-4731, F: 403-427-4347

Mortazavi, M.H.S., 53 Kilkenny Drive, Toronto, ON, M1W 1J9. T: 416-785-5110, F: 416-785-5120

Mouland, G.D., Fundy Engineering & Consulting, P.O. Box 6626, Saint John, NB, E2L 4S1. T: 506-635-1566, F: 506-635-0206

Muckle, J.G., Terraprobe Limited, 903 Barton Street, Stoney Creek, ON, L8E 5P5. T: 905-643-7560, F: 905-643-7559

Murphy, T.P., CBCL Limited, 1489 Hollis Street, P.O. Box 606, Halifax, NS, B3J 2R7. T: 902-492-6762, F: 902-423-3938

Murray, T.K., Klohn-Crippen Consultants Ltd., Ste. 114, 6815-8th St., NE, Calgary, AB, T2E 7H7. T: 403-274-3424, F: 403-274-5349

Musulak, S.C., 539 Woodside Place S.W., Calgary, AB, T2W 3J9

Naesgaard, E., Site I Box 38 Creek Road, Bowen Island, BC, V0N 1G0. T: 604-947-2637, F: 604-985-7725

Nakai, S.B., 2704 Byron Road, North Vancouver, BC, V7H 1M2. T: 604-684-4384

Nascimento, C.M., 7 Northey Drive, Willowdale, ON, M2L 2S8. T: 416-432-3600, F: 416-432-1926

Naylor, D.S., Naylor Engineering Associates Ltd., 353 Bridge St. E., Kitchener, ON, N2K 2Y5. T: 519-741-1313, F: 519-741-5422

Netherton, D.E., StreamFlow Power, 433 McIntyre Street East, North Bay, ON, P1B 1E1. T: 705-475-1994, F: 705-474-8095

Ng, J., 194 Risebrough Cir, Markham, ON, L3R 3E3. T: 416-474-9255

Ng, R.M.C., 54 Palomino Crescent, Willowdale, ON, M2K 1W3. T: 416-785-5110

Nicol, D., Ministry of Forests, Nelson Regional Office, 518 Lake Street, Nelson, BC, V1L 4C6. T: 250-354-6135, F: 250-354-6250

Nilsen, J.A., 641 Millwood Road, Toronto, ON, M4S 1L1. T: 416-225-9272, F: 461-225-0622

Noel, M., 7/1 Oakwood Way, Menai, NSW, 2234 Australia. T: 6129532-0996, F: 1.1613E+12

Nolin, J., Wardrop Engineering Inc., 400-386 Broadway, Winnipeg, MB, R3C 4M8

Noonan, D.K.J., Golder Associates, 2180 Meadowvale Blvd., Mississauga, ON, L5N 5S3. T: 905-567-4444, F: 905-567-6561

Northwood, R.P., 30 Park Drive, Woodbridge, ON, L4L 2H3. T: 416-213-1060, Ext. 228, F: 416-213-1070

Nowski, S.J., 24 Ferndell Circle, Unionville, ON, L3R 3Y8. T: 416-751-5230, F: 416-751-6745

O'Connor, M.J., O'Connor Associates, 1000-639 5th Avenue SW, Calgary, AB, T2P 0M9. T: 403-294-4200, F: 403-294-4240

O'Dwyer, T., C.T. Soil & Materials Eng. Inc., R.R. #1, 2000 Legacy Park Drive, Windsor, ON, N9A 6J3. T: 519-966-8863, F: 519-966-8870

O'Kane, M.A.T., R.R. #3, Box 62, Saskatoon, SK, S7K 3J6. T: 306-966-5154, F: 306-966-5154

Olson, T., 1062 Sherman Crescent, Pickering, ON, L1X 1P7. T: 416-495-8614, Ext. 226, F: 905-479-9326

O'Reilly, D., Trow Consulting Engineers, 1595 Clark Boulevard, Brampton, ON, L6T 4V1

O'Rourke, T.H., T.H. O'Rourke Structural Con. Inc., 184 Shorting Road, Scarborough, ON, M1S 3S7. T: 416-292-5502, F: 416-292-5525

Orpwood, T.G., 48 Trotters Lane, Brampton, ON, L6Y 1B6. T: 905-796-2650, F: 905-796-2250

O'Shaughnessy, V., 1504 Chomley Crescent, Ottawa, ON, K1G 0V8

Osho, P., 1404-907 Barberry Walk SW, Calgary, AB, T3C 2Y3. T: 403-508-2761

Ostiguy, J.-C., Labo S.M. Inc., 2111, boul. Fernand Lafontaine, Longueuil, QC, J4G 2J4. T: 450-651-0981

Oswell, J.M., 35 Hawkside Close NW, Calgary, AB, T3G 3K4. T: 403-235-8113, F: 403-248-2188

Ouarzidini, A., Laboratovic de materiaux de Quebec, (1987) Inc., 2445, Rue Dalton, Ste-Foy, QC, G1P 3S5. T: 418-659-5115, F: 418-659-3311

Ouimet, J.-M., 1670 Dumouchel, Chomedey-Laval, QC, H7S 1J5. T: 384-7970

Ozden, Z.S., 84 Fieldside Drive, Scarborough, ON, M1V 3C5. T: 416-213-1255, F: 416-213-1260

Pang, S., 4054 Trapper Crescent, Mississauga, ON, L5L 3A9. F: 905-567-6561

Pankratz, H.D., 48 Maplegrove Road, Winnipeg, MB, R2V 4J3. T: 204-488-2997, F: 204-489-8261

Papanicolas, D., Thurber Engineering Ltd., 200-9636 51st Avenue, Edmonton, AB, T6E 6A5. T: 403-438-1460, F: 403-437-7125

Paquette, J., 6510 Des Faucons, Laval, QC, H7L 4E9. T: 450-625-2373, F: 450-662-9959

Partsis, E.S., Geo- Ltd., 18-90 Nolan Court, Markham, ON, L3R 4L9. T: 416-474-9255

Pasqualoni, R., Inspec-Sol Inc., 228 Matheson Blvd. East, Mississauga, ON, L4Z 1X1. T: 905-712-4771, F: 905-712-0515

Patrick, R.A., EBA Engineering Consultants Ltd., 1-4373 Boban Drive, Nanaimo, BC, V9T 6A7. T: 250-756-2256, F: 250-756-2686

Pauls, G.J., AGRA Earth & Environmental Ltd., 4810-93rd Street, Edmonton, AB, T6E 5M4

Payne, J.G., 7 Black Willow Court, Richmond Hill, ON, L4E 2M7. T: 416-756-3400, F: 416-756-2266

Pazin, M., 42 White Oak Blvd., Toronto, ON, M8X 1J2. T: 416-240-0084, F: 416-248-0621

Peaker, K.R., 41 Valecrest Drive, Islington, ON, M9A 4P5. T: 416-213-1255, F: 416-213-1260

Peck, K.W., 3863 Route 102, Island View, NB, E3E 1H1. T: 506-454-7247, F: 506-454-6281

Peck, R.B., 1101 Warm Sands Drive S.E., Albuquerque, New Mexico, 87123 USA. F: 505-323-7760

Peggs, D.R.D., Acres International Inc., 4342 Queen Street, P.O. Box 1001, Niagara Falls, ON, L2E 6W1. T: 416-374-5200

Perret, D.H., Geological Survey of, 2535 Boul. Laurier Case, Postale 7500, Sainte-Foy, QC, G1V 4C7. T: 418-654-2686, F: 418-654-2615

Peterson, T.W., 2132 Nineth Avenue, Calgary, AB, T2N 1E4

Pheeney, P.E., 55 Neptune Cres., Hanwell, NB, E3C 1M7. T: 506-457-3200, F: 506-452-7652

Phillips, R., C-Core, Memorial University of Newfoundland, St. John's, NF, A1B 3X5. T: 709-737-8371, F: 709-737-4706

Piascik, A.M., 1455 Adamson Street, Mississauga, ON, L5C 2Z4. T: 905-567-4444, F: 905-567-6561

Pigeon, Y., 406 Avenue Lockhart, Mt. Royal, QC, H3P 1Y5. T: 514-287-8511, F: 514-287-8643

Pion, P. 2364 Fauteux, Chomedey, Laval, QC, H7T 2T1. T: 514-333-5151, F: 514-333-4674

Plewes, H.D., 5283 Ketch Place, Delta, BC, V4K 4Z4. T: 604-279-4337, F: 604-279-4387

Polan, B.J., 648 Sprucehurst Crescent, Waterloo, ON, N2V 2E3. T: 519-725-3313, F: 519-725-1394

Pollock, D.H., 35 Heritage Place, Regina, SK, S4S 2Z7. T: 306-780-5123

Pollock, J.B., 1911-10883 Saskatchewan Drive, Edmonton, AB, T6E 4S6. T: 780-436-2152, F: 780-436-8425

Polysou, N.C., 5483 Moriarty Crescent, Prince George, BC, V2N 4A5. T: 250-565-6195, F: 250-565-6928

Poot, S.E.M., Golder Associates Ltd., 2180 Meadowvale Blvd., Mississauga, ON, L5N 5S3. T: 416-429-5684

Popescu, R., 1 Dundas Street, St. John's, NF, A1B 1X1. T: 709-737-2591, F: 709-737-4706

Poschmann, A.S., Golder Associates, 2180 Meadowvale Blvd., Mississauga, ON, L5N 5S3. T: 905-567-4444 x301, F: 905-567-6561

Pottie, J., 323-11673 Seventh Avenue, Richmond, BC, V7E 4X3. T: 604-273-0311, F: 604-279-4387

Powell, D.P., Cullinane & Powell Consulting Inc., 2-953 Laval Crescent, Kamloops, BC, V2C 5P4. T: 250-374-1775, F: 250-374-4572

Powell, R.D., P.O. Box 1051, Woodbury, CT, 06798 USA

Prasad, G. Durga, 4100 Ponytrail Dr., Ste. 411, Mississauga, ON, L4W 2Y1

Priscu, C., Acres International Ltd., 6th Floor, 500 Portage Avenue, Winnipeg, MB, R3C 3H3. T: 204-786-8751, F: 204-786-2242

Proskin, S.A., 14911-39th Avenue NW, Edmonton, AB, T6R 1J6. T: 403-492-7244, F: 403-492-0249

Provencal, J., 40 rue Adrien D'Anjou, Notre-dame de, L'ile-Perrot, QC, J7V 9C7. T: 514-453-0684

Pudsey, E.G., 27 Chornick Drive, Winnipeg, MB, R2G 2N2. T: 204-474-3148, F: 204-474-4682

Pufahl, D.E., University of Saskatchewan, Dept. of Civil Engineering, Saskatoon, SK, S7N 0W0. T: 306-966-5374, F: 306-966-8710

Purdy, W.J., 4146 Garden Grove Dr., Burnaby, BC, V5G 4G6. T: 604-684-4384, F: 604-684-5124

Qayyum, A., Dept. of Highways & Transportation, 1610 Park Street, Regina, SK, S4P 3V7. T: 306-787-4935, F: 306-787-4582

Rahardjo, H., 45 Nanyang View, Singapore, 639641 Singapore. T: 65-799-6455, F: 65-791-0676

Rahman, M.G., Saskatchewan Water Corporation, 111 Fairford Street East, Moose Jaw, SK, S6H 7X9. T: 306-694-3956, F: 306-694-3944

Rainu, I., 139 Elmer Avenue, Toronto, ON, M4L 3R9. F: 416-690-7997

Ramage, R.G., AGRA Earth & Environmental Limited, 3017 Faithfull Avenue, Saskatoon, SK, S7K 8B3. T: 306-975-0444, F: 306-9554-2446

Ratcliffe, D.W., 70 Sprucegrove Way, Airdrie, AB, T4B 2E1

Rattue, D.A., 180 Townshend, St. Bruno, QC, J3V 1K8. T: 514-866-1000

Rennie, R.J., 782 Quinlan Road, Ottawa, ON, K1G 1S1. F: 613-521-7954

Riddoch, R.G., 46225 Larch Avenue, Chilliwack, BC, V2P 1E7. T: 604-792-2003, F: 604-792-2003

Robert, Y., 8394 Wilfrid Pelletier, Anjou, QC, H1K 1M4. T: 514-674-4901, F: 514-674-3370

Robertson, Andrew MacGregor, Robertson GeoConsultants Inc., 902-580 Hornby Street, Vancouver, BC, V6C 3B6. T: 604-684-8072, F: 604-681-4166

Robertson, P.K., University of Alberta, Industry Liaison Office, 222 Camput Tower, 8625-112 Street, Edmonton, AB, T6G 2E1. T: 780-492-8318, F: 780-492-7876

Robinson, K.E., Jacques Whitford & Associates Ltd., 1-3771 North Fraser Way, Burnaby, BC, V5J 5G5. T: 604-436-3014, F: 604-436-3074

Rodger, J.D., Golder Associates, 2825 Lauzon Parkway, Windsor, ON, N8T 3W5. T: 519-945-5500

Rodrigues, E., Ontario Ministry of the Environment, 2 St. Claire Avenue West, Toronto, ON, M4V 1L5. T: 416-314-7226

Rogers, B.T., Klohn-Crippen Consultants Ltd., 114, 6815-8th Street NE, Calgary, AB, T2E 7H7. T: 403-274-3424, F: 403-274-5349

Rorquist, W., Agriculture & Agri-Food, 603-1800 Hamiton Street, Regina, SK, S4P 4L2. T: 306-780-8705, F: 306-780-6683

Rose, S.V., 84 Beverley Street, Kingston, ON, K7L 3Y6. T: 613-548-3446, F: 613-548-7975

Rowe, R.K., Queen's University, Dept. of Civil Engineering, Ellis Hall, Kingston, ON, K7L 3N6. T: 613-533-3113, F: 613-533-6934

Roy, M., Universite Laval, Dept. de Genie Civil, Quebec, QC, G1K 7P4. T: 418-656-5936, F: 418-656-2928

Ruban, A.F., EBA Engineering Consultants Ltd., 14535-118 Avenue, Edmonton, AB, T5L 2M7. T: 780-451-2121, F: 780-454-5688

Ruffell, J.P., 4 Emerson Place, St. Albert, AB, T8N 5X3. T: 403-451-2121, F: 403-454-56888

Ryzuk, C.N., 28 Crease Ave., Victoria, BC, V8Z 1S3. T: 604-475-3131, F: 604-475-3611

San, E., Gartner Lee Limited, 102-140 Renfrew Drive, Markham, ON, L3R 6B3. T: 905-477-8400, Ext. 235, F: 905-477-1456

Sangiuliano, T., Ministry of Transportation of ON, 1201 Wilson Avenue, Room 223, Central Building, Downsview, ON, L6A 1V7. T: 416-235-5267, F: 416-235-5268

Sappal, U.S., 16 Bellside Drive, Unionville, ON, L3P 7B8. T: 905-479-5259, F: 905-479-5956

Sargent, D.W.J., 3982 West 31st Avenue, Vancouver, BC, V6S 1Y5. T: 604-872-6721, F: 604-872-3825

Sastry, V.V.R.N., St Marys University, Div. of Engineering, Halifax, NS, B3H 3C3. T: 902-420-5697, F: 902-420-5110

Schebesch, D., R.R. 3, 389 Irish Line, Cayuga, ON, N0A 1E0

Schellekens, F., 9821-89th Avenue, Edmonton, AB, T6E 2S3. T: 780-439-7904

Schlagintweit, M., 3536 Point Grey Road, Vancouver, BC, V6R 1A8. T: 604-685-0543, F: 604-685-0147

Schmidt, J.A., AGRA Earth & Environmental Ltd., 610 Richard Road, Prince George, BC, V2K 4L3. T: 604-564-3243, F: 604-562-7045

Schrank, J., HWA GeoSciences Inc., 19730-64th Avenue W. Suite 200, Lynnwood, WA, 98036 USA. T: 425-774-0106

Schreiner, B.T., Saskatchewan Research Council, 15 Innovation Blvd., Saskatoon, SK, S7N 2X8. T: 306-933-5404, F: 306-933-7896

Schulz, T.M., UMA Engineering Ltd., 17007-107th Avenue, Edmonton, AB, T5S 1G3. T: 403-486-7000, F: 403-486-7070

Scotton, S., 100 Eagle Crescent, Nanaimo, BC, V9S 2S6

Sedran, G., 358 Inverness Avenue East, Hamilton, ON, L9A 1H5. T: 604-684-4384

Sego, D.C., University of Alberta, Dept. of Civil Engineering, 220 Eng. Bldg., Edmonton, AB, T6G 2G7. T: 780-492-2059, F: 780-492-8198

Sellers, J.B., Geokon Inc., 48 Spencer Street, Lebanon, NH, 03766 USA. T: 603-448-1562, F: 603-448-3216

Seychuk, J.L., 38 Longfield Road, Toronto, ON, M9B 3G3. T: 905-567-4444, Ext. 159, F: 905-567-6561

Shang, J.Q., Dept. of Civil & Environmental Eng., University of Western Ontario, London, ON, N6A 5B9. T: 519-661-4218, F: 519-661-3942

Shields, D.H., 990 Terrace Avenue, Victoria, BC, V8S 3V3

Shirlaw, J.N., 134 Watten Estate Road, #02-02, Singapore, 287601 Republic of Singapore. T: 654639125, F: 653961150

Showers, D.K., 2126-5400 Preston Oaks, Dalas, TX, 75240 USA. T: 214-361-7900

Shtenko, V.W., 857 Strathaven Drive, North Vancouver, BC, V7H 2K1. T: 604-986-2072

Sierakowski, C., 98, Ste-Monique, Chicoutimi, QC, G7J 3W9. T: 418-696-6874, F: 418-696-6896

Silvestri, V., 380 Outremont, Outremont, QC, H2V 3M2. T: 514-340-4503, F: 514-340-5841

Singh, N.K.H., 14222-70th Avenue, Surrey, BC, V3W 0P4. T: 604-273-0311, F: 604-279-4300

Siu, D.Y., 10251 Hollymount Drive, Richmond, BC, V7E 4T5. T: 604-273-0311

Skaftfeld, K., 66 Brigantine Bay, Winnipeg, MB, R3P 1R1. T: 204-284-0580, F: 204-475-3646

Skibinsky, D.N., 9419-101st Street, Edmonton, AB, T5K 0W5. T: 436-2152

Skirrow, R., Alberta Infrastructure, 4999-98th Avenue, 1st Floor, Twin Atria Building, Edmonton, AB, T6B 2X3. T: 780-427-5578, F: 780-422-5426

Smith, D., 2000 Lloyd Avenue, North Vancouver, BC, V7P 2N8. T: 604-684-4384, F: 604-684-5124

Smith, G., 170 Highway 7, Locust Hill, ON, L0H 1J0. T: 905-294-7123

Smith, J.A., J.A. Smith & Associates Ltd., 8, 5555-2nd Street SE, Calgary, AB, T2H 2W4. T: 403-543-60803, F: 403-543-6083

Smith, N.J., Acres International Ltd., 6th Fl 500 Portage Avenue, Winnipeg, MB, R3C 3Y8. T: 204-786-8751, F: 204-786-2242

Smith, W.J., 844 Nicholson Avenue, Ottawa, ON, K1V 6N5. T: 613-995-5284, F: 613-995-1781

Snead, D.E., Cook Pickering & Doyle Ltd., 141 East 7th Avenue, Vancouver, BC, V5T 1M5. T: 604-879-0494, F: 604-879-6522

Snow, M., 24892 Avenida Avalon, Laguna Hills, CA, 92653 USA

Sobkowicz, J.C., 201 Sundown Way SE, Calgary, AB, T2X 2M2. T: 403-253-9217, F: 403-252-8159

Sowa, V.A., 5616 Timbervalley Road, Tsawwassen, BC, V4L 2E6. T: 604-436-3014, F: 604-436-3752

Staples, T.E., 411-125 Spruce Street, Fort McMurray, AB, T9K 1E2. F: 604-782-3430

Starke, B., R.R. 1, Goodwood, ON, L0C 1A0. T: 416-881-2882, F: 416-881-2564

Sterne, K.B., 210-4475 Viewmont Avenue, Victoria, BC, V8Z 6L8. T: 604-727-2201, F: 604-727-3710

Stewart, W.P., 107-585 Austin Avenue, Coquitlam, BC, V3K 3N2. T: 604-412-7424, F: 604-432-8973

St-Louis, M.W., G.M.M. Consultants, 624 Auguste-Mondoux, Aylmer, QC, J9H 5E1. T: 819-776-4446

Stolle, D.F.E., McMaster University, Dept. of Civil Engineering, 1280 Main St. W., Hamilton, ON, L8S 4L7. T: 416-525-9140, Ext. 24919, F: 416-529-9688

Stypulkowski, J.B., Geotechnical & Tunneling Division, Parsons Brinckeroff Quade & Douglas, One Penn Plaza, New York, New York, 10119 USA. T: 212-465-5247, F: 212-465-5592

Sully, J.P., MEG Consulting Ltd., 6211 Comstock Road, Richmond, BC, V7C 2X5. T: 604-277-2778, F: 604-619-6194

Sun, P.C., Peto MacCallum Ltd., 165 Cartwright Avenue, Toronto, ON, M6A 1V5. T: 416-785-5110, F: 416-785-5120

Sungaila, M.A., 152 The Westway, Etobicoke, ON, M9P 2C1. T: 416-477-8400, F: 416-477-1456

Sutherland, K., 14 MacDonnell Street, Box 393, Falconbridge, ON, P0M 1S0

Sy, A., 553 West 61st Avenue, Vancouver, BC, V6P 2B4. T: 604-279-4331, F: 604-279-4300

Szymanski, M.B., 6048 Camgreen Cir, Mississauga, ON, L5N 4M8. T: 905-568-2929, F: 905-568-1686

Tanos, M., Terraprobe Limited, 18-12 Bram Court, Brampton, ON, L6W 3V1. T: 416-793-2650, F: 905-796-2250

Taylor, B.B., 35 Newcastle Street, Dartmouth, NS, B2Y 3M6. T: 902-468-7777

Tedder, K.H., Petro-, 3275 Robecca Street, Oakville, ON, L6L 6N5. T: 905-469-3754, F: 905-496-3760

Tetreault, M., Department of Civil Engineering, Royal Military College of, P.O. Box 17000 STN Forces, Kingston, ON, K7K 7B4. T: 613-541-6000, Ext. 6350, F: 613-541-6599

Thomas, G., DST Consulting Engineers Inc., 22-5350 Canotek Road, Ottawa, ON, K1J 9E2. T: 613-748-1415, F: 613-748-1356

Thompson, G.A., 33 Westwood Crescent, P.O. Box 730, Hudson Heights, QC, J0P 1J0. T: 514-458-4925, F: 514-458-4925

Thurber, B.D., 15 Assiniboia Way West, Lethbridge, AB, T1K 6W2. T: 403-329-9009, F: 403-328-8817

Toombs, R.G., 1914 148th Street, White Rock, BC, V4A 6R3. T: 1-604-536-5872

Toth, P., 2204 Oakridge Cr., Burlington, ON, L7M 4C8. T: 416-592-7348, F: 416-592-4446

Touhidi-Baghini, A., AGRA Earth & Environmental Limited, 4810-93rd Street, Edmonton, AB, T6E 5M4. T: 780-436-2152

Touileb, B.N., Hydro Quebec, Barrages et Ouvrages de Genie Civil, 855 Sainte-Catherine est, 18eme eta, Montreal, QC, H2L 4P5. T: 514-840-3000, Ext. 4005, F: 514-840-5112

Tournier, J.P., DPPE/SEBJ – Hydro Quebec, 800 Est de Maisoncuve, Montreal, QC, H2L 4M8. T: 514-840-3000, F: 514-840-3085

Tovell, A., 171 Des Bouleaux, Gatineau, QC, J8R 2L6. T: 613-226-7381, F: 613-226-6344

Trainor, P.G., 606-1260 Nelson Street, Vancouver, BC, V6E 1J7. T: 604-488-2312, F: 604-488-2019

Treen, C., AGRA Earth & Environmental Ltd., 4810- 93rd Street, Edmonton, AB, T6E 5M4. T: 403-791-0848, F: 403-790-1194

Tremblay, M., Laboratoire de Ville de Montreal, 999 rue de Louvain est, Montreal, QC, H2M 1B3. T: 514-872-3926, F: 514-872-1669

Trew, J.D., 1494 Sandpiper Road, Oakville, ON, L6M 3R8. T: 905-567-4444, F: 905-567-6561

Trimble, J.R., EBA Engineering Consultants Ltd., Calcite Business Centre, 6-151 Industrial Road, Whitehorse, YT, Y1A 2V3. T: 867-668-3068, F: 867-668-4349

Tsang, R.H.-K., 336 Bowling Green Court, Mississauga, ON, L4Z 2T1. T: 416-207-6798, F: 416-234-1511

Turcotte, B.R., Groupe SM Inc., 3705 Boul. Industriel, Sherbrooke, QC, J1L 1X8. T: 819-566-8855, F: 819-566-0224

Tutkaluk, J.M., Box 7, Group 30, R.R. #1, Dugald, MB, R2C 0M8. T: 204-474-8072, F: 204-474-7513

Tweedie, R.W., Thurber Engineering Ltd., 200-9636- 51st Avenue, Edmonton, AB, T6E 6A5. T: 403-438-1460, F: 403-437-7125

Ulrich, B., Knight Piesold LLC, 1050 17th #500, Denver, CO, 80265 USA. T: 303-629-8788, F: 303-629-8789

Umadat, J., Manitoba Highways & Government Serv, Bridges & Structures, 6th Floor, 215 Garry Street, Winnipeg, MB, R3C 3Z1. T: 204-945-5206, F: 204-945-4456

Uthayakumar, M., 11131 Kingsbridge Drive, Richmond, BC, V7A 4T1. T: 604-922-0812, F: 604-922-9167

Vaid, Y.P., U.B.C., Dept. of Civil Engineering, 2324 Main Mall, Vancouver, BC, V6T 1Z4. T: 604-822-2204, F: 604-822-6901

Valliappan, P., 7202 Maywood Street, Niagara Falls, ON, L2E 5P5. T: 416-374-5200

Valsangkar, A.J., 99 Beechwood Crescent, Fredericton, NB, E3B 2S9. T: 506-453-4521, F: 506-453-3568

van Veen, W. (Walter), 36 Hillcrest Lane, Kitchener, ON, N2K 1S8. T: 519-725-3313

Vanapalli, S.K., Royal Military College of, Civil Engineering Department, P.O. Box 17000, Stn Forces, Kingston, ON, K7K 7B4. T: 613-541-6347

Virely, D., 1257 rue Sauvé, Ste-Foy, QC, G1W 3C9. T: 418-655-1114

Wade, N.H., 240 Valhalla Crescent N.W., Calgary, AB, T3A 2A1. T: 403-298-4629, F: 403-247-4601

Walker, A.J., 36 Russell Hill Road, R.R. #1, Grand Valley, ON, L0N 1G0. T: 905-567-4444

Walker, B.P., 306-400 Walmer Rd., West Tower, Toronto, ON, M5P 2X7

Walker, R., R.W.B. Engineering Ltd., 505-150 Consumers Road, Willowdale, ON, M2J 1P9. T: 416-756-3102, F: 416-756-3113

Wallis, D.M., Cook Pickering & Doyle Ltd., 141 East 7th Ave., Vancouver, BC, V5T 1M5. T: 604-879-0494, F: 604-879-6522

Walter, D.J., AGRA Earth and Environmental, 4810 93rd Street, Edmonton, AB, T6E 5M4. T: 780-436-2152

Wan, A., Flat 19G Block 8, 23 Greig Crescent, Quary Bay, Hong Kong. T: 85291500560

Wandschneider, H., HT & Associates, 657 Avondale Avenue, Kitchener, ON, N2M 2W4. T: 519-745-9178, F: 519-745-9847

Wang, B.W., 4405 Heathgate Crescent, Mississauga, ON, L5R 2C2. T: 416-252-5311, Ext. 479, F: 416-252-4376

Warith, M.A., 20 Monkland Avenue, Ottawa, ON, K1S 1Y9. T: 416-979-5000, Ext. 459

Watts, B.D., 2301 Oak Street, Vancouver, BC, V6H 2J8. T: 604-273-0311, F: 604-279-4300

Webb, G.S., 3 Burnbrook Crescent, Nepean, ON, K2H 9A6. T: 613-224-5864, F: 613-224-9928

Wedge, N.E., 203-3760 West 10th Avenue, Vancouver, BC, V6R 2G4. T: 604-298-6623, F: 604-298-5253

Wei, L.F., 52 Fairway Drive, Edmonton, AB, T6J 2C3. T: 780-420-5429

Weimer, N.F., Government of Alberta, Public Works Supply & Services, 3rd Floor, 6950-113th Street, Edmonton, AB, T6H 5V7. T: 403-422-7626, F: 403-422-9594

Welch, D.E., Golder Associates, 2180 Meadowvale Blvd., Mississauga, ON, L5N 5S3. T: 905-567-4444, F: 905-567-6561

Werbovetski, T.M., P. Machibroda Engineering Ltd., 2623B Faithfull Avenue, Saskatoon, SK, S7K. T: 306-665-8444, F: 306-652-2092

Westland, J., 3366 Trelawny Circle, Mississauga, ON, L5N 6N5. T: 905-567-4444, F: 905-567-6561

Wheeler, P.J., 303 Sunvale Dr. S.E., Calgary, AB, T2X 3B8. T: 403-254-1706

White, W.L., P.O. Box 1272, Fredericton, NB, E3B 5C8. T: 506-452-9000, F: 506-459-3954

Whitford, M.S., Jacques Whitford & Assoc. Ltd., 20 Broadview Avenue South, Saint John, NB, E2L 5C5. T: 506-634-2185, F: 506-634-8104

Wiesner, W.R., P.O. Box 62, La Salle, MB, R0G 1B0

Wightman, A., Klohn-Crippen Consultants Ltd., 10200 Shellbridge Way, Richmond, BC, V6X 2W7. T: 604-273-0311, F: 604-279-4300

Williams, D.R., O'Connor Associates, 1000-639 Fifth Aveue SW, Calgary, AB, T2P 0M9. T: 403-294-4204, F: 403-294-4240

Williams, N.K., Interior Testing Services Ltd., 1-1925 Kirschner Road, Kelowna, BC, V1Y 4N7. T: 250-860-6540, F: 250-860-5027

Windisch, E.J., Ecole de technologie superieure, 1100 Notre-Dame ouest, Montreal, QC, H3C 1K3. T: 514-396-8655, F: 514-396-8584

Wizniak, L.M., 80 MacTaggart Drive, P.O. Box 302, Nobleton, ON, L0G 1N0

Woeller, D.J., 9113 Shaughnessy St., Vancouver, BC, V6P 6R9. T: 604-327-4311, F: 604-327-4066

Wong, E.G., 17 Woodgrove Trail, Unionville, ON, L6C 2A3. T: 416-252-5831, F: 416-253-9202

Wong, K.K.C., Syncrude Ltd., P.O. Box 4009, M.D. 1070, Fort McMurray, AB, T9H 3L1. T: 403-790-4358, F: 403-790-4850

Wood, V.A., V.A. Wood Associates Ltd., 24-1080 Tapscott Road, Scarborough, ON, M1X 1E7. T: 416-292-2868, F: 416-292-5375

Wride, C.E., 2912-12th Avenue NW, Calgary, AB, T2N 1K8

Wright, S., 136 Gatesgill St., Brampton, ON, L6X 3S3

Wu, G., 20-7433 Sixteenth Street, Burnaby, BC, V3N 4Z5. T: 604-525-6826, F: 604-294-4664

Wu, P.L., Reinforced Earth Company Ltd., The Enterprise Centre, 229-1550 Enterprise Road, Mississauga, ON, L4W 4P4. T: 905-564-0896, F: 905-564-2609

Yanful, E., University of Western Ontario, Dept. of Civil Engineering, London, ON, N6A 5B9. T: 519-661-4068, F: 519-661-3942

Yang, D., 415-130 West Keith Road, North Vancouver, BC, V7M 1L5, F: 604-986-1222

Yarechewski, D., 30 Shadyside Drive, Winnipeg, MB, R2C 1L5. T: 204-284-0580, F: 204-475-3646

Yee, H.-C.D., 6330 Palace Place, Burnaby, BC, V5E 1Z7. T: 514-876-1900

Yip, E.C.C., 10777 Linden Court, Surrey, BC, V4N 1W3. T: 604-231-8929, F: 604-273-6177

Yoshida, R., Clifton Associates Ltd., 340 Maxwell Crescent, Regina, SK, S4N 5Y5. T: 306-721-7611, F: 306-721-8128

Zahursky, Al, AGRA Earth & Environmental Ltd., 3-600 Industrial Road #1, Cranbrook, BC, V1C 4C6. T: 250-426-7448, F: 250-426-5997

Zeleny, R.N., 12 Thornwood Place N.W., Calgary, AB, T2K 5N2. T: 403-274-9870

Zergoun, M., 301-2036 York Avenue, Vancouver, BC, V6J 1E6. T: 604-731-0145, F: 604-731-4710

Zhu, F., 407-89 MacDonald Drive, St. John's, NF, A1A 2K8. T: 709-737-8354, F: 709-737-4706

Zivkovic, A., 48 Chalet Road, North York, ON, M2J 3V4. T: 905-568-2929, F: 416-252-4376

CHILE

Secretary: Mr Ramón Verdugo, Sociedad Chilena de Geotecnia, IDIEM, Plaza Ercilla 883, Santiago.
F: 56 2 6718979, E: rverdugo@cec.uchile.cl

Total number of members: 53

Aguirre, M., Prof. Civil Engineering, Universidad de La Serena, Benavente 980, La Serena. T: (56-51)204227, F: (56-51)227020, E: maguirre@elqui.cic.userena.cl

Andrade, C., Civil Engineer, Golder Associates S.A., Av. 11 de Septiembre 2353, Piso 16, Providencia, Santiago. T: (56-9)8952091, E: candrade@golder.cl

Andrade, H., Civil Engineer, Knight Piésold Chile, Marchant Pereira 221 Piso 7, Providencia, Santiago. T: (56-2)341 7627, F: (56-2)3417628, E: handrade@kpsa.cl

Bard, E., Civil Engineer, Arcadis Geotécnica, Av. Eliodoro Yáñez 1893, Providencia, Santiago. T: (56-2)3816000, F: (56-2)3816001, E: edgar.bard@geotecnica.cl

Barrera, S., Civil Engineer, Arcadis Geotécnica, Av. Eliodoro Yáñez 1893, Providencia, Santiago. T: (56-2)3816000, F: (56-2)3816001, E: sergio.barrera@geotecnica.cl

Cano, C., Civil Engineer, Golder Associates S.A., Av.11 de Septiembre 2353, Piso 16, Providencia, Santiago. T: (56-9)7997103, E: ccano@golder.cl

Campaña, J., Civil Engineer, Arcadis Geotécnica, Av. Eliodoro Yáñez 1893, Providencia, Santiago. T: (56-2)3816000, F: (56-2)3816001, E: Jose.campana@geotecnica.cl

Carrasco, R., Civil Engineer, Almirante Manuel Señoret 1434, Punta Arenas. T: (56-61)223490, F: (56-61)204416

Carrasco, A., Civil Engineer, REG Estudios Ltda., Providencia 2330 Of. 61, Providencia, Santiago. T: (56-2)2337076, F: (56-2)2332572, E: redwards@rdc.cl

Dick, F., Civil Engineer, Arcadis Geotécnica, Av. Eliodoro Yáñez 1893, Providencia, Santiago. T: (56-2)3816000, F: (56-2)3816001, E: francisco.dick@geotecnica.cl

Ebensperger, E., Civil Engineer, EMPRO Ltda., Paicavi 3001, Concepción. T: (56-41)480090, F: (56-41)480092, E: empro@ctcreuna.cl

Echeverría, F., Civil Engineer, Arcadis Geotécnica, Av. Eliodoro Yáñez 1893, Providencia, Santiago. T: (56-2) 3816000, F: (56-2)3816001, E: fernando.echeverria@geotecnica.cl

Espinace, R., Prof. Civil Engineering, Universidad Católica de Valparaíso, Av. Brasil 2147 Casilla 4059, Valparaíso. T: (56-32)273611, F: (56-32)273808, E: respinac@ucv.cl

Feliú, R., Civil Engineer, REG Estudios Ltda., Providencia 2330 Of. 61, Providencia, Santiago. T: (56-2)2337076, F: (56-2)2332572, E: redwards@rdc.cl

Fernández, A., Civil Engineer, FCQ Geotecnia e Ingeniería Ltda., Domingo Faustino Sarmiento 136, Ñuñoa, Santiago. T: (56-2)3411785, F: (56-2)2695643, E: afs@tie.cl

Garcés, E., Civil Engineer, Edic Ingenieros Ltda., Alonso de Córdova 5151, Of. 1301, Las Condes, Santiago. T: (56-2)3785610, F: (56-2)3785656, E: edic@ingenieros.ltda.cl

Garrido, L., Civil Engineer, Av. Los Carrera 180 Depto. 22 B, Quilpué

Jamett, R., Civil Engineer, Knight Piésold Chile, Marchant Pereira 221 Piso 7, Providencia, Santiago. T: (56-2)3417627, F: (56-2)3417628, E: rjamett@kpsa.cl

Jaramillo, M., Civil Engineer, Ingendesa,Santa Rosa 76 Piso 12, Santiago. T: (56-2)6308187, F: (56-2)6354070, E: mjaramillo@ingendesa.cl

Karzulovic, A., Civil Engineer, A. Karzulovic y Asoc. Ltda., Brown Norte 476 Depto. 304, Santiago. T: (56-2)2255663, F: (56-2)2040548, E: akl@reuna.cl

Kort, I., President of Society, Civil Engineer, Issa Kort Kort, Valenzuela Castillo 1597, Providencia, Santiago. T: (56-2) 2360495, F: (56-2)2358407, E: ikort@rdc.cl

Lastrico, R., Civil Engineer, Arcadis Geotécnica, Av. Eliodoro Yáñez 1893, Providencia, Santiago. T: (56-2)3816000, F: (56-2)3816001, E: roberto.lastrico@geotecnica.cl

Lara, J., Civil Engineer, Arcadis Geotécnica, Av. Eliodoro Yáñez 1893, Providencia, Santiago. T: (56-2)3816000, F: (56-2)3816001, E: jose.lara@geotecnica.cl

Lobos, R., Civil Engineer, DICTUC U. Católica de Chile, Av. Vicuña Mackenna 4860, Santiago. T: (56-2)6864208, F: (56-2)5532268

Martínez, F., Civil Engineer, F. Martinez & Asoc., 11 de Septiembre 1480 Of. 41, Providencia, Santiago. T: (56-2) 2356485, F: (56-2)2356487

Morales, E., Av. Angamos 0610, Antofagasta

Musante, H., Civil Engineer, Geofun, María Luisa Santander 0231, Santiago. T: (56-2)3414800, F: (56-2)3414800

Noguera, G., Civil Engineer, Edic Ingenieros Ltda., Alonso de Córdova 5151, Of. 1301, Las Condes. T: (56-2)3785610, F: (56-2)3785656, E: edic@ingenieros.ltda.cl

Norambuena, C., Civil Engineer, Geovenor, Hernando de Aguirre 201 Of. 1101, Providencia, Santiago. T: (56-2)3353480, F: (56-2)3353483, E: geovenor@geovenor.cl

Ortigosa, P., Prof. Civil Engineering, Universidad de Chile, Plaza Ercilla 883, Santiago. T: (56-2)6784151, F: (56-2) 6718979, E: idiem@idiem.uchile.cl

Palma, A., Geotechnical Engineer, EMIN Ingeniería y Construcción Ltda., Félix de Amesti 90, Piso 3, Las Condes, Santiago. T: (56-2)2062815, F: (56-2)2062895

Palma, J., Geotechnical Engineer, U. Católica de Valparaíso, Av. Brasil 2147, Valparaíso. T: (56-32)273633, F: (56-32) 273802, E: jpalma@ucv.cl

Paredes, L., Civil Engineer, AMES Ingeniería y Proyectos, Las Bellotas 199 Of. 34, Providencia, Santiago. T: (56-2) 3351539, F: (56-2)2325077, E: molyp@cmet.net

Pérez, A., Civil Engineer, Geoprospec, Callao 2970 Of. 618, Santiago. T: (56-2)6391499, F: (56-2)2342306

Pérez, P., Civil Engineer, Golder Associates S.A., Av. 11 de Septiembre 2353, Piso 16, Providencia, Santiago. T: (56-9)8534374, E: pperez@golder.cl

Petersen, M., Prof. Civil Engineering, U. Técnica Federico Santa María, Avda. España 1680, Valparaíso. T: (56-32) 654385-654181, F: (56-32)654115, E: mpeterse@ociv.utfsm.cl

Pinilla, L., Civil Engineer, Ingendesa, Santa Rosa 76, Santiago. T: (56-2)6347266, F: (56-2)6354070

Poblete, M., Civil Engineer, M. P. Ingeniería, Miguel Claro 1084, Providencia, Santiago. T: (56-2)2259430, F: (56-2) 2258554, E: mpobl@ctcinternet.cl

Rebolledo, L., Civil Engineer, Knight Piésold Chile, Marchant Pereira 221 Piso 7, Providencia, Santiago. T: (56-2)3417627, F: (56-2)3417628, E: lrebolledo@kpsa.cl

Riveros, C., Civil Engineer, Arcadis Geotécnica, Av. Eliodoro Yáñez 1893, Providencia, Santiago. T: (56-2)3816000, F: (56-2)3816001, E: cecilia.riveros@geotecnica.cl

Rodríguez, F., Prof. Civil Engineering, Universidad Católica de Chile, Vicuña Mackenna 4860, Santiago. T: (56-2) 5522375, F: (56-2)5524054, E: Frroa@ing.puc.cl

Rojas, L., Civil Engineer, GEO Consultores. Calle 6½ Oriente 169. Viña del Mar. T: (56-32)976828, F: (56-32)697930, E: geoconsu@entelchile.net

Soto, V., Civil Engineer, Arcadis Geotécnica, Av. Eliodoro Yáñez 1893, Providencia, Santiago. T: (56-2)3816000, F: 3816001, E: victor.soto@geotecnica.cl

Tapia, C., Civil Engineer, Geosonda Ltda., Bustamante 32 Of. 31, Santiago. T: (56-2)2094830, F: (56-2)2094830, E: geosonda@geosonda.cl

Troncoso, J., Prof. Civil Engineering, Universidad Católica de Chile, Vicuña Mackenna 4860, Santiago. T: (56-2) 6864214, F: (56-2)2468082, E: troncoso@ing.puc.cl

Valdebenito, R., Civil Engineer, VST Ingenieros Ltda.,
Av. Antonio Varas 1803, Providencia, Santiago. T: (56-2)
3414128, F: (56-2)3414128
Valenzuela, L., Civil Engineer, Arcadis Geotécnica, Av.
Eliodoro Yáñez 1893, Providencia, Santiago. T: (56-2)
3816000, F: 3816001, E: luis.valenzuela@geotecnica.cl
Van Sint Jan, M., Prof. Civil Engineering, Universidad
Católica de Chile, Vicuña Mackenna 4860, Santiago.
T: (56-2)6864209, F: (56-2)6864243, E: vsintjan@ing.puc.cl
Veiga, J., Civil Engineer, Geovenor, Hernando de Aguirre 201
Of. 1101, Santiago. T: (56-2)3353480, F: (56-2)3353483,
E: jveiga@geovenor.cl

Velasco, L., Civil Engineer, Edic Ingenieros Ltda., Alonso de
Córdova 5151, Of. 1301, Las Condes. Santiago. T: (56-2)
3785610, F: (56-2)3785656, E: lvelasco@edic.cl
Ventura, H., Civil Engineer, Héctor Ventura y Asociados,
Biarritz 1953, Santiago. T: (56-2)2047543, F: (56-2)225
1608, E: venturaa@cmet.net
Verdugo, R., Prof. Civil Engineering, Universidad de Chile,
Plaza Ercilla 883, Santiago. T: (56-2)6784142, F: (56-2)
6718979, E: rverdugo@cec.uchile.cl
Villarroel, C., Laboratorio Nacional de Vialidad, Ministerio
de Obras Públicas. Morandé 59-71, Piso 2 y 3, Santiago.
T: (56-2)2211391, F: (56-2)2213745

CHINA/CHINE

Secretary: Zhang, Jian-Min, Prof., CISMGE-CCES, Dept. of Hydraulic Eng., Tsinghua University, Beijing 100084, PCR. T: 62785593 (O), F: 62785699, E: zhangjm@mail.tsinghua.edu.cn, cismge@tsinghua.edu.cn

Total number of members: 154

Bai, Ri-Sheng, Senior Engineer, Prof., The 4th Survey and Design Institute, Ministry of Railway, Yangyuan, Wuchang, Hubei 430063. T: (8627)6818619

Bao, Cheng-Gang, Prof., Yangtze River Scientific Research Institute, 23 Huangpu Road, Wuhan, Hubei 430010. T: (8627) 2829710, F: (8627)2829726, E: hub.geoengyz@hubei.shspt. china.mail

Chang, Lu, Senior Engineer, Shenzhen Building Engg. Foundation Co., Honggang, Shenzhen 518028. T: (86755) 9100272

Chen, Fan, Senior Engineer, Inst. of Foundation Engg., China Academy of Building Research, No. 30, Beisanhuan Dong Rd., Beijing 100013. T: (8610)64236191, F: (8610) 64213086

Chen, Huan, Prof., Department of Hydraulic Engg., Tianjin University, Tianjin 300072. T: (8622)7472074, F: (8622) 3358741

Chen, Liang-Sheng, Prof., Dept. of Hydraulic Engg., Tsinghua University, Beijing 100084

Chen, Long Zhu, Prof., Dept. of Civil Engg., Zhejiang University, Hangzhou 310027. T: (86571)7951398, F: (86571)7951846

Chen, Yu-Jiong, Senior Engineer, Prof., China Institute of Water Resources and Hydropower Research, 20 Chegong-zhuangxi Rd., Beijing 100044. T: (8610)68415522, Ext. 6900, F: (8610)68412316

Chen, Yun-Min, Dr, Prof., Dept. of Civil Engineering, Zhejiang University, Hangzhou 310027. T: (86571)5172244, Ext. 2319(O), (86571)7951846

Chen, Zheng-Han, Prof., Logistics Engineering College, Yuzhou Road, Chongqing 400016. T: (8623)68756762, F: (8623)68595773

Chen, Zhi-De, Senior Engineer, Beijing Geotechnical Institute, 15 Yang Fang Dian Road, Fu Xing Meng Wai, Beijing 100038. T: (8610)63967691, F: (8610)63967691

Chen, Zhong-Yi, Prof., Dept. of Hydraulic Engineering, Tsinghua University, Beijing 100084. T: (8610)62785803, F: (8610)62780265

Chen, Zu-Yu, Senior Engineer, Prof., China Institute of Water Resources and Hydropower Research, P.O. Box 366, Beijing 100044. T: (8610)68415522, Ext. 6205, F: (8610) 68412316, E: chenzy@mail.tsinghua.edu.cn

Cheng, Zhan-Lin, Senior Engineer, Geotechnical Engg. Divi-sion, Yangtze River Sc. Research Institute, 23 Huangpu Road, Wuhan 430010. T: (8627)2829710, F: (8627)282 9726

Cui, Jie, Institute of Engg. Mechanics, SSB, Xuefu Road, Harbin 150080. T: (451)6662901, Ext. 516, F: (451)6664755

Dai, Yun-Xiang, Dr, Senior Engineer, Shenzhen Investigation & Research Inst., Geotechnical Desigm & Res. Branch, Futian, Shenzhen 518026. T: (86755)3253392, F: (86755) 3328249

Feng, Guo-Dong, Prof., Wuhan University of Hydraulic & Electrical Engg., Wuhan 430027. T: (8627)722212, Ext. 246

Gong, Xiao-Nan, Prof., Dept. of Civil Engineering, Zhejiang University, Hangzhou, Zhejiang 310027. T: (86571)7894678, F: (86571)7651012

Gong, Yi-Ming, Senior Engineer, Prof., Fujian Institute of Building Research, No. 162 Yangqiaozhong Rd., Fuzhou 350002. T: (86591)3715646, F: (86591)3715748

Gu, An-Quan, Prof., Xian Highway University, 3 Cuihua Road, Xian 710064. T: (8629)7213192, F: (8629)5261532

Gu, Bao-He, Research Prof., Comprehensive Institute of Geotechnical Investigation & Surveying 177 Dongzhimennei Str., Beijing 100007. T: (8610)64013366

Gu, Xiao-Lu, Prof., Civil Engg. Dept., Tianjing University, Tianjing 300072. T: (8622)3359116

Gu, Xiao-Yun, Research Prof., Institute of Mechanics, Chinese Academy of Sciences, 15 Zhonguanchun Road, Beijing 100080. T: (8610)62554188, F: (8610)62561284, E: guxy@ cc5.imech.ac.cn

Gui, Ye-Kun, Shanghai Foundation Co., Yanan Dong Rd., Shanghai 200002

Guo, Le-Qun, Senior Engineer, Chief Engineer, Shenyang Institute of Geotechnical Investigation, Ministry of Metalur-gical Industry, Shenyang 110015

Guo, Xi-Rong, Senior Engineer, China Institute of Water Resources and Hydropower Research, 20 West Chegong-zhuang Road, Beijing 100044. F: (8610)68412316

Guo, Zeng-Yue, Prof., Geotechnical Res. Inst., Xi'an Univer-sity of Technology, P.O. Box 211, South Jinhua Rd., Xi'an 710048. T: (8629)3224670, F: (8629)3235545

He, Guang-Ne, Prof., Dept. of Civil Engg., Dalian University of Technology, Dalian, Lioanin 116023. T: (86411)4709778, F: (86411)4671009, E: dutpube@dlut.edu.cn

He, Huai-Jian, Research Prof., Institute of Rock & Soil Mechanics, Chinese Academy of Sciences, Xiao Hong Shan, Wuhan 430071

Hou, Xue-Yuan, Prof., Dept. of Geotechnics, Tongji Univer-sity, Shanghai 200092. T: (8621)65022243 (O), (8621) 65153857 (H), F: (8621)65025188

Hou, Wei-Sheng, Senior Engineer, Prof., Fujian Institute of Building Research, No. 162 Yangqiaozhong Rd., Fuzhou 350002. T: (86591)3715646, 3799249, E: hws@publ. Fz.Fj.cn

Hu, Ding, Prof., Geotechnical Engg. Division, Hydraulic Engg. Dept., Sichuan Union University, Chengdu, Sichuan 610065. T: (8628)5581554, F: (8628)5582670

Hu, Zhan-Fei, Associate Prof., Dept. of Geotechnical Engg., Tongji University, Shanghai 200092. T: (8621)65025080, Ext. 4011, F: (8621)65025188

Huang, Qiang, Senior Engineer, China Academy of Building Research, No. 30, Beisanhuan Dong Road, Beijing 100013. T: (8610)64214356 (O), 64202233, Ext. 3033 (H), F: (8610) 64221369

Huang, Shao-Ming, Senior Engineer, Prof., Shanghai Muni-cipal Institute of Architectural Design, 17 Guangdong Rd., Shanghai 200002. F: (8621)63290183

Huang, Xi-Ling, Member of Chinese Academy of Engg., Research Prof., China Academy of Building Research, No. 30 San Huan Dong Rd., Beijing 100013. T: (8610)64265641, F: (8610)64213086

Huang, Xin, Senior Engineer, Central Research Institute of Building & Construction of MMI, 33 Xi Tu Cheng Road, Beijing 100088. T: (8610)62015599, Ext. 3325, F: (8610) 62057789

Jiang, Guo-Cheng, Senior Engineer, Prof., China Institute of Water Resources and Hydropower Research, 20 West Chegongzhuang Road, Beijing 100044. T: (8610)68415522, Ext. 6420, F: (8610)68412316

Jiang, Peng-Nian, Senior Engineer, Prof., Nanjing Hydraulic Research Institute, 34 Hujuguan Road, Nanjing, Jiangsu 210024. T: (8625)3717041, F: (8625)3310321

Kong, Xian-jing, Prof., Dept. of Civil Engg., Dalian Univer-sity of Technology, Dalian 116024. T: (86411)4708501, F: (86411)4671009

Kuang, Jian-Zheng, Research Prof., Guangzhou Grouting Co., Xianlie Rd., Guangzhou 510070. T/F: (8620) 87768677

Li, Cheng-Jiang, Dr, Senior Engineer, Central Research Institute of Building & Construction of MMI, 33 Xi Tu Cheng Road, Beijing 100088. T: (8610)62015599, Ext. 4014, F: (8610)62013104

Li, Cheng-Quan, Senior Engineer, Guangdong Institute of Building Research, Guangzhou

Li, Fu-Min, Senior Engineer, Shenzhen Branch, China Academy of Railway Sciences, Huaming Bld., Yannan Rd., Shenzhen 518031. T: (86755)3233135, F: (86755)3233136

Li, Guang-Xin, Prof., Dept. of Hydraulic Engg., Tsinghua University, Beijing 100084. T: (8610)62785593, F: (8610) 62595699

Li, Guo-Xiong, General Manager, Engineer, Guangzhou Luban Water-Proofing Engineering. Co., 7-1# 4 Bld., Guangdongqian, Donghua Dong Rd., Guangzhou 510080. T: (8620)7676852

Li, Neng-Hui, Research Prof., Nanjing Hydraulic Research Institute, 34 Hujuguan, Nanjing 210024. T: (8625) 6636206 (O), (8625)3320432 (H), F: (8625)3310321

Li, Rong-Qiang, Dr, Senior Engineer, Shenzhen Investigation & Research Inst., Futian, Shenzhen 518026. T: (86755) 3357434, F: (86755)3328285

Li, Tze-Chun, Associate Research Prof., Geotechnical Division, China Academy of Railway Sciences, Beijing 100081. T: (8610)63224043

Li, Yao-Gang, Senior Engineer, Comprehensive Institute of Geotechnical Investigation & Surveying, Dongzhimen, Beijing 100007. T: (8610)64013391, (8610)64013189

Lin, Zai-Guan, Senior Engineer, Prof., Northwest Research Institute of Engg. Investigations & Design, No. 9 Xiwuyuan, Xi'an, Shaanxi 710003. T: (8629)7215401 (O), (8629) 7320283 (H), F: (8629)7321343

Liu, Guo-Bin, Associate Prof., Dept. of Geotechnical Engg., Tongji University, Shanghai 200092. T: (8621)65025080, Ext. 4011, F: (8621)65025188

Liu, Guo-Nan, Associate Prof., Geotechnical Division, China Academy of Railway Sciences, Beijing 100081. T: (8610) 63249415 (O), 62232661 (H), F: (8610)62256572

Liu, Han-Long, Prof., Geotechnical Engineering Research Institute, Hohai University, No. 1, Xikang Road, Nanjing 210098, China. T: (025)3713777, Ext. 51918 (O), (025) 3713777, Ext. 51884 (O), F: (025)3716161, E: hliu@pub. jlonline.com

Liu, Jian-Hang, Member of Chinese Academy of Engg., Senior Engineer, Shanghai Municipal Engg. Bureau, No. 193, Hangkou Rd., Shanghai 200002. F: (8621) 63230590

Liu, Jian-Hua, Senior Engineer, Vice-President, Beijing Geotechnical Institute, 15 Yang Fang Dian Road, Beijing 100038. T: (8610)63983608, F: (8610)63967691

Liu, Jin-Li, Research Prof., Institute of Foundation Engg., China Academy of Building Research, No. 30, Beisanhuan Dong Road, Beijing 100013. T: (8610)64236191, F: (8610) 64213086

Liu, Ming-Zhen, Prof., Xi'an University of Architecture & Technology, Yanta Road, Xi'an 710055. T: (029) 2201201 (H)

Liu, Song-Yue, Dr, Associate Prof., Geotechnical Engg. Inst., Southeast University Nanjing 210028. T: (8625) 3600598, F: (8625)7712719

Liu, Zu-De, Prof., Wuhan University of Hydraulic & Electrical Engg., Wuhan 430072. T: (8627)722212, Ext. 246

Liu, Zu-Dian, Prof., Institute of Water Resources & Hydro-Electric Engg., Xi'an University of Technology, Jinghua Road, Xi'an 710048. T: (8629)3232931, Ext. 3792, F: (8629)3235545

Lou, Zhi-Gang, Prof., Institute of Mechanics, Chinese Academy of Sciences, 15 Zhongguancun Road, Beijing 100080. T: (8610)62554188, F: (8610)62561284

Lu, Yao-Sheng, Senior Engineer, Shenzhen Branch, China Academy of Railway Sciences, Huaming Bld., Yannan Rd., Shenzhen 518031. T: (86755)3233135, F: (86755) 3233136

Lu, Yi-Jie, Senior Engineer, Shenzhen Branch, Central Res. Inst. of Building & Construction of MMI, No. 2-102 Longcheng Garden, Nanyou, Shenzhen 518054. T: (86755) 6400600

Lu, Zhao-Jun, Member of Chinese Academy of Sciences, Research Prof., Geotechnical Division, China Academy of Railway Sciences, Beijing 100081. T: (8610)63224396, F: (8610)62256572

Luan, Mao-Tian, Prof., Dept. of Civil Engg., Dalian University of Technology, Dalian 116024. T: (86411)4708513, F: (86411)4671009, E: mtluan@dlut.edu.cn

Luo, Shu-Xue, Associate Prof., Southwest Jiaotong University, Chengdu, Sichuan 610031. T: (8628)7524160, Ext. 49718, F: (8628)7784007

Luo, Yu-Sheng, Senior Engineer, Shanxi Institute of Building Sciences Research, 142 West Road, Xian, Shanxi 710082. T: (8629)862700, Ext. 316

Ma, Shi-Dong, Prof., Dept. of Civil Engg., Huachiao University, Quanzhou, Fujian 362011. T: (86595)2681797, F: (86595)2686969

Ma, Wei, Research Prof., Research Dept. of Frozen Soil Engineer, Cold and Arid Regions Environmental & Engineering Research Institute, China Academy of Sciences, 260# Donggangxi District, Lanzhou Gansu 730000. T: (0931)8279806 (O), (0931)8278139 (H)

Pan, Qiu-Yuan, Prof., Dept. of Civil Engineering, Zhejiang University, Hangzhou 310027. T: (8571)5172244, Ext. 2319 (O), (86571)7964509 (H), T: (86571)7951846

Pei, Jie, Chief Engineer, Shanghai Shenyuan Geotechnical Engineering Co., 18# 314 Nong Wulumoqinan Road, Shanghai 200031. T: (021)64337738, 13901647255

Peng, Da-Yong, Senior Engineer, Shanghai Engineering Consultants & Supervision Co., 1102 Room, No. 55 Haining Road, Shanghai 200080. T: (8621)63063773, F: (8621) 63067673

Pu, Jia-Liu, Prof., Dept. of Hydraulic Engg., Tsinghua University, Beijing 100084. T: (8610)62785457, F: (8610) 62785699, E: pjl-dhh@mail.tsinghua.edu.cn

Qian, Hong-Jin, Prof., Xi'an University of Architecture & Technology, 13 Yanta Road, Xi'an Shaanxi 710055. T: (8629)2202599, F: (8629)7215422

Qiao, Zheng-Shou, Senior Engineer, The 4th Surveying & Designing Institute, Ministry of Railway, Yangyuan, Wuhan 430063. T: (8627)6816139

Rao, Hong-Yan, Research Prof., Research Institute of Highway, The Ministry of Communication, No. 48 Beisanhuanzhong Rd., Beijing 100088. T: (8610)62013399, Ext. 2309, F: (8610)62014130

Sha, Xian-Lin, Senior Engineer, Hainan East Architectural Design Co. Ltd., D2-201 Nanxi Garden, Haikou, Hainan 570102. T: (86898)6767590

Shan, Chang, President, Comprehensive Institute of Geotechnical Investigation & Surveying, Dongzhimen, Beijing 100007

Shao, Long-Tan, Associate Prof., Dept. of Civil Engg., Dalian University of Technology, Dalian 116024. T: (86411) 4708501, F: (86411)4671009

Shen, Xiao-Ke, Senior Engineer, Beijing Geotechnical Institute, 15 Yang Fang Dian Road, Beijing 100038. T: (8610) 63983608, F: (8610)63967691

Shen, Zhu-Jiang, Member of Chinese Academy of Sciences, Research Prof., Dept. of Geotechnical Engg., Nanjing Hydraulic Research Institute, 34 Hujuguan Road, Nanjing 210024. T: (8625)6636206, F: (8625) 3310321

Shi, Chun-Lin, Associate Prof., Geotechnical Division, China Academy of Railway Sciences. T: (8610)63249415, F: (8610)62256572

Song, Er-Xiang, Prof., Dept. of Civil Engg., Tsinghua University, Beijing 100084. T: 62784979, E: songek@ tsinghua.edu.cn

Sun, Geng-Sheng, Senior Engineer, Prof., Shanghai City Planning Bureau, 193 Hankou Road, Shanghai 200002. T: (8621)63232923), F: (8621)63230590

Sun, Yu-Qi, Prof., Dept. of Civil Engg., Lanzhou Railway Institute, Anning District, Lanzhou, Gansu 730070. T: (86931)7661966, F: (86931)7666426

Teng, Yan-Jing, Research Prof., Institute of Foundation Engg., China Academy of Building Research, No. 30, Beisanhuan Dong Road, Beijing 100013. T: (8610) 64236191, F: (8610)64213086

Wang, Bao-Tian, Dr, Associate Prof., Geotechnical Engg. Institute, Hohai University, 1 Xikang Road, Nanjing 210089. T: (8625)3313648, F: (9625)3315375

Wang, Fu-Ming, Prof., Dept. of Hydraulic Engineering, Zhengzhou University, 97 Wenhua Road, Zhengzhou, Henan 450002. T: (0371)3887443 (O), (0371)3887447 (O), F: (0371)3886043

Wang, Gong-Xian, Research Prof., Northwest Branch, China Academy of Railway Sciences, 365 East Minzhu Road, Lanzhou, Gansu 730000. T: (86931)8934214

Wang, Ji-Wang, Senior Engineer, Central Research Institute of Building & Construction of MMI, 33 Xi Tu Cheng Road, Beijing 100088. T: (8610)62013104, F: (8610)62013104

Wang, Kun-Yao, Senior Engineer, Geotechnical Engg. Division, China Institute of Water Resources and Hydropower Research, Chegongzhuang, Beijing 100044. T: (8610) 68415522, Ext. 6701

Wang, Lan-Min, Research Prof., Lanzhou Institute of Seismology, China Seismological Bureau, 410 West Donggang Road, Lanzhou 730000

Wang, Ren, Research Prof., Wuhan Geotechnical Institute, Xiaohongshan, Wuhan 430071. T: (8627)87870516, (027) 87863386, E: rwang@dellwhrsm.ac.cn

Wang, Tie-Hong, Research Prof., Institute of Foundation Engg., China Academy of Building Research, No. 30, Beisanhuan Dong Road, Beijing 100013. T: (8610)64236191, F: (8610)64213086

Wang, Tie-Ru, Prof., Dept. of Civil Engg., Zhejiang University, Hanzhou, Zhejiang 310027. T: (86571)7984109, F: (86571)5171797

Wang, Wen-Shao, Member of Chinese Academy of Sciences, Research Prof., China Institute of Water Resources & Hydropower Research, P.O. Box 366, Beijing 100044. T: (8610)68415522, Ext. 6879

Wang, Xiao-Gang, Senior Engineer, Prof., Geotechnical Engg. Division, China Institute of Water Resources and Hydropower Research, Chegongzhuang, Beijing 100044. T: (8610)68415522, Ext. 6205, (8610)68438317, E: wangxg@IWHR.com

Wang, Xin-Jie, President, Beijing Urban Engg. Design & Research Institute, A-1 Xi'erhuan Road, Beijing100037. T: (8610)68318837, F: (8610)68312806

Wang, Yuan, Associate Research Prof., Research Institute of Highway, The Ministry of Communication, Xueyuan Rd., Beijing 100088. T: (8610)62079586

Wang, Zheng-Hong, Prof., Postgraduate School, North China Institute of Water Conservancy & Hydro-electric Power, Zizhuyuan, Beijing 100044. T: (8610)68417950

Wang, Zhong-Qi, Research Prof., Comprehensive Inst. of Geot. Engg. Investigat. & Surveying, 177 Dongzhimen St., Beijing 100007

Wei, Ru-Long, Research Prof., Geotechnical Engg. Dept., Nanjing Hydraulic Research Institute, 34 Hujuguan, Nanjing 210024. T: (8625)6636206, F: (8625)3310321

Wu, Chang-Yu, President, Senior Engineer, Yangtze River Scientific Research Institute, 23 Huangpu Road, Wuhan, Hubei 430010. T: (8627)82829818 (O)

Wu, Ju-Yuan, Senior Engineer, Beijing BUGG Metro and Foundation Engg. Co. Ltd., No. 15 Daliushu Rd., Beijing 100081. T: (8610)62185705, F: (8610)62185709

Wu, Shi-Ming, Prof., Institute of Geotechnical Engg., Zhejiang University, Hangzhou 310027. T: (86571)7951106, F: (86571)7951385, E: smwu@ldns.zju.edu.cn

Wu, Xiao-Ming, Research Prof., Geotechnical Division, China Academy of Railway Sciences, Beijing 100081. T: (8610)63249435, F: (8610)62780265, E: wuxmc@163bj.com

Xie, Ding-Yi, Prof., Xi'an University of Technology, P.O. Box 211, South Jinhua Rd., Xi'an 710084. T: (8629)3224670, F: (8629)3235545

Xie, Shi-An, Senior Engineer, Guangdong Institute of Electric Power Design, Guangzhou

Xie, Yong-Li, Dr, Associate Prof., Xian Highway University, 3 Cuihua Road, Xi'an 710064. T: (8629)7213192, F: (8629)5261532

Xu, Bang-Dong, Research Prof., Northwestern Branch, China Academy of Railway Sciences, Lanzhou 730000

Xu, Mei-Kun, Senior Engineer, Vice President, Research Institute of Harbor Engineering, Shanghai 200032. T: (8621)64696915

Xu, Wei-Yang, Senior Engineer, Prof., East China Architectural Design & Research Institute, 151 Hankou Road, Shanghai 200002. T: (8621)63217420, F: (8621)63214301

Yan, Ming-Li, Research Prof., Institute of Foundation Engg., China Academy of Building Research, No. 30, Beisanhuan Dong Road, Beijing 100013. T: (8610)64236191, F: (8610)64213086

Yang, Bin, Senior Engineer, Institute of Foundation Engg., China Academy of Building Research, 30 Beisanhuan Dong Rd., Beijing 100013. T: (8610)64202233, Ext. 2344, F: (8610)64213086

Yang, Guang-Hua, Senior Engineer, Guangdong Hydraulic Research Institute, Shougouling, Guangzhou 510610

Yang, Jun, Senior Engineer, Institute of Foundation Engg., China Academy of Building Research, No. 30, Beisanhuan Dong Road, Beijing 100013. T: (8610)64236191, F: (8610)64213086

Yang, Ming, Prof., Dept. of Geotechnical Engg., Tongji University, Shanghai 200092. T: (8621)65025080, Ext. 4011, F: (8621)65150683

Yang, Qing, Associate Prof., Dalian University of Technology, Dalian 116024. T: (0411)4708513, (0411)4674141, E: qyang@dlut.edu.cn

Yang, Zhi-Yin, Senior Engineer, Shenzhen Branch, Central Res. Inst. of Building & Construction of MMI, No. 2-102 Longcheng Garden, Nanyou, Shenzhen 518054. T: (86755)6400600

Yao, Yang-Ping, Prof., Civil Engg., Beijing University of Aeronautics and Astronautics, 37# Xueyuan Road, Haidian, Beijing 100083. T: (8610)82317538 (O), (8610)82316916 (H), F: (8610)82316916, E: ypyao@263.net

Ye, Guan-Bao, Associate Prof., Dept. of Geotechnical Engg., Tongji University, Shanghai 200092. T: (8621)65014339 (H), F: (8621)65025188

Ye, Yang-Sheng, Associate Research Prof., Geotechnical Division, China Academy of Railway Sciences, Beijing 100081. T: (8610)63224043, (8610)62256572

Yin, Zong-Ze, Prof., Hehai University, 1 Xikang Road, Nanjing 210098. T: (8625)3313648, F: (9625)3315375

Yuan, Jian-Xin, Research Prof., Institute of Rock & Soil Mechanics, Chinese Academy of Sciences, Xiao Hong Shan, Wuhan 430071. T: (8627)7881776, F: (8627)7862 413, E: jxyuan@dell.wursm

Yuan, Xiao-Ming, Research Prof., Institute of Engg. Mechanics, SSB, 9 Xuefu Road, Haibin 150080. T: (86451)6662901, Ext. 516 (O), Ext. 554 (H), F: (86451)6664755

Zai, Jin-Min, Prof., Nanjing Architectural & Civil Engineering Institute, 200# Zhongshanbei Rd., Nanjing 210009. T: (025)3239966, F: (025)3239888, E: zai@ilonline.com

Zeng, Guo-Xi, Prof., Dept. of Civil Engg., Zhejiang University, Hangzhou 310027. T: (86571)7981767, F: (86571)7951358, E: zadri@pub.zjptanet.cn

Zeng, Qing-Yi, Dr, Senior Engineer, Shenzhen Geotechnical Engg. Co., No. 4 Shangbu Zhong Rd., Shenzhen 518028. T: (86755)3352514

Zhang, Guo-Xia, Senior Engineer, Prof., Beijing Geotechnical Institute, 15 Yangfangdian Rd., Beijing 100038

Zhang, Hong-Ru, Prof., Northern Jiaotong University, Beijing 100044. T: (8610)63240217, (8610)63240044, E: Zhanghr@public.bta.net.cn

Zhang, Hong-Yu, Senior Engineer, Prof., Editorial Office, Chinese Journal of Geotechnical Engg. 34, Hujuguan Rd., Nanjing, 210024. T: (8625)6633662, Ext. 512, F: (8625)3310321

Zhang, Jian-Min, Prof., Dept. of Hydraulic Engg., Tsinghua University, Beijing 100084. T: (8610)62772074 (H), (8610)62783867 (O), F: (8610)62785566, E: zhangjm@tsinghua.edu.cn

Zhang, Jing, President, Tianjin Port Institute, Dagu Nan Rd., Tianjin 300222

Zhang, Qian-Li, Associate Research Prof., Geotechnical Division, China Academy of Railway Sciences, Beijing 100081. T: (8610)63249435

Zhang, Si-Ping, Vice President, Chongqing University of Architecture, Chongqing 630045

Zhang, Wei, Senior Engineer, Prof., Institute of Geotechnical Investigation & Surveying of MMI, No. 51 Xianning Zhong Rd., Xi'an 710043. T: (8629)3281275, F: (8629)3231254

Zhang, Wei-Min, Senior Engineer, Prof., Nanjing Hydraulic Research Institute, Hujuguan Road, Nanjing 210024. T: (8625)3738179, Ext. 541

Zhang, Wen-Zheng, Research Prof., China Institute of Water Resources & Hydropower Research, No. 20 Chegongzhuang Rd., Beijing 100044. T: (8610)68453859, (8610) 68453860, E: iwhr@mimi.cnc.ac.cn

Zhang, Yan, Senior Engineer, Institute of Foundation Engg., China Academy of Building Research, No. 30, Beisanhuan Dong Road, Beijing 100013. T: (8610)64236191, F: (8610) 64213086

Zhang, Yong-Jun, Research Prof., Institute of Foundation Engg., China Academy of Building Research, P.O. Box 752, 30 Beis San Huan Dong Rd., Beijing 100013. T: (8610) 64227010, F: (8610)64213086

Zhang, Yu-Fang, Senior Engineer, Shenzhen Branch, China Academy of Railway Sciences, Huaming Bld., Yannan Rd., Shenzhen 518031. T: (86755)3342227, F: (86755)3233136

Zhang, Zai-Ming, Senior Engineer, Prof., Beijing Geotechnical Institute, 15 Yang Fang Dian Rd., Beijing 100038. T: (8610)63964323, F: (8610)63967691

Zhao, Wei-Bing, Prof., Dept. of Geotechnical Engg., Hohai University, 1 Xikang Road, Nanjing 210024. T: (8625) 3323777, Ext. 50961, F: (9625)3315375

Zhao, Xue-Meng, Deputy Director, Xi'an Highway Research Institute, Xi'an 710061

Zhao, You-Ming, Senior Engineer, Shenzhen Branch, China Academy of Railway Sciences, Huaming Bld., Yannan Rd., Shenzhen 518031. T: (86755)3342227, F: (86755)3233136

Zheng, Ying-Ren, Prof., Dept. of Civil Engg., Logistic Engineering College, 79 Yuzhou Rd., Chongqing, Sichuan 630041. T: (86811)8756785

Zhou, De-Pei, Prof., Southwest Jiaotong University, Chengdu, Sichuan 610031

Zhou, Guo-Jun, Senior Engineer, Prof., Central Research Institute of Building & Construction of MMI, 33 Xi, Tu Cheng Road, Beijing 100088. T: (8610)62013104, F: (8610) 62057789

Zhou, Jing, Member of Chinese Academy of Engg., Research Prof., Geotechnical Division, China Academy of Railway Sciences, Beijing 100081. T: (8610)63224396, F: (8610)62256572

Zhou, Shen-Gen, Research Prof., Geotechnical Division, China Academy of Railway Sciences, Beijing 100081. T: (8610)63249425, F: (8610)62256572

Zhou, Shun-Hua, Prof., Institute of Geotechnical Engg., Shanghai Railway University, 450# Zhangnan Rd., Shanghai 200311. T: (021)56220585, (021)56220565, E: zhoushh@online.sh.cn

Zhu, Wei-Xin, Senior Engineer, Prof., Dept. of Geotechnical Engg., Nanjing Hydraulic Research Insititute, 34 Hujuguan Rd., Nanjing 210024. T: (8625)6634175, F: (8625)310321

Zhu, Xiang-Rong, Associate Prof., Dept. of Civil Engineering, Zhejiang University, Hangzhou 310027. T: (86571) 7985002 (H), F: (86571)7951846

COLOMBIA/COLOMBIE

Secretary: Eng. Jacobo Ojeda Moncayo, Sociedad Colombiana de Geotecnia, Apartado Aéreo 057045, Bogotá, D.C.
T: 57-1-2200287, 4123300, Ext. 465, F: 57-1-2220438, 4244592, E: jojeda@ingeomin.gov.co, scg1@colomsat.net.co

Total number of members: 29

Acosta, M.H.E., Ingeniero, Ingeniería y Geotecnia Ltda. Diagonal 127A No. 17-20, Bogotá, D.C. T: 2586676, 2746567, E: heacosta@geot.te.u-tokyo.ac.jp
Alvarez, P.J.A., Ingeniero. Cra. 29A #158-80 Int. 104, Bogotá, D.C. T: 6776578, E: jaa@col.net.co
Amórtegui, G.J.V., Ingeniero, Ingeniería y Geotecnia Ltda., Diagonal 127A No. 17-20, Bogotá, D.C. T: 2586676, 2166297, E: jamorteg@gaitana.interred.net.co
Angel, R.G., Ingeniero, IEH Grucon Ltda., Avenida 13 No. 118-30 Of. 703, Bogotá, D.C. T: 6294303, 6294303, E: gangelr@tutopia.com
Arias, C.L., Ingeniero, Ingeniería Oslo Ltda., Calle 39B Sur No. 37-27, Medellín. T: (094)3350540, 3314474, E: gol@epm.net.co
Caicedo, H.B., Ingeniero, Universidad de los Andes, Calle 122 A No. 14-71 Apto. 204 I.2, Bogotá, D.C. T: 3394949, 4141340
Carrizosa, G.A., Ingeniero, Eta S.A. Calle 29 No. 6-58 Of. 801, Bogotá, D.C. T: 2876693, 2872184, E: alcarri@netscape.net
Correa, V.E., Ingeniero, Sedic Ltda., Calle 49 No. 45-65 Piso 11, Medellín. T: (094)2511400, 2519418, E: geotecnia@sedic.com.co
García, L.M., Ingeniero, Ingeniería y Geotecnia Ltda., Diagonal 127A No. 17-20, Bogotá, D.C. T: 2586676, 2166297, E: igl@gaitana.interred.net.co
García, M.F., Ingeniero, 504 Emory Court Apt. 304, Salisbury, USA. T: 4105481591, E: fgarciag@worldnet.att.net
Gómez, A.G.A, Ingeniero, Inversiones Setema Ltda., Calle 166 #23-10, Bogotá D.C. T: 6717824, 6717568, E: ggustavo@uol.com.co
González, G.A.J., Ingeniero, A.G.C. Ltda., Cra. 54 A No. 172-50, Bogotá, D.C. T: 6720297, 6724251, E: ajgonzg@col1.telecom.com.co
Martínez, R.J.M., Ingeniero, Geoingeniería Ltda., Cra. 13 A No. 90-18 Piso 5, Bogotá, D.C. T: 6183000, 6183880, E: geoingenieria@geoingenieria.com
Mejía, G.L.M., Geóloga, Instituto Nacional de Vías, Calle 97 No. 57-47 Apto. 503, Bogotá, D.C. T: 4280400, Ext. 1192, 1492, E: lmejia@invias.gov.co
Montero, O.J., Geólogo, Ingeominas, Transv. 15A No. 130-80, Bogotá, D.C. T: 2165648, 2200246, E: jmo887@colomsat.net.co

Moya, B.J.E., Ingeniero, Ingeniería y Geotecnia Ltda., Diagonal 127A No. 17-20, Bogotá, D.C. T: 2586676, 2166297, E: igl@gaitana.interred.net.co
Oviedo, S.D.M., Ingeniero, Universidad Pontificia Bolivariana, Km. 7 Vía a Piedecuesta, Bucaramanga. T: (097) 6796220, Ext. 301, E: dm_oviedo@yahoo.com
Pabón, G.G., Ingeniero, Consultoría Colombiana S.A., Cra. 20 No. 37-28, Bogotá, D.C. T: 2875300, 3200941, Ext. 346, E: gpabon@concol.com
Parra, F.H.E., Ingeniero Ingeciencias S.A., Calle 72 No. 12-65 Of. 404, Bogotá D.C. T: 2124261, 4306, E: hecpageo@unete.com
Portilla, G.M.E., Geólogo, Universidad Nacional, Calle 1 D No. 31B-46 Piso 2, Bogotá, D.C. T: 2379393, 3165368, E: portilla@ciencias.ciencias.unal.edu.co
Rios, R.C.J., Escuela Colombiana de Ingeniería, Autopista Norte Km. 13, Bogotá, D.C. T: 6762666, Ext. 238, E: crios@escuelaing.edu.co
Rodríguez, G.E.E., Ingeniero, Georiesgos, Cra. 16 No. 96-64 Of. 313, Bogotá, D.C. T: 6369654, 87, E: ingeoriesgos@hotmail.com
Rodríguez, O.J.A., Ingeniero, Geoingeniería Ltda., Cra. 13 A No. 90-18 Piso 5, Bogotá, D.C. T: 6183000, 3175060, Ext. 116. E: jrodríguez@heingenieros.com
Sanabria, P.D., Ingeniero, Subsuelos S.A., Calle 166 No. 23-10, Bogotá, D.C. T: 6711277, 6717509, E: subsuelo@colomsat.net.co
Shuk, E.T., Ingeniero, Calle 111 No. 3A-50, Bogotá, D.C. T: 2133917
Sierra, M.J.M., Ingeniero, Transversal 62A No. 122-05, Bogotá, D.C. T: 2711313, 4139713, E: jmsierra@telesat.com.co
Uribe, S.A., Ingeniero, Alfonso Uribe S. y Cia. Ltda., Cra. 13 No. 96-82 Of. 301, Bogotá, D.C. T: 6184973, 6184997
Vesga, M.L.F., Ingeniero, Edifica Ltda., Cra. 32 No. 98-57, Bogotá, D.C. T: 2362121, 6122174, E: l-vesga@isis.uniandes.edu.co
Villegas, G.F., Ingeniero, Integral S.A., Cra. 46 No. 52-36 P.13, Medellín. T: (094)5115400, E: fvillegas@integral.com.co

CROATIA/CROATIE

Secretary: Dr. Vlasta Szavits-Nossan, Secretary of Croatian Society for Soil Mechanics and Geotechnical Engineering, Faculty of Geotechnical Engineering, University of Zagreb, Hallerova aleja 7, HR-42000 Varaždin.
T/F: +385 1 4833553, E: svlasta@grad.hr

Total number of members: 121

Arbanas, Ž., Civil Engineering Institute of Croatia, Rijeka, Vukovarska 10a, HR-51000 Rijeka. T: +385 51 330744, F: +385 51 330810, E: zarbanas@ri.igh.hr

Balija, H., Civil Engineering Institute of Croatia, Janka Rakuše 1, HR-10000 Zagreb. T: +385 1 6144111, F: +385 1 6144732, E: hbalija@zg.igh.hr

Barbalić, I., Civil Engineering Institute of Croatia, Split, Matice hrvatske 15, HR-21000 Split. T: +385 21 523939, F: +385 21 551152, E: ivo.barbalic@st.igh.hr

Benać, Č., Faculty of Civil Engineering, University of Rijeka. Viktora cara Emina 5, HR-51000 Rijeka. T: +385 51 352111, F: +385 51 332816, E: benac@master.gradri.hr

Benamatić, D., Moho, Svetog Mateja 127, HR-10000 Zagreb. T: +385 1 6602746, F: +385 1 683586, E: moho@zg.tel.hr

Bosančić, B., Geotehnika, Kupska 2, HR-10000 Zagreb. T: +385 1 6171812, F: +385 1 6171827, E: geotehnika@zg.tel.hr

Bradvica, I., Civil Engineering Institute of Croatia, Janka Rakuše 1, HR-10000 Zagreb. T: +385 1 6144111, F: +385 1 6144732, E: ibradvica@zg.igh.hr

Briški, G., Rijekaprojekt – Geotechnical Investigations, J. Polić Kamova 111, HR-51000 Rijeka. T: +385 51 436 613, F: +385 51 436 610

Brščić, Z., GeoKon, Starotrnjanska 16a, HR-10000 Zagreb. T: +385 1 6050055, F: +385 1 6050094, E: z.brscic@geokonzg.com

Bruncić, A., Civil Engineering Institute of Croatia, Janka Rakuše 1, HR-10000 Zagreb. T: +385 1 6144111, F: +385 1 6144732, E: abruncic@zg.igh.hr

Brunetta, I., Civil Engineering Institute of Croatia, Janka Rakuše 1, HR-10000 Zagreb. T: +385 1 6144111, F: +385 1 6144732, E: ibrunetta@zg.igh.hr

Čorko, D., Conex, Kalinovica 3/3, HR-10000 Zagreb. T: +385 1 3836500, F: +385 1 3836332, E: corko@conex.hinet.hr

Dašić, G., GeoKon, Starotrnjanska 16a, HR-10000 Zagreb. T: +385 1 6050055, F:+ 385 1 6050094, E: g.dasic@geokonzg.com

Drnjević, B., Civil Engineering Institute of Croatia, Janka Rakuše 1, HR-10000 Zagreb. T: +385 1 6144111, F: +385 1 6144732, E: bdrnjevic@zg.igh.hr

Dugić, M., Prizma, Bernarda Vukasa 22, HR-10000 Zagreb. T/F: +385 1 2346356, E: prizma@zg.tel.hr

Dujmić, D., Moho, Svetog Mateja 127, HR-10000 Zagreb. T: +385 1 6602746, F: +385 1 683586, E: moho@zg.tel.hr

Dusparić, S., Geotehnika, Kupska 2, HR-10000 Zagreb. T: +385 1 6171812, F: +385 1 6171827, E: geotehnika@zg.tel.hr

Filković, Z., Croatian Management for Roads, Vončinina 3, HR-10000 Zagreb. T: +385 1 4617422, F: +385 1 4617222

Frgić, L., Faculty of Mining, Geology and Petroleum Engineering, University of Zagreb, Pierottijeva 6, HR-10000 Zagreb. T: +385 1 4605224, F: +385 1 4836064, E: lfrgic@rgn.hr

Galić, Krešo, Elektroprojekt, Ulica grada Vukovara 37, HR-10000 Zagreb. T: +385 1 6307970, F: +385 1 6152685, E: kresimir.galic@elektroprojekt.tel.hr

Galić, Kruno, Elektroprojekt, Ulica grada Vukovara 37, HR-10000 Zagreb. T: +385 1 6307777, F: +385 1 6152685

Gjetvaj, V., Civil Engineering Institute of Croatia, Janka Rakuše 1, HR-10000 Zagreb. T: +385 1 6144111, F: +385 1 6144732, E: vjgetvaj@zg.igh.hr

Gotić, I., Faculty of Geotechnical Engineering, University of Zagreb, Hallerova aleja 7, HR-42000 Varaždin. T: + 385 42 212228, F: +385 42 313587, E: ivan.gotic@vz.tel.hr

Gotovac, B., Faculty of Civil Engineering, University of Split, Matice hrvatske 15, HR-21000 Split. T: +385 21 303335, F: +385 21 524162, E: gotovac@gradst.hr

Horvat, K., Elektroprojekt, Ulica grada Vukovara 37, HR-10000 Zagreb. T: +385 1 6307968, F: +385 1 6152685

Horvat, Z., Ministry of Environmental Protection and Physical Planning, Gajeva 30a, HR-10000 Zagreb. T: +385 1 4591931, F: +385 1 4591949, E: zlatko.horvat2@zg.hinet.hr

Hrešić, D., Faculty of Civil Engineering, University of Rijeka, Viktora cara Emina 5, HR-51000 Rijeka. T: +385 51 352111, F: +385 51 332816, E: dekanat@gradri.hr

Hudoletnjak, V., Faculty of Geotechnical Engineering, University of Zagreb, Hallerova aleja 7, HR-42000 Varaždin. T: +385 42 212228, F: +385 42 313587, E: geotehnicki-fak.vz@vz.tel.hr

Ivandić, K., Faculty of Civil Engineering, University of Zagreb, Kačićeva 26, HR-10000 Zagreb. T: +385 1 4561222, F: +385 1 4827008, E: ivandic@grad.hr

Ivanišević, I., Civil Engineering Institute of Croatia, Osijek, Drinska 18, HR-31000 Osijek. T: +385 31 274524, F: +385 31 274 400, E: ikolund@os.igh.hr

Ivoš, M., GeoKon, Starotrnjanska 16a, HR-10000 Zagreb. T: +385 1 6050055, F: +385 1 6050094, E: m.ivos@geokonzg.com

Ivšić, T., Faculty of Civil Engineering, University of Zagreb, Kačićeva 26, HR-10000 Zagreb. T: +385 1 4561222, F: +385 1 4827008, E: tom@grad.hr

Jagatić, I., Moho, Svetog Mateja 127, HR-10000 Zagreb. T: +385 1 6602746, F: +385 1 683586, E: moho@zg.tel.hr

Jardas, B., Civil Engineering Institute of Croatia, Rijeka, Vukovarska 10a, HR-51000 Rijeka. T: +385 51 330744, F: +385 51 330810, E: bjardas@ri.igh.hr

Jašarević, I., Faculty of Civil Engineering, University of Zagreb, Kačićeva 26, HR-10000 Zagreb. T: +385 1 4561222, F: +385 1 4827008

Jurman, A., Faculty of Civil Engineering, University of Zagreb, Kačićeva 26, HR-10000 Zagreb. T: +385 1 4561222, F: +385 1 4827008

Kandera, F., Civil Engineering Institute of Croatia, Osijek, Drinska 18, HR-31000 Osijek. T: +385 31 274524, F: +385 31 274 400, E: ikolund@os.igh.hr

Klarić, N., Civil Engineering Institute of Croatia, Split, Matice hrvatske 15, HR-21000 Split. T: +385 21 523939, F: +385 21 551152

Knežević, O., Geotehnika, Kupska 2, HR-10000 Zagreb. T: +385 1 6171812, F: +385 1 6171827, E: geotehnika@zg.tel.hr

Kolund, I., Civil Engineering Institute of Croatia, Osijek, Drinska 18, HR-31000 Osijek. T: +385 31 274524, F: +385 31 274 400, E: ikolund@os.igh.hr

Korpar, S., Faculty of Geotechnical Engineering, University of Zagreb, Hallerova aleja 7, HR-42000 Varaždin. T: +385 42 212228, F: +385 42 313587, E: geotehnicki-fak.vz@vz.tel.hr

Kovač, I., Faculty of Geotechnical Engineering, University of Zagreb, Hallerova aleja 7, HR-42000 Varaždin. T: +385 42 212228, F: +385 42 313587, E: geotehnicki-fak.vz@vz.tel.hr

Kovačević, M.S., Faculty of Civil Engineering, University of Zagreb, Kačićeva 26, HR-10000 Zagreb. T: +385 1 4561222, F: +385 1 4827008, E: msk@grad.hr

Kovačić, D., Conex BBR., Kalinovica 3/4, HR-10000 Zagreb. T: +385 1 3839220, F: +385 1 3839243, E: bbr-conex@zg.tel.hr

Krajnović, D., Conex, Kalinovica 3/3, HR-10000 Zagreb. T: +385 1 3836500, F: +385 1 3836332, E: conex@conex.hinet.hr

Kralj, N., Geotechnical Studio, Milivoja Matošeca 3, HR-10000 Zagreb. T: +385 1 3734323, F: +385 1 3735227, E: geotehnicki-studio@zg.tel.hr

Krsnik, M., Civil Engineering Institute of Croatia, Janka Rakuše 1, HR-10000 Zagreb. T: +385 1 6144111, F: +385 1 6144732, E: mkrsnik@zg.igh.hr

Kvasnička, P., Faculty of Mining, Geology and Petroleum Engineering, University of Zagreb, Pierottijeva 6, HR-10000 Zagreb. T: +385 1 4605152, F: +385 1 4836064, E: pkvasnic@rgn.hr

Lisac, Z., President of the Croatian Society for Soil Mechanics and Geotechnical Engineering, Civil Engineering Institute of Croatia, Janka Rakuše 1, HR-10000 Zagreb. T: +385 1 6144731, F: +385 1 6144732, E: zlisac@zg.igh.hr

Lovrenčić, D., Conex, Kalinovica 3/3, HR-10000 Zagreb. T: +385 1 3836500, F: +385 1 3836332, E: d.lovrencic@conex.hinet.hr

Lustig, R., Rijekaprojekt, Moše Albaharija 10a, HR-51000 Rijeka. T: +385 51 344 250, F: +385 51 344 195

Marenče, M., Vinogradi 53, HR-10000 Zagreb. T: +385 1 3703619

Marić, B., Conex, Kalinovica 3/3, HR-10000 Zagreb. T: +385 1 3836471, F: +385 1 3836332, E: bozica@conex.hinet.hr

Marjanović, P., Faculty of Geotechnical Engineering, University of Zagreb, Hallerova aleja 7, HR-42000 Varaždin. T: +385 42 212228, F: +385 42 313587, E: geotehnicki-fak.vz@vz.tel.hr

Marković, A., Rijekaprojekt, Moše Albaharija 10a, HR-51000 Rijeka. T: +385 51 344 192, F: +385 51 344 195

Marović, P., Faculty of Civil Engineering, University of Split, Matice hrvatske 15, HR-21000 Split. T: +385 21 303334, F: +385 21 524162, E: marovic@gradst.hr

Matešić, L., Faculty of Civil Engineering, University of Zagreb, Kačićeva 26, HR-10000 Zagreb. T: +385 1 4561222, F: +385 1 4827008, E: leomat@grad.hr

Matković, I., Civil Engineering Institute of Croatia, Janka Rakuše 1, HR-10000 Zagreb. T: +385 1 6144111, F: +385 1 6144732, E: imatkovic@zg.igh.hr

Matošević, V., Geo-5, Carera 59, HR-52210 Rovinj. T: +385 52 811380, F: +385 52 815492

Mavar, R., Civil Engineering Institute of Croatia, Janka Rakuše 1, HR-10000 Zagreb. T: +385 1 6144111, F: +385 1 6144732, E: rmavar@zg.igh.hr

Megla, T., Croatian Waters, Ulica grada Vukovara 220, HR-10000 Zagreb. T: +385 1 6307333, F: +385 1 6151793

Mesec, J., Faculty of Geotechnical Engineering, University of Zagreb, Hallerova aleja 7, HR-42000 Varaždin. T: +385 42 212228, F: +385 42 313587, E: jmesec@yahoo.com

Mihalinec, Z., Civil Engineering Institute of Croatia, Janka Rakuše 1, HR-10000 Zagreb. T: +385 1 6144111, F: +385 1 6144732, E: zmihalinec@zg.igh.hr

Mihovilović, M., Geo-5, Carera 59, HR-52210 Rovinj. T: +385 52 811380, F: +385 52 815492

Miljković, B., GeoKon, Starotrnjanska 16a, HR-10000 Zagreb. T: +385 1 6050055, F: +385 1 6050094, E: b.miljkovic@geokonzg.com

Mišćević, P., Faculty of Civil Engineering, University of Split, Matice hrvatske 15, HR-21000 Split. T: +385 21 303353, F: +385 21 524162, E: miscevic@gradst.hr

Mitrović, G., Civil Engineering Institute of Croatia, Janka Rakuše 1, HR-10000 Zagreb. T: +385 1 6144111, F: +385 1 6144732, E: gmitrovic@zg.igh.hr

Mlinarić, D., Geoexpert GTB, Majstora Radonje 12, HR-10000 Zagreb. T: +385 1 3843831, F: +385 1 3843836, E: geoexpert-gtb@zg.hinet.hr

Molk, N., Civil Engineering Institute of Croatia, Split, Matice hrvatske 15, HR-21000 Split. T: +385 21 523939, F: +385 21 551152, E: natasa.molk@st.igh.hr

Muhovec, I., Faculty of Geotechnical Engineering, University of Zagreb, Hallerova aleja 7, HR-42000 Varaždin. T: +385 42 212228, F: +385 42 313587, E: ivan.muhovec@zg.tel.hr

Mulabdić, M., Faculty of Civil Engineering, University of Osijek, Drinska 16a, HR-31000 Osijek. T: +385 31 274377, F: +385 31 274444, E: leta@zg.tel.hr

Nemet, Ž., Civil Engineering Institute of Croatia, Osijek, Drinska 18, HR-31000 Osijek. T: +385 31 274524, F: +385 31 274400, E: ikolund@os.igh.hr

Novosel, T., Civil Engineering Institute of Croatia, Janka Rakuše 1, HR-10000 Zagreb. T: +385 1 6144111, F: +385 1 6144732, E: tnovosel@zg.igh.hr

Ortolan, Ž., Laurenčićeva 18, HR-10000 Zagreb. T: +385 1 6144934

Pavlovec, E., Faculty of Civil Engineering, University of Rijeka, Viktora cara Emina 5, HR-51000 Rijeka. T: +385 51 352111, F: +385 51 332816, E: dekanat@gradri.hr

Pfeifer, D., Civil Engineering Institute of Croatia, Osijek, Drinska 18, HR-31000 Osijek. T: +385 31 274524, F: +385 31 274 400, E: ikolund@os.igh.hr

Plepelić, G., Prizma, Bernarda Vukasa 22, HR-10000 Zagreb, T/F: +385 1 2346356, E: prizma@zg.tel.hr

Polić, S., Faculty of Civil Engineering, University of Zagreb, Kačićeva 26, HR-10000 Zagreb. T: +385 1 4561222, F: +385 1 4827008

Prebeg, I., Civil Engineering Institute of Croatia, Janka Rakuše 1, HR-10000 Zagreb. T: +385 1 6144111, F: +385 1 6144732, E: iprebeg@zg.igh.hr

Radaljac, Ž.D., Conex, Kalinovica 3/3, HR-10000 Zagreb. T: +385 1 3836500, F: +385 1 3836332, E: conex@conex.hinet.hr

Reić, B., Civil Engineering Institute of Croatia, Janka Rakuše 1, HR-10000 Zagreb. T: +385 1 6144111, F: +385 1 6144732, E: breic@zg.igh.hr

Reljanović, Ž., Civil Engineering Institute of Croatia, Split, Matice hrvatske 15, HR-21000 Split. T: +385 21 523939, F: +385 21 551152, E: zeljko.reljanovic@st.igh.hr

Roje-Bonacci, T., Faculty of Civil Engineering, University of Split, Matice hrvatske 15, HR-21000 Split. T: +385 21 303341, F: +385 21 524162, E: Tanja.Roje-Bonacci@gradst.hr

Salković, A., Moho, Svetog Mateja 127, HR-10000 Zagreb. T: +385 1 6602746, F: +385 1 683586, E: moho@zg.tel.hr

Samardžija, I., Civil Engineering Institute of Croatia, Split, Matice hrvatske 15, HR-21000 Split. T: +385 21 523939, F: +385 21 551152, E: ivica.samardzija@st.igh.hr

Sapunar, N., Civil Engineering Institute of Croatia, Janka Rakuše 1, HR-10000 Zagreb. T: +385 1 6144111, F: +385 1 6144732, E: nsapunar@zg.igh.hr

Sesar, S., Civil Engineering Institute of Croatia, Janka Rakuše 1, HR-10000 Zagreb. T: +385 1 6144111, F: +385 1 6144732, E: ssesar@zg.igh.hr

Sever, K., GeoKon, Starotrnjanska 16a, HR-10000 Zagreb. T: +385 1 6050055, F: +385 1 6050094, E: k.sever@geokonzg.com

Sokolić, Ž., Geotechnical Studio, Milivoja Matošeca 3, HR-10000 Zagreb. T: +385 1 3734323, F: +385 1 3735227, E: geotehnicki-studio@zg.tel.hr

Soldo, B., Faculty of Geotechnical Engineering, University of Zagreb, Hallerova aleja 7, HR-42000 Varaždin. T: +385 42 212228, F: +385 42 313587, E: b_soldo@yahoo.com

Sorić, I., Geotechnical Studio, Milivoja Matošeca 3, HR-10000 Zagreb. T: +385 1 3734323, F: +385 1 3735227, E: geotehnicki-studio@zg.tel.hr

Stanić, B., Civil Engineering Institute of Croatia, Janka Rakuše 1, HR-10000 Zagreb. T: +385 1 6144111, F: +385 1 6144732, E: bstanic@zg.igh.hr

Stojković, B., Civil Engineering Institute of Croatia, Janka Rakuše 1, HR-10000 Zagreb. T: +385 1 6144111, F: +385 1 6144732, E: bstojkovic@zg.igh.hr

Strelec, S., Faculty of Geotechnical Engineering, University of Zagreb, Hallerova aleja 7, HR-42000 Varaždin. T: +385 42 212228, F: +385 42 313587, E: sstrelec@yahoo.com

Svirčev, S., Prizma, Bernarda Vukasa 22, HR-10000 Zagreb. T/F +385 1 2346356, E: prizma@zg.tel.hr

Szavits-Nossan, A., Faculty of Civil Engineering, University of Zagreb, Kačićeva 26, HR-10000 Zagreb. T: +385 1 4561272, F: +385 1 4827008, E: szavits@grad.hr

Szavits-Nossan, V., Faculty of Geotechnical Engineering, University of Zagreb, Hallerova aleja 7, HR-42000 Varaždin. T: +385 42 212228, F: +385 42 313587, E: svlasta@ grad.hr

Šepac, Z., Geotehnika, Kupska 2, HR-10000 Zagreb. T: +385 1 6171812, F: +385 1 6171827, E: geotehnika@ zg.tel.hr

Šestanović, S., Faculty of Civil Engineering, University of Split, Matice hrvatske 15, HR-21000 Split. T: +385 21 303329, F: +385 21 524162, Slobodan. E: sestanovic@ gradst.hr

Šilhard, V., Geoexpert GTB., Majstora Radonje 12, HR-10000 Zagreb. T: +385 1 3843831, F: +385 1 3843836

Škacan, B., Conex, Kalinovica 3/3, HR-10000 Zagreb. T: +385 1 3836500, F: +385 1 3836332, E: conex@conex. hinet.hr

Štefanek, Ž., Hidroinženjering, Okučanska 30, HR-10000 Zagreb. T/F: +385 1 258120, E: hidroinzenjering@zg.tel.hr

Štimac, D., Civil Engineering Institute of Croatia, Split, Matice hrvatske 15, HR-21000 Split. T: +385 21 523939, F: +385 21 551152, E: dragutin.stimac@st.igh.hr

Štuhec, D., Faculty of Geotechnical Engineering, University of Zagreb, Hallerova aleja 7, HR-42000 Varaždin. T: +385 42 212228, F: +385 42 313587, E: damir. stuhec@vz.tel.hr

Tomac, I., Conex, Kalinovica 3/3, HR-10000 Zagreb. T: +385 1 3836500, F: +385 1 3836332, E: conex@conex.hinet.hr

Tulić, T., Rijekaprojekt - Geotechnical Investigations, J. Polić Kamova 111, HR-51000 Rijeka. T: +385 51 436 613, F: +385 51 436 610

Tušar, Z., Cvijete Zuzorić 37, HR-10000 Zagreb. T: +385 1 6158589

Verić, F., Faculty of Civil Engineering, University of Zagreb, Kačićeva 26, HR-10000 Zagreb. T: +385 1 4561218, F: +385 1 4827008, E: veric@grad.hr

Višić, I., Center for Waters, Frankopanska 16, HR-10000 Zagreb. T: +385 1 4849072, F: +385 1 4849067

Vlahović, D., Hidroinženjering, Okučanska 30, HR-10000 Zagreb. T/F: +385 1 258120, E: hidroinzenjering@zg.tel.hr

Vrkljan, I., Civil Engineering Institute of Croatia, Janka Rakuše 1, HR-10000 Zagreb. T: +385 1 6144111, F: +385 1 6144732, E: ivrkljan@zg.igh.hr

Vrkljan, M., Civil Engineering Institute of Croatia, Janka Rakuše 1, HR-10000 Zagreb. T: +385 1 6144111, F: +385 1 6144732, E: mvrkljan@zg.igh.hr

Vukadinović, B., Geotechnical Studio, Milivoja Matošeca 3, HR-10000 Zagreb. T: +385 1 3734323, F: +385 1 3735227, E: bojanv@iname.com

Vukobrat, J., GeoKon, Starotrnjanska 16a, HR-10000 Zagreb. T: +385 1 6050055, F: +385 1 6050094, E: j.vukobrat@ geokonzg.com

Zelenika, M., Faculty of Geotechnical Engineering, University of Zagreb, Hallerova aleja 7, HR-42000 Varaždin. T: +385 42 212228, F: +385 42 313587, E: geotehnicki-fak.vz@vz.tel.hr

Zidar, M., Faculty of Geotechnical Engineering, University of Zagreb, Hallerova aleja 7, HR-42000 Varaždin. T: +385 42 212228, F: +385 42 313587, E: maczidar@yahoo.com

Zlatović, S., Faculty of Civil Engineering, University of Zagreb, Kačićeva 26, HR-10000 Zagreb. T: +385 1 4561222, F: +385 1 4827008, E: sonja@grad.hr

Znidarčić, D., University of Colorado at Boulder, Boulder, Colorado, 80309-0428, USA. T: +303 4927577, F: +303 4927317, E: znidarci@spot.Colorado.edu

Zvornik, D., Geotechnical Studio, Milivoja Matošeca 3, HR-10000 Zagreb. T: +385 1 3734323, F: +385 1 3735 227, E: geotehnicki-studio@zg.tel.hr

Žarković, V., Elektroprojekt, Ulica grada Vukovara 37, HR-10000 Zagreb. T: +385 1 6307746, F: +385 1 6152685, E: vladimir.zarkovic@elektroprojekt.tel.hr

Županić, Z., Civil Engineering Institute of Croatia, Osijek, Drinska 18, HR-31000 Osijek. T: +385 31 274524, F: +385 31 274 400, E: ikolund@os.igh.hr

CTGA/Comité Transnational des Géotechniciens d'Afrique

Secretary: Omar Chemaou Elfihri, CTGA, LPEE, 25 rue d'Azilal, 20 000 Casablanca, Marocco.
T: 212 2 54 75 17, 18, F: 212 2 45 01 49

Total number of members: 23

Abba, A., LNTP/B, BP 464 Naimey, Niger. T: 227 73 25 62

Adjati, A.M., Collège Polytechnique Universataire, 03 BP 2092, Cotonou, Benin. T: 229 33 11 62, F: 229 36 01 99, E: adjati@syfed.bj.refer.org

Ativon, Y., LNBTP, BP 20100 Lomé, Togo. T: 228 25 62 83, 25 92 68, F: 228 21 68 12

Chemaou, E.O., LPEE, 25 Rue d'Azilal, 20000 Casablanca, Maroc. T: 212 2 30 75 10 or 30 04 50 or 54 75 1718, F: 212 2 45 01 49

Cisse, I., Ecole Supérieur Polytechnique de Thiés, BP 10 Thies, Dakar, Sénégal. T: 221 8 95 14 22, F: 221 8 95 14 22, 95 14 76

Daillo, M.B., Laboratoire de Méchanique des Sols et Géotechnique, Université de Conakry, BP 1147, Conakry, Guinée. T: 224 11222448

Ekpini, K.G., Bureau National D'Etudes et de Development, 04 BP 945 Abidjan D4 Côte d'Ivore. T: 225 05 73, F: 225 44 68 03, E: gekpiri@bnetd.sita.net

Faye, M.A., CEREEQ, BP 189 Dakar, Sénégal. T: 221 8 32 35 18, F: 221 8 32 10 72

Gambin, M., 21 Quai d'Anjou, 9 Rue Poulletier 75004 Paris. T: 33 4 92 97 61 82, F: 33 4 92 97 61 82, E: mgambin@magic.fr

Guei, A., Institut Polytechnique Felix Houphuet Boigny, BP 1083 Yamoussoukro, Cote d'Ivoire. T: 225 64 19 88, F: 225 64 04 06, E: yoro@larina.Inyet.CI

Kpanou, S., CNERTP, BP:1270, Cotonou, Bénin. T/F: 229 33 0978

Menin, M., LBTP, 04 BP3 Abidjan, Cote d'Ivoire. T: 225 25 43 58, F: 225 25 33 69

Minkousse, D., LABOGENIE, BP 349 Yaounde, Cameroun. T: 237 30 65 72, F: 237 30 41 65

Moufo, J., Labogenie, BP 1094 Douala, Cameroun. F: 237 40 08 29, E: moufoi@cemnet.cm

N'Gouan, K.E., LBTP, 04 BP3 Abijan, Cote d'Ivoire. T: 225 25 43 58, F: 225 25 33 69

Olodo T.D., Directeur des Routes, BP 531, Ootonou, Bénin. T: 229 31 32 04, F: 229 33 09 78

Ould A.M.M., LNTP, BP 602, Nouakchott, Mauritanie. T: 222 25 11 03, F: 22 2 25 11 03

Saleh, A., LBTP, BP 104 Ndjamena, Chad. T: 235 52 20 36, F: 235 52 40 25

Sanfo S., Entreprise Travaux, BP 133, Ouagadougou, Burkina Faso. T: 226 30 23 09, F: 226 30 28 28

Sikali, F., Labogenie, BP 349, Yaoundé, Cameroun. T: 237 30 65 72, F: 237 30 41 65

Sissoko, S., CNREX/BTP, BP: 1398 Bamako, Mali. T: 223 20 21 43, F: 223 77 01 47, E: cnrex@spidu.tol.net.org

Tchehouali, D.A., CERC, 01 BP2009, Cotonou, Bénin. T: 229 36 67 23, F: 229 36 01 99

Toe, I., NBTP, 01 133 Ouagadougou, Burkina Faso 01. T: 226 34 05 93, F: 226 30 28 38

46

CZECH AND SLOVAK REPUBLICS/REPUBLIQUES TCHÉQUE ET SLOVAQUE

Secretary: Dr I. Herle, Czech and Slovak Committee for SMFE, Inst. of Theoretical and Applied Mechanics ASCR, Prosecká 76, 190 00 Praha 9, Czech Republic. T: +420-2-86882121, F: +420-2-86884634, E: herle@itam.cas.cz

Total number of members: 43

Barvínek, R., Na Dionysce 1553/1, 160 00 Praha 6. E: barvinek@iol.cz
Bažant, Z., Prof, Dept. of Geotechnics, Czech Technical University, Thákurova 7, 166 29 Praha 6. T: 02-24354557, F: 02-3114206, E: k135@fsv.cvut.cz
Boháč, J., Dr, Dept. of Engineering Geology, Charles University, Albertov 6, 128 43 Praha 2. T: 02-21952205, F: 02-21952180, E: bohac@natur.cuni.cz
Doležalová, M., Dr, Dolexpert-Geotechnika Nad Belvederem 3, 148 00 Praha 4. T: 02-7927426, F: 02-7927426, E: dolezalova@pha.pvtnet.cz
Ebermann, T., Kaňovského 6/1239, 182 00 Praha 8. T: 0603-528603, E: tomas_ebermann@post.cz
Feda, J., Assoc Prof, Institute of Theoretical and Applied Mechanics, Czech Academy of Sciences, Prosecká 76, 190 00 Praha 9. T: 02-86882121, F: 02-86884634, E: feda@itam.cas.cz
Frankovská, J., Dr, Geological Survey of the Slovak Republic, Mlynská dolina 1, 817 04 Bratislava. T: 07-60296207, F: 07-54771940, E: frankova@gssr.sk
Fussgaenger, E., Dr, Geofos, Velký diel 3323, 010 08 Žilina. T: 089-655249, F: 089-652747, E: geofos@vud.sk
Glisníková, V., Dr, Institute of Geotechnics, Veveří 95, 662 37 Brno. T: 05-41147234, F: 05-41147237, E: glisnikova.v@fce.vutbr.cz
Herle, I., Dr, Institute of Theoretical and Applied Mechanics, Czech Academy of Sciences, Prosecká 76, 190 00 Praha 9. T: 02-86882121, F: 02-86884634, E: herle@itam.cas.cz
Herle, V., SG-Geotechnika, Geologická 4, 152 00 Praha 5. T: 02-5818440, F: 02-5817995, E: herle@sggt.cz
Herštus, J., Dr, K Dolánkám 996, 282 01 Český Brod. T: 0203-622467, F: 0203- 620737, E: age-brod@comp.cz
Hořejší, V., SG-Geotechnika, Pekárenská 81, 372 13 České Budějovice. T: 038-24435, F: 038-24177, E: sggtcb@mbox.vol.cz
Hroch, Z., Dr, Tomanova 72, 169 00 Praha 6
Hulla, J., Prof, Dept. of Geotechnics, Slovak Technical University, Radlinského 11, 813 68 Bratislava. T: 07-5927 4666, F: 07-52925642, E: hulla@svf.stuba.sk
Jesenák, J., Prof, Dept. of Geotechnics, Slovak Technical University, Radlinského 11, 813 68 Bratislava. T: 07-59274668, F: 07-52925642
Jettmar, J., Assoc Prof, Dept. of Geotechnics, Czech Technical University, Thákurova 7, 166 29 Praha 6. T: 02-24354542, F: 02-3114206, E: jettmar@fsv.cvut.cz
Klablena, P., Prof, 569 72 Rohozná u Poličky č.p. 419. T: 0461-595228, F: 0461-595228, E: geotechnika@volny.cz
Klein, K., Dr, VUIS-Zakladanie stavieb, s.r.o, Dúbravská cesta 9, 842 37 Bratislava. T: 07-54776040, F: 07-54776040, E: vuis-zs@gtinet.sk
Koudelka, P., Dr, Ve svahu 25, 147 00 Praha 4. T: 02-8688 2121, F: 02-86884634, E: koudelka@itam.cas.cz
Kurka, J., Dr, Devonská 1/999, 152 00 Praha 5. E: kurka@barr.cz

Kuzma, J., Assoc Prof, Dept. of Geotechnics, Slovak Technical University, Radlinského 11, 813 68 Bratislava. T: 07-5927 4679, F: 07-52925642, E: kuzma@svf.stuba.sk
Kysela, Z., Dr, Na Dobešce 20/1102, 147 00 Praha 4.
Lamboj, L., Assoc Prof, Na šťáhlavce 1a, 160 00 Praha 6. T: 02-24354557, F: 02-3114206, E: lamboj@fsv.cvut.cz
Masarovičová, M., Dr, Dept. of Geotechnics, Slovak Technical University, Radlinského 11, 813 68 Bratislava. T: 07-5927 4673, F: 07-52925642, E: masarovm@svf.stuba.sk
Matys, M., Prof, Dept. of Engineering Geology PF UK, Mlynská dolina, 842 15 Bratislava. T: 07-60296638, E: matys@fns.uniba.sk
Procházka, P., Prof, Dept. of Structural Mechanics, Czech Technical University, Thákurova 7, 166 29 Praha 6. T: 02-24354480, F: 02-24310775, E: petrp@fsv.cvut.cz
Ravinger, R., Dr, Dept. of Geotechnics, Slovak Technical University, Radlinského 11, 813 68 Bratislava. T: 07-5927 4667, F: 07-52925642, E: ravinger@svf.stuba.sk
Rozsypal, A., Assoc Prof, SG-Geotechnika, Geologická 4, 152 00 Praha 5. T: 02-5818490, F: 02-5818590, E: sggtdir@mbox.vol.cz
Rybář, J., Assoc Prof, Institute of Rock Structure and Mechanics ASCR, V Holešovičkách 41, 182 09 Praha 8. T: 02-66009231, F: 02-6880105, E: rybar@irsm.cas.cz
Řičica, J., Soletanche ČR, s.r.o., K Botiči 6, 101 00 Praha 10. E: jindrich.ricica@soletanche.cz
Salák, J., Dr, Dept. of Geotechnics, Czech Technical University, Thákurova 7, 166 29 Praha 6. T: 02-24354908, F: 02-3114206, E: salak@fsv.cvut.cz
Sekyra, Z., SG-Geotechnika, Geologická 4, 152 00 Praha 5. T: 02-5818440, F: 02-5817995, E: sekyra@sggt.cz
Seyček, J., Dr, Xaveriova 11, 150 00 Praha 5
Slávik, I., Dr, Dept. of Geotechnics, Slovak Technical University, Radlinského 11, 813 68 Bratislava. T: 07-5927 4672, F: 07-52925642, E: slavik@svf.stuba.sk
Šimek, J., Prof, Dept. of Geotechnics, Czech Technical University, Thákurova 7, 166 29 Praha 6. T: 02-24354553, F: 02-3114206, E: k135@fsv.cvut.cz
Šťastný, J., Dr, Koněvova 242, 130 00 Praha 3. E: age@sro.cz
Turček, P., Prof, Dept. of Geotechnics, Slovak Technical University, Radlinského 11, 813 68 Bratislava. T: 07-5927 4665, F: 07-52925642, E: turcek@svf.stuba.sk
Vaníček, I., Prof, Dept. of Geotechnics, Czech Technical University, Thákurova 7, 166 29 Praha 6. T: 02-24354540, F: 02-3114206, E: vaniceki@fsv.cvut.cz
Weiglová, K., Assoc Prof, Institute of Geotechnics, Veveří 95, 662 37 Brno. T: 05-41147240, F: 05-41147237, E: weiglova.k@fce.vutbr.cz
Záleský, J., Dr, Dept. of Geotechnics, Czech Technical University, Thákurova 7, 166 29 Praha 6. T: 02-24354551, F: 02-3114206, E: zalesky@fsv.cvut.cz
Záruba, J., SG-Geotechnika, Geologická 4, 152 00 Praha 5. T: 02-5818440, F: 02-5817995, E: apgt@sggt.cz
Zavoral, J., Dr, AZ Consult, s.r.o., Klíšská 12, 400 01 Ústí nad Labem. E: azc@mbox.vol.cz

DENMARK/DANEMARK

Secretary: Dr. Jens Brink Clausen, Danish Technical Society, GEO, Maglebjergvej 1, P.O. Box 119, DK 2800 Lyngby, Denmark. T: +(45)45 88 44 44, F: (45)45 88 12 40, E: jens.brink.clausen@geoteknisk.dk

Total number of members: 345

Ahrentzen, P., Rishøjvej 9, Kattinge 4000 Roskilde
Alvin, D., Silkeborgvej 786 8220 Brabrand. T: 86 26 40 40, E: dra@cowi.dk
Andersen, A.T., P.P. Ørums Gade 8, st.tv. 8000 Århus C. T: 86 14 92 98, E: ata@aarsleff.com
Andersen, B.N., Langelandsvej 11, st.th. 2000 Frederiksberg C
Andersen, C., Stjernevænget 10 5592 Ejby. T: 64 46 21 54
Andersen, E., Gyden 1 6000 Kolding. T: 75 56 90 84
Andersen, I., Elmevej 31 4140 Borup. T: 57 52 18 12
Andersen, J.D., Nordmarksvej 2 A 9000 Aalborg. T: 98 11 41 71, E: jda@geoteknisk.dk
Andersen, J.J., Snoghøj Landevej 60 7000 Fredericia. T: 75 94 20 23
Andersen, L.L., Rugårdsvej 45, 2.sal 5000 Odense C. T: 66 14 45 71, E: la@geotek.dk
Andersen, S.B., Korsørgade 28, 1. Th. 2100 København Ø. T: 35 43 79 80, E: sia@cowi.dk
Andreasen, F., Helleskrænten 29 2860 Søborg. T: 39 66 23 30
Andreasen, M., Blichersvej 27 B 8620 Kjellerup
Andreasen, S., Egerupvej 105, Ørslevvester 4173 Fjenneslev Ringsted. T: 57 80 80 67
Arnung, J.P., Lighedsvej 4 2000 Frederiksberg. T: 38 86 13 88
Augustesen, A., Kollegievej 6, vær. 131 9000 Aalborg. T: 96 35 84 55, E: august96@civil.auc.dk
Avnstrøm, P.L., Vejgårdsvænget 33 3520 Farum. T: 44 95 61 35
Bagge, G., Ørbygade 12 3200 Helsinge. E: gunbag@labm.dtu.dk
Bai, W., Skovtoften 7 8700 Horsens. T: 75 66 52 32, E: wb@horsens.ih.dk
Bak, J.K., Bøgegårdsvej 58 5471 Søndersø. E: jkb@ramboll.dk
Balstrup, T., Maglegårdsvej 221 3480 Fredensborg. T: 48 48 42 31, E: dgi@geoteknisk.dk
Baumann, J., Søllerødgårdsvej 42 2840 Holte. T: 45 80 69 29, E: jeb@geoteknisk.dk
Bendixen, K., Stenløse Bygade 19 5260 Odense S. T: 66 15 06 40, E: kmb@geoteknisk.dk
Bielefeldt, B., Jord Teknik A/S, Generatorvej 21 2730 Herlev. T: 39 69 88 88
Bisgaard, A., Birkerød Parkvej 32, 2.th. 3460 Birkerød. T: 45 81 89 33, E: abi@cowi.dk
Bjerrum, A., Maridan, Agern Alle 3 2970 Hørsholm. T: Priv. 45890560, E: abj@maridan.dk
Blem, H., Hesselgårdsvej 47 3460 Birkerød. T: 45 81 74 32, E: hnb@ramboll.dk
Bliksted, T., Uldalsvej 20, 1. tv. 9400 Nørresundby. T: 98 17 98 79
Bock, M., Titangade 3M, 2.th. 2200 København N. T: 35 83 67 36
Boldsen, J., Krakesvej 75 8660 Skanderborg. T: 86 57 28 85
Bollerup, J., Østerbro 99, 3.th 9000 Aalborg. E: bollerup@bigfoot.com
Bonde, C., Langkærgårdsvej 36 3460 Birkerød. T: 45 81 94 64, E: crb@geoteknisk.dk
Bonde, O., Toftevænget 23 6000 Kolding. E: nbm@nbm-consult.dk
Brendstrup, J., Frederiksberg Allé 39, 1. 1820 Frederiksberg C. E: jbr@cowi.dk
Brødbæk, C., Langelandsvej 10B, 3.th. 2000 Frederiksberg. E: cxb@COWI.dk
Buhl, N.J., Kertemindevej 7, Himmelev 4000 Roskilde. T: 46 35 55 54
Bæk-Madsen, C., Hækkevej 17 2970 Hørsholm. T: 45 86 29 12, E: cbm@geoteknisk.dk
Bødker, K., Fortebakken 11 8240 Risskov. T: 86 17 97 28, E: kb@b.iha.dk

Bødker, L., Dronningensgade 16, 3. 9400 Nr. Sundby. T: 98 19 03 01, E: skaerm@12move.dk
Bøegh, L., Per, Bredkær Tværvej 53 8250 Egå. T: 86 22 23 67, E: pbl@cowi.dk
Bønding, N., Cedervangen 57 3450 Allerød. T: 42 89 34 47, E: nib@niras.dk
Christensen, A., Lyngby Hovedgade 44 2800 Lyngby. T: 45 93 18 93, E: ac@anderschristensen.dk
Christensen, A.N., Toftegårdsvej 7, Farstrup 9240 Nibe. E: anc@geoteknisk.dk
Christensen, B.S., Jægersborg Allé 86 2920 Charlottenlund. T: 31 63 72 44
Christensen, C.T., Boveskovvej 2 2800 Lyngby. T: 45 87 22 01
Christensen, J., Vimmelsbækløkken 6 5270 Odense N. E: jc@geotek.dk
Christensen, J.D., Skt. Annæ Gade 35, st.tv. 1416 København K. T: 32 96 69 63
Christensen, J.L., Drejøgade 35, st. 9 2100 København Ø. T: 26 24 28 34, E: jlc@geoteknisk.dk
Christensen, L., Klokkerfaldet 88 8210 Århus V. T: 86 15 83 40
Christensen, M.B., Vardevej 102 st.th. 7100 Vejle
Christensen, N., Vesterlundvej 43, Høm 6760 Ribe. T: 75 44 12 90, E: niels@jyskgeoteknik.dk
Christensen, N.B., Nannasgade 13, 4.th. 2200 København N. T: 35 85 49 30
Christensen, N.L., Hovmarksvej 21 8700 Horsens. T: 75 65 80 14, E: nlc@horsens.ih.dk
Christensen, O., Lunderødvej 55 4340 Tølløse. T: 53 48 60 44
Christensen, T., Kirkestien 14, Almind 8800 Viborg
Christensen, T., Elkjær, Jespervej 62 3400 Hillerød. T: 42 25 13 09
Christiansen, B.G., Stengårds Allé 11 2800 Lyngby. T: 45 93 16 35, E: bgc@cowi.dk
Christiansen, J.F., 10 Hammond Parkvej, Middleport New York 14105, USA
Christiansen, S., Kingosvej 9 3400 Hillerød
Clausen, E., Hvide Sande Havn, Fossannæsvej 22 6960 Hvide Sande. T: 75 11 64 84
Clausen, J.B., Bregnevej 15A 3500 Værløse. T: 44 48 47 57, E: jbc@geoteknisk.dk
Clausen, S.G., Vejdammen 11 2840 Holte. T: 45 80 18 87, E: sgc@monthor.dk
Dahlgren, S.I., Langs Hegnet 51 2800 Lyngby. T: 45 88 90 94, E: sid@cowi.dk
Dam, N.N., Hesteskoen 39 5250 Odense SV. T: 66 17 16 61, E: nnd@skude-jacobsen-v.dk
Damgaard, K., Kirkegårds Allé 33 5000 Odense C
Dannerfjord, S., Århusgade 29, 4.th. 2100 København Ø. T: 31 26 54 37
De Churruca, R., Skolevænget 3, Stevning 6430 Nordborg
Denver, H., Rådhusvej 47 2920 Charlottenlund. T: 39 63 12 37, E: hd@geoteknisk.dk
Du Thinh, K., c/o IGG, DTU, Bygning 373, Diplomvej 2800 Lyngby. T: 45 25 50 92, E: kdt@iabm.dtu.dk
Erichsen, L., Brandts Vænge 10 3460 Birkerød. T: 45 81 26 15, E: le@orestad.dk
Eriksen, F.S., Hesjegårdsvej 32 3460 Birkerød. T: 42 81 41 57, E: fse@kampsax.dk
Esbjerg, K., Teknik og Miljø Mohammad Ahmed Hussein, Frodesgade 30 6700 Esbjerg. T: Priv. 75 13 06 35
Eskesen, S.D., Biskop Monrads Vej 6 2830 Virum. T: 45 85 55 44, E: sde@cowi.dk
Federau, M., Langelinie 99 5230 Odense M. T: 66 13 62 71
Feld, T., Lyngskrænten 33 2840 Holte. T: 44 80 11 31, E: tof@ramboll.dk

Fjellerup, F.E., Sjælsøparken 34 3450 Allerød. T: 48 17 42 05
Flensted, A.M., Nyvej 17 7100 Vejle. T: 75 83 89 04, E: amf@ramboll.dk
Foged, B., Birkedommervej 12, st.th 2400 København NV. T: 3810 7290, E: bfoged@yahoo.dk
Foged, N., Gøngesletten 7 2950 Vedbæk. T: 45 89 43 52, E: foged@iabm.dtu.dk
Franck, B., Koralvej 15 8700 Horsens. T: 75 64 06 96, F: 75 61 70 61, E: bgf@geoteknik.dk
Franck, F., Industrivej 22 3550 Slangerup. T: 42 33 32 00
Fredericia, J., Elletoften 40 2800 Lyngby DK. T: 4444 2070, E: jfr@geus.dk
Frederiksen, F.F., Skanderborgvej 15 8370 Hadsten. E: 4AP@mail.tele.dk
Frederiksen, J., Johannevej 19 2920 Charlottenlund. T: 39 64 42 37, E: jrf@ramboll.dk
Frederiksen, K.M., Endrupvej 40 3480 Fredensborg. T: 42 28 54 16
Frederiksen, P., Vagtelvej 2 8660 Skanderborg. T: 87 69 90 80, E: 4AP@mail.tele.dk
Fredslund, K., Munkevænget 4 5492 Vissenbjerg
Friis, C., Kong Hans Allé 29 2860 Søborg. T: 39 69 93 73, E: miljo@miljocontractors.dk
Fuglsang, L.D., Krokusvej 2 3400 Hillerød. T: 48 28 74 40, E: fuglsang@iabm.dtu.dk
Gadeberg, D., Smedeengen 12, Torslunde 2635 Ishøj. T: 43 71 56 56, E: d.gadeberg@comet.dk
Galsgaard, J., Vestre Allé 10 2500 Valby. T: 36 46 85 57, E: jng@geoteknisk.dk
Gormsen, C., Svankærvej 19 2720 Vanløse. T: 38 71 80 86, E: clg@niras
Gormsen, K., Christoffers Allé 192 2800 Lyngby
Granhøj, S., Christoffers Alle 70, st.mf.th. 2800 Lyngby. T: 44 49 24 36
Graversen, E., Tværskiftet 15B 2730 Herlev
Gravgård, J., Nørretofte Alle 8 2500 Valby. T: 36 17 52 21, E: jgr@ramboll.dk
Gregersen, J., Hedeselskabet, Ringstedvej 20 4000 Roskilde
Grinsted, O., Tystrupvej 3 4250 Fuglebjerg. T: 53 64 95 09
Gronemann, G., Åløkkevej 32 A 2720 Vanløse. T: 31 86 94 21, E: gg@hcf.dk
Gudmundson, G., H.1, Drekahlid 9 550 Saudarkrokur Island 94 33 11
Gundorph, S., Dyrehavegårdsvej 50 C 2800 Lyngby. T: 42 88 90 97, E: sgu@ramboll.dk
Hammami, R., Holtegade 6, 3.th. 2200 København N. T: 35830248, E: rmh@ramboll.dk
Hanberg, F., Langesund 6, Vonsild 6000 Kolding. T: 75 56 60 66
Hansen, A., Bøgehegnet 71, 1.sal.tv. 2670 Greve. T: 43 69 19 58, E: aah@cowi.dk
Hansen, B., Valmuevej 18 2970 Hørsholm. T: 42 86 28 85
Hansen, C.N., Lille Østrupvej 15 9541 Suldrup. T: 98 37 86 64, F: 98 37 86 64, E: crh@vd.dk
Hansen, G., Granstuevej 39 2840 Holte. T: 42 42 39 65, E: glh@ramboll.dk
Hansen, G., Lundsbjergvej 51 5863 Ferritslev, Fyn. T: 65 98 10 20, E: gunnarhansen@vip.cybercity.dk
Hansen, H.H.D., Ribegade 126 6700 Esbjerg
Hansen, H.K., Hvidehusvej 28 3450 Allerød. T: 48 17 48 78, E: kh@geoteknisk.dk
Hansen, H.I., Ryvej 58, Gantrup 8752 Østbirk 75 78 11 9, E: hih@just.dk
Hansen, J.G., Roskildevej 53, 4-409 2000 Frederiksberg Dk. T: 36177175, E: jqh@cowi.dk
Hansen, P.B., Tokkekøbvej 39 3450 Allerød. T: 48 17 57 31, E: pbh@geoteknisk.dk
Hansen, S.A., Overstræde 8 5000 Odense C. T: 66 13 10 90, E: s-a-hansen@bret.iot.dk
Hansen, S.B., Borthigsgade 15, 1.sal 2100 København Ø. T: 35 35 97 45, E: srh@ramboll.dk
Hansen, A., 'blehaven 6 3450 Allerød. T: 42 27 30 14
Hansson, L., Jernbanegade 30, st.th. 3480 Fredensborg. E: lah@geoteknisk.dk
Hartvig, F., Bjerggårdsvej 6, Aale 7160 Tørrng. T: 75676399, Mobile: 40645875, E: fha@carlbro.dk

Hasbo, B., Vallerødgade 15A 2960 Rungsted Kyst. T: 45 76 33 88
Hauritz, P., Søndrevej 40 d 8700 Horsens. T: 75 63 13 18, F: 75 61 70 61, E: pbh@geoteknik.dk
Havmøller, O., Ekofiskveien 1 4056 Tananger Norge
Hededal, O., Jernbane alle 35 2720 Vanløse. T: 38 11 60 01
Hedegaard, J.R., Herluf Trolles Gade 31, 2.tv. 9000 Aalborg. T: 98 10 38 85
Henriksen, J.B., Skudstrupvej 13E 8541 Skødstrup. T: 86 99 34 05
Henriksen, P., Tjørneengen 20 2791 Dragør. T: 31 53 25 43, E: peh@geoteknisk.dk
Hessner, J., Strandvejen 32 B, 2.tv. 2100 København Ø. T: 31 29 15 60
Hoff, M., Krogagre 84 8240 Risskov. T: 86 21 96 60, E: mbh@geoteknisk.dk
Holm, E., Engtoftevej 16, Todbjerg 8530 Hjortshøj 86 99 93 9. T: 86 99 98 60
Holte, J., Baggesensgade 6, II.th. 2200 København N, E: jch@ramboll.dk
Horn, J., Rudersdalsvej 90, st.th. 2840 Holte. T: 45 41 37 10, E: jzh@cowi.dk
Hulgaard, E., Østerhegn 47 2880 Bagsværd. T: 42 98 29 16
Hulgaard, N., Ebberupvej 110 5631 Ebberup. E: nhh@kampsax.dk
Hunnerup, C.E., Dagmarvej 1 3060 Espergærde. T: 49 13 43 67
Hvam, T., Granbakken 48 3400 Hillerød. T: 45 26 71 57, E: th@geoteknisk.dk
Haahr, F., Dag Hammarskjölds Allé 27, 4. 2100 København Ø. T: 35 42 19 96, E: frh@carlbro.dk
Ibsen, L.B., Nordenshuse 9 9000 Aalborg. T: 98 16 45 45, E: i5lbi@civil.auc.dk
Ikkala, P.-I., Bellisvænget 9 5330 Munkebo. T: 65 97 77 57, E: pii@carlbro.dk
Jacobsen, H.J.A., Hjortevænget 9 4760 Vordingborg. T: 55 37 45 88, E: hjj@skude-jacobsen-v.dk
Jacobsen, S., Ringstedgade 533 4700 Næstved. E: sja@ncc.dk
Jakobsen, K.P., Usserød Kongevej 4, 2.tv. 2970 Hørsholm. T: 45 76 55 15, E: kpj@dhi.dk
Jakobsen, L., Sportsvej 41, 1.sal, dør 3 2600 Glostrup. T: 43 44 01 45
Jakobsen, P.-E., Bækkeskovvej 8 8722 Store Dalby, Hedensted. T: 75 89 04 87, E: poulerik@rotek.dk
Jansson, M., Gothersgade 93, 2.tv. 1123 København K. T: 33 32 93 20
Jensen, B.G., Enebakken 15 8520 Lystrup. T: 86 22 09 78
Jensen, B.S., Jens Chr. Skous Vej 9 8000 Århus C. T: 87 39 66 66, E: bes@cowi.dk
Jensen, J.H., Cedervangen 56 3450 Allerød. T: 4817 5729, E: jhj@geoteknisk.dk
Jensen, J.K., Otteshavevej 4 6600 Vejen. T: 75 58 85 45, Mobile: 407385, E: jkj@carlbro.dk
Jensen, J.L., Fasanhaven 17 2820 Gentofte. E: jlj@cowi.dk
Jensen, K.G., Øster Sundby Vej 39 9000 Aalborg. T: 98 10 40 42
Jensen, M.B., Norgesgade 13 9000 Aalborg. T: 98165912, E: maj@carlbro.dk
Jepsen, J.-E., Jættehøjen 29, Græse Bakkeby 3600 Frederikssund. T: 47 36 14 44, E: jnj@geoteknisk.dk
Jessen, G., Nivåvænge 224 2990 Nivå. T: 49 14 41 43
Johannesen, S.H., Kongensgade 29A, st.th. 7000 Fredericia. E: shj@jensjohanandersen.dk
Johansen, H., Møllevej 24 2640 Hedehusene. T: 46 59 02 99
Johansson, P., Vikans Industriväg 14 41268 Görteborg Sverige. E: peter.x.johansson@hercules.se
Jørck, M., Hjelmsgade 5 2100 København Ø. T: 31 18 06 61, E: mortenj@post.tele.dk
Jørgen, B.M., Rambøll, Bredevej 2 2830 Virum. T: 45 57 09 27, E: jorn@ltramboll.dk
Jørgensen, B., Hjortekærsvej 158 B 2800 Lyngby. T: 45 93 73 13, E: drilling.bjo@post.uni2.dk
Jørgensen, J., Solsortvej 14 2000 Frederiksberg. T: 38 19 00 41
Jørgensen, J., Gl. Skolevej 15, Vrold 8660 Skanderborg. E: jsj@geoteknisk.dk

Jørgensen, K.J., Bøgehøj 16 2900 Hellerup. T: 39 62 64 66, Mobile 20156214, E: kai@cowi.dk
Jørgensen, L.B., Nøragervej 2, Hornel 9850 Hirtshals. T: 96 56 99 98, E: lbj1@post11.tele.dk
Jørgensen, M., Nybro Vænge 63 2800 Lyngby
Jørgensen, M.B., Møllegårdsvej 5 9210 Aalborg SØ. T: 98 14 19 14, E: i5mbj@civil.auc.dk
Jørgensen, M., Kongsdalvej 26 2720 Vanløse. T: 38 74 72 18, E: moj@geoteknisk.dk
Jørgensen, P., Østermøllevej 13 9230 Svenstrup. T: 98 38 33 56, E: poj@carlbro.dk
Jørgensen, V.A., Humlebækgade 2, 1.tv. 2200 København N. T: 38 34 73 79, E: vjorgensen@comet.dk
Kalsmose, K., Birkebjergvej 1, Ougtved 4291 Ruds Vedby. T: 58 26 11 58
Kellezi, L., Svalegabet 14A 2850 Tværum DK. T: 4580 2578, E: lke@geoteknisk.dk
Kierkegaard, C., Kulsviervej 84A 2800 Lyngby. E: cki@ hoh-vand.dk
Kildsgaard, E., Alsædtægten 21, St. Darum 6740 Bramming. T: 75 17 91 99, E: ebbe@kildsgaard.dk
Kirkegaard, M., Forchhammers vej 11, 1.th 1920 Frederiksberg C. T: 33 31 41 72, E: mak@baene.dk
Kjeld, A., Unnerød Huse 10 4500 Nykøbing Sj
Kjeldsen, A., Greisvej 4 2300 København S. T: 32 52 50 03, E: ank@geoteknisk.dk
Kjeldsen, V., Grejsdals Vænge 4 7100 Vejle. T: 75 82 44 43, E: villy.kjeldsen@mail.tele.dk
Kjær, H., Hanstholmhavn, Auktionsgade 39 7730 Hanstholm. T: 96550710
Knudsen, B., Bushøjvænget 27 8270 Højbjerg. T: 86 27 72 10, E: bk@geoteknisk.dk
Knudsen, K.M., Aarsleff – Bachy Soletanche J.V. Tune Parkvej 5 Tune, 4000 Roskilde
Kofoed, J., Nivåpark 13 2990 Nivå. T: 49 14 76 12
Korshøj, J. Chr., Klokkerbakken 71 8210 Århus V
Krabbenhøft, S., Valmuevænget 27 6710 Esbjerg V
Kristensen, B.M., Thit Jensens Vej 1 9240 Nibe. T: 98 35 20 09
Kristensen, J.P., Gartnerkrogen 12 3500 Værløse. T: 42 48 30 18
Kristensen, P., Lehwaldsvej 3, 8F 2800 Lyngby. T: 45 87 15 56
Krogsbøll, A., Natravnevej 4 3660 Stenløse. T: 47 10 73 06, E: akr@byg.dtu.dk
Lade, P.V., Lille Borgergade 19, 3.tv. 9400 Nr. Sundby DANMARK. T: 98 17 83 99, E: i5pvl@civil.auc.dk
Lagoni, P., Strandvejen 300 a 2930 Klampenborg DANMARK. T: 45 39 64 71 80
Larsen, B., Thoravej 8 2400 København NV. T: 45 86 16 82
Larsen, F.A., Husbyvej4 ESB. 3230 Græsted. E: finn. anker.larsen@get2net.dk
Larsen, H., Rødager Allé 59 B 2610 Rødovre. T: 31 41 09 68, E: hl@geoteknisk.dk
Larsen, J., Solbjerg Hovedgade 154 8355 Solbjerg. T: 86 92 90 92, E: jql@geoteknisk.dk
Larsen, L.-H.N., Møgeltøndergade 3, 3.sal, th. 1755 København V. T: 33 25 85 30, E: lhl@geoteknisk.dk
Larsen, M., Oehlenschlägersgade 6, 2.th. 1663 København V. T: 33 31 19 01
Larsen, O., Frugthegnet 72 2830 Virum. T: 42 85 47 18
Larsen, O.S., Byvej 254 2650 Hvidovre. T: 36 78 81 28, E: baver@post4.tele.dk
Larsen, P., Kålmarken 17 2860 Søborg. T: 39 67 63 89, E: pol@geoteknisk.dk
Lauesen, L.M., Stegshavejvej 5750 Ringe
Laursen, P.B., Uggeløse Bygade 98 3540 Lynge. T: 42 42 19 39
Leeberg, R., Tornegade 4, 1. 3700 Rønne. T: 56 91 1032, F: 56 91 09 17, E: rlconsul@post4.tele.dk
Leth, C.T., Blegdamsgården, Blegdamsvej 74C, II. 2100 København Ø. T: 35 35 60 85
Liingaard, M., Vesterbrogade 20, 3th. 9400 Nørresundby. E: liin96@civil.auc.dk
Lindberg, C.B., Ericavej 20 2820 Gentofte. T: 39 69 71 83
Lisby, J., Spørringvej 34 8530 Hjortshøj. E: jos@aarsleff. com
Lollike, J., Rønnebærtoften 15 2950 Vedbæk. T: 42 89 31 05

Lorange, J., Frugthegnet 43 2830 Virum. T: 45 85 77 75, E: jlo@carlbro.dk
Luke, K., Sellerupskovvej 122 7080 Børkop. T: 75 86 92 95
Lund, P., Esbølvej 33 6893 Hemmet, Vestjylland. T: 97 37 50 73
Lund, W., Bentevej 21 9520 Skørping. E: i5wl@civil. auc.dk
Lundgren, H., Bernstorfflund Allé 14 B 2920 Charlottenlund. T: 39 63 22 27
Lundstrøm, M., Daddelvej 12 4700 Næstved. T: 53 72 12 38
Lyngby, D., Gullandsgade12, 4.tv. 2300 København S. T: 32 84 16 79, E: d.lyngby@comet.dk
Madsen, E.B., Vesterbro 93, 3.tv. 9000 Aalborg. T: 98 11 67 76
Madsen, J., Klintemarken 23 2860 Søborg. T: 39 69 41 65
Madsen, K.N., Stengårds Alle 53 2800 Lyngby. T: 44 98 72 02, E: kom@cowi.dk
Madsen, O.A., Bybækpark 2 3520 Farum. T: 42 95 30 15
Madsen, P.H., Gåsevænget 8 2791 Dragør. T: 32 53 47 78
Madsen, P., Mondrupsvej 11A 8260 Viby J. E: pem@ ramboll.dk
Madsen, P., Korshøj 112 3670 Veksø. T: 47 17 37 05
Mandrup, M., Bispeengen 76 5270 Odense N. T: 66 18 97 90, E: mma@optiroc.dk
Michaelsen, C., Statshavnsadministrationen, Postboks 2 6701 Esbjerg
Mikkelsen, H., Ellemosevej 39 2900 Hellerup. T: 26 27 54 80, E: hm@cowi.dk
Mikkelsen, J.B., Nørregade 7 5970 Ærøskøbing. T: 65 99 21 91, E: jmi@carlbro.dk
Milling, P., Reginsvej 22 8600 Silkeborg. T: 86 80 38 8
Mollerup, N., Finlandsgade 30 4690 Haslev. T: 56 31 50 00, F: 56 31 66 10, E: mollerup.aps@mobilixnet.dk
Mortensen, J.K., Højbjergvang 18 2840 Holte. T: 45 42 19 42, E: jkm@ramboll.dk
Mortensen, K., Bakkedraget 9 A 8270 Højbjerg. T: 86 27 23 62
Mortensen, N., Birkegade 27A, 2.th. 2200 København N. T: 35 39 87 86, E: nmo@cowi.dk
Mortensen, O., Rokhøj 8 8520 Lystrup. E: ole.mortensen@ skanska.dk
Mortensen, P.K., Stammen 63 9260 Gistrup. T: 98 32 30 52, E: pkm@kampsax.dk
Mouridsen, A.B., Hovmarksvej 33 8700 Hor-sens. T: 75 65 58 18, E: abm@aarsleff.com
Mousten, S., Kent, Forskerparken 8000 Århus C. T: 86202000, lok 4900, F: 86209788, E: dg@geofysik.dk
Munkholt, H., Hans Hedtoftsvej 19 9210 Aalborg SØ. T: 98 18 95 00
Mygind, M., Teglgardsparken 45C 5500 Middelfart. T: 64 41 97 88, E: mim@arkil-fundering.dk
Møller, B.H., Bjerggaardsvej 6, Åle 7160 Tørring. T: 75 67 65 87
Møller, L.H., Hedegårdsvej 17 B 2300 København S. T: 32 97 01 13
Møller, O., Henrik Hertz Vej 23 8230 Åbyhøj. T: 86 25 62 63, E: NOM@aarsleff.com
Møller, T.H., Hans Broges Gade 41, 5.sal 8000 Aarhus C. T: 86 18 19 36
Møller, T.D., Ny Sebberupvej 9 8723 Løsning. T: 75 65 16 16, E: tdm@sh-ing.dk
Mørch, C., Åboulevarden 88 A st.th. 8700 Horsens
Nielsen, B.N., Nordmarken 30, Vester Hassing 9310 Vodskov. T: 98 25 51 67, E: bnn@carlbro.dk
Nielsen, B., Kollensøvej 28, Slagslunde Skov 3550 Slangerup. T: 42 33 52 45
Nielsen, C.Q., Ledavej 100 9210 Aalborg SØ. T: 98 14 31 63
Nielsen, F.R., Frederikssundsvej 263, st. 2700 Brønshøj. T: 38 28 43 98
Nielsen, H.K., Lundeskovvej 22 2900 Hellerup. T: 39 62 42 84
Nielsen, H.L., Møllebakken 7 8660 Skanderborg. T: 86 51 05 58, E: hln@neg-micon.dk
Nielsen, J.M., Langballevej 104 8320 Mårslet
Nielsen, K.W., Bikivej 14 A 2970 Hørsholm. T: 42 86 60 31
Nielsen, K.K., Nørholmsvej 571 9420 Nibe. T: 98 31 42 72
Nielsen, L.M., Bygtoften 29 2800 Lyngby. T: 45 87 83 16, E: lmn@ncc.dk
Nielsen, O.B., Høeghsmindevej 56 2820 Gentofte. T: 39 65 19 72

Nielsen, P.M., Kvædevej 29 2830 Virum
Nielsen, R., Pile Allé 5B, lejlighed 18 2000 Frederiksberg. T: 31 39 39 57
Nielsen, S., Holstebrovej 26 6950 Ringkøbing. T: 97 32 04 16, E: soeren@ranfelt.dk
Nielsen, U.T., Ganehøj Alle 20, Ganløse 3660 Stenløse. T: 48 18 34 32, E: utn@ramboll.dk
Nygaard, P., Jean Miquel, Rebekkavej 44, 3.tv. 2900 Hellerup
Okkels, N., Stenrosevej 71 8330 Beder. T: 86 93 76 40, E: nio@geoteknisk.dk
Olesen, A.L., Søndergade 34, Assentoft 8900 Randers. T: 86 49 67 48, E: 4AP@mail.tele.dk
Olsen, M., Ville Heise Park 35 3450 Allerød. T: 48 17 57 05, E: moo@carlbro.dk
Olsen, N.J., Moesgårds Allé 14 8320 Mårslet. T: 86 27 60 06
Oppenhagen, R., Trollevænget 9 4400 Kalundborg. T: 59 560094
Ovesen, N.K., Sølvgade 22, 2.tv. 1307 København K. T: 33 32 45 21, E: nko@geoteknisk.dk
Pallesen, P.R., Dybbøl Bygade 15 C, Dybbøl 6400 Sønderborg. T: 74 48 78 44, E: prp@slothmoller.dk
Panduro, P., Vejteknisk Institut, P.O. Box 235 4000 Roskilde. T: 46 30 01 34
Pedersen, H., Jægervang 23 3460 Birkerød. T: 45 82 82 39, E: hgp@niras.dk
Pedersen, K., Gravquick A/S, Fabriksparken 16 2600 Glostrup. T: 42 45 16 00
Pedersen, M., Stærevej 5 4760 Vordingborg. T: 53 77 31 13
Pedersen, P.V., Evertvej 20 4040 Jyllinge. T: 46 73 25 40
Petersen, J.F., Vangedevej 224 A, 3.sal, tv. 2820 Gentofte. Mobile: 24231544
Petersen, K.H., Lille Blovstrødvej 55 3450 Allerød. T: 48 17 51 07, E: khp@geoteknisk.dk
Petersen, K., Bjergbakkevej 312 2600 Glostrup. T: 43 63 83 82, E: kim@copen.sgi.com
Petersen, S.B., P.B. Miljø A/S, Enebærvej 7 8850 Bjerringbro
Petersen, S.J., Enggårdsvej 9A 2610 Rødovre. T: 36 72 40 04, E: sjp@ramboll.dk
Petersen, T., Adamsminde 19 5462 Morud. E: tmp@ramboll.dk
Pihl, K.A., Nordgårdsvej 9, Tune 4000 Roskilde. T: 46 13 90 98
Porsvig, M., Hejrevej 10 2970 Hørsholm. T: 45 76 80 76, E: mop@geoteknisk.dk
Poulsen, H.S., Skovvej 9, Gram 8660 Skanderborg. T: 86 57 25 03, E: hsp@ramboll.dk
Primdahl, G., Kirkeby Sand 21 5771 Stenstrup. T: 62 26 19 17, F: 62 26 25 29, E: gpr@dge.dk
Qvist, S., Prisholmvej 26 2500 Valby. T: 36 16 89 74, E: sq@dynonobel.dk
Rabøl, J., Parcelvej 46 2840 Holte. T: 42 42 36 93
Rande, L., Alrøvej 320, Alrø 8300 Odder. T: 86 55 20 15, E: lra@aarsleff.com
Ranfelt, J., Odderskærvej 10, Boris 6900 Skjern. T: 97 36 63 10, F: 97 36 6610, E: jr@ranfelt.dk
Rasmussen, E., Vestermøllevej 31 8660 Skanderborg
Rasmussen, H.L., Røddinggade 10A 1735 København V
Rasmussen, I., Floraparken 40 4690 Haslev. T: 56 69 34 96
Rasmussen, J.L., Granparken 141 2800 Lyngby. T: 42 88 90 51, E: jlr@ramboll.dk
Rasmussen, K.F., Runebergs Alle 25 2860 Søborg. T: 39 66 24 39
Rasmussen, L., Viemosetoften 5 2610 Rødovre. E: lar@geoteknisk.dk
Rasmussen, M.S., Rødlundvej 292 8462 Harlev. E: msr@aarsleff.com
Rasmussen, P.K., Dalsø Park 40 3500 Værløse. T: 44 47 97 01
Rodevang, B., Løvbakken 9 3520 Farum. T: 42 48 17 23, E: br@iktmail.cph.ih.dk
Rosbæk, M.S., Færgevej 4 5700 Svendborg. T: 62 21 50 21
Rosenkvist, A., Stabelhjørnevej 36 8721 Daugaard. T: 75 89 51 51, F: 75 89 51 61, E: arne.rosenkvist@mail.tele.dk
Rossen, I.H.R.H., Skt. Pauls Kirkeplads 9, P.O. Box 96 8100 Aarhus C. E: ing-hrstofanet.dk
Rømhild, C.J., Bredebovej 35, st.tv. 2800 Lyngby. T: 45 87 06 50, amh. 4870822

Schiellerup, U., Thorsvej 4 B 3460 Birkerød. T: 45 82 44 81, E: uls@geoteknisk.dk
Schmidt, T., Sydvestvej 29 8700 Horsens. T: 75 63 07 06
Schnabl, W., Skude & Jacobsen, Næstvedvej 1 4760 Vordingborg. T: 55 37 16 00, E: ws@skude-jacobsen-v.dk
Schou, L.F., Ærøvej 13 6100 Haderslev. T: 74 53 04 15, F: 74 52 64 95
Schaarup-Jensen, A.L., P.P. Ørumsgade 8, st.tv. 8000 Århus C. T: 86 14 92 98, Mobile: 40442286
Sell, M., Dalvangen 20 8400 Ebeltoft
Simensen, M., Kirsebær Alle 9-11 3400 Hillerød. T: 42 26 06 66
Sjösten, G., Ved Hegnet 24 2960 Rungsted Kyst. T: 20 40 06 56
Skytte, J., Abildgaardsvej 95 2830 Virum. T: 45 85 28 58, F: 45 85 28 57, E: jan.skytte@skanska.dk
Slumstrup, E.H., Aasiaat Kommune, Teknisk forvaltning Postboks 220 3950 Aasiaat
Steen, O., Rørløkken 38 2730 Herlev. T: 44 84 93 20, E: ost@monthor.dk
Steenfelt, J.S., Moseskrænten 18 2860 Søborg. T: 39 56 50 99, E: jst@iabm.dtu.dk
Steensen-Bach, J.O., Ved Store Dyrehave 6, st.tv. 3400 Hillerød. E: osb@carlbro.dk
Steffensen, H.E., Elmegårdsvej 50 5250 Odense SV. E: hes@geoteknisk.dk
Steffensen, K., Vibevej 20 7000 Fredericia. T: 75 92 06 16
Stentsøe, S., Nørregårdsvej 13 2610 Rødovre. T: 3670 4452, F: 36704452, E: sns@cowi.dk
Strømann, H., Hans Bruunsvej 6 2920 Charlottenlund. T: 31 64 18 74, E: hls@geoteknisk.dk
Sundberg, M., Syv Holmevej 12S 4130 Viby Sj. Mobile: 40 30 04 71
Sundström, L., Geijersgatan 8 21 618 Malmö Sverige. T: 46 40 16 70 00, E: larsake.sundstrom@sweco.se
Systemteknik, Attn: Berit Carlson, Banestyrelsen, Pakhusvej 101 2100 København Ø. T: 82 34 51 31
Sørensen, C.S., Ridefogedvej 11 9000 Aalborg. T: 98 18 64 73, F: 98 18 65 22, E: i5css@civil.auc.dk
Sørensen, E., Solvænget 22 2800 Lyngby. T: 45 88 85 64, E: els@geoteknisk.dk
Sørensen, I., Ingeniørhøjskolen, Chr.M. Østergaardsvej 4 8700 Horsens. T: 75 68 42 45
Sørensen, J.L., Gartnerkrogen 25 3500 Værløse. T: 44 48 37 10
Sørensen, K.D., Baldursgade 28 6700 Esbjerg. T: 75 45 26 05
Sørensen, M., Engbakken 5 2830 Virum. T: 45 83 81 68
Sørensen, P.H., Nærum Hovedgade 76, 1.3 2850 Nærum. T: 45 80 45 76, E: pds@ramboll.dk
Sørensen, U.L., Højdevej 57, 2.th. 2300 København S. T: 32 55 99 59
Thamdrup, K., Amtsgården, Sorsigvej 35 6760 Ribe. T: 75 41 17 80, E: kt@ribeamt.dk
Therkildsen, C.G., Egehaven 55 8520 Lystrup. E: cgt@kampsax.dk
Thomsen, B.K., Haslundparken 84, V. Hassing 9310 Vodskov. E: bit@niras.dk
Thomsen, L., Løkkegårdsvej 77 6230 Rødekro. T: 74 66 16 30
Thorsen, G., Risbjergvej 24 9260 Gistrup. T: 98 31 49 82, F: 98 31 49 82, E: i5gt@civil.auc.dk
Thorsen, S., Risbjergvej 24 9260 Gistrup. T: 98 31 49 82, E: sgen.thorsen@compaqnet.dk
Thorsen, T., Rugbjergvej 84, Stautrup 8260 Viby J. T: 31 47 10 05, E: trt@geoteknisk.dk
Thøgersen, K., Urbansgade 21, II.tv. 9000 Aalborg. T: 98 18 82 32
Thøgersen, L., Midgaarden 27 9230 Svenstrup J. T: 98 38 26 09, E: i5lt@civil.auc.dk
Trankjær, H., Egegårdsvej 22 8900 Randers. T: 86 42 22 68, E: hlt@cowi.dk
Tychsen, H., Mårumvej 62 3230 Græsted. E: hty@bane.dk
van Deurs, G., Furesøvej 93 2830 Virum. T: 45 85 15 25
Vanggaard, M., Ravnsnæsvej 15 3460 Birkerød. T: 45 81 69 66, E: mva@geoteknik.dk
Vefling, G., Johan Wilmannsvej 44 2800 Kgs. Lyngby. T: 42 88 07 44

Vejrup, N.P., Lundholmvej 4 9900 Frederikshavn. T: 87
47 50 9
Vendelbo, N.J., Nordøgade 7 8200 Århus N
Vesterby, H., Nobisvej 8 3460 Birkerød. T: 42 81 17 00,
E: hv@geoteknisk.dk
Villumsen, A., Vipstjertvej 4 3400 Hillerød. T: 48 26 95 13,
F: 48 26 95 13, E: iggav@pop.dtu.dk

Vork, K.A., Holmstrupgårdsvej 192 8210 Århus V
Wendt, P.S., Westend 8, st.th. 1661 København V. T: 33
31 88 69
Wyrwik, B., Enebærkrogen 26, Abildøre 4560 Vig. T: 59 31
89 07, F: 59 31 00 04
Østergaard, H., Bredgade 11 6100 Haderslev. T: 74 53 40 44
Aarestrup, O., Iny Strandvej 3050 Humlebæk. T: 49 19 38 99

ECUADOR/EQUATEUR

Secretary: Ing. Renato Benavides Chica, Ecuadorian Society of Soil Mechanics and Rocks (SEMSIR), P.O. Box 9176, Guayaquil. T: (5934)28690, F: (5934)883433, E: semsir@telconet.net

Total number of members: 29

Andrango Loor, O., Consulting Engineering, Cdla. Modelo, calle 4ta. #304 y la 3ra.Guayaquil. T: 5934-399339

Cappelo Brito, J., Consulting Engineering, Cdla. Kennedy Vieja, 1ra. Peatonal y Calle 1ra. Oeste. P.O. Box 09-01-10923, Guayaquil. T: 5934-205006

Chávez, M.A., Prof. Escuela Superior Politécnica del Litoral, Consulting Engineering, Cdla. Kennedy Norte No. 702, P.O. Box 5863, Guayaquil. T: 5934-269131, F: 5934-852274, E: planific@espol.edu.ec

España Pico, H.E., Consulting Engineering, Calle 15 y Av. 10, esquina, 2do. Piso, Dept. #2, Manta. T: 05-611547, 622102, F: 05-627770/613553

González Moya, F., Consulting Engineering, Yogoslavia #143 y Rumipamba, P.O. Box 2759, Quito. T: 5932-529881, E: gonzalez@waccon.net.ec

Herbozo Alvarado, E., Consulting Engineering, Cdla. La Alborada XI Etapa Mz. 25 V. 13. T: 5934-234016, Guayaquil

León Toledo, V., Prof. Universidad de Guayaquil, Consulting Engineering, Cdla. La Garzota, 2da. Etapa, Mz. 135, Villa 16, Guayaquil. T: 5934-281037, 640828

Luque, C., Consulting Engineering, Avda. C.J. Arosemena, Km 2fi, Guayaquil. T: 5934-201009

Manchemo Orellana, A., Consulting Engineering, Cdla. La Alborada XI Etapa Albocentros Edif. C Piso 2 Ofic. 204, Guayaquil. T: 5934-248781, F: 5934-234197

Marín Estevez, D., Consulting Engineering, Cdla. nueva Kennedy, Calle 5ta. #103 y la E, Guayaquil. T: 5934-396597, E: desireemarin56@yahoo.com

Marín-Nieto, L., Prof. Universidad de Guayaquil, Consulting Engineering, Alianza entre Tercera y Cuarta, Urdesa, T: 5934-883433, P.O. Box 3699, Guayaquil. E: luismarin32@yahoo.com

Martínez Rephani, G., Consulting Engineering, Boyacá 1107 y P. Icaza 3er. Piso Dpto. 2, P.O. Box 11369, Guayaquil. T: 5934-304206

Maruri Díaz, R., Consulting Engineering, Calle Dolores Sucre #111 entre El Oro y Maracaibo, P.O. Box 03491, Guayaquil. T: 5934-255039, F: 5934-255064, E: cimentac@gye.satnet.net

Moreno Lituma, V., Prof. Universidad de Guayaquil, Consulting Engineering, Cdla. Las Acacias Mz. B3 V. 10, Guayaquil. T: 5934-281037

Moya, R., Consulting Engineering, Alborada IX Etapa Mz. 934 V. 4, P.O. Box 9879, Guayaquil. T: 5934-274531

Núñez Lomas, F., Consulting Engineering, Cdla. Las Acacias Mz. C-15 V. 17, Guayaquil. T: 5934-345205, E: fenl64@latinmail.com

Nuques Cobo, J., Consulting Engineering, Avda. C.J. Arosemena, Km 2fi, Guayaquil. T: 5934-201009

Pesantes, Pesantes, J., Consulting Engineering, Carchi #1526 entre Sucre y Colón, Guayaquil. T: 5934-451603, F: 5934-454385, E: aetecnicos@porta.net

Plaza Nieto, G., Prof. Escuela Politécnica Nacional, Consulting Engineering, Calle Isabel La Católica, P.O. Box 2759, Quito. T: 5932-507127

Ponce, O.M., Consulting Engineering, Cdla. La Florida, Mz. 403 V. 15, Guayaquil. T: 5934-260595

Ramos Olmedo, L., Consulting Engineering, Baquerizo Moreno #922 y V. M. Rendón, Guayaquil. T: 5934-304926

Rodríguez Torres, J., Consulting Engineering, Seis de Marzo 4301 y Rosendo Avilés, Guayaquil. T: 5934- 449429, F: 5934-441993

Ríos de Capello, G., Consulting Engineering, Cdla. Kennedy Vieja 1ra. Peatonal y Calle 1ra. Oeste, P.O. Box 10923, Guayaquil. T: 5934-205006

Rivera, G.J., Consulting Engineering, Sauces III Mz. 194, V. #567, Guayaquil. T: 5934-640846

Rivero Ordematt, C., Consulting Engineering, Las Riberas Mz. H V. 9, P.O. Box. 9176, Guayaquil. T: 5934-239119

Silva Sánchez, A., Consulting Engineering, Francisco de Marcos 330 5to. Piso, Guayaquil. T: 5934-400766, 402482

Terreros de Varela, C., Prof. Universidad de Guayaquil, Consulting Engineering, Cdla. Nueva Kennedy Séptima #103 y la E, Guayaquil. T: 5934-3989081

Velasco Filián, A., Consulting Engineering, Cdla. Guayacanes Mz. 36 V. 20, Guayaquil. T: 5934-331092

Yépez Intriago, L., Consulting Engineering, Coop. Juan Montalvo Mz. C-5, V. 10 (estación de la línea 10), Guayaquil. T: 5934-266502/266525

ESTONIA/ESTONIE

Secretary: Mr Johannes Pello, Estonian Geotechnical Society, Department of Structural Engineering, Tallinn Technical University, Ehitajate tee 5, 19086 Tallinn. T: +372 6202408, E: jpello@edu.ttu.ee

Total number of members: 25

Eller, E., Geologist, AS Maves, Marja tn 4D, Tallinn
Eskel, Ü., Executive director, AS GIB, Välja tn 8, Tallinn
Jaaniso, V., Ph.D., Asst. professor, Tallinn Technical University, Ehitajate tee 5, 19086 Tallinn
Järve, U., Executive director, OÜ REI Geotechnika, Rävala pst 8, Tallinn
Killar, E., Manager, AS Viacon Eesti, Madara tn 27
Leinsalu, T., Project manager, OÜ REI Geotechnika, Rävala pst 8, Tallinn
Lemberg, U., Manager, EKUK Geotehnikalabor, Suur-Sõjamäe 34, Tallinn
Liblik, T., Manager, Foundation of construction geology, Mustamäe tee 51, Tallinn
Mets, M., Ph.D., Science director, AS GIB, Välja tn 8, Tallinn
Oll, K., Senior engineer, Tallinn Technical University, Ehitajate tee 5, 19086 Tallinn
Parbo, A., Engineer-geologist, AS GIB, Välja tn 8, Tallinn
Pello, J., M.Sc., Lecturer, Tallinn Technical University, Ehitajate tee 5, 19086 Tallinn
Petermann, S., Senior geologist, Eesti Geoloogiakeskus, Kadaka tee 82, 12618 Tallinn

Piits, T., Project manager, REIB OÜ, Rävala pst 8, Tallinn
Riet, K.-H., Chief hydro-geologist, OÜ REI Geotehnika, Rävala pst 8, Tallinn
Sedman, P., Project manager, OÜ IPT Projektijuhtimine, Mustamäe tee 193-18, 12902 Tallinn
Talviste, P., Manager, OÜ IPT Projektijuhtimine, Mustamäe tee 193-18, 12902 Tallinn
Tammemäe, O., Manager, Säästva Eesti Instituut, SEI-Tallinn, Lai tn 34, 10502 Tallinn
Tang, H., Department manager, Eesti Keskkonnauuringute Keskus, Marja 4D, Tallinn
Trealt, A.-L., Director, OÜ Minaron, Seene tn 16-3, Tallinn
Tumm, A., Chairman of board, OÜ ESPAN, Rävala pst 8-C403
Tõevere, J., Engineer-geologist, AS GIB, Välja tn 8, Tallinn
Vallaots, T., Manager, Ehituskonstruktsioonide Tugevdamine OÜ, Pelguranna 21-43, Tallinn
Vilo, A., Consulting-engineer
Torn, H., Manager, AS GIB, Välja tn 8, Tallinn

FINLAND

Secretary: Mr Hans Rathmayer, Finnish Geotechnical Society, c/o VTT-RTE, P.O. Box 1800, FIN-02044 VTT.
T: +358-9-4564681, F: +358-9-463251, E: hans.rathmayer@vtt.fi

Total number of members: 390

Aalto, A., TKK, Rakennus- ja yhdyskuntatekn. os., Rakentajanaukio 4 A, FIN-02150 Espoo. T: 358-9-4513765, F: 358-9-4513826

Aarnio, T., Turun Viatek Oy, Linnankatu 3 a B, FIN-20900 Turku

Airaksinen, U., Finntrea Oy, FIN-71730 Kinnulanlahti

Aittapelto, O., Turun Viatek Oy, Linnakatu 3 aB, FIN-20110 Turku. T: 358-2-2733000, F: 358-2-2335560

Ajanko, S., MVR Juslenius Oy, Yliopistonkatu 6 B, FIN-20110 Turku. T: 358-2-336530

Akkanen, H., Ins. tsto Pohjatekniikka Oy, Nuijamiestentie 5 B, FIN-00400 Helsinki. T: 358-9-4777510, F: 358-9-47775111

Alinen, J., Geoinsinööri Oy, Gallen-Kallelankatu 8, FIN-28100 Pori. T: 358-2-6300800, F: 358-2-6300801

Ali-Runkka, T., Geopalvelu Oy, Tikkutehtaankatu 3, FIN-33250 Tampere. T: 358-3-2767200, F: 358-3-2767222

Antikainen, J., Pohjantie 8 A 18, FIN-02100 Espoo. T: 358-9-4552473

Anttikoski, U., Mahlarinne 12 B, FIN-02130 Espoo. T: 358-9-4524998

Anttila, H., Komeetankuja 4 D 22, FIN-02210 Espoo

Anttila, P., Vapaalantie 48 A, FIN-01650 Vantaa. T: 358-9-845118

Arjas, J., Kyläkirkontie 7 A 9, FIN-00370 Helsinki

Arkima, O., Ys – Yhdyskunta Oy, Westendintie 1 B, FIN-02160 Espoo. T: 358-9-43008110, F: 358-9-43008100

Asikainen, A., Kevätkuja 3 F, FIN-02200 Espoo. T: 358-9-8031790

Avellan, K., c/o Konsultointi Kareg Oy, Töölöntorinkatu 11 B, FIN-00260 Helsinki. T: 358-9-493411, F: 358-9-493032

Blomster, D., Imatran Voima Oy, Energialiiketoiminta, Annankatu 34-36 B, FIN-00019 IVO. T: 358-9-85616423, F: 358-9-85616764

Breilin, O., Geologian tutkimuskeskus, PL 1237, FIN-70211 Kuopio. T: 358-2055030, F: 358-2055013

Britschgi, R., Suomen ympäristökeskus, PL 140, FIN-00251 Helsinki. T: 358-9-40300444, F: 358-9-40300491

Eerola, L.O., Geosto Oy, Kulmakatu 10, FIN-15140 Lahti. T: 358-3-515501, F: 358-3-7521890

Eerola, M., Tielaitos, Tuotanto/Konsultointi, PL 157, FIN-00521 Helsinki. T: 358-204442543, F: 358-204442154

Ehrola, E., Syrjäkatu 1 B 3, FIN-90100 OULU. T: 358-8-336548

Eilu, P., Turun yliopisto, Geologian laitos, FIN-20014 Turku. T: 358-2-3335496, F: 358-2-3336580

Eklund, P., Mäntymäenkuja 8, FIN-02700 Kauniainen. T: 359-050-5053022

Engberg, F., Maa ja Vesi Oy, Itälahdenkatu 2, FIN-00210 Helsinki. T: 358-9-682661, F: 358-9-6826600

Eronen, K., Fasaanitie 18, FIN-04440 Järvenpää. T: 358-9-281739

Fagerholm, K., Oy VR-Rata Ab, Suunnitteluosasto, PL 488, FIN-00101 Helsinki. T: 358-9-7073240, F: 358-9-7073716

Filen, B., Studentvägen, S-95164 Luleå, Sweden

Fischer, A., Alkutie 32 J, FIN-00660 Helsinki. T: 358-9-742413

Forsman, J., SCC Viatek Oy, Piispanmäentie 5, FIN-02240 Espoo. T: 358-9-43011533, F: 358-9-4301560

Friberg, P., Puolarniitynkuja 4 A 4, FIN-02780 Espoo. T: 358-9-8054092

Gardemeister, R., IVO International Oy, Rajatorpantie 8, FIN-01019 IVO. T: 358-9-8561567, F: 358-9-5083402

Gers, L., AEL, Kaarnatie 4, FIN-00410 Helsinki. T: 358-9-5307458

Gulin, K., Fundus Oy, Melkonkatu 9, FIN-00210 Helsinki. T: 358-9-615810, F: 358-9-61581420

Gustavsson, H., Pelimannintie 24 H 76, FIN-00420 Helsinki. T: 358-9-5072737

Haavisto-Hyvärinen, M.-L., Katajaharjuntie 22 C 17, FIN-00200 Helsinki. T: 358-9-678036

Hailikari, T., Kaarikuja 1 I 75, FIN-00940 Helsinki. T: 358-9-304767

Hakkarainen, V., Piilopolku 3 I 54, FIN-02130 Espoo. T: 358-9-465162

Hakulinen, M., Julinintie 6 E, FIN-53200 Lappeenranta. T: 358-5-4517101

Halkola, H., Helsingin kaupunki, Geotekn. os., PL 2202, FIN-00099 Helsingin Kaupunki. T: 358-9-1694544, F: 358-9-1694555

Harjula, H., VR, PL 488, FIN-00101 Helsinki. T: 358-9-7072220, F: 358-9-7072330

Hartikainen, J., Espoon tekninen keskus/Geotekn. yksik., Virastopiha 2 C, FIN-02770 Espoo. T: 358-9-8065456, F: 358-9-8065457

Hartikainen, J., TTKK, Geotekniikan laboratorio, PL 600, FIN-33101 Tampere. T: 358-3-3652860, F: 358-3-3652884

Havukainen, J., Vesimyllyntie 7 A, FIN-00920 Helsinki. T: 358-9-3492988

Heikinheimo, R., Rusthollinkatu 2 G 61, FIN-20880 Turku. T: 358-2-2353635

Heikkilä, J., Arcus Oy, Maariankatu 4 C 63, FIN-20100 Turku. T: 358-2-2747050, F: 358-2-2330050

Heikkilä, J., SCC Viatek Oy, Piispanmäentie 5, FIN-02240 Espoo. T: 358-9-4301219, F: 358-9-4301560

Heikkinen, R., Jönsaksenkuja 3 D 31, FIN-01600 Vantaa. T: 358-9-539630

Heino, M., Korsutie 18, FIN-00370 Helsinki. T: 358-9-5652572

Heinonen, P., Junttan Oy, PL 1702, FIN-70701 Kuopio. T: 358-17-2874400, F: 358-17-2874411

Heinonen, J., Paavo Kolin katu 12 E 39, FIN-33720 Tampere

Hekkala, H., Kylmäniemenkuja 4, FIN-90540 Oulu. T: 358-8-512350

Helander, R., Itä-Suomen Viatek Oy, Kauppalankatu 14, Fin-45100 Kouvola. T: 358-5-7455400, F: 358-5-7455498

Herva, M., FIN-94400 Laurila

Herva, J., Keluveenväylä 16, FIN-90650 Oulu. T: 358-8-5565214

Hiekkanen, R., Pohjantähdenkuja 6, FIN-00740 Helsinki. T: 358-9-366136

Hietala, K., Seitasaarentie, FIN-95300 Tervola. T: 358-16-435911

Hietala, P., Seitasaarentie, FIN-95300 Tervola. T: 358-16-435911

Hilpi-Niemi, A., Kruununtie 8 E, FIN-02180 Espoo. T: 358-9-529425

Hinkkanen, H., Kuparitie 14 B 29, FIN-00440 Helsinki

Hirvonen, J., Itä-Suomen Viatek Oy, Vuorikatu 14 A, FIN-70100 Kuopio. T: 358-17-2631300, F: 358-17-2631104

Hoikkala, S., Hallainvuorentie 5 D 17, FIN-00920 Helsinki

Holmberg, H., FIN-02590 Lappers. T: 358-19-347225

Holopainen, P., Helsingin kaupunki, Geotekn. os., PL 2202, FIN-00099 Helsingin Kaupunki. T: 358-9-1631, F: 358-9-1634555

Holopainen, A., Suotie 31, FIN-04310 Tuusula. T: 358-9-248983

Holtari, M., Ylipalonkuja 3 B, FIN-00670 Helsinki. T: 358-9-746269

Honkaniemi, M., Rinnekuja 4, FIN-03400 Vihti. T: 358-9-2249383

Huhtala, M., VTT: RTE, PL 1800, FIN-02044 VTT. T: 358-9-4564960, F: 358-9-463251
Huokuna, M., Puolikuu 3 D 31, FIN-02210 Espoo. T: 358-9-8031991
Hurme, O., Geo-Suunnittelu O. Hurme Oy, Vesijärvenkatu 38, FIN-15140 Lahti. T: 358-3-7188600, F: 358-3-7188601
Husa, J., Visatie 12 J 85, FIN-04260 Kerava
Huttu, U., Länsi-Suomen ympäristökeskus, PL 156, FIN-60101 Seinäjoki. T: 358-6-3256511, F: 358-6-4141982
Hyrkkönen, A., Tampereen kaupunki, Tampereen kaupunki/TETO/SO PL 86, FIN-33211 Tampere. T: 358-3-2196048, F: 358-3-2196033
Hytti, K., Mäntysalontie 1 D, FIN-01820 Klaukkala. T: 358-400-410390
Hyvärinen, J., Etelä-Savon ympäristökeskus, Jääkärinkatu 14, FIN-50100 Mikkeli. T: 358-15-1913335, F: 358-15-363915
Hämäläinen, J., Kaitos Oy, Kytkintie 47, FIN-00770 Helsinki. T: 358-9-35070629, F: 358-9-35070610
Höynälä, H., Laurinmäenkuja 4 D 33, FIN-00440 Helsinki. T: 358-50-3530518
Ihalainen, P., Länsi-Uudenmaan vesi ja ympäristö ry, PL 51, FIN-08101 Lohja. T: 358-19-323623, F: 358-19-325697
Ikävalko, O., Espoon kaupunki/Geotekninen osasto, Virastopiha 2 C, FIN-02770 Espoo
Immonen, R., Pioneerikoulu, Puolustusvoimat, PL 1180, FIN-45101 Kouvola. T: 358-14-1812111
Isotalo, M., Takametsäntie 2 E, FIN-00620 Helsinki
Jaako, M., Geologintie 4, FIN-90570 Oulu. T: 358-8-5544645
Jaatinen, S., Oy VR-Rata Ab, Georyhmä, PL 488, FIN-00101 Helsinki. T: 358-9-7072292, F: 358-9-7073716
Jantunen, H., Koskenniskantie 5 E 63, FIN-48400 Kotka. T: 358-5-24010
Johansson, E., Ins.tsto Saanio & Riekkola Oy, Laulukuja 4, FIN-00420 Helsinki. T: 358-9-5666500, F: 358-90-5663354
Jokinen, J., Espoon kaupunki, Geotekniikkayks, Virastopiha 2 C, FIN-02770 Espoo. T: 358-9-8061, F: 358-9-8065566
Jokinen, E., Männistöntie 2 D 60, FIN-28120 Pori. T: 358-2-6327894
Jokiniemi, H., TTKK, Geotekniikan laboratorio, PL 600, FIN-33101 Tampere. T: 358-3-3652894, F: 358-3-3652884
Juhola, M.O., Lounaisväylä 2, FIN-00200 Helsinki. T: 358-9-677412
Jukarainen, O., Lujabetoni, FIN-71800 Siilinjärvi
Junnila, A., Innogeo Oy, Korppaanmäentie 22, FIN-00300 Helsinki. T: 358-9-47773732, F: 358-9-47773777
Juntunen, E., Viiriäisentie 7 as 8, FIN-28220 Pori. T: 0405139217
Jurmu, A., Anfallintie 37, FIN-02920 Espoo. T: 358-9-841552
Juvankoski, M., VTT: RTE, PL 1800, FIN-02044 VTT. T: 358-9-4564890, F: 358-9-463251
Jyllilä, H., Päivätie 16 A 1, FIN-02210 Espoo. T: 358-9-8845650
Jännes, E., Suomalainen Insinööritoimisto Oy, Pohjatie 12 A, FIN-02100 Espoo. T: 358-9-43781, F: 358-9-4378399
Järvelä, R., Helsingin kaupunki, Geotekn. os., PL 2202, FIN-00099 Helsinki Kaupunki. T: 358-9-35108230, F: 358-9-35108200
Järvenpää, H., Merivirta 7 B 24, FIN-02320 Espoo. T: 358-9-8136485
Järvi, I., Vanhatie 30 A 14, FIN-15240 Lahti
Järviö, E., Nervanderinkatu 7 B 17, FIN-00100 Helsinki. T: 358-9-444724
Jääskeläinen, R., Nurkkala, FIN-71570 Syväniemi. T: 358-17-541181
Kajander, S., Henrik Borgströmintie 5 E 36, FIN-00840 Helsinki. T: 358-9-6986007
Kalla, J., Kalamestarintie 1 E, FIN-04300 Tuusula
Kallio, T., IVO International Oy, PL 112, FIN-01601 Vantaa. T: 358-9-5081, F: 358-9-5083402

Kallionpää, T., Tiehallitus, Pasilan virastokeskus, Pl 33, FIN-00521 Helsinki. T: 358-204442144, F: 358-204442154
Kapanen, M., Isafjordinkatu 10, FIN-80140 Joensuu. T: 358-13-802488
Karstunen, M., Dep of Civil Engineering, Rankine Building, The University of Glasgow, Glasgow, G12 8LT, U.K. T: 44-141-3305208, F: 44-141-3304557
Karvonen, E., Fortum Power and Hea. Oy, Teknologia, PL 20, FIN-00048 Fortum. T: 358-104534650, F: 358-9-5632225
Kasari, T., Yrttikatu 7 A 1, FIN-33710 Tampere
Kaukonen, J., Postinkuja 6, FIN-21530 Paimio. T: 358-2-806051
Kauranne, K., Satukuja 1, F: 35, FIN-02230 Espoo. T: 358-9-8032740
Kaurila, M., Palotie, FIN-02880 Veikkola. T: 358-9-265141 (358-040-5454630)
Kejonen, A., Retkeilijäntie 9 C 22, FIN-70200 Kuopio
Kelkka, A., Jokikatu 33 D 5, FIN-06100 Porvoo. T: 358-19-5243544
Kemppinen, J., Perttulantie 8 C 43, FIN-00210 Helsinki. T: 358-9-675338
Keto, K., Fortum Engineering Oy, Rajatorpantie 8, FIN-00048 Fortum. T: 358-104532334
Kettunen, E., Itä-Suomen Viatek Oy, Kyminlinnantie 6, FIN-48600 Karhula. T: 358-5-2279500, F: 358-5-2279500
Kilpeläinen, M., Helsingin kaupunki, Geotekn. os., PL 2202, FIN-00099 Helsingin Kaupunki. T: 358-9-1694553, F: 358-9-1644555
Kinnunen, P., Jäkälätie 6, FIN-21600 Parainen. T: 358-2-4580027
Kinnunen, T., Limingantie 73 as 16, FIN-00560 Helsinki. T: 358-40-5410252
Kivekäs, L., Lohja Rudus Oy Ab, PL 49, FIN-00441 Helsinki. T: 358-9-5037357, F: 358-9-5037395
Kivelö, M., Murkelvägen 86, S-18434 Åkersberga, Sverige. T: 46-08-54067083
Kivikoski, H., VTT: RTE, PL 1800, FIN-02044 VTT. T: 358-9-4564840, F: 358-9-463251
Kivilaakso, E., Helsingin kaupunki, Kaavoitusosasto, Kansakoulukatu 3, FIN-00100 Helsinki. T: 358-9-1694271, F: 358-9-1694290
Kleemola, J., Terramare Oy, Laurinmäenkuja 3 A, FIN-00440 Helsinki. T: 358-9-613621, F: 358-9-61362700
Kohonen, E., Vesihydro Oy, Sentnerikuja 1, FIN-00440 Helsinki. T: 358-9-5650, F: 358-9-5650385
Kohtamäki, T., Rakennus Oy Lemminkäinen, PL 2, FIN-00241 Helsinki. T: 358-9-1599263, F: 358-9-1599227
Koistinen, T., Ins.tsto TJ Koistinen, Viljatie 7, FIN-15860 Hollola
Koivumäki, O., Näsiänkuja 5 A 3, FIN-36220 Suorama
Koivusalo, R., Turun kaupungin katurakennusosasto, Linnankatu 55, FIN-20100 Turku
Kolisoja, P., TTKK, Geotekniikan laboratorio, PL 600, FIN-33101 Tampere. T: 358-3-3652803, F: 358-3-3652884
Konttinen, A., Välitalontie 1 C, FIN-00660 Helsinki. T: 358-9-7546057
Koponen, H., Maa ja Vesi Oy, Itälahdenkatu 2, FIN-00210 Helsinki. T: 358-9-682661, F: 358-9-6826600
Korhonen, R., Vantaan kaupunki, rakennusvalvontavir, Kielotie 20 C, FIN-01300 Vantaa. T: 358-9-8394480
Korhonen, K.-H., Tuomaantie 10, FIN-02180 Espoo. T: 358-9-522561
Korhonen, O., Raitamaantie 11 F, FIN-00420 Helsinki. T: 358-9-5661689
Korjus, H., Lapinlahdenkatu 25 A 7, FIN-00180 Helsinki
Korkeakoski, P., Atomikatu 1 C 19, FIN-33720 Tampere. T: 358-40-5216455
Korkiala-Tanttu, L.-K., Venepellonmäki 6, FIN-02780 Espoo. T: 358-9-810456
Korkka-Niemi, K., Pampinkuja 2 as 2, FIN-20900 Turku. T: 358-2-2581902
Korpela, J., Kalkkipaadentie 2 G, FIN-00340 Helsinki. T: 358-9-488080

Korpinen, T., Geoinsinööri Oy, Gallen-Kallelankatu 8, FIN-28100 Pori. T: 358-2-6300806, F: 358-2-6300801
Korte, A., Koivutie 26 A 2, FIN-33096 Pirkkala. T: 358-3-3685786
Koskela, P., Hiidentie 2 C 36, FIN-90550 Oulu
Koskiahde, A., Imatran Voima Oy/IVO Teknologiakeskus, Rajatorpantie 8, FIN-01019 IVO. T: 358-9-85614651, F: 358-9-5632225
Koskinen, M., TTKK, Geotekniikan laitos, PL 600, FIN-33101 Tampere. T: 358-3-162861, F: 358-3-162884
Koskinen, J., Punatulkuntie 2 A 3, FIN-02660 Espoo. T: 358-9-57143210
Kosonen, S., Ristonmäenrinne 13, FIN-40500 Jyväskylä. T: 358-14-242921
Kotakorpi, J., Naavakatu 6 A 2, FIN-15950 Lahti
Kujala, K., Oulun yliopisto, Kasarmintie 8 E, FIN-90100 Oulu. T: 358-8-5534332, F: 358-8-3115781
Kulmala, H., Hämeen tiepiiri, PL 376, FIN-33101 Tampere. T: 358-204444196, F: 358-204444201
Kulman, M., TTKK, Geotekniikan laboratorio, PL 600, FIN-33101 Tampere. T: 358-3-3652857, F: 358-3-3652884
Kumila, K., VTT- RTE, PL 1800, FIN-02044 VTT. T: 358-9-4564697, F: 358-9-463251
Kuula-Väisänen, P., Elementinpolku 17 A 8, FIN-33720 Tampere
Kuusiniemi, R., Suomen ympäristökeskus, PL 140, FIN-00251 Helsinki. T: 358-9-40300556, F: 358-9-40300590
Kuusipuro, K., Nordkalk Oy Ab, FIN-21600 Parainen. T: 358-204556352, F: 358-204556083
Kuusisto, J., Etelätie 26 C, FIN-02710 Espoo. T: 358-9-5091251
Kuusola, J., Karhutie 3 C, FIN-02400 Kirkkonummi. T: 358-9-2989128
Kylänpää, H., Geotesti Oy, Satakunnankatu 24 A, FIN-33210 Tampere. T: 358-3-2468900, F: 358-3-2468914
Kämi, A., Hannuksenkuja 10 F:40, FIN-02270 Espoo
Kärkkäinen, M., LMK International, PL 18, FIN-53501 Lappeenranta. T: 7-812-3258795, F: 7-812-3150244
Laaksonen, R., VTT: RTE, PL 1805, FIN-02044 VTT. T: 358-9-4564692, F: 358-9-4567004
Lahtinen, P., SCC Viatek Oy/SGT, c/o SGT, FIN-36760 Luopioinen. T: 358-3-5361571, F: 358-3-5361584
Laihonen, E., Karvarinkatu 4 A 7, FIN-23800 Laitila. T: 358-2-855191
Laine, O., Purjehtijankuja 3 A, FIN-00570 Helsinki
Laitinen, T., Suomen Maarakentajien Keskusliitto r.y., Linnankatu 36 B 8, FIN-20100 Turku. T: 358-2-2505643, F: 358-2-2505643
Laitinen, M., Mahlatie 7, FIN-70150 Kuopio. T: 358-17-2833142
Laitinen, A., Tammipolku 8 C, FIN-36200 Kangasala
Lappalainen, K., Ajurinmäki 5 B 49, FIN-02600 Espoo
Lappalainen, E., Marjakatu 5, FIN-21200 Raisio. T: 358-2-782821
Lappalainen, V., Kotitontuntie 17 C, FIN-02200 Espoo. T: 358-9-423146
Lassila, J., Ins.tsto PSV Oy, Kalevankuja 8, FIN-90570 Oulu. T: 358-8-363222, F: 358-8-363772
Laurén, T., Kelohongantie 2 E 39, FIN-02120 Espoo. T: 358-9-466455
Lehtiniemi, R., Piisaminpolku 2, FIN-71800 Siilinjärvi. T: 358-17-426568
Lehtonen, J., Helsingin kaupungin rak.vir., Katuosasto, Kasarmikatu 21, FIN-00130 Helsinki. T: 358-9-1662612, F: 358-9-1662027
Lehtonen, J.L., Ruissalo 60, FIN-20100 Turku. T: 358-2-2589123
Leinonen, V., Geosto Oy, Porrassalmenkatu 1, FIN-50100 Mikkeli. T: 358-15-365911, F: 358-15-162227
Leinonen, J., Pikkusuontie 4 A, FIN-00670 Helsinki. T: 358-9-7542762
Leiskallio, A., Päijät-Hämeen Jätehuolto Oy, Vesijärvenkatu 21, FIN-15140 Lahti
Lempinen, T., Kesätuulentie 20, FIN-06100 Porvoo. T: 358-19-175633
Leppämäki, K., Lehmipolku 3 B 25, FIN-01360, Vantaa

Leppänen, M., Törmäniityntie 11 A 1, FIN-02710 Espoo
Leppänen, E., Taulutie 18 A, FIN-00680 Helsinki. T: 358-9-7281132
Leppänen, M., Hulaudentie 86, FIN-37500 Lempäälä. T: 358-3-3749733
Leskelä, A., IVO – International Oy, FIN-01019 IVO. T: 358-9-85612339, F: 358-9-5664156
Lindholm, J., YS – Yhdyskunta Oy, Mustionkatu 10, FIN-20750 Turku. T: 358-2-2766551, F: 358-2-2766550
Lindroos, P., Impivaarantie 10-12 A 1, FIN-02880 Veikkola. T: 358-40-5054526
Linna, H., Karstuntie 52 B 3, FIN-08100 Lohja. T: 358-19-25409
Lintu, Y., Golder Associates Oy, Ruosilankuja 3 E, FIN-00390 Helsinki. T: 358-9-5617210, F: 358-9-56172120
Liukas, J., Meripoiju 3 E 39, FIN-02320 Espoo. T: 358-9-8022457
Lojander, M., TKK, Rakennus- ja yhdyskuntatekn. os., PL 2100, FIN-02015 TKK. T: 358-9-4513733, F: 358-9-4513826
Lotvonen, S., PSV – Maa ja Vesi Oy, Kalevankuja 8, FIN-90570 Oulu. T: 358-8-5568222, F: 358-8-5569772
Loukola, E., Suomen ympäristökeskus, PL 140, FIN-00251 Helsinki. T: 358-9-40300555, F: 358-9-40300590
Luoma, R., Geobotnia Oy, Koulukatu 28, FIN-90100 Oulu. T: 358-8-373255, F: 358-8-373598
Lyytikäinen, J., GWM-Engineering Oy, Savilahdentie 6, FIN-70210 Kuopio. T: 358-17-5800818, F: 358-17-240211
Länsivaara, T., Isoistentie 11 D 8, FIN-02200 Espoo. T: 358-9-8843035
Maanpää, S., Ins.tsto Sauli Maanpää Ky, Itäpellontie 30 A, FIN-20300 Turku. T: 358-2-2395000, F: 358-2-2394078
Maijala, T., Viipurinkatu 4 A 9, FIN-00510 Helsinki
Maisala, M., Helsingin kaupunki, Geotekn. os., PL 2202, FIN-00099 Helsingin Kaupunki. T: 358-9-3510823, F: 358-9-35108200
Mali, J., Heikkiläntie 41, FIN-40270 Palokka. T: 358-14-783792
Manelius, M., Tielaitos, Tuotanto/Konsultointi, PL 157, FIN-00521 Helsinki. T: 358-20444150, F: 358-204442154
Markkanen, P., Kyyhkysmäki 10 A 10, FIN-02600 Espoo. T: 358-9-54891540
Matikainen, R., Geologian tutkimuskeskus, PL 96, FIN-02151 Espoo. T: 358-2055011, F: 358-2055012
Melander, K., Kytöpolku 6, FIN-00740 Helsinki. T: 358-9-360599
Meronen, J., Selkämerenkatu 16 B 33, FIN-00180 Helsinki. T: 358-9-6852764
Mettänen, O., Inkeroistentie 3, FIN-00950 Helsinki. T: 358-9-327886
Milén, E., Laaksokatu 7 A 8, FIN-15140 Lahti. T: 358-3-7341815
Mäkelä, H., Innogeo Oy, Korppmäentie 22, FIN-00300 Helsinki. T: 358-9-47773731, F: 358-9-47773777
Mäkeläinen, A., Oy Kreuto Ab, Vattuniemenkatu 12, FIN-00210 Helsinki. T: 358-9-6924115, F: 358-6927043
Mäkikyrö, M., Tilustie 4 C 1, FIN-90650 Oulu. T: 358-8-5304036
Mäkinen, R., Geotek Oy, PL 17, FIN-02211 Espoo. T: 358-9-613211, F: 358-9-8037715
Mäkinen, J., Tarates. T: Oy, Pilotinkatu 22, FIN-33900 Tampere. T: 358-3-3683322, F: 358-3-3683317
Määttänen, P., Temppelikatu 17 B 43, FIN-00100 Helsinki. T: 358-50-5868552
Natukka, A., Leirikatu 7 A 5, FIN-02600 Espoo. T: 358-9-4558656
Nauska, J., Revontuli 5 B, FIN-90450 Kempele
Niemelä, H., Laurinniityntie 6-8 C 25, FIN-00440 Helsinki
Niemi, O., Hirsikalliontie 15 C, FIN-02710 Espoo
Nieminen, P., FIN-36760 Luopioinen. T: 358-3-61148
Niini, H., Isonmastontie 4 A 3, FIN-00980 Helsinki. T: 358-9-314486
Niinimäki, R., Kiesitie 4, FIN-00750 Helsinki. T: 358-9-3464703
Nirhamo, J., Kalliolaaksontie 2, FIN-01200 Vantaa. T: 358-50-5379733

Nissinen, M., Lohjantie 34 B 3, FIN-03100 Nummela. T: 358-9-2223682

Nisula, J., Katajakuja 4 A, FIN-60100 Seinäjoki. T: 358-6-4140981

Niva, M., Vantaankallio 4 B 69, FIN-01730 Vantaa. T: 358-9-8786400

Norema, R., Rysätie 19, FIN-49210 Huutjärvi. T: 358-5-2201155

Norrbäck, S., Uudenkyläntie 4, FIN-02700 Kauniainen. T: 358-9-523340

Nuutilainen, O., Geobotnia Oy, Koulukatu 28, FIN-90100 Oulu. T: 358-8-373255, F: 358-8-373598

Nystén, T., Suomen ympäristökeskus, PL 140, FIN-00251 Helsinki. T: 358-9-40300441, F: 358-9-40300491

Näätänen, A., Joensuun kaupunki/Tekninen virasto, PL 148, FIN-80101 Joensuu. T: 358-13-2673500

Okko, O., VTT: RTE, PL 19041, FIN-02044 VTT. T: 358-9-4564873, F: 358-9-467927

Okko, V., Lahnaruohontie 3 B 15, FIN-00200 Helsinki. T: 358-9-676889

Oksanen, V., Geomap Oy, Mäkelänkatu 56, FIN-00510 Helsinki. T: 358-9-229011602, F: 358-9-229011610

Oksanen, J., Myrttitie 12 C 8, FIN-00720 Helsinki. T: 358-9-3511090

Olaste, A., Naalitie 8 A, FIN-01450 Vantaa. T: 358-9-8723268

Ollikainen, M., Osuuskunta Kuopion Avainosaajat, Päivärinteentie 31, FIN-70300 Kuopio. T: 358-17-158449, F: 358-17-158448

Onninen, H., VTT:- RTE, PL 1800, FIN-02044 VTT. T: 358-9-4564891, F: 358-9-463251

Orama, R., Tuulimyllyntie 1 F: 78, FIN-00920 Helsinki. T: 358-9-3493787

Oudman, A., Mätästie 3 K 109, FIN-00770 Helsinki. T: 358-9-387347

Paakkinen, I., Moinsalmentie 1854, FIN-57230 Savonlinna. T: 358-15-343108

Pajunen, H., Vantaan kaupungin rakennusvirasto, Kielotie 13, FIN-01300 Vantaa. T: 358-9-8392633, F: 358-9-8394341

Palmu, J.-P., Geologian tutkimuskeskus, PL 96, FIN-02151 Espoo. T: 358-2055011, F: 358-2055012

Palmu, H., Jukolantie 14 P 66, FIN-04200 Kerava. T: 358-9-2947860

Palmu, M., Puustellinrinne 3 C 27, FIN-00410 Helsinki

Palo, S., Ohjaajantie 11 C 26, FIN-00400 Helsinki. T: 358-9-574429

Palolahti, A., Hakarinne 2 M 165, FIN-02100 Espoo. T: 358-9-4521361

Partanen, M., SAVON TEKMI, Kiveläntie 5, FIN-70460 Kuopio. T: 358-17-2621125

Partanen, A., Lähderanta 8 H, FIN-02720 Espoo. T: 358-9-5863552

Partio, E., Putousrinne 1 D 27, FIN-01600 Vantaa. T: 358-9-531837

Patjas, E., Suomalainen Insinööritoimisto Oy, Pohjantie 12 A, FIN-02100 Espoo. T: 358-9-476111, F: 358-9-47611511

Peltomaa, P., Insinööritoimisto Perusfundamentti Oy, Askarkuja 5, FIN-00700 Helsinki. T: 358-9-3513920, F: 358-9-355822

Perttu, A., Pohjankyröntie 79, FIN-61500 Isokyrö

Petäjä, J., Vuorenpeikontie 12, FIN-06650 Hamari. T: 358-19-153223

Petäjä-Ronkainen, A.-K., Kaakkois-Suomen Ympäristökeskus, PL 1023, FIN-45101 Kouvola. T: 358-5-7544351, F: 358-5-3710893

Pihl, H., Partek Nordkalk Oy Ab, FIN-21600 Parainen. T: 358-204556539, F: 358-204556038

Pihlajamäki, J., VTT: RTE, PL 1800, FIN-02044 VTT. T: 358-9-4564688, F: 358-9-463251

Pipinen, K., Uudenmaan tiepiiri, PL 70, FIN-00521 Helsinki. T: 358-02044442807, F: 358-02044442717

Pitkäkoski, L., Ins. tsto Maatesti, Kuhatie 4 B, FIN-02170 Espoo

Pitkänen, I., Finnoontie 54 A 1, FIN-02280 Espoo 8039040

Pokki, E., Kallioruohonkuja 3, FIN-01300 Vantaa. T: 358-9-831560

Porkka, M., Ins. tsto. Pohjatekniikka Oy, Nuijamiestentie 5 B, FIN-00400 Helsinki. T: 358-9-4777510, F: 358-9-47775111

Pukkila, H., Ins.tsto Väylä Oy, Punamullantie 27, FIN-01900 Nurmijärvi. T: 358-9-2768022, F: 358-9-2768033

Punkari, S., Jäkärlän puistokatu 10 A 4, FIN-20460 Turku. T: 358-2-2536067

Puumalainen, N., Helsingin kaupunki, Geotekn. os , PL 2202, FIN-00099 Helsingin Kaupunki. T: 358-9-1694532, F: 358-9-1694555

Pylkkänen, K., Pysäkkitie 6, FIN-37550 Moisio. T: 358-3-3752832

Pyy, H., VTT: RTE, PL 1805, FIN-02044 VTT. T: 358-9-4561, F: 358-9-4567003

Pyyny, J., Kemijoki Oy, Valtakatu 9-11, FIN-96100 Rovaniemi. T: 358-16-32511, F: 358-16-3252325

Pöllä, J., Vehkatie 25-29 As. 19, FIN-04400 Järvenpää. T: 358-9-289574

Rahikainen, T., Pohjolantie 42 E, FIN-04230 Kerava. T: 358-9-2945352

Rahkonen, I., Maa ja Vesi Oy, Kenttäkatu 17 A, FIN-40700 Jyväskylä. T: 358-14-617270, F: 358-14-620924

Rajala, O., Päivänkaari 6 B, FIN-02210 Espoo. T: 358-9-880375

Rajala, M., Heinonkatu 14, FIN-53850 Lappeenranta. T: 358-5-26884

Ranta-aho, T., Ruutikuja 3 C 23, FIN-02600 Espoo. T: 358-9-5121871

Rantala, K., Helsingin kaupunki, Geotekn. os., PL 2202, FIN-00099 Helsingin Kaupunki. T: 358-9-1694529, F: 358-9-1694555

Rasmus, R., Kukkumäenraitti 4, FIN-37800 Toijala. T: 358-3-5423857

Rathmayer, H., VTT: RTE, PL 1800, FIN-02044 VTT. T: 358-9-4561, F: 358-9-463251

Raudasmaa, P., Laajakorvenkuja 2 As. 24, FIN-01620 Vantaa. T: 358-9-8783435

Rauhala, E., Signalistinkatu 19 C 21, FIN-20360 Turku. T: 358-2-481196

Rautavuoma, M., Alber. T: Perteliuksenkatu 5 D 38, FIN-01370 Vantaa

Rautio, T., Suomen Malmi Oy, Juvanteollisuuskatu 16, Pl 10, FIN-02921 Espoo. T: 358-9-8524010, F: 358-9-85240123

Ravaska, O., Teknillinen korkeakoulu, RY -osasto, Pl 2100, FIN-02015 TKK. T: 358-9-4513730, F: 358-9-4513826

Rekonen, R., Lintupiha 12 B, FIN-02660 Espoo. T: 358-9-5092979

Riekkola, R., Ins.tsto Saanio & Riekkola Oy, Laulukuja 4, FIN-00420 Helsinki. T: 358-9-5666500, F: 358-9-5663354

Riikonen, E., Sapilastie 4 b, FIN-00760 Helsinki. T: 358-9-383155

Rinne, M., Vaarinpiha 14, FIN-02400 Kirkkonummi. T: 358-9-2214110

Rinne, H., Karhekuja 6, FIN-01660 Vantaa. T: 358-9-847785

Rintala, J., Pähkinätie 7 D, FIN-00780 Helsinki

Ritola, J., VTT:- RTE, PL 19041, FIN-02044 VTT. T: 358-9-4566177, F: 358-9-467927

Roinisto, J., Kalliosuunnittelu Oy, Kellosilta 2 D, FIN-00520 Helsinki. T: 358-9-142244, F: 358-9-1454655

Roitto, J.-P., Stora Enso Oyj, Teollisuusrakentaminen, FIN-55800 Imatra. T: 358-204623196, F: 358-204624711

Roos, S., Santaharjunkuja 3, FIN-21360 Lieto As.

Ruohonen, K., Maa ja Vesi Oy, Itälahdenkatu 2, FIN-00210 Helsinki. T: 358-9-6826530, F: 358-9-6826600

Ruoppa, A., Veneentekijänkaari 3 A 4, FIN-00210 Helsinki. T: 358-9-633009, F: 358-9-633009

Rusanen, J., Kuopion kaupunki Tekninen virasto Mitt., Suokatu 42, PL 1097, FIN-70111 Kuopio. T: 358-17-185534

Ryynänen, S., Tähtikatu 30 B 28, FIN-45700 Kuusankoski. T: 358-5-3748515

Räihä, U., Saunalahdentie 11 A 4, FIN-00330 Helsinki. T: 358-9-486849

Räisänen, M., Helsingin yliopisto, Geologian laitos, PL 11, FIN-00014 Helsinki. T: 358-9-19124212, F: 358-9-19123466

Räisänen, E., Geologian tutkimuskeskus, PL 96, FIN-02151 Espoo. T: 358-2055011, F: 358-2055012

Rämö, P., Finnsementti Oy, FIN-21600 Parainen

Rönkä, E., Suomen ympäristökeskus, PL 140, FIN-00251 Helsinki. T: 358-9-40300437, F: 358-9-40300491

Saanio, T., Ins.tsto Saanio & Riekkola Oy, Laulukuja 4, FIN-00420 Helsinki. T: 358-9-5666500, F: 358-9-5663354

Saarela, J., Suomen ympäristökeskus, PL 140, FIN-00251 Helsinki. T: 358-9-40300557, F: 358-9-40300590

Saarelainen, S., VTT: RTE, PL 1800, FIN-02044 VTT. T: 358-9-4561, F: 358-9-463251

Saarelma, M., Parikkalantie 13, FIN-00920 Helsinki

Saarinen, L., Böle, FIN-02400 Kirkkonummi. T: 358-9-2984363

Sacklén, N., Ins.tsto Saanio & Riekkola Oy, Laulukuja 4, FIN-00420 Helsinki. T: 358-9-5666500, F: 358-9-5663354

Sahala, L., Geologian tutkimuskeskus, PL 96, FIN-02151 Espoo. T: 358-205502423, F: 358-2055012

Sahi, K., Ristikatu 8 A 10, FIN-33200 Tampere. T: 358-3-2148361, F: 358-3-2122089

Saksa, P., Ripusuontie 40 B, FIN-00660 Helsinki. T: 358-9-7542042

Salmelainen, J., Harmaahaikarankuja 3 N, FIN-00940 Helsinki

Salmenhaara, P., De Nee, F: Finland Oy, Itälahdenkatu 9 A, FIN-00210 Helsinki. T: 358-9-671063, F: 358-9-679311

Salo, P., Tiehallinto, Tie- ja liikennetekniikka, PL 33, FIN-00521 Helsinki. T: 358-204222145, F: 358-204222399

Salo, M., Tulustie 14, FIN-90420 Oulu. T: 358-8-5546485

Salo, R., Hernekuja 4 B 10, FIN-01300 Vantaa

Sandberg, E., Outokumpu Metals & Resources Oy, Kummunkatu 34, FIN-83500 Outokumpu. T: 358-13-5561, F: 358-13-556263

Sandin, P., Tiehallinto, Uudenmaan tiepiiri, Opastinsilta 12 B, FIN-00520 Helsinki. T: 358-9-14873616, F: 358-9-14873206

Sandström, H., Karakalliontie 14 L 75, FIN-02620 Espoo

Schmidt, H.-C., Borgmästaregatan 3, FIN-10600 Ekenäs. T: 358-19-711245

Schüller, M., Geomap Oy, Mäkelänkatu 56, FIN-00510 Helsinki. T: 358-9-229011601, F: 358-9-229011610

Selonen, O., Raskinpolku 8 D 71, FIN-20360 Turku

Seppälä, M., Lounais-Suomen Ympäristökeskus, Pl 47, FIN-20801 Turku. T: 358-2-2661794, F: 358-2-2661730

Simonen, A., Kekkurintie 1 K, FIN-15880 Hollola 3

Siren, A., Mäntykuja 8, FIN-60100 Seinäjoki. T: 358-6-134104

Sjöholm, M., Viapipe Oy, Vernissakatu 8, FIN-01300 Vantaa. T: 358-9-8230500, F: 358-9-8230501

Slunga, E., Toppelundintie 3 F: 79, FIN-02170 Espoo. T: 358-9-4128074

Smura, M., Tieliikelaitos, Konsultointi, PL 157, FIN-00521 Helsinki. T: 358-204442956, F: 358-204442929

Solovjew, N., Katajaharjuntie 20 A 3, FIN-00200 Helsinki. T: 358-9-678558

Sorkamo, M., Oy Talentek Ab, Hovioikeudenpuistikko 19 A, FIN-65100 Vaasa. T: 358-6-3208500, F: 358-6-3208501

Strandström, G., Nuijamiestentie 30 D 34, FIN-00400 Helsinki

Suhonen, S., Maa ja Vesi Oy, PL 50, FIN-01621 Vantaa. T: 358-400-412473, F: 358-9-6826600

Sundman, C., Ins.tsto Pohjatekniikka Oy, Nuijamiestentie 5 B, FIN-00400 Helsinki. T: 358-9-4777510, F: 358-9-47775111

Suokko, T., Suomen ympäristökeskus, PL 140, FIN-00251 Helsinki. T: 358-9-40281

Suominen, J., Geo-Master Oy, Fiskarsinkatu 7 C, FIN-20750 Turku. T: 358-2-530900, F: 358-2-530110

Suominen, V., Viherlaaksonranta 10 A 3, FIN-02710 Espoo. T: 358-9-593232

Taipale, H., Ruutikatu 10 B 47, FIN-02600 Espoo

Tammirinne, M., VTT: RTE, PL 1800, FIN-02044 VTT. T: 358-9-4564670, F: 358-9-463251

Tanska, H., Espoon kaupunki, Geotekniikkayks, Virastopiha 2 C, FIN-02770 Espoo. T: 358-9-8065477, F: 358-9-8065566

Tarkkala, J., Arentitie 14 as. 5, FIN-00410 Helsinki

Tarkkio, T., Valkamanpolku 4 L 5, FIN-05840 Hyvinkää. T: 358-19-434242

Tawast, I., Taivaanvuohentie 4 B A, FIN-01450 Vantaa. T: 358-9-8721567

Tenhola, M., Klovinrinne 10 A, FIN-02180 Espoo. T: 358-9-524030

Tikkanen, H., Jönsaksenkuja 3 C 22, FIN-01600 Vantaa. T: 358-9-5661175

Timonen, E., Matinraitti 7 H 99, FIN-02230 Espoo. T: 358-40-5123874

Toikka, K., Aleja Wojska Polskiego 75, PL-05520 Kostancin Jeziorna, Poland. T: 48-22-7542418

Toivanen, T., Suomalainen Insinööritoimisto Oy, Pohjantie 12 A, FIN-02100 Espoo. T: 358-9-47611605, F: 358-9-47611511

Tolla, P., Tieliikelaitos, Konsultointi, PL 157, FIN-00521 Helsinki. T: 358-204442146, F: 358-204442154

Tossavainen, M., Lohjan kaupunki, Tekn.virasto, Pl 39, FIN-08101 Lohja

Tuhola, M., VTT: RTE, PL 19041, FIN-02044 VTT. T: 358-9-4564689, F: 358-9-463251

Tuisku, T., Ins.tsto Pohjatekniikka Oy, Nuijamiestentie 5 B, FIN-00400 Helsinki. T: 358-9-4777510, F: 358-9-47775111

Tuohimaa, R., Innogeo Oy, Korppaanmäentie 22, FIN-00300 Helsinki. T: 358-9-47773733, F: 358-9-47773777

Turunen, A., Mirane. T: Oy, Huhtakoukku 3, FIN-02340 Espoo. T: 358-9-8019671, F: 358-9-8133415

Turunen, A., Linnanrakentajantie 11 C 42, FIN-00810 Helsinki. T: 358-9-7592195

Törönen, J., Vattulantie 57, FIN-34240 Kämmenniemi. T: 358-3-789550

Urmas, V., Maa ja Vesi Oy, Itälahdenkatu 2, FIN-00210 Helsinki. T: 358-9-682661, F: 358-9-6826600

Uusikartano, K., Kiertotie 27, FIN-33450 Siivikkala. T: 358-3-3466214

Uusinoka, R., TTKK, rakennusgeologian laitos, PL 600, FIN-33101 Tampere. T: 358-3-3162863, F: 358-3-3162884

Vaajasaari, M., Kesusmaa, FIN-59410 Kirjavala

Vahanne, P., VTT: RTE, PL 19041, FIN-02044 VTT. T: 358-9-4561, F: 358-9-467927

Valkeisenmäki, A., Särkiniementie 12 A 8, FIN-00210 Helsinki. T: 358-9-6925845

Valkonen, A., Pysäkkitie 6, FIN-37550 Moisio. T: 358-3-3752832

Vallius, P., Kaakkois-Suomen tiepiiri, Kauppamiehenkatu 4, FIN-45100 Kouvola. T: 358-204446373, F: 358-204446380

Valtonen, M., Ins.tsto Geotesti Oy, Satakunnankatu 24 A, FIN-33210 Tampere. T: 358-3-2146311, F: 358-3-2146319

Vanhoja, A., VR-Rata/Suunnitteluosasto, PL 488, FIN-00101 Helsinki. T: 358-9-7072280, F: 358-9-7073716

Varjo, V.-P., Taatasaarentie 11 I, FIN-23800 Laitila. T: 358-2-55768

Vaskelainen, J., Lahden tekninen virasto, PL 126, FIN-15141 Lahti. T: 358-3-8142387

Vehkaperä, H., Geopudas Oy, Oulun Geolab Oy, Eskonpolku 13, FIN-90820 Kello. T: 358-8-5402473, F: 358-8-3114915

Vehmas, H., Mikkelänahde 10, FIN-02770 Espoo. T: 358-9-8053929

Venhola, J., Suomalainen Insinööritoimisto Oy, Pohjantie 12 A, FIN-02100 Espoo. T: 358-9-43781, F: 358-9-4378399

Vepsäläinen, P., TKK, Rakennus- ja yhdyskuntatekn. os., PL 2100, FIN-02015 TKK. T: 358-9-4513731, F: 358-9-4513826

Vesa, H., Koivukuja 5, FIN-90460 Oulunsalo. T: 358-8-5212646

Vesala, E., Nuutintie 3, FIN-90650 Oulu. T: 358-8-5303925

Viitala, M., Keijuniityntie 5 C 13, FIN-02130 Espoo. T: 358-9-467116

Viitanen, P., Ruorimiehenkatu 5 B 9, FIN-02320 Espoo. T: 358-9-8033491

Wikström, R., Fundus Oy, Melkonkatu 9, FIN-00210 Helsinki. T: 358-9-615810, F: 358-9-61581420

Virtanen, P., Haavinkatu 3 B 6, FIN-21260 Raisio. T: 358-2-4350494

Volanen, N., Pakkastie 12 B, FIN-00700 Helsinki. T: 358-9-372059

Vunneli, J., Taavinharju 19 D, FIN-02180 Espoo. T: 358-9-5022630

Vuola, P., Laajaniitynkuja 1 D 54, FIN-01620 Vantaa

Vuorela, P., Hannusjärvenmäki 10, FIN-02360 Espoo. T: 358-9-8017848

Vuorimies, N., Teekkarinkatu 3 A 32, FIN-33720 Tampere

Vähäaho, I., Helsingin kaupunki, Geotekn. os., PL 2202, FIN-00099 Helsingin Kaupunki. T: 358-9-1694521, F: 358-9-1694555

Vähäkainu, P., Länsiportti 4 A 50, FIN-02210 Espoo

Vähäsarja, P., Honkavaarankuja 1 L 82, FIN-02710 Espoo. T: 358-9-590861

Wäre, M., Tammitie 8, FIN-00330 Helsinki

Väätäinen, R., Terramare Oy, PL 14, FIN-00441 Helsinki. T: 358-9-613621, F: 358-9-61362700

Ylinen, A., Tuohistanhua 4 B 7, FIN-02710 Espoo. T: 358-9-590139

Yrjänä, M., Kirjavaisenkatu 19 D 16, FIN-33560 Tampere

Ålander, C., Fundia Betoniteräkse. T: Oy, PL 24, FIN-02921 Espoo. T: 358 19 2213220, F: 358-9-8531957

Äikäs, T., Teollisuuden Voima Oy, Annankatu 42 C, FIN-00100 Helsinki

Öhberg, J., Karakalliontie 3 C 26, FIN-02620 Espoo. T: 358-9-593750

Öhberg, A., Vartiotuvantie 24, FIN-02430 Masala. T: 358-9-2977505

FRANCE

Secretary: Florence Altmayer, Comité Français de Mécanique des Sols et de Géotechnique, c/o Ponts Formation Edition, 28 rue des Saints-Pères, F-75343 Paris Cedex 07. T: +33 1 44 58 27 77, F: +33 1 44 58 27 06, E: cfms@mail.enpc.fr

Total number of members: 434

Absi, E., Bureau d'Etudes Et d'Expertise (B.E.E.), 44 Rue De Cronstadt, F-75015 Paris

Aggoun, S., IUP Génie Civil Et Infrastructures, Université De Cergy-Pontoise, 5 Mail Gay Lussac, Neuville-Sur-Oise, F-95031 Cergy Pontoise. T: +33 1 34 25 69 13, F: +33 1 34 25 69 41, E: Salima.Aggoun@iupgc.u-cergy.fr

Ah-Line, Clément, Alc Consultants, 18bis Rue Milius, F-97400 Saint-Denis, La Réunion. T: +0262 41 45 36, F: +0262 41 96 40, E: Alc.Consultants@wanadoo.fr

Albert, R., Coyne Et Bellier, 9 Allée Des Barbanniers, F-92632 Gennevilliers Cedex

Alimi Ichola, I., INSA De LYON, Lab URGC- Géotechnique, 20 Av. Albert Einstein, Bât. 304, F-69621 Villeurbanne Cedex. T: +33 4 72 43 84 63, F: +33 4 72 43 85 20

Al-Mukhtar, M., ESEM (Ec. Sup. Energie Et Matériaux), Université d'Orléans, F-45072 Orléans Cedex 2. T: +33 2 38 25 78 81, F: +33 2 38 41 70 63, E: Muzahim@cnrs-orleans.fr

Alonso, E., CEMAGREF, Groupement De Bordeaux, 50 Av. De Verdun, BP 3, F-33612 Cestas Cedex. T: +33 5 57 89 08 23, F: +33 5 57 89 08 01, E: Emmanuel.Alonso@cemagref.fr

Altmayer, F., BUREAU VERITAS, 18 Bd De l'Hôpital Stell, F-92563 Rueil-Malmaison Cedex. T: +33 1 47 52 49 27, F: +33 1 47 77 03 22, E: Florence.Altmayer@bureauveritas.com

Amar, S., LCPC (Laboratoire Central Des Ponts Et Chaussées), 58 Bd Lefebvre, F-75732 Paris Cedex 15. T: +33 1 40 43 52 62, F: +33 1 40 43 65 11, E: Samuel.Amar@lcpc.fr

Ammar-Boudjelal, A., IUP Génie Civil De La Rochelle, Univ. La Rochelle, Bât. Sciences Et Technologie, Av. Michel Crépeau, F-17042 La Rochelle. T: +33 5 46 45 82 79, F: +33 5 46 45 82 47, E: Aammar@univ-lr.fr

Andrei, A., Fondasol, Agence De Lille, 1 Rue Denis Papin, Parc Club Des Prés, F-59658 Villeneuve d'Ascq Cedex. T: +33 3 20 56 25 17, F: +33 3 20 56 20 94

Antea, Dépt Des Moyens Techniques, 3 Av. Claude Guillemin, BP 6119, F-45061 Orléans Cedex 2. T: +33 2 38 64 31 29, F: +33 2 38 64 36 43, E: E.Michalski@antea.brgm.fr

Apageo Segelm, ZA De Gomberville, BP 35, F-78114 Magny-Les-Hameaux. T: +33 1 30 52 35 42, F: +33 1 30 52 30 28, E: Apageo@compuserve.com

Arnould, M., 6 Carrière Marlé, F-92340 Bourg-La-Reine

Aste, J.P., J.P.A. Consultants, 503 Av. Du 8 Mai 1945, F-69300 Caluire. T: +33 4 78 23 22 52, F: +33 4 78 23 32 83, E: Jpaconsultants@compuserve.com

Auvinet, G., INSTITUTO DE INGENIERIA, UNAM, Ciudad Universitaria, Apdo Postal 70-472 Coyocan 4510 Mexico D.F., Mexique. T: +525 622 3500 À 04, F: +525 616 0784, E: Gauvinetg@pumas.iingen.unam mx

Averlan, J.-L., SOLS MESURES, 17 Rue Jean Monnet, ZA Des Côtes, F-78990 Elancourt. T: +33 1 30 50 34 50, F: +33 1 30 50 34 99, E: Sols.Mesures@wanadoo.fr

Ayadat, T., UNIVERSITE DE M'SILA, Institut De Génie Civil, BP 166, Chebilia, 28000 M'Sila, Algérie. T: +213 5 55 73 42, F: +213 5 55 18 36

Baguelin, F., FONDACONCEPT, 14 Rue Palestro, F-93500 Pantin. T: +33 1 41 83 19 28, F: +33 1 41 83 19 29, E: Fondaconcept.Idf@libertysurf.fr

Ballester, P., FONDASOL ETUDES, 55 Av. Louis Bréguet, Immeuble Apollo, F-31400 Toulouse Cedex. T: +33 5 61 20 55 16, F: +33 5 61 20 55 57

Bangratz, J.-L., LRPC De l'Est Parisien, DREIF-LREP, F-93351 Le Bourget. T: +33 1 48 38 81 36, F: 33 1 48 38 81 01, E: LREPBOURGET@wanadoo.fr

Bard, P.-Y., LCPC/LGIT, Observatoire De Grenoble, BP 53, F-38041 Grenoble Cedex. T: +33 4 76 82 80 61, F: +33 4 76 82 81 01, E: Pierre-Yves.Bard@obs.ujf-grenoble.fr

Bardot, F., Cabinet D'expertises BARDOT, 29 Rue Roger Bréchan, F-69003, Lyon. T: +33 4 72 33 69 61, F: +33 4 78 53 34 29, E: Bardot.Expert@wanadoo.fr

Barnoud, Fr. V., GEOTEC, 9 Bd De l'Europe, F-21800 Quetigny-Lès-Dijon. T: +33 3 80 48 93 31, F: +33 3 80 71 05 90

Barriere, A., CEBTP, CEE De Bordeaux, 105 Rue Jean Jaurès, F-33400 Talence. T: +33 5 56 80 36 11, F: +33 5 56 84 91 70, E: Cebtpso@easynet.fr

Barrois, B., SOLETANCHE BACHY, Dépt AMSOL, 6 Rue De Watford, F-92000 Nanterre Cedex. T: +33 1 47 76 54 27, F: +33 1 47 76 56 32, E: Bertrand.Barrois@soletanche-bachy.com

Barthe, O., SERVICES TECHNIQUES SEDCO-FOREX, 50 Av. Jean Jaurès, BP 599, F-92542 Montrouge Cedex. T: +33 1 47 46 68 38, F: +33 1 47 46 67 36, E: Barthe@montrouge.sedco-forex.slb.com

Bassal, J.-L., 9 Grand'rue, F-34400 Saturargues. T: +33 4 67 86 08 31, F: +33 4 67 86 08 40, E: Jean-Louis.Bassal@wanadoo.fr

Bastick, M., SPIE BATIGNOLLES, Pôle Magellan, F-95862 Cergy Pontoise Cedex. T: +33 1 34 24 36 04, E: Michel-Bastick@spiebatignolles.fr

Baudrillard, J., S.N. BAUDRILLARD-CONSULTANTS, BP 3, F-78125 La Boissière Ecole. T: +33 1 34 85 05 41, F: +33 1 34 85 08 43

Becue, J.-P., SAFEGE Ingénieurs Conseil, 15-27 Rue Du Port, BP 727, F-92007 Nanterre Cedex. T: +33 1 46 14 72 38, F: +33 1 46 14 72 31, E: Jbecue@safege.fr

Bedin, A., SORES, 12 rue des Cosmonautes, ZI du Palays, F-31400 Toulouse. T: +33 5 62 71 80 00, F: +33 5 62 71 80 05, E: soresmp@aol.com

Behnia, C., KHAKE MOSSALLAH IRAN (Terre Armée Iran), Jamalzadeh av., Shadid Gholamreza Toussi No. 7, F- Téhéran 14198, Iran. T: +98 21 920150, F: +98 21 920150

Bergin, J.-P., 9 Faubourg Sébastopol, F-31290 Villefranche-de-Lauragais. T: +33 5 61 27 06 30, F: +33 5 61 27 06 30

Bernardet, A., EEG SIMECSOL, 18 rue Troyon, F-92316 Sèvres Cedex. T: +33 1 46 23 77 95, F: +33 1 46 23 77 80, E: abernard@paris.simecsol.fr

Bernhardt, V., TERRASOL, Immeuble Hélios, 72 av. Pasteur, F-93108 Montreuil Cedex. T: +33 01 49 88 24 42, F: +33 01 49 88 06 66, E: v.bernhardt@terrasol.com

Bertaina, G., CETE de l'Ouest, LRPC d'Angers, 23 av. de l'Amiral Chauvin, BP 69, F-49136 Les Ponts de Cé Cedex. T: +33 2 41 79 13 01, F: +33 2 41 44 32 76, E: gilles.bertaina@equipement.gouv.fr

Berteaux, P., EURISK, 19 chemin de Prunay, BP 24, F-78431 Louveciennes Cedex. T: +33 1 30 78 18 62, F: +33 1 30 78 18 39, E: contact@eurisk.fr

Berthelot, P., BUREAU VERITAS, Alliance 2, 77 rue Samuel Morse, F-34000 Montpellier. T: +33 4 99 52 35 52/33 40, F: +33 4 99 52 32 50, E: patrick.berthelot@fr.bureauveritas.com

Berthou, F., TECHNOSOL S.A., route de la Grange aux Cercles, F-91160 Ballainvilliers. T: +33 1 69 09 14 51, F: +33 1 64 48 23 56, E: franck-berthou@yahoo.fr

Bescond, B., CETE Méditerranée, BP 37000, F- Aix-en-Provence Cedex 3. T: +33 4 42 24 78 53, F: +33 4 42 24 78 43, E: b.bescond@cete13.equipement.gouv.fr

Besson, C., INTRAFOR, 41-43 av. du Centre, F-78067 Saint-Quentin-Yv. Cedex. T: +33 1 39 44 85 85, F: +33 1 39 44 85 86

Biarez, J., ECOLE CENTRALE de PARIS, Grande Voie des Vignes, F-92295 Chatenay-Malabry Cedex. T: +33 1 41 13 13 50, F: +33 1 41 13 14 42

Bigot, G., LRPC de l'Est Parisien, DREIF, 319 av. Georges Clémenceau, Vaux-Le-Pénil, BP 505, F-77015, Melun Cedex. T: +33 1 60 56 64 30, F: +33 1 60 56 64 01, E: lb.lrep@wanadoo.fr

Biguenet, G., 3 rue du 19 mars 1962, F-38400 Saint-Martin d'Hères. T: +33 04 76 42 52 49

Billaux, D., ITASCA, Centre scientifique A. Moiroux, 64 chemin des Mouilles, F-69130 Ecully. T: +33 4 72 18 04 20, F: +33 4 72 18 04 21, E: d.billaux@itasca.fr

Binquet, J., COYNE et BELLIER, 9 allée des Barbanniers, F-92632 Gennevilliers Cedex. T: +33 1 41 85 03 40, F: +33 1 41 85 03 24, E: jean.binquet@coyne-et-bellier.fr

Blivet, J.-C., LRPC, CETE Normandie Centre, 10 Chemin de la Poudrière, BP 245, F-76121 Le Grand Quevilly Cedex. T: +33 02 35 68 81 63, F: +33 02 35 68 81 88

Blondeau, A., CETEN APAVE International, 191 rue de Vaugirard, F-75015 Paris. T: +33 1 45 66 17 60, F: +33 1 45 66 18 09, E: groupe@apave.com

Blondeau, F., 38 rue Boileau, F-75016 Paris

Blondeau, P., SOCOTEC, Les Quadrants, 3 av. du Centre, Guyancourt, F-78182 Saint-Quentin-Yv. Cedex

Bois, P., FONDASOL INTERNATIONAL, ZA des Amandiers, 35 rue des Entrepreneurs, F-78421 Carrières-sur-Seine Cedex. T: +33 1 39 14 77 00, F: +33 1 39 14 76 70

Bolle, G., 15 rue de Plélo, F-75015 Paris. T: +33 1 45 58 22 44, F: +33 1 45 58 22 44,

Bonaz, R., 17 square des Platanes, F-78870 Bailly

Bonin, J.-P., CAMPENON BERNARD SGE, 5 cours Ferdinand de Lesseps, F-92500 Rueil-Malmaison Cedex. T: +33 1 47 16 30 63, F: +33 1 47 16 33 80

Bonne Esperance, 11 rue des Gries, BP 4, F-67241 Bischwiller Cedex. T: +33 3 88 06 24 10, F: +33 3 88 53 90 19, E: Bonnesperance@wanadoo.fr

Bonnet, G., ENPC CERMMO, 6-8 av. Blaise Pascal, Champs-sur-Marne, F-77455 Marne-la-Vallée Cedex 2. T: +33 1 64 15 36 58, F: +33 1 64 15 37 48, bonnet@cermmo.enpc.fr

Bordes, J.-L., Expert près la Cour d'Appel de Paris, 20 rue de Madrid, F-75008 Paris. T: +33 1 42 93 87 90, F: +33 1 42 93 87 90, E: Jean-Louis.Bordes@wanadoo.fr

Borel, S., LCPC (Laboratoire Central des Ponts et Chaussées), 58 bd Lefebvre, F-75732 Paris Cedex 15. T: +33 1 40 43 52 76, F: +33 1 40 43 65 11, E: Serge.Borel@lcpc.fr

Bostvironnois, J.-L., BAUER Spezialtiefbau Gmbh, Geschäftsbereich Inland, Wittelsbacher Strasse 5, D-86529 Schrobenhausen, Allemagne. T: +49 82 52 97 16 47, F: +49 82 52 97 15 16

Botte Sade Fondations, 21 rue du Pont des Halles, Delta 112, F-94536 Rungis. T: +33 1 49 61 48 00, F: +33 1 49 61 48 01

Botte Sondages, 25 allée du Parc de Garlande, F-92220 Bagneux. T: +33 01 40 92 16 22, F: +33 01 40 92 07 14

Bouafia, A., UNIVERSITE de BLIDA, Dépt Génie Civil, route de Soumâa, BP 270, 9000 Blida, Algérie. T: +213 3 43 39 39, F: +213 3 43 39 39, E: univblida@hotmail.com

Bouchain, J., RATP/ITA/IDI/GEO/LAC P 61, 40bis rue Roger Salengro, LAC P 61, F-94724 Fontenay-sous-Bois Cedex. T: +33 1 41 95 37 98, F: +33 1 41 95 38 30, E: jean.bouchain@ratp.fr

Boucherie, M., SOCOTEC, Les Quadrants, 3 av. du Centre, F-78182 Saint-Quentin-Yv. Cedex

Boucraut, L.-M., 21 rue Bonnefoy, F-13006 Marseille. T: +33 4 91 37 19 00, F: +33 4 91 37 24 03

Boulon, M., LABORATOIRE 3S (Sols,Solides,Structures), BP 53, F-38041, Grenoble Cedex 9. T: +33 4 76 82 51 65, F: +33 4 76 82 70 43, E: boulon@hmg.inpg.fr

Bourdeau, Y., INSA de LYON, URGC Géotechnique, av. Albert Einstein, bât. 304, F-69621 Villeurbanne Cedex. T: +33 04 72 43 83 69, F: +33 04 72 43 85 20, E: bourdeau@gcu-geot.insa-lyon.fr

Bourgeois, E., LCPC (Laboratoire Central des Ponts et Chaussées), 58 bd Lefebvre, F-75732 Paris Cedex 15. T: +33 1 40 43 54 17, F: +33 1 40 43 65 11, E: Emmanuel.Bourgeois@lcpc.fr

Bouyge, B., A & M, Automates & Maintenance, Le Reclus, BP 10, F-34680 Saint-Georges d'Orques

Breysse, D., CDGA, Université Bordeaux 1, av. des Facultés, F 33405 Talence Cedex. T: +33 4 56 84 88 40, F: +33 4 56 80 71 38, E: denys.breysse@cdga.u-bordeaux.fr

Brossier, P., INGEROP, 174 bd de Verdun, F-92413 Courbevoie Cedex. T: +33 1 49 04 58 93, F: +33 1 49 04 58 91, E: paul.brossier@ingerop.com

Brucy, F., IFP (Institut Français du Pétrole), 1 & 4 av. de Bois-Préau, F-92852 Rueil-Malmaison Cedex. T: +33 1 47 52 61 43, F: +33 1 47 52 70 02, E: francoise.brucy@ifp.fr

Brun, P., EDF-CNEH, Savoie-Technolac, F-73373 Le Bourget du Lac Cedex. T: +33 4 79 60 60 28, F: +33 4 79 60 62 29, E: pierre.brun@edf.fr

Buet, EDF-DE-SQR, 905 av. du Camp de Menthe, BP 605, F-13093 Aix-en-Provence Cedex 02

Bustamante, M., LCPC (Laboratoire Central des Ponts et Chaussées), 58 bd Lefebvre, F-75732 Paris Cedex 15. T: +33 1 40 43 52 65, F: +33 1 40 43 65 11

Cambou, B., ECOLE CENTRALE de LYON, BP 163, F-69131 Ecully Cedex

Canepa, Y., LRPC de l'Est Parisien, Vaux-Le-Pénil, BP 505, F-77015 Melun Cedex. T: +33 1 60 56 64 29, F: +33 1 60 56 64 01, E: lb.lrep@wanadoo.fr

Canou, J., ENPC CERMES, 6-8 av. Blaise Pascal, Champs-sur-Marne, F-77455 Marne-la-Vallée Cedex 2. T: +33 1 64 15 35 46, F: +33 1 64 15 35 62, E: canou@cermes.enpc.fr

Capdepont, Y., D2I (Développement Ingénierie Industrie), 18 rue de Gabel, F-47300 Villeneuve-sur-Lot. T: +33 5 53 70 15 51, F: +33 5 53 41 91 93, E: yc_consulting@yahoo.fr

Carpinteiro, L., SOCOTEC, Les Quadrants, 3 av. du Centre, F-78182 Saint-Quentin-Yv. Cedex. T: +33 1 30 12 82 36, F: +33 1 30 12 83 90

Carrere, A., COYNE et BELLIER, 9 allée des Barban-niers, F-92632 Gennevilliers Cedex. T: +33 1 41 85 02 51, F: +33 1 41 85 03 74, E: alain.Carrere@coyne-et-bellier.fr

Carriere, M.-L., Avocat, 1 av. Franklin D. Roosevelt, F-75008 Paris. T: +33 1 56 59 20 00, F: +33 1 56 59 20 01, E: carriere@jas-avocats.com

Cassan, M., FONDASOL ETUDES, 290 rue des Galoubets, F-84140 Montfavet. T: +33 4 90 31 23 96

Cebtp (Centre Expérimental de Recherches et d'Etudes du Bâtiment et des Travaux Publics), Domaine de Saint-Paul, B.P. 37, 78470 Saint-Rémy-lès-Chevreuse. T: +33 1 30 85 24 00, F: +33 1 30 85 24 30, E: info@cebtp.fr

Centre D'etudes Techniques Maritimes et Fluviales, 2 bd Gambetta, BP 60039, F-60321, Compiègne Cedex. T: +33 3 44 92 60 16, F: +33 3 44 20 06 75, E: o.piet@cetmef.equipement.gouv.fr

Cetu (Centre d'Etudes des Tunnels), 25 av. François Mitterrand, Case n° 1, F-69674 Bron Cedex. T: +33 4 72 14 33 61, F: +33 4 72 14 34 30, E: pascal.dubois@cetu.equipement.gouv.fr

Chaillot, G., SNCF – Ligne nouvelle TGV Méditerranée, 1 bd Camille Flammarion, BP 22, F-13234 Marseille. T: +33 4 95 04 29 55, E: achaill@club-internet.fr

Challamel, N., GAZ DE FRANCE, Direction de la Recherche, 361 av. Président Wilson, BP 33, F-93211, Saint-Denis Cedex. T: +33 1 49 22 55 14, F: +33 1 49 22 58 91, E: noel.challamel@gazdefrance.com

Chapeau, C., LRPC de LYON, CETE de Lyon, 25 av. François Mitterrand, Case n° 1, F-69674 Bron. T: +33 4 72 14 32 66, F: +33 4 72 14 30 35

Chapuis, R.P., ECOLE POLYTECHNIQUE, Dépt. Génie Civil, Géologique et des Mines, CP 6979, succ. Centre ville, F- Montreal, Canada. T: +514 340 4711 ext 4427, F: +514 340 4477, E: rchapuis@mail.polymtl.ca

Chaput, D., CETE de l'Ouest, LRPC d'Angers, 23 av. de l'Amiral Chauvin, BP 69, F-49136 Les Ponts de Cé Cedex. T: +33 2 41 79 13 30, F: +33 2 41 44 32 76, E: daniel.chaput@equipement.gouv.fr

Charpentier, D., TECHNOSOL S.A., route de la Grange aux Cercles, F-91160 Ballainvilliers. T: +33 1 69 09 14 51, F: +33 1 64 48 23 56, E: TECHNOSOL@wanadoo.fr

Chassagne, P., ALIOS INGENIERIE, 1 rue de la Gravette, F-33320 Eysines. T: +33 5 56 28 10 07, F: +33 5 56 28 07 69, E: alios.ingenierie@wanadoo.fr

Chemali, S., SPIE FONDATIONS, 10 av. de l'Entreprise, Pôle Galilée, F-95865 Cergy Pontoise. T: +33 1 34 24 49 13, F: +33 1 34 24 37 56

Chevalier, P., SARETEC, 9 rue Thomas Edison, F-94025 Créteil. T: +33 1 49 56 84 54, F: +33 1 56 72 10 45, E: patrick.chevalier@wanadoo.fr

Chiappa, J., G F C (Géotechnique, Fondation, Contrôle), 20 chemin du Furet, F-31200 Toulouse. T: +33 5 34 25 01 50, F: +33 5 34 25 01 59

Chopin, M., MC Consulting, 50 rue du Rocher, F-75008 Paris. T: +33 1 42 93 84 30, F: +33 1 42 93 20 64, E: 101610.3637@compuserve.com

Cleaud, J.-J., GROUPE J, 124 av. Victor Hugo, F-69140 Rillieux-la-Pape. T: +33 4 78 88 75 83, F: +33 4 78 97 40 38, E: groupej@club-internet.fr

Clemenceau, P., 68 rue du Maréchal Murat, F-77340 Pontault-Combault, T/F: +33 1 64 40 52 82

Clerdouet, D., 48 bd de la Mission Marchand, F-92400 Courbevoie. T: +33 1 43 33 32 65, F: +33 1 43 33 32 65

Cognon, J.-M., MENARD SOLTRAITEMENT, 2 rue Gutenberg, F-91620 Nozay. T: +33 1 69 01 37 38, F: +33 1 69 01 75 05

Cojean, R., EMP – ENPC/CGI, 60 bd Saint-Michel, F-75272 Paris Cedex 06. T: +33 1 40 51 91 76, F: +33 1 43 26 36 56, E: cojean@cgi.ensmp.fr

Colin, J.-C., ANTEA Agence Lorraine, 1 rue du Parc de Brabois, F-54500 Vandoeuvre-lès-Nancy. T: +33 3 83 44 81 44, F: +33 3 83 44 45 36

Combarieu, O., LRPC de ROUEN, CETE Normandie, chemin de la Poudrière, BP 245, F-76121 Le Grand Quevilly. T: +33 2 35 68 81 54, F: +33 2 35 68 81 72

Compagnie Nationale Du Rhone (CNR), 2 rue André Bonin, F-69316 Lyon Cedex 04. T: +33 4 72 00 68 52, F: +33 4 72 10 66 62, E: G.Tratapel@cnr.tm.fr

Corte, J.-F., LCPC (Laboratoire Central des Ponts et Chaussées), Route de Bouaye, BP 4129, F-44341 Bouguenais Cedex. T: +33 2 40 84 58 15, F: +33 2 40 84 59 94, E: corte@lcpc.fr

Cosenza, P., UNIVERSITE PARIS VI, IST Géophysique – Géotechnique, 4 place Jussieu, Tour 22-32, 5ème étage, F-75232 Paris Cedex 05. T: +33 1 44 27 43 81/01 44 27 48 83, F: +33 1 44 27 45 88, E: cosenza@ccr.jussieu.fr

Costaz, J., COYNE et BELLIER, 9 allée des Barbanniers, F-92632, Gennevilliers Cedex. T: +33 01 41 85 03 10, F: +33 1 41 85 03 74

Coulet, C., IUT A, Dépt. Génie Civil, 43 bd du 11 Novembre 1918, F-69622 Villeurbanne Cedex. T: +33 4 72 69 21 21, F: +33 4 72 69 21 20, E: christian.coulet@iutagc.univ-lyon1.fr

Coyne et Bellier, 9 allée des Barbanniers, F-92632 Gennevilliers Cedex. T: +33 1 41 85 03 10, F: +33 01 41 85 03 74

Cui, Y.J., ENPC CERMES, 6-8 av. Blaise Pascal, Champs-sur-Marne, F-77455 Marne-la-Vallée Cedex 2. T: +33 1 64 15 35 50, F: +33 1 64 15 35 62, E: cui@cermes.enpc.fr

Cuvillier, A., CETE Nord Picardie, LRPC de Lille, 42 bis rue Marais – Sequedin, BP 99, F-59480 Haubourdin Cedex. T: +33 3 20 48 49 27, F: +33 3 20 50 55 09, E: Arnoult.Cuvillier@equipement.gouv.fr

Dagba, R., SETRA Ministère de l'Equipement, 46 av. Aristide Briand, BP 100, F-92223 Bagneux Cedex. T: +33 1 46 11 34 10, F: +33 1 46 11 31 69, E: rdagba@setra.fr

Dagnaux, J.-P., Le Galice Mirabeau – C, 11 rue Louise Colet, F-13090 Aix-en-Provence. T: +33 4 42 95 41 45, F: +33 4 42 95 41 45, E: J.P.Dagnaux@wanadoo.fr

Darve, F., LABORATOIRE 3S (Sols,Solides,Structures), BP 53, F-38041 Grenoble Cedex. T: +33 4 76 82 52 76, F: +33 4 76 82 70 00, E: Félix.Darve@hmg.inpg.fr

Dauvisis, J.-P., 4 rue Théophile Gautier, F-91600 Savigny-sur-Orge. T: +33 01 69 05 23 90

Debats, J.-M., VIBROFLOTATION S.A.R.L., 140 rue Serpentine, ZI Jalassières, F-13510 Eguyilles. T: +33 04 42 29 75 21, F: +33 04 42 59 01 53, E: jmdvibro@compuserve.com

De Buhan, P., ENPC CERMMO, 6-8 av. Blaise Pascal, Champs-sur-Marne, F-77455 Marne-la-Vallée Cedex 2

De Coninck, J., CETE Nord Picardie, LRPC de Lille, 42 bis rue Marais Sequedin, BP 99, F-59482, Haubourdin. T: +33 3 20 48 49 25, F: +33 3 20 50 55 09, E: lrpc.lille@wanadoo.fr

Decrion, M., COMPETENCE GEOTECHNIQUE, BP 94034, F-57040 Metz Cedex 1. T: +33 3 87 18 90 46, F: +33 3 87 21 00 00

Defer, G., SEFIA, 1 av. Sonia Delaunay, F-94500 Champigny-sur-Marne. T: +33 1 48 81 13 79

Degrugilliers, P., ECOLE des MINES de DOUAI, 941 rue Charles Bourseul, BP 838, F-59508 Douai Cedex. T: +33 3 27 71 24 28, F: +33 3 27 71 25 25, E: degrugilliers@ensm-douai.fr

De LA Chapelle, G., FRANKI FONDATION, 4ème rue n° 9, BP 2154, F-13847 Vitrolles. T: +33 4 42 89 09 58, F: +33 4 42 46 05 50

Delage, P., ENPC CERMES, 6-8 av. Blaise Pascal, Champs-sur-Marne, F-77455 Marne-la-Vallée Cedex 2. T: +33 1 64 15 35 42, F: +33 1 64 15 35 62, E: delage@cermes.enpc.fr

Delattre, L., LCPC (Laboratoire Central des Ponts et Chaussées), 58 bd Lefebvre, F-75732 Paris Cedex 15. T: +33 1 40 43 52 74/64, F: +33 1 40 43 65 11, E: delattre@lcpc.fr

Delmas, P., BIDIM GEOSYNTHETICS S.A., 9 rue Marcel Paul, BP 80, F-95873 Bezons Cedex. T: +33 1 34 23 53 95, F: +33 1 34 23 53 64, E: philippe.delmas@bidim.com

Demartinecourt, J.-P., HYDRO-GEOTECHNIQUE, Z.A. des Ormeaux, R.N. 6, F-71150 Fontaines. T: +33 3 85 46 70 92, F: +33 3 85 45 88 43, E: JeanPierreDemartinecourt@wanadoo.fr

Demonsablon, P., 66 rue Denfert Rochereau, F-92100 Boulogne. T: +33 1 48 25 19 62

Depardon, F., 3bis rue Le Bouvier, F-92340 Bourg-la-Reine. T: +33 1 45 36 93 92 /+33 6 85 93 20 14, E: FRANCOIS.DEPARDON@wanadoo.fr

Dervaux, M., CETE Nord Picardie, LRPC de Lille, 42 bis rue Marais Sequedin, BP 99, F-59482 Haubourdin Cedex. T: +33 3 20 48 49 19, F: +33 3 20 50 55 09

De Sloovere, P., ME2I, 4 allée des Jachères, BP 421, F-94263 Fresnes. T: +33 1 49 84 23 23, F: +33 1 49 84 23 00, E: me2i@calva.net

Desrues, J., LABORATOIRE 3S (Sols, Solides, Structures), BP 53, F-38041 Grenoble Cedex. T: +33 4 76 82 51 73, F: +33 4 76 82 70 00, E: jacques.desrues@hmg.inpg.fr

Desvarreux, P., SAGE (Société Alpine de Géotechnique), 2 rue de la Condamine, BP 17, F-38610 Gières. T: +33 4 76 44 75 72, F: +33 4 76 44 20 18, E: SAGE.INGENIERIE@wanadoo.fr

Detry, V., VERONIQUE DETRY CONSULTANT, 47 rue Raymond Marcheron, F-92170 Vanves

Dhouib, A., SOLEN ETUDES, 48-50 rue Eugénie Le Guillermic, BP 56, F-94290 Villeneuve-le-Roi. T: +33 1 45 97 17 40, F: +33 1 49 61 58 17

Diab, Y., UNIVERSITE de MARNE-LA-VALLEE, LGCU, 2 rue Einstein, F-77420 Champs-sur-Marne. T: +33 1 64 73 05 23, F: +33 1 64 73 05 21, E: ydiab@univ-mlv.fr

Di Benedetto, H., ENTPE, DGCB, rue Maurice Audin, F-69518 Vaulx-en-Velin. T: +33 4 72 04 70 65, F: +33 4 72 04 71 56, E: herve.dibenedetto@entpe.fr

Didier, G., INSA de LYON, URGC Géotechnique, 20 av. Albert Einstein, bât. 304, F-69621, Villeurbanne Cedex. T: +33 4 72 43 83 23, F: +33 4 72 43 85 20, E: geot@insa-lyon.fr

Dieudonne, A., PONTIGNAC SARL, 152 rue Henri Maurice, F-59494 Audry-du-Hainaut. T: +33 3 27 46 90 15, F: +33 3 27 46 43 85, E: pontignac@nordnet.fr

Di Nota, R., SOL PROGRES, 2 rue Louis Gousson, F-78120 Rambouillet
Dondaine, E., E.D.G., Agence Nord et Ouest, 6 rue de Watford, F-92000 Nanterre. T: +33 1 47 76 54 28, F: +33 1 49 06 97 34, E: edgparis@solétanche-bachy.cou
Dore, M., MECASOL, 43 rue Grosse Pierre, Silic 443, F-94593 Rungis Cedex. T: +33 1 46 87 20 40, F: +33 1 46 87 00 18, E: mecasol@wanadoo.fi
Dormieux, L., ENPC CERMMO, 6-8 av. Blaise Pascal, Champs-sur-Marne, F-77455 MARNE LA VALLEE Cedex 2
Dubie, J.-Y., EDF-SQR, TEGG, 905 av. du Camp de Menthe, BP 605, F-13093 Aix-en-Provence Cedex 02. T: +33 4 42 95 95 76, F: +33 4 42 95 95 00, E: jean-yves.dubie@edf.fr
Dubois, P., CETU (Centre d'Etudes des Tunnels), 25 av. François Mitterrand, Case n° 1, F-69674 Bron Cedex. T: +33 4 72 14 33 61, F: +33 4 72 14 34 30, E: pascal.dubois@cetu.equipement.gouv.fr
Dubreucq, LCPC (Laboratoire Central des Ponts et Chaussées), Centre de Nantes, BP 19, F-44340 Bouguenais
Duffaut, P., 130 rue de Rennes, F-75006 Paris. T: +33 1 45 48 91 39, F: +33 1 45 44 93 12
Dupas, J.-M., MECASOL, 43 rue de la Grosse Pierre, Silic 443, F-94593 Rungis Cedex. T: +33 1 46 87 20 40, F: +33 1 46 87 00 18, E: mecasol@wanadoo.fr
Dupla, J.-C., ENPC CERMES, 6-8 av. Blaise Pascal, Champs-sur-Marne, F-77455 Marne-la-Vallée Cedex 2. T: +33 1 64 19 39 93, F: +33 1 64 15 39 62, E: dupla@cermes.enpc.fr
Durand, L., SOL ESSAIS ETUDES, 49 rue des Sazières, F-92700 Colombes. T: +33 1 47 81 22 10, F: +33 1 47 82 52 25, E: SOL.ESSAIS.ETUDES@wanadoo.fr
Durot, D., RINCENT BTP INGENIERIE RECHERCHE, 4 rue d'Amsterdam, ZA Parisud 2 – Sénart, F-91250 Tigery. T: +33 1 69 13 80 20, F: +33 1 69 13 00 11
E.C.G.D. (Eiffage Construction, Gestion et Développement), 3 av. Morane Saulnier, F-78140 Vélizy-Villacoublay. T: +33 1 34 65 86 12, F: +33 1 34 65 87 07, E: mguerinet@fougerolle.aenix.fr
Ecole Centrale de Paris, Grande Voie des Vignes, F-92295 Chatenay-Malabry. T: +33 1 41 13 13 20, F: +33 1 41 13 14 42, E: fleureau@mss.ecp.fr
Edf-CNEH, S.-T., F-73373 Le Bourget du Lac Cedex. T: +33 4 79 60 61 78, F: +33 4 79 60 62 98, E: jean-jacques.fry@edf.fr
Edf-SQR, TEGG, 905 av. du Camp de Menthe, BP 605, F-13093 Aix-en-Provence Cedex 02. T: +33 4 42 95 95 76, F: +33 4 42 95 95 00, E: jean-yves.dubie@edf.fr
Eeg Simecsol, 18 rue Troyon, F-92316 Sèvres Cedex. T: +33 1 46 23 78 05, F: +33 1 46 23 78 07, E: direction@simecsol.fr
Eff (Entreprise Française de Fondations), 21 av. Jean Jaurès, F-69007 Lyon. T: +33 4 72 76 82 82, F: +33 4 78 61 10 88
Elmi, F., ECOLE CENTRALE de PARIS, Lab. MSS/ MAT, Grande Voie des Vignes, F-92295 Chatenay-Malabry. T: +33 1 41 13 15 62, F: +33 1 41 13 14 42, E: elmi@mss.ecp.fr
Emeriault, F., INSA de LYON, URGC Géotechnique, 20 av. Albert Einstein, bât. 304, F-69621 Villeurbanne Cedex. T: +33 4 72 43 79 26, F: +33 4 72 43 85 20, E: emeriaul@gcu-geot.insa-lyon.fr
Erg (Etudes et Recherches Géotechniques), ZI La Provençale, av. d'Estiennes d'Orves, F-83500 La Seyne-sur-Mer. T: +33 4 94 11 04 90, F: +33 4 94 30 29 71, E: erg@pacwan.fr
Erling, J.-C., Résidence Château Double, bât. 4, 3 rue Alexander Fleming, F-13090 Aix-en-Provence. T: +33 4 42 59 52 07, F: +33 4 42 59 52 07
Esta, J.-B., SOIL MECHANICS ASS, Sté ARABO EUROPEENNE TERRE ARMEE, rue Issa Maalarf, Imm Trad, Agrafieh Beyrouth, BP 165275, Beyrouth, Liban. T: +961 1 336 354, F: +961 1 330 657, E: jeanesta@cyberia.net.lb
Eurisk, 19 chemin de Prunay, BP 24, F-78431 Louveciennes Cedex. T: +33 1 30 78 18 62, F: +33 1 30 78 18 39, E: contact@eurisk.fr

Europrofil France, Division Palplanches, 17 rue Claude Chappe, F-57070 Metz. T: +33 3 82 59 11 20, F: +33 3 82 52 27 34, E: dominique.piault@profilarbed.lu
Faou, J., 96 rue Louis Rouquier, F-92300 Levallois-Perret. T: +33 1 40 87 00 21, E: Joel.Faou@wanadoo.fr
Faure, R., CETU (Centre d'Etudes des Tunnels), 25 av. François Mitterrand, F-69500 Bron. T: +33 4 72 04 34 81, F: +33 4 72 04 34 90, E: rene-michel.faure@cetu.equipement.gouv.fr
Favre, J.-L., ECOLE CENTRALE de PARIS, Grande Voie des Vignes, F-92295 Chatenay-Malabry Cedex. T: +33 1 41 13 11 22, F: +33 1 41 13 14 42, E: favre@mss. ecp.fr
Favre, M., GEOS INGENIEURS CONSEILS S.A., Bâtiment Athena, Parc d'Affaires International, F-74166 Archamps. T: +33 4 50 95 38 14, F: +33 4 50 95 99 36, E: info@geos.fr
Favreau, S., GAUDRIOT Ingénieurs Conseils, Bât. G, 7 av. Henri Becquerel, Parc Kennedy, F-33700 Mérignac. T: +33 5 56 47 88 33, F: +33 5 56 47 88 36, E: gaudriot-33@wanadoo.fr
Fayad, P., EDRAFOR, Imm. Sté Bancaire du Liban, 7ème étage, rue Hikmeh, Jdeidet El Metn, Liban. T: +961 1 878 313, F: +961 1 888 707, E: edrafor@dm.net.lb
Fayad, H., FOREX S.A.R.L., Imm. Sté Bancaire du Liban, rue de la Sagesse, F- Jdeidet El Metn, Liban. T: +961 1 878 313, F: +961 1 888 707, E: edrafor@dm. net.lb
Felix, B., ANDRA, 1-7 rue Jean Monnet, F-92298 Chatenay-Malabry. T: +33 1 46 11 80 91, F: +33 1 46 11 82 26, E: bernard-felix@andra.fr
Ferte, J.-C., QUILLE, 18 rue Henri Rivière, BP 1048, F-76112 Rouen Cedex. T: +33 2 35 14 49 40, F: +33 2 35 14 48 92, E: jean-claude.ferte@quille.fr
Flavigny, E., LABORATOIRE 3S (Sols,Solides,Structures), BP 53, F-38041 Grenoble Cedex 9. T: +33 4 76 82 51 45, F: +33 4 76 82 70 00, E: etienne.flavigny@hmg. inpg.fr
Fleureau, J.-M., ECOLE CENTRALE de PARIS, Lab. Méc. Sols Struct., Grande Voie des Vignes, F-92295 Chatenay-Malabry Cedex. T: +33 1 41 13 13 20, F: +33 1 41 13 14 42, E: fleureau@mss.ecp.fr
Florentin, P., INGEROP, 168-172 bd de Verdun, F-92413 Courbevoie Cedex. T: +33 1 49 04 56 80, F: +33 1 49 04 57 01, E: pierre.florentin@ingerop.com
Fondasol Etudes, 290 rue des Galoubets, F-84140, Montfavet. T: +33 4 90 31 23 96, F: +33 4 90 32 59 83, E: fondasol@wanadoo.fr
Fondasol International, ZA des Amandiers, 35 rue des Entrepreneurs, F-78421 Carrières-sur-Seine Cedex. T: +33 1 39 14 77 00, F: +33 1 39 14 76 70, E: fondasol@wanadoo.fr
Forni, M., 17 route de Marolles, F-94440 Santeny. T: +33 1 43 86 10 61, F: +33 1 43 86 09 14
Fosses, M., CRAMIF, 17/19 rue de Flandre, F-75935 Paris Cedex 19
Francq, J., COYNE et BELLIER, 9 allée des Barbanniers, F-92632 Gennevilliers Cedex. T: +33 1 41 85 03 64, F: +33 1 41 85 03 74, E: joël.francq@coyne-et-bellier.fr
Frank, R., ENPC CERMES, 6-8 av. Blaise Pascal, Champs-sur-Marne, F-77455 Marne-la-Vallée Cedex 2. T: +33 1 64 15 35 44, F: +33 1 64 15 35 62, E: frank@cermes.enpc.fr
Franki Fondation, 34 rue Charles Piketty, F-91170 Viry-Châtillon Cedex. T: +33 1 69 54 21 00, F: +33 1 69 54 21 10, E: Franki.Paris@wanadoo.fr
Fremond, M., LCPC (Laboratoire Central des Ponts et Chaussées), 58 bd Lefebvre, F-75732 Paris Cedex 15
Fruchart, A., SNCF – Direction de l'Ingénierie, Dépt. Etudes de Lignes Div. LGO, 122 rue des Poissonniers, F-75876 Paris
Fry, J.-J., EDF-CNEH, Savoie-Technolac, F-73373 Le Bourget du Lac Cedex. T: +33 4 79 60 61 78, F: +33 4 79 60 62 98, E: jean-jacques.fry@edf.fr
Gaboriaud, J.-M., FONDASOL, 290 rue des Galoubets, F-84035 Avignon. T: +33 4 90 31 23 96, F: +33 4 90 32 53 83, E: fondasol@wanadoo.fr
Gambin, M., 21 quai d'Anjou/9 rue Poulletier, F-75004 Paris. T: +33 01 43 54 86 46, F: +33 01 43 29 40 41, E: mgambin@magic.fr

Ganessane, B., Tour 22, Appt 135, 4 av. Henri Charon, F-91270 Vigneux-sur-Seine. T: +33 1 69 03 62 93, E: bganessane@voilà.fr

Gangneux, P., CEBTP, Dépt Géotechnique Ile-de-France, Domaine de Saint-Paul, BP 37, F-78470 Saint-Rémy-lès-Chevreuse. T: +33 1 30 85 23 88, F: +33 1 30 85 21 03

Garnier, J., LCPC (Laboratoire Central des Ponts et Chaussées) – Centre de Nantes, BP 4129, F-44341 Bouguenais Cedex. T: +33 2 40 84 58 19, F: +33 2 40 84 59 97, E: jacques.garnier@lcpc.fr

Gatmiri, B., ENPC CERMES, 6-8 av. Blaise Pascal, Champs-sur-Marne, F-77455 Marne-la-Vallée Cedex 2

Gaudin, B., SCETAUROUTE, Dépt Tunnels et Travaux Souterrains, Les Pléiades n° 35, Park Nord Annecy, F-74373 Pringy Cedex. T: +33 4 50 27 39 72, F: +33 4 50 27 39 40, E: b.gaudin@scetauroute.fr

Gaumy, Y., ALSTOM CENTRALES ENERGETIQUES S.A., 24–26 quai Alphonse Le Gallo, F-92512, Boulogne-Billancourt. T: +33 1 41 86 46 83, F: +33 1 41 86 47 14, E: yves.gaumy@energy.alstom.com

Gauthey, J.R., SPIE FONDATIONS, 10 av. de l'Entreprise, Pôle Galilée, F-95865 Cergy Pontoise Cedex

Geisler, J., SOL ESSAIS ETUDES, 49 rue des Sazières, F-92700 Colombes. T: +33 1 47 81 22 10, F: +33 1 47 82 52 25, E: Sol.Essais.Etudes@wanadoo.fr

Geolabo, 38 place de la Loire, Silic 413, F-94573 Rungis Cedex. T: +33 1 45 60 74 40, F: +33 1 45 60 74 44, E: geolabo@wanadoo.fr

Geomat Antilles, rue Ferdinand Forest, ZI de Jarry, BP 2292, 97198 Jarry Cedex, Guadeloupe. T: +33 5 90 26 83 30, F: +33 5 90 26 73 97

Gibon, P., GEOMECA, 18 place Albert Prévost, F-59175 Templemars. T: +33 3 20 60 24 00, F: +33 3 20 95 21 50

Gigan, J.-P., LRPC de l'Est Parisien, rue de l'Egalité, F-93350 Le Bourget. T: +33 1 48 38 81 17, F: +33 1 48 38 81 01

Gilbert, C., 36 route du Grand Pont, F-78110 Le Vésinet

Gilbert, Z., 36 route du Grand Pont, F-78110 Le Vésinet

Girard, H., CEMAGREF, 50 av. de Verdun, BP 3, F-33612 Cestas Cedex. T: +33 5 57 89 08 30, F: +33 5 57 89 08 01, E: hugues.girard@cemagref.fr

Givet, O., EEG SIMECSOL, 18 rue Troyon, F-92316 Sèvres Cedex

Golcheh, J., VIVALP, 27-29 bd du 11 Novembre, BP 2131, F-69603 Villeurbanne Cedex. T: +33 4 78 17 22 13, F: +33 4 72 82 04 46, E: info@vivalp.com

Gonin, H., 42 villa Brimborion, F-92190 Meudon. T: +33 1 46 26 43 14, F: +33 1 46 26 43 14

Gotteland, P., I.S.T.G., Dépt Géotechnique, BP 53, F-38041 Grenoble. T: +33 4 76 82 79 30-31, F: +33 4 76 82 79 01, E: philippe.gotteland@ujf-grenoble.fr

Goubet, R., EEG SIMECSOL, 18 rue Troyon, F-92310 Sèvres. T: +33 1 46 23 77 77, F: +33 1 46 23 77 80, E: rgoubet@paris.simecsol.fr

Gourc, J.-P., LIRIGM, Univ. Joseph Fourier – Grenoble 1, BP 53, F-38041 Grenoble Cedex 9. T: +33 4 76 82 80 90, F: +33 4 76 82 80 70, E: gourc@ujf-grenoble.fr

Gourves, R., CUST Réseau Eiffel Clermont-Ferrand, Institut des Sciences de l'Ingénieur, BP 206, F-63174 Aubière Cedex. T: +33 4 73 40 75 23, F: +33 4 73 40 74 94, E: gourves@lermes.univ-bpclermont.fr

Gouvenot, D., SOLETANCHE BACHY, 6 rue de Watford, BP 511, F-92000 Nanterre. T: +33 1 47 76 55 94, F: +33 01 49 6 97 34, E: d.gouvenot@soletanche-bachy.com

Granier, M., BUREAU VERITAS, 32-34 rue Rennequin, F-75850 Paris Cedex 17. T: +33 1 40 54 62 15, F: +33 1 47 63 19 42, E: marc.granier@fr.bureauveritas.com

Gress, J.-C., HYDRO-GEOTECHNIQUE S.A., Z.A. des Ormeaux, RN 6, F-71150 Fontaines. T: +33 3 85 45 88 44, F: +33 3 85 45 88 43

Guerinet, M., E.C.G.D. (Eiffage Construction, Gestion et Développement), 3 av. Morane Saulnier, F-78140 Vélizy-Villacoublay. T: +33 1 34 65 86 12, F: +33 1 34 65 87 07, E: mguerinet@fougerolle.aenix.fr

Gueroult, P., 6 rue Yvan Tourguenieff, F-78380 Bougival. T: +33 1 39 69 90 38

Guerpillon, Y., SCETAUROUTE, 3 rue du Dr Schweitzer, F-38180 Seyssins. T: +33 4 76 48 47 48, F: +33 4 76 48 44 47, E: y.guerpillon@scetauroute.fr

Guillaud, M., SOL EXPERT INTERNATIONAL, 285 av. Georges Clémenceau, BP 515, F-92005, Nanterre Cedex. T: +33 1 47 76 42 62, F: +33 1 47 73 92 76, E: mguillaud@sbc.fr

Guillermain, P., Cabinet GUILLERMAIN S.A., 6 rue Louis Pasteur, F-92774 Boulogne Cedex. T: +33 1 46 99 18 50, F: +33 1 46 03 00 68

Guilloux, A., TERRASOL, Immeuble Hélios, 72 av. Pasteur, F-93108 Montreuil Cedex. T: +33 1 49 88 24 42, F: +33 1 49 88 06 66, E: info@terrasol.com

Guiras, H., INSTITUT SUPERIEUR des ETUDES TECHNOLOGIQUES de NABEUL, El Merazka, 8000 Nabeul, Tunisie. T: +216 2 220 035, F: +216 2 220 033, E: skandaji@gnet.tn

Guitard, F., CEBTP, Région Sud-Ouest, 105 rue Jean Jaurès, F-33400 Talence. T: +33 5 56 80 36 11, F: +33 5 56 84 91 70, E: FrancisGuitard<cebtpso@easynet.fr>

Habib, P., LMS-X, ECOLE POLYTECHNIQUE, F-91128 Palaiseau Cedex. T: +33 1 69 33 33 81, F: +33 1 69 33 30 28, E: habib@g3s.polytechnique.fr

Haghgou, M., MECASOL, 43 rue de la Grosse Pierre, Silic 443, F-94593 Rungis Cedex

Haiun, G., SETRA-CTOA, Fondations-Soutènements, 46 av. Aristide Briand, BP 100, F-92223 Bagneux Cedex. T: +33 1 46 11 32 07, F: +33 1 46 11 33 52, E: hauin@setra.fr

Hamelin, J.-P., SOLETANCHE BACHY, 6 rue de Watford, BP 511, F-92000 Nanterre Cedex. T: +33 1 47 76 57 50, F: +33 1 49 06 97 34, E: jp.hamelin@soletanche-bachy.com

Hattab, M., LPMM-ISGMP, Ile du Saulcy, F-57045 Metz Cedex 01. T: +33 3 87 54 72 48, F: +33 3 87 31 53 66, E: hattab@lpmm.univ—metz.fr

Haza, E., LCPC Centre de Nantes, route de Bouaye, BP 4129, F-44341 Bouguenais Cedex. T: +33 2 40 84 58 21

Heintz, R., EURASOL S.A., bd Dr Charles Marx, Luxembourg 2130, Luxembourg. T: +352 48 94 42, F: +352 48 90 30

Henry, J.-Y., IUT Dépt Génie Civil, 1230 rue de l'Université, BP 819, F-62408 Béthune. T: +33 3 21 63 23 35

Holtzer, F., ARUP Geotechnic, 13 Fitzroy Street, London W1P 6BQ, Royaume Uni. T: +44 171 636 1531

Horaist, R., SAFEGE, 15-27 rue du Port, BP 727, F-92007 Nanterre. T: +33 1 46 14 72 33, F: +33 1 46 14 72 31

Hurtado, J., 6 rue Greffulhe, F-75008 Paris. T: +33 1 42 68 07 80

Hydro-Geotechnique S.A., Z.A. des Ormeaux, RN 6, F-71150 Fontaines. T: +33 3 85 45 88 44, F: +33 3 85 45 88 43 E: HYDROGEO_centre@compuserve.com

Ialynko, P., EEG SIMECSOL, 18 rue Troyon, F-92316 Sèvres Cedex. T: +33 1 46 01 24 06, F: +33 1 46 32 62 62, E: pialynko@plessis.simecsol.fr

Ialynko, P., GEOLABO, 38 place de la Loire, Silic 413, F-94573 Rungis Cedex. T: +33 1 45 60 74 40, F: +33 1 45 60 74 44, E: geolabo@wanadoo.fr

Insa de Lyon, URGC/Géotechnique, 20 av. Albert Einstein, bât. 307, F-69621 Villeurbanne Cedex. T: +33 4 72 43 83 21, F: +33 4 72 43 85 20, E: pierre.lareal@insa-lyon.fr

Intrafor, 41-43 av. du Centre, F-78067 Saint-Quentin-Yv. Cedex. T: +33 1 39 44 85 85, F: +33 1 39 44 85 86, E: intrafor.dt@wanadoo.fr

Iskandar, A., BOTTE SONDAGES, 25 allée du Parc de Garlande, F-92220 Bagneux. T: +33 1 40 92 16 22, F: +33 1 40 92 07 14

Isnard, A., 47bis rue Hippolyte Maindron, F-75014 Paris. T: +33 1 40 44 52 58, F: +33 1 40 44 52 58

Jacquard, C., SORES, Le Puech Radier – Lot n° 6, rue Montels Eglise, F-34979 Lattes Cedex 02. T: +33 4 67 06 04 90, F: +33 4 67 92 98 19, E: soresmp@aol.com

Jacquet, D., FRANKI FONDATION, Agence Rhône Alpes, 11 rue du Dôme, ZA, F-69630 Chaponost. T: +33 4 78 56 72 30, F: +33 4 78 56 72 39, E: FRANKI.Lyon@wanadoo.fr

Jardin, J., 7 rue de l'Orangerie, F-78000, Versailles

Jossinet, C., SICSOL Géotechnique S.A., Parc d' Activités Clément Ader, 19 rue Louis Bréguet, F-34830 Jacou. T: +33 4 67 59 40 10, F: +33 4 67 59 23 30, E: sicsol@wanadoo.fr

Jouanna, P., 765 chemin de la Tramontane, Montferrier sur Lez, F-34980 Saint-Gely du Fesc

Juillie, Y., B.E. ACCOTEC, 3 chemin des Passiflores, F-91190 Gif-sur-Yvette. T: +33 1 64 46 45 23, F: +33 1 69 07 53 74

Kamgueu, V., BP 10 019, F-Yaoundé, Cameroun. T: +237 20 72 22, F: +237 20 92 62

Kassab, T., 11 av. Garreau, F-92700 Colombes. T: +33 1 46 49 92 42, F: +33 1 46 49 92 42

Kazmierczak, J.-B., TERRASOL, 72 av. Pasteur, Immeuble Helios, F-93108 Montreuil Cedex. T: +33 1 49 88 24 42, F: +33 1 49 88 06 66, E: jb.kazmierczak@terrasol.com

Keller Fondations Speciales, Espace Plein Ciel, allée de l'Europe, BP 5, F-67960 Entzheim. T: +33 3 88 59 92 00, F: +33 3 88 59 95 90, E: marketing@keller-france.com

Kerisel, J., 28 av. d'Eylau, F-75116 Paris. T: +33 1 44 05 16 11, F: +33 1 45 53 30 71, E: jean.kerisel@wanadoo.fr

Khay, M., CETE Normandie Centre, Centre d'Expérimentation Routière, 10 Chemin de la Poudrière, BP 245, F-76121 Le Grand Quevilly Cedex. T: +33 2 35 68 82 12, F: +33 2 35 68 81 21, E: cetenc.cer@wanadoo.fr

Khemissa, M., Centre Universitaire Mohamed Boudiaf, BP 166, route d'Ichbilia, F-28003 M'Sila, Algérie. T: +213 5 55 18 36, F: +213 5 55 04 04

Kim, M.-S., Cap Massy 133, 5 rue Eric Tabarly, F-91300 Massy. E: msk.phoenix@magic.fr

Kovarik, J.-B., PORT AUTONOME de ROUEN, 34 bd de Boisguilbert, BP 4075, F-76022 Rouen Cedex. T: +33 2 35 52 54 20, F: +33 2 35 52 55 04

Kretz, A., SOLETANCHE BACHY, Agence AMSOL, 6 rue de Watford, F-92000 Nanterre. T: +33 1 47 76 42 62, F: +33 1 47 75 99 10, E: andre.kretz@soletanche-bachy.com

Labanieh, S., LABORATOIRE 3S (Sols, Solides, Structures), BP 53, F-38041 Grenoble Cedex 9. T: +33 4 76 82 51 57, F: +33 4 76 82 70 00, E: safwan.labanieh@hmg.inpg.fr

Lamotte, S., SOLS et FONDATIONS, 1038 rue de la Fontaine, F-45200 Amilly. T: +33 2 38 89 32 00, F: +33 2 38 89 30 90

Lanier, J., LABORATOIRE 3S (Sols, Solides, Structures), BP 53, F-38041 Grenoble Cedex 9. T: +33 4 76 82 52 88, F: +33 4 76 82 70 00, E: jack.lanier@hmg.inpg.fr

Lareal, P., INSA de LYON, URGC/Géotechnique, 20 av. Albert Einstein, bât. 307, F-69621 Villeurbanne Cedex. T: +33 4 72 43 83 21, F: +33 4 72 43 85 20, E: pierre.lareal@insa-lyon.fr

Launay, J., DUMEZ-GTM, 57 av. Jules Quentin, F-92022 Nanterre. T: +33 1 41 91 42 25, F: +33 1 41 91 45 87, E: jeanlaunay%dzgtm@dumez-gtm.fr

Lcpc (Laboratoire Central des Ponts et Chaussées), 58 bd Lefebvre, F-75732 Paris Cedex 15. T: +33 1 40 43 50 00, F: +33 1 40 43 54 98, E: Jacaues.Roudier@lcpc.fr

Lebegue, Y., 35bis rue Henri Barbusse, F-75005 Paris. T: +33 1 43 54 96 50

Leblais, Y., EEG SIMECSOL, 18 rue Troyon, F-92316, Sèvres Cedex. T: +33 1 46 23 77 90, F: +33 1 46 23 78 07, E: direction@simecsol.fr

Leblanc, J., Résidence les Marronniers, 6 av. de la Gare, F-91570 Bièvres. T: +33 1 69 41 12 54

Leca, E., SCETAUROUTE, Dépt. Tunnels et Travaux Souterrains, Les Pleïades n° 35, Park Nord Annecy, F-74373 Pringy Cedex. T: +33 4 50 27 39 54, F: +33 4 50 27 39 33, E: e.leca@scetauroute.fr

Ledoux, J.-L., CETE du Sud Ouest, LRPC de Bordeaux, 24 rue Carton, F-33019 Bordeaux. T: +33 5 56 70 63 61, F: +33 5 56 70 63 33, E: jledoux@cete33.equipement.gouv.fr

Lefebvre, A., MECASOL, 43 rue de la Grosse Pierre, Silic 443, F-94593 Rungis Cedex. T: +33 1 46 87 20 40, F: +33 1 46 87 00 18, E: mecasol@wanadoo.fr

Lefevre, F., EDF, Service Ingénierie Marseille, 82 av. de Hambourg, F-13008 Marseille. T: +33 4 91 74 91 04, E: Francis-E.Lefevre@edfgdf.fr

Lefevre, A., SNCF-Dir.Ingénierie, Dépt. des Ouvrages d'Art du Patrimoine, 122 rue des Poissonniers, F-75876 Paris Cedex 18. T: +33 1 55 31 16 35, F: +33 1 55 31 18 62

Legendre, Y., SOLETANCHE BACHY FRANCE, 6 rue de Watford, BP 511, F-92000 Nanterre. T: +33 1 47 76 42 62, F: +33 1 47 75 99 10, E: yves.legendre@soletanche-bachy.com

Le Guernic, J., SOGEO EXPERT, 3 rue Leclanché, F-86012 Poitiers. T: +33 5 49 37 92 86

Lejeune, J.-M., 15 bd Maréchal Leclerc, F-38000 Grenoble

Lerat, P., LMSGC – LCPC-CNRS, Cité Descartes, 2 allée Képler, F-77420 Champs-sur-Marne. T: +33 1 40 43 54 68, F: +33 1 40 43 54 50, E: patrick.lerat@lcpc.fr

Lerau, J., INSA de TOULOUSE, Complexe scientifique de Rangueil, F-31077 Toulouse Cedex 4. T: +33 5 61 55 99 01, F: +33 5 61 55 99 00, E: jacques.lerau@insa-tlse.fr

Le Tirant, P., 13 rue des Primevères, F-92500 Rueil-Malmaison. T: +33 1 47 51 15 46

Levacher, D., M2C-GRGC-UPRES A 6143 CNRS (Morphodynamique Continentale et Côtière), 24 rue des Tilleuls, F-14000 Caen. T: +33 2 31 56 57 09, F: +33 2 31 56 57 57, E: levacher@geos.unicaen.fr

Levillain, J.-P., EEG SIMECSOL, Agence de Nantes, 17 place Magellan, Le Ponant 2, Zone Atlantis, F-44812 Saint-Herblain Cedex. T: +33 2 40 92 76 21, F: +33 2 40 92 11 31, E: jplevill@nantes.simecsol.fr

Levy, R., 21 rue Labelonye, F-78400 Chatou. T/F: +33 1 30 53 23 96

Lewandowska, J., Laboratoire LTHE, Domaine Universitaire, BP 53, F-38041 Grenoble Cedex 9. T: +33 4 76 82 70 52, F: +33 4 76 82 52 86, E: Jolanta.Lewandowska@hmg.inpg.fr

Liausu, P., MENARD SOL TRAITEMENT, 2 rue Gutenberg, BP 28, F-91620 Nozay. T: +33 1 69 01 37 38, F: +33 1 69 01 75 05, E: menard.sol@wanadoo.fr

Lijour, P., PORT AUTONOME de NANTES/SAINT-NAZAIRE, 18 quai Ernest Renaud, BP 18609, F-44186 Nantes Cedex 04

Lizzi, F., Via Camillo De Nardis n° 7, F-80127 Naples, Italie. T: +39 81 560 43 91, F: +39 81 560 43 91, E: LIZZI@na.infn.it

Londez, M., MECASOL, 43 rue de la Grosse Pierre, Silic 443, F-94593 Rungis Cedex. T: +33 1 46 87 20 40, F: +33 1 46 87 00 18, E: mecasol@wanadoo.fr

Lossy, D., BONNE ESPERANCE, 11 rue des Gries, BP 4, F-67241 Bischwiller Cedex. T: +33 3 88 06 24 10, F: +33 3 88 53 90 19, E: Bonnesperance@wanadoo.fr

Loudiere, D., ENGEES, 1 quai Koch, BP 1039, F-67070 Strasbourg Cedex. T: +33 3 88 24 82 82, F: +33 3 88 37 04 97, E: loudiere@engees.u-strasbg.fr

Luong, M. P., CNRS – LMS-X, ECOLE POLYTECHNIQUE, F-91128, Palaiseau Cedex. T: +33 1 69 33 33 68, F: +33 1 69 33 30 26, E: luong@lms.polytechnique.fr

Magnan, J.-P., LCPC (Laboratoire Central des Ponts et Chaussées), 58 bd Lefebvre, F-75732 Paris Cedex 15. T: +33 1 40 43 52 60, F: +33 1 40 43 65 11, E: magnan@lcpc.fr

Mahe, A., ECOLE CENTRALE de NANTES, Lab. Mécanique des Solides, 1 rue de la Noë, BP 92101, F-44072 Nantes Cedex 3. T: +33 2 40 37 16 64, F: +33 2 40 74 74 06, E: amahe@oceanet.fr

Maiolino, S., LMS-X, ECOLE POLYTECHNIQUE, F-91128 Palaiseau Cedex. T: +33 1 69 33 36 96, F: +33 1 69 33 30 26, E: maiolino@lms.polytechnique.fr

Maleki, K., GIP G3S, ECOLE POLYTECHNIQUE, F-91128 Palaiseau Cedex. T: +33 1 69 33 35 30, F: +33 1 69 33 30 28, E: maleki@g3s.polytechnique.fr

Maleki, K., TECHNOSOL S.A., F-91160 Ballainvilliers. T: +33 1 69 09 14 51

Manojlovic, J., CEMAGREF, DEAN, Parc de Tourvoie, BP 44, F-92163 Antony. T: +33 1 40 96 60 44, F: +33 1 40 96 62 70, E: jovan.manojlovic@cemagref.fr

Marchal, J., TAI (Terre Armée International), Parc des Erables, bât. 4, 66 route de Sartrouville, F-78230 Le Pecq

Marcie, C., TECHNOSOL S.A., route de la Grange aux Cercles, F-91160 Ballainvilliers. T: +33 1 69 09 14 51, F: +33 1 64 48 23 56

Margarit, P., CEBTP Normandie, route des Gabions, BP 96, F-76700 Harfleur. T: +33 2 35 25 12 65, F: +33 2 35 25 12 64, E: Cebtpoue@easynet.fr

Mariotti, G., Chemin du Guazzore, F-20260 Calvi. T: +33 4 95 65 42 07, F: +33 4 95 65 42 07

Martial, G., Sté du Canal de Provence et d'Aménagement de la Région Provençale, Le Tholonet, BP 100, F-13603 Aix-en-Provence Cedex 1. T: +33 4 42 66 70 00, F: +33 4 42 66 70 80, E: SCP.Ingenierie@wanadoo.fr

Mascarelli, J.-P., SOL ESSAIS S.A., 460 av. Jean Perrin, F-13851 Aix-en-Provence Cedex. T: +33 4 42 39 74 85, F: +33 4 42 39 73 91

Masrouri, F., INSTITUT NAT. POLYTECHNIQUE DE LORRAINE, LAEGO, rue Doyen Marcel Roubault, BP 40, F-54501 Vandoeuvre-lès-Nancy. T: +33 3 83 59 63 04, F: +33 3 83 59 63 00, E: farimah.masrouri@ensg.u-nancy.fr

Massonnet, R., FONDASOL ETUDES, 290 rue des Galoubets, F-84140 Montfavet. T: +33 4 90 31 23 96, F: +33 4 90 32 59 83, E: fondasol@wanadoo.fr

Mathieu, P., INSA de LYON, URGC/Géotechnique, 20 av. Albert Einstein, bât. 304, F-69621 Villeurbanne Cedex. T: +33 4 72 43 83 70, F: +33 4 72 43 85 20, E: mathieu@gcu-geot.insa-lyon.fr

Mativat, F., 10 Oxford Road, Sidcup Kent DA14 6LW, Royaume Uni

Mattiuzzo, J.-L., SEGG (Société d'Etudes Géotechniques et Géophysiques, Savoie Technolac, BP 230, F-73375 Le Bourget du Lac. T: +33 4 79 25 35 80, F: +33 4 79 25 35 90, E: jl.mattiuzzo@segg.fr

Maurel, C., SETRA-CTOA, Fondations-Soutènements, 46 av. Aristide Briand, BP 100, F-92223 Bagneux Cedex. T: +33 1 46 11 31 93, F: +33 1 46 11 33 52, E: cmaurel@setra.fr

Mayeux, H., ERG (Etudes et Recherches Géotechniques), ZI La Provençale, av. d'Estiennes d'Orves, F-83500 La Seyne-sur-Mer. T: +33 4 94 11 04 90, F: +33 4 94 30 29 71, E: erg@pacwan.fr

Mazare, B., EEG SIMECSOL, Agence de Lyon, 17 rue Louis Guérin, F-69626 Villeurbanne Cedex. T: +33 4 78 89 81 18, F: +33 4 78 94 36 96, E: lyon@eeg-simecsol.com

Mecasol, 43 rue de la Grosse Pierre, Silic 443, F-94593 Rungis Cedex. T: +33 1 46 87 20 40, F: +33 1 46 87 00 18, E: mecasol@wanadoo.fr

Menard, J., CETE Méditerranée, BP 37000, F-13791 Aix-en-Provence Cedex 3

Menard Soltraitement, 2 rue Gutenberg, F-91620 Nozay. T: +33 1 69 01 37 38, F: +33 1 69 01 75 05 E: menard.sol@wabadoo.fr

Menasri, A., CERMES, 6-8 av. Blaise Pascal, Champs-sur-Marne, F-77455, Marne-la-Vallée Cedex 2. T: +33 1 64 15 35 66, F: +33 1 64 15 35 62

Meneroud, J.-P., CETE Méditerranée, Service Géologie Sols, 56 bd de Stalingrad, F-6300 Nice. T: +33 4 92 00 81 56, F: +33 4 92 00 81 99, E: jp.meneroud@cete13. equipement.gouv.fr

Mercieca, G., CEBTP, 109 rue du 1er Mars 43, BP 1032, F-69612 Villeurbanne. T: +33 4 72 33 08 46, F: +33 4 72 34 61 54, E: cebtprab@easynet.fr

Mercier, A., APAGEO SEGELM, ZA de Gomberville, BP 35, F-78114 Magny-les-Hameaux. T: +33 1 30 52 35 42, F: +33 1 30 52 30 28, E: apageo@compuserve.com

Mermet, J.-P., BIDIM GEOSYNTHETICS S.A., 9 rue Marcel Paul, BP 80, F-95870 Bezons Cedex. T: +33 1 34 23 53 63, F: +33 1 34 23 53 98

Merrien-Soukatchoff, V., ECOLE des MINES de NANCY., LAEGO, Parc de Saurupt, F-54042 Nancy Cedex. T: +33 3 83 58 42 92, F: +33 3 83 53 38 49, E: merrien@mines. u-nancy.fr

Mestat, P., LCPC (Laboratoire Central des Ponts et Chaussées), 58 bd Lefebvre, F-75732 Paris Cedex 15. T: +33 1 40 43 52 68, F: +33 1 40 43 65 11, E: Philippe.Mestat@lcpc.fr

Meunier, J., IFREMER Centre de Brest, DITI/GO/ MSG, BP 70, F-29280 Plouzané. T: +33 2 98 22 41 46, F: +33 2 98 22 46 50, E: jmeunier@ifremer.fr

Michalski, E.R., ANTEA, Agence Rhône Alpes, Le Parc Lyonnais, 392 rue des Mercières, F-69140 Rillieux-la-Pape. T: +33 4 37 85 19 90, F: +33 4 37 85 19 61, E: e.michalski@antea.brgm.fr

Michel, M.C., 98 rue de la Saussaie, F-94320 Thiais

Mieussens, C., LRPC de TOULOUSE., 1 av. du Colonel Roche, F-31400 Toulouse. T: +33 5 62 25 97 11, F: +33 5 62 25 97 98, E: cmieussens@cete33.equipement. gouv.fr

Millan, A., SETRA, 46 av. Aristide Briand, BP 100, F-92223 Bagneux Cedex

Modaressi, H., BRGM (Bureau de Recherches, Géologiques et Minières), av. Claude Guillemain, BP 6009, F-45060 Orléans la SOURCE. T: +33 2 38 64 30 73, F: +33 2 38 64 35 94, E: h.modaressi@brgm.fr

Monek, G., SOLETCO Nord-Est, Espace Valentin, BP 3053, F-25046 Besançon Cedex. T: +33 3 81 80 73 24, F: +33 3 81 85 03 33

Monnet, A., GEOCONSEIL, 10 av. Newton, F-92350 Le Plessis Robinson. T: +33 1 46 01 24 71, F: +33 1 46 01 24 73, E: geoconseil@wanadoo.fr

Monnet, J., LIRIGM, Université Joseph Fourier, BP 53, F-38041 Grenoble Cedex 9. T: +33 4 76 82 80 75, F: +33 4 76 82 80 70, E: jmonnet@ujf-grenoble.fr

Montenoise, J.J., SEFIA Ingénieurs Conseils, 1 av. Sonia Delaunay, F-94500 Champigny-sur-Marne

Morbois, A., SCETAUROUTE Ile-de-France et Ouest, 11 av. du Centre, St-Quentin-en-Yvelines, F-78286 Guyancourt Cedex. T: +33 1 30 48 46 03, F: +33 1 30 48 45 09

Morel, A., EDF-SEPTEN, Etat-Major, 12-14 av. Dutriévoz, F-69628 Villeurbanne Cedex. T: +33 4 72 82 74 31, F: +33 4 72 82 76 92, E: alain.morel@edf.fr

Mosse, J., 10 rue d'Avenay, F-51160 Germaine. T: +33 3 26 52 77 63, F: +33 3 26 59 24 31

Moulin, G., ECOLE CENTRALE de NANTES, 1 rue de la Noë, BP 92101, F-44321 Nantes Cedex 3. T: +33 2 40 37 16 63, F: +33 2 40 37 25 35

Moury, N., GEOCONSEIL, 10 av. Newton, F-92350, Le Plessis Robinson. T: +33 1 46 01 24 71, F: +33 1 46 01 24 73, E: geoconseil@wanadoo.fr

Moussouteguy, N., ALIOS INGENIERIE, Univ. Bx I – CDGA., 1 rue de la Gravette, F-33320 Eysines. T: +33 5 56 28 10 07, F: +33 5 56 28 07 69, E: nathalie.moussouteguy@wanadoo.fr

Musso, J., AGGLOCENTRE, 48 quai du Nouveau Port, BP 135, F-71305 Montceau-les-Mines Cedex. T: +33 3 85 58 58 84, F: +33 3 85 58 05 51

Nauroy, J.-F., IFP (Institut Français du Pétrole), 1 et 4 av. de Bois-Préau, F-92852 Rueil-Malmaison Cedex. T: +33 1 47 52 66 74, F: +33 1 47 52 70 63, E: j-francois.nauroy@ifp.fr

Neumann, C., FONDASOL INTERNATIONAL, ZA des Amandiers, 35 rue des Entrepreneurs, F-78421 Carrières-sur-Seine Cedex. T: +33 1 39 14 77 00, F: +33 1 39 14 76 70

Nguyen Thanh, L., LCPC (Laboratoire Central des Ponts et Chaussées), 58 bd Lefebvre, F-75732 Paris Cedex 15. T: +33 1 40 43 52 58, F: +33 1 40 43 54 98/65 16, E: long@lcpc.fr

Nibel, D., BOTTE SADE Fondations, 21 rue du Pont des Halles, Delta 112, F-94536 Rungis. T: +33 1 49 61 48 25, F: +33 1 49 61 48 01, E: dnibel@campenon.com

Olivari, G., ECOLE CENTRALE de LYON, BP 163, F-69131 Ecully

Oudin, M., ECOLE du BATIMENT et desTRAVAUX PUBLICS, 18 rue de Belfort, F-94307 Vincennes Cedex. T: +33 1 48 08 11 21, F: +33 1 43 98 96 87

Pal, O., FOREZIENNE D'ENTREPRISES, 7–9 rue Grangeneuve, BP 48, F-42002 Saint-Etienne. T: +33 4 77 43 37 68, F: +33 4 77 43 37 76, E: saint-etienne@forezienne-entreprises.fr

Panet, M., EDF DRD MTC, 1 route de Sens, F-77818 Moret-sur-Loing. T: +33 1 60 73 64 85, F: +33 1 60 73 65 59 ou 61 49, E: michel.panet@edf.fr

Panet, M., EEG SIMECSOL, 18 rue Troyon, F-92316 Sèvres Cedex. T: +33 1 46 23 77 90, F: +33 1 45 78 78 07, E: direction@simecsol.com

Pantet, A., ESIP (Ecole Supérieure d'Ingénieurs de Poitiers), 40 av. du Recteur Pineau, F-86022 Poitiers Cedex. T: +33 5 49 45 35 25, F: +33 5 49 45 44 44, E: anne.pantet@esip-univ.poitiers.fr

Parain, P., GEOMAT ANTILLES, rue Ferdinand Forest, ZI de Jarry BP 2292, F-97198 Jarry Cedex. T: +33 05 90 26 83 30, F: +33 05 90 26 73 97

Parez, L., 4 place de Mexico, F-75116 Paris. T: +33 1 47 27 75 79, F: +33 1 47 27 75 79

Pastor, J., ESIGEC Université de Savoie, Laboratoire LaMaCo, ESIGEC Chartreuse, F-73376 Le Bourget du Lac. T: +33 4 79 75 87 22, F: +33 4 79 75 86 65, E: joseph.pastor@univ-savoie.fr

Paumier, A., LRPC de l'Est Parisien, rue de l'Egalité prolongée, F-93351 Le Bourget Cedex

Pecker, A., GEODYNAMIQUE ET STRUCTURE, 157 rue des Blains, F-92220 Bagneux. T: +33 1 46 65 00 11, F: +33 1 46 65 58 54, E: pecker_geodyn@wanadoo.fr

Peignaud, M., 7 square Jeanne d'Arc, F-49100 Angers. T: +33 2 41 87 30 24, F: +33 2 41 87 30 24

Pellet, F., LABORATOIRE 3S (Sols, Solides, Structures), BP 53, F-38041 Grenoble Cedex 9. T: +33 4 76 82 70 23, F: +33 4 76 82 70 00, E: frederic.pellet@hmg.inpg.fr

Pernot, M., SOLETANCHE BACHY, 6 rue de Watford, F-92000, Nanterre. T: +33 1 47 76 42 62, E: michel.pernot@soletanche-bachy.com

Perrin, J., INGEVAL, Le Burizet, VEYSSILIEU, F-38460 Crémieu. T: +33 4 74 90 32 57, F: +33 4 74 90 21 72

Pfefer, D., SOBESOL, 10 av. Newton, BP 63, F-92354 Le Plessis Robinson. T: +33 1 46 01 81 10, F: +33 1 46 01 81 19, E: sobesol@easynet.fr

Philipponnat, G., Villa La Topia, 60 Val du Careï, F-6500 Menton, F: +33 4 93 35 83 83, E: latopia@club-internet.fr

Piault, D., EUROPROFIL FRANCE, Division Palplanches, 17 rue Claude Chappe, F-57070 Metz. T: +33 3 82 59 11 20, F: +33 3 82 52 27 34, E: dominique.piault@profilarbed.lu

Piault, D., INTERNATIONAL SHEET PILING CY (ISPC), 66 rue de Luxembourg, F-L-4009 Esch-sur-Halette, Luxembourg. T: +352 5313 3245, F: +352 5313 32 90, E: dominique.piault@profilarbed.lu

Piet, O., CENTRE D'ETUDES TECHNIQUES MARITIMES ET FLUVIALES, 2 bd Gambetta, BP 60039, F-60321 Compiègne Cedex. T: +33 3 44 92 60 16, F: +33 3 44 20 06 75, E: o.piet@cetmef.equipement. gouv.fr

Pilot, G., LCPC (Laboratoire Central des Ponts et Chaussées), 58 bd Lefebvre, F-75732 Paris Cedex 15. T: +33 1 40 43 50 28, F: +33 1 40 43 54 92, E: georges.pilot@lcpc.fr

Pincent, B., EEG SIMECSOL, 18 rue Troyon, F-92316 Sèvres Cedex. T: +33 1 46 23 78 02, F: +33 1 46 23 77 80, E: direction@simecsol.fr

Plas, F., ANDRA, Direction Scientifique, Parc de la Croix Blanche, F-92298 Chatenay-Malabry Cedex. T: +33 1 46 11 81 78, F: +33 1 46 11 84 08, E: Frederic.Plas@andra.fr

Plot, O., G D E, 3 rue Odilon Redon, F-91570 Bièvres

Plotto, P., IMS RN sarl, Parc d'Activités Pré Millet, bât. 9, F-38330 Montbonnot. T: +33 4 76 52 41 20, F: +33 4 76 52 49 09, E: ims-rn@alpes-net.fr

Plumelle, C., CNAM (Conservatoire National des Arts et Métiers) Chaire de Géotechnique, 2 rue Conté, F-75141 Paris Cedex 03. T: +33 1 40 27 28 78, F: +33 1 40 27 24 28, E: plumelle@cnam.fr

Poilpre, V., GTS (Géotechnique et Travaux Spéciaux), Agence de Paris, Continentale Square, 3 place de Londres, BP 10762, F-95727 Roissy Charles de Gaulle Cedex. T: +33 1 48 16 43 70, F: +33 1 48 16 43 71

Poitout, M.-J., SNCF-Dir.Ingénierie, 122 rue des Poissonniers, F-75018 Paris. T: +33 1 55 31 14 90

Popescu, L., IUT Génie Civil, Université de Marne-la-Vallée, 2 rue Albert Einstein, F-77420 Champs-sur-Marne. T: +33 1 64 73 05 13, F: +33 1 64 75 05 55, E: LiviuPopescu@univ-mlv.fr

Port Autonome du Havre, Terre plein de la Barre, F-76067 Le Havre Cedex. T: +33 2 32 74 74 21, F: +33 2 32 74 72 25

Post, G.R., 11 rue des Girondins, F-92210 Saint-Cloud. T: +33 1 46 02 35 93, F: +33 1 46 02 01 80

Poteur, M., GEOTRA, 1 chemin de la Baume, F-13740 Le Rove

Poudevigne, J., SOLEN, Parc d'Activités du Mirail, 23 av. du Mirail, F-33370 Artigues près Bordeaux. T: +33 5 57 77 01 60, F: +33 5 56 32 55 07, E: solen-bx@worldonline.fr

Pouget, L.R.P.C., 8 rue Bernard Palissy, Z.I. du Bréset, F-63014 Clermont-Ferrand

Pozzi, N., FRANKI FONDATION, 34 rue Charles Piketty, F-91170 Viry-Châtillon Cedex. T: +33 1 69 54 21 00, F: +33 1 69 54 21 10, E: Franki.Paris@wanadoo.fr

Proust, J.-L., SOLETANCHE BACHY, BP 511, F-92005 Nanterre Cedex

Puech, A., FUGRO France, 26 av. des Champs Pierreux, F-92000, Nanterre. T: +33 1 55 69 14 14, F: +33 1 55 69 14 15

Puech, J.-P., SCETAUROUTE, ZAC de la Condamine, Rond-Point de l'Europe, F-34436 Saint-Jean de Vedas Cedex. T: +33 4 67 69 07 00, F: +33 4 67 69 14 00, E: jp.puech@scetauroute.fr

Quenech, J.-L., ESITC de Caen, 1 rue Pierre et Marie Curie, F-14610 Epron. T: +33 2 31 46 23 00, F: +33 2 31 43 89 74, E: laboratoire@esitc-caen.fr

Quibel, A., CETE Normandie Centre, Centre d'Expérimentation Routière, Chemin de la Poudrière, BP 245, F-76121 Le Grand Quevilly Cedex. T: +33 2 35 68 82 09, F: +33 2 35 68 81 21, E: cetenc-cr@wanadoo.fr

Rajot, J.-P., LRPC d'AUTUN, bd de l'Industrie, BP 141, F-71405 Autun Cedex. T: +33 3 85 86 67 11, F: +33 3 85 86 90 83, E: jeanpierre.rajot@cetelyon.equipement. gouv.fr

Ramondenc, P., SNCF-Dir.Ingénierie, Dépt. des Ouvrages d'Art du Patrimoine, 122 rue des Poissonniers, F-75876 Paris Cedex 18. T: +33 1 55 31 17 46, F: +33 1 55 31 18 62

Ratp/ITA/IDI/GEO/LAC P 61, 40 bis rue Roger Salengro, LAC P 61, F-94724 Fontenay-sous-Bois Cedex. T: +33 1 41 95 37 98, F: +33 1 41 95 38 30, E: jean.bouchain@ratp.fr

Raynaud, D., AEROPORTS DE PARIS, BP 20102, F-95711 Roissy Charles de Gaulle. T: +33 1 48 62 11 75, F: +33 1 48 62 00 65, E: daniel.raynaud@adp.fr

Reiffsteck, P., LCPC (Laboratoire Central des Ponts et Chaussées), 58 bd Lefebvre, F-75732 Paris Cedex 15. T: +33 1 40 43 52 73, F: +33 1 40 43 65 11, E: philippe.reiffsteck@lcpc.fr ou refsteck@lcpc.fr

Reimbert, A., 67 bd de Reuilly, F-75012 Paris. T: +33 1 43 43 08 72

Riou, Y., ECOLE CENTRALE de NANTES, Dépt Génie Civil, 1 rue de la Noë, BP 92101, F-44321 Nantes Cedex 3. T: +33 2 40 37 16 64, F: +33 2 40 37 25 35, E: riou@ec-nantes.fr

Riviere, P., PORT AUTONOME du HAVRE, Terre plein de la Barre, F-76067 Le Havre Cedex. T: +33 2 32 74 74 21, F: +33 2 32 74 72 25

Riviere, M., 2779bis ch. Allo Marcellin, La Baronne, F-6610 La Gaude. T: +33 4 93 31 80 44, F: +33 4 93 14 65 50

Robert, J., EEG SIMECSOL, 18 rue Troyon, F-92316 Sèvres Cedex. T: +33 1 46 23 78 05, F: +33 1 46 23 78 07, E: direction@simecsol.fr

Robinet, J.-C., EURO-GEOMAT CONSULTING, 14 rue E. Biscara, F-45000 Orléans. T: +33 2 38 66 22 20, F: +33 2 38 66 24 70, E: robinetjc@3dnet.fr

Roche, M., SEFI, 2 av. du Général de Gaulle, BP 45, F-91170 Viry-Châtillon Cedex. T: +33 1 69 54 22 15, F: +33 1 69 96 92 03

Ropers, P., GEODYNAMIQUE ET STRUCTURE, 157 rue des Blains, F-92220 Bagneux. T: +33 1 46 65 00 11, F: +33 1 46 65 58 54, E: geodynamique@wanadoo.fr

Roudier, J., LCPC (Laboratoire Central des Ponts et Chaussées), 58 bd Lefebvre, F-75732 Paris Cedex 15. T: +33 1 40 43 50 00, F: +33 1 40 43 54 98, E: Jacques.Roudier@lcpc.fr

Rousseau, B., CEBTP – CEMEREX, 6 rue n° 14 – ZI, F-13127 Vitrolles. T: +33 4 42 10 98 10, F: +33 4 42 79 30 21

Rousseau, J., 9 rue de la Matauderie, F-86280 Saint-Benoit. T: +33 5 49 57 11 19, F: +33 5 49 88 76 38

Rouxel, N., LRPC de SAINT-BRIEUC, 5 rue Jules Vallès, F-22015 Saint-Brieuc Cedex. T: +33 2 96 75 93 00, F: +33 2 96 75 93 10, E: nicolas.rouxel@equipement. gouv.fr

Royet, P., CEMAGREF, Le Tholonet, BP 31, F-13612 Aix-en-Provence. T: +33 4 42 66 99 35, F: +33 4 42 66 88 65, E: paul.royet@cemagref.fr

Sadki, N., ECOLE CENTRALE de PARIS, Grande Voie des Vignes, F-92295 Chatenay-Malabry. T: +33 1 41 13 15 62, F: +33 1 41 13 14 42, E: sadki@mss.ecp.fr

Sage (Société Alpine de Géotechnique), 2 rue de la Condamine, BP 17, F-38610 Gières. T: +33 4 76 44 75 72, F: +33 4 76 44 20 18, E: SAGE.INGENIERIE@wanadoo.fr

Salaun-Vaujour, M.-D., PONTS FORMATION EDITIONS, 28 rue des Saints-Pères, F-75007 Paris. T: +33 1 44 58 27 32

Salençon, J., LMS-X, ECOLE POLYTECHNIQUE, F-91128 Palaiseau Cedex. T: +33 1 69 33 33 03, F: +33 1 69 33 30 67, E: bterrien@lms.polytechnique.fr

Samama, L., SCETAUROUTE, 11 av. du Centre, St-Quentin-en-Yvelines, F-78286 Guyancourt Cedex. T: +33 1 30 48 46 37, F: +33 1 30 48 49 79, E: l.samama@ scetauroute.fr

Sanglerat, G., 182bis av. Félix Faure, F-69003 Lyon. T: +33 4 78 54 42 90

Scherer, C., 1 rue des Peupliers, F-57050 Lorry-lès-Metz. T: +33 3 87 30 78 87/06 82 88 29 28

Schlosser, F., 7bis rue Brancas, F-92310 Sèvres. T: +33 1 45 34 08 32, F: +33 1 45 34 19 74

Schmitt, P., SOLETANCHE BACHY, 6 rue de Watford, BP 511, F-92000 Nanterre. T: +33 1 47 76 42 62, F: +33 1 40 90 94 87, E: 106241.2512@compuserve.com

Sefi, 2 av. du Général de Gaulle, BP 45, F-91170 Viry-Châtillon Cedex. T: +33 1 69 54 22 15, F: +33 1 69 96 92 93

Sejourne, M.-L., ROCSOL, 36 rue d'Estienne d'Orves, F-92120 Montrouge Cedex. T: +33 1 42 53 18 18, F: +33 1 42 53 53 20, E: rocsol@club-internet.fr

Semblat, J.-F., LCPC (Laboratoire Central des Ponts et Chaussées), 58 bd Lefebvre, F-75732 Paris Cedex 15. T: +33 1 40 43 50 94, F: +33 1 40 43 54 98, E: semblat@ lcpc.fr

Serratrice, J.-F., CETE Méditerranée, BP 37000, F-13791 Aix-en-Provence Cedex 3. T: +33 4 42 24 78 52, F: +33 4 44 24 78 18, E: jf.serratrice@cete13.equipement.gouv.fr

Services Techniques Sedco-Forex, 50 av. Jean Jaurès, BP 599, F-92542 MONTROUGE Cedex. T: +33 1 47 46 68 38, F: +33 1 47 46 67 36, E: barthe@montrouge.sedco-forex.slb.com

Setra-CTOA, Fondations-Soutènements, 46 av. Aristide Briand, BP 100, F-92223 Bagneux Cedex. T: +33 1 46 11 32 07, F: +33 1 46 11 33 52, E: hauin@setra.fr

Seve, G., CETE Méditerranée, Laboratoire de Nice, 56 bd de Stalingrad, F-6300 Nice. T: +33 4 92 00 81 82, F: +33 4 92 00 81 99, E: g.seve@cete13.equipement.gouv.fr

Sevestre, J.-J., SETEC Géotechnique, Tour Gamma D, 58 quai de la Rapée, F-75583 Paris Cedex 12. T: +33 1 40 04 69 54, F: +33 1 40 04 57 41

Shahrour, I., EUDIL (Ecole Univ. d'Ingénieurs de Lille), Département Génie Civil, F-59655 Villeneuve d'Ascq Cedex. T: +33 3 20 43 45 45, F: +33 3 20 43 45 83, E: isam. shahrour@eudil.fr

Sicsol Géotechnique S.A., Parc d'Activités Clément Ader, 19 rue Louis Bréguet, F-34830 Jacou. T: +33 4 67 59 40 10, F: +33 4 67 59 23 30, E: sicsol@wanadoo.fr

Sieffert, J.-G., ENSAIS, 24 bd de la Victoire, F-67084 Strasbourg Cedex. T: +33 3 88 14 47 66, F: +33 3 88 24 14 90, E: jean-georges.sieffert@ensais.u-strasbg.fr

Silleran, A., 121 rue de la Pompe, F-75116 Paris. T: +33 1 45 53 72 29, F: +33 1 53 70 91 95

Simon, B., TERRASOL, Immeuble Hélios, 72 av. Pasteur, F-93108 Montreuil Cedex. T: +33 1 49 88 24 42, F: +33 1 49 88 06 66, E: b.simon@terrasol.com

Sirieys, P., 13 rue de la République, F-38000 Grenoble. T: +33 4 76 44 71 63

Siwak, J.-M., ECOLE des MINES de DOUAI, Dépt Génie Civil, 941 rue Charles Bourseul, BP 838, F-59508 Douay Cedex. T: +33 3 27 71 24 20, F: +33 3 27 71 25 25, E: siwak@ensm-douai.fr

Sncf-Dir.Ingénierie, Dépt. des Ouvrages d'Art du Patrimoine, 122 rue des Poissonniers, F-75876 Paris Cedex 18. T: +33 1 55 31 16 35, F: +33 1 55 31 18 62

Sobesol, 10 av. Newton, BP 63, F-92354, Le Plessis Robinson. T: +33 1 46 01 81 10, F: +33 1 46 01 81 19, E: sobesol@ easynet.fr

Socotec, Les Quadrants, 3 av. du Centre, F-78182 Saint-Quentin-Yv. Cedex. T: +33 1 30 12 82 36, F: +33 1 30 12 83 90

Sol Essais Etudes, 49 rue des Sazières, F-92700 Colombes. T: +33 1 47 81 22 10, F: +33 1 47 82 52 25, E: SOL. ESSAIS.ETUDES@wanadoo.fr

Sol Essais S.A., 460 av. Jean Perrin, F-13851 Aix-en-Provence Cedex. T: +33 4 42 39 74 85, F: +33 4 42 39 73 91

Solen Geotechnique, 16 allée Prométhée, "Les Propylées III", BP 169, F-28003 Chartres Cedex. T: +33 2 37 88 03 30, F: +33 2 37 30 90 75, E: solengeo@aol.com

Soletanche Bachy, 6 rue de Watford, BP 511, F-92000 Nanterre. T: +33 1 47 76 55 94, F: +33 1 49 06 97 34, E: d.gouvenot@soletanche-bachy.com

Sols Mesures, 17 rue Jean Monnet, ZA des Côtes, F-78990 Elancourt. T: +33 1 30 50 34 50, F: +33 1 30 50 34 99, E: sols.mesures@wanadoo.fr

Soubra, A.H., ENSAIS, 24 bd de la Victoire, F-67084 Strasbourg Cedex. T: +33 3 88 14 47 00 poste 4809, F: +33 3 88 24 14 90, E: ahamid.soubra@ensais. u-strasbg.fr

Soyez, B., LCPC (Laboratoire Central des Ponts et Chaussées), 58 bd Lefebvre, F-75732 Paris Cedex 15

Spie Fondations, 10 av. de l'Entreprise, Pôle Galilée, F-95865 Cergy Pontoise. T: +33 1 34 24 49 13, F: +33 1 34 24 37 56

Sterenberg, J., COYNE et BELLIER, 9 allée des Barbanniers, F-92632 Gennevilliers

Stoehr, B., KELLER FONDATIONS SPECIALES, Espace Plein Ciel, allée de l'Europe, BP 5, F-67960, Entzheim. T: +33 3 88 59 92 00, F: +33 3 88 59 95 90, E: marketing@keller-france.com

Stoltz, D., CEBTP Nord-Est, 10 rue St-Hilaire, F-51100 Reims. T: +33 3 26 87 86 00, F: +33 3 26 87 86 01

Tabbagh, A., UNIVERSITE P. & M. CURIE, Dpt Géophysique Appliquée, Case 105, 4 place Jussieu, F-75252 Paris Cedex 05. T: +33 1 44 27 48 24, F: +33 1 44 27 45 88, E: alat@ccr.jussieu.fr

Taibi, S., UNIVERSITE du HAVRE, Fac. Sciences & Techniques, BP 540, F-76058 Le Havre Cedex

Tcheng, Y., 34 rue Duret, F-75116 Paris, F: +33 01 45 00 02 52

Tchocothe, F., LCPC (Laboratoire Central des Ponts et Chaussées) – BCAM., 58 bd Lefebvre, F-75732 Paris Cedex 15. T: +33 1 40 43 53 52

Technosol S.A., route de la Grange aux Cercles, F-91160 Ballainvilliers. T: +33 1 69 09 14 51, F: +33 1 64 48 23 56, E: TECHNOSOL@wanadoo.fr

Terrasol, I.H., 72 av. Pasteur, F-93108 Montreuil Cedex. T: +33 1 49 88 24 42, F: +33 1 49 88 06 66, E: info@ terrasol.com

Thiriat, D., BOTTE SADE Fondations, 21 rue du Pont des Halles, Delta 112, F-94536 Rungis. T: +33 1 49 61 48 00, F: +33 1 49 61 48 01

Thorel, L., LCPC, Centre de Nantes, route de Bouaye, BP 19, F-44340 Bouguenais. T: +33 2 40 84 58 08, F: +33 2 40 84 59 97, E: Luc.Thorel@lcpc.fr

Tisot, J.-P., ECOLE NATIONALE SUPERIEURE de GEOLOGIE, rue Doyen Marcel Roubault, BP 40, F-54501 Vandoeuvre-lès-Nancy. T: +33 3 83 59 64 00, F: +33 3 83 59 64 71, E: tisot@ensg.u-nancy.fr

Touquet, J.-L., SPIE FONDATIONS, Service Pieux, 10 av. de l'Entreprise, Pôle Galilée, F-95865 Cergy Pontoise. T: +33 1 34 24 48 93, F: +33 1 34 24 37 56

Toure-Bahloul, F., IUT, Service Génie Civil, 63 av. De Lattre de Tassigny, F-18020 Bourges Cedex. T: +33 2 48 23 80 67, F: +33 2 48 23 80 40

Tournery, H., SCETAUROUTE, Les Pléiades n 35, Park Nord Annecy, F-74373 Pringy Cedex. T: +33 4 50 27 39 62, F: +33 4 50 27 39 40

Trak, A., GEOFOND SARL, BP 2074, F- Libreville, Gabon.
T: +241 74 88 84, F: +241 74 88 84

Tran, V.N.J., SPIE, Pôle Magellan, Parc Saint-Christophe, F-95862 Cergy Pontoise. T: +33 1 34 24 38 40, F: +33 1 34 24 39 11

Tratapel, G., COMPAGNIE NATIONALE DU RHÔNE (CNR), 2 rue André Bonin, F-69316 Lyon Cedex 04. T: +33 4 72 00 68 52, F: +33 4 72 10 66 62, E: G.Tratapel@cnr.tm.fr

Ung, S.Y., FONDASOL INTERNATIONAL, ZA des Amandiers, 35 rue des Entrepreneurs, F-78421 Carrières-sur-Seine Cedex. T: +33 1 39 14 77 00, F: +33 1 39 14 76 70

Ursat, P., CETE de l'Est, LRPC de Strasbourg, rue Jean Mentelin, BP 9, F-67035 Strasbourg Cedex 2. T: +33 3 88 77 46 14, F: +33 3 88 77 46 40, E: paul.ursat@cete57.equipement.gouv.fr

Van De Graaf, H.C., LANKELMA Geotechniek Zuid b.v.i.o., BP 38, NL-5688 ZG Oirschot, Pays-Bas. T: +31 499 57 85 20, F: +31 499 57 85 73, E: post@inpijn-blokpoel.com

Varaksin, S., MENARD SOL TRAITEMENT, 2 rue Gutenberg, F-91620 Nozay. T: +33 1 69 01 37 38, F: +33 1 69 01 75 05

Vaseux, J., EN.OM.FRA, 6/8 av. Eiffel, F-77220 Gretz-Armainvilliers. T: +33 1 64 06 47 76, F: +33 1 64 06 47 59, E: ENOMFRA@wanadoo.fr

Venot, C., CETE de l'Est, LRPC de Nancy, 75 rue de la Grande Haie, BP 8, F-54510 Tomblaine. T: +33 3 83 18 41 13, F: +33 3 83 18 41 00, E: anne-marie.venot@cete57.equipement.gouv.fr

Verger, R., AFITEST, Les Courrières, BP 48, F-87170, Isle. T: +33 5 55 43 84 88, F: +33 5 55 43 84 81

Vergobbi, P., FUGRO FRANCE, 26 av. des Champs Pierreux, F-92000 Nanterre. T: +33 1 55 69 14 14, F: +33 1 55 69 14 15

Verity, A., GAUDRIOT Ingénieurs Conseils, ZI République II, 25 rue Victor Grignard, F-86060 Poitiers Cedex 9. T: +33 5 49 30 35 00, F: +33 5 49 30 35 35, E: gaudriot-86@wanadoo.fr

Vezole, P., SAE, 143 av. de Verdun, F-92442 Issy-les-Moulineaux Cedex. T: +33 1 41 08 38 19, F: +33 1 41 46 91 98

Virely, D., LRPC de TOULOUSE, Complexe Scient. de Rangueil, 1 av. du Colonel Roche, F-31400 Toulouse.

T: +33 5 62 25 97 12, F: +33 5 62 25 97 98, E: dvirely@cete33.equipement.gouv.fr

Vivalp, 27–29 bd du 11 Novembre, BP 2131, F-69603 Villeurbanne Cedex. T: +33 4 78 17 22 13, F: +33 4 72 82 04 46, E: info@vivalp.com

Volcke, J.-P., FRANKI FONDATION, 34 rue Charles Piketty, F-91170 Viry-Châtillon Cedex. T: +33 1 69 54 21 00, F: +33 1 69 54 21 10, E: Franki.Paris@wanadoo.fr

Vossoughi, K.-C., ECOLE CENTRALE de PARIS, MSS/Mat, Grande Voie des Vignes, F-92295 Chatenay-Malabry. T: +33 1 41 13 13 39, F: +33 1 41 13 14 42, E: vossough@mss.ecp.fr

Vossoughi-Jenab, B., ENPC – CERMES, 6-8 av. Blaise Pascal, Cité Descartes – Champs-sur-Marne, F-77455 Marne-la-Vallée Cedex. T: +33 1 64 15 35 24, F: +33 1 64 15 35 62, E: jenab@cermes.enpc.fr

Vuez, A., INSA de RENNES, Dépt Génie Civil et Urbanisme, 20 av. des Buttes de Coësmes, F-35043 Rennes. T: +33 2 99 28 65 38, E: alain.vuez@insa-rennes.fr

Vuillier, C.P., VIBRO-PILE (Aust.) Pty Ltd, No. 1 Steel Court, Elwood 3184 Victoria, Australie. T: +61 3 95 84 45 44, F: +61 3 95 83 86 29, E: vuillier@vibropile.com.au

Walter, J.-P., GEODYNAMIQUE ET STRUCTURE, 157 rue des Blains, F-92220 Bagneux. T: +33 1 46 65 00 11, F: +33 1 46 65 58 54, E: geodynamique@wanadoo.fr

Waschkowski, E., 17 rue du Point du Jour, F-41350 Vineuil

Welti, O., GEOMEDIA, 18 place de France, F-95200 Sarcelles. T: +33 1 39 94 46 38

Wojnarowicz, B., Cabinet GUILLERMAIN, 6 Louis Pasteur, F-92774 Boulogne Cedex. T: +33 1 46 99 18 50

Wojnarowicz, M., SEPIA, 26 rue Ampère ZI, F-91430 Igny. T: +33 1 69 33 22 90, F: +33 1 60 19 66 98, E: lacsepia@aol.com

Zaghouani, K., ETUDESOL, Z.A. du Moulin à Vent, rue des Mares Julienne, F-91380 Chilly Mazarin. T: +33 1 69 34 43 90, F: +33 1 69 34 41 45, E: ETUDESOL.ZAGHOUANI@wanadoo.fr

Zerhouni, M.I., SOLEN GEOTECHNIQUE, 16 allée Prométhée, "Les Propylées III", BP 169, F-28003 Chartres Cedex. T: +33 2 37 88 03 30, F: +33 2 37 30 90 75, E: solengeo@aol.com

Ziani, F., TRIADE, 195 av. du Général Leclerc, F-78220 Viroflay. T: +33 1 30 24 25 25

GERMANY/ALLEMAGNE

Secretary: Dr. rer. nat. Kirsten Laackmann, Deutsche Gesellschaft für Geotechnik e.V., Hohenzollernstr. 52, 45128 Essen. T: 0049/201/78 27 23, F: 0049/201/78 27 43, www.dggt.de, E: service@dggt.de

Total number of members: 885

Abel, H.-J., Dipl.-Ing., Byfanger Str. 98, 45257 Essen. T: 0201/482116(p)

Achmus, M., Dr.-Ing., Wissensch. Mitarbeiter, Meersmannufer 23, 30655, Hannover

Adam, E., Dipl.-Geol., Saarstr. 16a, 54455 Serring

Aicher, M., Dipl.-Ing., Max Aicher GmbH & Co., Teisenbergstr. 7, 83395 Freilassing. T: 08654/2011(d), F: 08654/3378(d)

Albers, K.-H., Dipl.-Ing., G. quadrat Geokunststoff-gesellschaft mbH, Kochstr. 44, 47805 Krefeld. T: 02151-368-241(d), F: 02151-368-243(d)

Alberts, D., Dipl.-Ing., Bundesanstalt f. Wasserbau, Wedeler Landstr. 157, 22559 Hamburg. T: 040/81908-310(d), F: 040/81908-373(d)

Albrecht, F., Dr. rer. nat., Selbst., Beratender Geologe, Baukauer Str. 46a, 44532 Herne. T: 02323/9274-0(d), F: 02323/9274-30(d), T: 02323/51325(p), F: 02323/57398(p)

Alexiew, D., Dipl.-Ing., HUESKER Synthetic GmbH & Co., Leiter Anwendungstechnik, Fabrikstr. 13–15, 48712 Gescher. T: 02542/701-290(d), F: 02542/701-499(d), T: 02542/2352(p)

Alexiew, N., Dipl.-Ing., HUESKER Synthetic GmbH & Co., Abt. Anwendungstechnik, Fabrikstr. 13–15, 48712 Gescher. T: 02542/701-291(d), F: 02542/701-499(d)

Alheid, H.-J., Dr., Bundesanstalt für Geowissenschaften & Rohstoffe, B 2.12, Stilleweg 2, 30655 Hannover. T: 0511/6432870(d), F: 0511/6432304(d)

Altmann, A., Dipl.-Ing., Hortensienweg 21, 85551 Kirchheim b. München. T: 089/9038283(p)

Amann, P., Prof. Dr.-Ing., AICON Amann Infutec Consult AG, O.-R. Str. 42, 64367 Mühltal/Darmstadt. T: 06151/1415-0(d), F: 06151/141514(d)

Andrä, G., Friedhofstr. 20, 63225 Langen. T: 06103/1512(p)

Armbruster-Veneti, H., Dipl.-Ing., Bundesanstalt für Wasserbau, Geotechnik, Kußmaulstr. 17, 76187 Karlsruhe. T: 0721/9726-386(d), F: 0721/9726-483(d)

Armstroff, L.O., Herderstr. 6b, 65239 Hochheim. T: 06146/84297(p)

Arslan, U., Univ.-Prof. Dr.-Ing., Technische Universität Darmstadt, Institut für Geotechnik, Bauingenieurwesen, Petersenstr. 13, 64287 Darmstadt. T: 06151/163749(d), F: 06151/166683(d)

Arz, P., Dipl.-Ing., Direktor, Schwanheimer Str. 18, 64683 Einhausen. T: 0611/708-0(d), F: 0611/708-236(d), T: 06251/587718(p), F: 06251/587718(p)

Asdecker, F., Dipl.-Ing., Ingenieurbüro H. Asdecker, Wolfsbacher Str. 32, 95448 Bayreuth. T: 09209/16289(d), F: 09209/1539(d)

Ast, W., Prof. Dipl.-Ing., Gluckstr. 6, 70195 Stuttgart. T: 07351/582-110(d), F: 07351/582-172(d), T: 0711/9960310(p), F: 0711/9960312(p)

Bacharach, W., Dipl.-Ing., Drees & Sommer AG Infra Consult & Management GmbH, Lautenschlagerstr. 2, 70173 Stuttgart. T: 0711/222933-22(d), F: 0711/222933-90(d), T: 07243/79976(p)

Bachmann, M., Dipl.-Geol., Wiss. Angest., Körner Str. 28, 31141 Hildesheim. T: 0531/3912325(d), 05121/869063(p)

Backes, W., Dr.-Ing., Am Irschberg 3, 66625 Nohfelden. T: 06875/1703(p)

Bahnsen, P., Dipl.-Ing., Geschäftsleitung, BBI Geo- und Umwelttechnik, Stormsweg 5, 22085 Hamburg. T: 040/229468-0(d), F: 040/229468-40(d), T: 040/7229596(p)

Balthaus, H., Dr.-Ing., DIC, M.S., Geschäftsf. Gesellsch., Vennstr. 162, 40627 Düsseldorf. T: 0211/6396-106(d),

F: 0211/6396-107(d), T: 0211/252246(p), F: 0211/2092070(p)

Barthel, N., Dipl.-Ing., öbuv Sachverständiger, Baugrundbüro Barthel, Magdeborner Str. 9, 04416 Markkleeberg. T: 034297/678-0(d), F: 034297/678-11(d)

Bärthel, U., Dipl.-Geotech., Talstr. 6, 07629 St. Gangloff

Batke, O., Dipl.-Ing., Baugrundlabor Batke GmbH, Kaufmannstr. 81a, 53115 Bonn. T: 0228/654611(d), F: 0228/657387(d)

Baudendistel, M., Prof. Dr.-Ing., Tunnel Consult, Tulpenstr. 20, 76275 Ettlingen. T: 07243/91923(d), F: 07243/99793(d)

Bauer, A., Dr.-Ing., Reichenaustr. 37, 81243 München. T: 089/28927149(d), F: 089/28927189(d), T: 089/83969991(p)

Bauer, F., Dipl.-Ing., Ingenieurbüro Bauer, Karl-Liebknecht-Str. 76, 03046 Cottbus

Bauer, K., Dr.-Ing. Dr.-Ing. E.h., Bauer Spezialtiefbau GmbH, Wittelsbacher Str. 5, 86529 Schrobenhausen. T: 08252/971219(d), F: 08252/971213(d)

Baumann, H.J., Dr.-Ing., Regierungsdirektor, Bayerisches Geologisches Landesamt, Heßstr. 128, 80797 München. T: 089/9214-2683(d), F: 089/9214-2647(d), T: 089/664352(p)

Bayer, E., Prof. Dr.-Ing., Bauberatung Zement Wiesbaden, Friedrich-Bergius-Str. 7, 65203 Wiesbaden. T: 0611/20042(d), F: 0611/24294(d)

Bechert, H., Prof. Dr.-Ing., Ingenieurbüro, Teckstr. 44, 70190 Stuttgart. T: 0711/1663422(d), F: 0711/1663500(d)

Beckefeld, P., Dr.-Ing., Postfach 10 03 44, 45003 Essen

Becker, A.W., Dipl.-Ing., Strobelallee 79, 66953 Pirmasens

Becker, J., Dipl.-Ing., Sachgebietsleiter, Rheinbraun AG, Abt. BT 4, Gebirgs- u. Bodenmechanik, Stüttgenweg 2, 50416 Köln. T: 0221/48022711(d), F: 0221/48022460(d)

Becker, S., Dipl.-Ing., Kiefernweg 14, 54516 Wittlich. T: 0651/14674-0(d), F: 0651/1467470(d), T: 06571/27927(p), F: 06571/260049(p)

Becker, T., Dipl.-Ing., Wissensch. Mitarbeiter, Grawertstr. 8, 54316 Pluwig. T: 0631/205-3807(d), F: 0631/205-3806(d), 0171/8128948(p)

Behrend, C., Yitzhat Rabin Str. 3, 70376 Stuttgart

Behrens, W., Dipl.-Ing., ISIS Gmbh, LüpertzenderStr. 6, 41061 Mönchengladbach

Beine, R., Dr.-Ing., Prof. Dr.-Ing. Jessberger & Partner, Am Umweltpark 5, 44793 Bochum

Bektas, Z., Parkstr. 42a, 65795 Hattersheim

Bennewitz, T., Dipl.-Bauing, Städtelner Str. 32, 04416 Markkleeberg. T: 0341/3581222(p), F: 0341/3581222(p)

Berg, J., Dipl.-Ing., Karlsbader Str. 44, 63150 Heusenstamm. T: 069/8051274(d), F: 069/8051221(d), T: 06104/61617(p)

Berner, U., Dr.-Ing., Heugäßle 11, 78465 Konstanz. T: 07531/59450(d), F: 07531/594550(d)

Bernsdorf, L., Dipl.-Ing., Kreisstr. 10b, 27711 Osterholz-Scharmbeck. T: 04791/3382(p)

Beutinger, P.J., Dipl.-Ing., Don-Carlos-Str. 36, 70563 Stuttgart. T: 0711/685 2439(d)

Beyer, H., Dipl.-Ing., Ingenieurbüro für Verkehrswegebau, An der Silberkuhle 8, 30655 Hannover. T: 0511/3885182(d), F: 0511/3885130(d)

Beyer, W., Prof., Paganinistr. 82a, 81247 München. T: 089/8116577(p), F: 089/8119921(p)

Biczok, E., Dr.techn., GTU Ingenieurgesellschaft mbH, Sahlkamp 149, 30179 Hannover. T: 0511/9089916(d), F: 0511/9089925(d)

Biedebach, S., Dipl.-Ing., Büroinhaber, Grundbauinstitut in Dortmund, Am Remberrg 180, 44269 Dortmund. T: 0231/441064(d), F: 0231/443665(d)

Bilz, P., Prof. Dr.-Ing.habil, Selbst. Sachverständiger, Jägerhofstr. 71, 01445 Radebeul

Binard-Kühnel, Christian-Rübsamenstr. 9, 35578 Wetzlar. T: 06441/92474-0(d), F: 06441/92474-20(d), T: 06441/782141(p), F. 06441/975965(p)

Bischoff, W., Dipl.-Geol., Bauingenieur, Ingenieurbüro für Baugrunduntersuchungen und Ingenieurgeologie, Bremer-Str. 75, 12623 Berlin. T: 030/5628428(d), F: 030/5633964(d)

Blankmeister, W., Dipl.-Geol., Dickebankstr. 36, 44866 Bochum. T: 02327/88224(p), F: 02327/15852(p)

Blessing, V.M., Dipl.-Ing., Zum Römerturm 10, 50127 Bergheim. T: 0221/824-2553(d), F: 0221/824-2597(d), T: 02271/66955(p)

Bloh von, G., Dr.-Ing., Grundbaulabor Bremen Dipl.-Ing. Dietrich Behnke, Kleiner Ort 2, 28357 Bremen. T: 0421/207700(d), T: 0441/58884(p)

Blume, K.-H., Dipl.-Ing., Zur Freiheit 18, 51491 Overath. T: 02204/43723(d), F: 02204/43673(d)

Blümel, W., Prof. Dr.-Ing., Hochschullehrer/Bauingenieur, Geotechnische Beratung und Begutachtung, Alabasterweg 15, 30455 Hannover. T: 0511/405706(d), F: 0511/7011219(d), T: 0172/4401800(p), F: 0511/405706(p)

Böck, R., Dipl.-Ing., Franki Grundbau GmbH, NL Süd, Albert-Roßhaupter-Str. 73, 81369 München. T: 089/7435040(d), F: 089/743504-10(d), T: 08177/8604(p)

Bockholt, G., Prof. Dipl.-Ing., Im Paßkamp 32, 45665 Recklinghausen

Böckmann, F., Dr.-Ing., BBI, Geo-u. Umwelttechnik, Geschäftsleitung, Stormsweg 5, 22085 Hamburg. T: 040/229468-0(d), F: 040/229468-40(d), T: 040/4505079(p)

Boenke, Dipl.-Ing., ISP Ingenieurbüro für Spezialtiefbau, Projekt-Management, Bertha-von-Suttner-Str. 79, 64846 Groß-Zimmern. T: 06071/48310(d), F: 06071/ 44428(d)

Boley, C., Dr.-Ing., Beratender Ingenieur, Windels-Timm-Morgen Beratende, Ingenieure im Bauwesen, Ballindamm 17, 22095 Hamburg. T: 040 35009 238(d), F: 040 35009 438

Borchert, C., Dipl.-Ing., Erdbaulabor Borchert & Lange, Finkenhof 12a, 45134 Essen. T: 0201/43555-0(d), F: 0201/43555-43(d)

Bordeaux, H., Baugrundsachverständiger, Grundbaubüro Bordeaux, Homburger Str. 17, 61118 Bad Vilbel. T: 06101/2214(d), F: 06101/7666(d)

Borrmann, C., Dipl.-Ing., Karl-Marx-Str. 11, 07774 Frauenprießnitz

Bosenick, J., Dipl.-Ing., ELE-Erdbaulaboratorium Essen, Susannastr. 31, 45136 Essen. T: 0201 8959-6, F: 0201 253733 (d)

Boss, N., Dipl.-Ing., Rilkestr. 23, 63165 Mühlheim. T: 069/8051272(d), 06108/76116(p)

Böttinger, C., Weinheimer Str. 5, 69493 Leutershausen. T: 06201/53573(p)

Brandl, H., Prof. Dr. techn., Institutsvorstand, Technische Universität Wien, Institut für Grundbau und Bodenmechanik, Karlsplatz 13, A-1040 Wien, Österreich. T: 0043/1/58801-22100(d)

Brandt, A., Knappenstr. 4, 59071 Hamm

Bräu, G., Dipl.-Ing., Agilolfingerstr. 16, 81543 München. T: 089/289-27139(d), F: 089/289-27189(d)

Brauch, H.F., Dr. rer. pol., Ingenieurbüro Kittelberger, Geschäftsleitung, Mundenheimer Str. 141–145, 67061 Ludwigshafen. T: 0621/5602-251(d), F: 0621/5602-233(d)

Breder, R., Dipl.-Ing., Selbständig, Ingenieurbüro für Geotechnik, Stadtstr. 66a, 79104 Freiburg. T: 0761/2021545(d), F: 0761/2021514(d), T: 0761/39099(p)

Brehm, J., Dipl.-Geol., Institut für Angewandte Geologie und Umweltanalytik Dipl.-Geol. J. Brehm GmbH, Bauhofstr. 18D, 63762 Großostheim. T: 06026/9733-0(d), F: 06026/9733-18(d)

Breinlinger, F., Dr.-Ing., Breinlinger + Partner Ingenieurgesellschaft mbH, Kanalstr. 1–4, 78532 Tuttlingen. T: 07461/184-0(d), F: 07461/184-100(d), T: 07461/163433(p)

Bretz, H., Dipl.-Ing., Beratender Ingenieur, Nuapliastr., 74a, 81545, München. T: 089 644751(d), F: 089 64248484(d)

Breymann, H., Dipl.-Ing. Dr. techn., Fehrenbachweg 8, 5550 Radstadt, Österreich. T: 0043/6452/7148(p), F: 0043/6452/7148(p)

Bruhn, D., Dipl.-Geol., Terra Control GmbH, Geschäftsführung, Postfach 15 61, 61215 Bad Nauheim. T: 06032/72855(d), F: 06032/72805(d)

Bruns, T., Dr.-Ing., Meiendorfer Mühlenweg 136a, 22159 Hamburg. T/F: 040/400264(p)

Bücherl, K., Dipl.-Geol., Geschäftsführer, Ingenieurbüro für Geotechnik und Umweltschutz GmbH, Im Gewerbepark D 60, 93059 Regensburg. T: 0941/46306-0(d), F: 0941/48741(d), T: 0941/32634(p)

Buchmaier, R.F., Prof. Dr.-Ing., Bussardweg 26, 71111 Waldenbuch. T: 07157/4206(p)

Buckow, G., Dipl.-Ing., Baugrundingenieur, Geotechnisches Ingenieurbüro Dipl.-Ing. Buckow, Wasserstr. 20, 06632 Freyburg. T: 034464/27690(d), F: 034464/61030(d), T: 034464/29100(p)

Bulk, D., Dipl.-Ing., selbst.-Gutachter, Wallstr. 3, 58313 Herdecke. T: 02330/800919(d), F: 02330/8112(d), T: 02330/84590(p)

Bumiller, B., Dipl.-Geophys., Frank & Bumiller & Kraft GmbH, Hofangerstr. 82, 81735 München. T: 089/684096(d), F: 089/6804425(d)

Burkhardt, M., Dr., Stettiner Str. 2, 48455 Bad Bentheim. T: 05922/4514(p)

Burmann, O., Dipl.-Ing., Burmann & Mandel, Gasstr. 18, Haus 6b, 22761 Hamburg. T: 040/6014548(d), F: 040/8901621(d)

Caballero Hoyos, J.F., Dipl.-Ing., P.O. Box 12351, - La Paz, Bolivien (S.A.)

Cantre, S., Patriotischer Weg 38, 18057 Rostock

Carrasco, R., Dr.-Ing., Ingenieurbau SKW GmbH, Hegauer Weg 25, 14163 Berlin. T: 030/80903257(d), F: 030/80903258(d), T: 030/42016544(p), F: 030/42016546(p)

Cejka, A., Dipl.-Ing., Sülldorfer Landstr. 78e, 22589 Hamburg. T: 040/8700310(p)

Chamier, H., Dipl.-Ing., Rungestr. 38, 18435 Stralsund. T: 03831/263526(p), F: 03831/294044(p)

Clasmeier, H.-D., Dr.-Ing., 3. Norderwieke 24, 26802 Moormerland. T: 04921/897-132(d), F: 04921/897-137(d), T: 04954/6474(p)

Clauer, C., Grafenstr. 15, 64283 Darmstadt

Clostermann, M., Dipl.-Ing., GB-Leiter, Deutsche Montan Technologie GmbH, Geschäftsbereich GUC Geo- und Bau Consult – Baugrundinstitut, Am Technologiepark 1, 45307 Essen. T: 0201/172-1900(d), F: 0201/172-1773(d)

Colombo, S., Remscheidenstr. 47, 42369 Wuppertal

Croissant, R., Dipl.-Ing. Ing. Civil, Mittelstr. 11, 07580 Thüringen, F: 0090/342 518 1111(d), T: 036602/22853(p)

Cunze, G., Dr.-Ing., Berat. Ingenieur, Agilolfingerweg 9, 30175 Hannover. T: 0511/853282(p)

Czapla, H., Dr.-Ing., Berat. Ingenieur, sgg Dr. Czapla, Am Gockert 69, 64354 Reinheim. T: 06162/912640(d), F: 06162/912641(d)

Czygan, C., Dipl.-Ing., Berat. Ingenieur, Ing.-Büro, ö.b.u.v., Birkenwaldstr. 20d, 63179 Obertshausen. T: 06104/75948(d), F: 06104/75940(d)

Damisch, A., Dipl.-Ing., Inhaber, Ingenieurbüro Damisch Hoch-, Tief- und Deponiebau Geo- und Umwelttechnik, Moltkestr. 12, 39576 Stendal. T: 03931/6481-51(d), F: 03931/6481-59(d)

Dannemann, H., Dipl.-Ing., Erdbaulaboratorium Ahlenberg, Am Ossenbrink 40, 58313 Herdecke. T/F: 02330/8009-252(d)

Dausch, G.L., Dipl.-Ing., Bergstr. 1, 76891 Niederschlettenbach

Dauwe, L., Dipl.-Ing., Frauenalber Str. 19, 76199 Karlsruhe. T: 07243/76320(d)

Deflorian, W., Bereichsleiter Grundbau, Badlochstr. 2b, A-6890 Lustehau, Österreich. T: 05577/88631(p), T: 0043/5574/79108(d)

Degen, W., Dr. sc. techn. ETH, Vibroflotation AG, Letzistr. 21, 8852 Altendorf, Schweiz. T: 0041/55/4519020(d), F: 0041/55/4519021(d)

Deman, F., Dr.-Ing., WPW Geoconsult GmbH, Hochstr. 61, 66115 Saarbrücken. T: 0681/9920-230(d), F: 0681/ 9920-239(d)

Demler, H.A., Dipl.-Kfm. Ing. grad., Geschäftsführer, Demler GmbH & Co., Lahnstr. 92, 57250 Netphen. T: 02738/ 6080(d), F: 02738/608130(d)

Denzer, K., Dr. Dipl.-Geol., Geotechnisches Büro Dr. Denzer GmbH, Untere Mühlgasse 5, 82497 Unterammergau. T: 08822/6114(d), F: 08822/4397(d)

Deppe, M., Dipl.-Ing., Freier Mitarbeiter, Zionskirchstr. 7, 10119 Berlin. T: 05222/59805(d), T: 030/44045174(p), F: 030/44045175(p)

Deutsch, R., Dr. rer. nat., Abteilungsleiter, Ruhrverband, Abt. Ingenieur- und Hydrogeologie, Kronprinzenstr. 37, 45128 Essen. T: 0201/178-2680(d), F: 0201/178-2605(d)

Dieler, H., Prof. Dr.-Ing., Bertholdstr. 7, 52066 Aachen. T: 0241/63751(p)

Dietrich, T., Prof. Dr.-Ing., Zu den Birken 4a, 64739 Höchst-Hassenroth

Dietz, K., Dipl.-Ing., Fuchsbergstr. 93, 40724 Hilden. T: 02103/23830(p)

Dillo, M., Dr.-Ing., Zentralschweizerisches Technikum Luzern Bauingenieurwesen, Technikumstr., 6048 Horw, Schweiz. T: 0041/413493454(d), F: 0041/413493460(d), T: 0041/326215467(p)

Dimmerling, A., Dipl.-Geol., Tannenweg 13, 36137 Großenlüder

Dirkes, M., Am Sonnenhang 33, 45289 Essen. T: 0201/ 8576396(p), F: 0201/8576397(p)

Döbbelin, J., Dr.-Ing., Bauingenieur, Rheinstr. 83a, 76185 Karlsruhe. T: 0721/400890(d), 0721/5966377(p)

Dölz, K., Nieder-Ramstädter-Str. 179/A 31, 64285 Darmstadt. T: 06151/423825(p)

Dommaschk, P., Dipl.-Ing., Referent, Sächsisches Landesamt für Unwelt und Geologie Ber. Boden und Geologie, Ref. IG, Halsbrücker Str. 31a, 09599, Freiborg. T: 03731 294153 (d), 03731 32234 (p), F: 03731 294115

Dressel, B., Prof. Dr.-Ing. habil., Berat. Ingenieur, Ingenieurbüro für Bautechnik Prof. Dr.-Ing. habil. B. Dressel, Hübnerstr. 27, 01187 Dresden. T: 0351/4711135(d), F: 0351/ 4724781(d)

Driemeier, D., Ringeler Str. 74, 49525 Lengerich. T: 02534/620028(d), F: 02534/620032(d), T: 05481/5844(p)

Dubbert, J.M., Dipl.-Ing., Öffentl. best. Verm. Ing. J.-M. Dubbert, Dorfstr. 7, Altes Gutshaus, 23968 Gramkow. T: 038428 6460 (d), F: 038428 64642 (d)

Dühmert, W., Im Wiesenring 12, 63150 Heusenstamm. T: 06104/921250(p), F: 06104/921256(p)

Düllmann, H., Prof. Dr.-Ing., Geotechnisches Büro Prof. Dr.-Ing. H. Düllmann, Neuenhofstr. 112, 52078 Aachen. T: 0241/92839-0(d), F: 0241/527762(d)

Dümcke, A., GF Dipl.-Ing., Baukontor Dümcke GmbH, Alfstr. 26, 23552 Lübeck

Dürrwang, R., Dipl.-Ing., Geschäftsführer Arcadis, Trischler und Partner GmbH, Berliner Allee 6, 64295 Darmstadt. T: 06151/388-220(d), F: 06151/388-994(d), T: 06155/ 5569(p)

Ebenhög, J.W., Dipl.-Ing., Rotkehlchenweg 96, 25421 Pinneberg. T: 0041 1 870620, F: 00411 8709111

Ebersbach, F., Dr.-Ing., Projektleiter, Krauthofstr. 13, 18439 Stralsund

Eberz-Schuster, U., Dr. rer. nat., Rohrbach Zement GmbH & Co KG, 72359 Dotternhausen. T: 07427/79225(d), F: 07427/79300(d)

Eckardt, H., Dr.-Ing., Im Hummerholz 54, 71397 Leutenbach – Weiler. T: 0711/88025060(d), F: 0711/ 8873328(d), T/F: 07195/3145(p)

Eckardt, M., Dipl.-Geol., Büro für Ing.- u. Hydrogeologie, Boden- und Felsmechanik, Umweltgeotechnik, Johanniter Str. 23, 52064 Aachen. T: 0241/402028(d), F: 0241/ 402027(d)

Eckert, W., Dipl.-Ing., Ingenieurbüro Eckert, Crusiusstr. 7, 09120 Chemnitz. T: 0371/53012-11(d), F: 0371/53012-10(d)

Eckstaller, W., Dipl.-Geol., Putzbrunnerstr. 1, 81737 München. T: 089 54272818(d), F: 089 54272855

Edelmann, L., Dr.-Ing., Bessunger Str. 3–5, 64285 Darmstadt. T: 06151/662171(p)

Effenberger, K., Dipl.-Geol., Victoriahain 6, 45141 Essen. T: 0201/211889(p)

Egey, Z., Dipl.-Ing., Ingenieurbüro Egey, Waseneckstr. 16, 77694 Kehl. T: 07854/489(d), F: 07854/9079(d), T: 07854/ 18704(p)

Egger, P., Prof. Dr.-Ing., Labor für Felsmechanik, Abtlg. Bauingenieurwesen, ETH Lausanne, CH-1015 Lausanne, Schweiz. T: 0041/21/6932328(d), F: 0041/21/6934153(d)

Egloffstein, T., Dr. rer. nat., Geschäftsführer, ICP Ingenieurgesellschaft Prof. Czurda & Partner mbH, Eisenbahnstr. 36, 76229 Karlsruhe. T: 0721 94477-0(d), F: 0721/94477-70(d), T: 07271/51466(p), F: 07271/ 950128(p)

Egner, G., Dipl.-Geol., Geschäftsführer, Egner + Partner, Ingenieurbüro für Angewandte Geologie und Umweltplanung, Waldhörnlestr. 18, 72072 Tübingen. T: 07071/ 78099(d), F: 07071/78022(d)

Ehlers, E.-H., Dr.-Ing., Nordlabor GmbH, Flensburger Str. 15, 25421 Pinneberg. T: 04101/78201-23(d), F: 04101/ 72737(d)

Ehlers, K.-D., Dr.-Ing., Vorstandsmitglied der Bilfinger + Berger Bauaktiengesellschaft, Carl-Reiss-Platz 1–5, 68165 Mannheim. T: 0621/459 2808(d), F: 0621/459 2814(d), T: 0621/4185 060(p)

Eickenbrock, C., Dipl.-Ing., Op'n Idenkamp 20, 22397 Hamburg. T: 040 227000-39(d), 040 60566490(p), F: 020 60566496 (p)

Eiermann, P., Prof. Dipl.-Ing., An der Egge 10a, 58093 Hagen. T: 02331/51756(p)

Eigenschenk, E., Dipl.-Geol., IFB Eigenschenk GmbH, Mettener Str. 33, 94455 Deggendorf. T: 0991/370150(d), F: 0991/33918(d)

Eisenbach, O., Dipl.-Ing., Briennerstr. 39a, 64560 Riedstadt

Eisenhardt, M., Dipl.-Ing., Windloh 49, 22589 Hamburg. T: 040/8702182(p), F: 040/874678(p)

Eitner, V., Dr., DIN Deutsches Institut für Normung e.V., NABau – Normenausschuß Bauwesen, Burggrafenstr. 6, 10787 Berlin. T: 030/2601-2526(d), F: 030/2601-42526(d)

El Mossallamy, Y., Dr.-Ing., Arcadis Trischler und Partner, Berliner, Allee 6, 64295 Darmstadt. T: 06151/388551(d), F: 06151/388996(d), T: 06151/782321(p)

Elsing, A., Dipl.-Ing., Huesker Synthetic GmbH & Co. Abt. Anwendungstechnik, Fabrikstr. 13–15, 48712 Gescher. T: 02542/701-299(d), F: 02542/701-499(d)

Emmrich, R., Dr.-Ing., Annabergweg 23, 30851 Langenhagen. T: 05341 823 272(d)

Engel, J., Müncher Platz 6, 01187 Dresden. T: 0351 4632314, F: 0351 4634131

Ennigkeit, A., Dipl.-Ing., Technische Universität Darmstadt Institut für Geotechnik, Petersenstr. 13, 64287 Darmstadt. T: 06151/163149(d)

Erban, P.-J. Prof. Dr.-Ing., Müllerwiese 17, 51491 Overath. T: 02206/82904(p), F: 02206/5063(p)

Erdemgil, M., Prof. Dr.-Ing., Cankaya Cad. 21/9, TR-06700 Ankara, Türkei. T: 0090/312/4468527(d), F: 0090/ 312/4468529(d), T: 0090/312/2666031(p)

Erichsen, C., Dr.-Ing., Geschäftsführer, WBI Prof. Dr.-Ing. W. Wittke Beratende Ingenieure für Grundbau und Felsbau GmbH, Henricistr. 50, 52072 Aachen. T: 0241/889870(d), F: 0241/8898733(d)

Estermann, U., Dipl.-Ing., ELE-Erdbaulaboratorium Essen Prof. Dr.-Ing. H. Nendza u. Partner, Susannastr. 31, 45136 Essen. T: 0201/26608-48(d), F: 0201/253733(d)

Faustmann, H., Dipl.-Ing., Ercosplan Ingenieurgesellschaft Geotechnik und Bergbau mbH, Arnstädter Str. 28, 99096 Erfurt. T: 0361/3810-0(d)

Fechner, D., Dr. rer. nat. Dipl.-Geol., Ingenieurgesellschaft Geo-Consult, An der Saline 31, 63654 Büdingen. T: 06042/4194(d), F: 06042/1382(d), T: 06031/92052(p)

Fedinger, H., Dipl.-Ing., Rhein-Ruhr Ingenieurges. mbH, Institut für Grundbau und Umwelttechnik, Burgwall 5, 44135 Dortmund. T: 0231/5482-0(d), F: 0231/575556(d)

Fehsenfeld, A., Karolingerstr. 9, 64823 Groß-Umstadt. T: 06078/8121(p)

Feix, W., Dipl.-Geol., Geotechnisches Büro Feix, Hopfengarten 18, 65795 Hattersheim

Felber, G.J., Dipl.-Ing. (FH), Schönbacher Str. 37, 86556 Kühbach. T: 08251/3505(p)

Fels, F., Dipl.-Ing., Geschäftsführer, TerraConsult GmbH, Beratende Ingenieure für Geo- und Umwelttechnik, Hobrechtstr. 34, 64285 Darmstadt. T: 06151/409922(d), F: 06151/409920(d)

Fenzl, O., Dipl.-Ing., Berat. Ing., Baugrundberatung Fenzl GmbH, Am Stühm Süd 112, 22175 Hamburg. T: 040/6400044(d), F: 040/6405766(d)

Ferron, D., 182 Rue de Parc, 3542 Dudelange, Luxenburg. T: 00352 510 780

Fesseler, W., Dr., Gewerbegebiet 8/1, 88213 Ravensburg. T: 0751/96961(p), F: 0751/96951(p)

Festag, G., Dipl.-Ing., Erlenweg 9, 64342 Seeheim-Jugenheim. T: 06257/904256(p), 06151/163249(d), F: 06151/166683(d)

Fiechter-Scharr, I., Dr.-Ing., Wiss. Angest., Hintere Gasse 21, 70825 Korntal-Münchingen. T/F: 07150/970806(p)

Filho, F.O, Ing., Q-15 Conjunto e Casa 55, BR-73000 Sobradinho Df, Brasilien

Fischer, C.M., Dipl.-Ing., Geschäftsführer, Scanrock GmbH, Schlossplatz 8, 29221 Celle. T: 05141/28200(d), F: 05141/6205(d)

Fischer, J., Ing., Saarwerdenstraße 41, 41541 Dormagen. T: 02133/49219(p), F: 02133/10781(p)

Fischer, T., Dipl.-Ing., Bereichsleiter, Keller Grundbau GmbH, Bereich Süd, Schwarzwaldstr. 1, 77871 Renchen. T: 07843/709-0(d), F: 07843/709-188(d), T: 07843/1283(p)

Fishman, Y., Dr.-Ing., Jenkelweg 16, 22119 Hamburg. T: 040/736-729-52(p)

Fliß, A., Kindermannstr. 144, 99867 Gotha. T: 03621/757305(p)

Florian, S., Ing., Fraunhoferstr. 16, 82152 Planegg (Martinsried). T: 089/8958044(p), F: 089/8958045(p)

Floss, R., Prof. Dr.-Ing. Dr.-Ing. E.h., Technische Universität München, Lehrstuhl und Prüfamt für Grundbau, Boden- und Felsmechanik, Baumbachstr. 7, 81245 München. T: 089/289-27132(d), F: 089/289-27189(d), T: 08024/48395(p), F: 08204/48195(p)

Flügge, F., Dr. agr., Huesker Synthetic GmbH & Co, Vertriebsbüro Rostock, Darwin Ring 4, 18059 Rostock

Foik, G., Dr.-Ing., Wulfsdorfer Weg 11c, 22949 Ammersbek

Franke, D., Univ.-Prof. Dr.-Ing. habil., Direkt. d. Inst. f. Geot., Plquenscher Ring 44, 01187 Dresden. T: 0351/4634248(d), F: 0351/4634131(d), T: 0351/4030594(p)

Franke, J., Dipl.-Ing., Wissenschaft. Mitarbeiter, TU Berlin, Grundbauinstitut Prof. Savidis, Straße des 17. Juni 135, 10623 Berlin. T: 030/31424-786(d), F: 030/31424-492(d), T: 030/3214924(p)

von Fransecky, U., Geschäftsführer, Huesker Synthetic GmbH & Co., Fabrikstr. 13–15, 48712 Gescher. T: 0254 701-0(d), F: 02542 701 490(d)

Friedewold, C., Dipl.-Ing, Bauingenieur, Posthof 9, 86609 Donauwörth

Friedrich, K., Dr., Ingenieurgeologe, Zedernweg 20, 65527 Niedernhausen. T: 06127/2477(p), F: 06127/78428(p)

Friese, M., Dipl.-Ing., Scherlingstr. 66, 58640 Iserlohn. T: 0171/1943812(p), F: 02304/953146(p)

Frisch, R., Dipl.-Ing., Suspa Spannbeton GmbH, Abt. Ankertechnik, Germanenstr. 8, 86331 Königsbrunn. T: 08232/960721(d), 08232/4367(p), F: 08232/730365(p)

Fröhlich, B., Dr.-Ing., Bauingenieur, gbm Gesellschaft für Baugeologie und Meßtechnik mbH, Robert Bosch Str. 7, 65549 Limburg/Lahn. T: 06431/9112-0(d), F: 06431/9112-10(d), T: 06479/1427(p)

Früchtenicht, H., Dr.-Ing., GeoIngenieure Früchtenicht GmbH, Wintergasse 24, 64832 Babenhausen. T: 06073/731146(d), F: 06073/731148(d)

Fuhrmann, D., Dipl.-Geol., Chefbauleiter, Theresienstr. 53, 80333 München. T/F: 089/528136(p)

Fülling, P., Dipl.-Geol., Fülling Beratende Geologen GmbH, In der Krim 42, 42369 Wuppertal. T: 0202/24649-0(d), F: 0202/24649-60(d)

Gaasch, F., Dipl.-Ing., Sales Manager, ISPC (International Sheet Pile Company) c/o Profil Arbed, 66 Rue de Luxembourg, L-4009 Esch/Alzette, Luxemburg. T: 00352/53133213(d), F: 00352/53133293(d), T/F: 00352/298981(p)

Gallinger, A., Dipl.-Geol., Streicher KG, Rohrleitungsbau, Schwaigerbreite 17, 94469 Deggendorf. T: 0991/330-247(d), F: 0991/330-260(d)

Gärber, R., Dipl.-Ing., Departement de Genie Civil, Ecole Polytechnique Federale de Lausanne, 1015 Lausanne - Ecublens, Schweiz

Garbrecht, D., Dr.-Ing., Geschäftsführer, Franken Consult, Gesellschaft für Ingenieurwesen mbH, Postfach 10 04 36, 95404 Bayreuth. T: 0921/8806-0(d), F: 0921/8806-88(d)

Gartung, E., Dr.-Ing., Leipziger Str. 4, 90518 Altdorf. T: 0911/6555561(d), F: 0911/6555599(d), T: 09187/2440(p), F: 09187/958509(p)

Gäßler, G., Prof. Dr.-Ing., Fachhochschule München, Fachbereich 02, Karlstr. 6, 80333 München. T: 089/1265-2665(d), F: 089/1265-2699(d)

Gattermann, J.H., Dr.-Ing., Paul-Gerhardt-Str. 28, 52072 Aachen. T: 0241/80-5249(d)

Gauglitz, E., Dr., Geschäftsführer, Erdbaulabor Göttingen GmbH, Erd- und Grundbau, Deponiebau, Umwelttechnik, Hans-Böckler-Str. 2, 37079 Göttingen. T: 0551/50540-0(d), F: 0551/50540-44(d)

Gebel, R., Dipl. Geogr., Geschäftsführer, Geoplan GmbH, Donau-Gewerbepark 5, 94486 Osterhofen. T: 09932/9544-0(d), F: 09932/9544-77(d)

Geil, M., Dr.-Ing., Von-Ossietzky-Ring 55, 45279 Essen. T: 0201/521973(p)

Geißler, J., Dipl.-Ing., Bauingenieur, Dr.-Otto-Nuschke-Str. 9, 08396 Waldenburg. T: 037608/21418(p)

Giszas, H., Staatsrat Prof. Dr.-Ing., Freie und Hansestadt Hamburg, Wirtschaftsbehörde, Alter Steinweg 4, 20459 Hamburg. T: 040/35041855(d), F: 040/35042818(d)

Glade, T., Dr., Geographisches Institut, Meckenheimer Allee 166, 53115 Bonn. T: 0228/739098(d), F: 0228/739099(d), T: 0228/797035(p)

Glawe, U., Dr. rer. nat., Projektleiter, Zürcherstr. 68, 8953 Dietikon, Schweiz. T: 0041/1/7411725(p)

Gneger, W., Dipl.-Ing., Jakob-Gross-Str. 47, 63879 Weibersbrunn

Gödecke, H.-J., Dr.-Ing., Geot. Büro Prof. Schuler und Dr.-Ing.-Gödecke GbR mbH, Salzmannstr. 29/1, 86163 Augsberg. T: 0821 62026(d), F: 0821 666630(d)

Gödecke, GbR mbH, Salzmannstr. 29/1, 86163 Augsberg. T: 0821/62026(d), F: 0821/666630(d)

Göhring, D., Johannestr. 56, 53225 Bonn. T: 0241/707249(d)

Goldscheider, M., Dr.-Ing., Universität Karlsruhe, Institut f. Boden- und Felsmechanik, Postfach 69 80, 76128 Karlsruhe. T: 0721/6083291(d), F: 0721/696096(d)

Gommeringer, K., Dipl.-Geol., Geschäftsführer, IUG – Ingenieurgesellschaft für Umweltgeologie und Geotechnik, In der Neckarhalle 127/1, 69118 Heidelberg. T: 06221/805862(d), F: 06221/805863(d), T: 06221/805324(p)

Gotschol, A., Pfortenteich 5, 99974 Mühlhausen

Gottheil, K.-M., Dipl.-Ing., Geschäftsführer, Steigenhohlstr. 11, 76275 Ettlingen. T: 07244/7013-13(d), F: 07244/7013-17(d)

Gräbe, B., Dipl.-Ing., Bauingenieur, Grüner Waldweg 45, 34121 Kassel

Grabe, J., Prof. Dr.-Ing., TU Hamburg-Harburg Arbeitsbereich 1–13 Geotechnik u. Bauproduktionstechnik, 21071 Hamburg. T: 040/428783762(d)

Grabe, M., Dipl.-Ing., Hans-Eidig-Str. 1, 21224 Rosengarten. T: 04105 58247(p), F: 04105 635936(p)

Graf, B., Dr.-Ing., Schneckenmannstr. 9, CH-8044 Zürich, Schweiz

Graf, H.J., Dipl.-Ing., Josephinenstr. 3, 81479 München

Grasshoff, H., Prof. Dr.-Ing., Baudirektor a.D., Riekestr. 2, 28359 Bremen. T: 0421/2386334(p)

Grauf, T., Dipl.-Geol., 21, rue le Verger Bonnet, F-86190 Beruges, Frankreich. T: 0033/549590710(p), 0033/549303500(d), F: 0033/549303535(d)

Griese, L., Dipl.-Ing., Ehnernstr. 52A, 26121 Oldenburg. T: 04152/8074-0(d), F: 04152/807420(d)

Griese, R., Firmeninhaber, Baugrund-Institut R. Griese, Lessingstr. 7, 93152 Nittendorf. T: 0171/8966917(d), F: 09404/5670(d), T: 09404/4995(p)

Grießl, D., Dr.-Ing., Geschäftsführer, Crimmitschaverstr. 108G, 08058 Zwichau

Grönemeyer, K., Dipl.-Ing., Rosenstr. 28, 21255 Tostedt. T: 040 229257212 (d), F: 040 229257299 (d), T: 04105/8690(p), F: 04105/869299(p)

Groß, U., Dr.-Ing., GEOTECHNIK PROJEKT, Gesellschaft für Geotechnik im Bauwesen, Bergbau und Umweltschutz bR, Diezmannstr. 67, 04207 Leipzig. T: 0341/4217800(d)

Grubert, M.Sc., Peter, Dr.-Ing. Dipl.-Ing., GGU Gesellschaft für Grundbau und Umwelttechnik mbH, Werner-vol-Siemens-Ring 13a, 39116 Magdeburg, T: 0391/6230243(d), F: 0391/6230244(d), T: 05331/31687(p)

Gruhn, H., Dipl.-Ing., Baugrundsachverständiger, Ingenieurbüro für Geotechnik (IbG), Wolfganggasse 12, 74072 Heilbronn. T: 07131/9959-0(d), F: 07131/9959-99(d)

Gründer, J., Dr. Dipl.-Geol., Geotechnisches Institut Dr. Gründer, Föhrenstr. 6, 90602 Pyrbaum. T: 09180/94040(d), F: 09180/940418(d)

Grundhoff, T., Dipl.-Ing., Selikumer Weg 23, 41464 Neuss. T: 040/389139-41(d), F: 040/38091/0(d), T: 02131/130327(p)

Grunwaldt, B., Hans-Leipelt-Str. 8, WG 15, 80805 München. T: 089/3231939(p)

Gülzow, H.-G., Prof. Dr.-Ing., Hochschullehrer FH, Osningstr. 71, 33605 Bielefeld. T: 0571/8385-0(d), F: 0571/8385-250(d), T: 0521/22052(p), F: 0521/22190(p)

Gündling, N., Dipl.-Ing., Ingenieurbüro für Geotechnik N. Gündling, Römerstr. 57, 64291 Darmstadt. T: 06151/373069(d), F: 06151/373079(d)

Günther, G., Dr., Geschäftsführer, ICP GmbH Leipzig, Fasanenweg 2, 04420 Markranstädt. T: 0341/94426-0(d), F: 0341/94426-15(d)

Günther, K., Dr.-Ing., IGB Ingenieurbüro für Grundbau, Bodenmechanik und Umwelttechnik, Heinrich-Hertz-Str. 116, 22083 Hamburg. T: 040/227000-31(d), F: 040/227000-34(d)

Günther, M., Dipl.-Ing., G. Puhlmann GmbH & Co KG, Potsdamer Str. 16–17, 14163 Berlin

Günther, U., Dipl.-Ing., Ingenieurbüro für Boden und Umwelttechnik, Im Heidewinkel 30, 40625 Düsseldorf. T: 0211/284150(d), F: 0211/298273(d)

Gürtler, J., Dipl.-Ing., Gürtler + Wiesel & Partner Ingenieurgesellechaft für Hochbau & Tiefbau, Karmeliter Str. 44, 53329 Bonn

Güttler, U., Dr.-Ing., August-Schmidt-Str. 24, 45739 Oer-Erkenschwick

Haberkorn, H., Alfred-Messel-Weg 6/91, 64287 Darmstadt. T: 06151/715443(p), F: 06151/715443(p)

Hackenbroch, W., Dipl.-Ing., Partner, Ingenieurbüro Domke Nachf., Mannesmannstr. 161, 47259 Duisburg. T: 0203/781080(d), F: 0203/750455(d), T: 0203/758400(p)

Häcker, H.-A., Dipl.-Geol., Niederlassungsleiter, GTU Ingenieurgesellschaft GmbH, NL Sangerhausen, Glück-Auf-Str. 41, 06526 Sangerhausen. T: 03464/622270(d), F: 03464/622272(d)

Hackl, L., Baurat h.c. Dipl.-Ing., Ing.-konsulent Bauwesen, Geoconsult ZT GmbH, Sterneckstr. 52, A-5020 Salzburg, Österreich. T: 0043/662/65965-0(d), F: 0043/662/65965-10(d)

Hähne, K., Dr.-Ing., Baubehörde der FHH, Amt für Wasserwirtschaft, Hochwasserschutz und Konstruktiver Wasserbau, Baumwall 3, 20459 Hamburg, T: 040/349131(d), F: 040/34913552(d)

Hahnen, C., Ströverstr. 9, 59427 Unna

Halbach, S., Dipl.-Ing., Garschager Heide 32, 42899 Remscheid. T: 02330/800-923(p)

Hallbauer, C., Dr.-Ing., Baugrundbüro Dr. Hallbauer + Dressel, Spiegelstr. 31, 08056 Zwickau. T: 0375/277048(d), 036624/20290(p)

Haller, H.-P., Dipl.-Ing. (FH), Haldenstr. 83, 71254 Ditzingen. T: 030/414723-0(d), 07156/959050(p), F: 07156/959052(p)

Haller, J., Erwin-von-Steinbach Weg 27a, 80937 München. T: 089/3137311(p)

Hangen, H., Huesker Synthetic GmbH & Co. Abt. Anwendungstechnik, Fabrikstr. 13–15, 48712 Gescher. T: 02542/701302(d), F: 02542/701499(d)

Hänsel, F., Dipl.-Ing., Am Pfannenstiel 19, 86152 Augsburg. T: 0821/7802172(d), 0821/5082509(p)

Hansel, H., Dipl.-Ing., Geschäftsleitung, Geotechnisches Institut, Prof. Dr. Kurt Magar, Winterhäuser Str. 9, 97084 Würzburg. T: 0931/61440(d), F: 0931/6144-200(d)

Hartmann-Linden, R., Dr.-Ing., Büchel 13, 51399 Burscheid. T: 02174 748288(p)

Hartung, M., Dipl. Ing.m Lehrstr. 29, 45356 Essen

Hartwig, U., Dipl.-Ing., Bruckenächer 5, 70565 Stuttgart. T: 0711/7803500(p)

Haschke, R., Elbestr. 6, 36041 Fulda. T: 0661/55324(p)

Hasel, K., Busehofstr. 29, 45144 Essen. T: 0201/7509524(p)

Haupt, W., Prof. Dr.-Ing., Landesgewerbeanstalt Bayern, Geotechnik – Versuchswesen, Tillystr. 2, 90431 Nürnberg. T: 0911/655-5531(d), F: 0911/655-5510(d), T: 09122/84998(p)

Hebener, H.-L., Dipl.-Ing., Gutzkowstr. 5, 10827 Berlin

Heerten, G., Dr.-Ing., Geschäftsf. Gesellsch., Naue Fasertechnik GmbH & Co. KG, Geschäftsleitung, Wartturmstr. 1, 32312 Lübbecke. T: 05741/4008-10(d), F: 05741/4008-83(d), T: 05741/20076(p)

Heibaum, M., Dr.-Ing., Primelweg 5, 76297 Stutensee-Friedrichstal. T: 0721/9726382(d), F: 0721/9726483(d)

Heil, H., Dr.-Ing., Berat. Ingenieur VBI, Ingenieurbüro IGB, Heinrich-Hertz-Str. 116, 22083 Hamburg. T: 040/227000-40(d), F: 040/227000-28(d)

Heil, M., Dipl.-Ing., Waldheimstr. 18, 78239 Rielasingen

Heimer, V., Dipl.-Ing., Erdbaulaboratorium Saar, Steigerstr. 51, 66292 Riegelsberg. T: 06806/440045(d), F: 06806/47213(d)

Hein, B.-R., Dipl.-Ing., Tulpenstr. 6, 79664 Wehr. T: 07762/2343(p)

Heine, A., Dipl.-Ing., selbständig, In den Hofwiesen 8, 45711 Datteln

Heineke, S.A., Dipl.-Ing., Wissensch. Mitarbeiter, Von-Ketteler-Str. 11, 64297 Darmstadt. T: 06151/162849(d), T: 06151/51254(p)

Heitfeld, M., Dr.-Ing., Geschäftsführer, Ingenieurbüro Heitfeld-Schetelig GmbH, Preusweg 74, 52074 Aachen. T: 0241/70516-0(d), F: 0241/70516-20(d), T: 0241/77652(p)

Heitzer, K., Dr.-Ing., A Schubert Beratendre Ingenieure für Geotechnik, Werner-v-Siemens Str. 17, 82140 Olching. T: 08142 49000(d), F: 08142 3795(d), T: 089 8344170(p)

Hellerer, H.-O., Dipl.-Geol., Mittenwalderstr. 37a, 81377 München. T: 089/289-27157(d), F: 089/289-27193(d), T: 089/7148786(p), F: 089/71019347(p)

Henning, D., Dr.-Ing. Dr.-Ing. E.h., Stresemannstr. 31, 52349 Düren. T: 0221/4801400(d), F: 0221/4801402(d), T: 02421/58616(p), F: 02421/502877(p)

Henzinger, J., Dr. Dipl.-Ing., Büro für Geotechnik, Plattach 5, A-6094 Grinzens, Österreich. T: 0043/52345533(d), F: 0043/523455334(d)

Heppner, B., Dipl-Ing., Baugrundingenieur, Ingenieurbüro B. Heppner, Hauptstr. 17, 18442 Zimkendorf. T: 038321/60052(d), F: 038321/60051(d)

Hermanns Stengele, R., Prof. Dr. sc. techn., Bauingenieur, Institut für Geotechnik ETH Zürich – Hönggerberg, 8093 Zürich, Schweiz

Herrmann, J., Dipl.-Geol., Abteilungsleiter, Institut für Baustoffprüfung und Umwelttechnik – IBE GmbH, Bössingerstr. 23, 74243 Langenbrettach 2. T: 07946/2001(d), F: 07946/2559(d), T: 0171/3772161(p)

Herrmann, R.A., Prof. Dr.-Ing., Universität – GH-Siegen, Fachbereich Bauingenieurwesen, Institut für Geotechnik, Paul-Bonatz-Str. 9–11, 57068 Siegen. T: 0271/740-2168(d), F: 0271/740-2572(d), T: 09825/353(p), F: 09825/ 358(p)

Hermann, S., Hans-Sachs-Str. 11, 76133 Karlsruhe. T: 0721/ 8305095

Hettler, A., Univ.-Prof. Dr.-Ing. habil., Universität Dortmund Lehrstuhl Baugrund – Grundbau, August – Schmidt – Str. 8, 44227 Dortmund. T: 0231/7553012(d), F: 0231/ 7555435(d)

Heun, T., Dipl.-Ing., 7, Les Etangs, F-44310 La Limouzinière, Frankreich. T: 0033/240787246(p), F: 0033/ 240787643(p)

Heusermann, S., Dr.-Ing., Wissensch. Direktor, Bundesanstalt für Geowissenschaften und Rohstoffe, Ref. B 2.6, Stilleweg 2, 30655 Hannover. T: 0511/6432429(d), F: 0511/ 6433694(d)

Heyer, D., Dipl.-Ing., TU München, Lehrstuhl und Prüfamt f. Grundbau, Boden- und Felsmechanik, Baumbachstr. 7, 81245 München. T: 089/289-27135(d), F: 089/289-27189(d), T: 040/847300(p)

Hochgürtel, T., Dr.-Ing., Wissensch. Angest., Zum Zehntgarten 20, 53894 Mechernich. T: 02443/31217(p)

Hock-Berghaus, K., Dr.-Ing., ö.b.u.v. Sachv. für Grundbau und Bodenmechanik, Ingenieurbüro für Geotechnik, Kreuzstr. 16, 42277 Wuppertal. T: 0202/5287770(d), F: 0202/ 5288989(d)

Hoemann, A., Dipl.-Geol., Freiber. Geologe, Geologie-Büro Hoemann, Alter Graben 2, 33034 Brakel-Gehrden. T: 05648/372(d), F: 05648/841(d)

Höfer, U., Dr.-Ing., Grundbauinstitut Dortmund, Am Remberg 180, 44269 Dortmund. T: 0231/441064(d), F: 0231/ 443665(d)

Hoffmann, H., Dipl.-Ing., Ingenieursozietät Prof. Dr.-Ing. Katzenbach gmbh Pfaffenwiese 14a, 65931 Frankfurt am Main. T: 069/9362230(d), F: 069/361049(d)

Höffner, W., Dipl.-Geol., Büro für Geotechnik Aalen, Robert-Bosch-Str. 59, 73431 Aalen. T: 07361/94060(d), F: 07361/940610(d)

Hofmann, E.H., Dipl.-Ing., Wiss. Mitarb., Universität-GH-Siegen, Fachbereich Bauingenieurwesen, Institut für Geotechnik, Paul-Bonatz-Str. 9–11, 57068 Siegen. T: 0271/ 740-2176(d), F: 0271/740-2572(d), T/F: 02733/ 51303(p)

Hofmann, H., Dr.-Ing., Großingesheimer Str. 39, 74321 Bietigheim – Bissingen. T: 07142/55988(p), F: 07142/ 56961(p)

Hohlrieder, R., Troxler GmbH, Gilchinger Str. 33, 82239 Alling. T: 08141/71063(d), F: 08141/80731(d)

Holzwarth, H.-G., Dipl.-Geol., Jettenburger Str. 29, 72127 Kusterdingen. T: 07071/369843(d), F: 07071/368933(d), T: 07071/31340(p)

Hönisch, K., Dr.-Ing., Rüsselheimer Str. 12c, 64560 Riedstadt. T: 0611/708731(d), F: 0611/708555(d), T: 06158/2850(p)

Hoof van, P., Dipl.-Ing., Grundermühlenweg 19, 51381 Leverkusen. T: 02171/731999(p)

Horn, A., Univ.-Prof. Dr.-Ing., Ahornstr. 1, 82152 Krailling. T: 089/8572793(p), F: 089/8575739(p)

Hornbruch, J., Dipl.-Ing., selbständig, Berkhopstr. 3, 30938 Burgwedel. T: 05139/891695(p), F: 05139/891696(p)

Hornig, E.-D., Dipl.-Ing. (Univ.), Wiss. Ang., FH Stuttgart, Augsburger Str. 281 D, 70327 Stuttgart. T: 0711/ 1212907(d), F: 0711/1212666(d)

Höwing, K.-D., Dr. Dipl.-Geol., Sammenheim 98, 91723 Dittenheim. T: 09833/95925(p)

Hu, Y., Dr.-Ing., Steinmetzstr. 18, 90431 Nürnberg. T: 0561/ 8042629(d), 0911/9657775(p)

Huber, G., Dr.-Ing., Universität Karlsruhe, Institut f. Boden- und Felsmechanik, Postfach 69 80, 76128 Karlsruhe. T: 0721/ 608-2221(d), F: 0721/696069(d)

Huch, T., Dipl.-Ing., Wissensch. Angestellter, TU Braunschweig, Institut für Grundbau und Bodenmechanik, Gaußstr. 2, 38106 Braunschweig. T: 0531/391-2730(d), T: 0531/73865(p)

Hügel, H.M., Dr.-Ing., Universität Kaiserslautern, Fachgebiet Bodenmechanik und Grundbau, Postfach 3049, 67653 Kaiserslautern. T: 0631/2052979(d)

Hundsdorf, H., Dipl.-Geol., Middle East Surveys, P.O. Box 27198, Safat 13132, Kuwait. T: 00965/4763481(d), F: 00965/5641369(d), T/F: 00965/5641369(p)

Hurler, H., Dr. rer. nat., Ingenieurbüro für Geotechnik Dr. Hurler + Partner, Sitzinger Str. 9–11, 67549 Worms. T: 06241/978210(d), F: 06241/9782118(d)

Huwe, H.-W., Dipl.-Geol., DMT IFP, Franz-Fischer-Weg 61, 45307 Essen. T: 0201/172-1411(d), T: 0209/71549(p)

Ickler, U., Dipl.-Ing., Taku-Fort-Str. 14, 81827 München. T/F: 089/4305919(p)

Illing, G., Dipl.-Ing., Prüfing. f. Baustatik, Ingenieurbüro für Baustatik und Tragwerksplanung, Glauchauer Str. 5, 09350 Lichtenstein. T/F: 037204/87707(d), T/F: 037204/ 2768(p)

Immig, H., Dipl.-Ing., Witt + Jehle Geotechnik GmbH, Zeisigstr. 4, 56075 Koblenz. T: 0261/95269-14(d), F: 0261/ 95269-20(d)

Ismail, T.H., 12 Rue Yehia El-Mazouni, B.P 66 Bis, DZ- Alger, Algerien. T: 002132/923893(p), F: 002132/ 923894(p)

Ittershaqen, M., Dipl. Ing., Darmastädterstr. 54, 64354 Reinheim/Odenwald. T: 06162 968967(p)

Jacobsen, N., Dipl.-Ing., Projektingenieur, WCI Umwelttechnik GmbH, Hochofenstr. 19/21, 23569 Lübeck. T: 0451/ 306022

Jagau, H., Prof. Dr.-Ing., Berat. Ingenieur, Prof. Dr.-Ing. H. Jagau & Partner, Geotechnik – Umwelttechnik, Leipziger Str. 19–21, 28857 Syke. T: 04242/50504(d), F: 04242/ 6112(d)

Jancsecz, S., Dipl.-Ing., Rückinger Str. 3, 63526 Erlensee. T: 06183/902077, F: 06183/901431(p)

Jebe, P., Dipl.-Ing., Freiber., Bethlehemplatz 2, 30451 Hannover. T: 0511/2106753(p)

Jechalik, R., Dipl.-Geol., Projektierungsgesellschaft für Geotechnik und Grundbau GmbH, Lise-Meitnerstr. 11, 70794 Filderstadt. T: 07127/88547(p)

Jehle, R., Dipl.-Ing., Witt + Jehle Geotechnik GmbH, Zeisigstr.4, 56075 Koblenz. T: 0261/952690(d), F: 0261/ 9526920(d)

Jennes, H., Dipl.-Ing., Selbst. Ingenieur, Forsthaus 1, 40883 Ratingen. T: 02102/68843(p)

John, I., Dipl.-Geol., Goldaper Str. 19, 48161 Münster. T: 0251/327909(d), F: 0251/327928(d), T: 02534/65127(p)

Jordan, P., Dr.-Ing., Geschäftsf. Gesellschaft., Prof. Dr.-Ing. Jessberger & Partner GmbH, Am Umweltpark 5, 44793 Bochum. T: 0234/68775-0(d), F: 0234/68775-10(d)

Jörger, R., Dipl.-Ing., Bilfinger + Berger Bauaktiengesellschaft Ingenieurbau Spezialtiefbau Technik, Carl-Reiß-Platz 1–5, 68165 Mannheim. T: 0621/4592382(d), F: 0621/4592433(d), F: 06222/51238(p)

Jossen, T.G., Dipl.-Geol., Spitzlei & Jossen, Ing.-Büro f. Bauwes. u. Geol. GmbH, Fichtenweg 1, 53721 Siegburg. T: 02241/9192-0(d)

Joswig, P., Dipl.-Ing., Ingenieurbüro Geyer & Joswig, Am Hubengut 4, 76149 Karlsruhe. T: 0721/97835-0(d), F: 0721/97835-99(d), T: 0721/402781(p)

Jovanovic, M., Dipl. Ing. Bauingenieurin, Bavariastr. 9a, 80336 München. T: 0721 608 2238(d), F: 0721 696096 (d), T: 089 76703922(p)

Jung, R.G., Dipl.-Ing., Baugrundinstitut – Ingenieurbüro, Fritz-Winter-Str. 11, 86911 Dießen. T: 08807/1883(d), F: 08807/6097(d)

Jurgk, D., Dipl.-Geol., Wiesenstr. 6, 38889 Blankenburg. T/F: 03944/5401(p)

Kaden, K., Dipl.-Ing., Geschäftsführer, G.U.B. Ingenieurgesellschaft Lausitz GmbH, Alte Coseler Str. 26, 02994 Bernsdorf. T: 035723/22801(d), F: 035723/22803(d)

Kahl, M., Dr.-Ing., Beratender Ingenieur, Grundbauingenieure Steinfeld und Partner GbR, Alte Königstr. 3, 22767 Hamburg. T: 040/38913934(d), F: 040/3809170(d)

Kämper, K., Dipl.-Ing., Busdorfer Str. 1, 24837 Schleswig. T: 04351/4031(d), F: 04351/41291(d), T: 04621/35638(p), F: 04621/35602(p)

Kany, M., Prof. Dr.-Ing., GEOTEC – Software, Vestnerstr. 5b, 90513 Zirndorf. T/F: 0911/606524(d), T/F: 0911/ 606524(p)

Karlsson, J.-A., Dipl.-Ing., Bilfinger + Berger Bauaktiengesellschaft, Gustav-Nachtigal-Str. 3, 65189 Wiesbaden. T: 0611/708-525(d), F: 0611/708-555(d)

Karstedt, J., Dr.-Ing., Geschäftsführer, Ingenieurbüro für Grundbau u. Bodenmechanik Dr.-Ing. R. Elmiger Dr.-Ing. J. Karstedt GmbH, Limastr. 25A, 14163 Berlin. T: 030/809926-0(d), F: 030/8011493(d)

Kast, K., Dr.-Ing., Dr.-Ing. K. Kast u. Partner Ing.-Gem. f. Umwelt- und Geotechnik, Erlerstr. 21, 76275 Ettlingen. T: 07243/39223(d), F: 07243/39224(d)

Katzenbach, R., Univ.-Prof. Dr.-Ing., Technische Universität Darmstadt, Direktor des Instituts und der Versuchsanstalt für Geotechnik, Petersenstr. 13, 64287 Darmstadt. T: 06151/162149(d), F: 06151/166683(d)

Kayser, J., Dr.-Ing., Bettina-Stieg 2, 22609 Hamburg. T: 040/81908368(d), F: 040/81908373(d), T: 040/82279803(p)

Keller, A., Dipl.-Ing., Sachverständiger, Allianz AG, Technische Beratung, Königinstr. 28, 80802 München. T: 089/38004868(d), F: 089/38004690(d)

Keller, J., Bau-Ing. ETH, Steinfeldstr. 12, CH-8153 Rümlang, Schweiz. T: 0041/1/8170751(p), F: 0041/1/8180468(p)

Kemnitz, M., Dipl.-Ing., Ing.-Büro R.-U. Wode, Büro für Geotechnik und Angewandte Umweltgeologie, Zur Felsenmühle 10, 02763 Mittelherwigsdorf. T: 03583/770410(d), F: 03583/770432(d)

Kempa, R., Prof. Dipl.-Ing., Geschäftsf. Gesellschaft, Ingenieurgesellschaft Kempa mbH, Falkenhorstweg 23, 81476 München. T: 089/7459357(d), F: 089/7459866(d)

Kempfert, H.-G., Univ.-Prof. Dr.-Ing., Universität GH Kassel, Fachgebiet Geotechnik, Mönchebergstr. 7, 34125 Kassel. T: 0561/804-2631(d), F: 0561/804-2651(d), T: 040/86646480 (p)

Kennepohl, T., Dipl.-Bauing., Betteläcker 9A, 69257 Wiesenbach. T: 06223/40959(p)

Kerreit, M., Dr.-Ing., Ing.-Büro Dr. Kerreit Baugrunderkundung/Gründungsberatung, Kurt-Huber-Weg 1, 04299 Leipzig. T/F: 0341/8612801(d)

Kief, H., Dipl.-Ing., Prokurist, Wiemer & Trachte Aktiengesellschaft, Märkische Str. 111, 44141 Dortmund. T: 0231/5413-395(d), F: 0231/5413-435(d)

Kiefer, P., Dipl.-Ing. ETH, Kiefer & Studer AG, Therwilerstr. 27, CH-4153 Reinach BL 1, SCHWEIZ. T: 0041/61/7119476(d), F: 0041/61/7119634(d)

Kiehl, J.R., Dr.-Ing. Dipl.-Phys., WBI Prof. Dr.-Ing. W. Wittke, Beratende Ingenieure für Grundbau und Felsbau GmbH, Henricistr. 50, 52072 Aachen. T: 0241/88987-0(d), F: 0241/88987-33(d)

Kinzel, J., Dipl.-Ing., Lichtenbergstr. 74, 64289 Darmstadt. T: 06151/163149(d), F: 06151/166683(d)

Kirac, M.K., Dr. rer. nat., Verantw, Geschäftsf., STFA Temel Investigation Inc., Tophanelioglu CAD No. 19, Altwniz., 81190 Istanbul, Türkei. T: 0090/216/3263546(d), F: 0090/216/3393300(d), T: 0090/2163886690(p)

Kirchberg, J., Dr.-Ing., GCE Geotechnisches Ing.-Büro Dr. Kirchberg & Ecke GmbH, Salbker Chaussee 17, 39116 Magdeburg. T: 0391/6355050(d), F: 0391/ 63550519(d)

Kirchner, S., Nieder-Ramstädter-Str. 181/B3, 64285 Darmstadt. T: 06151/423886(p)

Kirsch, K., Dipl. Ing. Rheinstr. 18, 61273 Wehrheim. T: 069/8051220, F: 069/8051229

Kittel, M., Dipl.-Ing., Friedensallee 26, 99334 Ichtershausen.

Klapperich, H., Univ.-Prof. Dr.-Ing., TU Bergakademie Freiberg Bodenmechanik, bergbauliche Geotechnik und Grundbau, Gustav-Zeuner-Str. 1, 09599 Freiberg. T: 03731/39-3614(d), F: 03731/39-3501(d)

Kleberger, J., Dr. phil. Geol., IC consulenten, ZT GmbH Salzburg Dr. Johannes Kleberger, Zollhausweg 1, 5101 Salzburg, Österreich. T: 0043/662/450 773 12(d), F: 0043/662/450 773 5(d)

Klee, J., Dr.-Ing., Freier Blick Nr. 13, 06618 Naumburg (Saale). T: 03445/762126(p), F: 03445/762162(p)

Kleen, H., Prof. Dr.-Ing., Ingenieurbüro für Geotechnik Kleen GmbH, Helmholtzstr. 13, 14467 Potsdam. T: 0331/2709266(d), F: 0331/2709268(d)

Klein, J., Dr.-Ing., Beratender Ingenieur, Deutsche Montan Technologie GmbH, Am Technologiepark 1, 45307 Essen. T: 0201 172-1432(d), F: 0201 172 1777(d)

Klein, H., Dipl.-Ing., Uhlandstr. 127, 10717 Berlin. T: 030/7532082(d), F: 030/75487445(d), T: 030/86409353(p), F: 030/86409355(p)

Kliesch, K., Dr.-Ing., An der Nachtweid 3, 68526 Ladenburg

Klingmüller, O., Dr.-Ing., Geschäftsführer, GSP Gesellschaft für Schwinguntersuchungen und dynamische Prüfmethoden mbH, Käfertaler Str. 164, 68167 Mannheim. T: 0621/331361(d)

Klobe, B., Dr.-Ing., Bauer Spezialtiefbau GmbH, Baukonstruktion, Wittelsbacherstr. 5, 86529 Schrobenhausen. T: 08252/97-1304(d), F: 08252/97-1516(d)

Klöckner, W., Dipl.-Ing., Schwanenstr. 1c, 68259 Mannheim. T: 0621/792309(d)

Klönne, H., Dipl.-Ing., Leiter Grund-u. Tunnelbau, Zerna, Köpper & Partner, Ingenieurgesellsch. f. Bautechnik, Industriestr. 27, 44892 Bochum. T: 0234/9204-123(d), F: 0234/9204-190(d)

Klüber, E., Dr.-Ing., Gambrinusstr. 28a, 64319 Pfungstadt. T: 0621/6055843(d), 06157/84332(p)

Knauber, J., Mühlenweg 38, 52531 Übach-Palenberg. T: 02451/909070(p), F: 02451/909072(p)

Kneißl, A., Dipl.-Ing., Badstr. 4, 92280 Kastl. T: 09188/9400-0(d), F: 09188/9400-49(d)

Knierim, H., Dipl.-Ing., Berat. Ingenieur, Das Baugrund Institut, Dipl.-Ing. Knierim GmbH, Wolfhager Str. 427, 34128 Kassel. T: 0561/96994-0(d), F: 0561/96994-55(d)

Knoll, P., Prof. Dr. sc. techn., Geschäftsführer, GEO-DYN GmbH Gesellschaft für Geophysikalisches Messen und Geotechnische Untersuchungen, Warthestr. 21, 14513 Teltow. T: 03328/3109-0(d), F: 03328/3109-20(d)

Knopf, S., Dipl.-Geol., Krebs & Kiefer, Beratende Ingenieure für das Bauwesen GmbH Karlsruhe, Karlstr. 46, 76133 Karlsruhe. T: 0721/3508-360(d), F: 0721/3508-211(d)

Koch, E., Dipl.-Ing., Koch Beratende Ingenieure, Dammstr. 26, 47119 Duisburg-Ruhrort. T: 0203/82811(d), F: 0203/85295(d)

Koerner, U., Dr. rer. nat. Dipl.-Geol., Ltd. Geol. Dir. i.R., Herlinstr. 6, 79312 Emmendingen. T: 07641/42563(p), F: 07681/25060(p)

Köhler, H.-J., Dipl.-Ing., Beiertheimer Allee 20a, 76137 Karlsruhe

Köhler, K., Dipl.-Ing., Janow-Podlaski-Str. 2, 16540 Hohen Neuendorf

Köhler, U., Dr.-Ing., Baugrundinstitut Dr. Köhler + Herold GmbH, Im Dorfe 6, 99428 Ulla b. Weimar. T: 03643/2429-0(d), F: 03643/242912(d)

Kohlhase, S., Prof. Dr.-Ing., Universität Rostock, FB Bauingenieurwesen, Institut für Wasserbau, Philipp-Müller-Str., 23966 Wismar. T: 03841/753-448(d), F:03841/753-306(d), T: 0381/4995-158(p), F: 0381/4995-159(p)

Kokemüller, E., Dipl.-Geol., selbständig, Ingenieurbüro Kokemüller & Möker, Merkurstr. 1D, 30419 Hannover. T: 0511/758098-3(d), F: 0511/758098-49(d)

Kolb, H., Dr.-Ing., Rasmussenweg 2, 70439 Stuttgart

Kölsch, T., Dipl.-Ing., Amyastr. 33, 52066 Aachen. T: 0241/911824(p)

Kolymbas, D., o. Prof. Dr.-techn. DI, Institut f. Geotechnik u. Tunnelbau, Universität Innsbruck, Technikerstr. 13, A-6020 Innsbruck, Österreich. T: 0043/512507-6670(d), F: 0043/512507-2996(d)

Könemann, F., Dr.-Ing., ELE-Erdbaulaboratorium Essen, Susannastr. 31, 45136, Essen. T: 0201 8959-6(d), F: 0201 253733(d)

König, D., Dr.-Ing., Akademischer Rat, Ruhr-Universität Bochum, Lehrstuhl für Grundbau und Bodenmechanik, Geb. IA 4/126, 44780 Bochum. T: 0234/7006135(d), F: 0234/7094150(d)

König, F.T., Prof. Dr.-Ing., Hasselholt 14, 23909 Ratzeburg. T: 04541/3909(p)

Kopelke, B., Dipl.-Ing., Geschäftsführer, Burmann, Mandel + Partner, Ingenieurbüro für Grundbau und Umwelttechnik, Gasstr. 18, Haus 6b, 22761 Hamburg. T: 040/896037(d), F: 040/8901621(d), T: 040/6787319(p)

Koppetsch, A., Dipl.-Ing., Weberstr. 81, 47918 Tönisvorst

Kordinand, R., Dipl.-Ing., Hinterm Vogelherd 30b, 22926 Ahrensburg

Kosack, A., Dipl.-Ing., Grundbauingenieure Steinfeld & Partner GbR, Alte Königstr. 3, 22767 Hamburg. T: 040/389139-38(d)

Köster, M., Dr.-Ing., Geschäftsf. Gesell., Baugrundinstitut Franke – Meißner u. Partner GmbH, Max-Planck-Ring 47, 65205 Wiesbaden-Delkenheim. T: 06122/51057(d), F: 06122/52591(d)

Kraft, H.-P., Dipl.-Ing., Frank + Kraft + Partner Geotechnik und Umwelttechnik GmbH, Hofer Str. 1, 81737 München. T: 089/670061-0(d), F: 089/670061-33(d), T: 089/6378999(p)

Krahl, W., Dipl.-Ing., Iländerweg 88a, 45239 Essen. T: 069/96227-416(d), T: 0201/403313(p)

Krajewski, W., Dr.-Ing., Grundbau-Institut Prof. Dr.-Ing. P. Amann Consult GmbH, Geschäftsleitung, Ober-Ramstädter Str. 42, 64367 Mühltal. T: 06151/1415-11(d), F: 06151/1415-14(d)

Kramer, H., Prof. Dr.-Ing., Hochschullehrer, Waldwinkel 31, 28759 Bremen. T: 0421/621361(p), F: 0421/621365(p)

Kramer, J., Obering., ELE-Erdbaulaboratorium Essen, Prof. Dr.-Ing. H. Nendza und Partner, Susannastr. 31, 45136 Essen. T: 0201/8959-6(d), F: 0201/253733(d)

Kratzenberg, G., Ing., Erd- u. Grundbaulaboratorium Kassel, Im Druseltal 134a, 34131 Kassel. T: 0561/36074(d), F: 0561/314811(d)

Krauß, E., Dr.-Ing., Leiter Tief- und Tunnelbau, IMS Ingenieurgesellschaft mbH, Stadtdeich 5, 20097 Hamburg. T: 040/32818-0(d), F: 040/32818-139(d)

Krauter, E., Prof. Dr., Ingenieurgeologe, geointernational Geo-Center Mainz, Mombacher Str. 49–53, 55122 Mainz. T: 06131/387071(d), F: 06131/387076(d)

Kretzer, D., Dr.-Ing., Am Steinbruch 14, 09557 Flöha

Kreuter, H., Dr. Dipl.-Geol., Hot Rock GmbH, Kaiserstr. 167, 76133 Karlsruhe. T: 0721/95789-0(d), F: 0721/95789-22(d)

Kriechbaum, J., Dipl.-Ing., Witt + Jehle Geotechnik GmbH, 56017 Koblenz

Krieg, S., Dipl.-Ing., Schurwaldstr. 23, 70771 Leinfelden-Echterdingen. T: 0711/13164-0(d)

Kroer, A., Fibertex AS, Postfach Box 80 29, DK-9220 Aalborg Ost, Dänemark. T: 0045/98/158600(d), F: 0045/98/158555(d)

Kron, G., Ingenieurbüro f. Geotechnik, Kölnstr. 144, 53111 Bonn. T: 0228/631315(d), F: 0228/652928(d)

Krüger, P., Prof., Ingenieurbüro GmbH, Giersdorf 12, 28870 Ottersberg-Posthausen. T: 04297/8188-0(d), F: 04297/8188-44(d)

Krüger, W., Dipl.-Ing., Ingenieurbüro für Bodenmechanik und Grundbau, Lutherstr. 32c, 06886 Wittenberg. T: 03491/401031(d), F: 03491/406266(d)

Krugmann, P., Dipl.-Ing., Ph. D., Brauhausstr. 9, 86732 Oettingen. T: 09082/1512(p), F: 09082/920933(p)

Krusche, J., Dipl.-Ing., Am Bahnhof 4, 64347 Griesheim. T: 06155/76746(p)

Kruse, K., Dr. sc. techn., Rheinallee 8a, 53579 Erpel

Kruse, T., Dr.-Ing., Am Kornfeld 43, 58239 Schwerte

Kudella, P., Dr.-Ing., Oberingenieur, Universität Karlsruhe, Lehrstuhl für Bodenmechanik und Grundbau, Postfach 69 80, 76128 Karlsruhe. T: 0721/608-3298(d), F: 0721/696096(d), T: 07262/4858(p)

Kudla, W., Univ. Prof. Dr. Ing., TU Bergakademie Freiberg Institut für Bodenmechanik und Grundbau, Gustav-Zeuner-Str. 1, 09596 Freiberg. T: 03731 39 2514(d), F: 03731 39 2524(d)

Kügler, M., Dipl.-Ing., Karl-Liebknecht-Str. 78, 03046 Cottbus

Kühn, G., Prof. Dr.-Ing., Universität Karlsruhe, Institut f. Maschinenwesen im Baubetr., Am Fasanengarten, 76131 Karlsruhe

Kuk, M., Westerfeldstr. 12, 30419 Hannover. T: 0511/759447(p)

Külzer, M.A., Dipl.-Ing., Brötzinger Str. 22, 75180 Pforzheim-Büchenbronn. T: 0721 6082227(d), F: 0721 696096(d), T: 07231 784253(p)

Kumbernuß, C., Ackerstr. 39, 10115 Berlin. T: 030/85909180(d), F: 030/85909120(d)

Kunz, E., Dipl.-Ing. f. Geot., Lemkestr. 198, 12623 Berlin. T: 030/4277769(p)

Kuschka, L., Halberstädter Str. 62, 37520 Osterode

Küster, J., Dübener Str. 17, 06774 Schwemsal. T/F: 034243/50788(d), T: 0341/4804076(p)

Kutzner, C., Prof. Dr.-Ing., Sachverständiger, Adolf-Kolping-Str. 6, 65719 Hofheim am Taunus. T: 06192/27643(p), F: 06192/24761(p)

Laabmayr, F., Dipl.-Ing., Ing.-Büro Laabmayr & Partner, Preishartlweg 4, A-5020 Salzburg/Liefering, Österreich. T: 0043/662/430703(d), F: 0043/662/430703/33(d)

Lächler, W., Dr.-Ing., Smoltczyk & Partner GmbH, Untere Waldplätze 14, 70569 Stuttgart

Landgraf, D., Dipl.-Ing., Mainstr. 39, 64732 Bad König. T: 06063/9306-12(d), F: 06063/9306-20(d), T: 06063/9306-0(p), F: 06063/9306-20(p)

Lange, W., Dipl.-Ing., Halbach + Lange Ingenieurbüro, Agetexstr. 6, 45549 Sprockhövel. T: 02339/9194-0(d), F: 02339/9194-99(d)

Lauber, C., Obere Wingert 24, 69437 Neckargerach. T: 02623/594(p)

Laue, J., Dr.-Ing., Ingenieurkontor Laue, Hofferhofer Str. 7, 51503 Rösrath

Lehmann, U., Dr.-Ing., Dietkircher Str. 9, 65555 Limburg. T: 06431/57358(p)

Lehners, C., Dr.-Ing., Selbständig, Dr.-Ing. Lehners + Partner, Schäferkamp 17, 23563 Lübeck. T: 0451/6926242(d), F: 0451/692012(d), T: 04503/889891(p)

Lehr, R., Dipl.-Geol., Geotec, Gartenstr. 25, 63263 Neu-Isenburg. T: 06102/1370(d)

Leiendecker, H., Dipl.-Ing., Unterer Eickeshagen 29, 42555 Velbert. T: 0234/68775-51(d), F: 0234/68775-10(d)

Lenz, J., Dipl.-Geol., SL-Geotechnik, Europastr. 11, 35394 Gießen. T: 0641/9433380(d), F: 0641/9433382(d)

Lepique, M., Dr.-Ing., Baugrundgutachter, DMT-GUC, Am Technologiepark 1, 45307 Essen. T: 0201/172-1895(d), 0202/435101(p)

Lesny, K., Dipl.-Ing., Schöne Aussicht 33, 45289 Essen. T: 0201/1832856(d), F: 0201/1832870(d), T: 0201/570347(p)

Liedtke, L., Dr.-Ing., Bundesanstalt für Geowissenschaften und Rohstoffe, Stilleweg 2, 30655 Hannover. T: 0511/6432418(d), F: 0511/6432304(d)

Linden, H., Dipl.-Ing., Inhaber, Linden – Grundbau, Hochstr. 27, 82024 Taufkirchen. T/F: 089/6122086(d), T/F: 089/6122086(p)

Linder, W.-R., Dr.-Ing., Leiter Betrieb u. Technik, Brückner Grundbau GmbH, Lüscherhofstr. 70, 45356 Essen. T: 0201/3108-246(d), F: 0201/3108-258(d)

Lippomann, R., Prof. Dr.-Ing. Dipl.-Geol., Berat. Ingenieur, Prof. Lippomann Dr. Kley GmbH, Händelstr. 14, 93128 Regenstauf

List, F., Dr., Pfeuferstr. 4, 81373 München. T: 089/7253133(p), F: 089/7469067(p)

Lizcano, A., Dipl.-Ing., Universidad de los Andes Dpto. de Ingenieria Civil y Ambietal, Carrera 1 A Este No 18 A-10, CO- Bogotá, Kolumbien. T: 57/1/4144503(d)

Lobin, M., Dipl.-Ing., Str. d. Einheit 8, 09599 Freiberg. T: 03731/365536(d), F: 03731/365537(d), T: 03731/697173(p)

Loer, U., Dipl.-Ing., Dyckerhoff & Widmann AG, Beim Strohhause 2, 20097 Hamburg. T: 040/24879-301(d), F: 040/24879-895(d), T: 02129/32893(p)

Löhmann, J., Dipl.-Ing., UMB Umwelt- und Risikomanagement- beratungs GmbH, John-F.-Kennedy-Str. 1, 65189 Wiesbaden. T: 069/595081(p)

Löschner, J., Dipl.-Ing., Abteilungsleiter, Arcadis Trischler & Partner GmbH Beratende Ingenieure Geotechnik – Umweltschutz, Berliner Allee 6, 64295 Darmstadt. T: 06151/388223(d), F: 06151/388996(d), T: 06151/534766(p)

Löw, M., Dipl.-Ing., Gustav-Adolf-Str. 11, 65195 Wiesbaden. T/F: 0611/9812878(p), T: 0611/537-318(d), F: 0611/537-327(d)

Lüdeling, R., Dr. Dipl.-Geol., Bundesanstalt f. Geowissenschaften und Rohstoffe, Technische Geologie, Umweltgeologie, Stilleweg 2, 30655 Hannover. T: 0511/6432864(d), F: 0511/6432304(d)

Lukas, H.-D., Dipl.-Ing., Hegerkamp 10, 45329 Essen. T: 0201/3108-02(d), F: 0201/367794(d)

Lüke, J., Dipl.-Ing. (USA), Heinrichsallee 60, 5206 Aachen. T: 0241 805254(d), F: 0241 8888-384(d)

Lüken, J., Dipl.-Ing., Beratender Ingenieur, Ingenieurbüro für Geotechnik, Winsener Str. 134c, 21077 Hamburg. T: 040/76411872(d), F: 040/76411871(d)

Luo, Z., Dipl.-Ing., Siemensstr. 14, 64289 Darmstadt. T/F: 06151/735382(p)

Lürssen, U., Dipl.-Ing. (FH), freier Ingenieur, Ing.-Büro für Geotechnik Problemschäden, Bauwerkssanierung, Hauptstr. 5, 06722 Wetterzeube. T/F: 036693/22500(d), T: 0177/5622500(p)

Lüthke, J., Dipl.-Geol., Am Mühlhügel 8, 07751 Jena-Wogau. T/F: 03641/448584(p)

Lux, K.-H., Prof. Dr.-Ing., Sundernstr. 58, 31224 Peine-Madanoglu, Tunc, Dr. Dipl.-Ing., KMT – Ingenieurgesellschaft mbH, Geschäftsführung, Erdkampsweg 4, 22335 Hamburg. T: 040/5009262(d), F: 040/5000077(d)

Madanoglu, T., Dr.Dipl.-Ing. KMT – Ingenieur-Gesellschaft gmbh Gesraftsführung, Erdlampsweg 4, 22335 Hamburg. T: 040/500/9262, F: 040/5000077

Magar, K., Prof. Dr., Geschäftsleitung, Geotechnisches Institut, Prof. Dr. Magar & Partner, Winterhäuser Str. 9, 97084 Würzburg. T: 0931/61440(d), F: 0931/6144200(d)

Malios, I., Dipl.-Ing., Malios and Assoc. Inc., Incorporation of Consulting Engineers, Management, 3rd Septembriou 43A, GR-10433 Athen, Griechenland. T: 00301/8217931(d), F: 00301/8212469(d)

Mallwitz, K., Prof. Dr.-Ing., FH-Neubrandenburg, Verkehrsbau, Grundbau und Bodenmechanik, Brodaer Str. 2, 17033 Neubrandenburg. T: 0395-5693-324(d), F: 0395-5693-399(d), T: 0395/5693-5825488(p)

Mandel, P.J., Dipl.-Ing., Burmann & Mandel Berat. Ing. für Grundbau u. Umwelttechnik, Gasstraße 18 Haus 6b, 22761 Hamburg. T: 040/896037(d), F: 040/8901621(d)

Mands, E., Dr., UBeG GbR, Zum Boden 6, 35580 Wetzlar. T: 06441/212910(d), F: 06441/212911(d), T: 06441/212912(p), F: 06441/212911(p)

Marth, E., Dipl.-Ing., Ingenieurbüro Eichenberger AG, Sumatrastr. 22, CH-8023 Zürich, Schweiz. T: 0041/1/2522077(d), F: 0041/1/2522025(d)

Massat, R., Dipl.-Ing., Bahnhofstr. 38, 64395 Brensbach. T: 06161/1667(p)

Massmeier, L., Dr.-Ing., ELE-Erdbaulaboratarium Essen, Susannastr. 31, 45136 Essen. T: 0201 8959-6, F: 0201 253733(d)

Mattersberger, C., Neufriedenheimer Str. 69, 81375 München. T: 089 71030466(p)

Matthesius, H.-J., Dr. rer. nat. Dipl.-Geol., gbm Bauingenieurwesen, Robert-Bosch-Str. 7, 65549 Limburg. T: 06431/9112-0(d), F: 06431/9112-10(d)

Matuschka, W., Dipl.-Ing., Geschäftsführer, Rhein-Ruhr Ingenieur-Gesellschaft mbH, Burgwall 5, 44135 Dortmund. T: 0231/5482-0(d), F: 0231/575556(d)

Meier, C.-H., Kleiststr. 16a, 69514 Laudenbach. T: 06201/74608(p)

Meier, K., Dipl.-Ing., Oberbauleiter, Bahnhofstr. 5, 23611 Bad Schwartau. T: 030/39708478(d), F: 030/39704404(d), T: 0451/24273(p)

Meinhardt, G., Dipl.-Ing., Projektingenieur, Schießhüttenweg 12, 88458 Indoldingen. T: 06151 388-433(d), F: 06151 388 998(d)

Meiniger, W., Dipl.-Ing., Forschungs- und Materialprüfungsanstalt Baden-Württemberg – Otto-Graf-Institut, Abt. Geotechnik, Pfaffenwaldring 4, 70569 Stuttgart. T: 0711/6853358(d), F: 0711/6856826(d)

Meißner, H., Prof. Dr.-Ing., Universität Kaiserslautern FB Bodenmechanik u. Grundbau, Postfach 30 49, 67653 Kaiserslautern. T: 0631/2052930(d), F: 0631/2053806(d)

Meißner, H., Dipl.-Ing. Dr. mont., Kurhausstr. 58, 65719 Hofheim/Ta. T: 06122/51057(d), F: 06122/52591(d)

Melzer, K.-J., Dr.-Ing., Consultant, KJM Industry Consult, Drosselweg 7a, 61440 Oberursel. T: 06172/390259(d), F: 06172/937596(d)

Mennings, B., Dipl.-Ing., Projektleiter, Stationsplein 4, NL-3800 Ae Amersfort, Niederlande. T: 0033/676594(p), F: 0033/676169(p)

Menzel, W., Dr.-Ing., Blumenstr. 29, 04420 Kulkwitz. T: 0341/336000(d), F: 0341/33600308(d), T: 034205/84857(p)

Mertens, M., Dipl.-Ing., CPB Cordes GmbH, Neustädter Passage 6, 06122 Halle (Saale). T: 0345/863836(d), F: 0345/8061405(d), T: 03461/501795(p)

Meseck, H., Dr.-Ing., Wiesenstr. 52, 40699 Erkrath

Meyer, J., Dipl.-Ing., ELE-Erdbaulaboratorium Essen, Susannastr. 31, 45136 Essen. T: 0201/8959-6(d), F: 0201/253733(d)

Meyer, N., Prof. Dr.-Ing., Hochschullehrer, Oberer Triftweg 14, 38640 Goslar. T: 05321/318233(p), F: 05321/318235(p)

Meyer-König, G., Dipl.-Ing., Hainbuchenweg 10, 70597 Stuttgart

Meyerhof, G.G., Prof. Dr. Dr.-Ing. E.h., Forschungsprofessor, Dalhousie University, Department of Civil Engineering, PO Box 1000, Halifax N.S. B3J 2X4. T: 001/902 420-7681(d), F: 001/902 420-2612(d)

Mikulitsch, V., Dipl.-Ing., Nonnengasse 5, 09599 Freiberg. T: 03731/39-3579(d), F: 03731/39-3581(d), T: 03731/218662(p)

Millgramm, H., Prof. Dipl.-Ing., Auf der Tenten 17, 53797 Lohmar

Minkley, W., Dr.-Ing., Projektleiter, IFG – Institut für Gebirgsmechanik GmbH, Friederikenstr. 60, 04279 Leipzig. T: 0341/33600216(d), F: 0341/33600308(d)

Mittag, J., Dipl.-Ing., TU Berlin, Grundbauinstitut, Sekr. B 7, Str. des 17. Juni 135, 10623 Berlin. T: 030(314-24915(d)

Möbius, H.-U., Dipl.-Ing., DMU Diederichs + Möbius Geotechnik, Kamsdorfer, Str. 5, 07334 Gosswitz. T: 03671/672011(d), F: 03671/672012(d), T: 08171/28643(p), F: 08171/22425(p)

Moore, D., Dipl.-Bauing., Zschokke Locher AG, Bahnhofstr. 24, 5001 Aarau, Schweiz. T: 0041/62 832 0400(d), F: 0041/62/832-0402(d), T: 0041 1 834 0654

Morgen, K., Dr.-Ing., Prüfingenieur f. Baustat., Dr.-Ing. R. Windels, Dr.-Ing. G. Timm, Dr.-Ing. K. Morgen Beratende Ingeniure VBI., Ballindamm 17, 20095 Hamburg. T: 040/350090(d), F: 040/35009100(d)

Morgner, J., Dipl.-Ing., Gerberstr. 10, 23966 Wismar. T: 03841/753305(d), F: 03841/753306(d), T: 03841/212502(p)

Moser, M., Prof. Dr., Hochschullehrer, Universität Erlangen Lehrstuhl für Angewandte Geologie, Schloßgarten 5, 91054 Erlangen. T: 09131/852697(d), F: 09131/859294(d), T: 09131/207607(p)

Moser, M.A., Dr.-Ing., Dr Ing. Bilfinger & Berger Bau AG, Postfach 15 09, 65005 Wiesbaden. T: 0611 7080, F: 0611 708499

Mößner, A., Dipl.-Geol., Rheinstr. 182, 50839 Wesseling. T: 02236 66097(d), F: 02236 68275(d)

Mühring, W., Dr.-Ing., Baudirektor a.D., Am Mittellandkanal 57, 49565 Bramsche. T: 05461/1470(p), F: 05461/71506(p)

Müller, H., Frankfurter Str. 146, 63263 Neu-Isenburg. T: 06102/39155(p)

Müller, K., Weidenstr. 35, 56470 Bad Marienberg. T: 02661/98780(p)

Müller, K., Prof. Dr.-Ing., Rotkleedamm 2, 21614 Buxtehude. T: 04161/62264(p), F: 04164/62264(p)

Müller, P., Dipl.-Ing., Brechtel Brunnenbau GmbH Brunnenbau und Spezialtiefbau, Industriestr. 11a, 67063 Ludwigshafen. T: 0621/69004-0(d), F: 0621/69004-24(d)

Müller, P.J., Dr., A-9711 Patermion 8, Österreich

Müller-Bornemann, H., Groffstr. 14, 80638 München

Müller-Kirchenbauer, A., Börtheidering 27, 31535 Neustadt. T: 05034/690(p)

Müller-Kirchenbauer, H., Prof. Dr.-Ing., Reichhelmstr. 1, 30519 Hannover

Müller-Ruhe, W., Dipl.-Ing., Bayernstr. 15, 34131 Kassel

Müllner, B., Dr.-Ing., Fuggerstr. 2, 90513 Zirndorf. T: 0911/6555582(p), F: 0911/6555599(p)

Muntzos, T., Dr., An der Alten Ziegelei 36, 48157 Münster. T: 0251/143870(d), F: 0251/143872(d), T: 05484/682(p)

Münzner, J., Dipl.-Ing., Etzelstr. 11, 70180 Stuttgart. T: 0711/64954-0(d), F: 0711/64954-10(d), T: 07152/26346(p)

Nacke, C., Dr.-Ing., Jörg von Schauenburgstr. 8, 77704 Oberkirch. T: 07802/50923(p)

Nagel, D., Am Schützeich 3, 09246 Pleißa

Natau, O., Prof. Dr.-Ing., Institutsleiter, Universität Karlsruhe, Institut f. Boden – u. Felsmechanik, Abt. Felsmechanik, Engler-Bunte-Ring, 76131 Karlsruhe. T: 0721/608-2228(d), F: 0721/693638(d), T: 0721/686553(p)

Natter, P., Dipl.-Ing., Leiter GF Verdichtung, Birkmannsweg 34, 45149 Essen. T: 02131/938-813(d), F: 02131/938-700(d), T/F: 0201/711775(p)

Nauhauser, F., Höhenweg 40, 52072 Aachen. T: 0241/8794257(p)

Neff, H.K., Dipl.-Ing., Berat. Ing., Brunnenstr. 66, 35460 Staufenberg. T: 06402/52260(d), F: 06402/5226-98(d), T: 06406/905008(p), F: 06406/905009(p)

Neher, H., Dipl.-Ing., Georg-Schurr-Str. 3, 70794 Filderstadt. T: 0711/685 2436/38(d), F: 0711/685 2439(d)

Neidhart, T., Prof. Dr.-Ing., Professor, Fachhochschule Regensburg Fachbereich Bauingenieurwesen, Prüfeninger Str. 58, 93049 Regensburg. T: 0941/943-1204(d), T: 0941/2803545(p)

Neubert, L., Ing.(grad.), Ostwender Str. 9, 30161 Hannover

Neuffer, E., Dipl.-Ing., Schloß Martinsburg, 56112 Lahnstein. T: 0261/5791015(p)

Neumann, R.B., Dipl.-Ing., Gebr. Neumann GmbH & Co. KG, Mackeriege 1, 26506 Norden. T: 04931/18070(d)

Ney, P., Dipl.-Ing., Kellergrund Weg 9, 61476 Kronberg. T: 06173/934815(d), 06173/79446(p), F: 06173/79446(p)

Nierste, A., Dipl.-Ing., Osdorfer Landstr. 311, 22589 Hamburg. T: 0172/4066968(d), T: 040/87082269(p)

Nies, D., Prof. Dipl.-Ing., Ziviling., Gf., M.P.T. Engineering GmbH Dipl.-Ing. Dieter Nies, Fischen 19, 4531 Kematen a.d. Krems, Österreich. T: 0043/7228/7400(d), F: 0043/7228/74004(d)

Nimmesgern, M., Prof. Dr.-Ing., Kellereistr. 4, 97199 Ochsenfurt. T: 0931/3511281(d), F: 0931/3511337(d), T: 09331/980240(p), F: 09331/980241(p)

Nöll, T., Dipl.-Geol., Ing.-geol. Beratungsbüro, Schellingstr. 8a, 30625 Hannover

Novy, A., Dipl.-Ing., Albert Schweitzer Str. 32, 02977 Hoyerswerda. T: 03573/782148(p)

Nowicki, P., Zu Niederndorf 23a, 55457 Horrweiler. T: 06721/942512(d), T: 06727/953187(p)

Nußbaumer, M.Sc., Manfred, Prof. Dr.-Ing. E.h., Vorstandsvorsitzender der Ed. Züblin AG, Albstadtweg 3, 70567 Stuttgart. T: 0711/7883616(d), F: 0711/ 7883668(d)

Obermanns, K., Dipl.-Geol., Milseburgstr. 4, 36093 Künzell-Dirlos. T: 0661/302441(p)

Obermeyer, L., Dr.-Ing., Geschäftsführer, Obermeyer Planungsgesellschaft f. Bau, Umwelt, Verkehr und Technische Ausrüstung mbH, Hansastr. 40, 80686 München. T: 089/5799600(d)

Ochmann, H., Dr.-Ing., Ochmann + Partner Geotechnik GmbH, Mendelssohnstr. 15F, 22761 Hamburg. T: 040/8100090(d), F: 040/8905665(d)

Odendahl, K., Dr.-Ing., UCON Ingenieurgesellschaft, Dr. Riedel – Dr. Odendahl GmbH, Kohlenstr. 70, 44795 Bochum. T: 0234/943620(d)

Oehme, H., Dipl.-Geol., Freiberufl., Ingenieurbüro Oehme Dipl.-Geol. Oehme, H. Beratender Ingenieur, Elite-Gewerbepark, Dammstr. 2-4, B1, 09618 Brand-Erbisdorf. T: 037322/41797(d), F: 037322/42977(d), T: 037322/3802(p)

Oehrl, M., Dipl.-Ing., Tenax Kunststoffe GmbH, Schloßstr. 13, 88131 Lindau. T: 08382/93040(d), F: 08382/930430(d)

Oesterreich, B., Dr. rer. nat. Dipl.-Geol., Geologisches Landesamt NRW, De-Greiffstr. 195, 47083 Krefeld. T: 02151/897410(d)

Oligmueller, L., Dipl.-Ing., Waldsaum 5, 45134 Essen. T: 0201/473834(p)

Opheys, S., Dr., Keltenstr. 2, 92339 Beilngries. T: 08461/601042(p)

Oswald, C., Dipl.-Ing., Prellerstr. 51, 04155 Leipzig

Oswald, K.-D., Dipl.-Ing., C & E Consulting und Engineering GmbH, FB Geotechnik, Jagdschänkenstr. 52, 09117 Chemnitz. T: 0371/8114279(d), F: 0371/8114577(d)

Ott, E., Weisse Breite 11a, 34130 Kassel. T: 0561/804-2197(d), F: 0561/804-2651(d), T: 0561/66124(p)

Otto, U., Dipl.-Ing., Berat. Ingenieur, Trischler & Partner GmbH, Berliner Allee 6, 64295 Darmstadt. T: 06151/388-0(d), F: 06151/388998(d)

Paehge, W., Dipl.-Geol., Institut für Geologie und Paläontologie, Abteilung für Ingenieurgeologie, Leibnizstr. 10, 38678 Clausthal-Zellerfeld

Palla, R., Dipl.-Ing., Gepacenter – HBPM, Alfred Ammon Str. 29, I-39042 Brixen, Italien. T: 0039/472/209912(d), F: 0039/472/201005(d), T: 0039/335/6265759(p)

Paproth, R., Dipl.-Ing., Friedrich-Ebert-Str. 325, 47800 Krefeld

Pasch, M., Peter-Wenzel-Weg 26, 69118 Heidelberg. T: 06221/800549(p)

Patron, J.A., Dipl.-Ing., Arcostr. 19, 15831 Mahlow

Paulus, M., Tiefbauing., ist Paulus Ingenieurplanung, Spezialtiefbau, Stolzenfelsstr. 4a, 81375 München. T: 089/74141272(d), F: 089/74141273(d)

Pause, H., Dr.-Ing. E.h. Dipl.-Ing., Tiefentalweg 20, 82402 Seeshaupt. T: 08801/2130(p), F: 08801/2139(p)

Peintinger, B., Prof. Dipl.-Ing., Berat. Ingenieur f. Geotechnik, Fachhochschule Deggendorf, FB Bauingenieurwesen, Ulrichsbergerstr. 17, 94469 Deggendorf. T: 0991/29054-35(d), F: 0991/29054-51(d), T: 08131/92512(p)

Pentzin, W., Dipl.-Ing., Erd- u. Grundbauinstitut Hamburg, Hinschenfelder Stieg 1, 22041 Hamburg. T: 040/6931314(d), F: 040/6931330(d)

Penzel, M., Dr.-Ing., GEOTECHNIK PROJEKT Gesellschaft für Geotechnik, im Bauwesen, Bergbau und Umweltschutz bR, Diezmannstr. 67, 04207 Leipzig. T: 0341/4217800(d), F: 0341/4217801(d)

Perau, E.W., Dr.-Ing., Rüttenscheider Str. 190, 45131 Essen. T: 0201 8417333 (p)

Peter, T., Dipl.-Ing., Geschäftsführer, Boley & Peter Beratung & Planung Ing.-Büro im Bauwesen GmbH, Hermann-Köhl-Str. 2, 93049 Regensburg. T: 0941/296060(d), F: 0941/2960611(d), T: 0941/31384(p)

Peters, M., Dipl.-Ing., Sen. Reserv. Eng., Finkenstr. 7A, 34225 Baunatal. T: 0561/3011916(d), F: 0561/3011195(d), T/F: 0561/4911318(p)

Peters, G., Dipl.-Ing., Berliner Str. 73a, 11467 Potsdam. T: 0331/2705791(p)

Petersen, H., Dr.-Ing., Wissensch. Assistent, Universität Karlsruhe Institut für Bodenmechanik und Felsmechanik, Postfach 69 80, 76128 Karlsruhe. T: 0721/608-2232(d), F: 0721/693638(d)

Petersen, U., Dr.-Ing., ISK Ingenieurgesellschaft für Bau- und Geotechnik mbH, An der Wiesenhecke 10, 63456 Hanau Klein-Auheim. T: 06181/96407-0(d), F: 06181/96407-77(d)

Philipp, H., Dr.-Ing., Geschäftsführer, BAUGEO GmbH Ingenieurbüro, Forststr. 10, 04229 Leipzig. T: 0341/487510(d), F: 0341/4875129(d), T: 0341/3385313(p)

Philipp, T., Dipl.-Ing., Str. des 18. Oktobers 10/9. Et., 04103 Leipzig

Pimentel, E., Dr.-Ing., Wissensch. Assistent, Universität Karlsruhe Institut für Bodenmechanik und Felsmechanik, Postfach 69 80, 76128 Karlsruhe. T: 0721/608-2232(d), F: 0721/693638(d)

Pingel, R.J., Dipl.-Ing., Ingenieurbüro für Geotechnik, Wiesenhöfen 2, 22359 Hamburg. T: 040/6037225(d), F: 040/6035829(d)

Pires, P., Francois, Konrad-Adenauer-Str. 17, 67663 Kaiserslautern. T: 0172/6806884(p)

Placke, A., Am Sonnenhang 7, 65719 Hofheim. T: 0032/495/585323(d), F: 06192/935241(p)

Placzek, D., Prof. Dr.-Ing., Geschäftsführung, ELE-Erdbaulaboratorium Essen, Susannastr. 31, 45136 Essen. T: 0201 8959 702(d), F: 0201 8959-795(d)

Plohmann, U., Dipl.-Ing., Niederlassungsleiter, Pfahlgründung Centrum Pfähle GmbH NL Karlsruhe, Hauptstr. 33, 76344 Eggenstein. T: 07247/963060(d), F: 07247/20555(d)

Pohlmann, H., Dipl.-Ing., Häbdekstr. 20, 48249 Dülmen Händelstr. T: 02542 701 0(d)

Porada, J., Dipl.-Geol., J. Porada – Geotechnik, Mädesüßweg 5a, 21698 Harsefeld. T: 04164/6767(d), F: 04164/6768(d), T: 04164/6767(p), F: 04164/6768(p)

Pötschick, L., Dorfstr. 25, 03238 Lieskau

Prinz, H., Prof. Dr., Ingenieurgeologe, Stromberger Str. 38, 55411 Bingen. T/F: 06721/35166(p)

Probst, U., Dipl.-Geol., Geschäftsführer, Ingenieurgesellschaft Fugro mbH, Valdenaire Ring 91, 54329 Konz. T: 06501/99116(d), F: 06501/2686(d), T: 06501/6211(p)

Pulsfort, M., Prof. Dr.-Ing., Kriemhildenstr. 6, 42287 Wuppertal. T: 0202/40646(d), F: 0202/401615(d), T: 0202/595349(p)

Purkert, P., Winklerstr. 20, 09599 Freiberg. T: 0177/3231955(p)

Quast, P., Dr.-Ing., Grundbauingenieure Steinfeld und Partner GbR, Alte Königstr. 3, 22767 Hamburg. T: 040/389139-0(d), F: 040/3809170(d)

Quick, H., Prof. Dipl.-Ing., Geschäftsführender Gesellschafter, Prof. Dipl.-Ing. H. Quick Ingenieure und Geologen GmbH, Gross-Gerauer Weg 1, 64295 Darmstadt. T/F: 06151/13036-0(d), T: 06204/740932(p), F: 06204/740933(p)

Raabe, E.-W., Dr.-Ing., Hinter Holtein 16, 44227 Dortmund

Rabus, E., Dipl.-Ing., Kurfürst-Max-Str. 23, 82377 Penzberg. T: 08856/934600(p), F: 08856/934601(p), T: 08856/934600(d), F: 08856/934601(d)

Rackwitz, F., Dipl.-Ing., Wissensch. Mitarbeiter, Technische Universität Berlin FG Grundbau und Bodenmechanik Sekr. B7, Straße des 17. Juni 135, 10623 Berlin. T: 030/314-22861(d), F: 030/314-24492(d), T: 0172/8841369(p)

Raeker, H.-M., Dipl.-Ing., Witt + Jehle Geotechnik GmbH, Zeisigstr. 4, 56075 Koblenz. T: 02241 390699(p)

Raithel, M., Dr.-Ing., Würzburg., Wölffelstr. 1, 97072 Würzburg

Raju, V., Dr.-Ing., Keller Grundbau GmbH, Overseas Divison, Kaiserleistr. 44, 63067 Offenbach. T: 069/8051-0(d), F: 069/8051-270(d)

Rauch, H.-J., Dipl.-Ing., Baugrundsachverständiger, IBES Baugrundinstitut GmbH, Postfach 10 06 30, 67406 Neustadt/Weinstraße. T: 06321/13081(d), F: 06321/15853(d)

Rausche, F., Dr., President, Goble Rausche Likins and Associates Inc., 4535 Renaissance Parkway, Cleveland OH 44128 Vereinig. Staaten von Amerike. T: 001 216 831 6131(d), F: 001 216 831-0916(d), T: 001 440 247 7905

Rauscher, W., Dr.-Ing., Ing.-Büro EDR GmbH, Hansastr. 28, 80686 München. T: 089/547112-0(d), F: 089/547112-50(d)

Rechlin, D., Dipl.-Geol., Regierungsdirektor, Schlesierstr. 18, 76275 Ettlingen. T: 0721/9726-378(d), F: 0721/9726-483(d), T: 07243/4701(p)

Rechtern, J., Dr.-Ing., Grundbauingenieure Steinfeld und Partner GbR, Alte Königstr. 3, 22767 Hamburg. T: 040/389139-0(d), F: 040/3809170(d)

Reder, K., Dr. rer. nat. Dipl.-Geol., Geschäftsführer, Bergstr. 53, 97638 Mellrichstadt. T: 03681/4490-0(p), F: 03681/4490-32(p)

Reetz, D., Dipl.-Ing., Gerhart-Hauptmann-Str. 32, 40699 Erkrath. T: 0211/204335(p), T: 02173/8501-50(d), F: 02173/8501-22(d)

Rehberg, H.-W., Dipl.-Ing., Abteilungsleiter, CES Consult. Eng. Salzgitter GmbH Geotechnik und Grundbau Infrastruktur, Nord-Süd-Str. 1, 38259 Salzgitter. T: 05341/823-440(d), T: 05123/8325(p)

Rehfeld, E., Dr.-Ing., Dürerstr. 119, 01309 Dresden

Rehm, K.H., Dr.-Ing. Dipl. Phys., Geotechnik Labor GmbH Prof. Riße, Dr. Rehm Rostock, Geschäftsführung, Am Hechtgraben 1B, 18147 Rostock. T: 0381/699333(d), F: 0381/699334(d)

Reich, S., Dipl.-Ing., Burmann, Mandel & Partner Ingenieurbüro für Grundbau und Umwelttechnik, Gasstr. 18, Haus 6b, 22761 Hamburg. T: 040/896037(d), F: 040/8901621(d), T/F: 040/3908275(p)

Reinders, P., Dr.-Ing., Franzstr. 15, 52064 Aachen. T: 0241/9290956(p)

Reiner, J., Dipl.-Ing., Prokurist, iwb Ingenieurgesellschaft mbH, Billwerder, Str. 44, 21033 Hamburg. T: 040/369854-0(d), F: 040/369854-99(d)

Reitmeier, W., Prof. Dr.-Ing., Eibenfeld 17, 84307 Eggenfelden. T: 07531/206224(d), F: 07531/206430(d), T: 08721/4864(p)

Rensing, D., Parkweg 3, 99735 Werther. T: 03631 654-212(d), F: 03631 600326, T: 03631 974403(p)

Rettenberger, G., Prof. Dipl.-Ing., Ingenieurgruppe RUK, Schockenriedstr. 4, 70565 Stuttgart. T: 0711/90678-0(d), F: 0711/90678-88(d)

Reul, O., Dipl.-Ing., Hamburger Str. 22a, 61130 Nidderau

De Reuter, W., Ingenieurbüro für Geotechnik und Baustofftechnologie, Lindenstr. 1, 48341 Altenberge. T: 02505/2010(d), F: 02505/3205(d)

Reuter, E., Dr.-Ing., Lindenweg 8, 31553 Auhagen

Richter, K., Dr.-Ing., Beratender Ingenieur, Ingenieurbüro Dr.-Ing. K. Richter, Demmeringstr. 92, 04177 Leipzig. T: 0341/9260230(d), F: 0341/9260231(d), T: 0341/9260232(p), F: 0341/9260231(p)

Richwien, W., Prof. Dr.-Ing., Universität GH Essen Fachbereich 10 Grundbau und Bodenmechanik, 45117 Essen. T: 0201/183-2858/57(d), F: 0201/1832870(d)

Riedel, H., Dr.-Ing., UCON Ingenieurgesellschaft Dr. Riedel – Dr. Odendahl GmbH, Kohlenstr. 70, 44795 Bochum. T: 0234/943620(d)

Riemer, W., Dr. Dipl.-Geol., Risch Haff, L-8529 Ehner/Redingen, Luxemburg. T: 00352/63518(p), F: 00352/639612(p)

Ries, W., Dipl.-Ing., Bautest, Mühlmahdweg 25a, 86167 Augsburg. T: 0821/72024-28(d), F: 0821/72024-0(d), T: 08222/6718(p)

Rilling, B., Dr.-Ing., Am Waldrand 6/6, 71111 Waldenbuch

Rinke, L., Dipl.-Ing., Grundbauingenieure Steinfeld und Partner GbR, Alte Königstr. 3, 22767 Hamburg

Rizkallah, V., Univ.-Prof. Dr.-Ing. habil., Universität Hannover, Bauingenieur- und Vermessungswesen, Appel Str. 9a, 30167 Hannover. T: 0511/7624155(d), F: 0511/7625105(d)

Röcke, R., Dipl.-Ing. (FH), Geschäftsführer, Geotechnisches Ingenieurbüro R. Röcke GmbH, Dorfstr. 49, 06862 Streetz. T: 034901/67621(d), F: 034901/67622(d), T/F: 034901/67046(p)

Röder, K., Dr.-Ing. habil., NL-Leiter, Preussag Spezialtiefbau GmbH, Abt. Umwelttechnik, Industriestr. 16, 04435 Schkeuditz. T: 034204/8260(d), F: 034204/82650(d)

Rodriguez, G., Carrera 53a, No. 135-49, Apto. 1104, Torre 2, Bogota, Kolumbien. T: 057 1 6137111(p), F: 057 1 6139450(p)

Rogall, M., Dr., Schusterstr. 12, 55116 Mainz

Rogge, T., Prof. Dr.-Ing., Postfach 12 04 61, 27518 Bremerhaven

Rogowski, E., Dr., Geol. Landesamt Baden-Württemberg, Urbanstr. 53, 70182 Stuttgart. T: 0711/2124818(d), F: 0711/21248363(d)

Rohrbach, J., Dipl.-Geol., Waldstr. 36a, 64521 Groß-Gerau

Rohwedder, P.-C., ö.b.u.v. Sachverständiger, Ing.-Büro für Spezialtiefbau VDI, Öffentlich bestellter und vereidigter Schverständiger, Ellingstedter Weg 19, 25767 Albersdorf. T: 04835/9400(d), F: 04835/9420(d), T: 04835/9410(p)

Rollberg, D., Dr.-Ing., Roßdörfer Str. 127, 64287 Darmstadt

Romberg, W., Dipl.-Ing., Berat. Ingenieur, Lindenweg 28, 64291 Darmstadt. T: 06151/374289(d), F: 06151/370169(d)

Rooney, P., Dipl.-Ing., Martin-Reck-Str. 15, 61118 Bad Vilbel. T: 06101/552307(d)

Rose, K., Dipl.-Ing., Heinr.-Zille-Str. 3, 99734 Nordhausen

Rothengatter, P., Dipl.-Ing., Ingenieurgeologie, Ingenieurgesellschaft für Baugeologie und MeßtechnikmbH (gbm), Baugrundinstitut, Pforzheimerstr. 126a, 76275 Ettlingen. T: 07243/763260(d), F: 07243/763250(d)

Röther, W.-A., Dipl.-Ing., Baukontor Dümcke GmbH, Alfstr. 26, 23552 Lübeck. T: 0451/1400260(d), F: 0451/1400254(d)

Rotter, K.-W., Dipl.-Ing., Geschäftsführer, Gartenstr. 11, 86510 Ried-Baindkirch. T: 08202/904607(d), F: 08202/904677(d), T/F: 08202/662(p)

Rübener, R.H., Univ.-Prof. i.R., Thingstr. 16, 45527 Hattingen. T: 02324/67478(p)

Ruch, C., Dr., Geologisches Landesamt Freiburg, Albertstr. 5, 79104 Freiburg. T: 0761/204-4433(d)

Rüdiger, A., Dipl.-Ing., Niedersachsenstr. 8, 41564 Kaarst

Rudolph, M., Dipl.-Ing., Ingenieurbüro für Tragwerksplanung, Prüfingenieur für Baustatik, Parkstr. 6, 08107 Saupersdorf. T/F: 037602/65219(d)

Rummel, F., Prof. Dr., Ruhr-Universität Bochum, Institut f. Geophysik, Universitätsstr. 150, 44780 Bochum. T: 0234/7003279(d), F: 0234/7094181(d)

Rumpelt, T., Dr.-Ing., Faching. f. Geotechnik, Smoltczyk & Partner GmbH, Geotechnik und Grundbau, Untere Waldplätze 14, 70569 Stuttgart. T: 0711/1316413(d), F: 0711/1316464(d)

Ruppert, F.-R., Dr.-Ing., BRP Consult, Beckmann, Ruppert und Partner GmbH, Inselwall 14, 38114 Braunschweig. T: 0531/48000-0(d), F: 0531/48000-48(d)

Rusert, K.-J., Dipl.-Ing., Friedrichstr. 10, 58135 Hagen. T: 02331/41834(p), F: 02331/42965(p)

Rütz, D., Dr.-Ing., Bauhaus-Universität Weimar, Fakultät Bauingenieurwesen, Prof. Bodenmechanik, Marienstr. 13–15, 99421 Weimar. T: 03643/584462(d)

Rux, W., An der Zschopau 29, 09487 Schlettau

Rychlovsky, E., Dipl.-Ing., Dyckerhoff & Widmann AG, Wilhelm-Weigand-Str. 5, 81925 München. T: 089/92553282(d), F: 089/92553464(d), T: 089/951223(p)

Saathoff, F., Dr.-Ing., Geschäftsführer, Bollstr. 4, 32312 Lübbecke. T: 05443/995661(d), F: 05443/995666(d), T: 05741/8153(p)

Sadgorski, W., Dr. techn. Dipl.-Ing., Rotbuchenstr. 73, 81547 München. T: 089/6907197(p)

Salden, D., Dr.-Ing., Erlenweg 2, 71272 Renningen 2. T: 07159/8228(p)

Salomo, K.-P., Prof. Dr.-Ing., Geschäftsführer, IGU Ingenieurgesellschaft für Geotechnik und Umweltmanagement mbH, St.-Viti-Str. 1, 29525 Uelzen. T: 0581/976050(d), F: 0581/9760599(d), T: 0581/75122(p)

Salveter, G., Dr.-Ing. (GF), Marktplatz 2, 57250 Netphen. T: 02738/698-0(p), F: 02738/4441(p)

Salveter, G., Dipl.-Ing., Bellingrathstr. 5, 01279 Dresden. T: 0511 7624152(d), T: 0351 2596800(p)

Salzer, K., Dr. rer. nat., Projektleiter, Institut für Gebirgsmechanik GmbH, Friederikenstr. 60, 04279 Leipzig. T: 0341/33600215(d), F: 0341/33600308(d)

Salzmann, H., Dr.-Ing., Geschäftsführer, IMS Ingenieurgesellschaft mbH, Stadtdeich 5, 20097 Hamburg. T: 040/328180(d), F: 040/32818139(d)

Sasse, U., Dipl.-Ing., Marotherstr. 29, 56269 Marienhausen. T: 02689/979031(p)

Sauer, M., Teufelsgrund 8B, 35580 Wetzlar. T: 06048/3571(p)

Savidis, S., o. Prof. Dr.-Ing. habil., Technische Universität Berlin Grundbau-Institut Sekr. B7, Straße des 17. Juni 135, 10623 Berlin. T: 030/31423321(d), F: 030/ 31424492(d)

Schad, H., Dr.-Ing. habil., Forschungs- und Materialprüfungsanstalt Baden-Würt. Otto-Graf-Institut Abt. Geotechnik, Pfaffenwaldring 4, 70569 Stuttgart

Schade, H.-W., Dr.-Ing., Dr. Schellenberg Ing. Ges., Maximilianstr. 15, 89340 Leipheim. T: 08221/72021(d), F: 08221/72554(d)

Schäfer, C., Adam-Rückert-Str. 30, 64372 Ober-Ramstadt. T: 06741/9209-0(d), F: 06741/9209-20(d), T: 06154/638995(p), F: 06154/638997(p)

Schanz, T., Univ. Prof. Dr. Sc. techn., Bauhaus Universität Weimar, FB Bodenmechanik, Paul Schneider Str. 36, 99425 Weimar. T: 03643/904279(d)

Scheffler, H., Dr.-Ing. habil., Meisenweg 15, 04416 Markkleeberg. T: 0341/3389912(d), F: 0341/3384000(d), T: 034299/76736(p), F: 034299/76761(p)

Scheid, Y., Dipl.-Ing., TU Graz Institut für Bodenmechanik und Grundbau, Rechbauerstr. 12, 8010 Graz, Österreich. T: 0043/316/873-6228(d), 0043/316/319704(p), F: 0043/316/873-6232(d)

Scherbatzki, A., Dipl.-Ing., Angestellter, Gregor-Mendel Str. 44, 14469 Potsdam. T: 0177 2355944(p)

Scherbeck, R., Dr.-Ing., Meisenweg 16, 45665 Recklinghausen. T: 02361/892263(p)

Scherzinger, T., Dr.-Ing., Wibel, Leinenkugel + Partner GbR Ingenieurgemeinschaft f. Geotechnik, Lindenstr. 12, 79199 Kirchzarten. T: 07661/9391-0(d), F: 07661/9391-75(d)

Schetelig, K., Prof. Dr. rer. nat., RWTH Aachen, Lehrst. f. Ingenieurgeologie u. Hydrogeologie, Lochnerstr. 4-20, 52064 Aachen. T: 0241/805740(d), F: 0241/8888-280(d)

Scheu, C., Dipl.-Ing., Ingenieurbüro Scheu & Co. GmbH, Bäckerstr. 33, 32312 Lübbecke. T: 05741/7044(d), F: 05741/20259(d)

Schick, A., August-Lämmle-Str. 4, 73635 Rudersberg-Stinenberg. T: 07183/7332(p)

Schick, P. Dr Ing. F. Kobell Str. 4, 85540 Haar. T: 089 6004 3475(d), T: 089 6883128(p)

Schiessl, S., Dipl.-Ing. (FH), Eberlingasse 2, 89312 Günzburg. T: 08221/906-19(d), F: 08221/906-40(d), T: 08221/250280(p), F: 08221/250281(p)

Schilling, D., Dr. rer. nat., Geologe, Büro für Geotechnik & Umweltschutz, Urlharting 79 1/3, 94081 Fürstenzell. T: 08506/922003(d), F: 08506/922004(d)

Schilling, S., Dipl.-Geol., Geotechnik und Umweltschutz S. Schilling, Büro für Grundbau u. Umweltgeologie, Langfuhr 11a, 56841 Traben-Trarbach. T/F: 06541/6972(d)

Schindler, R., Dipl.-Geol., Lahmeyer International GmbH, Friedberger Str. 173, 61118 Bad Vilbel. T: 0177/6672440(d)

Schirmer, R., Dipl.-Ing., Podbielskiallee 13, 14195 Berlin

Schlebusch, M., Dipl.-Ing., Bauingenieur, Gesellschaft für Baugeologie und -meßtechnik (gbm), Am Schlößchen 3, 65594 Runkel. T: 06431/911236(d), F: 06431/911210(d), T: 06482/6208(p)

Schleinig, J.-P., Dipl.-Ing., Kali und Salz GmbH, Fr.-Ebert-Str. 160, 34119 Kassel

Schlipköther, T., Dipl.-Ing., Kaarmannweg 21, 45239 Essen. T: 0201/402346(p)

Schlögl, F., Dipl.-Ing., Unteriglbach Erlengrund 13, 94496 Ortenburg. T: 089/28927136(d), T: 08542/1413(p)

Schlösser, T., Dipl.-Ing., Schlösser Spezialtiefbau, Schacher Str. 31, 88255 Baienfurt. T: 0751/41856(d)

Schlötzer, C., Dr. Ing., Alter Gutshof 2, 30419 Hannover. T: 0511 79922(p)

Schmid, c/o Schleiwies, Jürgen, Dipl. Ing., Hochkalterstr. 4, 81547 München, T: 089 6004-3475(d), F: 089/6004-3476(d), T/F: 089 6259129(p)

Schmidkunz, A., Dipl.-Ing., Hugo-von-Hoffmannsthal-Str. 18, 81925 München. T: 0171/5822200(d), T: 089/913406(p)

Schmidt, A., Am Bonneshof 12, 40474 Düsseldorf. T: 0211/452304(p)

Schmidt, H.-G., Dr.-Ing., Dr.-Ing. Wissensch, Mitarbeiter/ Lehrbeauftr., Lindenstr. 9, 99428 Niedergrunstedt, Weimar. T: 03643 584554(d), F: 03643 584564(d), T: 03643 503062(p), F: 03643 503067(p)

Schmidt, H.H., Prof. Dr.-Ing., Hochschullehrer/ Berat. Ing., Smoltczyk und Partner GmbH, Untere Waldplätze 14, 70569 Stuttgart. T: 0711/1316414(d), F: 0711/1316464(d)

Schmidt, J., Dr.-Ing., Berat. Ingenieur, Gesellschaft für Grundbau und Umwelttechnik mbH (GGU), Am Hafen 22, 38112 Braunschweig. T: 0531/312895(d), F: 0531/313074(d)

Schmidt, K.-U., Dr. rer. nat. Dipl.-Geol., Laborleiter, Bickhardt Bau AG, Industriestr. 9, 36275 Kirchheim. T: 06625/88-0(d), F: 06625/88-113(d), T: 02772/957680(p)

Schmidt, M.W., Dipl.-Geol., Fontaneweg 2, 38642 Goslar. T: 05336/89219(d)

Schmidt, N., Dipl.-Ing., Lahmeyer International Office Cairo, Masaken-Nasr City, ET- Cairo, Ägypten. T: 0202/4189211(d), F: 0202/4177507(d)

Schmidt, S., Dipl.-Geol., Brachterstr. 30, 35282 Rauschenberg. T: 06421/283465(d), T: 06425/6224(p)

Schmidt, T., Dipl.-Ing., IBS Ingenieurbüro Thomas Schmidt, Fuchsmühlenweg 7, 09599 Freiberg. T: 03731/31938(d), F: 03731/32210(d)

Schmidt-Schleicher, H., Dr.-Ing., Zerna, Köpper & Partner, Geschäftsführung, Industriestr. 27, 44892 Bochum. T: 0234/9204-0(d), F: 0234/9204-170(d)

Schmitt, A., Am Bonneshof 12, 40474 Düsseldorf. T: 0211 452304

Schmitt, G.-P., Dr.-Ing., Geschäftsführer, Björnsen Beratende Ingenieure, Maria Trost 3, 56070 Koblenz. T: 0261/8851-180(d), F: 0261/805725(d), T/F: 0261/52665(p)

Schmitt, J., Dipl.-Ing., Projektingenieur, Kelsterbacherstr. 118, 64546 Walldorf. T: 06105/951222(p)

Schmitt, V., Feldbergstr. 34, 64293 Darmstadt. T: 06151/894376(p)

Schmitz, I., Darmstädter Str. 29, 64409 Messel. T: 06159/5473(p), F: 06159/913365(p)

Schmitz, S., Dr.-Ing., Ing.-Büro Pirlet & Partner mbH, Cäcilienstr. 48, 50667 Köln. T: 0221/9257750(d), F: 0221/922577518(d)

Schnebele, H.P., Dipl.-Ing., Bundesanstalt für Wasserbau, Kußmaulstr. 17, 76187 Karlsruhe

Schneider, R., Dipl.-Ing., Crystal Geotechnik Beratende Ingenieure und Geologen GmbH, Hofstattstr. 28, 86919 Utting. T: 08806/480(d), F: 08806/2609(d)

Schnorr, F., Frauenbergstr. 14, 35039 Marburg. T: 06421/47087(p)

Schober, H.-D., Dipl.-Ing., Friedrichsfelder Weg 17, 68782 Brühl

Scholz, BRB Prüflabor GbR mbH, Albertshofer Chaussee, 16321 Bernau. T: 0338/396876(d), F: 0338/396888(d)

Scholz, M., Eißendorfer Str. 36, 21073 Hamburg-Hamburg

Scholz, W., Dipl.-Geol., Wildermuthstr. 3, 67065 Ludwigshafen. T: 0621/573614(p), F: 0621/573266(p)

Schramm, J.M., Univ.-Prof. Dr., Leonorenweg 20, A-5020 Salzburg, Österreich. T: 0043/662/8044-5410(d), F: 0043/662/8044-621(d)

Schran, M.Sc., U., Dipl.-Ing., Chausseestr. 32, 14542 Glindow. T: 03327/71570(p)

Schreyer, A., Dipl.-Ing., TU Bergakademie Freiberg, Institut für Geotechnik, Abt. Bodenmechanik, Gustav-Zeuner-Str. 1, 09596 Freiberg. T: 03731/392523(d), F: 03731/393501(d)

Schröder, H., Franz, Dr.-Ing., Projektgruppenleiter, Bundesanstalt f. Materialforschung und Prüfung (BAM), VI.1 Beständ. v. Polymerwerkstoffen, Unter den Eichen 87, 12205 Berlin. T: 030/81041611(d), T: 030/8023747(p), F: 030/8023747(p)

Schröder, P., Prof. Dr., Professor f. Bodenmech., Wilhelm-Busch-Ring 22 a, 63486 Bruchköbel. T: 06181/72897(p), F: 06181/72897(p)

Schröpfer, T., Dr.-Ing., Rosenberg 6, 38640 Goslar. T: 0531/2104111(d), F: 0531/313274(d)

Schubert, A., Dr.-Ing., Dr.-Ing. A. Schubert, Werner-v.-Siemensstr. 17, 82140 Olching. T: 08142/49000(d), F: 08142/3795(d)

Schuck, W., Dipl.-Ing., Bauingenieur, Deutsche Bahn AG Forschungs-und Technologiezentrum (TZ) Konstruktiver Ingenieurbau, Richelstr. 3, 80634 Mümchen. T: 089 1308-5616(d), F: 089 1308-5711(d)

Schuhmacher, I., Dipl.-Ing., Guntherstr. 61, 90461 Nürnberg. T: 0911/4719836(p)

Schuhmann, H., Dipl.-Ing., Büro für Baurealisierung, Weil am Rhein GmbH, Am Stammbachgraben 2, 79539 Lörrach. T: 07621/97790(d), F: 07621/793300(d)

Schulte, K., Dr.-Ing., Niederlassungsleiter, ELE-Erdbaulaboratorium Essen Niederlassung Berlin, Prinzessinnenstr. 8, 10969 Berlin. T: 030 616989-0(d), F: 030/61989-99(d)

Schultz, E.W., Dipl.-Ing., Ltr. Tiefbauabteilung, Dyckerhoff & Widmann AG, Walter-Kolb-Straße 5–7, 60594 Frankfurt. T: 069/96227-150(d), F: 069/96227-450(d), T: 06101/7951(p)

Schultz, H.G., Dr.-Ing., Prüfingenieur f. Baustat., CSK Ingenieurgesellschaft mbH, Westring 24, 44787 Bochum. T: 0234/96440-0(d), F: 0234/96440-40(d)

Schulz, E., Dr.-Ing., Baugrund Dresden Ingenieurgesellschaft mbH, Paul-Schwarze-Str. 2, 01097 Dresden. T: 0351/8241358(d), F: 0351/8030786(d)

Schulz, G., Dipl.-Ing., Arcadis Trischler und Partner GmbH Beratende Ingenieure Geotechnik — Umweltschutz, Berliner Allee 6, 64295 Darmstadt. T: 06151/388-0(d), F: 06151/388-999(d)

Schulz, H., Univ.-Prof. Dr.-Ing., Universität der Bundeswehr München, Institut für Bodenmechanik und Grundbau, 85577 Neubiberg. T/F: 089/6004-3476(d), T: 07245/937852(p), F: 07245/4073(p)

Schulz, T., Dipl.-Ing., Dames & Moore gmbh & co. KG, A division of URS Corp., Goernerstr. 32, 20249 Hamburg.

T: 040/460760-14(d), F: 040/460760-60(d), T: 0531/71743(p)

Schulze, R., Dipl.-Ing., Ingenieurbüro für Erd- und Grundbau, Doberkamp 17, 24223 Raisdorf. T: 04307/6850(d), F: 04307/7175(d)

Schuppener, B., Dr.-Ing., Abteilungsleiter, Geot., Bundesanstalt für Wasserbau, Kußmaulstr. 17, 76187 Karlsruhe. T: 0721/9726-380(d), T: 07247/22608(p)

Schürmann, A., Südhofstr. 11, 58099 Hagen. T: 02173/7902-72(d), T: 02331/632324(p)

Schütte, H., Ing., Am Richteich 2, 30916 Isernhagen

Schütz, H., Prof. Dipl.-Ing, Hohlenscheidterstr. 52, 42349 Wuppertal. T: 0202/403139(p), F: 0202/401615(p)

Schuwicht, A., Dipl.-Ing., Husenstieg 18, 37603 Holzminden. T: 05531/61215(p)

Schwab, A.J.M., Dipl.-Ing., Richard-Wagner-Str. 16, 67360 Lingenfeld

Schwab, R., Dr.-Ing., Wissensch. Angestellter, Bundesanstalt für Wasserbau, Abt. Geotechnik, Kußmaulstr. 17, 76187 Karlsruhe. T: 0721/9726-381(d), F: 0721/9726-483(d)

Schwan, K., Dipl.-Ing. (FH), Preihsl und Schwan Beraten und Planen im Bauwesen, Kreubergweg 1a, 93133 Burglengenfeld. T: 09471/7016-0(d), F: 09471/7016-17(d), T: 0941/51965(p), F: 0941/563799(p)

Schwarz, H., Dr.-Ing., Geschäftsführer, Stump Spezialtiefbau GmbH, Fränkische Str. 11, 30455 Hannover. T: 0511/9499910(d), F: 0511/499498(d)

Schwarz, W., Dr.-Ing., Bauer Spezialtiefbau GmbH, Bautechnik, Wittelsbacherstraße 5, 86529 Schrobenhausen. T: 08252/97-1401(d), F: 08252/97-1213(d), T: 08252/81478(p)

Schwerdt, S., Dipl.-Ing., Edderitzer Str. 30, 06366 Köthen. T: 03496/558191(p)

Schwing, E., Prof. Dr.-Ing., Hochschullehrer, Tilsiterstr. 2, 76139 Karlsruhe. T: 0721/689456(p)

Seeger, H., Dr.-Ing., Ing.-Büro für Geotechnik Dr.-Ing. Helmut Seeger, Steingrübenstr. 6, 71111 Waldenbuch. T: 07157/20184(d)

Sehrbrock, U., Dr.-Ing., Götte + Sehrbrock GmbH, Bevenroder str. 152, 38104 Braunschweig. T: 05131/3 78460(d), F: 05131/378371(d)

Seip, M., Kaupstr. 22, 64289 Darmstadt. T: 06151/783641(p)

Semprich, S., Prof. Dr.-Ing., Technische Universität Graz, Institut f. Bodenmechanik u. Grundbau, Rechbauerstr. 12, A-8010 Graz, Österreich. T: 0043/316/873-6230(d)

Sentko, M., Prof. Dr.-Ing., Berat. Ingenieur, Am Südhang 1, 49545 Tecklenburg. T: 05482/6312(p)

Seydel, E., Dipl.-Ing., Grundbaulabor München, St.-Martin-Str. 26, 81541 München. T: 089/6917147(d), F: 089/6927034(d)

Siebert, M., Dipl.-Ing. (U), Kohlenstr. 157, 44793 Bochum. T: 0234/685775-619(d), T: 0234/6407588(p)

Siebke, J., Baurat, Wasser- u. Schiffahrtsamt Eberswalde, Postfach 10 08 26, 16208 Eberswalde

Siegmund, M., Dipl.-Ing., Wissensch. Mitarbeiterin, Franz-Mehring-Str. 23, 99427 Weimar. T: 03643/421358(p)

Siegmundt, M., Dr.-Ing., TU Bergakademie Freiberg, Institut für Geotechnik, Gustav-Zeuner-Str. 1, 09596 Freiberg. T: 03731/392484(d), F: 03731/393638(d)

Siemer, C., Dipl.-Ing., Saarweystr. 3, 70191 Stuttgart. T: 0711/9371930(p)

Siemer, H., Prof. Dr.-Ing., Max-Liebermann-Str. 7, 40699 Erkrath. T/F: 0211/242683(p)

Siemer, J., Dr., Geschäftsführer, IBTB Ingenieurgesellschaft für Bautechnik mbH, An der Gehespitz 50, 63263 Neu-Isenburg. T: 06102/45-2044(d), F: 06102/45-2092(d), T: 06103/573348(p)

Simon, A., Dipl.-Geol., Geol. Oberrat, Landesamt für Geowissenschaften und Rohstoffe Brandenburg, Stahnsdorfer Damm 77, 14532 Kleinmachnow. T: 033203/36760(d), F: 033203/36702(d), T: 03329/615005(p), F: 03329/615003(p)

Simon, S., Dipl.-Ing., Alter Graben 24, 63571 Gelnhausen. T: 06051/883566(p)

Skalare, H., Dipl.-Geol., Atlas Copco MCT GmbH, Ernestinenstr. 155, 45141 Essen. T: 0201/8919-377(d), F: 0201/291412(d)

Sobe, A., Dieburger Str. 13, 64287 Darmstadt. T: 06151/719024

Smoltczyk, U., Prof. Dr.-Ing. Dr.-Ing. E.h., Berat. Ingenieur, Adlerstr. 63, 71032 Böblingen. T: 07031/273678(p), F: 07031/288677(p)

Sobolewski, J., Dr.-Ing., HUESKER Synthetic GmbH & Co., Abt. Anwendungstechnik, Fabrikstraße 13–15, 48712 Gescher. T: 02542/701-309(d), F: 02542/701-499(d), T: 02541/981247(p)

Sommer, H., Prof. Dr.-Ing., Ludwigstr. 142, 64367 Mühltal

Sommer, M., Dipl.-Ing., Lautenschlägerstr. 8, 64289 Darmstadt. T: 06151/300240(d), T: 06151/717194(p)

Sondermann, W., Dr.-Ing., Geschäftsführer, Keller Grundbau GmbH, Kaiserleistr. 44, 63067 Offenbach. T: 069/8051-220(d), F: 069/8051-224(d), T: 06106/647212(p)

Soos von, P., Dipl.-Ing., Ltd. Akad. Direktor i.R., Reußweg 30, 81247 München. T: 089/882738(p), F: 089/8205255(p)

Spang, K., Dr.-Ing., Projektmanager, Planungsgesellschaft Bahnbau Deutsche Einheit, Am Brauhaus 1, 01099 Dresden. T: 0351/808250(d), F: 0351/8082530(d), T: 0351/2690856(p)

Spang, R.M., Dr., Geschäftsführer, Dr. Spang GmbH, Westfalenstraße 5-9, 58455 Witten. T: 02302/91402-0(d), F: 02302/91402-20(d)

Spaun, G., Prof. Dr., TU München Lehrstuhl für Allgemeine, Angewandte und Ingenieur-Geologie, Arcisstr. 21, 80290 München. T: 089/289 25850(d), F: 089/289 25852(d)

Spielhoff, D., Dipl.-Ing., SPIBO Spielhoff-Bohrwerkzeuge GmbH, Kronprinzenstr. 26, 44135 Dortmund. T: 0231/528246(d), F: 0231/575237(d)

Spotka, J., Dr.-Ing., Ingenieurbüro für Grundbau, Finkenweg 3, 92353 Postbauer-Heng. T: 09188/94000(d), F: 09188/9400-49(d)

Srebak, B., Dipl.-Ing. (FH), Zugspitzenstr. 1, 85649 Brunnthal.

Stabel, B., Dipl.-Ing., Lahmeyer International GmbH, Friedberger Str. 173, 61118 Bad Vilbel

Stadtbäumer, F., Dr. rer. nat., Dr. Stadtbäumer Baugrunduntersuchung GmbH, Rotwandstr. 10, 85609 Aschheim. T: 089/991518-0(d), F: 089/991518-13(d)

Stahlhut, O., Dr.-Ing. Dipl.-Wirtsch.-Ing., iwb Ingenieursellschaft mbH, Stubbenhuk 10, 20459 Hamburg. T: 040/369854-16(d), F: 040/369854-99(d)

Stanke, C., Mittelstellberg 14, 36157 Ebersburg. T: 06656/6199(p)

Stapf, K., Dipl.-Ing. (FH), Berat. Ingenieur, Ingenieurgesellschaft Stapf und Sturny mbH, Mombacher Str. 93, 55122 Mainz. T: 06131/387689(d), F: 06131/385821(d)

Starke, C., Dipl.-Geol., Referatsleiter, Sächsisches Landesamt f. Umwelt u. Geologie Angew. Geologie/Ingenieurgeologie, Halsbrücker Str. 31a, 09599 Freiberg. T: 03731/2941-43(d), F: 03731/22918(d), T: 03731/459977(p), F: 03731/31731(p)

Staudt, A., Amyastr. 24, 52066 Aachen. T: 0221/405023(d), F: 0221/405527(d)

Staudt, H., Dipl.-Ing., Berat. Ing., Wiethaserstr. 5, 50933 Köln. T: 0221/405023(p)

Staudtmeister, K., Dipl.-Ing., An den Papenstücken 20, 30455 Hannover. T: 0511/7623044(d), F: 0511/7625367(d)

Steffen, H., Prof. Dr.-Ing., Dr.-Ing. Steffen, Ingenieursellschaft mbH, Abteilungsleitung, Im Teelbruch 128, 45219 Essen. T: 02054/1211-201(d), F: 02054/1211-216(d)

Stehling, K., Pappelweg 1, 26624 Südbrookmerland

Steiger, A., Dipl.-Bauing. ETH, Inhaber, Andreas Steiger & Partner AG, Beratende Ingenieure SIAASIC, Pilatusstr. 30, CH-6003 Luzern, Schweiz. T: 0041/41-2275101(d), F: 0041/41-2275102(d)

Steiger, H., Dipl.-Ing., Krebs und Kiefer, Hilpertstr. 20, 64295 Darmstadt. T: 06151/885143(d), F: 06151/885118(d), T: 06150/12369(p)

Steiner, H., Dipl.-Ing., Grundbauingenieure Steinfeld und Partner GbR, Alte Königstr. 3, 22767 Hamburg. T: 040/3891390(d), F: 040/3809170(d)

Steinhoff, J., Dr.-Ing., Geschäftsführer, MMS – Ingenieurgesellschaft für Bautechnik mbH, Gustavstr. 12,

42329 Wuppertal. T: 0202/27311-20(d), F: 0202/27311-22(d)

Stelzer, O., Heidenreichstr. 1, 64287 Darmstadt. T: 06151/423980(p)

Stelzig, S., Dipl.-Ing., Geschäftsführer, IGU Ingenieurgesellschaft für Grundbau und Umwelttechnik mbH, Bahnhofstr. 22, 87719 Mindelheim. T: 08261/3079(d), F: 08261/6956(d)

Stender, E.P., Dipl.-Ing., Tejastr. 4, 12105 Berlin

Sternath, R., Dipl.-Ing., DBProjekt GmbH Köln-Rhein/Main Fachprojektleitung Tunnelbau, Herriotstr. 5, 60528 Frankfurt/Main. T: 069/26529464(d), F: 069/26529469(d), T: 08024/1216(p)

Stiriz, K.H., Dipl.-Ing., Geschäftsführer, Wiesendamm 147, 22303 Hamburg. T/F: 040/2792446(d)

Stocker, M., Dr. Dipl.-Ing., Bauer Spezialtiefbau GmbH, Geschäftsbereichsleitung Technik, Wittelsbacher Str. 5, 86529 Schrobenhausen. T: 08252/97-1200(d), F: 08252/97-1213(d), T: 08252/7110(p)

Stöcker, T., Moritzstr. 26, 34127 Kassel. T: 0561/8042650(d), F: 0561/8042651(d)

Stockhammer, P., Dipl.-Ing., Geschäftsführer, Keller Grundbau GmbH, Mariahilferstr. 129, 1151 Wien, Österreich. T: 0043/1/838541(d), F: 0043/1/859813(d)

Stoewahse, C., Dipl.-Ing., Wissensch. Mitarbeiter, Universität Hannover Institut für Grundbau Bodenmechanik u. Energiewasserbau, Appelstr. 9A, 30167 Hannover. T: 0511/7624151(d), T: 0511/2108085(p)

Stoof, R., Dipl.-Ing., Geschäftsführer, Erdbaulaboratorium Dresden GmbH, Ing.-Büro f. Geotechnik u. Umwelt, Stolpener Str. 1, 01477 Arnsdorf/OT Fischbach. T: 035200/32930(d), F: 035200/32939(d)

Stötzner, U., Dr. rer. nat., freiberufl. Consultant, Dr. Ulrich Stötznor Geophysik-Consultant, Fasanenweg 25, 04316 Leipzig. T: 0341 6586240(d), F: 0341 6586239(d)

Stracke, M., Dipl.-Geol., Ingenieurgeologen Oberste Wilms & Stracke GbR OWS, Zum Wasserwerk 15, 48268 Greven. T: 02571/95288-0(d), F: 02571/95288-2(d), T: 02507/4182(p)

Strassl, C., Niederramstädter Str. 181, 64285 Darmstadt. T: 06151/43696(p)

Straußberger, F., 1, 91126 Schwabach. T: 0911/ 6555575(d), T: 09122/73607(p)

Streim, W., Dr., Soz. Grundbau & Ingenieurgeol., Salzschlirferstr. 16, 60386 Frankfurt/M. T: 069/414150(d), F: 069/417170(d)

Stroh, D., Dr.-Ing., Direktor, Hochtief Consult, Opernplatz 2, 45128 Essen. T: 0201/824-2421(d), F: 0201/824-2742(d)

Strohhäusl, S., Dipl.-Ing., Strohhäusl & Partner Ziviltechniker GmbH, Magazingasse 8/3, 4020 Linz, Österreich. T: 0043/732/771777(d), F: 0043/732/771777-10(d), T: 0043/72384721(p)

Strüber, S., Dipl.-Ing., Herdweg 60, 64285 Darmstadt. T: 06151/425766(p)

Struck, O., Dr.-Ing., Dipl.-Ing., Letteallee 90, 13409 Berlin. T: 030/34963-510(d), F: 030/34963-800(d), T: 030/4919083(p)

Stummer, H., Erlenweg 20, 64807 Dieburg

Stump, H., Dipl.-Ing. ETH, Stump Bohr AG Abt. Geschäftsleitung, Stationsstr. 57, CH-8606 Nänikon, Schweiz. T: 0041/1/9417777(d), F: 0041/1/9417800(d)

Stürmer, H., Dipl.-Ing., Bauingenieur, Im Uckerfeld 10, 53127 Bonn. T: 0228/282442(p)

Sturny, D., Dipl.-Ing. (FH), Ing.-Ges. Stapf & Sturny GmbH, Mombacher Str. 93, 55122 Mainz. T: 06131/387689(d), F: 06131/385821(d)

Suckow, H.-M., Dr.-Ing., BGA Suckow + Zarske GbR Beratende Geologen und Ingenieure, Hamelnweg 12, 38124 Braunschweig. T: 0531/691044(d), F: 0531/691046(d)

Tausch, N., Prof. Dr.-Ing., Professor, Fachhochschule Kaiserslautern, Morlauterer Str. 31, 67657 Kaiserslautern. T: 06131/3724505(d), F: 06131/3724555(d), T: 06131/77183(p), F: 06131/73585(p)

Teichmann, J., Dipl.-Ing., Heilit + Woerner Bau AG Verkehrswegebau NL Düsseldorf, Vogelsanger Weg 111, 40470 Düsseldorf. T: 0221/6104427(d)

Thaher, M., Dr.-Ing., Philipp Holzmann Bautechnik GmbH, Zentrales Geotechniklabor, An der Gehespitz 80, 63263 Neu-Isenburg. T: 06102/45-2701(d), F: 06102/45-2799(d)

Thiel, J., Dipl.-Ing., Dipl.-Ing. Joern Thiel Baugrunduntersuchung GmbH, Georg-Wilhelm-Str. 322, 21107 Hamburg. T: 040/3078804(d), F: 040/3078897(d)

Thiel, L., Dipl.-Ing., Kottwitzerstr. 18, 20253 Hamburg

Thiele, R., Dr.-Ing., leit. Mitarbeiter, Turmweg 6, 04277 Leipzig. T: 0341/9610906(p)

Thomä, M., Dr. rer. nat., Eichendorffstr. 57, 55122 Mainz. T: 06131/581589(p), F: 06131/593655(p)

Thuro, K., Dipl.-Geol. Dr. rer. nat., ETH Hönggerberg Ingenieurgeologie, 8093 Zürich, Schweiz. T: 0049/1/633-6818(d), F: 0049/1/633-1108(d)

Timm, G., Dr.-Ing., Prüfingenieur f. Baustat., Dr.-Ing. R. Windels, Dr.-Ing. G. Timm, Dr.-Ing. K. Morgen Beratende Ingenieure VBI, Ballindamm 17, 20095 Hamburg. T: 040/350090(d), F: 040/35009100(d)

Timmers, V., Krakaustr. 27a, 52064 Aachen. T: 0241/407915(p)

Tischner, B., Eichenweg 4, 14532 Kleinmachnow. T: 033203/70931(d), F: 033203/70935(d), T: 033203/70907(p)

Töniges, W., Dipl.-Geol., Töniges GmbH, Ingenieurgeologisches Büro, Kleines Feldlein 4, 74889 Sinsheim. T: 07261/92110(d), F: 07261/921122(d)

Tönnis, B., Dipl.-Ing., Hydroprojekt Ingenieurgesellschaft mbH, Monschauer Str. 1, 40549 Düsseldorf. T: 030/7877630(d), F: 030/78776310(d)

Tophinke, G., Dipl.-Ing., Am See 20, 24259 Westensee

Toth, L., Dr. Mag. rer. nat., Ingenieurkonsulent für Technische Geologie, Burwegstr. 3, 3032 Eichgraben, Österreich. T: 0043/2773/42806(d), F: 0043/2773/428064(d), T: 0043/2773/42806(p), F: 0043/2773/428064(p)

Trautner, O., Dipl.-Ing., Geschäftsführer/Mitinhab., Ingenieurbüro für Geotechnik GFPGbR, Bürgstr, 15, 47057 Duisburg. T: 0203 350539(d), F: 0203 350541(d), T: 02102 895598(p)

Triantafyllidis, T., Univ.-Prof. Dr.-Ing. habil., Ruhr-Universität Bochum Lehrst. f. Grundbau u. Bodenmechanik, Universitätsstr. 150/Geb. IA 4/126, 44801 Bochum. T: 0234/32-26135(d), F: 0234/32-14150(d), T: 0201/5024116(p), F: 0201/5024117(p)

Tropper, W., Dr. techn. Dipl.-Ing., Geotechnik Dr. Tropper GmbH, Fischnalerstr. 19, 6020 Innsbruck, Österreich. T: 0043/512/284192(d), F: 0043/512/282192-12(d)

Trunk, U., Dr.-Ing., Bauingenieur, Keller Grundbau GmbH, Veltener Str. 31, 16/67 Germendorf. T: 03301 5857-50(d), F: 03301 5857-20(d)

Turek, J., Dipl.-Ing., Lichtenbergstr. 74, 64289 Darmstadt

Ulrich, G., Dr. Ing. Baugrundinstitut Dr. Ing. Georg Ulrich Zun Brunnentobel 6, 88299 Leutkirch. T: 07561 9863-0(d), F: 07561 7571(d)

Vardoulakis, I.G., Prof. Dr.-Ing., National Technical Univers. Athens, Dept. of Engineering Science, 5 Heroes of Polytechnion Avenue, GR-157 73 Athen, Griechenland. T: 0030/1/777-2105(d), F: 0030/1/778-7272(d)

Vees, E., Prof. Dr.-Ing., Baugrundgutachter, Prof. Dr.-Ing. E. Vees und Partner Baugrundinstitut GmbH, Waldenbucher Str. 19, 70771 Leinfelden-Echterdingen. T: 0711/7979378(d), F: 0711/7979378(d), T: 0711/798006(p), F: 0711/7979378(p)

Veith, N., Schlüterbusch 20, 45289 Essen. T: 0211/472010(d), F: 0211/4180232(d), T: 0201/578026(p)

Vermeer, P.A., Prof. Dr.-Ing., Hochschullehrer, Universität Stuttgart, IGS Institut für Geotechnik, Pfaffenwaldring 35, 70569 Stuttgart. T: 0711/685 2436(d), F: 0711/685 2439(d), T/F: 07032/74314(p)

Verst, R., Dipl.-Ing., Stefan-George-Ring 9, 81929 München. T: 089/9267-291(d), F: 089/9267-366(d), T: 089/93940898(p)

Viedenz, U., Dipl.-Ing., Leiter Bautechnik, Siemens AG, Abt. EV HG 6, Bautechnik, Postfach 3220, 91050 Erlangen. T: 09131/733924(d), F: 09131/734397(d), T/F: 09131/994371(p)

Vittinghoff, T., Dipl.-Ing., TU Braunschweig, Institut für Grundbau und Bodenmechanik, Gaußstr. 2, 38106 Braunschweig. T: 0531/391-2730(d), F: 0531/391-4574(d)

Voge, c/o Andreas Zill, Michael, Geotechnica GmbH, Friedrichsruher Str. 26, 14193 Berlin. T: 030 6127531(d), F: 0172 3009157(p)

Vogel, H.-J., Dipl.-Ing., Geschäftsführer, Eilers & Vogel Beratende Ingenieure für Bauwesen GmbH, Sutelstr. 23, 30659 Hannover. T: 0511/90494-0(d), F: 0511/90494-44(d)

Vogel, H., Bachstr. 46, 76297 Stutensee. T: 06151/162271(d), T: 07244/740660(p)

Vogel, J., Dr.-Ing., Dr.-Ing. Meihorst u. Partner, Beratende Ingenieure für Bauwesen GmbH, Gehägestr. 46, 30655 Hannover. T: 0511/909560(d), F: 0511/9095611(d), T/F: 0511/871148(p)

Vogel, W., Dipl.-Ing., Lortzingstr. 17, 81241 München. T: 089/1308-5416(d), F: 089/1308-5711(d), T: 089/8203426(p)

Vogler, M., Dr.-Ing., Wachenbergstr. 13, 69469 Weinheim. T: 06151/363138(d), F: 06151/363130(d),

Vogt, C., Dr.-Ing. Projehtleiterin Wolfmahden-str. 42, 70563 Stuttgart. T: 0711/13164-32, F: 0711/13164-64

Vogt, J., Dipl.-Ing., Berat. Ing., Kirchstr. 38, 50181 Bedburg. T: 02272/3886(d), F: 02272/7330(d)

Vogt, N., Dr.-Ing., Prof. Dr.-Ing. Technische Universität Minchen, Lehrsturhl u. Prüfanet f. Grundbau, Boden – und Felsmechanik, Baumbachstr. 7, 81245 Minchen. T: 089/289-27132(d), F: 089/289-27189(d)

Voigt, T., Dr.-Ing., Spessartweg 9, 70794 Filderstadt. T: 0711/7800986(p), T: 0711 734386(p)

Voigtmann, H., Dipl.-Geol., Büro für Baugrund-Untersuchungen, Ingenieur-Geologie und Hydro-Geologie, Theodor-Heuss-Platz 3, 71364 Winnenden. T: 07195/9250-0(d), F: 07195/2622(d)

Völkl, J., Dipl.-Geol., Institut Dr.-Ing. Gauer Niederlassung Weiden Erdbau, Fasanenweg 24, 92721 Störnstein. T: 09602/94100(d), F: 09602/941031(d)

Vollenweider, U., Dr.-Ing. ETH., Hegarstr. 22, CH-8032 Zürich, Schweiz. T: 0041/1/3833844(p), F: 0041/1/3832676(p)

Vosteen, B., Baudirektor, Im Vogelsang 20, 35452 Heuchelheim. T: 0641/61611(p)

Vrettos, C., Priv.-Doz. Dr.-Ing., GuD Consult GmbH, Dudenstr. 78, 10965 Berlin. T: 030/789089-0(d), F: 030/789089-89(d)

Waas, G., Dr., Bergstr. 12a, 65817 Eppstein. T: 06198/8165(p)

Wagenhausen, J., Dipl.-Ing., Talstr. 5a, 02625 Bautzen. T: 035934/4488(d), F: 035934/4489(d), T: 03591/530630(p)

Waibel, P., Prof. Dipl.-Ing. Dr., Ziviling. für Bauwesen, BGG, Büro Prof. Dipl.-Ing. Dr. P. Waibel, Mariahilfer Str. 20, A-1070 Wien, Österreich. T: 0043/1/5242980-0(d), F: 0043/1/5242980-4(d)

Walbröhl, H.T., Dipl.-Ing., Nordstr. 75, 53111 Bonn

Waldhoff, P., Dr.-Ing., IGW Ingenieurgesellschaft für Geotechnik, IGW Ingenieurgesellschaft für Geotechnik, Uellendahl 70, 42109 Wuppertal

Wallner, M., Dr.-Ing., Wissensch. Direktor, Bundesanstalt für Geowissenschaften und Rohstoffe, Ref. B 2.1: Felsmech., Baugeologie, Stilleweg 2, 20655 Hannover. T: 0511/643-2422(d), F: 0511/643-3694(d), T: 0511/ 6479168(p)

Wälzel, E., Dipl.-Ing., Möhlstr. 19, 81675 München

Warmbold, U., Dipl.-Ing., Abt.-Ltr., Lausitzer Braunkohle AG (LAUBAG), Bodenmechanik – TG 3, Knappenstr. 1, 01968 Senftenberg. T: 03573/78-2159(d), F: 03573/78-782188(d), T: 0355/3819887(p)

Warnecke, W., Dr. Dipl.-Geol., T.A., Otto-Lilienthal-Str. 15, 31157 Sarstedt. T: 0511/3034413(d), T: 05066//4600(p)

Weber, E., Prof. Dr.-Ing. habil., Gutachterbüro Prof. Dr Weber, Bahnhofstr. 33, 03099 Kolwitz. T: 0355 287102(d), F: 0355 28619(d), T: 0355 287102(p), F: 0355 28619(p)

Weber, H., Dr.-Ing., Sachverständiger f. Geotechnik, Dr.-Ing. H. Weber Sachverstandiger f. Geotechnik Beratender Ingenieur, Dorfstr. 15, 06231 Kötzchau-Thalschütz. T: 03462 510-100(d), F: 03462 510-102(d), T: 03462 510 101(p)

Weber, T., Turnstr. 1a, 09235 Burkhardtsdorf. T: 03721/24380(p)

Weber, T., Dipl.-Ing., Sigmaringerstr. 5a, 10713 Berlin. T: 030/21304-147(d)

Wehr, M.Sc., Wolfgang, Dr.-Ing., Lämmerspieler Str. 83, 63165 Mühlheim/Main. T: 069 8051/257, F: 069 8051 244(d)

Weidenmüller, A., Kohlenstr. 13a, 08228 Rodewisch. T: 03744/31269(p)

Weidle, A., Wilhelminenstr. 15, 64283 Darmstadt. T: 06151/ 162349(d), F: 06151/166683(d), T: 06151/25581(p)

Weihrauch, S., Dr.-Ing., Grundbauingenieure Steinfeld und Partner GbR, Alte Königstr. 3, 22767 Hamburg. T: 040/ 389139-0(d), F: 040/3809170(d)

Weiler, H., Dr.-Ing., Bussardweg 2, 63303 Dreieich. T: 06103/699167(p)

Weinhold, H., Prof. Dr.-Ing., Ehrwalder Str. 81, 81377 München. T: 089/396500(p), F: 089/346956(p)

Weismann, A., Bruchwiesenstr. 6a, 64285 Darmstadt. T: 06151/663478(p)

Weiße, M., Dipl.-Ing., Berat. Ingenieur, Ing.-Büro Weiße, Kaiseritz Nr. 6, 18528 Bergen auf Rügen. T: 03838/ 23322(d), F: 03838/254773(d)

Weißenbach, A., Univ.-Prof. Dr.-Ing., Am Gehölz 14, 22844 Norderstedt. T: 040/5223321(p), F: 040/5358143(p)

Weissmann, R., Dipl.-Ing., Schürenbergstr. 19, 45139 Essen. T: 0201/239170(p)

Wellmer, F.-W., Prof. Dr.-Ing. Dr. h.c., Präsident der Bundesanstalt für Geowissenschaften und Rohstoffe, Stilleweg 2, 30655 Hannover. T: 0511/643-2243(d), F: 0511/643-3676(d)

Wendt, R., Dr.-Ing., Bauhaus-Universität Weimar Fakultät Bauingenieurwesen Professor Grundbau, Marienstr. 7, 99421 Weimar. T: 03643/504556(d), F: 03643/504564(d)

Wennmohs, K.H., Dipl.-Ing., Atlas Copco MCT GmbH, Projektabt., Ernestinenstr. 155, 45141 Essen

Wenz, K.-P., Prof. Dr.-Ing., Bismarckstr. 24, 66953 Pirmasens. T: 06331/96842(p), F: 06331/226687(p)

Werth, A., Dipl.-Ing., Obermeyer Planen & Beraten GmbH, Abt. Ingenieur-Tiefbau, Hansastr. 40, 80686 München. T: 089/5799-243(d), F: 089/5799-205(d)

Weß, S., Odilienstr. 1, 36124 Döllbach. T: 06656/1481(p)

Westhaus, T., Dr.-Ing., Baugrund-Institut Dr.-Ing. Westhaus GmbH, Rheinufer 11, 55252 Mainz-Kastel. T: 06134/ 180457(p), F: 06134/180458(p)

Westrup, J., Dr. rer. nat., Hessisches Landesamt für Bodenforschung, Leberberg 9, 65193 Wiesbaden. T: 0611/ 537316(d), F: 0611/537327(d)

Wichert, H.-W., Dr.-Ing., Dr. Hans- W. Wichert VDI Vermittlung, Beratung, Altlasten, Bodenschutz, Kölner Str. 21–25, 50171 Kerpen. T: 02237/925520(d), F: 02237/ 922717(d)

Wichter, L., Prof. Dr.-Ing., Brandenburgische TU Lehrstuhl für Bodenmechanik u. Grundbau/Geotechnik, Universitätsplatz 3–4, 03044 Cottbus. T: 0355/692602(d), F: 0355/ 692566(d)

Wiese, V., Dipl.-Ing., Zum Höfken 2, 58313 Herdecke

Wiesenmaier, H., Stuttgarter Str. 9, 71554 Weissach im Tal. T: 07191/51499(p)

Wiesiolek, B., Dipl.-Ing., Frank & Kraft + Partner, Geotechnik und Umwelttechnik GmbH, Institut für Erd- und Grundbau, Hofer Str. 1, 81737 München. T: 089/670061-0(d), F: 089/670061-33(d)

Wietek, B., Prof. Dipl.-Ing., Blumeserweg 290, A-6073 Sistrans b. Innsbruck, Österreich. T: 0043/512/378188-0(p), F: 0043/512/378188-4(p)

Wilden, U., Dipl.-Ing., Rheinbraun AG BT 4, Stüttgenweg 2, 50935 Köln. T: 0221/480-2789(d), T: 02234/84098(p)

Wilhelm, R., Dr. Dipl.-Geol., Chef-Geologe, Lahmeyer International GmbH GW-3, Friedberger Str. 173, 61118 Bad Vilbel. T: 06101/55-1414(d), F: 06101/55-1961(d), T/F: 06053/4575(p)

Wille, T., Dipl.-Min., Geschäftsführer, Wille Geotechnik GmbH

Wagenstieg, 8a, 37077 Göttingen. T: 0551/307520(d), F: 0551/3075220(d)

Wilmers, W., Dipl.-Geol. Dr. rer. nat., Techn. Angestellter, Berliner, Ring 72, 35576 Wetzlar. T: 06441/92474-22(d), F: 06441/92474-20(d), T: 06441/52334(p)

Winkelmann, O., Dipl.-Ing., Dipl.-Geol., Berat. Ing., Berat. Baugeol., Schulstr. 1, 48149 Münster. T: 0251/274930(d), F: 0251/298377(d), T: 0251/272154(p)

Winkelvoss, U., Dr.-Ing., Geschäftsführer, Baugrund – Institut Klein + Winkelvoss GmbH, Eichendorffstr. 35, 93138 Lappersdorf. T: 0941/82935(d), F: 0941/85977(d)

Witt, K.-J., Univ.-Prof. Dr.-Ing., Bauhaus-Universität Weimar, Fakultät Bauingenieurwesen, Professur Grundbau, Marienstr. 7, 99421 Weimar. T: 03643/584560(d), F: 03643/584564(d), T: 03643/517604(p)

Wittig, M., Dr.-Ing., BIUG GmbH, An der Bleiche 12, 09599 Freiberg. T: 03731/26010(d), F: 03731/260123(d), T: 03731/ 34773(p)

Wittke, W., Univ.-Prof. Dr.-Ing. Dr.-Ing. E.h., Geschäftsführer, WBI Prof. Dr.-Ing. W. Wittke, Beratende Ingenieure für Grundbau und Felsbau GmbH, Henricistr. 50, 52072 Aachen. T: 0241/889870(d), F: 0241/8898733(d)

Wittke-Gattermann, P., Dr.-Ing., WBI Prof. Dr.-Ing. W. Wittke, Beratende Ingenieure für Grundbau und Felsbau GmbH, Henricistr. 50, 52072 Aachen. T: 0241/889870(d), F: 0241/ 8898733(d)

Wittlinger, M., Dr.-Ing., Bauingenieur, Weidstr. 18, CH-8105 Watt, Schweiz. T: 0043/52/7202320(d), F: 0043/52/ 7201410(d), T: 0043/1/8400983(p)

Wölfel, H.P., Dr.-Ing, Wölfel Beratende Ingenieure GmbH + Co. Max-Planck-Str. 15, 97204 Höchberg. T: 0931 49708-0(d), F: 0931 49708-150(d)

Wolffersdorff, von, P.-A., Priv.-Doz. Dr.-Ing., Geschäftsführer, Baugrund Dresden Ingenieurgesellschaft mbH, Paul-Schwarze-Str. 2, 01097 Dresden. T: 0351/8241350(d), F: 0351/8030786(d), T: 0351 2656380(p)

Wollenhaupt, H., Dipl.-Ing., Thüringer Str. 91, 36208 Wildeck. T: 06678/679(p)

Woywod, C., Im Fuhlenbrock 190A, 46242 Bottrop. T: 02041/ 559519(p)

Wu, W., Dr.-Ing., Lahmeyer International GmbH, Friedberger Str. 173, 61118 Bad Vilbel

Wübbels, D., Rottstr. 9, 44793 Bochum. T: 0234/3252032(p)

Wullschläger, D., Prof. Dr.-Ing., Mozartstr. 47, 76307 Karlsbad

Wunsch, R., Dr.-Ing., Schöngasse 1, 67346 Speyer. T: 0621/459-2379(d), F: 0621/459-2219(d)

Wünscher, J., Dr.-Ing., Ingenieurgemeinschaft Baugrund und Grundbau Bad Langensalza, Büro Naumburg, Peter-Paul-Str. 22, 06618 Naumburg. T: 03445/7193-0(d), F: 03445/7193-12(d), T: 0341/8772150(p), F: 0341/ 8772150(p)

Würfel, M., Heinrich-Kröller-Str. 2, 81545 München. T: 089/ 647751(p)

Zacher, P., Dr.-Ing., Lilienweg 6, 64646 Heppenheim. T/F: 06252/72414(p)

Zangl, L.W., Dr.-Ing., Berat. Ingenieur, Ingenieurbüro für Geotechnik, Untere Hauptstr. 76, 67363 Lustadt. T: 06347/8642(d), F: 06347/6741(d)

Zarske, G., Dr., BGA Suckow + Zarske GbR Beratende Geologen und Ingenieure, Hamelnweg 12, 38124 Braunschweig. T: 0531/264160(d), F: 0531/2641677(d), T: 05302/ 7140(p)

Zäschke, T., Alfred-Messel-Weg 8–21, 64287 Darmstadt. T: 06151/713063(p)

Zeh, R., Thomas-Müntzer-Str. 30, 99423 Weimar. T: 03643 259002(p)

Zehetmaier, K.H., Ing., Dywidag Systems International GmbH, Erdinger Landstr. 1, 85609 Aschheim. T: 089/ 9267-227(d), F: 089/9267-366(d)

Ziegler, M., Univ.-Prof. Dr.-Ing., RWTH Aachen Lehrstuhl für Geotechnik im Bauwesen, Mies-van-der-Rohe-Str. 1, 52074 Aachen. T: 0241/80-5247(d), F: 0241/8888-384(d)

Zielasko, A., Dipl.-Geol., Gager 2, 99192 Neudietendorf. T: 0361/381397(d), T: 03602/81400(p)

Zimmer, C., Gardistenstr. 19, 64289 Darmstadt. T: 06151 719358(p)

Zimmermann, J., Dr.-Ing., Walter Bau-Aktiengesellschaft Vorstand, Böheimstr. 8, 86153 Augsburg. T: 0821/5582-306(d), F: 0821/5582-460(d)

Zirfas, J., Dr., Institut f. Geotechnik Dr. Jochen Zirfas, Egerländerstr. 46, 65556 Limburg. T: 06431/2949-0(d), F: 06431/2949-44(d)

Zöller, R., IGB Berlin Ingenieurgesellschaft für Grundbau, Bodenmechanik und Umwelttechnik mbH, Rheinstr. 46, 12161 Berlin. T: 030 85963856(d), F: 030 85963857(d)

Zsak, P., Alfred-Messel-Weg 10C 41, 64287 Darmstadt. T: 06151/79691(p)

Zweynert, M., Dipl.-Ing., selbständig, ibg Ingenieurbüro für Baugrund und Geotechnik, Johannisstr. 7, 54290 Trier. T: 0651/9941406(d), 0651/39962(p), F: 0651/9941408(d)

GHANA

Contact person: Dr. S.K. Ampadu, Ghana Geotechnical Society, Civil Engineering Department, University of Science & Technology, Kumasi-Ghana, Ghana. T: 051-60226, F: 051-60226, E: ampad@ighmail.com

Total number of members: 25

Acquaah, V.K.A., Mr., School of Engineering, Knust-Kumasi. T/F: 051-60226

Acquah, E.N.B., Mr., GHA Materials Division, P.O. Box 164, Accra. T: 021-772465, E: boniakwa@yahoo.com

Adams, C.A., Mr., School of Engineering, Knust-Kumasi. T/F: 051-60226

Amedzake, J.A., Mr., GHA Materials Division, P.O. Box 164, Accra. T: 021-772465

Ampadu, S.I.K., Dr., School of Engineering, Knust-Kumasi. T/F: 051-60226, E: ampad@ighmail.com

Andoh, M.B., Mr., B.R.R.I. Geotech Division, P.O. Box 40, Ust-Kumasi. T: 051-60064/051-60065, F: 051-60086

Ashiabor, S.Y.W., Mr., Structeng Const Eng., P.O. Box C2348, Cantonments. T: 021-220424, F: 225762

Donkor, C.B., Mr., G.H.A., P.O. Box 1641, Accra. T: 021-772465

Frempong, E.M., Mr., B.R.R.I. Geotech Division, P.O. Box 40, Ust-Kumasi. T: 051-60064/051-60065, F: 051-60086

Ghartey, E.B.H., Mr., B.R.R.I. Geotech Division, P.O. Box 40, Ust-Kumasi. T: 051-60064/051-60065, F: 051-60086

Gogo, J.O., Dr., B.R.R.I. Geotech Division, P.O. Box 40, Ust-Kumasi. T: 051-60064/051-60065, F: 051-60086

Hammond, F., Mr., G.H.A., P.O. Box 1641, Accra. T: 021-772862

Hazel, B.K., Mr., Conterra Ltd., P.O. Box 11658, Accra North. T: 021-712169

Kporku, T.K., Conterra Ltd., P.O. Box 11658, Accra North. T: 021-712169

Kumapley, N.K., Prof., Conterra Ltd., P.O. Box 11658, Accra North. T: 021-712169

Lamptey, L.L.L., Mr., G.H.A., P.O. Box 1641, Accra. T: 021-772465

Martin, C.N., Mr., Twum-Boafo & Partners, P.O. Box M156, Accra. T: 021-772465

Matrevi, E.K., Mr., G.H.A., P.O. Box 1641, Accra. T: 021-772465 Ext. 228

Oduroh, P.K., Mr., University, P.O. Box 40. T: 051-60064/051-60065, F: 051-60086

Oduro-Kunadu, E., Mr., G.H.A., P.O. Box 1641, Accra. T: 021-772862, E: konadu@ghanahighway.com

Oppong-Baah, T.K., Mr., G.H.A., P.O. Box 1641, Accra. T: 021-772465

Quansah, D.S., Mr., G.H.A., P.O. Box 1641, Accra. T: 021-772465

Tuffour, Y.A., Dr., School of Engineering, Knust-Kumasi. T/F: 051-60226

Yeboa, S.L., Dr., Twum-Boafo & Partners, P.O. Box M156, Accra. T: 021-228761/021-228776, F: 021-669504

Yeboah, J., Mr., G.H.A., P.O. Box 1641, Accra. T: 021-772465

GREECE/GRECE

Secretary: Dr. Á. Anagnostopoulos, Hellenic Society of SMFE, 42 Patission Str. (Polytechnion), 106 82 Athens. T: 01 7723434, F: 01 772 3428, E: geotech@central.ntua.gr

Total number of members: 173

Alamanis, N., Civil Engineer, 12 Panagouli Str., 412 22 Larissa. T: 041 226429, F: 041 226836

Alexandris, A., Civil/Geotechnical Engineer, 24 Evrou Str., 152 34 Chalandri. T: 01 6830950

Alkalais, E., Managing Director, 91 Ethnikis Antistasseos Av., 153 44 Pallini. T: 01 6032813, F: 01 6669019

Anagnostopoulos, A.G., Civil Engineer, Professor, National Technical University of Athens, 42 Patission Str., 106 82 Athens. T: 01 7723420, F: 01 7723428

Anagnostopoulos, C.T., Assoc. Prof., Soil Mechanics & Foundation Laboratory, Aristotle University of Thessaloniki, Greece, Civil Engineering Department, Geotechnical Engineering Division, P.O. Box 450, 540 06 Thessaloniki. T: 031 995715, F: 031 995619

Andrianis, E., Civil Engineer, 71 Platonos Str., 174 55 Kalamaki. T: 01 644 62 77

Andrikopoulou, K.P., Geotechnical Engineer, Scientific Associate, NTUA, 5 Avras Str., 145 65 Ekali. T: 01 8134290, F: 01 7723428

Antonopoulos, D., Civil Engineer, 181 Patission Str., 112 52 Athens. T: 01 8644563

Argyropoulos, V., Geologist, 25 Despotopoulou Str., 250 02 Vrahneika Patras. T: 061 670378

Athanassopoulos, G.A., Professor of Civil Engineering, University of Patras, Department of Geotechnical Engineering, 265 00 Rio Patras. T: 061 997677, F: 061 997274

Atmatzidis, D.K., Professor of Geotechnical Engineering, Department of Civil Engineering, University of Patras, 265 00 Rio Patras. T: 061 997677, F: 061 997274

Bardanis, M., Civil/Geotechnical Engineer, 29 Xanthipis Str., 104 44 Athens. T: 01 5123458, F: 01 7723428

Barounis, A.N., Mining Geotechnical Engineer, 10–12 Aristidou Str., 105 59 Kalithea Athens. T: 01 3220797

Belokas, G., Civil/Geotechnical Engineer, 94 Olympou Str., 152 34 Chalandri. T: 01 6854819 T/F: 01 6810915

Bouckovalas, G., Assist. Prof., NTUA, 1 Iktinou Str., 151 26 Marousi. T: 01 7723780, 01 8068393, F: 01 7723428

Brovas, D., Mining Eng., 1 Terpsitheas Str., 153 41 Ag. Paraskevi

Carydis, P. Gr., Prof., Laboratory for Earthquake Engineering, Polytechnic Campus, 9 Iroon Polytechniou, 157 73 Zografos. T: 01 7721185, F: 01 7721157

Cavounidis, S., Dr., Civil Engineer, Director, Edafos Ltd., 9 Iperidou Str., 105 58 Athens. T: 01 3222050, F: 01 3241607

Christaras, V., Dr., Engineering Geologist, Assist. Prof., Aristotle University of Thessaloniki, Geology Department, 54006 Thessaloniki. T: 031 998506

Christodoulias, J., Engineering Geologist, Eyde-Mede, 100 Pocrashou Street, 114 75 Athens. T: 01 6457492, F: 01 6400311

Chrysikos, D.A., Scientific Associate, University of Patras, Dept. of Geotechnical Engineering, Civil Engineer, 265 00 Patras Rio. T: 061 997677, 061 997274

Comodromos, C.E., Consulting Engineer, c/o Geognosis S.A., 11 Nikis Avenue, 546 24 Thessaloniki. T: 031 243170

Constantinidis, D., Geotechnical Engineer, Edrassis S.A., 5th km Peanias – Markopoulou, 194 00 Peania. T: 01 6020500, F: 01 6627748

Costopoulos, S., Assoc. Prof., Consulting Geotechnical Engineer (Dr. Sc. Tech.), 36 Skoufa Str., 106 72 Athens. T: 01 3600444, 01 3603225, F: 01 3624182

Coumoulos, D.G., Partner, Castor Ltd., 28Á Epidavrou Str., 152 33 Halandri, Athens. T: 01 6852483, F: 01 6894356

Daoutis, J.S., Consulting Engineer, 89 Kallistratous Str., 157 71 Athens. T: 01 7756510

Demiris, C.A., Prof., Alexandrou Mihailidou 5, 546 40 Thessaloniki. T: 031 822909

Didaskalou, G., Geotechnical Engineer, 2 Amazonon Str., 551 33 Kalamaria Thessaloniki. T: 031 431563, F: 031 430636

Doulis, G., Civil Engineer, 35 Perrevou Str., 124 61 Haidari. T: 01 5813682

Dounias, G.T., Director, Edafos Ltd., 9 Iperidou Str, 105 58 Athens. T: 01 3222050, F: 01 3241607

Draghiotis, T., Civil Geotechnical Engineer, 31 Andromedas Str., 162 31 Byron Athens. T: 01 7659551, F: 01 3221772

Economou, V., Civil Engineer M.Sc., Pangaea Consulting Engineer Ltd., 131 Kifissias Ave., 115 24 Athens. T: 01 6915926, 6921910, F: 01 6928137

Egglezos, D., Civil Engineer, 20 Papanikolaou Str., 173 42 Ag. Demitrios. T: 01 9927374

Exadaktylos, G., Dr., Assist. Prof. Technical University of Crete, 731 00 Chania. T: 0821 69561

Fegaras, G., Engineering Geologist, c/o ADF Ltd., 108 Kifissias Av., 115 26 Athens. T: 01 6981866

Fikiris, J., Civil Engineer, 48 Sokratous Str., 17672 Kallithea. T: 01 9597438

Fotiadis, D., Civil Engineer, 44 Avras Str., 14562 Kifissia. T: 01 8084040

Fourniotis-Pavlatos, C.F.P., Civil Engineer, 2 Sapfous Str., 151 26 Marousi. T: 01 8043881, F: 01 7723428

Fragidakis, D.J., Consulting Engineer, 89 Kallistratou Str., 157 71 Athens. T: 01 7756510, F: 01 7759977

Garatziotis, J., Civil Engineer, 8 Koumarianou Str., 114 73 Athens

Gargala, M., Civil Engineer, 3 Mela Str., 115 62 Holargos. T: 01 7220790

Gazelas, D., Civil Engineer M.Sc., 11 Knossou Str., 112 53 Athens. T: 01 8623205

Gazetas, G., Prof. NTUA, 36 Assimakopoulou Str., 153 42 Ag. Paraskevi. T: 01 6008578, F: 01 6007699

Georgiadis, M., Assoc. Prof., Aristotle University, School of Engineering, Department of Civil Engineering Geotechnical Engineering Div., Lab of Soil Mechanics & Foundation Engineering. P.O. Box 450, 54 006 Thessaloniki. T: 031 995684, F: 031 995619

Georgiannou, V., Civil/Geotechnical Engineer, Lecturer NTUA, Geotechnical Dept., 42 Patission Str., 106 82 Athens. T: 01 7723489, 01 7723408, F: 01 7723428

Giannopoulou, D., Civil Engineer M.Sc., 18 Miltiadou Str., 155 61 Holargos. T: 01 6512796

Gofas, Th C., Consulting Engineer, 21 Voukourestiou Str., 106 71 Athens. T: 01 3633664, F: 01 3612679

Hardaloupa, E., Civil Engineer, 84 Thoukididou Str., 174 55 Alimos

Hatzigogos, Th N., Assoc. Prof., Aristotle University of Thessaloniki, Lab of Soil Mechanics and Foundation Engineering, 1 N. Kasomouli Str., 546 55 Thessaloniki. T: 031 995713, F: 031 995619

Hatzigogou-Harissi, A., Research Assistant, Aristotle University, School of Engineering, Department of Civil Engineering, Geotechnical Engineering Div., Lab. of Soil Mechanics & Foundation Engineering, P.O. Box 450, 540 06 Thessaloniki. T: 031 995771, F: 031 995619

Ioannou, C.G., Consulting Engineer, 3 Lampsa Str., 115 24 Athens. T: 01 6816572, 01 6919484

Kallioglou, P., Civil Engineer, 15 K. Palaiologou Str., 546 22 Thessaloniki. T: 031 243920

Kalkani, E., Prof. NTUA, 50 Sevastoupoleos Str., 115 26 Athens. T: 01 7722830, F: 01 7723428

Kamariotis, A.M., Civil Engineer, 15 Stournara Str., 106 83 Athens. T: 01 3623440, F: 01 7723428

Kanakari, H., Civil Engineer, 45 Marasli Str., 106 76 Athens

Kantartzi, C., Civil/Geotechnical Engineer, c/o Pangaea Consulting Engineers Ltd., 131 Kifissias Av., 115 24 Athens. T: 01 6929484, F: 01 6928137

Kapenis, A., Mining Engineer, 21–23 Kyknon Str., 111 46 Athens. T: 01 2918547

Kappos, A., Civil Engineer, Managing Director, 91 Ethnikis Antistasseos Av., 153 44 Pallini. T: 01 6032813, F: 01 6669019

Karlaftis, A., Civil Engineer, 5 Paraschou Str., 154 52 Psychiko. T: 01 3303945, 01 6726147, F: 01 3815043, 01 6723174

Kavvadas, M.J., Assist. Prof. of Civil Engineering, 4 Ag. Annas Str., 145 63 Politia. T: 01 7723412. T: 01 8084059, F: 01 7723428

Klimis, K.N., Ph. D, Geotechnical Engineer, P.O. Box 53, 551 02 Finikas Thessaloniki. T: 031 476081-4, F: 031 476085

Kolias, V., Civil Engineer, DENCO Ltd., 16 Kifissias Av., 15125 Maroussi. T: 01 6854801-6, F: 01 6854800

Kollios, A.A., Dr. Civil Engineer, 9 Esperou Str., 175 61 Pal Faliro Athens. T: 01 6897568, F: 01 6897581

Koryalos, T.P., Partner, Castor Ltd., 28Á Epidavrou Str., 152 33 Halandri, Athens. T: 01 6852483, F: 01 6894356

Kotta, N.A., Civil Engineer, University Assistant, NTUA, 7 Valtetsiou Str., 154 51 N. Psychiko. T: 01 6729425, F: 01 6729525

Kotzias, P.C., Partner, Kotzias – Stamatopoulos, Geotechnical Consultants, 5 Isavron Str., 114 71 Athens. T: 01 3603911, 01 3624898, F: 01 3616919

Kountouris, P.J., Consulting Engineer, 81 Sof. Venizelou Str., 152 32 Halandri Athens. T: 01 6820509, F: 01 2237537

Lamaris, Ch, Civil Engineer, 13 Makedonias Str., 145 61 Kifissia

Laskaratos, P., Geotechnical – Consulting Engineer, 25–27 Ebedokleous Str., 116 35 Athens

Lepidas, I., Ph.D, Geotechnical Engineer, 23 Dinokratous Str., 106 75 Athens. T: 01 7223615

Lontzetidis, C., Civil Engineer, 16 Sarantaporou Str, 546 40 Thessaloniki. T: 031 816656

Makra, C., Civil Engineer, 4–8 Lachana Str., 55337 Thessaloniki. T: 031 942493, 031 995808, 031 995842

Makriyiannis, A., Civil Engineer, 19 Efpalinou Str., 112 53 Athens. T: 01 8677387, 01 8656823, 01 3614611

Malakis, N., Civil Engineer, 50 Iroon Polytechniou Str., 264 41 Patra. T: 061 426988

Manailoglou, J.T., Geotechnical Engineer, 41 Efroniou Str., 161 21 Athens. T: 01 7245904, F: 01 8235288

Manolopoulou-Papaliaga, S., Civil Engineer, Lecturer, Aristotle University of Thessaloniki, P.O. Box 450, 540 06 Thessaloniki. T: 031 995716

Manos, O., Civil Engineer, 55 Evelpidon Str., 11362 Athens. T: 01 8233961

Mantziaras, P., Civil Engineer, 98 El. Venizelou Str., 155 61 Holargos. T: 01 6535096

Maragos, Ch, Assist. Prof., Aristotle University of Thessaloniki. Dept. of Civil Engineering, Geotechnical Division, 540 06 Thessaloniki. T: 031 995694, F: 031 995619

Marinos, P., Prof. Engineering Geology, 23A Panetoliou Str., 117 41 Athens. T: 01 7723430, 01 9225835, F: 01 9242570, 01 772 2770

Markou, J., Civil Engineer, Lecturer, School of Civil Engineering, Demokritos University of Thrace, 15 Ilioupoleos Str., 671 00 Xanthi. T: 0541 64948, 0541 27952

Maronikolakis, S.S., Civil Engineer, 50 Pindou Str., 156 69 Papagos Athens. T: 01 6526972

Marselos, N., Civil Engineer, 23 Protis Str., 167 77 Elliniko. T: 01 8947415

Matsigos, N., Engineering Geologist, Pangaea Consulting Engineers Ltd., 4 Niovis Str., 112 52 Athens. T: 01 6915926, 01 6921910, F: 01 6928137

Mavridis, G., Civil Engineer, 3 Parou Str., 546 38 Thessaloniki. T: 031 948471

Metaxas, J.L., Geotechnical Engineer, c/o Castor Ltd., 28A Epidavrou Str., 162 33 Halandri. T: 01 6852483, F: 01 6894356

Michailidis, O., Civil Engineer, 10 G. Skokou Str., 2024 Nicosia, Cyprus. T: 02 31214902

Michalis, I., Civil/Geotechnical Engineer, 14 Karaiskaki Str., 155 62 Holargos. T: 01 6792382, F: 01 7723428

Milonas, H., Civil Engineer, VASIS Ltd., 6B Kalidopoulou Str., 546 42 Thessaloniki. T: 031 842533, F: 031 855828

Moutafis, N.J., Lecturer, NTUA, 7 Larnakos Str., 156 69 Papagou Athens. T: 01 6534641, F: 01 6520123

Nascos, N.N., Dr. Civil/Geotechnical Engineer, 85 Kritis Str., 546 55 Thessaloniki. T: 031 301097, F: 031 430710

Pachakis, M.D., Geotechnical Engineer, 15 Plastira Str., 151 27 Melissia Attikis. T: 01 8042688, F: 01 8235288

Panagiotou, G.N., Assist. Professor,10 Korai Str., 171 22 Nea Smyrni Athens. T: 01 9332600, F: 01 9335430

Panagopoulou, O., Civil Engineer, M.Sc., Public Building Soc., 235-9 Herakleous Str., 176 74 Kalithea Athens. T: 01 9423864

Pandis, C., Civil/Geotechnical Engineer, 21 Patriarchou Fotiou Str., 151 22 Maroussi. T: 01 8024234, F: 01 6928137

Pantazidou, M., Civil Engineer, 48 Nik. Litra Str., 114 74 Athens. T: 01 6432152

Pantelidis, P., Civil Engineer, 5 Rimini Str., 152 37 Philothei. T: 01 3222050, F: 01 3241607

Papadimitriou, A., Civil Engineer, 116 Aristotelous Str., 104 34 Athens. T: 01 8238183

Papadopoulos, V., Civil Engineer, Assist. Professor NTUA, 40 Zakinthinou Str., 156 69 Papagou Athens. T: 01 7728434, F: 01 7723428

Papageorgiou, H., Civil/Geotechnical Engineer, 8 Kyprou Str., 154 51 Neon Psychikon Athens. T: 01 6723666, 6723666

Papageorgiou, O., Consulting Engineer, 8 Kyprou Str., 154 51 Neon Psychikon Athens. T: 01 6756556, 01 6723666

Papakyriakopoulos, D., Consulting Engineer, 7 Aspasias Str., 156 69 Papagou Athens. T: 01 2515452, 01 6547992, F: 01 2520211

Papaliagas, Th, Civil Engineer Thessaloniki Technological Institute, P.O. Box 14561, 541 01 Thessaloniki

Papantonopoulos, C.I., Assist. Prof., University of Patras, Dept. of Civil Engineering, 265 00 Rio Patras. T: 061 997677, F: 01 997274

Papaspirou, S., Consulting Engineer, c/o Edafomichaniki Ltd., 5b Delfon Str., 151 25 Marousi. T: 01 6897568, F: 01 6897581

Pergantis, E., Civil Engineer, 73 Vassileos Pavlou, 154 52 Psychiko. T: 01 6715735

Pilitsis, S.P., Consulting Civil Engineer, 3 St. Gerodimon Str., 114 71 Athens

Pissakas, S., Civil Engineer, 25 Ragavi Str., 114 74 Athens

Pitilakis, K.D., Prof., Aristotle University – School of Engineering, Department of Civil Engineering, Geotechnical Engineering Div., Lab of Soil Mechanics & Foundation Engineering, P.O. Box 450, 540 06 Thessaloniki. T: 031 995693, F: 031 995619

Platis, A.D., M. Eng., Geotechnical Engineer, Managing Director, Geoconsult Ltd., 9 Salaminos Str., 153 43 Aghia Paraskevi Athens. T: 01 6004741, F: 01 6013044

Pliakas, F., Civil Engineer, Demokritio University of Thraki, 671 00 Xanthi. T: 0541 25373

Plitas, C., Dr. Civil Engineer, 22 Patron Str., 166 73 Voula. T: 01 9332136

Prassinou, H., Civil Engineer, M.Sc., 38 Siganeou Str., 111 42 Athens. T: 01 2931750, F: 01 6458809

Preza, A., Civil Engineer, 16–18 Armatolon Klefton Str., 114 71 Athens. T: 01 6462127, 01 6465547

Protonotarios, J., Civil Engineer, Assist. Prof., National Technical University of Athens, Geotechnical Dept., 42 Patission Str., 106 82 Athens. T: 01 7723488, F: 01 7723428

Protopappa, H., Civil Engineer, 19 Zefiron Str. 153 42 Ag. Paraskevi. T: 01 6013502

Proutsali, S.P., M.Sc., Civil Engineer, 42 Proxenou Koromila Str., 546 22 Thessaloniki. T: 031 469100, F: 031 469211

Psallidas, C.C., Dipl. Engineer – Managing Director. Edrassis Chr. Psallidas S.A., 5th km Peanias – Markopoulou, 194 00 Peania. T: 01 6680600, F: 01 6627748

Rachaniotis, N., Mining/Geotechnical Engineer, 60 Iasonidou Str., 551 33 Thessaloniki. T: 031 436226, 031 470310

Ritsos, A., Civil Geotechnical Engineer, 17 Artemidos Str., 156 69 Papagou. T: 01 6512132, 01 6197568, 01 6197581

Rizopoulos, G., Dr., Geotechnical Engineer Edrassis S.A., 5th km Peanias – Markopoulou, 194 00 Koropi. T: 01 6020500, F: 01 6627748

Sabatakakis, N., Engin. Geologist, 22 Valaoritou Str., 154 52 Psychiko. T: 061 996277

Sachpazis, C., Geotechnical Eng. M.Sc., Ph.D. Managing Director Geodomisi Ltd., Consulting Geotechnical Eng. – Eng. Geology, 32 Marni Str., 104 32 Athens. T: 30 10 5238127, F: 30 10 5231352, E: geodomisi@ath.forthnet.gr

Sakelariou, M., Civil Engineer, Assist. Professor NTUA, 25Á Amfissis Str., 155 62 Holargos. T/F: 01 7722621

Sarigiannis, D., Civil Engineer, 18 Kerasountos Str., 551 31 Thessaloniki. T: 031 417374, 031 445866

Schina, S., Civil Engineer, 5 Demonogianni Str., 104 45 Athens

Seferoglou, K.E., Civil Engineer M.Sc., 3 Ersis Str., 114 73 Athens. T: 01 8222127, F: 01 8233003

Sextos, A., Civil Engineer, 24 Fanariou Str., 55 133 Thessaloniki. T: 031 434052

Siahou, S., Civil Engineer M.Sc., 19 Voulgaroktonou Str., 152 33 Halandri. T: 01 6800277

Skias, S., Geologist, University of Thraki, Engin. Geology Dept, 671 00 Xanthi. T: 0541 78438

Sofianos, Á.É., Assoc. Professor NTUA, Dept. of Mining Eng., 9 Iroon Polytechniou Str., 157 80 Zografou. T: 01 7722200, F: 01 7722156

Somakos, L., Civil Engineer, Lazaridis Associates ATEM., 3 Evias Str., 151 25 Marousi. T: 01 8063741, F: 01 8055495

Sotiropoulos, E., Geomechaniki Ltd., Consulting Engineer, 91 Ethn. Adiastaseos Str., 153 44 Pallini. T: 01 6030181, F: 01 6669019

Sotiropoulos, V., 44 Mihalakopoulou Str., 115 28 Athens. T: 01 7252085, 01 7254086, F: 01 7251219

Soundias, E. Dr., Civil Engineer, 21 Harilaou Trikoupi Str., 185 36 Piraeus. T: 01 4523365, F: 01 4523367

Stamatopoulos, Á.C., Consulting Geotechnical Engineer, 5 Isavron Str., 114 71 Athens. T: 01 3624898, F: 01 3616919

Stamatopoulos, C.S., Civil Engineer, Kotzias – Stamatopoulos Co Ltd., 5 Isavron Str., 114 71 Athens. T: 01 3624898, 01 3622855, F: 01 3616919

Stavropoulou, M., Mining Engineer, 22 Doukissis Plakendias Str., 115 23 Athens. T: 01 6915163

Stefanidis, C., Civil Engineer, 9 Ag. Akindinon Str., 135 61, Ag. Anargiri Athens. T: 01 2618002

Stournaras, G., Assist. Professor Engin. Geology, 4 Hephaestou Str., 163 45 Ilioupolis Athens. T: 01 9940460, F: 01 9935981

Tassios, Th, Prof., NTUA, 4 Agias Lavras, 152 36 Palea Pendeli. T: 01 6139280

Theofili, Ch, Mining Engineer, 72–74 Korinthias Str., 115 27 Athens. T: 01 7701841

Tika, T., Assoc. Prof., Aristotle University, School of Engineering. Department of Civil Engineering, Geotechnical Engineering Div. Lab of Soil Mechanics & Foundation Engineering, P.O. Box 450, 540 06 Thessaloniki. T: 031 995735, F: 031 995619

Tolis, S., Civil Engineer, 3 2nd May Str., 171 21 Nea Smyrni

Tsakalides, C., Ph.D. Civil Engineer – Geotechnical Engineer, c/o Themeliodomi Ltd., Box 212, 570 01 Thermi Thessaloniki. T: 031 469241, F: 031 46921

Tsaltas, J., Civil Engineer, 38 Pericleous Str., 155 61 Holargos. T: 01 6522036

Tsamis, V., Geotechnical Engineer, Assistant Lecturer NTUA, 22 Ravine Str., 115 21 Athens. T: 01 7723746, F: 01 7723428

Tsatsanifos, C.P., Civil Engineer, Managing Director, Pangea Consulting Engineers Ltd., 131 Kifissias Ave., 115 24 Athens. T: 01 6915926, 01 6921910, F: 01 6928137

Tsatsanifou, F., Civil Engineer, Consultant, 131 Kifissias Ave., 115 24 Athens. T: 01 6915926, 01 6921910, F: 01 6928137

Tsatsos, N., Civil Engineer, 2 Spartis Str., 546 40 Thessaloniki. T: 031 849631, 031 469161, 031 469169

Tsiambaos, G., Engineering Geologist, Assist. Professor, NTUA, 25 Evmenous Str., 116 32 Athens. T: 01 7514983, 01 7723748, F: 01 7723428

Tsotsos, S.S., Prof., Aristotle University – School of Engineering, Department of Civil Engineering, Geotechnical Div., Lab. of Soil Mechanics & Foundation Engineering, P.O. Box 450, 540 06 Thessaloniki. T: 031 995771, F: 031 995619

Tsoutrelis, G.E., Professor of Mining Eng., 21 Parnithos Str., 154 52 P. Psychiko. T: 01 6713419

Tzirita, A., Civil Engineer, 51 Apollonos Str., 153 44 Pallini. T: 01 6659532

Vantolas, V., Civil/Geotechnical Engineer, 42 V. Tsitsani, 166 75 Glyfada. T: 01 9633826, 0291 52093

Vamvourellis, G., Geotechnical Engineer, M.Sc., DIC, 51 Theokritou Str., 811 00 Mitilini Lesvos. T/F: 0251 28633

Vardakastanis, D., Mining Engineer, 14 Eratous Str., 113 64 Athens. T: 01 8656054, 01 3302575, F: 01 3815043

Vardoulakis, J.I.G., Professor, Department of Mechanics, National Technical Unií. of Athens, 9 Iroon Polytechniou Ave., 157 73 Zografou Athens. T: 01 7721217, F: 01 7721218

Vatsellas, G., Civil Engineer, 43A 3rd Septembriou Str., 104 33 Athens. T: 01 8212263

Vettas, P., Civil Engineer, OTM S.A., 6–8 Koumarianou Str., 114 73 Athens. T: 01 8216432, F: 01 8235288

Vlachogiannis, J., Civil Engineer, 17 Drossini Str., 111 41 Athens. T: 01 2282984, F: 01 2237537

Vlavianos, G., Civil Engineer M.Sc., 41 Davaki Str., 156 69 Papagou. T: 01 643058, 01 6536462

Vouzaras, E., Civil Engineer, 72 Admitou Str., 104 46 Athens. T: 01 8670379, F: 01 7723428

Voyiatzoglou, C.J., Civil Engineer, Maunsell Hellas S.A., 42 Beikou and Drakon Str., 117 42 Athens

Vratsikidis, C., Civil Engineer, 44 Pavlou Mela Str., 54622 Thessaloniki. T: 031 281222

Xeidakis, G., Assist. Prof., University of Thraki, Engin. Geology Dept., 671 00 Xanthi. T: 0541 25373, F: 0541 20981

Xirotiri, A., Civil Engineer, 98 Halkidikis Str., 542 48 Thessaloniki

Yiagos, A.N., Civil Engineer, Geotechniki Ltd., 68 Orfanidou Str., 111 41 Athens. T: 01 2691421, F: 01 2520211

Zacas, M.K., Dr. Civil Engineer, Consulting Engineer, c/o Pangaea Consulting Engineers Ltd., 131 Kifissias Ave., 115 24 Athens. T: 01 6915926, 01 6921910, F: 01 6928137

Zairis, P., Civil Engineer, 19 Kolokotroni Str., 166 73 Ano Voula. T: 01 8955624

Zervogiannis, H., Civil Engineer, 44A Omirou & Pindou Str., 183 44 Moschato Athens. T: 01 9411441

Zouganelis, P., Civil Engineer M.Sc., 14 Samara Str., 15452 P. Phychiko. T: 01 6479710, 01 6000366

Zouridis, A.J., Civil Engineer, 33 Solomou Str., 106 82 Athens. T: 01 3833630, F: 01 3825881

HUNGARY/HONGRIE

President: Prof Peter Scharle, Head, Department of Engineering Structures, Széchenyi István College, 9007 Győr, P.O. Box 71. T: (96)503450, F: (96)503451, E: scharle@rs1.szif.hu

Total number of members. 94

Anka, M., 1075 Budapest, Károly körút 9c. T: (1)3521023
Asbóth, J., 1026 Budapest, Pasaréti út 111. T/F: (1)3931778
Ács, E., 1021 Budapest, Tárogató út 16. T: (1)2007521
Bagi, K., 1111 Budapest, Műegyetem rkp 3. T: (1)4631160, F: (1)4631099, E: kbagi@mail.bme.hu
Bakó, Gy., 1015 Budapest, Batthyány utca 46. T: (1)2141281, E: bako.gyula@interwave.hu
Balázs, F., 7635 Pécs, Nagyszkokói út 18. T: (72)310987, E: balazs@anyagtan.pmmfk.jpte.hu
Balázsy, B., 1133 Budapest, Véső utca 4d. T: (1)3498616
Batu, Á., 7100 Szekszárd, Napfény utca 19. T: (74)317282
Benák, F., 9023 Győr, Pusztaszeri utca 21. T: (96)413492, E: georam@dpg.hu
Berzi, P., 1093 Budapest, Lónyay utca 58. T: (1)2169420, F: (1)3652755
Biczók, E., 1111 Budapest, Budafoki út 32b. T: (49)(511) 7242429, E: biczok@gtu-online.de
Bogár, S., 1088 Budapest, Reviczky utca 4a. T: (1)3185641, F: (1)3384894, E: ftvrt@mail.matav.hu
Bojtár, I., 1521 Budapest, Műegyetem rkp 3. T: (1)4631160, F: (1)4631099, E: bojtar@mail.bme.hu
Boromisza, T., 1122 Budapest, Csaba utca 7a. T: (1)3555171, E: boromissza@mail.kozut.hu
Bólya, J., 1088 Budapest, Reviczky utca 4a. T: (1)2421694, F: (1)3384894, E: ftvrt@mail.matav.hu
Böröczky, Zs., 1145 Budapest, Varsó utca 23. T: (1)2511908, E: boezsolt@matavnet.hu
Buchberger, P., 7623 Pécs, Ungvár utca 14. T: (72)448078, E: bmi@freemail.hu
Bukics, T., 8900 Zalaegerszeg, Kinizsi út 14. T: (92)1313130
Cafaridu, P., 1021 Budapest, Tárogató út 87–89a. T: (1)275 1755, F: (1)4652323
Czap, Z., 1151 Budapest, Csobogós utca 1. T: (1)4632172, E: zczap@epito.bme.hu
Detre, Gy., 1165 Budapest, Kalitka utca 7. T: (1)4036909
Dékány, Cs., 1119 Budapest, Ormay Norbert utca 14. T: (1)205 8603, E: dekanycs@matavnet.hu
Domján, J., Sr, 1054 Budapest, Vértanúk tere 1. T: (1)3123944
Domján, J.Á., 7624 Pécs, Hunyadi út 75. T: (72)335013
Farkas, J., 1118 Budapest, Dayka Gábor utca 28. T: (1)463 1453, F: (1)4633006
Gabos, Gy., 1113 Budapest, Dávid Ferenc utca 2b. T: (1)466 2303
Gajdos, Gy., 1088 Budapest, Reviczky utca 4a. T: (1)3185641, F: (1)3384894
Gálos, M., 1089 Budapest, Biró Lajos utca 47. T: (1)3145327, F: (1)4632017, E: galos@goliat.eik.bme.hu
Greschik, Gy., Past President, 1126 Budapest, Orbánhegyi út 13. T: (1)3556182, E: h13250gre@helka.iif.hu
György, P., 1025 Budapest, Alsózöldmáli út 25b. T: (1)465 2346, E: gypmotah@elender.hu
Hajnal, I., 1027 Budapest, Frankel Leó utca 14. T: (1)3166202
Héjj, H., 1026 Budapest, Szilágyi Erzsébet fasor 45a. T: (1)2135377, E: hhejj@matavnet.hu
Horváth, Gy., 2092 Budakeszi, Makkosi út 170. T: (23)453142
Horváth, L., 1121 Budapest, Tállya utca 8–10. T: (1)3755745
Imre, E., Secretary, 1024 Budapest, Káplár utca 5. T: (1)463 1636, (1)3169838, E: imreemok@epito.bme.hu
Janitsáry, I., 1118 Budapest, Számadó utca 8. T: (1)385 1662, F: (1)2092059
Juhász, M., 1105 Budapest, Kápolna köz 1b. T: (1)3779309
Józsa, J., 1014 Budapest, Úri utca 36. F: (1)4631879, E: jozsa@vit.bme.hu
Kabai, I., 1142 Budapest, Ráskai Lea utca 42. T: (1)2525960, E: ikabai@epito.bme.hu
Karlóczai, P., 1125 Budapest, Kútvölgyi út 67. T: (1)3761856

Karvaly, E., 4029 Debrecen, Csapó utca 63. T: (52)324125, F: (52)418643
Kaszab, I., 6721 Szeged, Sóhordó utca 23. T: (62)544747, E: kaszab@jgytf.u-szeged.hu
Kaszás, F., 7635 Pécs, Bagoly dűlő 19c. T: (72)335370, F: (72) 310026
Kertész, P., 1122 Budapest, Tóth Lőrinc utca 14. T: (1)355 2823
Keszey, Zs., 1062 Budapest, Andrássy ut 88–90. T: (1)3022114, E: geohidro@elender.hu
Kleb, B., 1122 Budapest, Tóth L. utca 23. T: (1)3550321
Kondor, J., 2040 Budaörs, Eper utca 1. T: (23)417602, E: mavti@mail.matav.hu
Kovács, B., 3519 Miskolc, Bencések útja 111. T: (46)304193, E: hgkov@gold.uni-miskolc.hu
Kovács, G., 1119 Budapest, Major utca 7. T: (1)2059395
Kovács, M., 1111 Budapest, Műegyetem rkp 3. T: (1)4631492, E: mkovacs@epito.bme.hu
Kovácsházy, P., 1038 Budapest, Bebó Károly utca 2. T: (1)243 4328
Králik, B., 2335 Taksony, Dózsa György út 17. T: (24)377074
Lazányi, I., 1133 Budapest, Kárpát utca 48. T: (1)3406688, E: lazanyi@mail.matav.hu
Lőrincz, J., 2092 Budakeszi, Arany János utca 15. T: (1)465 2350
Mályusz, L., 1116 Budapest, Kisújszállás utca 30. T: (1)204 5871, E: malyusz@solaris.ymmf.hu
Máté, Gy., 2220 Vecsés, Fő utca 236. T: (29)353397, F: (1)251 1746
Mecsi, J., 1025 Budapest, Nagybányai utca 84b. T: (1)200 9226, E: h13773mec@helka.iif.hu
Mester, J., 2132 Göd, Duna utca 61. T: (27)330289, E: torokor@mail.datanet.hu
Molnár, J.P., 1194 Budapest, Léva utca 14. E: emab@mail.datanet.hu
Müller, M., 1124 Budapest, Fürj utca 29b. T: (1)3198145, F: (1)4633006, E: mmuller@epito.bme.hu
Nagy, J., 1021 Budapest, Kuruclesi út 12b
Nagy, L., 1106 Budapest, Pogány utca 7a. T: (1)2619331
Németh, G., 1028 Budapest, Vörösmarty utca 9. T: (1)376 8539
Paál, T., 1113 Budapest, Ulászló utca 62. T: (1)3656173, E: talaj@fomterv.hu
Parti, J., 8900 Zalaegerszeg, Lépcsősor utca 16. T: (92)316132
Pálossy, L., 1113 Vincellér utca 31. T: (1)3656748
Párdányi, J., 1034 Budapest, Zápor utca 17b. T: (1)3687608
Páti, Gy., 1016 Budapest, Berényi utca 9. T: (1)3868497
Petik, Á., 1215 Budapest, Táncsics Mihály utca 40. T: (1)322 1418, E: petikkft@matavnet.hu
Petrasovits, G., 2094 Nagykovácsi, Besenyőtelek utca 3. T: (26) 389409
Pőcz, B.I., 1119 Budapest, Albert utca 23. T: (1)2059274, E: ega@mgsz.hu
Prajczer, A., 1123 Budapest, Csörsz utca 15. T: (1)3754725, E: aprajczer@fomterv.hu
Rákóczi, L., 1021 Budapest, Labanc utca 6a. T: (1)2750323, E: lrakoczi@elender.hu
Rékai, J., 1112 Budapest, Brassó utca 5. T: (1)3196416
Richter, L., 1188 Budapest, Rákóczi út 43b. T: (1)2954018
Scharle, P., President, 1112 Budapest, Meredek utca 60. T: (1)3194649, E: h1281sch@helka.iif.hu
Schubert, J., 7633 Pécs, Építők útja 17a. T: (72)255067, F: (72)214682
Sigmond, A.E., 1122 Budapest, Maros utca 24. T: (1)356 2226
Soós, G., 2092 Budakeszi, Knáb János utca 39. T: (23)451879, F: (1)2042937, E: uvaterv@mail.datanet.hu

Szabó, Gy., 1026 Budapest, Balogh Ádám utca 7a. T: (1)200 0670

Szabó, I., 3515 Miskolc, Derkovits utca 54. T: (46)367451, F: (46)362972

Szalatkay, I., 1025 Budapest, Boróka utca 5a. T: (1)3257456

Szepesházi, R., Secretary, 9022 Győr, Liszt Ferenc utca 16. T: (96)503400, (96)315040, E: szepesr@rs1.szif.hu

Szijártó, A., 1029 Budapest, Kőfejtő utca 5. T: (1)3970224, F: (1)3663766

Szilágyi, G., 1088 Budapest, Reviczky utca 4a. T: (1)266 5410, F: (1)3384894, E: ftvrt@mail.matav.hu

Szilvágyi, I., 1118 Budapest, Minerva utca 2. T: (1)3664622

Szilvágyi, L., 1114 Budapest, Eszék utca 13/15. T: (1)209 1832, F: (1)2675782, E: geoplan@mail.datanet.hu

Szörényi, J., 1026 Budapest, Szilágyi Erzsébet fasor 85. T: (1)3565954

Telekes, G., 1114 Budapest, Bartók út 19. T: (1)3658628, E: gtelekes@solaris.ymmf.hu

Tóth, S., 2000 Szentendre, Móricz Zsigmond utca 16–18. T: (26)314018

Tőrös, E., 1142 Budapest, Kassai út 96. T: (1)3836533, F: (1)3843309, E: h5881tor@helka.iif.hu

Turányik, Gy., 1118 Budapest, Rétköz utca 43c. T: (1)246 1565, E: gyula.turanyik@mail.skanska.hu

Varga, L., 9023 Győr, Tihanyi Árpád utca 58. T: (96)423314

Zorkóczyné Szabó, M., 1157 Budapest, Zsókavár utca 62. T: (1)4177516

ICELAND

Secretary: Haraldur Sigursteinsson, The Icelandic Geotechnical Society, Vegagerdin, Borgartúni 7, IS-105 Reykjavik, Iceland.
T: +354 563 1400, F: +354 563 1450, E: has@vega.is

Total number of members: 10

Jónsson, B., Geological Engineer, University of Iceland, Engineering Department, Sudurgotu, IS-107 Reykjavik. T: +354 525 4000, F: +354 525 4632, E: bjonsson@hi.is

Stefánsson, B., Geotechnical Engineer, National Power Company, Háaleitisbraut 68, IS-103 Reykjavik. T: +354 515 9000, F: +354 515 9007, E: bjornst@lv.is

Sveindóttir, E.L., Geological Engineer, The Icelandic Building Research Institute Rb-Keldnaholt, IS-112 Reykjavik. T: +354 570 7300, F: +354 570 7311, E: els@rabygg.is

Rafnsson, E.Á., Geotechnical Engineer, Honnun h.f. Consulting Engineers, Sidumula 1, IS-108 Reykjavik. T: +354 510 4000, F: +354 510 4001, E: ear@honnun.is

Sigursteinsson, H., Engineer, Public Roads Administrator, Borgartuni 7, IS-105 Reykjavik. T: +354 563 2300, F: +354 562 4034, E: has@vegag.is

Sæmundsdóttir, I., Geotechnical Engineer, Technical College of Iceland, Höfðabakki 9, IS-110 Reykjavik. T: +354 577 1400, F: +354 577 1401, E: ingunn@ti.is

Skúlason, J., Geotechnical Engineer, Almenna Consulting Engineers, Fellsmula 26, IS-108 Reykjavik. T: +354 580 8100, F: +354 580 8101, E: js@almenna.is

Pálmason, P.R., Geotechnical Engineer, VST h.f. Consulting Engineers, Armula 4, IS-108 Reykjavik. T: +354 569 5000, F: +354 569 5010, E: prp@vst.is

Snorrason, S., Geologist, Líuhönnum, Sudurlandsbraut 4, IS-108 Reykjavik. T: +354 568 0181, F: +354 568 0681, E: sigfinnur@lh.is

Erlingsson, S., Geotechnical Engineer, University of Iceland, Engineering Department, Sudurgotu, IS-107 Reykjavik. T: +354 525 4000, F: +354 525 4632, E: sigger@verk.hi.is

INDIA/INDE

Secretary: Dr. K.S. Rao, Honorary Secretary, Indian Geotechnical Society, 206 Manisha, 75–76 Nehru Place, New Delhi 110 019, INDIA. T: 6211146, F: 6210361, E: igs@mantraonline.com

Total number of members: 235

Abhilash, T., 'Vijayasree', TC 27/1457, Rishimangalam Vanchiyoor, Trivandrum 695 035

Abhyankar, V.V., 'Surabhi', Plot No. 398, Mahatma Co-Op Hsg. Soc., Behind Gandhi Bhavan, Kothrud, Pune 411 029

Abraham, B.M., Reader in Civil Engg., School of Engineering, Cochin University of Science & Tech., Cochin 682 022

Agarwal, K.B., C/o Shri L. Agarwal, 3307, Sector D-3, Vasant Kunj, New Delhi 110 070

Aggarwal, P.K., N-505, Sector IX, R.K. Puram, New Delhi 110 022

Agarwal, V.K., Chairman, Railway Board & Ex-Officio, Principal Secretary to the Govt. of India, Flat No. 17-2A, Araval View Rail Vihar, Sector 26, Gurgaon 122003

Ali, S.S., Asst. Exec. Eng., Jeypore (R&B) Circle, Jeypore, Koraput (Dist.) 764 001

Ambare, P., 138, 14th Floor, R-Block, Budhwar Park, Colaba, Mumbai 400 005

Arora, V.K., Asst. Prof., C.E. Dept., Regl. Engg. College, Kurukshetra 132 119

Babu, D., 58/B, Paik Para Row, Calcutta 700 037

Babu Shanker, N., Principal, Kamala Inst. of Tech. & Science, Singapur, Huzurabad 505 468

Bagchi, J.K., Senior Manager, Engineers India Limited, A-15/16, Vasant Vihar, New Delhi 110 057

Baidya, D.K., Asst. Prof., Dept. of Civil Engg., I.I.T. Kharagpur, Kharagpur 721 302

Bhandari, N., Professor, Civil Engg. Dept., University of Roorkee, Roorkee 247 667

Bhandari, R.K., Head, International S&T Affairs Directorate, C.S.I.R. 2, Rafi Marg, New Delhi 110 001

Bhargava, S.N., Technical Officer 'B', Geot. Engineering Divn., C.B.R.I., Roorkee 247 667

Bhatia, S.K., Associate Professor, Dept. of Civil Engineering, Syracuse University, Syracuse, New York

Bibhas, K., B-7/302, PRDA Flats, South Srikrishnapuri, Patna 800 001

Bidasaria, A., Chief Executive, Ferro Concrete Co. (India) Ltd., Bhagirathpura, Indore 452 003

Bidasaria, M., Managing Director, Ferro Concrete Const. (I) Pvt. Ltd., Bhagirathpura, Indore 452 003. E: noble@bom4. vsnl.net.in

Boominathan, A., Associate Professor, Geotech. Engg. Divn., Dept. of Civil Engg., I.I.T. Madras, Chennai 600 036

Chacko, A. (Ms.), Senior Lecturer, Dept. of Civil Engg., M.A. College of Engg., Kothamangalam 686 666

Chandra, S., Professor, Civil Engg. Dept., I.I.T. Kanpur, Kanpur 208 016

Chang, N.-Y., Dept. of Civil Engineering, Univ. of Colorado at Denver, 1100, 14th Cross Denver, Colorado 80202

Chattopadhyay, K.K., Dept. of Civil Engg., Bengal Engg. College, Howrah 711 103

Chauhan, H.O., C-7/10, S.D.A., Hauz Khas, New Delhi 110 016

Chitra, R. (Ms.), Senior Research Officer, C.S.M.R.S., Olof Palme Marg, Hauz Khas, New Delhi 110 016

Chockalingom, S., 55, Meenakshi Illam, D.V.D. Colony, Kottar, Nagercoil 629 002

Chummar, A.V., Director, F.S. Engineers Pvt. Ltd., 109, Velachery Road, Guindy, Chennai 600 032

Dastidar, A.G., Geotech. Engg. Consultant, 12/1, Hungerford St., Calcutta 700 017

Datir, U.D., B/8, Sharam Safalya Soc., Near Tana Aptmt., Opp. Elora Park, Race Course, Baroda 390 007

Datta, M., Professor, Civil Engg. Dept., I.I.T. Delhi, Hauz Khas, New Delhi 110 016

Datye, K.R., Consulting Engineer, Ganesh Kutir, Ist Floor 68, Prarthana Samaj Road, Ville Parle (E), Mumbai 400 057

De, P., C/o B.D. Brothers, New Guwahati, P.O. Noonmati, Guwahati 781 020

De, P.K., BL-97, Sector-II, Salt Lake City, Calcutta 700 091

De, S.K., Superintending Engineer, M.P.P.W.D. E-1, Lalparade Ground, Guna 473 001

Desai, M.D., B-004, Heritage Appt., Behind Sarjan Society, Opp Ravidarsan, Umra, Surat 395 007

Deshmukh, A.M., Professor, C.E. Dept., College of Engineering, Farmagudi, Farmagudi 403 401

Deshpande, S.C., Managing Director, Insitu Geotech (P) Ltd, Apeejay Chambers, Wallace Street, Mumbai 400 001

Desousa, Proenca Pio, H. No. 4/77, Porbavaddo Calangute, Bardez 403 516

Devaraj, B., No. 121, Ramaswamy Street, Mannady, Chennai 600 001

Dewaikar, D.M., C.E. Dept., I.I.T. Powai, Mumbai 400 076

Dhinakaran, G., Senior Lecturer, Civil Engg. Dept., Shanmugha College of Engg., Tirumalaisamudram, Thanjavur 613 402

Dube, A.K., Emeritus Scientist, Central Road Research Institute, P.O. CRRI, Delhi–Mathura Road, New Delhi 110 020

Fotedar, S.K., Flat No. 2, Gr. Floor, Plot No. 189, Sector 28, Vashi, Navi Mumbai 400 703

Gandhi, N.S.V.V.S.J., Asst. Professor, Civil Engg. Dept., College of Engineering, Kakinada 533 003

Ganpule, V.T., C-4, Indrayani Complex G. Floor, J.K. Sawant Marg, Near Ruby Mill, Dadar (West)

Gharpure, A.D., Officine Maccaferri S.P.A., Liaison Office, 1408 Maker Chamber V, Nariman Point, Mumbai 400 021

Ghazvinian, A., Khiyabam-E-Nohom, Shaharak-E-Farhangiyan, Shahroud, IRAN

Ghosh, C., Visiting Research Fellow, Dept. of Urban & Civil Engg., Ibaraki University, 4-12-1, Hitachi, 316-8511

Ghosh, S., 7/D, Mrinal Park, P.O. Baranagore, Calcutta 700 036

Ghosh, T., Retd. Chief Engineer, PWD Bihar, 365/A, Ashok Nagar, Road No. 4B, Ranchi 834 002

Gill, B.S., S.D.O., S/o Late S. Mohinder Singh Nambardar, V.P.O. Gill, Dist. Ludhiana 141 116

Giri, P. (Ms.), Reader, Applied Mechanics Dept., 'Ashadeep', B-12, Shivam Tenaments, New Era High School Rd, Makarpura, Baroda

Gogoi, J.C., 196, Basantapth Beltola College Road, Bye Lane No. 2, P.O. Beltola, Guwahati 781 028

Golait, Y.S., Professor, C.E. Dept., Ramdeobaba Engg. College, Katol Road, Nagpur 440 013

Gopalarathnam, S., SGR Constructions Pvt. Ltd., 29, 2nd Street, Vinayakanagar, Karumandapam, Tiruchirappalli 620 001

Goswami, N.R., House No. D-7/40, Govt. Quarters Ram Nagar, Rander Road, Surat 395 009

Guha Niyogi, B.P., Executive Director, Pile Found. Const. Co. (I) (P) Ltd., 30, Chittaranjan Avenue, Calcutta 700 012

Guha Roy, S., Garrison Engineer, Chandigarh Airport Road, Chandigarh 160 003

Gulhati, S., Dept. of Civil Engg., Indian Institute of Tech., Delhi, Hauz Khas, New Delhi 110 016

Gupta, K.K., Associate Professor, Civil Engg. Dept., I.I.T., Hauz Khas, New Delhi 110 016

Gupta, S., Managing Director, Cengrs Geotechnica Pvt. Ltd., Cengrs House, B-3/87, Safdarjung Enclave, New Delhi 110 029. T: 011-6103774, 6105251, F: 011-6193985, E: cengrs@vsnl.com

Gupta, S., Flat 3A, 228, Mandeville Gardens, Calcutta 700 019

Gupta, T.N., Min. of Housing & Urban Dev., Nirman Bhavan, "G" Wing, I Floor, New Delhi 110 001

Gupta, V.K., Professor of Civil Engg., University of Roorkee, Roorkee 247 667

Jai Bhagwan, Scientist, C.R.R.I., Mathura Road, New Delhi 110 020. E: jbhagwan@cscrri.nic.in

Jain, R.N., E-3, Water Works Flats, Near Karol Bagh Terminal, New Delhi 110 005

Jakhanwal, M.P., Professor, C.E. Dept., Muzaffarpur Inst. of Tech., Muzaffarpur 842 003

Jawald, S.M.A., Kothi Nawab Saheb, House No 11A, Mohalla Bulaqipur, Gorakhpur 273 001

John, H.D.S.T., Principal, Geotechnical Consulting Group, 1A Queensberry Place, London SW7 2DL

Jose Babu, T., Prof. & Director, School of Engg., Cochin Univ. of Sci. & Tech., Cochin 682 022

Joshi, N.H., Applied Mech. Dept., Faculty of Engg. & Tech., M.S. University of Baroda, Vadodara 390 001

Joshi, V.H., Professor, E.Q. Engg. Dept., University of Roorkee, Roorkee 247 667

Joy, K.J., Superintending Engineer (Civil), 9-E, CMDA Bldg., O.N.G.C., Egmore, Chennai 600 008

Juneja, K.L., Consultant, B-2/10A, Rajouri Garden, New Delhi 110 027

Kadiwala, A.D., K.B.M. Engg. Res. Lab., Dhun House, Bhadra, Ahmedabad 380 001

Kalita, U.C., Head, Dept. of Civil Engg., Indian Institute of Tech., Institution of Engineers Bldg., Pan Bazar, Guwahati 781 039

Kapoor, R., Senior Railway Engineer, Konkan Railway Corpn. Ltd., Near Railway Overbridge, Rawanfond, Navelim, P.O. Margao 403707

Kapur, R., Director, Unitech Limited, 6, Community Centre, Saket, New Delhi 110 017

Karachiwala, Z.M., Maker Tower, "K" Block, Flat No. 151, Cuffe Parade, Colaba, Mumbai 400 005

Karandikar, A.V., C/o Descon Velho's Bldg., Near Mun. Garden, Panaji 403 001

Karandikar, D.V., Consulting Engineer, 7/85, Shivanand Co-Op. Housing Society, Play Ground Cross Road, Vile Parle (East), Mumbai 400 057

Karnik, A.B., Consulting Engineer, 41, Vishwak, Artek Aptmts., M. Kalekar Marg, Near Kala Nagar, Bandra (E), Mumbai 400 051

Kate, J.M., Associate Professor, Civil Engg. Dept., I.I.T., Hauz Khas, New Delhi 110 016

Katrak, F.E., B.S. Panthaky Baug, Bldg. No. 5, Flat No. 17, IVth Floor, H. Sanatorium Road, Andheri (E), Mumbai 400 069

Katti, A.R., D-63, Group 3, Sector 4, Airoli, New Mumbai 400 708

Katti, R.K., X-1, RH-4, Sector-9, CBD, Konkan Bhawan, New Mumbai 400 614

Khare, D., 1594, Napier Town, Jabalpur 482 001

Khare, N., 247, AWHO Colony, Ambabari, Jaipur

Khitoliya, R.K., Professor, Civil Engineering, Punjab Engineering College, Chandigarh 160 012

Krishnamurthy, D.N., E-8/38, "Besant Kunj", Arera Colony, Bhopal 462 039

Krishnamurthy, K.V., C/103, Staff Qrs, Bit Mesra, Ranchi 835 215

Kuberan, R., Consultant, 7-D, Surya Apartments, Sector 13, Rohini, New Delhi 110 085

Kulkarni, R.K., Kulkarni & Associates, 101, Dnyandeep Swanand Soc., Sahakar Nagar, No. 2, Parvati, Pune 411 009

Kumar, A., Dept. of Structural Engg. & Construction Management, Regional Engg. College, Jalandhar 144 011

Kumar, P., H. No. B-30, CBRI Colony, Roorkee 247 667

Kumar, S., Superintending Engineer (Civil), C-35, ONGC Colony, Nazira Dist., Sibsagar 785 685

Kumaresan, G., 27, Ram Raja Street, Pondicherry 605 001

Kurian, N.P., Professor of Geotechnical Engg., Dept. of Civil Engg., I.I.T. Madras, Chennai 600 036

Madhav, M.R., Professor, Civil Engg. Dept., I.I.T., Kanpur 208 016. T: 0512-597144, 597724, F: 0512-597395, 590260, E: madhav@civil.iitd.ernet.in

Mandal, A.K., 82, Khaluibill Math, First Lane, P.O. & Dist. Burdwan, Burdwan 713 101

Mandal, J.J., Assistant Professor, Dept. of Civil Engg., Jalpaiguri, Govt. Engg. College, Jalpaiguri 735 102

Manoranjani, G.M. (Ms.), C/o G. Thomas Henry, D. No. 54-11-51, H.B. Colony (Po), M.G. Officer's Colony, Visakhapatnam 530 022

Mittal, A., Proprietor Soils & Foundations, Mittal Colony, Sodala, Jaipur 302 006

Mohammadali, S.M., Italconsult, U.A.E., C/o Mr. Zafer F. Kadri 7/201, Jainar Manzil Mothwad Street, Navsari

Mohan, D., R-1/19, Raj Nagar, Ghaziabad 201 002

Mohan, K.S.R., Senior Lecturer, Dept. of Civil Engg., Shanmugha College of Engg., Tirumalaisamudram, Thanjavur 613 402

Moitra, S., P-194, Kalindi Housing Estate, Calcutta 700 089

Moza, K.K., 136, Charak Sadan, Vikas Puri, New Delhi 110 018

Mukerjee, S., Reader, Earthquake Engg. Dept., Roorkee University, Roorkee 247 667

Mukherjee, K., Senior Design Engineer, Stup Consultants Ltd., 301, Nirman Vyapar Kendra, Sector 17, Vashi, New Mumbai 400 705

Mukherjee, S., C.E. Dept., Jadavpur University, Calcutta 700 032

Mukherjee, S.K., 1, 42 Task Force, C/o Depot Bin, MEG-Centre, Bangalore 560042

Muthukameswaran, M., Principal, Shanmugha Polytechnic, Tirumalaisamudram, Thanjavur 613 402

Nagana, B.S., C/o Eureka Academy, I.G. Road, Masood Building, Beside Nagalaxmi Theatre, Chikmagalur 577 101

Nagaraj, T.S., Professor, C.E. Dept., Indian Inst. of Science, Bangalore 560 012. E: nagaraj@civil.iisc.ernet.in

Nanda, A., Engineers India Ltd., 9th Floor, Cama Place, R.K. Puram, New Delhi 110 066

Nandan, S., "Swasti Bhavan", Keota, Tyrebagan, P.O. Sahaganj, Hooghly 712 104

Narayanan, G., Chief General Engineer, No. 134-D, Railway Quarters, Thungabadra, Sterling Road, Nungambakkam, Chennai 600 034

Naresh, D.N., 4-C, DDA Flats, Sarai Julliana, New Delhi 110 025

Natarajan, K.R., Professor, Dept. of Civil Engg., Shanmugha College of Engg., Thanjavur 613 402

Nayak, N.V., Executive Director, Gammon India Limited, Gammon House, Prabhadevi, Mumbai 400 025

Pandey, G. (Ms.), D/o O.P. Pandey, B-18/121, Associated Society, Behind Akota Stadium, Akota, Baroda 390 015

Panesar, H.S., 343/7, Central Town, Jalandhar City 144 001

Paul, M., Marine Engg. Dept., Engineers India Ltd., EIB – 2nd Floor, 1, Bhikaji Cama Place, New Delhi 110 066

Pendse, M.D., Chartered Engineer, Madhu Malati Bldg, 3rd Floor, 9, Swastishri Coop. Hsng. Soc., Ganesh Nagar, Pune 411 029

Pingale, D.P., C2-5, 3:3, Sector-02, Vashi, Navi Mumbai 400 703

Pise, P.J., Professor, Civil Engg. Dept., I.I.T., Kharagpur 721 302

Poorooshasb, H.B., Dept. of Civil Engg., Concordia University, Montreal, Quebec H3G 1M8

Prakash, C., Scientist, Geotec. Engg. Divn., C.B.R.I., Roorkee 247 667

Prakash, S., Dept. of Civil Engg., School of Civil Engg., Univ. of Missouri, Rolla Bulter, Carlton, Civil Eng. Hall, Rolla, Missouri 65401

Prakasha, K.S., Dy. S.E. (Civil), Institute of Engg. & Ocean Tech., Oil and Natural Gas Corpn. Ltd., Phase-II, Panvel, Navi Mumbai 410 221

Raikar, R.N., Chairman & M.D., Structwel Designers & Consult., 1008/9, Raheja Centre, 10th Floor, Nariman Point, Mumbai 400 021

Rajagopal, K., Associate Professor, Civil Engg. Dept., Indian Inst. of Tech. Madras, Chennai 600 036

Raju, V.S., Dept. of Civil Engg., I.I.T. Madras, Chennai 600 036

Ramachandra, S., 72, Amarjyothi Layout, RMV-II Stage, Bangalore 560 094

Ramamurthy, T., Flat No. 42A, Pocket-A, Sukhdev Vihar, New Delhi 110 025

Ramamurthy, T.N., Dept. of Civil Engg., R.V. College of Engg., Bangalore 560 059

Ramasamy, G., Professor, Civil Engg. Dept., University of Roorkee, Roorkee 247 667

Ramaswamy, S.V., 37/4, Third Main Road, Gandhinagar, Adyar, Chennai 600 020

Ranjan, G., 5-A/46, Civil Lines, PWD Road, Roorkee 247 667

Rao, A.S., Principal, JNTU College of Engg., Kakinada 533 003

Rao, C.S., Professor, C.E. Dept., I.I.T., Kharagpur 721 302

Rao, K.S., Associate Professor, Dept. of Civil Engg., I.I.T., Hauz Khas, New Delhi 110 016. T: 011-6591206, 6861977 Ext. 4021, F: 011-6862037, E: raoks@civil.iitd.ernet.in

Rao, K.S., Plot No. 489, Vivekananda Nagar Colony, Kukatpally, Hyderabad 500 072

Rao, M.B., Superintending Engineer (Retd.), P.O. Mantada, Krishna District 521 256

Rao, M.R., Chief Engineer (Buildings), Nirmana Soudh, Bhubaneswar 751 001

Rao, M.R., Jayateertha, Professor & Head, Dept. of Civil Engg., Shanmugha College of Engg., Tirumalaisamudram, Thanjavur 613 402

Rao, P.J., Sector 19, House No. 399, Faridabad 121 002

Rao, R.R., Dept. of Civil Engg., Andhra University, Visakhapatnam 530 003

Rao, S., C5A/106, First Floor, Janakpuri, New Delhi 110 058

Rao, S.V., Research Officer, Soil Mech. Div. II, C.S.M.R.S., Olof Palme Marg, Hauz Khas, New Delhi 110 016

Rao, V.V.S., Nagadi Consultants (P) Ltd., 106/1D, Kishangarh, Vasant Kunj, New Delhi 110 070

Rao, V.S., Safe Geotechnics, 6, Citizen Apartments, 30th Road, Bandra, Mumbai 400 050

Ratnam, M.V., 44, Sunder Nagar, S.R. Nagar P.O., Hyderabad 500 038. E: ratnammalli@hotmail.com

Ravi Kumar, C., C-2-18/3, ONGC Colony, Panvel, Navi Mumbai 410 221

Ravindar, S., Dept. of Civil Engg., Shanmugha College of Engg., Tirumalaisamudram, Thanjavur 613 402

Raviskanthan, A., 15/16, Oxford Street Box Hill, Victoria 3128

Rawat, P.C., SFS, Flat No. 202, Pocket-C, Sheikh Sarai (Phase-I), New Delhi 110 017

Ray, S.K., 6/9, Gopal Ch Chatterjee Road, Cossipore, Calcutta 700 002

Rekhi, T.S., 3085, Phase II, Urban Estate, Dugri Road, Ludhiana 141 006

Sachdev, G., E-2419, Palam Vihar, Dist. Gurgaon 122 017

Saha, A., Assistant Engineer, Irrigation & Waterways Direct., Govt. of West Bengal, 14, Central Park, P.O. Bansdroni, Calcutta 700 070

Sahasi, K.M.S., H-502, Palam Vihar, Gurgaon 122 017

Sahu, A.K., Dept. of W.R., Govt. of Orissa (BBSR), AT: Duttatota (Bina Bhaban), P.O./Dist. Puri 752 001

Santhoshkumar, P.T., Assistant Engineer, P.W.D. (Irrigation), Kanjirappuzha, Irrigation Project Section No. IV/I, Ottappalam (P.O.), Palakkad 679 104

Saran, S., Professor, Civil Engg. Dept., University of Roorkee, Roorkee 247 667

Satyanarayana, C., 14-190, Old Gajuwaka, Visakhapatnam 530 026

Sengupta, S., Research Officer, C/o D. Sengupta, Adj. to Dr. C.S. Das, Child Specialist, Ambicapatty, P.O. Silchar 788004

Shah, D.L., 30, Amarjyoti Society, Sindhvaimata Road, Pratapnagar, Baroda 390 004

Shah, S.H., Shah Abad, Opposite Alumkulam, Near Metro Politan Hospital, Thrissur 680 007

Shanmugarajah, S., Civil Engineer, 15 IBC Road, Colombo 6

Sharma, A.K., Technical Officer, C.B.R.I., Roorkee 247 667

Sharma, B.V.R., 110A, Kanchanjunga Apartments, Sector-53, Noida 201 301

Sharma, H.K., C-127A, Malviya Nagar, Hari Marg, Jaipur

Sharma, J.C., Resident Engineer, M/s ITALCONSULT, U.A.E., E-62, Sector 55, Noida, Dist. Ghaziabad

Sharma, J.K., D-18, Staff Colony, Engg. College, Kota Campus, Rawatbhata Road, Akelgarh, Kota 324 010

Sharma, K.G., Professor, C.E. Dept., I.I.T., Hauz Khas, New Delhi 110 016

Sharma, S.K., Superintending Engineer, Indore Central Circle, Central P.W.D., B-Wing, Near White Church, Indore 452 001

Sharma, V.M., C-358, Vikas Puri, New Delhi 110 018

Shirke, D.P., Superintending Engineer, (MD) Ishavasyam, Purnawad Nagar, Gangapur Road, Akashvani 422005

Shivarudru, N.R., P.O. Box 355, Seri Complex, BA 1779 BSB, Brunei Darussalam, Via Singapore

Shroff, A.V., Professor, App. Mech. Dept., M.S. University of Baroda, Faculty of Tech. & Engg., P.O. Box 51, Kalabhavan, Baroda 390 001. T: 0265-553371

Singh, B., Assistant Professor, Civil Engg. Dept., Indian Institute of Technology, North Guwahati, Guwahati 781 031

Singh, D.N., Associate Professor, Dept. of Civil Engg., I.I.T. Bombay, Powai, Mumbai 400 076

Singh, J., School of Civil & Mining Engg., University of Sydney, Sydney NSW2006

Singh, J., Assistant Professor, B-10/196, Kamal Colony, Khanna Road, Samrala, Ludhiana 141 114

Singh, M., Assistant Professor, Civil Engg. Dept., M.N.R. Engg. College, Allahabad 211 004

Singh, R.B., Chief Research Officer, C.S.M.R.S., Olof Palme Marg, Hauz Khas, New Delhi 110 016

Sinha, A.K., R.N.T. Road (E), No. 3, Govt. Colony, P.O. Sodepur, Dist. 24 PGS (N) 743 178

Sinha, K.P., Area Devl. Comm. cum Chairman (Retired), Gada, 36, Mig House, Lohia Nagar, Patna 800 020

Sinha, P.K., Bombay Port Trust, 1, Anderson House, B.P.T. Qrs., Mazagaon (East), Mumbai 400 010

Sitharam, T.G., Associate Professor, Dept. of Civil Engg., Indian Institute of Science, Bangalore 560 012

Sivapullaiah, P.V., S.O.C.E. Dept., Indian Inst. of Science, Bangalore 560 012

Som, N., Professor & Head, Dept. of Civil Engg., Jadavpur University, Calcutta 700 032. T: 033-4138088, F: 033-4730687, E: nsom@cal2.vsnl.net.in

Sreekantiah, H.R., Professor, C.E. Dept., K.R.E. College Surathkal, Srinivasanagar 574 157

Sridharan, A., Advisor, Indian Institute of Science, Bangalore 560 012. T: 080-3600129, F: 080-3341683, E: asuri@vigyan.iisc.ernet.in

Srihari, V., Lecturer, Dept. of Civil Engg., Shanmugha College of Engg., Tirumalaisamudram, Thanjavur 613 402

Srivastava, A.K., Asst. Engineer, C/o Sri Nem Singh Chauhan (MLC), Bela Marg, Chauraha Vishnupuri, Aligarh 202 001

Srivastava, D.K., C/o Sri C.M. Srivastava, Advocate, Assam Road Carrier, G.T. Road, Maheshpur, Varanasi

Srivastava, R.K., Professor, Civil Engg. Dept., MNR Engg. College, Allahabad 211 004

Srivastava, V.K., 601/7, Dayanagar, Jabalpur 482 002

Sundaram, K.R., D-11/2002, Vasant Kunj, New Delhi 110 030

Sur, A., M/s Gannon Dunkerley & Co. Ltd., 23A, Netaji Subhas Road, 4th Floor, Calcutta 700 001

Swamee, P.K., Professor of Civil Engg., University of Roorkee, Roorkee 247 667

Swamy, G.P., Modinagar, At. & Post Murud, Tq. & Dist. Latur 413 510

Tata, C.R., 50, Hill Road, Bandra, Mumbai 400 050

Thilagavathi, S. (Ms.), Senior Lecturer (Civil), Shanmugha College of Engg., Tirumalaisamudaram, Thanjavur 613 402

Tiwari, R.P., Project Officer, Nodal Centre, NTIMIS, M.N.R. Engg. College, Allahabad 211 004

Tyagi, A.K., C-7/37, S.D.A., Hauz Khas, New Delhi 110 016

Uma, S.R. (Ms.), Senior Lecturer, Dept. of Civil Engg., Shamugha College of Engg., Tirumalaisamudram, Thanjavur 613 402

Uppal, R.S., 60, Manovikas Nagar, Secunderabad 500 009

Vaishampayan, V.V., Sohame Foundation Engg. (P) Ltd., C-2006, Station Plaza, Station Road, Bhandup (W), Mumbai 400 078

Varadarajan, A., Professor, Civil Engg. Dept., I.I.T., Hauz Khas, New Delhi 110 016

Vats, P.R., Akant Kuteer, 77, Ram Nagar, Delhi 110 051

Veletsos, A.S., Professor of Civil Engineering, Rice University Houston, Texas 77251

Venkatachalam, K., Director, C.S.M.R.S., Olof Palme Marg, Hauz Khas, New Delhi 110 016

Venkatasubramanian, C., Lecturer in Civil Engg., Shanmugha College of Engg., Tirumalaisamudram, Thanjavur 613 402

Verma, A., Acid Factory, 150, G.T. Road, Ghaziabad

Verma, N.C., Assistant Engineer, C/o (L) Dr. S.R. Sharma, 35, Circuit House Area (Old), P.O. Bistapur, Jamshedpur 831 001

Wadia, Z.H., "Manorath", 1, French Garden, Athwa Lines, Near Wadia Women's College, Surat 395 001

Institution members

Border Roads Organisation, Director General, Seema Sadak Bhawan, Ring Road, Delhi Cantt, New Delhi 110 010

Design & Consultancy, CME P.O., Pune 411 031

Highway Res. Station, Director, P.O. Box 2371, Guindy, Chennai 600 025

Jadavpur University, Registrar, Jadavpur, Calcutta 700 032

N.T.P.C. Ltd., Senior Manager (Library & Information), Engineering Office Complex, Plot No. A-8A, Sector-24, Dist. Ghaziabad, Noida 201 301

University of Roorkee, Head, Dept. of Civil Engg., Roorkee 247 667

Walchand Instt. of Tech., Principal, Walchand, Hirachand Marg, Solapur, Solapur 413 006

Associated members

M/s Afcons Infrastructure Limited, Afcons House, 16, Shah Indust. Est., Veera Desai Road, Azad Ngr P.O., Andheri Road, Mumbai 400 053

M/s Aimil Ltd., Naimex House, A-8, Mohan Coop. Ind. Est., Mathura Road, New Delhi 100 044

M/s Encardio-Rite Electronics Pvt. Ltd., A-5, Industrial Estate, Talkatora Road, Lucknow 226 011

M/s Engineers India Ltd., 1, Bhikaji Cama Place, New Delhi 110 066

M/s Kvaerner Cementation India, 5, Chowringhee Approach, P.O. Box 2119, Calcutta 700 072

M/s Simplex Concrete Piles (I) Ltd., 501-D, Poonam Chambers, Shiv Sagar Estate, Dr. Annie Besant Road, Worli, Mumbai 400 018

INDONESIA

Secretary: Ir Y.P. Chandra, M. Eng, Indonesian Society for Geotechnical Engineering, Basement Aldevco Octagon, Jl. Warung Jati Barat Raya No. 75 Jakarta 12740. T: 62 021 7981966, T/F: 62 021 7974795, E: hatti@indosat:net:id, Website: http://www.hatti.or.id

Total number of members: 20

Aspar, W.A.N., Dr. Ir, Jl. Papandayan J 298 Cinere Estate Cinere 16514. T: 62 021 3168433, F: 62 021 3141278, E: aspar@bppt.go.id

Chandra, Y.P. Ir M. Eng, P.T: Pondasi Kisocon Raya MidPlaza II 16th Floor Jl. Jend. Sudirman 10-11 Jakarta 12220. T: 62 021 5705145/6, F: 62 021 5720742, E: ypchan@indo.net.id

Firmansyah, I., Ir MSCE, Komplek Japos Graha Lestari Blok F5/9 Jurang Mangu Barat, Pondok Aren 15223. T/F: 62 021 7301735, E: irw13352@indosat.net.id

Hasan, L., Dr. M.S. Ir, Jl. Sawa CT: VIII/91 Karangayam Yogyakarta. T: 62 0274 560180, 896441, E: l.hasan@mailcity.com

Hutapea, B., Ir M.Sc. Ph.D., Perumahan Bumi Karang Indah, Jl. Karang Asri Raya C8/9 Jakarta Selatan. T: 62 021 7508425, (022) 2508125, E: bigman@geotech.pauir.itb.ac.id

Idrus, Ir M.Sc., Jl. Asmin No. 45 Susukan R. T: 005/02 Jakarta 13750. T/F: 62 021 87791126, E: idrus99@ indosat.net:id

Indarto, Dr. DEA Ir, Jl. Rungkut Barata VIII/11 Surabaya 60293. T: 62 031 8700207, 5947284

Irsyam, M., Ir MSCE Ph.D., Komplek PPR ITB No. B11 Pasirmuncang, Bandung. T: 62 022 2505217, 2534150, E: mi@soilmech.si.itb.ac.id

Laksita, B.H. Ir M.Sc., Jl. Nenas No. 308 B Blok A, Cinere 16514. T/F: 62 021 7544943, E: budiantari@ telkom.net

Makarim, C.A., Ir MSCE Ph.D., Jl. H. Naim III B No. 12 Cipete Utara Jakarta 12150. T: 62 021 7206512, 7269628, F: 63858253, E: geotech@indosat:net.id/ppsuntar@cbn.net.id

Muliadi, J., Ir Msi, Jl. Sangir No. 108 (baru), Ujung Pandang. T: 62 0411 315960, F: 62 0411 314617

Napitupulu, J., Dr. Ir, Jl. Masjdulhak No. 9A Medan 20152. T: 62 061 8453910, F: 62 061 8453911, E: pondasi@ indosat.net.id

Oemar, B., Ir M.S., Jl. Manunggal III/17 Palembang 30144. T: 62 0711 358188, E: bakrimar@mdp.co.id

Oetomo, K., Dipl. Ing, PT: Tetrasa Geosinindo Roxy, Mas Blok C4/18-20 Jakarta 10150. T: 62 021 6330150, 6330535, F: 62 021 6330540, E: office@geosinindo.co.id

Rahardjo, P.P., Prof. Dr. MSCE Ir, Jl. Lembahsuka Resmi I/15 Bandung 40162. T/F: 62 022 2032077, E: slawi@bdg.centrin.net.id

Sandjaja, G.S., Ir M.T., F.T: Sipil Universitas Tarumanegara Jl. Letjen. S. Parman No. 1 Grogol Jakarta Barat. T: 62 021 5672548, F: 62 021 5663277, E:ftsipil@cbn.net.id

Sengara, I.W., Ir MSEM MSCE Ph.D., Villa Bukit Mas Estate E-1 Bojongkoneng Bandung 40191. T/F: 62 022 7108157, E: iws@geotech.pauir.itb.ac.id

Soediro, Roeswan, Ir M.S., Jl. Cimandiri Raya No. 32 Semarang 50121. T: 62 024 3515849, F: 62 024 445152

Soepandji, Budi Susilo, Prof. Dr. Ir, F.T: Universitas Indonesia Kampus Baru Depok. T: 62 021 7863504, F: 62 021 7270050, E: budisus@eng.ui.ac.id

Soepriyono, Djoko, Ir M.T., Jl. Bratang Binangun IX/6 – Surabaya 60284. T: 62 031 5044414, F: 62 021 5045394, E: tmr@sby.mega.net.id

IRAN/IRAN

Secretary: Mr B. Gatmiri, Iranian Geotechnical Society, P.O. Box 19355-3719 Tehran. T: 98 21 220 62 16 and 98 21 220 64 17, F: 98 21 220 84 78, E: igs@sesce.com, icgesm@omran.net, gatmiri@daryakhak.com

Total number of members: 158

Amirsoleymani, T., Mandro Consulting Engineers, No. 46. Daman Afshar. North Vanak Sq., Tehran. T: 98 21 877 2901, F: 98 21 888 2809, E: mandro@jamejam.net

Abarham, R., Khak Pey Tehran Consulting Engineers, No. 21, 212 East Street, Tehran Pars Avenue, Tehran, Post Code 16557. T: 98 21 7886064, F: 98 21 776470

Abbasi, H., Unit 5, 9th floor, Sarv Business Complex, West Sarv St., Saadat Abad, Tehran 19818. T: 98 21 2075770, F: 98 21 2087373

Abdi, D., No. 154, Shhid Rahmandoust Street, 20m Street, Eslamabod, Zanjan

Abdoli, M., Geological Survey Organization of Iran, P.O. Box 13185-1494, Azadi Ave., Tehran. T: 98 21 9171, F: 98 21 6009338

Abedzadeh Anaraki, F., Soils Engineering Services, No. 135 Zhoubin Cul-de-sac, North Dastuor Street, Gheitarieh, Tehran 19317. T: 98 21 2206216, F: 98 21 2208478, E: ses@sesce.com

Abootalebi, A.R., AbNiroo Consulting Engineers, No. 18, Sink Alley, Sohrevardi Ave., Tehran 15779. T: 98 21 8754753, F: 98 21 8759345

Afkhami Aghda, A.R., No. 24, Afkhami Alley Corner, End of 2nd Baharestan, Bahaei St. Javidan St. Banihashem Sq., Resalat Ave. Tehran 16659. T: 98 21 2510639

Aftab-Roshad, M., Mandro Consulting Engineers, No. 46. Daman Afshar. North Vanak Sq., Tehran. T: 98 21 877 2901, F: 98 21 888 2809

Akhavan, Gh., Pey Kav Consulting Engineers, No. 4, Bahar Shiraz Street, Shariati Road, Tehran, Post Code 15649. T: 98 21 7500572, F: 98 21 761431

Alemzadeh, A.R., Darya Khak Pey Consulting Engineers, No. 17, Masoud Dead End, North of Sheykh bahaee Sq., Vanak St.,Tehran 19958-3391. T: 98 21 8055406 & 8059856, F: 98 21 8033599

Alizadeh, Sh., No. 5, Shahid Taimouri Alley Shahid Langari (Aghdasieh), Pasdaran, Tehran. T: 98 21 2586199

Amidi-rad, J., No. 3 Dabir Alaee Alley, Vali-E-Asr Ave. Tehran. T: 98 21 874 7765

Amigh, R., No. 137, 17th St., Sadeghieh 2nd Sq., Tehran. E: ramigh@ihug.co.nz, rambod@edc.co.nz

Amin, H., Mandro Consulting Engineers, No. 46. Daman Afshar. North Vanak Sq., Tehran. T: 98 21 877 2901, F: 98 21 888 2809, E: mandro@jamejam.net

Amini Hoseini, K., P.O. Box 11365/5766, Tehran. T: 98 21 39 17 97

Amini Asalami, A., Avang Sazeh Consulting Engineers, No. 40, Ebne Yamin Street, Sohrevardi Avenue, Tehran, Post Code 15568. T: 98 21 8762354, F: 98 21 8767892

Amini, B., Soils Engineering Services, No. 135 Zhoubin Cul-de-sac, North Dastuor Street, Gheitarieh, Tehran 19317. T: 98 21 2206216, F: 98 21 2208478, E: bamini@sesce.com

Arafati, N., Sako Consulting Engineers, No. 53, Atefi Gharbi Alley, Africa Avenue, Tehran Post Code 19679. T: 98 21 2053334, F: 98 21 2057648

Ardabili, N., Khak-O-Sang Company, No. 20 – West 182nd St., Tehran Pars, Tehran. T: 98 21 804 3698

AsgharBeik, M., Khakvaran Consulting Engineers, No. 125, Nastaran Alley, Ramsar Street, Enghelab Avenue, Tehran, Post Code 15818. T: 98 21 8848983-4

Ashrafzadeh, J., Block 3601, 5th Street, Karmandan Township, Zanjan, Post Code, 45139844765. T: 98 241 448922

Ashtari, M., Sazian Company, No. 32. Haraz Alley opposite to Africa Cinema, Vali-E-Asr, Ave., Tehran. T: 98 21 880 3628

Astaneh, S.M.F., Ab Va Khak Co, No. 86, Bozorgmeher St., Vali-e-Asr Ave., Tehran 14176. T: 98 21 6402824, F: 98 21 6410035

Atrchian, M.R., Khak-O-Sang Company, No. 39, 157 Mehr, Soheil St., Shariati Ave., Tehran. T: 98 21 223 37 33- 98 21 223 2446, F: 98 21 220 4396

Azharaddavami, N., Soils Engineering Services, No. 135 Zhoubin Cul-de-sac, North Dastuor Street, Gheitarieh, Tehran 19317. T: 98 21 2206216, F: 98 21 2208478, E: ses@sesce.com

Babakhanians, E., Zamiran Consulting Engineers, No. 48, Mehr Alley, Shadab Street, Nejatollahi Avenue, Post Code 15989. T: 9821 8907996, F: 98 21 8907126, E: Zamiran@dpi.net.ir

Babazadeh, H.A., P.O. Box 37185-3567 – Qom – Technical and Soil Mechanics Laboratory, Old – Tehran Rd, Qom – Tehran

Badeeifar, M., Baniyan Pey Co., No. 76. 8th St., Ameri Ave., Kooyeh Melat, Ahvaz, Tehran. T: 98 611 452875, F: 98 611 460817

Baghi, R., BandAb Consulting Engineers, No. 58, Chelsootoun St., Yousef Abad, Tehran,14318. T: 98 21 8712167, F: 98 21 8726886

Bahrami Samani, F., Tamavan Consulting Engineers, No. 3 Second Alley, South Kaj St., Fatemi Ave., Tehran 14147, P.O. Box 14565/161. T: 98 21 8951611, F: 98 21 8951610

Bahraminezhad, M.R., Soils Engineering Services, No. 135 Zhoubin Cul-de-sac, North Dastuor Street, Gheitarieh, Tehran 19317. T: 98 21 2206216, F: 98 21 2208478, E: ses@sesce.com

Bamani Yazdi, A., Pouyab Consulting Engineers, No. 49, Payam Boulevard, Mokhaberat Township, Saadat Abad, Tehran 19819. T: 98 21 2075453, F: 98 21 2062844

Baradaran Jamili, Sh., Soils Engineering Services, No. 135 Zhoubin Cul-de-sac, North Dastuor Street, Gheitarieh, Tehran 19317. T: 98 21 2206216, F: 98 21 2208478, E: ses@sesce.com

Basir, M., Iran Khak Consulting Engineers, No. 19, 20th Alley, Palizi Street, Sohrevardi Avenue, Tehran Post Code 15549. T: 98 21 8766163-5, F: 98 21 8768095

Behnia, K., Khak Mosallah Consulting Engineers, No. 7, Toosi Street, Jamalzadeh Avenue, Tehran Post Code 14198. T: 98 21 6431176, F: 98 21 920150

Behpoor, L., Shiraz University, Faculty of Engineering, Department of Civil Engineering, Shiraz. T: 98 711 303051, F: 98 711 52725

Bicheranloo, R., No. 111, Jami Street, Shirvan, Khorasan. T: 98 582422-6858, 98 21 6462616

Bloorchi, M.J., No. 106, Shaid Momen Nejad St., Pasdaran Ave., Tehran. T: 98 21 2581629

Daraee, H., No. 11Apt #3, 12th Alley, Nemati St., Dolat Ave., Tehran. T: 98 21 2001417, F: 98 21 2218364

Daraee, L., No. 11Apt #3, 12th Alley, Nemati St., Dolat Ave., Tehran. T: 98 21 2001417, F: 98 21 2218364

Davoodi, S.M.M., Soils Engineering Services, No. 135 Zhoubin Cul-de-sac, North Dastuor Street, Gheitarieh, Tehran 19317. T: 98 21 2206216, F: 98 21 2208478, E: ses@sesce.com

Ebrahimi Ghajar, Sh., No. 3, Noavar Bldg., 6th St., Shaid Shah Nazari St., Mirdamad, Tehran. T: 98 21 2227882, F: 98 21 2227883

Ebrahimi Medise, M.A., Soils Engineering Services, No. 135 Zhoubien Cul-de-sec, North Dastuor Street, Gheitarieh, Tehran 19317. T: 98 21 2206216, F: 98 21 2208478, E: ses@sesce.com

Ekrampour, V., Sakht azma Co., P.O. Box 71555-735, Shiraz: 98 711-7279435

Eslami, HR., Block 2868, Ghaem 10 Street, Zanjan. T: 98 241 446862

Fakher, A., Tehran University, Faculty of Technology, Department of Civil Engineering, P.O. Box 14155-6457, Tehran. T: 98 21 6112273, F: 98 21 6461024

Family, H., Kooban Kav Consulting Engineers, No. 90, 7th Street, Kargar shomali Avenue, Tehran Post Code 14396. T: 98 21 8002274, F: 98 21 8025146

Farahi, Sh., Mandro Consulting Engineers, No. 46. Daman Afshar. North Vanak Sq., Tehran. T: 98 21 877 290, F: 98 21 888 2809, E: mandro@jamejam.net

Farhang, M., Moshanir Co., Shahid Khidami St., Vanak Sq., Tehran 19944. T: 98 8776682, F: 98 21 8875010

Farzaneh, O., Tehran University, Faculty of Technology, Department of Civil Engineering, P.O. Box 14455-145, Tehran. T: 98 21 6112273, F: 98 21 6461024

Fatemi, Sh., Mandro Consulting Engineers, No. 46. Daman Afshar. North Vanak Sq., Tehran. T: 98 21 877 2901, F: 98 21 888 2809, E: mandro@jamejam.net

Fattahi, B., Soils Engineering Services, No. 135 Zhoubien Cul-de-sac, North Dastuor Street, Gheitarieh, Tehran 19317. T: 98 21 2206216, F: 98 21 2208478, E: ses@sesce.com

Garousi, M.R., No. 30 Eslami Alley, Malek Ashtar Ave., Tehran

Gatmiri, B., Darya Khak Pey Consulting Engineers, No. 17, Masoud Dead End, North of Sheykh bahaee Sq., Vanak St., Tehran 19958-3391. T: 98 21 8055406, 8059856, F: 98 21 8033599

Gha'edi, M., Saabir Company, 5th Block, Vanak Park, Hemmat Highway, Tehran P.O. Box 14155/5619. T: 98 21 8058314-6, F: 98 21 8058317

Ghahremani, A., Shiraz University, Faculty of Engineering, Department of Civil Engineering, Shiraz. T: 98 711 303051, F: 98 711 52725

Ghalibafian, M., Sano Consulting Engineers, No. 15, Tavanir Street, ValiAsr Avenue, Tehran Post Code 1434874881. T: 98 21 8770173, F: 98 21 8775520

Ghanbari, A., Ghods Niroo Consulting Engineers, No. 96, Motahari Ave. Tehran. T: 98 21 841149, 8417555, F: 98 21 8411704

Ghasemi, H., Emam Hosein University, Faculty of Engineering and Technology, Shaid Babaee Express way Tehran. T: 98 21 884 9481

Ghavaseh, A., Sako Consulting Engineers, No. 53, Atefi Gharbi Alley, Africa Avenue, Tehran Post Code 19679. T: 98 21 2053334, F: 98 21 2057648

Gholami, Gh., Khak-O-Sang Company, No. 264, Ghafari St., Kamali Ave., Karegr Jonoubi. Tehran. T: 98 21 504 4888, F: 98 21 541 205

Gholampoor, Gh., Pey Kav Consulting Engineers, No. 4, Bahar Shiraz Street, Shariati Road, Tehran Post Code 15649. T: 98 21 7500572, F: 98 21 761431

Habib Agahi, Gh., Shiraz University, Faculty of Engineering, Department of Civil Engineering, Shiraz. T: 98 711 303051, F: 98 711 52725

Hadad, A.H., No. 73. 16th St. Karegar Shomali Tehran. T: 98 21 800 2883

Hadji Mohamed, A.R., Khak-O-Sang Company. No. 86, West 118, Tehran Pars, Narmak, Tehran. T: 98 21 704 525

Hariri, Y., Pey Joo Iran Consulting Engineers, No. 63, Sazman AB Blvd., Sheikh Fazlollah Highway, Tehran Post Code 14546. T: 98 21 8230784-6, F: 98 21 8203235

Hatef, N., Shiraz University, Faculty of Engineering, Department of Civil Engineering, Shiraz. T: 98 711 303051, F: 98 711 52725

Havai, A.R., No. 941, Opposite Takhti Stadium, Takhti Cross Roads. Chaharbagh, Isfahan. T: 98 311-220329

Havaii, H.R., No. 14, Shahid Shojaee Doust Dead End, Shahid Ghazi Asgar Alley, Bozorgmehr Street, Isfahan. T: 98 311 679041, F: 98 311 677581

Heidari, S., Zamiran Consulting Engineers, No. 48, Mehr Alley, Shadab Street, Nejatollahi Avenue, Post Code 15989. T: 98 21 8907996, F: 98 21 8907126, E: Zamiran@dpi.net.ir

Heidari Fard, H., Sugar Cane Industries Development Company, Technical & Infra Structure office, Golestan Express way, Aahvaz. T: 98 611 36 4011

Hojabri, R., Khak Pey Tehran Consulting Engineers, No. 21, 212 East Street, Tehran Pars Avenue, Tehran, Post Code 16557. T: 98 21 7886064, F: 98 21 776470

Homayounfar, H., Mandro Consulting Engineers, No. 17, Mobasher St. Pol-E-Roomi – Shariati Ave., Tehran. T: 98 21 223 7908, E: Homayoun@chase-mandro.com

Hoseyni, M., Sano Consulting Engineers, No. 15, Tavanir Street, ValiAsr Avenue, Tehran Post Code 1434874881. T: 98 21 8770173, F: 98 21 8775520

Ilka, M., Khak-O-Sang Company-Air Force Blocks, Block No. 6, 1st floor, Iran Zamin, Cross Rd, Golestan Ave., Ponak Sq., Tehran

Imani, H., Ministry of Housing & Urban Development, Bldg., No. 2, 26, Shaid Mostafa hosseini St., Karim Khan Zand Ave., Tehran 15864. T: 98 21 8822064, F: 98 21 8822068

Jahanian, Sh., Mana Pey Gharb, Amin Building. Unit 2, 1st floor, Daneshgah Sq., Mirzadeh Eshghi Ave., P.O. Box: 65155-985 Hamadan. T: 98 81 8267574, F: 98 81 826 27 85

Jalili, M.M., Soils Engineering Services, No. 135 Zhoubin Cul-de-sac, North Dastuor Street, Gheitarieh, Tehran 19317. T: 98 21 2206216, F: 98 21 2208478, E: ses@sesce.com

Jazayeri, H., Mandro Consulting Engineers, No. 46, Daman Afshar. North Vanak Sq., Tehran. T: 98 21 877 2901, F: 98 21 888 2809

Karimi, M.R., Soils Engineering Services, No. 135 Zhoubien. Cul-de-sec, North Dastuor Street, Gheitarieh, Tehran 19317. T: 98 21 2206216, F: 98 21 2208478, E: ses@sesce.com

Kavianipoor, A., Iran Khak Consulting Engineers, No. 19, 20th Alley, Palizi Street, Sohrevardi Avenue, Tehran, Post Code 15549. T: 98 21 8766163-5, F: 98 21 8768095

Keyvanloo Nejad, B., Pey Joo Iran Consulting Engineers, No. 63, Sazman AB Blvd., Sheikh Fazlollah Highway, Tehran Post Code 14546. T: 98 21 8230784-6, F: 98 21 8203235

Keyvanpajouh, K., Darya Khak Pey Consulting Engineers, No. 17, Masoud Dead End, North of Sheykh bahaee Sq., Vanak St., Tehran 19958-3391. T: 98 21 8055406, 8059856, F: 98 21 8033599

Khalili Motlagh Kasmaee, A., Sabir Company, 5th Vanak Park Towers, 3rd Floor, Shaid Hemmat Express way P.O. Box 14155-5619. T: 98 21 8206880

Khameh Chian, M., Tarbiat Modaras University, College of Basic Sciences, Dept. of Geology. P.O. Box 14155-4838

Khodabandeh, C., Mandro Consulting Engineers, No. 46. Daman Afshar. North Vanak Sq., Tehran. T: 98 21 877 2901, F: 98 21 888 2809, E: mandro@jamejam.net

Khomami, M., Pey Joo Iran Consulting Engineers, No. 63, Sazman AB Blvd., Sheikh Fazlollah Highway, Tehran Post Code 14546. T: 98 21 8230784-6, F: 98 21 8203235

Khoramshahi, Gh., Baniyan Pey Co., No. 76. 8th St., Ameri Ave., Kooyeh Melat, Ahvaz. T: 98 611 452875, F: 98 611 460817

Kushesh, A., Tavan AB Consulting Engineers, No. 3, 2nd Street, Gisha Avenue, Tehran. T: 98 21 8268961

Litkouhi, S., Soils Engineering Services, No. 135 Zhoubin Cul-de-sac, North Dastuor Street, Gheitarieh, Tehran 19317. T: 98 21 2206216, F: 98 21 2208478, E: slitkouhi@sesce.com

Ma'afi, M., No. 83, Ershad Alley, Abouzar Avenue, Isfahan. T: 98311 237874

Maher, M.H. (Ali), Dept. of Civil Environmental Eng, Rutgers, the State University of New Jersey Piscataway, New Jersey 08855-0909 USA. T: 001 732 445 2232, F: 001 732 445 0577

Mahimani, S., Pey Joo Iran Consulting Engineers, No. 63, Sazman AB Blvd., Sheikh Fazlollah Highway, Tehran Post Code 14546. T: 98 21 8230784-6, F: 98 21 8203235

Mahmood, M., Deed Consulting Engineers, No. 138, Roodsar Street, Hafez Avenue, Tehran Post Code 15936. T: 98 21 6404862, F: 98 21 6404718

Makarchian, M., Engineering Faculty, Boualisina University Hamedan, P.O. Box 651 4161. T: 98 811 225046, F: 98 811 4221358

Malek-Zadeh, Z., P.O. Box 19585/763. T: 98 21 258 6199, F: 98 21 258 2799

Mir Karimi, S.M., Aygan Consulting Engineers, No. 18, Alizadeh Alley, Bahar Shomali St., Sadr Expressway, Tehran 19318. T: 98 21 2230004

Mir Moezi, S.M., Darya Khak Pey Consulting Engineers, No. 17, Masoud Dead End, North of Sheykh bahaee Sq., Vanak st. Tehran 19958-3391. T: 98 21 8055406, 8059856, F: 98 21 8033599

Mirmohamad Sadeghi, M., No. 29, Shahrokh Dead End, Hedayat Alley, Chahr Bag Bala St., Isphahan. T: 98 31-302 529, F: 98 31-302 061, Home T: 98 31-621 428

Moazzami, M., Geotechnical Division, Tehran Underground, Next to Dardasht Cross roads, Resalat Freeway. T: 98 21 702223, F: 98 21 7454890

Mohajeri, M., Jameh Iran Consulting Engineers, No. 9, Orchid Alley, Sheykh Bahaee, Mola Sadra, Tehran 19937. T: 98 21 8829171, F: 98 21 8842112

Mokhberi, M., No. 7, Shahid Faghihi Street, Estahban. T: 98 73242 22011,2

Molaei, Gh.A., No. 2, Varasteh Alley, North Bahar Street, Tehran

Mosayebi, A., Zayandab Company, No. 84, Naqsh-E-Jahan Malek Shahr, Isphahan. T: 98 311 243 921, F: 98 311 24 6770

Movassagnezhad, F., Soils Engineering Services, No. 135 Zhoubin Cul-de-sac, North Dastuor Street, Gheitarieh, Tehran 19317. T: 98 21 2206216, F: 98 21 2208478, E: ses@sesce.com

Nabavi, A., No. 54, Mohandes Alley, Salim street, Shariati Ave., Tehran 16137. T: 98 21 8550585

Nadi, B., Mandro Consulting Engineers, No. 46. Daman Afshar. North Vanak Sq., Tehran. T: 98 21 877 2901, F: 98 21 888 2809

Namadmalian Esfahani, A.R., Lika Co., No. 169, Beheshti Ave., Tehran. T: 98 21 8755031

Namazian, M., Soils Engineering Services, No. 135 Zhoubin Cul-de-sac, North Dastuor Street, Gheitarieh, Tehran 19317. T: 98 21 2206216, F: 98 21 2208478, E: ses@sesce.com

Negargar, S., No. 12, Nazaneen Alley, Foroughi St., Kooyeh Vali-e-Asr

Nevisi, B., Khak Pey Tehran Consulting Engineers, No. 21, 212 East Street, TehranPars Avenue, Tehran Post Code 16557. T: 98 21 7886064, F: 98 21 776470

Nourbakhash-Pir Bazari, M., No. 37, Parviz Roshan Alley, Adjacent to Ministry of Power, North Palestine.Tehran T/F: 98 21 880 1200

Nourzad, R., Sharif University, Civil Engineering College, Geotechniques Department, Azadi Ave. T: 98 21 6022733-49(15 lines), Ext. 4270, F: 98 21 6023201

Parvizi, F., Zamiran Consulting Engineers, No. 48, Mehr Alley, Shadab Street, Nejatollahi Avenue, Post Code 15989. T: 98 21 8907996, F: 98 21 8907126, E: Zamiran@dpi.net.ir

Pourakbar, M.A., Pars Kav Company, No. 24, shaid Gomnam – Fatemi Sq., Tehran. T: 98 21 440 4559

Rabanian, Sh., No. 2. 10th St., Jamshedieh, Shaid Bahonar Ave., Tehran. T: 98 21 2280213, M: 98 911 225 8568

Rafia, F., Kavoshgaran Consulting Engineers, No. 62, Bozorgmehr Street, Tehran 14168. T: 98 21 6468507, F: 98 21 6400053

Raha'ie, A., Abnieh Fanni Consulting Engineers, No. 1, 46th Street, Asadabadi Avenue, Tehran Post Code 14367. T: 98 21 8032181, F: 98 21 8032745

Rahimi, H., University of Agriculture, Department of Irrigation, Karaj

Rashidi, S.M.E., Ghir&Karzin, Fars Province. T: 98 711 305301 Ext.50

Ravadgar, B., Pey Kav Consulting Engineers, No. 4, Bahar Shiraz Street, Shariati Road, Tehran Post Code 15649. T: 98 21 7500572, F: 98 21 761431

Razavi Vokhshoorpoor, Gh., Moshanir Co., Shahid Khidami St., Vanak Sq., Tehran 19944. T: 98 8776682, F: 98 21 8875010

Rezvan, K., Khak Mosallah Consulting Engineers, No. 7, Toosi Street, Jamalzadeh Avenue, Tehran Post Code 14198. T: 98 21 6431176, F: 98 21 920150

Sadeghi, M., Khakvaran Consulting Engineers, No. 125, Nastaran Alley, Ramsar Street, Enghelab Avenue, Tehran, Post Code 15818. T: 98 21 8848983-4, F: 98 21 8828782

Sadrnejad, S.A., Khajeh Nasir University, Faculty of Civil Engineering, Tehran. T: 98 21 8779473, F: 98 21 8779476

Saeedi, A., Geological Survey Organization of Iran, P.O. Box 13185-1494, Azadi Ave., Tehran. T: 98 21 9171, F: 98 21 6009338

Samadzadeh Mahabadi, M.A., Soils Engineering Services, No. 135 Zhoubin Cul-de-sac, North Dastuor Street, Gheitarieh, Tehran 19317. T: 98 21 2206216, F: 98 21 2208478, E: ses@sesce.com

Sana'ee, Gh., Omran Iran Consulting Engineers, No. 4, Mir Motahhari Street, Seyd Khandan Bridge Tehran Post Code 15558, P.O. Box 15875/6151. T: 98 21 862144

Sani'ee, H., Pey Kav Consulting Engineers, No. 4, Bahar Shiraz Street, Shariati Road, Tehran Post Code 15649. T: 98 21 7500572, F: 98 21 761431

Sanjabi, R., Soils Engineering Services, No. 135 Zhoubin Cul-de-sac, North Dastuor Street, Gheitarieh, Tehran 19317. T: 98 21 2206216, F: 98 21 2208478, E: ses@sesce.com

Seraj, No. 139, Jahanial Cross-Road, Katanbaf St., Ahvaz. T: 98 611 213 7118

Seyd Karbasi, M., Mahab Ghods Cousulting Engineers, No. 9, 132 Street, Madani Avenue, Narmak, Tehran, Post Code 16349. T: (21)7838747

Seyedi, S.M., Darya Khak Pey Consulting Engineers, No. 17, Masoud Dead End, North of Sheykh bahaee Sq., Vanak St., Tehran 19958-3391. T: 98 21 805540, 8059856, F: 98 21 8033599

Shaaban Lari, P., Baniyan Pey Co., No. 76. 8th St., Ameri Ave., Kooyeh Melat, Ahvaz. T: 98 611 452875, F: 98 611 460817

Shahangian, Shk., Pey Avar Consulting Engineers, No 22, 19th Street, Gisha Avenue, Tehran, Post Code 14477. T: 98 21 8265829, F: 98 21 8276943

Shahangian, Shr., Pey Avar Consulting Engineers, No. 22, 19th Street, Gisha Avenue, Tehran, Post Code 14477. T: 98 21 8265829, F: 98 21 8276943

Shariati, A., Azarazma Consulting Engineers. Homa Building, 3rd Floor, South Park St., Pounak Sq., Tehran. T: 98 21 874 8296

Sharifi Soltani, A.R., Soils Engineering Services, No. 135 Zhoubin Cul-de-sac, North Dastuor Street, Gheitarieh, Tehran 19317. T: 98 21 2206216, F: 98 21 2208478, E: ses@sesce.com

Sherouyeh Araghi, M., Tamavan Consulting Engineers, No. 3 Second Alley, South Kaj St., Fatemi Ave., Tehran 14147, P.O. Box 14565/161. T: 98 21 8951611, F: 98 21 8951610

Shojaee, R., P.O. Box 13445-1136, Tehran. T: 98 21 4901415

Shokravi, Gh. R., No. 37 Hodhod Alley, Arasbaran Street, Seyed Khandan. T: 98 21 8080046, 8087800

Sinehsepehr, M., No. 17, 27th Street, North Kia Street, Shahrara, Tehran. T: 98 21 8779686, 8779119

Soleymanian, Ma., Khak Va Beton Iran Consulting Engineers, No. 42, 1st Mesagh St., Taleghani Cross Rd, Shariati Ave., Tehran 15638. T: 98 21 7501401, F: 98 21 766619

Soleymanian, Mo., Khak Va Beton Iran Consulting Engineers, No. 42, 1st Mesagh St., Taleghani Cross Rd, Shariati Ave., Tehran 15638. T: 98 21 7501401, F: 98 21 766619

Tabatabai-Yazdi, J., P.O. Box 13445-1136. T: 98 21 602 5874, F: 98 21 602 5706

Taherian, A.R., Lar Consulting Engineers, 3rd Floor No. 9, Kavous Alley, South Dibaji, Tehran. T: 98 21 255 2651

Taherian, M., Mandro Consulting Engineers, No. 46. Daman Afshar. North Vanak Sq., Tehran. T: 98 21 877 2901, F: 98 21 888 2809, E: mandro@jamejam.net

Tajedin, R., Mandro Consulting Engineers. No. 46. Daman Afshar. North Vanak Sq., Tehran. T: 98 21 877 2901, F: 98 21 888 2809

Tamannaee, H.R., BandAb Consulting Engineers, No. 58, Chelsootoun St., Yousef Abad, Tehran 14318. T: 98 21 8712167, F: 98 21 8726886

Taveezi, Gh., Ministry of Housing & Urban Development, Bldg., No. 2, 26, Shaid Mostafa hosseini St., Karim Khan Zand Ave., Tehran 15864. T: 98 21 8822064, F: 98 21 8822068

Toutounchi, M., P.O. Box 14155-1143, Tehran. T: 98 21
   8893927, F: 98 218809002
Valipoor, F., No. 6, Zamani Alley, Bahar Jonoobi Street, Sadr
   Highway, Tehran, Post Code 19398, P.O. Box 19395/3537
Vaziri, M., No. 66. Ehteshamieh St., 7th Niestan-Pasdaran-
   Tehran
Yazdanbakhsh, N., No. 8, To'hid Alley, Hafez Street,
   Ardestan 83817. T: 98 3242 3213
Yazdi, M., Avang Sazeh Consulting Engineers, No. 40,
   EbneYamin Street, Sohrevardi Avenue, Tehran Post Code
   15568. T: 98 21 8762354

Younesi, K., Mandro Consulting Engineers, No. 46. Daman
   Afshar, North Vanak Sq., Tehran. T: 98 21 877 2901, F: 98
   21 888 28 09
Zahari, Gh., Sano Consulting Engineers, No. 15, Tavanir
   Street, ValiAsr Avenue, Tehran Post Code 1434874881.
   T: (21)8770173, F: (21)8775520

IRELAND

Secretary: Dr. Ken Gavin, University College Dublin, Earlsfort Terrace, Dublin.
T: +353 1 7167292, F: +353 1 7167399, E: kenneth.gavin@ucd.ie

Total number of members: 34

Byrne, J., Mr., c/o Alpha Engineering Services, Unit 6, Crumlin Business Centre, Stanway Drive, Dublin 12. E: admin@alphaeng.ie
Callanan, F., Mr., c/o Ove Arrup & Partners, 10 Wellington Rd., Dublin. E: frank.callanan@arup.com
Cameron, D., Mr., 29 Royal Court Apartments, Lower Main St., Portrush, N. Ireland
Casey, B., Mr., c/o ESBI, Stephen Court, 18–21 St. Stephens Green, Dublin 2. E: bernard.casey esbi.ie
Creighton, R., Dr., c/o Geological Society of Ireland, Beggars Bush, Haddington Road, Dublin 4. E: creighton@tec.irlgov.ie
Cross, H., Ms., McCarthy Hyder Consultants, National Management Centre, Clonard, Sandyford Road, Dublin 16. E: helen.cross@hyder-con.co.uk
Crowley, D., Dr., Forramoyle West, Bearna, Co. Galway
Crowley, D., M. Mr., Forramoyle West, Barna, Galway
Deane, J., Mr., Springville House, Blackrock Rd, Co Cork. E: jdeane@egpettit.ie
Delehunty, M., Mr., Larsen-foundations, Cahergowen, Claregalway, Co. Galway. E: mdelatunty@larsen-foundations.ie
Duggan, D., Mr., Lisduff, Whitechurch, Co. Cork
Farrell, E., Dr., Trinity College, 7 Hainault Park, Foxrock, Dublin 18. E: efarrell@tcd.ie
Faulkner, A., Ms., c/o John Barrett & Assoc Ltd., Unit 7 Dundrum Business Park, Windy Arbour, Dublin 14. E: afaulkner@csa.ie
Galbrait, R., Mr., 6 Beechwood Ave Lower, Ranelagh, Dublin 6. E: ross.galbraith@dit.ie
Green, B., Mr., c/o IGSL, Newbridge Industrial Estate, Newbridge, Co. Kildare
Green, D., Mr., c/o Arup Consulting Engineers, 10 Wellington Road, Ballsbridge, Co. Dublin, Dublin 4. E: david.greene@arup.com
Johnston, T., Mr., 1 Milford Road, Leighlinbridge, Co. Carlow. E: tjohnston@agec.ie

Kenny, A., Mr., c/o Priority Drilling Ltd, 162 Clontarf Rd, Dublin 3
Lehane, B., Dr., c/o Dept. of Civil Engineering, Trinity College Dublin, Dublin 2. T: 01-6082389, E: blehane@tcd.ie
Leonard, N., Mr., c/o Geotech Specialists Ltd., Carenswood, Castlemartyr, Co. Cork. E: geotech@indigo.ie
Long, M., Mr., c/o Dept. of Civil Engineering, University College Dublin, Earlsfort Terrace, Dublin 2. E: mikelong@ucd.ie
Looby, M., Mr., c/o Alpha Engineering Services, Unit 12, Crumlin Business Centre, Stannaway Drive, Dublin 12. E: alphaeng@iol.ie
Luby, D., Mr., c/o John Barrett & Assoc Ltd., Unit 7 Dundrum Business Park, Windy Arbour, Dublin 14. E: dluby@csa.ie
Mc Donnell, K., Mr., c/o Descon Ltd., Monaghan Road, Co. Cork. E: desconk@indigo.ie
Murphy, B., Mr., c/o B.J. Murphy & Assoc., Kennedy Street, Carlow, Co. Carlow. E: bmacarlow@bma.ie
O'Connor, P., Mr.
Orr, T., Dr., c/o Dept. of Civil Engineering, Trinity College, Dublin 2. E: torr@tcd.ie
Ormsond, W., Mr., c/o MCOS, 01-288 4499. E: worsmond@dublin.mcos.ie
Parker, G., Mr., c/o K.T. Cullen, Bracken Business Park, Sandyford Industrial Estate, Sandyford, Dublin 18
Paul, T., Mr., Unit 7 Dundrum Business Park, Windy Arbour, Dublin 14
Russell, O., Mr., Palmerstown, Oldtown, Co. Dublin
Smith, A., Mr., c/o Clifton Scannel Emmerson Assoc., Seafort Lodge, Castledawson Ave., Blackrock, Co. Dublin. E: mailto:aidan.smith@csea.ie
Walsh, Ta., Mr., c/o Muir Assoc., 17 Fitzwilliam Place, Dublin 2
West, M., Mr., 8 Bayview Lawns, Killiney, Co. Dublin

ITALY/ITALIE

Secretary: Eng. Claudio Soccodato, Secretary General Associazione Geotecnica Italiana, Piazza Bologna 22, 00162 Roma.
T: +39 06 44249272-3, F: +39 06 44249274, E: agiroma@iol.it

Total number of members: 219

Albert, L., Soil S.r.l., Via Mercalli 11, 20122 Milano. T: 02 58322042, F: 02 58322356

Alongi, U.M., Via S. Agata 12, 94100 Enna. T: 0935/500613, F: 0935/500113

Ambrosini, F., Vipp S.p.A., Via Boscarola 41/a, 37050 Angiari (VR). T: 045 983015, F: 0442 97322

Amorosi, A., Dip. Ing. Strutt. e Geotecnica, Via Monte d'Oro 28, 00186 Roma. T: 06 49919611, F: 06 6878923, E: a.amorosi@poliba.it

Angeli, M.-G., IRPI/CNR, Via Madonna Alta 126, 06128 Perugia. T: 075/5006730-5054943, F: 075/5051325

Angelino, C., Via Millio 41, 10141 Torino. T: 011/3851904, F: 011/3850546

Aversa, S., Dip. Ingegneria Civile, Università di Pisa, Via Diotisalvi 2, 56126 Pisa. T: 081/7683474-7683475, F: 081/7683481, E: staversa@unina.it

Bagala', S., Via De Pieri 3, 36100 Vicenza. T: 0444 511397

Baldi, A.M., Studio di Geologia e Geofisica, Strada Massetana 56, 53100 Siena. T: 0577/49276, F: 0577/ 287254, E: baldi@sgg.it

Balossi Restelli, A., Via Montenapoleone 23, 20121 Milano. T: 02 76001006, F: 02 781518

Barla, A., Dip. Ingegneria Strutturale, Politecnico di Torino, Corso Duca degli Abruzzi 24, 10129 Torino. T: 011 5644824, F: 011 5644899, E: gbarla@polroc.polito.it

Battelino, D., Istituto di Idraulica, Via Valerio 10, 34127 Trieste. T: 04O/6763471, F: 040/572082, E: battelino@dic.univ.trieste.it

Beccati, A., Via del Gorgo 174/a, 44040 Gaibanella (FE). T: 0532 718115,

Berardi, R., DISEG – Dip. Ingegneria Strutt. e Geot., Via Montallegro 1, 16145 Genova. T: 010/3532506-525, F: 010/ 3532534

Bernard, R., I.Ve.Co.S. S.p.A., Via del Lavoro 35, 31014 Colle Umberto (TV). T: 0438/430290, F: 0438/430248

Bimbi, C., Via di Casabassa 29, 53034 Colle Val d'Elsa (SI). T: 0577 928776, F: 0577 920584

Bonini, M., Via Regione Noceto 1, 14030 Frinco (AT). T: 0141/904175, E: bonini.m@libero.it

Bruzzi, D., Sisgeo S.r.l., Via Filippo Serpero 4/F1, 20060 Masate (MI). T: 02 95764130, F: 02 95762011, E: info@sisgeo.com

Budetta, P., Via degli Etruschi 13, 84135 Salerno. T: 089/271017

Bulfamante, U., Via Liberta' 87, 90018 Termini Imerese (PA). T: 091/8114535, F: 091/8112590

Burghignoli, A., Dip. Ingegneria Strutturale e Geotecnica, Università "La Sapienza", Via Monte d'Oro 28, 00186 Roma. T: 06 49919604, F: 06 6878923

Burgio, S., P.zza Pantelleria, 92100 Agrigento. T: 0349 7436767, E: armandone@libero.it

Calabresi, G., Dip. Ingegn. Strutt. e Geot., Via Monte d'Oro 28, 00186 Roma. T: 06 7876133-6878886, F: 6878923

Caldara, M., Edison S.p.A., Foro Buonaparte 31, 20131 Milano. T: 02 62228929, F: 02 62227146

Calderaro, M., Via Roma 77, 07100 Sassari. T: 079 281520, F: 079 281520, E: geocald@tiscalinet.it

Callisto, L., Dip. Ingegn. Strutt. e Geotec., Via Monte d'Oro 28, 00186 Roma. T: 06 7876133-6878886, F: 6878923, E: luigi.callisto@uniroma1.it

Cancelli, A., Via Sansovino 23, 20133 Milano. T: 02 2666005, F: 02 2666005

Cancelli, P. Studio Cancelli Ass., Via Sansovino 23, 20133 Milano. T: 02 2666005, F: 02 2666005

Canetta, G., Ce.A.S. S.r.l., Viale Giustiniano 10, 20129 Milano. T: 02/29522923, F: 02/29512533

Canzoneri, V., Via Florestano Pepe 6, 90139 Palermo. T: 091/333278

Cargnel, G., Viale Fantuzzi 8, 32100 Belluno. T: 0437/943194, F: 0437/944610

Carrabba, E., Via F. Mitta 6, 23022 Chiavenna (SO)

Carraro, F., Via Volturno 80 -res. "I Cigni", 20047 Brugherio (MI). T: 039/882654, F: 039/2872014

Carraro, S., Via Mancinelli 10, 35132 Padova. T: 049/8648865, F: 049/8648865, E: simone.carraro@libero.it

Carruba, C., Via Liberta' 207, 90143 Palermo. T: 091/6259711, F: 091/6253129

Cazzola, A., PIRFS Inc., 4th Floor, Legaspi Tower 200, 107 Paseo de Roxas, Makati City (Philippines). T: 63 2 8176826, F: 63 2 8163491

Cazzuffi, D., Enel Hydro, Polo Idraulico e Strutturale, Via G. Pozzobonelli 6, 20162 Milano. T: 02 72243545, F: 02 72243640, E: cazzuffi.daniele@enel.it

Celotti, U.M., Studio Tecnico, Via Mincio 22, 20139 Milano. T: 02/5393977, F: 02/5392262

Cestari, F., Studio Geotecnico Italiano, Via Ripamonti 89, 20139 Milano. T: 02/5691841, F: 02/5691845

Cherubini, C., Ist. Geologia Applicata e Geotecnica, Politecnico di Bari, Via Orabona 4, 70125 Bari. T: 080 242363, F: 080 242529, E: c.cherubini@poliba.it

Chiara, N., Via Versilia 4, 90144 Palermo. T: 091/528950

Chioatto, P., St. Geologia Geomorfologia Applicata, Via Montorio 55/a, 37131 Verona. T: 045 8400347, F: 045 8400347, E: chioattopaul@geologi.it

Cividini, A., Dip. Ingegneria Strutturale, Politecnico di Milano, Piazza Leonardo da Vinci 32, 20133 Milano. T: 02 23994331, F: 02 23994220

Collotta, T., Gei S.r.l., Via P.L. da Palestrina 2, 20124 Milano. T: 02 66985943, F: 02 66986895

Colombo, C., Terra Company S.r.l., Viale Monza 16, 20127 Milano. T: 02/2610270, F: 02/26146646

Comastri, C., Thesis Engineering, Via Castello 7, 40037 Sasso Marconi (BO). T: 051/6192112, F: 051/561294, E: ccomast@tin.it

Cotecchia, F., Ist. Geologia Appl. e Geotecnica, Politecnico di Bari, Via Orabona 4, 70125 Bari. T: 080/ 5442338, F: 080/ 242529

Cotecchia, V., Studio Tecnico, Corso Alcide De Gasperi 384, 70125 Bari. T: 080/5650377, F: 080/353231, E: vcotecc@tin.it

Cusmano, G., Viale Regina Margherita 9, 90138 Palermo. T: 091 6521008, F: 091 6521006

De Sanctis, L., Via del Parco Margherita 59, 80121 Napoli. T: 081 415388, F: 081 7683481, E: lucads@engineer.com

D'Elia, B., Dip. Ingegn. Struttur. e Geotecnica, Via Gramsci 53, 00197 Roma. T: 49919175-5745178, F: 3221449, E: beniamino.delia@uniroma1.it

Di Francesco, R., Fraz. Casanova, 64040 Cortino (TE). T: 0861 286148, F: 0861 286148, E: romolodifrancesco@libero.it

Di Giovanna, M., Via Lido 9, 92019 Sciacca (AG). T: 0925 23157, E: mario@xacca.it

Di Landri, M., Geoservice S.a.s., Via Abate Conforti 30, 84090 S. Antonio di Pontecagnano (SA). T: 089 384835, F: 089 384835, E: geo.service@tiscalinet.it

Di Maio, S., Via Francesco Redi 5, 00161 Roma. T: 06 4403869, F: 06 44230858

Elitropi, M., Eurogeo S.n.c., Via Paglia 21, 24122 Bergamo. T: 035 271216, F: 035 271216

Esposito, A., Largo Temistocle Solera 7/10, 00199 Roma. T: 2251326, F: 2251335

Evangelista, A., Via Bonito 21, 80129 Napoli. T: 081/367004, E: evangeli@unina.it

Falcon, A., Geofondazioni Ingegneria e Lavori S.r.l., Via Boschi 30, 30030 Martellago (VE). T: 041 5403194, F: 041 5402870, E: geofond@tin.it

Fanfani, E., Consorzio Bonifica Muzza Bassa Lodigiana, Via Nino dall'Oro 4, 26854 Lodi. T: 0371/420189, F: 0371/50393, E: cmuzza@tin.it

Federico, A., Via Cesare Battisti 84, 73100 Lecce. T: 099/4733205, F: 099/4733229, E: federico@poliba.it

Ferraiolo, G., Istituto Giordano S.p.A., Via Rossini 2, 47814 Bellaria-Igea Marina (RN). T: 0541 343030, F: 0541 345540, E: gferraiolo@giordano.it

Ferrari, P., Nino Ferrari S.r.l., Via Tommaseo 8, 19100 La Spezia. T: 0187 36335, F: 0187 529043

Ferrari, S., Via della Stazione 11, 17029 Finale Ligure (SV). T: 019/698562, F: 019/698562

Filippi, S., Via Gastaldi 7, 12051 Alba (CN). T: 0173 440841, E: stefanofilippi@inwind.it

Fiumani, U., Studio Geognostico-Lab. Geot., Via Marco Polo 60, 60027 Osimo (AN). T: 071/716250, F: 071/716251

Fleres, V., Studio Fleres, Via Cesare Battisti 108, 98123 Messina. T: 090/716249, F: 090/774415

Focardi, P., Via N. Machiavelli 4, 50015 Grassina (FI). T: 055/641983-610947

Forlani, E., S.G.A.I. S.r.l., Via Mariotti 20, 47833 Morciano di Romagna (FO). T: 0541 988277-988972, F: 0541 987606, E: sgai@sgai.com

Fratalocchi, E., Dip. Scienze Materiali e Terra, Via Brecce Bianche, 60131 Ancona. T: 0734/810182, E: fratalocchi@popcsi.unian.it

Fuoco, S., S.Doppio.S., Via Malfatti 21, 38100 Trento. T: 0461/915577, F: 0461/915226, E: s.fuoco@enginsost.it

Gabbi, M., Via Villa Maria 6, 38050 Povo (TN). T: 0461/397394, F: 0461/390483

Gambino, C., Via Beato Angelico 101, 90145 Palermo. T: 091 204071, F: 091 6398521

Garrasi, A., Traversa Via Fanelli 227/10, 70125 Bari. T: 080/5025081, F: 080/5025081, E: alberto.garrasi@tin.it

Garzonio, C.A., Via Caterina Franceschi Ferrucci 2, 50135 Firenze. T: 055 602418, F: 055 587087

Giani, G.P., Via San Quintino 32, 10121 Torino. T: 011/512945, E: giani@parma1.eng.unipr.it

Giannini, R., D.R.E.A.M. Italia S.c.r.l., Via Enrico Bindi 14, 51100 Pistoia. T: 0573 365967, F: 0573 34714

Gonsalvi, L., Via Magenta 17, 26900 Lodi. T: 0371/66880, F: 0371/66880

Gottardi, G., DISTART, Univ. di Bologna, Viale Risorgimento 2, 40136 Bologna. T: 051 6443520, F: 051 6443527, E: ggottardi@distart.ing.unibo.it

Grasso, P., Geodata, Corso Duca degli Abruzzi 48/e, 10129 Torino. T: 011/503822, F: 011/597440

Gravino, S., Via Monte Carmelo, 00166 Roma. T: 06 66415526, E: salvatore.gravino@tiscalinet.it

Griffini, L., Via Pagliano 37, 20149 Milano. T: 02/ 4988607, F: 02/48193784

Grundler, D., Via G. Verdi 21, 40056 Crespellano (BO). T: 051 6722368,

Gulla, G., IRPI-CNR., Via Verdi 1, 87036 Rende (CS). T: 0984/839131-2-3, F: 0984/837382

Jamiolkowski, M., Studio Geotecnico Italiano, Via Ripamonti 89, 20141 Milano. T: 02/5220141, F: 02/5691845, E: sgi_jamiolkowski@studio-geotecnico.it

Jappelli, R., Dip. Ing. Civile, Univ. Tor Vergata, Via di Tor Vergata 110, 00133 Roma. T: 79597056, F: 72594586, E: rjappel@tin.it

Jommi, C., Politecnico di Milano, Piazza Leonardo da Vinci 32, 20133 Milano. T: 02/23994281, F: 02/23994220

La Terza, M., Nuova Domina S.c.ar.l., Casella Postale 33, 30019 Sottomarina Chioggia (VE). T: 041 5500887, F: 041 5507254

Lamperti, R., Via Ugo Foscolo 12, 20050 Lesmo (MI). T: 039/386608, E: romano.sales@sisgeo.com

Lavorato, A., Via Uccelli 3, 24123 Bergamo. T: 035/577579

Lizzi, F., Via C. De Nardis 7, 80127 Napoli. T: 081 5604391

Lo Presti, D.C.F., Dip. Ing. Strutturale, Politecnico di Torino, Corso Duca degli Abruzzi 24, 10129 Torino. T: 011/5644804, F: 011/5644899, E: diego.lopresti@polito.it

Lodini, A., Casoli & Lodini S.n.c, Via dei Mulini 11/a, 43030 Mamiano (PR). T: 0521/848110, F: 0521/848110

Madiai, Claudia, Viale Lavagnini 4, 50129 Firenze. T: 055/4796313-484249,

Madoni, Giancarlo, Studio Ingegneri Associati, Via Garian 64, 20146 Milano. T: 02/468646, F: 02/468454

Manassero, M., Via Pigafetta 4, 10129 Torino. T: 011 5611811, F: 011 5620568, E: manassero@polito.it

Mancuso, C., Via F. Cilea 265/b, 80127 Napoli. T: 081/5454056, F: 081/7683481, E: mancuso@unina.it

Mandolini, A., Via Domenico Fontana 25/e, 80128 Napoli. T: 081/7702768, E: alessandro.mandolini@unina2.it

Manfredini, G., Via Cremuzio Cordo 28, 00136 Roma. T: 85093701, F: 85092736

Mansueto, F., Via Mansueto 2A/9, 16159 Genova. T: 010/6443386, F: 010 567036, E: bizio.ge@tin.it

Marchetti, S., Via Bracciano 38, 00189 Roma. T: 06 30360107, F: 06 30367760, E: marchetti@flashnet.it

Marchi, G., En. Ser., Viale Baccarini 29, 48018 Faenza (RA). T: 0546 663423, F: 0546 663428, E: enser@mbox.dinamica.it

Margiotta, G., Via Liberta' 207, 90143 Palermo. T: 091 6259711, F: 091 6253129

Martinetti, S., Via Asmara 58, 00199 Roma. T: 0735 82068, F: 0735 655470, E: merlino@jth.it

Martis, M., Edilgeo S.r.l., Via P.L. da Palestrina 22, 09129 Cagliari. T: 070/497472, F: 070/497472

Mascardi, C., Studio Geotecnico Italiano, Via Ripamonti 89, 20141 Milano. T: 02/5691841, F: 02/5691845, E: sgi_mascardi@studio-geotecnico.it

Mascarucci, M., Via Gran Sasso 31, 66100 Chieti. T: 0871/69667, E: mamas@plugit.net

Maugeri, M., Corso Umberto 208, 95024 Acireale (CT). T: 095 338920, F: 095 894334, E: mmaugeri@ isfa.unict.it

Mazzetti, G.P., Piazza S. Quirino 6, 42015 Correggio (RE). T: 0522 641001, F: 0522 632162

Mazzucato, A., Dip. Costruzioni Marittime e Geotecnica, Universita' di Padova, Via Ognissanti 39, 35129 Padova. T: 049 8277980, F: 049 8277986, E: amg@geomar.ing.unipd.it

Meriggi, R., Dip. Georisorse e Territorio, Univ. di Udine, Via Cotonificio 114, 33100 Udine. T: 0432 558772, F: 0432 558700, E: meriggi@udgtls.dgt.uniud.it

Micolitti, G., Via Toscana 1, 04100 Latina. T: 0773 621038, F: 0773 621038, E: g.micolitti@libero.it

Migliorino, M., Via E. Hemingway 5, 90148 Palermo. T: 091 6910367, E: gugmigl@tin.it

Molinari, M., Via Bezzecca 6, 25128 Brescia. T: 0338 6932139

Monaco, P., DISAT-Fac. Di Ingegneria, Universitàdell'Aquila, Loc. Monteluco di Roio, 67040 Monteluco di Roio (AQ). T: 0862 434536, F: 0862 434548, E: p.monaco@ing.univaq.it

Monterisi, L., Ist. Geol. Appl. e Geot., Politecnico di Bari, Via Orabona 4, 70125 Bari. T: 080/242324-242362, F: 080/242529, E: l.monterisi@cstar.poliba.it

Monti, G., Tecnimont S.p.A., Viale Monte Grappa 3, 20124 Milano. T: 02 63139589, F: 02 63139155

Montrasio, L., Dip. Ing. Civile, Università di Parma, Viale delle Scienze, 43100 Parma. T: 0521/905932, F: 0521/905924, E: lorella.montrasio@unipr.it

Moraci, N., Dip. Meccanica e Materiali, Università di Reggio Calabria, Via Graziella Loc. Feo di Vito, 89060 Reggio Calabria. T: 0965 875263, F: 0965 875201, E: moraci@ing.unirc.it

Muscolino, S.R., Via Vincenzo Epifanio 3, 90146 Palermo. T: 091 6713404, E: sandrosauro@tin.it

Napoleoni, Q., Via Capo Peloro 10, 00141 Roma. T: 06 8272960, F: 06 44585016

Nesti, M., Piazza Domenico Ricci 12, 41027 Pievepelago (MO). T: 0536/71395, F: 0536/71395, E: maurizio.nesti@garassinosrl.it

Niccolai, A., Via G.B. Barinetti 1, 20145 Milano. T: 02/33603977

Nicotera, P., Via Manzoni 257, 80123 Napoli. T: 081/ 7693176

Nigrelli, F., Studio Ing. Nigrelli, Via Emilia 47, 90144 Palermo. T: 091/522477, F: 091/517806

Nocera, N., Via Zambrano 92, 84088 Siano (SA). T: 081/ 5181165 0338/3996862,

Nocilla, N., Via G. Cascino 18 (tra.Basile), 90128 Palermo. T: 091/486157, E: blunoc@neomedia.it

Nova, R., Dip. Ingegneria Strutturale, Piazza Leonardo da Vinci 32, 20133 Milano. T: 02/23994232, F: 02/ 23994220, E: nova@shu.polimi.it

Olcese, A., Via C. Cabella 22/b/28, 16122 Genova. T: 010 813451, F: 010 52035290

Ottaviani, M., Ist. di Geologia Applicata, Via Eudossiana 18, 00184 Roma. T: 461810,

Pasqualini, E., Dip. Scienze Materiali e Terra, Via Brecce Bianche snc, 60131 Ancona. T: 071/2204793, F: 071/ 2204714, E: pasqualini@popcsi.unian.it

Passalacqua, R., Via Puggia 21 – int. 5, 16131 Genova. T: 010/302212, F: 010/3532534

Patrizi, P., CERN, Chantier LHC P5, 01170 Cessy (FRANCIA). T: +33 450 403700, F: +33 450 419811

Pedroni, S., Enel S.p.A.-SRI-Polo Idraulico e Strutturale, Via G. Pozzobonelli 6, 20162 Milano. T: 02/ 72243510, F: 02/ 72243550-3540

Pelli, F., Via Ammiraglio G. Bettolo 14 1a, 16031 Bogliasco (GE). T: 010 3470208, F: 010 3470208

Pezzarossa, R., Siagi S.r.l., Piazza di Novella 1, 00199 Roma. T: 8312787-8390164, F: 8312787, E: siagi@ciaoweb.it

Pirollo, F., Via Collemeroni 2, 03040 S. Vittore del Lazio (FR). T: 0776/343038, F: 0776/343039

Pitullo, A., Via La Piscopia 24, 71014 S. Marco in Lamis (FG). T: 0882 833033

Placucci, A., Via Carducci 19, 06024 Gubbio (PG). T: 075 9221399, F: 075 9222306, E: aplacucci@cooprogtti.it

Popescu, M., Dept. of Civil and Architectural Engineering, Illinois institute of Technology, 3201 South Dearborn Street, IL 60616 Chicago (U.S.A.). T: +1 312 5673547, F: +1 312 5673519, E: mepopescu@usa.net

Porcheddu, A., Via Salvemini 8, 07100 Sassari. T: 079 233491, F: 079 233491

Prisco, R., Contrada Ponte, 93014 Mussomeli (CL). T: 0934/951254

Provenzano, P., Dip. Ingegneria Civile, Università Tor Vergata, Via di Tor Vergata 110, 00133 Roma. T: 06 72597057, F: 06 72597005, E: prvenzanopaola@libero.it

Rabagliati, U., Via Torricelli 4, 10128 Torino. T: 011 599561-5682718, F: 011 599561-5681342

Rampello, S., Dip. Ingegn. Strutt. e Geot., Via Monte d'Oro 28, 00186 Roma. T: 06/6878886, F: 06/ 6878923, E: sebastiano.rampello@uniroma1.it

Resnati, C., Studio Geoplan S.S., Via C. Rota 39, 20052 Monza (MI). T: 039/832781, F: 039/835750, E: studiogeoplan @iol.it

Ribacchi, R., Dip. Ingegn. Strutt. e Geotec., Via Monte d'Oro 28, 00186 Roma. T: 6878886, F: 6878923

Riccardi, F., Bocoge S.p.A., Via Alessandro Fleming 55, 00191 Roma. T: 5919141

Robotti, F., A.G.I.S.Co. S.r.l., Via A. Moro 2, 20060 Liscate (MI). T: 02 9587690, F: 02 9587381

Roda, C., Via Paolo Canciani 16, 33100 Udine. T: 0432/ 21845

Rodino, A., Igeas Engineering S.r.l., Via Reduzzi 9, 10134 Torino. T: 011 3181661-3181904, F: 011 318555

Ronco, F., Ingeo di Chendi M. e Ronco F., Via dei Lancieri 2, 36040 Torri Quartesolo (VI). T: 0444 581561, F: 0444 580896, E: roccoingeo@goldnet.it

Salvati, G., Via Vaccaro 113, 85100 Potenza. T: 0971/ 54789, F: 0971/472967

Savazzi, G., Via Marconi 32/U, 27040 Mezzanino (PV). T: 0385/716231, F: 0385/719063

Scalorbi, R., Via Bracciano 2, 00189 Roma. T: 3760027, F: 3710288, E: scalorbi@sieget.it

Scarascia Mugnozza, G., Dip. Scienze della Terra, P.le Aldo Moro 5, 00185 Roma. T: 06 49914445, F: 06 4454729

Scarpelli, G., Dip.Scienze Materiali e Terra, Via Brecce Bianche, 60131 Ancona. T: 071/2204421, F: 071/ 5893714, E: g.scarpelli@geotecnica.union.it

Scavia, C., Via Colle delle Finestre 4/bis, 10059 Susa (TO). T: 0122/2703, E: scavia@polito.it

Sciotti, M., Facolta' di Ingegneria, Via Eudossiana 18, 00184 Roma. T: 44585021, F: 44585016

Sembenelli, P., SC Sembenelli Consulting S.r.l., Via Santa Valeria 3/5, 20123 Milano. T: 02 72002689 72002705, F: 02 89010549, E: scsem@iol.it

Serafini, G., Via Mascagni 7, 41100 Modena. T: 059/ 211156

Sfondrini, G., Dip. Scienze della Terra, Via Mangiagalli 34, 20133 Milano. T: 02/23698234-23698216, F: 02/70638261

Silvestri, F., Dip. Difesa del Suolo, Università della Calabria, C.da S. Antonello, 87046 Montalto Uffugo (CS). T: 0984 839686, F: 0984 934245, E: f.silvestri@unical.it

Simeoni, L., Dip. Ing. Meccanica e Strutturale, Università degli Studi di Trento, Via Mesiano 77, 38050 Trento. T: 0461 882585, F: 0461 882599, E: lucia.simeoni@ing.unitn.it

Simonetti, G., Via San Gottardo 58, 20052 Monza (MI). T: 0721/7711

Simonini, P., Ist. Costruz. Marittime e Geot., Via Ognissanti 39, 35129 Padova

Soccodato, C., Via Caio Canuleio 121, 00174 Roma. T: 744018, E: csoccodato@iol.it

Speciale, G., Piazza Unita' d'Italia 4, 90144 Palermo. T: 091/302946, F: 091/307452

Stanzione, L.R., Studio Associato Stanzione & Stanzione, S.A.S.S., Via Leopardi 7, 23868 Valmadrera (LC). T: 0341 200641, F: 0341 200641, E: docgeolu@iol.it

Stefani, B., Edilfloor S.p.A., Via Leonardo da Vinci 14/15, 36066 Sandrigo (VI). T: 0444 750350, F: 0444 657246, E: lino@edilfloor.com

Susani, A., Via Calderai 42, 90133 Palermo. T: 091 6161511

Tamagnini, C., Ist. Ingegneria Ambientale, Univ. di Perugia, Via G. Duranti 1A/6, 06125 Perugia. T: 075 5852763, F: 075 5852756, E: tamag@unipg.it

Tassi, T., Favero e Milan Ingegneria s.r.l., Via Varotara 57, 30030 Zianigo Mirano (VE). T: 041/5785711, F: 041/ 4355933, E: fm@favero-milan.com

Tentor, A., L.G.T. S.n.c., Via Zorutti 84, 33030 Campoformido (UD). T: 0432 652490, F: 0432 652849, E: ellegiti@ tin.it

Togliani, G., Pedrozzi e Associati, Via Ligaino 20, 6963 Pregassona (SVIZZERA). T: +41 91 9412351, F: +41 91 9669974

Tommasi, P., Via Fossato di Vico 10, 00181 Roma. T: 06/ 44585005-76900984, F: 06/44585016, E: tommasi@dits. ing.uniroma1.it

Tornaghi, R., Via Guido d'Arezzo 8, 20145 Milano. T: 02/4690807

Totani, G., Dip. Ingegneria Strutture e Acque Territorio, Piazza Pontieri 2, 67040 Monteluco di Roio (AQ). T: 0862/434535, F: 0862/434548

Umilta`, G., Via Valdemone 57, 90144 Palermo. T: 091 519860, F: 091 524782, E: progeost@tin.it

Uzielli, M., Piazza Unganelli 3, 50125 Firenze. T: 328 6549144, F: 055 2478916, E: mazcanzi@tin.it

Valore, C., Via Rosina Anselmi 11, 90135 Palermo, F: 091/6568407, E: calogerovalore@tiscalinet.it

Van Impe, W.F., Ghent Univ.-Labo Voor Grondmec, Technologiepark 9, 9052 Zwijnaarde (BELGIUM). T: +32 9 2645723, F: +32 9 2645849

Vandoni, C., Viale Tunisia 13, 20124 Milano. T: 02/29405765

Vannucchi, G., Dip. Ingegneria Civile, Via S. Marta 3, 50139 Firenze. T: 055/52612, E: giovan@dicea.unifi.it

Vassallo, G.P., D'Appolonia S.p.A., Via S. Nazaro 19, 16145 Genova. T: 010 3367647, F: 010 367647

Venturi, M., Sidercem S.r.l., C.da Calderaro Z.I. Snc, 93100 Caltanissetta. T: 0937/568465, F: 0934/568388

Viggiani, C., Via Posillipo 281, 80123 Napoli. T: 081/662276, E: viggiani@hmg.inpg.fr

Wolf, E., Via San Valentino 28, 00197 Roma

Zani, O., Via Matteotti 50, 48012 Bagnacavallo (RA). T: 0545/26561-61733

Zaninetti, A., Enel Ricerca-Polo Idraulico e Strutturale, Via G. Pozzobonelli 6, 20162 Milano. T: 02/72243521, F: 02/72243550-3540

Zerbo, V., Via Vignicella 14, 90013 Castelbuono (PA). T: 0921 671759, E: vincenzozerbo@tiscalinet.it

A.L.G.I., Giovanni Dott. Geol. Rea, Via Frattina 35, 00187 Roma. T: 52200232, F: 52200232
Anisig, Franco Dott. Zucchi, Via G. Pascoli 58, 20133 Milano. T: 02 70637754
Aquater S.p.A., Antonio Dott. Ing. Carletti, Via Miralbello 53, 61047 S. Lorenzo in Campo (PS). T: 0721/7311, F: 0721/731308
Azienda Energetica S.p.A., Etschwerke AG, Giuliano Zamunaro, Via Dodiciville 8, 39100 Bolzano. T: 0471/924555, F: 0471/980419
Dip. Difesa del Suolo, Univ. Calabria, Contrada S. Antonello, 87040 Montalto Uffugo (CS)
Dip. Ing. Idraulica Marittima Geotecnica, Univ. di Padova, Via Ognissanti 39, 35129 Padova. T: 049/8071299
Dip. Ingegneria Geotecnica-Amministrazione, Univ. di Napoli, Via Claudio 21, 80125 Napoli. T: 081/7683481
Dip. Ingegneria Strutturale e Geotecnica, Univ. "La Sapienza", Via Monte d'Oro 28, 00186 Roma. T: 6878886-6876133, F: 6878923
DISTART, Univ. di Bologna, Raffaele Prof. Poluzzi, Viale Risorgimento 2, 40136 Bologna. T: 051/6443232-6443231, F: 051/6443236
Enel Hydro S.p.A., Gualtiero Dott. Ing. Baldi, Via Pastrengo 9, 24068 Seriate (BG). T: 035/307111, F: 035/302999
Enel S.p.A. Dir. Amministra.va, Viale Regina Margherita 137, 00198 Roma. T: 0685092834, F: 0685095905, E: romano.massimo@enel.it
Ferrovie dello Stato S.p.A., Istituto Sperimentale, Vittorio Dott. Misano, Piazza Ippolito Nievo 46, 00153 Roma. T: 47303182
Geosonda S.p.A., Lucio Dott. Ing. Diamanti, Via Girolamo da Carpi 1, 00196 Roma. T: 06 3221993, F: 06 3221755
Italferr S.p.A., Via Marsala 53/67, 00185 Roma. T: 06 49751, F: 06 49752437
Metropolitana Milanese S.p.A., Bruno Dott. Ing. Cavagna, Via del Vecchio Politecnico 8, 20121 Milano. T: 02/77471, F: 02/780033

Officine Maccaferri S.p.A., Francesco Dott. Ing. Ferraiolo, Via Agresti 6, 40123 Bologna. T: 051 6436000, F: 051 236507, E: utp.officine@maccaferri.com
Ordine Ingegneri Caserta, Ferdinando Dott. Ing. Luminoso, Via Botticelli 32, 81100 Caserta. T: 0832/326204-443365
Pagani Geotechnical Equipment, Ermanno Pagani, Loc. Santimento 44, 29010 Calendasco (PC). T: 0523 781520, F: 0523 781520, E: geo.pagani@tin.it
Pegaso S.c.r.l., Enrico Dott. Campa, Via F. Tovaglieri 17, 00155 Roma. T: 2305345, F: 2305134
Provincia di Oristano, Amministrazione, E. Sanna, Via Mattei s.n.c., 09170 Oristano. T: 0783/7391, F: 0783/793304
Regione Piemonte, Direzione Servizio Tecnico di Prevenzione, Giuseppe Ben, Via Pisano 6, 10152 Torino. T: 011 4324577-0, F: 011 4323535, E: direzione20@regione.piemonte.it
Rodio S.p.A., Jean Dott. Ing. Kissenpfennig, Via Pandina 5, 26831 Casalmaiocco (LO). T: 02/9834991, F: 02/9837556, E: rodio@rodiospa.com
Spea Ingegneria Europea S.p.A., Alessandro Dott. Ing. D'Amato, Via Vida 11, 20127 Milano. T: 02/88831, F: 02/8883355
Studio Geotecnico Italiano, Michele Prof. Ing. Jamiolkowski, Via Ripamonti 89, 20141 Milano. T: 02/5220141, F: 02/5691845
Tenax S.p.A., Via Industria 3, 23897 Vigano' (LC). T: 039 9219307, F: 039 9219200
Tensacciai S.p.A., Cesare Dott. Ing. Prevedini, Via F. Vegezio 15, 20149 Milano. T: 02 4300161, F: 02 48010726, E: tens@tensacciai.it
Trevi S.p.A., Davide Cav. Lav. Trevisani, Via Dismano 5819, 47023 Cesena (FO). T: 0547/319311, F: 0547/317395
Ajmone-cat, Alessio M., Via del Pigneto 133, 00176 Roma. T: 822393

108

JAPAN

Secretary: Prof. Hideo Sekiguchi, Secretary of the Japanese Geotechnical Society, Sugayama Building 4F, Kanda Awaji-cho 2-23, Chiyoda-ku, Tokyo 101-0063. T: 0332517661, F: 0332516688, E: jgs@jiban.or.jp

Total number of members: 1389

Abe, F., 2-4-3-716, Shirogane, Minato-ku, Tokyo 108

Abe, H., College Professor, Gunma National College of Technology, 580 Toribamachi, Maebashi, Gunma 371-0845. T: 0272549193, F: 0272549022, E: abe@cvl.gunmact.ac.jp

Abe, H., Geotechnical Director, 2641 Inaba, Minamimata, Nagano 380-0917. T/F: 0262232742, E: abe@chubugeotech.co.jp

Abe, H., Dr, Deputy Director, Kajima Technical Research Institute, 19-1, Tobitakyu 2, Chofu, Tokyo 182-0036. T: 0424893093, F: 0424897016, E: abeh@pub.kajima.co.jp

Abe, N., Dr, University Associate Professor, Osaka University, 2-1, Yamada-oka, Suita, Osaka 565-0871. T: 0668797624, F: 0668797629, E: abe@civil.eng.osaka-u.ac.jp

Aboshi, H., Representative, Prof. H. Aboshi Institute, 3-3-58, Takasu, Nishiku, Hiroshima 733-0871. T/F: 0822715722, E: aboshi@fukken.co.jp

Adachi, K., Dr, Professor, Shibaura Inst. of Technology, 3-9-14, Shibaura, Minato-ku, Tokyo 108-8548. T: 0354763048, F: 0354763166, E: adachi@sic.shibaura-it.ac.jp

Adachi, N., Research Engineer, Kajima Technical Research Institute, 19-1, Tobitakyu 2-chome, Chofu, Tokyo 182-0036. T: 0424897114, F: 0424897116, E: naohito@katri.kajima.co.jp

Adachi, T., University Professor, Dept. of Civil Engineering, Kyoto University, Yoshida Hon-machi, Sakyo-ku, Kyoto 606-01. T: 0757535104, F: 0757535104, E: adachi@chik.kuciv.kyoto-u.ac.jp

Adachi, T., Dr, University Professor, College of Science & Technology of Nihon University, 1–8, Surugadai, Kanda, Chiyoda-ku, Tokyo 101-8308. T: 0332590696, F: 0332938253, E: adachi@arch.cst.nihon-u.ac.jp

Aida, T., Professional Engineer, Nippon Koei Co., Ltd., Engineering Research Laboratory, 2304 Inarihara, Kukizakimachi, Inashiki-gun, Ibaraki 300-1259. T: 0298712065, F: 0298712021, E: a1380@n-koei.co.jp

Aikawa, A., Associate Professor, Oita National College of Technology, Maki, Oita 870-0152. T: 0975527652, F: 0975527949, E: aikawa@oita-ct.ac.jp

Akagi, H., Dr, University Professor, Waseda University, Dept. of Civil Engineering, 58-205, 3-4-1, Ohkubo, Shinjuku-ku, Tokyo 169-8555. T: 0352863405, F: 0352720695, E: akagi@mn.waseda.ac.jp

Akagi, T., Dr, University Professor, Dept. of Civil Engineering, Toyo University, Kujirai 2100 Kawagoe, Saitama 350-8585. T: 0492391394, F: 0492314482, E: akagi@eng.toyo.ac.jp

Akai, K., Prof. Emeritus, Geo-Research Institute, Osaka Science & Technology Centre Bldg.6F., 1-8-4, Utsubohonmachi, Nishi-ku, Osaka 550-0004. T: 0664438930, F: 0664439484, E: akai-k@geor.or.jp

Akamoto, H., Technical Research Institute, Toyo Construction Co., Ltd., 3-17-6, Naruohama, Nishinomiya, Hyogo 663. T: 0798435903, F: 0798435916

Akawa, T., c/o Hayakawa, Yamada Nishi-Machi, 1-51, Kita-ku, Nagoya 462. T: 0529811151

Akiyama, J., Professional Engineer, Japan Ground Water Development Co., Ltd., 777 Matsubara, Yamagata-shi, Yamagata 990-2313. T: 0236886000, F: 0236884122, E: jun_akiyama@jgd.co.jp

Akiyoshi, T., Dr, University Professor, Dept. of Civil Engineering & Architecture, Kumamoto University, 2-39-1, Kurokami, Kumamoto 860-8555. T: 0963423538, F: 0963423507, E: akiyoshi@kumamoto-u.ac.jp

Akutagawa, S., University Associate Professor, Dept. of Architecture and Civil Engineering, Kobe University, Rokkodai, 1-1, Nada, Kobe 657-8501. T: 0788036015, F: 0788036069, E: cadax@kobe-u.ac.jp

Ando, H., Consulting Engineer, Shingijutsu Keikaku Co., Ltd., 1-12, Minami-Ohi 6-chome, Shinagawa-ku, Tokyo 140-0013. T: 0337663211, F: 0337659700, E: jr1mlu@jarl.com

Ando, Y., Riverside 1-5-406, 5-5-7, Isehara-cho, Kawagoe-shi, Saitama 350

Anma, S., Dr, Engineering Geologist, Kensetsu Kiso Chosa Co., Ltd., 241-7, Kusunoki-shinden, Shimizu 424. T: 0543452415, F: 0543474669

Aoki, H., Geotechnical Engineer, Japan Railway Construction Public Corporation, 14-2, Nagata-cho 2-chome, Chiyoda-ku, Tokyo 100. T: 0335061861, F: 0335061891

Aoki, M., Chief Researcher, Takenaka Corporation, Research & Development Institute, 5-1, 1-chome Ohtsuka, Inzai, Chiba 270-1395. T: 0476471700, F: 0476473080, E: aoki.masamichi@takenaka.co.jp

Aoki, Y., Takizawa Dam Construction Office, Water Resources Development Public Corp., 10-18, Shimomiyaji, Chichibu

Aoyagi, T., Mr, Assistant Manager, Tokyo Electric Power Services Co., Ltd., 2-1-4, Uchisaiwai-cho, Chiyoda-ku, Tokyo 100-0011. T: 0345133153, F: 0345133017, E: aotaka@tepsco.co.jp

Aoyama, C., Dr, 2-11-10-919, Kemacho, Miyakogima, Osaka. T: 069257808

Aoyama, K., University Professor, Research Institute for Hazards in Snowy Areas, Niigata University, Igarashininocho 8050, Niigata 950-2181. T: 0252627053, F: 0252627050, E: aoyamak@gs.niigata-u.ac.jp

Arai, H., Research Engineer, EDM/RIKEN, 2465-1, Fukuimikiyama, Miki, Hyogo 673-0433. T: 0794836637, F: 0794836695, E: arai@miki.riken.go.jp

Arai, K., Dr, University Professor, Fukui University, 3-9-1, Bunkyo, Fukui-shi, Fukui 910-8507. T/F: 0776278594, E: katuhiko@anc.anc-d.fukui-u.ac.jp

Aramaki, N., University Lecturer, Sojo University, 4-22-1, Ikeda, Kumamoto 860-0082. T: 0963263111, F: 0963263000, E: aramaki@ce.sojo-u.ac.jp

Arioka, M., Officer & Joint General Manager, Kumagai Gumi Co., Ltd., Marketing Division, 2-1, Tsukudo-cho, Shinjuku-ku, Tokyo 162-8557. T: 0332358653, F: 0332353061, E: marioka@ku.kumagaigumi.co.jp

Asada, A., University Professor, Tohoku Institute of Technology, 35-1, Yagiyama-kasumicho, Taihaku-ku, Sendai. T: 0222291151, Ext. 448, F: 0222298393, E: asada@titan.tohtech.ac.jp

Asada, H., Hibarigaoka-ryo, 4-3, Nishi-Mannenzaka, Manganji, Kawanishi 655

Asai, K., Geotechnical Engineer, Takenaka Corporation, Research & Development Institute, 5-14, 2-chome Minamisuna, Koto-ku, Tokyo 136. T: 0336473161, F: 0336450911

Asakuma, M., 2-8-6-302, Momochi, Sawara-ku, Fukuoka 814

Asami, H., Consulting Engineer, Nikken Sekkei Ltd., 1-4-27, Kohoraku, Bunkyo-ku, Tokyo 112. T: 0338133361, F: 0338170518

Asano, Y., Project Manager, Meiyo Electric Co., Ltd., 485 Nanatsushinya, Shimizu 424-0066. T: 0543452213, F: 0543452215, E: y-asano@meiyoelc.co.jp

Asaoka, A., Dr, University Professor, Dept. of Civil Engineering, Nagoya University, Furo-cho, Chikusa-ku, Nagoya 464-8603. T: 0527894621, F: 0527894624, E: asaoka@soil.genv.nagoya-u.ac.jp

Aso, T., 3-20-24-301, Hakomatsu Higashi-ku, Fukuoka 812

Baba, K., Dr, Research Lecturer, Faculty of Engineering, Osaka University, Yamadaoka, Suita, Osaka 565. T: 068775111, F: 068771658

Babasaki, R., Geotechnical Engineer, Takenaka Technical Research Laboratory, 1-5-1, Ohtsuka, Inzai, Chiba 270-13. T: 0476471700, F: 0476473070

Bessho, M., Geotechnical Engineer, Onoda Chemico Co., Ltd., 2-17-4, Yanagibashi, Taito-ku, Tokyo 111-8637. T: 0338622253, F: 0338654078, E: m-bessho@nifty.ne.jp

Chai, J.C., Dr, University Professor, Dept. of Civil Engineering, Saga University, 1 Honjo, Saga 840-8502. T: 0952288580, F: 0952288190, E: chai@cc.saga-u.ac.jp

Chen, G., Associate Professor, Dept. of Civil & Structure Engineering, Kyushu University, 6-10-1, Hakozaki, Higashi-ku, Fukuoka 812-8581. T/F: 0926424406, E: chen@civil.kyushu-u.ac.jp

Chida, S., Dr, Managing Director, Public Works Research Center, 7-2, Taito 1-chome, Taito-ku, Tokyo 110. T: 0338353609, F: 0338327397

Cubrinovski, M., Senior Research Engineer, Kiso-Jiban Consultants Co., Ltd., 1-11-5, Kudan-kita, Chiyoda-ku, Tokyo 102-8220. T: 0352766737, F: 0352109405, E: misko-cubrinovski@kiso.co.jp

Daido, K., 2-11-7, Takemidai, Suita 565

Daikoku, A., Deputy Manager, Dept. of Technology Development, Nihon Shinko Co., Ltd., 800-9, Sawa, Kaizuka-shi, Osaka 597. T/F: 0724377306

Daito, K., Dr, Associate Professor, Dept. of Civil Engineering, Daido Institute of Technology, 40 Hakusui-cho, Minami-ku, Nagoya 457. T: 0526125571, Ext. 239, F: 0526125953, E: daito@daido-it.ac.jp

Deguchi, C., Dr, Associate Professor, Dept. of Civil Eng., Miyazaki University, Gakuenkibanadai Nishi, 1-1, Miyajaki 889-21. T: 0985582811, F: 0985582876

Denda, A., Deputy General Manager, Shimizu Corporation, Etchujima, 3-4-17, Koto-ku, Tokyo 135-8530. T: 0338205293, F: 0338205959, E: den@sit.shimz.co.jp

Echigo, Y., Managing Director, Kawasaki Steel Corp., Hibiya Kokusai Bldg., 2-3, 2chome Uchisaiwaicho, Chiyoda-ku, Tokyo 100. T: 0335973048, F: 0335974361

Eiki, A., Consulting Engineer, Token Geo Tech, Yushimatoukyu Bldg., 3-37-4, Yushima, Bunkyo-ku, Tokyo 113-0034. T: 033 8346191, F: 0338358903, E: eiki@tokengeotec.co.jp

Enami, A., Professor Emeritus of Nihon University, 3-4-20, Asagaya Kita, Suginami-ku, Tokyo 166-0001. T: 0333384046, F: 0333382969

Endo, M., Consulting Engineer, 1-490, Onari-cho, Omiya 331-0043. T: 0486657828, F: 0486602386, E: m-endou@muj.biglobe.ne.jp

Endo, O., 17-2-406, Satsukigaoka, Midori-ku, Yokohama 227

Enoki, M., Prof., Dept. of Civil Engineering, Tottori University, 4-101, Minami-cho, Tottori-shi, Tottori 680. T: 0857315289, F: 0857287899

Enomoto, M., Geotechnical Engineer, Kiso-Jiban Consultants Co., Ltd., 2-14-1, Ishikawa-cho, Ohta-ku, Tokyo 145. T: 0337276158, F: 0337276247

Enomoto, T., University Professor, Kanagawa University, 3-27-1, Rokkakubashi, Kanagawa-ku, Yokohama 221-8686. T: 0454815661, F: 0454917915, E: enomot01@kanagawa-u.ac.jp

Esaki, T., University Professor, Institute of Environmental Systems, Kyushu University, Fukuoka 812-8581. T: 0926423845, F: 0926423848, E: esaki@ies.kyushu-u.ac.jp

Fuchida, K., College Professor, Yatsushiro College of Technology, 2627 Hirayama-shinmachi, Yatsushiro 866-8501. T: 0965531346, F: 0965531349, E: fuchida@as.yatsushiro-nct.ac.jp

Fuchimoto, M., Research Engineer, Shimizu Corporation, Institute of Technology, 3-4-17, Etchujima, Koto-ku, Tokyo 135-8530. T: 0338205521, F: 0338205959, E: fuchi@sit.shimz.co.jp

Fudeyasu, T., Takahashi Bldg., 2-19, Kinya-cho, Minami-ku, Hiroshima 730

Fujii, H., University Professor, Faculty of Environmental Science and Technology, Okayama Universtiy, 1-1, Naka 3-chome, Tsushima, Okayama 700-8530. T: 0862518360, F: 0862518361, E: jhei0200@cc.okayama-u.ac.jp

Fujii, K., University Professor, Iwate University, 3-18-8, Ueda, Morioka, Iwate 020-8850. T: 0196216198, F: 0196216204, E: katfujii@iwate-u.ac.jp

Fujii, K., College Professor, Dept. of Civil Engineering, Anan National College of Technology, 265 Aoki, Minobayashi-cho, Anan-shi, Tokushima 774-0017. T: 0884237188, F: 0884237199, E: fujiik@ana-nct.ac.jp

Fujii, M., Consulting Engineer, Fujii Kiso Sekkei Co., Ltd., 1349 Higashitsuda-cho, Matsue-shi, Shimane 690. T: 0852216721, F: 0852252248

Fujii, M., Dr, Tokai University, 1117 Kitakaname, Hiratuka, Kanagawa 259-1292. T: 0463581211, F: 0463502024, E: fujii@keyaki.cc.u-tokai.ac.jp

Fujii, N., University Professor, Dept. of Civil Engineering, Chuo University, Kasuga, 1-13-27, Bunkyo-ku, Tokyo 112-8551. T: 0338171800, F: 0338171803, E: fujii@sip.civil.chuo-u.ac.jp

Fujii, S., Dr, General Manager, Taisei Building Research Institute, 344-1, Nase-cho, Totsuka-ku, Yokohama 245-0051. T: 0458147227, F: 0458147251, E: shunji.fujii@sakura.taisei.co.jp

Fujimoto, H., Dr, Technical Advisor, Soil Testing Center, 2043 Hongokitakata-Hirata, Miyazaki 880-0925. T: 0985522403, F: 0985544347, E: fuji9442@sweet.ocn.ne.jp

Fujimoto, K., University Professor, Kanazawa Institute of Technology, 7-1, Ohgigaoka, Nonoichimachi, Ishikawa-gun, Kanazawa 921-8501. T: 0762946712, F: 0762946713, E: fujimotk@neptune.kanazawa-it.ac.jp

Fujimura, H., Dr, Associate Professor, 101 4-chome Minami, Koyama, Tottori 680. T: 0857315296, F: 0857287899

Fujioka, T., Consulting Engineer, Chiyoda Corporation, 2-12-1, Tsurumichuo, Tsurumi-ku, Yokohama 230-8601. T: 0455067523, F: 0455069295, E: tfujioka@ykh.chiyoda.co.jp

Fujita, H., Dr, President, Institute of Slope Technology, 2-7-15, Ichigaya-tamachi, Shinjuku-ku, Tokyo 162-0843. T: 0332606682, F: 0332604820, E: fujita@ist-ls.co.jp

Fujita, K., University Professor, 5-37-4, Nishi-oizumi, Nerima-ku, Tokyo 178-0065. T: 0339233915, F: 0471239766

Fujiwara, H., College Professor, Tokuyama College of Technology, 3538 Kume, Tokuyama, Yamaguchi 745-8585. T: 0834296332, F: 0834289813, E: fujiwara@tokuyama.ac.jp

Fujiwara, K., Structural Engineer, Suzuki Architectural Design Office, 7-55, Aioichou, Yamagata 990-0055. T: 0236231778, F: 0236231779, E: gr2k-fjwr@asahi-net.or.jp

Fujiwara, T., Manager of Civil Engineering Department, Ohbayashi Corp., Technical Research Institute, Shimo-Kiyoto, 4-640, Kiyose, Tokyo 204. T: 0424950910, F: 0424950903

Fujiwara, T., 516-1-A 211, Maeda-cho, Totsuka-ku, Yokohama 244

Fujiwara, T., Geotechnical Engineer, Geo-Research Institute, 1-1-20, Kuise, Minami-Shinmachi, Amagasaki 660-0822. T: 0664888256, F: 0664887802, E: teruyuki@geor.or.jp

Fujiwara, Y., Senior Research Engineer, Geo-environment Engineering Section, Civil Engineering Research Institute, Technology Center, Taisei Corporation, 344-1, Nase-cho, Totsuka-ku, Yokohama 245-0051. T: 0458147217, F: 0458147257, E: yasushi.fujiwara@sakura.taisei.co.jp

Fujiwara, Y., College Associate Professor, Miyagi Agricultural College, Hatatate, 2-2-1, Taihaku-ku, Sendai 982-0215. T: 0222452211, F: 0222451534, E: fujifuji@mxh.mesh.ne.jp

Fujiyama, A., 4816-1, Oazakoujiro, Obatake-cho, Kuga-gun, Yamaguchi 749-01

Fukada, H., Geotechnical Engineer, Fudo Construction Co., Ltd., Taito, 1-2-1, Taito-ku, Tokyo 110. T: 0338376037, F: 0338376105

Fukagawa, R., Dr, University Professor, Ritsumeikan University, 1-1-1, Noji-higashi, Kusatsu, Shiga 525-8577. T: 0775612875, F: 0775612667, E: fukagawa@se.ritsumei.ac.jp

Fukasawa, T., Registered Consulting Engineer, TOA Corporation, 5 Yonban-cho, Chiyoda-ku, Tokyo 102-8451. T: 0332625102, F: 0332629536, E: PLA07898@nifty.ne.jp

Fukazawa, E., Soil Engineer, 14-1, Shinmei 2-chome, Hino-shi, Tokyo 191. T: 0425862807

Fukuda, F., Research Associate, Hokkaido University, Division of Structural and Geotechnical Engineering, Graduate School of Engineering, Kita-13, Nishi-8, Kita-ku, Sapporo 060-8628. T: 0117066194, F: 0117067204, E: fukuda@eng.hokudai.ac.jp

Fukuda, K., Dr, Geotechnical Engineer, Shimizu Corporation, Seavans South, 1-2-3, Shibaura, Minato-ku, Tokyo 105-8007. T: 0354410593, F: 0354410515, E: k-fukuda@civil.shimz.co.jp

Fukuda, M., 3-24-14-612, Tarume-cho, Suita-shi, Osaka 564

Fukuda, M., Dr, Managing Director, Mitsui Construction, 1-9-1, Nakase, Mihama-ku, Chiba 261. T: 0432127545, F: 0432127540

Fukuda, N., Dr, Managing Director, Fukken Co., Ltd., Consulting Engineers, 2-4-28, Hatabu, Shimonoseki, Yanmaguchi 751-0828. T: 0832524517, F: 0832540039, E: fukuda@fukken.co.jp

Fukue, M., University Professor, Tokai University, 3-20-1, Orido, Shimizu 424-8610. T: 0543370921, F: 0543349768, E: fukue@scc.u-tokai.ac.jp

Fukuhara, T., 2-21, Higashi Koraibashi, Chuo-ku, Osaka 540. T: 069411658, F: 069416264

Fukui, M., Dr, University Professor, Yamamotodai, 3-14-14, Takarazuka, Hyogo 665-0885. T: 0797881171, E: m-fukui@otemae.ac.jp

Fukui, S., Geotechnical Engineer, Osaka Municipal Government, 1-2-2-700, Umeda, Kita-ku, Osaka 530. T: 062089835, F: 063431570

Fukui, Y., Civil Engineer, Kansai Electric Power Co. Inc., 40 Okayamacho, Wakayam-Shi, Wakayama. T: 07059374251, F: 0734630639, E: k423410@kepco.co.jp

Fukumoto, T., Dr, University Professor, Dept. of Civil Engineering, Ritsumeikan University, 1-1-1, Nojihigashi, Kusatsu, Shiga 525. T: 0775661111, F: 0775612667

Fukuoka, M., Geotechnical Engineer, 5-15-12, Shimo, Kita-ku, Tokyo 115-0042. T: 0339014747, F: 0339032217

Fukushima, S., Dr, Geotechnical Engineer, Fujita Corporation, Ohdana-cho 74, Kohoku-ku, Yokohama 223. T: 0455913911, F: 0455925816

Fukushima, S., 1818-2-203, Takamatsu-cho, Takamatsu 761-01. T: 0878440583, F: 0878440583

Fukutake, K., Doctor Engineer, Shimizu Corporation, Fukoku-Seimei Bldg.27F, 2-2-2, Uchisaiwai-cho, Chiyoda-ku, Tokyo 100-0011. T: 0335088101, F: 0335082196, E: fukutake@ori.shimz.co.jp

Fukuwaka, M., Consulting Engineer, Kawasaki Steel Corp., Hibiya Kokusai Bldg.18F, 2-3, Uchisaiwaicho 2-chome, Chiyoda-ku, Tokyo 100. T: 0335974667, F: 0335974505

Fukuzumi, H., Geotechnical Engineer, Fudo Construction Co., Ltd., 2-16, 4-chome Hirano-machi, Chuo-ku, Osaka 541. T: 062019211, F: 062011130

Furukawa, K., Consulting Engineer, Nippon Koei Co., Ltd., 5-4, Koujimachi, Chiyoda-ku, Tokyo 102-8539. T: 0332388352, F: 0332388230, E: a3300@n-koei.co.jp

Furuya, T., Dr, Geotechnical Engineer, National Research Institute of Agricultural Engineering, 1-2, Kannondai 2-chome, Tsukuba-shi, Ibaraki 305. T: 0298387670, F: 0298387609

Futaki, M., Dr, Director of Production Engineering Dept., Ministry of Land, Infrastructure and Transport, Tachihara 1, Oho-machi, Tsukuba, Ibaraki 305. T: 0298646653, F: 0298646774, E: futaki@kenken.go.jp

Fuyuki, M., Associate Professor, Dept. of Civil Engineering, Tokai University, 1117 Kitakaname, Hiratsuka-shi, Kanagawa 259-12. T: 0463581211, F: 0463587184

Gibo, S., Dr, University Professor, University of the Ryukyus, 1 Senbaru, Nishihara, Okinawa 903-0213. T/F: 0988958787, E: gibo@eve.u-ryukyu.ac.jp

Gose, S., Consulting Engineer, Katahira Engineers International, Taiko Bldg., 4-2-16, Ginza, Chuo-ku, Tokyo 104-0061. T: 0335634053, E: Shingose@aol.com

Goto, K., Dr, Professor, Dept. of Civil Engineering, Nagasaki University, 1-14, Bunkyo-machi, Nagasaki 852. T: 0958471111, F: 0958483624

Goto, M., General Manager of Kansai Branch, Kiso-Jiban Consultants Co., Ltd., 1-11-14, Awaza, Nishi-ku, Osaka 550-0011. T: 0665361591, F: 0665361503, E: goto.masaaki@kiso.co.jp

Goto, S., Director Executive Specialist, Tokyo Gas Co., Ltd., 1-5-20, Kaigan, Minato-ku, Tokyo 105-8527. T: 0354007578, F: 0335788365, E: s_gotou@tokyo-gas.co.jp

Goto, S., Dr, University Professor, Dept. of Civil and Environmental Engineering, Yamanashi University, 4-3-11, Takeda, Kofu-shi, Yamanashi 400-8511. T: 0552208526, F: 0552208527, E: goto@ccn.yamanashi.ac.jp

Goto, S., Mr, Chief Geotechnical Research Engineer, Shimizu Institute of Technology, 3-4-17, Etchujima, Koto-ku, Tokyo 135-8530. T: 0338205245, F: 0338205489, E: goto@sit.shimz.co.jp

Goto, S., Dr, Emeritus Professor, Waseda University, (Home address) Otsuka, 5-33-11, Bunkyo-ku 112. T: 0339410796

Goto, T., Consulting Engineer, Chuken Consultant Co., Ltd., 7-1-55, Minamiokajima, Taisho-ku, Osaka 551-0021. T: 0665562380, F: 0665562389, E: goto@ccc.soc.co.jp

Gurung, S.B., Dr, Dept. of Civil and Environmental Engineering, Hiroshima University, 1-4, Kagamiyama 1-chome, Higashi-Hiroshima 739. T/F: 0824247785, E: sukh@ue.jpc-uc.ac.jp

Hachisu, S., Dr, College Professor, Maebashi City College of Technology, 460 Kamisadori-machi, Maebashi, Gumma 371. T: 0272650111

Hachiya, Y., Dr, Laboratory Chief, Port and Harbour Research Institute, Ministry of Land, Infrastructure and Transport, 1-1, Nagase 3, Yokosuka 239-0826. T: 0468445026, F: 0468444471, E: hachiya@cc.phri.go.jp

Hada, M., Consulting Engineer, Technical Research Institute of Fujita Corporation, 74 Ohdana-cho, Kohoku-ku, Yokohama 223. T: 0455913944, F: 0455925816, E: hada@giken.fujita.co.jp

Hada, M., 6-5-9, Tamadaira, Hino 191

Haga, Y., Dr, University Professor, Fukuyama University, Higashimura-cho, Fukuyama-shi, Fukuyama 729-0292. T: 0849362111, F: 0849362132

Hagino, Y., Geotechnical Engineeer, Fudo Construction Co., Ltd., 2-1, Taito 1-chome, Taito-ku, Tokyo 110-0016. T: 0338376140, F: 0338376024, E: hagino@fudo.co.jp

Hagiwara, T., Geotechnical Engineer, Nishimatsu Construction Co., Ltd., 2570-4, Shimotsuruma, Yamato, Kanagawa 242-8520. T: 0462857101, F: 0462857104, E: hagiwara@ri.nishimatsu.co.jp

Hamajima, R., 600, Sakae, Urawa 338

Hamasato, S., Geotechnical Engineer, Fudo Construction Co., Ltd., 10-40, Oosu 4-chome, Naka-ku, Nagoya 460. T: 0522617501, F: 0522615139

Hanai, T., Dr, Associate Professor, Dept. of Architecture, Nagasaki Institute of Applied Science, 536 Abamachi, Nagasaki 581-01. T: 0958393111

Hanamura, T., University Professor, Okayama University, Dept. of Environmental and Civil Engineering, 3-1-1, Tsushima-Naka, Okayama 700-8530. T/F: 0862518152, E: hanamura@cc.okayama-u.ac.jp

Hanazato, T., Dr, Senior Research Engineer, Tajimi Engineering Services Ltd., Tachibana Shinjuku Bldg.3F, 3-2-26, Nishishinjuku, Shinjuku-ku, Tokyo 160. T: 0333458431, F: 0333458434

Hanzawa, H., Soil Laboratory, Toa Harbor Works Co., Ltd., Anzen-cho, 1-3, Tsurumi-ku, Yokohama 230. T: 0455033741, F: 0455021206

Hara, H., Assistant, Ryukyu University, 1 Senbaru, Nishihara, Okinawa 903-01. T: 0988958672, F: 0988958672, E: soil@tec.u-ryukyu.ac.jp

Hara, K., Geotechnical Engineer, 2-25-12, Daishin, Kooriyama, Fukushima 963-8852. T/F: 0249355958, E: khararose.ocn.ne.jp

Hara, T., Consulting Engineer, CTI Engineering Co., Ltd., 9th Chuo Bldg., 4-9-11, Honcho Nihonbashi, Chuo-ku, Tokyo

103-8430. T: 0336680451, F: 0356951885, E: t-hara@ctie.co.jp

Harada, K., Geotechnical Engineer, Fudo Construction Co., Ltd., 1-2-1, Taito, Taito-ku, Tokyo 110-0016. T: 0338376034, F: 0338376158, E: h-kenji@fudo.co.jp

Harumoto, S., Professor, Takamatsu National College of Technology, 355 Chokushi-cho, Takamatsu 761

Hasegawa, A., Dr, University Professor, Dept. of Civil Eng., Hachinohe Institute of Technology, 88-1, Ohbiraki, Myo, Hachinohe 031-8501. T: 0178258075, F: 0178250722, E: hasegawa@hi-tech.ac.jp

Hasegawa, H., Geosurvey Engineer, Chiken Co., Ltd., Wakitahon-cho, 11-27, Kawagoe, Saitama 350-11. T: 0492456419, F: 0492456432

Hasegawa, S., 2-19-50-2-214, Izumi, Okegawa 363

Hasegawa, T., Dr, University Professor, Div. of Environmental Science & Technology, Kyoto University Graduate School, Kitashirakawa, Sakyo-ku, Kyoto 606. T: 0757536151, F: 0757536346, E: hasegawa@emeu.kais.kyoto-u.ac.jp

Hashiguchi, K., Dept. Agr. Eng., Kyushu University, Hakozaki, 6-10-1, Higashi-ku, Fukuoka 812-8584. T: 0926422927, F: 0926422932, E: khashi@agr.kyushu-u.ac.jp

Hashimoto, M., Engineer, Kawasaki Steel Corp., Construction Materials Center, Hibiya Kokusai Bldg., 2-3, Uchisaiwai-cho 2-chome, Chiyoda-ku, Tokyo 100. T: 0335974518, F: 0335974948

Hashimoto, N., 1-39-73-208, Sakae-cho, Higashimurayama 189

Hashimoto, O., Manager, Kawasaki Steel Corp., Dojima-Avanza Bldg.10F, 1-6-20, Dojima, Kita-ku, Osaka 530-8353. T: 0663420738, F: 0663420724, E: o-hashimoto@kawasaki-steel.co.jp

Hashimoto, S., Kumagai Gumi Co., Ltd

Hashimoto, T., Geotechnical Engineer, Geo-Research Institute, 4-3-2, Itachibori, Nishi-ku, Osaka 550-0012. T: 0665392977, F: 0665786256, E: hasimoto@geor.or.jp

Hatanaka, M., University Professor, Chiba Institute of Technology, Dept. of Architecture, 17-1, Tsudanuma 2-chome, Narashino, Chiba 275-0016. T/F: 0474780479, E: munenori.hatanaka@pf.it-chiba.ac.jp

Haya, H., 1-45-4-4-405-C-201, Hikari-cho, Kokubunji 185

Hayakawa, K., University Professor, 1-1-1, Noji-higashi, Kusatsu, Shiga 525-8577. T/F: 0775612789, E: kiyoshi@se.ritsumei.ac.jp

Hayashi, I., Consulting Engineer, Pacific Consultants Co., Ltd., Railway Dept., Daiichi Seimei Bldg.25F, 2-7-1, Nishishinjuku, Shinjuku-ku, Tokyo 163-0730. T: 03334 40583, F: 0333440805, E: Ichirou.Hayashi@tk.pacific.co.jp

Hayashi, K., Consulting Engineer, Chuo Fukken Consultants Co., Ltd., 1-8-29, Nishimiyahara, Yodogawa-ku, Osaka 532. T: 063931190, F: 063931146, E: hayashik@osk.threewebnet.or.jp

Hayashi, M., Research Engineer, NKK Corporation, 1-1, Minamiwatarida-cho, Kawasaki-ku, Kawasaki 210. T: 0443226221, F: 0443226519

Hayashi, M., Design Engineer, Honshu-Shikoku Bridge Authority, 5-1-5, Toranomon, Minato-ku, Tokyo 105. T: 0334347281, F: 0335789298

Hayashi, M., Honorary Research Advisor, Central Research Institute of Electric Power Industry, Ootemachi Bldg.7F, 1-6-1, Ootemachi, Chiyoda-ku, Tokyo 100-8126. T/F: 0471826612, E: hayashim@amy.hi-ho.ne.jp

Hayashi, N., Constru. Geotechnical Engineer, 1534-1, Yonku-cho, Nishinasuno-machi, Nasu-gun, Tochigi 329-2746. T: 0287392116, F: 0287392133, E: Norio.Hayashi@mail.penta-ocean.co.jp

Hayashi, S., Dr, Professor, Maebashi City College of Technology, 460 Kamisadori-machi, Maebashi, Gumma 371. T: 0272650111, F: 0272653837

Hayashi, S., Dr, Professor, Institute of Lowland Technology, Saga University, Honjo-machi 1, Saga 840. T: 0952288582, F: 0952288189, E: hayashis@cc.saga-u.ac.jp

Hayashi, Y., Geotechnical Engineer, Ohta Geo Research Co., Ltd., 3-19-5, Kamisakabe, Amagasaki-shi, Hyogo 661. T: 064948939, F: 064948938, E: ohta-gz@mars.dtinet.or.jp

Hayashi, Y., University Research Associate, Kumamoto University, 2-31-1, Kurokami, Kumamoto 860-8555. T/F: 0963423550, E: yhayashi@gpo.kumamoto-u.ac.jp

Hazarika, H., University Professor, Dept. of Civil Engineering, Kyushu Sangyo University, 2-3-1, Matsukadai, Higashi-ku, Fukuoka 813-8503. T: 0926735050, F: 0926735699, E: hazarika@ip.kyusan-u.ac.jp

Hibino, S., Geotechnical Engineer, Tenox Corporation, 13-7, Akasaka 6-chome, Minato-ku, Tokyo 107. T: 0335825168, F: 0335824714

Higaki, K., Geotechnical Engineer, Taisei Corporation, 344-1, Nase-cho, Totsuka-ku, Yokohama 245-0051. T: 0458147236, F: 0458147253, E: kanji.higaki@sakura.taisei.co.jp

Higashi, S., 3-4-9-301, Sengendai-nishi, Koshigaya 343

Higuchi, Y., Geotechnical Engineer, Taisei Corporation, 344-1, Nase-cho, Totsuka-ku, Yokohama 245-0051. T: 0458147217, F: 0458147257, E: yuichi.higuchi@sakura.taisei.co.jp

Hira, M., University Assistant Professor, Kagoshima University, 21-24, Korimoto 1, Kagoshima 890-0065. T/F: 0992858690, E: hira@bio2.agri.kagoshima-u.ac.jp

Hirai, Y., Geotechnical Engineer, Takenaka Corporation, 5-1, 1-chome Ohtsuka, Inzai-shi, Chiba 270-13. T: 0476471700, F: 0476473080

Hirakawa, S., Dr, University Professor, Dept. of Civil Engineering, Fukuyama University, 1 Sanzo, Gakuen-cho, Fukuyama 729-02. T: 0849362111, F: 0849362023

Hirama, K., General Manager, Obayashi Corporation, Technical Research Institute, Shimokiyoto, 4-640, Kiyose, Tokyo 204. T: 0424950911, F: 0424950903

Hirano, K., Researcher, 5724 Taumi, Omi-machi, Nishikubikigun, Niigata 949-03. T: 0255624461

Hirano, Y., Senior Managing Director, Itogumi Construction Co., Ltd., 4-1, Kita-Shijo-Nishi, Chuo-ku, Sapporo 060-8554. T: 0112616111, F: 0112225398

Hirao, K., Assistant Professor, Dept. of Civil Engineering, Nishinippon Institute of Technology, 1-11, Aratsu, Kanda, Miyako-gun, Fukuoka 800-0394. T: 0930231491, F: 09302 47900, E: hirao@nishitech.ac.jp

Hirata, T., Dr, College Professor, Kagoshima National College of Technology, 1460-1, Sinko, Hayato-cho, Aira-gun, Kagoshima 899-5193. T: 0995429121, F: 0995432584, E: hirata@kagoshima-ct.ac.jp

Hirata, T., University Professor, Dept. of Environmental Systems, Wakayama University, 930 Sakaedani, Wakayama 640-8510. T: 0734578372, F: 0734578373, E: hirata@sys.wakayama-u.ac.jp

Hirayama, H., Dr, Research Geotechnical Engineer, Geotop Corporation, 2-1-10, Koraibashi, Chuo-ku, Osaka 541. T: 062260871, F: 062260992

Hirayama, M., Consulting Engineer, Taisei Foundation Design & Research Co., Ltd., Taisei Bldg., 5-6, Sendagaya 4-chome, Shibuya-ku, Tokyo 151. T: 0334784111, F: 0334784140

Hiro-oka, A., Dr, University Professor, Dept. of Civil Engineering, Kyushu Institute of Technology, 1-1, Sensui-cho, Tobata-ku, Kitakyushu-shi, Fukuoka 804. T: 0938843113, F: 0938843113, E: ahirooka@civil.kyutech.ac.jp

Hisanaga, K., Consulting Engineer, Geo-Tech Co., Ltd., Miyauchi, 4244-1, Hatukaichi, Hiroshima. T: 0829398316, F: 0829394796, E: rock@dear.ne.jp

Hisatake, M., Dr, University Professor, Civil Engineering, Kinki University, 3-4-1, Kowakae, Higashi-Osaka, Osaka 577-8502. T: 067212332, F: 0729955192, E: hisatake@civileng.kindai.ac.jp

Hokugo, H., Dr, Professor, Nippon Institute of Technology, 4-1, Gakuendai, Miyashiro-machi, Minami-Saitama-gun, Saitama 345. T: 0480344111, F: 0480342941

Honda, M., Consulting Engineer, Nikken Sekkei, Civil Engineering, 4-6-2, Koraibashi, Chuo-ku, Osaka 541-8528. T: 0662033694, F: 0662271534, E: hondam@nikken.co.jp

Honda, R., University Research Associate, Disaster Prevention Research Institute, Kyoto University, Gokasho, Uji, Kyoto 611-0011. T: 0774384067, F: 0774384070, E: honda@catfish.dpri.kyoto-u.ac.jp

Honda, S., Geotechnical Engineer, Nikken Soil Research Ltd. 4-5-22, Doshomachi, Chuo-ku, Osaka 541-0045. T: 0662026093, F: 0662027090, E: hondas@nikken.co.jp

Honda, T., University Research Associate, University of Tokyo, Dept. of Civil Engineering, University of Tokyo, 7-3-1, Hongo, Bunkyo-ku, Tokyo 113-8656. T: 0358416137, F: 0358418504, E: honda@geot.t.u-tokyo.ac.jp

Hong, Z., Research Engineer, Soil Mechanics Laboratory, Port and Harbour Research Institute, Nagase, Yokosuka 239-0826. T: 0468445021, F: 0468444577, E: hong@cc.phri.go.jp

Hongo, H., Alpusu Testing Laboratory, Mita, 1828-4, Horigane, Minamiazumi, Nagano 399-82. T: 0263725222

Honjo, Y., University Professor, Dept. of Civil Eng., Gifu University, 1-1, Yanagido, Gifu 501-1193. T: 0582932435, F: 0582301891, E: honjo@cc.gifu-u.ac.jp

Horiguchi, T., President, A&G Horiguchi Consultant Office, 4-45-21-110, Izumi, Suginami, Tokyo 168-0063. T/F: 0333136283, E: tak-hori@aioros.ocn.ne.jp

Horii, H., Professor, The University of Tokyo, Dept. of Civil Engineering, 7-3-1, Hongo, Bunkyo-ku, Tokyo 113-8656. T: 0358416090, F: 0358417496, E: horii@ohriki.t.u-tokyo.ac.jp

Horii, K., Geotechnical Engineer, Chuo Kaihatsu Corporation, 13-5, Nishiwasada 3-chome, Shinjuku-ku, Tokyo 160. T: 0332083111

Horii, N., Senior Research Officer, Construction Safety Research Division, National Institute of Industrial Safety, Ministry of Health Labour and Welfare, 1-4-6, Umezono, Kiyose, Tokyo 204-0024. T: 0424914512, F: 0424917846, E: horii@anken.go.jp

Horikoshi, K., Ph.D., Geotechnical Engineer, Taisei Corporation, 344-1, Nase-cho, Totsuka-ku, Yokohama 245-0051. T: 0458147236, F: 0458147253, E: kenichi.horikoshi@sakura.taisei.co.jp

Horiuchi, S., Dr, Senior Research Engineer, Institute of Technology, Shimizu Corporation, 3-4-17, Etchujima, Koto-ku, Tokyo 135-8530. T: 0338205437, F: 0338205955, E: horiuchi@sit.shimz.co.jp

Horiuchi, T., Dr, University Professor, Meijo University, 1-501, Siogamaguchi, Tenpaku-ku, Nagoya 468-8502. T: 0528321151, F: 0528321179, E: horiuchi@meijo-u.ac.jp

Hosaka, Y., University Research Associate, Dept. of Civil Engineering and Architecture, Niigata University, 8050 Ikarashi-2nocho, Niigata-shi, Niigata 950-2181. T: 0252627032, F: 0252627021, E: hosaka@eng.niigata-u.ac.jp

Hoshikawa, T., University Research Fellow, Nagoya Institute of Technology, Gokiso-cho, Showa-ku, Nagoya, Aichi 466-8555. T/F: 0527357157, E: hoshi@tuti1.ace.nitech.ac.jp

Hoshino, Y., Senior Researcher, 3523-1-102, Minami-ogishima, Koshigaya 343. T: 0489791592, E: fuhd4522@mb.infoweb.or.jp

Hosoi, T., General Manager, Nishimatsu Construction Co., Ltd., Toranomon, 1-20-10, Minato-ku, Tokyo 105. T: 0335027575, F: 0335020228

Huang, Y., Consulting Engineer, Kiso-jiban Consultants Co., Ltd., 2-14-1, Ishikawa-cho, Ohta-ku, Tokyo 145-0061. T: 0337276158, F: 0337276247, E: huang@kiso.co.jp

Hyodo, M., Dr, University Professor, Dept. of Civil Engineering, Yamaguchi University, Tokiwadai, 2-16-1, Ube, Yamaguchi 755-8611. T: 0836859343, F: 0836859301, E: hyodo@po.cc.yamaguchi-u.ac.jp

Iai, S., Dr, Laboratory Director, Port and Harbour Research Institute, Nagase, 3-1-1, Yokosuka 239-0826. T: 0468445030, F: 0468440839, E: iai@cc.phri.go.jp

Ibaraki, T., Dr, University Professor, Dept. of Civil Engineering, Chuo University, Kasuga, 1-13, Bunkyo-ku, Tokyo. T: 0339554380

Ichihara, M., Professor Emeritus, Kamenoi, 2-25, Meito-ku, Nagoya 465. T: 0527020847

Ichii, K., Research Engineer, Port & Harbour Research Institute, 3-1-1, Nagase, Yokosuka 239-0831. T: 0468445028, F: 0468440839, E: ichiikoji@aol.com

Ichikawa, K., Dr, Managing Director, Chiken Consultants Co., Ltd., Soken Bldg., 11-27, Wakitahoncho, Kawagoeshi, Saitama 350-1123. T: 0492456800, F: 0492456442, E: ichikawa@chikencon.co.jp

Ichikawa, T., Assistant Professor, Dept. of Civil Engineering, Tokai University, Akama Munakata, Fukuoka 811-41. T: 0940323311

Ichikawa, Y., University Professor, Division of Environmental Engineering and Architecture, Graduate School of Environmental Studies, Nagoya University, Nagoya 464-8603. T: 0527893829, F: 0527891176, E: YIchikawa@nucc.cc.nagoya-u.ac.jp

Idoguchi, K., Wakashi-ryo, 1-6-39, Hagurazaki, Izumisano 598

Igarashi, H., Geotechnical Engineer, Kajima Corporation, 19-1, Tobitakyu 2-chome, Chofu-shi, Tokyo 182. T: 0424897069, F: 0424897034

Igarashi, M., Consulting Engineer, 3-1-2, Ikebukuro, Toshima-ku, Tokyo 171-0014. T: 0339866186, F: 0339869408, E: M.Igarashi@diaconsult.co.jp

Iida, T., Consulting Engineer, Nikken Sekkei Ltd., 1-4-27, Koraku, Bunkyo-ku, Tokyo 112. T: 0338133361, F: 0338170517

Iizuka, A., University Associate Professor, Dept. of Civil Engineering, Kobe University, 1-1, Rokkodai, Nada-ku, Kobe 657-8501. T/F: 0788036029, E: iizuka@kobe-u.ac.jp

Ikeda, M., Consulting Engineer, Tokyu Construction Co., Ltd., 1-16-14, Shibuya, Shibuya-ku, Tokyo 150-8340. T: 0354665824, F: 0337977547, E: m_ikeda@hd.tokyu-cnst.co.jp

Ikegami, S., Research Associate, Dept. of Civil and Environmental Engineering, Hiroshima University, 1-4-1, Kagamiyama Higashi-Hiroshima, Hiroshima 724. T/F: 0824247783

Ikeuchi, K., 1-63, Tosaki, Ageo 362

Ikeura, Prof I., Tomakomai National College of Tech., 422 Nishikioka, Tomakamai. T: 0144670211, F: 0144670814

Ikuta, Y., Professional Engineer, 3309 Naka, 1-10-101, Kouyou-cho, Higashinada-ku, Kobe 658-0032

Imai, G., Dr, University Professor, Yokohama National University, Tokiwadai, Hodogaya-ku, Yokohama 240-8501. T: 0453394037, F: 0453311707, E: imaigo@cvg.ynu.ac.jp

Imaizumi, S., Dr, University Professor, Graduate School of Engineering, Utsunomiya University, 7-1-2, Yoto, Utsunomiya, Tochigi 321-8585. T/F: 0286896217, E: imaizumi@cc.utsunomiya-u.ac.jp

Imamura, S., Geo-Environmental Engineer, Technology Center, Taisei Corporation, 344-1, Nasemachi, Totsuka-ku, Yokohama 245-0051. T: 0458147217, F: 0458147257, E: satoshi.imamura@sakura.taisei.co.jp

Imamura, S., Geotechnical Engineer, Nishimatsu Construction Co., Ltd., Aikawa Technical Research Institute, 4054 Nakatsu, Aikawa-cho, Aikoh-gun, Kanagawa 243-0303. T: 0462857101, F: 0462857104, E: imamura@ri.nishimatsu.co.jp

Imamura, Y., Dr, Dept. of Civil Engineering, Science University of Tokyo, 2641 Yamazaki, Noda, Chiba 278. T: 0471241501, F: 0471239766

Imanishi, H., Director of Kyushu Office, Geo-Research Institute, Daiichi-hoki Bldg.5F., 3-5-1, Otemon, Chuo-ku, Fukuoka 810-0074. T: 0927628650, F: 0927263877, E: imanishi@kyushu.geor.or.jp

Imano, M., Associate Professor, Dept. of Civil Engineering, Nihon University, 2-1, Izumicho 1-chome, Narashino-shi, Chiba 275. T: 0474742455, F: 0474742449

Inaba, T., Managing Director, Yamato-shi, shimoturuma, 2570-4, Kanagawa 242-8520. T: 0462756793, F: 0462756796, E: inaba@ri.nishimatsu.co.jp

Inada, M., Dr, Prof. Emeritus, 9-4, Chigusadai, Midori-ku, Yokohama 227. T: 0459731207

Inagaki, H., Geotechnical Engineer, Kankyo Chisitsu Co., Ltd., Ishihara Bldg.5F, 1-9-7, Watarida, Kawasaki-ku, Kawasaki 210. T: 0443554313, F: 0443553809, E: inagaki@kankyo-c.com

Ineoka, S., 3-2-6, Sumiyoshi-cho, Shibata 957

Inoue, K., Managing Director, Kansai Soil Research Center, 1-3-3, Higashibefu, Settu, Osaka 566-0042. T: 0668278833, F: 0668292257, E: inoue@ks-dositu.or.jp

Inoue, T., Dr, Consulting Engineer, Fukken Co., Ltd., 2-10-11, Hikarimachi, Higashi-ku, Hiroshima 732. T: 0825061826, F: 0825061894

Iseda, T., Dr, University Emeritus Professor, Inage Skytown 4-407, 488-1, Konakadai-cho, Inage-ku, Chiba, F: 0432842175

Isemoto, N., Geotechnical Engineer, Toda Corporation, Research & Development Department, Shin Yaesu Bldg., Kyobashi, 1-7-1, Chuo-ku, Tokyo 104. T: 0335626111, F: 0332067180

Isemura, K., Consulting Engineer, Nikken Sekkei, 2-33-14-407, Senkawa, Toshima-ku, Tokyo 171-0041. T/F: 0359665843, E: isemura@nifty.com

Ishibashi, H., Consulting Engineer, Suimon Research Inc., Isono Bldg., 3-11-19, Matsunami, Chuo-ku, Chiba 260-0044. T: 0432567121, F: 0432568834, E: h-sr@courante. plala.or.jp

Ishida, H., Dr, 41-36, Nishiaoyama 3-chome, Morioka 020-0132. T: 0196476147

Ishida, M., Consulting Engineer, Kawatetsu Engineering, Ltd., 1-1-5, Koraku, Bunkyo-ku, Tokyo 112-0004. T: 0338170675, F: 0338170178, E: ishida@kel. kawatetsu.ne.jp

Ishida, T., Consulting Engineer, 19-17, 10-jo 2-chome, Fushiko, Higashi-ku, Sapporo 065. T: 0117820257

Ishida, T., Dr, University Associate Professor, Toyo University, 2100 Kujirai, Kawagoe-shi, Saitama 350-8585. T/F: 0492391409, E: ishida@eng.toyo.ac.jp

Ishiguro, T., 2-13-8-102, Hikawadai, Nerima-ku, Tokyo 179

Ishihara, A., University Professor, Science Univ. of Tokyo, 2641 Yamazaki, Nada-shi, Chiba. T: 0471241501 Ext. 4006, F: 0471239766

Ishii, T., Consulting Engineer, Nikken Sekkei Ltd., 2-1-3, Koraku, Bunkyo-ku, Tokyo 112-8565. T: 0338133361, F: 0338170517, E: ishii@nikken.co.jp

Ishii, Y., Chief Research Engineer, Obayashi Corporation, 640 Shimokiyoto 4-chome, Kiyose-shi, Tokyo 204-8558. T: 0424951024, F: 0424959401, E: ishii@tri.obayashi.co.jp

Ishikawa, K., Dr, Consulting Engineer, Chuo Kaihatsu Corporation, Osaka Br., 3-34-12, Tarumicho, Suita, Osaka 564. T: 063863691, F: 063865082

Ishikawa, Y., Geotechnical Engineer, Institute of Technology, Shimizu Corporation, 3-4-17, Etchujima, Koto-ku, Tokyo 135-8530. T: 0338205570, F: 0336437260, E: yutaka@sit. shimz.co.jp

Ishimaru, S., Structural Engineer, Hokkaido Nikken Sekkei, Kita-1 Higashi-2, Chuo-ku, Sapporo 060. T: 0112419438, F: 0112417598

Ishimoto, H., Research Engineer, NTT. Telecommunication Field Systems R&D Center, 1-7-1, Hanabatake, Tsukuba, Ibaraki 305. T: 0298522555, F: 0298522593

Ishiyama, K., Consulting Engineer, KI Construction Consultant Co., Ltd., 1-7-22, Kaminomiya, Turumi-ku, Yokohama. T: 0455721659/0333601805, F: 0333608447

Ishizaki, H., Dr, Geotechnical Engineer, Sumitomo-osaka Cement Co., Ltd., 7-1-55, Minamiokajima, Taisho-ku, Osaka 551. T: 065562260

Ishizaki, T., Head of Physics Section, Tokyo National Research Institute of Cultural Properties, 13-43, Ueno-Park, Taito-ku, Tokyo 110-8713. T: 0338234880, F: 0338223247, E: ishizaki@tobunken.go.jp

Isobe, K., Consulting Engineer, Raito Kogyo Co., Ltd., 4-2-35, Kudan-Kita, Chiyoda-ku, Tokyo 102-8236. T: 0332652458, F: 0332652678, E: kisobe@raito.co.jp

Isoda, T., 2-15-102, Nirenokidai, Asahigaoka-machi, Hanamigawa-ku, Chiba 262

Isozaki, S., Consulting Engineer, Docon Co., Ltd., 1-5-4-1, Atsubetsu-chuo, Atsubetsu-ku, Sapporo 004-8585. T: 0118011570, F: 0118011571, E: si1273@mb.docon. co.jp

Itabashi, K., Dr, University Professor, Meijo University, Shiogamaguchi, 1-501, Tenpaku-ku, Nagoya 468-8502. T: 0528321151, F: 0528321178, E: itabashi@meijo-u.ac.jp

Itakura, M., 6-21-13-201, Kikuna, Kohoku-ku, Yokohama 222. T: 0454331583

Itaya, K., Manager of Technical Department, Chiyoda Civil Engineering Co., Ltd., 940 Kamiko-cho, Ōmiya-shi, Saitama 331-0853. T: 0486425252, F: 0486430866, E: wing-x@pop16.odn.ne.jp

Ito, A., Dr, University Professor, Kansai University, 3-35, Yamate-cho 3, Suita, Osaka 564-8680. T: 0663680785, F: 0663397720, E: ito@ipcku.kansai-u.ac.jp

Ito, H., Dr, Research Fellow, 1646 Abiko, Abiko-shi, Chiba 270-11. T: 0471821181, F: 0471833182

Ito, K., Hanenodai, 800-331, Haneno Asa, Tonemachi, Kita-Soumagun, Ibaraki 270-12. T: 0297682937

Ito, M., Dr, General Manager, Civil Engineering Dept., Maeda Corporation, 10-26, Fujimi 2-chome, Chiyoda-ku, Tokyo 102. T: 0352769420, F: 0352769432, E: itoma@ jcity.maeda.co.jp

Ito, T., Dr, Professor, Akita National College of Technology, 1 Bunkyo, Iijima, Akita 011-8511. T/F: 0188476077, E: taito@ipc.akita-nct.ac.jp

Ito, T., University Professor Emeritus, (Home Address) 11-21, Suehiro-cho, Neyagawa-shi, Osaka 572-0009. T: 0728312150

Ito, T., University Professor, Tohoku Institute of Technology, 35-1, Kasumi-cho, Yagiyama, Taihaku-ku, Sendai 982-8577. T: 0222291151, Ext. 440, F: 0222298393, E: tito+@titan. tohtech.ac.jp

Ito, T., Professor, Gifu National College of Technology, Shinsei-cho, Motosu-gun, Gifu 501-04. T: 0583241101, F: 0583232709

Ito, Y., Dr, Assistant Manager, Kumagai Gumi Co., Ltd., Nuclear & Energy Dept., 2-1, Tsukudo-cho, Shinjuku-ku, Tokyo 162. T: 0332602111, F: 0352619350

Ito, Y., Associate Professor, Dept. of Civil Engineering, Setsunan University, 17-8, Ikeda-Nakamachi, Neyagawa, Osaka 572-8508. T/F: 0728399701, E: cito@civ.setsunan.ac.jp

Iwabe, T., Technical College Lecturer, Yatsushiro National College of Technology, 2627 Hirayama-shinmachi, Yatsushiro, Kumamoto 866-8501. T: 0965531332, F: 0965531349, E: iwabe@as.yatsushiro-nct.ac.jp

Iwamatsu, A., University Professor, Inst. Earth Sci., Fac. Sci., Kagoshima University, 1-21-35, Korimoto, Kagoshima 890. T: 0992858145, F: 0992594720

Iwamoto, H., Researcher, Takenaka Corporation, 5-1, 1-chome Ohtsuka, Inzai, Chiba 270-1356. T: 0476470065, F: 0476473080, E: iwamoto.hiroshi@takenaka.co.jp

Iwao, Y., Dr, Professor, Dept. of Civil Engineering, Faculty of Science & Engineering, Saga University, 1 Honjyo-machi, Saga 840. T: 0952245191, F: 0952298867

Iwasaki, K., Managing Director, Kiso-Jiban Consultants Co., Ltd., 4-13-25, Nagatsuka, Asaminami-ku, Hiroshima 731-0135. T: 0822387227, F: 0822387949, E: iwasaki. kimitoshi@kiso.co.jp

Iwasaki, T., Dr, President, CRL, Kyodo Bldg.4F, 1–18, Kanda Suda-cho, Chiyoda-ku, Tokyo 101-0041. T: 0332549481, F: 0332549448, E: iwasaki@crl.or.jp

Iwasaki, Y., Director, Geo-Research Institute, 4-3-2, Itachi-bori, Nishi-ku, Osaka 550-0012. T: 0665392976, F: 0665786255, E: iwasaki@geor.or.jp

Iwashita, T., Senior Research Engineer, Public Works Research Institute, Ministry of Land, Infrastructure and Transport, 1 Asahi, Tsukuba, Ibaraki 305-0804. T: 0298642211, F: 0298640164, E: iwashita@pwri.go.jp

Jiang, G.L., Dr, Senior Geotechnical Research Engineer, Integrated Geotechnology Institute, Fukugen Bldg.7F, 2-15-16, Akasaka, Minato-ku, Tokyo 107-0052. T: 0335823373, F: 0335823509, E: jiang@igi.co.jp

Jiang, J.C., Assistant Professor, Dept. of Civil Engineering, The University of Tokushima, 2-1, Minami-josanjima-cho, Tokushima 770-8506. T: 0886567346, F: 0886567347, E: jiang@ce.tokushima-u.ac.jp

Jiang, J.Q., University Professor, College of Civil Engineering and Architecture, Zhejiang University, Yuquan, Hangzhou 310027, China. T/F: 865718920983, E: dr,eamwh@mail. hz.zj.cn

Kadowaki, K., Consulting Engineer, Kokusai Kogyo Co., Ltd., 2 Rokuban-cho, Chiyoda-ku, Tokyo 102-0085.

T: 0332626221, F: 0332640657, E: kiyoshi_kadowaki@kkc.co.jp

Kaga, M., University Associate Professor, Dept. of Civil Engineering, Toyo Univ., 2100 Kujirai, Kawagoe-shi, Saitama 350. T: 0492391406, F: 0492314482, E: kaga@krc.eng.toyo.ac.jp

Kaide, T., Geotechnical Engineer, Minamiyama-cho, 3-95, Seto 489. T: 0561835404

Kaino, T., Managing Director, JR East Japan Consultants Co., Ltd., 2-2-6, Yoyogi, Shibuya-ku, Tokyo 151. T: 0353713371, F: 0353713374

Kakurai, M., Deputy General Manager, Takenaka Research & Development Institute, 5-1, 1-chome Ohtsuka, Inzai, Chiba 270-1395. T: 0476471700, F: 0476473080, E: kakurai.masaaki@takenaka.co.jp

Kamada, H., Consulting Engineer, 27-6, Hiraki-cho, Uji-shi, Kyoto 611. T: 0774448710

Kamao, S., University Lecturer, Nihon University, 1–8, Kanda-Surugadai, Chiyoda-ku, Tokyo 101-8308. T: 0332590667, F: 0332933319, E: kamao@civil.cst.nihon-u.ac.jp

Kamata, M., Dr, Vice President, Kosaka Giken Co., Ltd., 56-2, Kamiikarida, Ohaza, Naganawashiro, Havhinohe, Aomori 039-1103. T: 0178273444, F: 0178273445, E: kosaka05@infoaomori.ne.jp

Kamei, T., Dr, University Professor, Dept. of Geoscience, Shimane University, 1060 Nishikawatsu, Matsue, Shimane 690. T: 0852326460, F: 0852326469

Kamiura, M., University Professor, Hokkai Gakuen University, South 26 West 11, Chuo-ku, Sapporo 064-0926. T: 0118411161, F: 0115512951, E: kamiura@cvl.hokkai-s-u.ac.jp

Kamiya, K., Research Associate, Dept. of Civil Engineering, Gifu University, 1-1, Yanagido, Gifu 501-11. T: 0582932421, F: 0582301891, E: kkamiya@cc.gifu-u.ac.jp

Kamiya, M., Dr, University Professor, Hokkaido Institute of Technology, 4-1, 7-15, Maeda, Teine-ku, Sapporo 006-8585. T: 0116812161, F: 0116840522, E: kamiya@hit.ac.jp

Kamiyama, M., Dr, University Professor, Dept. of Civil Engineering, Tohoku Institute of Technology, 35-1, Yagiyama-Kasumicho, Sendai, Miyagi 982. T: 0222291151, F: 0222298393

Kamon, M., University Professor, Disaster Prevention Research Institute, Kyoto University, Gokasho, Uji, Kyoto 611-0011. T: 0774384090, F: 0774333521, E: kamon@geotech.dpri.kyoto-u.ac.jp

Kanatani, M., Geotechnical Engineer, Central Research Institute of Electric Power Industry, 1646 Abiko, Abiko-shi, Chiba 270-1194. T: 0471821181, F: 0471842941, E: kanatani@criepi.denken.or.jp

Kanda, Y., Geotechnical Engineer, Fudo Construction Co., Ltd., 4-2-16, Hiranomachi, Chuo-ku, Osaka 541. T: 062019210, F: 062011130

Kaneda, K., Research Associate, Nagoya University, Dept. of Civil Engineering, Furo-cho, Chikusa-ku, Nagoya 464-8603. T/F: 0527894624, E: kaneda@soil.genv.nagoya-u.ac.jp

Kaneko, O., Geotechnical Engineer, Toda Corporation, 4-6-1, Hacchobori, Chuo-ku, Tokyo 104. T: 0332067186, F: 0332067180

Kang, M.S., Dr, Research Engineer, Geotechnical Engineering Div., Port and Harbour Research Institute, 3-1-1, Nagase, Yokosuka 239-0826. T: 0468445021, F: 0468444577, E: mskang@ipc.phri.go.jp

Kani, Y., General Manager of Nagoya Branch, Nippon Concrete Industries Co., Ltd., 3-11-22, Meieki, Nakamura-ku, Nagoya 450-0002. T: 0525810666, F: 0525412530, E: kani@star.ncic.co.jp

Kano, K., Research Associate, Hiroshima University, 4-1, Kagamiyama 1-chome, Higashi-hiroshima, Hiroshima 739-8527. T/F: 0824247785, E: skano@hiroshima-u.ac.jp

Karasawa, M., 1366-66, Nakagami-cho, Akishima-shi, Tokyo 196

Karkee, M.B., University Professor, Akita Prefectural University, Dept. of Architecture and Environment System, Honjo, Akita 015-0055. T: 0184272047, F: 0184272086, E: karkee@akita-pu.co.jp

Karube, D., Prof., University Professor, Dept. of Civil Engineering, Kobe University, Rokkodai-cho, Nada-ku, Kobe 657. T: 0788031020, F: 0788825478

Kasai, Y., Chief, Soil and Foundation Section, Building Structural Engineering Dept., Kajima Technical Research Institute, 19-1, Tobitakyu 2-chome, Chofu-shi, Tokyo 182. T: 0424897097, F: 0424892020, E: kasai@katri.kajima.co.jp

Kasama, K., Research Associate, Graduate School of Engineering, Kyushu University, 6-10-1, Hakozaki, Higashi-ku, Fukuoka 812-8581. T/F: 0926424406, E: kasama@civil.kyushu-u.ac.jp

Kashiwagi, A., Engineering Advisor, Maeda Corporation, 2-10-26, Fujimi, Chiyoda-ku, Tokyo 102. T: 0352765214

Kasuda, K., Consulting Engineer, Kiso-Jiban Consultants Co., Ltd., Kanto Branch, Axis Bldg.6F, 3-22-6, Toyo, Koto-ku, Tokyo 135-0016. T: 0356326824, F: 0356326816, E: kasuda.kinichi@kiso.co.jp

Katagiri, M., Dr, Geotechnical Engineer, Nikken Sekkei Nakase Geotechnical Institute, 4-11-1, Minamikase, Saiwai-ku, Kawasaki 212-0055. T: 0445991151, F: 0445999444, E: katagiri@nikken.co.jp

Katakami, N., Consulting Engineer, Nikken Sekkei Nakase Geotechnical Institute, 4-11-1, Minamikase, Saiwai-ku, Kawasaki 212-0055. T: 0445991151, F: 0445999444, E: katakami@nikken.co.jp

Kataoka, K., Managing Director, Chuken Consultant Co., Ltd., 7-1-55, Minami Okajima, Taisho-ku, Osaka 551-0021. T: 0665562380, F: 0665562389, E: kataoka@ccc.soc.co.jp

Kato, M., Dr, University Professor, Tokyo Univ. of Agriculture and Technology, 3-5-8, Saiwai-cho, Fuchu, Tokyo 183. T: 0423675756, F: 0423665391, E: mkato@cc.tuat.ac.jp

Kato, S., Research Associate, The Graduate School of Science & Technology, Kobe University, Rokkodai-cho, Nada-ku, Kobe 657. T: 0788030147, F: 0788030147, E: 1skato1@icluna.kobe-u.ac.jp

Katsube, Y., Consulting Engineer, Every Plan Corporation Ltd., Asahimachi 489, Matsue-shi, Shimane 690. T: 0852552100, F: 0852552101, E: BYB02470@niftyserve.or.jp

Katsumata, M., Manager, Maeda Corporation, 1-39-16, Asahi-cho, Nerima-ku, Tokyo 179-8914. T: 0339772584, F: 0339772251, E: katumatm@jcity.maeda.co.jp

Katsumi, T., Associate Professor, Dept. of Civil Engineering, Ritsumeikan University, Kusatsu, Shiga 525-8577. T: 0775614987, F: 0775612667, E: tkatsumi@se.ritsumei.ac.jp

Katsura, Y., 2-482-22-D-104, Hasama-cho, Funabashi 274

Kawabata, S., University Research Associate, Hokkaido Institute of Technology, 4-1, 7-15, Maeda, Teine-ku, Sapporo, Hokkaido 006-8585. T: 0116812161, F: 0116840522, E: kawabata@hit.ac.jp

Kawabata, T., Dr, University Associate Professor, Kobe University, Faculty of Agriculture, Dept. of Agriculture and Environmental Engineering, 1-1, Rokkodai, Nada-ku, Kobe 657-8501. T/F: 0788035902, E: kawabata@eng.ans.kobe-u.ac.jp

Kawabe, K., Geotechnical Engineer, 202, 1-1-5, Hara-machida, Machida, Tokyo 194-0013. T: 0427396848, F: 0427376847, E: kawabekz@infoweb.ne.jp

Kawachi, T., University Professor, 384 Kohjiya, Tokorozawa, Saitama 359-1166

Kawaguchi, T., College of Technology, Research Associate, Hakodate National College of Technology, 14-1, Tokura-cho, Hakodate 042-8501. T/F: 0138596481, E: kawa@hakodate-ct.ac.jp

Kawai, K., University Research Associate, Dept. of Civil Engineering, Kobe University, 1-1, Rokkoudai-cho, Nada-ku, Kobe 657-0013. T: 0788036281, F: 0788036069, E: kkawai@kobe-u.ac.jp

Kawai, T., Research Engineer, Central Research Institute of Electric Power Industry (CRIEPI), 1646 Abiko, Abiko-shi, Chiba 270-1194. T: 0471821181, F: 0471842941, E: t-kawai@criepi.denken.or.jp

Kawai, Y., 15C4 Mitkorn Mansion, 153/1 Soi Mahadelkluang 1, Rajdamri Road, Bangkok 10330, Thailand. T/F: 6622551877, E: yoji@kt.rim.or.jp

Kawaida, M., Project Manager, Kameyama Construction Office, Japan Highway Public Corporation, 558 Nishimachi, Kameyama, Mie 519-0153. T: 0595835368, F: 0595834800, E: Minoru_Kawaida@gw.japan-highway.go.jp

Kawakami, H., Dr, Geotechnical Engineer, 1-37-17, Isemiya, Nagano 380-0958. T/F: 0262275601

Kawakami, T., University Associate Professor, 3-20-1, Orido, Shimizu, Shizuoka 424-8610. T: 0543340411, F: 0543349768, E: kawakami@scc.u-tokai.ac.jp

Kawama, I., Consulting Engineer, Nikken Sekkei Ltd., Kouraku, 2-1-2, Bunkyo-ku, Tokyo 112. T: 0338133361, F: 0338170517, E: kawama_cts@nics.nikken.co.jp

Kawamoto, K., University Research Associate, Saitama University, Shimo-Okubo 255, Urawa, Saitama 338-0825. T: 0488583572, F: 0488587374, E: kawamoto@dice.dr5w. saitama-u.ac.jp

Kawamoto, T., University Prof., Dept. of Civil Engineering, Aichi Institute of Technology, 1247 Yachigusa, Yagusa-cho, Toyota 470-03. T: 0565488121, F: 0565480277

Kawamura, H., Managing Director, Kaiyoho Kogyo Co., Ltd., 12 Funa-cho, Shinjuku-ku, Tokyo 160. T: 0333598792

Kawamura, K., Dr, University Professor, Dept. of Civil Engineering, Kanazawa Institute of Technology, Nonoichi-machi, Ishikawa-gun, Ishikawa 921-8501. T: 0762481100, Ext. 2401, F: 0762941480, E: kawamura@neptune. kanazawa-it.ac.jp

Kawamura, M., University Professor, Toyohashi University of Technology, 1-1, Tenpaku-cho, Toyohashi 441-8580. T: 0532446847, F: 0532446830, E: kawamura@acserv. tutrp.tut.ac.jp

Kawamura, M., University Professor, Nihon University, College of Industrial Technology, 1-2-1, Izumi-cho, Narashio, Chiba 275-8575. T: 0474742490, F: 0474742499, E: kawamura@arch.cit.nihon-u.ac.jp

Kawamura, S., Research Associate, Muroran Institute of Technology, 27-1, Mizumoto-cho, Muroran 050-8585. T: 0143465282, F: 0143465283, E: kawamura@news3. ce.muroran-it.ac.jp

Kawamura, T., University Research Associate, Shinshu University, Dept. of Civil Engineering, 4-17-1, Wakasato, Nagano 380-8553. T: 0262695289, F: 0262234480, E: t_kawa@gipwc.shinshu-u.ac.jp

Kawanabe, O., 2-3-3-503, Yokodai, Isogo-ku, Yokohama 235

Kawano, K., Geotechnical Engineer, Fudo Construction Ltd., Kamiyama Bldg.5F, 1, kita3-jyo nishi2-cho, Chuo-ku, Hokkaido 060-0003. T: 0112816771, F: 0112224727, E: kawano@fudo.co.jp

Kawasaki, K., Consulting Engineer, Taisei Co., 1-25-1, Nishi-Shinjuku, Shinjuku-ku, Tokyo 163-0606. T: 0353815206, F: 0333449476, E: koji.kawasaki@sakura.taisei.co.jp

Kawasaki, R., Chief Geologist, Sanyu Consultants, Tech-Laboratory, Nakajima 121, Yahata, Chita, Aichi 478. T: 0562321351, F: 0562332633

Kawasaki, S., Research Associate, Dept. of Global Architecture, Graduate School of Engineering, Osaka University, 2-1, Yamada-oka, Suita, Osaka 565-0871. T: 0668797622, F: 0668797617, E: kawasaki@ga.eng.osaka-u.ac.jp

Kawasaki, T., Dr, Adviser, Tokyo Soil Research Co., Ltd., 2-11-16, Higashigaoka, Meguro-ku, Tokyo 152-0021. T: 0334101711, F: 0334181494

Kawase, Y., Manager, Fudo Construction Co., Ltd., 2-1, Taito 1-chome, Taito-ku, Tokyo 110

Kazama, H., Dr, University Lecturer, Hydroscience & Geo-technology Laboratory, Saitama University, 255 Shimo-Okubo, Urawa 338. T: 0488583546, F: 0488559361, E: kazama@dice.dr5w.saitama-u.ac.jp

Kazama, M., University Professor, Tohoku University, Graduate School of Engineering, Dept. of Civil Engineering, Aoba06, Ara-maki, Aoba-ku, Sendai 980-8579. T: 0222177434, F: 0222177435, E: kazama@mechanics.civil.tohoku.ac.jp

Kazama, M., Managing Director, Kajima Corporation, 3-15, Awaza 1-chome, Nishi-ku, Osaka 550-0011. T: 0665363311, F: 0665367806

Kazama, S., Dr, University Professor, Waseda University, Kikuicho 17, Shinjuku-ku, Tokyo 162-0044. T/F: 0332039445, E: kazama@mn.waseda.ac.jp

Kidera, S., General Manager, West Japan Engineering Consultants, Inc., Road Department, 1-1, 1-chome Watan-abe-Dori, Chuo-ku, Fukuoka 810-0004. T: 0928432965, F: 0927246529, E: kidera@civil.wjec.co.jp

Kiku, H., Dr, Geotechnical Engineer, Sato Kogyo Co., Ltd., 4-12-20, Nihonbashi-honcho, Chuo-ku, Tokyo 103-8639. T: 0358232350, F: 0358232358, E: kiku@satoko-gyo.co.jp

Kikuchi, Y., Research Engineer, Geotechnical Engineering Div., Port & Harbour Research Institute, 3-1-1, Nagase, Yokosuka 239-0826. T: 0468445024, F: 0468440618, E: kikuchi@cc.phri.go.jp

Kim, H.K., Manager, 13 Goong-Dong, Dong-ku, Kwangju 501-040, Korea. T: 82622222201, F: 82622261515, E: choonchoo@hananet.net

Kim, M., Consulting Engineer, CTI Engineering Co., Ltd., 2-1-10, Watanabedori, Chuo-ku, Fukuoka 810. T: 0927142211, F: 0927118316

Kimata, T., University Assistant Professor, Osaka Prefecture University, 1-1, Gakuen-cho, Sakai, Osaka 599-8531. T/F: 0722549435, E: kimata@envi.osakafu-u.ac.jp

Kimura, K., Associate Professor, Dept. of Civil Engineering, Aichi Inst. of Technology, Yakusa-cho 1247, Toyota, Aichi 470-03. T: 0565488121

Kimura, K., Director of Construction, Kanto Regional Office of Japan Railway Construction Public Corporation, 1-10-14, Kita-Ueno, Taito-ku, Tokyo 110-0014. T: 0338457046, F: 0338458845, E: koh.kimura@jrcc.go.jp

Kimura, M., Dr, University Associate Professor, Dept. of Civil Engineering, Kyoto University, Kyoto 606-8501. T: 0757535105, F: 0757535104, E: kimura@toshi.kuciv. kyoto-u.ac.jp

Kimura, T., President, National Institution for Academic Degrees, 2-1-2-11F, Hitotsubashi, Chiyoda-ku, Tokyo 101-8438. T: 0342128202, F: 0342128210, E: tkimura@ niad.ac.jp

Kinoshita, F., Senior Research Engineer, Kajima Inst. of Const.Technology, 19-1, Tobitaktu 2-chome, Chofu, Tokyo 182. T: 0424851111, F: 0424892020

Kishi, N., Dr, Professor, Civil Engineering, Muroran Institute of Technology, Muroran 050. T: 0143473168, F: 0143473127

Kishida, H., Dr, Professor, Science Univ. of Tokyo, (Home address) 1-61-17, Kitasenzoku, Ohta-ku, Tokyo 145-0062. T: 0337234334, F: 0337237447, E: kishida@ maple.ocn.ne.jp

Kishida, K., Research Associate, Dept. of Civil Engineering, Kyoto University, Yoshida Hon-machi, Sakyo-ku, Kyoto 606-8501. T: 0757535106, F: 0757535104, E: kishida@ toshi.kuciv.kyoto-u.ac.jp

Kishida, T., Dr, Chief Director, Toa Corporation, Technical Research Institute, 1-3, Anzen-cho, Tsurumi-ku, Yokohama 230-0035. T: 0455033741, F: 0455021206, E: t_kishida@ toa-const.co.jp

Kishino, Y., Dr, University Professor, Dept. Civil Engineering, Tohoku University, Aobayama06, Sendai 980-8579. T: 0222177421, F: 0222177423, E: kishino@civil. tohoku.ac.jp

Kita, D., Dr, Adviser, Environmental Dept., EAC Corp., Okochi Bldg., 52-8, Ikebukuro 2-chome, Toshima-ku, Tokyo 171. T: 0339872181, F: 0339870562

Kita, K., Dr, Dept. of Marine Civil Engineering, Tokai University, 3-20-1, Orido, Shimizu, Shizuoka 424-8610. T: 0543340411, Ext. 2263, F: 0543349768, E: kita@scc.u-tokai.ac.jp

Kitagawa, T., Dr, Geotechnical Engineer, Nishimatsu Construction Co., Ltd., 20-10, 1-chome Toranomon, Minato-ku, Tokyo 105. T: 0335027637, F: 0335020228, E: JDA02111@niftyserve.or.jp

Kitago, S., Dr, Professor Emeritus, 6-4, Yamanote, Nishi-ku, Sapporo 063. T: 0116216426

Kitamura, R., Dr, University Professor, Dept. of Ocean Civil Engineering, Kagoshima University, 1-21-40, Korimoto,

Kagoshima 890-0065. T: 0992858473, F: 0992581738, E: kitamura@oce.kagoshima-u.ac.jp

Kitamura, Y., Geotechnical Engineer, Japan Highway Public Corporation, 1-4-1, Tadao, Machida, Tokyo 194-8508. T: 0427911621, F: 0427912380, E: yoshinori_kitamura@gw.japan-highway.go.jp

Kitazono, Y., University Professor, Kumamoto University, 2-39-1, Kurokami, Kumamoto 860-8555. T/F: 0963423540, E: kitazono@kumamoto-u.ac.jp

Kitazume, M., Chief of Soil Stabilization Laboratry, Port and Harbour Research Institute, 3-1-1, Nagase, Yokosuka 239-0826. T: 0468445023, F: 0468418098, E: kitazume@cc.phri.go.jp

Kiyama, H., University Professor, Tottori University, Dept. of Civil Engg., 4-101, Koyama-minami, Tottori-shi, Tottori 680-8552. T: 0857315295, F: 0857287899, E: kiyama@cv.tottori-u.ac.jp

Kiyono, T., Dr, University Associate Professor, Faculty of Engineering, Yamaguchi University, Tokiwadai, Ube 755. T: 0836359484, F: 0836359484

Kiyota, Y., Daisan-midoriso 15, 2-8-1, Koyanagi-cho, Fuchu-shi, Tokyo 183

Kobayashi, A., University Associate Professor, Kyoto University, Kitashirakawa, Sakyo, Kyoto 606-8502. T: 0757536152, F: 0757536346, E: kobadesu@relief.kais.kyoto-u.ac.jp

Kobayashi, I., Research Associate, Tokyo Institute of Technology, Oh-okayama, 2-12-1, Meguro-ku, Tokyo 152-8552. T: 0357343583, F: 0357343577, E: koba@geotech.cv.titech.ac.jp

Kobayashi, M., Research Engineer, Soil Mech. Lab. Soil Div., Port & Harb. Res. Inst., Min. of Transport, 1-1, Nagase 3-chome, Yokosuka 239. T: 0468445020, F: 0468444577

Kobayashi, S., Research Associate, Dept. of Civil Engineering, Kyoto University, Sakyo, Kyoto 606-8501. T/F: 0757534794, E: koba@baseball.kuciv.kyoto-u.ac.jp

Kobayashi, T., University Assistant Professor, Tokyo Denki University, College of Science and Engineering, Dept. of Civil and Environmental Engineering, Hatoyamacho, Hiki-gun, Saitama 350-0394. T: 0492962911, F: 0492966501, E: kobo-t@g.dendai.ac.jp

Kobayashi, T., Technical Adviser, Dia Consultant Co., Ltd., Towa No. 2 Bldg., 34-5, Toshima-ku, Tokyo 171. T: 0339865191, F: 0339865192

Kobayashi, Y., Drain Kogyo Co., Ltd., Tenjinyama-cho, 1-106, Nishi-ku, Nagoya 451. T: 0525221316, F: 0525221317

Kobayashi, Y., University Lecturer, Dept. of Civil Engineering, Tokyo Metropolitan University, 1-1, Minamiosawa, Hachioji-shi, Tokyo 192-0363. T: 0426771111, Ext. 4526, F: 0426772772, E: kobayashi-yoshio@c.metro-u.ac.jp

Kobayashi, Y., Chief Research Engineer, Technology Research Center for River Front Development, 3-8, Sanban-cho, Chiyoda-ku, Tokyo 102. T: 0332657121, F: 0332657456

Kocho, T., Geotechnical Engineer, Kiso-Jiban Consultants Co., Ltd., 1-11-5, Kaminagoya, Nishi-ku, Nagoya, Aichi 451. T: 052522317, F: 0525242729

Kodaka, T., Associate Professor, Dept. of Civil Engineering, Kyoto University, Yoshida Hon-machi, Sakyo-ku, Kyoto 606-8501. T: 0757535085, F: 0757535086, E: kodaka@nakisuna.kuciv.kyoto-u.ac.jp

Koga, Y., Director, Onoda Chemico Co., Ltd., 2-17-4, Yanagibashi, Taito-ku, Tokyo 111-8637. T: 0338622253, F: 0338654078, E: LEI01427@nifty.ne.jp

Kogai, Y., Fudokensetsu Tatsubuse-ryo, 1-38-2, Ooka, Minami-ku, Yokohama 365

Kogure, K., Academy Professor, Dept. of Civil Engineering, National Defense Academy, 10-20, Hashirimizu 1-chome, Yokosuka, Kanagawa 239. T: 0468413810, F: 0468445913

Kohata, Y., University Associate Professor, Muroran Institute of Technology, 27-1, Mizumoto-cho, Muroran, Hokkaido 050-8585. T: 0143465281, F: 0143465283, E: kohata@news3.ce.muroran-it.ac.jp

Kohchi, T., Consulting Engineer, Kohchi Sekkei Jimusho, 1-8-32, Hirafuku, Okayama-shi, Okayama 702. T: 0862644812, F: 0862645258

Kohda, M., Geotechnical Engineer, Railway Technical Research Institute, 2-8-38, Hikari-cho, Kokubunji-shi,

Tokyo 185-8540. T: 0425737261, F: 0425737248, E: mkoda@rtri.or.jp

Kohgo, Y., Ph.D., Geotechnical Engineer, National Research Institute of Agricultural Engineering, 2-1-2, Kannondai, Tsukuba, Ibaraki 305-8609. T/F: 0298387570, E: kohgo@nkk.affrc.go.jp

Kohno, E., Dr, Associate Professor, Nihon University, 1866 Kameino, Fujisawa 252. T: 0466816241, F: 0466821310

Kohno, F., Geotechnical Engineer, Maruyama-urasando-city-house 701, 1-253, S-1, W-16, Chuo-ku, Sapporo 060-0061. T/F: 0116167051

Kohno, I., Dr, University President, Okayama University, 1-1-1, Tsushima-Naka, Okayama 700-8530. T: 0862517000

Kohno, T., Geotechnical Engineer, R&D Institute, Takenaka Corp., 1-5-1, Ootuka, Inzai, Chiba 270-1395. T: 0476471700, F: 0476473080, E: koono.takao@takenaka.co.jp

Koike, M., Geotechnical Engineer, Taisei Corporation, Advanced Analytical Technology Section, Design Dept. I., Civil Engineering Div., 1-25-1, Nishi Shinjuku, Shinjuku-ku, Tokyo 163-0606. T: 0353815296, F: 0333451914, E: koike-m@ce.taisei.co.jp

Koike, Y., Geotechnical Engineer, OYO Corporation, 2F Rokubancho Kyodo Bldg., 6 Rokubancho, Chiyoda-ku, Tokyo 102-0085. T: 0352115183, F: 0352115184, E: koike-yutaka@oyonet.oyo.co.jp

Koizumi, K., Consulting Engineer, Dia Consultants Co., Ltd., 1-297, Kitabukuro-cho, Omiya, Saitama 330. T: 0486448385, F: 0486472810, E: koi@ins.diaconsult.co.jp

Koizumi, T., Tokai University, 1117 Kitakaname, Hiratsuka, Kanagawa 259-12. T: 0463581211, F: 0463587485

Kojima, J., Consulting Engineer, Tokai Technology Center, 2-710, Inokoshi, Meito-ku, Nagoya 465-0021. T: 0527715161, F: 0527715164, E: j_kojima@zttc.or.jp

Kojima, K., Dr, University Associate Professor, Fukui University, 3-9-1, Bunkyo, Fukui 910-8507. T: 0776278592, F: 0776278746, E: Keisuke@anc.anc-d.fukui-u.ac.jp

Kojima, K., Research Engineer, Railway Technical Research Institute, 2-8-38, Hikari-cho, Kokubunji-shi, Tokyo 185. T: 0425737261, F: 0425737248, E: kojima@rtri.or.jp

Kokusho, T., Dr, Professor, Civil Engineering Dept., Science & Engineering Faculty, Chuo University, 1-13-27, Bunkyo-ku, Tokyo 112. T: 0338171798, F: 0338171803, E: kokusho@civil.chuo-u.ac.jp

Komatsu, T., University Professor, Graduate School of Engineering, Hiroshima University, 1-4-1, Kagamiyama, Higashi-Hiroshima 739-8527. T/F: 0824247824, E: toshiko@hiroshima-u.ac.jp

Komatsuda, S., Dr, Consulting Engineer, 3-27-3, Araisono, Sagamihara 228. T: 0462530726

Komine, A., 28-578, Kohata, Hirano, Uji 611

Komine, H., Dr, Geotechnical Engineer, Central Research Institute of Electric Power Industry, 1646 Abiko, Abiko-shi, Chiba 270-1194. T: 0471821181, F: 0471842941, E: komine@criepi.denken.or.jp

Komiya, K., Dr, University Lecturer, Chiba Institute of Technology, Tsudanuma, 1-17-1, Narashino-shi, Chiba 275. T: 0474780449, F: 0474780474, E: komiya@ce.it-chiba.ac.jp

Komiya, Y., University Lecturer, College of Agriculture Univ. of the Ryukyus, 1 Senbaru, Nishihara-cho, Okinawa 903-01. T: 0988952221, F: 0988952864

Konami, T., Geotechnical Engineer, Okasanlivic Co., Ltd., 5-5, Shibadaimon 2-chome, Minato-ku, Tokyo 105-0012. T: 0334360700, F: 0334360850, E: kanami@okasanlivic.co.jp

Kondo, K., Consulting Engineer, Kajitani Engineering Co., Ltd., 3-13-14, Nishiazabu, Minato-ku, Tokyo 106-0031. T: 0334783181, F: 0334783380, E: kondou-kouichi@kajitani.co.jp

Kondo, T., Professor Emeritus, 4-11-4, Minamigaoka, Tsu 514-0822. T/F: 0592264606, E: ta-kondo@nifty.com

Kondo, T., Consulting Engineer, OYO Corporation, 2-6, Kudan-Kita 4-chome, Chiyoda-ku, Tokyo 102. T: 0332340811, F: 0332396425, E: ty6t-kndu@asahi-net.or.jp

Konishi, J., University Professor, Dept. of Civil Engineering, Faculty of Engineering, Shinshu University, 4-17-1,

JAPAN 117

Wakasato, Nagano 380-8553. T: 0262695288, F: 0262234480, E: junkoni@gipwc.shinshu-u.ac.jp

Koreeda, K., Kiso-Jiban Consultants Co., Ltd., 2-16-7, Hara, Sawara-ku, Fukuoka 814. T: 0928312511

Koseki, J., Associate Professor, Institute of Industrial Science, University of Tokyo, 4-6-1, Komaba, Meguro-ku, Tokyo 153-8505. T: 0354526421, F: 0354526423, E: koseki@iis.u-tokyo.ac.jp

Kotake, N., Geotechnical Engineer, Toyo Construction Co., Ltd., 3-7-1, Kandanishiki-cho, Chiyoda-ku, Tokyo 101-8463. T/F: 0332964623, E: kotake-nozomu@toyo-const.co.jp

Kotera, H., Dr, Chief of Civil Engineering Division, Nittodaito Construction Co., Ltd., 1-14-32, Maruyamadai, Kohnan-ku, Yokohama. T/F: 0458439619, E: h.kotera@lax.allet.ne.jp

Koumoto, T., Dr, University Professor, Saga University, 1 Honjo-machi, Saga 840-8502. T: 0952288759, F: 0952288709, E: koumotot@cc.saga-u.ac.jp

Koyama, S., University Professor, Osaka Prefecture University, Dept. of Regional Environmental Science, 1-1, Gakuen-cho, Sakai, Osaka 599-8531. T/F: 0722549439, E: koyama@envi.osakafu-u.ac.jp

Koyama, Y., Managing Director, Railway Technical Research Institute, 22-8-38, Hikari-cho, Kokubunji, Tokyo 185-8540. T: 0425737481, F: 0425737372, E: koyama@rtri.or.jp

Koyamada, K., Research Engineer, Kajima Corporation, KI Bldg.5F, 6-5-30, Akasaka, Minato-ku, Tokyo 107-8502. T: 0355612425, F: 0355612431, E: koyamada@krc.kajima.co.jp

Kubo, H., Dr, University Professor, Hokkaigakuen University, Faculty of Engineering, S-26, W-11, 1-1, Chuo-ku, Sapporo, 064-0926. T: 0118411161, Ext. 722, F: 0115512951, E: kubo@cvl.hokkai-s-u.ac.jp

Kudo, K., 5-2-8, Tojo-cho, Kashiwa-shi, Chiba 277

Kuki, M., 4-7-10, Sendagi, Bunkyo-ku, Tokyo 113

Kumada, T., Engineer, Hirose & Co., Ltd., Toyo-Central Bldg., 1-13, Toyo 4-chome, Koto-ku, Tokyo 135. T: 0356344508, F: 0356340268

Kumagai, K., University Professor, Dept. of Civil Engineering, Hachinohe Institute of Technology, 88-1, Myo Ohbiraki, Hachinohe, Aomori 031-8501. T: 0178258079, F: 0178250722, E: kumagaik@hi-tech.ac.jp

Kumamoto, N., Geotechnical Engineer, Mitsubishi Heavy Industries, Ltd., 5-1, Eba-Oki-machi, Naka-ku, Hiroshima 730-8642. T: 0822943626, F: 0822918310, E: kumamoto@eba.hrdc.mhi.co.jp

Kumazaki, I., Chief, Chubu Electric Power Co., Inc., 20-1, Kitasekiyama, Odaka-cho, Midori-ku, Nagoya 459-8522. T: 0526216101, F: 0526235117, E: Kumazaki.Ikutarou@chuden.co.jp

Kumota, K., 5-36-11, Mukono-sho, Amagasaki 661

Kunito, T., Executive Managing Director, Seiko Kogyo Co., Ltd., 1-7-24, Otemae, Chuo-ku, Osaka 540-0008. T: 0669103111, F: 0669103131, E: gijutsu@seikou-knet.co.jp

Kuno, G., University Professor, Dept. of Civil Engineering, Chuo University, 13-27, Kasuga 1-chome, Bunkyo-ku, Tokyo 112. T: 0338171798, F: 0338171803

Kurachi, Y., Research Engineer, Shiraishi Co., 1-14, Kanda-iwamotocho, Chiyoda-ku, Tokyo 101-8588. T: 0332539111, F: 0332537427, E: kurachiy@shiraishi.com

Kurata, N., Civil Engineer, 3-10-29, Wakamiya, Ichihara, Chiba 290. T: 0436413320

Kuribayashi, E., Dr, University Professor, Dept. of Civil Eng., Regional Planning, Toyohashi Univ. of Tech., Toyohashi 440. T: 0532470111, Ext. 714 & 721, F: 0532482830/450480

Kurihara, H., Dr, Manager, Civil Engineering Dept. 2, Kajima Technical Research Institute, 19-1, Tobitakyu 2-chome, Chofu, Tokyo 182. T: 0424897082, F: 0424897034

Kurisu, H., Engineer, Meishin Construction Consultants Co., 4-12, Kanon-Honmachi 1-chome, Hiroshima 733. T: 0822913141, F: 0822951065

Kuroda, K., Dr, Professor, Dept. of Civil & Environmental Engineering, Kumamoto University, 2-39-1, Kurokami, Kumamoto 860. T: 0963442111, F: 0963445063

Kuroda, S., Consulting Engineer, Chuo Kaihatsu Co., Ltd., 2-16, Ushidadori, Nakamura-ku, Nagoya 453. T: 0524816266, F: 0524828777

Kurose, M., Professional Engineer, Taiyo Gijutsu Kaihatsu Co., Ltd., 9-9, Ieno-machi, Nagasaki 852-8136. T: 0958481211, F: 0958478144, E: headoffice@n-tgk.co.jp

Kuroshima, T., Manager, Mitsui Construction Co., Ltd., 1-9-1, Nakase, Mihama-ku, Chiba 261-0023. T: 0432127545, F: 0432127540, E: IchiroKuroshima@mcc.co.jp

Kurumada, Y., Geotechnical Engineer, Penta-Ocean Construction Co., Ltd., 1534-1, Yonku-cho, Nishinasuno-machi, Nasu-gun, Tochigi 329-2746. T: 0287392116, F: 0287392133, E: Yoshinori.Kurumada@mail.penta-ocean.co.jp

Kusabuka, M., University Professor, Hosei University, Kajino-cho, Koganei-shi, Tokyo 184-8584. T: 0423876268, F: 0423876124, E: m.kusa@k.hosei.ac.jp

Kusakabe, F., Consulting Engineer, Fudo Construction Co., Ltd., Kita 3 jyo Nishi 2-chome 1, Chuo-ku, Sapporo 060. T: 0112816771, F: 0112224727

Kusakabe, K., University Professor, Dept. of Architecture and Civil Engineering, Kobe University, Rokkodai-cho, Nada-ku, Kobe 657-8501. T/F: 0788036023, E: kusakabe@kobe-u.ac.jp

Kusakabe, O., University Professor, Tokyo Institute of Technology, Dept. of Civil Engineering, O-okayama, Meguro-ku, Tokyo 152-8552. T: 0357342798, F: 03523743578, E: kusakabe@geotech.cv.titech.ac.jp

Kusakabe, S., Geotechnical Engineer, Okumura Corporation, TRI., 387 Ohsuna, Tsukuba-shi, Ibaraki 300-2612. T: 0298651521, F: 0298651522, E: oku05937@gm.okumuragumi.co.jp

Kusakabe, Y., Professional Engineer, Hokkaido Development Bureau, Civil Engineering Research Institute, Hiragishi, 1-3, Toyohira-ku, Sapporo 062-8602. T: 0118411775, F: 0118429173, E: kusakabe@ceri.go.jp

Kusano, K., Research Engineer, Tokyo Institute of Civil Engineering, Tokyo Metropolitan Government, 1-9-15, Shinsuna, Koto-ku, Tokyo 136. T: 0356831520

Kusumi, H., Dr, Associate Professor, Dept. of Civil Engineering, Kansai University, Suita, Osaka 564-8680. T/F: 0663680837, E: kusumi@ipcku.kansai-u.ac.jp

Kutsuzawa, S., Consulting Engineer, Chuo Kaihatsu Corporation, 2-16, Ushidadori, Nakamura-ku, Nagoya 453. T: 0524816261, F: 0524828777

Kuwabara, F., Dr, University Professor, Nippon Institute of Technology, 4-1, Gakuen-dai, Miyashiro-machi, Minamisaitama-gun, Saitama 345-8501. T: 0480344111, F: 0480337715

Kuwabara, M., Manager, Fudo Construction Co., Ltd., 1-2-1, Taito, Taito-ku, Tokyo 110-0016. T: 0338376035, F: 0338376158, E: mkuwa@fudo.co.jp

Kuwahara, S., Consulting Engineer, Japan Irrigation & Reclamation Consultants Co., Ltd., 34-4, Shinbashi 5-chome, Minato-ku, Tokyo 105. T: 0334343831, F: 0334590642

Kuwano, J., Dr, Associate Professor, Dept. of Civil Engineering, Tokyo Institute of Technology, 2-12-1, O-okayama, Meguro, Tokyo 152-8552. T: 0357342593, F: 0357343577, E: jkuwano@geotech.cv.titech.ac.jp

Kuwano, R., University Lecturer, University of Tokyo, Dept. of Civil Engineering, 7-3-1, Hongo, Bunkyo-ku, Tokyo 113-8656. T: 0358416122, F: 0358418504, E: rkuwano@geot.t.u-tokyo.ac.jp

Kuwayama, T., Professor, Daido Institute of Technology, 40 Hakusui-cho, Minami-ku, Nagoya 457-8532. T: 0526125571, F: 0526125953, E: kuwayama@daido-it.ac.jp

Kyoya, T., University Associate Professor, Dept. of Civil Engineering, Tohoku University, Aramaki-Aoba06, Aoba, Sendai 980-8579. T: 0222177422, F: 0222177423, E: kyoya@civil.tohoku.ac.jp

Li, L., Geotechnical Engineer, Nippon Koei Co., Ltd., R&D Center, 2304 Inarihara, Kukizaki-machi, Inashiki-gun, Ibaraki 300-1259. T: 0298712064, F: 0298712021, E: a5109@n-koei.co.jp

Mae, I., 5-487-79, Hanakoganei, Kodaira, Tokyo 187

Maeda, K., Assistant Professor, Nagoya Institute of Technology, Gokiso-cho, Showa-ku, Nagoya 466-8555. T/F: 0527355497, E: maeda@doboku2.ace.nitech.ac.jp

Maeda, Y., Consulting Engineer, Konakadaimachi, 567-1-109, Inage-ku, Chiba 263. T: 0432878305

Maeda, Y., University Professor, Kyushu Kyoritsu University, 1-8, Jiyugaoka, Yahatanishi-ku, Kitakyushu, Fukuoka 807-8585. T/F: 0936933229, E: maeda@kyukyo-u.ac.jp

Maegawa, F., Consulting Engineer, NEWJEC Inc., 1-20-19, Shimanouti, Chuo-ku, Osaka 542-0082. T: 0662454901, F: 0662452246, E: maegawaft@osaka.newjec.co.jp

Maejima, T., Geotechnical Engineer, Asahi Chemical Industry Co., Ltd., 6-46, Kagurazaka, Shinjuku-ku, Tokyo. T: 0332676671, F: 0332676654

Maekawa, H., Associate Professor, Nature and Construction Systems Core, Kanazawa Inst. of Technology, 7-1, Ogigaoka, Nonoichi-machi, Ishikawa 921. T: 0762481100, F: 0762946713

Makihara, Y., Geotechnical Engineer, Alpha Geo Co., Ltd., 4-9-6, Nakagawa, Tsuzuki-ku, Yokohama 224-0001. T: 0459137420, F: 0459137421, E: YRF01037@nifty.ne.jp

Makiuchi, K., Dr, University Professor, Dept. of Transportation Engineering, College of Science and Technology, Nihon University, 7-24-1, Narashinodai, Funabashi, Chiba 274-8501. T/F: 0474695217, E: makiuchi@trpt.cst.nihon-u.ac.jp

Maruoka, M., Chief Researcher, Takenaka Corporation, 5-14, 2-chome Minamisuna, Koto-ku, Tokyo 136. T: 0336473161, F: 0336450911

Maruyama, T., 2-36-1, Sunagawa-cho, Tachikawa 190

Masago, S., Geotechnical Engineer, Kokusai Kogyo Co., Ltd., 3-6-1, Asahigaoka, Hino, Tokyo 191. T: 0425833611, F: 0425841784

Matsubara, M., Consulting Engineer, OYO Corporation, Technical Center, 61-5, Toro-cho 2-chome, Omiya, Saitama 330-8632. T: 0486651811, F: 0486679250, E: matsubara-mikio@oyonet.oyo.co.jp

Matsuda, H., Dr, University Professor, Yamaguchi University, 2-16-1, Tokiwadai, Ube, Yamaguchi 755-8611. T: 0836859324, F: 0836859301, E: hmatsuda@po.cc.yamaguchi-u.ac.jp

Matsuda, T., Chief Research Engineer, Obayashi Corp., Technical Research Institute, 4-640, Shimokiyoto, Kiyose-shi, Tokyo 204. T: 0424950954, F: 0424950903, E: matsuda@tri.obayashi.co.jp

Matsui, K., Consulting Engineer, CTI Engineering Co., Ltd., 2-24-1, Koyodai, Fukuma, Munakata-gun, Fukuoka 811-3223. T/F: 0940343053, E: KenjiJP@aol.com

Matsui, T., Dr, University Professor, Dept. of Civil Engineering, Osaka University, 2-1, Yamadaoka, Suita, Osaka 565-0871. T: 0668797623, F: 0668797626, E: t-matsui@civil.eng.osaka-u.ac.jp

Matsukura, Y., Dr, University Professor, University of Tsukuba, Ibaraki 305-8571. T: 0298534460, F: 0298519764, E: matukura@atm.geo.tsukuba.ac.jp

Matsumoto, H., Deputy Head of Planning Division, 3-2-1-301, Ushidawaseda, Higashi-ku, Hiroshima-shi, Hiroshima 732. T: 0822113889

Matsumoto, J., 3-1-45-105, Kashiwagi, Aoba-ku, Sendai 981

Matsumoto, K., Consulting Engineer, OYO Corporation, Ichigaya Bldg.4F, 2-6, Kudan kita 4-chome, Chiyoda-ku, Tokyo 102. T: 0332340811, F: 0332396425

Matsumoto, K., Geotechnical Engineer, Hazama Corporation, 515-1, Karima, Tsukuba, Ibaraki 305-0822. T: 0298588813, E: matsu@hazama.co.jp

Matsumoto, T., General Manager, Kajima Corporation, Dept. of Intellectual Property & License, 1-2-7, Motoakasaka, Minato-ku, Tokyo 107-8388. T: 0337467099, F: 0334023178, E: matumoto@pub.kajima.co.jp

Matsumoto, T., Dr, University Professor, Dept. of Civil Engineering, Kanazawa Univ., 2-40-20, Kodatsuno, Kanazawa 920-8667. T: 0762344625, F: 0762344632, E: matsumot@t.kanazawa-u.ac.jp

Matsunaga, Y., Geotechnical Engineer, Fudo Construction Co., Ltd., 3-11, Oomachi 2-chome, Aoba-ku, Sendai, Miyagi 980. T: 0222623415, F: 0222627009

Matsuo, K., Dr, Chief of Research Section, 4th Research Center, Technical Research and Development Institute, Japan Defense Agency (TRDI JDA), Fuchinobe, 2-9-54, Sagamihara, Kanagawa 229-0006. T: 0427522941, F: 0427522940, E: matsuoz@jda-trdi.go.jp

Matsuo, M., President, Nagoya University, Furo-cho, Chikusa-ku, Nagoya 464-8601. T: 0527892000, F: 0527892005, E: t5730430@post.jimu.nagoya-u.ac.jp

Matsuo, O., Geotechnical Engineer, Public Works Research Institute, Asahi-1, Tsukuba 305-0804. T: 0298642933, F: 0298642576, E: matsuo@pwri.go.jp

Matsuoka, H., University Professor, Dept. of Civil Engineering, Nagoya Institute of Technology, Gokiso-cho, Showa-ku, Nagoya 466-8555. T/F: 0527355483, E: matsuoka@doboku2.ace.nitech.ac.jp

Matsushima, T., Assistant Professor, Institute of Engineering Mechanics and Systems, University of Tsukuba, 1-1-1, Tennodai, Tsukuba, Ibaraki 305-8573. T/F: 0298535269, E: tmatsu@kz.tsukuba.ac.jp

Matsuzawa, H., Dr, Geotechnical Engineer, OYO Corporation, Chubu Branch, 102 Nakashima, Seko, Moriyama-ku, Nagoya 463-8541. T: 0527938321, F: 0527948477, E: matsuzawa-hirosh@oyonet.oyo.co.jp

Michi, Y., Registered Consulting Engineer, Yoshimitsu Corporation, 118-ko Nagasaki-machi, Komatsu, Ishikawa 923-0004. T: 0761245151, F: 0761245152, E: yosimite@plum.ocn.ne.jp

Michihiro, K., University Professor, Setsunan University. T/F: 0728399126, E: michi@civ.setsunan.ac.jp

Midorikawa, S., University Professor, Tokyo Institute of Technology, 4259 Nagatsuta, Midori-ku, Yokohama 226. T: 0459245602, F: 0459223840

Mihara, M., Geotechnical Engineer, Hazama Corporation, 2-5-8, Kita-Aoyama, Minato-ku, Tokyo 107-8658. T: 0334054052, F: 0334058372, E: mihara@hazama.co.jp

Mikami, H., Chief Research Engineer, Institute of Technology & Development, Sumitomo Construction Co., Ltd., 1726 Niragawa, Minamikawachi-machi, Kawachi-gun, Tochigi 329-0432. T: 0285482611, F: 0285482655, E: hmikami@sumiken.co.jp

Mikasa, M., University Professor Emeritus, Soil Engineering Co., Ltd., 14-13, 2-chome Sakuragawa, Naniwa-ku, Osaka 556-0022. T: 0665683587, F: 0665681631

Miki, G., University Professor Emeritus, (Home address) 2-21-3, Midori-cho, Inage-ku, Chiba 263-0023. T/F: 0432467080

Miki, H., Dr, Managing Director, Public Works Research Institute, 1 Asahi, Tsukuba, Ibaraki 305-0804. T: 0298642211, F: 0298640564, E: miki@pwri.go.jp

Miki, K., Dr, Consulting Engineer, Kawasakichisitsu, 1-11-1, Omorikita, Ohta-ku, Tokyo 143. T: 0337637721

Mimura, M., Associate Professor, Disaster Prevention Research Institute of Kyoto University, Gokasho, Uji, Kyoto 611-0011. T: 0774384091, F: 0774384094, E: mimura@geotech.dpri.kyoto-u.ac.jp

Mimuro, T., Consulting Engineer, A-tic Corporation, 6-1, 1-jyo 5-chome, 24-ken, Nishi-ku, Sapporo 063-0801. T: 0116442807, F: 0116442892, E: mimuro@a-tic.co.jp

Minagawa, S., 2-7-1-303, Namiki, Kanazawa-ku, Yokohama 236

Minami, K., Managing Director, Osaka Port Corporation, Osaka W.t.c. Bldg.31F, 1-14-16, Nanko, Suminoe-ku, Osaka 559-0034. T: 0666157211, F: 0666157210, E: kminami@optc.or.jp

Minegishi, K., Research Associate, Dept. of Transportation Engineering & Socio-Technology, College of Science & Technology, Nihon University, 7-24-1, Narashinodai, Funabashi-shi, Chiba 274-8501. T: 0474695217, F: 0474692581, E: kmine@trpt.cst.nihon-u.ac.jp

Mineta, K., Consulting Engineer, Docon Co., Ltd., 4-1, Atsubetsu-Chuo, 1-jo 5-chome, Atsubetsu-ku, Sapporo, Hokkaido 004-8585. T: 0118011570, F: 0118011571, E: km698@mb.docon.co.jp

Mise, T., Dr, Professor Emeritus, Dept. of Civil Engineering, Osaka City University, (Home address) 13-22, Kitakatahoko-cho, Hirakata-shi, Osaka. T: 0720575205

Misumi, K., Dr, University Associate Professor, Dept. of Ocean Civil Engineering, Kagoshima University, 1-21-40, Korimoto, Kagoshima 890-0065. T/F: 0992858474, E: misumi@oce.kagoshima-u.ac.jp

Mitachi, T., Dr, Professor, Graduate School of Engineering, Hokkaido University, North 13, West 8, Kita-ku, Sapporo 060-8628. T: 0117066192, F: 0117067204, E: mitachi@eng.hokudai.ac.jp

Mitsunari, T., Consulting Engineer, Toda Corporation, 4-6-1, Hatyobori, Chuo-ku, Tokyo 104-0032. T: 0332067186, F: 0332067185, E: takashi.mitsunari@toda.co.jp

Miura, K., Dr, Associate Professor of Geomechanics, Geomechanics Group, Faculty of Engineering, Hokkaido University, N-13, W-8, Kita-ku, Sapporo, Hokkaido 060. T: 0117066202, F: 0117262296, E: miura-k@eng.hokudai.ac.jp

Miura, K., University Professor, Structural Engineering, Faculty of Engineering, Hiroshima University, Kagamiyama, 1-4-1, Higashi-Hiroshima 739-8527. T/F: 0824247794, E: miurak@hiroshima-u.ac.jp

Miura, N., Dr, University Professor, Saga University, Honjo 1, Saga 840-8502. T: 0952288576, F: 0952288190, E: miuran@cc.saga-u.ac.jp

Miura, S., Dr, University Professor, Hokkaido University, Kita 13, Nishi 8, Sapporo 060-8628. T: 0117066201, F: 0117066202, E: s-miura@eng.hokudai.ac.jp

Miura, T., Geotechnical Manager, Asia Air Survey Co., Ltd., 5-42-32, Asahi-cho, Atsugi, Kanagawa 243. T: 0462290794, F: 0462296482

Miwa, S., Registered Consulting Engineer, Tobishima Corporation, 5472 Kimagase, Sekiyado-machi, Higashikatsushika-gun, Chiba 270-02. T: 0471987553, F: 0471987585

Miyajima, M., Dr, University Professor, Kanazawa University, 2-40-20, Kodatsuno, Kanazawa, 920-8667. T: 0762344656, F: 0762344644, E: miyajima@t.kanazawa-u.ac.jp

Miyajima, S., Senior Research Engineer, Port and Harbour Research Institute, Nagase, 3-1-1, Yokosuka, Kanagawa 239-0826. T: 0468445023, F: 0468418098, E: miyajima@cc.phri.go.jp

Miyake, M., Dr, Chief Research Engineer, Soil Laboratory Toyo Construction Co., Ltd., 17-6, Naruohama 3-chome, Nishinomiya, Hyogo 663. T: 0798435903, F: 0798435916

Miyamori, T., University Professor, Dept. of Transportation Engineering, Nihon University, 7-24-1, Narashinoda, Funabashi, Chiba 274-8501. T: 0474695228, F: 0474692581

Miyamoto, K., 5-25-13-209, Nishi-Gotanda, Shinagawa-ku, Tokyo 141

Miyamoto, T., Registered Consulting Engineer, Shimizu Corporation, 4-17, Etchujima 3-chome, Koto-ku, Tokyo 135-8530. T: 0338205520, F: 0338205959, E: miyamoto@sit.shimz.co.jp

Miyashita, T., 18-2-101, Nagamori-honmachi 1-chome, Gifu 500

Miyata, K., University Professor, Dept. of Forestry, Faculty of Agriculture, Tottori University, 101 Minami 4-chome, Koyama-cho, Tottori 680. T: 0857280321, F: 0857286293

Miyata, M., Senior Research Engineer, Port and Harbour Research Institute, 3-1-1, Nagase, Yokosuka 239-0826. T: 0468445029, F: 0468440839, E: miyata@cc.phri.go.jp

Miyata, Y., University Assistant Professor, National Defense Academy, 1-10-20, Hashirimizu, Yokosuka 239-8686. T: 0468413810, Ext. 3527, F: 0468445913, E: miyamiya@cc.nda.ac.jp

Miyazaki, K., Research Engineer, Technical Research Institute, Nishimatsu Construction Co., Ltd., 2570-4, Shimotsuruma, Yamato, Kanagawa 242. T: 0462751135, F: 0462756796

Miyazaki, M., Dr, Consulting Engineer, Nippon Geophysical Prospecting Co., Ltd., 2-12, Naka-Magome 2-chome, Ohta-ku, Tokyo 143. T: 0337743161, F: 0337743180

Miyazaki, S., 10-20-705, Kawahigashi-cho, Nishinomiya 662

Miyazaki, T., 4-6-3-101, Kita-Koiwa, Edogawa-ku, Tokyo 133

Miyazaki, Y., Dr, University Professor, 2-1-1, Miyake Saekiku, Hiroshima. T: 0829213121, F: 0829237083, E: miyazaki@cc.it-hiroshima.ac.jp

Miyoshi, A., 1758-202, Kasama-cho, Sakae-ku, Yokohama 247

Mizukami, J., Senior Research Engineer, 1-1, Nagase 3-chome, Yokosuka, Kanagawa 239. T: 0468445021, F: 0468444577

Mizuno, H., Research Director for Advanced Technology, Building Research Institute, Ministry of Land Infrastructure and Transport, 1 Tachihara, Tsukuba, Ibaraki 305-0802. T: 0298646760, F: 0298646772, E: mizuno21@kenken.go.jp

Mizuno, H., Geotechnical Engineer, Ooi, 7-4-21, Shinagawa-ku, Tokyo 140-0014. T: 0337724796, E: hiroshi_mizuno@kkc.co.jp

Mizuno, Y., Fudokensetsu Hibarigaoka-ryo, 4-3, Manganji, Nishimannenzaka, Chuo-ku, Osaka 665

Mizutani, T., Research Engineer, Port and Harbour Research Institute, Ministry of Land, Infrastructure and Transport, 3-1-1, Nagase, Yokosuka 239-0826. T: 0468445024, F: 0468440618, E: mizutani@ipc.phri.go.jp

Mochida, S., Group Manager, Building Engineering Department, Kajima Technical Research Institute, 19-1, Tobitakyu 2-chome, Chofu-shi, Tokyo. T: 0424893157, F: 0424892020, E: mochida@katri.kajima.co.jp

Mochinaga, R., Consulting Engineer, Dohyu Enji Ando Bldg.7F, 7-1, Taito-2, Taito-ku, Tokyo 110-0016. T: 0338335998, F: 0338336455, E: motinaga@dohyou.co.jp

Mochizuki, A., University Professor, Dept. of Civil Engineering, Faculty of Engineering, The University of Tokushima, 2-1, Minamijosanjima-cho, Tokushima 770-8506. T/F: 0886569721, E: motizuki@ce.tokushima-u.ac.jp

Mochizuki, Y., 1-23-38, Kuriki, Isogo-ku, Yokohama 235

Monden, H., Dr, Professor, Environmental Studies, Hiroshima Inst. of Technology, 1-1, Miyake 2-chome, Saeki-ku, Hiroshima 731-51. T: 0829213121, F: 0829221480

Mori, A., Dr, Emeritus Professor, (Home address) 1-5-5, Ehara-cho, Nakano-ku, Tokyo 165-0023. T/F: 0339513187

Mori, I., Geotechnical Engineer, Kyushu Kyoritsu University, 1-8, Jiyugaoka, Yahatanishiku, Kitakyushu, Fukuoka 807-8585. T: 0936933195, F: 0936933225, E: i-mori@kyukyo-u.ac.jp

Mori, K., 2-14-6, Tsukushino, Abiko 270-11

Mori, K., Dr, President, Kiso-Jiban Consultants Co., Ltd., 1-11-5, Kudan-Kita, Chiyoda-ku, Tokyo 102-8220. T: 0332633611, F: 0332611810, E: mori.kenji@kiso.co.jp

Mori, M., University Professor, Dept. of Civil Engineering, Meisei University, 2-1-1, Hodokubo, Hino, Tokyo 191. T: 0425915111

Mori, S., University Professor, Dept. of Civil and Environmental Engineering, Ehime University, 3 Bunkyo, Matsuyama, Ehime 790-8577. T: 0899279818, F: 0899279845, E: mori@coe.ehime-u.ac.jp

Mori, Y., Dr, University Professor, Nihon University, Tamuramachi, Koriyama, Fukushima 963. T: 0249568718, F: 0249568858

Mori, Y., Civil Engineer, Nishi Sinkoiwa, 5-16-4-416, Katushika-ku, Tokyo 124. T: 0356988072

Mori, Y., Deputy General Manager, Tokyo Electric Power Services Co., Ltd., 3-3, Higashiueno 3-chome, Taito-ku, Tokyo 110-0015. T: 0344645238, F: 0344645190, E: yoshiaki@tepsco.co.jp

Morii, T., University Professor, Faculty of Agriculture, Niigata University, Niigata 950-2181. T/F: 0252626652, E: morii@agr.niigata-u.ac.jp

Morikawa, Y., Research Engineer, Port and Harbour Research Institute, Nagase, 3-1-1, Yokosuka, Kanagawa 239-0826. T: 0468445022, F: 0468440618, E: morikawa@cc.phri.go.jp

Morio, S., Geotechnical Engineer, Okumura Corporation, 2-2-2, Matsuzaki-cho, Abeno-ku, Osaka 545-8555. T: 0666253772, F: 0666237699, E: oku05988@gm.okumuragumi.co.jp

Morisaki, H., Consulting Engineer, Pacific Consultants Co., Ltd., Shin-Osaka Kimura Daisan Bldg., 4-3-24, Nishinakajima, Yodogawa-ku, Osaka-shi, Osaka 532-0011. T: 0668868460, F: 0668868489, E: Hiroshi.Morisaki@os.pacific.co.jp

Morita, K., Engineer, Oyo Corporation, 4-2-6, Kudan-Kita, Chiyoda-ku, Tokyo 102. T: 0332340811

Morita, T., Researcher, Port & Harbour Research Inst., 3-1-1, Nagase, Yokosuka, Kanagawa 257. T: 0468445028, F: 0468444095

Morita, Y., Consulting Engineer, Kiso-Jiban Consultants Co., Ltd., 1-11-5, Kudan-kita, Chiyoda-ku, Tokyo 102-8220. T: 0352766599, F: 0332347439

Moriwaki, T., Dr, University Associate Professor, Hiroshima University, Graduate School of Engineering, 1-4-1, Kagamiyama, Higashi-Hiroshima 739-8527. T/F: 0824247784, E: tmori@hiroshima-u.ac.jp

Moriya, M., Consulting Engineer, Aoi Engineering Co., Ltd., 9-3, Kugenumakaigan 1-chome, Fujisawa, Kanagawa 251. T: 0466364854

Morohashi, T., Managing Director, Koa Kaihatsu Co., Ltd., Shasoku Bldg., 5-3-13, Kotobashi, Sumida-ku, Tokyo 130. T: 0336337351, F: 0336337356

Moroto, N., Dr, University Professor, Hachinohe Institute of Technology, Oobiraki, 88-1, Hachinohe, Aomori 031. T: 0178253111, F: 0178250722

Muguruma, H., University Professor, Disaster Prevention Research Institute, Kyoto University, Uji, Kyoto 611. T: 0774323111, F: 0774326065

Mukaitani, M., Assistant Professor, Dept. of Civil and Environmental Engineering, Takamatsu National College of Technology, 355 Chokushi-cho, Takamatsu 761-8058. T: 0878693921, F: 0878693929, E: mitsu@takamatsu-nct.ac.jp

Muraishi, H., Consulting Engineer, Railway Technical Research Institute, 2-8-38, Hikari-cho, Kokubunji, Tokyo 185-8540. T: 0425737263, F: 0425737398, E: muraishi@din.or.jp

Murakami, A., University Professor, Graduate School of Natural Science & Technology, Okayama University, Okayama 700-8530. T/F: 0862518361, E: akira_m@cc.okayama-u.ac.jp

Murakami, H., Manager, Asahi Chemical Construction Materials Co., Ltd., 1-1, Uchisaiwaicho 1-chome, Chiyoda-ku, Tokyo 100. T: 0335077524, F: 0335077536

Murakami, S., Geotechnical Engineer, Tobishima Corporation, 5472 Kimagase, Sekiyado-machi, Chiba 270-0222. T: 0471981101, F: 0471987586, E: seiki_murakami@tobishima.co.jp

Murakami, S., 6-23-10, Shikawa, Mitaka 181

Murakami, S., Geotechnical Engineer, 4-25, Fukuro-machi, Naka-ku, Hiroshima 730-0036. T: 0822480138, F: 0822496826, E: shigehir@fudo.co.jp

Murakami, S., University Research Assistant, Ibaraki University, 4-12-1, Nakanarusawa-cho, Hitachi, Ibaraki 316-8511. T: 0294385174, F: 0294385268, E: murakami@civil.ibaraki.ac.jp

Murakami, Y., Dr, Professor, Yamanashi University, 3-11, Takeda 4-chome, Kofu, Yamanashi 400. T: 0552208528, F: 0552208527, E: mura@ccn.yamanashi.ac.jp

Muramatsu, M., 3-2-302, Honmokuhara, Naka-ku, Yokohama 231

Murao, Y., Consulting Engineer, Murao Chiken Co., Ltd., Tsukahara 150, Toyama 939. T: 0764292511, F: 0764292403

Murata, H., University Professor, Yamaguchi University, Tokiwadai, Ube 755-8611. T/F: 0836859342, E: hmurata@po.cc.yamaguchi-u.ac.jp

Murata, M., 4-1-1, 2 jyo, Miyanomori, Chuo-ku, Sapporo 064

Murata, O., General Manager, Railway Technical Research Institute, 2-8-38, Hikari-cho, Kokubunji, Tokyo 185. T: 0425737260, F: 0425737472, E: o-murata@rtri.or.jp

Murata, S., Dr, University Professor, Sojo University, Ikeda, 4-22-1, Kumamoto 860-0082. T: 0963263111, F: 0963263000, E: murata@ce.sojo-u.ac.jp

Murata, S., University Professor, Dept. of Civil Engineering, Kanto-Gakuin University, Mutsuura-cho 4834, Kanazawa-ku, Yokohama 236. T: 0457812001

Murata, Y., Geotechnical Engineer, Chubu branch of OYO Corporation, Aza Nakazima102, Oaza Seko, Moriyama-ku, Nagoya, Aichi 463-8541. T: 0527933221, F: 0527948477, E: murata-yoshinobu@oyonet.oyo.co.jp

Muro, T., University Professor, Ehime University, 3 Bunko-cho, Matsuyama 790-8577. T: 0899279814, F: 0899279845, E: muro@dpc.ehime-u.ac.jp

Muromachi, T., Dr, Geotechnical Engineer, Sakata Denki Co., Ltd., 1-20-8, Hon-cho, Kichijouji, Musashino, Tokyo 180-0004. T: 0422205522, F: 0422209444

Muto, A., University Professor, Dept. of Material Sic. and Eng., Muroran Inst. of Technology, 27-1, Mizumoto-cho, Muroran-shi, Hokkaido 050. T: 0143444181

Nabeshima, Y., University Assistant Professor, Dept. of Civil Engineering, Osaka University, 2-1, Yamadaoka, Suita, Osaka 533-0023. T: 0668797625, F: 0668797626, E: nabesim@civil.eng.osaka-u.ac.jp

Nagai, K., Prof., Setsunan Univ., 17-8, IkedaNakamachi, Neyagawashi, Osaka 572-8508. T: 0728399133, F: 0728386599, E: nagai@arc.setsunan.ac.jp

Nagai, O., Goyokensetsu Toride-ryo 6, 1-5-11, Toride, Toride 302

Nagaoka, H., Dr, University Professor, Kyoto University, Graduate School of Engineering, Yoshida-honmachi, Saky o-ku, Kyoto 606-8501. T: 0757535758, F: 0757535748, E: nagaoka@archi.kyoto-u.ac.jp

Nagasaka, Y., Consulting Engineer, SLS Corp., Daimaru Bldg. 4F, 1-11-6, Uchikanda, Chiyoda-ku, Tokyo. T: 0352823330, F: 0352823116, E: nagasaka@muf.biglobe.ne.jp

Nagase, H., Dr, University Associate Professor, Kyushu Institute of Technology, 1-1, Sensui-cho, Tobata-ku, Kitakyushu, Fukuoka 804-8550. T: 0938843111, F: 0938843100, E: nagase@civil.kyutech.ac.jp

Nagataki, Y., Research Engineer, Technology Research Center, Taisei Corp., 344-1, Nase-machi, Totsuka-ku, Yokohama 245. T: 0458121211, F: 0458142139

Nagaya, J., Consulting Engineer, Geo-Research Institute, 4-3-2, Itachibori, Nishi-ku, Osaka 550-0012. T: 0665392971, F: 0665786560, E: nagaya@geor.or.jp

Nagura, K., 388-2-B-104, Tomioka-cho, Kanazawa-ku, Yokohama 236

Nagura, M., Consulting Engineer, Oyo Corporation, 4-2-6, Kudan-Kita, Chiyoda-ku, Tokyo 102. T: 0332340811, F: 0332625169

Naito, K., Consulting Engineer, 7-4-16, Onodai, Sagamihara- shi, Kanagawa 229. T: 0427585521, F: 0427585521

Nakabori, K., Consulting Engineer, Nakabori Soil Corner Co., Ltd., 3-4-11, Esaka-cho, Suita, Osaka 564-0063. T: 0663849069, F: 0663868772, E: nsc@mva.biglobe.ne.jp

Nakagawa, K., Graduate School Professor, Graduate School of Science, Osaka City University, Sugimoto, 3-3-138, Sumiyoshi-ku, Osaka 558-8585. T: 0666052588, F: 0666052522, E: knaka@sci.osaka-cu.ac.jp

Nakagawa, K., Dr, Senior Research Geohydrologist, Central Research Institute of Electric Power Industry, 1646 Abiko, Abiko-shi, Chiba 270-1194. T: 0471821181, F: 0471833182, E: nakagawa@criepi.denken.or.jp

Nakahira, A., Consulting Engineer, C.T.I.Engineering Co., Ltd., Osaka Branch Office, 2-15, Otemae 1-chome, Chuo-ku, Osaka 540

Nakai, T., Dr, University Professor, Dept. of Systems Management and Engineering, Nagoya Institute of Technology, Gokiso-cho, Showa-ku, Nagoya 466-8555. T/F: 0527355485, E: nakai@tuti1.ace.nitech.ac.jp

Nakajima, H., Consulting Engineer, 19-3, Nishikubo 1-chome, Musashino, Tokyo 180-0013. T: 0422371222, F: 0422371344, E: geo-nakajima@mvd.biglobe.ne.jp

Nakajima, M., Geotechnical Engineer, Diaconsultants Co., Ltd., 2-272-3, Yoshino-cho, Omiya, Saitama 330-8660. T: 0486543011, F: 0486543833, E: M.Nakajima@diaconsult.co.jp

Nakajima, S., Consulting Engineer, Hazama Corporation, 515-1, Nishimukai, Karima, Tsukuba, Ibaraki 305. T: 0298588841, F: 0298588848

Nakamura, H., Dr, Associate Professor, Dept. of Environment and Natural Resources, Tokyo University of Agriculture and Technology, 3-5-8, Saiwai-cho, Fuchu-shi, Tokyo 183. T: 0423643311, F: 0423608830

Nakamura, J., 5-12-8, Akashiadai, Sanda 669-13

Nakamura, K., 4-29-23, Takadanobaba, Shinjuku-ku, Tokyo 169

Nakamura, K., Kochinda 804, Kochinda-cho, Shimajiri-gun, Okinawa 901-04

Nakamura, K., Registered Civil Engineer, Kindai-Sekkei Consultant Inc., 1-9-16, Kaji-cho, Chiyoda-ku, Tokyo 101. T: 0332558961, F: 0332513783

Nakamura, K., Geotechnical Engineer, Kiso-Jiban Consultants Co., Ltd., Chugoku-Branch, 4-13-25, Nagatuka, Asaminamiku, Hiroshima 731-01. T: 0822387227, F: 0822387949

Nakamura, K., Consulting Engineer, Geodesign Co., Ltd., Hamamatsu-cho, TS Bldg., 2-8-14, Hamamatsu-cho, Minato-ku, Tokyo. T: 0354257356, F: 0354257353, E: nakamra@geodesign.co.jp

Nakamura, M., Consulting Engineer, Fudo Construction Co., Ltd., Geo-Engineering Research Division, 2-1, Taito 1-chome, Taito-ku, Tokyo 110. T: 0338376131, F: 0338395469

Nakamura, R., Chief Research Engineer, Kyushu Nat. Agric. Exp. Sta., Min. of Agric., For. & Fisheries, Nishigoshi, Kumamoto 861-11. T: 0962421150, F: 0962423919

Nakamura, S., Consulting Engineer, Sakura, 1-9-17-303, Koganei, Tokyo 184-0005. T/F: 0353704552, E: zaf96805@rose.zero.ad.jp

Nakamura, S., University Associate Professor, Dept. of Civil Engineering, College of Engineering, Nihon University, Tokusada aza, Nakagawara 1, Tamura, Koriyama, Fukusima 963-8642. T: 0249568712, F: 0249568858, E: s-nak@civil.ce.nihon-u.ac.jp

Nakamura, T., University Associate Professor, Dept. of Marine Civil Engineering, Tokai University, 3-20-1, Orido, Shimizu 424. T: 0543340411, F: 0543345095

Nakamura, T., Dr, 24-8, Shinmachi 1-chome, Setagaya-ku, Tokyo 154. T: 0337013371

Nakamura, T., Dr, Emeritus Professor of Ehime University, (Home address) 1-7-43, Shoenji, Matsuyama, Ehime 790-0904. T: 0899771378

Nakamura, T., Research Engineer, Port and Harbour Research Institute, Ministry of Land, Infrastructure and Transport, 3-1-1, Nagase, Yokosuka 239-0826. T: 0468445023, F: 0468418098, E: nakamurat@cc.phri.go.jp

Nakamura, T., Research Associate, Tomakomai National College of Technology, 443 Nishikioka, Tomakomai, Hokkaido 059-1275. T: 0144678058, F: 0144678028, E: tsutomu@civil.tomakomai-ct.ac.jp

Nakane, A., Chief Research Engineer, Technology Research Center, Tekken Corporation, 9-1, Shinsen, Narita, Chiba 286. T: 0476362357, F: 0476362380, E: TIS04759@niftyserve.or.jp

Nakano, M., Dr, Professor, Faculty of Agriculture, Tokyo University, 1-1-1, Yayoi, Bunkyo-ku, Tokyo 113. T: 0338122111

Nakano, M., University Professor, Dept. of Civil Engineering, Nagoya University, Furo-cho, Chikusa-ku, Nagoya 464-8603. T: 0527894622, F: 0527894624, E: nakano@soil.genv.nagoya-u.ac.jp

Nakano, R., Dr, University Professor, 25-66, Gotsubo-cho, Gifu-shi, Gifu 501-8157. T/F: 0582476030, E: ry-nakano@mui.biglobe.ne.jp

Nakano, T., University Associate Professor, Dept. of Agricultural Engineering, Niigata University, Igarashi-Nino-cho 8050, Niigata 950-21. T: 0252626656

Nakanodo, H., Consulting Engineer, Fukken Co., Ltd., 2-10-11, Hikari-machi, Higashi-ku, Hiroshima 732-0052. T: 0822865211, F: 0822628132, E: nakanodo@fukken.co.jp

Nakasaki, H., 1199-203, Oji-cho, Midori-ku, Chiba, Chiba 267. T: 0432944712

Nakasato, A., 81-10, Tatesawashinden, Tomisato-machi, Inba-gun, Chiba 286-02

Nakase, H., Section Chief, Tokyo Electric Power Services Co., Ltd., 3-3, Higashi-Ueno 3-chome, Taito-ku, Tokyo 110-0015. T: 0358187602, F: 0358187608, E: nakase@tepsco.co.jp

Nakata, K., Dr, Associate Professor, Kinki University, 1-39, Katamachi 1-chome, Miyakojim-ku, Osaka 534. T: 063514322

Nakata, S., Geotechnical Engineer, Kawasaki Steel Co. Ltd., 2-2-3, Uchisaiwaicho, Chiyoda-ku, Tokyo 100. T: 0335974511

Nakata, Y., Research Associate, Yamaguchi University, Tokiwadai 2557, Ube, Yamaguchi 755. T: 0836315100, Ext. 3823, F: 0836359429, E: nakata@geotech.civil.yamaguchi-u.ac.jp

Nakayama, J., Consulting Engineer, 1598-11, Nogaya-machi, Machida, Tokyo 194-01. T/F: 0427345441

Nakayama, Y., Manager, 3-22-1, Nagao Nishi-cho, Hirakata, Osaka 573-0162. T: 0728670184, E: nakayama@ks-dositu.or.jp

Nakazawa, A., Manager, Advanced Geotechnical Eng., Dept., Raito Kogyo Co., Ltd., 4-2-35, Kudankita, Chiyoda-ku, Tokyo 102-8265. T: 0332652551, F: 0332952678, E: nakaz@angel.ne.jp

Nakazawa, S., 1-9, Asahigaoka, Kawachinagano 586

Nakazumi, I., Nirenokidai, 1-2-404, 3-1-16-1103, Yatsu, Narashino 275

Namikawa, K., Inhouse Engineer, Besshyo, 1-32-2-403, Hachioji, Tokyo. T/F: 0426753964, E: namikawa.k@mex.go.jp

Namikawa, T., Research Engineer, Takenaka Research & Development Institute, 1-5, Otsuka, Inzai, Chiba 270-13. T: 0476471700, F: 0476473070

Narahashi, H., Dr, University Professor, Kyushu Sangyo University, 2-3-1, Matsukadai, Higashi-ku, Fukuoka 813-8503. T: 0926735783, F: 0926735094, E: narahashi@ip.kyusan-u.ac.jp

Narita, K., Dr, University Professor, Aichi Institute of Technology, Yachigusa 1247, Yagusa, Toyota, Aichi 470-0392. T: 0565488121, F: 0565483749, E: narita@aitech.ac.jp

Nasu, M., University Professor, Dept. of Civil Engineering, Maebashi Institute of Technology, 460-1, Kamisadori-cho, Maebashi, Gunma 371-0816. T/F: 0272657342, E: nasu@maebashi-it.ac.jp

Nasu, T., 3-3-27-402, Inagekaigan, Mihama-ku, Chiba 281

Niibori, T., Consulting Engineer, East Japan Railway Company, Construction Department., Yoyogi, 2-2-2, Shibuya-ku, Tokyo 151-8578. T: 0353341288, F: 0353341289, E: niibori@head.jreast.co.jp

Niimura, K., Civil Engineer, Kawasaki Steel Corp., Engineering Division, Hibiya Kokusai Bldg., 2-2-3, Uchisaiwai-cho, Chiyoda-ku, Tokyo 100. T: 0335973111

Niiro, T., Chief Geotechnical Engineer, 505-2, Shinsei-machi, Kajiki-cho, Kagoshima 899-5223. T: 0995624804, E: niiro@mint.ocn.ne.jp

Nishi, K., Deputy Director, Abiko Research Laboratory, Central Research Institute of Electric Power Industry, 1646 Abiko, Abiko-shi, Chiba 270-1194. T: 0471821181, F: 0471829417, E: nishi@criepi.denken.or.jp

Nishi, M., University Professor, Reclamation Engineering Research Institute, Faculty of Engineering, Kobe University, Rokkodaimachi, Nada, Kobe 657. T: 0788811212, F: 0788024157

Nishida, K., University Professor, Kansai University, Yamate-cho, 3-3-35, Suita, Osaka 564. T: 0663681121, F: 0663680898, E: nishida@ipcku.kansai-u.ac.jp

Nishida, K., Research Engineer, Obayashi Corporation T.R.I., 640 Shimokiyoto 4-chome, Kiyose, Tokyo 204-8558. T: 0424950910, F: 0424950903, E: nishida@tri.obayashi.co.jp

Nishie, S., Director of Geotechnical Development, Chuo Kaihatsu Corp., 3-13-5, Nishi-Waseda, Shinjuku-ku, Tokyo 169-8612. T: 0332085252, F: 0332089915, E: shunsaku_nishie@ckc-unet.ocn.ne.jp

Nishigaki, M., Dr, University Professor, Faculty of Environmental Science and Technology, Okayama University, 3-1-1, Tsushimanaka, Okayama 700-8530. T/F: 0862518164, E: n_makoto@cc.okayama-u.ac.jp

Nishigaki, Y., Dr, Managing Director, Kiso-Jiban Consultants Co., Ltd., Rock Engineering Center, 3-12-21, Tatsunominami, Sango, Ikoma-gun, Nara 636-0822. T: 0745326486, F: 0745325616, E: nishigaki.yoshihiko@kiso.co.jp

Nishigata, T., Dr, University Associate Professor, Kansai University, 3-3-35, Yamate-cho, Suita, Osaka 564-8680. T/F: 0663680898, E: nisigata@ipcku.kansai-u.ac.jp

Nishihara, A., Dr, Fukuyama Univ., 985 Gakuen-cho, Fukuyama 729-02. T: 0849362111, F: 0849362213

Nishihata, S., Consulting Engineer, Nippon Koei Co., Ltd., Corporate Development Administration, Information Management Office, 5-4, Kojimachi, Chiyoda-ku, Tokyo 102-8539. T: 0332388195, F: 0332388326, E: a03299@n-koei.co.jp

Nishikawa, J., Dr, Research Engineer, Civil Engineering Research Institute of Hokkaido, 1-3, Hiragishi, Toyohira-ku, Sapporo 062-8602. T: 0118415288, F: 0118417333, E: nishikaw@ceri.go.jp

Nishikawa, K., Geotechnical Engineer, Kisojiban Consultants Co., Ltd., 1-11-14, Awaza, Nishiku, Osaka 550-0011. T: 0665361781, F: 0665362124, E: nishikawa.katsuhiro@kiso.co.jp

Nishimura, A., Dr, Laboratory Director, Railway Technical Research Institute, 2-8-38, Hikaricho, Kokubunji-shi, Tokyo 185. T: 0425737262, F: 0425737248

Nishimura, M., Consulting Engineer, Kaihatsu Koei Co., Ltd., Asty 45, North 4, West 5-1, Chuo-ku, Sapporo, Hokkaido 060-0004. T: 0112073666, F: 0112185777, E: nishimura@kai-koei.co.jp

Nishimura, S., Dr, University Professor, Dept. of Environmental Management Engineering, Okayama University, 3-1-1, Tsushima-naka, Okayama 700-8530. T/F: 0862518353, E: theg1786@cc.okayama-u.ac.jp

Nishimura, S., Director, Fugro Geoscience Co., Ltd., 4-3-16, Kudan-Kita, Chiyoda-ku, Tokyo 102-0073. T: 0332882936, F: 0332882984, E: nishimura@fugro.co.jp

Nishimura, T., Research Associate, Tottori University, 4-101, Minami, Koyama, Tottori-shi, Tottori 680-8552. T: 0857315297, F: 0857287899, E: tnishi@cv.tottori-u.ac.jp

Nishimura, T., Dr, University Associate Professor, Ashikaga Institute of Technology, Dept. of Civil Engineering, 268 Omae, Ashikaga, Tochigi 326-8558. T: 0284620605, F: 0284641061, E: tomo@ashitech.ac.jp

Nishimura, T., Consulting Engineer, 1-27-25, Maioka, Totsuka-ku, Yokohama 244-0814. T: 0458235792, F: 0458212331, E: nishitel@nifty.com

Nishimura, Y., 1-6-39, Hakurazaki, Izumisano 598

Nishio, S., Dr, Senior Research Engineer, Institute of Technology, Shimizu Corporation, 3-4-17, Etchujima, Koto-ku, Tokyo 135-8530. T: 0338205268, F: 0338205959, E: nishio@sit.shimz.co.jp

Nishiyama, S., Research Associate, Kyoto University, School of Civil Engineering, Sakyo-ku, Kyoto 606-8501. T/F: 0757535129, E: nisiyama@geotech.kuciv.kyoto-u.ac.jp

Nishiyama, T., Research Associate, Okayama University, 3-1-1, Tsushima-naka, Okayama 700-8530. T: 0862518362, F: 0862518361, E: nisiyama@cc.okayama-u.ac.jp

Nitao, H., Geotechnical Engineer, Fudo Construction Co., Ltd., 2-11-16, Hakataekimae, Hakata-ku, Fukuoka 812-0011. T: 0924514179, F: 0924117088, E: nitao@fudo.co.jp

Noda, S., Executive Adviser, Mitsubishi Heavy Industries Ltd., 5-1, Marunouchi 2-chome, Chiyoda-ku, Tokyo 100-8315. T: 0332123111, F: 0332129833, E: Q09884@hq.mhi.co.jp

Noda, T., University Professor, Dept. of Geotechnical & Environmental Engineering, Nagoya University, Furo-cho, Chikusa-ku, Nagoya 464-8603. T: 0527893833, F: 0527893836, E: noda@soil.genv.nagoya-u.ac.jp

Noguchi, K., Dr, University Professor, Waseda University, 3-4-1, Ohkubo, Shinjuku-ku, Tokyo 169-8555. T: 0352863325, F: 0352863491, E: knog@mn.waseda.ac.jp

Noguchi, T., General Manager, Railway Technical Research Institute, 2-8-38, Hikari-cho, Kokubunji, Tokyo 185-8540. T: 0425737304, F: 0425737398, E: noguchi@rtri.or.jp

Nojiri, A., Dr, General Manager, Intellectual Property & License Department, Kajima Co., 2-7, Motoakasaka 1-chome, Minato-ku, Tokyo 107. T: 0337466900, F: 0334023178, E: nojiriak@pub.kajima.co.jp

Noto, S., Dr, Director of Structures Division, Civil Eng. Research Inst. of HDB., 1-3, Hiragishi, Toyohira-ku, Sapporo 062. T: 0118411111, F: 0118241226

Notsukitaira, Y., Geotechnical Engineer, Geo Consultant Co., Ltd., 1-15-13, Oyaba, Urawa, Saitama 336. T: 0488837575, F: 0488837576

Nozawa, D., Dr, 2-20-12-116, Senjyu-Azuma, Adachi-ku, Tokyo 120. T: 0338825447

Nozu, A., Research Engineer, Port and Harbour Research Inst., 3-1-1, Nagase, Yokosuka 239-0826. T: 0468445030, F: 0468440839, E: nozu@ipc.phri.go.jp

Nozu, M., Geotechnical Engineer, Fudo Construction Co., Ltd., 1-2-1, Taito, Taito-ku, Tokyo 110-0016. T: 0338376034, F: 0338376158, E: nozu@fudo.co.jp

Numakami, K., Consulting Engineer, Tokyu Construction Co., Ltd., 1-16-14, Shibuya, Shibuya-ku, Tokyo 150. T: 0354665224, F: 0334066715, E: numakami@tokyu-cnst.co.jp

Numata, A., Geotechnical Engineer, Tobishima Corporation, 5472 Kimagase, Sekiyado-machi, Higashikatsushika-gun, Chiba 270-0222. T: 0471987553, F: 0471987586, E: atsunori_numata@tobishima.co.jp

Ochi, K., Manager, Tokyu Construction Co., Ltd., 2-17-18, Marunouchi, Naka-ku, Nagoya 460-0002. T: 0522321214, F: 0522018575, E: ochi@na.tokyu-cnst.co.jp

Ochiai, H., Dr, University Professor, Dept. of Civil Engineering, Kyushu University, 6-10-1, Hakozaki, Higashi-ku, Fukuoka 812-8581. T: 0926423283, F: 0926423285, E: ochiai@civil.kyushu-u.ac.jp

Oda, E., Dr, Professor Emeritus, Tokushima University, (Home address) 6-25, Fujiyamadai 10-chome, Kasugai, Aichi 487. T: 0568921198

Oda, K., University Research Associate, Geotechnical Division, Dept. of Civil Engineering, Graduate School of Engineering, Osaka University, 2-1, Yamadaoka, Suita, Osaka 565-0871. T/F: 0668797626, E: oda@civil.eng.osaka-u.ac.jp

Oda, M., University Professor, Saitama University, Dept. of Civil and Environmental Engineering, Faculty of Engineering, Saitama 338-8570. T/F: 0488583542, E: oda@dice.dr5w.saitama-u.ac.jp

Ogata, N., Senior Research Fellow, Central Research Institute of Electric Power Industry, 1646 Abiko, Abiko-shi, Chiba 270-11. T: 0471821181, F: 0471825934

Ogawa, S., Dr, President, Nagaoka National College of Technology, Nishikatakai-888, Nagaoka 940-8532. T: 0258326435, F: 0258349700, E: shoji@nagaoka-ct.ac.jp

Ogawa, Y., Dr, Researcher, Institute of Civil Engineering, TMG 1-9-15, Shinsuna, Koutou-ku, Tokyo 136. T: 0356831530, F: 0356831515

Ogisako, E., Dr, Geotechnical Engineer, Shimizu Corporation, 4-17, Etchujima 3-chome, Koto-ku, Tokyo 135-8530. T: 0338205533, F: 0338205959, E: ogisako@sit.shimz.co.jp

Ogura, H., Chief Research Engineer, GEOTOP Corporation, 1-16-3, Shinkawa, Chuo-ku, Tokyo 104-0033. T: 0355344601, F: 0355434610, E: ogura@eng.geotop.co.jp

Ohba, S., Dr, Professor, Dept. of Architecture, Osaka Institute of Technology, 5-16-1, Omiya, Asahi-ku, Osaka 535-8585. T: 0669544213, F: 0669572132, E: ohba@archi.oit.ac.jp

Ohbayashi, J., Consulting Engineer, Fudo Construction Co., Ltd., 1-2-1, Taito, Taito-ku, Tokyo 110. T: 0338376034, F: 0338376158

Ohfuka, N., Consulting Engineer, ARS Consultants Co., Ltd., 2-1-1, Izuminodemachi, Kanazawa 921-8116. T: 0762484004, F: 0762484174, E: ohfuka@po.incl.ne.jp

Ohhara, J., Research Manager, Tokyo Soil Research Co., Ltd., 15-3, Ikejiri 3-chome, Setagaya-ku, Tokyo 154. T: 0334107281, F: 0334246416

Ohhashi, M., 7-8, Minami Tokiwadai 1-chome, Itabashi-ku, Tokyo 174. T: 0339565357

Ohhashi, T., 5-206, Tomyoan 1, Toshin-cho, Iwakura 482

Ohkawa, H., Dr, University Professor, Faculty of Engineering, Niigata University, Ikarashi 2, Niigata 950-2181. T: 0252626793, F: 0252627021, E: bigriver@eng.niigata-u.ac.jp

Ohki, N., Senior Chief Researcher, Takenaka Corporation, 5-1, 1-chome Ohtsuka, Inzai-shi, Chiba 270-13. T: 0476471700, F: 0476473080, E: ohki@ccm.rdi.takenaka.co.jp

Ohmachi, T., Dr, University Professor, Tokyo Institute of Technology, 4259 Nagatsuta, Midori-ku, Yokohama 226-8502. T: 0459245605, F: 0459245574, E: ohmachi@enveng.titech.ac.jp

Ohmaki, S., Dr, Geotechnical Engineer, National, Research Institute of Fisheries Engineering, Fisheries Research Agency, Ebidai, Hazaki-machi, Kashima-gun, Ibaraki 314-04. T: 0479445940, F: 0479441875, E: omaki@nrife.affrc.go.jp

Ohmine, K., Dr, Dept. of Civil Engineering, Kyushu University, 6-10-1, Hakozaki, Higashi-ku, Fukuoka 812-8581. T/F: 0926423285, E: oomine@civil.kyushu-u.ac.jp

Ohmori, K., Geotechnical Engineer, Research Institute of Digital Geo-Environment (RIDGE), Ro 86-1, Kannondo-machi, Kanazawa, Ishikawa 920-0352. T/F: 0762687953, E: ohmori-10@safedraw.co.jp

Ohne, Y., Dr, Professor, Dept. of Civil Engineering, Aichi Institute of Technology, 1247 Yachigusa, Yagusa-cho, Toyota, Aichi 470-03. T: 0565488121, F: 0565483749

Ohnishi, T., Consulting Engineer, Hanshin Consultants Co., Ltd., Shouwa Bldg., 2-5-24, Nishihon-machi, Nishi-ku, Osaka 550. T: 065430201, F: 065430253

Ohnishi, Y., Dr, Professor, School of Civil Engineering, Kyoto University, Kyoto 606-8501. T: 0757535127, F: 0757535129, E: ohnishi@geotech.kuciv.kyoto-u.ac.jp

Ohno, M., Managing Director, Hazama Corporation, 515-1, Karima, Tsukuba, Ibaraki 305-0822. T: 0298588803, F: 0298588809, E: ohno@hazama.co.jp

Ohno, S., University Research Associate, Dept. of Civil Engineering, Kinki University, 3-4-1, Kowakae, Higashi-Osaka, Osaka 577-8502. T: 0667212332, Ext. 4656, F: 0729955192, E: ohno@civileng.kindai.ac.jp

Oh-Oka, H., Institute Professor, Niigata Institute of Technology, Fujihashi 1719, Kashiwazaki, Niigata 945-1195. T/F: 0257228171, E: oh-oka@abe.nilt.ac.jp

Ohshima, A., University Lecturer, Dept. of Civil Engineering, Osaka City University, 3-3-38, Sugimoto, Sumiyoshi-ku, Osaka 558-8585. T: 0666052996, F: 0666052726, E: oshima@civil.eng.osaka-cu.ac.jp

Ohshita, T., Chief Researcher, Takenaka Corporation Research & Development Institute, 3-1-8, Mokuzai-dori, Mihara-cho, Minamikawachi-gun, Osaka 587. T: 0723626110, F: 0723623851

Ohta, H., University Professor, Tokyo Institute of Technology, 2-12-1, O-okayama, Meguro-ku, Tokyo 152-8552. T: 0357343583, F: 0357343577, E: ohta@geotech.cv.titech.ac.jp

Ohta, S., Managing Director, Chemical Grouting Co., Ltd., Anzen Bldg., 1-6-4, Motoakasaka, Minato-ku, Tokyo 107-8309. T: 0334750201, F: 0334751545, E: ohtasozo@pub.kajima.co.jp

Ohtani, J., University Professor, Dept. of Civil & Environmental Engineering, Kumamoto University, Kumamoto 860-8555. T/F: 0963423535, E: junotani@gpo.kumamoto-u.ac.jp

Ohtani, K., Director, Disaster Prevention Research Division, National Research Institute of Earth Science and Disaster Prevention, Science & Technology Agency, Tennodai, 3-1, Tsukuba-shi, Ibaraki 305. T: 0298511611, F: 0298528512

Ohtani, Y., Professional Engineer, Hirose & Co., Ltd., 1-12-19, Minami-horie, Nishi-ku, Osaka 550-0015. T: 0665326923, F: 0665332423, E: yotani@pearl.ocn.ne.jp

Ohtomo, H., Senior Geophysicist, Oyo Corp., 4-2-6, Kudan-Kita, Chiyoda-ku, Tokyo 102. T: 0332340811, F: 0332625169

Ohtsuka, S., Dr, University Professor, Nagaoka University of Technology, Dept. of Civil and Environmental Engineering, Nagaoka, Niigata, 940-2188. T: 0258479633, F: 0258479600, E: ohtsuka@vos.nagaokaut.ac.jp

Ohuchi, M., Engineer, Shiraishi Corp., SN-Iwamoto-cho Bldg.6F, 3-2-10, Iwamoto-cho, Chiyoda-ku, Tokyo 101. T: 0356876585, F: 0356876587

Ohya, S., President, Oyo Corporation, 4-2-6, Kudan-Kita, Chiyoda-ku, Tokyo 102-0073. T: 0332340811, F: 0332340383, E: ohya-satoru@oyonet.oyo.co.jp

Oikawa, H., Dr, Associate Professor, Dept. of Civil Engineering, Mining College, Akita University, 1-1, Tegata, Gakuencho, Akita 010. T: 0188892360, F: 0188370407

Oka, A., Geotechnical Engineer, 765 Nonaka-machi, Kurume-shi, Fukuoka 839-0862. T/F: 0942333314

Oka, F., University Professor, Dept. of Civil Engineering, Division of Soil Mechanics, Graduate School of Engineering, Kyoto University, Yoshida-Honmachi, Sakyo-ku, Kyoto 606-8501. T: 0757535084, F: 0757535086, E: foka@nakisuna.kuciv.kyoto-u.ac.jp

Okabayashi, I., Dr, Associate Director, 6-24-4, Minamirinkan, Yamato-shi, Kanagawa 242

Okabayashi, K., Associate Professor of National College of Technology, Kochi National College of Technology, Monobe, 200-1, Nankoku, Kochi 783. T: 0888645589, F: 0888645581, E: oka@ce.kochi-ct.ac.jp

Okabe, H., Consulting Engineer, Nikken Sekkei Ltd., 1-4-27, Koraku, Bunkyo-ku, Tokyo 112. T: 0338133361, F: 0338170517

Okada, F., Lecturer, Dept. of Civil Engineering, Meijyo University, 1-501, Shiogamaguchi, Tenpaku-ku, Nagoya 468. T: 0528321151

Okada, J., Consulting Engineer, Nakabori Soil Corner, 4-11, Esaka-cho 3-chome, Suita 564. T: 063849069, F: 063868772

Okada, K., Dr, University Professor, Faculty of Engineering, Kokushikan University, 4-28-1, Setagaya, Setagaya-ku, Tokyo 154-8515. T/F: 0354815862, E: okadak@kokushikan.ac.jp

Okamoto, S., Dr, 1-4-18, Anagawa, Inageku, Chiba 263. T: 0432515573

Okamoto, S., Consulting Engineer, 1-25, N-24 E-7, Higashi-ku, Sapporo 065-0024. T/F: 0117313543, E: okamotos@rainbow.ne.jp

Okamoto, T., Dr, Manager, Central Research Institute of Electric Power, 1646 Abiko, Abiko-shi, Chiba 270-11. T: 0471821181, F: 0471832916

Okamura, M., Geotechnical Researcher, Public Works Research Institute, 1 Asahi, Tsukuba 305-0804. T: 0298644969, F: 0298642576, E: okamura@pwri.go.jp

Okimura, T., Dr, University Professor, Kobe University, Rokkodai-cho, Nada-ku, Kobe 657-8501. T: 0788036010, F: 0788036394, E: okimura@kobe-u.ac.jp

Okino, Y., Manager, Toa Grout Kogyo Co., Ltd., Shin-Osaka Bldg.5F, 7-3, Nishi-Nakajima 6-chome, Yodogawa-ku, Osaka 532-0011. T: 0663070880, F: 0663077567, E: yusuke.okino@toa-g.co.jp

Okuda, S., 41-1-813, Suzugamine-cho, Nishi-ku, Hiroshima 733

Okumura, F., Dr, General Manager, Railway Technical Research Institute, 2-8-38, Hikari-cho, Kokubunji-shi, Tokyo 185-8540. T: 0425737352, F: 0425737370, E: okumura@rtri.or.jp

Okumura, I., 1-1-MW-105, Oguradai, Inzai-cho, Inba-gun, Chiba 270-13

Okumura, T., Dr, Professor, Dept. of Environmental & Civil Eng., Okayama University, 2-1-1, Tsushima-naka, Okayama 700. T: 0862518160, F: 0862532993, E: okumura@cc.okayama-u.ac.jp

Okumura, T., Dr, University Professor, Dept. of Civil Engineering, Aichi Institute of Technology, 1247 Yachigusa, Yagusa-cho, Toyota, Aichi 470-0392. T: 0565488121, F: 0565483749, E: okumura@ce.aitech.ac.jp

Okuno, T., Geotechnical Engineer, Institute of Technology, Shimizu Corporation, 3-4-17, Etchujima, Koto-ku, Tokyo 135-8530. T: 0338205537, F: 0338205959, E: okuno@sit.shimiz.co.jp

Onitsuka, K., Dr, University Professor, Dept. of Civil Engineering, Faculty of Science and Engineering, Saga University, 1 Honjyo, Saga 840-8502. T/F: 0952288690, E: onitukak@cc.saga-u.ac.jp

Ono, H., Consulting Engineer, Chuo Kaihatsu Co., Ltd., 289-15, Toyoshiki, Kashiwa, Chiba 277-0863. T: 0471462028, E: hideono@fa2.so-net.ne.jp

Ono, S., Geotechnical Engineer, Chuo Kaihatsu Co., Ltd., Osaka Branch, 34-12, Tarumi-cho 3-chome, Suita 564. T: 063863691, F: 063863020

Ono, T., University Professor, Dept. of Civil Engineering, Hokkai-Gakuen University, S-26, W-11, Chuo-ku, Sapporo 064-0926. T: 0118411161, Ext. 754, F: 0115512951, E: ono@cvl.hokkai-s-u.ac.jp

Onoue, A., Dr, Professor, Nagaoka National College of Technology, 888 Nishikatagai, Nagaoka-shi, Nagaoka 940-8532. T/F:0258349439, E: onoue@nagaoka-ct.ac.jp

Orense, R.P., Associate Professor, Dept. of Civil Engineering, University of Tokyo, 7-3-1, Hongo, Bunkyo-ku, Tokyo 113-8656. T: 0358417452, F: 0358418504, E: orense@geot.t.u-tokyo.ac.jp

Ozaki, E., University Professor Emeritus, 239-7, Kitanoda, Sakai, Osaka 599-8123. T/F: 0722376080

Ozaki, M., Dept. of Civil Engineering, Maebashi Municipal Junior College of Techn., Kamisadori-machi 460, Maebashi 371. T: 0272650111, F: 0272653837

Ozawa, H., Managing Director, Demip Giken Co., Ltd., 30-8, Ichinoe 7-chome, Edogawa-ku, Tokyo. T: 0356072011, F: 0356072014

Ozawa, Y., Dr, Technical Advisor, Nikken Sekkei Ltd., 2-1-3, Koraku, Bunkyo-ku, Tokyo 112-8565. T: 0338133361, F: 0338147594, E: ozawa@nikken.co.jp

Park, J.K., University Professor, Seoul National University of Technology, Dept. of Civil Engineering, 172 Gongneung-dong, Nowon-gu, Seoul 139-743. T: 82029706514, F: 82029480043, E: jokpark@duck.snut.ac.kr

Park, S.Z., University Professor, Dept. of Civil Engineering, Pusan National University, #30, Changjeon-dong, Kumjeong-ku, Pusan 609-735. T: 820515102349, F: 820515156810, E: szpark@hyowon.cc.pusan.ac.kr

Rito, F., General Manager, OYO Corporation, 61-5, Toro-cho 2-chome, Omiya 330-8632. T: 0486651811, F: 0486679250, E: ritou-fusao@oyonet.oyo.co.jp

Ryokai, K., Dr, Senior Research Engineer, Shimizu Corporation Institute of Technology, 4-17, Etchujima 3-chome, Koto-ku, Tokyo 135. T: 0336434311, F: 0336437260

Ryu, S., Nagoya Institute of Technology

Saeki, E., Managing Director of Eco-Pile, Nippon Steel Corporation, Otemachi, 2-6-3, Chiyoda-ku, Tokyo 100-8071. T: 0332756626, F: 0332755978, E: saeki.eiichiroh@eng.nsc.co.jp

Saiki, K., Geotechnical Consultant, Saiki Research Studio, 1-30-31-301, Seta, Setagaya-ku, Tokyo 158-0095. T: 0337005032, F: 0337005079, E: saikirs@aol.com

Saito, K., University Professor, Chuo University, 1-13-27, Kasuga, Bunkyo-ku, Tokyo 112-8551. T: 0338171804, F: 0338171803, E: saitoh@civil.chuo-u.ac.jp

Saito, M., Dr, Consulting Engineer, 2-25-12, Ikego, Zushi, Kanagawa 249. T: 0468718379

Saito, M., Consulting Engineer, Onoda Chemico Co., Ltd., 2-17-4, Yanagibashi, Taito-ku, Tokyo 111. T: 0338622250, F: 0338665928

Saito, T., Dr, 188-1, Kawabehorinouchi, Hino-shi, Tokyo 191. T: 0425837465

Saito, Y., Managing Director, Kiso-Jiban Consultants Co., Ltd., Kyushu Branch, 2-16-7, Hora, Sawara-ku, Fukuoka 814-0022. T: 0928312511, F: 0928222393, E: saito.yoshinori@kiso.co.jp

Saji, S., Geotechnical Engineer, Tokyu Construction Co., Ltd., Chikatetsu Bldg.5F, 1-16-14, Shibuya, Shibuya-ku, Tokyo 150-8340. T: 0354665402, F: 0354665573, E: saji@hd.tokyu-cnst.co.jp

Sakaba, Y., Consulting Engineer, KYOSEI Inc., 1-23-1, Shinjuku, Shinjuku-ku, Tokyo 160-0022. T: 0332259912, F: 0332259914, E: sakaba@kyosei-kk.co.jp

Sakaguchi, O., Dr, University Professor of Architecture & Bldg. Science, University of Kinki, 3-4-1, Kowakae, Higashi Osaka-shi, Osaka 577. T: 067212332, F: 067301320

Sakai, A., Dr, University Associate Professor, Saga University, 1 Honjo, Saga 840-8502. T: 0952288572, F: 0952288190, E: sakaia@cc.saga-u.ac.jp

Sakai, H., Researcher, Railway Technical Research Institute, Hikari-cho, 2-8-38, Kokubunji, Tokyo 185-8540. T: 0425737265, F: 0425737398, E: water@rtri.or.jp

Sakai, K., Managing Director, 7-20-408, Shimonoge 1-chome, Takatsu-ku, Kawasaki 213. T: 0448119440

Sakai, S., Consulting Engineer, Hanshin Consultants Co., Ltd., 2-860-1, Shijo-ooji, Nara 630-8014. T: 0742360211, F: 0742361989, E: sakai@hanshin-consul.co.jp

Sakai, S., Chief Research Engineer, Fudo Construction Co., Ltd., 1-2-1, Taito, Taito-ku, Tokyo 110-0016. T: 0338376032, F: 0338376158, E: shiesa@fudo.co.jp

Sakai, T., Consulting Engineer, Applied Research Co., Ltd., 904-1, Tohigashi, Tsukuba, Ibaraki 300-2633. T: 0298481095, F: 0298481096, E: sakai@applied.co.jp

Sakai, T., Associate Professor, Ehime University, 3-5-7, Tarumi, Matsuyama, Ehime 790-8566. T: 0899469885, F: 0899469883, E: sakai@agr.ehime-u.ac.jp

Sakaino, N., 2-1-1-205, Nishi-kashiwadai, Kashiwa 277

Sakajo, S., Dr, Managing Director, Kiso-Jiban Consultants Co., Ltd., 1-11-5, Kudan-kita, Chiyoda-ku, Tokyo 102-8220. T: 0352766231, F: 0332347439, E: sakajo.saiichi@kiso.co.jp

Sakamoto, Y., Director, Kiso-Jiban Consultants Co., Ltd., 11-5, Kudan-Kita 1-chome, Chiyoda-ku, Tokyo 102. T: 0332633611

Sakaue, T., University Professor Emeritus, Sumikawa, 3-1-9-61, Minami-ku, Sapporo 005-0003. T/F: 0118310311

Sakurazawa, M., Assistant Manager, Fukuda Corporation, 2-778, Nishibori-dori, Niigata, Niigata 951-8061. T: 0252273531, F: 0252273522, E: sakurazawa@fkd.co.jp

Sandanbata, I., Chief Research Engineer, Hazama Corporation, 515-1, Karima, Tsukuba, Ibaraki 305-0822. T: 0298588813, F: 0298588819, E: sandan@hazama.co.jp

Sano, I., University Associate Professor, Dept. of Civil Engineering, Faculty of Engineering, Osakasangyo University, 3-1-1, Nakagaito, Daito, Osaka 574-8530. T: 0728753001, Ext. 3738, F: 0728755044, E: sano@ce.osaka-sandai.ac.jp

Sano, S., Consulting Engineer, Toa Grout Kogyo Co., Ltd., 10-3, Yotsuya 2-chome, Shinjuku-ku, Tokyo 160-0004. T: 0333554457, F: 0333551532, E: sano@ringnet.dion.ne.jp

Sano, Y., College Professor, Hakodate National College of Technology, 14-1, Tokura-cho, Hakodate, Hokkaido 042-8501. T/F: 0138596483, E: sano@hakodate-ct.ac.jp

Sasaki, S., Professor, Wakayama National College of Technology, 37 Noshima, Nada, Gobo, Wakayama 644-0023. T: 0738298448, F: 0738298469, E: sasaki@wakayama-nct.ac.jp

Sasaki, T., Research Engineer, Public Works Research Institute, 1 Asahi, Tsukuba, Ibaraki 305-0804. T: 0298644283, F: 0298642688, E: sasaki@pwri.go.jp

Sasaki, Y., Dr, University Professor, Hiroshima Univ., 1-4-1, Kagamiyama, Higashi-Hiroshima 739-8527. T/F: 0824247783, E: ysasaki@hiroshima-u.ac.jp

Sasakura, T., Geotechnical Research Engineer, Kajima Technical Research Institute, 2-19-1, Tobitakyu, Chofu-shi, Tokyo 182-0036. T: 0424897067, F: 0424897034, E: taksas@katri.kajima.co.jp

Sassa, K., University Professor, Disaster Prevention Research Institute, Kyoto University, Uji, Kyoto 611-0011. T: 0774384110, F: 0774325597, E: sassa@SCL.kyoto-u.ac.jp

Sassa, S., Research Fellow of the Japan Society for the Promotion of Science, Disaster Prevention Research Institute, Kyoto University, Uji, Gokasho, Kyoto 611-0011. T: 0774384309, F: 0774384180, E: sassa@rcde.dpri.kyoto-u.ac.jp

Satake, M., University Professor, 201, 1-5-1, Yagiyama-Honcho, Taihaku-ku, Sendai 982-0801. T/F: 0222295309, E: satake@tjcc.tohoku-gakuin.ac.jp

Sato, A., Research Associate, Dept. of Civil Engineering, Kumamoto University, Kurokami, 2-39-1, Kumamoto 860-8555. T/F: 0963423694, E: asato@alpha.msre.kumamoto-u.ac.jp

Sato, D., Sumitomo Electric Industries, Ltd., Special Steel Wire Division, 1-3-12, Motoakasaka, Minato-ku, Tokyo 104-8468. T: 0334235141, F: 0334235001, E: sato-daigo@sei.co.jp

Sato, E., Consulting Engineer, Takenaka Corporation, Takenaka Research & Development Institute, 5-1, 1-chome Ohtsuka, Inzai, Chiba 270-1395. T: 0476471700, F: 0476473080, E: satou.eiji@takenaka.co.jp

Sato, H., University Professor, Japan National Defense Academy, 1-10-20, Hashirimizu, Yokosuka, Kanagawa 239-8686. T: 0468413810, F: 0468445913, E: satoh@cc.nda.ac.jp

Sato, H., Senior Researcher, Tokyo Electric Power Company, Civil Engineering Group, Power Engineering R&D Center, 4-1, Egasaki-cho, tsurumi-ku, Yokohama 230-8510. T: 0456133365, F: 0456133399, E: t0526186@pmail.tepco.co.jp

Sato, K., 4-3-1001, Shinkotoni, 9-jyo 1-chome, Kita-ku, Sapporo 001

Sato, K., Geotechnical Engineering Director, West Japan Railway Company, 3-8-21, Futabanosato, Higashi-ku, Hiroshima 732. T: 0822630761, F: 0822611258

Sato, K., Dr, College Professor, Dept. of Civil Engineering, Nagaoka College of Technology, 888 Nishikatakai, Nagaoka 940. T: 0258349273, F: 0258349284

Sato, K., Research Associate, Dept. of Civil Engineering, Fukuoka University, 8-19-1, Nanakuma, Jo-nanku, Fukuoka 814-80. T: 0928716631, F: 0928656031, E: sato@tsat.fukuoka-u.ac.jp

Sato, M., 8-10, Fujimi-cho, Hachioji 192

Sato, M., 46-2-15-101, Higashi-naruse, Isehara-shi, Kanagawa 259-11

Sato, T., Dr, University Professor, Disaster Prevention Research Institute, Kyoto University, Gokasho, Uji 611-0011. T: 0774384065, F: 0774384070, E: sato@catfish.dpri.kyoto-u.ac.jp

Sato, T., Dr, Dept. of Civil Engineering, Gifu University, Yanagido, Gifu 501-11. T: 0582301111, F: 0582301891

Sato, T., Geotechnical Engineer, Toyo Construction Co., Ltd., Higashikantoh Branch, Dai-2 Yamazaki, Bldg.4F, 2-13-1, Fujimi, Chuoh-ku, Chiba 260. T: 0432243625, F: 0432246728

Sato, T., Senior Planner for Regional Air Transport Infrastructure, Ministry of Land, Infrastructure and Transport, 2-1-3, Kasumigaseki, Chiyoda-ku, Tokyo 100-8989. T: 0352538717, F: 0352531658, E: satou-t2tj@mlit.go.jp

Sato, Y., Geotechnical Engineer, JDC Corporation, Technical Research Institute, 4036-1, Nakatsu, Aikawa-cho, Aikoh-gun, Kanagawa 243-03. T: 0462854924, F: 0462861642

Sawa, K., Dr, College Professor, Akashi College of Technology, 679-3, Nishioka, Uozumi-cho, Akashi Hyogo 674-8501. T: 0789466170, F: 0789466184, E: sawa@akashi.ac.jp

Sawada, K., Research Associate, Dept. of Civil Engineering, Gifu Univ., 1-1, Yanagido, Gifu 501-11. T: 0582932422, F: 0582301891

Sawada, S., University Professor, Disaster Prevention Research Institute, Kyoto University, Gokasho, Uji, Kyoto 606-0011. T: 0774384066, F: 0774384070, E: sawada@catfish.dpri.kyoto-u.ac.jp

Sawada, S., Geotechnical Engineer, OYO Corporation, Technical Center, 2-61-5, Toro-cho, Omiya, Saitama 330-8632. T: 0486679141, F: 0486679275, E: sawada-shun@oyo-net.oyo.co.jp

Sawada, T., Professor Emeritus, Kyoto University, (Home address)5-7, Ooharano, Kamizato, Ojika-cho, Nishikyo-ku, Kyoto 610-11

Sei, H., Chief Research Engineer, Obayashi Corporation, Simokiyoto, 4-640, Kiyose, Tokyo 204-8558. T: 0424951225, F: 0424959401, E: h.sei@tri.obayashi.co.jp

Sekiguchi, H., Dr, Professor, Disaster Prevention Research Institute, Kyoto University, Uji, Kyoto 611-0011. T: 0774384176, F: 0774384180, E: sekiguch@rcde.dpri.kyoto-u.ac.jp

Sekiguchi, K., Senior Research Engineer, NKK Corporation, 1-1, Minami-watarida-cho, Kawasaki-ku, Kawasaki 210-0855. T: 0443226222, F: 0443226519, E: ksekigut@lab.keihin.nkk.co.jp

Sekine, E., Research Engineer, Railway Technical Research Institute, 2-8-38, Hikari-cho, Kokubunji-shi, Tokyo 185. T: 0425737261, F: 0425737248

Seko, T., Managing Director, Chuo Kaihatsu Corporation, 3-13-5, Nishi-waseda, Shinjuku-ku, Tokyo 169. T: 0332083111, F: 0332083127

Sento, N., University Research Assistant, Tohoku University, 06 Aramaki-aza, Aoba, Aoba-ku, Sendai, Miyagi 980-8579. T: 0222177436, F: 0222177435, E: nsentoh@civil.tohoku.ac.jp

Shamoto, Y., 299-74, Takatsu, Yachiyo-shi, Chiba 276

Shibata, A., Director of Geotechnical Laboratory, Kowa Co., Ltd., 4-7-22, Toyano, Niigata-shi, Niigata. T: 0252815135, F: 0252810258, E: a-shibata@kowa-net.co.jp

Shibata, H., University Professor, Dept. of Civil Engineering, Kokushikan University, 4-28-1, Setagaya-ku, Tokyo 154. T: 0354813277, F: 0354813277

Shibata, T., University Professor, 6-Iga, Momoyama-cho, Fushimi-ku, Kyoto 612-8026. T/F: 0756017769, E: kyoto.SHIBATA@nifty.ne.jp

Shibazaki, M., 21-23, Akatsuka Shinmachi 3-chome, Itabashiku, Tokyo 175. T: 0339304622

Shibutani, O., Geotechnical Engineer, 1-179-19, Wagahara, Tokorozawa, Saitama 359-1162. T/F: 0429494478, E: os2@bh.mbn.or.jp

Shibuya, S., Dr, University Professor, Division of Structural and Geotechnical Engineering, Graduate School of Engineering, Hokkaido University, Sapporo 060-8628. T: 0117066193, F: 0117067204, E: shibuya@eng.hokudai.ac.jp

Shigemura, T., Dr, University Professor, National Defense Academy, 1-10-20, Hashirimizu, Yokosuka, Kanagawa 239-8686. T: 0468413810, Ext. 2363, F: 0468445913, E: shigemra@cc.nda.ac.jp

Shigeno, Y., Research Engineer, Takenaka Corporation, 5-14, 2-chome Minamisuna, Koto-ku, Tokyo 136. T: 0336473161, F: 0336450911

Shima, H., Dr, Research Manager, OYO Corporation, 2-2-19, Daitakubo, Urawa, Saitama 336. T: 0488825374, F: 0488828386

Shima, S., Associate Professor of University, Hiroshima Institute of Technology, 2-1-1, Miyake, Saeki-ku, Hiroshima 731-5193. T: 0829213121, F: 0829237083, E: shima@cc.it-hiroshima.ac.jp

Shimada, K., University Professor, Tokyo University of Agriculture and Technology, Faculty of Agriculture, 3-5-8, Saiwai-cho, Fuchu, Tokyo 183-8509. T/F: 0423675760, E: shimadak@cc.tuat.ac.jp

Shimazu, A., Consulting Engineer, 3-15-503, Tsukushino, Abiko, Chiba 270-1164. T/F: 0471848580, E: shimazu@mxh.mesh.ne.jp

Shimizu, E., Dr, University Professor, Dept. of Civil Engineering, Chiba Institute of Technology, 17-1, Tsudanuma 2-chome, Narashino 275. T: 0474780451, F: 0474780474, E: shimizu@ce.it-chiba.ac.jp

Shimizu, H., Geotechnical Engineer, Maeda Corporation, 1-39-16, Asahi-cho, Nerima-ku, Tokyo 179-8914. T: 0339772568, F: 0339772251, E: shimizuh@jcity.maeda.co.jp

Shimizu, K., Dr, University Professor, Kyushu Institute of Technology, 1-1, Sensui-cho, Tobata-ku, Kitakyushu 804-8550. T: 0938843105, F: 0938843100, E: shimizu@civil.kyutech.ac.jp

Shimizu, M., Dr, Associate Professor, Dept. of Civil Engineering, Faculty of Engineering, Tottori University, 4-101, Koyama-minami, Tottori 680-8552. T: 0857315290, F: 0857287899, E: mshimizu@cv.tottri-u.ac.jp

Shimizu, M., Consulting Engineer, AOI Engineering Co., Ltd., 154 Hutasemachi, Nakamura-ku, Nagoya 453. T: 0524131871, F: 0524131890

Shimizu, Y., Dr, University Associate Professor, Meijo University, Shiogamaguchi, 1-501, Tenpaku-ku, Nagoya 468-8502. T: 0528321151, F: 0528321178, E: shimuzuy@meijo-u.ac.jp

Shimizu, Y., 849-2, Oobano, Tabuse-cho, Kumage-gun, Yamaguchi 742-15

Shimobe, S., Dr, Full-time Lecturer, Faculty of Junior College, Nihon University, 24-1, Narashinodai 7-chome, Funabashi, Chiba 274. T: 0474695442, F: 0474617342

Shimodaira, T., Taisei Corporation

Shimokawa, E., University Professor, Faculty of Agriculture, Kagoshima University, 1-21-24, Korimoto, Kagoshima 890-0065. T: 0992858703, F: 0992853667, E: sabos@env.agri.kagoshima-u.ac.jp

Shimokura, H., Consulting Engineer, Nippon Koei Co., Ltd., 2304 Inanihara, Kukizaki-machi, Inashiki-gun, Ibaraki 300-1259. T/F: 0298712071

Shimomura, Y., Senior, Advisor, Hazama Corp., 2-5-8, Kita-Aoyama, Minato-ku, Tokyo 107-8658. T: 0334051111, F: 0334755817, E: shimomura@hazama.co.jp

Shimura, S., University Professor, Dept. of Civil Engineering, Meisei University, 2-1-1, Hodokubo, Hino, Tokyo 191. T: 0425919793, F: 0425919632

Shinjo, T., Dr, University Professor, University of the Ryukyus, Senbaru 1, Nishihara-cho, Okinawa 903-0213. T/F: 0988958780, E: shinjot@agr.u-ryukyu.ac.jp

Shinkawa, N., Fudokensetsu Kounosu-ryo, 1-1-34, Matsubara, Kounosu 365

Shino, K., Dr, University Professor, Faculty of Agriculture, Kochi University, Otsu 200, Monobe, Nankoku, Kochi 783. T: 0888645162, E: shino@cc.kochi-u.ac.jp

Shinomiya, H., General Manager, Mizushima Office, Kawatetsu Techno-Construction Co., Ltd., 1 Kawasaki-Dori, Mizushima, Kurashiki, Okayama 712-8074. T: 0864474510, F: 0864474513, E: shinomiya@ktc.co.jp

Shinomiya, K., 3-31-17-101, Isehara, Isehara 259-11

Shinsha, H., Manager, Penta-Ocean Construction Co., Ltd., Institute of Technology, 1534-1, Yonku-cho, Nishinasuno-machi, Nasu-gun, Tochigi 329-2746. T: 0287392100, F: 0287392132, E: Hiroshi.Shinsha@mail.penta-ocean.co.jp

Shinshi, Y., Groundwater Engineer, Konoike Construction Co., Ltd., 3-6-1, Kitakyuhoji-machi, Chuo-ku, Osaka 541-0057. T: 0662443617, F: 0662443676, E: shinshi_yh@konoike.co.jp

Shintani, N., Manager, Civil Engineering Dept., The Chugoku Electric Power Co., Inc., 4-33, Komachi, Naka-ku, Hiroshima 730-91. T: 0822410211, F: 0822425989

Shioi, Y., Dr, College Professor, Hachinohe Institute of Technology, 88 Ohbiraki, Myo, Hachinohe, Aomori 031-8501. T: 0178258081, F: 0178250722, E: shioiyu@hi-tech.ac.jp

Shiomi, T., Chief Researcher, Research and Development Institute, Takenaka Corp., 1-5-1, Ohtsuka, Inzai, Chiba 270-1395. T: 0476471700, F: 0476473060, E: shiomi.tadahiko@takenaka.co.jp

Shirai, K., Geotechnical Engineer, Takenaka Technical Research Laboratory, 5-1, Ohtsuka 1-chome, Inzai-shi, Chiba 270-13. T: 0476471700, F: 0476473070

Shiraishi, S., Dr, Chairman, Tashi Fudosan Co., Ltd., 8 Minamimotomachi, Shinjuku-ku, Tokyo 160-0012. T: 0333508560, F: 0333416455, E: XMB00215@nifty.ne.jp

Shirakawa, K., Geotechnical Engineer, Hanshin Consultants Co., Ltd., 2-5-24, Nishihommachi, Nishi-ku, Osaka 550. T: 065430203, F: 065438575

Shiraki, W., University Professor, Kagawa University, Dept. of Reliability-based Information Systems Engineering, 2217-20, Hayashi-cho, Takamatsu, Kagawa 761-0396. T/F: 0878642243, E: shiraki@eng.kagawa-u.ac.jp

Shirako, H., Consulting Engineer, Construction Project Consultants, No. 8 Matsuda Bldg.8F, 2-1-9, Okubo, Shinjuku-ku, Tokyo 169. T: 0332028122, F: 0332044839

Shiwakoti, D.R., Geotechnical Engineer, Geotechnical Investigation Laboratory, Port and Harbour Research Institute, 3-1-1, Nagase, Yokosuka 239-0826. T: 0468445025, F: 0468445058, E: shiwakoti_dinesh@hotmail.com

Shogaki, T., Dr, University Professor, National Defense Academy, 1-10-20, Hashirimizu, Yokosuka 239-8686. T: 0468413810, F: 0468445913, E: shogaki@cc.nda.ac.jp

So, E.K., Dr, University Professor, Dept. of Civil Engineering, Kanto Gakuin University, Mutsuura-cho, Kanazawa-ku, Yokohama 236. T: 0457812001, F: 0457848153

Sodekawa, M., 47-41, Sakasai, Kashiwa-shi, Chiba 277

Someya, T., Chief Engineer, Okumura Corporation, Tokyo Office, Civil Engineering Dept., 7-12-6, Nishikasai, Edogawa-ku, Tokyo 134. T: 0338041520, F: 0336754711

Song, D., Consulting Engineer, Kiso-Jiban Consultants Co., Ltd., Chubu Branch, kikui, 2-14-24, Nishi-ku, Nagoya 451-0044. T: 0525891058, F: 0525891275, E: song.dejun@kiso.co.jp

Song, M.Y., University Professor, Chungnam National University, Dept. of Geology, Taejon 305-764, Korea. T: 82428216423, F: 82428233722, E: mysong@cnu.ac.kr

Sramoon, A., Research Associate, Civil and Environmental Engineering Department, Nagaoka University of Technology, 1603-1, Kamitomioka, Nagaoka, Niigata 940-2188. T: 0258479620, F: 0258479600, E: aphichat@vos.nagaokaut.ac.jp

Suemasa, N., Associate Professor, Dept. of Civil Engineering, Musashi Institute of Technology, 1-28-1, Tamazutsumi, Setagaya-ku, Tokyo 158-8557. T: 0337033111, Ext. 3245, F: 0357072202, E: nsuemasa@eng.musashi-tech.ac.jp

Sueoka, T., Dr, Geotechnical Engineer, Taisei Research Center, Taisei Corporation, 344-1, Nase-cho, Totsuka-ku, Yokohama 245-0051. T: 0458147221, F: 0458147250, E: toru.sueoka@sakura.taisei.co.jp

Suetsugu, D., University Research Associate, National Defense Academy, 1-10-20, Hashirimizu, Yokosuka 239-8686. T: 0468413810, Ext. 3522, F: 0468445913, E:suetsugu@cc.nda.ac.jp

Sugahara, M., President, Maruhachi Co., Ltd., 2-29, Tonya-machi, Fukui-shi, Fukui 918-8231. T: 0776240808, F: 0776220582, E: maruhati@lilac.ocn.ne.jp

Sugano, T., Dr, Chief, Structural Dynamics Lab., Port & Harbour Research Institute, 3-1-1, Nagase, Yokosuka, Kanagawa 239. T: 0468445029, F: 0468440839

Sugawara, A., 1-43-6-40, Satato-cho, Moriguchi 570

Sugawara, N., General Manager, Tsukuba Technical R&D Center of OYO Corporation, Miyukigaoka, 34 Tsukuba, Ibaraki 305-0841. T: 0298516621, F: 0298515450, E: sugawara-noriaki@oyonet.oyo.co.jp

Sugie, S., Dr, Geotechnical Engineer, Obayashi Corporation, 640 Shimokiyoto 4-chome, Kiyose-shi, Tokyo 204-8558. T: 0424951097, F: 0424950903, E: sugie@tri.obayashi.co.jp

Sugii, T., Dr, Associate Professor, Dept. of Civil Engineering, Chubu University, 1200 Matsumoto-cho, Kasugai, Aichi 487-8501. T: 0568511111, F: 0568511495, E: nanto@isc.chubu.ac.jp

Sugimoto, M., Dr, Associate Prof., Dept. of Civil Engineering, Nagaoka Univ. of Technology, 1603-1, Kamitomioka-cho, Nagaoka, Niigata 940-2188. T: 0258479618, F: 0258479600, E: sugimo@nagaokaut.ac.jp

Sugimoto, T., Dr, Head of Department, Institute of Civil Engineering in Tokyo Metropolitan Government, 1-9-15, Minamisuna, Koto-ku, Tokyo 136-0075. T: 0356831522, F: 0356831515, E: takasugi@sc4.so-net.ne.jp

Sugimura, Y., University Professor, Dept. of Architecture and Building Science, School of Engineering, Tohoku University, Aoba 06, Aramaki, Sendai 980-8579. T: 0222177867, F: 0222177869, E: sugi@strmech.archi.tohoku.ac.jp

Sugito, M., Dr, University Professor, Gifu University, Yanagido, 1-1, Gifu 501-11. T: 0582932420, F: 0582301891, E: sugito@cive.gifu-u.ac.jp

Sugiyama, T., Chief Engineer, Railway Technical Research Institute, 2-8-38, Hikari-cho, Kokubunji-shi, Tokyo 185. T: 0425737265, F: 0425837398

Sumioka, N., Dr, Consulting Engineer, Chuden Engineering Consultants Co., Ltd., 2-3-30, Deshio, Minami-ku, Hiroshima 734-8510. T: 0822563351, F: 0822561968, E: sumioka@cecnet.co.jp

Sun, D.A., Research Associate, Dept. of Civil Engineering, Nagoya Institute of Technology, Gokiso-cho, Showa-ku, Nagoya 466. T/F: 0527355497, E: sun@doboku2.ace.nitech.ac.jp

Sunaga, M., Planning Manager, Railway Technical Research Institute, 2-8-38, Hikari-cho, Kokubunji, Tokyo 185-8540. T: 0425737320, F: 0425737209, E: makoto@rtri.or.jp

Sunami, S., Chief of Engineering Research, Nikken Sekkei Nakase Geotechnical Institute, 4-6-2, Kouraibashi, Chuo-ku, Osaka 541-8528. T: 0662296372, F: 0662012433, E: sunami@nikken.co.jp

Suwa, S., Managing Director, Geo-Research Institute, 4-3-2, Itachibori, Nishi-ku, Osaka 550-0012. T: 0665393135, F: 0665786255, E: suwa@geor.or.jp

Suzuki, A., Dr, Professor, Dept. of Civil and Environmental Engineering, Kumamoto University, Kurokami, 5-39-1, Kumamoto 860-8555. T/F: 0963423539, E: a-suzuki@kumamoto-u.ac.jp

Suzuki, H., Dr, Associate Professor, Dept. of Civil Engineering, The University of Tokushima, 2-1, Minami, Jousanjima, Tokushima 770. T: 0886232311, F: 0886549632

Suzuki, H., Electric Power Civil Engineer, Chubu Electric Power Corporation, 1 Toshin-cho, Higashi-ku, Nagoya-shi, Aichi 461-91. T: 0529518211, F: 0529733173

Suzuki, K., Dr, University Professor, Dept. of Civil & Environmental Engineering, Saitama Univ., 255 Shimo-Okubo, Urawa, Saitama 338-8570. T: 0488589572, F: 0488587374, E: suzuki@dice.dr5w.saitama-u.ac.jp

Suzuki, M., Dr, Researcher, Shimizu Corporation, Izumi Research Institute, Fukoku Seimei Bldg., 2-2-2, Uchisaiwai-cho, Chiyoda-ku, Tokyo 100-0011. T: 0335088101, F: 0335082196, E: suzuki@ori.shimz.co.jp

Suzuki, M., University Research Associate, Dept. of Civil Engineering, Faculty of Engineering, Yamaguchi University, 2-16-1, Tokiwadai, Ube 755-8611. T: 0836859303, F: 0836859301, E: msuzuki@jim2.civil.yamaguchi-u.ac.jp

Suzuki, O., Dr, Managing Director, Komoto Construction Inc., 2544 Kita-Narushima-machi, Tatebayashi, Tokyo 374-0057. T: 0276723321, F: 0276751500, E: o-suzuki@komoto.co.jp

Suzuki, S., Geotechnical Engineer, OYO Corporation, WBG Malib West 30F, 2-6, Nakase, Mihama-ku, Chiba-shi, Chiba 261-7130. T: 0432994111, F: 0432994113, E: suzuki-shinji@oyonet.oyo.co.jp

Suzuki, T., Dr, University Professor, Kitami Institute of Technology, 165 Kouen-cho, Kitami, Hokkaido 090-8507. T: 0157269475, F: 0157239408, E: suzuki@rock.civil.kitami-it.ac.jp

Suzuki, Y., Deputy General Manager, Dept. of Energy & Nuclear Engineering, Takenaka Corporation, 21-1, 8-chome Ginza, Chuo-ku, Tokyo 104. T: 0335427100, F: 0335426855

Suzuki, Y., Geotechnical Engineer, Kajima Corporation, 19-1, Tobitakyu 2-chome, Chofu, Tokyo 182-0036. T: 0424897097, F: 0424892020, E: suzuki@katri.kajima.co.jp

Suzuki, Y., General Manager, Takenaka Corporation, 1-5-1, Ohtsuka, Inzai-shi, Chiba 270-1395. T: 0476471700, F: 0476473080, E: suzuki.yoshio@takenaka.co.jp

Suzuki, Y., Structural Engineer, 65-5, Yoshino, Yamagata-shi, Yamagata 990-0074. T: 0236414367, F: 0236414700, E: yu1_suzuki@nifty.com

Tachikawa, T., Dr, Meiji University, School of Science and Engineering, Architecture Dept., 1-501, Shiogamaguchi, Tenpakuku, Nagoyashi. T: 0528321151, Ext. 5216, F: 0528519221

Taga, N., University Professor, Fukuoka University, 8-19-1, Nanakuma, Jonann-ku, Fukuoka 814-0180. T: 0928716631, Ext. 6511, F: 0928656031, E: taga@fukuoka-u.ac.jp

Tagaya, R., Dr, General Manager, Aratani Civil Engineering Consultants Co., Ltd., General Engineering Department, 14-15, Funairiminami 4-chome, Naka-ku, Hiroshima 730. T: 0822345660, F: 0822344961

Taira, K., Registered Engineer, 6-21-4, Oowada-machi, Hachioji, Tokyo 192-0045. T/F: 0426428916, E: k.taira@tokyo.kajima.co.jp

Taji, Y., 3-28-1-407, Tokiwadaira, Matsudo 270

Tajiri, N., Dr, Consulting Engineer, Fukken Co., Ltd., 2-10-11, Hikarimachi, Higashi-ku, Hiroshima 732-0052. T: 0825061825, F: 0825061897, E: tajiri@fukken.co.jp

Takaba, N., Chief Civil Engineer, Design, Nikken Sekkei Ltd., Civil Engineering Office, 1-4-27, Koraku, Bunkyo-ku, Tokyo 112. T: 0338133361, F: 0338170515

Takada, N., University Professor, Dept. of Civil Engineering, Osaka City University, Sugimoto, Sumiyoshi-ku, Osaka 558-8585. T: 0666052724, F: 0666052769, E: takada@civil.eng.osaka-cu.ac.jp

Takahashi, A., University Research Associate, Civil Engineering Course, Dept. of International Development Engineering, Tokyo Institute of Technology, 2-12-1, O-okayama, Meguro-ku, Tokyo 152-8552. T: 0357342592, F: 0357343577, E: takihiro@geotech.cv.titech.ac.jp

Takahashi, K., Dr, Research Engineer, Geotechnical Engineering Division, Port & Harbour Research Institute, Ministry of Transport, 1-1, Nagase 3-chome, Yokosuka 239. T: 0468445022, F: 0468440618, E: takahashik@phri.go.jp

Takahashi, K., Consulting Engineer, Suimon Gijyutsu Consultant Co., Ltd., 1-24-2, Takasu, Mihama-ku, Chiba,

Chiba 261. T: 0432792311, F: 0432700654, E: takahashi@hydrology-tec.com

Takahashi, M., Senior Researcher, Civil Engineering & Architecture Dept., Power Engineering R&D Center, Tokyo Electric Power Co., 4-1, Egasaki-cho, Tsurumi-ku, Yokohama 230. T: 0455858600

Takahashi, S., Chief Research Engineer, Geo-technical Engineering Dept., Technical Research Institute, Obayashi Corporation, 4-640, Shimokiyoto, Kiyose, Tokyo 204. T: 0424950910, F: 0424950903, E: s-taka@tri.obayashi.co.jp

Takami, Y., Consulting Engineer, 2-7-17, Shimohara, Nishi-tokyo, Tokyo 202-0004. T: 0424222464, E: u-takami@ananet.or.jp

Takayama, M., Geotechnical Engineer, 3-22-10, Nishi-nooka, Nishi-ku, Fukuoka 819-0046. T/F: 0928851700

Takeda, S., General Manager, Docon Co., Ltd., 4-1, Atsubetsu-cho, 1-5, Atsubetsu-ku, Sapporo 004-8585. T: 0118011570, F: 0118011571, E: st671@mb.docon.co.jp

Takeda, T., Managing Director, Onoda Chemico Asia Pte. Ltd., 15 West Coast Highway #02-01/02, Pasir Panjang Bldg., Singapore 117861. T: 657742274, F: 657746614, E: chemasia@mbox5.singnet.com.sg

Takehana, K., 1-9, Takamatsu 1-chome, Toshima-ku, Tokyo 171. T: 0339572065

Takehara, N., Geotechnical Engineer, Tokyo Soil Research Co., Ltd., Tsukuba General Laboratory, 2-1-12, Umezono, Tsukuba, Ibaraki 305. T: 0298519501, F: 0298519559, E: LEDO2635@niftyserve.or.jp

Takehara, Y., Geotechnical Engineer, Fudo Construction Co., Ltd., 1-2-1, Taito, Taito-ku, Tokyo 110. T: 0338376097, F: 0338376158

Takei, A., 192 Kinugasa, Kaiboshi-cho, Kamigyo-ku, Kyoto 603. T: 0754633976

Takei, M., Geotechnical Engineer, Nishimatsu Construction Co., Ltd., 2570-4, Shimotsuruma, Yamato-shi, Kanagawa 242-8520. T: 0462751135, F: 0462756796, E: takei@ri.nishimatsu.co.jp

Takeichi, K., Dr, University Professor, Faculty of Engineering, Hokkai Gakuen University, South 26, West 11, Chuo-ku, Sapporo 064-0926. T: 0118411161, F: 0115512951, E: takeichi@cvl.hokkai-s-u.ac.jp

Takemura, J., Dr, Associate Professor, Tokyo Institute of Technology, 2-12-1, O-okayama, Meguro-ku, Tokyo 152-8552. T: 0357342592, F: 0357343577, E: jtakemur@geotech.cv.titech.ac.jp

Takeshita, S., University Professor, Ritsumeikan University, Kusatsushi, Nojicho 1916, Kusatsu 525. T: 0775661111, F: 0775612667, E: takesita@bkc.ritsumei.ac.jp

Takeshita, Y., Dr, University Associate Professor, Okayama University, 1-1, Tsushimanaka 3-chome, Okayama 700-8530. T/F: 0862518153, E: yujitake@cc.okayama-u.ac.jp

Takesue, K., Geotechnical Engineer, Kajima Technical Research Institute, 19-1, Tobitakyu 2-chome, Chofu-shi, Tokyo 182-0036. T: 0424897098, F: 0424892020, E: takesue@katri.kajima.co.jp

Taki, M., Project Engineer, Fukken Co., Ltd., Yokohama Branch Office, 6-83, Onoe-cho, Naka-ku, Yokohama 231-0015. T: 0456649551, F: 0452244877, E: taki@fukken.co.jp

Tamaki, S., University Teacher, The Faculty of Science and Technology, Nihon University, Kanda Surugadai, 1-8, Chiyoda-ku, Tokyo 101. T: 0332933251

Tamano, T., Dr, University Professor, Dept. of Civil Engineering, Osaka Sangyo Univ., Nakagaito, 3-1-1, Daito, Osaka 574-8530. T: 0728753001, F: 0728755044, E: tamano@ce.osaka-sandai.ac.jp

Tamari, Y., Professional Engineer, Tokyo Electric Power Services, 3-3-3, Higashi-ueno, Taito-ku, Tokyo 110-0015. T: 0344645472, F: 0344645490, E: etamari@tepsco.co.jp

Tamate, S., Senior Researcher, National Institute of Industrial Safety, 1-4-6, Umezono, Kiyose, Tokyo 204-0024. T: 0424914512, F: 0424917846, E: tamate@anken.go.jp

Tamura, A., 2-5-1-801, Satsukidaira, Misato 341

Tamura, M., Dr, Senior Research Engineer, Building Research Institute, 1 Tatehara, Tsukuba, Ibaraki 305. T: 0298642151, F: 0298642989

Tamura, S., University Associate Professor, Shinshu University, Faculty of Engineering, Architecture and Civil Engineering, 4-17-1, Wakasato, Nagano 380-8553. T/F: 0262695362, E: tamura@gipwc.shinshu-u.ac.jp

Tamura, T., Dr, University Professor, Kyoto University, Yoshida-honmachi, Sakyo-ku, Kyoto 606-8501. T: 0757534793, F: 0757534794, E: tamura@baseball. kuciv.kyoto-u.ac.jp

Tamura, Y., Consulting Engineer, Integrated Geotechnology Institute Ltd., Akasaka Fukugen Bldg.7F, 2-15-16, Akasaka, Minato-ku, Tokyo 107-0052. T: 0335823373, F: 0335823509, E: y_tamura@igi.co.jp

Tanabashi, Y., Dr, University Professor, Dept. of Civil Engineering, Faculty of Engineering, Nagasaki University, 1-14, Bunkyo-machi, Nagasaki 852-8521. T: 0958479356, F: 0958483624, E: tanabasi@net.nagasaki-u.ac.jp

Tanahashi, H., University Professor, Kyoto Prefectural University, Hangi-cho, Simogamo, Sakyo-ku, Kyoto, 606-8522. T/F: 0757035428, E: tana@kpu.ac.jp

Tanaka, A., 6-63-1, Nakanohigashi, Aki-ku, Hiroshima 739-03

Tanaka, H., Dr, Research Engineer, Port & Harbour Research Institute, 3-1-1, Nagase, Yokosuka 293-0826. T: 0468445025, F: 0468445058, E: tanakah@cc.phri.go.jp

Tanaka, H., Dr, President of Cons. Engineering, Tokusyu Kozo Sekkei Co., 1813-1, Takasu, Kochi 780. T: 0888610369, F: 0888610258

Tanaka, I., Consulting Engineer, OYO International Corporation, Rokubancho Kyodo Bldg.2F, Rokubancho, Chiyoda-ku, Tokyo 102-0085. T: 0352115181, F: 0352115184, E: ichiro-t@kk.iij4u.or.jp

Tanaka, M., 2-24-4, Fujigaoka, Midori-ku, Yokohama 227

Tanaka, M., Senior Research Engineer, Port and Harbour Research Institute, Ministry of Land, Infrastructure and Transport, 3-1-1, Nagase, Yokosuka, Kanagawa 239-0826. T: 0468445025, F: 0468445058, E: tanakam@cc.phri.go.jp

Tanaka, M., Consulting Engineer, Geo-Research Institute, 4-3-2, Itachibori, Nishi-ku, Osaka 550-0012. T: 0665392971, F: 0665786560, E: makoto@geor.or.jp

Tanaka, O., Manager of Tanaka Research Office Technological Consultancy Division, Fujita Corporation, 4-6-15, Sendagaya, Shibuya-ku, Tokyo 151. T: 0337962451, F: 0337962365

Tanaka, R., Director, Chuo Fukken Consultants Co., Ltd, 8-29-32, Nishimiyahara 1-Chome, Yodogawa-ku, Osaka 532-0004. T: 0663931132, F: 0663931148, E: tanaka_r@ cfk.co.jp

Tanaka, S., Director, OYO Corporation, 4-2-6, Kudankita, Chiyoda-ku, Tokyo 102-0073. T: 0332340811. F: 0332396425, E: tanaka-soichi@oyonet.oyo.co.jp

Tanaka, S., University Professor Emeritus, Construction Engineering Research Institute Foundation, 1-3-10, Tsurukabuto, Nada-ku, Kobe 651-0011. T: 0788511850, F: 0788515454, E: kensetsu@pearl.ocn.ne.jp

Tanaka, S., Geotechnical Engineer, Kajima Technical Research Institute, 19-1, Tobitakyu 2-chome, Chofu-shi, Tokyo 182. T: 0424897098, F: 0424892020

Tanaka, T., University Professor, Dept. of Agriculture, Meiji University, 1-1-1, Higashi-mita, Tama-ku, Kawasaki 214. T: 0449347160, F: 0449347902, E: tanakat@isc.meiji.ac.jp

Tanaka, T., Dr, University Professor, Faculty of Agriculture, Kobe University, Rokkodai 1, Nada, Kobe 657-8501. T/F: 0788035901, E: ttanaka@kobe-u.ac.jp

Tanaka, Y., University Professor, Research Center for Urban Safety and Security(RCUSS), Kobe University, Nada, Kobe 657-8501. T: 0788036058, F: 0788036394, E: ytgeotec@ kobe-u.ac.jp

Tanaka, Y., Consulting Engineer, Daitou Boring Sakusen Kogyosho, 2-8-12, Dogashiba, Tennoji-ku, Osaka 543-0033. T: 0667711362, F: 0667723901, E: daitou1@vesta.ocn.ne.jp

Tanaka, Y., Dr, Professor Emeritus, 5-23-13, Sagamiono, Sagamihara, Kanagawa 228. T/F: 0427424511

Tanaka, Y., Dr, Research Engineer, Central Research Institute of Electric Power Industry, 1646 Abiko, Abiko-shi, Chiba 270-1194. T: 0471821181, F: 0471842941, E: yu-tanak@ criepi.denken.or.jp

Tang, X.W., University Professor, Institute of Lowland Technology, Saga University, Honjo-machi 1, Saga 840-8502. T: 0952288695, F: 0952264998, E: tangxw@cc. saga-u.ac.jp

Tang, Y.X., Geotechnical Engineer, Keihin Office, Kanmon Kowan Kensetsu, Co., Ltd., 3-10, Honmoku Juniten, Naka-ku, Yokohama 231-0803. T: 0456239061, F: 0456239062, E: yxtang@konmon-const.co.jp

Tani, K., Dr, Research Engineer, Criepi., 1646 Abiko, Abiko-shi, Chiba 270-11. T: 0471821181, F: 0471832962

Tani, S., Dr, Geotechnical Engineer, National Research Institute of Agricultural Engineering, 2-1-2, Kannondai, Tsukuba-shi, Ibaraki 305. T: 0298387575, F: 0298387609

Tanifuji, S., Consulting Engineer, 5-35-4, Sakurajosui, Setagaya-ku, Tokyo 156-0045. T: 0333020263, F: 0333043939

Tanigawa, S., Consulting Engineer, OYO Corporation, 2-1, Ohtsuka 3-chome, Bunkyo-ku, Tokyo 112

Tanikawa, M., Civil Engineer, Nittoc Construction Co., Ltd., Technical Division Tsukuba 1, 5-5, Tohkodai, Tsukuba, Ibaraki 300-26. T: 0298478670, F: 0298478664

Tanimoto, C., University Professor, Dept. of Civil Engineering, Kyoto University, Yoshida-Honmachi, Sakyo-ku, Kyoto 606. T: 0757535105, F: 0757610646

Tanimoto, K., 4-25-6, Uzumoridai, Higashi-nada-ku, Kobe 658. T: 0788416300

Tanizawa, F., Chief Engineer, 440-3-212, Morooka-cho, Kohoku-ku, Yokohama 222-0002. T: 0455441465, E: fusao.tanizawa@sakura.taisei.co.jp

Tano, H., Dr, University Professor, College of Engineering, Nihon University, Tamura-machi, Koriyama, Fukushima 979-66. T: 0249441300, F: 0249432894

Tanoue, Y., Geotechnical Engineer, Kiso-Jiban Consultants Co., Ltd., 2-16-7, Hara, Sawara-ku, Fukuoka 814. T: 0928312511, F: 0928315445

Tarumi, H., Dr, General Manager, Railway Technical Research Institute, 2-8-38, Hikari-cho, Kokubunji, Tokyo 185-8540. T: 0425737480, F: 0425737372, E: tarumi@ rtri.or.jp

Tatebe, H., Dr, Professor, Dept. of Civil Engineering, Aichi Institute of Technology, Yakusacho 1247, Toyota 470-03. T: 0565488121, F: 0565483749

Tateyama, K., Dr, Lecturer, Dept. of Civil Engineering, Kyoto University, Yoshida-Honmachi, Sakyo-ku, Kyoto 606-01. T: 0757535106, F: 0757610646

Tateyama, M., Geotechnical Engineer, Railway Technical Research Institute, Kokubunji-shi, Tokyo 185. T: 0425737261, F: 0425737248

Tatsui, T., 2-2-8-304, Yanaka, Adachi-ku, Tokyo 120

Tatsuoka, F., Dr, Professor, University of Tokyo, 7-3-1, Hongo, Bunkyo-ku, Tokyo 113-8656. T: 0358416120, F: 0358418504, E: tatsuoka@geot.t.u-tokyo.ac.jp

Tayama, S., Assistant Section Chief, Research Institute, Japan Highway Public Corporation, 1-4-1, Tadao, Machida-shi, Tokyo 194. T: 0427911621, F: 0427928650, E: tayama@jhri.japan-highway.go.jp

Terada, M., Geotechnical Engineer, Technical Research Institute, Okumura Corporation, 387 Ohsuna, Tsukuba, Ibaraki 300-2612. T: 0298651714, F: 0298650782, E: oku06250@gm.okumuragumi.co.jp

Terashi, M., Dr, Principal, Nikken Sekkei Geotechnical Institute, 4-11-1, Minami-kase, Saiwai-ku, Kawasaki 212-0055. T: 0445991151, F: 0445999444

Tobita, J., Associate Professor, Dept. of Civil and Architecture, Nagoya University, Furo-cho, Chikusa-ku, Nagoya 464-8603. T: 0527893754, F: 0527893768, E: tobita@ sharaku.nuac.nagoya-u.ac.jp

Tobita, Y., University Professor, Dept. of Civil and Environmental Engineering, Tohoku-Gakuin University, Tagaiyo 985-8537. T: 0223687396, F: 0223687070, E: tobita@tjcc. tohoku-gakuin.ac.jp

Todo, H., Geotechnical Engineer, Kiso-Jiban Consultants Co., Ltd., 1-11-5, Kudan-kita, Chiyoda-ku, Tokyo 102-8220. T: 0332394451, F: 0332394597, E: todo.hiroaki@kiso.co.jp

Todoroki, Y., 19-3-2-105, Takahama, Chiba 260

Togari, A., Researcher, East Japan Railway Company, Tokyo Kotsu Kaikan 7F, 2-10-8, Yuraku-cho, Chiyoda-ku, Tokyo

100-0006. T: 0332111118, F: 0352198678, E: ota@head. jreast.co.jp

Tohda, J., Dr, University Associate Professor, Dept. of Civil Engineering, Faculty of Engineering, Osaka City University, 3-3-138, Sugimoto, Sumiyoshi-ku, Osaka 558-8585. T/F: 066052725, F: tohda@civil.eng.osaka-cu.ac.jp

Tohno, I., Dr, Section Leader, National Institute for Environmental Studies, 16-2, Onogawa, Tsukuba, Ibaraki 305-0053. T: 0298502484, F: 0298502576, E: tohno@nies.go.jp

Toki, S., Doctor of Engg., Hokkaido Institute of Technology, 7-15, Maeda, Teine-ku, Sapporo 006-8585. T: 0116812161 (319), F: 0116840522, E: toki@hit.ac.jp

Tokimatsu, K., Dr, Professor, Tokyo Institute of Technology, O-okayama, Meguro-ku, Tokyo 152-8552. T: 0357343160, F: 0357342925, E: kohji@o.cc.titech.ac.jp

Tominaga, K., Dr, University Prof., Graduate School for IDEC., Hiroshima Univ., 1-7-2, Kagamiyama, Higashi-hiroshima 739. T: 0824247797, F: 0824227194, E: tomi@ue.ipc.hiroshima-u.ac.jp

Tominaga, M., Professional Engineer, JP, 2-18-10, Nishishiba, Kanazawa-ku, Yokohama 236-0017. T/F: 0457824613, E: tominaga@mtc.biglobe.ne.jp

Tominaga, S., Managing Director, Kajitani Engineering Co., Ltd., 1-29-5, Shinmachi, Nishi-ku, Osaka 550-0013. T: 0665324165, F: 0665324169, E: tominaga-shohei@kajitani.co.jp

Tomita, R., Manager of Chiba Branch, Koa Kaihatsu Co., Ltd., 970-9, Miyako-cho, Chuo-ku, Chiba 260. T: 0432324891, F: 0432327981

Tomohisa, S., Dr, College Professor, Akashi College of Technology, 679-3, Nishioka, Uozumi, Akashi, Hyogo 674-8501. T: 0789466172, F: 0789466184, E: tomohisa@akashi.ac.jp

Tomosawa, H., Managing Director, Fudo Construction Co., Ltd., Taito, 1-2-1, Taito-ku, Tokyo 110. T: 0338376033, F: 0338347396

Tonosaki, A., Dr, Associate Professor, Dept. of Civil Engineering, Kanazawa Inst. of Technology, Oogigaoka, 7-1, Nonoichi-machi, Kanasawa 921. T: 0762481100

Tonouchi, K., Consulting Engineer, OYO Corporation, Ichigaya Bldg., 2-6, Kudan Kita 4-chome, Chiyoda-ku, Tokyo 102. T: 0332340811, F: 0332625169

Torii, T., Consulting Engineer, Construction Project Consultants Inc., 3-5-25, Utsubo-honmachi, Nishi-ku, Osaka 550-0004. T: 0664414614, F: 0664483686, E: torii@cpcinc.co.jp

Toriihara, M., Geotechnical Engineer, Obayashi Corporation, Geotechnical Engineering Dept., Technical Research Institute, 640 Shimokiyoto 4-chome, Kiyose-shi, Tokyo 204-8588. T: 0424950913, F: 0424950903, E: m.tori@tri.obayashi.co.jp

Towhata, I., Professor, Dept. of Civil Engineering, University of Tokyo, 7-3-1, Hongo, Bunkyo-ku, Tokyo 113-8656. T: 0358416121, F: 0358418504, E: towhata@geot.t.u-tokyo.ac.jp

Toyosawa, Y., Research Officer, National Institute of Industrial Safety, Ministry of Health, Labour and Welfare, 1-4-6, Umezono, Kiyose-shi, Tokyo 204-0024. T: 0424914512, F: 0424917846, E: toyosawa@anken.go.jp

Toyota, H., University Associate Professor, Nagaoka University of Technology, 1603-1, Kamitomioka, Nagaoka, Niigata 940-2188. T: 0258479619, F: 0258479600, E: toyota@vos.nagaokaut.ac.jp

Tsubakihara, Y., Dr, Geotechnical Engineer, Takenaka Corporation, 5-1, 1-chome Ohtsuka, Inzai, Chiba 270-1395. T: 0476471700, F: 0476473080, E: tubakihara.yasunori@takenaka.co.jp

Tsuboi, H., Doctoral Engineer & Manager, Fudo Construction Co., Ltd., 1-2-1, Taito, Taito-ku, Tokyo 110-0016. T: 0338376170, F: 0338376158, E: htsuboi@fudo.co.jp

Tsubota, K., Branch Manager, Kiso-Jiban Consultants Co., Ltd., 11-1-5, Kaminagoya, Nishi-ku, Nagoya 451. T: 0525223171, F: 0525242729

Tsuchida, H., Geotechnical Engineer, Nippon Steel Corporation, 6-3, Otemachi 2-chome, Chiyoda-ku, Tokyo 100-8071. T: 0332755894, F: 0332755648, E: tsuchida.hajime@eng.nsc.co.jp

Tsuchida, T., Dr, Geotechnical Engineer, Port and Harbour Research Institute, 3-1-1, Nagase, Yokosuka, Kanagawa 239-0826. T: 0468445021, F: 0468444577, E: tsuchida@cc.phri.go.jp

Tsuchikura, T., Dr, University Associate Professor, Maebashi Institute of Technology, 460-1, Kamisadori-cho, Maebashi, Gunma 371-0816. T: 0272657305, F: 0272653837, E: tsuchi@maebashi-it.ac.jp

Tsuchiya, H., Dr, Geotechnical Engineer, Kiso-Jiban Consultants Co., Ltd., 2-14-1, Ishikawa-cho, Ohta-ku, Tokyo 145-0061. T: 0337276158, F: 0337276247, E: tsuchiya.hisashi@kiso.co.jp

Tsuchiya, T., Dr, University Professor, Muroran Institute of Technology, 27-1, Mizumoto-cho, Muroran, Hokkaido 050-8585. T/F: 0143465215, E: tsuchi@mmm.muroran-it.ac.jp

Tsuchiya, T., Geotechnical Engineer, Takenaka Corporation, Research & Development Institute, 5-1, Ohtsuka 1-chome, Inzai-shi, Chiba 270-1395. T: 0476471700, F: 0476473080, E: tsuchiya.tomio@takeneka.co.jp

Tsukahara, J., Geotechnical Engineer, Chuo Kaihatsu Co., Ltd., Osaka Branch, 3-34-12, Tarumi-cho, Suita, Osaka 592-0062. T: 0663863691, F: 0663863020, E: jun_tsukahara@ckc-unet.ocn.ne.jp

Tsukakoshi, H., Consulting Engineer, 694-5, Tokoro-Kuki, Kuki 346. T: 0480233186

Tsukamoto, H., Consulting Engineer, Construction Project Consultants Inc., 3-5-25, Utsubohonmachi, Nishi-ku, Osaka 550-0004. T: 0664414721, F: 0664483683, E: tsukamoto@cpcinc.co.jp

Tsukamoto, T., Fudo kawagoe-ryo, 100-1, Aza-Nishiyashiki, Ooaza-Toyoda, Kawagoe-cho, Mie-gun, Mie 512

Tsukamoto, Y., University Lecturer, Dept. of Civil Engineering, Science University of Tokyo, 2641 Yamazaki, Noda, Chiba 278-8510. T: 0471241501, Ext. 4004, F: 0471239766, E: ytsoil@rs.noda.sut.ac.jp

Tsunematsu, S., Dr, President, Northern Japan Soil General Laboratory Co., Ltd., West 5-1, Nakanuma, Higashi-ku, Sapporo 065. T: 0117911651, F: 0117915241, E: tunematu@nj-soil.co.jp

Tsushima, M., Dr, Professor, Akita National College of Technology, 1-1, Iijima, bunkyo-cho, Akita 011-8511. T/F: 0188476073, E: tsushima@ipc.akita-nct.ac.jp

Uchida, A., Geotechnical Engineer, Hazama Corp., 2-12-3, Numakage, Urawa, Saitama. T: 067677275, F: 067677278

Uchida, A., Research Engineer, Research & Development Institute, Takenaka Corporation, 1-5-1, Ohtsuka, Inzai, Chiba 270-1395. T: 0476471700, F: 0476473080, E: uchida.akihiko@takenaka.co.jp

Uchida, A., Geotechnical Engineer, Kiso-Jiban Consultants Co., Ltd., 1-11-20, Chuou, Matsuyama, Ehime 791-8015. T: 0899275808, F: 0899275812, E: uchida.atsushi@kiso.co.jp

Uchida, K., Dr, University Professor, Dept. of Ag. & Env. Eng., Kobe University, 1-1, Rokkodai-cho, Nada-ku, Kobe 657-8501. T/F: 0788035900, E: uchidak@kobe-u.ac.jp

Uchida, K., Managing Director, Trans Tokyo Bay Highway Corp., Ichiban-cho, NN Bldg., 15-5, Ichiban-cho, Chiyoda-ku, Tokyo 102. T: 0332396582

Uchimura, T., University Research Associate, Dept. of Civil Engineering, University of Tokyo, 7-3-1, Hongo, Bunkyo-ku, Tokyo 113-8656. T: 0358416124, F: 0358418504, E: uchimura@civil.t.u-tokyo.ac.jp

Uchiyama, K., Consulting Engineer, Urban & Ground Engineering Laboratory Co., Ltd., 174-2, Mameguchidai, Naka-ku, Yokohama 231. T: 0456225667

Udaka, T., Chief Executive, Jishin Kogaku Kenkyusho, 2-19-108, Tomihisa-cho, Shinjuku-ku, Tokyo 162-0067. T: 0332268733, F: 0332268735, E: udaka@flush.co.jp

Ue, S., College of Technology Associate Professor, Tokuyama College of Technology, 3538 Takajo, Kume, Tokuyama, Yamaguchi 745-8585. T/F: 0834296321, E: ue@tokuyama.ac.jp

Ueda, T., Dr, Manager, Office of LNG Facilities, Takenaka Corporation, 28-21-1, Ginza, Chuo-ku, Tokyo 104-8182. T: 0335427728, F: 0335450974, E: ueda.takeo@takenaka.co.jp

Uehara, H., Director, Uehara Geotechnical Research Institute, Tsubokawa Bldg.3F, 165 Tsubokawa, Naha-shi, Okinawa 900-0025. T: 0988368485, F: 0988361137, E: ugeotech@lime.ocn.ne.jp

Uematsu, T., Dr,. Eng., Hokkaido Prefectual Cold Region Housing and Urban Research Institute, 4-jo, 1-chome, 24ken, Nishi-ku, Sapporo 063-0804. T: 0116214211, F: 0116214215, E: uematsu@hri.pref.hokkaido.jp

Ueno, K., University Lecturer, Dept. of Engineering, Faculty of Engineering, The University of Tokushima, 2-1, Minami-Jyosanjima-cho, Tokushima 770-8506. T/F: 0886567342, E: ueno@ce.tokushima-u.ac.jp

Ueno, S., Engineering Geologist, OYO Co., 61-5, Torocho 2-chome, Omiya-shi, Saitama 330. T: 0486651811, F: 0486679340

Ueshita, K., Dr, University Professor, 614, 3-chome, Inokoshi, Meito-ku, Nagoya 465-0021. T/F: 0527724385, E: ueshita@bronze.ocn.ne.jp

Ueta, K., Denryoku Chuokenkyusho, Kashiwa-ryo, 3-8-15, Akehara, Kashiwa 277

Ueta, Y., Consulting Engineer, Hanshin Consultants Co., Ltd., 2-860-1, Shijo-ooji, Nara 630-8014. T: 0742360212, F: 0742361989, E: ueta@hanshin-consul.co.jp

Ugai, K., Dr, Prof., Gunma University, Tenjin 1, Kiryu, Gunma 376. T/F: 0277301620

Ukaji, F., Consulting Engineer, Japanese Civic Consulting Company, Nippori UC Bldg.6F, 26-2, Nippori 2-chome, Arakawa-ku, Tokyo 116-0013. T: 0356047500, F: 0356047555

Umehara, Y., Dr, Executive Director, Yachiyo Engineering Co., Ltd., 10-21, 1-chome Nakameguro, Meguro-ku, Tokyo 153-8639. T: 0337151231, F: 0337105910, E: Umehara@yachiyo-eng.co.jp

Umemura, J., University Lecturer, Nihon University, Tamura-machi, Nakagawara, Koriyama, Fukushima 963-8642. T: 0249568709, F: 0249568858, E: umemura@ce.nihon-u.ac.jp

Umezaki, T., Dr, Associate Professor, Dept. of Civil Engineering, Shinshu University, 4-17-1, Wakasato, Nagano 380-8553. T/F: 0262695291, E: umezaki@gipwc.shinshu-u.ac.jp

Uno, K., Dr, University Professor, Kyushu Kyoritsu University, 1-8, Jiyugaoka, Yahatanishi-ku, Kitakyushu, Fukuoka 807-8585. T/F: 0936933226, E: uno@kyukyo-u.ac.jp

Uno, T., Dr, Professor, Gifu University, 1-1, Yanagido, Gifu-shi, Gifu 501-1193. T/F: 0582932415, E: unotakao@cc.gifu-u.ac.jp

Ushiro, T., Consulting Engineer, Daiichi-Consultants Co., Ltd., 3-1-5, Takasushinmachi, Kochi 780-8122. T: 0888852123, F: 0888852136, E: usiro@daiichi-c.co.jp

Uto, K., University Professor, 22-24, Bunkyo 1-chome, Sagamihara 228. T: 0427440319

Uwabe, T., Deputy General Manager, Service Center of Port Engineering, SCOPE Management System, 3-1, 3-chome Kasumigaseki, Chiyoda-ku, Tokyo 100-0013. T: 0335032280, F: 0355126922, E: uwabe@apricot.ocn.ne.jp

Uzuoka, R., Research Engineer, Earthquake Disaster Mitigation Research Center, 2465-1, Mikiyama, Miki, Hyogo 673-0433. T: 0794836637, F: 0794836695, E: uzuoka@miki.riken.go.jp

Wada, S., University Professor, Faculty of Agriculture, Kyushu University, Fukuoka 812-8581. T: 0926422844, F: 0926422845, E: wadasi@agr.kyushu-u.ac.jp

Wakai, A., University Professor, Dept. of Civil Engineering, Gunma University, 1-5-1, Kiryu, Gunma 376-8515, T/F: 0277301624, E: wakai@ce.gunma-u.ac.jp

Wakamatsu, K., University Research Fellow, Institute of Industrial Science, University of Tokyo, 4-6-1, Komaba, Meguro-ku, Tokyo 153-8505. T: 0354526388, F: 0354526389, E: wakamatu@iis.u-tokyo.ac.jp

Wako, T., Consulting Engineer, Japan Port Consultants Ltd., TK Gotanda Bldg.7F, 8-3-6, Nishi-Gotanda, Shinagawa-ku, Tokyo 141-0031. T: 0354345309, F: 0354345393, E: tatsuo_wako@jportc.co.jp

Wang, L., Ph.D., Chuo Kaihatsu Corporation, 3-13-5, Nishiwaseda, Shinjuku-ku, Tokyo 169-8612. T: 0332083111, F: 0332089915, E: lin_wang@ckc-unet.ocn.ne.jp

Watabe, Y., Senior Researcher, Port & Harbour Research Institute, 3-1-1, Nagase, Yokosuka, Kanagawa 239-0826. T: 0468445021, F: 0468444577, E: watabe@ipc.phri.go.jp

Watahiki, K., Dr, Professor, Dept. of Civil Engineering, Tokai University, 1117 Kita-kaname, Hiratsuka-shi, Kanagawa 259-12. T: 0463581211, F: 0463502045

Watanabe, H., 19-15, Tomio-Kita 3-chome, Nara 631

Watanabe, H., Dr, Professor, Dept. of Civil and Environmental Engineering, Saitama University, Shimo-Ohkubo 255, Urawa-shi, Saitama 338. T/F: 0488583541

Watanabe, K., Consulting Engineer, Koa-Kaihatsu Co., Ltd., Shasoku Bldg.6F, 5-3-13, Kotobashi, Sumida-ku, Tokyo 130. T: 0336337351, F: 0336337356, E: kunihiro.watanabe@koa-kaihatsu.co.jp

Watanabe, T., 3-24-A-201, Shonan-machi, Higashikatsushika-gun, Chiba 277

Watanabe, T., University Emeritus Professor, (Home address) 4-35-8, Daita, Setagaya-ku, Tokyo 155-0033. T/F: 0333287053

Watari, Y., Dr, Chairman, Mizuno Institute of Technology, Miura Bldg., 2-22, Nakajima-cho, Naka-ku, Hiroshima 730. T: 0822463311, F: 0822463370

Wu, J., Leader of R&D Group, Central Giken Co., Ltd., 1-2-13, Motoyokoyama-cho, Hachioji, Tokyo 192-0063. T: 0426458276, F: 0426458307, E: wujiaycg@coral.ocn.ne.jp

Wu, Z., Associate Professor, Dept. of Urban and Civil Engineering, Ibaraki University, 4-12-1, Nakanarusawa, Hitachi 316-8511. T: 0294385179, F: 0294385268, E: zswu@ipc.ibaraki.ac.jp

Xu, G.L., Consulting Engineer, Daichi Consultants Co., Ltd., Koi Honmachi, 2-20-16, Nishi-ku, Hiroshima 733-0812. T: 0822731471, F: 0822737644, E: xugl@cdaichi.co.jp

Xu, T., University Research Associate, Hiroshima University, Kagamiyama, 1-5-1, Higashi-hiroshima-shi, Hiroshima 739-8529. T/F: 0824246930, E: ting@hiroshima-u.ac.jp

Yabuuchi, S., President, Geotop Corporation, 2-1-10, Koraibashi, Chuo-ku, Osaka 541-0043. T: 0662260451, F: 0662260873, E: sadao_yabuuchi@mail.geotop.co.jp

Yagi, K., Construction Engineer, Chizaki Kogyo, S4 W7, Chuo-ku, Sapporo, Hokkaido 064-8588. T: 0115118114, F: 0115112660, E: 1754@chizaki.co.jp

Yagi, N., University Professor, 30-1, Higashiiori-cho, Kitashirakawa, Sakyo-ku, Kyoto 606-8251. T/F: 0757110388, E: nyagi@hh.iij4u.or.jp

Yagiura, Y., Consulting Engineer, Kiso-Jiban Consultants Co., Ltd., 1-11-14, Awaza, Nishi-ku, Osaka 550-0011. T: 0665361591, F: 0665361503, E: yagiura.yoshiyuki@kiso.co.jp

Yajima, J., Research Engineer, Tokyu Construction Co., Ltd., 3062-1, Soneshita, Tana, Sagamihara 229-11. T: 0427639511, F: 0427639504, E: yajima@etd.tokyu-const.co.jp

Yajima, J., Geotechnical Engineer, Tekken Corporation, 2-5-3, Misaki-cho, Chiyoda-ku, Tokyo 101-8366. T: 0332212185, F: 0332212161, E: juichi-yajima@tekken.co.jp

Yamabe, S., Geotechnical Engineer, Araigumi Co., Ltd., 2-26, Tsutonishiguchi-cho, Nishinomiya, Hyogo 663-8231. T: 0798268360, F: 0798368104, E: yamabe@mbd.sphere.ne.jp

Yamabe, T., Dr, University Associate Professor, Saitama University, Shimo-Ohkubo 255, Urawa, Saitama 338-8570. T: 0488583544, F: 0488587374, E: yamabe@rock.civil.saitama-u.ac.jp

Yamada, E., Assistant Professor, Nagoya University, Dept. of Geotechnical and Environmental Engineering, Furo-cho, Chikusa-ku, Nagoya, 464-8603. T: 0527893834, F: 0257893836, E: e-yamada@soil.genv.nagoya-u.ac.jp

Yamada, H., Manager, Engineering Div., NITTOC Construction Co., Ltd., 8-14-14, Ginza, Chuo-ku, Tokyo 104. T: 0335429120, F: 0335429133

Yamada, K., 432-128-204, Terada-machi, Hachioji 193

Yamada, K., Dr, Professor, Dept. of Civil Engineering, College of Engineering, Chubu University, 1200 Matsumoto-cho, Kasugai, Aichi 487. T: 0568511111

Yamada, A., Dr, University Professor, Nihon University, 1-8, Kanda-Surugadai, Chiyoda-ku, Tokyo 101-8308.

T: 0332590667, F: 0332933319, E: yamada@civil.cst.
nihon-u.ac.jp

Yamada, K., Dr, Consulting Engineer, 531-16, Teshiro-cho,
Soka, Saitama 340-0021. T/F: 0489248173

Yamada, K., Research Engineer, Fudo Construction Company,
Hirano-machi, 4-2-16, Chuo-ku, Osaka 541. T: 062019211,
F: 062011130

Yamada, T., Geotechnical Engineer, Fudo Constrution
Co., Ltd., 1-2-1, Taito, Taito-ku, Tokyo 110. T: 0338376035,
F: 0338376158

Yamada, T., Research Engineer, Takenaka Research &
Development Institute, 5-14, 2-chome Minamisuna,
Koto-ku, Tokyo 136. T: 0336473161, F: 0336450911

Yamada, Y., University Professor, Institute of Engineering
Mechanics and Systems, University of Tsukuba, Tsukuba
305-8573. T: 0298535146, F: 0298535207, E: yamada@
kz.tsukuba.ac.jp

Yamagami, T., University Professor, Dept. of Civil Engineering, Faculty of Engineering, The University of Tokushima,
2-1, Minamijosanjima-cho, Tokushima 770-8506.
T/F: 0886567345, E: takuo@ce.tokushima-u.ac.jp

Yamaguchi, H., Dr, University Professor, The National Defense
Academy, 1-10-20, Hashirimizu, Yokosuka, Kanagawa
239-8686. T: 0468413810, F: 0468445913, E: hareyuki@
oregano.ocn.ne.jp

Yamaguchi, H., Geotechnical Engineer, Fudo Construction
Co., Ltd., Geo-Engineering Division, 2-1, Taito 1-chome,
Taito-ku, Tokyo, 110-0016. T: 0338376097, F: 0338376158,
E: hirohisa@fudo.co.jp

Yamaguchi, M., Consulting Engineer, Nippon Giken Inc., 2-16-
10, Chiyoda, Naka-ku, Nagoya 460-0012. T: 0522611321,
F: 0522611655, E: masahiro-yamaguchi@npgk.co.jp

Yamaguchi, Y., Dr, Geotechnical Engineer, Public Works
Research Institute, Ministry of Land, Infrastructure
and Transport, 1 Asahi, Tsukuba, Ibaraki 305-0804.
T: 0298642413, F: 0298642688, E: yamaguti@pwri.go.jp

Yamakado, A., Emeritus Professor, 1-55-4, Yoyogi, Shibuya-
ku, Tokyo 151. T: 0333736140

Yamakami, Y., Alpha Hydraulic Engineering Consultants
Co., Ltd., 9-14-516-336, Hassamu, Nishi-ku, Sapporo 063

Yamakawa, O., Consulting Engineer, NITA Soil, Rock and
Water Research Co., Ltd., 38-2, Kawauchi-cho, Suzuen-
ishi, Tokushima-shi, Tokushima 771-01. T: 0886653618,
F: 0886653962

Yamamoto, H., University Assoc. Professor, Hiroshima Univ.,
IDEC, 1-5-1, Kagamiyama, Higashi-Hiroshima 739-8529.
T/F: 0824246928, E: yamamoto@idec.hiroshima-u.ac.jp

Yamamoto, K., Research Associate, Dept. of Ocean Civil
Engineering, Kagoshima University, 1-21-40, Korimoto,
Kagoshima 890-0065. T: 0992858475, F: 0992581738,
E: yamaken@oce.kagoshima-u.ac.jp

Yamamoto, M., 3-3-3-8-301, Sengendai-Nishi, Koshigaya 343

Yamamoto, T., Research Assistant, Faculty of Architecture,
Kanagawa University, 3-27, Rokkakubashi, Kanagawa-ku,
Yokohama 221. T: 0454815661

Yamamoto, T., Dr, Assistant Director, Kajima Technical
Research Institute, 19-1, Tobitakyu 2-chome, Chofu, Tokyo
182. T: 0424851111, F: 0424897016

Yamamoto, T., Dr, University Professor, Yamaguchi University, 2-6-1, Tokiwadai, Ube, Yamaguchi 755-8611.
T: 0836859302, F: 0836859301, E: tyamamot@jim2.civil.
yamaguchi-u.ac.jp

Yamamoto, Y., Consulting Engineer, Kiso-Jiban Consultants
Co., Ltd., 16-7, Hara 2-chome, Sawara-ku, Fukuoka 814.
T: 0928312511, F: 0928222393

Yamamura, K., Dr, University Professor, 6-33-14-602,
Jingumae, Shibuya-ku, Tokyo 150

Yamanouchi, T., Professor Emeritus, 4-3-9, Ropponmatsu,
Chuo-ku, Fukuoka 810-0044. T/F: 09275100191

Yamaoka, K., Consulting Engineer, Pacific Consultants Co.,
Ltd., Shinjuku Daiichi Seimei Bldg., P.O. 5073, 2-7-1,
Nishi-Shinjuku, Shinjuku-ku, Tokyo 163-07. T: 0333441304,
F: 0333441366, E: yamaoka.kazumasa@pacific.co.jp

Yamashita, J., Research Engineer, Engineering Research Dept.,
Taisei Corporation, Shinjuku Center Bldg., 25-1, Nishshinjuku
1-chome, Shinjuku-ku, Tokyo 163. T: 0333481111

Yamashita, K., Consulting Engineer, Takenaka Technical
Research Laboratory, 5-14, Minamisuna 2-chome, Koto-
ku, Tokyo 136. T: 0336473161, F: 0336450911

Yamashita, S., Dr, Associate Professor, Kitami Institute of
Technology, 165 Koen-cho, Kitami, Hokkaido 090-8507.
T: 0157269480, F: 0157239408, E: yamast@king.cc.
kitami-it.ac.jp

Yamato, T., University Professor, Dept. of Civil Engineering,
Fukuoka Univ., 8-19-1, Nanakuma, Jyonan-ku, Fukuoka
814-80. T: 0928716631, F: 0928648901

Yamauchi, H., Dr, Civil Engineer, Penta-Ocean Construction
Co., Ltd., 145-1, Sarumaru Uetanbo, Shokowa-mura,
Ohno-gun, Gifu 501-54. T: 0576922281, F: 0576923048

Yamazaki, H., Chief of Soil Dynamics Laboratory, Port & Harbour Research Institute, 3-1-1, Nagase, Yokosuka 239-0826.
T: 0468445022, F: 0468440618, E: yamazaki@cc.phri.go.jp

Yamazaki, M., Research Assistant, Dept. of Architecture
and Architectural Systems, Kyoto University, Yoshida
Hon-machi, Sakyo-ku, Kyoto 606-8501. T: 0757535737,
F: 0757535748, E: yama@archi.kyoto-u.ac.jp

Yamazaki, T., Civil Engineer, The Tokyo Electric Power
Co., Inc., 1-1-3, Uchisaiwai-cho, Chiyoda-ku, Tokyo 100.
T: 0335018111, F: 0335968527, E: t0561175@pmail.
tepco.co.jp

Yamazaki, T., Senior Researcher, National Institute of
Advanced Industrial Science and Technology, 16-1,
Onogawa, Tsukuba, Ibaraki 305-8569. T: 0298618721,
F: 0298618709, E: p1810@nire.go.jp

Yanagida, T., Geotechnical Engineer, Fudo Construction
Co., Ltd., 2-16, 4-chome Hirano-machi, Chuo-ku, Osaka
541. T: 062019211, F: 062011130

Yanagisawa, E., University Professor, Dept. of Civil
Engineering, Tohoku, Univ., Aramaki, Aoba, Sendai 980.
T: 0222177433, F: 0222177435, E: yana@mechanics.civil.
tohoku.ac.jp

Yanai, E., Dr, Professor Emeritus, Yamanashi Univ., (Home
address) 4-18-2, Mejirodai, Hachioji-shi, Tokyo 193.
T/F: 0426632047

Yanase, S., Consulting Engineer, Koa Kaihatsu Co., Ltd.,
3-13, Kohtohbashi 5-chome, Sumida-ku, Tokyo 130.
T: 0336337351

Yang, J., University Research Associate, Hokkaido University,
North 13, West 8, Sapporo 060-8628. T: 0117066203,
F: 0117066202, E: yangij@eng.hokudai.ac.jp

Yang, J., Senior Researcher, Disaster Prevention Research
Institute, Kyoto University, Gokasho, Uji, Kyoto 611-0011.
T: 0774384068, F: 0774384070, E: yang@catfish.dpri.
kyoto-u.ac.jp

Yano, K., Dr, Technical Advisor, Ocean Consultant, Japan
Co., Ltd., 27-4, Ichibancho, Chiyoda-ku, Tokyo 102-0082.
T: 0335661577, F: 0332392050

Yano, T., 2-31, Yamanawate, Terada-cho, Mukho-shi, Kyoto 617

Yao, S., University Professor, Kansai University, 3-3-35,
Yamato-cho, Suita, Osaka 564-8680. T: 0663681121,
F: 0663303770, E: yao@ipcku.kansai-u.ac.jp

Yashima, A., University Professor, Dept. of Civil Engineering, Gifu University, 1-1, Yanagido, Gifu 501-1193.
T/F: 0582932419, E: yashima@cc.gifu-u.ac.jp

Yasuda, S., University Professor, Tokyo Denki University,
Hatoyama, Hiki-gun, Saitama 350-0394. T: 0492962911,
F: 0492966501, E: yasuda@g.dendai.ac.jp

Yasue, T., Dr, Geotechnical Engineer, 2-19-7, Masuodai,
Kashiwa-shi, Chiba 277

Yasufuku, N., Dr, University Associate Professor, Dept.
of Civil Engineering, Kyushu University, 6-10-1,
Hakozaki, Higashi-ku, Fukuoka 812-8581. T: 0926423284,
F: 0926423285, E: yasufuku@civil.kyushu-u.ac.jp

Yasuhara, K., Dr, Dept. of Urban and Civil Engineering,
Ibaraki University, 4-12-1, Nakanarusawa, Hitachi, Ibaraki
316. T: 0294356101, F: 0294358146

Yasui, M., Geotechnical Engineer, Toda Corporation, 4-6-1,
Hacchobori, Chuo-ku, Tokyo 104-0032. T: 0332067186,
F: 0332067185, E: mitoshi.yasui@toda.co.jp

Yasukawa, I., Teacher, Dept. of Civil Engineering, Fushimi
Technical High School, Fukakusa, Fushimi-ku, Kyoto 612.
T: 0756415121, F: 0756415950

Yasunaka, M., Government Official, Agriculture, Forestry and Fisheries Research Council Secretariat, Ministry of A.F.F., Kasumigaseki, 1-2-1, Chiyoda-ku, Tokyo 100-8950. T: 0335072811, Ext. 5187, F: 0335078794, E: yasu@s.affrc.go.jp

Yatabe, R., Dr, University Professor, Ehime University, Bunkyo 3, Matsuyama, Ehime 790-8577. T: 0899279817, F: 0899279820, E: yatabe@dpc.ehime-u.ac.jp

Yokohama, S., Lecturer, Hokkaido College Senshu University, 1610-1, Bibai, Hokkaido 079-0197. T: 0126630246, F: 0126623666, E: yokohm@senshu-hc.ac.jp

Yokoo, F., Geotechnical Engineer, OYO Corporation Singapore Branch, 32 Maxwell Road, #01-04 Whitehouse, Singapore 069115. T: 652270023, F: 652270352, E: yokoo@singnet.com.sg

Yokoo, Y., Dr, Minamimizocho 10, Yamashina-ku, Kyoto 607. T: 0755812565

Yokose, H., Dr, Professor, Dept. of Agricultural Engineering, Kagawa University, Ikenobe 2393, Mikmi-cho, Kita-gun, Kagawa 761-07. T: 0878989788, F: 0878987295

Yokota, H., Dr, University Professor, Dept. of Civil & Environmental Engineering, Miyazaki University, Gakuen Kibanadai, Miyazaki 889-2192. T: 0985587330, F: 0985587344, E: yokota@civil.miyazaki-u.ac.jp

Yokoyama, Y., Consulting Engineer, 13-15, Kita-Ichunosawa-machi, Utsunomiya, Tochigi 320-0048. T/F: 0286254135, E: yokouki@ucatv.ne.jp

Yonekura, R., Dr, University Professor Emeritus, 4-5-3, Miyamae, Suginami-ku, Tokyo 168-0081. T/F: 0359302711, E: yoneryo@jcom.home.ne.jp

Yonezawa, T., Soil Mechanics & Foundation Engineering, Japan Railway Construction Public Corporation, Sanno Bldg.6F, 2-14-2, Nagata-cho, Chiyoda-ku, Tokyo 100. T: 0335061861, F: 0335061891

Yoshida, H., University Associate Professor, Kagawa University, 2217-20, Hayashi-cho, Takamatsu, Kagawa 761-0396. T: 0878642187, F: 0878642188, E: yoshida@eng.kagawa-u.ac.jp

Yoshida, I., Dr, President, Honshu-Shikoku Bridge Engineering Co., 1-13, 3-chome Irifune, Chuo-ku, Tokyo 104. T: 0335526711, F: 0335526710

Yoshida, N., 1-45-15, Kamo, Sawara-ku, Fukuoka 814

Yoshida, N., Managing Researcher, Sato Kogyo, Co., Ltd., Nihonbashi-honcho, 4-12-20, Chuo-ku, Tokyo 103-8639. T: 0358232350, F: 0358232358, E: Nozomu.Yoshida@satokogyo.co.jp

Yoshida, N., Dr, Associate Professor, Research Center for Urban Safety and Security, Kobe University, 1-1, Rokkodai, Nada-ku, Kobe 657-8501. T: 0788036031, F: 0788036394, E: nyoshida@kobe-u.ac.jp

Yoshida, S., Dr, University Professor, Dept. of Agricultural Engineering, Niigata University, 8050 Nino-cho, Ikarashi, Niigata 950-21. T: 0252626655, F: 0252631659

Yoshida, S., Deputy General Manager, Tenox Corporation, 6-13-7, Akasaka, Minato-ku, Tokyo 107-8533. T: 0335825168, F: 0335824714, E: yoshida-s@tenox.co.jp

Yoshida, T., Okumura Corp., Architectural Engineering Section, 2-22, Matsuzaki-cho, Abeno-ku, Osaka 545. T: 066253704

Yoshida, T., Dr, Research Engineer, Civil Engineering Dept. 2nd, Kajima Technical Research Institute, Tobitakyu, 2-19-1, Chofu-shi, Tokyo 182. T: 0424897067, F: 0424897034

Yoshida, Y., Assistant Professor, Dept. of Civil Engineering, Tokyo Denki University, Hatoyama, Saitama 350-03. T: 0492962911, F: 0492966501

Yoshida, Y., Geotechnical Engineer, Central Research Service Co., Ltd., Kamakura No.3 Bldg., 2-11-23, Sibasakidai, Abiko-shi, Chiba 270-1176. T: 0471835711, F: 0471835691, E: JDX04715@nifty.ne.jp

Yoshifuku, T., 2-9-101, Koshien-Rokkokucho, Nishinomiya 663

Yoshikoshi, H., Dr, Associate Director, Tokyo Electric Power Company, 1-1-3, Uchisaiwai-cho, Chiyoda-ku, Tokyo 100-0011. T: 0335018111, F: 0335968534, E: T0429308@pmail.tepsco.co.jp

Yoshikuni, H., Dr, University Professor, Hiroshima Institute of Technology, 2-1-1, Miyake, Saeki-ku, Hiroshima 731-5143. T: 0829213121, F: 0829237083, E: yosikuni@cc.it-hiroshima.ac.jp

Yoshimi, Y., Professor Emeritus, 4074 Shimotsuruma, Yamato 242-0001, E: yoshiaki.yoshimi@nifty.ne.jp

Yoshimine, M., University Lecturer, Tokyo Metropolitan University, Minami-Osawa, 1-1, Hachioji, Tokyo 192-0397. T: 0426772773, F: 0426772772, E: yoshmine-mitsutoshi@c.metro-u.ac.jp

Yoshimoto, N., University Research Associate, Dept. of Civil Engineering, Faculty of Engineering, Yamaguchi University, 2-16-1, Tokiwadai, Ube 755-8611. T: 0836589344, F: 0836859301, E: nyoshi@po.cc.yamaguchi-u.ac.jp

Yoshimura, H., Research Engineer, Konoike Construction Co., Ltd., Research Institute of Technology, 1-20-1, Sakura, Tsukuba, Ibaraki 305-0003. T: 0298572000, F: 0298572123, E: yoshimura_hs@konoike.co.jp

Yoshimura, M., Fukuoka Prefectural Government

Yoshimura, Y., Dr, College Associate Professor, Dept. of Civil Engineering, Gifu National College of Technology, Shinsei-cho, Motosu-gun, Gifu 501-04. T: 0583201401, F: 0583201409, E: yuji@gifu.nct.ac.jp

Yoshinaka, R., University Professor, Dept. of Civil and Environmental Engineering, Saitama University, 255 Shimo-Okubo, Urawa, Saitama 338. T: 0488583540, F: 0488559361

Yoshioka, M., Structural Consulting Engineer, Park-Hights Yokohama D-911, 3-1, Kawabe-cho, Hodaogaya-ku, Yokohama 240-0001. T: 0453402547, E: mktyoshi@cds.ne.jp

Yoshitake, S., Geotechnical Engineer, Nihon Kensetu Gijutsu, 1417-1, Tokusue, Kitahatamura, Higashimatsuura-gun, Saga 847-1201. T: 0955642528, F: 0955644255, E: mkg.con@topaz.ocn.ne.jp

Yoshitake, Y., University Professor, Dept. of Rural Engineering, Ehime Univ., 3-5-7, Tarumi, Matsuyama 790. T: 0899469891, F: 0899774364

Yoshizawa, H., Research Engineer, JDC Corporation, Technical Research Institute, Nakatsu, 4036-1, Aikawa-cho, Aikoh-gun, Kanagawa 243-03. T: 0462854924, F: 0462861642

Yoshizawa, M., Consulting Engineer, Seiko Kenkyusyo Co., Ltd., 2-18-20, Kakazu, Ginowan, Okinawa 901-2226. T: 0988702334, F: 0988702321, E: yoshizawa@seikoken.co.jp

Yukitomo, H., Consulting Engineer, 1330-43, Koshigoe, Kamakura, Kanagawa 248-0033. T/F: 0467325955, E: yukitomoh@pop06.odn.ne.jp

Yunoki, Y., Civil Engineer, Nippon Koei Co., Ltd., 5-4, Kojimachi, Chiyoda-ku, Tokyo 102. T: 0332388267, F: 0332656469

Zaoya, K., Consulting Engineer, 2-24-23, Minami-yawata, Ichikawa, Chiba 272-0023. T: 0473784720

Zen, K., Dr, University Professor, Dept. of Civil Engineering, Kyushu University, 6-10-1, Hakozaki, Higashi-ku, Fukuoka 812-8581. T/F: 0926423316, E: zen@civil.kyushu-u.ac.jp

Zhang, F., University Professor, Dept. of Civil Engrg., Gifu University, Yanagido, 1-1, Gifu 501-1193. T/F: 0582932465, E: zhang-f@cive.gifu-u.ac.jp

Collective members

Central Research Institute of Electric Power Industry, Abiko Research Lab., 1646 Abiko, Abiko-shi, Chiba 270-1194. T: 0471821181, F: 0471841336

Chuden Engineering Consultants Inc., 3-30, Deshio-cho 2-chome, Minami-ku, Hiroshima 734-8510. T: 0822555501, F: 0822510302

Chuo Fukken Consultants Co., Ltd., 5-26, Higashimikuni 3-chome, Yodogawa-ku, Osaka 532-0004. T: 063931121, F: 063950677

Chuo Kaihatsu Corporation, 3-13-5, Nishiwaseda, Shinjuku-ku, Tokyo 169-8612. T: 0332085251, F: 0332083572

Chuo Kaihatsu Co., Ltd., Osaka Branch, 34-12, Tarumi-cho 3-chome, Suita 564-0062. T: 063863691, F: 063865082

Dia Consultants Co., Ltd., Touwa No.2 Bldg., 34-5, Minami Ikebukuro 2-chome, Toshima-ku, Tokyo 171-0014. T: 0339865191, F: 0339865192

Electric Power Development Co., Ltd., Engineering & Research Institute, 9-88, Chigasaki 1-chome, Chigasaki, Kanagawa 253-0041. T: 0467871211, F: 0467824003

FUJITA Corporation, Ono, Atsugi, Kanagawa 243-0125. T: 0462507095, F: 0462507139

INA Corporation, Sekiguchi, 1-44-10, Bunkyo-ku, Tokyo 112-0014. T: 0352615711, F: 0332682776, E: ina@pa2.so-net.or.jp

JGC Corporation, 14-1, Bessho 1-chome, Minami-ku, Yokohama 232. T: 0457121111, F: 0457217305

Kajima Corporation Co., Ltd., 2-7, Motoakasaka 1-chome, Minato-ku, Tokyo 107-0051. T: 0334043311, F: 0334701444

Kawasaki Geological Engineering Co., Ltd., Kansai Headquarters Office, Shikizu-nishi, 2-1-12, Naniwa-ku, Osaka 556-0015. T: 066492215, F: 066492240

Keisoku Research Consultant Co., 1-665-1, Fukuda, Higashi-ku, Hiroshima 735. T: 0828995471, F: 0828995479

Kiso-Jiban Consultants Co., Ltd., 11-5, Kudan-Kita 1-chome, Chiyoda-ku, Tokyo 102-8220. T: 0332633611, F: 032611810

Konoike Construction Co., Ltd., Technical Research Div., 6-1, 3-chome Kitakyuhoji-machi, Chuo-ku, Osaka 541-0057. T: 062443600, F: 062443633

Meiji Consultant Co., Ltd., Branch of Sapporo, No.2 Hiroyasu Bldg., S-7, W-1-13, Chuo-ku, Sapporo 064-0807. T: 0115623066, F: 01156231991

Nihon Suiko Sekkei Co., Ltd., 3-12-1, Kachidoki, Chuo-ku, Tokyo 104-0054. T: 0335345511, F: 0335345510

Nikken Consultants Co., Ltd., 353 Mamedo-cho, Kouhoku-ku, Yokohama, Kanagawa 222-0032. T: 0454331611, F: 0454340185

Nippon Koei Co., Ltd., Urban Civil Engineering Dept., 5-4, Kojimachi, Chiyoda-ku, Tokyo 102-8539. T: 0332388354, F: 0332388379

Obayashi Corporation, Technical Research Institute, 4-640, Shimokiyoto, Kiyose, Tokyo 204-0011. T: 0424951111, F: 0424950901

Ohmoto Gumi Co., Ltd., 1-13, Uchisange 1-chome, Okayama 700-8550. T: 0862255131, F: 0862275174

Okumura Corporation, 2-2-2, Matsuzaki-cho, Abeno-ku, Osaka 545-8555. T: 0666211101, F: 0666237692, W: http://www.okumura.co.jp

Oriental Consultants Co., Ltd., 16-14, Shibuya 1-chome, Shibuya-ku, Tokyo 150-0002. T: 0334097551, F: 0334090208

OYO Corporation, Ichigaya Bldg., 4-2-6, Kudan-Kita, Chiyoda-ku, Tokyo 102-0073. T: 0332340811, F: 0332636854, E: prosight@oyonet.oyo.co.jp

Taiheiyo Cemento Corporation, 2-4-2, Osaku, Sakura, Chiba 285-8655. T: 0434983817, F: 0434983821

Taisei Technology Center, 344-1, Nase-cho, Totsuka-ku, Yokohama, Kanagawa 245-0051. T/F: 0458147254

Takenaka Corporation Research & Development Institute, 1-5-1, Ohtsuka, Inzai, Chiba 270-1395. T: 0476471700, F: 0476473080

Tokyo Gas Co., Ltd., Fundamental Technology Research Laboratory, 16-25, Shibaura 1-chome, Minato-ku, Tokyo 105-0023. T: 0354844632, F: 0334520915

Tokyo Gas Co., Ltd., Pipeline & Facilities Engineering Dept., 1-5-20, Kaigan, Minato-ku, Tokyo 105-8527. T: 0334332111, F: 0334379177

Tokyu Construction Co., Ltd., Institute of Technology, 3062-1, Soneshita, Tana, Sagamihara 229-1124. T: 0427639511, F: 0427639503

KAZAKHSTAN/KAZAQUIE

Secretary: Associate Professor, Dr. T.M. Baytasov, Kazakhstan Geotechnical Society, the L.N. Gumilyov Eurasian National University, 5, Munaitpassov Str., Astana, Kazakhstan, 473021. T: 7-3172-353740, F: 7-3172-353740, E: askarz@nets.kz

Total number of members: 25

Ahmetov, A.S., Principal of Aktau University named after Sh. Esenov, Aktau, 24 district, Kazakhstan, 466200. T: 7-3292-338432, F: 7-3292-337814

Aldungarov, M.M., Head of Dept., Geotechnical Engineering of KazGASA, Professor, Dr.Ph., 28 Ryskulbekova Str., Almaty, Kazakhstan. T: 7-3272-200322

Aytaliev, Sh.M., Dr.Sc., Professor, Academic of National Academy of Science of the Republic of Kazakhstan, Kazakhstan Academy of Transport and Communications, 97 Shevchenko Street, Almaty, Kazakhstan, 480012. T: 7-3272-923434

Bakenov, B.B., Head of Dept. Civil Engineering, the L.N. Gumilyov Eurasian National University, 5, Munaitpassov Str., Astana, Kazakhstan, 473021. T: 7-3172-355667

Baytasov, T.M., Associate Professor, Dr.Ph., General Secretary of Kazakhstan Geotechnical Society, the L.N. Gumilyov Eurasian National University, 5, Munaitpassov Str., Astana, Kazakhstan, 473021. E: askarz@nets.kz

Bazarov, B.A., Associate Professor, Dr.Ph., Karaganda Metallurgical Institute, 34 Lenin Avenue, Temirtau, Kazakhstan, 472300. T: 7-3213-914872, F: 7-3213-914872, E: askar@ada.kz

Bekenov, T.N., Head of Transport Engineering Dept., Professor, Dr.Sc., the L.N. Gumilyov Eurasian National University, 5, Munaitpassov Str., Astana, Kazakhstan, 473021. T: 7-3172-356244, F: 7-3172-353740, E: askarz@nets.kz

Bittibaev, S.M., Head of Department Applied Mechanics, Kazakhstan Academy of Transport and Communications, 97 Shevchenko Str., Almaty, Kazakhstan, 480012

Bizhanov, K.S., Dr.Ph., General Director of "Arka-Kurylys" Ltd., 88 flat, 43 Building, Al-Farabi district, Astana, Kazakhstan, 437000. T: 7-3172-282519

Bozhanov, E.T., Professor, Dr.Sc., 13 Markov Street, 6 flat, Almaty, Kazakhstan. T: 7-3272-923954

Dussembaev, I.N., Professor, Dr.Ph., Vice-principal of Aktau University named after Sh. Esenov, Aktau, 24 district, Kazakhstan, 466200. T: 7-3292-338432, F: 7-3292-337814

Filatov, A.V., Head of Dept. Civil Engineering, Professor, Dr.Sc., Karaganda Metallurgical Institute, 34 Lenin Avenue, Temirtau, Kazakhstan, 472300

Gofshtein, F.A., Dr.-Ing., President of JSE "Karagandinsky Promstroiproekt", Kazakhstan, 470032, Karaganda, N. Abdirov, 3. T: 7 3213412260, F: 7-3212-412260, E: gofr@nursat.kz

Isakhanov, E.A., Professor of Dept. Civil Engineering, Dr.Sc., Kazakhstan Academy of Transport and Communications, 97 Shevchenko Str., Almaty, Kazakhstan, 480012. T: 7-3272-923434

Kudaikulov, U.A., Dr.Ph., Student, Kazakhstan Academy of Transport and Communications, 97 Shevchenko Str., Almaty, Kazakhstan, 480012

Kusainov, A.A., Professor, Dr.Sc., President of Kazakhstan Leading Academy of Architecture and Civil Engineering, Riskulbekova, 28, Almaty, Kazakhstan. T: 7-3272-294611

Kusainov, M.K., Professor, Dr.Ph., Dept. Civil Eng., the L.N. Gumilyov Eurasian National University, 5, Munaitpassov Str., Astana, Kazakhstan, 473021. T: 7-3172-355667

Nuguzhinov, Z.S., Head of Civil Engineering Laboratory, Karaganda Technical University, 56 Bulvar Mira, 470075, Karaganda, Kazakhstan

Popov, V.N., Dr.Ph., President of Karaganda GIIZ and Co., Ltd., 29 Gogol Avenue, Karaganda, Kazakhstan. T: 7-3212-520795, F: 7-3212-520795

Sonin, A.M., Assoc. Prof., Dr.Ph., Dept. Civil Engineering, the L.N. Gumilyov Eurasian National University, 5, Munaitpassov Str., Astana, Kazakhstan, 473021

Teltaev, B.B., Head of Dept. Civil and Transport Engineering, Kazakhstan Academy of Transport and Communications, 97 Shevchenko Str., Almaty, Kazakhstan, 480012. T: 7-3272-923434

Unaibaev, B.Zh., Assoc. Prof., Dr.Ph., Dept. Civil Eng., Karaganda Technical University, 56 Bulvar Mira, 470075, Karaganda, Kazakhstan. T: 7-3212-527068, F: 7-3172-353740, E: askarz@nets.kz

Zaharov, N.I., Head of Dept. Geotechnical Eng., Eastern Kazakhstan Technical University, Assoc. Prof., Dr.Ph., Building 35, Lugovaya Str., 492034, Ustkamenogorsk, Kazakhstan. T: 7-3232-446412

Zhakulin, A.S., Dean of Civil Engineering Faculty, Professor, Dr.Ph., Karaganda Technical University, 56 Bulvar Mira, 470075, Karaganda, Kazakhstan

Zhusupbekov, A.Zh., Director of Geotechnical Institute, President of Kazakhstan Geotechnical Society, Professor, Dr.Sc., the L.N. Gumilyov Eurasian National University, 5, Munaitpassov Str., Astana, Kazakhstan, 473021. T: 7-3172-353740, F: 7-3172-353740, E: askarz@nets.kz

KENYA

Secretary: Dr. Bernard N. Njoroge, Kenya Geotechnical Society, c/o Faculty of Engineering, Department of Civil Engineering, P.O. Box 30197, Nairobe, Kenya. T: 254 3 248079, E: bmknjoroge@uonbi.ac.ke

Total number of members: 28

Bruno, I., Eng., Norconsult A.S., P.O. Box 48176, Nairobi. T: +254 -2-225580, F: +254 -2-337703, E: norcon@form-net.com

Carrington, R., Eng., P.O. Box 13886, Nairobi. T: +254-2-37443 39/351191, F:+254-2-3748503, E: geotechnics@caagroup.com

Getuno, J., Eng., H.P. Gauff Consulting Engineers, P.O. Box 49817, Nairobi. T: +254-2-445288/445966, F: +254-2-446124, E: jbgnbo@africaonline.co.ke

Gichaga, J.F., Prof., University of Nairobi, P.O. Box 30197, Nairobi. T: +254-2-334244, F: +254-2-212604, E: vc@uonbi.ac.ke

Kago, J.K., Eng., El Nino Emergency Project (Rural Roads & Bridges), P.O. Box 40213, Nairobi. T: +254-2-253640, F: +254-2-253637

Karanja, P. C., Mr., University of Nairobi (Civil Eng. Dept.), P.O. Box 30197, Nairobi. T: +254-2-334244, F: +254-2-336885

Maina, J.N., Eng., MOR&PW, P.O. Box 11873, Nairobi, F: +254-2-540117, E: aterials@roadsnet.go.ke

Maithulia, C., Mr., Construction Projects Consultants Inc., P.O. Box 34842, Nairobi. T: +254-2-710442/712678, F: +254-2-713681, E: cpc@africaonline.co.ke

Mambo, Eng., Gibb(EA) Ltd., P.O. Box 30020, Nairobi. T: +254-2-338992/250577, F: +254-2-210694/244493, E: gibbea@gibb.co.ke

Mugambi, Eng., MOR&PW, P.O. Box 11873, Nairobi. F: +254-2-540117, E: materials@roadsnet.go.ke

Mukabi, J.N., Dr., Construction Projects Consultants Inc., P.O. Box 34842, Nairobi. T: +254-2-710442/712678, F: +254-2-713681, E: cpc@africaonline.co.ke

Murunga, P.A., Eng., Norconsult A.S., P.O. Box 48176, Nairobi. T: +254-2-225580/226883, F: +254-2-337703, E: norcon@form-net.com

Mutonyi, Eng., Wajohi Consulting Engineers, P.O. Box 21714, Nairobi. T: +254-2-576690/1/2/3, F: +254-2-576693, E: wce@form-net.com, ikwmutonyi@hotmail.com

Mwangi, N.S., Mr., MOR&PW, P.O. Box 51413, Nairobi. F: +254-2-713681, E: saykigs@yahoo.com

Mwea, K.S., Eng., University of Nairobi (Civil Eng. Dept.), P.O. Box 30197, Nairobi. T: +254-2-334244, F: +254-2-336885

Ndemi, J., Eng., Norconsult A.S., P.O. Box 48176, Nairobi. T: +254-2-225580, F: +254-2-337703, E: norcon@form-net.com

Njoroge, B.N., Dr., University of Nairobi (Civil Eng. Dept.), P.O. Box 30197, Nairobi. T: +254-2-248079, F: +254-2-336885, E: bnknjoroge@uonbi.ac.ke

Nyandika, Eng., Norconsult A.S., P.O. Box 48176, Nairobi. T: +254-2-225580, F: +254-2-337703, E: norcon@form-net.com

Nyangweso, Eng., Norconsult A.S., P.O. Box 48176, Nairobi. T: +254-2-225580, F: +254-2-337703, E: norcon@form-net.com

Omolo, A., Mr., Construction Projects Consultants Inc., P.O. Box 34842, Nairobi. T: +254-2-710442/712678, F: +254-2-713681, E: yomollo@yahoo.com, cpc@africaonline.co.ke

Shoona, S.J., Mr., Mass Labs Ltd., P.O. Box 10782, Nairobi. T: +254-2-571985

Tibwitta, B.W., Eng., H.P. Gauff Consulting Engineers, P.O. Box 49817, Nairobi. T: +254-2-445288/445966, F: +254-2-446124, E: jbgnbo@africaonline.co.ke

Too, J.K., Eng., Jomo Kenyatta University of Agriculture & Technology, P.O. Box 62000, Nairobi. T: +254-2-0151-52711, E: araptoojk@yahoo.com

Wainaina, J.N.B., Eng., Otieno Odongo Consulting Engineers, P.O. Box 54021, Nairobi. T: +254-2570022/570032, F: +254-2-570236/570103, E: oop@form-net.com

Wambura, J.H.G., Eng., MOR&PW, P.O. Box 11873, Nairobi. F: +254-2-540117, E: materials@roadsnet.go.ke

Wamburu, A.N., Eng., Gibb(EA) Ltd., P.O. Box 30020, Nairobi. T: +254-2-338992/250577, F: +254-2-210694/244493, E: gibbea@gibb.co.ke

Wangusi, B., Mr., MOR&PW, P.O. Box 11873, Nairobi F: +254-2-540117, E: materials@roadsnet.go.ke

Waweru, S.G., Eng., Kajima Corporation, P.O. Box 58615, Nairobi. T: +254-2-725023/4, F: +254-2-729493, E: kajimake@swiftkenya.com

KOREAN REPUBLIC

Secretary: Dr. Seung Rad Lee, Secretary, The Korean Geotechnical Society, Rm. 1201, Seocho World Officetel, 1355-1 Seocho-dong, Seocho-gu, Seoul, Korea. T: 822 3474 4428, F: 822 3474 7379, E: kgssmfe@chollian.net

Total Number of members: 171

Ahn Sang-Ro, Mr., Korea Infrastructure Safety & Technology Corporation, 2311 Daewha-dong, Ilsan-gu Goyang-shi Kyunggi-do, 411-410. T: 82-31-910-4186, F: 82-31-910-4179

Ahn Tae-Bong, Prof., WooSong Univ., San 7-6., Jayang-dong, Dong-gu, Taejeon-shi 300-718. T: 82-42-630-9735, F: 82-42-631-2346

An Jin-Ho, Mr., Dooho Engineering Co., Ltd., 530-28, Bangbae-dong, Seocho-gu, Seoul 137-064. T: 822-599-7343, F: 82-594-7346, E: dooho@mail.hitel.net

Bae Gyu-Jin, Dr., Korea Institute of Construction Technology 2311, Daewha-dong, Ilsan-gu Goyang-shi Kyunggi-do 411-710. T: 82-31-9100-212, F: 82-31-9100-211

Baek Kyung-Jong, Mr., Dasol Consultant Co., Ltd., Rlo Bldg., 4F #790-2., Yuksam-dong, Kangnam-Ku, Seoul 135-080. T: 822-508-2290, F: 822-508-2297, E: dasolcon@hananel.net

Bang Yoon-Kyung, Prof., Daewon Science College 599., Shinwol-dong, Jecheon-Shi, Chung Buk 390-230. T: 82-43-649-3268

Chae Seung-Ho, Dr., SamSung Corporation 428-5., Gongse-Ri, Giheung Eup Yongin-City Kyunggi-Do 449-900. T: 82-31-289-6755, F: 82-31-289-677

Chae Young-Su, Prof., Dept. of Civil Eng. The University of Suwon San 2-2, Bongdam-eup, Wawoo-ri, whasung-gun Kyunggi-do 445-743. T: 82-31-220-2318, F: 82-31-220-2494, E: yschae@mail.suwon.ac.kr

Chang Soo-Ho, Mr., Dept. of Civil Eng. Seoul National Univ., San 56-1, Shinlim-dong, Kwanak-gu, Seoul 151-742. T: 822-880-7232, F: 822-877-0925

Chang Yong-Chai, Prof., Mokpo National Maritime Univ., 571-2, Chukyo-dong, Mokdo-si Chonnam 530-729. T: 82-61-240-7218, F: 82-61-240-7284, E: geo@mail.mmu.ac.kr

Cho Chun-Whan, Mr., Piletech Consulting Engineers Co., Rm 412 Midas Officetel, 775-1, Janghang-dong, llsangu, Goyangsi, Kyunggi-do 411-380. T: 82-31-908-5992, F: 82-31-908-5994, E: cwcho@piletech.co.kr

Cho Hyun-Tae, Mr., Korea Land Cooperation 217, Jongja-dong, Bundang-gu Sungnam-shi, Kyungi-do 463-010. T: 82-31-738-7529

Cho Nam-Jun, Prof., 253 13, Kangnam-gu, Nonheyn-dong, Seoul 135-010. T: 822-547-3478

Cho Sam-Deok, Dr., Korea Institute of Construction Technology 2311, Daewha-dong, Ilsan-gu Goyang-shi, Kyunggi-do 411-710. T: 82-31-9100-214, F: 82-31-9100-211, E: sdcho@kict.re.kr

Cho Seong-Ha, Mr., Dasan Consulteuts Co., Ltd., 1004, Sammi Bldg., Daechi-dong Kangnam-gu, Seoul 135-283. T: 822-2222-4107, F: 822-539-1361

Cho Sung-Eun, Mr., Kaist 373-1, Gusung-dong, Yusung-gu Taejeon-shi, 305-701. T: 82-42-869-3657, F: 82-42-869-3610

Cho Sung-Min, Dr., Korea Highway Corporation, 293-1, Kumto-dong, Sujong-Gu, Songnam-Shi Kyunggi-do 461-380. T: 822-2230-4657, F: 822-2230-4183, E: chosmin@freeway.co.kr

Cho Sung-Won, Mr., Dept. of Civil Eng., Seoul National Univ., San 56-1, Shinlim-dong, Kwanak-gu, Seoul 151-742. T: 822-880-8701, F: 822-875-6933

Choe Myong-Jin, Mr., Korea Telecom 62-1, Whaam-dong, Yusung-gu, Taejeon-shi 305-384. T: 82-42-866-3091, F: 82-42-866-3095

Choi Byeong-Seong, Mr., Dept. of Civil Eng., Seoul National Univ., San 56-1, Shinlim-dong, Kwanak-gu, Seoul, 151-742. T: 822-880-7354, F: 822-875-6933

Choi Hang Gill, Prof., Daebul Univ., 72, Samho-ri Samho-Myun Youngam-Gun Chunnam 526-890. T: 82-61-469-1328, F: 82-61-469-1338

Choi Jin-O, Mr., Dept. of Civil Eng., Seoul National Univ., San 56-1, Shinlim-dong, Kwanak-gu, Seoul, 151-742. T: 822-880-7354, F: 822-875-6933

Choi Jung-Bum, Mr., Saegil Engineering Co., Ltd., Hyun Min B/D, #65-3 Bangyi-dong, Songpa-gu, Seoul 138-050. T: 822-416-0904, F: 822-418-0904

Choi Kwang-Chul, Mr., Korea Telecom 62-1, Whaam-dong, Yusung-gu, Taejeon-shi, 305-384. T: 82-42-866-3091, F: 82-42-866-3095

Choi Kye-Shik, Mr., SSangyong Engineering & Construction Co., Ltd. (Institute of Construction Technology). T: 82-31-750-6405, F: 82-31-750-6690

Choi Yong Kyu, Prof., Dept. of Civil & Environmental Engineering 110-1, Daeyeon-dong, Nam-Gu, Pusan 603-736. T: 82-51-620-4753, F: 82-51-621-0729

Choo Jae-Keon, Mr., Dasan Eng. Co., Ltd., Sajo B/D, #507, Daechi-dong, Kangnam-gu, Seoul, 135-280. T: 822-3452-1700, F: 822-3452-9003, E: dasangeo@hitel.net

Chun Byung-Sik, Prof., Dept. of Civil Eng., Hanyang Univ., 17, Haengdang-dong, Sungdong-gu, Seoul, 133-791. T: 822-2290-0326, F: 822-2298-3270

Chung Choong-Ki, Prof., Dept. of Civil Eng., Seoul National Univ., San 56-1, Sinlim-dong, Kwanak-gu, Seoul, 151-742. T: 822-880-7347, F: 822-887-0349

Chung, Hyung-Sik, Prof., Dept. of Civil Eng., Hanyang Univ., 15-186, Seongbuk-dong, Seongbuk-gu, Seoul 136-020. T: 822-2298-1770, F: 822-2293-9977

Chung In-Jun, Prof., Dept. of Civil Eng. Seoul National Univ., San 56-1, Shinlim-dong, Kwanak-gu, Seoul 151-742. T: 822-880-7347

Chung Kyung-Whan, Mr., Dong-A Geological Engineering Co., Ltd., #1033-2, Guseo-Dong, GeumJeong-Gu, Busan, 609-420. T: 82-51-583-5500, F: 82-51-583-5505, E: ghjeong@dage.co.kr

Chung Moon-Kyung, Dr., Korea Institute of Construction Technology, Geotechnical Eng. Division 2311, Daewha-dong, Ilsan-gu

Chung Sang-Seom, Prof., Dept. of Civil Eng., Yonsei University Civil Eng., College of Engieering, Yonsei Univ., Seoul 120-749. T: 822-2123-2807, F: 822-364-5300, E: soj9081@yonsei.ac.kr

Goyang-shi, Kyunggi-do, 411-710. T: 82-31-9100-097, F: 82-31-9100-091, E: mkchung@kict.re.kr

Gu Ho-Bon, Mr., Korea Institute of Construction Technology 2311, Daewha-dong, Ilsan-gu Goyang-shi, Kyunggi-do 411-710. T: 82-31-9100-0217, F: 82-31-9100-211, E: hbkoo@kict.re.kr

Heo Yol, Prof., Chungbuk Univ., San 48, Gaeshin-dong, ChongJu-Si, Chungbuk 360-763. T: 82-43-261-2405, F: 82-43-261-2405

Hong Sung-Wan, Dr., Korea Institute of Construction Technology 2311 Daewha-dong, Ilsan-gu Goyang-shi, Kyonggi-do 411-410. T: 82-31-9100-485, F: 82-31-9100-441, E: swhong@smtppc.kict.re.kr

Hong Won-Pyo, Prof., Chung-Ang Univ., 221, Heuk-Suk dong Dong-Jak Gu, Seoul 156-756. T: 822-820-5258, F: 822-817-8050

Hwang Dae-Jin, Dr., SamSung Corporation 428-5, Gongse-Ri, Giheung Eup Yongin-City Kyunggi-do 449-900. T: 82-31-289-6763, F: 82-31-289-6768

Hwang Seon-Keun, Dr., Korea Railroad Research Institute, 374-1, Woulam-Dong, Uiwang-city, Kyonggi-do 437-050. T: 82-31-461-8531, F: 82-31-461-8536

Hwang Seong-Chun, Prof., Kyongju Univ., 42-1, Hyohyun-dong, Kyongju-city Kyongbuk 780-210. T: 82-54-770-5172, F: 82-54-748-2825, E: yeons@kyongju.ac.kr

Ihm Chol-Woong, Mr., Sangjee Menard. Co., Ltd., 6F, Seowon Bldg., #1515-4, Seocho-dong, Seocho-Ku, Seoul 137-070. T: 82?-587-9286, 822-587-9285, E: cwihm@sangjee-menard.co.kr

Im Jong-Chul, Prof., Dept. of Civil Eng. Pusan National Univ., #30, Jangjeon-dong, Gumjeong-gu, Pusan 609-735. T: 82-51-510-2442, F: 82-51-518-3084, E: imjc@hyowon. pusan.ac.kr

Im Soo-Bin, Mr., Dong-Il Engineering Consultants Co., Ltd., 107-7 MunJeong-Dong, Songpa-Gu, Seoul, 138-200. T: 822- 3400-5630, F: 82-431-4903, E: soobeen@dongileng.co.kr

Jang Hak-Sung, Mr., Yooshin Engineering Corporation 832-40, Yoksam-dong, Kangnam-Gu, Seoul 135-936. T: 822-555-7132, F: 822-538-1931, E: y12159@yooshin.co.kr

Jang Hyo-Wan, Mr., 551-7., Dogok-dong, Kangnam-Gu, Seoul 135-270. T: 822-3463-4766

Jang Won-Gil, Mr., Korea Land Cooperation 217, Jongja-dong, Bundang-gu, Sungnam-shi, Kyung-do, 463-010. T: 82-31-220-0352

Jang Youn-Soo, Prof., Dept. of Civil Eng. Dongguk Univ., 3-26, Phil-dong, Chung-gu, Seoul, 100-715. T: 822-2269-3265, F: 822-2266-8753, E: ysjang@cakra.dongguk.ac.kr

Jeon Jun-Soo, Mr., Daewoo Engineering Co., Ltd., Bundang P.O. Box 20, 9-3, Sunae-dong, Bundang-gu, Seongnam-Shi, Kyunggido 463-020. T: 82-31-738-0771, F: 82-31-738-0802

Jeon Mong-Gag, Prof., 602-111. Namhyun-dong, Seoul 151-080. T: 822-587-0560

Jeon Seok-Won, Dr., Dept. of Civil Eng., Seoul National Univ., San 56-1, Shinrim-dong, Kwanak-gu, Seoul 151-742. T: 822-880-8807, F: 822-871-8938, E: sjeon@rockeng.snu.ac.kr

Jeong Soon-Yong, Mr., Kaist 373-1, Gusung-dong, Yusung-gu Taejeon-shi 305-701. T: 82-42-869-5657, F: 82-42-869-3610

Joh Sung-Ho, Prof., Chung-Ang Univ., Kyunggi-do, 156-756. T: 82-31-675-7143, F: 82-31-675-1387

Joo Soo-Il, Mr., Dong-Il Engineering Consultants Co., Ltd., 107-7 MunJeong-Dong, Songpa-Gu, Seoul, 138-200. T: 822-3400-5504, F: 822-431-3153

Ju Jae-Woo, Prof., Sunchon National Univ., Dept. of Civil Eng., Sunchon National Univ., 315, Maegoc-dong, Sun-chon 540-742. T: 82-61-750-3515, F: 82-62-750-3510, E: woo3310@sunchon.ac.kr

Jung Du-Hwoe, Prof., PuKyong National Univ., San 100, Yongdang-Dong, Nam-Gu, Pusan 608-739. T: 82-51-620-1453, F: 82-51-628-2231, E: dhjung@pine.pknu.ac.kr

Kang Byng-Hee, Prof., Inha University, 253, Yonghyun-dong, Nam-ku, Inchon 402-751. T: 82-32-860-7564, F: 82-32-873-7560, E: bhkang@inha.ac.kr

Kang In-Kyu, Mr., Vniel, Consultant Co., Ltd., 807 Milleana 2nd 79-5 Garakbon-dong, Songpa-gu, Seoul 138-169. T: 822-3452-9130, F: 822-3452-9132, E: kangik@vnicl.co.kr

Kang Ki-Young, Mr., Nae Kyung Engineering Co., Sewon Bldg., 5-1, Yangjae-Dong Seocho-Ku, Korea, Seoul 137-130. T: 82-32-860-7564, F: 82-32-873-7560, E: bhkang@inha.ac.kr

Kang Kyung-Hun, Mr., GG Eng. & Const. Co., Ltd., 3F, Jaehyun Bldg., 88-8, Nonhyun-Dong, Kang Nam-Ku, Seoul 135-010. T: 822-541-5113, F: 822-3444-9990

Kang Shin-Chu, Mr., Boram Engineering Co., SamHwan Bldg., 3F, 214-4 Poy-Dong Kang, Nam-Gu, Seoul, 135-260. T: 822-529-5383, F: 822-529-5385

Kang Yea-Mook, Prof., Dept. of Agricultural Eng., Chungnam National Univ., 220, Gung-dong, Yuseong-gu, Taejeon-shi 305-764. T: 82-42-824-0467, F: 82-42-824-0467, E: ymkang@hanbat.chungnam.ac.kr

Kim Chang-Youb, Mr., Dept. of Civil Eng. Seoul National Univ., San 56-1, Shinlim-dong, Kwanak-gu, Seoul 151-742. T: 822-880-7348, F: 822-887-0349

Kim Dong-Soo, Prof., Korea Advanced Institute of Science and Technology (KAIST) 373-1, Kusung-dong, Yusung-gu Taejan, 305-701. T: 82-42-869-3619, F: 82-42-869-3610, E: dskim@Kaist.ac.kr

Kim Hak-Moon, Prof., Dept. of Civil & Environmental Eng. Dankook Univ., 8, Hannam-dong, Yongsan-gu, Seoul 140-714. T: 822-709-2555, E: Khm1028@anseo.dankook. ac.kr

Kim Hong-Jung, Mr., Korea Highway Corporation, 293-1, Kumto-dong, Sujong-Gu, Songnam-Shi, Kyunggi-do 461-380. T: 822-2230-4658, F: 822-2230-4183

Kim Hong-Taek, Prof., Dept. of Civil Eng., Hongik University 72-1, Sangsu-dong, Mapo-gu, Seoul 121-791. T: 822-320-1624, F: 822-336-2656

Kim Hyun-Gi, Mr., Dept. of Civil Eng. Seoul National Univ., San 56-1, Shinlim-dong, Kwanak-gu, Seoul 151-742. T: 822-880-7348, F: 822-887-0349

Kim Hyung-Joo, Prof., Kunsan National Univ., 68, Miryong-don Kunsan 573-701, Jeonbuk 573-360. T: 82-63-469-4760, F: 82-63-469-4760

Kim In-Kuin, Mr., Daewoo Engineering Co., Ltd., Bundang P.O. Box 20, 9-3, Sunae-dong, Bundang-gu, Seongnam-Shi, Kyunggido 463-020. T: 82-31-738-0300, F: 82-31-738-0115

Kim Jae-Soo, Mr., Han Ah Engineering Co., Ltd., Chowon APT. 105-104, 898, Pyongchon-dong, Dongan-gu, Anyang-shi Kyunggi-do, 431-070. T: 822-517-9052, F: 822-517-9055

Kim Jeong-Hwan, Mr., SamSung Corporation 428-5, Gongse-Ri, Giheung Eup Yongin-City, Kyunggi-Do 449-900. T: 82-31-289-6762, F: 82-31-289-6768, E: jnghwan@samsung.co.kr

Kim Jin-Man, Dr., Korea Institute of Construction Technology 2311, Daewha-dong, Ilsan-gu, Goyang-shi, Kyunggi-do, 411-710. T: 82-31-9100-222, F: 82-31-9100-21, E: jmkim@kict.re.kr

Kim Joo-Bum, Mr., Korea Construction Safety Engineering Association 58-1, SamSung-Dong, KangNam-Gu, Seoul 135-090. T: 822-512-0808, F: 822-547-7083

Kim Kyo-Won, Prof., Dept. of Geology Kyungpook Univ., 1370, Sankyuk-dong, Buk-gu, Taegu 702-701. T: 82-53-950-5357, F: 82-53-950-5362

Kim Kyung-Suk, Mr., KOREA Highway Corporation 293-1, Kumto-dong, Sujong-Gu, Songnam-Shi, Kyunggi-do 461-380. T: 822-2230-4658, F: 822-2230-4183, E: kskim@freeway.co.kr

Kim Man-Goo, Mr., Taejon Metropolitan City 1420, TunSan-dong So-gu Taejon 302-120. T: 82-42-600-3151, F: 82-42-600-2149, E: kmgoo9@hanamail.net

Kim Myung-Mo, Prof., Dept. of Civil Eng., Seoul National Univ., San 56-1, Shinlim-dong, Kwanak-gu, Seoul, 151-742. T: 822-880-7348, F: 822-887-0349, E: geotech@gong.snu.ac.kr

Kim Oon-Young, Prof., Jukong A.P.T 314-1004, dwunchon-dong, Kangdong-Ku, Seoul 134-773. T: 822-488-4335

Kim Pal-kyu, Prof., Chungnam National Univ., 220, Gung-dong, Yuseong-gu, Taejeon-shi 305-764. T: 82-42-824-0467, F: 82-42-824-0467

Kim Sang-Kyu, Prof., Dong Pusan College 640, Pansong-dong, Haeundae-gu, Pusan 612-715. T: 82-51-540-3701, F: 82-51-542-7188, E: skkim@sb.dpc.ac.kr

Kim Seong-Ho, Mr., Suseong Engineering Co., Ltd., 4F, Beumyang B/D, 7-14, Dongbinggo-dong, Yongsan-gu, Seoul 140-230. T: 822-796-8206, F: 822-749-8697

Kim Seung-Ryul, Dr., ESCO Engineers & Consultants Co., Ltd., 3F, Syspol Bldg., 869-9, Bangbae-Dong, Seocho-Gu, Seoul 137-060. T: 822-3461-4160, F: 822-3461-4167, E: srkim@escoeng.com

Kim Soo-Il, Prof., Dept. of Civil Eng., Yonsei Univ., 134, Shinchan-dong, Seodaemun-gu, Seoul 120-749. T: 822-361-2730, F: 822-312-9592, E: geo0273@challian.net

Kim Soo-Sam, Prof., Chung-Ang Univ., 221, Heuk-Suk dong Dong-Jak Gu, Seoul 156-756. T: 822-820-5259, F: 822-816-8987, E: kimss@cau.ac.kr

Kim Yong-Ku, Mr., Dohwa Consulting Engineers Co. Ltd., Seoul. T: 822-2056-0747, F: 822-3444-3584

Kim Young-Chin, Dr., Korea Institute of Construction Technology 2311, Daewha-dong, Ilsan-gu, Goyang-shi, Kyunggi-do 411-710. T: 82-31-9100-215, F: 82-31-9100-211

Kim Young-gu, Mr., SSangyong Engineering & Construction Co., Ltd. (Institute of Construction Technology), 7-23, Shinchon-Dong

Kim Young-Jin, Mr., Dasan Eng. Co., Ltd., Sajo B/D., #507, Daechi-dong, Kangnam-gu, Seoul 135-280. T: 822-3452-9002, F: 822-3452-9003, E: dasangeo@hitel.net

Koh li-Yong, Prof., Chodang University Sungnam-Ri 419, Muan-Eup, Muan-Goon Chullanam-Do 530-804. T: 82-61-450-1304, F: 82-61-453-4696, E: yikoh@chodang.ac.kr

Koo Ho-Bon, Mr., Korea Institute of Construction Technology 2311, Daewha-dong, Ilsan-gu, Goyang-shi, Kyunggi-do 411-710. T: 82-31-9100-217, F: 82-31-9100-211, E: hbkoo@kict.re.kr

Kwon Ki-Tae, Mr., 90-8., Yeonhy-Dong Seodaemun-Gu, Seoul 120-110. T: 822-334-7951

Kwon Oh-Kyun, Mr., Keimyung Univ., 1000, Shindang-dong, Dalsu-Ku Taegu 704-200. T: 82-53-580-5280, F: 82-53-580-5165, E: ohkwon@kmucc.keimyungac.kr

Kwon Oh-Soon, Mr., Coastal & Harbour Engineering Research Laboratory ANSAN P.O. Box 29, Seoul 425-170. T: 82-31-400-6311, F: 82-31-408-5823

Lee Byung-Sik, Prof., Kongju Univ., 182., Shinkwan-dong., Kongju-Shi ChungNam 314-701. T: 82-41-850-8634, F: 82-41-856-9388

Lee Chang-Ho, Mr., Korea Telecom 62-1. Whaam-dong, Yusung-gu, Taejeon-shi 305-384. T: 82-42-866-3092, F: 82-42-866-3065

Lee Chong-Kyu, Prof., Dept. of Civil & Enviromental Eng. Dankook Univ., 8, Hannam-dong, Youngsan-gu, Seoul 140-714. T: 822-709-2552, 2770, F: 822-796-8285, E: leeck@dankook.ac.kr

Lee Chung-In, Prof., College of Eng. Seoul National Univ., San 56-1, Shinrim-dong, Kwanak-gu, Seoul 151-742. T: 822-880-7221, F: 822-880-8708, E: cilee@plaza.snu.ac.kr

Lee Chung-In, Prof., Dept. of Civil Eng., Seoul National Univ., San 56-1, Shinlim-dong, Kwanak-gu, Seoul 151-742. T: 822-880-7221, F: 822-880-8708

Lee Dal-Won, Prof., Dept. of Agricultural Eng., Chungnam National Univ., 220, Gung-dong, Yuseong-gu, Taejeon-shi, 305-764. T: 82-42-821-5793, F: 82-42-824-0467, E: dwlee@hanbat.Chungnam.ac.kr

Lee Do-Seop, Mr.

Lee Eun-Soo, Mr., Korea Institute of Construction Structural Safety, 913-11, Daechi-dong, Kangnam-gu, Seoul 135-280. T: 822-554-4482, F: 822-538-1127, E: kicss@chollian.dacom.co.kr

Lee In-Keun, Dr., Seoul Metropolitan Government, 37, Seosomun-dong, Chung-gu, Seoul 100-110. T: 822-3707-8010, F: 822-3707-8029, E: iklee@mail.catholic.or.kr

Lee In-Mo, Prof., Korea Univ./Civil Eng. Dept. 1, 5-Ro, Anam-dong, Sungbuk-gu, Seoul 136-701. T: 822-3290-3314, F: 822-928-7656, E: inmolee@kuccnx.korea.ac.kr

Lee Jae-Hoon, Mr., Woodae Engineering Consultants Co., Ltd., Speciality Construction Center Bldg., 395-70, Shin-daebang-dong Dongjak-gu, Seoul 156-010. T: 822-3284-2977, F: 822-3284-2900

Lee Jae-Hyun, Mr., Pyeong Won Engineering Co., Ltd., Duk CHANG Bldg., 1044-1, Sadang-dong, Dongjak-Gu, Seoul 156-091. T: 822-522-1161, F: 822-585-4514, E: baekcheon@hanmil.net

Lee Kwang-Yeol, Prof., DongSeo Univ., San 69-1, Jurey-dong, Sasang-Gu, Pusan 617-012. T: 82-51-320-1819, F: 82-51-3123-1046

Lee Kyeong-Jun, Mr., Dong-A Geological Engineering Co., Ltd., #1033-2, Guseo-Dong, GeumJeong-Gu, Busan, 609-420. T: 82-51-512-1151, F: 82-51-513-8612

Lee Man-Soo, Mr., Hyundai #102-4, Mabjk-r, Goosung-Myun, Yongin-s, Kyunggi-do 449-710. T: 82-31-280-7268, F: 82-31-280-7070, E: ms'ee@hdae.co.kr

Lee Moon-Soo, Prof., Chonnam National Univ., 300, Yong-bong-dong, Buk-gu Kwangju-shi 500-757. T: 82-62-530-2153, F: 82-62-530-0653

Lee Myung-Whan, Dr., Piletech Consulting Engineers Co., Rm 412 Midas Officetel, 775-1, Janghang-dong, llsangu, Goyangsi, Kyunggi-do 411-380. T: 82-31-908-5992, F: 82-31-908-5994

Lee Sang-Duk, Prof., Ajou Univ., Wonchon-dong, Paldal-gu, Suwon, Kyunggi-do 442-749. T: 82-31-219-2503, F: 82-31-213-2801

Lee Sang-Tae, Mr., Korea Speciality Contractors Association 395-70., Sindaebang-Dong., Dongjak-Gu, Seoul 156-010. T: 822-3284-1001, F: 822-3284-1044

Lee Seong-Ho, Prof., Sangji Univ., 660, Woosan-dong, Wonju-Si kangwon-do 220-702. T: 82-33-730-0473, F: 82-33-730-0403, E: shlee@mail.sangji.ac.kr

Lee Seung-Hyun, Mr.,

Lee Seung-Rae, Prof., Kaist 373-1, Gusung-dong, Yusung-gu, Taejeon-shi 305-701. T: 82-42-869-3617, F: 82-42-869-3610, E: srlee@kaist.ac.kr

Lee Song, Prof., Seoul City Univ., of Technology 8-3, Jeonnong-dong, Dongdaemun-gu, Seoul, 130-743. T: 822-2210-2515, F: 822-248-9406

Lee Su-Hyung, Mr., Dept. of Civil Eng. Seoul National Univ., San 56-1, Shinlim-dong, Kwanak-gu, Seoul 151-742. T: 822-880-7354, F: 822-875-6933

Lee Sung-Ki, Mr., Sambo Engineering Co., Ltd., HanMih-Bldg., 45 Bangyi-dong, Songpa-gu, Seoul 138-050. T: 822-3433-3009, F: 822-554-4473

Lee Wan- Ho, Mr.

Lee Woo-Jin, Prof., Korea Univ./Civil Eng. Dept., 1, 5-Ro, Anam-dong, Sungbuk-gu, Seoul 136-701. T: 822-3290-3319, F: 822-921-2086

Lee Yang-Hee, Mr., Sye Gee Engineering Co., Ltd., HONG SEO Bldg., 2nd Floor, 699-15, Yoksam-dong, Kangnam-gu, Seoul 135-080. T: 822-562-5935, F: 822-562-6255, E: geojohn@unitel.Co.Kr

Lee Yeong-Saeng, Prof.

Lee Yong-Soo, Dr., Korea Institute of Construction Technology 2311, Daewha-dong, Ilsan-gu, Goyang-shi, Kyunggi-do 411-710. T: 82-31-9100-223, F: 82-31-9100-211

Lee Young-Huy, Prof., Dept. of Civil Eng., Yeungnam Univ., 214-1, Dae-dong, Gyongsan, Kyungbuk 712-749. T: 82-53-810-2417, F: 82-53-811-9640, E: younghuy@ynucc.yeungnam.ac.kr

Lim Jong-Seok, Prof., Mokpo National Univ., 61, Dorim-ri, Chungkye-myeon, Muan-gun, Jeonnam 534-729. T: 82-61-450-2474, F: 82-61-452-6468, E: jslim@chungkye.mokpo.ac.kr

Lim Tae-Kwan, Mr., Kyungin Engineering, Co., Ltd., (Geotechnical Engineering Dept) Yoksam-Dong 773-4, Kangnamgu, Seoul 151-080. T: 822-508-3211, F: 822-508-3214, E: tklim@kyungineng.co.kr

Lim Yu-Jin, Prof., Pai Chai Univ., 439-6, Doma2-dong, Seo-Ku Taejon 302-735. T: 82-42-520-5402, F: 82-42-864-1141

Ma Sang-Joon, Dr., Korea Institute of Construction Technology 2311, Daewha-dong, Ilsan-gu Goyang-shi Kyunggi-do 411-710. T: 82-31-9100-222, F: 82-31-9100-211, E: sjma@kict.re.kr

Min Tuk-Ki, Prof., Dept. of Civil Eng., Univ., of Ulsan 29, Mugae-dong, Dept. of Civil Eng., Univ., of Ulsan, Ulsan 680-749. T: 82-52-259-2259, F: 82-52-259-2629

Moon Jong-Kyu, Mr., Dong Myung Eng. Co., Ltd., 250, Maewang-li, Yoju-eup Yoju-gun, Kyunggido, 469-800

Moon Sang-Jo, Mr., Yooshin Engineering Corporation 832-40. Yoksam-dong, Kangnam-Gu, Seoul 135-936. T: 822-555-7132, F: 822-558-1050, E: musim96@yooshin.co.kr

Na Kyung-Joon, Mr., Korea Diagnosis & Reinforcement Co., Ltd., 2nd FL., Shallom Bldg., 231-13, P'oi-Dong, Kangnam-Gu, Seoul 135-260. T: 822-571-9270, F: 822-571-9274

Nam Jung-Man, Prof., JeJu Univ., 1. Ara1-dong., Jeju-si JeJudo 690-121. T: 82-64-754-3454, F: 82-64-725-2519

Paik Young-Shik, Prof., Kyung Hee Univ., 1 Seocheon-Ri, Kihung-Eup, Yongin-City, Kyunggi-do 449-701. T: 82-31-201-2550, F: 82-31-202-8854, E: yspaik@nms.kyunghee.ac.kr

Park Byung-Kee, Prof., Chonnam National Univ., 300, Yongbong-dong, Buk-gu Kwangju-shi, 500-757. T: 82-62-530-1651, F: 82-62-530-1659

Park Chan-Ho, Mr., YouGu Geotechnical Eng. Co., Ltd., 312-8. YangJae-Dong, Seocho-Ku, Seoul 137-130. T: 822-529-5974, F: 822-578-9225

Park Chi-Myeon, Mr., Esco Engineers & Consultants Co., Ltd., 3F, Syspol Bldg., 869-9, Bangbae-Dong, Seocho-Gu, Seoul 137-060. T: 822-3461-4160, F: 822-3461-4167

Park Dug-Keun, Prof., National Institute for Disaster prevention 253-42, Gongdeok-dong, Mapo-gu, Seoul 121-719, T: 822-3274-2204, F: 822-3274-2209, E: dr_park@mogaha.go.kr

Park Hee-Sauk, Mr.

Park Hyeong-Dong, Prof., Dept. of Civil Eng., Seoul National Univ., San 56-1, Shinlim-dong, Kwanak-gu, Seoul 151-742. T: 822-880-8808, F: 822-871-8938

Park Hyuck-Jin, Dr., Korea Institute of Construction Technology, Geotechnical Eng., Division, 2311, Daewha-dong, Ilsan-gu, Goyang-shi, Kyunggi-do 411-710. T: 82-31-9100-561, F: 82-31-9100-211

Park Jong Ho, Mr., DaeWon Soil Co., Ltd., 53, Samsung-dong, Kangnam-gu, Seoul 135-090. T: 822-547-1940, F: 822-588-0230, E: jongho2000@hanmail.net

Park Jun-Bom, Prof., Dept. of Civil Eng., Seoul National Univ., San 56-1, Shinlim-dong, Kwanak-gu, Seoul 151-742. T: 822-880-8356, F: 822-887-0349

Park Nam-Seo, Mr., Daeduk Consulting & Construction Co., 1337-3, Seocho-dong, Seocho-gu, Seoul 137-070. T: 822-3474-2146, F: 822-581-1841, E: daedukcc@elim.net

Park Seong-Jae, Mr., Pusan National Univ., #30, Changjeon-dong, Kumjeong-Ku, Pusan 609-735. T: 82-51-510-2349, F: 82-51-515-6810, E: szpark@hyowon.cc.pusan.ac.kr

Park Tae-Soon, Prof., Seoul National Univ., of Technology, 172 Gongreung-dong, Dept. of Civil Engineering, Nowon-gu Seoul 139-743. T: 822-970-6506, F: 822-948-0043, E: tpark@duck.snut.ac.kr

Park Tae-Soon, Prof., Seoul National Univ., of Technology, 172, Gongreung-dong, Dept. of Civil Engineering Nowon-gu, Seoul, 139-743. T: 822-970-6506, F: 822-948-0043

Park Yeon-Jun, Prof., Dept. of Civil Eng., The University of Suwon, San 2-2, Bongdam-eup, Wawoo-ri, whasung-gun, Kyunggi-do 445-743. T: 82-31-220-2580, F: 82-31-220-2522

Park Yong-Seob, Mr.

Park Yong-Won, Prof., Dept. of Civil Eng., Univ., of Myongji San 38-2, Nam-dong, Yongin-Shi, Kyunggi-Do, Yongin-shi, Kyunggi-do 449-728. T: 82-31-330-6350, F: 82-31-335-9998, E: ywpark@wh.Myongji.ac.kr

Park Young-Ho, Mr., Korea Highway Corporation, 293-1, Kumto-dong, Sujong-Gu, Songnam-Shi, Kyunggi-do 461-380. T: 822-2230-4658, F: 822-2230-4183

Ro Hyung-Don, Mr., Korea Infrastructure Safety & Technology Corporation, 2311 Daewha-dong, Ilsan-gu Goyang-shi, Kyunggi-do 411-410. T: 82-31-910-4108, F: 82-31-910-4178

Ryu Chi-Hyob, Prof., Hanlyo Univ., 199-4, Dokryori, Kwangyang, Chonam 545-800. T: 82-61-760-1182, F: 82-61-761-6709, E: chryu@hlu.hanlyo.ac.kr

Ryu Jeong-Soo, Mr., Vniel, Consultant Co., Ltd., 807 Milleana 2nd 79-5 Garakbon-dong Songpa-gu, Seoul 138-169. T: 822-3452-9130, F: 822-3452-9132

Shin Bang-Woong, Prof., Chungbuk Univ, San 48, Gaeshin-dong, ChongJu-Si, Chungbuk 360-763. T: 82-43-261-2379, F: 82-43-262-8686

Shin Min-Ho, Mr., Korea Railroad Rescarch Institute 374-1, Woulam-Dong, Uiwang-city, Kyonggi-Do 437-050. T: 82-31-461-8374, F: 82-31-461-8374, E: mhshin@krri.re.kr

Song Byong-Mu, Prof., Kyle Enterprise, 86-506, Hyundae APT, Apgujung-dong, Kangnam-gu, Seoul 135-110. T: 822-515-3967

Songpa-Gu, Seoul, 138-726. T: 822-3433-7754, F: 822-3433-7759, E: kp09@chollion.net

Woo Kee-Hyeung, Mr., Anam Consultants Co., Ltd., 695-1, Jayang-Dong, Kwangjin-gu, Seoul 143-192. T: 822-453-2365, F: 822-453-2378

Yang Hyun-Sung, Mr., Woodae Engineering Consultants Co., Ltd., Speciality Construction Center Bldg., 395-70, Shindaebang-dong, Dongjak-gu, Seoul, 156-010. T: 822-3284-2800, F: 822-3284-2900

Yang Ku-Seung, Mr., SamSung Corporation 428-5., Gongse-Ri, Giheung Eup Yongin-City, Kyunggi-do 449-900. T: 82-31-289-6764, F: 82-31-289-6768

Yi Seung-Weon, Mr.

Yoo Chung-Sik, Prof., Dept. of Civil Eng., SungKyunkwan Univ., 300, Chunchun-dong, Suwon 440-746. T: 82-31-290-7518, F: 82-31-290-7549

Yoo Han-Kyu, Prof., Dept. of Civil Eng., Hanyang Univ., 1271, Sa1-dong., Ansan-si, Kyunggi-do 425-791. T: 82-31-400-5147, F: 82-31-400-4029

Yoo Kun-Sun, Prof., Halla Univ., San 66, Heungup, Wonju, Kangwondo 220-712. T: 82-33-760-1265, F: 82-33-762-6704, E: ksyoo@hit.halla.ac.kr

Yoo Kwang-Ho, Prof., Dept. of Civil Eng., The University of Suwon San 2-2, Bongdam-eup, Wawoo-ri, whasung-gun, Kyunggi-do 445-743. T: 82-31-220-2566, F: 82-31-220-2522

Yoo Nam-Jae, Prof., Dept. of Civil Eng, Kangwoon National Univ., 192-1. H yuJa-dong Chuncheon, Kangwondo 200-701. T: 82-33-250-6237, F: 82-33-250-7237, E: njyoo@ce.kangwoon.ac.kr

Yoo Tae-Sung, Mr., DaeBon Engineering Co., Ltd., 71-2, Nonhyun-dong, Kangnam-Gu, Seoul, 135-701. T: 822-546-3800, F: 82-546-7141, E: tsyoo@DaebonEng.co.kr

Yoon Yeo-Won, Prof., Inha Univ., 253, Yonghyun-dong Nom-gu Inchon 402-751. T: 82-32-860-7568, F: 82-32-874-8392, E: yoonyw@inha.ac.kr

Yoon Yoon-Mo, Mr., Dae Han Ind.Mat'l. Co., Ltd., 4 Floor Sejin BD. 54-6, Kui-2Dong, Kwanjin-Ku, Seoul, 143-202. T: 822-456-4900, F: 822-456-7210, E: daehan@aeoskorea.com

Yun Ji-Sun, Prof., Inha Univ., 253, Yonghyun-dong, Nam-gu Inchon, 402-751. T: 82-32-860-7557

LATVIA

Secretary: V. Markvarts, Latvian Geotechnical Society, Skolas Street 21, Riga PDP-1306, Latvia.
T: +371 2 273417, E: celminabpb@vide.lv

Total number of members: 31

Ališauskas, Ķ., A/S "Geoserviss", LV-1073. Rīga, Rencēnu iela 6. T: +371 7248916

Apškalējs, V., A/S "Ceļuprojekts", LV-1011. Rīga, Stabu iela 63. T: +371 2272567

Ārgalis, G., A/S "Agroprojekts", LV-1050. Rīga, Dzirnavu iela 140

Buks, I., SIA "Celmiņa būvkonstrukciju projektēšanas birojs", LV-1010. Rīga, Skolas iela 21-404. T: +371 7369824

Celmiņš, V., SIA "Celmiņa būvkonstrukciju projektēšanas birojs", LV-1010. Rīga, Skolas iela 21-404. T/F: +371 7273417

Dāvids, H., A/S "Geoserviss", LV-1073. Rīga, Rencēnu iela 6. T: +371 7248916

Dišlere, S., Latvian University, LV-1010. Rīga, Alberta iela 10

Dzilna, I., Dr., ZPGC "Junikons", LV-1045. Rīga, Rūpniecības iela 32b. T: +371 7287835

Kadišs, F., Dr., Rīga Technical University, LV-1048. Rīga, Āzenes iela 16. T: +371 7616984

Kausa, A., Rīga, LV-1010, Vīlandes iela 20-1. T: +371 7331578

Līduma, I., Rīga, LV-1013, Kr.Valdemāra iela 147/1-50. T: +371 7361027

Ļakmunds, L., Dr., Rīga, LV-1069, Kurzemes pr.110-57. T: +371 7416457

Markvarts, V., SIA "CM GIB", LV-1010. Rīga, Skolas iela 21-304. T: +371 9420177

Miķelsons, B., Rīga, LV-1039, Raunas iela 46-47. T: +371 7563780

Moldāne, L., A/S "Geoserviss", LV-1073. Rīga, Rencēnu iela 6. T: +371 7248916

Ņikitins, A., SIA "SINUSS", LV-1010. Rīga, Skolas iela 21-401C. T: +371 7277487

Ošiņa, J., Rīga Technical University, LV-1048. Rīga, Āzenes iela 16. T: +371 7089229

Pedčenko, A., A/S "Siltumelektroprojekts", LV-1001. Rīga, Kr.Barona iela 98. T: +371 7369214

Priede, V., A/S "Geoserviss", LV-1073. Rīga, Rencēnu iela 6. T: +371 7249879

Rasa, H., A/S "Agroprojekts", LV-1050. Rīga, Dzirnavu iela 140. T: +371 7213006

Ratnieks, L., Rīga, LV-1048, Uzvaras bulv.5-56. T: +371 7614720

Reķis, I., A/S "Siltumelektroprojekts", LV-1001. Rīga, Kr.Barona iela 98. T: +371 7369257

Roga, I., VDI, LV-1010. Rīga, Kr.Valdemāra iela 38. T: +371 7021741

Skrastiņš, M., A/S "Pilsētprojekts, LV-1010, Skolas ielā 21

Šnore, A., Rīga, LV-1013, Kr.Valdemāra 97-10. T: +371 7378598

Švēde, A., SIA "ATW", LV-1010. Rīga, Kr.Valdemāra 38-603. T: +371 7369827

Terentjeva, S., ZPGC "Junikons", LV-1045. Rīga, Rūpniecības iela 32b. T: +371 7323548

Treimanis, V., SIA "BALT OST GEO", LV-1039. Rīga, Struktoru iela 14a. T: +371 7552209

Tužilkins, E., Rēzekne, LV-4600, Skolas iela 8a-31. T: +371 4624334

Vikaušs, R., ZPGC "Junikons", LV-1045. Rīga, Rūpniecības iela 32b. T: 7-323548

Zavickis, J., A/S "Ceļuprojekts", LV-1011. Rīga, Stabu iela 63. T: +371 2276008

LITHUANIA

Secretary: Vincentas Stragys, Associate Professor, Lithuanian Geotechnical Society, Vilnius Gediminas Technical University, Saulėtekio al.11, 2040 Vilnius. T: 370 2 766739, F: 370 2 760434, E: Vincentas.Stragys@st.vtu.lt

Total number of members: 44

Alikonis, A., Professor, Geotechnical Department of Vilnius Gediminas Technical University, Saulėtekio al.11, 2040 Vilnius. T: 370 2 766739

Amšiejus, J., Assoc. Professor, Geotechnical Department of Vilnius Gediminas Technical University, Saulėtekio al.11, 2040 Vilnius. T: 370 2 766739

Andrijaitis, P., Geotechnical Engineer, a/b Geostatyba, Durpių 1/35, Vilnius. T: 370 2 733344

Astrauka, A., Director, a/b Geostatyba, Durpių 1/35, Vilnius. T: 370 2 733344

Audronis, K., Geologist, a/b Geostatyba, Durpių 1/35, Vilnius. T: 370 2 733344

Baikštys, R., Director, a/b Požeminiai darbai, Lazdijų 20, 3018 Kaunas. T: 370 7 298313

Bendoravičius, Š., MEng., a/b NCC Industry, Verkių 29, 2600 Vilnius. T: 370 2 300116, 370 2 300117

Dagys, A., MEng., Geotechnical Laboratory of VGTU, Saulėtekio al.11, 2040 Vilnius. T: 370 2 767868

Dundulis, K., Head of Department of Hydrology and Engineering Geology, Vilnius University, Čiurlionio 21, 2009 Vilnius. T: 370 2 233091

Furmonavičius, L., Head of Geotechnical Laboratory of VGTU, Saulėtekio al.11, 2040 Vilnius. T: 370 2 767868

Gabrielaitis, L., Dr., Dept. of Eng. Architecture, of VGTU, Saulėtekio al.11, 2040 Vilnius. T: 370 2 767880

Gadeikis, S., Dr., Department of Hydrology and Engineering Geology, Vilnius University, Čiurlionio 21, 2009 Vilnius. T: 370 2 331012

Gavelis, A., Structural Engineer, a/b Rentinys, Savanorių pr. 174 A, Vilnius. T: 370 2 322031

Jonavičius, R., Geotechnical Engineer, a/b Geostatyba, Durpių 1/35, Vilnius. T: 370 2 733344

Juknius, A., Geotechnical Engineer, a/b Pramprojektas, Donelaičio 60, 3000 Kaunas. T: 370 7 322403

Krutinis, A., Assoc. Professor, Structural Department of VGTU, Saulėtekio al. 11, 2040 Vilnius. T: 370 2 766734

Kuliešius, V., Dr., Geotechnical Laboratory of VGTU, Saulėtekio al 11, 2040, Vilnius. T: 370 2 767868

Kutenis, N., Dr., a/b Geoprojektas, P.O. Box 552, 5813 Klaipėda. T: 370 6 431335

Mackevičius, R., Assoc. Professor, Geotechnical Department of VGTU, Saulėtekio al.11, 2040 Vilnius. T: 370 2 766739

Medzvieckas, J., Dr., Geotechnical Department of VGTU, Saulėtekio al. 11, 2040 Vilnius. T: 370 2 766739

Milvydas, R., Geotechnical Engineer, Geotechnical Laboratory of VGTU, Saulėtekio al. 11. T: 370 2 767868

Monstvilas, K., Geotechnical Engineer, a/b Krašto Projektai, Ukmergės 41, Vilnius. T: 370 2 724804

Paškauskas, S., Director, a/b Geosintetika, 5800 Klaipėda, P.O. Box 165. T: 370 98 24080

Pelakauskas, A., Director, a/b Rapasta, Donelaičio 60-503, 3000 Kaunas. T: 370 7 208672

Pranaitis, V., Associate Professor, Geotechnical Department of VGTU, Saulėtekio al.11, 2040 Vilnius. T: 370 2 766739

Remeika, V., Civil Eng., KUB Stafonas, Dailidžių 16, LT-4580 Alytus. T: 370 35 25921

Sidauga, B., Professor, Geotechnical Department of VGTU, Saulėtekio al.11, 2040 Vilnius. T: 370 2 766739

Sližyte, D., Lecturer, Geotechnical Department of VGTU, Saulėtekio al.11, 2040 Vilnius. T: 370 2 766739

Skrinskas, S., Dr., a/b NCC Industry, Verkių 29, 2600 Vilnius. T: 370 2 300116, 370 2 300117

Sobolevskij, D., Professor, Belorussian State Polytechnical Academy, Pushkin Av. 39, office 5–6, Belorus, 220092 Minsk. T: 375 17 2757416, 375 17 2102045

Šcesnulevičius, J. Geotechnical Engineer, a/b Rapasta, Donelaičio 60, 3000 Kaunas. T: 370 7 208672

Šimkus, J., Professor, Geotechnical Department of VGTU, Saulėtekio al.11, 2040 Vilnius. T: 370 2 766739

Šlauteris, A., Director, a/b Geoprojektas, 5813 Klaipėda, P.O. Box 552. T: 370 6 431335

Šulskis, R., Eng., a/b NCC Industry, Verkių 29, 2600 Vilnius. T: 370 2 300116, 370 2 300117

Tumosa, K., Geotechnical Engineer, a/b Rentinys, Savanorių 174 A, Vilnius. T: 370 99 49829

Trumpis, G., Geotechnical Engineer, a/b Geostatyba, Durpių 1/35, Vilnius. T: 370 2 733344

Urbonas, J., Civil Engineer, a/b Kausta SV-1, 9 Forto pl. 60, Kaunas. T: 370 7 231510

Užpolevičius, B., Assoc. Professor, Structural Department of VGTU, Saulėtekio al. 11, 2040 Vilnius. T: 766724

Vaičaitis, J., Assoc. Professor, Baltų 35-3, 3008 Kaunas. T: 370 7 231050

Vapšys, A., Civil Engineer, a/b Statybos Sandoriai, 2000 Vilnius. Goštauto 8, T: 370 2 227511.1

Zubrickas, K., Director, a/b Klaipėdos Hidrotechnika, Nemuno str.42, LT-5804, Klaipėda. T: 370 6 355237

Zykus, J., Director, a/b Rentinys, Savanorų 174 A, Vilnius. T: 370 2 653148

Ždankus, N., Head, Department of Geoengineering, KTU, Donelaičio 73, 3006 Kaunas. T: 370 7 202346

Žutautas, A., Geotechnical Engineer, a/b Geoprojektas, 5813 Klaipėda, P.O. Box 552. T: 370 6 341335

THE FORMER YUGOSLAV REPUBLIC OF MACEDONIA

Contact person: Vlatko SESOV, Institute for Earthquake Engineering and Engineering Seismology, P.O. Box 101, 1000, Skopje, The Former Yugoslav Republic of Macedonia. T: 389 2 176 155, F: 389 2 112 163, E: vlatko@pluto.iziis.ukim.edu.mk

Total number of members: 72

Aleksovski, D., Institute for Earthquake Engineering and Engineering Seismology, P.O. Box 101, Skopje. T: 389 2 176 155, E: dalek@pluto.iziis.ukim.edu.mk

Andonov, V., Geoproekt, blvd. ASNOM 8-II/5, Skopje. T: 389 2 434 966

Andreevski, B., REK Bitola – Development and Investments, Partizanska 99/6, Bitola

Bagasov, U., ADG Mavrovo, EE Geomavrovo, Blagoja Toska 21/8, Tetovo. T: 389 2 458 352

Cakarovski, L., PIKP "Cakar – Partners", Zeleznicka 37, Skopje. T: 389 2 112 361

Cekorova, R., DGR "Geotehnika"–ADG "Pelagonija", blvd. Jane Sandanski 37/III-8, Skopje. T: 389 2 162 104

Celebieva-Stojanoska, U., Institute for studies and designing, "Beton", Jurij Gagarin 15, Skopje. T: 389 2 380 888, Ext. 263

Crcarevski, J., ADG Mavrovo, EE Geomavrovo, blvd. Srbija b.b., Skopje. T: 389 2 458 352

Cubrinovski, M., Kiso-Jiban Consultants Co., Ltd., 1-11-5 Kudan-kita, Chiyoda-ku, Tokyo 102-8220, JAPAN. T: 81 3 5276 6737, E: misko.cubrinovski@kiso.co.jp

Delipetrev, T., Faculty of mining and geology – Stip, RGF – Stip

Dimitrievski, Lj., Faculty of Civil Engineering – Skopje, blvd. Partizanski Odredi b.b., Skopje. T: 389 2 116 066, Ext. 147, F: 389 2 117 367, E: dimitrievski@stobi.ga.ukim.edu.mk

Dubrovski, P., ADG Mavrovo, EE Geomavrovo, blvd. Srbija b.b., Skopje. T: 389 2 458 352, F: 389 2 458 364

Gadza, V., Institute for Earthquake Engineering and Engineering Seismology, P.O. Box 101, Skopje. T: 389 2 176 155

Gapkovski, N., Faculty of Civil Engineering – Skopje, blvd. Ilinden 87/11, Skopje. T: 389 2 231 034

Gligorijevic, Lj., Geohidroproekt, Ivan Cankar 109 a, Skopje. T: 389 2 340 582

Gligorijevic, M., Geological Institution, Ivan Cankar 109 a, Skopje. T: 389 2 340 582

Gorceski, Lj., EMO – HEP, Naroden front 54-5/1, Skopje. T: 389 2 125 389

Gorgevski, S., Faculty of Civil Engineering – Skopje, blvd. Partizanski Odredi b.b., Skopje. T: 389 2 116 066

Gorgevski, H., GEOING, Naroden front 31/4, Skopje. T: 389 2 135 254, F: 389 2 135 254

Gurkov, K., Institute for Civil Engineering "Makedonija", Drezdenska 52, Skopje. T: 389 2 366 816

Ilievski, D., D.G.P.U. "GEING", Borka Taleski 24, Skopje. T: 389 2 110 205, F: 389 2 132 369, E: geing@mkinter.net

Jovanovski, M., Faculty of Civil Engineering – Skopje, blvd. Partizanski Odredi b.b., Skopje. T: 389 2 116 066

Karagovski, G., Geoproekt, blvd. AVNOJ 42, Skopje. T: 389 2 434 966

Kirovakova, B., Petar Pop Arsov 19/I/18, Skopje. T: 389 2 215 890

Kitanovski, Lj., GP Granit A.D., OE Laboratory, Makedonsko-Kosovska Brigada b.b., Skopje. T: 389 2 622 481

Koneski, S., GP Granit A.D., OE Laboratory, Makedonsko-Kosovska Brigada b.b., Skopje. T: 389 2 614 494, 389 2 622 481

Kostovski, K., AD Granit, Daskal Kamce 3, Skopje. T: 389 2 368 039

Krango, N., D.G.P.U. "GEING", Borka Taleski 24, Skopje. T: 389 2 110 205, F: 389 2 132 369, E: geing@mkinter.net

Manasiev, J., REK Bitola, PE Mine "Suvodol", Rudnicka naselba, v. Sopotnica

Marina, M., Geohidroproekt, Bihacka 6, Skopje. T: 389 2 237 742

Marinov, V., D.G.P.U. "GEING", Borka Taleski 24, Skopje. T: 389 2 110 205, F: 389 2 132 369, E: geing@mkinter.net

Markovski, V., DGR Geotehnika, ADG Pelagonija, Vera Jocic 19/1-9, Skopje. T: 389 2 218 840

Medarski, N., Institute for Civil Engineering "Makedonija", Drezdenska 52, Skopje. T: 389 2 363 040

Micevski, E., Geohidroproekt, Bihacka 6, P.O. Box 196, Skopje. T: 389 2 237 742, F: 389 2 114 760

Mihailov, V., Institute for Earthquake Engineering and Engineering Seismology, P.O. Box 101, Skopje. T: 389 2 176 155, E: mihailov@pluto.iziis.ukim.edu.mk

Mihajlovski, S., Emo-Ohrid, Institute for energetics "Hidroelektro-proekt", Nikola Vapcarov 11, Skopje. T: 389 2 220 766, E: emoskp@lotus.mpt.com.mk

Miladinov, V., Institute for Civil Engineering "Makedonija", Drezdenska 52, Skopje. T: 389 2 366 829

Milovanovic, S., Faculty of Civil Engineering – Skopje, blvd. Partizanski Odredi b.b., Skopje. T: 389 2 116 066

Mirakovski, G., Institute for Earthquake Engineering and Engineering Seismology, P.O. Box 101, Skopje. T: 389 2 176 155, E: mirak@pluto.iziis.ukim.edu.mk

Mitevski, T., ADG Mavrovo, EE Mavrovoproekt, blvd. Partizanski Odredi 27-4/3, Skopje. T: 389 2 165 088

Mitrov, T., blvd. Ilinden 57, Skopje. T: 389 2 228 638

Moslavac, D., Faculty of Civil Engineering – Skopje, blvd. Partizanski Odredi b.b., Skopje. T: 389 2 116 066, Ext. 139, E: moslavac@stobi.ga.ukim.edu.mk

Nanus, A., Institute for Civil Engineering "Makedonija", Drezdenska 52, Skopje. T: 389 2 363 040

Neceski, S., D.G.P.U. "GEING", Borka Taleski 24, Skopje. T: 389 2 110 205, F: 389 2 132 369, E: geing@mkinter.net

Nikolov, S., Institute for Civil Engineering, Drezdenska 52, Skopje. T: 389 2 363 040

Nikoloska, V., D.G.P.U. "GEING", Borka Taleski 24, Skopje. T: 389 2 110 205, F: 389 2 132 369, E: geing@mkinter.net

Novaceski, T., REK Bitola, REK "Suvodol". T: 389 097 206 626

Paskalov, A., Institute for Earthquake Engineering and Engineering Seismology, P.O. Box 101, Skopje. T: 389 2 176 155, E: paskal@pluto.iziis.ukim.edu.mk

Petkovski, Lj., Faculty of Civil Engineering – Skopje, blvd. Partizanski Odredi b.b., P.O. Box 560, Skopje. T: 389 2 116 066, Ext. 236, E: petkovski@stobi.ga.ukim.edu.mk

Petreski, S., Institution for materials research "Skopje" – Skopje, Rade Koncar 16, Skopje. T: 389 2 313 125

Petreski, Lj., REK Bitola, REK Bitola, PE Mine "Suvodol"

Petrovska, M., D.G.P.U. "GEING", Borka Taleski 24, Skopje. T: 389 2 110 205, F: 389 2 132 369, E: geing@mkinter.net

Petrusev, E., AD Granit, Krste Asenov 7/I-23, Skopje. T: 389 2 203 590

Pjevic, S., D.G.P.U. "GEING", Borka Taleski 24, Skopje. T: 389 2 110 205, F: 389 2 132 369, E: geing@mkinter.net

Ribeski, D., Institution for materials research "Skopje" – Skopje, Rade Koncar 16, Skopje

Siderovski, K., Institute for Earthquake Engineering and Engineering Seismology, P.O. Box 101, Skopje. T: 389 2 176 155

Sendov, B., blvd. Kliment Ohridski 15/3-13, Skopje. T: 389 2 225 371

Sesov, N., Institute for Earthquake Engineering and Engineering Seismology, P.O. Box 101, Skopje. T: 389 2 176 155

Sesov, V., Institute for Earthquake Engineering and Engineering Seismology, P.O. Box 101, Skopje. T: 389 2 176 155, E: vlatko@pluto.iziis.ukim.edu.mk

Spirovski, D., DGR Geotehnika, ADG Pelagonija, Pandil Siskov 24, Skopje. T: 389 2 162 104

Talaganov, K., Institute for Earthquake Engineering and Engineering Seismology, P.O. Box 101, Skopje. T: 389 2 176 155, E: kotal@pluto.iziis.ukim.edu.mk

Tanatarec, Lj., REK Bitola I, Minela Babukovski 3/5, Bitola

Tancev, Lj., Faculty of Civil Engineering – Skopje, blvd. Partizanski Odredi b.b., P.O. Box 560, Skopje. T: 389 2 237 972, E: tancev@stobi.ga.ukim.edu.mk

Taneski, B., Institution for materials research "Skopje" – Skopje, Rade Koncar 16, Skopje. T: 389 2 221 363

Tanevski, G., GP Granit A.D., Teodosie Gologanov, Skopje. T: 389 2 374 143

Terzioski, D., Institute for Civil Engineering "Makedonija", Drezdenska 52, Skopje. T: 389 2 363 040

Trajkovski, V., D.G.P.U. "GEING", Borka Taleski 24, Skopje. T: 389 2 110 205, F: 389 2 132 369, E: geing@mkinter.net

Turnovaliev, Z., Republic Institution for court expertise, Zeleznicka b.b., Skopje. T: 389 2 375 419

Velevski, A., Institute for Civil Engineering "Makedonija", Drezdenska 52, Skopje. T: 389 2 363 040

Vitanov, V., Faculty of Civil Engineering – Skopje, blvd. Partizanski Odredi b.b., Skopje. T: 389 2 116 066

Veljanovska, K., D.G.P.U. "GEING", Borka Taleski 24, Skopje. T: 389 2 110 205, F: 389 2 132 369, E: geing@mkinter.net

Zafirova, I., Institute for Earthquake Engineering and Engineering Seismology, P.O. Box 101, Skopje. T: 389 2 176 155

MAROCCO/MAROC

Secretary: Mr H. Ejjaaouani, Comité Marocain de la Mécanique des Sols et des Roches, lpee Centre Expérimental des Sols, Km 7 Route d'El Jadida, Casablanca, Oasis BP 8066. F: 022 23 41 88, E: Ejjaouan@open.met.ma

Total number of members: 12

Khalid, R., President, Ecole Mohannedia, 56 Rue Sebou, Agdal, Rabat. F: 037 77 88 53, E: Diremi@emi.bc.ma

Ejjaaouani, H., Secretaire General, LPEE-CES, 97 Rue 60 Hay Tarik, Sidi Bernoussi. F: 022 23 41 88, E: Ejjaouan@open.net.ma

Akenkou, H., Tresorier, LPEE – CES, 586 Bd Abou Choib Doukkali Hay Idraissia 1, Casablanca. F: 061 23 41 88, E: Akenkou@caramail.com

Bouchaqour, Y., Ministere Equipment – Dat, Rabat, Chellah. F: 037767827, E: Boucha@rntpnet.gov.ma

Choukali, A., TESCO, 19 Rue al Kadi Bakker, Quartier Burger, Casablanca. F: 022 98 42 84, E: TESCO@open.net.ma

Essaadaoui, El M., Forasol Maroc, 11 Rue 17 Lotis Tazi, Mloudi Californie, Casablanca. F: 022 33 58 13, E: Forasolmaroc@Emi.ac.ma

Lhansali, Y., ONCF, Rue Abderahim Rafiki, Rabat, Agdal. F: 037 77 44 80, CMMSR@caramail.com

Louridi, M., ADM, BP 6526, Hay Riad, Rabat. F: 037 71 10 56, E: CMMSR@caramail.com

Sahli, M., EHTP, Residence Rami, 1 Rue Aboutou 20100 Maarif, Casablanca. F: 022 23 42 61

Lahsiba, D., ODEP, Casablanca. F: 022 25 74 66, E: D-LHASIBA@ODEP.org.ma

Rais, A., DRCR, Residence EC Boustane IMM

Refass, A., NBR Centre, 19 Rue Kazouini Chautilly, Casablanca 20100. F: 022 34 33 35, E: Labonbr@casanet.net.ma

MEXICO

Secretary: Célicia Chávez Jaimes, Sociedad Mexicana de Mecánica de Suelos, A.C. Valle de Bravo 19, Vergel de Coyoacán, Mexico, D.F. T: 56 77 37 30, F: 56 79 36 76, E: smms@prodigy.net.mx

Total number of members: 333

Aburto, V.R., Pico de Verapaz 449-106, torre 1 piso 6, Col. Jardines de la montaña, c.p. 14210, Tlalpan, México, D.F. T: 55102032, F: 55100573, E: raburto@tolsa.mineria. unam.mx

Adaya, T.P., Norte 70-A No. 5128, Col. Bondojito, c.p. 07850, G.A. Madero, México, D.F. T: 55514756, F: 55514026, E: geoconsa@prodigy.net.mx

Aguirre, M.L.M., Américas No. 225, Parque San Andrés, c.p. 04030, México, D.F. T: 55446602, F: 56893513, E: geosol@mpsnet.com.mx

Alanis, D.M.J., Arcos de Zacatecas 1230, Fraccionamiento Los arcos, c.p. 21079, Mexicali, Baja California. T: 0165641650, 0165575226, F: 0165575226

Alberro, A.J., Apartado postal 70-472, Col. Coyoacán, c.p 4510, Coyoacán, México, D.F. T: 5622-34-26, F: 5616-1514, E: ja@pumas.iingen.unam.mx

Alcerreca, C.E., Serafín Olarte 250, col. Vertíz Narvarte, c.p. 3600, Benito Juárez, México, D.F. T: 0124623793, F: 0124626363, E: ealcerre@sct.gob.mx

Alemán, V.J. de D., Augusto Rodin No. 265, Col. Nochebuena, c.p. 3720, Benito Juárez, México, D.F. T: 2309310, Ext. 8625, F: 55631005

Alonso, M.J., Calle Agua No. 144, col. Satélite Fovisste, c.p. 76110, Querétaro, Querétaro. T: 0142182923

Alvarez, M.A.A., Tabachin No. 17, Col. Alamos 2ª. Sección, c.p. 76160, Querétaro, Queré-taro. T: 0142127209, F: 0142241540, E: aamanilla@hotmail.com

Alvarez, S.M., Ocampo 72-D Sur, c.p. 76000, Querétaro, Querétaro. T: 0142125828

Aquino, M.C., Calle 56 norte No. 1629, Edificio E. Depto. 8, Col. San José Villaverde, c.p. 73210, Puebla de los Angeles, Puebla. T: 0122826219

Arean, C.F., Pico de Verapaz No. 461, torre 1 piso 6, Jardines en la montaña, c.p. 14210, México, D.F. T: 55247365, F: 55247444

Arias, E.F.J., Donizetti No. 158, Col. Vallejo, c.p. 07870, Gustavo A. Madero, México, D.F. T: 52729991, F: 52729991, Ext. 2509, E: ariasf@ica.com.mx

Astudillo, S.L., Heliotropos No. 27, Col. Rancho Alegre I, c.p. 0956, Coatzacoalcos, Veracruz. T: 0192187783, F: 0192186960

Auvinet, G.G.I.A., Francia 131, casa 10, Col. Florida, c.p. 01030, Alvaro Obregón, México, D.F. T: 56223500, F: 56160784, E: gaug@pumas.iingen.unam.mx

Avelar, C.J.A., Vía Dr. Gustavo Baz No. 300, Fraccionamiento Bosque de Echegaray, c.p. 53310, Naucalpan de Juárez, Edo. De México. T: 53734500, F: 53734599, E: iecsacv@ netservice.com.mx

Avelar, C.R., Vía Dr. Gustavo Baz No. 300, Fraccionamiento Bosque de Echegaray, c.p. 53310, Naucalpan de Juárez, Edo. De México. T: 53734500, F: 53734599, E: iecsacv@net service.com.mx

Avelar, L.R., Vía Dr. Gustavo Baz No. 300, Fraccionamiento Bosque de Echegaray, c.p. 53310, Naucalpan de Juárez, Edo. De México. T: 53734500, F: 53734599, E: iecsacv@ netservice.com.mx, E: ati_opalo@infosel.net.mx

Azomoza, P.J.G., 15 Poniente No. 4309, Col. Belisario Domínguez, c.p. 72160, Puebla, Puebla. T: 0122497076, F: 0122497096, E: Azomozac@pue1.telmex.net.mx

Bajo, S.J.L., Alfonso Vicens Saldivar 1306, Fraccionamiento Cedros, Col. Gaviotas, c.p. 86090, Villa Hermosa, Tabasco. T: 0193156212, F: 0193156330

Baranda, E.A., Tokio 503 Bis, Col. Portales, c.p. 03000, Benito Juárez, México, D.F. T: 56053506

Barredes, E.J.L., Calzada del hueso 160, ediF: G 13-404, Col. Coyoácan, c.p. 04850, Exhacienda coapa, México, D.F. T: 56713961

Barriga, L.P., Calle F., edificio 30 depto. 23, Unidad H. alianza popular revolucionaria, c.p. 04800, Coyoácan, México, D.F. T: 56779400

Bello, M.A., Martín L. Guzmán 273, Col. Iztacihuatl, c.p. 03520, Benito Juárez, México, D.F. T: 56719603

Benamar, I., M. No. 145, Col. Escandon, c.p. 11800, Escandón, México, D.F. T: 52729991, Ext. 3249, F: 52729991, Ext. 2509, E: benamari@ica.com.mx

Benavente, L.J., La lonja 755, Col. Bolivar, c.p. 78330, San Luis Potosí, San Luis Potosí. T: 0148127139, 0148204643

Bonilla, C.H., Patriotismo No. 683, Col. Mixcoac, c.p. 03910, Benito Juárez, México, D.F. T: 55985218, 55983863, F: 56889752, E: hbonilla@ser-df,imt.mx

Borja, N.G., Av. María No. 13, Col. del Carmen Coyoacán, c.p. 04000, Coyoacán, México, D.F. T: 55549265, F: 56887608, E: gilbor@df1.telmex.net.mx

Bosco, R.R.J., Adolfo Prieto No. 1238, Col. del Valle, c.p. 31000, Benito Juárez, México, D.F. T: 55599055, F: 55751556

Bravo, G.G., Rancho el Ciprés 34-103, Col. Prado Coapa, c.p. 14350, Tlalpan, México, D.F. T: 56792815, 53581264

Bueno, M.J.L., 265 Cozumel C.T., Laguna Beach, c.p. 92651, California, USA. T: (949)5832700, Ext. 109, F: (949)5832700, E: jbueno@golder.com

Caballero, A.T., Ayocuan No. 141, Col. Nezahualcoyotl, c.p. 68140, San Martín Mexicapan, Oaxaca, Oaxaca. T: 0195122076, F: 0195170400

Cacho, V.A., Av. Desierto de los leones 4938, Tetelpan, c.p. 01700, Alvaro Obregón, México, D.F. T: 55958814, F: 55811795

Calderón, F.L., España No. 2099, Col. Moderna, c.p. 44150, Guadalajara Jalisco. T: 0136158343

Calderón, G.J., Manuel Balbontin No. 446, Chapultepec Oriente, c.p. 58260, Morelia Michoacán. T: 0143167205

Calderón, H.H.M., Av. Revolución No. 595, San Pedro de los Pinos, c.p. 3800, Benito Juárez, México, D.F. T: 55433800

Calles, R.G., Calle 24 No. 128, Col. Florida, c.p. 86040, Villahermosa Tabasco. T: 151559 y 154328

Canales, F.A., Guadalupe Victoria No. 602, Col. Residencial Periférico, c.p. 66420, San Nicolás de los Garza, Nuevo León

Cancino, L.F., Augusto Rodin No. 265, Col. Nochebuena, c.p. 3720, Benito Juárez, México, D.F. T: 55294400, Ext. 8628, F: 55631005, E: fcl19880@cfe.gob.com

Capallera, C.J.F., Alaminos No. 297, Col. Reforma, c.p. 91919, Veracruz, Veracruz. T: 0129316868, F: 0129316868

Cárdenas, C.B., Puesta del Sol No. 1785, Col. Cerro del Tesoro, c.p. 45081, Tlaquepaque, Jalisco. T: 6693445, F: 8114355, E: cardben@cianet.com.mx

Carmona, R.J. Sergio, Monte Carmelo No. 97, Col. Independencia, Guadalajara, Jalisco. T: 6513688

Carranza, E.R., Tekit manzana 127, lote 6, Col. Ejidos de Padierna, c.p. 14200, Tlalpan, México, D.F. T: 55241050

Carreola, N.J., Privada del Relox No. 16, Col. Copilco el Bajo, c.p. 04340, Alvaro Obregón, México, D.F. E: eambriz@sct.gob.mx

Carreón, F.D.C., Vertíz No. 490, Col. Narvarte, c.p. 03020, Benito Juárez, México, D.F

Casales, G.C., Monrovia No. 504, Col. Portales, c.p. 03300, Benito Juárez, México, D.F

Casales, L.V., Monrovia No. 504, Col. Portales, c.p. 03300, Benito Juárez, México, D.F. T: 56118446

Castañeda, M.H., Primo Verdad No. 625, Col. centro, c.p. 20000, Aguascalientes, Aguascalientes. T: 0149161068

Castellanos, H.G.Y., Av. Hidalgo No. 1311, Col. Centro, c.p. 68000, Oaxaca de Juárez, Oaxaca. T: 0195162904, F: 0195140621, E: esmic@spersaoaxaca.com.mx

Castellanos, H.R.E., Av. Hidalgo No. 1110, Col. Centro, c.p. 68000, Oaxaca de Juárez, Oaxaca. T: 0195140007, F: 0195140621, E: esmic@spersaoaxaca.com.mx

Castilla, C.J., Augusto Rodin No. 265, primer piso, Col. Nochebuena, c.p. 03720, Benito Juárez, México, D.F. T: 52294400, Ext. 8610, F: 55631005

Castillo, G.H., 1ª. Cerrada Benito Juárez No. 30, Col. Castillo chico, Cuautepec Bajo, c.p. 07220, Gustavo A. Madero, México, D.F. T: 55444776

Castillo, S.J.L., Ignacio Zaragoza No. 324, Col. Chapultepec Oriente, c.p. 58260, Morelia, Michoácan. T: 0143144666

Castro, A., 1ero. De Mayo No. 2, Col. santa Lucia del camino Oaxaca, c.p. 71228, Oaxaca, Oaxaca. T: 0195175019, F: 0195157364

Castro, V.J.A., Av. Reforma No. 530, Col. La piragua, c.p. 68300, Tuxtepec Oaxaca. T: 0128751044, 0128751880

Cervantes, M.M.A., 35 norte No. 3023 Unidad Aquiles Serdan, Col. Unidad Aquiles Serdan c.p. 72070, Puebla, Puebla. T: 0122315836, F: 0122315836, E: lacocs@ prodigy.net.mx

Chávez, J.C., Hacienda Santa Cecilia No. 206, Col. Villa quietud, c.p. 04960, Coyoacán, México, D.F. T: 55245798, 55249106, 55244918, 55249265, Ext. 315, F: 55249265, E: zelikha1@hotmail.com

Chávez, R.L. Efren, Augusto Rodin No. 265, Col. Nochebuena, c.p. 03720, Benito Juárez, México, D.F. T: 52309090, F: 52309272, E: lchavez@cfe.gob.mx

Collazo, T.A., Jardín del estudiante No. 1 Col. Centro, c.p. 20000, Aguascalientes, Aguascalientes. T: 0149123345, Ext. 518

Cordova, A.J.M., Blvd. García de León No. 586, Col. Nueva Chapultepec, c.p. 58280, Morelia, Michoacán, T: 0143145227, 0143264628

Corro, C.S., Apartado postal 70-472, Ciudad Universitaria, c.p. 04510, Coyoacán, México, D.F. T: 55500388, F: 56162894, E: scc@pumasiingen.unam.mx, pscc@ prodigy.net.mx

Corte, N.A., 15 poniente 4309, Col. Belisario Domínguez, c.p. 72160, Puebla, Puebla. T: 0122497076, F: 0122497096

Couttolenc, E.O.R. Medicina No. 68-3, Col. Copilco, c.p. 04360, Coyoacán, México, D.F. T: 55437044, F: 55235061

Covarrubias, V.S., Apdo. postal 60-depto. 613, San Pedro de los pinos c.p. 03800, Benito Juárez, México, D.F. T: 55986177, F: 55983729

Crespo, V.C., Pedro Quintanilla No. 390, Col. Chepe Vera, c.p. 64030, Monterrey, Nuevo León. T: 0183463371, F: 0183284213

Cruickshank, G.G., J. Loreto Favela No. 850, Col. San Juan de Aragón, c.p. 07950, Gustavo A. Madero, México, D.F. T: 57600066, 57600283

Cruz, C.B., Casilla Postal No. 09-01-5477, Col. Del. Ciudadela 9 de Octubre, Ecuador. E: ahidrovo@porta.net

Cruz, C.M., Calzada Tecnológico S/N, Col. Centro, c.p. 68000, Oaxaca, Oaxaca. T: 0195161722

Cruz, C.P., Adolfo Duclos S. No. 47, Col. Santa Martha Acatitla, c.p. 09510, Iztapalapa, México, D.F. T: 57335521, F: 57335521

Cruz, M.S., Salud B-29, Col. 19 de Mayo, C.P. 70070, Juchitán de Zaragoza, Oaxaca. T: 018007152114, 0197113492

Cuanalo, C.O.A., 19 Norte No. 3621, Col. San Miguel Hueyotlipan, c.p. 72050, Puebla, Puebla

Cuevas, O.H., Santa Rosa No. 11, Col. San Bartolo Atepehuacan Lindavista, c.p. 07730, Gustavo A. Madero, México, D.F. T: 55677528, 55874162

Cuevas, O.J. Manuel, Unidad la patera, ediF: 11 departamento 18, Col. Industrial Vallejo, c.p. 07800, Gustavo A. Madero, México, D.F. T: 53650323, F: 53628751, E: aci@mpsnet.com.mx

Cuevas, R.A., Av. Miguel Hidalgo No. 77, San Lucas Tepetlacalco, c.p. 54055, Tlalnepantla, Edo. De México. T: 53650321, 53650323, F: 53628751, E: aci@mpsnet.com.mx

De Cervantes, Padilla Humberto Leonel, Franz Halss No. 139, Col. Alfonso XIII, c.p. 01460, Alvaro Obregón, México, D.F. T: 55635956, F: 55635479, E: eta3@prodigy.com.mx, etacu@hotmail.com

De la Cruz, M.C., Roberto Gayol No. 1255, desp. 402, Col. del Valle, c.p. 03100, Benito Juárez, México, D.F. T: 55590160, F: 55753449

De la Rocha, P.R., Apartado postal No. 1818, Santo Domingo, República Dominicana

Demeneghi, C.A., Cruz Verde No. 133-4, Toriello Guerra, c.p. 14050, Tlalpan, México, D.F. T: 56228003, F: 56228007

Díaz, C.J.L., Av. Melchor Ocampo No. 26, Col. San Mateo 1ª. Sección, c.p. 42850, Tepeji del río de Ocampo, Hidalgo. T: 0177330115

Díaz, C.M., Paseo del pedregal No. 1200, Col. Jardines del pedregal c.p. 01900, Alvaro Obregón, México, D.F. T: 55684296, F: 56533589

Díaz, G.R., Sur 28 No. 27, Col. Agrícola Oriental, c.p. 08500, Iztacalco, México, D.F. T: 55245699, F: 55246389

Díaz, R.J.A., Apartado Postal 70-256, c.p. 04510, Coyoacán, México, D.F. T: 56223230, F: 56223229, E: jadrdiaz@ servidor.unam.mx

Díaz-Infante, S.G., Hacienda de peñuelas No. 348, Fraccionamiento hacienda echegaray, c.p. 53300, Naucalpan, Edo. De México. T: 55605605, F: 55605605

Dimas, J.G., Av. Cuauhtémoc No. 45, condominio 2B casa 87, Unidad San Jacinto, Ixtapaluca, Edo. De México. T: 56799785, F: 56799100

Dumas, G.G.C., Onimex 5 lote 3 casa 51, Col. El potrero residencial los laureles, c.p. 55090, San Cristobal Ecatepec, Edo. De México. T: 52294400, Ext. 8792, F: 55630624

Elizondo, R.A.M., Rancho Altamira No. 116, Fraccionamiento los Sauces, c.p. 04940, Coyoacán, México, D.F. T: 56889803, F: 56887608, E: elizondo@citlali.imt.mx

Ellstein, R.A., Presa Temascal No. 10, Col. Electra, c.p. 54060, Tlalnepantla, Edo. De México. T: 53972022, F: 53615076

Espinosa, G.L., Calle 5 No. 63, reparto Dolores Patron, Col. García Gineres, c.p. 97070, Merida Yucatán. T: 0199250714

Esquivel, D.R. Francisco, Pasadena No. 23, Col. del Valle, c.p. 03100, Benito Juárez, México, D.F. T: 56690969, F: 56690769, E: raúlesq@df1.telmex.net.mx

Estrada, G.I.J., 38 Sur No. 4037, Fraccionamiento Castro Green La Mesa, c.p. 22650, Tijuana, Baja California. T: 0166890534, F: 0166890534, 55906938

Estrada, R.J.C., Calle Prueba No. 6, Col. Industrial, c.p. 07800, Gustavo A. Madero, México, D.F. T: 55775630, F: 55775630

Farías, L.G., Gonzalo Curiel No. 405, Col. Lomas del roble, c.p. 66450, San Nicolas de los Garza, Nuevo León. T: 0183522696, F: 0183522696

Farjeat, P.D., Zacatepetl No. 345, Col. Parques del pedregal, c.p. 14010, Tlalpan, México, D.F. E: farjeat@mail.internet.com.mx

Farjeat, P.E., Pico de Somosierra No. 27, Col. Jardines de la montaña, c.p. 14210, Tlalpan, México, D.F. T: 0174826038, F: 56452132

Favela, L.F., Roberto Gayol No. 55, Col. del Valle, c.p. 03100, Benito Juárez, México, D.F. T: 55591769

Felix, V.R., 1ª. Cerrada de Alborada No. 8, Col. Parques del pedregal, c.p. 14010, Tlalpan, México, D.F

Fernández, R.J. Francisco, Cerro del Chinaco No. 115, Col. Campestre Churubusco, c.p. 04200, Coyoacán, México, D.F. T: 52294400, Ext. 8779, F: 55631005, E: jfr81001@cfe.gob.mx

Ferrer, T.H.O., Benito Juárez No. 802, Col. San Felipe Hueyotlipan, c.p. 72030, Puebla, Puebla. T: 0122880760, E: ferrer@upaep.mx, hugo.ferrer@upaep.mx

Figueroa, V.G.E., La loma No. 13, col. Lomas de San Angel Inn, c.p. 01790, A. Obregon, México, D.F. E: ficuasa@hotmail.com

Flores, B.R., Matamoro No. 211, Col. Tlalpan,c.p. 14000, Tlalpan, México, D.F. T: 01731940, Ext. 705, 55737728, F: 0173208725, E: rflores@tlaloc.imta.mx

Flores, G.P., Iglesia No. 485, Col. Jardines del pedregal, c.p. 01900, A. Obregón, México, D.F. T: 56750486

Flores, N.J., Cañada No. 46, Col. Plazas de la Colina, c.p. 54080, Tlalnepantla, Edo. De México. T: 53973516

Flores, O.J.E., Llama No. 117, Col. Jardines del pedregal, c.p. 01900, A. Obregón, México, D.F. T: 55590160, F: 55753449

Franco, C.F.J., Electra 32, Aeropuerto, Col. Mesa de Otay, c.p. 22300, Tijuana, Baja California. T: 0166825464

Fuentes, S.F.R., Lomas de los altos No. 2039, Lomas de Atemajac, c.p. 45170, Zapopan, Jalisco. T: 0138237928, F: 0138332000, E: ffuentes@quantum.ucting.udg.mx

Galindo, G.L., Cerrada de Clarin mza. 12 lote 19-3, Fraccionamiento Rinconada de Aragón, c.p. 55140, Ecatepec, Edo. De México. T: 57755242, F: 57784343

Galindo, S.A., Av. Coyoacán No. 1895, Col. Acacias, c.p. 03420, Benito Juárez, México, D.F. T: 55249265, F: 55249265

Gallo, A.G., Cascada 1141-105, Col. Las reynas, c.p. 36660, Irapuato, Guanajuato. T: 0146241977, F: 0146241977

García, A.C., Gob. Rafael Rebollar 119-5, Col. San Miguel Chapultepec, c.p. 11850, Miguel Hidalgo, México, D.F. T: 55944572, F: 56732992

García, F.R., Paseo Ensenada No. 3005, sección costa hermosa, Col. Playas de Tijuana, c.p. 22240. T: 0166806988, F: 0166806988, E: geoing@telnor.net

García, G.A., Privada de Rayon No. 104, Col. centro, c.p. 68000, Oaxaca de Juárez, Oaxaca. T: 0195147030, F: 0195147030

García, R.G., Mar Baffil, ediF: 2-504, Col. Casas lindas Infonavit, c.p. 52947. Atizapan de Zaragoza, Edo. De México. T: 57523174, 57523157

García, R.L.A., Antiguo Camino a Xochimilco No. 5099, raudal No. 101, Ampliación Tepepan, c.p. 16029, Xochimilco, México, D.F. T: 56755167, F: 56754288

García, T.E., Presidentes No. 511 inT: 18, Col. Portales, c.p. 03300, Benito Juárez México, D.F. T: 56055376, E: garciaer@df1telmex.net.mx

Garnica, A.P., Km. 12+000 carretera Querétaro-Galindo, Col. Sanfandila, c.p. 76700, Pedro Escobedo, Querétaro. T: 0142169777, F: 0142169671, E: Paul.Garnica@imt.mx

Girault, D.L.P., Volcán No. 120, Col. Lomas de Chapultepec, c.p. 11000, Miguel Hidalgo, México, D.F. T: 55404329, F: 55204408

Goddard, E. J., Cerro del Agua No. 114, Col. Fraccionamiento Colinas del Cinatario, c.p. 76090, Querétaro, Querétaro. T: 0142235431, F: 0142235431, E: goddard@sisenet.net.mx

Gómez, C.P., Paseo del quetzal No. 70, Col. Lomas Verdes, c.p. 53120, Naucalpan, Edo. De México. T: 55349565

Gómez, L.J.A., Km. 12+000 carretera Querétaro-Galindo, Col. Sanfandila, c.p. 76700, Pedro Escobedo, Querétaro. T: 0142169777, F: 0142169671

Gómez, P.T., Segunda Sección 100, Unidad modelo, c.p. 68100, Oaxaca, Oaxaca. T: 0195136369

González, V.F., Augusto Rodin No. 265, Col. Nochebuena, c.p. 03720, Benito Juárez, México, D.F. T: 52294400, Ext. 8685, F: 55631005, E: fgonval@cfe.gob.mx

Guerrero, S.A., Abasolo No. 161, Col. Juárez Pantitlan, c.p. 57460, Nezahualcoyotl, Edo. De México. T: 57970872, F: 55248665

Gutierrez, A.E., Av. Del Imán 660, lote 8, departamento 602, Col. Pedregal del maurel, c.p. 04720, México, D.F. T: 56660598, F: 56660598

Gutierrez, S.C.E., Av. Miguel Hidalgo No. 77, San Lucas Tepetlacalco, c.p. 54055, Tlalnepantla, Edo. De México. T: 53650321, F: 53652917

Hass, M.H.S., Jalisco No. 14, Col. Héroes de Padierna, c.p. 10700, Magdalena Contreras, México, D.F. T: 56228003, F: 56228003

Hedron, Jr, A.J., 2230C Newmark Civil Engineer, c.p. 61801, Urbana Illinois, USA

Hernández, A.M., Llovizna No. 330, Col. La herradura, c.p. 29038, Tuxtla Gutierrez, Chiapas. T: 0196152808

Hernández, B.R., Colina de las ventiscas No. 81, Fraccionamiento Boulevares, c.p. 53140, Naucalpan, Edo. De México. T: 55608172, F: 55608172

Hernández, I.R., Av. Coyoacán No. 1895, Col. Acacias, c.p. 03240, Benito Juárez, México, D.F. T: 55246389, F: 55246389

Hernández, R.A., Carlos Pereira no 1 departamento 9, Col. Viaducto piedad, c.p. 08200, Iztacalco, México, D.F. T: 57941100, Ext. 2086, F: 57600049,

Hernández, S.C., Virginia No. 48, Col. Nativitas, c.p. 03500, Benito Juárez, México, D.F. T: 55902514

Hernández, S.R., Tacoteno No. 253, Col. Las Flores, c.p. 68354, Tuxtepec, Oaxaca. T: 0128751880, F: 0128751880

Hernández, S.R.H., Benito Juárez No. 203, Col. San Felipe del Agua, c.p. 68020, Oaxaca, Oaxaca. T: 0195158072, F: 0195140755

Herrera, A.A., Libertad No. 1468, Col. Centro, c.p. 68300, Tuxtepec, Oaxaca. T: 0128751044, F: 0128751880

Herrera, C.S. Raúl Congreso No.88, casa 4, Col. Tlalpan, c.p. 14000, Tlalpan, México, D.F. T: 56556935, F: 55500040

Herrera, P.T.S., Volcán Fujiyama No. 80, col. Mirador, c.p. 14449, Tlalpan, México, D.F. T: 55156551, F: 55161917

Hiriart, B.F., Zaragoza No. 21, Col. Barrio de Santa Catalina, c.p. 04000, Coyoacán, México, D.F. T: 55435461, F: 55435461

Hjort, D.E., Minería No. 145, EdiF: F, Col. Escandón, c.p. 11800, Miguel Hidalgo, México, D.F

Hristov, V.V., Paseo Cuauhnahuac No. 8532, Col. Progreso, c.p. 62550, Jiutepec, Morelos

Hurtado, M.D., Jacarandas No. 13, Col. del Bosque, c.p. 76209, Hércules de Querétaro, Querétaro. T: 2155236, F: 042155236, E: dariohm@sunserver.uaq.mx

Iriarte, V.J., San Pedro No. 122, Col. del Carmen, c.p. 04100, Coyoacán, México, D.F. T: 58493041, F: 58493041

Izabal, G.R., Aquiles Serdan No. 566-2, Col. Miguel Alemán, c.p. 80200, Culiacán, Sinaloa. T: 0167151520, F: 0167155642

Jaime, P.A., Cruz Verde No. 119-15, Col. San Jerónimo Lomas Quebradas, c.p.10600, Magdalena Contreras, México, D.F. T: 52374023, F: 56615430

Jaimes, P.L.R., Av. Acueducto No. 620 casa 1-E, Santiago Tepecatlalpan, c.p. 16090, Xochimilco, México, D.F. T: 56731249, 56731358, E: rjaimes@compaq.net.mx

Jara, L.M., Gorostiza No. 35, Circuito Diplomáticos, Col. Ciudad Satélite, c.p. 53100, Naucalpan, Edo. De México. T: 55622985, F: 53989679

Jímenez, S.J.A., Av. De Burgos No. 25, Col. Iberinsa, c.p. 28036, Madrid España. T: 983413023741

Jímenez, S.M., Calle 18 No. 22, depto. 102, Fraccionamiento San José Vista Hermosa, c.p. 72190, Puebla, Puebla. T: 0122320266, F: 0122325251

Juárez, B.E., Tepanco No. 32, Col. Barrio de San Lucas, c.p. 04030, Coyoacán, México, D.F. T: 56597054, F: 56161073, E: suelos@servidor.unam.mx

Juárez, T.J., Ixayac No. 91, Col. Benito Juárez, c.p. 57000. Nezahualcoyotl, Edo. De México. T: 55242197, F: 55242197

Lagunas, T.A.I., 15 poniente No. 4309, Col. Belisario Domínguez, Puebla, Puebla. T: 0122497076, F: 0122497096, E: icd@mail.g_networks.net

Landazuri, S.C.R., Jorge Drom Lote 29 y Gaspar de Villaroel, Quito, Ecuador

Lara, A.J.L., Dr. Mariano Romero No. 84, Col. Chapultepec, c.p. 80040, Culiacán, Sinaloa. T: 0167151520, F: 0167155642

Lara, M.E., Allende 78 casa B, col. centro de Tlalpan, c.p. 14000, Tlalpan, México, D.F. T: 55942727, F: 55942727

Laris, A.E., Paseo de la reforma No. 2229, Col. Lomas de Chapultepec, c.p. 11000. Miguel Hidalgo, México, D.F. T: 52297540

Lartigue, G.G., Indianapolis No. 35, Col. Napoles, c.p.03810, Benito Juárez, México, D.F. T: 55363488, F: 55363488

Lau, N.H., 4 Sur No. 104, Fraccionamiento Villa Satélite la Calera, c.p. 72500, Puebla, Puebla. T: 0122454866

Lazcano, D.C. Salvador, Cauda 964-1, Col. Jardines del Bosque, c.p. 44520, Guadalajara, Jalisco. T: 0136477981, F: 0131223557, E: slazcano@micronet.com.mx

Legorreta, C.H.A., Pedro de Alba No. 251, Col. Villa de Cortés, c.p. 03530, Benito Juárez, México, D.F. T: 56220879, F: 56228007

Li Liu Xiangyue, Paseo Cuauhnahuac No. 8532, Col. Progreso, c.p. 62550, Jiutepec, Morelos. T: 0173194000, Ext. 721, F: 0173194361, E: fareast@mail.giga.com

Limón, L.J. de J., Av. Alcalde No. 564, Col. Sector Hidalgo, c.p. 44760, Guadalajara, Jalisco

Limón, M.J.M., Sebastian Bach No. 4774, Col. Prados de Guadalupe, c.p. 45030, Zapopan, Jalisco. T: 0136297975, F: 0136297975

Lira, J.E., Av. Del Bosque 6310, Col. Bugambilias, c.p. 72580, Puebla, Puebla. T: 0122335529, F: 0122335528

López, B.R., Río Colorado No. 5919, Col. San Manuel, c.p. 72570, Puebla, Puebla. T: 0122450981, F: 0122456068

López, C.R., M. No. 81 casa 5, Col. Tlalpan, c.p. 14000, Tlalpan, México, D.F. T: 55735025

López, L.T., Xicotencatl No. 64, Col. Niños Héroes, c.p. 76010, Santiago, Guanajuato, T: 0142163599, F: 0142150898, E: lolte@sunserver.dsi.uaq.mx

López, P.J.L., Cerrada 21, poniente No. 310-A, casa 6, el Carmen, Col. El Carmen, Centro, c.p. 72000, Puebla, Puebla. T: 0122454866, F: 0122454866

López, P.R., Rancho San Mateo No. 120, Col. Cecilia, c.p. 04930, Coyoacán, México, D.F. T: 56710820, F: 57261700, Ext. 14477

López, P.V.M., Santiago No. 453, casa 11-B, Col. San Jerónimo Lídice, c.p. 10200, Magdalena Contreras, México, D.F. T: 52729991, Ext. 3370, F: 52275066

López, R.G., Martín Luis Guzmán No. 267, Col. Villa de Cortés, c.p. 03530, Benito Juárez, México, D.F. T: 55549624, F: 55791207

López, R.R., Insurgentes Sur No. 1871 despacho 402, 4°. Piso, Col. Guadalupe Inn, c.p. 01020, Alvaro Obregón, México, D.F. T: 56610414, F: 56617263, E: lopezr@prodigy.net.mx

López, R.V.M., Othon Almada 119, entre Reyes y Naranjo, Col. Balderrama, c.p. 83000, Hermosillo, Sonora. T: 0162147046, F: 0162147076

López, V.D.B., Km. 12+000 Carretera Querétaro-Galindo, Col. Sanfandila, C.P. 76700, Pedro Escobedo, Querétaro. T: 0142169777, F: 0142169671

López, Z.R.A., Ejército Nacional No. 505-10 depto. 1002, Col. Granada, c.p. 11520, Miguel Hidalgo, Edo. De Mexico. T: 52542710, F: 52036510, E: ralopez@keller.com.mx

Lowe III, J., 26 Grandiview Blvd., c.p. 10710, Yonkers, Nueva York, USA

Magaña, del T.R., Gabino Barreda No. 96-203-A, Col. San Rafael, c.p. 06470, Cuauhtemoc, México, D.F. T: 56223500, F: 56160784

Márquez, R.R., Volcan Ceboruco 616, casa 6 2ª. Privada de real de San Javier, Col. Azteca, c.p. 50180, Toluca, Edo. De México. T: 0148114830, F: 0148114830

Marquina, B.J. Othon, Retorno 38 No. 19, Col. Avante,c.p. 04460, Coyoacán, México, D.F. T: 53650321, F: 53652917

Martínez, A.J.J., Santa Barbara No. 102, Col. Lindavista, c.p. 07300, Gustavo A. Madero, México, D.F. T: 57606549, F: 57606549

Martínez, C.A., Fraccionamiento Las Alamedas No. 206, Col. Las Alamedas, c.p. 52500, Atizapan de Zaragoza, Edo. De México. T: 52294400, Ext. 6193

Martínez, de la P.C., Ejército Nacional No. 505, despacho 1002, Col. Granada, c.p. 11520, Miguel Hidalgo, México, D.F. T: 52542710, F: 52950020

Martínez, F.J.A., Avenida Don Bosco Quinta Urupagua No. 19 Urb. La Florida No. 61.621, Caracas 1050, Venezuela, c.p. 10710, Caracas Venezuela. T: 7314408, F: 7316866, E: ingeosolum@cantv.net

Martínez, M.J.A., Londres No. 44, Col. del Carmen Coyoacán, c.p. 04100, Coyoacán, México, D.F. T: 56882368, F: 56048206, E: mamj4811@servidor.unam.mx

Martínez, P.E., Club Cuicalli No. 86, Col. Circuito Cronistas, c.p. 53100, Cd. Satélite Naucalpan, Edo. De México. T: 55723948

Martínez, R.G., San Ernesto 3755, Col. Jardines de San Ignacio, c.p. 45040, Zapopan, Jalisco. T: 0131216471

Martínez, V.J.J., Apartado Postal 36-003, c.p. 11641, Miguel Hidalgo, México, D.F

Mascareño, J.D., Blvd. R. Sánchez Taboada No. 10403-306, Col. Zona Río, c.p. 22320, Tijuana, Baja California. T: 0166343506, F: 0166346163

Mayoral, R.P.A., Jesús Galindo y Villa 3259, Col. Jardines de la Paz, c.p. 44860, Guadalajara, Jalisco. T: 0136291975, E: alonsomy@hotmail.com

Medina, C.M.A., 148 Oriente No. 172, Col. Moctezuma, Segunda Sección, c.p. 15500, Venustiano Carranza, México, D.F. T: 55719133, F: 55719133

Medina, D.Z., 21 Diagonal No. 942 entre 21 y 21-A, Col. Fraccionamiento jardines de Mérida, c.p. 97135, Mérida, Yucatán

Medina, M.J., Avenida Potam No. 1316, Col. Fraccionamiento Camino Real, c.p. 83178, Hermosillo, Sonora. T: 0162195574, F: 0162197417, E: prolas@rtn.uson.mx Coyoacán, México, D.F. T: 56566678, F: 56075694, E: wpaniagua@hotmail.com

Megchun, L.C. del R., 17ª. Sur Poniente No. 203, Col. San Francisco, c.p. 29000, Tuxtla Gutierrez, Chiapas. T: 0196141835, F: 0196141835

Mejía, R.J., Manuel González Ureña No. 16, Col. Nueva Chapultepec Sur, c.p. 58260, Morelia, Michoácan. T: 0143120418, F: 56160784

Melara, R.E., Senda No. 1 casa No. 11, Col. Residencial Brumas de la Escalon, San Salvador. T: 98503841028

Menache, V.A., Chicago No. 20, Col. Napoles, c.p. 03810, Alvaro Obregón, México, D.F. T: 55231985, F: 55233816

Méndez, G.C., Puente del Poniente No. 115, Fraccionamiento El Bosque, c.p. 29047, Tuxtla Gutierrez, Chiapas. T: 0196120025

Mendoza, L.M.J., Pachuca s/n retorno 3 No. 13, Condominio retornos del pedregal, c.p. 14140, Tlalpan, México, D.F. T: 56223500, F: 56160784

Merrefield, C.C.C., Mar Blanco No. 93, Col. Popotla, c.p. 11400, Miguel Hidalgo, México, D.F

Mesa, J.C.M., Apartado Postal 1003-094, Col. Zona Arraiján, Panamá, República de Panamá. T: 2638000, Ext. 2145, F: 2641401, E: cmesa@utp.ac.pa

Meyer, J., Litorales No. 44, Col. Las Aguilas, c.p. 01710, Alvaro Obregón, México, D.F. T: 56805239, F: 55934167

Meza, C.M., Andador Escritores No. 7 manzana "F", Infonavit 1º De Mayo, 7ª. Etapa, c.p. 68020, Oaxaca de Juárez, Oaxaca. T: 5131694, F: 5138250

Minaburo, C.J., Norte 71 No. 2924, Col. Obrero Popular, c.p. 02840, Azcapotzalco, México, D.F. T: 53967977, 53967824

Miranda, M.S.M., Francisco Díaz Covarrubias No. 43-101, Col. San Rafael, c.p. 06470, Cuauhtemoc, México, D.F. T: 55663813, F: 55663813

Monforte, O.A., Torres de Mixcoac EdiF: E-6-23, Col. Mixcoac, c.p. 01460, Alvaro Obregón, México, D.F. T: 56826662, F: 56826888

Montalvo, G.J., Circunvalación No. 127, Col. Atlantida, c.p. 04370, Coyoacán, México, D.F. T: 55492315, F: 55492315

Montañez, C.L.E., Melchor OcampoCopilco No. 300, edificio 8, depto. 302, Col. Copilco Universidad, c.p. 04360, México, D.F. T: 52294400, Ext. 44100, F: 52294400, Ext. 44007, T: 56659932, F: 56651135

Montellano, M.F., Canova No. 32 PH, Col. Mixcoac, c.p. 03910, Benito Juárez, México, D.F. T: 55630244, F: 55428175

Montoya, A.J.G., Hacienda de los Portales No. 97, Col. Prados del Rosario, c.p. 02410, Azcapotzalco, México, D.F. T: 55444776, F: 55444776

Mooser, H.F., Av. Volcanes No. 176, Col. Carretera México-Toluca Km. 16.5, c.p. 01310, Cuajimalpa, México, D.F. T: 52920752, F: 52920752

Morales, D.V. Juventino, Carteros No. 106, Col. Postal, C.P. 68080, Oaxaca de Juárez, Oaxaca. T: 0195155828, 0195150088

Morales, G.A. Jorge, Monte Tabor 34, Col. Independencia, c.p. 44240, Guadalajara, Jalisco. T: 0136385877, E: aquilesm@prodigy.net.mx

Morales, V.L.R., Avenida Coyoacán No. 614-101, Col. del Valle, c.p. 03100, Benito Juárez, México, D.F. T: 54811218, F: 54811263, E: ctcna@supernet.com.mx

Morales, y M.R., apartado postal 74-089, Administración de correos No. 7404230, c.p. 09081, Coyoacán, México, D.F. T: 56892906, 55446190, F: 56893643

Moreno, F.A., Fuente de Vulcano No. 32, Col. Lomas de Tecamachalco, c.p. 53950, Huixquilucan, Edo. De México. T: 55751188, F: 55752130

Moreno, H.J., Privada Tabasco No. 411, Col. del Carmen, c.p. 72000, Puebla, Puebla. T: 0122374996, F: 0122374996, 56228007

Moreno, P.G., Larga distancia No. 62, Col. Ampliación Sinatel, c.p. 09470, Iztapalapa, México, D.F. T: 56228007, F: 56228007

Muñoz, C.J., Periférico Echeverría No. 1320-3er. Piso, Col. Centro, c.p. 25000, Saltillo, Coahuila. T: 0184301316, F: 0184301316

Murillo, F.R., Cruz del Sur No. 66, Col. Prados Churubusco, c.p. 04230, Tlalpan, México, D.F. T: 52374133, Ext. 2606, F: 56615430

Navarro, G.M.C., Pradera No. 47, Col. Pedregal de San Angel, c.p. 01900, Alvaro Obregón, México, D.F. T: 55149020, F: 56520101

Nieto, R.J.A., Insurgentes Sur No. 2140, Col. Ermita San Angel, c.p. 01070, Alvaro Obregón, México, D.F. T: 54811238, F: 54811263

Ontañon, L.J.L., Río Grijalva No. 5310, Col. San Manuel, c.p. 72570, Puebla, Puebla. T: 0122448244, F: 0122448244

Ordoñez, R.J., Av. Jaina No. 240, Col. Maya, c.p. 29000, Tuxtla Gutierrez, Chiapas. T: 0196150527, F: 0196150322

Orendain, O.A., Libertad No. 1966, Col. Americana, c.p. 44140, Guadalajara, Jalisco, T: 8257188, F: 8251447, E: gruposci@infosel.net.mx

Orozco, C.M. Jorge, Circuito Le Mans No. 107, Col. Villa Verdum, c.p. 01810, Alvaro Obregón, México, D.F. T: 52729991, Ext. 2250, F: 52275066, E: orozcoj@ica. com.mx

Orozco, S.R.V., Hacienda Chapa No. 5, Col. Prado Coapa, c.p. 14350, Tlalpan, México, D.F. T: 56719540, F: 56719540

Orozco, y O.C.I., Holbein No. 69, Col. Ciudad de los deportes, c.p. 03710, Benito Juárez, México, D.F. T: 0138113765, 55634239, F: 0138110144

Orozco, y O.J.I., Manuel M. Ponce No. 321-102, Col. Guadalupe Inn, c.p. 01020, Alvaro Obregón, México, D.F. T: 56889803, F: 56887608

Orozco, y O.J.M., Av. Coyoacán No. 1895, 1er. Piso, Col. Acacias, c.p. 03240, Benito Juárez, México, D.F. T: 55247325, 55243481, F: 55245426, E: jorozco@sct.gob.mx

Ortíz, H.G., Av. Insurgentes esquina Oaxaca, c.p. 63140, Tepic, Nayarit, T: 0132136844, F: 0132130188

Ortíz, R.R., 1ª. Privada 13 de septiembre No. 107, Col. Vicente Suárez, c.p. 68030, Oaxaca, Oaxaca. T: 0195140061, F: 0195140061

Ovando, S.E., Instituto de Ingeniería, Ciudad Universitaria, UNAM, Col. Coyoacán, c.p. 04510, Coyoacán, México, D.F. T: 56223500, F: 56160784, E: eovs@pumas.iingen. unam.mx

Padilla, C.E., Lope de Vega 264, Col. Arcos del Sur, c.p. 45040, Guadalajara, Jalisco, T: 0136159010, 56159010

Padilla, V.R.R., Insurgentes Sur 4411, edificio 15, inT: 304, Col. Tlalcoligia, c.p. 14430, Tlalpan, México, D.F. T: 56228003, F: 56228007

Paniagua, E.W., Cerro del embarcadero No. 17, Col. Pedregal de San Francisco, c.p. 04320, Coyoacán, México, D.F. T: 56566678, F: 56075694

Paniagua, Z.W.I., Cerro del embarcadero No. 17, Col. Pedregal de San Francisco, c.p. 04320. T: 0142163599, Ext. 15, F: 0142150898, E: erg@sunserver.uaq.mx

Peck, R., 101 Warm Sands Drive, S.E., c.p. 87123, Aburquerque, New Mexico, USA. T: 955052932484, F: 0196252388

Pérez, A.J.L., Márquez de Aguayo No. 11, Circuito Fundadores, c.p. 53100, Ciudad Satélite, Edo. De México. T: 55627052, F: 0173115002, Ext. 4803

Pérez, G.A., B. Quintana E-4, Col. Viveros, c.p. 76140, Querétaro, Querétaro. T: 0142243628, F: 01422152512, E: alfre@sunserver.uaq.mx

Pérez, G.N., Km. 12+000 Carretera Querétaro-Galindo, Col. Sanfandila, c.p. 76700, Pedro Escobedo, Querétaro. T: 0142169777, F: 0142169671, E: Paul.Garnica@imt.mx

Pérez, M.A., Carretera México-Piedras Negras, Km. 265 + 680, Tramo Matehuala-Saltillo, c.p. 25299, Saltillo, Coahuila. T: 0184301544, F: 0184301662

Pérez, R.M. de la Luz, H. Colegio Militar No. 109, Col. Niños Heroes, c.p. 76010, Querétaro, Querétaro. T: 0142163599, F: 0142155236, E: perea@sunserver. uaq.mx

Pérez, S.A., Km. 12+000 Carretera Querétaro-Galindo, Col. Sanfandila, c.p. 76700, Pedro Escobedo, Querétaro. T: 0142169777, F: 0142169671, E: Paul.Garnica@imt.mx

Pérez, S.M.E., Abasolo No. 206, Col. Centro, c.p. 68000, Oaxaca, Oaxaca. T: 0195145812, F: 0195161954

Pilatowsky, V.A., Bruselas 117, Col. del Carmen, c.p. 04100, Coyoacán, México, D.F. T: 55245609

Pineda, V.G., Avenida Sinaloa No. 400, Col. Isaac Arriaga, c.p. 58210, Morelia, Michoácan. T: 0143150207, F: 0143150207

Ponce, S.J.A., Barranca de Tarango No. 80 casa 6, privada 1, Col. San Antonio Tarango, c.p. 01620, A. Obregón, México, D.F. T: 52729991, Ext. 3322, F: 52275066

Porras, L.A., Calle 73 No. 89, depto. 9, Col. Puebla, c.p. 15020, Venustiano Carranza, México, D.F. T: 57588660, F: 57004049

Portillo, M.G.M., Celsa Vírgen No. 278, Fraccionamiento Prados de la Villa, c.p. 28970, Villa de Alvarez, Colima. T: 0133112040, F: 013113040, E: bssjportillo@yahoo.com

Poucell, P.R., Hacienda Totoapan No. 60, Col. Prado Coapa, c.p. 14350, Tlalpan, México, D.F

Preciado, C.H.F., Pasteur Norte No. 27, Col. Centro, c.p. 76000, Querétaro, Querétaro. T: 0142121031, F: 0142126979

Puebla, C.M., Cruz Verde No. 133-4, Col. Toriello Guerra, c.p. 14050, Tlalpan, México, D.F. T: 56228003, F: 56228007

Pujalte, P.A., Zeus No. 4, despacho 2, Col. Crédito Constructor, c.p. 03940, Benito Juárez, México, D.F. T: 56615347, 56618225

Quintana, I.B., Minería No. 145, edificio central 4º. Piso, Col. Escandón, c.p. 11800, Miguel Hidalgo, México, D.F. T/F: 52729991, Ext. 2000

Rabago, M.A., Magisterio Nacional, 151-8, Col. Tlalpan, c.p. 14000, Tlalpan, México, D.F. T: 56589906

Ramírez, G.M.A., Cerrada de 23 Poniente Sur No. 250, Col. Santa Elena, c.p. 29000, Tuxtla Gutierrez, Chiapas. T: 0127520149

Ramírez, M.S.O., Prolongación de Sauces No. 203, Col. Antiguo Aeropuerto, c.p. 68050, Oaxaca, Oaxaca. T: 0195151736, F: 55733969

Ramírez, R.A., Insurgentes Sur 4411, edificio 25-202, Col. Tlalcoligia, c.p. 14430, Tlalpan, México, D.F. T: 56552531, F: 56552531, Ext. 3969, E: arascon@prodigy.net.mx

Ramírez, R.E., Martín Mendalde No. 837-402, Col. del Valle, c.p. 03100, Benito Juárez, México, D.F. T: 56118446, Ext. 13, F: 56118380

Ramírez, R.M., Guerrero No. 31, Col. San Angel, c.p. 01080, Alvaro Obregón, México, D.F. T: 54811218, 54811100, Ext. 4369, F: 54811263, E: geotecnia@supernet. com.mx

Ramos, B.A., Tlacotal "P" No. 2517, Col. Ramos Millan, c.p. 08720, Iztacalco, México, D.F. T: 56503312, 53734599

Ramos, P.C., Avenida Alemania No. 1639, col. Moderna, c.p. 44190, Guadalajara, Jalisco. T: 0138113765, F: 0138113356

Rangel, N.J.L., Av. La Garita No. 7, departamento 5-1, col. Villa Coapa, c.p. 14390, Tlalpan, México, D.F. T: 55947308, E: jnrangel@prodigy.net.mx

Reséndiz, N.D., Brasil No. 31, of. 306, piso 2, col. Centro, c.p. 06029, Cuauhtémoc, México, D.F. T: 53281028, F: 53281034, 012421646

Reyes, M.R., Felix Romero No. 528, col. Unidad ISSSTE, c.p. 68040, Oaxaca de Juárez, Oaxaca. T: 0195152143, F: 0195157812O1

Reyes, S.A., Casas Grandes No. 18-2, col. Narvarte, c.p. 03020, Benito Juárez, México, D.F. T: 55307730

Reyes, S.L., Lago de Patzcuaro No. 6533, Fraccionamiento Lagulena, c.p. 72580. Puebla, Puebla. T: 0122334510, 0122622495, F: 0122334510, E: reyescca@prodigy.net.mx

Rincón, V.R., Buharros No. 13, Col. Lomas de las Aguilas, c.p. 01730, Alvaro Obregón, México, D.F. T: 56353053, F: 52275066

Ríos, A.M.A., Miguel Angel de Quevedo 627, departamento 3, Col. Romero de Terreros, c.p. 04310, Coyoacán, México, D.F. T: 55543285, 56115120, E: isora@prodigy.net.mx

Ríos, L.N., Ramón Beteta No. 8, circuito economistas, Col. Ciudad Satélite, c.p. 53100, Naucalpan, Edo. De México. T: 55721015

Rivas, L.J., Av. Hidalgo No. 1311, Col. Centro, c.p. 68000, Oaxaca de Juárez, Oaxaca. T: 0195162904, F: 0195140621, E: esmic@spersaoaxaca.com.mx

Rivera, C.R., Cerrada Flor de Yuca manzana 81, lote 10, Col. Ejidos de San Pedro Martir, c.p. 14640, Tlalpan, México, D.F. T: 56557775, F: 56557775, E: riverac@servidor.unam.mx

Rocher, P.J.A., Watteau No. 6 esquina Holbein, Col. Mixcoac, c.p. 03910, Benito Juárez, México, D.F. T: 56117626, F: 56114311

Rodal, C.J.A., Oriente 4, No. 530, c.p. 94300, Orizaba, Veracruz. T: 0127210484, F: 0127211465, E: constru@yoshnet.net.mx

Rodríguez, G.L.B., Bosques de Olinala No. 27-A, Fraccionamiento Bosques de la herradura, c.p. 52784. T: 56621444, F: 56621425, E: luisber@webtelmex.net.mx

Rodríguez, P.C.E., Calle 7 num. 1223 bajos, Ext. San Agustín-río Piedras, c.p. 00926, San Juan, Puerto Rico. T: 988097471780

Rodríguez, S.R., Avenida 2 oriente mza. 6, lote 3, Puerto industrial pesquero "Laguna Azul", Ciudad del Carmen, c.p. 24140, Cd. del Carmen, Campeche. T: 0193821784, F: 0193829411

Rojas, G.E., Mar mediterráneo No. 100-28, Col. Las hadas, c.p. 76160, Querétaro, Querétaro

Rojo, I.M.F., Avenida Arenal No. 11, Col. Pueblo la Magdalena Petlacalco, c.p. 14480, Tlalpan, México, D.F. T: 58463197, F: 58463197, E: tiisa@tutopia.com

Roman, C.G., Estudiante No. 1, Col. Centro, c.p., 62790, Xochitepec, Morelos. T: 01733612119, F: 0173434141

Romo, O.M.P., Instituto de Ingeniería de la UNAM, Sección Geotecnia, edificio 4, Apdo. Postal 70-472, c.p. 04510, Coyoacán, México, D.F. T: 56223500 al 04, F: 56160784, E: mromo@pumas.iingen.unam.mx

Rosales, O.V.A., P. de los jardines No. 38, Col. Paseos de Taxqueña, c.p. 04250, Coyoacán, México, D.F. T: 56893307

Ruíz, F.D., Picacho No. 610, Col. Jardines del pedregal, c.p. 01900, A. Obregón, México, D.F. T: 56615555, F: 56617965

Ruíz, G.G., Juan Cayetano Portugal 3605, depto. 10, Col. Residencial poniente, Zapopan, Jalisco. T: 0138613940

Saborio, U.J., Rubén Dario No. 1242, Col. Providencia, c.p. 33620, Guadalajara, Jalisco. T: 0138120497, F: 0138120497

Sahab, H.E., Cerrada de providencia No. 6 esq. San Bernabe, Col. San Jerónimo Lídice, c.p. 10200, Magdalena Contreras, México, D.F. T: 56522958, F: 56522958

Salazar, E.J.H., Calle de la concordia No. 11, Col. Villa Satélite. T: 0122497096, E: azomozac@pue1.telmex.net.mx

Salazar, R.A., Constitución No. 50, Col. Escandon, c.p. 11800, México, D.F. T: 55440515, F: 53360543, E: sabma@terra.com.mx

Salcedo, V.D., Km. 12+000 Carretera Querétaro-Galindo, Col. Sanfandila, c.p. 76700, Pedro Escobedo, Querétaro. T: 0142169777, F: 0142169671, E: Paul.Garnica@imt.mx

Sámano, A.A. Andrés, Lago San José No. 13, Col. Ampliación Granada, c.p. 11520, Miguel Hidalgo, México, D.F. T: 52729991, Ext. 2514, F: 52275055, E: samanoa@ica.com.mx

Sánchez, G.A., Esperanza No. 706-2, Col. Narvarte, c.p. 03020, Benito Juárez, México, D.F. T: 52729991, Ext. 4707

Sánchez, M.I., Av. Coyoacán, No. 1895, Col. Acacias, c.p. 03240, Benito Juárez, México, D.F. T: 55244561, F: 55247005

Sánchez, R.R., Andador Minería No. 11, manzana H., 6ª. Etapa primero de mayo INFONAVIT, c.p. 68020, Oaxaca de Juárez, Oaxaca. T: 0195134929, F: 0195161937

Sangines, G.H., José María Rico No. 538-A, Col. del Valle, c.p. 03100, Benito Juárez, México, D.F. T: 56228003, F: 56228007

Santoyo, V.E., Av. México No. 1256, Casa 310, col. Santa Teresa, c.p. 10710, Magdalena Contreras, México, D.F. T: 55599055, F: 55653150, E: tgc2000@infosel.net.mx

Schmitter, M. del C. J. J., Av. México No. 1256, casa 303, Col. Santa Teresa, c.p. 10710, Magdalena Contreras, México, D.F. T: 52729991, Ext. 2517, F: 522750555, E: schmittj@ica.com.mx

Sedano, L.S., Agapando No. 7, Col. Jardines de coyoacán, c.p. 04890, Coyoacán, México, D.F. T: 56770916, F: 56770916

Senties, A.E., Planicie No. 25, Parques del pedregal, c.p. 14010, Tlalpan, México, D.F. T: 56528635, 56650235

Serrano, L.F., Paseo de Tullerias No. 129, Col. Lomas Verdes 3ª. Sección, c.p. 53120, Naucalpan, Edo. De México. T: 56693418, F: 56875873

Sierras, F.F., León Tolstoi No. 775, Fraccionamiento Justo Sierra, c.p. 21230, Mexicali, Baja California. T: 0165616689, F: 0165617535

Sifuentes, V.A., Ursulo Galván No. 70, Col. Presidentes ejidales, c.p. 04480, Coyoacán, México, D.F. T: 56077096, F: 56077096, E: ansif@prodigy.net.mx

Silva, Z.B., Mario No. 16, Col. Pavón, c.p. 57610, Nezahualcoyotl, Edo. De México. T: 52729991, Ext. 2514, F: 52275055, E: silvaben@hotmail.com

Silva-González, P.G., Avenida Copilco No. 390, depto. 4, Col. Copilco-Universidad, c.p. 04360, Coyoacán, México, D.F. T: 56228138, F: 56228137

Simpser, D.B., Avenida Constituyentes 345, depto. 101-103, Col. Daniel Garza, c.p. 11830, Miguel Hidalgo, México, D.F. T: 52729969

Sosa, G.A., 4 norte No. 1206, despacho 304, Col. Centro, c.p. 72000, Puebla, Puebla. T: 012421646, Querétaro, Querétaro. T: 0142167659, F: 0142167659, E: lolte@sunserver.dsi.uaq.mx

Soto, G.A., José Azueta No. 1, manzana 2, Col. Guerrero 200, c.p. 39090, Chilpancingo, Guerrero. T: 0174727152

Soto, Y.E., Ejido Huipulco No. 69, Ejidos de San Francisco Culhuacán, c.p. 04420, Coyoacán, México D.F. T: 56588901, F: 56952547, E: soye@prodigy.net.mx

Springall, C.G., Londres No. 44, Col. del Carmen, c.p. 04100, Coyoacán, México, D.F. T: 56882368, F: 56048206, E: gspringall@adetel.net.mx

Springall, C.J., Londres No. 44, Col. del Carmen c.p. 04100, Coyoacán, México, D.F. T: 56882368, F: 56048206

Streu, C.W., Avenida Río Churubusco 407, edificio B-101, Col. Xoco, c.p. 03330, Benito Juárez, México, D.F. T: 6756054217, F: 56054217

Taboada, U.V.M., Apartado postal 70-472, Col. Coyoacán, c.p. 04510, Coyoacán, México, D.F. T: 56223500, F: 56160784, E: vtau@pumas.iingen.unam.mx

Tamez, G.E., Explanada No. 1615, Col. Lomas de Chapultepec, c.p. 11000, Miguel Hidalgo, México, D.F. T: 55204006, F: 55400373

Terés, F.J.R., Cisne 1486-2, Col. Morelos, c.p. 44440, Guadalajara, Jalisco. T: 0138103962, F: 0138110144, E: jrteres@hotmail.com, jrteres@yahoo.com

Tinajero, S.J., Lluvia No. 450, Casa 11, Col. Jardines del pedregal, c.p. 01900, Alvaro Obregón, México, D.F. T: 55206206, F: 55205684

Torales, T.E., Francisco Villa No. 54, Col. Vista Hermosa, c.p. 39050, Chilpancingo Guerrero. T: 0174413767, F: 0174413767

Trejo, M.A., Allende Norte No. 81, Col. Centro, c.p. 76000, Querétaro, Querétaro. T: 0142124988, F: 0142124988

Trueba, L.V., Av. Toluca No. 441, Col. Olivar de los padres, c.p. 01780, Alvaro Obregón, México, D.F. T: 52298603, F: 52298630, E: vtrueba@gsmn.cna.gob.mx

Ulloa, G.C., Aida No. 21, Col. San Angel, c.p. 01000, Alvaro Obregón, México, D.F. T: 55503595, F: 55503595, E: culloa@iwm.com.mx

Valle, C.J.A., Arenal No. 419, edificio Nardo No. 001, Col. Tepepan, c.p. 16020, Tlalpan, México, D.F. T: 56566678, F: 56075894

Valverde, L.H., Calle 3 No. 4, Col. Espartaco, c.p. 04870, Coyoacán, México, D.F. T: 56799785, F: 56799100, E: iecsc97@ienlaces.com.mx

Vázquez, V.A., Periférico Sur No. 3672, casa 51, Col. Jardines del pedregal, c.p. 01900, A. Obregón, México, D.F. T: 52275001

Vela, T.P., Allende No. 29, Col. Primera Sección, c.p. 70110, Ixtepec, Oaxaca. T: 0197111042, F: 0197112555

Vera, A.A., 15 poniente No. 4309, Col. Belisario Domínguez, c.p. 72180, Puebla, Puebla. T: 0122497076, F: 0122497096, E: azomozac@pue1.telmex.net.mx

Victoria, O.E., 2°. Callejón Independencia No. 31, Col. San Andrés Tetepilco, c.p. 09440, Ixtapalapa, México D.F. T: 55358756, 56720805

Vieitez, U.L., Juan Racine No. 120-402, Col. Casco Morales, c.p. 11510, Miguel Hidalgo, México, D.F. T: 56454834, F: 56455087

Villicaña, C.J. Herminio, Xallan No. 274, Col. Villa Izcalli, c.p. 28979, Villa de Alvarez, Colima. T: 10832

Wong, R.A., Rinconada Fresales No. 6, Col. Villa del Puente, c.p. 14335, Tlalpan, México, D.F. T: 55244561, 55245798, F: 55249265

Yañez, S.D., Las Flores No. 72, edificio G, Depto. PH1, Col. Santa Ursula Coapa, c.p. 04650, Coyoacán, México, D.F. T: 52729991, Ext. 7215, F: 52729991, Ext. 2509, E: yanezd@ica.com.mx

Zamora, M.F., Puente de piedra No. 47, Col. Toriello Guerra, c.p. 14050, Tlalpan, México, D.F. T: 55437044, F: 55235061

Zárate, A.M., Antigua Taxqueña No. 174, Col. Barrio de San Lucas, c.p. 04030, Coyoacán, México, D.F. T: 55446602, 55446603, F: 56893513, E: geosol@infosel.net.mx

Zárate, C.O., Av. Lázaro Cárdenas No. 708, Col. Ixcotel, c.p. 68000, Oaxaca, Oaxaca. T: 019517296000, Ext. 2704

Zeevaert, W. A., Eje Central Lázaro Cárdenas 2-2506, Col. Centro, c.p. 06007, Cuauhtémoc, México, D.F. T: 55109903, F: 551280018

Zeevaert, W.A., Torre Latinoamericana No. 2506, Col. Centro, c.p. 06000, Cuauhtémoc, México, D.F. T: 55109903, F: 55128018

Zeevaert, W.L., Isabel La Católica No. 67 altos, Col. Centro, c.p. 06080, Cuauhtémoc, México, D.F. T: 57094208, F: 57094208

Zepeda, G.J.A., Prof. José Guadalupe Velázquez No. 10, Col. Acueducto, c.p. 76020

Zueck, R.R., Paseo de las Palmas No. 29, Col. Jardines de la Florida, c.p. 53130, Naucalpan, Edo. de México. T: 53604512, F: 53604512

NEPAL

Secretary: Dr. Sukh B. Gurung, Nepal Geotechnical Society, P.O. Box 20360, Kathmandu, Nepal.
E: ngsxcom@yahoogroups.com, sukhgurung@hotmail.com

Total number of members: 17

Chaudhary, Dr. S.K., M98026-I 405 Wagon Road, Gallup, NM 87301, USA. E: sushilchaudhary@hotmail.com

Dhakal, Dr. A.S., M95009, Civil Engineer, Department of Irrigation, HMG/Nepal, Panipokhari, Kathmandu. T: 977-1-411-496, F: 81-298-53-4897, E: amoddhakal@hotmail.com

Ghimire, Mr. H.N., M00033 Doctoral Student, Rock Mechanics Laboratory, Division of Structural & Geotechnical Engineering, Graduate School of Engineering, Hokkaido University, N13W8, Sapporo 060 8628, JAPAN. T: 81-11-706-6301

Gurung, Dr. S.B., F95015-I, Opposite: Nepal Bank Limited, Taxi Chouk, Bagar, Pokhara. E: sukhgrung@hotmail.com

Gurung, Dr. N., M97021-I, Research Fellow, School of Civil Eng., Queensland University of Technology (QUT), 2 George St., Brisbane, Queensland 4001, AUSTRALIA. T: 61-7-3864-2776, F: 61-7-3864-1515

Humagain, Dr. I.R., M00034, Lecturer of Engineering Geology, Department of Civil Engineering, Tribhuvan University Pulchowk Campus, Lalitpur, Nepal Postal address: G.P.O. Box 9578, Kathmandu. T: 977-1-525477, F: 977-1-521985, E: ihumagain@wlink.com.np

Jha, Mr. K.Kr., M98025-I, Graduate Student, Department of Civil Engineering, Saitama University, Saitama, JAPAN. E: kaushal@dice.dr5w.saitama-u.ac.jp

Karkee, Dr. M.B., F97020-I, President of Napal Geotechnical Society also Professor of Structural Mechanics, Dept. of Architecture & Environment Systems, Faculty of System Science & Technology, Akita Prefectural University, 84-4 Tsuchiya, Honjo, Akita 015-0055, JAPAN. T: 81-184-27-2047, F: 81-184-27-2186, E: karkee@akita-pu.ac.jp

Karna, Mr. A.K., M99029, 1-36-1-101, Musashidai, Fuchushi, Tokyo, JAPAN. E: akhilesh_karna@hotmail.com

Lohani, Dr. T.N., M97022-I, Civil Engineer, Post-Doctoral Fellow, Geotechnical Laboratory, Department of Civil Engineering, University of Tokyo, JAPAN. E: tlohani@geot.t.u-tokyo.ac.jp

Neaupane, Dr. K.M., M97024-I, Asst. Prof., Civil Engineering program, Sirindhorn International Institute of Technology, Thammasat University, P. Box-22, Klong-luang, Pathumthani 12121 Bangkok, THAILAND. E: krishna@siit.tu.ac.th

Pokharel, Dr. G., M94003-I, Bidhyalaya Chowk, Chhinedanda, Pokhara-18, Kaski District (currently at Gifu University, Japan). T: 977-61-25951, E: gyaneswor@geotechnical.org

Sah, Mr. A.K., M97023-I, Land Resource Specialist, Project Administration and Management Dept. PASCO Int'l Inc., 2-8-10 Higashiyama, Meguro-ku, Tokyo 154, JAPAN. T: 81-3-3794-9961, F: 81-3-3794-9607, E: sah@pascointl.co.jp

Shahi, Dr. B.R., M00032, Central Road Research Laboratory, Department of Roads, Lalitpur. T: 977-1-521605, E: dshahi@wlink.com.np

Shiwakoti, Dr. D.R., M96019-I, Executive Geotech Engineer, Tao Corporation, Pulau Ubin & Tekong Reclamation Project, 23 Pandan crescent, SINGAPORE 128472. T: 65-5468455, F: 65-5468141, E: shiwakoti_dinesh@hotmail.com

Tamrakar, Mr. S.B., M98027-I, Graduate School of Engineering, Division of Structural and Geotechnical Engineering, Hokkaido University, North-13, West-8, Kita-ku, Sapporo, Hokkaido, JAPAN. E: tamrakar@eng.hokudai.ac.jp

Tiwari, Mr. B., M99031, Terao Kita 1-1-4-423, Niigata 950-2061, JAPAN. T: 025-267-2220 (Res.), 025-223-6161, Ext. 7937 (lab.), E: rihsa-17@gs.niigata-u.ac.jp

THE NETHERLANDS

Secretary: Mr Ir G.J.C.M. Hoffmans (Gijs), c/o Ministry of Transport, Public Works and Water, Road and Hydraulic Engineering Division, P.O. Box 5044, 2600 GA Delft, The Netherlands. T: 31 15 251 84 16, F: 31 15 251 85 55, F: g.j.c.m.hoffmans@dww.rws.minvenw.nl

Total number of members: 733

Aartsen, R.J., Ir, Oostblok 47, 2612 KL Delft, Holland Railconsult, Utrecht
Adrichem, T.S.C. van, Ir, Willem Frederikstraat 48, 3136 BR Vlaardingen, N.V. Koninklijke Nederlandsche Petroleum Maatschappij, 's-Gravenhage
Afman, H.B., Dr, Roelofslaantje 8, 1411 HA Naarden, OMEGAM, Amsterdam
Akker, C. van den, Prof Dr Ir, Kerkewijk 254, 3904 JL Veenendaal, TUD Communicatie en Marketing Groep, Delft
Akker, J.J.H. van den, Ir, Goudenregenstraat 15, 5213 HM 's-Hertogenbosch, DLO-Staring Centrum, Instituut V.Onderz. V/H Landelijk Gebied (SC-DLO), Wageningen
Allaart, A.P., Dr Ir, Tomatenstraat 164, 2564 CX 's-Gravenhage, Holland Railconsult, Utrecht
Alsem, D.M., Ir, Zesweg 53, 6601 HA Wijchen, Koninklijke Haskoning Groep BV, Nijmegen
Altink, H., Ir, Piet Heinlaan 14, 2121 XB Bennebroek, De Weger Architecten- en Ingenieursbureau BV, Rotterdam
Amesz, A.W., Ir, Hoeveweg 29, 8395 TD Steggerda, HBG Civiel Grondtechniek, Gouda
Andringa, R.J., Ir, Flensburghof 26, 3067 PT Rotterdam, Multiconsult Den Haag BV, 's-Gravenhage
Asin, M., Dr Ir, Fregelaan 64, 1062 KL Amsterdam, Ingenieursbureau Amsterdam, Amsterdam Zuidoost
Asin, A., Ir, Brigantijn 5, 3448 KB Woerden, Holland Railconsult, Utrecht
Baardewijk, A.P.H. van, Ir, Van Maerlantpark 6, 2902 BS Capelle aan den IJssel
Baaren, J.P. van, Ir, Oude Leedeweg 147, 2641 NP Pijnacker, Petro-Logica BV, Pijnacker
Baas, R.J., Ir, Van Eijdenhof 49, 3833 JX Leusden, Grontmij Verkeer & Infrastructuur, De Bilt
Bakker, M.L., Ir, Knuttelstraat 34, 2613 XX Delft, Oranjewoud Capelle a/d IJssel, Capelle aan den IJssel
Bakker, J.G.M., Dr, Cissy van Marxveldtstr 31, 6708 SJ Wageningen, Synoptics Integrated Remote Sensing & GIS Applications, Wageningen
Baldee, A., Ir, Hossenbosdijk 8, 3237 KD Vierpolders, NS Railinfrabeheer BV, Utrecht
Barel, F.L., Ir, Wulpstraat 9, 3815 HK Amersfoort, DHV Milieu en Infrastructuur BV, Amersfoort
Barendregt, L., Mr Ir, Meijerskade 7, 2313 EG Leiden
Barends, Prof Dr Ir F.B.J., Boomkleverstraat 56, 2623 GW Delft, GeoDelft, Delft
Bazelmans, F.M.M., Ir, Julianalaan 76, 2628 BJ Delft, De Straat Milieu-adviesurs BV, Delft
Bazuin, A.J., Ir, Freesiastraat 11, 2651 XL Berkel en Rodenrijs, Interbeton BV, Rijswijk Zh
Beekkerk, B.W., Ir, van Ruth, Dam 4, 3111 BD Schiedam, Grootint BV, Zwijndrecht
Beest, J. van, Ir, Graaf Floris 5, 1276 XA Huizen, Adviesburo van Beest, Huizen
Beetstra, G.W., Ir, Munnikenweg 4, 3214 LK Zuidland, Geo-Delft, Delft
Bellis, J.C.G., Ir, Westerstraat 138, 2613 RL Delft, NBM Beton- en Industriebouw, Breda
Bend, L.M. van der, Mrs Ir, Graswinckelstraat 23, 2613 PT Delft, Gemeentewerken Rotterdam, Rotterdam
Bennenk, H.W., Prof Ir, Elger 13, 1141 CC Monnickendam, Technische Universiteit Eindhoven Fac. Bouwkunde, Eindhoven
Berg, E.M. van den, Ir, Graswinckelstraat 21, 2613 PT Delft, IWACO BV Adviesbureau voor Water en Milieu, Rotterdam
Berg, P. van den, Dr Ir, Heesterlaan 47, 2803 BJ Gouda, Geo-Delft, Delft

Berg, P. van den, Ir, Kornelis van Tollaan 25, 3065 DA Rotterdam, Hoogheemraadschap van Rijnland, Leiden
Beringen, F.L., Ir, Oostering 24, 7933 TM Pesse, MCT, Wassenaar
Berkelaar, R., Ir, Grote Beer 14, 2665 WL Bleiswijk, Gemeentewerken Rotterdam, Rotterdam
Besteman, M.A., Ms Ing, Herenweg 144, 2101 MT Heemstede, Arcadis Advies BV, Hoofddorp
Bielefeld, M.W., Drs Ir, Koevordermeerstraat 36, 8531 RP Lemmer, IJsselmeerbeton Funderingstechnologie BV, Lemmer
Bijloo, A., Ir, Stilpot 23, 5708 GW Helmond, Smits Bouwbedrijf Someren BV, Someren
Bijnagte, J.L., Ir, Van Naeldwijcklaan 50, 2651 GN Berkel en Rodenrijs, GeoDelft, Delft
Bik, E.A., Ir, Achter de Kamp 55, 3811 JE Amersfoort, Ingenieursbureau Amsterdam, Amsterdam Zuidoost
Blanker, A.M., Ir, Noorderhavenkade 108-B, 3038 XR Rotterdam, Aveco BV, Utrecht
Blokpoel, C., Ir, Australielaan 5, 5691 JK Son, Raadgevend Ingenieursbureau Inpijn-Blokpoel BV, Son
Blom, J., Ir, Kloosterkoolhof 28-B, 6415 XT Heerlen, Coman Raadgevende Ingenieurs BV, Heerlen
Blommaart, P.J.L., Ir, Rivierpad 3, 2614 XB Delft, Rijkswaterstaat Dienst Weg- en Waterbouwkunde, Delft
Boer, T.J. de, B.Sc., M.Sc., Ir, Spiegelstraat 1-B, 2011 BN Haarlem, Arcadis Advies BV, Hoofddorp
Boer, F. de, Ir, Brugakker 65-24, 3704 RL Zeist, Holland Railconsult, Utrecht
Boer, A. de, Ir, Magnoliaplein 3, 6823 NM Arnhem, Rijkswaterstaat Bouwdienst, Utrecht
Boer, P. den, Ir, Klarinetstraat 133, 6922 JX Duiven, Koninklijke Haskoning Groep BV, Nijmegen
Boere, P.W., Ir, c/o Boskalis International BV, Far East Shopping Centre, 545 Orchard Road #16-01, Singapore 238882, Singapore, Baggermaatschappij Boskalis BV, Papendrecht
Bokkel, Huinink, Ir, S.A. ten, Jacob Mosselstraat 75, 2595 RG 's-Gravenhage, PricewaterhouseCoopers NV Technology Consultants, 's-Gravenhage
Bolhuis, J.A., Ir, Riddersborch 83, 3992 BH Houten, Ingenieursbureau H.K. Boorsma BV, Drachten
Bondt, A.H. de, Dr Ir, Graaf Diederiklaan 2, 3434 ST Nieuwegein, Ooms Avenhorn BV, Avenhorn
Bonhomme, G.M.N., Ir, 's-Gravesandelaan 24, 1222 SZ Hilversum, Hogeschool Alkmaar, Alkmaar
Bonnier, P.G., Dr Ir, Latijns Amerikalaan 78, 2622 BD Delft, PLAXIS BV, Delft
Boo, A.J. de, Ir, Dokter W.v.d. Horstlaan 1, 2641 RT Pijnacker
Boomgaard, D.J. van den, Ir, Via Gorizia 25, 23900 Lecco, Italy, Ecodeco Spa, Giussago (Pv), ITALY
Boon, T.J., Ir, Cort van der Lindenlaan 1, 4384 KH Vlissingen, Rijkswaterstaat Directie Zeeland, Middelburg
Boorsma, K., Ir, Vlaslaan 38, 9244 CH Beetsterzwaag, Ingenieursbureau H.K. Boorsma BV, Drachten
Borst, W.G., Ir, Latijns Amerikalaan 92, 2622 BD Delft, Netherlands Dredging Consultants BV, Delft
Bos, F. van den, Drs, Gravin Catharinalaan 24, 2263 TN Leidschendam, Wintershall Noordzee BV, 's-Gravenhage
Bosman, P.G.F., Ir, De Braak 2, 8101 GJ Raalte, Witteveen + Bos Raadgevende Ingenieurs BV, Deventer
Bot, A.P., M.Sc., Ir, Söderblomplaats 348, 3069 SL Rotterdam, Ir. A.P. Bot Raadgevend Ingenieur Water/Grond/ Milieu, Rotterdam
Boumans, A.J.M.O., Ir, Leharlaan 9, 2625 ZE Delft, Heijmans Infrastructuur en Milieu/van der Linden Verkeerstechniek BV, Uden

Bouwers, J., Ir, Parkweg 7, 2585 JG 's-Gravenhage, Architecten en Ingenieursbureau Friedhoff en van Heerde, 's-Gravenhage
Boven, P. van, Ir, De Klepelslag 10, 8455 JX Katlijk, Ingenieursbureau Oranjewoud BV, Heerenveen
Boxsel, W.A. van, Ir, Baronielaan 148, 4818 RE Breda, Bouwtechnisch Adviesburo Ir, W.A. van Boxsel BV, Oosterhout Nb
Brakel, H., Ir, Postbus 4906, 2003 EX Haarlem, Pieters Bouwtechniek Delft BV, Delft
Bremer, H.H.J., Ir, Ing, Hoefkensestraat 12, 6996 DS Drempt, Bremer Engineering, Drempt
Breur, A.A., Ir, Mevrouw Stoelstraat 6, 8017 HE Zwolle, Christelijke Hogeschool Windesheim, Zwolle
Brink, K.A., Ir, Prinses Beatrixstraat 40, 6942 JG Didam, ABT Adviesbureau voor Bouwtechniek BV, Arnhem
Brinkgreve, R.B.J., Dr Ir, Copernicuslaan 6, 3204 CJ Spijkenisse, Mos Grondmechanica BV, Rhoon
Broek, W.L.A.H. van den, M.Sc., Ir, Van Aerssenstraat 248, 2582 JX 's-Gravenhage, Van Oord ACZ, Gorinchem
Broer, J., Ir, Noordzijde 63, 2977 XB Goudriaan, BV Grint- & Zandexploitatie Maatschappij, Roermond
Broersma, A., Ir, Groot Hertoginnelaan 33, 2517 EB 's-Gravenhage, Broersma BV Adviesbureau voor Beton- en Staalconstructies, 's-Gravenhage
Brouwer, R., Prof Ir, Praetoriumstraat 25, 6522 HN Nijmegen, TUD Communicatie en Marketing Groep, Delft
Bruijn, V.M.P.J. de, Mrs Ir, Herengracht 414, 1017 BZ Amsterdam, Price Waterhouse Coopers, Amsterdam
Bruijn, P.G.A. de, Ir, Damastroos 13, 3068 BW Rotterdam, Ballast Nedam Beton en Waterbouw BV, Amstelveen
Bruyn, L. de, Ir, c/o Ballast Nedam Caribbeau, P.O. Box 960, St Maarten, Netherlands Antilles, Ballast Nedam International BV, Amstelveen
Buijs, C.E.H.M., Ir, Tjalk 3, 2991 PM Barendrecht, Gemeentelijk Havenbedrijf Rotterdam, Rotterdam
Bunschoten, C.J., Ir, Havezathelaan 80, 9472 PX Zuidlaren, Grontmij Groningen, Haren Gn
Burger, P., Ir, Aagje Dekenlaan 49, 3768 XR Soest, DHV Consultants BV, Amersfoort
Burger, H., Ir, Parkhaven 3, 8242 PE Lelystad, DHV Milieu en Infrastructuur BV, Amersfoort
Buth, L.J., Drs Ir, Zjoekowlaan 97, 2625 PL Delft, Nederlands Normalisatie-instituut, Delft
Büttner, J.H., Ir, Prins Alexanderstraat 125, 7009 AJ Doetinchem
Buykx, S.M., 's-Gravendijkwal 38-A, 3014 EC Rotterdam, Witteveen + Bos Raadgevende Ingenieurs BV, Deventer
Calle, E.O.F., Ir, Nesciohove 26, 2726 BH Zoetermeer, Geo-Delft, Delft
Calveen, H.C.A. van, Ir, Gombertstraat 121, 8031 LD Zwolle, Calveen Consult, Zwolle
Carels, K.W., Ir, Burgemeester Vd Heideln 26, 3451 ZT Vleuten
Carree, P., Ir, Korvezeestraat 355, 2628 DR Delft, Witteveen + Bos Raadgevende Ingenieurs BV, Deventer
Carree, G.J., Ir, Benedendorpsweg 43, 6862 WC Oosterbeek, ABT Adviesbureau voor Bouwtechniek BV, Arnhem
Catshoek, J.W.D., Ir, Herman Gorterhof 62, 2624 XE Delft, Rijkswaterstaat Directie Zuid Holland, Rotterdam
Cirkel, R.J., Ir, Graaf Bentincklaan 25, 6712 GS Ede, Rijkswaterstaat Bouwdienst, Utrecht
Cloo, H.H., Graaf Adolflaan 82, 3818 DD Amersfoort, DHV Milieu en Infrastructuur BV, Amersfoort
Coers, E.J., Ir, Spoorsingel 84, 2613 BB Delft, Coers Project en Interim Management BV, Delft
Couperus, B., Ir, Wilgenlaan 17, 9363 CS Marum, Raadgevend Ingenieursbureau Wiertsema & Partners, Tolbert
D'Angremond, K., Prof Ir, Hogedijk 16-A, 2435 ND Zevenhoven, TUD Faculteit Civiele Techniek & Geowetenschappen/Civiele Techniek, Delft
Dalen, J.H. van, Ir, Nieuwstraat 226, 3011 GM Rotterdam, Gemeentewerken Rotterdam, Rotterdam
Dalen, J.W. van, Ir, Padangstraat 8, 3531 TC Utrecht, Visser & Smit Bouw BV, Papendrecht

Debets, N.T., Ir, Karrenstraat 53, 5211 EH 's-Hertogenbosch, Raadgevend Ingenieursbureau Inpijn-Blokpoel BV, Son
Deen, J.K. van, Dr, Groot Hertoginnelaan 24, 2517 EG 's-Gravenhage, GeoDelft, Delft
Deest, F. van, Ir, Bodegraafsestraatweg 159, 2805 GN Gouda, Gouda Damwand BV, Gouda
Degen, B.T.A.J., Drs, Cantaloupenburg 66, 2514 KM 's-Gravenhage, GeoCom BV, 's-Gravenhage
Degenkamp, G., Eur Ing Ir, Oosteinde 21, 2271 EA Voorburg, Vryhof Anchors BV, Krimpen aan den IJssel
Dijk, H.A. van, Ir, Zijdeweg 40, 2244 BE Wassenaar, Züblin AG, Zürich, SWITZERLAND
Dirks, W.G., Ir, 25 Namly Drive, Singapore 267439, Singapore, Ballast Nedam Baggeren BV, Zeist
Dirksen, J., Ir, In de Korenmolen 1, 1115 GN Duivendrecht, Ingenieursbureau Amsterdam, Amsterdam Zuidoost
Dirkzwager, N., Ir, Kuringen 28, 4761 VA Zevenbergen, Interbeton BV, Rijswijk Zh
Dirven, B.P.J., Ir, Lange Slagen 30, 4823 LJ Breda, Heijmans Beton en Waterbouw BV, 's-Hertogenbosch
Doef, M.R. van der, Ir, Burgemeester V Beugenstr 31, 4904 LS Oosterhout, Rijkswaterstaat Bouwdienst, Utrecht
Dolman, J.A., Ir, Biezenkuilen 64, 5502 PE Veldhoven, DHV AIB BV, Eindhoven
Don, W.M.W., Ir, Achtersloot 110, 3401 NZ IJsselstein, Koninklijke Wegenbouw Stevin BV District Utrecht, De Meern
Dongen, A.J.C. van, Ir, Wagenmaker 42, 2353 WJ Leiderdorp, Heeremac, Kuala Lumpur, MALAYSIA
Doornhein, M., Ir, Smedenweg 2, 1261 BH Blaricum, D & S Civil Engineering, Blaricum
Dorp, R.F. van, Ir, Doornstraat 8, 3581 TT Utrecht, IFCO Funderingsexpertise BV, Waddinxveen
Dresken, S.J.B., Ir, Tetterodestraat 44-ROOD, 2023 XP Haarlem, Koninklijke Boskalis Westminster NV, Papendrecht
Druenen, L.W.M. van, Ir, Neuhuyskade 116, 2596 XM 's-Gravenhage, Interbeton BV, Rijswijk Zh
Dubbeldam, J.W., Ir, Berkenlaan 14, 2224 EJ Katwijk, Holland Railconsult, Utrecht
Dubbers, R.A.W., Ir, Marineblauw 163, 2718 KE Zoetermeer, Shell International Exploration and Production BV-Res.& Tech.Services, Rijswijk Zh
Duskov, M., Dr Ir, Frieslandlaan 29, 3137 GE Vlaardingen, Intrastructural Engineering Delft bv
Ebels, L.J., Ir, c/o Uhlmann, Witthaus & Prins, Southernwood, P.O. Box 11170, 5213 East London/Eastern Cape, South-Africa, Uhlmann, Witthaus & Prins, East London/Eastern Cape, South-Africa
Eckhardt, W., Ir, Hooikade 57, 2514 BK 's-Gravenhage, PGS-Marine Services, Walton-on-Thames, ENGLAND
Ee, G. van, Ir, Jan Ligthartplein 9, 3706 VC Zeist, Grontmij Advies-& Techniek BV Houten, Houten
Elias, J.M., M.Sc., Ir, Albert Neuhuysstraat 16, 6813 GX Arnhem, Colbond Geosynthetics, Arnhem
Elsman, M.A., Ir, Korenmolen 70, 3481 AW Harmelen, Rijkswaterstaat Bouwdienst, Utrecht
Elzen, M.J.M. van den, Ir, Heirust 9, 5242 JJ Rosmalen, Infratech BV, Rosmalen
Ende, K.C.J. van den, Ir, Van Blankenburgstraat 7, 2517 XL 's-Gravenhage, Rijksinstituut voor Kust en Zee Rws, 's-Gravenhage
Enserink, E.W., Ir, Weimarstraat 216, 2562 HP 's-Gravenhage, De Weger Architecten- en Ingenieursbureau BV, Rotterdam
Epskamp, C.J., Ir, Alpenroos 5, 2317 EX Leiden, Van Splunder Funderingstechniek BV, Rotterdam
Es, H. van, Ir, Hurstwood Lane, Haywards Heath, England RH17 7SZ, Balfour Beatty Ltd., Thornton Heath, ENGLAND
Eussen, A.M. van, Mrs Ir, Duizendblad 58, 2201 SR Noordwijk, Witteveen + Bos Raadgevende Ingenieurs BV, 's-Gravenhage
Everts, H.J., Ing, Beelaertspark 7, 2731 AE Benthuizen, GeoDelft, Delft
Ewijk, R.J. van, Ir, Sportlaan 39, 3135 GR Vlaardingen, Iv-Consult BV, Papendrecht
Faassen, T.F., Ir, Leliestraat 27-BIS-A, 3551 AS Utrecht, Rijkswaterstaat Bouwdienst, Utrecht

Feijter, J.W. de, Ir, Groendaal 10, 2641 LN Pijnacker, GeoDelft, Delft

Flórián, G.J., Ir, Plevierstraat 1, 6883 CG Velp

Franken, L.P.M., Ir, Kastanjedreef 9, 2920 Kalmthout, Belgium, CFE Beton- en Water BV, Rotterdam

Frenay, J.W.I.J., Dr Ir, Van Lidth de Jeudelaan 13, 6703 JA Wageningen, ENCI Marketing, 's-Hertogenbosch

Galle, R., Ir, Beverodelaan 153, 6952 JE Dieren, Hogeschool van Arnhem en Nijmegen, Nijmegen

Geel, J.W.A., Ir, Eekhoornlaan 33, 3734 GW Den Dolder, Ballast Nedam Engineering BV, Amstelveen

Geertse, J.J.A., Ir, Bastiaansblok 47, 4613 GB Bergen op Zoom, Geomet BV, Alphen aan den Rijn

Geraedts, J.J.A., Ir, Asseltsestraat 60, 6071 BT Swalmen, Bouwbedrijf Geraedts BV, Swalmen

Gerritsen, J., Ir, Korengracht 31, 3262 CD Oud Beijerland, Gemeentewerken Rotterdam, Rotterdam

Gijt, J.G. de, Ir, Sonderdankstraat 14, 2596 SE 's-Gravenhage, Gemeentewerken Rotterdam, Rotterdam

Ginkel, N. van, Ir, Vlaskamp 18, 2353 HT Leiderdorp, HAM, Hollandsche Aanneming Maatschappij BV, Rotterdam

Gisbergen, N.W.M. van, Ing, Lepelaar 14, 4822 RL Breda, Ingenieursbureau Snellen, Meulemans en van Schaik BV/ DHV-Zuid, Breda

Gitz, H.J.J., Ir, Jacob van Ruisdaellaan 26, 3401 NL IJsselstein, Aveco BV, Utrecht

Glas, H.J.M., Ir, Torenakker 16, 5268 BX Helvoirt, P2 Managers, Rosmalen

Goedemoed, S.S., Ir, Conradlaan 21, 2627 BT Delft, Fugro Engineers BV, Leidschendam

Goossen, G.R., Ir, Wilhelminalaan, #3 Curaçao, Netherlands Antilles, Ingenieurs en Archictectenbureau Ascon NV, Curaçao, Curaçao Netherlands Antilles

Gorter, J.H., Ir, Grote Beerstraat 21, 1973 ZP Ijmuiden, Stork Alpha Engineering BV, Beverwijk

Goudswaard, J.J., Ir, Lijsterbeslaan 147, 2282 LK Rijswijk, Ingenieursbureau Den Haag, 's-Gravenhage

Graaf, H.J. van der, Ir, Grootslag 57, 3991 RB Houten, Rijkswaterstaat Bouwdienst, Utrecht

Graaf, F.F.M. de, Ir, Amaliastein 20, 4133 HC Vianen, Rijkswaterstaat Bouwdienst, Utrecht

Graaf, P. de, Ir, Bouriciuslaan 70, 9203 PG Drachten, Ingenieursbureau H.K. Boorsma BV, Drachten

Greeve, J.M.M., Ir, Lisdodde 1, 2631 DJ Nootdorp, Dura Bouw BV, Amsterdam

Groen, R.A., Ir, Wittgensteinlaan 238, 1062 KE Amsterdam, De Boer Techniek + Methode BV, Baarn

Grootenboer, M., Ir, c/o HAM Singapore, P.O. Box 8574, 3009 AN Rotterdam, HAM, Hollandsche Aanneming Maatschappij BV, Rotterdam

Grote, B.J.H., Ir, Lindelaan 2, 2803 SJ Gouda, Strukton Groep NV, Maarssen

Gun, J.H.J. van der, Ir, Wickenburghselaan 2-A, 3998 JW Schalkwijk, Bodembeheer BV, Schalkwijk

Haan, W. de, Ir, Graaf Florisweg 131, 2805 AK Gouda, Grontmij Advies & Techniek BVm, Vestiging Zuid Holland V, Waddinxveen

Haan, K.E. de, Ir, Sparrenlaan 12, 8051 BB Hattem, De Haan Dredging Consultancy, Hattem

Haan, F.S. de, Ir, Oosterlaan 7, 8746 ND Schraard, Ingenieursbureau de Marne, Bolsward

Haeren, J.C.M. van, Ir, Voorstraat 15, 2611 JJ Delft, Van Haeren Advies- en Constructieburo, Delft

Haitjema, H.M., Prof Dr Ir, 2738 Brig's Bend, Bloomington IN 47401, USA, Indiana University, Bloomington, USA

Haitsma, W., Ir, Midlumerlaan 79, 8861 JJ Harlingen, Haitsma Management Assistance BV, Harlingen

Halter, W.R., Ir, Jan van Scorelstraat 29, 3583 CJ Utrecht, Fugro Ingenieursbureau BV, Nieuwegein

Ham, N.A.V.M. van, Ir, Zaagmolen 50, 2906 RK Capelle aan den IJssel, Gemeentewerken Rotterdam, Rotterdam

Hamilton-Huisman, M.J., Mrs Ir, Beeklaan 14, 6865 VH Doorwerth, Provincie Gelderland, Arnhem

Harmelen, A.H.J. van, Ir, Meerstraat 52, 1411 BJ Naarden, Bravenboer & Scheers BV Raadgevend Ingenieurs, Almere

Hart, R. 't, Ir, Volkerak 2, 2641 SW Pijnacker, Rijkswaterstaat Dienst Weg- en Waterbouwkunde, Delft

Hart, R.P. 't, Ir, Postbus 211, 8160 AE Epe, Swot-Groep Apeldoorn, Apeldoorn

Hartjes, A.G., Ir, Judith Leysterlaan 10, 2104 SM Heemstede, Stichting Bouwresearch, Rotterdam

Hartmans, D.C., Ir, Ransuillaan 119, 2261 DB Leidschendam

Hassel, C.J. van, Ir, De Bosch Kemperlaan 4, 3818 HC Amersfoort, Witteveen + Bos Raadgevende Ingenieurs BV, Deventer

Heemstra, J., Ir, Jacob van Lennepkade 25-HS, 1054 ZE Amsterdam, GeoDelft, Delft

Heeres, O.M., Ir, Dennenlaan 33, 3843 BW Harderwijk, TUD Communicatie en Marketing Groep, Delft

Heering, R.P., Ir, Elisabeth Hoeve 29, 2804 HM Gouda, NACO Nederlands Ontwerpbureau voor Luchthavens BV, 's-Gravenhage

Heidweiller, C.M., Mrs Ir, Glenn Millerstraat 36, 2324 LM Leiden, Iv-Consult BV, Papendrecht

Heijden, J.H.A. van der, Ir, Gerrit Rietveldlaan 47, 2343 MA Oegstgeest, Gemeentewerken Rotterdam, Rotterdam

Heijmans, R.W.M.G., Ir, Spijkerstraat 32, 3513 SL Utrecht, Arcadis Bouw/Infra, Amersfoort

Heijnen, W.J., Ir, Buitenberg 62, 4707 SX Roosendaal

Heins, W.F., Ing, Frederik Hendriklaan 19, 2582 BR 's-Gravenhage, Deltaland BV Ingenieursbureau Grondmechanica en Funderingstechniek, 's-Gravenhage

Helbo, T., Ir, Kazernestraat 68-B, 2514 CW 's-Gravenhage, Van Oord ACZ, Gorinchem

Hemmelder, E.H.M., Ir, Biltstraat 140, 3572 BL Utrecht, Ingenieursbureau Amsterdam, Amsterdam Zuidoost

Hemmen, B.R., Ir, Nieuwe Binnenweg 96-1L, 3015 BD Rotterdam, GeoDelft, Delft

Hendriks, J.C.F., Ir, Norbertushof 22, 4133 TA Vianen, Rijkswaterstaat Bouwdienst, Utrecht

Hendriks, J.M.G., Ir, Binnenweg 27, 4824 ZE Breda, Aronsohn Raadgevende Ingenieurs BV, Rotterdam

Herbschleb, J., Ir, Insulindeweg 976, 1095 DX Amsterdam, De Weger Architecten- en Ingenieursbureau BV, Rotterdam

Hergarden, H.J.A.M., Ir, Molierezijde 15, 2725 NG Zoetermeer, GeoDelft, Delft

Heukelom, D., Ir, Bloeme 21, 2811 AN Reeuwijk, TBI Beton en Waterbouw/Haverkort BV, Apeldoorn

Hjelde, H.G., Ir, 's-Gravenweg 542, 3065 SG Rotterdam, van Hattum en Blankevoort BV, Woerden

Hoefnagels, A.A.J.V., Ir, Groenhovenweg 397, 2803 DK Gouda, I-Bureau Gemeentewerken Rotterdam, Rotterdam

Hoff, J. van 't, Ir, Kersbergenplein 11, 3703 AR Zeist, Ingenieursbureau Van't Hoff BV, Zeist

Hollander, J.M., Ir, Reinout 14, 2202 PN Noordwijk, Gemeente Voorhout, Voorhout

Holtus, R.K.M., Ing, Hoofdstraat 16, 6881 TH Velp, Acordis/ Akzo Nobel, Arnhem

Hoogcarspel, P.A.H., Ir, Buerweg 33, 1861 CG Bergen

Hoogendoorn, A., Ir, Groenedijk 60, 3544 AB Utrecht, HD Consult, Utrecht

Hospers, B., Ir, Apollolaan 27, 2324 BR Leiden, NV Koninklijke Nederlandsche Petroleum Maatschappij, 's-Gravenhage

Hudig, E.P., Ir, Van Alkemadelaan 1080, 2597 BK 's-Gravenhage, Shell International Exploration en Production BV, 's-Gravenhage

Hulsbergen, J.G., Ir, Heilweg 57, 6932 KG Westervoort, ABT Adviesbureau voor Bouwtechniek BV, Arnhem

Huyskes, E.J., Ir, Magnolialaan 7, 5342 HG Oss, Koninklijke Haskoning Groep BV, Nijmegen

Iperen, J. van, Ir, Verdipad 37, 2912 XL Nieuwerkerk Aan de IJssel, Fugro Engineers BV, Leidschendam

Ittersum, D. van, Ir, Villapark 6, 7491 BV Delden, Hogeschool Enschede, Enschede

Jadoenathmisier, G.A., Ir, Huntum 68, 1102 JC Amsterdam Zuidoost, Ballast Nedam Engineering BV, Amstelveen

Jagher, J.T. de, Ir, Wethouder in T Veldstr 205, 1107 BB Amsterdam Zuidoost, GCEI, Amsterdam Zuidoost

Janse, E., Ir, Munnikenland 9, 2716 BV Zoetermeer, GeoDelft, Delft

Jansen, M.H.P., Ir, Queridostraat 65, 2274 XD Voorburg, Ingenieursbureau Svasek BV, Rotterdam

Jansen, H.L., Ir, Pijpenpad 19, 2802 BG Gouda, Fugro Ingenieursbureau BV, Leidschendam

Jansen, G.P.J., Ir, Wilgenlaan 36, 3831 XV Leusden, BMC-Bodemconsult, Amersfoort

Janssen, N., Ir, Van Leeuwenhoeksingel 8, 2611 AA Delft, Hollandse Aannemingsmaatschappij BV, Capelle aan den IJssel

Janssen, W.P.S., Ir, Kastanjelaan 21, 6584 CN Molenhoek, Koninklijke Haskoning Groep BV, Nijmegen

Jas, H.A., Ir, Goudhoekweg 5, 3233 AM Oostvoorne, Tensar Grids BV, Oostvoorne

Jelgersma, F., Ir, c/o Sternmesschen 22, 9403 ZZ Assen, Nederlandse Aardolie Maatschappij, Schoonebeek

Jessen, G.R.M., Ir, Bilt 10, 6107 BM Stevensweert, Geoconsult Geotechniek BV, Hoensbroek

Jong, P. de, Ir, Postbus 26, 2280 AA Rijswijk, Adviesbureau Ir J.G. Hageman BV, Rijswijk Zh

Jong, R.J. de, Ir, Fijnjekade 35, 2521 CR 's-Gravenhage, Arcadis Bouw/Infra, 's-Gravenhage

Jong, R.G.A. de, Ir, Sint Olofsstraat 29, 2613 EK Delft, Mos Grondmechanica BV, Rhoon

Jong, J. de, Ir, Lopikerweg Oost 118, 3411 LX Lopik, van Hattum en Blankevoort BV, Woerden

Jong, J.A. de, Ir, Prinses Beatrixweg 9, 7433 DB Schalkhaar, Witteveen + Bos Raadgevende Ingenieurs BV, Deventer

Jonge, H. de, Ir, Van Kleffenslaan 86, 4334 HK Middelburg, Heijmans Beton- en Waterbouw BV, Terneuzen

Jongert, P.N., Ir, Asfilstraat 12, 9031 Afsnee, BELGIUM

Jongh, J. de, Ir, Vleutenseweg 381, 3532 HH Utrecht, Grontmij Advies-& Techniek BV Houten, Houten

Jonker, A., Ing, Schijfmos 80, 2914 AM Nieuwerkerk Aan de IJssel, Stichting CUR, Gouda

Jonker, J.H., Ing, Harderwijkerweg 45, 3852 AA Ermelo, Holland Railconsult, Utrecht

Jonker, T.T., Ir, Laurenspark 47, 4835 GW Breda, Heijmans Beton en Waterbouw BV, 's-Hertogenbosch

Joustra, K., Ir, Voorburgseweg 54, 2264 AH Leidschendam, Joustra Consult BV, Leidschendam

Karim, U.F.A., Dr Ir, Pijpenstraat 9-A, 7511 GM Enschede, Universiteit Twente Civiele Technologie en Management, Enschede

Kayser, T.H., Ir, Kaag 3, 3121 XE Schiedam, Unilever Research, Vlaardingen

Kedde, H.J., Ir, Kwinkenplein 59, 9712 GX Groningen, Ingenieursbureau Dijkhuis BV, Groningen

Kerkhof, F.I.C.M., Ir, Vivaldilaan 37, 5242 HD Rosmalen, Raadgevend Ingenieursbureau van Nunen, Rosmalen

Keusters, A.C.A.M., Ir, Achillesstraat 89, 4818 BM Breda, Heijmans Beton en Waterbouw BV, 's-Hertogenbosch

Kieft, P., Ir, Mereveldlaan 55, 3454 CB De Meern, Rijkswaterstaat Hoofdkantoor, 's-Gravenhage

Kirstein, A.A., Ir, De Duiker 12, 3461 HN Linschoten, van Hattum en Blankevoort BV, Woerden

Klap, C.Q., Ir, Tintlaan 91, 2719 AM Zoetermeer, Hexa Consult, Amstelveen

Klaver, K.J., Ir, Burgemeester Hendrixstr 50, 2651 JV Berkel en Rodenrijs, Europe Combined Terminals BV, Rotterdam

Kleef, H. van, Ir, Max Havelaarburg 41, 2907 HH Capelle aan den IJssel, Technisch Adviesbureau Klebaver BV, Rotterdam

Kleinman, C.S., Prof Ir, Schutteheiweg 5, 6343 AG Klimmen, Technische Universiteit Eindhoven, Eindhoven

Klerks, L.F.G., Ir, Dennenlaan 16, 3739 KN Hollandsche Rading, Grontmij Advies-& Techniek BV Houten, Houten

Kliffen, H.W., Ir, Emmapark 22, 2641 EL Pijnacker, VROM/Rijksgebouwendienst, 's-Gravenhage

Knibbeler, A.G.M., Ir, Boezemweg 23, 2741 MV Waddinxveen, Mos Grondmechanica BV, Rhoon

Kock, R.A.J. de, Ir, Van Alkemadelaan 38, 2597 AM 's-Gravenhage, Provincie Zuid-Holland, 's-Gravenhage

Koelewijn, A.R., Ir, Latijns Amerikalaan 147, 2622 BB Delft, GeoDelft, Delft

Koers, G., Ir, Bingerdensedijk 10, 6987 EA Giesbeek, Bouwadviesbureau Strackee BV, Amsterdam

Koeyer, D.M. de, Ir, Prins Hendriklaan 36, 2635 JG Den Hoorn, Heerema Marine Contractors V.O.F., Leiden

Kok, M.B.A.M. de, Ir, c/o Ballast Nedam, P.O. Box 149, 3700 AC Zeist, Ballast Nedam Baggeren BV, Zeist

Kolk, H.J., Ir, Bartokhof 13, 2402 GC Alphen aan den Rijn, Fugro Engineers BV, Leidschendam

Koning, J. de, Prof Ir, Soetendaal 20, 1081 BP Amsterdam, Kodreco Dredging Consultants, Amsterdam

Koning, A. de, Ir, Driehoek 15, 3328 KG Dordrecht, Geka Bouw BV, Dordrecht

Koning, J.S. de, Ir, Ingenieur Menkolaan 61, 3761 XK Soest, Arcadis Bouw/Infra, Amersfoort

Kooi, D.A. van der, Ir, Roland Holstlaan 1080, 2624 JR Delft, SGS EcoCare BV, Dordrecht

Kool, A.M., Mrs Ir, Schiedamsedijk 8-E, 3011 EC Rotterdam, GeoDelft, Delft

Koolwijk, W., Ir, Kamgras 13, 2851 ZS Haastrecht, Heeremac, Kuala Lumpur, MALAYSIA

Kort, D.A., Ir, Koornmarkt 73-A, 2611 EC Delft, TUD Communicatie en Marketing Groep, Delft

Kort, P.J.C.M. de, Ir, Sikkelstraat 35, 4904 VA Oosterhout, Ballast Nedam Funderingstechnieken BV, Dordrecht

Kort, J. de, Ir Ing ir, Luiten Ambachtstraat 21, 4944 AS Raamsdonk, Akzo Nobel NV, Arnhem

Kortsmit, L.L.J., Ir, Zuiderbaan 25, 4386 CK Vlissingen, Baggermaatschappij Boskalis BV, Papendrecht

Koster, G., Drs Ir, Warmoesland 35, 2635 LB Den Hoorn, Heigroep Durieux, Delft

Koten, H. van, Ing, Tweede Stationsstraat 232, 2718 AC Zoetermeer, Flow Engineering BV, Rijswijk Zh

Kraaijeveld, P.A., Ir, Leerbroekseweg 2, 4245 KT Leerbroek, N. Kraaijveld BV, Sliedrecht

Kramer, G.J. de, Ir, Heimstede, Woestduinlaan 11, 3941 XA Doorn, Fugro NV, Leidschendam

Krause, E.R., Ing, c/o Turbigas Solar of. 22, J. Salguero 2745-piso 2, 1425 Buenos Aires, Argentinia, TURBIGAS SOLAR, Buenos Aires, ARGENTINIA

Kremer, R.H.J., Ir, H.A. Lorentzweg 37, 1402 CB Bussum, Ingenieursbureau Amsterdam, Amsterdam Zuidoost

Kruijff, H. de, Ing, Landzichtweg 26, 4105 DP Culemborg, Holland Railconsult, Utrecht

Kruisman, R.H., Ir, Dokter W Nijestraat 71, 2064 XB Spaarndam, Valkenberg Raadgevend Ingenieursburo BV, Haarlem

Kruizinga, J., Ir, Hoog Barge 21, 7211 DT Eefde, KOAC.WMD Instituut voor Materiaal- en Wegbouwkundig Onderzoek, Apeldoorn

Kruse, G.A.M., Drs, Wasstraat 60, 2313 JK Leiden, Geo-Delft, Delft

Kuiper, J.C., Ir, Rietdekkershoek 21, 3981 TN Bunnik, Holland Railconsult, Utrecht

Kuipers, J.J., Ir, Vuurdoornstraat 23, 6903 CJ Zevenaar, Grontmij Advies-& Techniek BV, Arnhem

Kwast, E.A., Ing, Braamgaarde 21, 3436 GL Nieuwegein, Holland Railconsult, Utrecht

Laan, W.A.M. van der, Ir, Koningsspil 45, 3642 ZN Mijdrecht, Architecten- en Ingenieursbureau H.W. van der Laan BV, Mijdrecht

Lambrechtsen, W.J.C., B.Eng, Ing, c/o BKH Adviesbureau, P.O. Box 5094, 2600 GB Delft, BKH Adviesbureau, Delft

Landwehr, J.C., Ir, Albert Schweitzerlaan 76, 3223 WG Hellevoetsluis, GeoDelft, Delft

Lange, W.M.J. de, Ir, Velserdijk 30, 1981 AA Velsen Zuid, Hogeschool Alkmaar, Alkmaar

Lange, M.W.P. van, Ir, Valkenbergerweg 7, 6991 JA Rheden, MTI Holland BV, Kinderdijk

Lavooij, H.A., Ir, Anker 50, 3904 PM Veenendaal, Haskoning Ingenieurs- en Architectenbureau, Capelle aan den IJssel

Leendertse, W.L., Ir, Prof. J.W. Dieperinklaan 20, 3571 WK Utrecht, Rijkswaterstaat Bouwdienst, Utrecht

Leepel, G.A., Ir, Vliegeniersweg 112, 2171 PE Sassenheim, Allseas Engineering BV, Delft

Leeuw, J.C.M. de, Ir, Opwijckstraat 3, 2272 BC Voorburg, Fugro Ingenieursbureau BV, Leidschendam

Leijsen, P.J.M. van, Ir, Watermolenberg 4, 4707 MA Roosendaal, Ingenieursbureau Oranjewoud BV, Oosterhout Nb

Leupen, F.D., Ir, Pinkenbergseweg 28, 6881 BE Velp, Grontmij Verkeer & Infrastructuur, De Bilt

Lijke, T.P. van der, Ir, Kinkerstraat 142-C-3, 1053 EG Amsterdam, Grontmij Advies-& Techniek BV, Houten

Limbergen, R. van, Ir, Veenhof 2347, 6604 DR Wijchen, Koninklijke Haskoning Groep BV, Nijmegen

Linden, A.M. van der, Ir, Naaierstraat 20, 2801 NG Gouda, DHV Consultants BV, Amersfoort

Lindenberg, J., Ir, Floris V Wevelikhovenstr 3, 8325 GD Vollenhove, Rijkswaterstaat Dienst Weg- en Waterbouwkunde, Delft

Lindhout, P.H., Ir, Henri Polaklaan 5-II, 1018 CP Amsterdam, Arcadis Bouw/Infra, 's-Gravenhage

Linssen, L.P.M., Ir, Spoordonkseweg 4-A, 5688 KD Oirschot, DHV Milieu en Infrastructuur BV, Amersfoort

Lohuizen, H.P.S. van, Prof Ir, Amsterdamsestraatweg 19, 1398 BS Muiden

Looff, J.C. de, Ir, De Savornin Lohmanlaan 14, 3362 CD Sliedrecht, Puri Nederland BV, Sliedrecht

Loxham, M., B.Sc., Eur Ing, Prof Dr Ir, Nieuwe Haven 45, 4301 DJ Zierikzee, GeoDelft, Delft

Lubking, P., Ir, Rode Klaver 24, 2923 GH Krimpen aan den IJssel, GeoDelft, Delft

Luger, H.J., Ir, Saffier 49, 2651 SZ Berkel en Rodenrijs, GeoDelft, Delft

Luipen, P. van, Ir, Moortwiete 29, 25479 Ellerau, Germany, MENCK, Ellerau, GERMANY

Luiten, G.T., Dr Ir, 2e de Carpentierstraat 226, 2595 HM 's-Gravenhage, HBG Hollandsche Beton Groep NV, Rijswijk Zh

Maas, G.J., Prof Ir, Esscheweg 167, 5262 TZ Vught, Technische Universiteit Eindhoven, Eindhoven

Maas, P.A.M., Ir, Schoolstraat 10, 5541 EG Reusel, Silidur Nederland BV, Rotterdam

Maase, T., Ir, Aegidiusstraat 106, 3061 XP Rotterdam, Visser & Smit Hanab BV, Papendrecht

Maatje, F., Ir, Ter Leedelaan 33, 2172 JL Sassenheim, ECCS BV Engineering CAD CAM Steel, Hoofddorp

Maertens, J.F.A., Ir, H. Consciencestr. 4, 2340 Beerse, Belgium, Jan Maertens BVBA, Beerse, BELGIUM

Maijers, A.H., Ir, Ten Oeverstraat 32, 8012 EV Zwolle, Allart Maijers Consulting, Zwolle

Majoor, R., Ir, Frankenslag 47, 2582 HD 's-Gravenhage, Ballast Nedam Baggeren BV, Zeist

Mans, D.G., Ir, Tintlaan 77, 2719 AM Zoetermeer, Arcadis Bouw/Infra, 's-Gravenhage

Markslag, R.B.G.H., Ing, c/o adv.bur.R.Markslag BV, Aa of Weerijs 106, 5032 BE Tilburg, Adviesbureau Markslag BV, Tilburg

Maurenbrecher, P.M., M.Sc., B.Sc., Groot Hertoginnelaan 15, 2517 EA 's-Gravenhage, TUD Faculteit Civiele Techniek & Geowetenschappen/Tech. Aardwetenschap, Delft

Meek, J., Ir, Treebord 8, 2811 EA Reeuwijk, Heeremac, Kuala Lumpur, MALAYSIA

Meel, R.L.B. van der, Ir, Wouwsestraatweg 63, 4622 AB Bergen op Zoom, Mos Grondmechanica BV, Rhoon

Meer, E.A. van der, Ir, Gorterstraat 3, 2761 LL Zevenhuizen, Strukton Groep NV, Maarssen

Meer, M.T. van der, Ir, Benedictijnenhove 5, 3834 ZA Leusden, Fugro Ingenieursbureau BV, Nieuwegein

Meeuse, J.C., Ir, Julianalaan 14, 3135 JJ Vlaardingen, Gemeente Den Haag, Dienst Stadsbeheer, 's-Gravenhage

Meijden, H.C.A.L. van der, Ir, Molengraafseweg 16, 5281 LN Boxtel, IWACO BV, 's-Hertogenbosch

Meijer, K.L., Dr Ir, 't Klooster 3, 8355 AR Giethoorn, Meyer Technisch-Wetenschappelijke Dienstverlening, Giethoorn

Meijers, P., Ir, Colombiahof 6, 2622 AC Delft, GeoDelft, Delft

Meisner, E., Ir, Stuijvesantplein 20-ROOD, 2023 KV Haarlem, Ingenieursbureau Amsterdam, Amsterdam Zuidoost

Meulemans, A.C., Ir, Liesbosdreef 12, 4841 JL Prinsenbeek, Ingenieursbureau Snellen, Meulemans en van Schaik BV/DHV-Zuid, Breda

Meurs, G.A.M. van, Dr Ir, Gaelstraat 46, 2291 SJ Wateringen, GeoDelft, Delft

Meyer, M.A., Ir, Coppenamestraat 6, Paramaribo, Surinam, SUNECON (Suriname Engineering Consultants), Paramaribo, Surinam

Meyvogel, I.J., Ir, Prins Alexanderlaan 29, 2224 XN Katwijk, GeoDelft, Delft

Middendorp, P., Ir, Koningin Wilhelminalaan 516, 2274 BL Voorburg, TNO Bouw, Delft

Molenaar, A.A.A., Prof Dr Ir, Schuylenburgh 7, 2631 CN Nootdorp, TUD Communicatie en Marketing Groep, Delft

Molenbroek, L.D., Ir, Marinus Batenburgplein 43, 3065 KL Rotterdam, Ingenieursbureau Molenbroek BV, Rotterdam

Molenkamp, F., Prof Dr Ir, 25 Bickerton Road, Altrincham, England WA14 4UN, University of Manchester Institute of Science and Technology, Manchester, ENGLAND

Mom, R.A.J.M., Ir, Csardasstraat 12, 6544 RX Nijmegen, Adams Bouwadviesbureau BV, Druten

Monster, H.B., Ir, Everwijnlaan 1, 6816 RC Arnhem, ABT Adviesbureau voor Bouwtechniek BV, Arnhem

Mooij, H.G.M., Ir, Iordensstraat 17, 2012 HA Haarlem, BV Adviesbureau TJaden voor Technisch Bodemonderzoek, Haarlem

Mooijman, P.J.C., Ir, Goudmos 36, 2914 AH Nieuwerkerk Aan de IJssel, Gemeentewerken Rotterdam, Rotterdam

Moons, J.M., Ir, Grotestraat 9, 5141 JM Waalwijk, Ingenieursbureau Moons, Waalwijk

Mulder, H.J., Ir, Batavierenplantsoen 37, 2025 CJ Haarlem, VROM/Rijksgebouwendienst, 's-Gravenhage

Muller, T.K., Ir, Marga Klompéhoeve 17, 2743 HV Waddinxveen, IFCO Funderingsexpertise BV, waddinxveen

Mynett, A.E., Prof, Lausbergstraat 9, 2628 LA Delft, Waterloopkundig Laboratorium/Delft Hydraulics, Delft

Naafs, H.J., Ir, Marcus Samuelstraat 28, 7468 BP Enter, de Bondt Rijssen BV, Rijssen

Niekerk, W.J. van, Ir, Kreekraklaan 30, 3544 WB Utrecht, Ballast Nedam Grond en Wegen BV, Amstelveen

Niese, M.S.J., Ir, Akkerwinde 15, 8131 GP Wijhe, Conewel BV, Alphen aan den Rijn

Nieuwaal, A., Ir, Moerheimstraat 43, 7701 CA Dedemsvaart, Grondboorbedrijf Haitjema BV, Dedemsvaart

Nieuwenhuis, J.D., Prof Dr, Roucooppark 27, 2251 AX Voorschoten, TUD Communicatie en Marketing Groep, Delft

Nieuwenhuis, T.J., Ir, Prins Hendriklaan 9, 3832 CN Leusden, DHV Milieu en Infrastructuur BV, Amersfoort

Nieuwenhuis, J.K., Ir, Broedheuvel 19, 5685 AM Best, EO Hydro, Eindhoven

Nieuwjaar, M.W.C., Ir, Nieuwe Kerksplein 20-A, 2011 ZT Haarlem, Provincie Flevoland, Lelystad

Nijhuis, H.W., Ir, Middenweg 113, 9468 GL Annen, Hanzehogeschool van Groningen Faculteit Techniek, Groningen

Nijs, R.E.P. de, Ir, Haarlemmerstraat 46, 1013 ES Amsterdam, IFCO Funderingsexpertise BV, Waddinxveen

Nijs, P.J.M. den, Ir, Churchilllaan 11, 2012 RM Haarlem, Wareco Amsterdam BV, Amsterdam Zuidoost

Nonneman, D.J., Ir, Geervliet 133, 1082 NN Amsterdam, Aveco BV, Utrecht

Oene, H.B. van, Ir, Barchemseweg 8, 7241 JD Lochem, Koninklijke Haskoning Groep BV, Nijmegen

Ogilvie, R.J., Ir, Van Deventerlaan 45, 2271 TV Voorburg, Heeremac, Kuala Lumpur, MALAYSIA

Oldeniel, L.B.J. van, Ir, Simon van Collemstraat 38, 1325 NB Almere, Geomet BV, Alphen aan den Rijn

Olthof, M., Ir, Rubenslaan 18, 3116 BN Schiedam, CAP Gemini Nederland BV, Utrecht

Omtzigt, H.A.M., Ir, Tine van Dethstraat 53, 2331 CD Leiden, Geo Mil Equipment BV, Alphen aan den Rijn

Ooms, K., Ir, Loevestein 51, 2403 JD Alphen aan den Rijn, Demas Dredging Consultants BV, Waddinxveen

Oosterman, J., Ir, c/o DHV consultants, P.O. Box 1399, 3800 BJ Amersfoort, DHV Consultants BV, Amersfoort

Oostlander, L.J., Ir, Dorpsstraat 149, 2995 XG Heerjansdam, Netherlands Pavement Consultants BV, Utrecht

Oostveen, J.P., Ir, Arthur van Schendelplein 135, 2624 CV Delft, Mos Grondmechanica BV, Rhoon

Ophem, J.I.M. van, Ir, Koperslagershoek 2, 3981 SB Bunnik, Ingenieursbureau Amsterdam, Amsterdam Zuidoost

Otte, L.J., Ir, Groepsekom 3, 3831 RG Leusden, Aannemingsmaatschappij Sikking BV, Hilversum
Oud, E.L.C.M., Ir, Prinsenlaan 25, 2104 AD Heemstede, van Rossum Infra BV, Almere
Paap, H.A., Ir, Carolinenburg 3, 8925 CA Leeuwarden, Waterschap Friesland, Leeuwarden
Paesschen, J.C.L., Ir, Prins Bernhardlaan 77, 4615 BB Bergen op Zoom, Arbed Damwand Nederland BV, Moerdijk
Parent, M.G., Ir, Florastraat 21, 2931 TB Krimpen aan de Lek, Gemeentewerken Rotterdam, Rotterdam
Pauwels, J., Ir, Dopheidestraat 62, 2165 VS Lisserbroek, Broersma BV Adviesbureau voor Beton- en Staalconstructies, 's-Gravenhage
Peerdeman, H.C., CI, Ir, Lichtegaarde 58, 3436 ZW Nieuwegein, Bruijn, Molenaar, Ter Riet en Parners BV, Nieuwegein
Peerlkamp, K.P., Ir, Anjerlaan 18, 4286 CR Almkerk, Van Oord ACZ, Gorinchem
Pelt, W.C.A. van, Ir, Rodenrijseweg 63, 2651 BN Berkel en Rodenrijs, Ballast Nedam NV, Amstelveen
Pereboom, D., Ir, Prinsenhof 54, 2641 RS Pijnacker, GeoDelft, Delft
Pichel, G., Ir, Mecklenburglaan 42, 3843 BP Harderwijk, DHV Consultants BV, Amersfoort
Piggelen, H.P. van, Ir, Meijendellseweg 24, 2243 GN Wassenaar, NACO Nederlands Ontwerpbureau voor Luchthavens BV, 's-Gravenhage
Pijls, C.G.J.M., Ir, Lange Maat 119, 6932 AC Westervoort, Tauw BV, Deventer
Pijpers, J.P., Ir, Scheveningsebos 53, 2716 HW Zoetermeer, Kvaerner BV, Zoetermeer
Pilarczyk, K.W., Ir, Nesciohove 23, 2726 BJ Zoetermeer, Rijkswaterstaat Dienst Weg- en Waterbouwkunde, Delft
Plasman, Z.J., Ir, c/o Rodelaan 283, 2272 SG Voorburg, Fugro Engineers BV, Leidschendam
Ploeg, T.J., Ir, Dalweg 56, 6865 CV Doorwerth, Kema Transmission & Distribution, Arnhem
Poel, J.T. van der, Ir, Watermolen 32, 3995 AP Houten, Geo-Delft, Delft
Poel, P. van der, Ir, Bennekomseweg 35, 6866 DB Heelsum, Ingenieursbureau v.d. Poel, Heelsum
Poelman, A., Ir, De Steen 9, 3931 VK Woudenberg, Grontmij Advies-& Techniek BV Houten, Houten
Pool, J.H., Ir, Graaf van Lyndenlaan 2, 3771 JC Barneveld, Grontmij Advies-& Techniek BV Houten, Houten
Postma, H., Ir, Willem de Zwijgerlaan 5, 2252 VN Voorschoten, Baggermaatschappij Boskalis BV, Papendrecht
Postma, M.H., Ir, Harddraversweg 16, 8501 CM Joure, Ingenieursbureau Oranjewoud BV, Heerenveen
Profittlich, M.J., Ir, Westplantsoen 26-B, 2613 GL Delft, Fugro Ingenieursbureau BV, Leidschendam
Pruijssers, A.F., Dr Ir, Abdij 40, 3335 DJ Zwijndrecht, Dirk Verstoep BV, Rotterdam
Pruisken, H.J.A., Ir, Kijckerweg 42, 2678 AD De Lier, NBM West, 's-Gravenhage
Put, J.L. van der, Ir, Waterpeper 2, 2804 PR Gouda, Rijkswaterstaat Bouwdienst, Utrecht
Quelerij, L. de, Ir, Kleine Woerdlaan 48, 2671 AS Naaldwijk, Fugro Ingenieursbureau BV, Leidschendam
Raadt, P.J. van den, Ir, Roemer Visscherstraat 227, 2026 TT Haarlem, BV Adviesbureau TJaden voor Technisch Bodemonderzoek, Haarlem
Raadt, J.P. de, Ir, Werve 12, 3155 GM Maasland, Technip Benelux BV, Zoetermeer
Raalte, G.H. van, Ir, Beekenstein 61, 3328 ZC Dordrecht, Boskalis/Hydronamic, Papendrecht
Radersma, A.L., Ir, Valkenkamp 640, 3607 MP Maarssen, Instituut Geotechniek Nederland BV, Hardinxveld Giessendam
Ramadhin, D., Ir, Jean Sibeliusstraat 27, 3069 MJ Rotterdam, Holland Railconsult, Utrecht
Ramaekers, J.J.F., Dr Ir, Raar 25, 6231 RP Meerssen, Nederlands Instituut voor Toegepaste Geowetenschappen TNO, Haarlem
Raming, M.P., Ir, Voetboogstraat 12, 1012 XL Amsterdam, Arcadis Bouw/Infra, 's-Gravenhage

Ramler, J.P.G., Ir, Vijverlaan 49, 1851 ZV Heiloo, Bam Infrabouw, Bunnik
Ratingen, R.C.C. van, Ir, Roerzicht 17, 6041 XV Roermond, Ingenieursbureau R.F.W. van Ratingen BV, Roermond
Ree, H.J. van, Ir, Peelstraat 139, 1079 RN Amsterdam, Witteveen + Bos Raadgevende Ingenieurs BV, Deventer
Ree, P.J. van der, Ir, Bilderdijklaan 42, 3743 HS Baarn
Reincke, H.J.J., Ir, Cor Bruijnweg 8, 1521 MB Wormerveer, Reincke Beheer BV, Wormerveer
Remmelts, W.R., Ir, Westhaven 18, 2801 PH Gouda, Interbeton BV, Rijswijk Zh
Ridder, C.A.H.H. de, Mrs Ir, Top Naeffstraat 19, 4207 MT Gorinchem, Dakmerk, Nieuwegein
Rigter, B.P., Ir, Professor Gerbrandylaan 19, 3445 CW Woerden, Rijkswaterstaat Bouwdienst, Utrecht
Ritt, H.P., Ir, Grubbenweg 2, 6343 CB Klimmen, Ritt Consult-Geo-en Funderingstechniek, Klimmen
Roelands, P.A.A., Ir, Pruimengaarde 7, 3992 JK Houten, NS Railinfrabeheer, Utrecht
Roep, V.D.L., B.Sc., Vijf Meilaan 79, 2321 RJ Leiden, Deloitte & Touche ICT Services, Amsterdam
Roesink, A.A.W., Ir, Catharijnesingel 100-DE, 3511 GV Utrecht, Berenschot-Osborne BV, Utrecht
Roetink, J.H., Ir, Goethelaan 22, 3533 VS Utrecht, Ingenieursbureau J.H. Roetink, Utrecht
Rol, A.H., Ir, Ter Wadding 3, 2253 LW Voorschoten, Holland Railconsult, Utrecht
Roo, M.E. de, Ir, Fichtestraat 25, 3076 RA Rotterdam, Kajima Europe BV, Amstelveen
Roosdorp, W.B.M., Ir, Mozartlaan 20, 4837 EK Breda, NV Koninklijke Nederlandsche Petroleum Maatschappij, 's-Gravenhage
Rosendahl, C., Ir, c/o Postbus 1149, 2260 BC Leidschendam, Fugro Sdn Bhd, Bandar Seri Begawan, BRUNEI
Rövekamp, N.H., Ir, Groenedijk 31, 3311 DB Dordrecht, HBW Hollandsche Beton- en Waterbouw BV, Gouda
Royen, S.H. van, MBA, Ir, Parkstraat 34, 3581 PL Utrecht, ProCap Projectmanagement BV, Utrecht
Ruinen, R.M., Ir, Hoflaan 55, 2926 RC Krimpen aan den IJssel, Vryhof Anchors BV, Krimpen aan den IJssel
Ruitenburg, C. van, Ir, Schumannlaan 23, 3055 HS Rotterdam, Adviesbureau voor Bouwconstructies Ir, C.V. Ruitenburg BV, Rotterdam
Salazar Rivera, J.R., M.Sc., Merellaan 227, 2903 GE Capelle aan den IJssel, Gemeentewerken Rotterdam, Rotterdam
Sanden, P.G.F.J. van der, Ir, Wandelhei 20, 5685 GM Best, Holland Railconsult, Utrecht
Sas, P.C., Ir, Van Neckstraat 12, 2597 SG 's-Gravenhage, Besix Nederland BV, Breda
Scheele, J.J., Ir, Albertus Perkstraat 79, 1217 NP Hilversum, Boskalis Baggermaatschappij, Papendrecht
Schellekens, J.P., Ir, Asterdkraag 68, 4823 GB Breda
Schellingerhout, A.J.G., Dr Ir, Torenstraat 14, 2821 BJ Stolwijk, IFCO Funderingsexpertise BV, Waddinxveen
Schokking, F., Dr Ir, Bentveldsweg 102-A, 2111 ED Aerdenhout, GeoConsult BV, Haarlem
Scholten, N.P.M., Ir, Waalsteeg 7, 1011 ER Amsterdam, TNO Bouw, Delft
Schothorst, K.H., Ir, Nedreveien 16, 4018 Stavanger, Norway, Mobil Exploration Norway Inc., Stavanger, NORWAY
Schouten, C.P., Ir, Laan van Klarenbeek 57, 6824 JM Arnhem, Arcadis Bouw/Infra, Amersfoort
Schram, A.R., Ir, Chico Mendesring 35, 3315 RL Dordrecht, De Weger Architecten- en Ingenieursbureau BV, Rotterdam
Schreuders, C., Ir, Meijlingsgaarde 9, 7622 HC Borne, Schreuders Bouwtechniek BV, Hengelo Ov
Schrieck, G.L.M. van der, Ir, Burgemeester den Texlaan 43, 2111 CC Aerdenhout, HAM, Hollandsche Aanneming Maatschappij BV, Rotterdam
Schunk, M.P., Ir, Kasteel Bleienbeekstraat 1-C, 6222 XG Maastricht, Fugro Ingenieursbureau BV, Maastricht
Schweig, J.C., Ir, Koppellaan 2, 3721 PE Bilthoven, Rijkswaterstaat Directie Utrecht, Nieuwegein
Segaar, M.J., Ir, Ganimedesstraat 6, 1562 ZL Krommenie, Hogeschool Alkmaar, Alkmaar

Sies, E.M., Ir, Goeverneurlaan 136, 2523 BN 's-Gravenhage, Rijkswaterstaat Bouwdienst, Utrecht

Sip, J.W., Ir, Zwartewater 11, 6741 EN Lunteren, DHV Milieu en Infrastructuur BV, Amersfoort

Six, G.J.P., Ing, De Scheifelaar 204, 5463 HX Veghel, Raadgevend Ingenieursbureau Inpijn-Blokpoel BV, Son

Slaa, G. te, Ir, Kastanjelaan 13, 6584 CN Molenhoek, Koninklijke Haskoning Groep BV, Nijmegen

Sluimer, G., Ir, Van Boetzelaerstraat 65, 2406 BD Alphen aan den Rijn, Ingenieursbureau Amsterdam, Amsterdam Zuidoost

Sluis, H.R. van der, Ir, Hof van Azuur 44, 2614 TB Delft, CIHR Adviezen BV, Delft

Sluis, J. van der, Ing, Menkemaborgstraat 34, 3223 WD Hellevoetsluis, Mos Grondmechanica BV, Rhoon

Smit, S.B., Ir, Oostdijk 25, 3077 CP Rotterdam, Koninklijke Boskalis Westminster NV, Papendrecht

Smit, M.G., Ir, Wickenburgstraat 138, 3077 TL Rotterdam, SIMTECH Automatisering BV, Rotterdam

Smit, F., Ir, Middelstebaan 34, 5263 EW Vught, Visser & Smit Hanab BV, Papendrecht

Smits, E.Th.J.H., Ir, Mecklenburg 38, 2171 DV Sassenheim, Fugro Ingenieursbureau BV, Leidschendam

Smulders, F.M.G., Ir, Lissevenlaan 7, 5582 KB Waalre, Jonckheere Bus & Coach NV, Roeselare, BELGIUM

Snijders, L.E.M.J., Ir, Postbus 292, 6430 AG Hoensbroek, Geoconsult Geotechniek BV, Hoensbroek

Snip, D.W., Ir, Sabangstraat 14, 2612 BK Delft, HAM, Hollandsche Aanneming Maatschappij BV, Rotterdam

Soehoed, A.R., Ir, 3 Jalan Maluku, Jakarta 10350, Indonesia, Pt Puri Fadjar Mandiri, Jakarta, INDONESIA

Sombroek, J.W.M., Ir, Oosterzijweg 46, 1851 PC Heiloo, De Weger Architecten- en Ingenieursbureau BV, Rotterdam

Sonke, E.J., Ir, Lemmerhengst 4, 4617 GL Bergen op Zoom, Bouwdienst Rijkswaterstaat Tilburg, Tilburg

Spelt, C., Drs Ir, Lester Pearsonweg 14, 3731 CD De Bilt, Elsenburg Beleidsontwikkeling & Projectmanagement, De Bilt

Spiekhout, J., Ir, Hoofdweg 115-C, 9761 ED Eelde, NV Nederlandse Gasunie, Groningen

Splunder, T.H.W. van, Ir, Prinsenhof 36, 3481 HB Harmelen, Strukton Groep NV, Maarssen

Stam, J.L.P., Ir, Leidseweg 395, 2253 JD Voorschoten, OMEGAM, Amsterdam

Stam, T., Ir, Margriethof 27-29, 3355 EJ Papendrecht, Bouwtechnisch Adviesbureau Ir. T. Stam, Papendrecht

Staveren, J.W.B. van, Ir, Grevelingen 82, 2401 DP Alphen aan den Rijn, Ten Hagen en Stam BV, 's-Gravenhage

Steenbergen-Kajabová, J., Mrs Ir, Satijnvlinder 8, 3723 RG Bilthoven, Grontmij Advies-& Techniek BV Houten, Houten

Steenwijk, W. van, Ir, Zwaardvegersstraat 5, 9646 CJ Veendam, Scheldebouw BV, Veendam

Stienstra, P., Dr, Mijnbouwplein 3, 2628 RT Delft, Zuiveringsschap Hollandse Eilanden en Waarden, Dordrecht

Stijn, M.G.E. van, Ir, Geistingen, Letterveld 26, 3640 Kinrooi, Belgium, Boormaatschappij Maasland NV, Maaseik, BELGIUM

Stoelhorst, P.A.G., Ir, Maandenweg 50, 1335 KR Almere, Handelmaatschappij Gooimeer BV, Almere

Stoevelaar, R., Ir, Westerweg 74, 1906 EH Limmen, Geo-Delft, Delft

Stuit, H.G., Dr Ir, Maria Rutgerslaan 93, 2135 PB Hoofddorp, Holland Railconsult, Utrecht

Stuurman, E.H., Ir, Burgemeester de Zeeuwstr 48, 3281 AJ Numansdorp, Ingenieursbureau Stuurman-Numansdorp BV, Numansdorp

Suiker, A.S.J., Ir, Doctor Schaepmanstraat 43, 2612 PJ Delft, TUD, Lucht- en Ruimtevaarttechniek, Delft

Taat, J., Ir, Emmastraat 72, 2595 EK 's-Gravenhage, Geo-Delft, Delft

Taselaar, F.M., Eur Ing Drs, Milletstraat 56-1HOOG, 1077 ZG Amsterdam, Adviesbureau Taselaar, Amsterdam

Tegelberg, E.R., Ir, Prof. Lorentzlaan 101, 3707 HC Zeist, Strukton Railinfra Services, Dordrecht

Temmink, H.E.M., Ir, Seinwachterstraat 59, 1019 TD Amsterdam, Ingenieursbureau De Koster & Van Brussel, 's-Gravenhage

Termaat, R.J., Ing, Kerklaan 506, 2903 HK Capelle aan den IJssel, Rijkswaterstaat Dienst Weg- en Waterbouwkunde, Delft

Terpstra, J.D., Ir, Paltrokmolen 10, 2906 SE Capelle aan den IJssel, De Weger Architecten- en Ingenieursbureau BV, Rotterdam

Teunissen, E.A.H., Ir, Mozartstraat 100, 7391 XM Twello, Witteveen + Bos Raadgevende Ingenieurs BV, Deventer

Thijs, J.M.J.F., Ir, Marie Koenenlaan 5, 5044 ND Tilburg, Iv-Consult BV, Papendrecht

Thoenes, P.D., Drs, Boven Zevenwouden 5, 3524 CK Utrecht, Nederzand BV, Beuningen GLD

Thorborg, B.B.W., Ir, De Meij V Streefkerkstr 73, 2313 JM Leiden, Ministerie Verkeer en Waterstaat, Directie Telecom en Post, 's-Gravenhage

Tiemessen, N.T.M., Ir, Henry Dunantlaan 18, 3862 XD Nijkerk, Grontmij Verkeer & Infrastructuur, De Bilt

Timmerman, J.I.H., Ir, De Bruynestraat 31-B, 2597 RD 's-Gravenhage, SGS Redwood Nederland BV, Spijkenisse

Timmerman, J., Ir, Koning Davidlaan 26, 6564 AD Heilig Landstichting, Koninklijke Haskoning Groep BV, Nijmegen

Tjaden, J.H., Ir, Arnoldlaan 1, 2061 AL Bloemendaal, BV Adviesbureau TJaden voor Technisch Bodemonderzoek, Haarlem

Tol, F.A. van, Ir, Beijerdstraat 2, 4112 NE Beusichem, Managementgroep Betuweroute (NS-RIB)

Tongeren, H. van, Prof Ir, Sophialaan 52, 3743 CX Baarn, Ballast Nedam Engineering BV, Amstelveen

Tonnisen, J.Y., Rembrandt van Rijnlaan 24, 2343 SV Oegstgeest, Ballast Nedam Engineering BV, Amstelveen

Toppler, J.F., Helmlaan 7, 2244 AZ Wassenaar, Ingenieursbureau Toppler, Wassenaar

Turkstra, G., Ir, Coebelweg 69, 2324 KX Leiden, Koninklijke Boskalis Westminster NV, Papendrecht

Twillert, J.A. van, Ir, Aspergedreef 1, 4614 HE Bergen op Zoom, GeoDelft, Delft

Uffink, A.C.M., Ir, Floraweg 21, 6542 KA Nijmegen, Hogeschool 's-Hertogenbosch, 's-Hertogenbosch

Uijting, B.G.J., Ir, Schouwtjesplein 3, 2012 KP Haarlem, Railpro BV, Hilversum

Uijttewaal, F.J., Ir, Handelstraat 24, 2651 VE Berkel en Rodenrijs, GeoDelft, Delft

Unger, M.K., Ir, Diakenhuisweg 97, 2033 AP Haarlem, Toornend & Partners, Haarlem

Vambersky, J.N.J.A., Prof Dipl-Ing, Rosa Spierstraat 8, 2642 BZ Pijnacker, Corsmit Raadgevend Ingenieursbureau BV, Rijswijk Zh

Van, M.A., Dr Ir, Hof van Zilverlicht 8, 2614 TW Delft, Geo-Delft, Delft

van de Griend, A.A. van de, Ir, Winstongaarde 26, 3824 BZ Amersfoort, Tauw BV, Utrecht

van de Velde, A.H. van de, Ir, Stadionkade 138-2, 1076 BT Amsterdam, Ingenieursbureau Oranjewoud BV, Almere

van de Waal, W.W. van de, Ir, Borgesiusstraat 7-A, 3038 TA Rotterdam, Geo Tech, Rotterdam

van de Wouw, A.W.J. van de, Ir, Biezelaar 4, 5121 NR Rijen, Bouwtechnisch Adviesburo Ir, W.A. van Boxsel BV, Oosterhout Nb

Veer, A., Ir, Ecodusweg 33, 2614 WS Delft, BKH Adviesbureau, Delft

Vegt, E., Ir, Schuilenburg 14, 3328 CH Dordrecht, Tebodin BV Consultants & Engineers, Eindhoven

Vellekoop, L., Ir, Koerillen 6, 2904 VL Capelle aan den IJssel, Koninklijke Boskalis Westminster NV, Papendrecht

Velthorst, H.W., Ir, Bloemkeshof 98, 5301 WT Zaltbommel, Dywidag-Systems International BV, Zaltbommel

Venmans, A.A.M., Ir, St. Josephstraat 4, 4847 SE Teteringen, Rijkswaterstaat Dienst Weg- en Waterbouwkunde, Delft

Vergeer, G.J.H., Ir, Berlageplan 28, 2728 EG Zoetermeer, Ingenieursbureau Oranjewoud BV, Rotterdam

Verhaar, D.W., Ing, Dorpszicht 1, 1391 LW Abcoude

Vermeer, Th W.M., Ir, Pottenbakkerstraat 42, 2984 AX Ridderkerk, EMN BV, Ridderkerk

Verruijt, A., Prof Dr Ir, Tweede Stationsstraat 229, 2718 AB Zoetermeer, TUD Communicatie en Marketing Groep, Delft

Verstappen, J.K.J., Ir, Boldert 10, 5667 AT Geldrop, JV2 Bouwadvies BV/JV2 Staaladvies BV, Nuenen

Verwijs, W., Ir, Appelgaard 37, 6662 HL Elst, Feenstra Architekten en Ingenieurs, Arnhem

Viergever, M.A., Ir, Zoom 16-02, 8225 KK Lelystad, Geo-Delft, Delft

Viersma, R.H., Ir, Korte Dreef 28, 9752 JN Haren, HAM Bagger- en Waterbouwkundige Werken, Rotterdam

Vinkenvleugel, J.J.M., Ir, De Meeren 48, 4761 SL Zevenbergen, Gemeente Dordrecht, Dordrecht

Vinks, T.J.N., Ing, Aalbespad 6, 2995 TA Heerjansdam, Dirk Verstoep BV, Rotterdam

Visser, G.H., Ir, Schoolstraat 49, 2282 RB Rijswijk, Adviesburo Zevenbergen voor Bouw en Waterbouwkunde, Rijswijk Zh

Visser, J.H.M., Mrs Dr, Anna Beijerstraat 25, 2613 DP Delft, TNO Bouw, Delft

Vlagsma, J.J., Ir, Klam 22, 8603 DS Sneek, Ingenieursbureau Wiersma-Vlagsma, Heerenveen

Vlasblom, W.J., Prof Ir, Kerverslaan 1-A, 1911 SH Uitgeest, TUD Faculteit Ontwerp, Constructie & Productie/Maritieme Techniek, Delft

Vlasblom, E., Ir, Weresteyn 3, 3363 BK Sliedrecht, Instituut Geotechniek Nederland BV, Hardinxveld Giessendam

Vollebregt, J.A.A.M., Ir, Thorbeckelaan 34, 2811 CE Reeuwijk, HBM Utiliteitsbouw BV Regio West, Rotterdam

Vonk, J.A., Ir, Dollardlaan 5, 1784 BE Den Helder

Vonk, R.A., Dr, Vossiuslaan 25-40, 2353 BC Leiderdorp, De Weger Architecten- en Ingenieursbureau BV, Rotterdam

Vorm, P. van der, Ir, Charloisse Lagedijk 908, 3088 LA Rotterdam, Gemeentewerken Rotterdam, Rotterdam

Vos, J. de, Ing, Neel Gijsenpad 5, 3043 MB Rotterdam, Geomet BV, Alphen aan den Rijn

Vos, O.M.T., Ir, Jupiterburg 2, 3437 GP Nieuwegein, Strukton Integrale Projecten, Maarssen

Voskamp, W., Ir, Maasoord 27, 3448 BM Woerden, Akzo Nobel Colbond Geosynthetics, Arnhem

Vreeburg, I.C., Ir, Reijerskoop 129, 2771 BH Boskoop, HBW Hollandsche Beton- en Waterbouw BV, Gouda

Vreugdenhil, R., Ir, Benedictastraat 8, 5331 BM Kerkdriel, Geofox BV Adviesbureau voor Bodem, Water & Milieu, Oldenzaal

Vriend, A.C., Ir, Pruylenborg 141, 3332 PC Zwijndrecht, Betonson BV, Son

Vries, A. de, Ir, Driehuizen 19, 2641 LE Pijnacker, Stork Engineers & Contractors BV, Schiedam

Vries, J.T. de, Ir, Boomstede 234, 3608 AL Maarssen, Rijkswaterstaat Bouwdienst, Utrecht

Waal, V.J. de, Dr Ir, Achtergracht 35, 1017 WN Amsterdam, Walinco puls- en heibedrijf BV, Amsterdam

Waal, J.C.M. de, Ir, c/o Thorbeckelaan 174, 4708 KS Roosendaal, Shell International Oil Products BV, 's-Gravenhage

Waverijn, C.G., Ir, Borchsatelaan 5, 3055 ZB Rotterdam, Gemeentewerken Rotterdam, Rotterdam

Weele, A.F. van, Ir, Woubrechterf 40, 2743 HM Waddinxveen, IFCO Funderingsexpertise BV, Waddinxveen

Weele, A.F. van, Prof Ir, Hofstede 12, 2821 VX Stolwijk, IFCO Funderingsexpertise BV, Waddinxveen

Weeren, C. van, Prof Ir, Lange Geer 6, 2611 PV Delft, TUD Faculteit Bouwkunde/Bouwkunde, Delft

Weesep, B. van, Ing, Verduinsbos 36, 4661 NW Halsteren, Fundamentum BV, Sliedrecht

Weijers, J.B.A., Ir, Lange Hille 3, 3261 TK Oud Beijerland, Rijkswaterstaat Dienst Weg- en Waterbouwkunde, Delft

Wendrich, J.H., Ir, Werdorperwaard 22, 3984 PR Odijk, Adviesbureau D3BN Civiel Ingenieurs, Rotterdam

Wentink, J.J., Ing, Betje Wolfflaan 2, 3723 DL Bilthoven, GeoDelft, Delft

Wesselius, P.M., Ir, IJsselstraat 21-B, 9725 GA Groningen, NV Nederlandse Gasunie, Groningen

Wessels, H.B.J., Ir, Amelterhout 120, 9403 EJ Assen, Christelijke Hogeschool Windesheim, Zwolle

Westerveld, F.M., Ir, Abraham Crijnssenstraat 27, 1411 DV Naarden, Ballast Nedam Beton en Waterbouw BV, Amstelveen

Weytingh, K.R., Ir, Nieuwe Markt 29, 7411 PC Deventer, Ingenieursbureau Oranjewoud BV, Heerenveen

Wielenga, J.C., Ir, De Kringloop 77, 2614 WJ Delft, KIWA Management Consultants, Rijswijk Zh

Wijnen, P.A. van, Ir, De Kunning 14, 5258 KR Berlicum, Hogeschool 's-Hertogenbosch, 's-Hertogenbosch

Wijngaarden, H. van, Ing, Meeuwenlaan 24, 3411 BE Lopik, Milieudienst Zuidoost Utrecht, Zeist

Wijtzes, J., Dr Ir, c/o Q-Consult, Adviesbureau, Populierendreef 874, 2272 HS Voorburg, Q-Consult Adviesbureau voor Bodem en Water, Voorburg

Wilschut, D., Ir, Burg D Josselin D Jongln 4, 3042 NH Rotterdam, Gemeentewerken Rotterdam, Rotterdam

Wiltjer, R.H., Ir, Adrianalaan 184, 3053 MH Rotterdam, Adviesbureau D3BN Civiel Ingenieurs, 's-Gravenhage

Wingerden, C.R. van, Ir, Bieslookveld 5, 3124 VB Schiedam, Gemeentewerken Rotterdam, Rotterdam

Winkel, T.H., Buiter, Ir, Henri Dunantstraat 11, 4273 EX Hank, Adviesbureau Winkel Buiter, Hank

Wit, J. de, Ing, Wilde Wingerdlaan 13, 2803 VR Gouda, J. de Wit Advies-& Expertise BV, Gouda

Wit, J.C.W.M. de, Ir, Korte Nieuwstraat 2, 5161 GP Sprang-Capelle, De Weger Architecten- en Ingenieursbureau BV, Rotterdam

Wolfs, P.P.A., Ir, Voorstraat 26, 2611 JR Delft, Visser & Smit Bouw BV, Papendrecht

Wolfs, H.G.J.W.A., Ir, Minahasastraat 6, 2612 GM Delft, Fugro Ingenieursbureau BV, Leidschendam

Wolk, M. van der, Ir, van Bleyswijckstraat 58, 2613 RT Delft, Fugro Engineers BV, Leidschendam

Wolters, W., Dr Ir, Prunusstraat 5, 6706 CA Wageningen, International Institute For Land Reclamation and Improvement ILRI, Wageningen

Woudenberg, J.W. van, Ir, De Blouwel 1, 4761 XX Zevenbergen, NBM Beton- en Industriebouw, Breda

Zandbergen, H. van, Ir, Trintel 109, 1141 DM Monnickendam, Ingenieursbureau Amsterdam, Amsterdam Zuidoost

Zandwijk, C. van, Ir, Herbarenerf 13, 2743 HA Waddinxveen, Heerema Marine Contractors V.O.F, Leiden

Zanten, D.C. van, Ir, Colenbranderstraat 5, 3221 BD Hellevoetsluis, Gemeentewerken Rotterdam, Rotterdam

Zegers, J.G.W., Ir, Hof van Berle 8, 5258 CD Berlicum, De Weger Architecten- en Ingenieursbureau BV, Rotterdam

Zelfde, W.D.A. van 't, Ir, Gershwinstraat 14, 2807 SG Gouda, De Weger Architecten- en Ingenieursbureau BV, Rotterdam

Zigterman, W., Ir, Balistraat 21, 2612 ED Delft, ITC-Delft/Enschede, Delft

Zijl, F. van, Ir, Jos Lussenburglaan 40, 8072 HX Nunspeet, Gamay Gelro Flevo Aanneembedrijf BV, Zutphen

Zijlmans, R.F.G.M., Ir, Brikkenoven 5, 6247 BG Gronsveld, Internationale Maascommissie, Liège, BELGIUM

Zonderland, H.E., Ir, Pegasus 29, 8531 MZ Lemmer, Ingenieursbureau Oranjewoud BV, Heerenveen

Zoun, J.M.C., Ir, Winterpad 8, 2614 WE Delft, Mos Grondmechanica BV, Rhoon

Zuidberg, H.M., Ing, Koningin Wilhelminalaan 2, 2231 VV Rijnsburg, Fugro Engineers BV, Leidschendam

Zwaag, G.L. van der, Ir, Tollenaersingel 11, 2352 JL Leiderdorp, Fugro Engineers BV, Leidschendam

Zwart, A.A., Ir, Hilhorstpoort 10, 3833 HT Leusden, Bouw- en Aannemingsmaatschappij Van Zwol BV, Amersfoort

Zweers, W., Ir, De Grippen 1212, 6605 TB Wijchen, Van Looy Group BV, 's-Hertogenbosch

Zwol, B. van, Ir, Vlamingstraat 35, 2611 KS Delft, TUD Faculteit Ontwerp, Constructie & Productie/Maritieme Techniek, Delft

NEW ZEALAND/NOUVELLE ZEALANDAE

Secretary: Mrs D.L. Fellows, Management Secretary, New Zealand Geotechnical Society, P.O. Box 12 241, Wellington, New Zealand. T: +64 9 817 7759, F: +64 9 817 7035, E: dfellows@xtra.co.nz

Total number of members: 298

Abdulla, K.M.N., Mr, 1/12 Waipuna Road, Mt. Wellington, Auckland, 1006

Adams, B.M., Mr, URS New Zealand Limited, P.O. Box 4479, Christchurch, 8001

Ahmed-Zeki, A., Mr, 24 Davey Place, South Hobart, Tasmania, 7004, Australia

Al-Alusi, N.M.M., Ms, MWH NZ Ltd., P.O. Box 13249, Christchurch, 8020

Alexander, G.J., Mr, 86 Croft Lane, Coatesville, RD 3, Albany

Ampualam, K.Y., Mr, Kompleks Banjar Wijaya, Block A8/7, JL Raya Cipondoh, Tangerang 15148, INDONESIA

Anderson, K.R., Mr, Singlair Knight Merz Ltd., P.O. Box 9806, Newmarket, Auckland, 1031

Anderson, S.J., Mr, 22 Nui Mana Place, Te Atatu South, Auckland

Anwar, M.S., Mr, 2/6 Ambrico Place, New Lynn, Auckland, 1007

Avdjiev, H.G., Mr, 2/34 Heathcote Road, Milford, Auckland, 1309

Bailey, R., Dr, 17B Jasmine Grove, Maungaraki, Lower Hutt

Barker, P.R., Mr, 19 Washington Avenue, Brooklyn, Wellington, 6002

Bartlett, P.E., Mr, Caltex New Zealand Ltd., P.O. Box 31246, Lower Hutt, 6315

Bartley, F.G., Mr, 1 George Laurenson Lane, Mt Roskill, Auckland, 1004

Bauld, C.J., Mr, 22 First Avenue, Kingsland, Auckland, 1003

Baunton, P.H., Mr, 45 Matahiwi Road, RD 2, Te Puna, Tauranga, 3021

Beetham, R.D., Mr, IGNS Ltd., P.O. Box 30368, Lower Hutt, 6315

Berrill, J.B., Mr, C/o School of Engineering, University of Canterbury, Private Bag 4800, Christchurch, 8030

Bevin, J.E., Mr, 3 Karami Crescent, Matamata, 1463

Billings, L.J., Mr, 1 Berescourt Place, Mount Maunganui, 3002

Bindra, M.S., Mr, Permathene Limited, Civil Engineering Division, P.O. Box 71015 Avondale, Auckland, 1230

Black, B.A., Mr, Riley Consultants Limited, P.O. Box 100253, North Shore Mail Centre, Auckland, 1333

Blakeley, G.L., Mr, 29 St Albans Avenue, Mt Eden, Auckland, 1003

Blakeley, J.P., Mr, 3/24 Sylvia Road, St Heliers, Auckland, 1005

Bloxam, M.J., Mr, Bloxam Burnett & Olliver Ltd., P.O. Box 9041, Hamilton, 2015

Borlase, O.M., Mr, Borlase & Associates Ltd., P.O. Box 8111, Dunedin, 9030

Bosselmann, P.B.C., Mr, Foundation Engineering Limited, P.O. Box 74549, Market Road, Auckland, 1130

Boswell, G.B., Mr, 18A Sonia Avenue, Remuera, Auckland, 1003

Brabhaharan, P., Mr, Opus International Conslts Ltd., P.O. Box 12003, Wellington, 6015

Bramley, A.M., Mr, Professional Pursuits Ltd., P.O. Box 12111, Napier, 4015

Brockliss, C.C., Mr, P.O. Box 12536, Penrose, Auckland, 1006

Brown, B.J.B., Mr, Fraser Thomas Limited, P.O. Box 23273, Papatoetoe, Auckland, 1030

Browne, A.A., Mr, Opus International Conslts Ltd., P.O. Box 5848, Auckland, 1030

Burr, J.P., Dr, Tonkin & Taylor Limited, P.O. Box 12152, Wellington, 6038

Burt, M.J., Mr, Beca Carter Hollings & Ferner, P.O. Box 6345, Auckland, 1036

Cameron, B.J., Mr, 15 Sunny Downs Drive, Tauranga, 3001

Carter, D.P., Dr, Beca Carter Hollings & Ferner, P.O. Box 6345, Auckland, 1030

Carter, R.P., Sir, Beca Group Limited, P.O. Box 6345, Auckland, 1030

Chand, V., Mr, 55 Burswood Drive, Pakuranga, Auckland, 1706

Chapman, J.J., Mr, P.O. Box 585, New Plymouth, 4615

Cheenikal, L.A., Mr, Tonkin & Taylor Limited, P.O. Box 5271, Wellesley Street, Auckland, 1030

Cheung, K.C., Dr, Peters and Cheung Ltd., P.O. Box 251154, Pakuranga, 1730

Chisholm, B.I., Mr, Hanlon & Partners, 219 High Street, Dunedin, 9001

Chisnall, J.W.S., Mr, CSP Pacific, P.O. Box 33 268, Petone, 6340

Christensen, S.A., Mr, 3 Wendover Street, Bishopdale, Christchurch, 8005

Clark, C.R., Mr, 7 Ridvan Grove, Ngaio, Wellington, 6004

Clark, W.D.C., Mr, 92 Hill Road, Belmont, Lower Hutt, 6009

Clayton, P.J., Mr, 173 Portland Road, Remuera, Auckland, 1005

Cobb, R.R., Mr, 1 Waimarie Road, Whenuapai Village, Auckland, 1008

Cocks, J.H.C., Mr, Montgomery Watson NZ Ltd., P.O. Box 4, Dunedin, 9015

Coleman, D.N., Mr, P.O. Box 46081, Herne Bay, Auckland, 1030

Collen, M.N., Mr, 3 Beacon Avenue, Campbells Bay, Auckland, 1310

Convery, D.J., Mr, Soil & Rock Conslts Ltd., P.O. Box 15233, New Lynn, Auckland, 1232

Cook, P.J., Mr, 27 A Anzac Road, Whangarei, 0101

Cooper, K.J., Mr, 26 Hedge Row, Pakuranga, Auckland, 1706

Cope, M.S., Mr, Geotek Services Limited, P.O. Box 39015, Howick, Auckland, 1705

Corlett, R.H., Mr, 30 Earls Road, St Clair Park, Dunedin, 9001

Costello, T.N., Mr, Cook Costello Limited, 19 Lower Tarewa Road, Whangarei, 0101

Couch, R.L., Mr, Opus International Conslts Ltd., Private Bag 6019, Napier, 4020

Coutts, D.R., Mr, 25 Premier Avenue, Pt Chevalier, Auckland, 1002

Cowbourne, A.J., Mr, 3 Waikowhai Road, Hillsborough, Auckland, 1004

Cowley, R.C., Mr, 1a Jesmond Dene, 2 Lithos Road, London NW3 6EG, England

Crawford, S.A., Mr, 24B 13th Avenue, Tauranga, 3001

Croft, S.F., Mr, Tonkin & Taylor Limited, P.O. Box 5271, Wellesley Street, Auckland, 1030

Crook, S.D., Mr, Meritec, P.O. Box 4241, Newmarket, Auckland, 1031

Crow, T.W., Mr, Transit NZ, P.O. Box 1459, Auckland, 1015

Dale, J.T., Dr, P.O. Box 13057, Hamilton, 2030

Davidge, R.J., Mr, 15 Christchurch Avenue, Kelson, Lower Hutt, 6009

Davidson, D.C., Mr, 117 Paihia Road, Onehunga, Auckland, 1006

Dawn, J.D., Mr, 27 A. Speddings Road, Whangarei, 0101

Dennison, D., Mr, Opus International Conslts Ltd., Private Bag 3057, Hamilton, 2030

Diprose, P.L., Mr, Diprose Consultants Ltd., P.O. Box 77038, Mt Albert, Auckland, 1030

Dooley, L.G., Mr, 40 Te Kawa Road, One Tree Hill, Auckland, 1003

Duske, G.C., Mr, 10 Weatherly Road, Torbay, Auckland, 1311

East, G.R.W., Mr, Opus International Consultants, P.O. Box 5848, Auckland, 1030

Edger, N.J., Mr, 38 Sunnyhills Avenue, Hamilton, 2001

Fairclough, A., Mr, 168-11-5 Kiara Park, Condominiums, Jalan Burhanhddin Helmi, Jaman Tun Dr Ismail, 6000 Kuala Lumpur, Malaysia

Fairless, G.J., Dr, 11 Golf Grove, St Andrews, Hamilton, 2001

Farquhar, G.B., Mr, Meritec Ltd., P.O. Box 4241, Newmarket, Auckland, 1015

Fellows, D.L., Ms, 6 Sylvan Valley Avenue, Titirangi, Auckland, 1007

Fendall, H.D.W., Dr, Hugh Fendall Consultants Ltd., 6 A Montel Avenue, Henderson, Auckland, 1008

Finlay, P.T., Mr, 62 Woodside Road, Massey, Auckland, 1008

Fitch, N.R., Mr, Riley Consultants Limited, P.O. Box 100253, North Shore Mail Centre, 1333

Fitchett, P.F.R., Mr, Sigma Consultants Limited, P.O. Box 553, Rotorua, 3215

Flower, C.E., Mr, McBreen Jenkins Const Ltd., P.O. Box 48, Whangarei, 0115

Fong, J.S.W., Mr, 34 Quona Avenue, Mt Roskill, Auckland, 1004

Franklin, J.T., Mr, 3 Stott Avenue, Birkdale, Auckland, 1310

Freer, C.J., Mr, Tonkin & Taylor Limited, P.O. Box 5271, Wellesley Street, Auckland, 1030

Galloway, G.A., Ms, Flat 2, 8 Davis Street, Thorndon, Wellington, 6001

Geddes, P.M.L., Mr, Hawthorn Geddes Limited, P.O. Box 575, Whangarei, 0115

Gibson, R.B., Mr, Cameron Gibson & Wells Ltd., 26 Nile Street, Nelson, 7001

Gilbertson, T.E., Mr, 2 Kimbolton Lane, Broadoaks, Cashmere, Christchurch, 8002

Giles, E.L., Mr, URS New Zealand Limited, P.O. Box 821, Auckland, 1030

Gillard, R.G., Mr, 4 Dawn View Place, RD 6, Minden, Tauranga, 3021

Gillespie, K.H., Mr, 27 Hollyhock Place, Browns Bay, Auckland, 1331

Gillon, M.D., Mr, DamWatch Services Limited, P.O. Box 1549, Wellington, 6015

Glanville, R.C., Mr, 96 Kauri Point Road, Laingholm, Auckland, 1007

Glassey, P.J., Mr, IGNS, Private Bag 1930, Dunedin, 9020

Goldsmith, P.R., Dr, Fraser Thomas Limited, P.O. Box 23273, Papatoetoe, Auckland, 1030

Graham, C.J., Dr, Meritec Limited, P.O. Box 4241, Auckland, 2015

Gray, W.J., Mr, Opus International Conslts Ltd., Private Bag 6019, Napier, 4020

Green, H.R., Mr, Transit New Zealand, P.O. Box 154, Kawakawa, 0291

Gribben, G.W., Mr, Duffill Watts & King Limited, P.O. Box 481, Whangarei, 0115

Gulati, S.K., Mr, Waste Disposal Services, P.O. Box 150, Whitford, Auckland, 1730

Guy, E.D., Mr, 224 Church Street, Onehunga, Auckland, 1006

Hadfield, G.J., Mr, Beca Carter Hollings & Ferner, P.O. Box 6345, Auckland, 1030

Hall, R.C., Mr, BHC Consulting Ltd., P.O. Box 8054, Havelock North, 4230

Hall, R.J., Mr, 78 Beverley Road, Timaru, 8601

Halton, R.P., Mr, 21 Saleyards Road, Puhoi, 1240

Hawley, J.G., Dr, 4 Gordon Craig Place, Algies Bay, Warkworth, 1241

Hedley, M.R., Mr, 7 Wharf Road, Te Atatu Peninsula, Auckland, 1008

Henderson, F.G., Mr, 29 Jasmine Place, Mt Maunganui, 3002

Hendy, I.T., Mr, 2/38 Ayton Drive, Glenfield, Auckland, 1310

Higginbotham Jr, J.W., Mr, 7 Harper Street, Nelson, 7001

Higgins, G., Mr, 27 Stanley Avenue, Palmerston North, 5301

Hill, T.M., Mr, TMH Consulting Limited, P.O. Box 5320, Mt Maunganui, 3030

Holtrigter, M., Mr, Soil Engineering Limited, P.O. Box 104073, Lincoln North, Auckland, 1230

Holyoake, R.J., Mr, Beca Carter Hollings & Ferner, P.O. Box 13960, Christchurch, 8020

Horrey, P.J., Mr, 45 Exeter Street, Lyttleton, 8012

Howell, C.W., Mr, Chris W Howell & Assocs. Ltd., Reg. Consulting Engineers, P.O. Box 100006, North Shore Mail Centre, Auckland 1333

Hughes, J.M.O., Dr, Suite 804 938 Howe Street, Vancouver, British Columbia, CANADA V6Z 1N9

Hughes, M.W., Mr, P.O. Box 231, Tauranga, 3030

Hurley, A.J.C., Mr, Montgomery Watson NZ Ltd., P.O. Box 13249, Christchurch, 8030

Hutchison, R.J., Mr, Keith Gillespie & Associates Ltd., P.O. Box 35876, Browns Bay, Auckland, 1330

Jacka, N.K., Mr, URS New Zealand Limited, P.O. Box 821, Auckland, 1036

Jairaj, V.P., Mr, 66 Sunnynook Road, Forrest Hill, Auckland, 1309

Jamieson, G.H., Mr, Bloxam Burnett & Olliver Ltd., P.O. Box 9041, Hamilton, 2015

Jennings, D.N., Mr, Opus International Conslts Ltd., Private Bag 3057, Hamilton, 2020

Jewell, D.G., Mr, Fletcher Const. Ltd., (Eng), Private Bag 92059, Auckland, 1030

Kay, A.N.P., Mr, Sinclair Knight Merz Limited, P.O. Box 9806, Newmarket, Auckland, 1030

Kayes, T.J., Mr, Tonkin & Taylor Limited, P.O. Box 5271, Wellesley Street, Auckland, 1030

Kell, G.N., Mr, Michael Newby & Associates, P.O. Box 904, Hastings, 4201

Kelsey, P.I., Mr, Earthtech Consulting Limited, P.O. Box 721, Pukekohe, South Auckland, 1730

Kennedy, H.J.F., Mr, 453 Te Moana Road, Waikanae, 6454

Kerslake, F.D., Mr, Tse Taranaki and Associates, P.O. Box 237, New Plymouth, 4615

King, J.A., Mr, 25 Charters Street, Glenross, Dunedin, 9001

King, R.A., Mr, P.O. Box 626, Tauranga, 3030

Kiryakos, M.D., Mr, 10 Holland Road, Hillcrest, Auckland, 1310

Knocker, M.G., Mr, 9 Gerontius Glade, Torbay, North Shore City, 1311

Knowles, R.B., Mr, RB Knowles & Associates Ltd., 150 Lichfield Street, Christchurch, 8001

Knowles, R.J., Mr, Foundation Engineering Limited, P.O. Box 74549, Market Road, Auckland, 1130

Ko, H.B.J., Mr, 76 Gosford Drive, Howick, Auckland, 1705

Korte, N., Mr, 58 Shetland Street, Titirangi, Auckland, 1007

Kortegast, A.P., Mr, Tonkin & Taylor Limited, P.O. Box 5271, Wellesley Street, Auckland, 1030

Kumar, P., Mr, PK Engineering Ltd., P.O. Box 464, Kerikeri, 0470

Land, J.W., Mr, 46 Lloyd Avenue, Mt Albert, Auckland, 1003

Lander, S.G., Mr, Meritec Ltd., P.O. Box 4241, Newmarket, Auckland, 1036

Langbein, A.C., Mr, 11 John Stokes Terrace, Newmarket, Auckland, 1001

Lapish, E.B., Mr, Lapish Enterprises Limited, 84 Stamford Park Road, Hillsborough, Auckland, 1004

Larkin, T.J., Dr, Civil Engineering Department, University of Auckland, Private Bag 92019, Auckland, 1030

Laws, M.J., Mr, 155 Paritai Drive, Orakei, Auckland, 1005

Lee, C.K., Ms, C/-Bushsider, RDI, Ashburton, 8300

Lewis, W.R., Mr, Lewis & Barrow Limited, P.O. Box 13282, Christchurch, 8030

Lim, S.H., Mr, 53 Woodbury Street, Avonhead, Christchurch, 8004

Linton, A.J., Mr, 19 Lyren Place, Bucklands Beach, Auckland, 1704

Litherland, J.W., Mr, 10 Liley Place, Remuera, Auckland, 1005

Liu, S.S.F., Mr, 41K Universal Mansion, 52 Hillwood Road, TST, Hong Kong

Loney, G.A., Mr, P.O. Box 37599, Parnell, Auckland, 1033

Lust, E., Mr, P.O. Box 21121, Edgeware, Christchurch, 8030

Luxford, N.S., Mr, Babbage Consultants Limited, P.O. Box 2027, Auckland, 1030

Lyons, C.D., Mr, Tonkin & Taylor Limited, P.O. Box 12152, Wellington, 6015

MacLeod, R.K., Mr, 35 Claremont St, Maori Hill, Dunedin, 9001

Malan, P., Mr, 2/15 Puriri Avenue, Greenlane, Auckland, 1005

Manktelow, C.D., Mr, Cutfield Consulting, P.O. Box 37, Whakatane, 3080

Manning, P.A., Mr, C/O Freeport Indonesia Co, P.O. Box 616, Cairns Qld 4870, AUSTRALIA

Marchant, P.G.D., Mr, 54 A Symonds Street, Royal Oak, Auckland, 1003

Marsh, E.J., Dr, 74 Calgary Street, Balmoral, Auckland, 1003

Martin, M.J., Mr, Babbage Consultants Limited, P.O. Box 2027, Auckland, 1030

Matuschka, T., Dr, Director, Engineering Geology Ltd., P.O. Box 33426, Takapuna, Auckland, 1332

McCahon, I.F., Mr, 29 Norwood Street, Christchurch, 8001

McFadden, D.R., Mr, Montgomery Watson (NZ) Limited, P.O. Box 13 249, Christchurch, 8015

McGuigan, D.M., Mr, 86 Grants Road, Paranui, Christchurch, 8005

McKay, D.R., Mr, 7 Hunter Place, New Plymouth, 4601

McLarin, M.W., Dr, 17 Wairere Road, Lower Hutt, 6009

McLintock, J.G., Mr, JG McLintock & Associates, P.O. Box 65037, Mairangi Bay, Auckland, 1330

McManus, K.J., Dr, Dept of Civil Engineering, University of Canterbury, Private Bag 4800, Christchurch, 8030

McNaughton, H.M., Mr, 23 Marshall Laing Ave, Mt Roskill, Auckland, 1004

McNulty, D.W., Mr, 70 Sale Street, Hokitika, 7900

McPherson, I.D., Mr, Connell Wagner Ltd., P.O. Box 1591, Wellington, 6015

McShane, J.B., Mr, 5 Oakfield Avenue, Mt Albert, Auckland, 1003

Melville-Smith, R.W., Mr, Foundation Engineering Limited, P.O. Box 74549, Market Road, Auckland, 1130

Millar, P.J., Mr, Tonkin & Taylor Limited, P.O. Box 5271, Wellesley Street, Auckland, 1030

Mitchell, M.T., Mr, P.O. Box 9123, Hamilton, 2031

Monk, R.B., Mr, Sigma Consultants Limited, P.O. Box 553, Rotorua, 3215

Moody, K.E., Mr, 12 Varlene Terrace, Forrest Hill, North Shore City, 1309

Moon, S.K., Mr, 59 A Deep Creek Road, Torbay, Auckland, 1311

Moon, V.G., Dr, 43 Edinburgh Road, Hillcrest, Hamilton, 2001

Moss, P.J., Dr, Dept of Civil Engineering, University of Canterbury, Private Bag 4800, Christchurch, 8030

Murashev, A., Dr, BCHF Ltd., P.O. Box 3942, Wellington, 6015

Murray, J.G., Mr, Sinclair Knight Merz Limited, P.O. Box 9806, Newmarket, Auckland, 1031

Mutton, R.B., Mr, Montgomery Watson NZ Ltd., P.O. Box 3455, Richmond, Nelson, 7031

Napier, D.G., Mr, David Napier & Assoc Ltd., P.O. Box 7197, Palmerston North, 5315

Nelson, A.H., Mr, EarthTech Consulting Ltd., P.O. Box 721, Pukekohe, 1730

Newby, G., Mr, 98 A Astley Avenue, New Lynn, Auckland, 1007

Newton, C.J., Mr, Meritec Limited, P.O. Box 4241, Auckland, 1030

Norfolk, P.D., Mr, Tonkin & Taylor Limited, P.O. Box 5271, Wellesley Street, Auckland, 1030

Norrie, J.A., Mr, Cheal Consultants, P.O. Box 165, Taupo, 2730

Northey, R.D., Dr, 127 Knights Road, Lower Hutt, 6009

O'Brien, M., Mr, 42 No Street, Wellington, 6001

O'Halloran, M., Dr, P.O. Box 11098, Papamoa Domain, Tauranga, 3030

Okada, W., Mr, 7/21 Kipling Avenue, Epsom, Auckland, 1003

O'Rourke, K., Mr, 9 Welsh Place, Richmond, Nelson, 7002

Olsen, A.J., Mr, Meritec Ltd., P.O. Box 4241, Auckland, 1030

Osborne, R.M., Mr, 31 Pitoitoi Avenue, Greenhithe, Auckland, 1008

Palmer, S.J., Mr, 41 Kano Street, Karori, Wellington, 6005

Park, R.J.T., Dr, 36 Cranwell Street, Churton Park, Wellington, 6004

Pender, M.J., Prof. Dept. Civil & Resource Eng., University of Auckland, Private Bag 92019, Auckland, 1030

Peploe, R.J., Dr, 12 A Appleyard Crescent, Meadowbank, Auckland, 1005

Pere, V.H., Mr, 254 Blenheim Road, Riccarton, Christchurch, 8004

Perry, R.C., Mr, 3 Hutchinson Cres, Durie Hill, Wanganui, 5001

Petrie, S.A., Mr, 24 Michie Street, Belleknowes, Dunedin, 9001

Pickens, G.A., Mr, 28 A Glover Road, Auckland, 1001

Pickford, A.J., Mr, 34 Ford Road, Opawa, Christchurch, 8002

Pike, T.J., Mr, Richardson Stevens Consultants, 2 Seaview Road, Whangarei, 0101

Plutecki, J.W., Mr, Plutecki Consulting Group Ltd., P.O. Box 186, Hamilton, 2015

Pranjoto, S., Mr, 2 Orton Street, Glenfield, Auckland, 1310

Preston, D.R., Mr, 11 Taylors Ave, Christchurch, 8001

Preston, M.J., Mr, Civil Engineering Services, P.O. Box 240, Te Aroha, 2971

Price, W.C., Mr, 396 Riddell Road, Glendowie, Auckland, 1005

Proffitt, G.T., Dr, Pattle Delamore Partners Ltd., P.O. Box 6136, Wellington, 6015

Ranchhod, S., Mr, 16 Dallas Place, New Windsor, Auckland, 1007

Rayudu, D.N.P., Mr, Unit 21, 72 Kitchener Road, Milford, Auckland

Richards, L.R., Dr, Rock Eng., Consultant, The Old School House, Swamp Road Ellesmere, R D 5 Christchurch, 8021

Ridgley, C.L., Mr, 8 A Raeben Avenue, Hillcrest, North Shore City, 1310

Ridgley, N.A., Mrs, 8 A Raeben Avenue, North Shore, Auckland, 1001

Roberts, R.W., Mr, 572 Waitomo Valley Road, RD 7, Otorohanga, 2564

Robinson, A.J., Mr, C/o Martin McCaulay, Morton Ltd., P.O. Box 1009, Mt Manganui, 3030

Rogers, N.W., Mr, 25 C Prospect Terrace, Milford, Auckland, 1309

Russell, P.F., Mr, Montgomery Watson NZ Ltd., P.O. Box 3455, Richmond, Nelson, 7031

Salt, G.A., Mr, P.O. Box 1319, Dunedin, 9015

Saul, G.J., Mr, 222 Cockayne Rd, Ngaio, Wellington, 6004

Schubert, D.R., Mr, 39 Rawhitiroa Road, Kohimarama, Auckland, 1005

Scott, C.R., Mr, Riley Consultants Limited, P.O. Box 2281, Christchurch, 8030

Scott, J.W., Mr, 99 Grange Road, Tauranga, 3001

Shaw, G.T., Mr, 325 Hibiscus Coast Hwy, Orewa, 1461

Shores, K.D.T., Mr, 127 Long Drive, St Heliers, Auckland, 1005

Sickling, J.B., Mr, 165 Mauku Road, RD 3, Pukekohe, 1730

Sinclair, M.D., Mr, Eliot Sinclair & Partners Ltd., P.O. Box 4597, Christchurch, 8030

Sinclair, M.R., Mr, c/-98 Park Rise, Campbells Boy, Auckland, 1311

Sinclair, T.J.E., Mr, 158 Victoria Avenue, Remuera, Auckland, 1005

Smith, M.A., Mr, 2 Bowenvale Avenue, Cashmere, Christchurch, 8002

Smith, R.P., Mr, 25 Fishermans Cove, Army Bay, Whangaparaoa, Auckland, 1463

Sole, W.J., Mr, 99 Redoubt Road, Manukau, Auckland, 1001

Soric, D., Mr, Soil and Rock Consultants, P.O. Box 15233, New Lynn, Auckland, 1232

Speight, N.I., Mr, 32 Mappin Place, Birkenhead, Auckland, 1310

Spicer, M.B., Mr, 31 Vanessa Crescent, Glendowie, Auckland, 1005

Spiers, R.J., Mr, 20 Forrester Drive, Welcome Bay, Tauranga, 3001

Stapleton, M.J.D., Mr, Babbage Consultants Limited, P.O. Box 2027, Auckland, 1030

Starke, W., Mr, 133 Astley Avenue, New Lynn, Auckland, 1007

Stevens, G.R., Mr, 24C Charles Dickens Drive, Mellons Bay, Auckland, 1704

Stewart, D.L., Mr, Duffill Watts and Tse Ltd., P.O. Box 6643, Wellington, 6030

Stiles, A.P., Mr, Tonkin & Taylor Limited, 19 Morgan Street, Newmarket, Auckland, 1001

Strayton, G., Mr, 43A Vincent Street, Howick, Auckland, 1705

Strong, R.B., Mr, 10 Seddon Terrace, Newtown, Wellington, 6001

Suchanski, A., Mr, SH6 Pahautane, P.O. Box 57, Punakaiki, West Coast, 7850

Sullivan, R.D., Mr, P.O. Box 21185, Christchurch, 8030

Sutherland, R.D., Mr, Property & Land Mgmt Services, P.O. Box 751, Blenheim, 7315

Swanney, J.J., Mr, 1 Melrose Road, Wandsworth, London SW18 1ND, UK

Syme, W.R., Mr, 59 Palmerston Road, Birkenhead, Auckland, 1310

Symmans, B., Mr, Tonkin & Taylor Limited, P.O. Box 12152, Wellington, 6015

Tate, D.R., Mr, 34 Gavin Street, Ellerslie, Auckland, 1005

Taylor, N.A., Mr, Verstoep & Taylor Limited, P.O. Box 69215, Glendene, Auckland, 1230

Taylor, D.K., Mr, 25 Market Road, Remuera, Auckland, 1005

Taylor, Prof P.W., 15 Melwood Drive, Northwood Park, Warkworth, 1241

Terzaghi, S., Mr, 131 Taylors Road, Mt Albert, Auckland, 1003

Thomas, M.R., Mr, Harrison Grierson Consultants, P.O. Box 1199, Tauranga, 3030

Thomas, N.J., Mr, 8 Travolta Street, Stafford Heights, Brisbane 4053, AUSTRALIA

Thompson, P.J., Mr, 100 S Sunrise Way #153, Palm Springs, CA 92262, United States of America

Thomson, D.M., Mr, 68 Harbour View Road, Northland, Wellington, 6005

Thomson, W.G., Mr, Argo Thomson Limited, P.O. Box 507, Auckland, 1015

Thorburn, R.J., Mr, Thorburn Consultants Limited, P.O. Box 99669, Newmarket, Auckland, 1031

Thorp, Y.F., Miss, 29 Towle Place, Remuera, Auckland, 1005

Toan, D.V., Dr, Beca Carter Hollings & Ferner, P.O. Box 6345, Auckland, 1030

Traylen, N.J., Mr, Geotech Consulting Ltd., 18 Dyers Pass Road, Cashmere, Christchurch, 8002

Twose, G.R., Mr, 70 Gills Road, Bucklands Beach, Auckland, 1311

Tyndall, A.E., Mr, Tyndall & Hanham, Consulting Engineers, P.O. Box 13117, Christchurch, 8030

Van Barneveld, J.H., Mr, Transit New Zealand, P.O. Box 5084, Wellington, 6015

Vaughan, S.A., Mr, Riley Consultants Limited, P.O. Box 100253, North Shore Mail Centre, 1333

Vautier, E.W., Mr, 18 Wairere Road, R D 2, Henderson, Auckland, 1008

Wakeman, G.C., Mr, 701b Beach Road, Browns Bay, Auckland, 1330

Wallace, G.I., Mr, 16 Steyne Avenue, Plimmerton, 6006

Wallis, A.J., Mr, 23 Wolseley Street, Morningside, Auckland, 1003

Watson, N.G., Mr, Meritec Limited, P.O. Box 4241, Auckland, 1030

Watt, A.J., Mr, 375 East Coast Road, Mairangi Bay, Auckland, 1311

Watt, R.J., Mr, Airey Consultants Limited, P.O. Box 177, Pukekohe, 1730

Wells-Green, P.S., Mr, 50 Connaught St, Blockhouse Bay, Auckland, 1007

Wesley, L.D., Dr, Civil Engineering Department, University of Auckland, Private Bag 92019, Auckland, 1030

Wesseldine, M.A., Mr, Belgravia Building Cons Ltd., P.O. Box 28766, Remuera, Auckland, 1136

Wilcox, P.J., Mr, 79 Campbell Street, Karori, Wellington, 6005

Williams, P.W.M., Mr, Harrison Grierson Conslts Ltd., P.O. Box 5760, Wellesley Street, Auckland, 1030

Williams, R.L., Mr, Opus International Conslts Ltd., Private Bag 3057, Hamilton, 2020

Wilshere, D.S., Mr, Secretary – NZSOLD, P.O. Box 12313, Wellington, 6015

Woodward, S.J., Mr, Geotek Services Ltd., P.O. Box 39015, Howick, Auckland, 1730

Yang, L.L., Mr, 42 Littlejohn Street, Hillsborough, Auckland, 1004

Yeoman, P., Mr, 54 Totara St, Christchurch, 8001

Yetton, M.D., Dr R D 1, Chateris Bay, Christchurch, 8021

Yonge, J.C., Mr, P.O. Box 13613, Onehunga, Auckland, 1132

Young, R., Mr, 29 Southwood Court, Wynyatt Street, London EC1 V7HX

NIGERIA

Secretary: Engr. Ronald Aborowa, Nigerian Geotechnical Association, (postal address) P.O. Box 72667, Victoria Island, Lagos.
Secretariat: National Engineering Centre, 1 Engineering Close, Victoria Island, Lagos.
T/F: 234 1 4936988, E: progress@alpha.linkserve.com

Total numbers of members: 32

Aborowa, R.O., Eng., Earth Surveys Ltd., 39 John Olugbo Street, P.O. Box 3191, Ikeja. T: 01 4936968
Aigboduwa, J.O., Eng., Trevi Foundations (NIG) Ltd., Plot 1292 Akin Adesola Street, P.O. Box 70621, Victoria Island, Lagos. T: 01 2622097, 2619383, 2616257, F: 01 2617509
Akinwale, B., Eng., Ove-Arup & Partners Nig., 25 Boyle Street, Lagos. T: 01 2600520 9
Akpokiniovo, R.O., Earthcore Engineering Ltd., 6 Ayanboye Street, Anthony Village, Lagos. T: 01 4972499
Efretuei, O.E., Eng., Geotech. Engineering Services Ltd., 33 Edgerley Road, Calabar, P.O. Box 918, Calabar. T: 087 231514, F: 087 231441
Elumese, P.O., Eng., Epon Associates, 11 Adeyinka Street, Ilupeju Estate, Lagos
Fasehun, E.O.O., Eng., P.O. Box 413, aba, Lagos. T: 01 4964333
Fatokun Olaposi, S., Eng., Trevi Foundations (NIG) Ltd., Plot 1292 Akin Adesola Street, P.O. Box 70621, Victoria Island, Lagos. T: 01 2622097, 2619383, 2616257, F: 01 2617509
Folayan, J.K., Eng., Progress Engineers, 4 Babatola Drive, Ikeja, P.O. Box 6450, Marina, Lagos
George, E.A.J., Eng., 61 Moorehouse Street, P.M.B. 6168, Port Harcourt. T: 084 235748
Ifaturoti, P.O., Eng., Trevi Foundations (NIG) Ltd., Plot 1292 Akin Adesola Street, P.O. Box 70621, Victoria Island, Lagos. T: 01 2622097, 2619383, 2616257, F: 01 2617509
Ifedayo, A., Eng., Profen Consultants, P.O. Box 6331, Agodi Gate, Ibadan. T: 02 715447, 713331, 710210
Katchy, N.P., Eng., 17 Olufemi Street, Off Nathan Street, Ojuelegba, Surulere, Lagos
Kuye, C.O., Colak Geotechnics, 323 Herbert Macaulay, Alagomeji, Yaba. T: 01 288 0868
Madeor, A.O., Eng., Progress Engineers, 4 Babatola Drive, Ikeja, P.O. Box 6454, Marina, Lagos. T: 01 4976841, 4976888
Meshida, S.O., Eng., Nigerian Soil Engineering Co. Ltd., m 43 Chief Okupe Estate, Maryland. T: 01 496 3717

Meshida, E.A., Dr., 3 Akindele Street, New Garage, Lagos
Obayelu, B.O., Eng., OMPADEC, 22 William Jumbo Street, P.M.B 5253, Port Harbourt. T: 084 237665
Odedairo, E., Eng., 12 Bola Ajibola Street, Off Allen Avenue, Ikeja. T: 087 231514, F: 087 231441
Ofojebe, A.P., Eng., Lagos Anglican Diocesan Seminary, 17 Broad Street, Marina, Lagos
Ogunbanjo, A.O., Eng., Trevi Foundations (Nigeria) Ltd., Plot 1292 Akin Adesola Street, P.O. Box 70621, Victoria island, Lagos. T: 01 2622097, 2619383, 2616257, F: 01 2617509
Okorie, I.E., Eng., Trevi Foundations (NIG) Ltd., Plot 1292 Akin Adesola Street, P.O. Box 70621, Victoria Island, Lagos. T: 01 2622097, 2619383, 2616257, F: 01 2617509
Olatunji, O.A., 2nd. Avenue, 22 Road, D Close, Block 4, Flat 2 Festac Town (postal) P.O. Box 142, Festac Town, Lagos
Ologuntoye, E.O., c/o Geotechnical Engineering Nig. Ltd., 3 Akindele Street, New Garage, Gbagada, Lagos
Olusola, J.M., Ove-Arup & Partners Nig., 25 Boyle Street, Lagos. T: 01 2600520 9
Oni, O.A., Labion Geotechnics Ltd., 56 Seriki Aro Avenue, P.O. Box 6762, Ikeja. T: 01 4930489, 4970296
Opanuga, A.O., New 36 Cardoso Street, Mushin, Lagos
Osanmor, J., Eng., Trevi Foundations (Nig.) Ltd., Trevi Foundations (NIG) Ltd., Plot 1292 Akin Adesola Street, P.O. Box 70621, Victoria Island, Lagos. T: 01 2622097, 2619383, 2616257, F: 01 2617509
Otoko, G.R., Eng., Head of Civil Engineering Department, Rivers State University of Science and Technology, P.M.B. 5080, Port Harcourt
Popoola, K.A., Eng., Total Nigeria Plc., 4 Afribank Street, Victoria Island, Lagos. T: 01 2621780-9
Teme, S.C., Prof., Rivers State University of Science and Technology, P.O. Box 25, Port Harbourt. T: 084 238585
Uche, A.N., Kotai Engineering Ltd., 7 Ibezim Obiajulu Street, Off Adelabu Street, P.O. Box 3228, Surulere. T: 01 834027

166

NORWAY

Secretary: Geraldine Sørum, Norwegian Geotechnical Society, c/o Norwegian Geotechnical Institute, P.O. Box 3930 Ullevaal Stadion, 0806 Oslo. T: (47) 22 02 30 00, F: (47) 22 23 04 48, E: ngf@ngi.no

Total number of members: 290

Andenæs, E., Civil Eng., Aker Marine Contractors a.s, Lilleakerveien 8, Pb. 248, 0216 Oslo
Andersen, K.H., Civil Eng., Norwegian Geotechnical Institute, Sognsvn. 72, Pb. 3930 Ullevål Stadion, 0806 Oslo. T: 22 02 30 00, F: 22 23 04 48, E: kha@ngi.no
Andersen, R.E., Cand. Scient, Bragerhagen 16, 3012 Drammen. T: 32 83 90 10, E: rolfeand@os.telia.no
Andersen, Ø., Civil Eng., Norwegian Public Roads Administration, Pb. 8142 Dep., 0033 Oslo. T: 22 07 35 00, F: 22 07 37 68
Andersson, H., Civil Eng., Geotechnical Peak Technology, Austadveien 44, 3034 Drammen. T: 32 81 71 00, F: 32 81 31 20, E: helem_mk_andersson@hotmail.com
Andresen, A.A., Andresen, Skausnaret 3, 1262 Oslo. T: 22 61 12 60
Andresen, L., M.Sc., Norwegian Geotechnical Institute, Sognsvn. 72, Pb. 3930 Ullevål Stadion, 0806 Oslo. T: 22 02 30 00, F: 22 23 04 48, E: la@ngi.no
Antonsen, P., Civil Eng., Norwegian Pollution Control Authority, Strømsveien 96, Pb. 8100 Dep., 0032 Oslo. T: 22 57 34 00, F: 22 67 67 06, E: per.antonsen@sftospost.md.dep.telemax.no
Arnesen, K., M.Sc., Det Norske Veritas AS, Veritasvn. 1, 1323 Høvik. T: 67 57 72 50, F: 67 57 74 74
Athanasiu, C.M., Dr, NOTEBY A/S Consulting Engineers, Hoffsveien 1, Pb. 265 Skøyen, 0212 Oslo. T: 22 51 54 00, F: 22 51 54 01, E: athanasi@online.no
Backer, L., Civil Eng., Norwegian Geotechnical Institute, Sognsvn. 72, Pb. 3930 Ullevål Stadion, 0806 Oslo. T: 22 02 30 00, F: 22 23 04 48
Bakkan, M.E., Civil Eng., BanePartner, Stortorget 7, Pb. 1162 Sentrum, 0107 Oslo. T: 22 45 61 50, F: 22 45 61 10, E: mb@jbv.no
Barton, N., Dr, Fjordveien 65c, 1363 Høvik. T: 67 53 15 06
Bellis, J., Civil Eng., Kampebråten 7B, 1300 Sandvika. T: 67 54 55 17
Berger, A., M.Sc., NOTEBY A/S Consulting Engineers, Hoffsveien 1, Pb. 265 Skøyen, 0212 Oslo. T: 22 51 54 00, F: 22 51 54 01
Bergersen, N., M.Sc., NOTEBY A/S Consulting Engineers, Hoffsveien 1, Pb. 265 Skøyen, 0212 Oslo. T: 22 51 54 00, F: 22 51 54 01
Berggren, A-L., Dr. Ing., Geofrost Engineering A.S, Grinidammen 10, 1359 Eiksmarka. T: 67 14 73 50
Bergsås, A., Civil Eng., Norwegian National Rail Administration, Stortorget 7, Pb. 1162 Sentrum, 0107 Oslo. T: 22 45 61 50, F: 22 45 61 10
Berre, T., Civil Eng., Norwegian Geotechnical Institute, Sognsvn. 72, Pb. 3930 Ullevål Stadion, 0806 Oslo. T: 22 02 30 00, F: 22 23 04 48
Bertnes, J.H., Civil Eng., NOTEBY A/S Consulting Engineers, Hoffsveien 1, Pb. 265 Skøyen, 0212 Oslo. T: 22 51 54 00, F: 22 51 54 01
Bjørnstad, B., Norwegian Pollution Control Authority, Strømsveien 96, Pb. 8100 Dep., 0032 Oslo. T: 22 57 34 00, F: 22 67 67 06, E: bjorn.bjornstad@sft.telemax.no
Bogen, L.O., Civil Eng., NOTEBY A/S Consulting Engineers, Hoffsveien 1, Pb. 265 Skøyen, 0212 Oslo. T: 22 51 54 00, F: 22 51 54 01, E: lob@noteby.no
Bollingmo, P., Civil Eng., NOTEBY A/S Consulting Engineers, Hoffsveien 1, Pb. 265 Skøyen, 0212 Oslo. T: 22 51 54 00, F: 22 51 54 01
Bostrøm, B., M.Sc., SINTEF Civil and Environmental Engineering, Klæbuvn. 153, 7465 Trondheim. T: 73 59 46 00, F: 73 59 53 40, E: bard.bostrom@civil.sintef.no
Bostrøm, T., Dr. Ing., Reinertsen Engineering ANS, Postuttak, 7005 Trondheim. T: 73 52 60 40, F: 73 52 13 21

Brattensborg, G., Civil Eng., Aker Engineering a.s., Tjuvholmen, 0250 Oslo. T: 22 94 50 00, F: 22 94 73 38
Brattlien, K., M.Sc., Norwegian Geotechnical Institute, Sognsvn. 72, Pb. 3930 Ullevål Stadion, 0806 Oslo. T: 22 02 30 00, F: 22 23 04 48, E: kb@ngi.no
Breedveld, G., Ph.D., Norwegian Geotechnical Institute, Sognsvn. 72, Pb. 3930 Ullevål Stadion, 0806 Oslo. T: 22 02 30 00, F: 22 23 04 48, E: gbr@ngi.no
Brekke, R., Civil Eng., Postboks 6881 St. Olavs plass, 0130 Oslo
Brennodden, H., Dr. Ing., GeoPartner Marin as, Tungasletta 2, 7485 Trondheim. T: 73 80 16 80, F: 73 80 16 90
Bruun, H., Civil Eng., Norwegian Public Roads Administration, Gaustadalléen 25, Pb. 8142 Dep., 0033 Oslo. T: 22 07 39 00, F: 22 07 34 44
Brylawski, E., Civil Eng., Geonor Inc., 8 Greenwood Hills, P.O. Box 903, Milford PA 18337, USA. T: 001 717 296 4884, F: 001 717 296 4886, E: geonorus@ptd.net
Brønstad, G., Senior Consultant, NOTEBY A/S Consulting Engineers, Hoffsveien 1, Pb. 265 Skøyen, 0212 Oslo. T: 22 51 54 00, F: 22 51 54 01
Braaten, A., Civil Eng., Norwegian Public Roads Administration, Gaustadalléen 25, Pb. 8142 Dep., 0033 Oslo. T: 22 07 39 00, F: 22 07 34 44, E: anne.braaten@vegdir.vegvesen.no
By, T., Civil Eng., DSND Subsea AS, Hoffsveien 17, 0275 Oslo. E: trondby@online.no
Bye, A., Civil Eng., NOTEBY A/S Consulting Engineers, Hoffsveien 1, Pb. 265 Skøyen, 0212 Oslo. T: 22 51 54 00, F: 22 51 54 01
Baardvik, G., Civil Eng., Norwegian Public Roads Administration Akershus, Lørenvn. 57, Pb. 8166 Dep., 0034 Oslo. T: 22 52 72 00
Caprona, G. de, Dr, Mgr Int. Sales, Geotech AB, Dataveien 53, S-43632 Gøteborg, Sverige. T: 0046 31 28 99 20, F: 0046 31 68 16 39, E: caprona@geotech.se
Christensen, S., Civil Eng., SINTEF Civil and Environmental Engineering, Klæbuvn. 153, 7465 Trondheim. T: 73 59 46 00, F: 73 59 53 40, E: stein.christensen@civil.sintef.no
Clausen, B., M.Sc., Hoffsveien 1, Postboks 265 Skøyen, 0213 Oslo. T: 22 51 51 51, F: 22 51 51 52, E: bjc@multiconsult.no
Clausen, C.J.F., Civil Eng., Carl J. Frimann Clausen, Cidex 424 Bis, F-06330 ROQUEFORT-les PINS, Frankrike. T: 0033 493 775 275, F: 0033 493 771 979
Dadasbilge, B., Civil Eng., Sivilingeniør Baykara Dadasbilge, Bakkerovn 2, Pb. 475, 1401 SKI. T: 64 87 36 26
Dahle, T., Civil Eng., NOTEBY A/S Consulting Engineers, Hoffsveien 1, Pb. 265 Skøyen, 0212 Oslo. T: 22 51 54 00, F: 22 51 54 01
Dahlén, O., Civil Eng., NVK AS Norsk Vandbygningskontor, Holtevegen 5, Pb. 280, 1401 SKI. T: 64 85 55 00, F: 64 85 55 55
Dalberg, R.G., Dr, Det Norske Veritas AS, Veritasvn. 1, 1323 Høvik. T: 67 57 72 50, F: 67 57 74 74, E: rda@dnv.no
DiBiagio, E., Dr, DiBiagio, Bj. Bjørnsonsvei 46, 1412 Sofiemyr. T: 66 80 27 96
Duncumb, R.W., Civil Eng., Groener AS, Fornebuvn 11, Pb. 400, 1327 Lysaker. T: 67 12 80 00, F: 67 12 82 12, E: duncumb@online.no
Dyvik, R., Ph.D., Norwegian Geotechnical Institute, Sognsvn. 72, Pb. 3930 Ullevål Stadion, 0806 Oslo. T: 22 02 30 00, F: 22 23 04 48, E: rd@ngi.no
Eek, E., M.Sc., Norwegian Geotechnical Institute, Sognsvn. 72, Pb. 3930 Ullevål Stadion, 0806 Oslo. T: 22 02 30 00, F: 22 23 04 48, E: ee@ngi.no
Eggen, A., M.Sc., Norwegian Geotechnical Institute, Sognsvn. 72, Pb. 3930 Ullevål Stadion, 0806 Oslo. F: 22 23 04 48

Eggestad, Å., Civil Eng., Eggestad, Hamang terrasse 71, 1336 Sandvika. T: 67 54 12 57

Eikslund, G.R., Dr. Eng., SINTEF Civil and Environmental Engineering, Klæbuvn. 153, 7465 Trondheim. T: 73 59 46 00, F: 73 59 53 40, E: gudmund.eiksund@civil.sintef.no

Eklund, T., M.Sc., Det Norske Veritas AS, Veritasvn. 1, 1323 Høvik. T: 67 57 72 50, F: 6/ 5/ 74 74, E: tek@dnv.no

Ellefsen, V., Civil Eng., Bragerhagen 16, 3012 Drammen. T: 32 83 90 10, E: vellefse@online.no

Ellingbø, O., Civil Eng., Norconsult International A.S, Vestfjordgt 4, 1338 Sandvika. T: 67 57 10 00, E: oe@norconsult.no

Ellingsen, K.E., M.Sc., ETH, Entreprenørservice A/S, Rudsletta 24, Pb. 4, 1351 RUD. T: 67 17 30 00, F: 67 17 30 01

Emdal, A., Asst. Prof., Norwegian University of Science and Technology, Høgskoleringen 6, 7491 Trondheim. T: 73 59 50 00, F: 73 59 46 09, E: arnfinn.emdal@bygg.ntnu.no

Engen, A., Civil Eng., Norwegian Geotechnical Institute, Sognsvn. 72, Pb. 3930 Ullevål Stadion, 0806 Oslo. T: 22 02 30 00, F: 22 23 04 48, E: an@ngi.no

Engh, E.A., SCC Abel Engh AS, Torgeir Vraas pl 4, 3044 Drammen. T: 32 83 19 00, E: eae@abel.engh.scc.no

Enlid, E., Civil Eng., Scandiaconsult AS, Øvre Flatåsvei 10, Pb. 6032, 7493 Trondheim. T: 72 58 17 66, F: 72 58 00 50, E: ey.enlid@scc.no

Eriksen, K., Eng., Norwegian Public Roads Administration Nord-Trøndelag, Byavn. 21, 7700 Steinkjer. T: 74 16 85 00, F: 74 16 86 06

Espedal, K., Civil Eng., NVK Terraplan a.s., Tollbugt. 49, Pb. 2345, 3003 Drammen. T: 32 20 62 70, F: 32 20 62 71

Falstad, B.A., Civil Eng., Norwegian National Rail Administration, Stortorget 7, Pb. 1162 Sentrum, 0107 Oslo. T: 22 45 61 50, F: 22 45 61 10

Fergestad, S., Civil Eng., Dr. Ing., A. Aas-Jakobsen A/S, Lilleakervn. 4, 0283 Oslo. T: 22 50 50 44, F: 22 52 00 83

Finborud, B., Civil Eng., Norconsult International A.S., Vestfjordgt 4, 1338 Sandvika. T: 67 57 10 00, E: bfi@norconsult.no

Finstad, J.A., Civil Eng., NOTEBY A/S Consulting Engineers, Hoffsveien 1, Pb. 265 Skøyen, 0212 Oslo. T: 22 51 54 00, F: 22 51 54 01

Finstad, K., Bjørn Kynningsrud AS, Vallehellene 3, Postboks 13, 1662 Rolvsøy. T: 69 30 97 00, F: 69 30 97 01

Flaten, G., Eng., Flaten, Sørkedalsvein 99, 0376 Oslo. T: 22 14 83 60

Flaate, K., Dr, Flaate, Bernh. Herresv. 6, 0376 Oslo. T: 22 49 10 87

Foss, I., Civil Eng., Ivar Foss Quality Management, Nedre Vollgt 9, 0158 Oslo. T: 22 41 66 93

Fredriksen, K.S., Civil Eng., Norwegian Public Roads Administration Aust-Agder, Skydebanevn 1, Pb. 173, 4801 Arendal. T: 37 01 98 00

Fredriksen, U., Civil Eng., Oslo Municipality, Water and Sewage Works, Herslebsgt. 5, Pb. 4704 Sofienberg, 0506 Oslo. T: 22 66 43 10, F: 22 66 40 81

Fremstad, A., Civil Eng., Oslo Byggeadministrasjon A/S, Postboks 30 Bekkelagshøgda, 1109 Oslo. T: 22 76 95 00, F: 22 76 95 01

Fritzvold, H.K., M.Sc., Norconsult International A.S., Vestfjordgt 4, 1338 Sandvika. T: 67 57 10 00, E: hkf@norconsult.no

Frydenlund, T.E., Civil Eng., Norwegian Public Roads Administration, Gaustadalléen 25, Pb. 8142 Dep., 0033 Oslo. T: 22 07 39 00, F: 22 07 34 44, E: tor-erik.frydenlund@vegdir.vegvesen.no

Funderud, P.Ø., Civil Eng., Asker kommune, Pb. 353, 1371 ASKER. E: per-oystein.funderud@akershus.telemax.no

Furuberg, T., Civil Eng., SINTEF Civil and Environmental Engineering, Klæbuvn. 153, 7465 Trondheim. T: 73 59 46 00, F: 73 59 53 40, E: tone.furuberg@civil.sintef.no

Færgestad, O., Civil Eng., NOTEBY A/S Consulting Engineers, Hoffsveien 1, Pb. 265 Skøyen, 0212 Oslo. T: 22 51 54 00, F: 22 51 54 01

Føyn, T., Civil Eng., Norconsult International A.S., Vestfjordgt 4, 1338 Sandvika. T: 67 57 10 00, E: tf@norconsult.no

Giske, S., Civil Eng., Norwegian Public Roads Administration, Gaustadalléen 25, Pb. 8142 Dep., 0033 Oslo. T: 22 07 39 00, F: 22 07 34 44

Gjelsvik, V., Civil Eng., SCC Kummeneje AS, Engebrets vei 5, Postboks 450 Skøyen, 0212 Oslo. E: vidar.gjelsvik@kummeneje.scc.no

Gregersen, O.S., Civil Eng., Norwegian Geotechnical Institute, Sognsvn. 72, Pb. 3930 Ullevål Stadion, 0806 Oslo. T: 22 02 30 00, F: 22 23 04 48

Gulliksen, C.C., Civil Eng., Grunn-Teknikk AS, Peder Bogensgt. 2 A, 3200 Sandefjord. T: 33 46 37 70

Gutierrez, M., Dr, Norwegian Geotechnical Institute, Sognsvn. 72, Pb. 3930 Ullevål Stadion, 0806 Oslo. T: 22 02 30 00, F: 22 23 04 48, E: mg@ngi.no

Guttormsen, T.R., Dr Scient, Norsk Hydro ASA, 0246 Oslo. E: tom.guttormsen@hydro.com

Hagberg, R., Export Sales Manager, Geonor AS, Grinidammen 10, Pb. 99 Røa, 0701 Oslo. T: 67 14 75 50, F: 67 14 58 46

Haldorsen, K., Norsk Hydro asa, Pb. 200, 1321 Stabekk. T: 22 73 81 00, F: 22 73 74 30

Hamre, L., Civil Eng., NOTEBY A/S Consulting Engineers, Hoffsveien 1, Pb. 265 Skøyen, 0212 Oslo. T: 22 51 54 00, F: 22 51 54 01, E: lha@noteby.no

Handberg, A., GeoSafe as, Varmbuvn. 2, 7084 Melhus. T: 72 87 00 70, F: 72 87 00 75

Hansen, B., Civil Eng., Norwegian Geotechnical Institute, Sognsvn. 72, Pb. 3930 Ullevål Stadion, 0806 Oslo. T: 22 02 30 00, F: 22 23 04 48

Hansen, E.L., Civil Eng., Siv.ing. Erik Lindbo Hansen, Frogs vei 15, 3600 Kongsberg. T: 32 76 81 25, E: eriklh@online.no

Hanson, Y.A., Civil Eng., Noteby A/S, Fiolvn. 13, 9016 Tromsø. T: 77 65 53 00, F: 77 65 75 40

Hansteen, O.E., Civil Eng., Hansteen, Ostadalsveien 68, 0753 Oslo. T: 22 14 54 30

Hauan, R., Civil Eng., SCC Bonde & Co. AS, Engebrets vei 5, Pb. 439 Skøyen, 0212 Oslo. T: 22 51 80 00, F: 22 51 80 02

Haug, T., Civil Eng., Topaas og Haug A/S, Olav Ingstads vei 11, Pb. 94, 1351 Rud. T: 67 13 52 90, F: 67 13 48 96

Hauge, A., Civil Eng., Norwegian Geotechnical Institute, Sognsvn. 72, Pb. 3930 Ullevål Stadion, 0806 Oslo. T: 22 02 30 00, F: 22 23 04 48, E: ah@ngi.no

Hauge, J., Cand.scient, Norsk Leca a.s., Brobekkvn. 84, Pb. 216 Alnabru, 0614 Oslo. T: 22 88 77 00, F: 22 64 54 54

Hauge, Kj., Civil Eng., Norwegian Geotechnical Institute, Sognsvn. 72, Pb. 3930 Ullevål Stadion, 0806 Oslo. T: 22 02 30 00, F: 22 23 04 48, E: kjh@ngi.no

Hauge, P.F., Civil Eng., Entreprenørservice A/S, Rudsletta 24, Pb. 4, 1351 Rud. T: 67 17 30 00, F: 67 17 30 01

Haugen, C., Civil Eng., GeoVita as, Lilleakerveien 4, 0283 Oslo. T: 22 50 82 50, F: 22 50 83 20

Haugerud, H.J., Civil Eng., SCC Abel Engh AS, Torgeir Vraas pl 4, 3044 Drammen. T: 32 83 19 00

Heiberg, S., Civil Eng., Statoil, Forusbeen 50, 4035 Stavanger. T: 51 80 80 80, E: shei@statoil.no

Heimli, P., Civil Eng., NOTEBY A/S Consulting Engineers, Hoffsveien 1, Pb. 265 Skøyen, 0212 Oslo. T: 22 51 54 00, F: 22 51 54 01

Helgesen, K.H., Civil Eng., Norsk Leca a.s., Brobekkvn. 84, Pb. 216 Alnabru, 0614 Oslo. T: 22 88 77 00, F: 22 64 54 54, E: kjell-hakon.helgesen@norskleca.scancem.com

Helgaas, O.-A., Norwegian Public Roads Administration Troms, Mellomvn. 40, 9005 Tromsø. T: 77 61 70 00, F: 77 61 76 66

Hennum, G., Civil Eng., Norconsult International A.S., Vestfjordgt 4, 1338 Sandvika. T: 67 57 10 00, E: guh@norconsult.no

Hermann, S., Civil Eng., Norwegian Geotechnical Institute, Sognsvn. 72, Pb. 3930 Ullevål Stadion, 0806 Oslo. T: 22 02 30 00, F: 22 23 04 48, E: sth@ngi.no

Hermstad, J., Civil Eng., Aker Technoligy A/S, Pb. 249 Lilleaker, 0216 Oslo. T: 22 94 50 00, F: 22 94 73 38

Hernes, S., Civil Eng., Norconsult International A.S., Vestfjordgt 4, 1338 Sandvika. T: 67 57 10 00, E: shy@norconsult.no

Heyerdahl, H., Norwegian Geotechnical Institute, Sognsvn. 72, Pb. 3930 Ullevål Stadion, 0806 Oslo. T: 22 02 30 00, F: 22 23 04 48

Hjorteset, A., Civil Eng., GeoVita as, Lilleakerveien 4, 0283 Oslo. T: 22 50 82 50, F: 22 50 83 20

Hoel, P.K., Eng., Nordisk Kartro AS, Holmavn. 20, Pb. 54, 1313 Vøyenenga. T: 67 17 36 00, F: 67 17 36 01

Hoem, O.I., Civil Eng., Gjemnes kommune, Kommunehuset, 6631 Batnfjordsøra

Hollerud, B., Civil Eng., Norconsult International A.S., Vestfjordgt 4, 1338 Sandvika. T: 67 57 10 00, E: birger. hollerud@cable.statnett.no

Hovem, S.G., Civil Eng., Groener AS, Fornebuvn 11, Pb. 400, 1327 Lysaker. T: 67 12 80 00, F: 67 12 82 12, E: sgh@ groner.no

Hylen, S., M.Sc., Advanced Production and Loading AS, Kittelsbuktvn 5, Pb. 53, 4801 Arendal. T: 37 00 25 00, F: 37 02 41 28, E: apl@sn.no

Høeg, K., Prof. Dr, Norwegian Geotechnical Institute, Sognsvn. 72, Pb. 3930 Ullevål Stadion, 0806 Oslo. T: 22 02 30 00, F: 22 23 04 48, E: kh@ngi.no

Håland, G., Civil Eng., Statoil, Forusbeen 50, 4035 Stavanger. T: 51 80 80 80, E: gish@statoil.no

Hårvik, L., Civil Eng., Norwegian Geotechnical Institute, Sognsvn. 72, Pb. 3930 Ullevål Stadion, 0806 Oslo. T: 2 02 30 00, F: 22 23 04 48, E: lh@ngi.no

Haavardsholm, B., Civil Eng., Selmer ASA, St. Olavsgt. 25, Pb. 1175 Sentrum, 0107 Oslo. T: 22 03 06 00, F: 22 20 88 30

Ihler, H., M.Sc., Norwegian Public Roads Administration, Gaustadalléen 25, Pb. 8142 Dep., 0033 Oslo. T: 22 07 39 00, F: 22 07 34 44, E: harald.ihler@vegvesen.no

Ilstad, T., Civil Eng., SINTEF Civil and Environmental Engineering, Klæbuvn. 153, 7465 Trondheim. T: 73 59 46 00, F: 73 59 53 40, E: Trygve.Ilstad@civil.sintef.no

Instanes, A., Dr. Eng., Norwegian Geotechnical Institute, Sognsvn. 72, Pb. 3930 Ullevål Stadion, 0806 Oslo. T: 22 02 30 00, F: 22 23 04 48, E: ai@ngi.no

Janbu, N., Dr, Janbu, Prestegårdsv 5B, 7016 Trondheim. T: 73 59 45 92

Jensen, H.R., Civil Eng., SCC Kummeneje AS, Prosjektsentret, Ringveien 26, 2800 Gjøvik. T: 61 17 59 20, F: 61 17 69 37

Jensen, H-P., Civil Eng., Hans Petter Jensen Geotechn. Consultant, Dronningen, Bygdøy, Pb. A Bygdøy, 0211 Oslo. T: 22 43 75 75, F: 22 44 79 85, E: hpj@internet.no

Jensen, T.G., Civil Eng., Norwegian Geotechnical Institute, Sognsvn. 72, Pb. 3930 Ullevål Stadion, 0806 Oslo. T: 22 02 30 00, F: 22 23 04 48, E: tgj@ngi.no

Johansen, P.M., Civil Eng., Norconsult International A.S., Vestfjordgt 4, 1338 Sandvika. T: 67 57 10 00

Johansen, T., Civil Eng., GeoVita as, Lilleakerveien 4, 0283 Oslo. T: 22 50 82 50, F: 22 50 83 20

Jonassen, H.M., M.Sc., Norwegian Geotechnical Institute, Sognsvn. 72, Pb. 3930 Ullevål Stadion, 0806 Oslo. T: 22 02 30 00, F: 22 23 04 48, E: hj@ngi.no

Jostad, H.P., Civil Eng., Norwegian Geotechnical Institute, Sognsvn. 72, Pb. 3930 Ullevål Stadion, 0806 Oslo. T: 22 02 30 00, F: 22 23 04 48

Jullum, R., Civil Eng., Directorate of Public Construction and Property, Stensberggaten 25, Pb. 8106 Dep., 0032 Oslo. T: 22 24 28 00

Julsheim, D.E., M.Sc., NOTEBY A/S, Christianslund allé 2, 1613 Fredrikstad. T: 69 31 28 04, F: 69 31 28 08

Jørve, J., Civil Eng., NOTEBY A/S Consulting Engineers, Hoffsveien 1, Pb. 265 Skøyen, 0212 Oslo. T: 22 51 54 00, F: 22 51 54 01

Karal, K., Dr. Ing., UMOE Anchor Contracting A.S., Holtet 45, Pb. 150, 1321 Stabekk. T: 67 10 90 00, F: 67 10 95 83

Karlsen, K., Civil Eng., Norwegian National Rail Adm., Southern Region, Strømsø veg 1, 3006 Drammen

Karlsen, V., Civil Eng., Norsk Hydro asa, Pb. 200, 1321 Stabekk. T: 22 73 81 00, F: 22 73 74 30

Karlsrud, K., M.Sc., Norwegian Geotechnical Institute, Sognsvn. 72, Pb. 3930 Ullevål Stadion, 0806 Oslo. T: 22 02 30 00, F: 22 23 04 48

Kavli, A., Civil Eng., Geovest A/S, Sandveien 12A, 6400 Molde. T: 71 20 59 20, F: 71 20 59 30, E: arne.kavli@ civil.sintef.no

Kaynia, A., Dr. Eng., Norwegian Geotechnical Institute, Sognsvn. 72, Pb. 3930 Ullevål Stadion, 0806 Oslo. T: 22 02 30 00, F: 22 23 04 48, E: amk@ngi.no

Kilen, A.K., Graduate Eng., SCC Kummeneje AS, Engebrets vei 5, Postboks 450 Skøyen, 0212 Oslo. E: Ann. Karin.Kilen@kummeneje.scc.no

Kirkebø, S., Dr. Ing., GeoVita as, Lilleakerveien 4, 0283 Oslo. T: 22 50 82 50, F: 22 50 83 20, E: signe.kirkeboe@ aas-jakobsen.no

Kjekstad, O., Deputy Man. Dir., Norwegian Geotechnical Institute, Sognsvn. 72, Pb. 3930 Ullevål Stadion, 0806 Oslo. T: 22 02 30 00, F: 22 23 04 48, E: ok@ngi.no

Klem, J.H., Civil Eng., Norconsult International, Saint Louis Mansion Co. Ap. 1002, 138 Saint Louis Soi, 2, 10120 South Sathorn Rd. Bangkok 10330, THAILAND

Klevar, Ø., MSBA, Geonor AS, Grinidammen 10, Pb. 99 Røa, 0701 Oslo. T: 67 14 75 50, F: 67 14 58 46, E: ok@ geonor.no

Kleven, A., Civil Eng., Norwegian Geotechnical Institute, Sognsvn. 72, Pb. 3930 Ullevål Stadion, 0806 Oslo. T: 22 02 30 00, F: 22 23 04 48

Knudsen, S., Civil Eng., Norwegian Geotechnical Institute, Sognsvn. 72, Pb. 3930 Ullevål Stadion, 0806 Oslo. T: 22 02 30 00, F: 22 23 04 48, E: sk@ngi.no

Kolstad, P., Civil Eng., Norwegian Geotechnical Institute, Sognsvn. 72, Pb. 3930 Ullevål Stadion, 0806 Oslo. T: 22 02 30 00, F: 22 23 04 48

Korbøl, B., M.Sc., 2260 Kirkenær. T: 62 94 76 33, F: 62 94 77 77

Kummeneje, O., Civil Eng., Kummeneje, Broddes vei 14, 7021 Trondheim. T: 72 55 85 90

Kurås, J.K., Eng., Norsk Stål AS, Nye Vakås vei 80, Pb. 123, 1360 Nesbru. T: 66 84 28 00, F: 66 84 28 50

Kvalstad, T.J., Civil Eng., Norwegian Geotechnical Institute, Sognsvn. 72, Pb. 3930 Ullevål Stadion, 0806 Oslo. T: 22 02 30 00, F: 22 23 04 48, E: tk@ngi.no

Kvalsvik, H.J., Civil Eng., Groener AS, Fornebuvn 11, Pb. 400, 1327 Lysaker. T: 67 12 80 00, F: 67 12 82 12

Kviterud, A., Civil Eng., Norwegian Petroleum Directorate, Prof. Olav Hansens vei 10, Pb. 600, 4001 Stavanger. T: 51 87 60 00, F: 51 55 15 71, E: arne.kvitrud@ npd.no

Lacasse, S., Dr, Norwegian Geotechnical Institute, Sognsvn. 72, Pb. 3930 Ullevål Stadion, 0806 Oslo. T: 22 02 30 00, F: 22 23 04 48, E: sl@ngi.no

Ladehammerkaia, 7041 Trondheim. T: 73 50 03 50, F: 73 94 38 61, E: jon@geoprobing-tech.no

Landva, A., Ph.D., Landva, 16 Silverwood Ct., Fredericton, New Brunswick E3C 1K2, CANADA

Langø, H., Dr. Ing., GeoPartner Marin as, Tungasletta 2, 7485 Trondheim. T: 73 80 16 80, F: 73 80 16 90

Lau, J., Civil Eng., Siv.ing. Jack Lau, Erik Bues vei 6, 2600 Lillehammer. T: 61 25 08 27, E: jlau@online.no

Laugesen, J., Civil Eng., Det Norske Veritas AS, Veritasvn. 1, 1323 Høvik. T: 67 57 72 50, F: 67 57 74 74, E: jens. laugesen@dnv.com

Lauritzsen, R.A., Civil Eng., Norwegian Geotechnical Institute, Sognsvn. 72, Pb. 3930 Ullevål Stadion, 0806 Oslo. T: 22 02 30 00, F: 22 23 04 48, E: ral@ngi.no

Lefstad, O., Civil Eng., Scandiaconsult AS, Øvre Flatåsvei 10, Pb. 6032, 7493 Trondheim. T: 72 58 17 66, F: 72 58 00 50

Leirvik, K., Export Sales Manager, Leirvik, Almelien 25, 1358 JAR. T: 67 15 92 83

Lerfaldet, M.A., Civil Eng., Norwegian Public Roads Administration Østfold, Verftsgt. 7, Pb. 310, 1501 MOSS. T: 69 27 45 00, F: 69 27 47 78

Liavaag, T., Civil Eng., Selmer ASA, St. Olavsgt. 25, Pb. 1175 Sentrum, 0107 Oslo. T: 22 03 06 00, F: 22 20 88 30

Lied, K., Norwegian Geotechnical Institute, Sognsvn. 72, Pb. 3930 Ullevål Stadion, 0806 Oslo. T: 22 02 30 00, F: 22 23 04 48

Lieng, J.T., Dr. Ing., GeoProbing Technology AS, Ormen Langes vei 12, 7041 Trondheim. T: 73500350, F: 73500349, E: jon.lieng@geoprobing-tech.no

Lundqvist, K.E., Civil Eng., Delko International AS, Sandakerveien 110C, Pb. 38 Grefsen, 0409 Oslo. T: 22 15 00 45, E: lundqvis@telepost.no

Lunne, T.A., Civil Eng., Norwegian Geotechnical Institute, Sognsvn. 72, Pb. 3930 Ullevål Stadion, 0806 Oslo. T: 22 02 30 00, F: 22 23 04 48, E: tlu@ngi.no

Lyche, E.B., Civil Eng., Scandiaconsult AS, Øvre Flatåsvei 10, Pb. 6032, 7493 Trondheim. T: 72 58 17 66, F: 72 58 00 50

Lyngroth, T., M.Sc., Norconsult International A.S., Vestfjordgt 4, 1338 Sandvika. T: 67 57 10 00, E: tly@norconsult.no

Løken, T., Geologist, Norwegian Geotechnical Institute, Sognsvn. 72, Pb. 3930 Ullevål Stadion, 0806 Oslo. T: 22 02 30 00, F: 22 23 04 48, E: tl@ngi.no

Løset, F., Cand.real, Norwegian Geotechnical Institute, Sognsvn. 72, Pb. 3930 Ullevål Stadion, 0806 Oslo. T: 22 02 30 00, F: 22 23 04 48

Løvholt, J., M.Sc., Norwegian Geotechnical Institute, Sognsvn. 72, Pb. 3930 Ullevål Stadion, 0806 Oslo. T: 22 02 30 00, F: 22 23 04 48, E: jhl@ngi.no

Løvlien, P., M.Sc., Løvlien Georåd AS, Grønnegt. 83, 2317 HAMAR. T: 62 52 05 50, F: 62 52 05 55, E: ploevlie@online.no

Madshus, Chr., M.Sc., Norwegian Geotechnical Institute, Sognsvn. 72, Pb. 3930 Ullevål Stadion, 0806 Oslo. T: 22 02 30 00, F: 22 23 04 48, E: cm@ngi.no

Moe, A., Eng., Norwegian Public Roads Administration Oslo, Pb. 8037 Dep., 0030 Oslo. T: 22 57 55 00, F: 22 57 55 15

Moe, D., Vice President, Norconsult International A.S., Vestfjordt 4, 1338 Sandvika. T: 67 57 10 00, E: drm@norconsult.no

Mokkelbost, K.H., M.Sc., Norwegian Geotechnical Institute, Sognsvn. 72, Pb. 3930 Ullevål Stadion, 0806 Oslo. T: 22 02 30 00, F: 22 23 04 48, E: khm@ngi.no

Morseth, B.R., Eng geologist, NOTEBY A/S Consulting Engineers, Hoffsveien 1, Pb. 265 Skøyen, 0212 Oslo. T: 22 51 54 00, F: 22 51 54 01

Mortensen, S-E., Civil Eng., Norwegian Council for Building Standardization, Forskningsvn. 3 B, Pb. 129 Blindern, 0314 Oslo. T: 22 96 59 50, F: 22 60 85 70

Motzfeldt, E., Civil Eng., Norwegian National Rail Adm., Eastern Region, Stenersgt 1A, 0048 Oslo. T: 23 15 34 74, E: else.motzfeldt@jbv.no

Mowinckel, A., Civil Eng., Bertil O. Steen, Økernveien 147, Pb. 52. 1471 Skårer. T: 23 37 92 90, E: amund.mowinckel@bosteen.no

Murvold, M., Civil Eng., NOTEBY A/S Consulting Engineers, Hoffsveien 1, Pb. 265 Skøyen, 0212 Oslo. T: 22 51 54 00, F: 22 51 54 01, E: mhm@noteby.no

Myklebust, S., Eng., Sverre Myklebust A/S, Nyvn. 23 B, 1320 Stabekk

Müller, I-M., Secr., Norwegian Geotechnical Institute, Sognsvn. 72, Pb. 3930 Ullevål Stadion, 0806 Oslo. T: 22 02 30 00, F: 22 23 04 48

Myrvoll, F., M.Sc., Norwegian Geotechnical Institute, Sognsvn. 72, Pb. 3930 Ullevål Stadion, 0806 Oslo. T: 22 02 30 00, F: 22 23 04 48, E: fm@ngi.no

Mørk, L., Director, Norwegian National Rail Administration, Wessels vei 3, 2005 RÆLINGEN. T: 22 45 61 50, F: 22 45 61 10, E: lm@jbv.no

Nadim, F., Dr, Norwegian Geotechnical Institute, Sognsvn. 72, Pb. 3930 Ullevål Stadion, 0806 Oslo. T: 22 02 30 00, F: 22 23 04 48, E: fna@ngi.no

Narjord, H., M.Sc., NOTEBY A/S, Sverresdalen 26, Pb. 1139 Sverresborg, 7002 Trondheim. T: 72 55 25 00, F: 72 55 26 61

Nergaard, S., Civil Eng., NOTEBY A/S Consulting Engineers, Hoffsveien 1, Pb. 265 Skøyen, 0212 Oslo. T: 22 51 54 00, F: 22 51 54 01

Nerland, O., Civil Eng., Thomassen Byggteknikk a.s., Fløtmannsgt. 2, 6413 Molde. T: 71 25 18 44, F: 71 21 66 91

Nerland, Ø., Civil Eng., Norwegian Geotechnical Institute, Sognsvn. 72, Pb. 3930 Ullevål Stadion, 0806 Oslo. T: 22 02 30 00, F: 22 23 04 48, E: on@ngi.no

Ness, M., Norwegian Geotechnical Institute, Sognsvn. 72, Pb. 3930 Ullevål Stadion, 0806 Oslo. T: 22 02 30 00, F: 22 23 04 48

Nøstvold, J.T., Civil Eng., Scandiaconslult AS, Øvre Flatåsvei 10, Pb. 6032, 7493 Trondheim. T: 72 58 17 66, F: 72 58 00 50

Nordal, R., Professor, Nordal, Oscar Wistingsvei 18B, 7020 Trondheim. T: 73 59 47 07

Nordal, S., Professor, Norwegian University of Science and Technology, Høgskoleringen 6, 7491 Trondheim. T: 73 59 50 00, F: 73 59 46 09, E: steinar.nordal@bygg.ntnu.no

Norem, H., Dr. Ing., Dr. Ing., Harald Norem A/S, Schøyensgt. 6, 7030 Trondheim

Nowacki, E.H.F., Dr, Norwegian Geotechnical Institute, Sognsvn. 72, Pb. 3930 Ullevål Stadion, 0806 Oslo. T: 22 02 30 00, F: 22 23 04 48, E: fn@ngi.no

Nyberg, P., Civil Eng., Norconsult International A.S., Vestfjordt 4, 1338 Sandvika. T: 67 57 10 00

Ocunsov, E., Dr, UNICONE Ltd., Rupniecibas 34, P.O. Box 241, Riga, LV – 1047, Latvia. T: 0037 17 28 78 35, F: 37 17 28 02 44

Olsen, T.S., M.Sc., NOTEBY A/S Consulting Engineers, Hoffsveien 1, Pb. 265 Skøyen, 0212 Oslo. T: 22 51 54 00, F: 22 51 54 01

Olsson, R., Ph.D., Groener AS, Fornebuvn 11, Pb. 400, 1327 Lysaker. T: 67 12 80 00, F: 67 12 82 12, E: rol@groner.no

Oset, F., Senior Eng., Norwegian Public Roads Administration, Gaustadalléen 25, Pb. 8142 Dep., 0033 Oslo. T: 22 07 39 00, F: 22 07 34 44, E: frode.oset@.vegvesen.no

Otter, R., Geologist, Norwegian Geotechnical Institute, Sognsvn. 72, Pb. 3930 Ullevål Stadion, 0806 Oslo. T: 22 02 30 00, F: 22 23 04 48, E: ro@ngi.no

Palmstrøm, A., Civil Eng., Norconsult International A.S., Vestfjordgt 4, 1338 Sandvika. T: 67 57 10 00

Paulsen, G., M.Sc., BorgTotal AS, Pb. 385, 1601 Fredrikstad. T: 22 50 82 50, F: 22 50 83 20, E: gisle.paulsen@bigfoot.com

Pharo, K.T., Senior Soils Engineer, Norwegian Public Roads Administration Hordaland, Spelhaugen 12, Pb. 3645, 5845 Fyllingsdalen. T: 55 17 38 35, F: 55 17 38 33

Ramlo, G.A., Civil Eng., Reinertsen Engineering, Thunes vei 2, Pb. 415, Skøyen, 0212 Oslo. T: 22 56 10 20, F: 22 56 10 19, E: ramlo@reinertsen.no

Refsdal, G.A., Senior Eng., Norwegian Public Roads Administration Buskerud, Pb. 2265 Strømsø, 3003 Drammen. T: 32 20 25 00, F: 32 83 06 05, E: refsdal.geir@vegvesen.buskerud.no

Riise, P.K., Civil Eng., Riise, Övre Axåsvägen 4, S-44332 Lerum., Sverige. T: (46) 31 7714 928, P: 46 302 12317, E: per.riise@swipnet.se

Robsrud, A.J., Civil Eng., BanePartner, Pb. 1162 Sentrum, 0107 Oslo. T: 22 45 62 39, F: 22 45 61 10, E: arr@jbv.no

Rognlien, B., Sen.Found.Eng., Aker Engineering a/s, Tjuvholmen, 0250 Oslo. T: 22 94 50 00, F: 22 94 73 38

Rohde, J.K.G., Sen. Civil Eng., Groener AS, Fornebuvn 11, Pb. 400, 1327 Lysaker. T: 67 12 80 00, F: 67 12 82 12, E: jkr@groner.no

Rose, R., Eng., Ing., R. Rose Oppmåling/Geoteknikk, Strømsbuveien 44, 4800 Arendal

Roti, D., M.Sc., NOTEBY A/S, Fiolvn. 13, 9016 Tromsø. T: 77 65 53 00, F: 77 65 75 40, E: dir@noteby.no

Rye, O.A., Civil Eng., Scandiaconsult AS, Øvre Flatåsvei 10, Pb. 6032, 7493 Trondheim. T: 72 58 17 66, F: 72 58 00 50

Rystad, V., Civil Eng., NOTEBY A/S Consulting Engineers, Hoffsveien 1, Pb. 265 Skøyen, 0212 Oslo. T: 22 51 54 00, F: 22 51 54 01, E: vr@noteby.no

Røe, O., Civil Eng., NOTEBY A/S, Sverresdalen 26, Pb. 1139 Sverresborg, 7002 Trondheim. T: 72 55 25 00, F: 72 55 26 61, E: trheim@noteby.no

Rønning, S., Civil Eng., NOTEBY A/S Consulting Engineers, Hoffsveien 1, Pb. 265 Skøyen, 0212 Oslo. T: 22 51 54 00, F: 22 51 54 01

Sagbakken, A.K., Eng., Norwegian Public Roads Administration Akershus, Lørenvn. 57, Pb. 8166 Dep., 0034 Oslo. T: 22 52 72 00

Saghaug, T., Civil Eng., Pilestredet 56, 0167 Oslo. T: 22 56 40 60, F: 22 56 42 40, E: tor.saghaug@rif.no

Sand, K., Civil Eng., Trondheim Municipality, Holtermannsvei 1, 7005 Trondheim. T: 72 54 60 65

Sande, O.Kr., Civil Eng., Statoil, Forusbeen 50, 4035 STAVANGER. T: 51 80 80 80, E: odks@statoil.no

Sandersen, F., M.Sc., Norwegian Geotechnical Institute, Sognsvn. 72, Pb. 3930 Ullevål Stadion, 0806 Oslo. T: 22 02 30 00, F: 22 23 04 48

Sandven, R., Dr. Ing., Norwegian University of Science and Technology, Høgskoleringen 6, 7491 Trondheim. T: 73 59 50 00, F: 73 59 46 09, E: rolf.sandven@bygg.ntnu.no

Schjetne, K., M.Sc., Norwegian Geotechnical Institute, Sognsvn. 72, Pb. 3930 Ullevål Stadion, 0806 Oslo. T: 22 02 30 00, F: 22 23 04 48, E: ks@ngi.no

Schrøder, K., Civil Eng., Norwegian Geotechnical Institute, Sognsvn. 72, Pb. 3930 Ullevål Stadion, 0806 Oslo. T: 22 02 30 00, F: 22 23 04 48, E: ksc@ngi.no

Seim, T., Civil Eng., Norwegian Public Roads Administration Akershus, Lørenvn. 57, Pb. 8166 Dep., 0034 Oslo. T: 22 52 72 00

Senneset, K., Professor, Senneset, Sundlandskrenten 37, 7032 Trondheim. T: 73 93 76 51, F: 73 59 46 02

Sharma, R., Cand.Scient., Norwegian Public Roads Administration Akershus, Lørenvn. 57, Pb. 8166 Dep., 0034 Oslo. T: 22 52 72 00

Simonsen, A.S., Civil Eng., NOTEBY A/S Consulting Engineers, Hoffsveien 1, Pb. 265 Skøyen, 0212 Oslo. T: 22 51 54 00, F: 22 51 54 01, E: ass@noteby.no

Sindre, H.S., Civil Eng., Entreprenørservice A/S, Rudsletta 24, Pb. 4, 1351 Rud. T: 67 17 30 00, F: 67 17 30 01, E: hallvard. sindre@entreprenorsevice.no

Skauerud, S.E., Civil Eng., NOTEBY A/S, Pb. 161 Vågsbygd, 4602 Kristiansand. T: 38 01 39 00, F: 38 01 27 46

Skomedal, E., Civil Eng., Statoil, Forusbeen 50, 4035 Stavanger. T: 51 80 80 80, E: eis@statoil.no

Skotheim, A.Å., Dr. Eng., Geovest A/S, Sandveien 12A, 6400 MOLDE. T: 71 20 59 20, F: 71 20 59 30, E: geovest@sn.no

Skuggedal, H., Civil Eng., Skuggedal, Kirkehaugsvn. 1, 0283 Oslo

Skulason, J., Civil Eng., Almenna Verkfrædistofan H.F., Fellsmuli 26, 108 Reykjavik, island. T: 00354 167 8100, F: 00354 168 0284

Slapgård, S., Civil Eng., Aas-Jakobsen A/S, Lilleakerveien 4, 0283 Oslo. T: 22 50 50 44

Slungaard, J., Civil Eng., Groener AS, Fornebuvn 11, Pb. 400, 1327 Lysaker. T: 67 12 80 00, F: 67 12 82 12, E: js@groner.no

Solheim, O.M., M.Sc., Consulting Engineers, Overlege Bratts vei 15, 7026 Trondheim. T: 72 55 53 21, E: oms@noteby.no

Solhjell, E., Civil Eng., Norwegian Geotechnical Institute, Sognsvn. 72, Pb. 3930 Ullevål Stadion, 0806 Oslo. T: 22 02 30 00, F: 22 23 04 48, E: es@ngi.no

Sparrevik, M., M.Sc., Norwegian Geotechnical Institute, Sognsvn. 72, Pb. 3930 Ullevål Stadion, 0806 Oslo. T: 22 02 30 00, F: 22 23 04 48, E: msp@ngi.no

Sparrevik, P., Civil Eng., Norwegian Geotechnical Institute, Sognsvn. 72, Pb. 3930 Ullevål Stadion, 0806 Oslo. T: 22 02 30 00, F: 22 23 04 48, E: psp@ngi.no

Stenhamar, P., Vice President, Statkraft Grøner AS, Fornebuveien 11, Pb. 400, 1327 Lysaker. T: 67 12 80 00

Stensrud, O.M., Civil Eng., Kroghs vei 4 A, Postboks 63 Ellingsrudåsen, 1006 Oslo

Stordal, A.D., Civil Eng., NOTEBY A/S, Hopsnesvn. 21, Pb. 153, 5040 Paradis. T: 55 91 07 00, F: 55 91 05 74

Strandvik, S.O., Cand.Scient, Norwegian Geotechnical Institute, Sognsvn. 72, Pb. 3930 Ullevål Stadion, 0806 Oslo. T: 22 02 30 00, F: 22 23 04 48, E: sts@ngi.no

Straumsnes, A.O., Civil Eng., NOTEBY A/S, Alexander Kiellandsgt 24, 3716 Skien. T: 35 52 31 15, E: straums@online.no

Strout, J.M., Dr, Norwegian Geotechnical Institute, Sognsvn. 72, Pb. 3930 Ullevål Stadion, 0806 Oslo. T: 22 02 30 00, F: 22 23 04 48, E: jms@ngi.no

Strøm, B., Civil Eng., Siviling. Bjørn Strøm, Andebuvn. 23, 3170 SEM. T: 33 33 33 77, F: 33 33 30 60

Strøm, M., Siviling. Bjørn Strøm, Andebuvn. 23, 3170 Sem. T: 33 33 33 77, F: 33 33 30 60

Strøm, P.J., M.Sc., Det Norske Veritas AS, Veritasvn. 1, 1323 Høvik. T: 67 57 72 50, F: 67 57 74 74

Strøm, T., Siviling. Bjørn Strøm, Andebuvn. 23, 3170 Sem. T: 33 33 33 77, F: 33 33 30 60

Sture, S., Ph.D., University of Colorado, Dept of Civ, Env & Archit Engr., Campus Box 428, Boulder, Boulder, COLORADO 80309-0428, USA. T: 001 303 491 7651, F: 001 303 4927 315, E: sture@grieg.colorado.edu

Sundholm, M., Civil Eng., Statoil, Eyvind Lyches vei 10 Pb. 110, 1301 Sandvika. T: 67 57 30 00, E: mesu@statoil.no

Svanø, G., Chief Research Eng., SINTEF Civil and Environmental Engineering, Klæbuvn. 153, 7465 Trondheim. T: 73 59 46 00, F: 73 59 53 40, E: geir.svano@civil.sintef.no

Systad, H., Civil Eng., NOTEBY A/S, Hopsnesvn. 21, Pb. 153, 5040 PARADIS. T: 55 91 07 00, F: 55 91 05 74

Sætre, A., Manager, Geogrunn A/S, Pb. 1813, 3703, Skien. T: 35 53 60 02, F: 35 53 77 75, E: ar-saetr@online.no

Sørlie, J.E., Civil Eng., Norwegian Geotechnical Institute, Sognsvn. 72, Pb. 3930 Ullevål Stadion, 0806 Oslo. T: 22 02 30 00, F: 22 23 04 48, E: jes@ngi.no

Sørum, G.A., Civil Eng., Norwegian Geotechnical Institute, Sognsvn. 72, Pb. 3930 Ullevål Stadion, 0806 Oslo. T: 22 02 30 00, F: 22 23 04 48, E: gas@ngi.no

Terjesen, L.B., Civil Eng., Selmer ASA, St. Olavsgt. 25, Pb. 1175 Sentrum, 0107 Oslo. T: 22 03 06 00, F: 22 20 88 30

Thorn, E., Civil Eng., NOTEBY A/S Consulting Engineers, Hoffsveien 1, Pb. 265 Skøyen, 0212 Oslo. T: 22 51 54 00, F: 22 51 54 01

Thorvaldsen, T., Civil Eng., Siv. ing. Trond Thorvaldsen A/S, Rudsletta 90, Pb. 63, 1351 Rud. T: 67 13 99 94

Tilrem, K., Civil Eng., Norwegian National Rail Administration, Stortorget 7, Pb. 1162 Sentrum, 0107 Oslo. T: 22 45 61 50, F: 22 45 61 10

Tokheim, O., Dr Techn, MBA, Feedback Research AS, Rådhusgt 7B., 0151 Oslo. T: 22 00 47 00, F: 22 33 61 66, E: oddvin.tokheim@feedback.no

Torblaa, I., Civil Eng., Torblaa, Langmyrgrenda 39C, 0861 Oslo. T: 22 23 31 39

Tuft, P., Eng., Norwegian Geotechnical Institute, Sognsvn. 72, Pb. 3930 Ullevål Stadion, 0806 Oslo. T: 22 02 30 00, F: 22 23 04 48, E: pt@ngi.no

Tvedt, G., Civil Eng., NOTEBY A/S Consulting Engineers, Hoffsveien 1, Pb. 265 Skøyen, 0212 Oslo. T: 22 51 54 00, F: 22 51 54 01, E: grt@noteby.no

Tyssebotn, R., Civil Eng., NOTEBY A/S, Hopsnesvn. 21, Pb. 153, 5040 Paradis. T: 55 91 07 00. T: 55 91 05 74

Valstad, T., Civil Eng., Norwegian Geotechnical Institute, Sognsvn. 72, Pb. 3930 Ullevål Stadion, 0806 Oslo. T: 22 02 30 00, F: 22 23 04 48, E: tva@ngi.no

Vaslestad, J., Dr. Eng., ViaCon AS, Strandveien 3 B, 2005 Rælengen. T: 63 83 02 97, F: 66 91 96 01

Vinjerui, O., Eng., Civil Aviation Administration, Norway, Wergelandsveien 1, Pb. 8124 Dep., 0032 Oslo. T: 22 94 20 00, F: 22 94 23 90

Vold, R.Chr., Civil Eng., Vold, Jon Østensens vei 12, Pb. 101, 1378 Nesbru, http//ourworld.compuserve.com/homepages/GEODATA. E: geodata@compuserve.com

Wabakken, A., Civil Eng., Wabakken, Bergknappen 5, 1352 KOLSÅS. T/F: 67 13 11 24

Watn, A., Civil Eng., SINTEF Civil and Environmental Engineering, Klæbuvn. 153, 7465 Trondheim. T: 73 59 46 00, F: 73 59 53 40, E: arnstein.watn@civil.sintef.no

Westerdahl, H., M.Sc., Norwegian Geotechnical Institute, Sognsvn. 72, Pb. 3930 Ullevål Stadion, 0806 Oslo. T: 22 02 30 00, F: 22 23 04 48, E: hw@ngi.no

Westerlund, G.J., Dr.Ing., Hardanger Vekst AS, Eitrheimsvn. 32, Pb. 204, 5751 Odda. T: 53 67 10 30, F: 53 67 10 40

Wetterling, S., B.Sc., Soilex AB, Bjurholmsplan 26, S-116 63 Stockholm, Sverige. T: 0046 86 41 88 26, F: 0046 86 40 18 21, E: soilex.aktiebolag@mbox200.swipnet.se

Wiig, T.W., Civil Eng., Norconsult International A.S., Vestfjordgt 4, 1338 Sandvika. T: 67 57 10 00, E: twi@norconsult.no

Williams, P.J., Prof, Carleton University, Department of Geography, 1125 Colonel By Drive, Ottawa, ON K1S 5B6, CANADA. T: 001 613 520 2852, F: 001 613 520 3640, E: peter_williams@carleton.ca

Ødemark, E., Eng., Statkraft Engineering, Veritasveien 26, Pb. 191, 1322 Høvik. T: 67 57 70 10, F: 67 57 70 11

Øiseth, E.Ø., SINTEF Civil and Environmental Engineering, Klæbuvn. 153, 7465 Trondheim. T: 73 59 46 00, F: 73 59 53 40, E: even.oiseth@civil.sintef.no

Øiseth, T., Eng., Entreprenørservice A/S, Rudsletta 24, Pb. 4, 1351 Rud. T: 67 17 30 00, F: 67 17 30 01

Ølnes, A., Civil Eng., Albert Ølnes, Fagerstrand Brygge, Pb. 65, 1464 Fagerstrand. T: 22 50 44 70, F: 22 50 26 25

Østmoe, O.Ø., Civil Eng., NOTEBY A/S Consulting Engineers, Hoffsveien 1, Pb. 265 Skøyen, 0212 Oslo. T: 22 51 54 00, F: 22 51 54 01

Aabøe, R., Civil Eng., Norwegian Public Roads Administration, Gaustadalléen 25, Pb. 8142 Dep., 0033 Oslo. T: 22 07 39 00, F: 22 07 34 44, E: roald.aabøe@vegdir.vegvesen.no

Aas, G., Civil Eng., Groener AS, Fornebuvn 11, Pb. 400, 1327 Lysaker. T: 67 12 80 00, F: 67 12 82 12, E: gaa@groner.no

Aas, P.M., Civil Eng., Norwegian Geotechnical Institute, Sognsvn. 72, Pb. 3930 Ullevål Stadion, 0806 Oslo. T: 22 02 30 00, F: 22 23 04 48

Aastorp, J.I., Civil Eng., Aastorp, Rødkleivfaret 12, 0788 Oslo. T: 67 53 49 36, F: 67 59 09 59

PAKISTAN

Pakistan Geotechnical Engineering Society (PGES)
Secretary: Dr. T. Masood, 6K, Block H, Gulberg II, Lahore, Pakistan. E: hrlcivil@wol.net.pk

Total number of members: 78

Abeer, N., E-in-C's Branch GHQ Rawalpindi. T: 9271253, F: 9271288

Afzal, J., (NESPAK) T: 5160500, 28-A Bankers Street Pir Ghazi Road, Ichhra, Lahore

Agha, A.M., Chief Executive, Pakistan Hydro Consultants, 22 A.E/2, Gulberg III, Lahore. T: 5760150-2

Ahmad, M., Ex. Vice President NESPAK (Geotech Division), 109 (A) E-1, Hali Road, Gulberg III, Lahore

Ahmad, S., 68/B Near Boharwala Chowk, A.I. Road, Railway Colony, Lahore

Ahmad, S.N., V.C. Sir Sayed, University of Engg. & Tech University Road, Karachi

Ahmed, M., Director (Geology), Project Monitoring Office Tarbela Dam

Ahmed, M.M., Senior Engineer NESPAK (G.T. & G.E. Division) St.No.6 Kot Khadim Ali Shah, Sahiwal

Ajaz, A., Vice President NESPAK (M & E Division), House 1-CN-Block Model Town Extension, Lahore

Akhtar, H.S., Consulting Engineer, 58/I, 11th Street, Phase V Defence, Karachi

Akhter, R.B., c/o Pakistan Engineering Services (PES), 188-Y, Phase III Defence Housing Society, Lahore

Akhter, M.A., PMO Tarbela Dam, Vill.294 G.B. Siakpu Tobateksingh

Ali, M., Associate Professor. Civil Dept. UET, Lahore

Ali, S., Chairman Civil Engg. Dept. UET, Taxila

Ali, S.J., ARO DSO Wapda, 19-Birdwood Road, Lahore

Ansari, M.S., PEMCON, B-264, Block-10, Gulshan-e-Iqbal, Karachi

Ansari, Y.S., Senior Geophysicist NESPAK (G.T. & G.E. Division) E-5/37, Rehmania St. Cavalry Ground, Officers Colony, Lahore

Anwar, J., Dy. Managing Director, PES 188-Y Phase-III Defence Housing Scheme, LCCHS, Lahore

Anwar, M., Senior Engineer CE (O & M) Water Tarbela, Ch.No. 4. G.D. Via Rawala Khurd, Okara

Asghar, M.A., Superintending Research Officer Mangla Dam

Aziz, A., Director I.C.C., 65 Main Gulberg II Lahore 36 Tariq Block, New Garden Town, Lahore

Chaudhry, M.M., Director PMO Wapda, Tarbela Dam Project Distt., Hazara

Dodhi, N.I., NESPAK T: 5160500, 175 Ahmed Block, New Garden Town, Lahore

Durrani, I.N.K., Chief Engineer Highway & Transportation Engg. Division (NESPAK). T: 5160500, 93-D-1, Darul Fazal, College Road, NESPAK Society, Lahore

Faruqi, N.A., Director Dso Lahore, 3/18 Block N, Jinnah Park Gulberg II, Lahore

Gilani, S.H., 319-Karim Block, Allama Iqbal Town, Lahore

Goraya, M.A., Assistant Research Officer, DSO Office Wapda, 19-Birdwood Road, Lahore

Haleem, A., Senior Engineer NESPAK (G.T. & G.E. Division) 128, St. No. 3 Cavalry Ground, Lahore

Haq, I., General Manager Tarbela Wapda

Hashmi, A.G., 373-B, Faisal Town 54700, Lahore

Hayat, T.M., Principal Engineer Dams (G.T. & G.E. Division) 1-c N-Block NESPAK 6/8 Mohammed Nagar Allama Iqbal Road, Lahore

Husain, S.S., Senior Engineer (Structures), Associated Consulting Engineers (ACE), 1-C/2 M.M. Alam Road Gulberg III, Lahore. T: 5759417-9

Ilyas, M., General Manager (Geology) NESPAK (G.T. & G.E. Division) 220 P, Model Town, Lahore

Jamal, S.Q., 525 Shadman Colony, Lahore

Jamil, M., Dy. Director Wapda, Khanpur Dam Project

Javed, F., H. No.14-D Officers Colony Narc, Park Road, Islamabad

Khalid, M.M., Director Operations FWO 40-A Chaklala Scheme No. 3, Rawalpindi. T: 590069, 590685

Khaliq, A., Technical Adviser to Member Water Wapda 708-Wapda House, Lahore

Khan, A., D-77 Sanober Colony Tarbela Dam

Khan, A.H., 24-D-1 LDA Officers Colony New Muslim Town, Lahore. T: 5867767

Khan, K., Ex. V.P. (NESPAK). T: 5160500, House 1-C N-Block Model Town Extension, Lahore

Khan, A.S., Principal Engineer Geotechnical PHC Lahore, 342, Street No. 7 Cavalry Grounds Ext., Lahore

Khan, M.A., SDO Wapda Thermal Power Station, Muzaffargarh

Khan, S.M.T., Member Wapda Wapda, Project Monitoring Office, Tarbela Dam Project

Khan, B.A., 11-A, St#13 Main Bazaar Farooq Gunj, Lahore

Kibria, S., Principal Engineer NESPAK (G.T. & G.E. Division) House 1-C N-Block Model Town Extension, Lahore. E: nesp@paknet4.ptc.pk, gtnesp@brain.net.pk

Kunert, N.R.E.W., 57462, 0lpe, Imbergstabe 5 Germany. T: 02761-5016, 02761-5111

Latif, W., Senior Engineer (Geotech), Associated Consulting Engineers (ACE), 1-C/2 M.M. Alam Road, Gulberg III, Lahore. T: 5759417-9, 021-4588750, E: wlatif@usa.net

Majeed, N., H. No-2/3 C Mangla Colony Mangla

Malik, J.R., Director Design (Geotech) CDO Room-804, Wapda House Lahore. T: 9202211, Ext. 2818

Malik, P., 31-T., Gulberg II Lahore. T: 5760799

Munir, F., Civil Dept. UET Lahore, H. No. 40/6 Jalalpur Jathan, Gujrat

Mushtaque, M., (C & W), 96 N, Samanabad, Lahore

Naeem, A., Senior Research Officer (SRO),Village Do-Burji Aryan P.O. Sialkot City

Naseer, S., Sr. Engr. Mangla Dam Project Wapda, H.No: 32/c Shadman Street, Ahmad Munir Road, Ichhra, Lahore

Patel, N.S., CRESOFT, 37 Q LCCHS Lahore Cant

Qadir, M.A., Chief Engineer GTZ., 158-B., Phase III, Govt. Employees Cooperative Housing Authority, Link Road, Lahore

Qayyum, T.I., Professor Civil Engineering Department, L-74 Staff Colony Engineering University, Lahore

Qureshi, M.S., Professor, Foundation Engineering Division, Civil Engg. Dept. University of Engg. & Technology, Lahore

Qureshi, M.A., Senior Research Officer (SRO) CMTL Wapda, 48 Shahdin Scheme Ichhra, Lahore

Rasheed, S., 447-N., Samanabad, Lahore

Rehman, S., 139 E LCCHS Phase-I Lahore 54792

Sadiq, M.M., Quality System Manager GBHP Barotha 67-M, Gulberg III, Lahore. T: 856809

Saeed, I., Regional Director (North) Associated Consulting Engineers (ACE), 1-C/2 M.M. Alam Road, Gulberg III, Lahore. T: 5759417-9, E: aceron@brain.net.pk

Samoon, M.I., Associated Consulting Engineers (ACE).1-C/2 M.M. Alam Road Gulberg III, Lahore. T: 5759417-9, 5760150-1, E: aceron@brain.net.pk

Shabbir, M.J., Chief Engineer (NESPAK) Structural Division T: 5160500 House 1-C N-Block Model Town Extension, Lahore

Shahid, M.A., Civil Dept. UET Lahore, 177/25 Ward No.7 Railway Road, Jhang City

Shamsi, K.A., 347-H, Gulshane Ravi, Lahore. T: 7411015

Sharif, M., SRO Power Complex Ghazi Barotha Hydropower Project Barotha, Atttock

Sheikh, M.S., General Manager G.T. & G.E. Div. NESPAK House 1-C N-Block Model Town Extension, Lahore

Sheikh, A.S., Dean of Civil Engineering, Civil Engg. Dept. University of Engg. & Technology, Lahore

Siddiqi, M.K., ARO Wapda, H/18 Street/3 Kot Shahab Din Shahdara, Lahore

Tariq, M.A., 25-D/1,Gulberg III, Lahore

Victoria, J.J., Wapda/HEPO/GTZ, 56-Gulberg Road, Lahore

Younus, M., HEPO Wapda Sunny View Lahore, 157-C Model Town, Lahore

Zafar, M.S., Site Engineer (ACE), 23-Y-7, Madina Town, Faisalabad

Zaheer, M.M., Senior Engineer NESPAK (G.T. & G.E. Division) 13/60 Askari St.Muhammadi Mohallah Wasanpura, Lahore

Zaidi, S.I.A., H-6 St-4 Chowk Muhammad Pura Ichra, Lahore

PARAGUAY

Secretary: Eng. Rodrigo Labbe, Sociedad Paraguaya de Geotecnia, Avda. España 959, Casilla Correo 336, Asunción, Paraguay. T: (59521)202-424, F: (59521)205-019, E: cpi@conexion.com.py, jbareiro@logos.com.py

Total number of members: 22

Andrada, M., Eng., Teniente Jara Méndez Nr 540, Asunción. T: 59521491968, F: 59521211641, E: maac@pla.net.py
Arce Otaño, O., Eng. Pilotes & Fundaciones S.R.L., Roma Nr. 489, Asunción. T: 59521371682, F: 59521371682, E: guard@conexion.com.py
Bareiro Montiel, J.C., Eng. LOGOS S.R.L., Nuestra Señora de la Asunción Nr 930, Asunción, 11 de Setiembre Nr 1326, Fernando de la Mora. T: 59521445269, F: 59521445277, E: jbareiro@logos.com.py
Bellasai, C., Eng. TAF S.R.L., Piribebuy Nr 621, Ingavi Nr 2639, Asunción. T: 59521443998, F: 59521445130, E: taf@supernet.com.py
Bosio Ciancio, J.J., Eng. Freelance Consultor, Bélgica Nr 1315, Asunción. T: 59521606993, F: 59521606 117, E: bosio@rieder.net.py
Came, O., Eng., Avenida Mariscal López c/ Senador Long, Avenida Aviadores del Chaco Nr 34, Asunción. T: 59521608623, F: 59521608623, E: ocame@pol.com.py
Cantero, N., Eng. TAF S.R.L., Piribebuy Nr 621, Asunción, Inmaculada Concepción Nr 788, Lambare. T: 59521443998, F: 59521445130, E: taf@supernet.com.py
Figueredo, M., Eng. Yegros Nr 32, San Lorenzo. T: 595 21585510, F: 59521585510
Gavilán, E., Eng. ANDE, Washington Nr 439, Julio Rivas Nr 1636, Asunción. T: 59521200954, F: 59521200695
Jiménez, Z., Eng. GEO-STAN S.R.L., Sicilia Nr 941, Corrales Nr 804 esq. Dr. Molinas, Asunción. T: 59521 420592, F: 59521481747, E: geostan@pla.net.py
López Bosio, C., Eng. LOGOS S.R.L., Nuestra Señora de la Asunción Nr 930, Ercilia López Nr 5641, Asunción. T: 59521445269, F: 59521445 277, E: clopez@logos.com.py
Maidana, G., Eng., 25 de Mayo y Kubitschek, Asunción, Las Residentas y Rosendo Benítez Nr 737, Fernando de la Mora. T: 595212172050, F: 595212172050

Mena, F., Eng., Mariscal Estigarribia Nr 2264, Teniente Zavala Nr 921, Asunción. T: 59521200 910, F: 59521200 910, E: menaraul@hotmail.com
Mieres, R., Eng., Luis Alberto del Paraná c/ 12 de Junio, Fernando de la Mora. T: 59521680852, F: 59521680852, E: rdm@telesurf.com.py
Morales, S., Eng., Chile Nr 1229-2°piso-of.7, Darío Gómez Serrato Nr 1225, Asunción. T: 59521490840, F: 59521 490840, E: simonmor@rieder.net.py
Ocampos, C., Eng. Ocampos Construcciones, Acahay c/ Capitán Britos, Comandante Aguirre Nr 791, Asunción. T: 59521292849, F: 59521293048, E: cocampos@conexion.com.py
Ramírez, F., Eng. HARZA, Engineering Company, INC., Independencia Nacional Nr 811, Asunción. T: 59521 440316, F: 59521072315, E: harza@itunet.com.ar
Recalde, A., Eng. ANTELCO, El Mesías y el Salto Nr 1434, Villa Policial, Lambare. T: 59521907 911, F: 59521907 911, E: ingrecalde@hotmail.com
Restaino, C., Eng., Casianov c/ Bertoni Nr 471, Dr. Morra Nr 474 c/ Del Maestro, Asunción. T: 59521601095, F: 59521601095, E: ece@quanta.com.py
Stanichevsky, M., Eng. GEO-STAN S.R.L., Sicilia Nr 941, Papa Juan XXIII Nr 1837, Asunción. T: 59521420592, F: 59521481747, E: geostan@pla.net.py
Tiozzo, A., Eng. Ministerio de Obras Públicas y Comunicaciones (M.O.P.C), Oliva y Alberdi, Profesor Samaniego y Sacramento Nr 1448, Asunción. T: 59521 444140
Torres, P., Eng. CONAVI., Independencia Nacional y Manuel Domínguez, Diego Velásquez Nr 848, Asunción. T: 59521 444340 int. 128, F: 59521444340, E: 436247@telesurf.com.py

PERU

Secretary: Eng. Oscar Neyra Garcia, Societad Peruana de Geotechnica, P.O. Box 679, Lima 18, Peru.
T: +511 433 9591, F: +511 0299, E: arnaldo@amauta.rcp.net.pe

Total number of members: 25

Agreda, Jr, G.L., Iquique 145-103, Lima 05. T: 4339184
Atala, Jr, A.C.A., Arnaldo Marquez 1153, 3 piso, C, Lima 11.
T: 4244107
Crdenas, P.J.L., Tipas 174, BellAv.ista, Callao 02. T: 4339591
Carrillo, D.C., Tipas 174, BellAv.ista, Callao 02. T: 4339591
Carrillo, D.E.F., Tipas 174, BellAv.ista, Callao 02. T: 433
9591, F: 4330299
Carrillo, G.A., Emilio Fernandez 296-703, Lima 01. T: 433
9591, F: 4330299, E: arnaldo@amauta.rcp.net.pe
Carrillo, A.A., Emilio Fernandez 296-703, Lima 01. T: 433
9591, F: 4330299, E: arnaldo@amauta.rcp.net.pe
De La Torre, S.M., Parque Rospigliosi 190-A, Lima 21.
T: 4206340
Donayre, Jr, C.L.E., Los Corales 378-A, Lima 13. T: 4253036
Dongo, I.M., Cayalli 536, Urb. CC Monterrico, Lima 33.
T: 4357550
Harman, I.J., Gervasio Santillana 336, Lima 18. T: 4407126
Hernandez, A.R., Mz. C Lote 21, Urb. Villa Alegre, Lima 33.
T: 2570728

Humala, A.G., Av. Marcona 190, Lima 33. T: 448176328
Madrid, S.M., Av. Velazco Astete 216, Lima 41. T: 4356628
Menendez, G.J., Av. Centenario 637, Cusco, Cusco
Meza, C.L., Jacaranda 773, Lima 33. T: 4484514
Michelena, C.R., Calle Boccioni 229, Lima 41. T: 4231572
Neyra, G.C.O., Av. Santa Catalina 582, Lima 13. T: 4723884,
F: 2653969, E: nisa@terra.com.pe
Neyra, T.J.E., Av. Santa Catalina 582, Lima 13. T: 4723884,
F: 2653969, E: nisa@terra.com.pe
Ordoñez, H.E., Las Jacarandas, Manzana N., Lote 18, Lima 33.
T: 3451501
Rios, B.R., Martin Napanga 214, Lima 18. T: 2415222
Shuan, Jr, L.L.E., Santa Leonor, Lima 31. T: 5369742,
E: Luisashuan@terra.com.pe
Velarde, S.M.J., Matamoros 196, Lima 41. T: 4353643
Tong, M.J., Paseo de la Republica 3905, Lima 34. T: 4410694
Zegarra, P.J., Clemente X 110-701, Lima 17. T: 4626277

POLAND/POLOGNE

Secretary: Prof. Zbigniew Młynarek, Polish Committee on Geotechnics, c/o Akademia Rolnicza, Katedra Geotechniki, ul/Str Piątkowska 94, 61-691 Poznań. T: +48-61-846-64-49, F: +48-61-846-64-51, E: hebo1@wp.pl, mlynhebo@poznan.tpnet.pl

Total number of members: 325

Bałachowski, L., Dr Eng., University Teacher, Technical University of Gdańsk, Geotechnical Department, ul Narutowicza 11/12, 80-952 Gdańsk. T: (+48-58)347-23-50, F: (+48-58) 341-98-14, E: abal@pg.gda

Barański, T., Dr Eng., University Teacher, Department of Geotechnical Engineering, Warsaw Agricultural University, GEOTEKO Geotechnical Consultants Ltd., Wałbrzyska 3/5, 02-739 Warszawa. T: 22-6459518, F: 22-6459519, E: info@geoteko.com.pl

Bartoszewicz, A., Dr Eng., University Teacher. University of Warmia and Mazury, Institute of Building and Sanitary Engineering, ul. Prawocheńskiego 19,10-720 Olsztyn. T: (+48-89) 523-47-98, F: (+48-89)523-47-59

Batog, A., M.Sc., Technical University of Wrocław, ul. Pautscha 5-7/2, 51-642 Wrocław. T: 3202299

Bella, M., Dr, University Teacher, Silesian Technical University Gliwice, ul. Arkońska 7/3, 41-100 Gliwice. T: 32-316886

Biedrowski, Zb, Dr, P-ń Technical Univ., Civil Eng. Institute, Geotechnical Eng. and Geology Division, ul. Piotrowa 5, 61-138 Poznań. T: 61-86652428, H:ul. Św. Rocha 6B m.7, 61-142 Poznań. T: 61-8774429

Biajgo, R., Dr Eng., ul. Jedności Narodowej 123 a/11, 50-301 Wrocław. T: 3228799

Bojanowski, W., Dr Eng., University Teacher, Częstochowa Technical University, ul. Myśliwska 5 m.37, 93-519 Łódź. T: 42-815217

Bolt, A., Dr Habil. Eng, University Professor, Technical University of Gdańsk, Geotechnical Depart., ul. Narutowicza 11/12, 80-952 Gdańsk. T/F: (+48-58)347-29-03, E: abolt@pg.gda.pl

Boniecki, K., M.Sc., ul. Piotrowska 183/7 m.46 90-456 Łódź

Borowczak, P., Dr Eng., Technical Univ., Civil Eng. Institute, Geotechnical Eng. And Geology Dvision, ul. Piotrowo 5, 61-138 Poznań. T: 61-86652123, H:ul. Szubińska 16, 60-114 Poznań. T: 61-8306743

Borowczyk, M., Dr Eng., Consulting Engineer, Sewastopolska 1 m. 36, 02-758 Warszawa. T: 22-8424397

Bracha, J., M.Sc., Meksykańska 3 m.30, 03-948 Warszawa. T: 22-8171335

Broniatowska, M., M.Sc., Eng., Geotechnical Institute, Cracow Technical University, Os. Dywizjonu 303, 46/129, 31-875 Kraków

Broś, B., Prof, Dr Eng., ul. Wyczółkowskiego 6, 51-639 Wrocław

Brzeska, M., M.Sc., Eng., University Teacher, Łódź Technical University, ul. Cieszkowskiego 11 m.1, 93-504 Łódź. T: 42- 486923

Brzesko, Z., Dr Eng., ul. Lipowa 9 m. 10, 02-496 Warszawa

Brząkała, W., Dr Habil. Technical University of Wrocław, ul. Zaolziańska 39/7, 50-566 Wrocław

Buca, B., M.Sc., Consulting Engineer, Geoprojekt Company, ul. Piastowska 74/E/7, 80-363 Gdańsk. T: 58-310418, F: 58-315838

Buca, B., M.Sc, Consulting Eng. GEOPROJEKT Gdańsk, ul. Dyrekcyjna 6, 80-363 Gdańsk. T: (+48-58)301-04-18, F: (+48-58)301-58-38

Bukowski, M., Dr Eng., Opaczewska 33b m. 3, 02-372 Warszawa. T: 22-6584001

Burda, H., M.Sc., Eng., Department of Geotechnics and Earth Structure Agricultural Academy Cracow, 30-040 Kraków

Byczkowski, M., M.Sc., Civ. Eng. Technical University of Gdańsk, Geotechnical Department, ul. Narutowicza 11/12, 80-952 Gdańsk. T: (+48-58)347-28-82, F: 58-341-98-14, E: mbycz@pg.dga.pl

Byrski, M., Eng., WSI Opole, ul. Licealna 2/9, 45-714 Opole. T: 77-745944

Bzówska, J., M.Sc., Eng., University Teacher, Silesian Technical University, ul. Wodna 8 ,44-178 Przyszowice. T: 32-1357212

Cała, M., Dr Eng., Dept. of Mining Geomechanics and Geotechnics, Univ. of Mining and Metallurgy, ul. Wysockiej 2c/59, 31-580 Kraków

Chaciński, Z., Dr Eng., University Teacher, Warsaw University of Technology, Warszawska 24, 05-807 Podkowa Leœna. T: 22-589637

Chmielniak, S., Dr, University Teacher, Silesian Technical University, ul. Kochanowskiego 29/18, 44-100 Gliwice. T: 32-1302313

Chrzanowski, A., Dr Eng., ul. Kraljevska 24/15, 65-052 Zielona Góra. T: 0601 4666281

Cichy, W., Dr Eng., University Teacher, Technical University of Gdańsk, Geotechnical Department, ul. Narutowicza 11/12, 80-952 Gdańsk. T: (+48-58)347-28-09, F: 58-341-98-14, E: wcic@pg.gda.pl

Cielenkiewicz, T., M.Sc., Research Engineer, Research Centre "HYDROBUDOWA", Batuty 1 m. 1104, 02-743 Warszawa. T: 22-8471619

Cieślak, J., M.Sc., Civil Engineer, ul. Batalionów Chłopskich 53/5, 70-770 Szczecin

Coufal, R., D.Sc., Ph.D., Institute of Civil Engineering, Technical University of Szczecin, ul. Jana Styki 2/6,71-138 Szczecin

Czaplicki, J., M.Sc., Engineer, "HYDROBUDOWA BIS" Co Ltd., Bitwy Warszawskiej 1920r, 02-300 Warszawa

Czarnota-Bojarski, R., Dr Eng., Pluga 1/36, 02-047 Warszawa. T: 22-8222627

Damicz, J., Dr Eng., University Teacher, University of Warmia and Mazury, Institute of Building and Sanitary Engineering, ul. Prawocheńskiego 19, 10-720 Olsztyn. T: (+48-89)523-47-98, F: (+48-89)523-47-59

Data, J., M.Sc., Geologist, SALIX, ul. Towarowa 12 m. 61, 15-007 Białystok. T: 85-324039

Dąbrowski, H., Dr Eng., University Teacher, Warsaw University of Technology, Petötiego 4 m. 46, 01-917 Warszawa. T: 22-6638039

Dembicki, E., Prof., Technical University of Gdańsk, Geotechnikal Department, ul. Narutowicza 11/12. T: (+48-58) 347-27-01, F: (+48-58)341-98-14, E: edemb@pg.gda.pl

Dłużewski, J., Prof., Faculty of Civil Enginerering, Warsaw University of Technology, Armii Ludowej 16, 00637 Warszawa. T: 22-8253933, E: Janusz.Dluzewski@il.pw.edu.pl

Dłużewski, R., Dr Eng., ul. Bukowa 1, 05-830 Nadarzyn. T: 22-6213370

Dłużewski, W., M Eng., Academy of Technology and Agriculture Faculty of Environmental and Civil Engineering, Geotechnical Department, ul. S. Kaliskiego 7, 85-796 Bydgoszcz. T: 52-3408157, 52-3408408, H:ul. Żmudzka 43/75, 85-028 Bydgoszcz. T: 52-3420737

Dmitruk, St, Prof., Dr Eng., ul. Legionów 12/13, 50-045 Wrocław

Dmowski, M., ul. Chocimska 116/118, 00-801 Warszawa. T: 22-243332

Dmowski, M., M.Sc., Engineer, Majorki 33a, 02-030 Białołęka Dworska

Dobrowolski, K., M.Sc., Eng., Consulting Engineer, Stalowa 9 m. 17, 03-425 Warszawa. T: 22-6196752

Domski, J., Dr Eng., University Teacher, Geotechnical Institute, Technical University of Cracow, ul. Biernackiego 5/3, 30-4- Kraków. T: 32-335873

Drązkiewicz, J., M.Sc., Consulting Engineer, Designing Office for Maritime Structure, 'PROJMORS.,'ul. Kuruczkowskiego 2, 80-288 Gdańsk. T: 58-520-33-06, F: 58-520-33-04

Dudek, J., Dr, ul. Księgowa 33, 53-335 Wrocław

Dudek, J., M.Sc., Engineer, Przędsiębiorstwo Geologiczne, ul. Mała Góra 16/1/0, 30-864 Kraków

Dudzikowski, R., M.Sc., Consulting Engineer, CONECO, ul. Wyczółkowskiego 17a, Gdańsk. T: 58-302-55-19, F: 58-341-94-11

Dunaj, P., Dr, University Teacher, Faculty of Architecture, Białystok Technical University, ul. Waszyngtona 22 m.5, 15-269 Białystok. T: 85-428345

Dymek, P., M.Sc., Civil Engineer, Poznań Technical University, ul. Głogowska 45 m.1, 60-735 Poznań. T: 616 50867

Dymek, P., Mgr, Eng., Pracownia Inżyniersko-Usługowa GEOPERITUS., ul. Arciszewskiego 29/33, 60-271. T: 0-501-616-088, dom: ul. Pokrzywno 39 B, 61-315 Poznań. T: 0-501-616-088

Dzik, G., M.Sc., Consulting Engineer, KELLER Polska Ltd., Zaciszańska 15, 03-284 Warszawa. T: 22-6755102, F: 22-6757529, E: Keller-Polska@keller.com.pl

Erdini, M., Dr Eng., ul. Łagiewnicka 183, 91-863 Łódź

Fedczuk, P., Dr, University Teacher, WSI Opole, ul.Skałtów Opolskich13/207, 45-286 Opole. T: 77-39212

Fedorowicz, M., M.Sc., Engineer, Geotechnical Institute, Cracow Technical University, ul. Grota Roweckiego 15/74, 30-348 Kraków

Fiedler, K., Dr Eng., ul. Lużycka 10 m.32, 02-732 Warszawa. T: 22-416978

Flisiak, D., Dr Eng., Dept. of Mining Geomechanics and Geotechnics, Univ. of Mining and Metallurgy, ul. Długa 8/42, 32-065 Krzeszowice

Flisiak, J., Dr Eng., Dept. of Mining Geomechanics and Geotechnics, Univ. of Mining and Metallurgy, ul. Długa 8/42, 32-065 Krzeszowice

Florkiewicz, A., Dr Hab. Eng., Prof PP, Technical Univ., Civil Eng. Institute, Geotechnical Eng. and Geology Division, ul. Piotrowo 5. T: 61-86652148, H:ul. Piękna 6 m.1, 60-591 Poznań. T: 61-8474523

Florkowski, J., M.Sc., Eng., HYDROPROJEKT- Bureau of Studies and Design for Hydraulic Eng. in Cracow, ul. Szafera 9/42, 31-543 Kraków

Foltyn, P., M.Sc., Consulting Engineer, GEOTEKO Geotechnical Consultants Ltd., Wałbrzyska 3/5, 02-739 Warszawa. T: 22-6459518, F: 22-6459519, E: info@geoeko.com.pl

Frankowski, Z., Dr Eng., ul. Sekocińska 13 m.6, 02-313 Warszawa. T: 22-236620

Fuks, G., M.Sc., 'Geotest' Tychy, ul. Junaków 17, 43-100 Tychy. T: 32-1177434

Fürstenberg, A., Dr Eng., Consulting Engineer, GEOTEKO Geotechnical Consultants Ltd., Wałbrzyska 3/5, 02-739 Warszawa. T: 22-6459518, F: 22-6459519, E: info@geoteko.com.pl

Gajewski, K., PhD., Dr, M.Sc., Koszalin Technical University, ul. Racławicka 15-17, 75-620 Koszalin

Gałczyński, St., Prof., Dr Eng., Wrocław Technical University, ul. Bacciarellego 81, 51-649 Wrocław

Gamończyk, S., M.Sc., Eng., OBRBG 'Budokop', ul. Niepodległości 134/27, 43-100 Tychy. T: 32-1277644

Garbulewski, K., Dr Eng., University Teacher, Department of Geotechnical Engineering, Warsaw Agricultural University, Nowoursynowska 166, 02-787 Warszawa. T: 228439041, Ext. 11753, F: 22-8470013, E: garbulewski@alpha.sggw.waw.pl

Głażewski, M., M.Sc., Research Engineer, IBMER, Rakowiecka 32, 02-532 Warszawa. T: 22-8493231, Ext. 285

Garlikowski, D., Dr Eng., Wrocław Academy of Agriculture, ul. Rogowska 11/17, 54-440 Wrocław

Gaszyńska-Freiwald, G., M.Sc., Eng., Geotechnical Institute, Cracow Technical University, ul. Halszki 29/5, 30-611 Kraków

Gaszyński, J., Dr, Geotechnical Institute, Cracow Technical University, Ul. Łokietka 57/26, 21-279 Kraków

Glazer, Z., Prof., Dr, ul. Okrężna 68, 02-925 Warszawa. T: 22-424632

Glinicka, M., Dr, University Teacher, Faculty of Building and Environmental Eng., Białystok Technical University, ul. Bielska 9, 15-852 Białystok. T: 85-523255

Glinicki, S.P., Assoc Prof., Faculty of Building and Environmental Eng., Białystok. T: 85-523255

Gołębiewska, A., Dr Eng., University Teacher, Department of Geotechnical Engineering, Warsaw Agricultural University, Nowoursynowska 166, 02-787 Warszawa. T: 22-8439041, Ext. 11753, F: 22-8470013, E: golebiewska@alpha.sggw.waw.pl

Grabowski, Z., Prof., Faculty of Civil Engineering, Warsaw University of Technology, Marszałkowska 7 m. 4, 00-626 Warszawa. T: 22-8253933

Gryczmański, M., Prof., Dr, Silesian Technical University, ul. Góry Chełmskiej 25/15, 44-100 Gliwice. T: 32-316089

Gwizdała, K., Dr Habil. Eng., University Professor. Technical University of Gdańsk, Geotechnical Department, ul. Narutowicza 11/12 80-952 Gdańsk. T : (+48-58) 347-11-32, F: (+48-58)341-98-14, E: kgwiz@pg.gda.pl.

Hauptmann, J., M.Sc., Consulting Engineer, 'BUDMORS'- Consulting, ul. Pilotów 3, 80-460 Gdańsk. T/F: 58-557-43-01, 557-43-02, E: budmors@itnet.pl

Hauryłkiewicz, J., Dr Hab. Eng., Koszalin Technical University, ul. Racławicka 15, 75-620 Koszalin. T: 94-3427881 w. 379, 365, H:ul. Gostyńska 58, 60-103 Poznań. T: 61-8306953

Hawrysz, M., Dr Eng., Wrocław Technical University, ul. Oleszkowskiego 75/2, 51-642 Wrocław

Hellman, M., M Eng., PP Pozprojekt, ul. Zielona 8, Poznań. T: 61-8527581, H: Os. Kosmonautów 20 m. 80, 61-245 Poznań. T: 61-8206318

Herzig, J., Dr Eng., Dept. of Eng. Geology and Environmental Geotechnics, Univ. of Mining and Metallurgy, ul. Witosa 19/78, 30-612 Kraków

Inerowicz, M., M.Sc., Eng., Maritime Institute Gdańsk, ul.Długi Targ 41/42, 80-830 Gdańsk. T: 58-552-03-68, F: 58-552-46-13

Izbicki, R., J., Prof. Dr, Technical Unicersity of Wrocław, ul. Bora-Komorowskiego 61A, 51-210 Wrocław. T: 3202628

Jagosz, Zb, M.Sc., ul. Na Ostatnim Groszu 64/2, 54-207 Wrocław

Jańecki, W., Dr Eng., ul.Drobuta 10, 51-511 Wrocław

Janiński, S., Dr Eng., Technical Univ., Civil Eng. Institute, Geotechnical Eng. and Geology Division, ul. Piotrowo 5. T: 61-86652417, H: Os. Rusa 4 m. 64, 61-245 Poznań. T: 61-8769767, 0-602-433-943

Jaremski, J., Dr, University Teacher, WSI Opole, ul. Katowicka 48, 45-310 Opole. T: 77-39212

Jastrzębska, M., M.Sc., Engineer, University Teacher, Silesian Technical University, ul. Pszczyńska 68, 43-190 Mikołów. T: 32-1263414

Jastrzębski, L., P.h.D., ul. Powst. Wielkopolskich 16/11, 75-100 Koszalin

Jaworska, K., M.Sc., Consulting Engineer, General Board of Subway Construction, Marszałkowska 77-79, 00-683 Warszawa. T: 22-6130058

Jeske, T., Dr Eng., University Teacher, Łódź Technical University, ul. Tokarzewskiego 25a m. 6, 91-842 Łódź. T: 42-570980

Jeż, J., Dr Hab. Eng., Prof. PP., Technical Univ., Civil Eng. Institute, Geotechnical Eng. and Geology Division, ul. Piotrowo 5, 61-138 Poznań. T: 61-86652418, 86652134, H: Os. Wichrowe Wzgórza 8 m. 97, 61-674 Poznań. T: 61-8207093

Jędryka, G., Dr Eng., University Teacher, Department of Geotechnical Engineering, Warsaw Agricultural University, Nowoursynowska 166, 02-787 Warszawa. T: 22-8439041, Ext. 11276

Juszkiewicz-Bednarczyk, B., M.Sc., Eng., Martime Institute Gdańsk, ul. Długi Targ 41/42, 80-830 Gdańsk. T: 58-552-03-68, F: 58-552-46-13

Kaczyński, R., Prof., Department of Geology, Warsaw University, Korsykańska 5 m.13, 02-763 Warszawa. T: 22-6421824

Kajewski, I., Dr Eng., Academy of Agriculture, ul. Stalowa 63/28, 53-440 Wrocław

Kania, M., Dr Eng., Prof. PP., Technical Univ., Civil Eng. Institute, Geotechnical Eng. and Geology Division, ul. Piotrowo 5, 61-138 Poznań. T: 61-86652425, 86652138, H: Os. Stefana Batorego 24 c m.23, 60-687 Poznań. T: 61-8217882, E: kania@sol.put.poznan.pl

Kański, T., Dr Eng., Chocimska 33 m. 3a, 00-791 Warszawa. T: 22-8494171

Kapołka, M., M.Sc., Eng., Drogowa Trasa Średnicowa S.A., ul. Mieszka I., 40-873 Katowice. T: 32-1545003

Karski, J., M.Sc., Engineer, Wałowa 4 m.48, 00-211 Warszawa. T: 22-8318603

Kawalec, D., Dr, University Teacher, Silesian Technical University, ul., Płk. Kiełbasy 62, 43-190 Mikołów. T: 32-1260056

Kawulok, M., Dr, University Teacher, Instytut Techniki Budowlanej Oddział Gliwice, ul. Jasnogórska 14/20, 44-100 Gliwice. T: 32-316481

Kłębek, A., M.Sc., Al. Niepodległości 35 m.30, 02-653 Warszawa. T: 22-432276

Kleczewska-Pawlak, B.M., Geotechnical and Geological, Office 'GEOPROJEKT'-Poznań, ul. Ratajczaka 10/12, 61-815 Poznań. T: 61-8525030. T/F: 8523767, H: Os. Zwycięstwa 9 m.88, 61-646 Poznań

Klich, St, M.Sc., Geoprojekt Kraków, ul. Brodowicza 5/36, 31-518 Kraków

Kłosek, K., Prof., Silesian Technical University, ul. Konarskiego 23B/13, 44-100 Gliwice. T: 32-310232

Kłosiński, B., Dr Eng., Research Engineer, Road and Bridge Research Institute, Jagiellońska 80, 03-301 Warszawa. T: 22-6754375, F: 22-8111792, E: bklosinski@ibdim.edu.pl

Knabe, W., Prof., Dr, Institute of Hydro-Eng., Polish Academy of Sciences, ul. Szara 5/21, 80-116 Gdańsk. T: 58-522011, F: 58-524211

Kociszewski, R., M.Sc., Geoprojekt, Batorego 37 m.57, 02-591 Warszawa. T: 22-8256223

Koda, E., Dr Eng., University Teacher, Department of Geotechnical Engineering, Warsaw Agricultural University, Nowoursynowska 166, 02-787 Warszawa. T: 22-8439041, Ext. 11753, F: 22-8470013, E: koda@alpha.sggw.waw.pl

Koda, F., Dr Eng., University Teacher, Department of Geotechnics, Agricultural University of Warsaw, ul. Bernardyńska 3 m.70, 02-904 Warszawa. T: 22-439041, Ext. 1753, F: 22-470013, X: 814790 sggw pl

Kokowski, J., Dr Eng., H: Os. Przyjaźni 13A m.3, 61-687 Poznań. T: 61-8224139

Koluch, Zb, M.Sc., Consulting Engineer, Hydrogeo Cracow, 1 Box 674, 10-960 Kraków

Kondek, A., M.Sc., Eng., Biuro Projektów Kolejowych Katowice, ul. Jana Galla 11, 41-800 Zabrze. T: 32-1575719

Konderla, H., Dr Eng., Wrocław Technical University, ul. Zatorska 26, 51-215 Wrocław

Koniecko, M., M.Sc., Eng., University Teacher, Częstochowa Technical University, ul. Czapskiego 11, 42-200 Częstochowa. T: 42-643113

Kopański, A., M.Sc., Eng., 'Geobud' Katowice, ul. Sikorskiego 34/36, 40-282 Katowice. T: 32-1554530

Kopka, Z., M.Sc., Eng., University Teacher, Silesian Technical University, ul. Konarskiego 13/4, 44-100 Gliwice. T: 32-315760

Korczyński, J., Eng., PRI Katowice, ul. Misjonarzy Oblatów 26/9, 40-129 Katowice. T: 32-538592

Korzeniowska, E., Dr, Geotechnical Institute, Technical University Kraków, ul. Składowa 12/17, 30-010 Kraków

Kosecki, M., M.Sc., Civil Engineer, ul. Stoisława 5/7, 70-223 Szczecin

Kosmala-Kot, W., Dr Eng., University Teacher, Częstochowa Technical University, ul. Dekabrystów 9/17 m.56, 42-200 Częstochowa. T: 42-255309

Kostrzewski, W., Dr Eng., Technical Univ., Civil Eng. Institute, Geotechnical Eng. and Geology Division, ul. Piotrowo 5, 61-138 Poznań. T: 61-86652427, H: ul. Zagrodnicza 31A., 61-645 Poznań, T: 61-8206538

Kościk, P., Mgr, Eng., Arcadis Ecoconrem, ul. Saperów 49/2, 53-151 Wrocław. T: 3382979

Koszela, J., Dr Eng., ul. Morelowskiego 7, 52-429 Wrocław

Kowalski, J., Prof. Dr, Agricultural University of Wrocław, ul. Buska 10/15, 53-326 Wrocław

Kowalski, K., M.Sc., Consulting Engineer, 'KAPPA' Pracownia Projektów, Ekspertyz i Wdrożeń Budownictwa Specjalistycznego, ul. Kołobrzeska 47A/1, 80-391 Gdańsk. T: 58-536822

Kowalski, W., M.Sc., Eng., ul. Abramowskiego 6/10, 90-355 Łódź. T: 42-372212

Kozielska-Sroka, E., M.Sc., Eng., Department of Geotechnics, Agricultural University of Cracow, ul. Armii Krajowej 89/48, Kraków

Kozłowski, T, M.Sc., Civil Engineer, Szczecin Technical University, al. Piastów 50a, 70-311 Szczecin

Kozubal, J., Mgr, Eng., Technical University of Wrocław, ul. Lakiernicza 20, 53-205 Wrocław.

Krasiński, A., Dr Eng., University Teacher, Technical University of Gdańsk, Geotechnical Depart., ul. Narutowicza 11/12, 80-952 Gdańsk. T: 58-347-22-09, F: 58-341-98-14, E: akra@pg.gda.pl

Krokoszyński, P., Dr Eng., Dept. of Eng. Geology and Geotechnics, Univ. of Mining and Metallurgy, ul. Armii Krajowej 7/117, 30-150 Kraków

Król, P., Dr Eng., University Teacher, Department of Geotechnical Engineering, Warsaw Agricultural University, Nowoursynowska 166, 02-787 Warszawa. T: 22-8439041, Ext. 11751, F: 22-8470013

Królikiewicz, A., M.Sc., Engineer, Krakowskie Przedsię biorstwo Robót Drogowych, Os, 1000-Lecia, Kraków

Krysiak, S., M.Sc., Consulting Engineer, GEOTEKO Geotechnical Consultants Ltd., Wałbrzyska 3/5, 02-739 Warszawa. T: 22-6459518, F: 22-6459519, E: info@geoteko.com.pl

Krzywiec, M., M.Sc., Consulting Engineer, ROLEX., ul. Wasilkowska 10 m.47, 15-137 Białystok. T: 85-751003

Kubiczek, M., M.Sc.,Eng., Regional Manager, 'Contest Melbourne Weeks', ul. Gombrowicza 1 m.139, 42-200 Częstochowa. T: 42-623971

Kuchler, A., Dr Eng., University Teacher, Warsaw University of Technology, Kossutha 8 m. 45, 01-315 Warszawa. T: 22-8369075

Kudlaszyk, M., TRANSPROJEKT Poznań, ul. Chłapowskiego 29, T: 62-8320044, 61-8331901, H: ul. Raszyńska 37 m.36, 60-135 Poznań. T: 61-8619194

Kumor, M.K., Prof. Dr Hab. Eng., Academy of Technology and Agriculture Faculty of Environmental and Civil Engineering, Geotechnical Department, ul. Kaliskiego 7, 85-796 Bydgoszcz. T: 52-3408408, H: ul. Krotoszyńska 3, 85-376 Bydgoszcz. T: 52-3797781

Kurałowicz, Z., Dr, University Teacher Technical University of Gdańsk, Faculty of Hydro and Environmental Engineering, ul. Narutowicza 11/12, 80-952 Gdańsk. T: 58-347-22-20, F: 58-347-24-13, E: zkur@pg.gda.pl

Laskowska, J., Dr Eng., University Teacher, Department of Geotechnical Engineering, Warsaw Agricultural University, Nowoursynowska 166, 02-787 Warszawa. T: 22-8439041, Ext. 11751, F: 22-8470013

Lechowicz, Z., Prof., University Teacher, Department of Geotechnical Engineering, Warsaw Agricultural University, Nowoursynowska 166, 02-787 Warszawa. T: 22-8439041, Ext. 11719, F: 22-8470013, E: lechowicz@alpha.sggw.waw.pl

Lewandowska, J., Dr Eng., University Teacher, Faculty of Environmental Engineering, Gdańsk Technical University, ul.Startowa 17C/4, 80-461 Gdańsk. T: 58-347-28-09, F: 58-341-98-14, E: jlew@pg.gda.pl

Leźnicki, J., Dr Eng., 74 Pease Avenue, Verona, New Jersey 07044

Lichwierowicz, J., M.Sc., Eng., Consulting Engineer, 'Polgeol', ul. 1 Maja 24/26 m.27, 90-178 Łódź. T: 42-338313

Lipiński, M., M.Sc., Research Engineer, Department of Geotechnics, Agricultural University of Warsaw, ul. Króla Maciusia 7a/100, 04-526 Warszawa. T: 22-439041, Ext. 1751, F: 22-470013, X: 814790 sggw pl

Lipiński, M., Dr Eng., University Teacher, Department of Geotechnical Engineering, Warsaw Agricultural University, ul. Nowoursynowska 166, 02-787 Warszawa. T: 22-8439041, Ext. 11752, F: 22-8470013, E: lipinski@alpha.sggw.waw.pl

Lisowska, J., Mgr, Eng., Agricultural University, ul. Bacciarellego 23/5, 51-649 Wrocław. T: 3205544

Liszkowski, J., Prof. Dr Hab., Department of Cainozoic Geology Institute of GeologyAdama Mickiewicza University, ul. Maków Polskich 16, 61-606 Poznań

Loska, F., M.Sc., Consulting Engineer, ul. Batorego 39/1, 80-251 Gdańsk. T: 58-416615

Lossman, W., M.Sc., Consulting Engineer, GEOPROJEKT-Olsztyn, ul. Mickiewicza 17/6, 10-509 Olsztyn. T: 89-527-57-68, F: 89-527-49-86

Lubieniecki, J., Dr Eng., ul. Barycka 3/33, 50-325 Wrocław. T: 3277439

Łacheta, St, Dr Eng., Department of Geotechnics, Agricultural University of Cracow, ul. Pallacha 3/34, Kraków

Łęcki, P., M.Sc., Civil Engineer, ul. Św. Jerzego 15c m.6,61-546 Poznań. T: 22-431471

Łęcki, P., M Eng., PPB CONDIX., ul. Świerzawska 1, 60-321 Poznań. T: 61-8614-008, H: Złotniki, ul. Koźlarzowa 29, 62-002 Suchy Las. T: 0-601-593-879

Łukasik, St, Dr Eng., Building Research Institute, Ksawerów 21, 02-656 Warszawa. T: 22-8431471, Ext. 254, E: S_lukas@itb.pl

Łuszczyński, W., M.Sc., Fabryczna 44 m. 9, 00-446 Warszawa. T: 22-8299634

Madej, J., Prof., Dr, Koszalin Technical University, ul. Pancerniaków 4, 75-644 Koszalin

Malesza, M., M.Sc., Senior Lecturer, Faculty of Building and Environmental Engineering, Białystok Technical University, ul. Porzeczkowa 19 m.48, 15-815 Białystok. T: 85-532896

Mamok, B., Mgr, Eng., Agricultural University, ul. Baccoarellego 23/5, 51-649 Wrocław. T: 3205544.

Mazur, R., M.Sc., ul. Drukarska 11/8, 53-312 Wrocław

Mazurek, J., Dr Eng., Dept. of Mining Geomechanics and Geotechnics, Univ. of Mining and Metallurgy, ul. Duza Góra 35/6, 30-857 Kraków

Mazurkiewicz, B., Prof. Technical University of Gdańsk, Marine Civil Engineering Department, ul. Narutowicza 11/12, 80-952 Gdańsk. T: 58-347-26-11, F: 58-347-14-36, E: bmazur@pg.gda.pl

Meyer, Z., Prof. Dr. Szczecin Technical University, Al. Piastów 50a, 70-311 Szczecin

Michalski, P., Dr Eng., Department of Geotechnics, Agricultural University of Cracow, ul. Prusa 15/4, Kraków

Milancej, P., Dr Eng., Technical University of Gdańsk, Geotechnical Department, ul. Narutowicza 11/12, 80-952 Gdańsk. T: 58-347-12-03, F: 58-341-98-14, E: pmil@pg.gda.pl

Mirecki, J., Dr Eng., University Teacher, Department of Geotechnical Engineering, Warsaw Agricultural University, Nowoursynowska 166, 02-787 Warszawa. T: 22-8439041, Ext. 12003, F: 22-8470013

Miturska, H., M.Sc., Okocimska 5 m. 4. 01-114 Warszawa. T: 22-8405148, Ext.139

Mlynarek, Zb, Prof., Dr Hab., Eng., August Cieszkowski Agricultural University, Department of Geotechnics, ul. Piątkowska 94, 61-691 Poznań. T: 61-8466449, HEBO Poznań, Sp.z.o.o., ul. Podolska 27a, 60-615 Poznań. T: 61 8669030. T/F: 61-8483317, H: ul. Kępińska 9, 61-115 Poznań. T: 61-8305912, E: mlynhebo@POZNAN.TPNET.PL

Molski, T., Dr Eng., Agricultural University of Wrocław, ul. Baciarellego 47/3, 51-649 Wrocław

Motak, E.D., Eng., Institute of Geotechnics, Cracow Technical University, ul. Kryniczna 19/43, 31-463 Kraków

Mrozek, W., Dr Eng., Senior Lecturer, Faculty of Building and Environmental Engineering, Białystok Technical University, ul. Mazowiecka 37A m. 4, 15-301 Białystok. T: 85-442188

Musiał, M., Eng., Institute of Geotechnics, Cracow Technical University, ul. Chelców15-17/6, 31-148 Kraków

Musiał, R., M.Sc., Consulting Engineer, AARSLEFF Ltd., Lambady 6, 02-830 Warszawa. T: 22-6488835, F: 22-6488836, E: aarsleff@aarsleff.com.pl

Naborczyk, J., Dr Eng., Institute of Geotechnics, Cracow Technical University, ul. Armii Krajowej 17/15, 30-150 Kraków

Nadybski, J., M.Sc., Eng., Biuro Projektów Kolejowych w Krakowie, Aleja Prazmowskiego 71, 31-514 Kraków

Niedzielski, A., Dr Hab., Eng., August Cieszkowski Agricultural University, Department of Geotechnics, ul. Piątkowska 94, 61-691 Poznań. T: 061-8466452, H: Os. Zwycięstwa 13 m.44, 61-647 Poznań. T: 61-8208886

Nitecki, T., M.Sc., Civil Engineer, Koszalin Technical University, Ul., Racławicki 15-17, 75-620 Koszalin

Nowacki, J., Eng., Geodrom-Kraków, ul. Św. Jana 2/17, Kraków

Nowak, K., M.Sc., Eng., WODEKO Water Well Drilling and Environmental Engineering Company, ul. Flanka 3a, 30-898 Kraków

Nowak, H., Dr Eng. T: 61-8430329, 30, ul. Dąbrowskiego 97E m.3, 60-574 Poznań. T: 61-8473959

Nowak, J., M Eng., Technical Univ., Civil Eng. Institute, Geotechnical Eng. and Geology Division, ul. Piotrowo 5, 61-138 Poznań. T: 61-86652429, H: Os. Wichrowe Wzgórza 33 m.46, 61-699 Poznań. T: 61-8200274

Odrobiński, W., Prof. Technical University of Gdańsk, Geotechnical Department, ul. Narutowicza 11/12, 80-952 Gdańsk. T: 58-347-13-48, F: 58-341-98-14 , E: wodr@pg.gda.pl

Okruszek, Z., M.Sc., Eng., Research Engineer, Łódź Technical University, al. Wyszyńskiego 63 m.23, 04-047 Łódź

Olejnik, W., Dr Eng., Technical Univ., Civil Eng. Institute, Geotechnical Eng. and Geology Division, ul. Piotrowo 5, 61-138 Poznań. T: 61-86652429, H: ul. Lisa Witalisa 34, 60-195 Poznań. T: 61-8681818

Oprzyński, P., M.Sc., GEOPROJEKT Olsztyn, ul. Mickiewicza 17/6 10-509 Olsztyn. T/F: 89-527-49-86

Orzeszyna, H., Dr Eng., Agricultural University of Wrocław, ul. Baciarellego 32/9, 51-649 Wrocław

Pabian, Zb, M.Sc., Eng., Institute of Geotechnics, Cracow Technical University, Os.Dywizjonu 303 14-142, Cracow

Pachowski, J., Prof., Road and Bridge Research Institute, Œwiêtojañska 17 m. 4, 00-266 Wałszawa. T: 22-8313510

Pallado, J., M.Sc., Eng., Polkon Ltd., ul.Drozdów 48, 40-530 Katowice. T: 32-517752

Pałka, J., Prof., Dr, Institute of Geotechnics, Cracow Technical University, 30-138 Kraków

Paprocki, P., M.Sc., Consulting Engineer, GEOTEKO Geotechnical Consultants Ltd., Wałbrzyska 3/5, 02-739 Warszawa. T: 22-6459518, F: 22-6459519, E: info@geoteko.com.pl

Parylak, K., Dr Eng., Agricultural University of Wrocław, ul. Baciarellego 40/6, 51-649 Wrocław

Pawelczak, M., M.Sc., Os. Rzeczypospolitej 19 m.26, 61-397 Poznań. T: 61-775595

Pawelczak, M., Przedsiebiorstwo Wiertniczo-Geologiczne GEOSONDA., ul. Wilczak 45/47, Poznań. T: 61-8202081 w 276, 274

Pawłowski, A., Dr Eng., Agricultural University of Wrocław, ul. Powstańców Sląskich 122/4., 53-333 Wrocław

Petri, N., Eng., Consulting Engineer, Metroprojekt, Żelazna 34/38 m. 104, 00-832 Warszawa. T: 22-6209284

Piaskowski, A., Prof., Puławska 107a/1, 02-595 Warszawa

Piechowski, N., M.Sc., Civil Engineer, ul. Wrońskiego 2/1, 71-302 Szczecin

Pieczyrak, J., Dr, University Teacher, Silesian Technical University, ul. Kokoszki 6/39, 44-114 Gliwice. T: 32-322030

Pilecki, K., Mgr, Eng., Arcadis Ecoconrem, ul. Legnicka 30/19. 53-673 Wrocław. T: 55721

Pietruszniewicz, S., M.Sc., ZUG 'Geotechnika', ul. Studzińskiego 78 m.15, 91-521 Łódź. T: 42-584393

Pietrzyk, K., Prof., Dr, Geotechnical Institute, Cracow Technical University, ul. Skarwińskiego 2, Kraków

Pinińska, J., Prof., University Professor, Faculty of Geology, Warsaw University, Woziwody 36, 02-908 Warszawa

180 POLAND/POLOGNE

Pisarczyk, S., Prof., Faculty of Civil Engineering, Warsaw University of Technology, Bruna 34 m. 14, 02-594 Warszawa. T: 22-259107
Plichta, Z., M.Sc., ul. Engelsa 41, 75-361 Koszalin
Podgórski, K., Prof., Silesian Technical University, ul. Czestochowska 21/5, 44-100 Gliwice. T: 32-311444
Popiołek, St, M.Sc., Hydrogeo-Kraków, ul. Nadwiślańska 17-19/9, Kraków
Porębski, R., Dr Eng., Technical Univ., Civil Eng. Institute, Geotechnical Eng. and Geology Division, ul. Piotrowo 5, 61-138 Poznań. T: 61-86652485, H: OS. Polan 50 m.7, 61-253 Poznań. T: 61-8750273, 0-601-961-845
Poziemski, J., M.Sc., Eng., ul. Wróblewskiego 25, 51-627 Wrocław
Priebe-Piechowska, M., Ph.D., Civil Engineer, Szczecin Technical University, ul. Wrońskiego 2/1, 71-302 Szczecin
Przedecki, T., Prof., Dr, ul. Piotrowska 257 m.32, 90-456 Łódź. T: 42-375333
Przewłócki, J., Dr Habil. Eng., University Teacher Technical University of Gdańsk, Faculty of Architecture, Department of Building Technology, ul. Narutowicza 11/12, 80-952 Gdańsk. T: 58-347-17-77, F: 58-347-13-15, E: jprzew@pg.gda.pl
Przybyłowicz, W., Dr Eng., Dept. of Geotechnics, Kielce Technical Univ. Ul. Manifestu lipcowego 129/20, 25-432 Kielce
Przystański, J., Prof., Dr Eng., Technical Univ., Civil Eng. Institute, Geotechnical Eng. and Geology Division, ul. Piotrowo 5, 61-138 Poznań. T: 61-86652426, H: ul. Poznańska 37, 62-040 Puszczykowo. T: 61-8133023
Puła, O., Dr Eng., Wrocław Technical University, ul. Morelowskiego 3, 52-429 Wrocław
Puła, W., Dr Eng., Technical University of Wrocław, ul. Ojca Beyzyma 20/8, 53-204 Wrocław. T: 3632685
Rafalski, L., Prof., Road and Bridge Research Institute, Jagiellońska 80, 03-301 Warszawa. T: 22-8110383, F: 22-8111792, E: lrafalski@ibdim.edu.pl
Rawski, H., M.Sc., WSI Opole, ul. Chabrów 98/101, 45-221 Opole. T: 7732021
Rola, S., M.Sc., Consulting Engineer, ROLEX, ul. Witosa 8 m.12, 15-660 Białystok. T: 85614334
Roman, G., M.Sc., ZUG 'Geotechnika', ul. Telewizyjna 41, 91-164 Łódź. T: 42-560247
Rosikoń, A., Prof., ul. Drzymały 17/16, 40-059 Katowice
Rybak, J., Mgr Eng., Technical Unicersity of Wrocław ul. Darmrota 41/8, 50-306 Wrocław. T: 3202841
Rybucki, St, Prof., Dr, Zakład Hydrogeologii i Geologii Inzynierskiej AGH w Krakowie, ul. Mazowiecka 44/17, 30-019 Kraków
Rymsza, B., Dr Eng., University Teacher, Faculty of Civil Engineering, Warsaw University of Technology, Bełska 28 m.3, 02-638 Warszawa. T: 22-8482593
Ryżyński, W., Dr, University Teacher, Faculty of Bulding and Environmental Engineering, Białystok Technical University, ul. Nowy Swiat 13 m.20, 15-453 Białystok. T: 85-421934
Rzeźniczak, J., Dr Eng., Technical Univ., Civil Eng. Institute, Geotechnical Eng. and Geology Division, ul. Piotrowo 5, 61-138 Poznań. T: 61-86652137, H: ul. Kołłataja 150, 61-421 Poznań. T: 61-8323973
Sachajko, S., M.Sc., Eng., ul. 11 Listopada 53 m.25, 91-371 Łódź. T: 42-341172
Sala, A., D Eng., Geotechnical Institute, Cracow Technical University, ul. Brodowicza 5b/28, 31-518 Kraków
Sanecki, L., D Eng., Geotechnical Institute, Cracow Technical University, ul. Boh. Wietnamu 5/30, 31-475 Kraków
Sanowski, W., M.Sc., Eng., Przedsiśbiorstwo Geologiczne w Krakowie, ul. Fałata 9/2, 30-118 Kraków
Sas, W., M.Sc., Research Engineer, Department of Geotechnical Engineering, Warsaw Agricultural University, Nowoursynowska 166, 02-787 Warszawa. T: 22-8439041, Ext. 120-03, F: 22-8470013
Satanowski, L., M Eng., ul. Kościuszki 1 A, 62-800 Kalisz. T: 62-73051, H: ul. Asnyka 45 m.5, 62-800 Kalisz. T: 62-35831
Satkiewicz, Zb, M.Sc., Civil Engineer, ul. Franciszka Gila 6 m.2, 71-457 Szczecin

Sękowski, J., Dr, University Teacher, Silesian Technical University, Ul. Sztabu Powstańczego 4/3, 44-100 Gliwice. T: 32-1792825
Siewczyński, L., Prof., Dr Hab., Eng., Prof. PP, Railway Structure Division, 61-138 Poznań, ul. Piotrowo 5. T: 61-86652431, 86652135, H: ul. Wieczorynki 14, 60-193 Poznań, 61-8681765
Sikora, Zb, Dr Habil., University Professor, Technical University of Gdańsk, Geotechnical Department, ul. Narutowicza 11/12, 80-952 Gdańsk. T: 58-347-20-75, F: 58-341-98-14, E: zbig@pg.gda.pl
Siwiec, S., M.Sc., Eng., Dept. of Geotechnics and Hydrotechnics, Rzeszów Technical Univ. ul. Dominikańska 5/5, 35-210 Rzeszów
Skarżyńska, K., Prof., Dr, Department of Geotechnics, Agricultural University of Cracow, ul. Biskupia 11/9, 31-144 Kraków
Skutnik, Z., M.Sc., Research Engineer, Department of Geotechnical Engineering, Warsaw Agricultural University, Nowoursynowska 166, 02-787 Warszawa. T: 22-8439041, Ext. 12003, F: 22-8470013, E: skutnikz@alpha.sggw.waw.pl
Sławiński, M., M.Sc., ul. Wiązowa 20/8, 53-127 Wrocław
Smoktunowicz, E., M.Sc., Daniłowiczowska 11/10, 00-084 Warszawa. T: 22-8275816
Sobkowiak, J., Dr Eng., Technical Univ., Civil Eng. Institute, Geotechnical Eng. and Geology Division, ul. Piotrowo 5, 61-138 Poznań. T: 61-86652149, H: Os. Stare Żegrze 55 m.7, 61-249 Poznań. T: 61-8797185
Soczawa, A., Dr, University Teacher, Silesian Technical University, ul. Klonowa 8/30, 41-800 Zabrze. T: 32-1717113
Sołtys, K., M.Sc., Morion, ul. 1 Maja 46/41, 41-300 Dabrowa Górnicza. T: 32-1625168
Sorbjan, P., M.Sc., Consulting Engineer, GEOTEKO, Geotechnical Consultants Ltd., Wałbrzyska 3/5, 02-739 Warszawa. T: 22-6459518, F: 22-6459519, E: info@geoteko.com.pl
Stachoń, M., Dr Eng., Technical University of Wrocław, ul.Ułańska 3a/3, 58-100 Świdnica. T: 3203669
Stankowski, Zb, Eng., Railway Structure Division, 61-138 Poznań, ul. Piotrowo 5. T: 61-86652407, H: Os.Przyjaźni 10 m.99, Poznań. T: 61-8200077
Steckiewicz, R., Assoc Prof., Faculty of Building and Environmental Engineering, Białystok Technical University, ul. Mickiewicza 17 m.3, 15-257 Białystok. T: 850327704
Sternik, K., M.Sc.,Eng., University Teacher, Silesian Technical University, Ul. Kosmonautów 3d/9, 47-220 Kędzierzyn-Koźle
Stępkowska, E., Prof. Institute of Hydro-Engineering, Polish Academy of Sciences Gdańsk, ul. Kościerska 7, 80-328 Gdańsk. T: 58-552-20-11, F: 58-552-42-11
Stopa, M., Ph.D., M.Sc., Civil Engineer, Szczecin Technical University, Ul. Domańskiego 2a/3, 71-312 Szczecin
Strzelecki, T., Prof., Dr, ul. Jagodowa 3, 53-007 Wrocław
Stilger-Szydło, W., Prof., Dr, Habil, Technical University of Wrocław, ul. Horbaczewskiego 35/5, 54-130 Wrocław. T: 3515509, 3287217
Subotowicz, W., Prof. Technical University of Gdańsk, Department of Hydrogeology and Engineering Geology, ul. Narutowicza 11/12, 80-952 Gdańsk. T: 58-347-24-92, F: 58-341-24-13, E: wsub@pg.gda.pl
Suchnicka, H., Prof., Dr, Wrocław Technical University, ul. Sudecka 12/21, 50-088 Wrocław.
Sukowski, T., Dr Eng., Consulting Engineer, Hydrogeological Modelling Department, ul. Doroszewskiwgo 18/2, 80-307 Gdańsk. T/F: 58-524482
Sulewska, M., Dr Eng., University Teacher, Faculty of Building and Environmental Engineering, Białystok Technical University, ul. Dziesieciny 41 m.1,15-806 Białystok. T: 85-541033
Surma-Gaczoł, J., M.Sc., Eng., Biuro Projektów Kolejowych w Krakowie, ul. Brak, Kraków
Surowiecki, A., Prof. Dr, Habil, Technical University of Wrocław, ul. Bacciarellego 19/1, 51-649 Wrocław. T: 3489257

Szpakowski, K., M Eng., Academy of Technology and Agriculture Faculty of Environmental and Civil Engineering, Geotechnical Department, ul. Kaliskiego 7, 85-796 Bydgoszcz. T: 52-3408157, H: ul. Stroma 13 a m.325, 85-158 Bydgoszcz. T: 0501-021997

Szpikowski, M., ul. Horbaczewskiego 7 m.20, 03-966 Warszawa

Szwech, Z., M.Sc., Żywnego 16 m.146, 02-701 Warszawa. T: 22-8442526

Szymański, A., Prof, Department of Geotechnical Engineering, Warsaw. Agricultural University, Nowoursynowska 166, 02-787 Warszawa. T: 22-8439041, Ext. 11428, F: 22-8470013, E: szymanski@alpha.sggw.waw.pl

Szymański, L., Dr Eng., University Teacher. University of Warmia and Mazury, Institute of Building and Sanitary Engineering, ul. Prawocheńskiego 19, 10-720 Olsztyn. T: 89-523-47-98, F: 89-523-47-59

Szyszka, J., M.Sc., Eng., Dept. of Geotechnics and Hydrotechnics, Rzeszów Technical Univ. ul. Króla Augusta 33/67, 35-210 Rzeszów

Świdziński, W., Dr Eng., Institute of Hydro-Engineering, Polish Academy of Sciences Gdańsk, ul. Podwale Przedmiejskie 24/12, 80-824 Gdańsk. T: 58-522011, F: 58-524211

Świeca, M., Dr Eng., Research Centre "HYDROBUDOWA", Promyka 5 m.39, 01-604 Warszawa. T: 22-8397645

Świniański, J., Dr Eng., KELLER Polska, ul. Sojowa 2d/30, 81-589 Gdynia. T: 58-629-75-10, F: 58-629-74-70, E: Keller-Gdynia@keller.com.pl

Szcześniak, K., Dr Eng., pl. Muzealny 11/20, 50-035 Wrocław. E: Keller-Gdynia@keller.com.pl

Szkup, St., M.Sc., ul. Klonowa 35 m.37, 91-036 Łódź. T: 42531220

Sztromajer, Z., Dr, University Teacher, ŁódŸ Technical University, ul. Piotrowska 78 m.1, 90-423 Łódź. T: 42-332757

Szwech, Z., ul. Żywnego 16 m.146, 02-701 Warszawa. T: 22-442526

Szymański, A., Prof. University Teacher, Department of Geotechnics, Agricultural University of Warsaw, ul. Pstrowskiego 12, 05-075 Wesoła. T: 22-734863, F: 22-470019, X: 814790 sggw pl

Szymański, L., Eng., University Teacher, Agricultural University of Olsztyn, ul. Iwaszkiewicza 23/17, 10-089 Olsztyn. T: 5234798

Szypcio, Z., Dr, University Teacher, Faculty of Building and Environmental Engineering, Białystok Technical University, ul. Gniła 72, 16-003 Kozińce. T: 85-197134

Tajdus, A., Prof., Dr Hab. Eng. Dept. of Mining Geomechanics and Geotechnics, Univ. of Mining and Metallurgy, ul. Gołaśka 7/14, 30-619 Kraków

Tarnowski, M., Dr, Geoprojekt Szczecin, ul. Chopina 52/8, 71-450 Szczecin

Tejchman, A., Prof. Technical University of Gdańsk, Geotechnical Department, ul. Narutowicza 11/12, 80-952 Gdańsk, T/F: 58-347-15-76, E: atej@pg.gda.pl

Tomaszewski, M., M.Sc., ul. Chlapowskiego 27 m.21, 63-100 Śrem. T: 36022

Tondel, Z., M.Sc., Consulting Engineer, GEOPROJEKT Gdańsk ul. Dyrekcyjna 6, 80-363 Gdańsk. T: 58-301-39-63, F: 58-301-58-38 Tatrzańska

Topolnicki, M., Prof., Dr Habil. Eng., Technical University of Gdańsk, Marine Civil Engineering Department, ul. Narutowicza 11/12, 80-952 Gdańsk. T: 58-347-10-85, F: 58-347-14-36, E: mtopol@pg.gda.pl

Troć, M.M., Technical Univ., Civil Eng. Institute, Geotechnical Eng. and Geology Division, ul. Piotrowo 5, 61-138 Poznań. T: 61-86652428, H: ul. Miśnieńska 30, 60-169 Poznań. T: 61-8678436, 090329761

Tschuschke, W., Dr Eng., August Cieszkowski Agricultural University, Department of Geotechnics, ul. Piątkowska 94, 61-691 Poznań. T: 61-8466453, HEBO Poznań, Sp. z. o.o., ul. Podolska 27a, 60-615 Poznań. H: Os. Wichrowe Wzgórza 13 m.118, 61-675 Poznań. T: 61-8205954, 0601-914-788

Ukleja, K., Prof., Dr Eng., Technical University of Wrocław ul. Głogowczyka 36/2, 51-604 Wrocław. T: 3488176

Ukleja, J., Dr Eng., ul. Rubinowa 13, 52-215 Wrocław. T: 488176

Waliński, K.M., Geotechnical and Geological, Office 'GEOPROJEKT'-Poznań, ul. Ratajczaka 10/12, 61 815 Poznań. T: 61-8525030. T/F: 8523767, H: ul. Szczepana 50, 61-465 Poznań. T: 61-8321954

Werno, M., Prof. Maritime Institute Gdańsk, ul. Długi Targ 41/42, 80-830 Gdańsk. T: 58-552-03-68, F: 58-552-46-13

Wierzba, A., Eng., Institute Geotechnical, Cracow Technical University, ul. Strzelców 15a/26, 31-422 Kraków

Wierzbicki, St, M.Sc., Kruczkowskiego 14 m.15, 00-386 Warszawa

Winskiewicz, M., M.Sc., GEOPROJEKT -Olsztyn, ul. Mickiewicza 17/6 10-509 Olsztyn. T/F: 89-527-49-86

Włodrczyk, M., M.Sc., Eng., ul. Tuszyńska 103/105, 93-376 ŁódŸ

Wnorowska, J., M.Sc., Senior Lecturer, Faculty of Building and Environmental Engineering, Białystok Technical University, ul. Białostoczek 11 m.32, 15-869 Białystok. T: 85-515069

Wojnicka-Janowska, E., M Eng., Zielona Góra Technical University, Civil Eng. Institute Geotechnical and Geodesy Department, ul. Podgórna 50, 65-246 Zielona Góra. T: 68-3282276, 3282286, H: ul. Rasia 9 m.3, 65-520 Zielona Góra. T: 68-3243779

Wojnicki, J., Zb, Dr Eng., Zielona Góra Technical University, Civil Eng. Institute Geotechnical and Geodesy Department, ul. Podgórna 50, 65-246 Zielona Góra. T: 68-3282276, 3282286, H: ul. Wypoczynek 22/18, 65-519 Zielona Góra. T: 68-3263990

Wojtasik, A., Dr Eng., Technical Univ., Civil Eng. Institute, Geotechnical Eng. and Geology Division, ul. Piotrowo 5, 61-138 Poznań. T: 61-86652417, H: ul. Porażińskiej 37, 60-195 Poznań. T: 61-8684069, 0-502-649-120

Wojtowicz, M., Mgr, Eng., ul. P. Jasienicy 51, Wrocław

Wojtuszewicz, A., M.Sc., ul. Odziezowa 26/11, 71-502 Szczecin

Wolski, W., Prof., Department of Geotechnical Engineering, Warsaw Agricultural University, ul. Londynska 14 m.10, 03-921 Warszawa. T: 22-8176888, F: 22-8470013, E: info@geoteko.com.pl

Woziwodzki, Zb, Dr Eng., Academy of Technology and Agriculture Faculty of Environmental and Civil Engineering, Geotechnical Department, ul. Kaliskiego 7, 85-796 Bydgoszcz. T: 52-3408407, H: ul.Wyzwolenia 98 m.7, 85-796 Bydgoszcz

Woźniak, H., Dr Eng., Dept. of Engineering Geology and Environmental Geotechnics, Univ. of Mining and Metallurgy, ul. Kwiatkowskiego 8, 30-135 Kraków

Woźniak, M., Prof., Łódź Technical University, ul. Międzynarodowa 58 m.63, 03-992 Warszawa. T: 22-617954

Wyjadłowski, M., Dr Eng., Technical University of Wrocław, ul. Dźwirzyńska 24/6, 54-320 Wrocław. T: 3203348

Wysokiński, L., Prof., Building Research Institute, Dept. of Geotechnics, Zdrojowa 23, 02-925 Warszawa. T: 22-8426366, 22-8488668, E: grunty_itb@zigzg.pl

Zbielska-Adamska, K., M.Sc., University Teacher, Faculty of Building and Environmental Engineering, Białystok Technical University, ul. Krupniki 32, 15-623 Białystok. T: 85-424545

Zaborowski, M., Dr Eng., ul. Ciołkowskiego 25, 01-980 Warszawa

Zadroga, B., Prof. Technical University of Gdańsk, Geotechnical Department, ul. Narutowicza 11.12, 80-952 Gdańsk. T: 58-347-27-01, F: 58-341-98-14, E: bzad@pg.gda.pl

Zapał, A., M.Sc., Eng., Geotechnical Institute, Cracow Technical University, ul. Powiśle 2/4, 31-101 Kraków

Zarucki, T., M.Sc., Consulting Engineer, Design Office, Agricultural Building Engineering, ul. Rolna 85, 10-804 Olsztyn. T/F: 89-527-49-86

Zawalski, A., M.Sc., Eng., Academy of Technology and Agriculture Faculty of Environmental and Civil Engineering, Geotechnical Department, ul. Kaliskiego 7, 85-796 Bydgoszcz. T: 52-3408407, H: ul. Andersena 8 m.47, 85-792 Bydgoszcz. T: 52-3435479

Zawisza, E., D Eng., Department of Geotechnics, Agricultural University of Cracow, ul. Włoska 15/23, 30-681 Kraków

Zych, J., Prof., Silensian Technical University, ul. Opawska 18/16, 44-100 Gliwice. T: 32-320014

Zychowicz, R., Engineer, ul. Wiolinowa 11 m.16, 02-789 Warszawa

Żmudziński, Z., Prof., Dr, Geotechnical Institute, Cracow Technical University, ul. Malawskiego 60, 31-471

183

PORTUGAL

Secretary: Fernando Pardo de Santayana, Sociedade Portuguesa de Geotecnia (SPG), Laboratório Nacional de Engenharia
Civil, Avenida do Brasil, 101, 1700-066 Lisboa. T: +351 21 844 3321, F: +351 21 844 3021,
E: fpardo@lnec.pt, spg@lnec.pt

Total number of members: 223

Abreu, R.M.M. de, Estrada do Desvio, 28-3° Esq°, 1750-080 Lisboa

Abrunhosa, M.J.F.G., Museu Lab. Mineralogia e Geologia, Faculdade de Ciências do Porto, Praça Gomes Teixeira, 4050-290 Porto

Alfaro Lopes, M. da G.D., Instituto Superior de Engenharia de Lisboa, Rua Conselheiro Emídio Navarro, 1900 Lisboa

Almeida, I.M.B.M., Av. João XXI, 6-3° Esq°, 1000-031 Lisboa

Almeida, L.M.F., Hidroprojecto, Av. Marechal Craveiro Lopes, 6, 1749-010 Lisboa

Alvarez, N.M.C.M., Rua Cidade da Beira, 50-5° Dt°, 1800 Lisboa

Alves, Á. de O.F., Rua. Eugénio de Castro, 238, Habitação 79, 4100-225 Porto

Amaral, A.R., Bairro São João da Carreira, Rua Mimosas, 13, Travassos de Cima, 3500 Viseu

Amaral, F.J.C., Rua Fialho de Almeida, 21-2°, Apartamento 6, 4400 Vila Nova de Gaia

Amaral, J.A. dos S.S. do, Travessa das Mercês, 50-3° Fte, 1200 Lisboa

Anahory, C.I.F.G.R., Rua Washington, 9-3°, 1170 Lisboa

Andrade, J.J. dos S. de, Av. do Uruguai, 8-R/c Dt°, 1500-613 Lisboa

Andrade, R.J. da S., Alameda Jardins da Arrábida, 436-3° B, 4400-478 V. Nova de Gaia

Antão, A.M. de M.C., Rua Comandante Salvador do Nascimento, 78-4° Esq°, 6300 Guarda

Antunes, J.A.R., Av. Miguel Bombarda, 15, 2710 Sintra

Araújo, P.A.M. de, Praça do Condestável, 117-5° Dt°-Tras., 4700 Braga

Baptista, M. de L.P., Av. Afonso de Albuquerque, 203-7E, S. António da Caparica, 2825 Costa da Caparica

Barbosa, P.J.N.F., Vale de Lagar, Lote 22, C. P. 52, 8500 Portimão

Barroso, J.A., Rua Dr. Celso Dias Gomes, 134, Piratininga 24300, Niteroi RJ, BRASIL

Basto, D.C.S.T., Rua do Molhe, 24-3° Esq°, 4150-498 Porto

Bastos, M.J.N., Praceta Dionísio Matias, 12-1° Dt°, 2780 Paço de Arcos

Begonha, F.M.H. Bacelar de, Praceta Rainha Santa, 3-5° Dt°, 1600-687 Lisboa

Beiro, N.J.M., Rua Marechal Gomes da Costa, Lote 57, R/c, Famões, 1675 Pontinha

Bento, F.S.G., Rua do Paraiso, 42-4°, 4000 Porto

Bonito, F.A.B., Correios Universidade, Ap. 2006, 3810 Aveiro

Borges, M.A.A., Rua República da Bolívia, 91-2° Dt°, 1500 Lisboa

Brito, J.A.M. de, Urb. da Portela, Lote 144-10° Esq°, 2685 Sacavém LRS

Cabanas, C.M.M., Travessa do Pote de Água, 4-R/c, 1700 Lisboa

Caldeira, L.M.M.S., LNEC-DG, Av. do Brasil, 101, 1700-066 Lisboa

Câmara, M.I.C. de Aguiar, Av. Praia da Vitória, 34-2°, 1000-248 Lisboa

Cardoso, A.J.M.S., Faculdade de Engenharia do Porto, Rua Dr Roberto Frias, 4200-465 Porto

Cardoso, J.C.P., Campus de Santiago – Universidade de Aveiro, Departamento de Engenharia, 3850 Aveiro

Cardoso, M.A.L.O., Rua Lagares D'el Rei, 21-2° Esq°, 1700-268 Lisboa

Carlos, A. da Palma, Rua. Diogo Bernardes, 18-5° Dt°, 1700-129 Lisboa

Carmo, C.S.N.E., Rua Vila do Seixal, 19-R/c Esq., 2610 Feijó

Carneiro de Barros, R.M.M., Rua Eugénio de Castro, 370-Hab. 66, 4100 Porto

Carreto, J.M.R., Praça de Londres, 8-5° Esq°, 1000 Lisboa

Carvalho, A.T.P. de, Largo Costa Pinto, 11R/c Dt°, 2800 Pragal

Carvalho, J.A.R. de, Rua Prof. Prado Coelho, 27-3° Dt°, 1600-651 Lisboa

Carvalho, R.C. de, Rua Casal do Sobreiro, 15, Livramento, 2765-368 Estoril

Caspurro, I.M. dos Santos, R. do Calvário, 270, Águas Santas, 4445 Ermesinde

Castilho, A.M.de Aguiar, Rua Gonçalves Zarco, 29, 3030-053 Coimbra

Catarino, A.E.L., Rua Marchal Saldanha, 724-2°A, 4150 Porto

Cavaleiro, V.M.P., Universidade da Beira Interior, Av. Marquês-D'Ávila e Bolama, 6200 Covilhã

Cavilhas, J.L.A., Largo Cristovão da Gama, 10-9° Frente, Damaia de Baixo, 2720-154 Amadora

Centro de Estudos de Geologia e Geotecnia de Santo André, Lugar da Galiza, 7500 Santo André

Cerejeira, J.M.F.G., Alameda Fernão Lopes, 29-8° Dt°, 1495-136 Algés

Chaves, J.A.B., Rua António Stromp, 5-3° Dt°, 1600 Lisboa

Claro, E.M. da Costa, Rua Santo Estêvão, 5-R/c Dt°, Algueirão, 2725-128 Mem-Martins

Coelho, A.M.L.G., LNEC-DG, Av. do Brasil, 101, 1700-066 Lisboa

Coelho, S.A., E.S.T.D., Av. das Forças Armadas, 125-4°A, 1600-079 Lisboa

Constantino, M. da S.M., Hidrotécnica Portuguesa-Edifício H.P., Rua. David Sousa, 9-1°, 1000-105 Lisboa

Correia, A.G., Instituto Superior Técnico-DEC, Av. Rovisco Pais, 1052-001 Lisboa

Correia, J.E.C., Rua Prof. Moisés Amzalak, 8-1° C, 1600-648 Lisboa

Correia, P.A.G.D., Praceta da Concórdia, Lote B-1° Dt°, 2675-603 Odivelas

Correia, R.M.B.P., LNEC-Direcção, Av. do Brasil, 101, 1700-066 Lisboa

Coutinho, A.G.F.S., R. Eugénio Salvador, Lote 10, 5° Esq°, 1600-448 Lisboa

Couto, T.R., Quinta de Sernado, 4760 Vila Nova de Famalicão

Cruz, A.A.M., Quintela, Tetraplano, Rua Tobis Potuguesa, Lote 10-2° Esc.9, 1750-292 Lisboa

Cruz, F.I.N. e Santos, Rua Sá de Miranda, 49-2° Esq°, 3000-353 Coimbra

Cruz, N.B. de Faria, Praceta Amorim Carvalho, 163-6° Dt°, 4460 Senhora da Hora

Cunha, M.J.F., Rua Júlio Dinis, 206-9° Centro, 4050 Porto

Dias, S.C.P.G., Estrada do Zambujal, V.J.F.G.D, Zambujal, 2640 Mafra

Diniz, R.M.R., Casal Carvalho Velho, 2600 S. João dos Montes

Duarte, P.P.C.T., Fundações Teixeira Duarte, Av. da República, 42-8°, 1050-194 Lisboa

Espinoza, M.T., Tomas de Bemur, 254 Y Fco Urrutia, Quito – Equador

Estrada, Jr, J.A.G., Andres Ramon Mejia, N. 793, Huaraz – Ancash, Perú

Falcão, J.M.C.B., LNEC-DB, Av. do Brasil, 101, 1700-066 Lisboa

Falorca, I.M. da Conceição Fonseca Gonçalves, Urbanização Quinta das Palmeiras, Eixo TCT, Lote 5-5° Esq°, 6201-907 Covilhã

Faria, P.A.G., Estrada de Benfica, 452-3° Dt°, 1500 Lisboa
Felix, C.M. da Silva, Rua Dr³ Maria Manuela Moreira de Sá, 230, 4465 S.Mamede de Infesta
Fernandes, B.P.A., Rua Dr. António Elvas, 49-1° Dt°, 2810-167 Almada
Fernandes, M.A. de Matos, Gabinete de Estruturas – Faculdade, de Engenharia – Rua Dr Roberto Frias, 4200-465 Porto
Ferreira, A.M.G. dos Santos, Rua Alegre, 41-1° Esq°, 1495 Algés
Ferreira, A.M. de Sousa C. Liberal, Rua Franco Nogueira, 11-1° Esq°, 2780 Oeiras
Ferreira, C.A.M., Rua Barreiros Cunha, 80-2° Dt°, Argacosa, Meadela, 4900 Viana do Castelo
Ferreira, H.N., Laboratório Engenharia Civil de Macau, Rua da Sé, 22, Macau (Via Hong-Kong)
Ferreira, M. e O.Q., Universidade de Coimbra, Departamento de Ciências da Terra, Apart. 3014, 3049 Coimbra – Codex
Ferreira, M.R.P.V., Praceta do Cidrel, 3, 3000 Coimbra
Ferreira, P.J.A., Av. Prof. Dr. Egas Moniz, 7-4° Esq°, Massamá, 2745 Queluz
Ferreira, S.C.P.C., Rua General Ferreira Martins, 1-5° Dt°, 1495 Algés
Ferreira, V.A.V.P.M., Av. Infante Santo, 64 C-7° Esq°, 1350-180 Lisboa
Flôr, A.F.T., Calçada do Moinho de Vento, 18-1A, 1150-236 Lisboa
Folque, J. de B., Alameda D. Afonso Henriques, 13-3° Esq°, 1900 Lisboa
Fonseca, A.J.P.V., Rua Nova da Ribeira, 168, S. Pedro de Avioso, 4470 Maia
Freitas, R.J.F. da Luz, Rua Cidade da Bolama, Lote 21, 11°D Frente, 1800 Lisboa
Furtado, R.J. de Almeida, Rua Antero de Quintal, 42-3° Dt°, 3000 Coimbra
Galhano, A.C.G.A., Rua da Marafusa – Lote 11 – Arades, 3810 Aveiro
Garcia e Costa, C.M.R. de Carvalho, Largo Dr. Francisco Sá Carneiro, 61-6° Esq°, 8000 Faro
Geosolve-Soluções de Engenharia, Geotecnia e Topografia, Ld³, Apartado 358, 2776-904 Carcavelos
Gomes, A.M.T., Padrão-Vila Maior-Santa Maria da Feira, 4535 Vila Maior
Gomes, C.M.B.C., R. do Penedo, Casa do Sobreiro, 4405 Valadares
Gomes, R.P.C., Rua Frederico George, Lote 23-1°Esq°, Telheiras Norte, 1600 Lisboa
Gonzalez, J.M.V., Av. Afonso de Albuquerque, 203-7° Esq°, S. António da Caparica, 2825 Costa da Caparica
Grossmann, N.F., LNEC-DB, Av.do Brasil, 101, 1700-066 Lisboa
Guedes, J.F., Rua Viana da Mota, 87, 4470 Maia
Guerra, N.M. da Costa, Largo Pedro Correia Marques, 3-2° Dt°, 1500 Lisboa
Guerreiro, H.J.P., Rua Joaquim de Almeida, 4-2° Esq°, Botequim – Quinta da Barriga, 2815 Charneca
Guimarães, J.M. da Silva, Edifício Vale Marinha, 2°Q, Caniço de Baixo, 9125 Caniço
Hadjadji, T., LNEC-DVC, Av. do Brasil, 101, 1700-066 Lisboa
Horta, J.C.C.S., Av. Bombeiros Voluntários, 42-3° F., Algés, 1495 Lisboa
Jorge, C.M.M., Rua Domingos de Paiva, Lote 4-1° Dt°, 2735 Cacém
Jorge, S., Av. da Liberdade, 18-Paivas, 2840 Seixal
Kaidussis, F.R.N., Rua Dr. José António Serrano, Lote 99-8° Esq° - Massamá, 2745 Queluz
Laboratório de Engenharia de Macau, Rua da Sé, 22, Macau (Via Hong Kong)
Laboratório de Ensaios de Materiais de Oeiras, Câmara Municipal de Oeiras, 2780 Oeiras
Ladeira, F.L., Rua Cândido dos Reis, 41-2° Esq°, 3800-357 Aveiro
Leite, A.G.M., Praça Professor Santos Andrea, 13-2° Dt°, Benfica, 1500 Lisboa

Lemos, L.J.L., Urbanização Quinta D. João, Lote 12-4° Andar, 3030 Coimbra
Lima, D.C., Condomínio Acamari, 98-36570-000, Viçosa, Minas Gerais, BRASIL
Lopes, A.J.G. dos Santos, Vivenda Santos Lopes, Rua. Josefina do Livramento, Sobreda da Caparica, 2825 Monte da Caparica
Lopes, J.J.A.B., Av. do Brasil, 182-1°A, 1750-618 Lisboa
Lopes, M.J.F. de Pinho, Universidade de Aveiro, Secção Autónoma de Engenharia, Campus, Universitário de Santiago, 3810 Aveiro
Lopes, M. de L. da Costa, Faculdade de Engenharia. da Universidade do Porto, Dep. de Eng³ Civil, Secção Geotecnia, Rua Dr. Roberto Frias, 4200-465 Porto
Lucas, F.J.F., Escola Superior de Tecnologia, Av. do Empresário, 6000 Castelo Branco
Luis, J.A.M.A., Rua Feliciano Ramos, 32-12° C, 4700 Braga
Machado, A.P.G., Rua D. Nuno Rodrigues de Almeida, 7-2° Dt, 2300 Tomar
Machado, F.A., Rua Mariano Pina, Lote 15097-7° C, 1500 Lisboa
Machado, J.M. da Costa, Av. 1° de Maio, 299-2° Esq°, 5000 Vila Real
Magalhães, C.L. dos Santos Filipe, Av. Estados Unidos da América, 118-9° Dt°, 1700-179 Lisboa
Malvar, W.E., Rua Manuel Teixeira Gomes, 13-4° Dt°, 2735 Carnaxide
Manuel, B.P., Rua Rei Katiavala, 30-4°Andar B, 10867 Luanda, Angola
Marques, J.M.M.C., Faculdade de Engenharia da Universidade do Porto-Estruturas, Rua Dr. Roberto Frias, 4200-465 Porto
Marques, M.A.L., Rua Oliveira Martins 15, 1° Esq°, Feijó, 2800 Almada
Marques, S.H.M., Rua Florentino Pedro Lopes, 17, S. Romão, 2410 Leiria
Martins, F.F., Rua Nascente, 211-2° Dt°, 4710 Braga
Martins, J.B., Rua Padre António Vieira, 62-6° Esq°, 4710 Braga
Mata, P.M.H.P. da, Rua Vencedores 3 Quintinhos, 2815-354 Charneca da Caparica
Matos, A.M.B.C.E., Rua. Oliveira Martins, 181 Hab. 13, 4200 Porto
Melaneo, F.F.S., Rua Maria Veleda, 3-5° G, 1500 Lisboa
Melo, F.A.G., Rua Tenente Coronel Ribeiro dos Reis, 6-3° Esq°, 1500 Lisboa
Menezes, J.E.T.Q. de, Rua Dr. Joaquim Pires de Lima, 308, 4200 Porto
Mineiro, A.J.C., Av. Barbosa Du Bocage, 130-4° Dt°, 1050 Lisboa
Mira, M.J.M.M.F., Rua Cidade do Rio Maior, 1, 2040-052 Azambujeira
Monteiro, B.P.B., Travessa. do Possolo, 4, 1350 Lisboa
Monteiro, M.J.R., Rua Aval de Cima, 137-4° Dt°, 4200 Porto
Moreira, A.P. de Amorim M., Rua Padre António Lino de Sousa Vale, 201-2° Esq°, 4440 Valongo
Morgado, J.M. dos Santos, Rua Água das Maias, 34, 2300 Tomar
Nascimento, Ú. da Fonseca, Av. de Roma, 14-5° Dt°, 1000 Lisboa
Neves, A.S.F., Tecnovia Madeira, Estrada da Eira do Serrado, 42, Santo António, 9000 Funchal
Neves, E.J.L. Maranha das, Rua Prof. Queiroz Veloso, Lote 15, 1600 Lisboa
Neves, J.M.C. das, Rua D. Maria I., 39-1° Esq°, 2734 Cacém
Nogueira, J.N.V. e Sá, Ministério das Obras Públicas, Laboratório de Engenharia da Guiné, C. P.14, República da Guiné-Bissau
Novatecna Europa Lda, Consolidações e Construções Lda, Rua dos Correeiros, 92-1° Esq°, 1100-167 Lisboa
Nunes, A.J. de Carvalho, Rua Bombeiros Voluntários Lote 12 A-2° Esq°, 2560 Torres Vedras
Oliveira, R.A.M. de, Coba, Av. 5 de Outubro, 323, 1600 Lisboa
Pacheco, J.A.L.D., Rua Maria Veleda, Torre 2-15° C, 1500 Lisboa

Pais, L.J.A., Av. S. Salvador, 15-R/c Dt°, 6200 Teixoso
Pascoa, J.M.S.C., Rua José Rodrigues Migueis, 3-2°B, Benfica, 1500-378 Lisboa
Paula, J.A.A. de, Praça João do Rio, 2-2° Dt°, 1000-180 Lisboa
Pereira, A. da Silva Costa, Rua Belo Horizonte, 9-2° Dt°, Paço D'arcos, 2780 Oeiras
Pereira, C. dos Santos, Rua. Carvalhão Duarte, 8-5 D, 1600 Lisboa
Pereira, L.I.L. da Silva, Rua Frei Fortunato São Boaventura, 47 R/c, 1900-242 Lisboa
Pereira, O.J.P., Rua Gonçalo Cristovão, 294-7° Dt°, 4000 Porto
Pimentel, V.M.J., Quinta do Marquês, Lote 10-3° A, 2780 Oeiras
Pinelo, A.M.S., LNEC-DVC, Avenida do Brasil, 101, 1700-066 Lisboa
Pinho, A.B. de, Universidade de Évora, Departamento de Geociências, 7000 Évora
Pinho, M.A.M. de, Rua Prof. Reinaldo dos Santos, 12-12° Dt°, 1500 Lisboa
Pinto, A. da Luz, Rua Marquês de Soveral, 9-3° Dt°, 1700 Lisboa
Pinto, A.A.V., LNEC-DG, Av. do Brasil, 1700-066 Lisboa
Pinto, A.D., Rua Inocêncio Francisco da Silva, 18-3° Esq°, 1500 Lisboa
Pinto, M.I.M., Urbanização Vale das Flôres, Lote 7-2° Esq°, 3000 Coimbra
Pinto, P.S.S. e, LNEC-DG, Av. do Brasil, 101, 1700-066 Lisboa
Pinto, S.L.D.L., Rua 11, Lote 212-3° Esq°, Tapada das Mercês, 2725 Algueirão
Pinto de Sousa, Ricardo de Oliveira Carneiro, Rua Mouzinho de Albuquerque, 182-1°, 4435 Rio Tinto
Portugal, J.M.C.S., LNEC-DG, Av. do Brasil, 1700-066 Lisboa
Projectope-Gabinet, de Topografia e Projectos Lda, Av. Fontes Pereira de Melo, 35-3° F/J., 1050-118 Lisboa
Quadros, C.A., Rua Maria Teles Mendes, 6-10° Dt°, 2780-659 Paço de Arcos
Quaresma, L.M.T., LNEC-DVC, Av. do Brasil, 101, 1700-066 Lisboa
Quelhas, J.A. da Silva, Metro do Porto, Av. dos Aliados, 133-2°, 4000 Porto
Ramos, A.N.M., Rua Bombeiros Voluntários, 25, 4° Esq°, 3885 Esmoriz
Rebelo, V.M. de Albuquerque, Urbanização da Quinta da Caldeira, Unidade 8A, Torre 3 (Ed.Jardim) 14° B Cidade Nova, 2670 Loures
Rêgo, M.M. Isabel Lopes do, Rua Visconde de Almeidinha, 12, 3830 Ílhavo
Resende, A.J.C.N., Av. Dr. Fernando Aroso, 878-4° Dt°, Leça da Palmeira, 4450 Matosinhos
Ribeiro, M.E. de Carvalho Alves, Rua 28, 576-3°, 4500 Espinho
Ribeiro, M.A., Apartado 82, Machico Codex, 9201 Madeira
Rodrigues, A.J.V.M., Sociedade Panificação Marão Lda, Av. Madame Brouillard, 5000 Vila Real
Rodrigues, C.F.S., Urbanização dos Moinhos, Lote 7, Casa 3 – Sitio dos Moinhos-Assomada, 9125 Caniço – Funchal
Rodrigues, C.M.G., Qta dos Bentos, Rua das Barreiras, 1-2° Esq°, 6300 Guarda
Rodrigues, F.M. da C.P., Rua do Parque, 73 L., Bairro de Santa Cruz, 1500 Lisboa
Rodrigues, J.D., LNEC-DG, Av. do Brasil, 101, 1700-066 Lisboa
Rodrigues, L.G., Rua Actor Taborda, 27-1° Esq°, 1000-007 Lisboa
Rodrigues, R. da Costa, Rua da Soenga, 142-Coimbrões, Santa Marinha, 4400 Vila Nova de Gaia
Rodrigues, V.A.C., Rua Antero de Figueiredo, Lote 56, 3°A, 2795-016 Linda a Velha

Sacadura, C.J.G., Praceta Rui de Pina, 1, 2855 Corroios
Salgado, F.M.G.A., LNEC-DG, 1700-066 Lisboa
Sanches, R., Rua Soldados da India, 30, 1400 Lisboa
Santayana, F.P. de, LNEC- DG, Av. do Brasil, 101, 1700-066 Lisboa
Santos, J.A. dos, Praceta Cristóvão Falcão Lote 53-9° Dt°, Massamá, 2745 Queluz
Saraiva, A.L. de Almeida, Departamento de Mineralogia e Geologia, Universidade de Coimbra, 3001-453 Coimbra
Secil Britas, S.A., Termas de São Vicente, 4575 Paredes PNF
Serra, J.P.B., LNEC-DG, Av. do Brasil, 101, 1700-066 Lisboa
Serrasqueiro Rossa, Sara Rita Louro Guerreiro, Rua Capitão Alfredo Guimarães, P. Belo Horizonte, 6° D, 4800 Guimarães
Silva, V.M.B.C.E., Rua Pedro Nunes, 27-1°, 1050 Lisboa
Silva, A.F.F. da, Rua José Joaquim Ribeiro Teles, 780-R/C, Apart. 2, 4445-554 Ermesinde
Silva, E.M. de Jesus da, Rua Serpa Pinto, Blocos Alegria, R/c 2, 8150 São Brás de Alportel
Silva, H.M.C. de Sá F.L., Quinta do Romão, Lote C 8, Apartado 555 Vila Moura, 8125 Quarteira
Silva, J.M.M.M. da, LNEC-DG, Av. do Brasil, 101, 1700-066 Lisboa
Silva, J.E.C. de Matos E., Av. das Túlipas, 16-6° Esq°, Miraflores, 1495 Algés
Silva, M.M.C.C., Rua Dom João I., 87-Aviais Fala, 3040-024 Coimbra
Silva, M.G.M. da, Rua Carvalhão Duarte, 2 -2° Dt°, Telheiras, 1600 Lisboa
Somague, S. de Construções S.A. Rua Dr António Loureiro Borges, Edifício 9, Miraflores, 1495-131 Algés
Sondagens Ródio, Av. dos Combatentes, 52, Apartado 112, Abrunheira, 2710 Sintra
Sopecate, S.A., -Sociedade de Pesquisa e Captações de Água e Transportes, Rua do Arsenal, 146-2, 1100-041 Lisboa
Sousa, M.J.L.E., Mineralogia e Geologia – Faculdade de Ciências, Praça Gomes Teixeira, 4099-002 Porto
Sousa, A.M.P.M. de Oliveira, Rua 9, 723-2° Esq°, 4500 Espinho
Sousa, H.J.A.R. de, Monte da Granja, Santa Maria, 7100 Estremoz
Sousa, J.M.R.D. de, Praceta Anibal Faustino, 6 B, Quinta da Piedade, 2625 Póvoa de Santa Iria
Sousa, J.N.V.A.E., Praceta Machado de Castro, Lote 4-2A, 3000 Coimbra
Sousa, L.M.R. e, LNEC-DG, Av. do Brasil, 101, 1700-066 Lisboa
Sousa, M.L. de A.A., Rua João de Paiva, 13, 1400 Lisboa
Tavares, A.S.M., Rua da Vitória, 73-3° Esq°, 1100 Lisboa
Tecnasol-Fundações e Geotecnia, Rua das Fontaínhas, 58, 2700-891 Venda Nova
Teixeira Duarte,-Engenharia e Construções S.A., Av. das Forças Armadas, 125,- 4A, 1600-079 Lisboa
Triede-Consultoria e Projectos de Engenharia Civil, S.A., Rua Margarida Palla, 9A, 1495-143 Algés
Universidade de Aveiro – Serviços de Documentação, Departamento de Engenharia Civil, Campus de Santiago, 3800 Aveiro
Vale, M. de L.A. do, Urbanização. da Portela, Lote 119-6° Esq°, 2685 Sacavém
Varatojo, A.P.C., Rua João de Santarém, 1, 1400 Lisboa
Vaz, E.A., Engenharia, Gestão e Consultoria Lda., C.P. 1349 Maputo, República. Popular de Moçambique
Vaz, J.G.C.O., Rua Mestre Guilherme Camarinha, 94-4 Hab. 4, 4200 Porto
Vieira, A.M.C. Pinheiro, LNEC-DG, Av. do Brasil, 101, 1700-066 Lisboa
Vieira, C.M.R.D., Rua General Ferreira Martins, 10-R/c A, 1485 Algés
Vieira, C.F. da Silva, Rua Dr Albino dos Reis, 41, 3720-241 Oliveira de Azeméis

186

ROMANIA/ROUMANIE

Secretary: Professor Nicoleta Radulescu, Technical University of Civil Engineering Bucharest,
124 Lacul Tei Blvd, 72302 Bucharest, P.O. Box 38-71, F: (40)-1-2420866, E: nicoleta@hidro.utcb.ro, manoliu@hidro.utcb.ro

Total number of members: 109

Abramescu, T., 2 Poienari Street, bl. 12, ap. 110, 75128 Bucharest. T: (40)-1-6756991

Albert, C., Technical University "Gh. Asachi" Iasi, 43 Mangeron D. Street, 6600 Iasi. T: (40)-32-278683, F: (40)-32-233368, E: calbert@ce.tuiasi.ro

Andrei, S., Technical University of Civil Engineering Bucharest, 124 Lacul Tei Blvd, 72302 Bucharest. T: (40)-1-2421208, Ext. 263, F: (40)-1-2420866, E: geoffice@hidro.utcb.ro

Antonescu, I., Technical University of Civil Engineering Bucharest, 124 Lacul Tei Blvd, 72302 Bucharest. T: (40)-1-2421208, Ext. 263, F: (40)-1-2420866, E: geoffice@hidro.utcb.ro

Balaj, V., SC PROIECT BIHOR SA ORADEA, Geotechnical Section, 23 Magheru Blvd, 3700 Oradea. T: (40)-59-415359, F: (40)-59-415353

Balan, D.M.A.S., 19 Domeasca Street, bl. C, sc. 1, et. 1, ap. 5, 6200 Galati. T: (40)-36-417051

Bally, R.J., 70 Popa Nan Street, 73117 Bucharest. T: (40)-1-3207019

Barariu, A., Consitrans, 23 General Macarovici Street, 76329 Bucharest. T: (40)-1-4113225, F: (40)-1-410 7400

Batali, L., Technical University of Civil Engineering Bucharest, 124 Lacul Tei Blvd, 72302 Bucharest. T: (40)-1-2421208, Ext. 280, F: (40)-1-2420866, E: loretta@hidro.utcb.ro

Boboc, J., ISPE, 4 Lunca Cernei Aleea, bl. D47, sc. H., ap. 109, 77409 Bucharest. T: (40)-1-2107080, Ext. 1478, F: (40)-1-3242443, E: iboboc@ispe.ro

Bogdan, I.A., "Politehnica" University of Timisoara, 2 Victoriei Square, 1900 Timisoara. T: (40)-56-192971, F: (40)-56-193110

Borsaru, I., ICIM, 294 Independentei Spl., 77703 Bucharest. T: (40)-1-2210975, Ext. 189

Boti, N., Technical University "Gh. Asachi" Iasi, 1 Pinului Street, 6600 Iasi. T: (40)-32-113155, F: (40)-32-233368, E: iboti@ce.tuiasi.ro

Botu, N., Technical University "Gh. Asachi" Iasi, 4 Aleea Decebal, sc. 4, ap. 10, 6600 Iasi. T: (40)-32-237761, F: (40)-32-233368, E: nbotu@ce.tuiasi.ro

Brad, I., SC Sinergic Sistem SRL, 1 Ioan Vidu Street, 2900 Arad. T: (40)-56-192971, F: (40)-56-193110

Braniste, Fl., SC RomGeoSlope, 12 Povernei Street, 71124 Bucharest. T: (40)-1-6503973, E: rgs@hades.ro

Breaban, V.G., "Ovidius" University Constanta, 124 Mamaia Blvd, 8700 Constanta. T: (40)-41-616399, F: (40)-41-511512, E: Breaban@univ_ovidius.ro

Burilescu, T., 38 Dinicu Golescu Blvd, 77113 Bucharest. T: (40)-1-6387862, F: (40)-1-3121416

Cazacu, G.B., SC Proiect SA Constanta, 143A, Tomis Blvd, 8700 Constana. T: (40)-41-55505, Ext. 211

Chioveanu, Gh., SC Proiect Bucuresti SA, 4 Vasile Alecsandri Street, 71122 Bucharest. T: (40)-1-6505040, Ext. 292, F: (40)-1-3129504

Chiriac, P., OSN SRL, 19 Impacarii Street, bl. 913, tr. I., et. 3, ap. 11, 6600 Iasi. T: (40)-32-123928

Chirica, A., Technical University of Civil Engineering Bucharest, 124 Lacul Tei Blvd, 72302 Bucharest. T: (40)-1-2421208, Ext. 176, F: (40)-1-2420866, E: achirica@hidro.utcb.ro

Ciortan, R., IPTANA SA Hydraulic Dept., 38 Dinicu Golescu Blvd, 79684 Bucharest. T: (40)-1-2231206

Coman, M.L., ISPIF SA, 35-37 Oltenitei Sos., 75501 Bucharest. T: (40)-1-3321845

Cosovliu, I.O., Project-Galati, 3 Navelor Street, 6200 Galati. T: (40)-36-417541

Culita, C., Agisfor SRL, 35 Costache Sibiceanu Street, 78261 Bucharest. T: (40)-1-6662611, F: (40)-1-6662611

Cvaci, R., Constrantions State Inspection Bucharest, 10 Unirii Square, bl. B5, 70404 Bucharest. T: (40)-1-3301001

Dan, E.S., SC Gemenii Proiect SRL, 34-36 Andrei Saguna Street, 3400 Cluj-Napoca. T: (40)-64-138553, F: (40)-64-192055

Dimitriu, D.V.S., AMEC Earth & Environmental Ltd., 3096 Devon Drive, Windsor, Ontario N8 4L2, Canada. E: vicand@netrover.com

Dumitrescu, A., ENPC-CERMES, Cité Descartes, 6/8 Blaise Pascal Blvd., Champs-sur-Marne, 77455 Marne-la-Vallée Cedex 2, France. E: geoffice@hidro.utcb.ro

Dumitrescu, V., IPTANA SA, 11-13 Iani Buzoiani Street, ap. 4, 78223 Bucharest. T: (40)-1-2232538, F: (40)-1-31-21416

Dumitriu, L.D., SC Consas SA, 27 Oasului Street, 3400 Cluj-Napoca. T: (40)-64-195699, F: (40)-64-192055

Farcas, V.S., Technical University Cluj-Napoca, 15 Daicoviciu C. Street, 3400 Cluj-Napoca. T: (40)-64-195699, F: (40)-64-192055

Feodorov, V., Iridex Group Constructii SRL, 6-8 Stefanesti Street, 72902 Voluntari. T: (40)-1-2404041, E: iridex@com.pcnet.ro

Fetea, L., Technical University Cluj-Napoca, 15 Daicoviciu C. Street, 3400 Cluj-Napoca. T: (40)-64-195699, F: (40)-64-192055

Floriansics, E., SC Salto Edila SA, 1 Observatorului Street, 3400 Cluj-Napoca. T: (40)-64-195699, F: (40)-64-192055

Fosti, V., Technical University Cluj-Napoca, 15 Daicoviciu C. Street, 3400 Cluj-Napoca. T: (40)-64-195699, F: (40)-64-192055

Gadea, A., "Politehnica" University of Timisoara, 2 Victoriei Square, 1900 Timisoara. T: (40)-56-192971, F: (40)-56-193110

Gadinceanu, C., Geo Construct Design SRL, 4 Constantin Brancoveanu Blvd, bl. 12A, sc. 2, et. 4, ap. 45, 75531 Bucharest. T: (40)-1-6363740

Galer, M., Technical University of Civil Engineering Bucharest, 124 Lacul Tei Blvd, 72302 Bucharest. T: (40)-1-2421208, Ext. 176, F: (40)-1-2420866, E: geoffice@hidro.utcb.ro

Gall, F., Technical University "Gh. Asachi" Iasi, 43 Mangeron D. Street, 6600 Iasi. T: (40)-32-233368

Grecu, V., Technical Univ. "Gh. Asachi" Iasi, 43 Costache Negri Street, bl. T1, ap. 5, 6600 Iasi. T: (40)-32-214079, F: (40)-32-233368, E: vgrecu@ce.tuiasi.ro

Gruia, A., "Politehnica" University of Timisoara, 2 Victoriei Square, 1900 Timisoara. T: (40)-56-192971, F: (40)-56-193110

Haida, V., "Politehnica" University of Timisoara, 2 Victoriei Square, 1900 Timisoara. T: (40)-56-192971, F: (40)-56-193110

Has, I., 188 Calea lui Traian, bl. 4, sc. D, ap. 4, 1000 Ramnicu Valcea. T: (40)-50-747393

Hogea, D., Technical University of Civil Engineering Bucharest, 124 Lacul Tei Blvd, 72302 Bucharest. T: (40)-1-2421208, Ext. 150, F: (40)-1-2420866, E: geoffice@hidro.utcb.ro

Ilnitchi, R., 3 Turnu Magurele Street, bl. C2, sc. 6, ap. 248, 75546 Bucharest. T: (40)-1-6838005, F: (40)-1-3121416

Juravle, S., GEOSIT PROIECT SRL, 27 Primaverii Street, sc. B, ap. 5, 6800 Botosani. T: (40)-31-511288, F: (40)-31-511288

Juravle, V., GEOSIT PROIECT SRL, 27 Primaverii Street, sc. B, ap. 5, 6800 Botosani. T: (40)-31-511288, F: (40)-31-511288

Klein, R., 9 Simon Bolivar Street, 72583 Bucharest. T: (40)-1-2125858

Lacatus, F.Gh., SC Consas SA, 27 Oasului Street, 3400 Cluj-Napoca. T: (40)-64-195699, F: (40)-64-192055

Liciu, M., SC Olt-Proiect SA Slatina, 2 Sevastopol Street, 0500 Slatina. T: (40) 49-422145, F: (40)-49-423596

Luca, E., USAMV, Land Reclamation and Environment Engineering Faculty, 59 Marasti Blvd, 71551 Bucharest. T: (40)-1-3242443, F: (40)-1-3242443

Lungu, I., Technical University "Gh. Asachi" Iasi, 53 Ion Creanga Street, bl. T3, et. 8, ap. 33, 6600 Iasi. T: (40)-32-177219, F: (40)-32-233368, E: ilungu@ce.tuiasi.ro

Manea, S., Technical University of Civil Engineering Bucharest, 124 Lacul Tei Blvd, 72302 Bucharest. T: (40)-1-2421208, Ext. 280, F: (40)-1-2420866, E: smanea@hidro.utcb.ro

Manescu, I.G., 5 Romancierilor Street, bl. C14, sc. C, ap. 82, 77396 Bucharest. T: (40)-1-7785092

Manoliu, I., Technical University of Civil Engineering Bucharest, 124 Lacul Tei Blvd, 72302 Bucharest. T: (40)-1-2421208, Ext. 139, F: (40)-1-2420866, E: manoliu@hidro.utcb.ro

Marchidanu, E., Technical University of Civil Engineering Bucharest, 124 Lacul Tei Blvd, 72302 Bucharest. T: (40)-1-2421208, Ext. 259, F: (40)-1-2420866, E: geoffice@hidro.utcb.ro

Marcu, A., Technical University of Civil Engineering Bucharest, 124 Lacul Tei Blvd, 72302 Bucharest. T: (40)-1-2421208, Ext. 259, F: (40)-1-2420866, E: amarcu@hidro.utcb.ro

Marin, M., "Politehnica" University of Timisoara, 2 Victoriei Square, 1900 Timisoara. T: (40)-56-192971, F: (40)-56-193110

Mihu, P., "Politehnica" University of Timisoara, 2 Victoriei Square, 1900 Timisoara. T: (40)-56-192971, F: (40)-56-193110

Mirea, M., "Politehnica" University of Timisoara, 2 Victoriei Square, 1900 Timisoara. T: (40)-56-192971, F: (40)-56-193110

Mitrofan, I., Primaria Municipiului Iasi, 45 Stefan cel Mare Street, 6600 Iasi. T: (40)-32-279121

Muresanu, D.F., Technical University Cluj-Napoca, 15 Daicoviciu C. Street, 3400 Cluj-Napoca. T: (40)-64-195699, F: (40)-64-192055

Musat, V., Technical University "Gh. Asachi" Iasi, 24 Independentei Blvd, bl. Y3, ap. 27, 6600 Iasi. T: (40)-32-218165, F: (40)-32-233368, E: musat@ce.tuiasi.ro

Nicola, G., Stizo – Bauer SA, 2 Bodesti Street, bl. 29b, sc. C, et. 4, ap. 105, 73516 Bucharest. T: (40)-1-2112964

Nicoras, V., "Politehnica" University of Timisoara, 2 Victoriei Square, 1900 Timisoara. T: (40)-56-192971, F: (40)-56-193110

Nicuta, A., Technical University "Gh. Asachi" Iasi, 49 Sfantu Lazar Street, bl. A3, ap. 19, 6600 Iasi. T: (40)-32-136487, F: (40)-32-233368

Olinic, E.D., Technical University of Civil Engineering Bucharest, 124 Lacul Tei Blvd, 72302 Bucharest. T: (40)-1-2421208, Ext. 280, F: (40)-1-2420866, E: ernest@hidro.utcb.ro

Olteanu, A., Technical University of Civil Engineering Bucharest, 124 Lacul Tei Blvd, 72302 Bucharest. T: (40)-1-2421208, Ext. 261, F: (40)-1-2420866, E: geoffice@hidro.utcb.ro

Orha, I.D., SC Proiect Bucuresti SA, 4 Vasile Alecsandri Street, 71122 Bucharest. T: (40)-1-6505040, Ext. 293, F: (40)-1-2503991

Pantea, P., "Politehnica" University of Timisoara, 2 Victoriei Square, 1900 Timisoara. T: (40)-56-192971, F: (40)-56 193110

Paunescu, D., Technical University of Civil Engineering Bucharest, 124 Lacul Tei Blvd, 72302 Bucharest. T: (40)-1-2421208, F: (40)-1-2420866, E: Paunescu@hidro.utcb.ro

Perlea, M.P., U.S. Army Corps of Engineers, 212 Raintree Drive, Lee's Summit, MO 64082, USA. E: perlea@compuserve.com

Perlea, V., U.S. Army Corps of Engineers, 212 Raintree Drive, Lee's Summit, MO 64082, USA. E: perlea@compuserve.com

Pirlea, H., Institut Proiect Cluj SA, 6-8 Motilor Street, 3400 Cluj-Napoca. T: (40)-64-195699, F: (40)-64-192055

Platica, D., Technical University "Gh. Asachi" Iasi, 5 Pacurari Street, bl. 540, tr. III, ap. 4, 6600 Iasi. T: (40) 32 162427, F: (40)-32-233368, E: platica@ce.tuiasi.ro

Pogany, A., "Politehnica" University of Timisoara, 2 Victoriei Square, 1900 Timisoara. T: (40)-56-192971, F: (40)-56-193110

Popa, A., Technical University Cluj-Napoca, 15 Daicoviciu C. Street, 3400 Cluj-Napoca. T: (40)-64-195699, F: (40)-64-192055

Popa, H., Technical University of Civil Engineering Bucharest, 124 Lacul Tei Blvd, 72302 Bucharest. T: (40)-1-2421208, Ext. 150, F: (40)-1-2420866, E: horatiu@hidro.utcb.ro

Popescu, M., Technical University of Civil Engineering Bucharest, 124 Lacul Tei Blvd, 72302 Bucharest. T: (40)-1-2421208, Ext. 176, F: (40)-1-2420866, E: mihpop@hidro.utcb.ro

Privighetorita, C., 7 Serdarului Intr., bl. 46A, et. 9, ap. 20, 71308 Bucharest. T: (40)-1-6667294

Radea, M., SC Mistar Proiect SRL, 74 Sondelor Street, 2000 Ploiesti. T: (40)-44-125679

Radulescu, N.M., Technical University of Civil Engineering Bucharest, 124 Lacul Tei Blvd, 72302 Bucharest. T: (40)-1-2421208, Ext. 139, F: (40)-1-2420866, E: nicoleta@hidro.utcb.ro

Raileanu, P., Technical University "Gh. Asachi" Iasi, 51 Sfantu Lazar Street, bl. A2, ap. 10, 6600 Iasi. T: (40)-32-130175, F: (40)-32-233368, E: raileanu@ce.tuiasi.ro

Rebeleanu, V., SC Consas SA, 27 Oasului Street, 3400 Cluj-Napoca. T: (40)-64-195699, F: (40)-64-192055

Roman, F., Technical University Cluj-Napoca, 15 Daicoviciu C. Street, 3400 Cluj-Napoca. T: (40)-64-195699, F: (40)-64-192055

Rotar, V., Noua Street, bl. N9, ap. 3, 4700 Zalau. T: (40)-56-192971, F: (40)-56-193110

Rotaru, A., Technical Univ. "Gh. Asachi" Iasi, 75 Eternitate Street, 6600 Iasi. T: (40)-32-278683, F: (40)-32-233368, E: arotaru@ce.tuiasi.ro

Schein, T., "Politehnica" University of Timisoara, 2 Victoriei Square, 1900 Timisoara. T: (40)-56-192971, F: (40)-56-193110

Scordaliu, I., INCERC Timisoara, 2 Traian Lalescu Street, 1900 Timisoara. T: (40)-56-192971, F: (40)-56-193110

Serban, T., SC INSPET Ploiesti, 15 Democratiei Street, 2000 Ploiesti. T: (40)-44-129367

Serbulea, M., Technical University of Civil Engineering Bucharest, 124 Lacul Tei Blvd, 72302 Bucharest. T: (40)-1-2421208, Ext. 261, F: (40)-1-2420866, E: mserb@hidro.utcb.ro

Sima, N.D., SC GEOSOND SA, 294 Independentei Spl., 77703 Bucharest. T: (40)-1-3123224

Siminea, I., USAMV, Land Reclamation and Environment Engineering Faculty, 59 Marasti Blvd, 71551 Bucharest. T: (40)-1-2223719

Stanciu, A., Technical University "Gh. Asachi" Iasi, 5 Basota Street, bl. D9, tr. I, ap. 7, 6600 Iasi. T: (40)-32-217362, F: (40)-32-233368, E: astanciu@ce.tuiasi.ro

Stanculescu, I., Technical University of Civil Engineering Bucharest, 124 Lacul Tei Blvd, 72302 Bucharest. T: (40)-1-2421208, Ext. 263, F: (40)-1-2420866, E: geoffice@hidro.utcb.ro

Stanescu, R., Technical University of Civil Engineering Bucharest, 124 Lacul Tei Blvd, 72302 Bucharest. T: (40)-1-2421208, Ext. 259, F: (40)-1-2420866, E: geoffice@hidro.utcb.ro

Stefanica Nica, M., INCERC Bucharest, 19 Elena Cuza Street, 75146 Bucharest. T: (40)-1-6336807

Stoica, R.I., Geo Construct Design SRL, 4 Constantin Brancoveanu Blvd, bl. 12A, sc. 2, et. 4, ap. 45, 75531 Bucharest. T: (40)-1-6363740

Tache, D., Constructions State Inspection Ilfov, 21 Nicolae Balcescu Blvd, 70112 Bucharest. T: (40)-1-6863434

Tanasoiu, C., Military Technical Academy, 81-83 George Cosbuc Blvd, 75275 Bucharest. T: (40)-1-6653129

Tirla, O., SC Consas SA, 27 Oasului Street, 3400 Cluj-Napoca. T: (40)-64-195699, F: (40)-64-192055

Tokes, T., SC Consas SA, 27 Oasului Street, 3400 Cluj-Napoca. T: (40)-64-195699, F: (40)-64-192055

Toma, H.R., Project-Galati, 3 Navelor Street, 6200 Galati. T: (40)-36-417541

Trifan, L., Geo Construct Design SRL, 4 Constantin Brancoveanu Blvd, bl. 12A, sc. 2, et. 4, ap. 45, 75531 Bucharest. T: (40)-1-6363740

Tripa, A.A.C., Technical University Cluj-Napoca, 15 Daicoviciu C. Street, 3400 Cluj-Napoca. T: (40)-64195699, F: (40)-64-192055

Tripa, I.I., SC Arconex Consulting SRL, 104, 21 Decembrie 1989 Blvd, 3400 Cluj-Napoca. T: (40)-64-195699, F: (40)-64-192055

Zaharia, C.S., SC Aris SA, 5 Culturii Street, 4800 Baia Mare. T: (40)-64-195699, F: (40)-64-192055

RUSSIA/RUSSIE

Secretary: I.V. Kolybine, Ph.D., Russian Society for Soil Mechanics, Geotechnics and Foundation Engineering (RSSMGFE), NIIOSP, 2-nd Institutskaya St., 6, Moscow 109428. T: 007 095 171 22 40, F: 007 095 170 27 67, E: rssmgfe@m9.ru

Total number of members: 269

Abelev, M., Yu, Dr, Prof, Dept Hd, GASIS Inst, Trifonovskaya St., 57, Moscow 129272. T/F: 007 095 288 8733

Abbasov, P.A., Dr, Prof, Dir, Eng Centre, DalNIIS Inst, Borodinskaya St., 14, Vladivostok 690049. T/F: 007 4232 461498

Abukhanov, A.Z., Ph.D., Dept Hd, Novocherkassk St Land-Reclamation Academy, Pushkinskaya St., 3, Novocherkassk 346409. T: 007 86352 55768

Aleynikov, S.M., Ph.D., Ass Prof, Voronezh St Arch and Civ Eng Acad, 20 Let Oktyabrya St., 84, Voronezh 394006. T: 007 0732 715362

Alexandrovsky, Yu V., Ph.D., Lead Res, Niiosp, 2-nd Institutskaya St., 6, Moscow 109428. T/F: 007 095 170 2715

Alexeev, S.I., Ph.D., Ass Prof., St-Petesburg St Arch-Civ Eng Univ, 2-nd Krasnoarmeyskaya St., 4, St-Peterburg 198005. T/F: 007 812 3166118, 3162120

Amaryan, L.S., Dr, Prof, Dept Hd, PNIIIS Inst, Okruzhnoy Pr.18, Moscow 105052. T: 007 095 430 7410

Anikiev, A.V., Ph.D., Techn Dir, Fundamentstroyproject Comp., 2-nd Institutskaya St., 6, Moscow 109428. T: 007 095 170 28 28, F: 007 095 170 32 68

Akhpatelov, D.M., Ph.D., Ass Prof, Moscow St Civ Eng Univ, Yaroslavskoe Shosse, 26, Moscow 129337. T/F: 007 095 2615988

Aripov, N.F., Ph.D., Chief Geot Eng, Inst Hydroproject, Volokolamskoe Shosse, 2, Moscow 125812. T: 007 095 1553614

Astrakhanov, B.N., Ph.D., Chief Eng, Comp Hydrospetsfundamentstroy, Dobroslobotskaya St., 6, Moscow 107066. T/F: 007 095 2652777

Badeev, A.N., Ph.D., Lead Res, Central Inst of Const, Mintransstroy, Kosskaya St., 1, Moscow 129329. T: 007 095 189 5126, 189 3871

Bagdasarov, Yu A., Ph.D., Lab Hd, NIIOSP, 2-nd Institutskaya St., 6, Moscow 109428. T: 007 095 1702749

Bakenov, Kh.Z., Ph.D., Deputy Sec Gen, Interparliamentary Assembly of CIS, Shpalernaya St., 47, St-Petersburg 193015. T: 007 812 2791848

Balykov, B.I., Ph.D., Lab Hd, VNIIG-TEST Centre, Gzhatskaya St., 21, St-Petersburg 195220. T: 007-812-5358868

Baramidze, Zh.I., Ph.D., Sen Res, Lenmorniiproject Inst, Mezhevoi Canal, 5a, St-Petersburg. T: 007 812 251 5565

Barvashov, V.A., Ph.D., Sen Res, NIIOSP, 2-nd Institutskaya St., 6, Moscow 109428. T: 007 095 1706941

Bartolomey, A.A., Dr, Prof, Dept Hd, Perm St Techn Univ, Komsomolsky Pr., 29, Perm 616600. T: 007 3422 391573, F: 007 3422 391496

Bakholdin, B.V., Dr, Prof, Lab Hd, NIIOSP, 2-nd Institutskaya St., 6, Moscow 109428. T/F: 007 095 1706967

Baulin, V.V., Ph.D., Dir, PNIIIS Inst, Okruzhnoi Pr., 18, Moscow 105058. T: 007 095 366 2573, 366 3189, F: 007 095 366 3190, 366 3485

Bellendir, E.N., Ph.D., Deputy Dir Gen, VNIIG Inst, Gzhatskaya St., 21, St-Petersburg 195220. T: 007 812 535 2807, F: 007 812 535 6720

Belov, D.V., Ph.D., Deputy Dir, Road Educ-Eng Centre, Prof. Molchanov St., 23, Pavlovsk, St-Petersburg 189623. T: 007 812 465 1171, F: 007 812 465 1097

Betelev, N.P., Ph.D., Civ Eng, NIIOSP, 2-nd Institutskaya St., 6, Moscow 109428

Bezrodnyi, K.P., Dr, Prof, Deputy Dir Gen, Inst Lenmetrogiprotrans, Bolshaya Moscovskaya St., 2, St-Petersburg 191002. T: 007 812 312 7811, F: 007 812 251 2536

Bobylev, L.M., Ph.D., Chairman, "BOS" Comp, Petrozavodskaya St., 15, bld.2, apart.229, Moscow 125502. T/F: 007 095 455 0233

Bogomolov, A.N., Dr, Prof, Dept Hd, Volgograd State Arch-Civ Eng Academy, Academician St., 1, Volgograd 400074. T: 007 8442 44 8183, 44 1957

Boldyrev, G.G., Dr, Prof, Dept Hd, Penza State Arch-Civ Eng Academy, Titov St., 28, Penza 440028. T: 007 8412 62 05 01

Bolshakov, V.F., Chief Eng, "Stav Ltd.," Comp, Rubstovskaya Emb., 3, Moscow 107082. T/F: 007 095 265 2980, 265 0327

Bondarev, Yu.V., Ph.D., Chief Specialist, Geostroy Comp, Zagorodnyi Prospect, 27/21, St-Petersburg 191180. T: 007 812 164 8783, F: 007 812 311 7339

Bondarenko, G.I., Ph.D., Lab Head, Niiosp, 2-nd Institutskaya St., 6, Moscow 109428. T: 007 095 170 2821

Borozenets, L.M., Ph.D., Toljatti Higher Military Civ Eng School, Samara Region, Gsp- 681, Toljatti 445025. T: 007 8482 32 55 66

Bortkevich, S.V., Ph.D., Lab Hd, Niies Centre, Stroitelnyi Pr., 7a, Moscow 123362. T: 007 095 492 97 15, 492 97 22

Broid, I.I., Ph.D., Dept Head, Remhydrospetsstroy Comp, 1-st Parkovaya St., 1/53, Moscow. T: 007 095 367 83 91, F: 007 095 237 41 70

Bronin, V.N., Dr, Prof, St-Petersburg St Arch-Civ Eng Univ, 2-nd Krasnoarmeyskaya St., 4, St-Petersburg 198005. T: 007 812 586 09 36

Bugrov, A.K., Dr, Prof, Dept Hd, St-Petersburg St Tech Univ, Polytechnicheskaya St., 29. T: 007 812 247 5961

Budin, A. Ya., Dr, Prof, Dept Hd, St-Petersburg St Agro Univ, Pr Slava, 43/49, St-Petersburg 192286. T: 007 812 476 0431

Bulatov, V.V., Dr, Prof, Moscow Inst of Steel and Alloyes (Branch), Pervomayskaya St., 7, Electrostal 144000. T: 007 257 431 34

Bulychev, N.S., Dr, Prof, Dept Hd, Tula St Techn Univ, Lenin Pr. 92, Tula 300600. T: 007 0872 252113

Burenkova, V.V., Ph.D., Sen Res, Vodgeo Inst, Komsomolsky Pr., 42, Moscow G-48. T: 007 095 245 9788

Buslov, A.S., Dr, Prof, Dept Hd, Moscow St Open Univ, Pavel Korchagin St. 22, Moscow 129805. T: 007 095 283 8797

Bykov, B.I., Ph.D., Sc Manager, Novatia Comp, Volodarsky St., 41, apart.1, Orenburg 460000. T: 007 3532 77 60 28

Chernyshev, S.N., Dr, Prof, Moscow St Civ Eng Univ, Yaroslavskaya Highway, 26, Moscow 129337. T: 007 095 2619120

Chetyrkin, N.S., Ph.D., Lab Head, SKTB "Tunnelmetrostroy", Krasnaya Presnya St., 288, Moscow 123022. T: 007 095 253 7742, F: 007 095 252 2149

Chikishev, V.M., Dr, Prof, Rector, Tjumen St Arch-Civ Eng Acad, Lunacharsky St., 2, Tjumen 625001. T: 007 3452 261010

Dashko, R.E., Dr, Prof, St-Petersburg St Inst of Mines, 21 Line of Vasilievski Island, 2, apart.113, St-Petersburg 199026. T: 007 812 328 82 88

Demin, A.M., Dr, Prof, Ipkon Inst, Russian Acad of Sc, Krakowsky Tup. 4, Moscow 111020. T: 007 095 360 4211, F: 007 095 360 8960

Demin, V.F., Ph.D., Dept Hd, VNIIG Inst, Gzhatskaya St., 21, St-Petersburg 195220. T: 007 812 535 50 89, F: 007 812 535 88 61

Demkin, V.M., Ph.D., Dir, "Saratovselinvestproject" Inst, Pugachev St., 159, Saratov 410005. T: 007 8452 24 78 73, 24 78 68, F: 007 8452 51 28 20, 99 05 51

Denisov, O.L., Ph.D., Ass Prof, Ufa St Oil Tech Univ, Mendeleev St., 197, Ufa 450080. T: 007 3472 282400

Dimov, L.A., Dr, Prof, Dir, Inst "NIPIneftegazstroydiagnostika", Glavpochtamp, P.O. Box 56, Ukhta 169400. T: 007 82147 345 68

Didukh, B.I., Dr, Prof, Dept Hd, Russian People's Friendship Univ, Ordzhonikidze St., 3, Moscow 117419. T: 007 095 955 0948, 955 0805

Dobrov, E.M., Dr, Prof, Hd of Dept, Moscow Motor-Road Techn Univ, Leningrad Pr., 4, Moscow 125829. T: 007 095 155 08 39

Doroshkevich, N.M., Ph.D., Ass Prof, Moscow St Civ Eng Univ, Yaroslavskaya Highway, 26, Moscow 129337. T/F: 007 095 261 5988

Dudler, I.V., Ph.D., Ass Prof, Moscow St Civ Eng Univ, Yaroslavskaya Highway, 26, Moscow 129337. T: 007 095 261 8120

Dyba, V.P., Ph.D., Ass Prof, Novocherkassk St Techn Univ, Prosveschenie St., 132, Novocherkassk 346428. T: 007 86352 55 416

Dzagov, A.M., Ph.D., Sen Res, NIIOSP, 2-nd Institutskaya St., 6, Moscow 109428. T/F: 007 095 170 69 67

Dzhantimirov, Kh.A., Ph.D., Lab Hd, NIIOSP, 2-nd Institutskaya St., 6, Moscow 109428. T: 007 095 170 69 31, F: 007 095 170 27 57

Egorov, A.I., Ph.D., Dir, "Restavrator G3R" Comp, Kozhevnicheskaya St., 13, Moscow 103104. T/F: 007 095 235 26 10, 235 34 23

Eisler, L.A., Eng, VNIIG Inst., Gzhatskaya St., 21, St-Petersburg 195220

Eppel, D.I., Ph.D., Chief Geologist, Mosoblgeotrest, Novinsky Bul, 27/10, Moscow 123242. T: 007 095 255 48 47

Evtushenko, S.I., Ph.D., Dept Hd, Novocherkassk St Techn Univ, Prosveschenie St., 132, Novocherkassk 346428. T: 007 86352 5 54 16, E: evtushen@novoch.ru

Fadeev, A.B., Dr, Prof, St-Petersburg St Arch-Civ Eng Univ, Krasnoarmeiskaya St., 4, St-Petersburg 198005. T: 007 812 316 4806, 316 6988, 113 5714, F: 007 812 113 5209

Fedorovsky, V.G., Ph.D., Lab Hd, NIIOSP, 2-nd Institutskaya St., 6, Moscow 109428. T: 007 095 170 6941, F: 007 095 170 2757

Feklin, V.I., Dr, Prof, Dept Hd, Toljatti Polytechn Inst, Belorusskaya St.14, Toljatti 445667. T: 007 8482 29 66 91

Fotieva, N.N., Dr, Prof, Dept Hd, Tula St Tech Univ, Lenin Pr., 92, Tula 300600. T: 007 0872 332298

Garagash, B.A., Ph.D., Dir Gen, Sc-Techn Centre "Intellekt", Kanunnikov St., 13, Volgograd 400001. T/F: 007 8442 94 48 99

Glagovsky, V.B., Ph.D., Lab Hd, VNIIG Inst, Gzhatskaya St., 21, St-Petersburg 195220. T: 007 812 5359843

Glozman, L.M., Ph.D., Hd of Dept, "Georeconstruction" Comp; Sen Res, Inst. VNIIGS, Karlovskaya St., 4, St-Petersburg 193148. T/F: 007 812 316 61 18

Glukhov, V.S., Ph.D., Hd of Dept, Penza Arch-Civ Eng Acad.; Dir Gen, "Novotex" Comp, Titov St., 28, Penza 440028. T/F: 007 8412 62 42 30, E: novotex@sura.com.ru

Goldin, A.L., Dr, Prof, VNIIG Inst, Gzhatskaya St., 21, St-Petersburg 195220. T: 007 812 535 73 95

Goldfeld, I.Z., Ph.D., Lab Hd, NIIOSP, 2-nd Institutskaya St., 6, Moscow 109428. T: 007 095 170 6930

Golli, A.V., Ph.D., Ass Prof, St-Petersburg St Arch-Civ Eng Univ, 2-nd Krasnoarmeiskaya St., 4, St-Petersburg 198005. T: 007 812 259 53 98

Goncharov, B.V., Dr, Prof, Dept Hd, Ufa St Oil Tech Univ, Kosmonavts St., 1, Ufa 450062. T: 007 3472 28 24 00

Gotman, A.L., Dr, Prof, Hd of Dept, "BashNIIstroy" Inst, Constitution St., 3, Ufa 450064. T: 007 3472 425285, F: 007 3472 429955

Gotman, N.Z., Ph.D., Sen Res, "BashNIIstroy" Inst, Constitution St., 3, Ufa 450064. T: 007 3472 424254, F: 007 3472 429955

Grachov, Yu.A., Ph.D., Lab Hd, NIIOSP, 2-nd Institutskaya St., 6, Moscow 109428. T: 007 095 170 27 76, F: 007 095 170 27 57

Grebenets, V.I., Ph.D., Dept Hd, NIIOSP, 2-nd Institutskaya St., 6, Moscow 109428. T: 007 095 170 27 14, F: 007 095 170 27 57

Grechischev, S.E., Dr, Prof, Lab Hd, "SoyuzdorNII" Inst, Balashikha-6, 143000 Moscow Region. T/F : 007 095 521 00 47, E: cryodor@balashikha.x400.rosprint.ru

Grib, S.I., Ph.D., Krasnoyarsk St Arch-Civ Eng Acad, Svobodnyi Pr., 82, Krasnoyarsk 660062. T: 007 3912 45 66 69, 45 58 92

Grigoryan, A.A., Dr, Prof, Hd of Dept, NIIOSP, 2-nd Institutskaya St., 6, Moscow 109428. T: 007 095 170 70 08, F: 007 095 170 27 57

Gugnin, A.A., Ph.D., Dir, Ural-Siberian Branch of NIIOSP, Ostrovski St., 60, Perm 614007. T: 007 3422 31 09 18

Guryanov, I.E., Ph.D., Lab Hd, Inst of Permafrost, Russian Acad of Sc, Yakutsk 677010. T: 007 4112 927135, 927259

Ignatova, O.I., Ph.D., Lead Res, NIIOSP, 2-nd Institutskaya St., 6, Moscow 109428. T: 007 095 170 27 45, F: 007 095 170 27 57

Ikonin, S.V., Ph.D., Ass Prof, Voronezh St, Arch & Civ Eng Acad, 20 Let Oktyabrya St., 84, Voronezh 394680. T: 007 0732 57 54 00, F: 007 0732 57 59 05

Ilyichev, V.A., President of Society, Dr, Prof, Dir, NIIOSP, 2-nd Institutskaya St., 6, Moscow 109428. T: 007 095 171 22 40, F: 007 095 170 27 57, E: root@niiosp.m9.ru

Inozemtsev, V.K., Ph.D., Dir Gen, "Podzemstroyreconstruction" Comp, Lazaretnyi Per, 2, St-Petersburg 191180. T: 007 812 113 57 14, 315 69 88, F: 007 812 113 52 09

Isaev, O.N., Ph.D., Sen Res, NIIOSP, 2-nd Institutskaya St., 6, Moscow 109428. T: 007 095 170 27 39, F: 007 095 170 27 57

Ivanov, V.N., Dr, Prof, Dir, Inst Aeroproject, Leningradskaya Highway 7, Moscow A-171 125171. T: 007 095 150 0222

Ivanov, M.N., Ph.D., Deputy Dir, Res Dept, Moscow St Civ Eng Univ, Yaroslavskaya Highway, 26, Moscow 129377. T: 007 095 188 94 01, 183 33 74, F: 007 095 183 53 10

Kabantsev, O.V., Eng, Chief Designer, Hd of Dept, Central Design Inst of Min of Defence, Moscow k-160. T/F: 007 095 296 02 73

Kagan, G.L., Ph.D., Ass Prof, Vologda Polytechn Inst, Lenin St., 15, Vologda 160600. T: 007 81722 250 03

Kalinkevich, D.A., Eng, Vice-President, "Hydrospetsstroy" Corp; Dir Gen, Mosgorhydrospetsstroy Comp, Kadashevskaya Emb, 6/1, Moscow 109017. T: 007 095 237 74 70

Kandaurov, I.I., Dr, Prof, St-Petersburg St Railway Eng Univ, Moscow Pr., 9, St-Petersburg 190031. T: 007 812 310 3472

Karapetov, G.Ya., Ph.D., Dept Hd, GASIS Inst, Trifonovskaya St., 57, Moscow 129272. T/F: 007 095 281 05 18

Karaulov, A.M., Ph.D., Lab Hd, Siberian St Railway Eng Acad, Dusya Kovalchuk St., 191, Novosibirsk 630023. T: 007 3832 28 76 00

Karimov, R.M., Ph.D., Ass Prof, Magnitogorsk Min & Metallurg Acad, Lenin Pr., 38, Magnitogorsk 455000. T: 007 3511 32 84 41

Kazantsev, V.S., Ph.D., Lab Hd, Chelyabinsk St Techn Univ, P.O. Box 8116, Chelyabinsk 454080. T: 007 3512 39 90 00

Kazarnovsky, V.D., Dr, Prof, Lab Hd, "SoyuzDorNII" Inst, Balashikha-6 143900, Moscow Region. T: 007 095 521 0065, 524 0350, 521 0111, F: 007 095 521 1892

Khamov, A.P., Ph.D., Ass Prof, Russian St Open Railway Eng Techn Univ, Chasovaya St., 22/2, Moscow GSP 125808. T: 007 095 156 58 46, 156 57 78

Khanin, R.E., Ph.D., Chief Expert, "Fundamentproject" Inst, Volokolamskaya Higway, 1, Moscow GSP 125843. T: 007 095 158 1281, F: 007 095 158 3078

Khazanov, M.I., Ph.D., Dir, Expedition Centrgeolnerud, Taganskaya St. 24, Moscow 109004. T: 007 095 911 0276, T/F: 007 095 911 2240

Khrustalev, L.N., Dr, Prof, Lab Hd, Moscow State University, Vorobiovy Gory, Moscow 119899. T: 007 095 939 14 53

Kim, M.S., Ph.D., Sen Teach, Voronezh St Arch & Civ Eng Acad, 20 Let Oktyabrya St., 84, Voronezh 394006. T: 007 0732 71 54 00

Kitaikina, O.V., Ph.D., Vniintpi Inst, Vernadsky Pr., 29, Office 503, Moscow 117331. T: 007 095 133 9235, F: 007 095 133 12 81

Klyachko, M.A., Ph.D., Dir, Sc-Techn Centre for Antiseismic Const, Pobeda Pr., 9, Petropavlovsk-Kamchatsky 683006. T: 007 415 225 87 74, F: 007 415 229 22 50

Kolybin, I.V., Ph.D., Sen Res, NIIOSP, 2-nd Institutskaya St., 6, Moscow 109428. T: 007 095 170 2826, F: 007 095 170 27 67, E: root@niiosp.m9.ru

Konovalov, P.A., Dr, Prof, Lab Hd, NIIOSP, 2-nd Institutskaya St., 6, Moscow 109428. T/F: 007 095 170 1927

Koreneva, E.B., Dr, Prof, Moscow St Civ Eng Univ, Yaroslavskaya Highway, 26, Moscow 129337

Koroliov, M.V., Ph.D., Deputy Prorector, Moscow St Civ Eng Univ, Yaroslavskaya Highway, 26, Moscow 129337. T: 007 095 183 33 74, F: 007 095 183 53 10

Kozmin, D.D., Ph.D., Ass Prof, Arkhangelsk St Techn Univ, North Dvina Emb, 17, Arkhangelsk 163007. T: 007 812 44 93 23, E: kosmin@agtu.ru

Kostin, S.V., Civ Eng, Dept Hd, JS "Trust GPR-3", Dobrolyubov St., 16/4, Moscow. T: 007 095 2102982

Krasilnikov, N.A., Ph.D., Sen Res, Inst NIIES, Stroitelnyi Pr., 7a, Moscow 123362. T: 007 095 492 9722

Kronik, Ya.A., Ph.D., Moscow St Civ Eng Univ, Yaroslavskaya Highway, 26, Moscow 129337. T/F: 007 095 261 59 88

Kryzhanovsky, A.L., Ph.D., Moscow St Civ Eng Univ, Shluzovaya Emb, 8, Moscow 113114. T/F: 007 095 261 59 88

Kubetsky, V.L., Dr, Prof, Russian St Open Railway Eng Univ, Chasovaya St., 22/2, Moscow 125808 GSP 47. T: 007 095 156 56 22, 156 57 51

Kulachkin, B.I., Dr, Lab Hd, NIIOSP, 2-nd Institutskaya St., 6, Moscow 109428. T: 007 095 170 27 68, F: 007 095 170 27 57

Kulchitsky, G.B., Dr, Prof, Hd of Dept, Surgut St Univ, St of Energetikov, 14, Surgut 626400, Tjumen Region. T: 007 3462 21 41 96

Kuprin, V.M., Ph.D., Dir, Comp "Soyuz Severnykh Gorodov", Moscow. T: 007 095 921 9769

Kurillo, S.V., Ph.D., Lead Res, NIIOSP, 2-nd Institutskaya St., 6, Moscow 109428. T: 007 095 170 69 41

Kutvitskaya, N.B., Ph.D., Lead Res, NIIOSP, 2-nd Institutskaya St., 6, Moscow 109428. T: 007 095 170 2821

Kushnir, S.Ya., Dr, Prof, Tjumen St Oil – Gaz Univ, Melnikaite St., 72, Tjumen 625001. T: 007 3452 22 50 28

Kushnir, L.G., Civ Eng, Dir Gen, "Rosstroyizyskaniya" St Union, Spasski Tup, 6, bld.1, Moscow 129090. T: 007 095 975 32 56

Kuzakhmetova, E.K., Dr, Lead Res, "Soyuzdornii" Inst, Moscow Region, Balashikha-6 143900. T: 007 095 524 0347

Kuzevanov, V.V., Ph.D., Deputy Dir, "Spetsfundamentstroy" Comp, D.Bednyi St., 4, Kemerovo 650026. T: 007 3842 360212

Kuznetsov, G.I., Dr, Prof, Krasnoyarsk St Techn Univ, Kirensky St., 26, Krasnoyarsk 660017. T: 007 3912 49 74 79

Lavrov, S.N., Civ Eng, Dir Gen, "Stroyizyskaniya" Comp, Frunze St., 14, Novosibirsk-99 630099. T: 007 3832 24 45 92

Lapin, S.K., Ph.D., Chief Expert, "Leningradsky Promstroyproject" Comp, Leninsky Pr., 160, St-Petersburg 196247. T: 007 812 290 98 10, F: 007 812 295 92 15

Larina, T.A., Ph.D., Deputy Dir, Pniiis Inst, Okruzhnoy Pr., 18, Moscow 105848, GSP-118. T: 007 095 366 31 89, F: 007 095 366 31 90, 366 34 85

Lezonenko, M.V., Ph.D., Chief Geophysician, St Comp "Antikarst and Coast Protection", Mayakovsky St., 33, Nizhegorodsky Region, Dzerzhinsk 606023. T/F: 007 8313 259 801, E: karst@kis.ru

Leonychev, A.V., Dr, Prof, Dept Hd, Moscow St Railway Eng Univ, Obraztsov St., 15, Moscow 101475. T: 007 095 978 72 74, 288 33 38

Lerner, V.G., Ph.D., The First Deputy Dir Gen, "Mosingstroy" Comp, M.Bronnaya St., 15-B, Moscow 103104. T: 007 095 290 43 94

Likhovtsev, V.M., Ph.D., Dir, "Aurora" Comp, Rozhdestvenka St., 11, Markhi, room 313, Moscow 103754. T: 007 095 923 8504

Lipovetskaya, T.F., Civ Eng, Sen Res, VNIIG Inst, Gzhatskaya St., 21, St-Petersburg 195220. T: 007 812 535 38 01

Lushnikov, V.V., Dr, Prof, Dept Hd, Ural Res Arch and Civ Eng Inst, Blyukher St., 26, Ekaterinburg 620137. T: 007 3432 74 83 92, F: 007 3432 49 16 38

Lychko, Yu.M., Ph.D., Lead Res, NIIOSP, 2-nd Institutskaya St., 6, Moscow 109428

Lyalin, Ya.D., Ph.D., Ass Prof, Volgograd St Arch and Civ Eng Acad, Academician St., 1, Volgograd 400074. T: 007 8442 44 06 55, 44 91 65

Lvovich, L.B., Civ Eng, Dir Gen, "Hydrotekhstroy" Comp, Mnevniki St., 1, Moscow 123308. T: 007 095 191 66 56, 191 35 11, F: 007 095 191 40 54

Maximenko, E.S., Ph.D., Dir, North Branch of NIIOSP, Yanovsky St., 1, Vorkuta 169900. T: 007 82151 4 48 03, 4 48 79

Maximyak, R.V., Ph.D., Sec, Journal "Bases, Foundations and Soil Mechanics", 2-nd Institutskaya St., 6, Moscow 109428. T: 007 095 170 27 53, 135 20 49

Maltsev, A.V., Ph.D., Ass Prof, Samara St Arch & Civ Eng Academy, Molodogvardeyskaya St., 194, Samara 443001. T: 007 8462 39 14 69

Malyshev, M.V., Dr, Prof, Moscow St Civ Eng Univ, Yaroslavskaya Highway, 26, Moscow 129337. T: 007 095 261 5988

Mangushev, P.A., Dr, Prof, St-Petersburg St Arch-Civ Eng Univ, 2-nd Krasnoarmeyskaya St., 4, St-Petersburg 197005. T: 007 812 259 5398

Margolin, V.M., Ph.D., Prof, Moscow St Open Tech Univ, Pavel Korchagin St., 22, Moscow 129805. T: 007 095 283 87 97, F: 007 095 283 80 71

Mariupolsky, L.G., Ph.D., Hd of Dept, NIIOSP, 2-nd Institutskaya St., 6, Moscow 109428. T/F: 007 095 170 2735

Melnik, B.G., Dr, Prof, Lab Hd, "Vodgeo" Inst, Komsomolsky Pr., 42, Moscow 119826. T: 007 095 245 95 71

Merzlikin, A.E., Ph.D., Lead res, "SoyuzDorNII" Inst, Moscow Region, Balashikha-6 143000. T: 007 095 524 03 42, E: 7453.g23@g23relcom.ru

Mescheryakov, Ya.M., Ph.D., Ass Prof, Moscow Arch Inst., Rozhdestvenka St., 11, Moscow 103754. T: 007 095 923 8504

Minaev, O.P., Ph.D., Deputy Dir Gen, "Promyshlenno-Stroitelnaya Comp", B.Monetnaya St., 26, St-Petersburg 197061. T/F: 007 812 113 05 04

Minkin, M.A., Dr, Prof, Hd of Dept, "Fundamentproject" Inst, Volokolamskoe Shosse, 1, Moscow 125843. T: 007 095 158 95 38

Mikheev, V.V., Ph.D., Lead Expert, NIIOSP, 2-nd Institutskaya St., 6, Moscow 109428. T: 007 095 170 2704

Mirenburg, Yu S., Ph.D., Sen Res, St-Petersburg St Mining Inst, Branch, Lenin St., 44, Vorkuta 169900. T: 007 2151 7 36 25

Mironov, V.S., Ph.D., Novosibirsk St Acad of Bld, Leningradskaya St., 113, Novosibirsk-8 630008. T: 007 3832 668360

Mishakov, V.A., Ph.D., Sc Dir, "Spetsstroyservice" Comp, Zaozernaya St., 1, bld. 2, St-Petersburg 196084. T: 007 812 298 30 36, F: 007 812 298 26 91

Mitkina, G.V., Ph.D., Lead Res, "BashNIIstroy" Inst, Constitution St., 3, Ufa 450064. T: 007 3472 424254, F: 007 3472 429955

Mulyukov, E.I., Dr, Prof, Dept Hd, Bash St Agro Univ, P.O. Box 2009, Ufa-1 450001. T: 007 3472 28 07 13, 28 07 17

Murzenko, Yul.N., Dr, Prof, Novocherkassk St Techn Univ, Prosveschenie St., 132, Novocherkassk GSP-1, 346428. T: 007 86352 55 416, F: 007 86352 28 463

Murzenko, A.Yul., Ph.D., Ass Prof, Novocherkassk St Techn Univ, Prosveschenie St., 132, Novocherkassk GSP-1, 346428. T: 007 86352 55 416, F: 007 86352 42 396

Narbut, R.M., Dr, Prof, Dept Hd, St-Petersburg St Univ of Water Communications, Dvinskaya St., 5/7, St-Petersburg 198035. T: 007 812 2595766

Nezamutdinov, Sh.R., Dr, Prof, Ufa St Oil Tech Univ, Mendeleev St., 197, Ufa 450000. T: 007 3472 28 24 00

Nevzorov, A.L., Ph.D., Hd of Dept, Arkhangelsk St Techn Univ, Emb of North Dvina, 17, Arkhangelsk 163007. T: 007 8182 449323

Nikiforova, N.S., Ph.D., Lead Res, NIIOSP, 2-nd Institutskaya St., 6, Moscow 109428. T/F: 007 095 170 27 00

Nizovkin, G.A., Ph.D., Expert, Inst of Railway Transport, Rusanov Pr., 2, Moscow 129323. T: 007 095 180 10 57

Nuzhdin, L.V., Ph.D., Lab Hd, Novosibirsk St Arch-Civ Eng Univ, Leningradskaya St., 113, Novosibirsk 630008. T: 007 3832 66 83 60, F: 007 3832 66 09 91, E: nugdin@ngas.nsk.su

Oding, B.S., Ph.D., Ass Prof, Voronezh St Arch & Civ Eng Acad, 20 Let Oktyabrya St., 84, Voronezh 394006. T: 007 0732 715400

Ofrikhter, V.G., Ph.D., Chief Eng, Uralenergostroy Corp, "Permtetsstroy" Bld Com, Kamchatovskaya St., 26, Perm 614016. T: 007 3422 497105, F: 007 3422 34 47 76

Omelchak, I.M., Ph.D., Dir, "Space" Comp, Komsomolsky Pr., 85, apart. 34, Perm 614010. T: 007 3422 64 04 33

Osokin, A.I., Ph.D., Dir Gen, "Geostroy" Comp, Zagorodnyi Pr., 27/21, St-Petersburg 191180. T: 007 812 315 02 36, 312 09 36, F: 007 812 311 73 39

Ostyukov, B.S., Civ Eng, Hd of Dept, NIIOSP, 2-nd Institutskaya St., 6, Moscow 109428. T: 007 095 170 27 15

Panov, S.I., Dr, Prof, Dept Hd, Krasnoyarsk St Arch & Civ Eng Acad, Svobodnyi Pr., 82, Krasnoyarsk 660041. T: 007 3912 446940

Paramonov, V.N., Dr, Prof, Techn Dir, "Georeconstruction" Comp, Izmailovsky Pr., 4, St-Petersburg 198005. T/F: 007 812 316 61 18

Pavilonsky, V.M., Ph.D., "Vodgeo" Inst., Komsomolsky Pr., 42, Moscow 119826

Pavlik, G.N., Ph.D., Lab Hd, Rostov State University, Prospect Stachki, 200/1, Rostov/Don 344104. T: 007 8632 28 58 88, F: 007 8632 34 87 38

Paushkin, G.A., Dr, Prof, Dept Hd, Moscow St Civ Eng Univ, Yaroslavskaya Highway, 26, Moscow 129337. T: 007 095 261 8120

Perepelkin, I.Z., Eng, Deputy Dir, "Karkas" Comp, Sokolovaya St., 18/40, Saratov 410030. T: 007 8452 64 99 26, 64 96 87, F: 007 8452 64 95 41

Pertsovsky, M.I., Civ Eng, "Stav Ltd" Comp, Rubtsovskaya Emb., 3, Moscow 107082. T: 007 095 265 29 80

Pertsovsky, S.M., Civ Eng, Dir Gen, "Stav Ltd" Comp, Rubtsovskaya Emb., 3, Moscow 107082. T/F: 007 095 265 29 80, 265 03 27, 956 66 39, E: secretary@stav.edunet.ru

Petrashenj, A.S., Ph.D., Ass Prof, Dalnevostochnyi St Techn Univ, Krasnoe Znamya Prospect, 66, Vladivostok 690014. T: 007 4232 25 29 91

Petrov, M.V., Ph.D., Hd of Dept, Orel St Agro Academy, General Rodin St., 69, Orel 302019. T: 007 08600 65152

Petrovich, P.P., Ph.D., Ass Prof, Moscow Motor-Car-Road Inst (Techn Univ), Leningradsky Pr., 64, Moscow GSP-47 125829. T: 007 095 155 03 22, 155 03 43

Petrukhin, V.P., Dr, Prof, Deputy Dir, NIIOSP, 2-nd Institutskaya St., 6, Moscow 109428. T: 007 095 170 57 92, F: 007 095 170 27 57

Pilyagin, A.V., Dr, Prof, Hd of Dept, Mari St Techn Univ, Square of Lenin, 3, Ioshkar-Ola 424025. T: 007 83625 59 60 44

Pivnik, N.P., Eng, Chief Designer, "Krasnodargrazhdan-project" Inst, Ordzhonikidze St., 41, Krasnodar 350000. T: 007 8612 62 28 89, F: 007 8612 62 48 37

Polischuk, A.I., Dr, Prof, Hd of Dept, Tomsk St Arch & Civ Eng Acad, Solyanaya Square, 2, Tomsk 634003. T: 007 3822 72 18 24, F: 007 3822 72 44 22

Ponomaryov, A.B., Dr, Prof, Ass Prof, Perm St Techn Univ, Komsomolsky Pr., 29-A, Perm GSP-45 614600. T: 007 3422 39 15 73, F: 007 3422 39 14 96

Postoev, G.P., Dr, Prof, Lead Res, Vsegingeo Inst, Moscow Region, "Zelenyi Posiolok". T: 007 095 529 11 01 add. 22 73

Potapov, A.D., Ph.D., Hd of Dept, Moscow St Civ Eng Univ, Yaroslavskaya Highway, 26, Moscow 129337. T: 007 095 188 01 02

Prikhodchenko, O.E., Dr, Prof, Hd of Dept, Rostov St Civ Eng Univ, Socialistichiskaya St., 162, Rostov/Don 344022. T: 007 8632 65 98 88

Pshenichkin, A.P., Dr, Prof, Dept Hd, Volgograd St Arch & Civ Eng Acad, Academician St., 1, Volgograd 400074. T: 007 8442 449165

Rabinovich, M.V., Ph.D., Deputy Hd of Dept, Yakutsk St Univ, Belinsky St., 58, Yakutsk 677000. T: 007 4112 26 33 44

Radkevich, A.I., Ph.D., Sen Res, NIIOSP, 2-nd Institutskaya St., 6, Moscow 109428. T: 007 095 170 2768

Rasskazov, L.N., Dr, Prof, Dept Hd, Moscow St Civ Eng Univ, Yaroslavskaya Highway, 26, Moscow 129337. T: 007 095 261 49 56

Razbegin, V.N., Civ Eng, Editor-in-Chief, Journal "Bases, Foundations and Soil Mechanics"; Dir, Rus Branch Corp "Transcontinental", 2-nd Institutskaya St., 6, Moscow 109428. T: 007 095 170 27 53, T/F 007 095 135 35 39, E: smfe@mail.ru, icirtp@aha.ru

Revenko, V.V., Ph.D., Ass Prof, Novocherkassk St Techn Univ, Prosveschenie St., 132, Novocherkassk 346428. T: 007 86352 55 416, F: 007 86352 42 396

Roman, L.T., Dr, Prof, Lead Res, Moscow State University, Vorobjevy Gory, Moscow 119899. T: 007 095 939 14 53

Ryazanov, N.S., Ph.D., Ass Prof, Tomsk St Arch & Civ Eng Acad, Solyanaya Sq., 2, Tomsk 634003. T: 007 3822 72 31 34

Ryzhkov, I.B., Dr, Prof, Bash St Agro Univ, 50 Let Oktyabrya St., 34, Ufa. T: 007 3472 28 07 13, 28 07 17

Rukin, V.V., Ph.D., Lab Hd, "Orgenergostroy" Inst, Mintopen-ergo RF., Varshavskaya Highway, 17, Moscow 113105. T: 007 095 952 2906

Sadovsky, A.V., Ph.D., Chief Expert, NIIOSP, 2-nd Institut-skaya St., 6, Moscow 109428. T: 007 095 170 27 14, F: 007 095 170 27 57, E: root@niiosp.m9.ru

Sakharov, I.I., Dr, Prof, St-Petersburg St Arch-Civ Eng Univ, Krasnoarmeiskaya St., 4, St-Petersburg 198005. T: 007 812 259 53 98

Salnikov, B.A., Ph.D., Lab Hd, "TsNIIS" Inst, Kolskaya St., 1, Moscow 129329. T: 007 095 189 5126

Sapegin, D.D., Dr, Prof, Deputy Dir, VNIIG Inst, Gzhatskaya St., 21, St-Petersburg 195220. T: 007 812 535 2807

Savinov, A.V., Ph.D., Ass Prof, Saratov St Techn Univ; Dir Gen, "Geotechnics" Comp, Politekhnicheskaya St., 77, Saratov 410016. T: 007 8452 25 73 43, 25 77 33, 25 73 57, 007 8452 98 40 89

Sazhin, V.S., Dr, Prof, Lab Hd, "MosgiproNIIselstroy" Inst, Obruchev St., 46, Moscow 117863. T: 007 095 333 41 41, 334 71 69

Scherbina, V.I., Ph.D., Deputy Dir Gen, "NIIES" Centre, Stroitelnyi Pr., 7a, Moscow 123362. T/F: 007 095 492 7612

Scherbina, E.V., Ph.D., Moscow St Civ Eng Univ, Yaroslav-skaya Highway, 26 Moscow 129337. T/F: 007 095 188 05 03

Seleznev, A.F., Ph.D., Dept Hd, "Rostov PromstroyNIIpro-ject" Inst, Lenin St., 66, Rostov/Don 344038. T: 007 8632 31 03 08, 31 09 90

Semkin, V.V., Ph.D., Lead Res, NIIOSP; Dir Gen, "BALTYI" Comp, 2-nd Institutskaya St., 6, Moscow 109428. T/F: 007 095 170 15 58

Shadunts, K.Sh., Dr, Prof, Dept Hd, Kubanj St Agro Univ, Kalinin St., 13, Krasnodar 350044. T: 007 8612 50 64 65, 56 09 39

Shaposhnikov, A.V., Ph.D., Lead Res, NIIOSP, 2-nd Institut-skaya St., 6, Moscow 109428. T/F: 007 095 170 28 18

Shapiro, D.M., Dr, Prof, Deputy Chief Eng, "GiprodorNII" Inst, Moskovsky Pr., 4, Voronezh 394068. T: 007 0732 52 67 59, F: 007 0732 52 68 67

Shashkin, A.G., Ph.D., Dir, "Georeconstruction" Comp, Izmailovsky Pr., 4, office 411, S-Petersburg 198005. T/F: 007 812 316 61 18, 316 21 20, 251 81 82, 259 54 72

Sheinin, V.I., Dr, Prof, Lab Hd, NIIOSP, 2-nd Institutskaya St., 6, Moscow 109428. T: 007 095 170 27 24, F: 007 095 170 27 57, E: root@niiosp.m9.ru

Shemenkov, Yu.M., Ph.D., Lab Hd, "BashNIIstroy" Inst, Constitution St., 3, Ufa 450064. T: 007 3472 42 42 54

Shiryaev, R.A., Ph.D., Sen Res, VNIIG Inst, Gzhatskaya St., 21, St-Petersburg 195220. T: 007 812 535 21 46

Shishkin, V.Ya., Ph.D., Lead Res, NIIOSP, 2-nd Institutskaya St., 6, Moscow 109428. T: 007 095 170 28 28, T/F: 007 095 170 32 88

Shishkov, Yu.A., Civ Eng, Chief Designer, Novosibirsk St Design Inst, Sovetskaya St., 52, Novosibirsk-104 630104. T: 007 3832 21 81 43, F: 007 3832 21 67 10

Shkolnikov, I.E., Ph.D., Sen Res, "TsNIIS" Inst of Mintranstroy, Kolskaya St., 1, Moscow 129329. T: 007 095 189 54 71

Shmatkov, V.V., Ph.D., Ass Prof, Novocherkassk St Techn Univ; President, Rus-Germ Build Comp "Top-Design", Budennovskaya St., 146/36, apart.2, Novocherkassk. T: 007 86352 423 96, E: topd@novoch.ru

Shulyatjev, O.A., Ph.D., Sen Res, NIIOSP, 2-nd Institutskaya St., 6. T: 007 095 170 27 39, F: 007 095 170 27 57, E: root@niiosp.m9.ru

Shvetsov, G.I., Dr, Prof, Deputy Rector, Dept Hd, Altaj St Tech Univ, Lenin Pr., 46, Barnaul 656099. T: 007 3852 36 85 35, 26 16 05

Sirota, Yu.L., Ph.D., Sen Res, VNIIG Inst, Gzhatskaya St., 21, St-Petersburg 195220. T: 007 812 535 41 82

Skachko, A.N., Ph.D., Hd of Dept, NIIOSP, 2-nd Institut-skaya St., 6, Moscow 109428. T: 007 095 170 69 71

Skalnyi, V.S., Ph.D., Prof, Orel St Agro Academy, General Rodin St., 69, Orel 302019. T: 007 09222 6 51 52

Skrylev, G.E., Ph.D., Deputy Hd of Lab, Moscow St Civ Eng Univ, Spartakovskaya St., 2/1, Moscow 107066. T: 007 095 261 87 83

Smolin, Yu.P., Ph.D., Ass Prof, Siberian St Railway Eng Acad, Dusya Kovalchuk St., 191, Novosibirsk 630023. T: 007 3832 28 74 69, E: agul@mail.ru

Sobolev, N.I., Civ Eng, Chief Expert, "Teploenergoproject" Inst, Spartakovskaya St., 2a, Moscow 107066. T: 007 095 263 38 26, F: 007 095 261 73 69

Sokolov, V.S., Ph.D., Hd of Dept, GSPI Inst, Minatom of Russia, Novoryazanskaya St., 8a, Moscow 107014. T: 007 095 261 86 48

Sokolov, N.S., Ph.D., Hd of Dept, Chuvash St Univ, Cheboksary 428000. T: 007 8350 22 00 75

Solomin, V.I., Dr, Prof, Dept Hd, Chelyabinsk St Techn Univ, Lenin Pr., 76, Chelyabinsk 454080. T/F: 007-3512-399000, 399348, 399359

Soloviev, Yu.I., Ph.D., Ass Prof, Siberian St Railway Eng Acad, Dusya Kovalchuk St., 191, Novosibirsk 630023. T: 007 3832 29 47 84

Sorochan, E.A., Dr, Prof, Lab Hd, NIIOSP, 2-nd Institutskaya St., 6, Moscow 109428. T: 007 095 170 2763

Sorokin, Yu.M., Min Eng, Dir, "STAV Ltd.," Comp, Rubtsovskaya Emb., 3, Moscow 107082. T/F: 007 095 265 29 80, 265 03 27, E: secretary@stav.edunet.ru

Sotnikov, S.N., Dr, Prof, Dept Hd, St-Petersburg St Arch-Civ Eng Univ, 2-nd Krasnoarmeyskaya St., 4, St-Petersburg 198005. T: 007 812 259 53 98, F: 007 812 316 58 72

Spektor, Yu.I., Dr, Prof, Chief Eng, "Gidrotruboprovod" Comp, Dobrolyubov St., 16, office 2, Moscow 127254. T: 007 095 174 25 42, 219 01 48, F: 007 095 219 82 79

Stavnitser, L.R., Dr, Prof, Lab Hd, NIIOSP, 2-nd institutskaya St., 6, Moscow 109428. T: 007 095 170 70 12, F: 007 095 170 27 57, E: root@niiosp.m9.ru

Svetinsky, E.V., Ph.D., Lead Res, NIIOSP, 2-nd Institutskaya St., 6, Moscow 109428. T: 007 095 170 69 67

Svirsky, S.I., Civ Eng, Dir Gen, "Mosingstroy" Comp, M.Bronnaya St., 15-B, Moscow 103104. T: 007 095 202 56 48, 202 16 48, F: 007 095 924 71 26

Ter-Martirosyan, Z.G., Dr, Prof, Lab Hd, Moscow St Civ Eng Univ, Yaroslavskaya Highway, 26, Moscow 129337. T/F: 007 095 261 5988

Tikhomirova, L.K., Ph.D., Ass, St-Petersburg Arch-Civ Eng Univ, 2-nd Krasnoarmeiskaya St., 4, St-Petersburg 190005. T: 007 812 259 53 98

Timofeeva, L.M., Dr, Prof, Dir, "Transport communications and bridges" Res Lab, Korolev St., 12, apart.123, Perm 614013. T/F: 007 3422 39 15 37

Tolmachev, V.V., Ph.D., Dir, St Comp "Anticarst and Coast Protection", Mayakovsky St., 33, Nizhny Novgorod Region, Dzerzhinsk 606023. T: 007 8313 55 58 01, F: 007 8313 25 05 08

Trofimenkov, Yu.G., Ph.D., Chief Expert, NIIOSP, 2-nd Institutskaya St., 6, Moscow 109428. T: 007 095 170 6071

Tseitlin, M.G., Ph.D., Chief Res, VNIIGS Inst, Pinegin St., 4, St-Petersburg 193148

Tseeva, A.N., Ph.D., Hd of Dept, Yakutsk Design Res Inst, Dzerzhinsky St., 20, Yakutsk 677000. T: 007 4112 45 21 25

Trubin, V.M., Civ Eng, Dir Gen, "Hydrospetsfundamentstroy" Comp, Dobroslobodskaya St., 6, Moscow 107066. T/F : 007 095 265 27 77

Trusov, G.V., Civ Eng, Dir, "Vysotspetsstroy" Comp, F. Engels St., 6, bld.2, Moscow 107005. T: 007 095 267 93 35

Ukhov, S.B., Dr, Prof, Moscow St Civ Eng Univ, yaroslavskaya Highway, 26, Moscow 129337. T/F: 007 095 261 5988

Ulitsky, V.M., Dr, Prof, St-Petersburg St Arch-Civ Eng Univ, 2-nd Krasnoarmeyskaya St., 4, St-Petersburg 198005. T/F: 007 812 316 6118, 316 2120, 251 8182, 259 5472

Uzdin, A.M., Dr, Prof, St-Petersburg St Railway Eng Univ, Moscovsky Pr., 9, St-Petersburg 190031. T: 007 812 168 89 25, 168 82 49, F: 007 812 113 86 12, 524 85 26, E: uzdin@cendr.spb.ru

Ushkov, S.M., Ph.D., Ass Prof, Chuvash St Univ, Lenin Pr., 6, Cheboksary 428000. T: 007 8350 22 00 75

Vasilkov, G.V., Dr, Prof, Dept Hd, Rostov St Civ Eng Univ, Socialisticheskaya St., 162, Rostov/Don 344022. T: 007 8632 65 51 00

Verstov, V.V., Dr, Prof, Dept Hd, St-Petersburg St Arch-Civ Eng Univ, 2-nd Krasnoarmeyskaya St., 4, St-Petersburg 198005. T: 007 812 3161609

Vilkov, I.M., Ph.D., Prof, Volgograd St Arch & Civ Eng Acad, Tsiolkovsky St., 1, Volgograd 400074. T: 007 8442 44 06 55

Vinogradov, V.V., Dr, Prof, Deputy Prorector, Moscow St Railway Eng Univ, Obraztsov St., 15, Moscow 101475. T: 007 095 284 21 10

Vorobiev, E.A., Ph.D., Chief Expert, PNIIIS Inst, Okruzhnoy Proezd, 18, Moscow 105848. T: 007 095 369 75 23

Vorontsov, G.I., Ph.D., Dir, VNIINTPI Inst, Vernadsky Pr., 29, office 2202, Moscow 117331. T/F: 007 095 133 07 71, T: 007 095 131 39 65, E: NTPI@mail.ru

Voskresensky, S.M., Ph.D., Dir Gen, "Hydrospetsproject" Inst; President "Hydrospetsstroy" Corp, Kadashevskaya Emb., 6/1, Moscow 109017. T: 007 095 237 58 15, 237 68 05, F: 007 095 237 58 65

Yakovleva, T.G., Dr, Prof, Moscow St Railway Eng Univ, Obraztsov St., 15, Moscow 101475. T: 007 095 281 1913

Yarovoy, Yu.I., Dr, Prof, Hd of Dept, Ural St Railway Eng Univ, Kolmogorov St., 66, Ekaterinburg 620034. T: 007 3432 45 47 67, F: 007 3432 45 31 88

Yastrebov, P.I., Ph.D., Lead Res, NIIOSP, 2-nd Institutskaya St., 6, Moscow 109428. T: 007 095 170 69 67, F: 007 095 170 27 57

Yurkin, R.G., Ph.D., Vereskovaya St., 3, bld.2, apart.30, Moscow 129329. T: 007 095 180 0278

Zaitsev, A.K., Ph.D., Hd of Dept, St-Petersburg Military Civ Eng Univ, Zakharievskaya St., 22, St-Petersburg. T: 007 812 278 03 03

Zaretsky, Yu.K., Dr, Prof, Dir, Intern Inst of Geomech & Hydrotech, Kadashevskaya Emb., 6/1, Moscow 109017. T/F: 007 095 237 63 49, T: 007 095 151 13 51

Zekhniev, F.F., Ph.D., Lead Res, NIIOSP, 2-nd Institutskaya St., 6, Moscow 109428. T/F: 007 095 170 19 27

Zertsalov, M.G., Dr, Prof, Hd of Dept, Moscow St Civ Eng Univ, Yaroslawskaya Highway, 26, Moscow 129377. T/F: 007 095 261 81 88

Zhivoderov, V.N., Ph.D., Deputy Dept Hd, "Transhydrostroy" Comp, Krzhizhanovsky St., 20/30, bld.1, Moscow 117218. T/F: 007 095 120 14 62, 128 83 41

Ziangirov, R.S., Dr, Prof, Chief Geologist, "Mosgorgeotrest" Comp, Leningradsky Pr., 11, Moscow 125040. T: 007 095 458 65 10, F: 007 095 458 86 37

Zinoviev, A.V., Ph.D., Chief Eng, "Teploproject" Inst, Komintern St., 7, bld.2, Moscow 129344. T: 007 095 184 1681

Zlatoverkhovnikov, L.F., Ph.D., Sen Res, "LenmorNIIproject" Inst, Mezhevoy Canal, 3, bld.2, St-Petersburg 198035. T: 007 812 259 8220

Znamensky, V.V., Ph.D., Ass Prof, Moscow St Civ Eng Univ, Yaroslavskaya Highway, 26, Moscow 129337. T: 007 095 261 59 88

SOUTHEAST ASIAN GEOTECHNICAL SOCIETY

Secretary: Prof. D.T. Bergardo, Southeast Asian Geotechnical Society, Asian Institute of Technology, P.O. Box 4, Klong Luang, Pathumthani 12120, Thailand. T: 66-2-524-5519, 516-0130-44
F: 66-2-524-5509, 516-2126, E: seags@ait.ac.th

Total number of members: 261

Adachi, K., Professor, Shibaura Institute of Technology, 3-9-14 Shibaura, 3 Chome, Minatoku, Tokyo 108-8548, JAPAN. T: 03-5476-3048, F: 03-5476-3166, E: adachi@sic.shibaura-it.ac.jp

Akagi, T., Professor, Dept. of Civil Engineering, Toyo University, Kujirai 2100, Kawagoe City, Saitama 350-8585, JAPAN. T: 81-492-39-1394, F: 81-492-31-4482, E: akagi@eng.toyo.ac.jp

Al-Saoudi, N.K.S., Prof., Head of Highway and Bridges Section, Building & Construction Engineering Department, University of Technology, 12906 Baghdad, IRAQ. E: techuni@uruklink.net

Alawaji, H.A., Dr., Asso. Professor Geotech. Engrg., Civil Engineering Department, College of Engineering King Saud University, P.O. Box 800, Riyadh 11421, SAUDI ARABIA. E: alawaji@ksu.edu.sa

Al-Gharieb, M., Sakr., Dr. M., Faculty of Engineering, Tanta University, Tanta, EGYPT

Anandasiri, K.A.R., Geotechnical Engineer, Geotechnical Engineering Office, 14F Civil Engineering Building, 101 Princess Margaret Rd, Homantin, Kowloon, Hong Kong, China. T: 852-2890-8104 (Off), 2716-8622 (Off), F: 852-2715-7572, 2795-9611, E: gesoil2_lab@ced.gov.hk

Arenicz, R.M., Dr., Dept. of Civil & Mining Engineering, University of Wollongong, P.O. Box 1144, Wollongong, NSW 2500, AUSTRALIA

Arulrajah, A., 11, Jalan 36/70A, Desa Sri Hartamas, 50480 Kuala Lumpur, MALAYSIA. E: ryanroy@pd.jaring.my

Azam, T., Chief Executive Officer, Ranhill Civil Sdn., Bhd., No. 7 Jalan Kelawar 6/4H, Section 6, Shah Alam, 40000 Selangor, DE, MALAYSIA. T: 603-4260-3170, F: 603-4260-3169, E: tarique@recsb.po.my

Balasubramaniam, A.S., Prof., Chair Professor, School of Civil Engineering, Asian Institute of Technology, P.O. Box 4, Klong Luang, Pathumthani 12120, THAILAND. T: 66-2-524-5508, F: 66-2-524-5509, E: bala@ait.ac.th

Barry, A.J., Tony & Mariyana Barry, Jacob & Denis, 11-3-2 Menara Hartamas, Jalan Sri Hartamas, 50480 Kuala Lumpur, MALAYSIA. E: tbarry@evalueco.com

Becker, D., Editor, Canadian Geotechnical Journal, 2180 Meadowvale Blvd., Mississauga, Ontario L5N 5S3, CANADA. E: dbecker@golder.com

Bellis, J.A., Kampebraten 7B, N-1300 Sandvika, NORWAY

Bergado, D.T., Professor, School of Civil Engineering, Asian Institute of Technology, P.O. Box 4, Klong Luang, Pathumthani 12120, THAILAND. T: 66-2-524-5512, F: 66-2-524-5509, E: bergado@ait.ac.th

Botelho, O.F., Civil Engineer, Head of Department, Laboratorio de Engenharia Civil de Macau (LECM), Rua da Se, 30, Macau, Hong Kong, China. F: 853-578930, E: oib@lecm.org.mo

Brand, E.W., Dr., Consultant Carpenter's Cottage, Nether Westcote OX7 6SD, UK

Brenner, R.P., Dr., Rosentrasse 8, CH-8570, Weinfelden, SWITZERLAND. E: rpb.g-de@weinfelden.ch

Budhihartanto, G., Jl. Mimosa K-11, Buncit Indah, Jakarta Selatan 12510, INDONESIA

Budihardjo, S., Jl. Ciwulan 19, Bandung 40114, INDONESIA. E: mettana@indo.net.id

Chaidrata, S., Consulting Managers, PT Putra Satria Prima, Citra Garden II, Block 06, No 25, Jakarta Barat 11830, INDONESIA. F: 62-21-690-1530, E: suheri@cbn.net.idpsp@psprima.com

Champa, S., Managing Director, GMS Power (Public) Co, 244 Sukumvit Soi 115, Sukhumvit Road, Theparaks, Smutarakarn 10270, THAILAND. E: 66-2-394-551, F: 66-2-394-0551

Chan, V.H.C., 3/F, Flat A, 22 Graupian Road, Kowloon, Hong Kong, China. T: 2337-8322, F: 2337-1497

Chan, S.F., Dr., 33-22-3 Sri Penaga, Jalan Medang Serai, Bukit Bandar Raya, 59100 Kuala Lumpur, MALAYSIA. E: sfchan@pc.jaring.my

Chang, S.Y., Sinotech Engineering Consultants Ltd., 171, Nan King E. Road, Sec. 5, Taipei 10572, R.O.C. E: evp@sinotech.org.tw

Chang, C.T., Prof., Vice President, Sinotech Engineering Consultants Ltd., No. 171, Nan King East Road, Sec. 5, Taipei 105, Taiwan, R.O.C. E: changct@sinotech.com.tw

Chang, D.-W., Prof., Dept. of Civil Engineering, Tamkang University, Tamsui, Taiwan 25137, R.O.C. T: +886 2-2623-4224, F: +886 2-2623-4224, E: dwchang@mail.tku.edu.tw

Chang, H.W., Dr., Professor, National Central University, Department of Civil Engineering, 38, Wu-chuan Li, Chungli, Taiwan 320, R.O.C. T: 886-3-4227151, Ext. 4121, F: 886-3-4252960, E: hueiwen@cc.ncu.edu.tw

Chang, M.F., Dr., Nanyang Technological University, School of Civil & Structural Engineering, Nanyang Avenue, Singapore 639798, SINGAPORE. E: CMFCHANG@ntu.edu.sg

Chantawarang, K., Dept. of Civil Engineering, Faculty of Engineering, Kasetsart University, Bangkok 10900, THAILAND

Chau, N.N., Manager (S.I. Department), VIBRO (HK) Limited, 287/F New World Tower, 16-18 Queen's Road Central, Hong Kong, China. F: 852 2524 9724, E: cn_ngan@hiphing.com.hk

Chen, M.S., Water Resources Department, Mini. of Econo. Affairs, 2F 48 Chan-An West Road, Taipei, Taiwan 104, R.O.C.

Chen, C.H., Prof., Department of Civil Engineering, National Taiwan University, 1 Roosevelt Rd, Sec 4, Taipei, Taiwan, R.O.C.

Chen, R.C., Deputy Controller/Building Control Unit, JTC Corporation, 7, Sunset Close, Singapore 597522, SINGAPORE. E: renchung@jtc.gov.sg

Cheng, Y., Chairman, Board of Director, Sinotech Foundation for Research & Development of Engineering Sciences & Technology, 5F, No. 75-4, Chung Hsiao Rd, Sec. 4, Taipei 106, Taiwan, R.O.C. T: 886-2-27789988, F: 886-2-27789911, E: sinotecf@m532.hinet.net

Cheng, F.C., Project Manager, Cluna Engineering Consultants Inc., 3rd Floor, No: 5-1, Lane 46, Shih-Tung Rd, Shin-Lin, Taipei, Taiwan, R.O.C.

Chern, J.C., Geotechnical Research Center, Sinotech Engineering Consultants Inc., Sinotech Building, 171 Nanking East Road, Sec. 5, Taipei 105, R.O.C. E: sinotech@sinotech.org.tw

Cheung, P.C.T., Geotechnical Engg. Office, 13/F, Civil Engineering Bldg., 101 Princess Margaret Rd, Homantin, Kowloon, Hong Kong, China. E: paulcheu@asiaonline.net

Chew, R.Y.C., Robert Y. Chew Geotech. Inc., 26062 Eden Landing Rd., Suite #7, Hayward, CA 94545, USA

Chew, S.H., Dr., Department of Civil Engineering, National University of Singapore, 10 Kent Ridge Crescent, Singapore 0511, SINGAPORE. E: cvecsh@nus.edu.sg

Chiam, T.T., 16 Jalan SS 2/63, 47300 Petaling Jaya, MALAYSIA

Chien, L.K., Prof., National Taiwan Ocean University, Dept. of Harbour & River Engineering, No: 2 Peining Road, Keelung, Taiwan 202-24, R.O.C.

Chin, C.H., Group General Manager, Pengurusan Lebuhraya Berhad, 9F, Menara 2, Faver Towers, Jalan Desa Bahagia, Taman Desa, Off Jalan Klang Lama, 58100 Kuala Lumpur, MALAYSIA. T: 603-7627-2788, F: 603-7625-6212, E: chinch@plb.kkellas.com.my

Chin, C.T., Dr., Moh & Associates Inc., 22/F, No. 112, Shin-Tai-Wu Road, Sec. 1, Oriental Technopolises Tower A, Hsi Chih Town, Taipei Hsien 221, Taiwan, R.O.C. E: maagroup@ms.maa.com.tw

Chin, E.C.H., Nylex (Malaysia) Berhad, Persiarah Selangor, Shah Alam Industrial Estate, Sec. 6, 40910, P.O. Box 7033, Kuala Lumpur, MALAYSIA. E: nylex@tm.net.my

Chin, J.C.P., P.O. Box 597, Tg. Aru, 88858 Kota Kinabalu, Sabah, MALAYSIA. E: jcpchin@tm.net.my

Chin, M.C., 18 Matthew Street, Merrylands, New South Wales 2160, AUSTRALIA

Chin, Y.P.R., Prof., Department of Civil Engineering, National Cheng Kung University, Tainan, Taiwan 70101, R.O.C. T: 886-6-209-4508, F: 886-6-235-8542, E: ypchin@mail.ncku.edu.tw

Chiong, S.K., Director, Geospec Sdn., Bhd., P.O. Box 644 MPC, Old Airport Berakas 3577, Negara Brunei Darussalam, BRUNEI. E: skchiong@brunet.bn

Chong, C.P.P., Consulting Engineer, No. 25, Kenyalang Park Shopping Centre, 93300 Kuching, Sarawak, MALAYSIA. T: 082-333622, F: 082-243718, E: iemsb@po.jaring.my

Chong, M.K., Dr., 1 Marine Vista, #16-77, Neptune Court, Singapore 449025, SINGAPORE

Chong, P.S., Soil and Foundation Pte. Ltd., 608-C East Coast Road, Singapore 459004, SINGAPORE

Chong, Y.W., KP CHAI, Engg. & Management Consultants, Block 204, Hougang Street 21, #03-117, Singapore 530204, SINGAPORE

Choon, L.E., Dr., Nanyang Tech. University, School of Civil & Struc. Engineering, Nanyang Ave., Singapore 639798, SINGAPORE

Chow, Y.K., Dr., Department of Civil Engineering, National University of Singapore, 10 Kent Ridge Crescent, Singapore 119260, SINGAPORE

Chu, J., Dr., Associate Professor, School of Civil & Structural Engineering, Nanyang Technological University, #01A-10, 50 Nauyang Ave., Singapore 639798, SINGAPORE. T: 65-790-4563, F: 65-792-1650, E: cjchu@ntu.edu.sg

Chua, S.H., Managing Director, Autoways Holdings Berhad, Lot 17, Section 92A, Jlan Dua, Off Jalan Chan Sow Lin, 55200 Kuala Lumpur, MALAYSIA. T: 603-9221-881, F: 603-9221-8880, E: chua@autoways.com.my

Coleman, R.A., P/T Lecturer & Consultant, University of Technology, Sydney, 80 Wakehurst Parkway, North Narrabeen, NSW 2101, AUSTRALIA. T/F: 61-2-9913 7724, E: cdemanra@hotmail.com

Contini, M., Sales Manager, Pagani Geotechnical Equipment, Loc. Santechnical Equipment, 29010 Calendasco, Piacenza, ITALY. T: 39-0335-6160503, F: 39-0523-771535, E: contini.pagani@tin.it

Cowland, J.W., 38 Magazine Gap Road, 2F, Hong Kong, China. E: cowland@asiaonline.net

Daito, Dr., K., Assistant Professor, Dept. of Civil Engineering, Daido Institute of Technology, 40 Hakusui-cho, Minami-ku, Nagoya 457, JAPAN

Daley, P., 2 Arundel Road, Tunbridge Wells, Kent, TN1 1TB, UK

Domanski, T., Managing Director, Bauer (Malaysia) Sdn., Bhd., 40 (2nd & 3rd) Jalan 52/4, 46200 Petaling Jaya, Selangor Darul Ehsan., MALAYSIA. E: domanski@po.jaring.my

Dutton, C., Director, Binnie Black & Veateh HK Ltd., 11th Floor, New Town Tower, Pak Hok Ting St., Shatin, NT, Hong Kong, China

Fan, S.L., Principal Soil & Foundation Engineer, Sinotech Engineering Consultants Ltd., No.171, Nanking E. Road, Sec. 5, Taipei, R.O.C. F: 886-2-2763 4555

Fang, Y.S., Prof., Department of Civil Engineering, National Chiao Tung University, 1001 Ta Hsueh Rd., Hsinchu, Taiwan 30010, R.O.C. T: 886-3-571-8636, F: 886-3-571-6257, E: ysfang@cc.nctu.edu.tw

Fellenius, B.H., Dr., Urkkada Technology Ltd., 735 Ludgate Court, Ottawa, Ontario K1J 8K8, CANADA. E: urkkada@attcanada.ca

Freer, C., Senior Geotechnical Engineer, c/o Tonkin & Taylor Ltd., P.O. Box 5271, Auckland 1, NEW ZEALAND. E: cfreer@tonkin.com

Fujioka, M.R., MASA Fujioka & Associates, 99-1205 Halawa Valley Street, Suite 302, Aiea, Hawaii 96701-3281, USA

Fujita, K., Dr., Dept. of Civil Engineering, Science University of Tokyo, 2641 Yamazaki, Noda-shi, Chiba 278-8510, JAPAN. T: 81-471-24-1501, Ext. 4011, F: 81-471-23-9766

Gasaluck, W., Department of Civil Engineering, Khon Kaen University, Faculty of Engineering, Khon Kaen 40002, THAILAND. E: watgas@kku1.kku.ac.th

Golombek, S., Rua Baronesa de Itu, 858 Paulo, Brazil CEP 01231-000, BRAZIL. E: consultrix@trycomm.com.br

Goughnour, R., Dr., Managing Engineering, Nilex Corporation, 705 Duft Road, N.E., Leesburg, Virginia 20176, USA. T: 703-771-0135, F: 703-771-0975, E: bgoughnour@nilex.com

Gue, S.S., Dr., Level 6, Wisma SSP, 1, Jalan SR 8/3, Serdang raya Seksyen 8, 43300 Seri Kembangan, Selangor Darul Ehsan, MALAYSIA

Gui, M.W., Dr., Assistant Professor, National Taipei University of Technology, Civil Engineering Department, No 1, Sec. 3, Chung-Hsiao E. Rd, Taipei 10643, Taiwan, R.O.C. E: mwgui@ntut.edu.tw

Guilford, C.M., Retired Consoltant, Flat 8C, Block 18, Village Gardens, 44 Fa Po Street, Kowloon, Hong Kong, China. T: 852-2778-8613, F: 852-2778-8613

Guo, Y.Y., ARUP Jururunding Sdn., Bhd., 72-76, Jalan 3/62, Bandar Menjalara, 52200 Kuala Lumpur, MALAYSIA. E: guo.yeang.yang@arupkl.po.my

Han, D., 103-1004, Kangbyon Apt., Manyondong, Seoku, Taejon 302-150, KOREA

Handali, S., Dr., Immanuel Christran University, (UKRIM), JL. Solo Km. 11, 1, P.O. Box 4/YKAP, Yogyakarta, INDONESIA. E: safe@hand.wlink.com.np

Handidjaja, B.P., Blk. 131, Tampines St. 11, #04 242, Singapore 1852, SINGAPORE. E: paulhandidjaja@pacific.net.sg

Harnpattanpanich, T., Dr., Geotechnical Engineer, Royal Irrigation Department, Geotechnical Division, Samsen Road, Dusit, Bangkok 10300, THAILAND

Hassan, C.A., Dr., Syarikat Pembinaan Sri Kumbang Sdn., Bhd., Lot 4-05, 4th Floor, Plaza Prima, Batu 41/2, Jalan Kelang Lama, 58200 Kuala Lumpur, MALAYSIA

Hj, O., H.K.B., Director, Jurutera Konsultant (M) Sdn., Bhd., 435, Wisma Jk, Jalan Ipoh, 51200 Kuala Lumpur, MALAYSIA. T: 603-4043-5122, F: 603-4043-3784, E: khalitjkc@hotmail.com

Ho, Y.M., MAA Engineering Consultants (H.K.) Ltd., Unit C, 11/F, China Overseas Building, 139 Hennessy Road, Hong Kong, China

Ho, S.T.M., 16th Floor, Stage 5, 16A Nassau Street, Mei Foo Sue, Sun Chuen, Kowloon, Hong Kong, China

Ho, C.E., 284 Vassar Street, Apt. G2, Cambridge, MA. 02139, USA. T: 617-577-5891, E: chueuho@mit.edu

Hoffmann, G.L., Operations Director, Keiler (M) Sdn., Bhd., 35 Lorong B., Kampung Pakar Batu 5, Jalan Sungei Besi, 5700 Kuala Lumpur, MALAYSIA. F: 603-7800349, E: kellerm@po.jaring.my

Hong, L.P., 40-2 Jalan 2/109E, Desa Business Park, Taman Desa, 58100 Kuala Lumpur, MALAYSIA. E: edasu@po.jaring.my

Hong, S.W., Dr., Korea Institute of Construction Technology, 2311, Daehwa-dong, Ilsan-gu, Koyang-shi, Kyonggi-do 410-411, KOREA. E: swhong@kict.re.kr

Hongsnoi, M., 301 Soi Ruamsirimitr, Superhighway, Bangkhen, Bangkok 10900, THAILAND

Hou, P. C., c/o Sinotech Engineering Consultants Ltd., No. 171, Nanking East Rd., Sec. 5, Taipei 10572, Taiwan, R.O.C. F: 886-2-2763-4555

Hsieh, C.S., Sinotech Engineering Consultants Ltd., Sinotech Building, 14th Floor, 171, Sec. 5, Nanking East Road, Taipei 105, Taiwan, R.O.C. F: 886-2-276-34555

Hsu, C.I., Vice President, CTCI Corporation, 12th Floor, CTCI Tower 77, Sec. 2, Tun Hwa South Road, Taipei, Taiwan, R.O.C. F: 886-2-709-9303, E: c.I.hsu@ctci.com.tw

Huang, C., Senior Tunnelling Engineer Deputy Project Driector/TANEEB 3rd Proj. Office, 3F. No. 7, Lane 79, Sec. 2, Mu Hsin Road, Mu Cha 11646 Taipei, Taiwan, R.O.C. E: chih@taneeb03.gov.tw

Huong, T.I., Director, Geospec Sdn., Bhd., Lot 587, 1st & 2nd Fl, Lorong 2B, Off Jalan Blacksmith, 93100 Kuching, Sarawak, MALAYSIA. E: darujob@pc.jaring.my

Hussein, A.N., Dr., Road Design Unit, Roads Branch, JKR Malaysia, Jalan Sultan Salahuddin, 50582 Kuala Lumpur, MALAYSIA. E: hanadzri@hg.jkv.gov.my

Hussein, M., Vice President, GRL and Associates, 8000 South Orange Avenue, Orlando, Florida 32809, USA. T: 407-826-9539, F: 407-826-4747, E: mhgrlfl@aol.com

Hwang, R.N., Moh & Associates Inc., 22/F, No. 112, Shin-Tai-Wu Road, Sec. 1, Oriental Technopolises Tower A, Hsi Chih Town, Taipei Hsien 221, Taiwan R.O.C. E: maagroup@ms.maa.com.tw

Hwang, Z.M., President, Asia World Engineering & Construction Co., 7F-2 No.81, Sec. 1, Shin Tai Wu Rd., Hsichih, Taipei County, Taiwan, R.O.C. E: awg4988@ms25.hinet.net

Indraratna, B., Dr., Professor of Civil Engineering, Faculty of Engineering, University of Wollongong, Wollongong, NSW 2522, AUSTRALIA. E: Indra@uow.edu.au

Jamaludin, A., Director, Khairi Consult Sdn., Bhd., No 88A, Jalan SG 4/8, Taman Seri Gombak, 68100 Batu Caves, Selangor Darul Ehsan, MALAYSIA. T: 03-6841652/3, F: 03-6841654, E: azmikc@tm.net.my

Jeng, C.J., Vice President/Adjunct Assistant Professor, Chung-chi Technical Consultant Co., Ltd., National I-Lan Institute of Technology, 12F-1, No. 431, Kuang-Fu South Rd, Taipei 110, Taiwan, R.O.C. E: jcj46380@tptsi.seed.net.tw

K. Engineering Consultants Co. Ltd., Managing Director, 136 Soi Intamara 18, Din Daeng, Bangkok 10400, THAILAND. F: 66-2-691-8366

Kamon, M., Prof., Disaster Prevention Research Institute, Kyoto University, Gokasho, Uji, JAPAN. E: kamon@geotech.dpri.kyoto-u.ac.jp

Kang, B.H., Professor, Dept. of Civil Engineering, Inha University, No. 253, Yonghyun-dong, Nam-ku, Incheon, Korea 402 751, KOREA. E: bhkang@inha.ac.kr

Kar, W., Senior of Geotechnical Engineer, Soil & Foundation P/L, No. 7 Harrison Rd, #02-00, Harrison Industrial Bldg., Singapore 369650, SINGAPORE. T: 65-3801703 (off), 65-9743-4704 (Home), F: 65-288-3451, E: karwinn@pacific.net.sg

Katzir, M., Foundation and Soil Limited, 44 Revivim St., Zahala, Tel-Aviv 69274, P.O. Box 10107, ISRAEL

Kazi, A., Prof., Vice Chancellor, Isra University, Hala Road, P.O. Box 313, Hyderabad-Sindh, PAKISTAN. T: 92-221-620177, F: 92-221-620180, E: israhd@hyd.paknet.com.pk

Khan, A.M., Dr., Lecturer, Geotechnical Group, Faculty of Engineering, Civil Engineering Department, King Abdulaziz University, P.O. Box 80204, Jeddah 21589, SAUDI ARABIA. E: amksk@hotmail.com

Khanchanusthiti, R., 120/1 Soi 5, Moo Ban navathanee, Sukhaphiban 2 Rd., Bungkum, Bangkok 10230, THAILAND

Kim, S.K., Prof., President, Dong-Pusan College, 640 Pansong-Dong, Haeundae-Gu, Pusan 612-715, KOREA. T: 82-51-540-3701, F: 82-51-542-7188, E: skkim@sb.dpc.ac.kr

Kim, S.S., Prof., Dept. of Civil Engineering, Chung-Ang University, #229, Heuk-Suk dong, Dong-Jak-ku, Seoul, KOREA. E: kimss@cau.ac.kr

Kim, H.T., Prof., Dept. of Civil Engineering, College of Engineering, Hong-Ik University, 72-1, Sangsu-Dong, Mapo-ku, Seoul 121-791, KOREA. E: htaek@wow.hongik.ac.kr

Kim, M.M., Prof., Seoul National University, Civil Engineering Department, Gwanak-ku 15, Seoul, KOREA. E: geotech@gong.snu.ac.kr

Kulhawy, F.H., Prof., Cornell University, Hollister Hall, Ithaca, NY 14853-3501, USA

Kurzeme, M., Dr., Golder Associates Pty. Ltd., P.O. Box 6079, Hawthorn West Vic 3122, AUSTRALIA. E: melbourne@golder.com.au

Kuwano, J., Dr., Dept. of Civil Engineering, Tokyo Institute of Technology, O-okayama, Meguro-ku, Tokyo 152, JAPAN

Lai, H.W., Dr., President Geostructure – Hydrotechnology, A-12, 1250, Eglinton Avenue West, Suite 228, Ontario, L5V 1N3, CANADA. T: 905-542-2610, F: 905-542-1278, E: hla10@attglobal.net

Lai, W.M., Operations Director, Keiler (M) Sdn., Bhd., 35 Lorong B., Kampung Pakar Batu 5, Jalan Sungei Besi, 5700 Kuala Lumpur, MALAYSIA. T: 03-7802894, F: 03-7800349

Lam, Y.C., Minconsult Sdn., Bhd., No. 14, Lorong 20/16A, 46300 Petaling Jaya, Selangor, MALAYSIA

Lau, J.C.W., Dr., Managing Director, Fong On Construction & Engineering Co., Ltd., 17/F, Vulcan House, 21-23 Leighton Rd, Hong Kong, China. T: 2893-0332, F: 2838-0011, E: jcl@fong-on.com.hk

Lau, Y.S., Dr., Bukit Timah Shopping Centre, 170 Upper Bukit Timah Road, #11-04 Singapore 2158, SINGAPORE. F: 65 468 6813

Lawrance, C., Senior Research Officer, Transport and Road Research Laboratory, Engineering Geology Section, Oversea Unit, Crowthorne, Berkshire RG45 6AU, UK

Lawson, C.R., CEO, Royal Ten Cate Regional Office, 11th Floor, Menara Glomac, Kelana Business Centre, 97 Jalan, SS 7/2, 47301 Petaling Jaya, MALAYSIA. F: 60-3-7492-8285, E: royal@po.jaring.my

Lee, S.H., Dr., National Taiwan Institute of Technology, Department of Construction Engineering, P.O. Box 90-130, Taipei, Taiwan 106, R.O.C. F: 886-2-7376606

Lee, C.J., Associate Professor, National Central University, Department of Civil Engineering, Chungli, Taiwan 32054, R.O.C. E: cjleeciv@cc.ncu.edu.tw

Lee, S.L., Prof., Department of Civil Engineering, National University of Singapore, 10 Kent Ridge Crescent, Singapore 0511, SINGAPORE. T: 65-772-2149, F: 65-779-1635, E: cveleesl@nus.edu.sg

Lee, F.H., Dr., Department of Civil Engineering, National University of Singapore, 10 Kent Ridge Crescent, Singapore 119260, SINGAPORE. E: cveleefh@nus.edu.sg

Lekhak, B.M., Kumagai Gumi Co. Ltd., (Thailand), 5/F Chaiyo Bldg., 91/1 Rama 9 Rd., Huay Kwang, Bangkok 10320, THAILAND

Leong, P.C.S., Apt. Blk. 469, Jurong West St. 41, #04-489, Singapore 640469, SINGAPORE

Leung, C.F., Dr., Dept. of Civil Engineering, National University of Singapore, 10 Kent Ridge Crescent, Singapore 119260, SINGAPORE

Li, F.H., 199 Tung Choi St., 3rd Floor, Kowloon, Hong Kong, China

Li, C.K.K., Room 1205, Tung On Building, 171 Prince Edward Road, Kowloon, Hong Kong, China

Li, D.W., Prof., Institute of Structural Engineering, Lanzhou Railway College, Lanzhou, Gansu 730070, P.R. China. E: rewnx@lzri.edu.cn

Li, J.C.C., Dr., Department of Civil Engineering, National Central University, Chung-li, Tao-yuan, Taiwan 320, R.O.C. T: 886-2711-9300, 4227251, Ext. 4195, F: 886-2-2741-8437(H), 2776-3238(O), E: chienchungli@yahoo.com.tw

Liew, Y.Y., No. 7, Jln 9/37, Taman Bukit Maluri, 52100 Kuala Lumpur, MALAYSIA. E: ktaski@po.jaring.my

Likins, G., President, Pile Dynamics Inc., 4535 Renaissance Parkway, Cleveland, Ohio 44128, USA. T: 216-831-6131, F: 216-831-0916, E: garland@pile.com

Lin, M.L., Prof., Department of Civil Engineering, National Taiwan University, Taipei, Taiwan 106, R.O.C. T: 886-2-23626281, F: 886-2-23626281, E: mllin@ce.ntu.edu.tw

Lin, H.D., Dr., Professor, Department of Construction Engineering, National Taiwan Institute of Technology, P.O. Box 90-130, Taipei, Taiwan 10672, R.O.C. T: 886-2-27376559, F: 886-2-27376606, E: hdlin@hp.ct.ntust.edu.tw

Liu, W.T., Prof., #1821 Chun-chan Road, Lu-chu Town, Kaohsiung County, Taiwan, R.O.C. E: wtliu@cc.kyit.edu.tw

Lo, K.H., Senior Geotechnical Engineer, MAA Engineering Consultants (HK) Ltd., Flat C., Block 1, 13/F., Woodview Court, 75 Kung Lok Rd, Kwun Tong, Kowloon, Hong Kong, China. T: 852-234-28507, F: 852-2861-2801, E: kanhunglo@yahoo.com.hk

Loke, K.H., Dr., Polyfelt Geosynthetics Sdn., Bhd., (Company No. 180781-W), 14, Jalan Seksyen 27, 40400 Shah Alam, Selongor Darul Ehsan, MALAYSIA. F: 603 512 8575, E: lokekh@polyfelt.com.my

Low, K.F., NRIC No: 5, Chapel Close, Singapore 429562 SINGAPORE. E: lkf@beca-asia.com

Lwin, T., Soil Laboratory Engineer (Supervisor), Reclamation at Changi East Phase Area "A" North, Hyundai Engineering & Construction Co., Ltd., Changi Airport, P.O. Box 0310, Singapore 918151, SINGAPORE. F: 65-545-5611, E: tun_lwin@pacific.net.sg

Madhav, M.R., Prof., Department of Civil Engineering, Indian Institute of Technology, Kanpur-208016, INDIA

Madjar, H., President, Terratech, 455 Rene-levesque Blvd. West, Montreal (Quebec), Canada H2Z 1Z3, CANADA. T: 514-393-1000, Ext. 7710, F: 514-393-9540, E: madjh@snc-lavalin.com

Mah, G.S., Minconsult Sdn., Bhd., P.O. Box 12263, 50772 Kuala Lumpur, MALAYSIA. T: 603-752-5757, F: 603-754-7373

Malone, A.W., Prof., Department of Earth Science, The University of Hong Kong, Pokfulam Road, Hong Kong, China. T: 852-2559-2555, F: 852-2517-6912, E: awmalone@netvigator.com

Martin, R.P., Dr., Geotechnical Engg. Office, 10/F Civil Engineering Bldg., 101 Princess Margaret Rd., Kowloon, Hong Kong, China

Marto, A.B.T., Dr., Faculty of Civil Engineering, Universiti Teknologi Malaysia, Locked Bag 791, 80990 Johor Bahru, MALAYSIA. E: aminaton@fka.utm.my

Matsubara, K., Taisei Corporation, A509 2-8-3 Somechi, Chofu City, Tokyo 182, JAPAN

Matsuda, H., Dr., Associate Professor, Dept. of Civil Engineering, Yamaguchi University, 2-16-1, Tokiwadai, Ube, Yamaguchi, 755-8611, JAPAN. T: 81-836-85-9324, F: 81-836-85-9301, E: hmatsuda@po.cc.yamaguchi-u.ac.jp

Maugeri, M., Prof., Corso Umberto, 208 95024 Acireale, ITALY

Maung, T., Principal Engineer Materials, Department of Works, P.O. Box 1108, Boroko, N.C.D. 111, PAPUA NEW GUINEA. T: 675-324-1244, F: 675-324-1230, E: tmaung@datec.com.pg

Miura, N., Prof., Department of Civil Engineering, Faculty of Science & Engineering, Saga University, 1 Honjo, Saga 840, JAPAN

Moe, K.-W.S., Geotechnical Engineer, Blk-508, #03-377, Bedok North Ave. 3, Singapore 460508, SINGAPORE. E: soemoe@pacipic.net.sg

Moh, Z.C., Dr., President, Moh & Associates Inc., 22/F. No. 112, Shin-Tai-Wu Road, Sec.1, Oriental Technopolises, Tower A, Hsi Chih Town, Taipei Hsien 221, Taiwan, R.O.C. T: 886-2-2696-7888, F: 886-2-2696-8688, E: zcmoh@ms.maa.com.tw

Myint, W.B., Blk. 402, #02-223, Fajar Road, Singapore 670402, SINGAPORE

Nakashima, Y., Kiso Jiban Consultants Co., Ltd., 1-11-5 Kudan Kita, Chiyoda-ku, Tokyo 102, JAPAN

Narayanan, R., GCS Consultants Sdn., Bhd., 128 MJln. SS21/35 Damansara Utama, 47400 Petaling Jaya, Selangor, MALAYSIA. E: harayan@pc.jaring.my

Nelson, J.D., Prof., Department of Civil Engineering, Colorado State University, Fort Collins, Colorado 80523-1372, USA. T: 970-491-6081, F: 970-491-3584, E: john@engr.colostate.edu

Ng, C.W.W., Dr., Associate Professor, Department of Civil Engineering, Hong Kong University of Science & Technology, Clear Water Bay, Kowloon, Hong Kong, China. E: cecwwng@ust.hk

Ng, K.B., 13 A, Komplek Sukanmu'adzam Shah, Lebuhraya Sultan Abdul Halim, 05400 Alor Setar, Kedah 05400, MALAYSIA

Nolan, Jr., T.A., Vice President, c/o Hogentogler & Co. Inc., P.O. Box 2219, Colombia MD 21045, USA

Oh, Y.N., Assistant Professor, Chinese Undersea Technology Association, 6F-3, 160, Jian-Kuo South Rd., Sec.1,Taipei 106, Taiwan, R.O.C. T: 886-2703-3422/2462-2192, Ext. 6169, F: 886-2703-3432, E: ohyanam@ms52.url.com.tw

Ohta, H., Prof., Dept. of Civil Engineering, Kanazawa University, Kodatsuno 2-40-20, Kanazawa 920, JAPAN. E: ohta@geotech.cu.titech.ac.jp

Ooi, L.H., Soil & Foundation Sdn., Bhd., No. 23, Jalan Desa, Taman Desa, Off Jalan Kelang Lama, 58100 Kuala Lumpur, MALAYSIA. F: 603-781-2767

Ooi, K.Y., Principal, Engineering Consultant, 16B, Jalan Perang, Taman Pelangi, 80400 Johor Bahru, MALAYSIA. T: 07-3311 377, F: 07-3311 378, E: kaoyang@pc.jaring.my

Ooi, T.A., Dr., Transfield, 15F Menara Multi-Purpose, Capital Square, 8 Jalan Munshi Abdullah, 50100 Kuala Lumpur, MALAYSIA. T: 603-4296-1668, F: 603-4295-2742, E: ooihy@pop.jaring.my

Ou, C.D., Dr., Deputy Mayor, Taipei City Government, 11 Fl., No. 1, Shih-Fu Road, Taipei 110, Taiwan, R.O.C. F: 886-2-2725-5354, E: oucd@serv2.tcg.gov.tw, oucd@ms1.tcg.gov.tw

Paisely, J.M., Managing Director, Fugro Singapore Pte. Ltd., 159 Sin Ming Road, Amtech Building, #06-07, Singapore 575625, SINGAPORE. T: 65-552-8600, F: 65-552-8955, E: jmpfugro@singnet.com.sg

Pan, Y.W., Prof., Department of Civil Engineering, National Chiao Tung University, 1001 Ta Hsueh Rd, Hsinchu, Taiwan 30010, R.O.C. F: 886 3 571 6257, E: ywpan@cc.nctu.edu.tw

Pang, P.L.R., Dr., Geotechnical Engineering Office, 12/F, Civil Engineering Bldg., 101 Princess Margaret Rd, Homantin, Kowloon, Hong Kong, China. E: cge_spd@ced.gov.hk

Peng, M.K., No.1 Jalan TS6/7, Taman Industri Subang, 47510 Subang Jaya, Selangor Darul Ehsan, MALAYSIA

Petchgate, K., Department of Civil Engineering, King Mongkut's Institute of Technology, Thonburi, Bangmod, Ras Burana, Bangkok 10140, THAILAND

Phienwej, N., Dr., Associate Professor, Geotechnical Engineering Program, Asian Institute of Technology, P.O. Box 4, Klong Luang, Pathumthani 12120, THAILAND. T: 66-2-524-5507, F: 66-2-524-5509, E: noppadol@ait.ac.th

Pitakdumrongkit, B., 140/2, Soi Sahamitr, Mahapruktaram Road, Bangrak, Bangkok 10500, THAILAND

Piyaboon, S., Regional Manager, Polyfelt Geosynthetics (Thailand) Ltd., 169/98, 3rd Floor, Serm Srap Building, Ratchadapisek Road, Din-daeng, Bangkok 10320, THAILAND. T: 66-2-692-6680-2, F: 66-2-692-6679, E: somsak@polyfelt.co.th

Pollard, J., Director, Meinhardt (Thailand) Ltd., 15th Fl, Grand Amarin Tower, 1550 New Petchburi Rd, Makkasan, Ratchtevee, Bangkok 10310, THAILAND. F: 66-2-207-0573

Poopath, V., Dr., Associate Professor, 38/1 Senaruam, Paholyothin 11, Phya Thai Road, Bangkok 10400, THAILAND. T: 66-2-278-3129

Porrazo, V.F., P.O. Box 7444, Domestic Airport Post Office, Domestic Airport Road, Pasay City, PHILIPPINES. E: vporrazzo@aol.com

Poulos, H.G., Prof., Senior Principal, Coffey Geoscienes Pty. Ltd., 142 Wicks Road, North Ryde NSW 2118, AUSTRALIA. T: 61-2-9888 7444, F: 61-2-9888 9977, E: harry_poulos@coffey.com.au

Prakash, S., Prof., Department of Civil Engineering, University of Missori Rolla, Carlton Civil Engineering Bldg., Missouri 65401, USA

Prawono, S., Lecturer, Petra Christian University, Prapen Indah Blok I/12A, Surabaya 60299, INDONESIA. T: 62-31-841-0630, F: 62-31-849-3751, E: supra@peter.petra.ac.id

Premchitt, Dr., J., Senior Geotechnical Engineer, HKSAR Government, Special Duties Office, Civil Engineering Dept., 3rd Floor, Civil Engineering Building, 101 Princess Margaret Rd., Homantin, Kowloon, Hong Kong, China. T: 852-2762 5638, F: 852-2714 0103, E: jprem@hkstar.com

Qiu, Y.J., Prof., School of Civil Engineering, Southwest Jiaotong University, Chengdu 610031, P.R. China, E: qiuyanjun@hotmail.com

Rafiquddin, A.K.M., Managing Director, Development Design Consultants Ltd., DDC Centre, 47, Mohakhali Commercial Area, Dhaka - 1212, BANGLADESH. E: ddcon@bangla.net

Rahardjo, H., Dr., Director, NTU-PWD Geotechnical Research Centre, School of Civil & Structural Engineering, Nanyang Technological University, Nanyang Ave., Singapore 639798, SINGAPORE. E: chrahardjo@ntu.edu.sg

Raju, P.S., Keiler (M) Sdn., Bhd., 35 Lorong B., Kampung Pakar Batu 5, Jalan Sungei Besi, 5700 Kuala Lumpur, MALAYSIA. T: 03-7802894, F: 03-7800349

Raju, V.R., Dr., Keiler (M) Sdn., Bhd., 35 Lorong B., Kampung Pakar Batu 5, Jalan Sungei Besi, 5700 Kuala Lumpur, MALAYSIA. T: 03-7802894, F: 03-7800349

Raman, J., Dr., Associate Director, Minconsult Sdn., Bhd., No. 6, Jalan 51A/223, 46100 Petaling Jaya, Selangor, MALAYSIA. T: 603-7954-5757, F: 603-7954-7373, E: minco@po.jaring.my

Roongthanee, J., State Railway of Thailand, Research & Analysis Division, Civil Engineering Department, Bangkok, THAILAND

Ruengkrairergsa, T., Dr., Materials & Research Division, The Department of Highways, Phayathai, Sri Ayutthaya Road, Bangkok 10400, THAILAND. T: 66-2-247-6908, Ext. 2224, 246-1122-33, Ext. 2325, F: 66-2-247-9425

Rujivipat, V., Managing Director, Infinity Services Co., Ltd., 100/176 Chonlada 25, Bangkrui-Sai Noi Road, Bangbuathong, Nonthaburi 11110, THAILAND. T: 66-1-813-2861

Sae-Tia, W., Department of Civil Engineering, Thamamasat University – Rangsit Campus, Klong Luang, Pathumthani 12121, THAILAND. T: 662-564-3001-09, Ext. 3166, F: 66-2-564-3010, E: sweeraya@engr.tu.ac.th

Sagae, T., Kiso-Jiban Consultants (M) Sdn., Bhd., 115-1, Jalan Mega Mendung, Kompleks Bandar, Batu 5, Jalan Kelang Lama, 58200 Kuala Lumpur, MALAYSIA

Sam, M.K., Lecturer, Civil Engineering & Building Department, Singapore Polytechnic, 500 Dover Road, Singapore 139651, SINGAPORE. T: 65-772-1166, F: 65-772-1973, E: sammk@sp.edu.sg

Sambhandharaksa, S., Prof., 49/7 Sukhumvit 60, Bangkok 10250, THAILAND. T: 66-2-218-6462, 311-2262, F: 66-2-311-2556

Santoso, D., Prof., Professor of Applied Geophysics, Department of Geophysical Engineering, Institute of Technology Bandung, Jl.Ganesa 10, Badung-40132, INDONESIA. T: 62-22-423-4413, 250-9168, F: 62-22-423-4413, 250-9168, E: dsantoso@indo.net.id

Scott, J.C., Woodward-Clyde 4582S., Ulster Street, Suite 1000, Denver, Colorado 80237, USA

Serajuddin, M., House No. 25, Block A., Road No. 18, Banani Model Town, Dhaka 1213, BANGLADESH. E: smseraj@bangla.net

Shieh, B.J., Section Manager, CTCI Corporation, 11th Floor, CTCI Tower 77, Sec. 2, Tun Hwa South Road, Taipei, Taiwan, R.O.C. F: 886 2 709 9303, E: soil@ctci.com.tw

Shimizu, N., Department of Civil Engineering, Yamaguchi University, Tokiwadai 2557, Ube 755-8611, JAPAN

Shin, U.T., Prof., Department of Civil Engineering, National Taiwan University, Taipei, Taiwan 10617, R.O.C. E: ueng@ce.ntu.edu.tw

Shinjo, T., Dr., College of Agriculture, University of the Ryukyus, 1 Senbaru Nishihara cho Nakagami gun, Okinawa, 903-01, JAPAN

Shiraishi, S., Dr., Chairman, Tashi Fudohsan Co., Ltd., Tashi Bldg., 8 Minamimotomachi, Shinjuku-ku, Tokyo 160-0012, JAPAN. T: 81-3-3350-8560, F: 81-2-3341-6455, E: XMB00215@nifty.ne.jp

Shiu, Y.K., Geotechnical Engineering Office, Civil Engineering Bldg., 101 Princess Margaret Rd., Homantin, Kowloon, Hong Kong, China

Shivashankar, R., Dr., Asst. Professor, Department of Civil Engineering, Karnataka Regional Engineering College, Surathkal, P.O. Srinivasnagar 574157 (D.K.), Karnataka State, INDIA

Sikam, E., Head of Geotechnical Branch, Department of Works, P.O. Box 1261, Lae 411, M, PAPPUA NEW GUINEA. T: 675-472-4389, F: 675-472-1349, E: esikam@Global.net.pg

Silva, F., Dr., 12 Baskin Road, Lexington, Massachusetts 02421, USA

Silver, M.L., Prof., Vietnam Representative Office, 3 Ngo Quyen Street, Hanoi, VIETNAM. E: msilver@hn.vnn.vn

Snodgrass, R.B., Engineers 9000 Pte., Ltd., 809 French Road, #04-164 Singapore 200809, SINGAPORE

Srisuppachaiya, P., Chief Executive Officer, S. Power Construction Co., Ltd., LPN Tower, 10th Floor, Linchee Road, Chongnonsee, Yannawa, Bangkok 10120, THAILAND. T: 66-2-285-4334, F: 66-2-285-4340, E: s_power_con@yahoo.com

Sugimoto, M., Dr., Associate Professor, Dept. of Civil Engineering, Nagaoka University of Tech., 1603-1 Kamitomioka-Chou, Nagaoka, Niigata, 940-21, JAPAN

Suwono, J.I., Director, Data Persada, Margorejo Indah D-507, Surabaya 60235, INDONESIA. T: 62-31-843 7296, F: 62-31-843 7302, E: data@mitra.net.id

Taesiri, Y., Dr., 28/128 Chicha Country Club, Rama II Road, Bangkhunthien, Bangkok 10150, THAILAND

Takada, N., Prof., Civil Engg. Department, Osaka City University 3-3-138, Sugimoto Sumiyoshi-ku, Osaka 558-8585, JAPAN. T: 81-6-6605-2724, F: 81-6-6605-2769, E: takada@civil.eng.osaka-cu.ac.jp

Tan, B.K., 8, Jalan SS21/36, Damansara Utama, 47400 Petaling Jaya, MALAYSIA. E: bktan@pkrisc.cc.ukm.my

Tan, Y.K., EK Consultants, 476 Lorong Stampin 20, Jalan Stampin 93350 Kuching, Sarawak, MALAYSIA. T: 082-451476, F: 082-463609, E: tanyk@pc.jaring.my

Tan, T.S., Department of Civil Engineering, National University of Singapore, Kent Ridge Crescent, Singapore 119260, SINGAPORE

Tanaka, Y., Dr., Research Centre for Urban Safety and Security, Kobe University, Nada, Kobe, JAPAN. E: ytgeotec@kobe_u.ac.jp

Tang, T.S., 5, Jalan 17/62, 46400 Petaling Jaya, Selangor, MALAYSIA

Tang, S.K., Blk. 98 Toa Payoh Lorong 1, #12-307, Singapore 1231, SINGAPORE

Tashiro, N., c/o Neutrino Inc., (Taisei Kensetsu), Takahashi Bldg., 1-44-3, Fuda, Chofu-shi, Tokya 182-0024, JAPAN. F: 0424-84-5556, E: book@neutrino.ca.jp

Teh, K.O., Test Technical Laboratory Sdn., Bhd., No. 23, Jalan Desa, Taman DesaOff Jalan Kelang Lama, 58100 Kuala Lumpur, MALAYSIA. F: 603 781 2767

Teng, K.S., Engineering Manager, Geocon (B) Sdn., Bhd., P.O. Box 653, Seri Complex BA1779, BRUNEI. T: 6732-660 702, F: 6732-663 747, E: geocon@brunet.bn

Teong, E.G., Block 117, #04-135, Jurong East St. 13., Singapore 60017, SINGAPORE. E: tegroup@cyberway.com.sg

Thasnanipan, N., SEAFCO Company Limited, 26/10 Rarm Intra 109 Road, Bangchan, Klong Sam Wah, Bangkok 10501, THAILAND. T: 66-2-919-0090-7, F: 66-2-518-3088, 919-0098, E: seafco@seafco.co.th

Theramast, N., Ph.D Student, Dept. of Geotechnical & Environmental Engineering, Nagoya University, Furo-gho, Chikusa-ku, Nagoya 464-8603, JAPAN. F: 81 52 789 3837, E: theramast@genv.nagoya-u.ac.jp

Thomson, R.R., Director, c/o Scott Wilson Kirkpatric (Thailand) Ltd., 17th Floor One Pacifi Place140 Sukhumvit Road, Bangkok 10110, THAILAND. T: 66-2-254-4205, F: 66-2-254-4202-4, E: swkt@mozart.net.co.th

Ting, H.W., Dr., Zaidun-Leeng S/B, 6th Floor, Bangunan Ming, Jalan Bukit Nanas, 50250 Kuala Lumpur, MALAYSIA. T: 603-230-3033, F: 603-201-8964, 230-9576, E: tingwh@pop.jaring.my

Tirtotjondro, R., Jalan Sindangsirna No. 25, Bandung 40153, INDONESIA

Tjandra, S., Managing Director, PT. Engitama Nusa Geotestindo, Jl. Tebet Barat IV No. 33 Jakarta 12810, INDONESIA. T: 62-21-830-1646, F: 62-21-829-0163, 830-0369, E: pteng@indo.net.id

Tjhoa, H.S., P.O. Box 3229, Jakarta 10032, INDONESIA. F: 62-21-626-5383, E: tjhoal@indosat.net.id

Todo, H., Geotechnical Engineer, Kiso-Jiban Consultants Co., Ltd., Overseas Department, 3F, Mori Kaikan Bldg., 1/11/2005 Kudan-Kita, Chiyoda-ku, Tokyo 102-8220, JAPAN. T: 81-3-3239-4451, F: 81-3-3239-4597, E: todo.hiroaki@kisa.co.jp

Toh, C.T., Dr., Consultant, 34-3, Medan Setia 2, Plaza Damansara, Bukit Damansara, Kuala Lumpur 50490, MALAYSIA

Towhata, I., Dr., Dept. of Civil Engineering, University of Tokyo 7-3-1, Hongo, Bunkyo-ku, Tokyo 113 113, JAPAN

Tsai, K.A., 56 Choa Chu Kang North 6, #06-33 Yew Mei Green, Singapore 689577, SINGAPORE

Tungboonterm, Assoc. Prof. P., Department of Civil Engineering, King Mongkut's of Technology, Thonburi, Sukswasd 48 Rd, Bangmod, Thungkru, Bangkok 10140, THAILAND. E: pinit.tum@kmutt.ac.th

Tusgate, S., Department of Energy Development & Promotion, 17 Rama 1 Road, Pathwun, Bangkok 10330, THAILAND. F: 66-2-221-8855

Uddin, M.K., Dr., Asst. Professor, Institute of Appropriate Technology, Bangladesh University of Engineering & Technology, Dhaka-1000, BANGLADESH. E: kamal@iat.buet.edu

Vasinvarthana, V., STS Engineering Consultants Co., Ltd., 196/10-12, Soi Pradipat, Pradipat Road, Samsennai, Phayathai, Bangkok 10400, THAILAND. F: 66-2-618-5401

Wada, A., Director, Asia Georesearch Agency Corporation Pte. Ltd., 43 Carpenter Street #03-00 Greatwood Building, Singapore 059922, SINGAPORE. T: 65-538-0400, F: 65-538-0422, E: wada@agacorp.com.sg

Wan, K.Y., 49 Jalan Selampit 27, Kaw 3, Taman Klang Jaya, 41200 Klang, Selangor, MALAYSIA

Wesley, L.D., Dr., Senior Lecturer, Department of Civil & Resource Engineering, University of Auckland, Private Bag, 92019 Auckland, NEW ZEALAND. T: 64-9-373-7999, F: 64-9-373-7462, E: l.wesley@auckland.ac.nz

Wieland, M., Dr., Electrowatt Engineering Services Ltd., Bellerivestrasse 36, P.O. Box, CH-8034, Zurich, SWITZERLAND

Wong, K.W.A., Reinforced Earth (S.E.A.) Pte., Ltd., 511 Kampong Bahru Road, 05-06, Keppel Distripark, Singapore 099447, SINGAPORE. T: 65-272-0035, F: 65-276-9353, E: reco_singapore@pacific.net.sg

Wong, K.S., Prof., School of Civil & Structural Engineering, Nanyang Technological University, Nanyang Avenue, Singapore 639798, SINGAPORE. F: 65-791-0676, E: ckswong@ntu.edu.sg

Wong, G.C.Y., Dr., Consulting Engineers & Project Managers, Greg Wong & Associates Ltd., Unit D, 8/Fl., Seabright Plaza, 9-13 Shell Street, North Point, Hong Kong, China. T: 852-2512-8003, F: 852-2571-3105, E: gwalhome@gwal.com.hk

Woo, S.M., Dr., President, Trinity Foundation Engineering Consultants Co. Ltd., 3 Floor, No. 28, Lane 102, Section

1An-Ho Rd, Section 1, Taipei 106, Taiwan, R.O.C. T: 886-2-2755-7828, F: 886-2-2705-5338, E: trinity@ms4.hinet.net

Yamagami, T., Prof., Dept. of Civil Engineering, Faculty of Engineering, University of Tokushima, 2-1 Minamijosanjima-cho, Tokushima 770-8506, JAPAN. T: 81-88-656-7345, F: 81-88-656-7345, E: takuo@ce.tokushima-u.ac.jp

Yang, L.W., Dr., I-Tarng Engineering Consultant Co., Ltd., Rm. 1, 7th Floor, 46 Tung-Hsing East Street, Taichung, Taiwan 408, R.O.C. T: 886-4-376 6500, F: 886-4-376-5459, E: chiuyang@ms9.hinet.net

Yang, Z.Y., Prof., Department of Civil Engineering, Tamkang University, Tamsui, Taipei 25137, Taiwan, R.O.C. T: 886-2-2621-5656, Ext. 2671, F: 886-2-2623-4215, E: yang@mail.tku.edu.tw

Yap, H.J., Dr., Principal Engineer, Treadwell & Rollo, Inc., 555 Montgomery Street, Suite 1300, San Francisco, California 94111, USA. T: 415-955-9040, F: 415-955-9041, E: hadiyap@home.com

Yee, K.S., c/o Menard Geosystems Sdn., Bhd., No. 27B (2nd Floor) Jalan USJ 10/1A, UEP Subang Jaya, Selangor 47620, MALAYSIA

Yin, J.H., Associate Professor, Dept. of Civil and Structural Engineering, Hong Kong Polytechnic University, Hung Hom, Kowloon, Hong Kong, China. T: 852-2766-6065, F: 852-2334-6389, E: cejhyin@polyu.edu.hk

Yokoo, F., Managing Director, OYO International Pte. Ltd., No. 2, Boon Leat Terrace, #08-02, Harbourside Industrial Bldg. II, Singapore 119844, SINGAPORE

Yong, A.F.K., Apt 8-12-2 Endah Villa Condominium, Jalan 2/149 B, Taman Sri Endah, 57000 Kuala Lumpur, MALAYSIA. E: y141300@pc.jaring.my

Yong, K.Y., Dr., Head, Department of Civil Engineering, National University of Singapore, 10 Kent Ridge Crescent, Singapore 0511, SINGAPORE. F: 65-779-1635, E: cvehead@nus.sg

Yue, K.F., Q.C. Manager, Industrial Concrete Products Berhad, 2nd Fl., Wisma IJM, Jalan Yong Shook Lin, 46050 Petalig Jaya, Selangor, MALAYSIA. T: 03-7955 8888, F: 03-7958 1111, E: icpb@po.jaring.my

Yung, C.Y.P., Dr., Flat 7, 19/Fl., Block B., King Lai Court, Ngau Chi Wan, Kowloon, Hong Kong, China

Yung, C.Y.P., Dr., Geotechnical Engineer, Flat c, 30/Fl., Tower 9, Island Harbourview, 11, Hoi Fai Road, Kowloon, Hong Kong, China. T: 2326 48237. T: 2326 4837, E: pyungkwu@netvigator.com

SLOVENIA/SLOVENIE

Secretary: Dr. Janko Logar, Faculty of Civil and Geodetic Engineering, Jamova 2, 1000 Ljubljana, Slovenia.
T: +386 1 476 85 26, F: +386 1 425 06 81, E: jlogar@fgg.uni-lj.si

Total number of members: 116

Ajdič, I., M.Sc., DDC, Blagovica, 1223 Blagovica. T: +386 1 723 49 21, F: +386 1 723 43 30, E: igor.ajdic@dd-ceste.si, igor.ajdic@siol.net

Bajželj, U., Ph.D., Professor, NTF-Oddelek za geotehnologijo in rudarstvo, Aškerčeva 12, 1000 Ljubljana. T: +386 1 470 46 11, F: +386 1 252 41 05, E: uros.bajzelj@ntfgam.uni-lj.si

Battelino, L., M.Sc., Vodnogospodarski inštitut, Hajdrihova 28c, 1000 Ljubljana. T: +386 1 477 53 36, F: +386 1 477 53 43, E: lilian.battelino@guest.arnes.si

Bebar, M., IGMAT, Polje 351c, 1260 Ljubljana Polje. T: +386 41 64 49 43, F: +386 1 586 26 01, E: marko.bebar@igmat.si

Bertoncelj, M., M.Sc., Gradbeni inštitut ZRMK, Dimičeva 12, 1000 Ljubljana. T: +386 2 604 02 30, F: +386 2 604 02 31, E: milica.bertoncelj@guest.arnes.si

Birsa, M., GRADING, Obrežna ul.1, 2000 Maribor. T: +386 2 420 55 41, F: +386 2 420 55 41, E: grading.mb@siol.net

Bizilj, P., Geoinženiring, Dimičeva 14, 1000 Ljubljana. T: +386 1 234 56 33, F: +386 1 234 56 10, E: p.bizilj@geo-inz.si

Blažič, A., M.Sc., Rudnik lignita Velenje, Partizanska 78, 3320 Velenje. T: +386 31 32 00 78, F: +386 3 586 91 31, E: andrej.blazic@rlv.si

Bohar, F., IGMAT, Polje 351c, 1260 Ljubljana Polje. T: +386 1 586 26 10, F: +386 1 586 26 01, E: feri.bohar@igmat.si

Borec-Merlak, J., GZL Geoprojekt, Letališka 27, 1000 Ljubljana. T: +386 1 524 90 80, F: +386 1 524 90 84

Breznik, M., Ph.D., Professor, Vrtača 8, 1000 Ljubljana. T: +386 1 425 72 09

Brinšek, R., Savske elektrarne Ljubljana, Gorenjska c. 46, 1215 Medvode. T: +386 1 474 91 34, F: +386 1 474 92 72, E: rudi.brinsek@savske-el.si

Brožič, D., GEOT, Dimičeva 12, 1000 Ljubljana. T: +386 1 280 82 57, F: +386 1 280 81 91, E: dbrozic@gi-zrmk.si

Crnkovič, I., GEOINVEST, Dimičeva 16, 1000 Ljubljana. T: +386 1 300 43 71, F: +386 1 300 43 80, E: ivan.geoinvest@boter.net

Crnkovič-Klanjšek, J., IBT Nizke gradnje Trbovlje, Gimnazijska 21, 1420 Trbovlje. T: +386 3 563 41 10, F: +386 3 563 41 11

Čarman, M., M.Sc., Geološki zavod Slovenije, Dimičeva 14, 1000 Ljubljana. T: +386 1 436 75 98, F: +386 1 436 75 96, E: magda.carman@geo-zs.si

Čuček, D., GEOINVEST, Dimičeva 16, 1000 Ljubljana. T: +386 1 300 43 70, F: +386 1 300 43 80, E: draga.geoinvest@boter.net

Demšar, V., ZAG Ljubljana, Dimičeva 12, 1000 Ljubljana. T: +386 1 280 42 61, F: +386 1 280 42 64, E: vladimir.demsar@zag.si

Dešman, K., DDC, Tržaška 19a, 1000 Ljubljana. T: +386 1 478 84 36, F: +386 1 478 83 98, E: karmen.desman@dd-ceste.si

Dokl, S., Geoing Maribor, Primorska 10, 2000 Maribor. T: +386 2 320 38 80, F: +386 2 320 38 81

Dolinar, B., M.Sc., Fakulteta za gradbeništvo, Smetanova 17, 2000 Maribor. T: +386 2 229 43 25, F: +386 2 252 41 79, E: bojana.dolinar@uni-mb.si

Dular, A., M.Sc., GEOT, Dimičeva 12, 1000 Ljubljana. T: +386 1 280 82 10, F: +386 1 280 81 91, E: geot@gi-zrmk.si

Dvanajščak, D., M.Sc., SCT, Slovenčeva 22, 1000 Ljubljana. T: +386 1 428 76 40, F: +386 1 366 62 48, E: drago.dvanajscak@sct.si

Ečimović, V., GEOT, Dimičeva 12, 1000 Ljubljana. T: +386 1 280 84 56, F: +386 1 280 84 51, E: vecimovi@gi-zrmk.si

Fašalek, M., GEOT, Dimičeva 12, 1000 Ljubljana. T: +386 1 280 81 81, F: +386 1 280 81 91, E: mfasalek@gi-zrmk.si

Fila, S., Gradis T.E.O., Šmartinska 134a, 1000 Ljubljana. T: +386 1 540 41 02, F: +386 1 524 24 19, E: sabine.fila@gradis-teo.si

Fras, B., Sinteza Lining, Kidrieva 3, 3000 Celje. T: +386 3 426 72 70, F: +386 3 426 72 73, E: sinteza-lining@celje.eurocom.si

Gaberc, A., M.Sc., Fakulteta za gradbeništvo in geodezijo, Jamova 2, 1000 Ljubljana. T: +386 1 476 85 27, F: +386 1 425 06 81, E: agaberc@fgg.uni-lj.si

Gostič, B., Gradbeni inštitut ZRMK, Dimičeva 12, 1000 Ljubljana. T: +386 1 280 82 64, F: +386 1 280 81 91, E: bgostic@gi-zrmk.si

Grubišič, Z., ZAG Ljubljana, Dimičeva 12, 1000 Ljubljana. T: +386 1 280 44 80, F: +386 1 280 42 64, E: zoran.grubisic@zag.si

Hoblaj, R., Geoinženiring, Dimičeva 14, 1000 Ljubljana. T: +386 1 234 56 30, F: +386 1 234 56 10, E: r.hoblaj@geo-inz.si

Hrast, K., IRGO, Slovenčeva 93, 1000 Ljubljana. T: +386 1 560 36 58, F: +386 1 534 16 80, E: nina.hrast@i-rgo.si

Huis, M., IRGO, Slovenčeva 93, 1000 Ljubljana. T: +386 1 560 36 70, F: +386 1 534 16 80, E: melanija.huis@i-rgo.si

Isakovič, S., Gradbeni inštitut ZRMK, Dimičeva 12, 1000 Ljubljana. T: +386 1 280 82 03, F: +386 1 280 84 51, E: sisakovi@gi-zrmk.si

Janžeković, B., GMG, Kersnikova 4, 2250 Ptuj. T: +386 2 748 18 14, E: bozo.janzekovic@siol.net

Jerman, J., Geoinženiring, Dimičeva 14, 1000 Ljubljana. T: +386 1 234 56 12, F: +386 1 234 56 10, E: j.jerman@geo-inz.si

Jezeršek, J., Rudnik lignita Velenje, Partizanska 78, 3320 Velenje. T: +386 3 587 14 65, F: +386 3 586 91 31, E: janez.jezersek@rlv.si

Jovičič, V., Ph.D., IRGO, Slovenčeva 93, 1000 Ljubljana. T: +386 1 560 36 21, F: +386 1 534 16 80, E: vojkan.jovicic@i-rgo.si

Kalanj, J., GEOKAL, Sokolska ulica 44-46, 2000 Maribor. T: +386 2 429 61 60, F: +386 2 420 01 74, E: geokal@siol.net

Kalanj, T., GEOKAL, Sokolska ulica 44-46, 2000 Maribor. T: +386 2 429 61 60, F: +386 2 420 01 74, E: geokal@siol.net

Kink, J., IGMAT, Polje 351c, 1260 Ljubljana Polje. T: +386 1 586 26 00, F: +386 1 586 26 01, E: jasmina.kink@igmat.si

Klemenc, I., M.Sc., ZAG Ljubljana, Dimičeva 12, 1000 Ljubljana. T: +386 1 280 44 26, F: +386 1 436 74 49, E: iztok.klemenc@zag.si

Klinc, M., GZL Geoprojekt, Letališka 27, 1000 Ljubljana. T: +386 1 524 90 80, F: +386 1 524 90 84

Kočevar, M., Geoinženiring, Dimičeva 14, 1000 Ljubljana. T: +386 1 234 56 26, F: +386 1 234 56 10, E: m.kocevar@geo-inz.si

Koren, V., Geoinženiring, Dimičeva 14, 1000 Ljubljana. T: +386 1 234 56 11, F: +386 1 234 56 10

Korpič, P., GEOT, Dimičeva 12, 1000 Ljubljana. T: +386 41 28 76 61, F: +386 1 280 84 51, E: pkorpic@gi-zrmk.si

Košir, M., IGMAT, Polje 351c, 1260 Ljubljana Polje. T: +386 1 586 26 00, F: +386 1 586 26 01, E: marko.kosir@igmat.si

Kovačič, A., GRACEN, Ptujska 7, 1000 Ljubljana. T: +386 1 436 75 65, F: +386 1 436 77 23

Kraljič Kenk, M., Geoinženiring, Dimičeva 14, 1000 Ljubljana. T: +386 1 234 56 08, F: +386 1 234 56 10, E: m.kenk@geo-inz.si

Kuščer, P., Ph.D., Professor, Gotska 9, 1000 Ljubljana. T: +386 1 519 28 54

Leben, B., ZAG Ljubljana, Dimičeva 12, 1000 Ljubljana. T: +386 1 280 45 06, F: +386 1 280 42 64, E: bojan.leben@zag.si

Lenart, S., ZAG Ljubljana, Dimičeva 12, 1000 Ljubljana. T: +386 1 280 44 00, F: +386 1 436 74 49, E: stanislav.lcnart@zag.si

Lesjak, I., SLP, Šmartinska 32, 1000 Ljubljana. T: +386 1 540 44 65, F: +386 1 540 44 65, E: slp@siol.net

Likar, J., Ph.D., Assistant Professor, IRGO, Slovenčeva 93, 1000 Ljubljana. T: +386 1 560 36 41, F: +386 1 534 16 80, E: jakob.likar@i-rgo.si

Ločniškar, A., DDC, Tržaška 19a, 1000 Ljubljana. T: +386 1 478 83 86, F: +386 1 478 83 78, E: andrej.locniskar@dd-ceste.si

Logar, J., Ph.D., Assistant Professor, Fakulteta za gradbeništvo in geodezijo, Jamova 2, 1000 Ljubljana. T: +386 1 476 85 26, F: +386 1 425 06 81, E: jlogar@fgg.uni-lj.si

Macuh, B., M.Sc., Fakulteta za gradbeništvo, Smetanova 17, 2000 Maribor. T: +386 2 229 43 32, F: +386 2 252 41 79, E: borut.macuh@uni-mb.si

Maček, R., DDC, Blagovica, 1223 Blagovica. T: +386 41 39 28 23, F: +386 1 723 43 30, E: roman.macek@dd-ceste.si

Majes, B., Ph.D., Professor, Fakulteta za gradbeništvo in geodezijo, Jamova 2, 1000 Ljubljana. T: +386 1 476 85 22, F: +386 1 425 06 81, E: bmajes@fgg.uni-lj.si

Markič, S., IRGO, Slovenčeva 93, 1000 Ljubljana. T: +386 1 560 36 57, F: +386 1 534 16 80, E: simona.markič@i-rgo.si

Medved, S., LINEAL, Ul. Pohorskega bataljona 49, 2000 Maribor. T: +386 2 429 27 25, F: +386 2 429 27 11, E: samo.medved@lineal.si

Mikoš, M., Ph.D., Professor, Fakulteta za gradbeništvo in geodezijo, Hajdrihova 28, 1000 Ljubljana. T: +386 1 425 43 80, F: +386 1 251 98 97, E: mmikos@fgg.uni-lj.si

Muhič, D., GPRO, Sokolska 22, 2103 Maribor. T: +386 2 429 58 50, F: +386 2 429 58 51

Muršec, B., GEOKAL., Sokolska ulica 44-46, 2000 Maribor. T: +386 2 429 61 60, F: +386 2 420 01 74, E: geokal@siol.net

Oblak, R., DDC, Snebrska 120b, 1000 Ljubljana. T: +386 1 528 50 76, F: +386 1 528 29 60, E: robert.oblak@dd-ceste.si

Ocepek, D., Geoinženiring, Dimičeva 14, 1000 Ljubljana. T: +386 1 234 56 24, F: +386 1 234 56 10, E: d.ocepek@geo-inz.si

Osrečki, I., FLOWTEX Geoprojekt, C. Dolomitskega odreda 10, 1000 Ljubljana. T: +386 1 423 99 20, F: +386 1 256 21 14, E: flowtex.ivan@siol.net

Peček, D., Geoinženiring, Dimičeva 14, 1000 Ljubljana. T: +386 1 234 56 25, F: +386 1 234 56 10, E: d.pecek@geo-inz.si

Pernovšek, I., Sinteza Lining, Kidričeva 3, 3000 Celje. T: +386 3 426 72 70, F: +386 3 426 72 73, E: sinteza-lining@celje.eurocom.si

Petkovšek, A., Gradbeni inštitut ZRMK, Dimičeva 12, 1000 Ljubljana. T: +386 1 280 82 87, F: +386 1 280 84 51, E: apetkovs@gi-zrmk.si

Petkovšek, B., Ph.D., ZAG Ljubljana, Dimičeva 12, 1000 Ljubljana. T: +386 1 280 43 20, F: +386 1 280 42 64, E: borut.petkovsek@zag.si

Petrica, R., ZAG Ljubljana, Dimičeva 12, 1000 Ljubljana. T: +386 1 280 44 68, F: +386 1 280 42 64, E: rajko.petrica@zag.si

Plajnšek, L., GEOKAL, Sokolska ulica 44-46, 2000 Maribor. T: +386 2 429 61 60, F: +386 2 420 01 74, E: geokal@siol.net

Poglajen, J., PINO, Gradbeni inženiring, Ul. M. Pregljeve 4, 1270 Litija. T: +386 1 898 01 50, F: +386 1 898 39 09, E: pino.doo@siol.net

Popovič, Z., TERRAS, Metelkova 1, 1000 Ljubljana. T: +386 1 438 21 30, F: +386 1 438 21 34, E: popovicn@siol.net

Prokop, M., M.Sc., GEOT., Dimičeva 12, 1000 Ljubljana. T: +386 1 280 82 98, F: +386 1 280 81 91, E: bprokop@gi-zrmk.si

Pulko, B., Ph.D., Fakulteta za gradbeništvo in geodezijo, Jamova 2, 1000 Ljubljana. T: +386 1 476 85 23, F: +386 1 425 06 81, E: bpulko@fgg.uni-lj.si

Ravbar, E., Gradis inženiring, Letališka 33, 1000 Ljubljana. T: +386 1 540 40 52, F: +386 1 540 40 62, E: projekt@gradis.si

Ravnikar-Turk, M., ZAG Ljubljana, Dimičeva 12, 1000 Ljubljana. T: +386 1 280 43 93, F: +386 1 280 42 64, E: mojca.turk@zag.si

Resanovič, I., Geockspert, Ob Koprivnici 57, 3000 Celje. T: +386 3 492 20 40, F: +386 3 492 20 40

Ribičič, M., Ph.D., Assistant Professor, Gradbeni inštitut ZRMK., Dimičeva 12, 1000 Ljubljana. T: +386 1 280 85 38, F: +386 1 280 84 51, E: mribicic@gi-zrmk.si

Rihtarič, N., Apros, Sokolska 44-46, 2000 Maribor. T: +386 2 420 01 55, F: +386 2 429 56 16, E: nevenka.rihtaric@mbapros.si

Rijavec, B., Geoinženiring, Dimičeva 14, 1000 Ljubljana. T: +386 1 234 56 02, F: +386 1 234 56 10, E: b.rijavec@geo-inz.si

Robas, A., Fakulteta za gradbeništvo in geodezijo, Jamova 2, 1000 Ljubljana. T: +386 1 476 85 28, F: +386 1 425 06 81, E: arobas@fgg.uni-lj.si

Schrott, T., Geoinženiring, Dimičeva 14, 1000 Ljubljana. T: +386 1 234 56 06, F: +386 1 234 56 10, E: t.schrott@geo-inz.si

Skok, J., GEOT, Dimičeva 12, 1000 Ljubljana. T: +386 1 280 84 57, F: +386 1 280 81 91, E: jskok@gi-zrmk.si

Skrinar, M., Ph.D., Assistant Professor, Fakulteta za gradbeništvo, Smetanova 17, 2000 Maribor. T: +386 2 229 43 58, F: +386 2 252 41 79, E: skrinar@uni-mb.si

Sovinc, A., Vodnogospodarski inštitut, Hajdrihova 28, 1000 Ljubljana. T: +386 1 425 64 58, F: +386 1 426 41 62, E: andrej.sovinc@guest.arnes.si

Stopinšek, M., Slovenske železnice, Sekcija vzdrževanja prog Celje, Ul. 14. divizije 2, 3000 Celje. T: +386 3 293 21 00, F: +386 3 293 38 01, E: marjan.stopinsek@amis.net

Strniša, G., SLP, Šmartinska 32, 1000 Ljubljana. T: +386 1 540 44 65, F: +386 1 540 44 65, E: slp@siol.net

Svetličič, S., DDC, Einspielerjeva 6, 1000 Ljubljana. T: +386 31 65 84 92, F: +386 1 478 83 78, E: suzana.svetlicic@dd-ceste.si

Škrabl, M., GRADING, Obrežna ul.1, 2000 Maribor. T: +386 2 420 55 41, F: +386 2 420 55 41, E: grading.mb@siol.net

Škrabl, S., Ph.D., Professor, Fakulteta za gradbeništvo, Smetanova 17, 2000 Maribor. T: +386 2 229 43 27, F: +386 2 252 41 79, E: stanislav.skrabl@uni-mb.si

Šterk, V., Šterk Grafika, Geomehanika, Zaloška 143, 1000 Ljubljana. T: +386 1 540 83 30, F: +386 1 540 83 30, E: sterk@siol.net

Štern, K., Geoinženiring, enota Maribor, Gorkega 1, 2000 Maribor. T: +386 2 330 07 50, F: +386 2 330 37 57, E: geoinzeniring.mb@amis.net

Štrukelj, A., Fakulteta za gradbeništvo, Smetanova 17, 2000 Maribor. T: +386 2 229 43 36, F: +386 2 252 41 79, E: andrej.strukelj@uni-mb.si

Tancer, M., Apros, Sokolska 44-46, 2000 Maribor. T: +386 2 429 56 15, F: +386 2 429 56 16, E: marjan.tancer@mbapros.si

Trajkovski, S., ZAG Ljubljana, Dimičeva 12, 1000 Ljubljana. T: +386 1 280 44 00, F: +386 1 436 74 49, E: sebastian.trajkovski@zag.si

Trauner, L., Ph.D., Professor, Fakulteta za gradbeništvo, Smetanova 17, 2000 Maribor. T: +386 2 229 43 01, F: +386 2 252 41 79, E: trauner@uni-mb.si

Umek, U., GEOT, Dimičeva 12, 1000 Ljubljana. T: +386 1 280 83 48, F: +386 1 280 81 91, E: uumek@gi-zrmk.si

Venturini, S., INI, Bravničarjeva 20, 1000 Ljubljana. T: +386 1 436 68 82, F: +386 1 436 68 83, E: geo@siol.net

Vidic, F., GEOTEC LJ, Miličinskega 77, 1000 Ljubljana. T: +386 1 505 81 86, F: +386 1 505 81 86

Vidmar, S., Ph.D., Professor, Arharjeva cesta 38, 1000 Ljubljana. T: +386 1 512 63 57

Vogrinčič, G., Ph.D., Assistant Professor, FMF – Oddelek za matematiko in mehaniko, Lepi pot 11, 1000 Ljubljana. T: +386 1 425 00 52, F: +386 1 425 00 72, E: geza.vogrincic@uni-lj.si

Volf, S., Geoinženiring, Dimičeva 14, 1000 Ljubljana. T: +386 1 234 56 07, F: +386 1 234 56 10

Vrabec, M., Podjetje za urejanje hudournikov, Hajdrihova 28, 1000 Ljubljana. T: +386 1 477 52 24, F: +386 1 251 00 30, E: martin.vrabec@puh.si

Vrecl, H., Fakulteta za gradbeništvo, Smetanova 17, 2000 Maribor. T: +386 2 229 43 27, F: +386 2 252 41 79, E: helena.vrecl@uni-mb.si

Wohinz, L., DDC, Tržaška 19a, 1000 Ljubljana. T: +386 1 478 84 29, F: +386 1 478 83 98, E: ladi.wohinz@dd-ceste.si

Zarnik, B., Občina Litija, Jerebova ul. 14, 1270 Litija. T: +386 1 898 12 11, F: +386 1 898 38 35, E: blaz.zarnik@litija.si

Zavšek, S., M.Sc., Rudnik lignita Velenje, Partizanska 78, 3320 Velenje. T: +386 3 587 14 65, F: +386 3 586 91 31, E: simon.zavsek@rlv.si

Zorič, Z., M.Sc., Fakulteta za gradbeništvo, Smetanova 17, 2000 Maribor. T: +386 2 229 43 05, F: +386 2 252 41 79, E: zdenko.zoric@uni-mb.si

Zrim, S., Gradis TEO – Gradbeni laboratorij, Letališka 33, 1000 Ljubljana. T: +386 41 63 00 84, F: +386 1 524 24 19, E: simona.zrim@gradis-teo.si

Železnik, D., SCT, Slovenčeva 22, 1000 Ljubljana. T: +386 1 589 86 56, F: +386 1 589 86 80, E: drago.zeleznik@sct.si

Žigman, F., IRGO, Slovenčeva 93, 1000 Ljubljana. T: +386 1 560 36 53, F: +386 1 534 16 80, E: franc.zigman@i-rgo.si

Žigman, U., IRGO, Slovenčeva 93, 1000 Ljubljana. T: +386 1 560 36 73, F: +386 1 534 16 80, E: urska.zigman@i-rgo.si

Žlender, B., Ph.D., Professor, Fakulteta za gradbeništvo, Smetanova 17, 2000 Maribor. T: +386 2 229 43 28, F: +386 2 252 41 79, E: bojan.zlender@uni-mb.si

SOUTH AFRICA

Geotechnical Division of the South African Institute for Civil Engineers
http://www.ee.up.ac.za/civil/saice

Total number of members: 339

Allen, P., c/o Peter Allen & Associates, P.O. Box 844, New Germany, 3620
Anthony, N., 313 The Rand Street, Menlo Park, 0081
Austin, J.L., P.O. Box 289, Roodepoort, 1725
Badenhorst, D.B., c/o BKS Ing., P.O. Box 3173, Pretoria, 0001
Badenhorst, P.J., P.O. Box 914 442, Wingate Park, 0153
Badenhorst, P.J., P.O. Box 38268, Garsfontein, 0042
Barker, W.R., P.O. Box 68421, Bryanston, 2021
Barnard, M.C., P.O. Box 12735, Hatfield, 0028
Barrett, A.J., 1121 Gordon Avenue West Vancouver BC Canada V7T IP8
Becker, P.C., P.O. Box 1347, Cape Town, 8000
Bergmann, T., P.O. Box 6503, Roggebaai, 8012
Berry, A.D., P.O. Box 10055, Vorna Valley, 1686
Bester, S.W.J., P.O. Box 21, Ceres, 6835
Blight, G.E., c/o Civil Engineering Dept., Univ. of The Witwatersrand, PBag 3, Wits, 2050
Bloem, W.F., P.O. Box 2446, Kimberley, 8300
Boniface, A., IREG 3rd Floor, Taiwan High Speed Rail Corp. 100 Section 5, Hsin Yi Road Taipei, Taiwan
Boonzaaier, E.R.W., 114th Avenue, Walmer, Port Elizabeth, 6070
Boswell, J.E.S., c/o Bohlweki, P.O. Box 11784, Vorna Valley, 1686
Botha, P.S., P.O. Box 2566, Pietersburg, 0700
Botha, D.B., Postnet Suite 366, PBag X 09, Weltevreden Park, 1715
Bowen, A.J., 193 Frances Street East, Observatory, 2198
Braatvedt, I.H., P.O. Box 72296, Parkview, 2122
Braithwaite, E.J., 150-5th Avenue, Edenvale, 1609
Bredenkamp, H.P., P.O. Box 67424, Bryanston, 2021
Breytenbach, F.J., P.O. Box 73478, Lynnwood Ridge, 0040
Brink, D., c/o Jones & Wagener, P.O. Box 1434, Rivonia, 2128
Brink, A.B.A., P.O. Box 67193, Bryanston, 2021
Brown, R.A.M., P.O. Box 2201, Randburg, 2125
Burger, S.W., P.O. Box 3239, Bloemfontein, 9300
Burger, A.C., P.O. Box 604, Oranjemund, Namibia
Buttrick, D.B., P.O. Box 2790, Rivonia, 2128
Cameron-Williger, R.C., P.O. Box 4733, Cape Town, 8000
Cattanach, J.A., P.O. Box 6731, Weltevreden Park, 1715
Chamberlain, G.W., P.O. Box 1277, Northcliff, 2115
Chrystal, T.B., P.O. Box 189, Richards Bay, 3900
Clanahan, C.R.H., P.O. Box 5495, Nelspruit, 1200
Clark, D.L.B., 321 Cato Road, Glenwood, 4001
Clayton, R.A., 27 Templeman Street, Vergesig, 7550
Cloete, J.P.L., P.O. Box 2345, Cramerview, 2060
Cockcroft, T.N., c/o VKE, P.O. Box 1462, Pinegowrie, 2123
Cohen, A., P.O. Box 3417, Durban, 4000
Collins, L.E., 20 Fairbridge Road, Roosevelt Park, 2195
Colman, P.A., c/o Franki Africa (Pty) Ltd., P.O. Box 693, Pinetown, 3600
Comninos, T.M., 11 Dale Brook Crescent, Victory Park, Johannesburg, 2195
Conradie, R.V., P.O. Box 3040, Durbanville, 7551
Cook, B.J., P.O. Box 585, Glenvista, 2058
Copeland, A.M., P.O. Box 2005, Fourways, 2055
Corbin, P.G., P.O. Box 799, Worcester, 6849
Courtney, T.A., P.O. Box 1700, Fourways, 2055
Cousins, B.A., P.O. Box 339, Umbogintwini, 4120
Cowburn, S.J., 22 Tullyallen Crescent, Rondebosch, 7700
Cridlan, B., P.O. Box 619, Rivonia, 2128
Cross, B.N., c/o Kantey & Templer, P.O. Box 399, Port Shepstone, 4240
Dale, S.G., c/o Knight Piesold, P.O. Box 1710, Northgate Hornsby, NSW 2077 Australia
Danzfuss, F.M., P.O. Box 55530, Arcadia, 0007
Davies, P., 25 Kloof, 3610lands Road, Kloof, 3610

Day, P.W., c/o Jones & Wagener, P.O. Box 1434, Rivonia, 2128
De Beer, A., P.O. Box 78125, Sandton, 2146
De Bruin, C.G., P.O. Box 912-387, Silverton, 0127
De Jager, C.J., P.O. Box 25733, Monumentpark, 0105
De Sousa Vinagre, T.J.V., Rua Antònio Palha 67 Vila Franca de Xira 260 V.F. de Xira Portugal
De Wet, L.F., P.O. Box 2005, Somerset West, 7129
De Wet, H., P.O. Box 781510, Sandton, 2146
De Wet, M., Berg Avenue 15, Stellenbosch, 7600
De Witt, M.J., P.O. Box 69971, Bryanston, 2021
Donaldson, G.W., Private Bag X10001, Garsfontein, 0042
Dorren, D.I., P.O. Box 72292, Lynnwood Ridge, 0040
Drennan, J.A., 20 Glenart Road, Kloof, 3610
Druyts, F.H.W., P.O. Box 13824, Sinoville, 0129
Du Preez, L.M., P.O. Box 1434, Rivonia, 2128
Dunbar, R.M., c/o BCP Engineers, P.O. Box 37379, Overport, 4067
Dunn, G.C., P.O. Box 50361, Colleen Glen, 6018
During, D.J., c/o BKS International, P.O. Box 3173, Pretoria, 0001
Durow, B.P., P.O. Box 94, Pietermaritzburg, 3200
Elges, H.F.W., Private Bag X313, Pretoria, 0001
Ellis, P.J., P.O. Box 509, Gramamstown, 6140
Ellmer, A., c/o Ellmer Partnership, P.O. Box 2608, Rand-burg, 2125
Engelbrecht, J.C., Rupertlaan 35, Marina-Landgoed, Somerset West, 7130
Erikson, F.V., P.O. Box 1031, Bedfordview, 2008
Errera, L.A., 14 Ludlow Road, Vredehoek, 8001
Esterhuizen, L.C., P.O. Box 2042, Pretoria, 0001
Everett, J.P., P.O. Box 521, Umhlali, 4390
Everett, P.R., 78 Forsdick Road, Carrington Heights, Durban, 4001
Fakudze, I.M., P.O. Box 5106, Mbabane, Swaziland
Ferreira, B.G., 52 Erskine Street Alva, Clackmannanshire FK12 5LU, Scotland UK
Fick, B.A., P.O. Box 65967, Benmore, 2010
Fitzgerald, P.P., c/o Keeve Steyn, P.O. Box 180, Sunninghill, 2157
Foden, A.J., P.O. Box 1342, Rivonia, 2128
Fourie, I.S., 11 Linda Road, Constantiakloof, Florida, 1709
Fourie, C.C., P.O. Box 25725, Langenhovenpark, 9330
Fourie, D.E., P.O. Box 28892, Kensington, 2101
Fourie, A.B., 21 Cardiff Road, Parkwood, 2193
Frame, J.A., c/o Gammon Construction Ltd., 28/F Devon House, Tai Noo Place 979 Kings Road Quarry Bay, Hong Kong
Friedlaender, E.A., 28A Day Road, Cheltenham NSW, 2119, Australia
Geertsema, A.J., P.O. Box 74319, Lynnwood Ridge, 0040
Gerber, J., c/o BKS KZN Structural Division, P.O. Box 56, Westville, 3630
Gertenbach, J.J., P.O. Box 39075, Bramley, 2018
Gibberd, J.P., 40 St Thomas Road, Claremont, 7700
Goldstein, A.E., P.O. Box 91904, Auckland Park, 2006
Gouws, P.J., P.O. Box 11770, Bendor, 0699
Gowan, M.J., c/o Golder Associate, P.O. Box 1734 Milton BC, Brisbane Australia, 4064
Grieve, G.R.H., 344 Delphinus Street, Waterkloof Ridge, 0181
Groenewald, F.W., P.O. Box 37150, Faerie Glen, 0043
Hanekom, A.C., P.O. Box 3173, Pretoria, 0001
Hansel, V.E., 23 Richmond Park, Tekwane Place, Kloof, 3610
Harrison, D.M., P.O. Box 35660, Menlo Park, 0102
Harrison, B.A., P.O. Box 2174, Halfway House, 1685
Hattingh, G.M., c/o G M Hattingh & Vennote, P.O. Box 2201, Randburg, 2125

Hatz, H., P.O. Box 23014, Windhoek, Namibia
Heinz, W.F., c/o Rodio S A Pty. Ltd., P.O. Box 1533, Halfway House, 1685
Hemingway, M.S., P.O. Box 887, Ladysmith, 3370
Hendry, R.W., 16 Doordrift Village, Doordrift Road, Constantia, 7800
Heymann, G., P.O. Box 100627, Moreletapark, 0044
Hicks, R.L., P.O. Box 15485, Lambton, 1414
Holland-Muter, L.M., P.O. Box 1450, Faerie Glen, 0043
Holmes, R.J., c/o B.S. Bergman & Partners Inc., P.O. Box 15654, Vlaeberg, 8018
Holub, J.A., P.O. Box 1092, Saxonwold, 2132
Honiball, H.P., Kristalweg 154, Lyttelton Manor, Centurion, 0157
Howell, G.C., P.O. Box 15502, Panorama, Parow, 7506
Hugo, F., 7 Keet Road, Stellenbosch, 7600
Hugo, P.L., P.O. Box 33804, Glenstantia, 0010
Huisman, I.L., 120 Cape Road, Mill Park, Port Elizabeth, 6001
Hutchison, P.N., P.O. Box 456, Randpark Ridge, 2156
Ippel, J., 297 Albertus Avenue, Erasmusrand, 0181
Jacobs, J.J.H., P.O. Box 12079, Brandhof, 9324
Jaros, M.B., 16 Broadlands, 105 Waterkant Road, Durban North, 4051
Jefferiss, I.G., 26 Palmiet Drive, Westville, 3630
Jermy, C.A., c/o Dept. of Geology, University of Natal, King George V Avenue, Durban, 4041
Jewaskiewitz, S.M., 12 Joubert Street, Rynfield, Benoni, 1501
Johnson, D.J.H., P.O. Box 3033, Durban, 4000
Jones, B.P., P.O. Box 1263, Wandsbeck, 3631
Jones, G.A., P.O. Box 1238, Northcliff, 2115
Jooste, N.J.J., P.O. Box 74786, Lynnwooddrif, 0040
Jordaan, T.B., P.O. Box 1731, Kroonstad, 9500
Joubert, J.W., P.O. Box 13817, Hatfield, 0028
Kabeya, K.K., P.O. Box 32683, Glenstantia, 0010
Kartsounis, M., P.O. Box 1798, Houghton, 2041
Kasunzuma, M., P.O. Box 18168, Nelspruit, 1200
Kawanga, C., Hanley Pepper Owenstown House, Forsters Avenue, Blackrock Co Dublin, Ireland
Kempe, J.A., P.O. Box 1434, Rivonia, 2128
Kerst, E., P.O. Box 486, Trichardt, 2300
Keyter, G.J., c/o SRK Consulting, P.O. Box 55291, Northlands, 2116
King, W.T., 72 Deane Road, Glenmore, Durban, 4001
Kirsten, H.A.D., c/o SRK Consulting, P.O. Box 55291, Northlands, 2116
Kitsopoulos, S., P.O. Box 28578, Danhof, 9310
Klintworth, W.J., 404 Meadow Lane, Iffley Oxford OX4 4ED United Kingdom
Klomp, F.J., P.O. Box 36868, Menlo Park, 0102
Knight, K., 45 Sandown Village, 27 Harvey Road, Pinetown, 3610
Knoetze, M.G., P.O. Box 21211, Parow, 7499
Knottenbelt, E.C., P.O. Box 2055, George, 6530
Koelman, A., P.O. Box 192, Allen's Nek, 1737
Krepelka, E.J.A., 30 Chestnut Drive, Hout Bay, 7800
Kreuiter, A., P.O. Box 818, Lonehill, 2062
Krone, B., c/o Esor (Pty) Ltd., P.O. Box 6478, Dunswart, 1508
Laros, P.A., 44 Orient Road, Lakeside, 7945
Lathleiff, E.P., 628 A Stella Road, Escombe, Queensburgh, 4093
Le Roux, G.J.R., P.O. Box 11036, Wierdapark Suid, 0149
Legg, P.A., 28-8th Avenue, Northmead, Benoni, 1501
Legge, K.R., 64 Jean Avenue, Doringkloof, Centurion, 0157
Legge, T.F.H., P.O. Box 68271, Bryanston, 2021
Leibnitz, A.M., 57 Collins Road, Hayfields, 3201
Lerch, H.E., P.O. Box 768, Windhoek, Namibia
Liebenberg, A.C., 10 The Avenue, Silverhurst Estate, Constantia, 7800
Lloyd, T., 227 Columbine Avenue, Mondeor, 2091
Lombard, H.A., P.O. Box 72885, Lynnwood Ridge, 0040
Lord, S.M., P.O. Box W193, Waterfalls, Harare, Zimbabwe
Lorio, R., c/o Uhlmann Witthaus & Prins, P.O. Box 7627, Roggebaai, 8012
Lötter, J.B.Z., Private Bag X66, Bryanston, 2021
Lourens, J.P., P.O. Box 3173, Pretoria, 0001

Luker, I., P.O. Box 30686, Braamfontein, 2017
Lund, B.G.A., P.O. Box 1935, Witkoppen, 2068
Lyell, K.A., P.O. Box 62545, Marshalltown, 2107
Maas, N.F., P.O. Box 849, Fourways, 2055
Macfarlane, N.G., P.O. Box 2947, Honeydew, 2040
Mackellar, D.C.R., c/o Ninham Shand Inc., Private Bag X136, Centurion, 0157
Manferdini, L.A., P.O. Box 72192, Parkview, 2122
Maree, L., P.O. Box 33060, Glenstantia, 0010
Marshall, T., P.O. Box 1772, Pinegowrie, 2123
Marx, F.N., P.O. Box 22600, Windhoek, Namibia
Masterson, P.S., 35 Royal Avenue, Hout Bay, 7800
Maud, C.J., c/o Keeve Steyn (Pty) Ltd., P.O. Box 37841, Overport, 4067
Mazibuko, S.V., 1034-42nd Hill, Intabazwe Township, Harrismith, 9880
McCarter, E.A., P.O. Box 522, Ladybrand, 9745
McDonald, B.D., P.O. Box 693, Pinetown, 3600
McDonald, I.A.C., 21 Herschel Road, Fish Hoek, Cape Town, 8000
McDonald, D.C.F., P.O. Box 3186, George Industria, 6536
McIlwraith, R.I., 47 Northgrove, 118 Prospect Hall Road, Durban North, 4051
McKelvey, J.G., c/o Matsoku Diversion Partnership, Private Bag A476, Maseru 100 Lesotho
McPhail, G.I., P.O. Box 1596, Cramerview, 2060
Meintjes, H.A.C., P.O. Box 258, Fontainebleau, 2032
Melvill, A.L., c/o Ninham Shand, P.O. Box 760, Pietermaritzburg, 3200
Mills, A.J., P.O. Box 256, East London, 5200
Molenaar, P.J., P.O. Box 495, Newlands, 0049
Montgomery, D., P.O. Box 652692, Benmore, 2010
Moore, L.W., P.O. Box 1263, Wandsbeck, 3631
Moores, A.K., Castaway, Beaumont Hill Great Dunom, Essex CM6 2AW, United Kingdom
Morgan, A.T.M., P.O. Box 548, Cramerview, 2060
Moss-Morris, A.L., 1 Hodford Road, London NW11 8NL, England
Mountain, M.J., c/o M J Mountain & Partners, P.O. Box 18959, Wynberg, 7824
Mouton, D.J., P.O. Box 1252, Garsfontein, 0042
Msutwana, S.B., P.O. Box 8117, Johannesburg, 2000
Muller, H.S., c/o I.D.C., P.O. Box 784055, Sandton, 2146
Netterberg, F., 79 Charles Jackson Street, Weavind Park, 0184
Newsome, C.R., 14 Harry Stodel Road, Kenilworth, 7700
Nortje, J.H., P.O. Box 39361, Moreletapark, 0044
Norton, H.G., P.O. Box 6478, Dunswart, 1508
Notcutt, S., 332 Moore Road, Durban, 4001
Ntshumaelo, K.M., P.O. Box 177, Ermelo, 2350
Odell, P.L., 25 Rietbok Road, Robin Hills, Randburg, 2194
Ofer, Z., P.O. Box 891528, Lyndhurst, 2106
Oosthuizen, A.P.C., P.O. Box 650235, Benmore, 2010
Oosthuizen, P.H., P.O. Box 39540, Faerie Glen, 0043
Paige-Green, P., 46 Peffer Street, Elardus Park, 0181
Paola, J.C., P.O. Box 95198, Waterkloof, 0145
Parrock, A.L., c/o ARQ Associates, 66 Ingersol Road, Lynnwood Glen, 0081
Parry-Davies, R., P.O. Box 26895, Hout Bay, 7872
Partridge, T.C., 13 Cluny Road, Forest Town, Johannesburg, 2193
Pass, J.C., P.O. Box 641, Westville, 3630
Paulsen, D.M.R., 44 William Trollip Road, Heuwelsig, 9301
Pavlakis, M., 7 Danya Road, Victory Park, 2195
Pidgeon, J.T., c/o Waffmark, P.O. Box 95011, Waterkloof, 0145
Pienaar, P.A., P.O. Box 35158, Menlo Park, 0102
Pistorius, J.J., P.O. Box 36304, Menlo Park, 0102
Plant, G.W., Londolozi 3 Acacia Grove Berkhamsted HP4 3AJ Herts, United Kingdom
Plantema, G.F.W., P.O. Box 606, Suider-Paarl, 7624
Pollock, I.G., P.O. Box 73245, Fairlands, 2030
Pretorius, F.J., P.O. Box 608, Empangeni, 3880
Pretorius, A.F., P.O. Box 556, Secunda, 2302
Prinsloo, S., P.O. Box 510, Silverton, 0127
Raaf, A.J., P.O. Box 18897, Dalbridge, 4014
Rademeyer, I.L., P.O. Box 1435, Port Elizabeth, 6000
Retief, A.C.G., P.O. Box 54173, Ninapark, 0156

Richardson, B.W., P.O. Box 3276, Cape Town, 8000
Robins, M., P.O. Box 62117, Marshalltown, 2107
Robinson, B.C.S., 121 Limpopo Avenue, Lyttelton Manor, Ext. 5, 0157
Robinson, A.W., P.O. Box 2861, Northcliff, 2115
Robinson, G.B., PO Box 107, Kloof, 3640
Rohde, A.W., P.O. Box 35280, Menlo Park, 0102
Rohm, H.B., PBag 9083, Cape Town, 8000
Rohrs, M.G., P.O. Box 753, Sabie, 1260
Rose, D.A., 4 Avenue Bordeaux, Fresnaye, Cape Town, 8001
Rosenthal, G.N., 8 Sunningdale Road, Kenilworth, 7700
Rossouw, A.K., 5 Blesbok Road, Bethal, 2310
Roux, P., P.O. Box 11380, Queenswood, 0121
Rozowsky, J., c/o Rozowsky Ass, P.O. Box 4969, Cape Town, 8000
Rust, E., P.O. Box 39108, Moreletapark, 0044
Sahli, C.L.W., P.O. Box 1164, Louis Trichardt, 0920
Sankar, A., P.O. Box 19392, Dormerton, 4015
Saunders, A.W., P.O. Box 19702, Tecoma, 5214
Saunderson, R., P.O. Box 779, Westville, 3630
Savage, R.H., 23 Copson Close, Chase Valey, 3201
Scheurenberg, R.J., c/o Knight Piésold Consultores SA, San Borja Sur 143, San Borja Lima 41, Peru
Schreiner, H.D., c/o Dept. of Civil Engineering, University of Natal, Private Bag X10, Dalbridge, 4014
Schultz, E.A., No. 49 North Dragon Hill High Miaoli Miaoli City Taiwan R.O.C.
Schulze-Hulbe, A., P.O. Box 905-866, Garsfontein, 0042
Sellers, E.J., P.O. Box 263, Auckland Park, 2006
Sellick, E.C.H., 37 Crescent Road, Waterkloof Ridge, 0181
Silberman, M.I., P.O. Box 34885, Glenstantia, 0010
Simpson, A.M., P.O. Box 1509, Cramerview, Sandton, 2060
Smit, G.J.N., c/o Klomp Consult Africa (Pty) Ltd., P.O. Box 2779, Ermelo, 2350
Smit, A.W., 222 South Road, Vryheid, 3100
Smith, W.H., P.O. Box 1164, Bethlehem, 9700
Smith, M.E., P.O. Box 732, Johannesburg, 2000
Smith, A.C.S., 28 St Albans Avenue, Craighall Park, 2196
Sparks, A.D.W., 4 Orkney Street, Rondebosch, 7701
Speers, C.R., Postnet Suite #347, Private Bag X18, Rondebosch, 7700
Staphorst, M.S.J., 7 Heron Road, Durban, 4001
Steffen, O.K.H., c/o SRK Consulting, P.O. Box 55291, Northlands, 2116
Sterianos, B., P.O. Box 744, Green Point, 8051
Stern, E.J., P.O. Box 65331, Benmore, 2010
Steyn, N., P.O. Box 1, Vryheid, 3100
Steyn, G.P., P.O. Box 11003, Silver Lakes, 0054
Steyn, J.D., P.O. Box 5198, Cresta, 2118
Steyn, W.J.V., P.O. Box 908052, Montana, 0151
Still, D.A., P.O. Box 11431, Dorpspruit, 3206
Stocken, R.W., 76 Cambridge Street, Farrarmere, Benoni, 1501
Strauss, A., P.O. Box 14186, Hatfield, 0028
Swallow, J.W., P.O. Box 411340, Craighall, 2024
Swart, J.J., P.O. Box 100872, Brandhof, 9324
Sydney, M.R.T., 19 Linnet Road, Woodhaven, 4004
Taitz, M.R., P.O. Box 11019, Vorna Valley, 1686
Tanner, A.W., P.O. Box 1718, Bromhof, 2154
Taute, A., c/o Van Niekerk Kleyn & Edwards, P.O. Box 72927, Lynnwood Ridge, 0040
Taylor, S.J., 1348 E Alice Avenue, Phoenix AZ 85020
Taylor, P.F., 45 First Street, Abbotsford, 2192
Tedder, B.N., 59 Vernon Road, Berea, Durban, 4001
Terblanche, E.H., P.O. Box 48921, Roosevelt Park, 2129
Theron, E., c/o SRK Consulting, P.O. Box 6824, Roggebaai, 8012

Thobejane, M.H., P.O. Box 17090, Doornfontein, 2028
Thompson, G.D., P.O. Box 67036, Bryanston, 2021
Tluczek, H.J., Cluster Box 1527, Forest Drive, Forest Hills, 3610
Toussaint, W.B., P.O. Box 73646, Lynnwood Ridge, 0040
Trebicki, D.D.P., c/o Rand Water, P.O. Box 1127, Johannesburg, 2000
Triebel, D.R., 179 Retha Road, Northcliff, Ext. 2, 2195
Tromp, B.E., P.O. Box 4233, Randburg, 2125
Van der Leij, R.S., P.O. Box 411606, Craighall, 2024
Van der Merwe, W.J., c/o Kantey & Templer, P.O. Box 3132, Cape Town, 8000
Van der Toorn, H., 8 Wald Avenue, Dunvegan, Edenvale, 1609
Van der Vlugt, R., P.O. Box 115, Kalk Bay, 7990
Van Huyssteen, R.J., P.O. Box 72727, Lynnwood Ridge, 0040
Van Rooyen, M., P.O. Box 72962, Lynnwood Ridge, 0040
Van Schalkwyk, A., c/o Department of Earth Sciences, University of Pretoria, Pretoria, 0001
Van Tonder, C.P.G., P.O. Box 12350, Bendor Park, 0699
Van Wieringen, M., P.O. Box 32224, Camps Bay, 8040
Van Wyk, L.C., P.O. Box 25048, Monumentpark, 0105
Van Wyk, F.J.V., P.O. Box 16591, Atlasville, 1465
Van Zyl, W.S., P.O. Box 8, Kimberley, 8300
Venter, J.S.M., P.O. Box 961, Germiston, 1400
Venter, J.P., P.O. Box 7271, Hennopsmeer, 0046
Venter, I.S., P.O. Box 36114, Menlo Park, 0102
Verbeek, T., 122 Pendoring Avenue, Wonderboom, 0182
Vermeulen, N.J., c/o Dept. Siviele Ingenieurs, University of Pretoria, Pretoria, 0001
Viljoen, D.P., c/o Keeve Steyn Inc., P.O. Box 37841, Overport, 4067
Viljoen, B.C., P.O. Box 382, North Riding, 2162
Viljoen, J.H., P.O. Box 1131, Kimberley, 8300
Vogt, S.J., P.O. Box 3973, Randburg, 2125
Von Geusau, C.G., P.O. Box 23022, Claremont, 7735
Vuba, B., P.O. Box 9212, Queenstown, 5320
Wagener, F.V.M., P.O. Box 1434, Rivonia, 2128
Walters, D.E., c/o Walters & Associates, 1 Northview Crescent, Vincent, 5247
Wardle, G.R., c/o Jones & Wagener, P.O. Box 1434, Rivonia, 2128
Watermeyer, C.F., P.O. Box 784506, Sandton, 2146
Wates, J.A., P.O. Box 6001, Halfway House, 1685
Webb, D.L., P.O. Box 50129, Musgrave, 4062
Webb, M.C., c/o SSIS Consulting Engineers, P.O. Box 74819, Lynnwood Ridge, 0040
Weir, I.J.R., P.O. Box 210, Jukskei Park, 2153
Wesseloo, J., P.O. Box 14503, Lyttelton, Centurion, 0140
Westerberg, A., P.O. Box 13946, Humewood, Port Elizabeth, 6013
White, A.C., c/o SRK Consulting, P.O. Box 55291, Northlands, 2116
Wilkinson, K.F.D., c/o KFD Wilkinson & Partners, 12th Floor-Picbel Parkade, Strand Street, Cape Town, 8001
Williams, A.A.B., 141 Dorado Street, Waterkloof Ridge, 0181
Williams, M.R., P.O. Box 1570, Vereeniging, 1930
Williams, D.A., 76 Naturaliste Boulevard, Iluka, Western Australia, WA6028
Williamson, J.R.G., P.O. Box 221, Rivonia, 2128
Wilson, C., 14 Palm Springs, 155 Ridge Road, Durban, 4001
Wolff, H., P.O. Box 231, Westville, 3630
Wright, D.F.H., 25 Daphne Street, Hout Bay, 7800
Zaaiman, G.T., P.O. Box 817, Florida Hills, 1716
Zietsman, J.F.W., P.O. Box 26426, Hout Bay, 7872

SPAIN/ESPAGNE

Secretary: E. Dapena, Sociedad Española de Mecánica del Suelo e Ingeniería Geotecnica, Laboratorio de Geotecnia. CEDEX. Alfonso XII, 3, Madrid 28014, Spain. T: 34-91-3357357, F: 34-91-3357322, E: e.dapena@cedex.es

Total number of members: 322

Abad, G., Guerrero y Mendoza, 16 (Chalet), 28002 Madrid. T: 91-4138984

Aceseg, Asoc. Catalana de Empresas de Sondeos y Estudios Geotecnicos, Rosellon, 340, 08025 Barcelona. T: 93-245221

Adalid, J.L., Arapiles, 21-3° Dcha., 28015 Madrid. T: 91-4458093

Adell, F. Avda. Ferrol, 37-3° Dcha., 28029 Madrid. T: 91-3294477

Aepo, B., 22, 28029 Madrid. T: 91-3789660

Afonso, C., Grupo Terratest Cimyson, I.CO.S, S.A., Alcalá, 65-4°, 28014 Madrid. T: 91-4237500

Alain, D., Grupo Terratest, Cimyson, I.CO.S, S.A., Alcalá, 65-4°, 28014 Madrid. T: 91-4237500

Alameda, J.G., P.B. Ing. Consultores, c/Caleruega, 67-7ª plta. 28033 Madrid. T: 91-4025062

Alcaide, A., Doctor Ingeniero Caminos, Arzobispo Morcillo, 42-10°A, 28029 Madrid

Alocen, J.R., Rodio Cimentaciones Especiales, P° de la Castellana, 130-6ª, 28046 Madrid. T: 91-5624610

Alonso Nieto, S., UTE.Eptisa-Inteinco-Seg, Arrollo de las Pilillas, 8-3°C, 28030 Madrid. T: 91-4399980

Alonso Pérez de Agreda, E+, Catedrático de Ingeniería del Terreno, E.T.S. Ingenieros de Caminos, Gran Capitán, s/n, Modulo D2, 08034 Barcelona. T: 93-4016866

Alvarez Martínez, A., c/San Juan Bautista, 5-4°A (Las Rozas), 28230 Madrid. T: 91-6375232

Alvarez Martínez, A., Divina Pastora, 5-7°, E-18012 Granada

Andrés de, M., ECM, Raimundo Fernández Villaverde, 45-2° D, 28003 Madrid. T: 91-5542129

Aniceto, J.Mª., Ferrivual-Agroman, General Moscardó, 21-D, 3° D, 28020 Madrid. T: 91-5860255

Antelo, J., Kronsa Internacional, General Ramírez de Madrid, 8-10-3°, 28029 Madrid. T: 91-4252890

Anton, J.L., Hermosilla, 70-5°C, 28001 Madrid. T: 91-5029000

Arcones Torrejón, A., Henares, 18, 28220 Majadahonda, Madrid

Arcos, J.L., Kronsa Internacional, General Ramirez de Madrid, 8-10-3°, 28029 Madrid. T: 91-4252890

Ariza, J., Grupo Terratest, Cimyson, I.CO.S, S.A., Alcalá, 65-4°, 28014 Madrid. T: 91-4237500

Arroyo, R., Barrio El Somo, 9-C, San Román, 38012 Santander. T: 942-343934

Asanza, E., Porto Lagos, 7-6°B Izq. 28924 Alcorcón. Madrid. T: 91-6107858

Aventín, A., Kronsa Internacional, General Ramirez de Madrid, 8-10-3°, 28029 Madrid. T: 91-4252890

Ayuga, F., E.T.S. de Ingenieros Agrónomos, Ciudad Universitaria, s/n. 28040 Madrid. T: 91-5274193

Ayuso, J., Apartado de Correos, n° 3048, 14080 Córdoba. T: 957-295464

Azcoiti, J.I., c/Sancho Ramirez, 7-6°B, 31008 Pamplona. T: 948-277037

Azkue, P.Mª., P° de Salamanca, 12, 20003 San Sebastián

Ballester Mercader, N., Moianes, 16-18, 3°-4ª dcha., 98014 Barcelona. T: 93-4319567

Ballester Muñoz, F., Director de Departamento de Transportes y Tecnología de P. y P., Av. de los Castros, s/n, Santander. T: 942-201750, F: 942-201703

Ballesteros, L.A., Limón, 41, 13300 Valdepeñas (Ciudad Real). T: 926-320626

Barba, J., Empresa Municipal de Aguas de Huelva, c/Alonso Sánchez, 3, 21003 Huelva. T: 959-280001, F: 959-259614

Barceló, L., San Isidro, 6-1°C, 28007 Alcalá de Henares, Madrid. T: 91-8787136

Barco, J.A., Depart. Geotecnia, Univ. De Cantabria. Jorge Vigón, 49-3°D, 26003 Logroño, La Rioja. T: 941-234403

Bartesaghi, A., San Sebastian 2967, Dpto. 304. Las Condes, Santiago, Chile. T: 59826044293

Barra, R., Diputación de Cádiz, Plaza de España, s/n, 11004 Cádiz. T: 956-274502

Basabe, J., Hermanos Pinzón, 4-7°, 28036 Madrid. T: 91-3198250

Bayón, E., Arturo Soria, 112, 28027 Madrid

Berganza, F., Ingeniero de Caminos, Canales y Puertos, Junta de Castilla y León-Carreteras, Glorieta de Bilbao, s/n, 09006 Burgos. T: 947-281536, F: 947-239735

Bermejo, J., Oficina Técnica de proyectos, La Paz, 20, 28100 Alcobendas, Madrid. T: 91-651855, F: 91-6518555

Bernal, A., Consulting Geotechnical Engineer, Ferrer del Rio, 14, 28028 Madrid. T: 91-7258315, F: 91-7669303

Beteta, M., c/Cristobal Bordiu, 8-4°B, 28003 Madrid. T: 91-3475854

Bianqui, G., Farmacéutico Angel Establiel, 8-3 Bloque-3°B, 03008 Alicante. T: 96-5103632

Biosca, F., Cimentaciones Especiales Rodio, S.A. Paseo de la Castellana, 130-6°, 28046 Madrid. T: 91-5624610

Bisbal, L. c/Alfonso de Córdoba, 6-1° -3ª, 46010 Valencia. T: 96/3600915

Borrego, J.A., Ingeniero de Caminos, Canales y Puertos, Plaza del Perú, 5-4°E (Oficina), Apolonio Morales, 21-4° Dcha. (Particular), 28036 Madrid. T: 91-2508615, F: 91-2504780

Boyarizo, E., Pilotes Santiago Sánchez. Plaza de Morenos, 5. 19001 Guadalajara. T: 949-222558

Bravo, G., Conf. Hidrográfica del Guadalquivir, Avda. de Madrid, 7, 18012 Granada. T: 958-245000

Burbano, G., Construcciones Especiales y Dragados, S.A., Ubanización Monteclaro, c/Magnolias, 11, 28023 (Pozuelo de Alarcón) Madrid. T: 91-5832000, F: 91-5833834

Calavera, J., Intemac, S.A., Monte Esquinza, 30-4°D, 28010 Madrid. T: 91-3105158

Calle, A.J., Sanchez Barcaiztegui, 23-4° Dcja. B, 28007 Madrid. T: 91-5514709

Canalda, A., Ingeniero Civil, Bonetero, 6-2° Dcha. 28016 Madrid. T: 91-5777140

Cancela, Mª.D., Av. de Francia, 1, portal 1-1°B, 28224 (Pozuelo de Alarcón) Madrid

Candela, J., Grupo Terratest Cimyson, I.CO.S, S.A., Alcalá, 65-4°, 28014 Madrid. T: 91-4237500

Cano, H., Laboratorio de Geotecnia-CEDEX, Alfonso XII-3, 28014 Madrid. T: 91-3357300, F: 91-3357322

Cañedo, C., Avda. de Burgos, 12, 28020 Madrid. T: 91-3025280

Cañizal, J., Urbanización Las Lindes, 11, 39600 Santander. T: 942-201813

Cañizo del, L., Alonso Quijano, 31, 28034 Madrid. T: 913597878

Casado, R., c/Doctor Esquerdo, 183, 28007 Madrid. T: 91-5521167

Casero, L., Ingeniero de Caminos, Canales y Puertos, Azalez, 572, Urb. El Soto, 28109 (Alcobendas) Madrid

Castanedo, F.J., Paseo Marqués de Zafra, 38 bis 1°A, 28028 Madrid

Castillo, E., Escuela de Ing. de Caminos, Avda. de los Castros s/n, Santander

Castro de, J.C., Inteinco, c/Serrano, 85-2°, 28006 Madrid. T: 91-5641512

Casuso, C., Grupo Terratest Cimyson, I.CO.S, S.A., Alcalá, 65-4°, 28014 Madrid. T: 91-4237500

Cea de, J.C., c/Santa María Magdalena, 3 Bajo D, 28016 Madrid

Ceico, S.L., Carretera Nacional 301 P.K. 397.900, 30100 Espinardo, Murcia. T: 968-260511, F: 968-341352

Celemín, M., Condesa de Sagasta, 36, 24001 Leon. T: 98-7291819

Celma, J.J., Avda. Barón de Carcer, 40-plta. 9, 46009 Valencia. T: 96-3877583

Centro de Estudios y Experimentación de Obras Públicas, Alfonso XII, 3 y 5, 28014 Madrid. T: 91-3357515, F: 91-3357222

Cimentaciones Especiales RODIO, S.A., Pº de la Castellana, 130-4ª planta, 28046 Madrid. T: 91-5624610

Conde, J.I., Grupo Terratest Cimyson, I.CO.S, S.A., Alcalá, 65-4º, 28014 Madrid. T: 91-4237500

Consec, S.A., Avenida de Cervantes, 43, pabellón 2, entreplanta, Departamento 9, 48970 Basauri (Vizcaya). T: 94/6156497

Cónsola, M., Plaza Creu del Terme, 2-2ºA, Bell-lloc d'Urgell, 25220 Lérida. T: 973-560735

Construcciones Especiales y Dragados, S.A., Avda. de Tenerife, 4-6, Edif. Flores 2ª planta, 28700 San Sebastian de los Reyes, Madrid. T: 91-5833381, F: 91-5832276

Cortacans, J.A., Ingeniero de Caminos, Entrecanales y Tavora, c/Juan de Mena, 8, 28014 Madrid. T: 91-7660758, F: 91-7661225

Cubero, A., Carabela La Niña, 24-bis, 7º-1ª, 08017 Barcelona. T: 93-2040918

Cuéllar, V., c/ Duque de Sesto, 30, 28009 Madrid

Cuenca Lorenzo, J.L., Constitución, 113-4º H, 28100 Alcobendas, Madrid. T: 91-6516359

Cuenca Paya, A., Laboratorio de Carreteras, Carretera de Ocaña, 4, 03007 Alicante. T: 96-5100939, F: 96-5101804

Da Costa, A., E.T.S. Ing. de Caminos, Canales y Puertos. Avda. de los Castros, s/n, 39005 Santander, Cantabria. T: 942-201813

Dal-Re, R., Profesor Doctor Ingeniero Agrónomo, c/Nuñez de Balboa, 12, 28001 Madrid. T: 91-5444807, F: 91-5437849

Dapena, J.E.*+, Dr. Civil Engineer, Laboratorio de Geotecnia-CEDEX, Alfonso XII, 3 y 5, 28014 Madrid. T: 91-3357357, F: 91-3357322

De Diego, J., Cimentaciones Especiales Rodio S.A., Paseo de la Castellana 139-6º, 28046 Madrid. T: 91-5624610

De Juan, M.A., Kronsa Internacional, General Ramirez de Madrid, 8-10-3º, 28029 Madrid. T: 91-4252890

De la Guardia, F., Grupo Terratest Cimyson, I.CO.S, S.A., Alcalá, 65-4º, 28014 Madrid. T: 91-4237500

De Santiago, J.C., Kronsa Internacional, General Ramirez de Madrid, 8-10-3º, 28029 Madrid. T: 91-4252890

Delgado, C+, Dr. Ingeniero de Caminos, Canales y Puertos, Contratas, Industrias y Materiales S.A., c/Hermano Garate, 8-5ºD, 28020 Madrid. T: 91-5708702

Delgado Marchal, J., Complejo Vistahermosa, Torre 2-9ºB, 03016 Alicante. T: 96-5903400

Diez, F., O'Donnell, 6, Torre B, Piso 13, 28009 Madrid. T: 91-5752941

DM Iberia, S.A., Empresa de Ingeniería, Pedro Muguruza, 8-2ºA, 28036 Madrid. T: 91-3503804

Domenech, C., Instituto Técnico de la Construccion, Avenida de Elche, 164, 03008 Alicante. T: 96-5104600

Domingo, J.A., Demarcación de Carreteras del Estado de Andalucia. Ministerio de Fomento. Avda. de Madrid, 7, Granada 18012. T: 958-271700

Dumont, F., Grupo Terratest Cimyson, I.CO.S, S.A., Alcalá, 65-4º, 28014 Madrid. T: 91-4237500

Edificios Gema, S.L., Nicolas Salmerón, 5-1º, 39009 Santander. T: 942-224000

Echave, J.M., Ingeniero de Caminos, Grupo Terratest, Alcalá, 65, 28014 Madrid. T: 91-5777140, F: 91-5768858

Eptisa, Arapiles, 14-4º, 28015 Madrid. T: 91-4450300

Escario, V., Dr. Civil Engineer, Geotechnical Consultant, Orense, 33, 28020 Madrid. T: 91-5557040, F: 91-5557040

Escolano, J.L., Avda. Andalucía, 23-11 D, 29006 Málaga

Esparcia, A., Kronsa Internacional, General Ramirez de Madrid, 8-10-3º, 28029 Madrid. T: 91-4252890

Espinace, R., Dr. Ingeniero, Universidad Catolica de Valparaiso, Av. Brasil 2147, Valparaiso, Chile. T: 32-273611

Estaire, J., Laboratorio de Geotecnia, Alfonso XII, 3y 5, 28014 Madrid. T: 91-3357300

Euro Geotecnica, Parc. Tecnologic del Valles. 08290 Cerdanyola, Barcelona. T: 93-5820160

Euroconsult Geotecnia, S.A., Avda. Montes de Oca, s/n, 28700 San Sebastian de los Reyes (Madrid). T: 91-6524899

Euroestudios, Empresa de Ingeniería, Castelló, 128-7º, 28006 Madrid. T: 91-4113212

Fernández, J.L., Cimentaciones Especiales Rodio, S.A., Paseo de la Castellana 139-6º, 28046 Madrid. T: 91-5624610

Fernández Blanco, S+, Dr. Ingeniero de Caminos, C. y P., Comision Comunidades Europeas, Rue de la Loi, 200 1049 Bruxelles

Fernández Bogas, D., Grupo Terratest Cimyson, I.CO.S, S.A., Alcalá, 65-4º, 28014 Madrid. T: 91-5777140, F: 91-5768858

Fernández Moya, J., Valdebernardo, 10, 28030 Madrid

Fernánez Ontivero, J., Narvaez, 43-3º puerta 4, 28009 Madrid

Fernández Renau, L., Dr. Civil Engineer, M.S. University of IOWA (USA), Avenida de San Luis, 95-9ºF, 28033 Madrid. T: 91-6501240

Fernández Salso, J.L., Geotecnia Estructural, S.L., Valderribas, 48-3ª Esc. 5ºC, 28007 Madrid. T: 91-4337796

Fernández Toledano, D., Grupo Terratest Cimyson, I.CO.S, S.A., Alcalá, 65-4º, 28014 Madrid. T: 91-5777140, F: 91-5768858

Ferrovial-Agroman, S.A., Empresa Constructora, Principe de Vergara, 108, 28002 Madrid. T: 91-5862500, F: 91-5862677

Florez, J., Dr. Ingeniero de Caminos, Escuela Tec. Sup. Ing. Caminos, Canales y Puertos, Ciudad Universitaria, s/n, 28040 Madrid. T: 91-3366719

Forcat, A., Construcciones Especiales y Dragados, S.A., Orense, 81, 28020 Madrid

Fuente de la, P., Dr. Ingeniero de Caminos, Canales y Puertos, Pza. Doctor Laguna, 10, 28009 Madrid. T: 91-4093277

García de la Oliva, J.L., Marqués de Lozoya, 12-10ºE, 28007 Madrid

García Díaz, M.A.*+, Ingeniero de Caminos, Canales y Puertos, Almagro, 11-6º-10, 28010 Madrid. T: 91-4106601/91-6548725

García Gamallo, A.Mª., Estudio de Arquitectura, Cobre,1, 28791 Soto del Real, Madrid. T: 91-8477005

García Girones, M., Grupo Terratest Cimyson, I.CO.S, S.A., Alcalá, 65-4º, 28014 Madrid. T: 91-5777140, F: 91-5768858

García Gonzalo, E., Cimentaciones Especiales Rodio, S.A., Paseo de la Castellana 139-6º, 28046 Madrid. T: 91-5624610

García Guirao, J., Cimentaciones Especiales Rodio, S.A., Paseo de la Castellana 139-6º, 28046 Madrid. T: 91-5624610

García Leiva, Mª.T., Cimentaciones Especiales Rodio, S.A., Paseo de la Castellana 139-6º, 28046 Madrid. T: 91-5624610

García Mañes, M., Fuestespina, 14-7ºA, Santa Eugenia, 28031 Madrid

García Mina, J., Grupo Terratest Cimyson, I.CO.S, S.A., Alcalá, 65-4º, 28014 Madrid. T: 91-5777140, F: 91-5768858

García Olguera, J., Grupo Terratest Cimyson, I.CO.S, S.A., Alcalá, 65-4º, 28014 Madrid. T: 91-5777140, F: 91-5768858

García Prado, J.A., Santa Virgilia, 7-9ºA, 28033 Madrid. T: 91-7642361

García-Roselló, J., Dr. Civil Engineer, Arturo Soria, 92, 28027 Madrid. T: 91-4088866

Gens, A*+, Profesor, Universidad de Ingenieria del Terreno, Universidad Politécnica Cataluña, Gran Capitán s/n, Edificio D-2, 09034 Barcelona. T: 93-4016867/4016866, F: 93-4016504

Gestión de Infraestructura de Andalucía, S.A., Rioja, 14–16, 2ª planta, 41001 Sevilla. T: 95-4211555

Gil Lablanca, D., Kronsa Internacional, General Ramirez de Madrid, 8-10-3º, 28029 Madrid. T: 91-4252890

Gil Mejias, F., Industrial de Sondeos, S.A., Pº de Goya, 15, 28032 Móstoles, Madrid. T: 91-6132700

G.O.C., S.L., Ingeniería Geotécnica y Consulting, Rio Bibey, 16 bajo, 32001 Orense. T: 988-214357, F: 988-214331

Gómez Corral, F.J., I.C.C.P. Consultor de Idom, Avda. Lendakari Aguirre, 3-4°, 48014 Bilbao. T: 94-4479600, F: 94-4761804

Gómez García, J.F., Luis Misson, 5-1°, 28039 Madrid. T: 91-3111809/8885714

Gómez Hermoso, J., Ingeniero de Caminos, Canales y Puertos, Alfredo Marquerie, 29-6°D, 28034 Madrid. T: 91-7390354

González Abril, E., Lacoex, S.L., Avda. Virgen de la Montaña, 1-1°, 10004 Cáceres. T: 927-226285

González de Vallejo, L.I.*+, Professor of Engineering Geology, Universidad Complutense, Fac. Ciencias Geológicas, 28060 Madrid. T: 91-5567693, F: 91-5567615

González Fernández, J., Instituto Técnico de la Cosntrucción, Avda. de Elche, 164, 03008 Alicante. T: 96-5104600, F: 96-5104819

González González, C., Escuela Tec. Sup. Ing. Caminos, Canales y Puertos, Avda. de los Castros, s/n, 39005 Santander. T: 942-201813

González Nicieza, C., Catedrático de Universidad, Independencia, 13, 33004 Oviedo (Asturias). T: 985-104267

Guedán, G., NECSO Entrecanales Cubiertas, Avda. Europa, 18, Parque Empresarial La Moraleja, 28108 Alcobendas, Madrid. T: 91-3483333

Guinda, P., Dr. Civil Engineer, Avda. del Puerto, 1, 46021 Valencia. T: 96-3528727, F: 96-3517094

Gurrera, J., Civil Engineer, Consulting Geotechnical Engineer, Germandat St. Sebastia, 46, 25700 La Seu D'Urgell (Lérida). T: 973-351790

Guerra, A., Geoestudios, S.A., Avda. Alberto Alcocer, 7-7° Centro, 28036 Madrid. T: 91-3598485

Gutierrez Blanco, F., P° de la Habana, 84-5° D, 28046 Madrid. T: 91-4570505

Gutierrez Manjón, J.M., Cardenal Herrera Oria, 167-6°A, 28034 Madrid

Gutierrez Pereda, J.Mª., Cimentaciones Especiales Rodio, S.A., Paseo de la Castellana 139-6°, 28046 Madrid. T: 91-5624610

Hacar, M.A., Dr. Ingeniero de Caminos, Canales y Puertos, Lic. en C.C. Exactas y Físicas, P° San Fco. de Sales, 2, 28003 Madrid. T: 91-5442070

Heras de las, C., Ingeniero de Caminos, Laboratorio de Geotecnia, CEDEX, Alfonso XII, 3 y 5, 28014 Madrid. T: 91-3357242/3357300, F: 91-3357322

Hermosilla, C., Kronsa Internacional, General Ramirez de Madrid, 8-10-3°, 28029 Madrid. T: 91-4252890

Hermoso, J.A.+, Ingeniero de Caminos, Condesa de Venadito, 12, 28027 Madrid. T: 91-4053845

Hernández Muñoz, J.M., Ingeniero de Caminos, Consultor Estructuras y Cimentaciones, Constancia, 10-1° Dcha., 29002 Málaga. T: 95-2351412, F: 95-2351401

Hernández Rebollo, J., Grupo Terratest Cimyson, I.CO.S, S.A., Alcalá, 65-4°, 28014 Madrid. T: 91-5777140, F: 91-5768858

Herrador, J.Mª., Kronsa Internacional, General Ramirez de Madrid, 8-10-3°, 28029 Madrid. T: 91-4252890

Heymo Ingenieria, S.A., Arequipa, 1 – planta 3ª, 28043 Madrid. T: 91-3810072, F: 91-3810664

Hinojosa, J.A., Dr. Ingeniero de Caminos, Canales y Puertos, Don Ramón de la Cruz, 40, 28001 Madrid. T: 91-5977822

Industrial de Sondeos, P° de Goya, 15, 28932 Móstoles, Madrid. T: 91-6132700

Ingeomat, S.L., Ingeniería Geotécnica y Medio Ambiente, Doctor Federico Rubio y Gali, 19 bajoF, 28003 Madrid. T: 91-5537239

Intecom, Avda. del Mar, s/n, 03187 Los Montesinos, Alicante. T: 96-6721218, F: 96-6721247

Jevenois, J., Almendro del Paular, 6, 28224 Madrid. T: 91-7153109/3365864

Jugo, I., Ikerlur, S.L., Avda. Zarautz, 82-3°, 20009 San Sebastian. T: 943-310471

Jurado, C., Corsan, E.C., S.A., Avda. de la Independencia, 34-4°A, 28700 San Sebastian de los Reyes, Madrid. T: 91-6531302, F: 91-4424387

Justo, J.L., Head of Department, E.T.S. Arquitectura, Av. Reina Mercedes, 2, 41012 Sevilla. T: 95-4622561, F: 95-4556534

Keller Técnicas del Suelo, S.L., Josefa Valcarcel, 8-2° planta, 28027 Madrid. T: 91-3202884

Laboratorio de Ensayos Navarra, S.A., Polígono Landaben. c/L y B, Pamplona, 31012 Navarra. T: 945-276500

Laboratorio de Ensayos Técnicos, S.A., Polígono Argualas, Nave, 52 A, c/Argualas, s/n, 50012 Zaragoza. T: 976-566875, F: 976-566612

Laque, A., Cimentaciones Especiales Rodio, S.A., Paseo de la Castellana 139-6°, 28046 Madrid. T: 91-5624610

Lama, J.L., Narvaez, 86-4°A, 28009 Madrid

LDH Geotecnia y Estructuras, S.L., Playa de Louro, 2, Esc.Centro 3°C, 28400 Villalba, Madrid. T: 91-8502467

Ledesma, A., Prof. Titular Universidad, E.T.S. Ingenieros de Caminos, Gran Capitán s/n, Modulo D-2, 08034 Barcelona. T: 93-4016864, F: 93-4016504

Leira, J.A., Fernandez Shaw, 2 portal 4-9°C, 28007 Madrid. T: 91-5390197

Lesarri, J.I., Castilla, 7-2° Izq., 39002 Santander. T: 942-360782

Liria, J., c/Juan de la Cierva, 6-3°, 28006 Madrid. T: 91-5618871

Lobato, A., Arcotecnos, S.A., Poligono Molino del Pilar, Camino de los Molinos, s/n, calle B, Nave 68, 50015 Zaragoza. T: 976/731710, F: 976-733934

López Guarda, R., Ingeniero de Caminos, Canales y Puertos, Zaragoza, 13-4° Izda. 22002 Huesca. T: 974-242845

López Martos, J.J., Las Mercedes, 2, 18009 Granada. T: 958-2822400

López-Pelaez, A., Ingeniero de Caminos, Canales y Puertos, Plaza Maestro Odón Alonso, 4-7°C, 24002 León. T: 987-242076

López Rodríguez, R., Kronsa Internacional, General Ramirez de Madrid, 8-10-3°, 28029 Madrid. T: 91-4252890

López Serrano, M., Cimentaciones y Obras Oúblicas, S.L., Carretera del Valle, 23, 30150 La Alberca, Murcia. T: 968-841109

Lorenzo, F., Arte, 31-5°C, 28033 Madrid. T: 91-3838455

Lucas, C., Sequillo, 2, Bajo B, 28017 Madrid. T: 91-4081117

Llorens, M., c/Eraso, 6-3°C, 28028 Madrid

Lloret, A., E.T.S.I.C.C.P., Jorge Girona Salgado, 31, 08034 Barcelona. T: 93-2566111

Maestre, M.V., Grupo Terratest Cimyson, I.CO.S, S.A., Alcalá, 65-4°, 28014 Madrid. T: 91-5777140, F: 91-5768858

Maldonado, J.Mª., San Gerardo, 37-3° A, 28035 Madrid. T: 91-3163402

Manzanares, J.L., Agua y Estructuras, S.A. (Ayesa), Pabellón de Checoslovaquia, Isla de la Cartuja, 41092 Sevilla. T: 95-4461300

Manzanas, J., Laboratorio de Geotecnia, Alfonso XII, 3 y 5, 28014 Madrid. T: 91-3357300, F: 91-3357322

Mañueco, Mª.G., INTECSA, c/Orense, 70-5° Dcha., 28020 Madrid. T: 91-5832810

Marrero, P.M., Consulting Engineer, Dr. Apolinario Macias, 35, 35011 Las Palmas de Gran Canarias. T: 928-257609, F: 928-250588

Marsal, R., José Abascal, 36-5° Izda., 28003 Madrid. T: 91-4461161

Martín, T., ASEFA, S.A., Orense, 58-7 A y B, 28020 Madrid. T: 91-5561002

Martínez Benito, A., Santa Brígida, 20, Urb. Los Olmos, Pta. 7-2°D, Majadahonda, 28220 Madrid

Martínez Carretero, R., Kronsa Internacional, General Ramirez de Madrid, 8-10-3°, 28029 Madrid. T: 91-4252890

Martínez-Cattaneo, F., Ingeniero de Caminos, Canales y Puertos, Alvarez de Baena, 5, 28006 Madrid. T: 91-5643022, 91-5628161(part.), F: 91-5634532

Martínez Santamaría, J.M., Laboratorio de Geotecnia-CEDEX, Alfonso XII, 3, 28014 Madrid. T: 91-3357300, F: 91-3357322

Martínez Torres, A., Grupo Terratest Cimyson, I.CO.S, S.A., Alcalá, 65-4°, 28014 Madrid. T: 91-5777140, F: 91-5768858

Matesanz, P., Dr. Ingeniero de Caminos, Canales y Puertos, c/Nuñez de Balboa, 91-7° Izda., 28006 Madrid

Maza, I., Kronsa Internacional, General Ramirez de Madrid, 8-10-3°, 28029 Madrid. T: 91-4252890

Megia, M., Azhuma, 10-3°, 18005 Granada. T: 958-261832
Minguez, J.A., Cimentaciones Especiales Rodio, S.A., Paseo de la Castellana 139-6°, 28046 Madrid. T: 91-5624610
Minguillon, F.J., REGSA, c/General Brito, 6-5ª planta, 25007 Lérida. T: 973/222838
Miranda, J., Civil Mining Engineer. Geotechnical Consulting. Labein, Cuesta de Olabeaga, 16, 48013 Bilbao. T: 94-4892400, F: 94-4411749
Molinero, J., Grupo Terratest Cimyson, I.CO.S, S.A., Alcalá, 65-4°, 28014 Madrid. T: 91-5777140, F: 91-5768858
Montemayor, J., Grupo Terratest Cimyson, I.CO.S, S.A., Alcalá, 65-4°, 28014 Madrid. T: 91-5777140, F: 91-5768858
Monte Ramos del, E., Técnica y Proyectos, S.A., Pza. Liceo, 3, 28043 Madrid. F: 91-3881686
Monte Saez, J.L., Civil Engineer, M.S., Ph.D., Laboratorio de Geotecnia-CEDEX, Alfonso XII, 3, 28014 Madrid. T: 91-3357358, F: 91-3357322
Mora-Rey, C., Grupo Terratest Cimyson, I.CO.S, S.A., Alcalá, 65-4°, 28014 Madrid. T: 91-5777140, F: 91-5768858
Moreno Galdo, J., c/Castillo de Utrera, 2-Esc 1-Izda, 41013 Sevilla
Moreno García, R., Kronsa Internacional, General Ramirez de Madrid, 8-10-3°, 28029 Madrid. T: 91-4252890
Moreno Robles, J., Valle de Pas, 24, 2ºE, 28023 Madrid. T: 91-3573208
Mozas, D., Grupo Terratest Cimyson, I.CO.S, S.A., Alcalá, 65-4°, 28014 Madrid. T: 91-5777140, F: 91-5768858
Mozas, F.J., Grupo Terratest Cimyson, I.CO.S, S.A., Alcalá, 65-4°, 28014 Madrid. T: 91-5777140, F: 91-5768858
Muelas, A., Santa Engracia, 19-6° Izda., 28010 Madrid. T: 91-4477132
Muñoz Armañac, J.J., Bauer Cimentaciones y Equipos, S.A., c/de la Pelaya, 6, Pol. Ind, rio de Janeiro, 28110 Algete, Madrid. T: 91-4237500
Muñoz Díaz, V., Cimentaciones Especiales Rodio, S.A., Paseo de la Castellana 139-6°, 28046 Madrid. T: 91-5624610
Murillo, T., Sondeos Inyecciones y Trabajos Especiales, Ramón Power, 58, 28043 Madrid. T: 91-7473444
Muro, M., Grupo Terratest Cimyson, I.CO.S, S.A., Alcalá, 65-4°, 28014 Madrid. T: 91-5777140, F: 91-5768858
Muzás, F., c/Pedro Texeira, 16-6ºA, 28020 Madrid
Nageli, C., Sant Salvador, 3, E-43500 Torotosa, Tarragona
Narbaiza, E., Ingeniero de Caminos, Canales y Puertos, Hirigintza, S.A., Pescadores de Terranova, 21-Bajo, 20014 San Sebastian (Guipuzcoa). T: 943-470904, F: 943-468083
Navarro, V., ETSICCP, de la Coruña, Campos de Elviña, s/n, 15192 La Coruña. T: 981-167000
Necso Entrecanales Cubiertas, Avda. Europa, 18, Parque Empresarial La Moraleja, 28108 Alcobendas, Madrid. T: 91-6632900
Nestares, E., Icyfsa, Travesia de la Almenara, 9, Tres Cantos, 28760 Madrid. T: 91-8031301
Nieto, F., Marqués de Lozoya, 3 Esc. C, 2ºB, 28007 Madrid
Nocito J., Geonoc, Martin Iriarte, 7, Las Matas, 28290 Madrid. T: 91-6301313
Novillo, A., Cimentaciones Especiales Rodio, S.A., Pº de la Castellana, 130-6ª planta, 28046 Madrid. T: 91-5624610
Olalla, C., Dr. Civil Engineer, Arganda, 24, 28005 Madrid. T: 91-4747784
Oliden, F.J., Via Augusta, 59, 08006 Madrid
Oliveros, M.A., Uriel y Asociados, S.A., c/Oslo 1, Clun azata-Somosaguas, Chalet 5, 28224 Pozuelo de Alarcón (Madrid). T: 91-3524820
Olmos, P.J., c/Lira, 2, 1ºB, 47003 Valladolid. T: 983-259913
Oñoro, J.M., Ingeniero de Caminos, Canales y Puertos, General Asensio Cabanillas, 35, 28003 Madrid. T: 91-5337034
Ordozgoiti, S., c/Juan Vigón, 9-3° I., 28003 Madrid. T: 91-5537980
Ortega, J., Grupo Terratest Cimyson, I.CO.S, S.A., Alcalá, 65-4°, 28014 Madrid. T: 91-5777140, F: 91-5768858
Ortuño, L., Consulting Engineer, c/o Uriel & Asociados, Oslo 1, Club Azata, chalet 5, 28224 Pozuelo de Alarcón (Madrid). T: 91-3524820, F: 91-3524980

Oteo, C., Prof. Dr. Ingeniero de Caminos, Canales y Puertos, E.T.S.I.C.C y P., Universidad Complutense, 28040 Madrid. T: 91-3453881, F: 91-3506088
Palma, J.H., Universidad Catolica de Valparaiso, Escuela de Ing. de Construccion, Avda. Brasil, 2147, Valparaiso, Chile
Pardo, F., Arrieta, 4, 28013 Madrid
Parrondo, J., Plaza do Olavide, 10-6ºF, 28010 Madrid
Pastor Pérez, M., Pº de la Castellana, 251, 28046 Madrid
Pedraza, L., DM Iberia, S.A., c/Pedro Muguruza, 8-2ºA, 28036 Madrid. T: 91-3503804
Pedrero, A., Maestro Ripoll, 3-2º Dcha., 28006 Madrid. T: 91-5615133
Peña, A., Batalla del Salado, 26-4°-4ª, 28045 Madrid
Pérez Arenas, R.A., Ingeniero de Caminos, Canales y Puertos, Soto Hidalgo, 10-4ºB, Urbanización Embajada, 28042 Madrid. T: 91-4350700, F: 91-5763285
Pérez Cecilia, R., c/Colombia, 2-3B, 47014 Valladolid. T: 983-473907
Pérez Ruiz, J.I., Avda. Conde Lumiares, 30-4ºD, 03010 Alicante. T: 965-252329
Pérez Sainz, A., Ofiteco, Guzman el Bueno, 133-1ª planta, Edificio Britania, 28003 Madrid. T: 91-3612433
Perucho, A., Laboratorio de Geotecnia, Alfonso XII, 3 y 5, 28014 Madrid. T: 91-3357300, F: 91.3357322
Pestaña, J.A., Grupo Terratest Cimyson, I.CO.S, S.A., Alcalá, 65-4°, 28014 Madrid. T: 91-5777140, F: 91-5768858
Picazo, I., Unidad de Carreteras de Albacete, c/Alcalde Conangla, 4, 02071 Albacete. T: 967-219604
Piferrer, J., Balmes, 18, 5ª-2ª, 08007 Barcelona. T: 93-4144446
Pilotes Posada, S.A., Carretera de Bayona, 44 interior, 36213 Vigo, Pontevedra. T: 986-293500
Pinillos, L.M., Juan Pantoja, 28-3ºE, 28039 Madrid. T: 91-5533026
Plaza, P., Arturo Soria, 353-4ºB, 28033 Madrid. T: 91-7669822
Polo, J., ATICA, S.L., Pº Fernando el Católico, 32, Pral. Dhca. 50009 Zaragoza. T: 976-559248, F: 976-567121
Porcellinis de, P., Cimentaciones Especiales Rodio, S.A., Pº de la Castellana 139-6°, 28046 Madrid. T: 91-5624610
Prada, L.C., c/Caballeros, 27-29 portal 1-6ºD, 15005 La Coruña. T: 981-153790
Prefabricados y Suelos Reforzados, Orense, 10, 28020 Madrid. T: 91-5556266
Prieto, L., Cimentaciones Especiales Rodio, S.A., Pº de la Castellana 139-6°, 28046 Madrid. T: 91-5624610
Puerta, F., Ingeniero de Caminos, Canales y Puertos, c/Ferrol del Caudillo, 3-9°, 28029 Madrid. T: 91-3233702
Quiles, E., Gines Navarro Construcciones, S.A., c/ Basauri, 5, 28023 Madrid. T: 91-3077944
Ramos, C., Grupo Terratest Cimyson, I.CO.S, S.A., Alcalá, 65-4°, 28014 Madrid. T: 91-5777140, F: 91-5768858
Rebollo, A., Raset, 25, 08021 Barcelona. T: 93-2013005
Reig, Mª.I., San Graciano, 7-21C. Esc. Dcha., 28026 Madrid. T: 91-4760850/91-6167062
Ripoll, J.L., Arturoa Soria, 201, 28043 Madrid
Rodríguez Ballesteros, F., c/Acanto, 22, 28045 Madrid
Rodríguez Monteverde, P., Oña, 109-8º-1, 28050 Madrid. T: 91-3839203
Rodríguez-Ortiz, J.Mª*+, Prof. Soil Mechanics. Consulting Engineer, Gamma, S.L., profesor Waksman, 5-9ºC, 28036 Madrid, T: 91-3453400, F: 91-3509171
Rodríguez Roa, F., Casilla 2983, RCH-Santiago, Chile
Rodríguez Sánchez, M., Gasómetro, 11 portal 8, 2ºC, 28005 Palma. T: 91-5345447
Rojo, J.L.+, Ingeniero de Caminos, Canales y Puertos. Dtor. Producción, Cimentaciones Especiales Rodio, S.A., Pº de la Castellana, 130-4ºP, 28046 Madrid. T: 91-5624610, F: 91-5613013
Romana, M+, Ingeotec, c/Ponzano, 54, 28003 Madrid. T: 91-4422288, F: 91-4416120
Romero Morales, E., Universidad Politecnica de Cataluña, Jordi Girona, 1-3 Módulo D-2, 09034 Barcelona. T: 93-4016888
Romero Ramos, E., Grupo Terratest Cimyson, I.CO.S, S.A., Alcalá, 65-4°, 28014 Madrid. T: 91-5777140, F: 91-5768858

Rossique, F., Conde de Cartagena, 5, 28007 Madrid. T: 91-5514412

Ruiwamba, F.J., Esteyco, S.A., Menendez Pidal, 17, 28036 Madrid. T: 91-3597878

Ruiz, F., Cimentaciones Especiales Rodio, S.A., P° de la Castellana, 130-6ª planta, 28046 Madrid. T: 91-5624610, F: 91-5613013

Sagaseta, C., La Mina, 52-A, Puente Arce, 39478 Cantabria. T: 942-275600, F: 942-201703

Sainz de, B., Grupo Terratest Cimyson, I.CO.S, S.A., Alcalá, 65-4°, 28014 Madrid. T: 91-5777140, F: 91-5768858

Sánchez Alciturri, J.M., Universidad de Cantabria, Hernán Cortés, 23-3° Izq., 39003 Santander. T: 942-226412

Sánchez Caro, F.J., Camino de Leganés, 19-2°A, Móstoles, 28931 Madrid. T: 91-6465365

Sánchez Díaz, G., Cimentaciones Especiales Rodio, S.A., P° de la Castellana, 130-6ª planta, 28046 Madrid. T: 91-5624610, F: 91-5613013

Sánchez Díaz, C.J., Polifelt Geosynthetics Iberia, S.L., Azalea, 1, Edif. B-2ª, 28109 Soto de la Moraleja, Madrid. T: 91-6506461

Sánchez García, J.F., Ancha, 32, 23400 Ubeda, Jaen. T: 953-751798

Sánchez del Río, F.J., Dragados y Construcciones, S.A., Avda. de Tenerife, 4-6, Edif. Flores 2ª plta., 28700 San Sebastian de Los Reyes, Madrid

Sánchez Lavín, R., Torrelaguna, 81-41C, 28043 Madrid

Sanroma, A., Cimentaciones Especiales Rodio, S.A., P° de la Castellana, 130-6ª planta, 28046 Madrid. T: 91-5624610, F: 91-5613013

Santamera, J.J., Ruiz Perelló, 3-4°B, 28028 Madrid. T: 91-3568094

Santos, A., Sándalo, 5-2ªG (La Piovera), 28042 Madrid

Sanz Lanzuela, D., Suelotest, S.L., Pi i Margal, 25, Esc. B-Entresuelo 1ª, 08024 Barcelona. T: 93-2841049

Sanz Saracho, J.Mª., Potosí, 4, 28016 Madrid. T: 91-4586732

Serrano González, A., Avda. de Baviera, 13-7° Izda., 28028 Madrid

Serrano Petterson, C., Civil Engineer. M.S. Geotechnical Eng., c/Jujuy, 1479-1°, 5000 Córdoba, Barrio Cofico, (Argentina)

Sierra, A., Grupo Terratest, Cimyson, I.CO.S, S.A., Alcalá, 65-4°, 28014 Madrid. T: 91-4237500

Simic, D., Director del Area de Ferrovial-Agroman, S.A., Principe de Vergara, 108, 28002 Madrid. T: 91-5869038

Sistemas de Cimentación, S.A., Via Augusta, 15-D112, 08006 Barcelona. T: 93-2173208

Site, c/Febrero, 36, 28022 Madrid. T: 91-7473444

Sola, P., Ingeniero de Caminos, Canales y Puertos, c/Hernani, 54-5°B, 28020 Madrid. T: 91-5720510

Soler, A., c/Nou, 22-2ª, 17600 Figueras, Gerona. T: 972-671191

Sopeña, L., Urb. Las Viñas, Bloque A-Portal 1, 3°C, 28220 Majadahonda, (Madrid)

Soriano, A., Hilarión Eslava, 27, 28015 Madrid, F: 91-5940595

Tamés, P., Diputación Foral de Guipuzcoa, Plaza de Guipuzcoa, s/n, 20004 San Sebastian. T: 943-423511

Tarraga, J.L., c/Tomás López Torregrosa, 11, 03002 Alicante. T: 96-5204080, F: 96-5666900

Tapia, J., Intemac, S.A., Monte Esquinza, 30-4°D, 28010 Madrid. T: 91-4105158

Tierra Armada, S.A., c/Melchor Fernández Almagro, 23, 28029 Madrid. T: 91-3239500, F: 91-3239511

Tissera, L.E., Ingeniero de Caminos. Jefe Dpto. Est. Geotécnicos, Tecnos, S.A., c/del Pino, 16, 28411 Moralzarzal, Madrid. T: 91-8578626

Torres, E., Cimentaciones y Sondeos, S.A., Alcalá, 65-4ª planta, 28014 Madrid. T: 91-5777150

Trebisa, Avda. Co-Princep Francées, 106, Parroquia d'Encamp, Principado de Andorra

Ucar, R., Apartado Postal 289, Mérida 5101ª, Estado de Mérida, Venezuela. T: 662970

Universidad de Almería, Biblioteca, Cañada de S. Urbano, s/n, 04071 Almeria. T: 950-215336

Universidad de Cádiz, Biblioteca del Campus de Algeciras, Avenida Ramon Pujol, s/n, 11002 Algeciras, Cádiz. T: 956-667649

Uriel Romero, S., Pedro Texeira, 18, 28020 Madrid

Valerio, J., Laboratorio de Geotecnia-CEDEX, Alfonso XII, 3, 28014 Madrid. T: 91-3357300, F: 91-3357322

Velasco, M., DM Iberia, S.A., Pedro Muguruza, 8-2°, 28036 Madrid. T: 91-3503804

Villar, Mª.V., CIEMAT-Técnicas Geológicas, Avda. Complutense, 22, 28040 Madrid. T: 91-6386610

Villarroel, J.Mª., Federico Rubio, 106, (Moderno 4A), 28040 Madrid

Vizcayno, J.Mª., Cimentaciones Especiales Rodio, S.A., P° Castellana, 130-6ª, 28046 Madrid. T: 91-5624610

Vorsevi, S.A., c/Cordel de Tomares, 2, La Pañoleta-Camas, 41900 Sevilla. T: 954-394305

Zapico, A., Tecnicas Especiales de Perforación, Torpederp Tucumán, 6 Bajo C, 28016 Madrid. T: 91-3459659

Zavala, R., Urb. Virgen de Iciar, 16-3°A, 28220 Majadahonda, Madrid. T: 91-3728433

Zeballos, M.E., c/Lavalleja, 2174, 5000, Barrio Alta, Córdoba, Argentina. T: 5451710855

Zegarra, J.V., Pontificia Universidad Católica del Peru, Clemente X, 110 Dpto. 701, Lima 17, Peru. T: 5114604510

SRI LANKA

Secretary: Dr. S.A.S. Kulathalika, National Building Research Organisation, 99/1 Jawatte Road, Colombo – 5, Sri Lanka.
T: 94 1 650657, F: 94 1 650622, E: sas@civil.mrt.ac.lk

Total number of members: 44

Arambepola, N.M.S.I., Mr., Head – Landslide Division, NBRO, 99/1 Jawatta Road, Colombo – 5

Bulathsinhala, V., Mr., No. 20, Rajagiriya Road, Rajagiriya

Deepthi, U., Dr. (Ms.), Senior Lecturer, Department of Civil Engineering, University of Peradeniya, Peradeniya

De Silva, S. Dr., Director, Soil Engineering and Deepwells Ltd., 736/1 A, Negambo Road, Mattumagala, Ragama

Edirisinghe, A.G.H.J., Dr., Senior Lecturer, Department of Civil Engineering, University of Peradeniya, Peradeniya. E: jayalath@civil.pdn.ac.lk

Fernando, H.Y., Mr., Deputy Director – RDA, No. 6, Sri Wijaya Road, Walana, Panadura

Fernando, B.P., Mr., 231, Delduwa, Wadduwa

Fernando, T.M., Mr., 802/1, Hill Street, Dehiwala

Fernando, T.R.P., Mr., No. 45, Circular Lane, Sapumal Place, Rajagiriya

Fernando, T.R.J., Mr., 45, Circular Lane, Sapumal Place, Rajagiriya

Fernando, G.S.K., Dr., Senior Lecturer, Department of Civil Engineering, University of Peradeniya, Peradeniya. E: sarath@soils.pdn.ac.lk

Godage, D., Mr., Chief Engineer – Ports Authority, 96, Jayanthipura Road, Battaramulla

Gunasekara, K.K.J., Mr., No. 63, Parakum Mawatha, Banadrawatta, Gampaha

Gunatilaka, I.R.P., Mr., 21/2, Kurunduhena Road, Weediyawatta, Udugampola

Hathurusinghe, R.L., Ms., 400/52 A, Mahaweli Housing Complex, Sarana Road, Colombo – 7

Herath, N., Mr., Director, Soil and Foundation (Pvt) Ltd., 950 H, Pannipitiya Road, Battaramulla

Jagath, H.P., Mr., Rupasinghe, 10/7, Kawiraja Mawatha, Wallawaththa, Wekada, Panadura

Jayasinghe, J.M.L.R., Mr., 53, Wathumulla, Udugampola

Jayawardana, U. De S., Dr., Senior Lecturer, Department of Civil Engineering, University of Peradeniya, Peradeniya

Jayawardane, W.A.D.K., Mr., 132/10, Nawala Road, Nugegoda

Karunarathna, S.A., Mr., 43/120, Poorwarama Mawatha, Colombo – 5. E: stems1@sltnet.lk

Kulathilaka, S.A.S., Dr., Senior Lecturer, Department of Civil Engineer, University of Moratuwa, Moratuwa. E: sas@civil.mrt.ac.lk

Kurukulasuriya, L.C., Dr., Department of Civil Engineering, University of Peradeniya, Peradeniya. E: chandana@soils.pdn.ac.lk

Landage, P.S., Mr., C/o J.T. Rajapaksa, Ambalanwatta, Pallewela

Mallawarachchi, D.P., Mr., General Manager – RDA, 84, 5th Lane, Colombo – 3

Peiris, T.A., Dr., Senior Lecturer, Department of Civil Engineering, University of Moratuwa, Moratuwa. E: ashok@civil.mrt.ac.lk

Perera, K.W., Mr., Deputy Director – Irrigation Department, 97/1, Madiwela Road, Embuldeniya, Nugegoda

Perera, M.G.E., Mr., 27, Mihindu Mawatha, Gampaha

Puswewala, U.G.A., Dr., Senior Lecturer, Department of Civil Engineering, University of Moratuwa, Moratuwa. E: ugap@civil.mrt.ac.lk

Rajapaksa, L., Mr., 154 A, Stanley Thilakarathna Mw., Nugegoda

Rajapaksa, J.T., Mr., Ambalanwatta, Pallewela

Rathnaweera, H.G.P.A., Dr., Senior Lecturer – Open University of Sri lanka, 7/2, Muhandiram E D Dabare Road, Colombo – 5

Sahabandu, K.L.S., Mr., Senior Design Engineer, C.E.C.B, 594, Galle Road, Colombo – 3

Selvarajah, S., Mr., F/8, Anderson Flats, 215, Park Road, Colombo – 5

Senanayaka, K.W.W.K., Mr., 280/1, Kandy Road, Wattegama

Senerath, R.M.A., Mr., 245/2, Pahalawela Road, Pelawatta, Battaramulla. E: stems1@sltnet.lk

Seneviratne, H.N., Prof., Department of Civil Engineering, University of Peradeniya, Peradeniya. E: nimals@soils.pdn.ac.lk

Sumanarathna, I.H.D., Mr., 46, School Lane, Nawala, Rajagiriya

Tennekoon, B.L., Prof., Department of Civil Engineering, University of Moratuwa, Moratuwa. E: blt@civil.mrt.ac.lk

Thawfeek, M.A., Mr., 44/28, Pamandada Road, Colombo – 6

Theivasagayam, K., Mr., No. 12, Daniel Avenue, Off Prathiba Road, Narahenpita, Colombo – 5

Tilakasiri, H.S., Dr., Senior Lecturer, Department of Civil Engineering, University of Moratuwa, Moratuwa. E: saman@civil.mrt.ac.lk

Viraj, A.A.D., Mr., No. 1072/1, Liyanagoda, Pannipitiya

Wijesundara, A.W., Mr., 49/4, Railway Avenue, Nugegoda. E: stems1@sltnet.lk

SUDAN/SOUDAN

Secretary: Dr. Ahmed Mohamed Elshareif, Sudanese Society for Soil Mechanics and Geotechnical Engineering (SSSMGE), Building and Road Research Institute, University of Khartoum, P.O. Box 321, Sudan. E: asharief@hotmail.com

Total number of members: 33

Abou, S.H., Dr.
Abd-Elnabi, O.
Abd-Elrahman, E.
Abdo, A.M.
Abou, S.A.-E.M.
Ali, E.M., Dr.
Ali, G.A.-A.
Ahmed, E.O., B.R.R.I, P.O. Box 321. E: elfatoman@usa.net
Elamin, A.-E.M., Dr.
Elmanan, W.A.
Ebrahim, A.B.M.
Elsawi, O.
Elhassan, E.M., Dr.
Elobeid, H., Dr.
Eltaib, E.M., Dr.
Fadel, M.O., Dr.
Gassim Elseid, K.M.
Hamadto, M.E. E: hamadto@arriyadh.net
Mustafa, E.

Mohamedzein, Y.E.A., Dr., Building and Road Research Institute University of Khartoum, P.O. Box 321, Khartoum
Mohamedzein, A.E.M., Dr. B.R.R.I., University of Khartoum, P.O. Box 321, Khartoum
Mohamed, H.E.
Mohamed, O.E.
Mokhtar, M.
Mostafa, A.E., Dr.
Mohamed, A.A.
Omer, O.M., Dr., Univrsity of Khartoum, Faculty of Eng, P.O. Box 321, Khartoum
Osman, M.A., Dr., Building and Road Research Institute, University of Khartoum, P.O. Box 321, Sudan. E: drmao@hotmail.com
Osman, M.K.
Omer, B.A.-A.
Saad, A.-A.B.
Yousif, H.O.
Yamen, T.

SWEDEN

Secretary: Ms Christina Berglund, Swedish Geotechnical Society, SE-581 93 Linköping, Sweden.
T: 46 13 20 18 67, F: 46 13 20 19 14, E: info@sgf.net

Total number of members: 750

Aarnio, T., M.Sc., Turun Viatek Oy, Slottsgatan 3 a B, FI-20100 ÅBO. T: +358-2-2733000, E: tomi.aarnio@viatek.fi
Abedinzadeh, R., Geotechnical Eng., Helsingborgsgatan 15, 2 tr, 164 78 KISTA
Adding, L., Eng., Myrvägen 5, 184 38 ÅKERSBERGA
Adestam, L., M.Sc., J&W Samhällsbyggnad, Box 13052, 600 13 NORRKÖPING. T: 011-139530, F: 011-124552, E: lennart.adestam@jw.se
Adolfsson, K., Eng., Ingenjörsfirma KADO, Tvetgatan 117, 442 33 KUNGÄLV. T: 0303-12587, F: 0303-12587, E: kado@gamma.telenordia.se
Ahlqvist, A.T., Tråggränd 49, 906 26 UMEÅ
Albertsson, A., M.Sc., Skanska Teknik AB, Lilla Bommen 2, 405 18 GÖTEBORG. T: 031-7711358, F: 031-153815, E: annelie.albertsson@teknik.skanska.se
Albertsson, B., Eng., J&W Samhällsbyggnad, 121 88 STOCKHOLM-GLOBEN. T: 08-6886421, F: 08-6886914, E: bjorn.albertsson@jw.se
Albertsson, K., Banverket, Box 43, 971 02 LULEÅ. T: 010-2012526
Alén, C., Head of Office, Swedish Geotechnical Institute, Chalmers Vasa, hus 5, 412 96 GÖTEBORG. T: 031-7786565, F: 031-7786575, E: claes.alen@swedgeo.se
Alenius, U., Civil Eng., Tyréns Infrakonsult AB, 118 86 STOCKHOLM. T: 08-4290000, F: 08-6448850, E: ulf.alenius@tyrens.se
Allard, K., Eng., Geometrik i Stockholm AB, Box 4010, 102 61 STOCKHOLM. T: 08-7430900, F: 08-7430908, E: kent_allard@geometrik.se
Allenius, I., B.Sc., Golder Grundteknik KB, Box 20127, 104 60 STOCKHOLM. T: 08-50630600, F: 08-50630601, E: torbjorn_allenius@golder.se
Alm, J., Geologist, KM Geoteknik, Rullagergatan 6, 415 26 GÖTEBORG. T: 031-7272500, F: 031-7272501, E: johan.alm@km.se
Alskär, J., J&W Samhällsbyggnad, 121 88 STOCKHOLM-GLOBEN. T: 08-6886000, F: 08-6886914, E: john.alskar@jw.se
Alström, J., M.Sc., Scandiaconsult Sverige AB, Box 454, 851 06 SUNDSVALL. T: 060-663600, F: 060-614984, E: jambns@scc.se
Alte, B., M.Sc., Bo Alte Geosenior, Måns Torbjörnsvägen 9, 417 29 GÖTEBORG. T: 031-239291, F: 031-239291
Alvarsson, T., Eng., Swedish National Road Administration, Box 601, 301 16 HALMSTAD. T: 035-151565, F: 035-128197, E: tore.alvarsson@vv.se
Anderberg, M., M.Sc., Per Aarsleff A/S, Postboks 1216, DK-4000 ROSKILDE. T: +45-46-163333, F: +45-46-163549
Anderson, T., Design Coordinator, Swedish National Rail Administration, Gyllerogatan 1, 233 51 SVEDALA. T: 040-406844, F: 040-406898, E: torbjorn.anderson@svedala.com
Andersson, B., Eng., VBB VIAK AB, Box 34044, 100 26 STOCKHOLM. T: 08-6956000, F: 08-6956010
Andersson, C.-A., M.Sc., c/o Björn Corlin, VBB Anläggning AB, Box 34044, 100 26 STOCKHOLM. T: 08-6956000, F: 08-6956060, E: carl_andersson@interar.com.ar
Andersson, H., M.Sc., Sycon Teknikkonsult AB, Österängsvägen 2 A, 554 63 JÖNKÖPING. T: 036-342886, F: 036-342899
Andersson, J.-E., Geologist, Tyréns Infrakonsult AB, 118 86 STOCKHOLM. T: 08-4290000, F: 08-4290020, E: jan-erik.andersson@tyrens.se
Andersson, L.J.E, M.Sc., Vägverket Konsult, Box 1062, 551 10 JÖNKÖPING. T: 036-151457, E: lars-vvk.andersson@vv.se

Andersson, M., Byggindustrin, Narvavägen 19, 114 60 STOCKHOLM
Andersson, P., Civil Engineer, Skanska Teknik AB, 405 18 GÖTEBORG. T: 031-7711000, F: 031-153815, E: pia.andersson@teknik.skanska.se
Andersson, T., Byggförstärkarna, Aröds Industriväg 74, 422 43 HISINGS BACKA. T: 031-518904, F: 031-229794
Andersson, Ö., Eng., Swedish National Road Administration, Box 4202, 171 04 SOLNA. T: 08-7576600, F: 08-283707, E: osten.andersson@vv.se
Andréasson, A., Civil Eng., J&W, Samhällsbyggnad, Rullagergatan 6, 415 26 GÖTEBORG. T: 031-7272795, F: 031-7272501, E: annika.andreasson@jw.se
Andreasson, B.O., M.Sc., Swedish Fed of Architects & Consulting Engineers, Box 16105, 103 22 STOCKHOLM. T: 08-7626702, F: 08-7626710, E: bengt.o.andreasson@ai-foretagen.se
Andréasson, B., Senior Consultant, J&W, Box 382, 401 26 GÖTEBORG. T: 031-614300, F: 031-614370, E: bo.andreasson@jw.se
Andrén, A., M.Sc., Banverket, 781 85 BORLÄNGE. T: 0243-445643, F: 0243-445617, E: anna.andren@banverket.se
Andrén, S., M.Sc., VBB VIAK AB, Box 34044, 100 26 STOCKHOLM. T: 08-6956259, F: 08-6956240, E: susanna.andren@sweco.se
Aradi, J., M.Sc., Scandiaconsult Sverige AB, Box 5343, 402 27 GÖTEBORG. T: 031-3353345, F: 031-400571, E: JAIMVG@scc.se
Aréen, L.G., Jacobson & Widmark, Byggmästargatan 5, 803 24 GÄVLE. T: 026-663550, F: 026-663560
Arnell, V., Ph.D., Vårgårdsvägen 28, 582 70 LINKÖPING
Arnér, E., Scandiaconsult Sverige AB, Box 4205, 102 65 STOCKHOLM. T: 08-6156318, F: 08-7021934, E: erik.arner@scc.se
Aronsson, I., M.Sc., Nordic Construction Company AB, 405 14 GÖTEBORG. T: 031-7715410, F: 031-151188, E: inge.aronsson@ncc.se
Aronsson, S., Geotechnician, VBB VIAK AB, Box 34044, 100 26 STOCKHOLM. T: 08-6956557, E: stefan.aronsson@sweco.se
Arteus, M., M.Sc., Golder Grundteknik KB, Box 20127, 104 60 STOCKHOLM. T: 08-50630623, F: 08-50630601, E: martin_arteus@golder.se
Askmar, B., M.Sc., GF Konsult AB, Box 5056, 402 22 GÖTEBORG. T: 031-3355000, F: 031-3358955, E: bengt.askmar@gfkonsult.se
Axelsson, G., Geotechnical Eng., Skanska Teknik AB, 169 83 SOLNA. T: 08-50435380, F: 08-7536048, E: Gary.Axelsson@teknik.skanska.se
Axelsson, J., M.Sc., J&W Samhällsbyggnad, Box 382, 401 26 GÖTEBORG. T: 031-614300, F: 031-614370, E: jonas.axelsson@jw.se
Axelsson, K., M.Sc., Skanska Berg och Bro, 169 83 SOL NA. T: 08-504 352 70, F: 08-753 47 90, E: karin.axelsson@skanska.se
Axelsson, K., Professor, Ingenjörshögskolan i Jönköping, Box 1026, 551 11 JÖNKÖPING. T: 036-157535, F: 036-302884, E: axke@ing.hj.se
Axelsson, M., M.Sc., Skanska Teknik AB, 169 83 SOLNA. T: 08-50435338, F: 08-7536048, E: morgan.axelsson@teknik.skanska.se
Back, B., M.Sc., SwedPower AB, Box 800, 771 28 LUDIKA. T: 0240-87200, F: 0240-87350, E: birger.back@swedpower.vattenfall.se
Bahrekazemi, M., Ph.D. Student, KTH Kungliga Tek-niska Högskolan, 100 44 STOCKHOLM. T: 08-7907925, F: 08-7907928, E: kazemi@ce.kth.se

Baker, S., Ph.D. Student, Chalmers University of Technology, 412 96 GÖTEBORG. T: 031-7722119, F: 031-7722070

Bard, G., J&W Samhällsbyggnad, 121 88 STOCKHOLM-GLOBEN. T: 08-6886423, F: 08-6886914, E: goran.bard@jw.se

Barth, C., Tyréns Infrakonsult AB, Box 8062, 700 08 ÖREBRO. T: 019-174435, F: 019-183710, E: christer.barth@tyrens.se

Bengtsson, J., Eng., Skanska, Box 112, 234 22 LOMMA. T: 040-144988, F: 040-411630, E: Jim.Bengtsson@Skanska.se

Bengtsson, P.-E., M.Sc., PEAB Sverige AB, Box 1934, 581 18 LINKÖPING. T: 013-249934, F: 013-249930, E: per-evert.bengtsson@peab.se

Benkert, J.-P., Geologist, Dr. Benkert AG, Zähringerstrasse 12, CH-3315 BÄTTERKINDEN. T: +41-32-6653737, F: +41-32-6653737

Bennet, T., M.Sc., Tyréns Infrakonsult AB, Drottninggatan 62, 252 21 HELSINGBORG. T: 042-289820, F: 042-289816, E: tomas.bennet@tyrens.se

Berg, J., Eng., J&W Samhällsbyggnad, Byggmästargatan 5, 803 24 GÄVLE. T: 026-663550, F: 026-663560, E: jerry.berg@jw.se

Berg, K.-O., M.Sc., KM Geoteknik, Box 758, 851 22 SUNDSVALL. T: 060-149564, F: 060-159738, E: kjell-ola.berg@jw.se

Bergdahl, T., Eng., Harpungränd 36, 175 47 JÄRFÄLLA

Bergdahl, U., Chief Engineer, Swedish Geotechnical Institute, 581 93 LINKÖPING. T: 013-201835, F: 013-201914, E: ulf.bergdahl@swedgeo.se

Berger, H., M.Sc., Swedish National Rail Administration, Box 1070, 172 22 SUNDBYBERG. T: 08-7623452, F: 08-7623388, E: hakan.berger@banverket.se

Bergfelt, A., Prof., Chalmers University of Technology, 412 96 GÖTEBORG. T: 031-7721000

Berggren, B., Head of Department, J&W Samhällsbyggnad, 121 88 STOCKHOLM-GLOBEN. T: 08-6886424, 013-70075, F: 08-6886914, E: bo.berggren@jw.se

Berglund, C., M.Sc., Swedish Geotechnical Institute, 581 93 LINKÖPING. T: 013-201867, F: 013-201914, E: christina.berglund@swedgeo.se

Berglund, H., M.Sc., Tyréns Infrakonsult AB, 118 86 STOCKHOLM. T: 08-4290303, F: 08-4290020, E: helena.berglund@tyrens.se

Bergman, E., Consulting Eng., Odenvägen 10, 181 32 LIDINGÖ

Bergman, J., B.A., MB Envirotech AB, Rullagergatan 9, 415 26 GÖTEBORG. T: 031-7275190, F: 031-7275191, E: jonny@mb-envirotech.com

Bergman, M., M.Sc., Ringvägen 45, 133 35 SALT-SJÖBADEN. T: 08-7175505, F: 08-7177365, E: magnus.bergman@mbox3.swipnet.se

Bergström, J., M.Sc., Göteborgs Gatubolag AB, Box 1086, 405 23 GÖTEBORG. T: 031-628357, F: 031-468825, E: anders.bergstrom@gatubolaget.goteborg.se

Bergwall, A., M.Sc., Golder Grundteknik KB, Box 20127, 104 60 STOCKHOLM. T: 08-50630624, F: 08-50630601, E: annika_bergwall@golder.se

Bernander, S., Chief Eng., CONGEO AB, Tegelformsgatan 10, 431 36 MÖLNDAL. T: 031-871104, F: 031-879532, E: stig.bernander@telia.com

Bernstone, C., Dr. Sc., Vattenfall Utveckling AB, 814 26 ÄLVKARLEBY. T: 026-83827, F: 026-83630, E: christian.bernstone@utveckling.vattenfall.se

Berntson, J.A., Head, Geotechn. Dept, Tyréns Infrakonsult AB, Stora Badhusgatan 6, 411 21 GÖTEBORG, E: jan.berntson@tyrens.se

Bjerin, L., M.Sc., Swedish National Road Administration, Box 4202, 171 04 SOLNA. T: 08-7576600, F: 08-283080, E: lars.bjerin@vv.se

Bjulemar, L., Geophysicist, Scandiaconsult Sverige AB, Box 1932, 791 19 FALUN. T: 023-84158, F: 023-19983, E: lars.bjulemar@scc.se

Björk, L., Eng., Swedish National Road Administration, Kaserngatan 10, 551 91 JÖNKÖPING. T: 036-192146, F: 036-192018, E: lars.bjork@vv.se

Blom, L., M.Sc., Swedish Geotechnical Institute, Chalmers Vasa, hus 5, 412 96 GÖTEBORG. T: 031-7786562, F: 031-7786575, E: lennart.blom@swedgeo.se

Blomberg, S., M.Sc., Statens geotekniska institut, Hospitalsgatan 16 A, 211 33 MALMÖ. T: 040-356771, F: 040-938685, E: sara.blomberg@swedgeo.se

Blomqvist, L., Eng., GEO-STUDIO, Ågatan 17, 614 34 SÖDERKÖPING. T: 0121-15321, F: 0121-15107, E: lars.blomqvist@mailbox.swipnet.se

Blumenberg, J., Consulting Eng., Statens geotekniska institut, Fabriksgatan 32, 851 71 SUNDSVALL. T: 060-162154, F: 060-162151, E: jan.blumenberg@swedgeo.se

Bodare, A., D.Sc., Royal Institute of Technology, 100 44 STOCKHOLM. T: 08-7908024, F: 08-7907928, E: bodare@ce.kth.se

Bodin, C.-A., Krokvägen 11, 131 41 NACKA

Bohlin, Å., Marketing Director, Kernfest-Webac AB, Box 22, 125 21 ÄLVSJÖ. T: 08-4471050, F: 08-4471066, E: ashland-suedchemie@kernfest-webac.se

Bohm, H., Chief, AB Stockholm Konsult, Box 9611, 117 91 STOCKHOLM. T: 08-7858505, F: 08-7858504, E: hakan.bohm@stockholmkonsult.se

Bono, N., M.Sc., c/o Alfredsson, Båtsman Stens väg 70, 163 41 SPÅNGA, E: nabono@aol.com

Book, J., Geologist, KM Geoteknik, Box 17, 781 21 BOR-LÄNGE. T: 0243-88180, F: 0243-84250, E: jenny.book@km.se

Boox, C., Bjerking Ingenjörsbyrå AB, Box 2006, 750 02 UPPSALA. T: 018-651159, F: 018-710485, E: connie.boox@bjerking.se

Borchardt, P., M.Sc., Citytunneln, Box 4012, 203 11 MALMÖ. T: 040-321443, F: 040-321500, E: peter.borchardt@citytunneln.com

Borgstig, L.-E., VBB VIAK, Box 50120, 973 24 LULEÅ. T: 0920-35500, F: 0920-26376

Borgström, Å., M.Sc., Ljungåsvägen 1 B, 296 38 ÅHUS

Bramstång, O., Eng., Kärret 1519, 440 07 SJÖVIK

Brandhammar, U., UBA-Kontroll AB, Gilleskroken 10, 222 47 LUND. T: 046-147895, F: 046-125905

Brandt, S., Eng., Gator och Grönt i Linköping AB, Box 1937, 581 18 LINKÖPING. T: 013-208383, F: 013-208016, E: soren.brandt@tekniskaverken.se

Bredenberg, H., Ph.D., Bredenberg Geoteknik HB, Igeldammsgatan 22 C, 112 49 STOCKHOLM. T: 08-6505134, F: 08-6505137, E: hakan.bredenberg@telia.com

Brink, L.-G., AB Jacobson & Widmark, Box 502, 901 09 UMEÅ. T: 090-717300, E: lars-goran.brink@jw.se

Brink, R., M.Sc., Drottningvägen 24, 181 32 LIDINGÖ. T: 08-7652086, F: 08-7673023, E: rolf.brink@mbox.pi.se

Broms, B.B, Professor, Strindbergsgatan 36, 115 31 STOCKHOLM. T: 08-6643937, F: 08-6644942, E: b.broms@geo.se

Brorsson, I., Office Manager, Swedish Geotechnical Institute, Hospitalsgatan 16 A, 211 33 MALMÖ. T: 040-356772, F: 040-938685, E: inge.brorsson@swedgeo.se

Brorsson, J., Geophysicist, Scandiaconsult Sverige AB, Stora Varvsgatan 11 N:3, 211 19 MALMÖ. T: 040-105400, F: 040-126650, E: jorgen.brorsson@scc.se

Bylin, K., Hudiksvalls Kommun, Box 149, 824 80 HUDIKSVALL. T: 0650-19498, F: 0650-38155, E: kalle.bylin@hudiksvall.se

Bågevik, K., Nitro Consult AB, Björklund 1631, 762 91 RIMBO. T: 08-6814323, F: 08-6814336, E: kare.bagevik@eu.dynoasa.com

Båtelsson, O., Civil Engineer, VBB VIAK AB, Box 34044, 100 26 STOCKHOLM. T: 08-6956349, F: 08-6956350, E: olle.batelsson@sweco.se

Bähr, G., FB Engineering, Box 12076, 402 41 GÖTEBORG. T: 031-7751000

Böhmer, W., Eng., Tyréns Projektledarna AB, 118 86 STOCKHOLM. T: 08-4290000, F: 08-4290010, E: wolf.bohmer@tyrens.se

Börgesson, L., Dr., Clay Technology, Ideon, 223 70 LUND. T: 046-2862573, F: 046-134230, E: lb@claytech.se

Börjeson, H., Eng., Marknadsundersökningar på Ingarö AB, Källvreten, 134 65 INGARÖ

Cadling, L.O.T., Consulting Eng., Furusundsgatan 16, 115 37 STOCKHOLM. E: lyman.cadling@swipnet.se

Carlberg, G., Skanska Berg och Bro AB, 163 89 SOLNA. E: goran.carlberg@skanska.se

Carling, M., Research & Consulting Eng., Swedish Geotechnical Institute, 581 93 LINKÖPING. T: 013 201805, F: 013-201909, E: maria.carling@swedgeo.se

Carlring, J.-E., M.Sc., J&W Samhällsbyggnad, Box 8094, 700 08 ÖREBRO. T: 019-178950, F: 019-133200, E: jan-eric. carlring@jw.se

Carlsson, B., M.Sc., Miljoteknik, Repslagaregatan 19, 582 22 LINKÖPING. T: 013-357273, F: 013-357271, E: bo.carlsson @envipro.se

Carlsson, E., M.Sc., KM Geoteknik, Box 6540, 906 12 UMEÅ. T: 090-103282, F: 090-103158, E: Eric.Carlsson@ km.se

Carlsson, K.-A., Eng., Rådhusgatan 54, 571 32 NÄSSJÖ

Carlsson, M., M.Sc., Skanska Sverige AB Division Berg och Bro, 169 83 SOLNA. T: 08-504 361 54, F: 08-753 47 90, E: mats.r.carlsson@skanska.se

Carlsson, P., Eng., VBB VIAK AB, Box 34044, 100 26 STOCKHOLM. T: 08-6956355, F: 08-6956360, E: per. carlsson@sweco.se

Carlsson, R., M.Sc., Färgspelsgatan 86, 421 63 VÄSTRA FRÖLUNDA

Carlsten, P., M.Sc., Scandiaconsult Sverige AB, Teknikringen 1C, 583 30 LINKÖPING. T: 013-212260, F: 013-210460, E: peter.carlsten@scc.se

Cedersjö, H., M.Sc., SWEPRO Training AB, Brovägen 1, 182 76 STOCKSUND. T: 08-4465484, F: 08-7534295, E: helena.cedersjo@swepro.se

Cederwall, K., Professor, Royal Institute of Technology, 100 44 STOCKHOLM. T: 08-7908053, F: 08-208946, E: klasc@aom.kth.se

Christensen, G., B.A., J&W Samhällsbyggnad, 121 88 STOCKHOLM-GLOBEN. T: 08-6886425, F: 08-6886914, E: goran.christensen@jw.se

Claesson, P., M.Sc., Skanska Teknik AB, 405 18 GÖTEBORG. T: 031-7711000, F: 031-153815, E: peterc@vsect. chalmers.se

Clifford, F., Skanska Teknik AB, 169 83 SOLNA. T: 08-50435317, F: 08-7536048, E: fredrik.clifford@teknik. skanska.se

Corlin, B., Principal Engineer Hydropower, VBB VIAK AB, Box 34044, 100 26 STOCKHOLM. T: 08-6956000, F: 08-6956060, E: bjorn.corlin@sweco.se

Creutz, M., M.Sc., Golder Grundteknik KB, Box 20127, 104 60 STOCKHOLM. T: 08-50630600, F: 08-50630601

Dahlberg, A., M.Sc., Tyréns Infrakonsult AB, Drottninggatan 62, 252 21 HELSINGBORG. T: 042-289818, F: 042-289816, E: anders.dahlberg@tyrens.se

Dahlberg, H., M.Sc., J&W Samhällsbyggnad, Byggmästargatan 5, 803 24 GÄVLE. T: 026-663572, F: 026-663560, E: helena.dahlberg@jw.se

Dahlberg, R., Senoir Principal Engineer, Det Norske Veritas AS, Veritasveien 1, NO-1322 HÖVIK. T: +47-67577450, F: +47-67579911, E: rune.dahlberg@dnv.com

Dahlin, T., Ph.D., Lund Institute of Technology, Box 118, 221 00 LUND. T: 046-2229658, F: 046-2229127

Dannewitz, N., M.Sc., Golder Grundteknik KB, Box 20127, 104 60 STOCKHOLM. T: 08-50630626, F: 08-50630601, E: niklas_dannewitz@golder.se

Davidsson, L., M.Sc., KM Miljöteknik AB, Kastanjeallen 1, 302 31 HALMSTAD. T: 035-219025, F: 035-219902, E: lars.davidsson@km.se

de Caprona, G., Ph.D., Ingenjörsfirman Geotech AB, Datavägen 53, 436 32 ASKIM. T: 031-289920, F: 031-681639, E: caprona@geotech.se

Dehlbom, B., M.Sc., Scandiaconsult Sverige AB, Box 1932, 791 19 FALUN. T: 023-84170, F: 023-19983, E: bjorn. dehlbom@scc.se

Delblanc, F., Fil. Lic., Delblanc Miljöteknik AB, Alnarpsvägen 54, 232 53 ÅKARP. T: 040-464310, F: 040-464312, E: fredrik@dmab.com

Delfs, W., Civ. Eng., KM Geoteknik, Rullagergatan 6, 415 26 GÖTEBORG. T: 031-7272500, F: 031-7272501

Diener, A., A D Markkonsult, Klostergatan 35, 567 92 VAGGERYD. T: 0393-22109, F: 0393-22109, E: admarkkonsult@ swipnet.se

Du, T.K., D.Sc., Scandiaconsult Sverige AB, Box 5343, 402 27 GÖTEBORG. E: kdtmvg@scc.se

Dueck, A., Doctoral student, Clay Technology AB, Ideon Research Center, 223 70 LUND. T: 046-2852570, F: 046 134230, E: ad@claytech.se

Edlund, L., Engineer, J&W Samhällsbyggnad, 121 88 STOCKHOLM-GLOBEN. T: 08-6886000, F: 08-6886914

Edstam, T., Ph.D., J&W Samhällsbyggnad, Box 382, 401 26 GÖTEBORG. T: 031-614300, F: 031-614305, E: torbjorn. edstam@jw.se

Edvardsson, T., Sjöfartsverket, 601 78 NORRKÖPING. T: 011-191189, F: 011-126791

Egnelöv, G., M.Sc., 43 Epirus road, LONDON SW6 7UR

Ehnbåge, G., Project Engineer, VBK Konsulterande Ingenjörer, Mölndalsvägen 85, 412 85 GÖTEBORG. T: 031-7033534, F: 031-7033501, E: gunnar.e@vbk.se

Ehnström, H., Managing Director, KB, Ehnen Fastighetsförvaltning, Länsmansstigen 1, 182 36 DANDERYD. T: 070-6329988, F: 08-7559620

Ekdahl, U., Technical Manager, PEAB Sverige AB, St Varvsgatan 11 B, 211 19 MALMÖ. T: 040-357515, F: 040-977905, E: ulf.ekdahl@peab.se

Ekenberg, M., M.Sc., Skanska Teknik AB, 405 18 GÖTEBORG. T: 031-7711000, F: 031-153815, E: Mats. Ekenberg@teknik.skanska.se

Ekevärn, G., Viltorpsbacken 81, 162 70 VÄLLINGBY

Eklund, C., Managing Director, AB BORRCENTER, Antennvägen 10, 135 48 TYRESÖ. T: 08-7982040, F: 08-7423020, E: borrcenter@borrcenter.se

Eklund, S., Swedish National Road Administration, Box 1008, 901 20 UMEÅ. T: 090-172782, F: 090-172610, E: stefan.eklund@vv.se

Eklund, T., Eng., J&W Samhällsbyggnad, Östra Storgatan 67, 553 21 JÖNKÖPING. T: 036-304320, F: 036-165875, E: torbjorn.eklund@jw.se

Ekstedt, K.-A., Eng., Bjerking Ingenjörsbyrå AB, Box 2006, 750 02 UPPSALA. T: 018-651132, F: 018-710485, E: karl-arne.ekstedt@bjerking.se

Ekström, I., M.Sc., Swedpower AB, Box 527, 162 16 STOCKHOLM. T: 08-7397219, F: 08-7395332, E: ingvar. ekstrom@swedpower.vattenfall.se

Ekström, J., Jord & Bild, Sandvägen 28, 240 12 TORNAHÄLLESTAD. T: 046-530 17, F: 046-530 17

Ekström, J., D.Sc., Vägverket, 405 33 GÖTEBORG. T: 031-635049, F: 031-635212, E: jan.ekstrom@vv.se

Elander, P., M.Sc., Envipro Miljöteknik AB, Repslagaregatan 19, 582 22 LINKÖPING. T: 013-357276, F: 013-357271, E: par.elander@envipro.se

Eldh, T., Bjerking Ingenjörsbyrå AB, Box 2006, 750 02 UPPSALA. T: 018-651100, F: 018-710485, E: thomas.eldh@ bjerking.se

Eliasson, A., M.Sc., Trångsundsvägen 14, 135 52 TYRESÖ

Eliasson, B., Manager Civil Design Dept. Sycon Teknikkonsult AB, Box 332, 631 05 ESKILSTUNA. T: 016-154250, F: 016-154241, E: bo.eliasson@confortia.se

Eliasson, K., Eng., Terravia, Skjulvägen 25, 120 48 ENSKEDE GÅRD. T: 08-6002260, F: 08-6002260, E: Terravia@telia.com

Elliot, A.L., M.Sc., GF Konsult AB, Box 5056, 402 22 GÖTEBORG. T: 031-3355000, F: 031-3358955, E: annlouise.elliot@gfkonsult.se

Elmgren, K., Managing Director, Environmental Mechanics AB, Kungegårdsgatan 7, 441 57 ALINGSÅS. T: 0322-17101, F: 0322-636756, E: kjell@envi.se

Elvelind, K., J&W Samhällsbyggnad, Byggmästargatan 5, 803 24 GÄVLE. T: 026-663585, F: 026-663560, E: katrin.elvelind@jw.se

Emmelin, A., M.Sc., Rättviksvägen 8, 167 75 BROMMA. E: ann.emmelin@svebefo.se

Enestedt, K.-O., M.Sc., Swedish National Road Administration, Box 4202, 171 04 SOLNA. T: 08-6274122, F: 08-283707, E: kjell-olof.enestedt@vv.se

Eng, T., Managing Director, Engtex AB, 565 22 MULLSJÖ. T: 0392-10676, F: 0392-10185, E: torbjoern.eng@engtex.se

Engdahl, M., Senior Geologist, Geological Survey of Sweden, Guldhedsgatan 5 A, 413 20 GÖTEBORG. T: 031-7082651, F: 031-7082675, E: Mats.Engdahl@sgu.se

Englund, P., RGW Mark & Dataprojektering AB, Skärfsta 443, 881 30 SOLLEFTEÅ. T: 0620-17997, F: 0620-17997, E: peder@rgw.se

Englöv, P., M.Sc., VBB VIAK AB, Geijersgatan 8, 216 18 MALMÖ. T: 040-167000, F: 040-154347, E: peter.englov@sweco.se

Engman, S., Senior Research Eng., Gripgatan 2, 582 43 LINKÖPING

Engström, M., J&W Samhällsbyggnad, Box 382, 401 26 GÖTEBORG. T: 031-614383, F: 031-614370, E: michael.engstrom@jw.se

Engström, P., Consulting Eng., VBB Viak, Box 34044, 100 26 STOCKHOLM. T: 08-6956000, F: 08-6956350, E: per.engstrom@sweco.se

Engström, Å., M.Sc., HydroTerra Ingenjörer AB, Box 1031, 651 15 KARLSTAD. T: 054-567430, F: 054-567433, E: ake.engstrom@hydroterra.se

Engvall, L., Consulting Eng., VBB Viak, Box 34044, 100 26 STOCKHOLM. T: 08-6956000, F: 08-6956010

Envang, S., Eng., Milstensvägen 19, 187 33 TÄBY. E: geoteknik-.stig.envang@swipnet.se

Eresund, S., D.Sc., Stenöregatan 63, 230 44 BUNKE-FLOSTRAND. F: 040-510602

Ericson, G., M.Sc., Scandiaconsult Sverige AB, Stora Varvsgatan 11 N, 211 19 MALMÖ. T: 040-105410, F: 040-105510, E: gosta.ericson@scc.se

Ericson, J., Eng., Åhusvägen 4, 3 tr, 121 51 JOHANNESHOV

Ericson, M., Eng., J&W Samhällsbyggnad, Byggmästargatan 5, 803 24 GÄVLE. T: 026-663550, F: 026-663560, E: mats.ericson@jw.se

Ericsson, E., Eng., J&W Samhällsbyggnad, Box 71, 581 02 LINKÖPING. T: 013-253171, F: 013-125082, E: ewald.ericsson@jw.se

Ericsson, J., M.Sc., Tibnor AB, Box 1073, 581 10 LINKÖPING. T: 013-357800, F: 013-104550, E: johan.ericsson@tibnor.se

Eriksson, A., Eng., KM Geoteknik, Box 92093, 120 07 STOCKHOLM. T: 08-55523166, F: 08-55523010, E: alf.eriksson@km.se

Eriksson, Å., J&W Energi och Miljö, 121 88 STOCKHOLM-GLOBEN. T: 08-6886405, F: 08-6886922, E: anders.eriksson@jw.se

Eriksson, B., Maratongatan 6, 871 51 HÄRNÖSAND

Eriksson, H., M.Sc., Böö 3401, 471 91 KLÖVEDAL. E: hakan.eriksson@operamail.com

Eriksson, H., Ph.D., Hercules Grundläggning AB, Box 1283, 164 29 KISTA. T: 08-7503315, F: 08-7506057, E: hakan.l.eriksson@hercules.se

Eriksson, L.G., Lic Eng., MRM Konsult AB, Box 63, 971 03 LULEÅ. T: 0920-60460, F: 0920-60474, E: lars.eriksson@mrm.se

Eriksson, L., Head of Department, Swedish Geotechnical Institute, 581 93 LINKÖPING. T: 013-201870, F: 013-201914, E: leif.eriksson@swedgeo.se

Eriksson, L., M.Sc., Skanska, 205 13 MALMÖ. T: 040-144348, F: 040-237047

Eriksson, M., M.Sc., J&W Samhällsbyggnad, 121 88 STOCKHOLM-GLOBEN. T: 08-6886428, F: 08-6886914, E: maria.e.eriksson@jw.se

Eriksson, P., M.Sc., NCC AB, 405 14 GÖTEBORG. T: 031-7715065, F: 031-151188, E: per.e.eriksson@ncc.se

Eriksson, T., Eng., Vägverket Konsult, 781 87 BORLÄNGE. T: 0243-94319, F: 0243-83927, E: tobias.eriksson@vv.se

Eriksson, T., Hifab Byggprojektledaren AB, Portalgatan 2 A, 754 23 UPPSALA. T: 018-246090, F: 018246045, E: torbjorn.eriksson@hifab.se

Eriksson, U.B., M.Sc., NCC AB, 170 80 SOLNA. T: 08-6553345, F: 08-6553360, E: ulf.eriksson@ncc.se

Eriksson, Å., Project Hydrogeologist, Golder Grundteknik KB, Anders Perssonsgatan 12, 416 64 GÖTEBORG. T: 031-800840, F: 031-154950, E: ake_eriksson@golder.se

Fagerström, H., M.Sc., Björnbo 21, 181 46 LIDINGÖ

Falge, B., Eng., Swedish National Rail Administration, Box 1070, 172 22 SUNDBYBERG. T: 08-7622000, F: 08-7625755, E: bo.falge@banverket.se

Falk, J., M.Sc., BAT Cofra AB, Box 1060, 186 26 VALLEN-TUNA. T: 08-51170600, F: 08-51173361, E: Johan@batab.se

Falksund, M., M.Sc., Hercules Grundläggning AB, Förrådsvägen 16, 901 32 UMEÅ. T: 090-134576, F: 090-134578, E: mattias.falksund@hercules.se

Fall, B., Eng., Geofall, Norrtullsgatan 25 E, 113 27 STOCK-HOLM

Fallsvik, J., Research & Consulting Eng., Swedish Geotechnical Institute, 581 93 LINKÖPING. T: 013-201845, F: 013-201914, E: jan.fallsvik@swedgeo.se

Fellenius, B.H., Professor, Urkkada Technology Ltd, 735 Ludgate Court, OTTAWA, Ontario, K1J 8K8. T: +1-613-7415071, E: bfellenius@achilles.net

Filipson, H., B.A., Swedish National Rail Administration, Box 1346, 172 27 SUNDBYBERG. T: 08-7623465, F: 08-7623388, E: hans.filipson@banverket.se

Flodin, L.-E., Skandinaviska Geoteknik AB, Roslagsvägen 54-56, 113 47 STOCKHOLM. T: 08-6121130, F: 08-4589620, E: stockholm@eurotema.se

Fogelström, R., Eng., Kapellgatan 9, 291 33 KRISTIANSTAD. T: 044-123752, F: 044-202015, E: rolf.fogelstrom@telia.se

Forsberg, L.

Forsberg, T., M.Sc., Hercules Grundläggning AB, Box 1283, 164 29 KISTA. T: 08-7503317, F: 08-7506057, E: tony.forsberg@hercules.se

Forsgren, I., Eng., Chalmers University of Technology, 412 96 GÖTEBORG. T: 031-7722113

Forssblad, L., D.Sc., Gustaf Lundbergs väg 4, 168 50 BROMMA

Forssman, G., Field Eng., AB Stockholm Konsult, Box 9611, 117 91 STOCKHOLM. T: 08-7858658, F: 08-7442210, E: goran.forssman@stockholmkonsult.se

Franzén, G., M.Sc., Scandiaconsult Sverige AB, Box 5343, 402 27 GÖTEBORG. T: 031-3353394, F: 031-400571, E: gunilla.franzen@scc.se

Franzén, T., Research Director, Swedish Rock Engineering Research, Box 47047, 100 74 STOCKHOLM. T: 08-6922280, F: 08-6511364, E: tomas.franzen@svebefo.se

Fredbäck, H.E., Chief Engineer, Fortifikationsverket, Kungsgatan 43, 631 89 ESKILSTUNA. T: 016-154817, F: 016-133702, E: hkfk@fortv.se

Fredriksson, A., D.Sc., Golder Grundteknik KB, Box 20127, 104 60 STOCKHOLM. T: 08-50630600, F: 08-50630601, E: anders_fredriksson@golder.se

Friberg, A.B.A., Engineer, J&W Samhällsbyggnad, 121 88 STOCKHOLM-GLOBEN. T: 08-6886000, F: 08-6886914, E: arne.friberg@jw.se

Fridh, R., Eng., Street Council of Malmö, 205 80 MALMÖ. T: 040-341316, F: 040-341358

Friman, S., Eng., GF Konsult AB, Box 5056, 402 22 GÖTE-BORG. T: 031-3355320, F: 031-3358955, E: sven.friman@gfkonsult.se

Frumerie, L., M.Sc., M5 Consulting AB, Östermalmsgatan 11, 3 tr, 114 24 STOCKHOLM. T: 08-141817, F: 08-141817, E: lars.frumerie@swipnet.se

Fröidh, J., Tyréns Infrakonsult AB, 118 86 STOCKHOLM. T: 08-4290000, F: 08-6448850

Fällman, A.-M., Research & Consulting Eng., Swedish Geotechnical Institute, 581 93 LINKÖPING. T: 013-201807, F: 013-201909, E: annmarie.fallman@swedgeo.se

Gabrielsson, A., Research & Consulting Eng., Swedish Geotechnical Institute, 581 93 LINKÖPING. T: 013-201815, F: 013-201912, E: anna.gabrielsson@swedgeo.se

Gabrielsson, Å., Eng., Swedish National Road Administration, Box 4202, 171 04 SOLNA. T: 08-7576600, F: 08-983030, E: ake.gabrielsson@vv.se

Garin, H., M.Sc., Scandiaconsult Sverige AB, Kaj 24, Stora Varvsg 11 N:3, 211 19 MALMÖ. T: 040-105431, F: 040-126650, E: hgnsm@scc.se

Gereben, L., D.Sc., ELGETE Consulting AB, Box 1451, 171 28 SOLNA. T: 08-7358890, F: 08-7354257, E: laszlo@elgete.se

Girdo, D., Eng., Multrågatan 144, 162 55 VÄLLINGBY
Gjers, A., M.Sc., Banverket, Box 1070, 172 22 SUNDBY-BERG. T: 08-7623475, F: 08-7623387, E: Anna.Gjers@banverket.se
Graad, M., Geotechnical Eng., Vägverket Konsult, Box 164, 371 22 KARLSKRONA. T: 0455-25509, F: 0455-25030, E: magnus.graad@vv.se
Graeffe, H., Hydrogeologist, Vägverket Konsult, Box 1733, 751 47 UPPSALA. T: 018-167018, F: 018-167070, E: heidi.graeffe@vv.se
Granström, M., Consulting Eng., J&W Samhällsbyggnad, Byggmästargatan 5, 803 24 GÄVLE. T: 026-663550, F: 026-663560, E: mats.granstrom@jw.se
Green, M., M.Sc., VBB VIAK AB, Box 33, 721 03 VÄSTERÅS. T: 021-14 11 00
Grén, A., Eng., GM Consult, Box 17071, 104 62 STOCKHOLM. T: 08/7208401, E: gren@gmconsult.se
Griwell, F., VBB VIAK AB, Geijersgatan 8, 216 18 MALMÖ. T: 040-167176, F: 040-154347, E: fredrik.griwell@sweco.se
Grundfelt, B., Managing Director, Kemakta Konsult AB, Box 12655, 112 93 STOCKHOLM. T: 08-6176755, F: 08-6521607, E: bertil@kemakta.se
Grävare, B., Managing Director, Europile/BACHY, Äsperedsgatan 9, 424 57 GUNNILSE. T: 031-943610, F: 031-943315, E: bengt.europile@bachy.se
Grävare, C.-J., Consulting Eng., Pålanalys AB, August Barks gata 13 C, 421 32 VÄSTRA FRÖLUNDA. T: 031-454307, F: 031-459980, E: carl-john@palanalys.se
Grävare, M., Construction Eng, Pålanalys AB, August Barks gata 13 C, 421 32 VÄSTRA FRÖLUNDA. T: 031-454307, F: 031-459980, E: office@palanalys.se
Gullbrandsson, P.A., Eng., PG Geoteknik AB, Upplagsvägen 17, 117 43 STOCKHOLM. T: 08-7442233, F: 08-7440560, E: pggeoteknik@swipnet.se
Gunnarsson, A., M.Sc., Vägverket, Box 494, 581 06 LINKÖPING. T: 013-289033, F: 013-289010, E: agne.gunnarsson@vv.se
Gustafson, G., Geohydrologist, Chalmers Tekniska Högskola, 412 96 GÖTEBORG. T: 031-7721928, F: 031-189705, E: g2@geo.chalmers.se
Gustafsson, B.-O., Ph.D., J&W Energi och Miljö, Slagthuset, 211 20 MALMÖ. T: 040-108200, F: 040-108201, E: bjorn-olof.gustafsson@jw.se
Gustafsson, J.-O., Eng., Geo-gruppen AB, Marieholmsgatan 122, 415 02 GÖTEBORG. T: 031-438450, F: 031-489450, E: janove.gustafsson@geogruppen.se
Gustafsson, L., Eng., Flygfältsbyrån, Box 12076, 402 41 GÖTEBORG. T: 031-7751118, F: 031-7751122, E: lg@fbe.se
Gustafsson, M., M.Sc., Chalmers Tekniska Högskola, 412 96 GÖTEBORG. T: 031-7722104, F: 031-7722107, E: maling@vsect.chalmers.se
Gustafsson, S., Head of Division, Scandiaconsult Sverige AB, Kaj 24, Stora Varvsg 11 N:3, 211 19 MALMÖ. T: 040-105430, F: 040-126650, E: sgnsm@scc.se
Gustavsson, L., M.Sc., VBK Konsulterande Ingenjörer KB, Mölndalsvägen 85, 412 85 GÖTEBORG. T: 031-7033500, F: 031-7033501, E: leif.g@vbk.se
Gustavsson, M., Banverket, Box 43, 971 02 LULEÅ. T: 0920-35261, F: 0920-35446, E: mattias.gustavsson@banverket.se
Gustavsson, S., AB Jacobson & Widmark, Box 8094, 700 08 ÖREBRO. T: 019-178950, F: 019-133200, E: stig.gustavsson@jw.se
Gustavsson, T., Luleå Kommun, 971 85 LULEÅ. T: 0920-93070, E: torbjorn.gustavsson@tekn.lulea.se
Guve, R., VBB VIAK AB, Box 34044, 100 26 STOCKHOLM. T: 08-6956354, F: 08-6956350, E: robert.guve@sweco.se
Gyllfors, L.-G., Björkhagsvägen 17, 186 35 VALLENTUNA
Göransson, B., Geologist/Eng., Tyréns Infrakonsult AB, Box 27, 291 21 KRISTIANSTAD. T: 044-106800, F: 044-111326
Göransson, L., M.Sc., Boverket, Box 534, 371 23 KARLSKRONA. T: 0455-353276, F: 0455-353221, E: lars.goransson@boverket.se
Haag, Ö., Managing Director, Welltech Systems AB, Box 2011, 433 02 PARTILLE. T: 031-7956100, F: 031-7957662

Hagberg, T., M.Sc., GEO, Box 119, DK-2800 LYNGBY. T: +45-45204122, F: +45-45881240, E: tora.hagberg@geoteknisk.dk
Hagblom, J., Project Leader, Skanska Berg och Bro, Industrivägen 6-8, 137 37 VÄSTERHANINGE. T: 08-50073865, F: 08-50073815, E: johan.hagblom@skanska.se
Hagman, M., M.Sc., J&W Samhallsbyggnad, Slagthuset 8 D, 211 20 MALMÖ. T: 040-108221, F: 040-108201, E: mats.hagman@jw.se
Hall, L., D.Sc., FB Engineering AB, Box 12076, 402 41 GÖTEBORG. T: 031-7751565, F: 031-7751122, E: lhl@fbe.se
Hallberg, J., Constructional Eng., Golder Grundteknik KB, Box 20127, 104 60 STOCKHOLM. T: 08-50630600, F: 08-50630601
Hallberg, O., M.Sc., GeoNordic AB, Reimersholmsgatan 23, 2 tr, 117 40 STOCKHOLM. T: 08-6699630, F: 08-6699621, E: ove.hallberg@geonordic.se
Halldén, B., M.Sc., Ripstigen 5, 170 74 SOLNA
Hallin, A., President, Golder Grundteknik KB, Box 20127, 104 60 STOCKHOLM. T: 08-50630600, F: 08-50630601, E: axel_hallin@golder.se
Hallingberg, A., M.Sc., Swedish National Rail Administration, Box 1014, 405 21 GÖTEBORG. T: 031-103296, F: 031-103203, E: anders.hallingberg@banverket.se
Hansbo, S., Prof., Lyckovägen 2, 182 74 STOCKSUND. E: sven.hansbo@swipnet.se
Hansen, F., Eng., Skanska Teknik AB, 405 18 GÖTEBORG. T: 031-7711000, F: 031-153815, E: fredhy.hansen@teknik.skanska.se
Hansson, A., M.Sc., Skanska Teknik AB, 405 18 GÖTEBORG. T: 031-7711326, F: 031-153815, E: anders.hansson@teknik.skanska.se
Hansson, B., M.Sc., Tyrens Infrakonsult AB, Lilla Badhusgatan 4, 411 21 GÖTEBORG. T: 031-60 63 12, F: 031-60 63 01, E: bengt.hansson@tyrens.se
Hansson, P.-H., Swedish National Rail Administration, Box 43, 971 02 LULEÅ. T: 010-2584592
Hansson, P., Eng., Geotech AB, Datavägen 53, 436 32 ASKIM. T: 031-289920, F: 031-681639, E: peter.hansson@geotech.se
Hansson, T., M.Sc., Hercules Grundläggning AB, Vikans Industriväg 14, 418 78 GÖTEBORG. T: 031-530110, F: 031-546085, E: tobias.hansson@ic24.net
Hansson, Ö., Georådgivning AB, Åkerts gård, 240 35 HARLÖSA. T: 046-62050, F: 046-62050
Harling, R., Production manager, Skanska Berg och Bro, Industrivägen 6-8, 137 37 VÄSTERHANINGE. T: 08-50073885, F: 08-50073815, E: risto.harling@skanska.se
Hartlén, J., Ph.D., JH GeoConsulting AB, Järavallsgatan 31 C, 216 11 LIMHAMN. T: 040-164538, F: 040-152770, E: jan.hartlen@swipnet.se
Hartzen, L.-O., Civil Engineer, VBB VIAK AB, Geijersgatan 8, 216 18 MALMÖ. T: 040-167000, F: 040-154347, E: lars-olof.hartzen@sweco.se
Hasselvall, P., Construction Eng., Stångjärnsvägen 3, 168 68 BROMMA
Hedar, P.A., D.Sc., Storuddens gård, 635 17 NÄSHULTA
Hedberg, B., Swedish National Road Administration, Box 1023, 651 15 KARLSTAD. T: 054-140036, F: 054-140051, E: bjorn.hedberg@vv.se
Hedberg, S., M.Sc., Skanska Anläggning AB, Kemivägen, 705 97 GLANSHAMMAR. E: susanne.hedberg@skanska.se
Hedlund, S., Senior Engineer, VBB VIAK AB, Box 34044, 100 26 STOCKHOLM. T: 08-6956000, F: 08-6956480
Hedström, L., Eng., Skanska Teknik AB, 405 18 GÖTEBORG. T: 031-7711000, F: 031-153815, E: Lennart.Hedstrom@teknik.skanska.se
Heiner, A., D.Sc., VBB VIAK AB, Box 34044, 100 26 STOCKHOLM. T: 08-6956223, F: 08-6956350
Helander, J., Eng., Granbacken 5, 141 31 HUDDINGE
Helgesson, H., Research & Consulting Eng., Swedish Geotechnical Institute, 581 93 LINKÖPING. T: 013-201814, F: 013-201912, E: helena.helgesson@swedgeo.se
Hellgren, A., Chief Eng., Rådmansgatan 23, 114 25 STOCKHOLM

Hellgren, M., M.Sc., Golder Grundteknik KB, Box 20127, 104 60 STOCKHOLM. T: 08-50630600, F: 08-50630601, E: martin_hellgren@golder.se

Hellgren, N., Eng., Sth Geomek AB, Upplagsvägen 17, 117 43 STOCKHOLM. T: 08-7440500, F: 08-7442210, E: nils@geomek.com

Henningsson, B.-M., M.Sc., GF Konsult AB, Box 5056, 402 22 GÖTEBORG. T: 031-3355000, F: 031-3358955, E: britt-marie.henningsson@gfkonsult.se

Henricsson, L., Engineer, J&W Samhällsbyggnad, 121 88 STOCKHOLM-GLOBEN. T: 08-6886429, F: 08-6886914

Henrikson, P., M.Sc., Scandiaconsult Sverige AB, Box 5343, 402 27 GÖTEBORG. T: 031-3353381, F: 031-400571, E: per.henrikson@scc.se

Henriksson, A., Eng., J&W Byggprojektering, Rullagergatan 6, 415 26 GÖTEBORG. T: 031-7272662, F: 031-7272501, E: arne.henriksson@jw.se

Henriksson, C., Eng., J&W Samhällsbyggnad, Box 503, 391 25 KALMAR. T: 0480-449205, F: 0480-27642, E: christer.henriksson@jw.se

Henriksson, M., M.Sc., Rovågersvägen 2, 907 88 TÄTEÅ

Hermansson, C., Project Manager, Europile/BACHY, Äsperedsgatan 9, 424 57 GUNNILSE. T: 031-943610, F: 031-943315, E: chermansson@europile.se

Hermansson, I., Civ. Eng., Pålanalys AB, August Barks gata 13 C, 421 32 VÄSTRA FRÖLUNDA. T: 031-454307, F: 031-459980, E: ingemar@palanalys.se

Hintze, S., Ph.D., Skanska Teknik AB, 169 83 SOLNA. T: 08-50435493, F: 08-7536048, E: staffan.hintze@teknik.skanska.se

Holm, B., B.A., Bjerking Ingenjörsbyrå AB, Box 2006, 750 02 UPPSALA. T: 018-651100, F: 018-110485

Holm, G., Director, Swedish Geotechnical Institute, 581 93 LINKÖPING. T: 013-201861, F: 013-201914, E: goran.holm@swedgeo.se

Holmberg, R., Eng, Nitro Consult AB, Dunkersgatan 5, 903 27 UMEÅ. T: 090-132840, F: 090-132848, E: ralf.holmberg@nitroconsult.se

Holmqvist, L., M.Sc., Kopparåsvägen 56, 427 38 BILLDAL

Holmström, L., M.Sc., Hercules Grundläggning AB, Box 1283, 164 29 KISTA. T: 08-7503320, F: 08-7506057, E: lars.holmstrom@hercules.se

Holtz, R.D., Prof. of C.E., University of Washington, Box 352700, SEATTLE, WA 98195-2700. T: +1-206-5437614, F: +1-206-685386, E: holtz@u.washington.edu

Hornö, B., Sales Representative, Ingenjörsfirman Geotech AB, Datavägen 53, 436 32 ASKIM. T: 031-289925, F: 031-681639, E: bill@geotech.se

Hugner, A., M.Sc., J&W Samhällsbyggnad, 121 88 STOCKHOLM-GLOBEN. T: 08-6886431, F: 08-6886914, E: anders.hugner@jw.se

Hult, G., VBB VIAK AB, Box 385, 651 09 KARLSTAD. T: 054-141730, F: 054-158926

Hultén, C., M.Sc., KM Geoteknik, Rullagergatan 6, 415 26 GÖTEBORG. T: 031-7272500, F: 031-7272501, E: carina.hulten@km.se

Hultsjö, S., M.Sc., Golder Grundteknik KB, Box 20127, 104 60 STOCKHOLM. T: 08-50630600, F: 08-50630601, E: sven_hultsjo@golder.se

Hydén, G., Tekniska Verken i Linköping AB, Box 1500, 581 15 LINKÖPING. E: gosta.hyden@tekniskaverken.se

Hylander, J., MSc., Hylanders Geobyrå AB, Fabriksgatan 12, 602 23 NORRKÖPING. T: 011-108950, F: 011-105930, E: hgb@telia.com

Hübinette, P., M.Sc., Golder Associates AB, AndersPerssonsgatan 12, 416 64 GÖTEBORG. T: 031-800840, F: 031-154950, E: per_hubinette@golder.se

Hågeryd, A.-C., Geologist, Swedish Geotechnical Institute, 581 93 LINKÖPING. T: 013-201848, F: 013-201914, E: annchristine.hageryd@swedgeo.se

Håkansson, H., Geotechnical Eng., Bjerking Ingenjörsbyrå AB, Box 2006, 750 02 UPPSALA. T: 018-651149, F: 018-710485, E: henrik.hakansson@bjerking.se

Håkansson, S., M.Sc., Vägverket Konsult, Box 1398, 801 38 GÄVLE. T: 026-149363, F: 026-129944, E: stefan.hakansson@vv.se

Håkansson, U., Regional Manager, ATLAS COPCO, P.O. Box 69407, KWUN TONG, HONG KONG. T: +852-23171262, F: +852-23171161

Hägerstrand, S., M.Sc., Vibratec Isolation AB, Norrsund 1859, 760 17 BLIDÖ. T: 0176-82240, F: 0176-82773, E: svante.hagerstrand@telia.com

Hässler, L., D.Sc., Vattenfall Hydropower AB, Box 533, 162 16 STOCKHOLM. T: 08-7396000, F: 08-7396983

Högberg, M., Eng., Kolvedsvägen 2, 831 73 ÖSTERSUND

Höglander, B., Swedish Institute of Building Documentation, Sankt Eriksgatan 46, 112 34 STOCKHOLM. T: 08-6177450, F: 08-6177460, E: borje.hoglander@byggdok.se

Ingers, C.E., Consulting Engineer, Hammarlunda 1, 240 35 HARLÖSA

Ingimarsson, R., Islands Geotekniska Förening, Hjardarhaga 2-6, IS-107 REYKJAVIK

Ismael, E., Chief Engineer, Bronsåldersgatan 80, 421 63 VÄSTRA FRÖLUNDA

Ittner, T., Hydrogeologist, Scandiaconsult Sverige AB, Box 1403, 801 38 GÄVLE. T: 026-149533, F: 026-127150, E: tirbng@scc.se

Iwanowski, T., Ph.D., TADYNA AB, Åbergssonsvägen 6, 170 77 SOLNA. T: 08-850080

Iwers, L.-G., Eng., Tyréns Infrakonsult AB, Tångringsgatan 2, 784 30 BORLÄNGE. T: 0243-88640, E: larsgoran.iwers@tyrens.se

Jacobsson, J.Å., Hydrogeologist, JAC-konsult, Puckvägen 23, 740 30 BJÖRKLINGE. T: 073-9892551, F: 018-4906589, E: jackonsult@x-stream.se

Jakobsson, C., Eng., Golder Grundteknik KB, Box 20127, 104 60 STOCKHOLM. T: 08-50630642, F: 08-50630601, E: carina_jakobsson@golder.se

Jakobsson, M., M.Sc., VBB VIAK AB, Box 50120, 973 24 LULEÅ. T: 0920-35581, F: 0920-35545, E: marten.jakobsson@sweco.se

Janson, N., Dr. Sc., Golder Grundteknik KB, Box 20127, 104 60 STOCKHOLM. T: 08-50630600, F: 08-50630601, E: thomas_janson@golder.se

Jansson, B., Eng., J&W Management, Vasagatan 34, 722 15 VÄSTERÅS. E: bo.l.jansson@jw.se

Jansson, S., Principal Eng., VBB Anläggning AB, Box 34044, 100 26 STOCKHOLM. T: 08-6956000, F: 08-6956110

Jansson, S., M.Sc., VA-Projekt AB, Ribbingsgatan 11, 703 63 ÖREBRO. T: 019-105415, F: 019-6113011, E: soren.jansson@vap.se

Janzon, A., M.Sc., Citytunneln, Box 4012, 203 11 MALMÖ. T: 040-321452, F: 040-321500, E: anders.janzon@citytunneln.com

Jelisic, D., M.Sc., Tyréns Infrakonsult AB, Drottninggatan 62, 252 21 HELSINGBORG. T: 042-289822, F: 042-289816, E: drago.jelisic@tyrens.se

Jelisic, N., Lic. Eng., Swedish National Road Administration, Box 186, 871 24 HÄRNÖSAND. T: 0611-44186, F: 0611-44054, E: nenad.jelisic@vv.se

Jendeby, L., Chief Geotech. Eng., NCC, 405 14 GÖTEBORG. T: 031-7715000, F: 031-151188, E: Leif.Jendeby@ncc.se

Jensen, K., Consultant, Box 22, 178 21 EKERÖ. T: 08-56030919, F: 08-56030919

Jerbo, A., Ph.D., JERBOL AB, Norra Sandåsgatan 39, 386 32 FÄRJESTADEN. T: 0485-30791, F: 0485-34704, E: jerbol@telia.com

Joelson, K.-G., Research Engineer, Swedish Geotechnical Institute, 581 93 LINKÖPING. T: 013-201887, F: 013-201913, E: karlgustav.joelson@swedgeo.se

Johannesson, L.-E., M.Sc., Clay Technology AB, 223 70 LUND. T: 046-2862577, F: 046-134230, E: lej@claytech.se

Johansson, B., M.Sc., NCC Bygg-Väst AB, 405 14 GÖTEBORG. T: 031-7715035

Johansson, H.G., Senior Researcher, Swedish Road and Transport Research Institute, 581 95 LINKÖPING. T: 013-204279, F: 013-141436, E: hans.g.johansson@vti.se

Johansson, J., Eng., Miljökonsult, Kemigatan 16, 275 38 SJÖBO. T: 0416-19520, F: 0416-19520

Johansson, J., M.Sc., Swedish National Road Administration, Box 1062, 551 10 JÖNKÖPING. T: 036-151468, F: 036-151453, E: jan-p.johansson@vv.se

Johansson, L.E., Licenciate in Engineering, VBB VIAK AB, Geijersgatan 8, 216 18 MALMÖ. T: 040-167250, E: lars. erik.johansson@sweco.se

Johansson, L.O., M.Sc., J&W Samhällsbyggnad, Box 8094, 700 08 ÖREBRO. T: 019-178950, F: 019-133200, E: lars.o. johansson@jw.se

Johansson, L.O., M.Sc., VBB VIAK AB, Box 1848, 581 17 LINKÖPING. T: 013-252728, F: 013-142860, E: lars.o. johansson@sweco.se

Johansson, L., Göteborgs Kommun, Box 2554, 403 17 GÖTEBORG. T: 031-611745, F: 031-611710, E: lennart. johansson@stadsbyggnad.goteborg.se

Johansson, L., M.Sc., Vägverket Konsult, Södra Hamngatan 5, 411 14 GÖTEBORG. T: 031-3396129, F: 031-3396111, E: lennart.johansson@vv.se

Johansson, P., M.Sc., Hercules Grundläggning AB, Vikans Industriväg 14, 418 78 GÖTEBORG. T: 031-647867, F: 031-546085, E: peter.x.johansson@hercules.se

Johansson, R., J&W Samhällsbyggnad, Box 502, 901 09 UMEÅ. T: 090-125560, E: roger.johansson@jw.se

Johansson, S., Eng., Swedish National Road Administration, Box 213, 871 25 HÄRNÖSAND. T: 0611-44200, F: 0611-44203, E: stefan-p.johansson@vv.se

Johansson, T., Eng., J&W Byggprojektering, Box 503, 391 25 KALMAR. T: 0480-449207, F: 0480-27642, E: torbjorn. johansson@jw.se

Johansson, Å., Research & Consulting Engineer, Swedish Geotechnical Institute, Chalmers Vasa, hus 5, 412 96 GÖTEBORG. T: 031-7786561, F: 031-7786575, E: ake. johansson@swedgeo.se

Johansson, Ö., Engineer, J&W Samhällsbyggnad, 121 88 STOCKHOLM-GLOBEN. T: 08-6886000, F: 08-6886914, E: orjan.johansson@jw.se

Johnson, U., M.Sc., Ulf Johnson Geo AB, Karlbergsvägen 33, 113 62 STOCKHOLM. T: 08-304338, F: 08-304338, E: uj.geo@swipnet.se

Jonell, P., M.Sc., Ingenjörsfirman Geotech AB, Datavägen 53, 436 32 ASKIM. T: 031-289920, F: 031-681639

Jons, A.-K., M.Sc., SWECO VIAK, Box 34044, 100 26 STOCKHOLM. T: 08-6956465, F: 08-6956350, E: anna karin.jons@sweco.se

Jonsson, A., M.Sc., Blegdamsvej 80, DK-2100 KÖPEN-HAMN

Jonsson, A., M.Sc., J&W Samhällsbyggnad, Byggmästargatan 5, 803 24 GÄVLE. T: 026-663550, F: 026-663560, E: annika.jonsson@jw.se

Jonsson, E., M.Sc., Swedish National Road Administration, Box 375, 831 25 ÖSTERSUND. T: 063-194937, F: 063-194881, E: eva.jonsson@vv.se

Jonsson, G., Eng., Geo-Projektering AB, Box 53, 751 03 UPPSALA. T: 018-152290, F: 018-149520

Jonsson, J., M.Sc., NCC AB, 170 80 SOLNA. T: 08-6553368, F: 08-6553360, E: jonas.jonsson@ncc.se

Jonsson, R., Ragnar Jonsson GEO-Rådgivning AB, Ella-hagsvägen 27 A., 183 38 TÄBY. T: 08-7584974

Jonsson, R., Eng., Scandiaconsult Sverige AB, Bergsgatan 13, 852 36 SUNDSVALL. T: 060-663600, F: 060-614984, E: rjnbns@scc.se

Josefson, A., M.Sc., Swedish National Road Administration, Box 1023, 651 15 KARLSTAD. T: 054-140032, F: 054-140051, E: axel.josefson@vv.se

Josefsson, K., VBB VIAK AB, Box 385, 651 09 KARLSTAD. T: 054-141700, F: 054-158926, E: kent.josefsson@sweco.se

Junkers, C., M.Sc., J&W Samhällsbyggnad, Box 382, 401 26 GÖTEBORG. T: 031-614300, F: 031-614370, E: charlotte. junkers@jw.se

Jämsä, M., Tyréns Infrakonsult AB, Västra Norrlandsgatan 10, 903 27 UMEÅ. T: 090-702950, E: markku.jamse@tyrens.se

Jönsson, C., Eng., Tyréns Infrakonsult AB, 118 86 STOCK-HOLM. T: 08-4290000, F: 08-6448850

Jönsson, S.-S., M.Sc., AB Jacobson & Widmark, Östra Storgatan 67, 553 21 JÖNKÖPING. T: 036-304320, F: 036-165875, E: sten-sture.jonsson@jw.se

Jönsson, Å., M.Sc., VBB VIAK AB, Box 34044, 100 26 STOCKHOLM. T: 08-6956487, F: 08-6956350, E: asa. jonsson@sweco.se

Karlefors, T., AB Jacobson & Widmark, Box 502, 901 10 UMEÅ. T: 090-125560, F: 090-774600, E: torbjorn. karlefors@jw.se

Karlsson, G., M.Sc., Bygg-och Geokonstruktioner AB, Torsgatan 10, 561 31 HUSKVARNA. T: 036-139060, F: 036-139060, E: bygg-geo@telia.com

Karlsson, G., Engineer, J&W Samhällsbyggnad, 121 88 STOCKHOLM-GLOBEN. T: 08-6886000, F: 08-6886914, E: gunnar.karlsson@jw.se

Karlsson, M., M.Sc., Banverket, 781 85 BORLÄNGE. T: 0243-445498, F: 0243-445617, E: magnus.karlsson@ banverket.se

Karlsson, R., Tyréns Infrakonsult AB, Västra Norrlandsgatan 10, 903 27 UMEÅ. T: 090-702950, F: 090-702959, E: roger. karlsson@tyrens.se

Kers, A., M.Sc., Lillsandsvägen 8, 893 40 KÖPMAN-HOLMEN

Kilnes, L., M.Sc., Lilldalsvägen 51, 144 64 RÖNNINGE

Kivelö, M., Ph.D., Skanska Berg och Bro, 169 83 SOLNA. T: 08-50436858, F: 08-7534790, E: matti.kivelo@ skanska.se

Kjellberg, K., Sales Eng., Rautaruukki (Sverige) AB, Box 127, 781 22 BORLÄNGE. T: 0243-88744, F: 0243-84210, E: Kjell.Kjellberg@rautaruukki.fi

Klingenberg, H., M.Sc., KFS AnläggningsKonstruktörer AB, Industrivägen 5, 171 48 SOLNA. T: 08-4700562, F: 08-4700561, E: hans.klingenberg@kfs.se

Knutsson, S., Associate Professor, Luleå University of Technology, 971 87 LULEÅ. T: 0920-91332, F: 0920-72075, E: Sven.Knutsson@ce.luth.se

Knutz, Å., Geologist, National Swedish Road Administration, 781 87 BORLÄNGE. T: 0243-94320, F: 0243-94310, E: ake.knutz@vv.se

Komarek, R., Consulting Engineer, Swedish National Road Administration, Box 4107, 171 04 SOLNA. T: 08-4041002, F: 08-4041001, E: rudolf.komarek@vv.se

Kovanen, L., M.Sc., Reduttvägen 52, 187 68 TÄBY

Kristensson, O., M.Sc., Centerlöf & Holmberg AB, Box 29, 201 20 MALMÖ. T: 040-258928, F: 040-258901, E: ok@coh.se

Kullingsjö, A., M.Sc., Skanska Teknik AB, Lilla Bommen 2, 405 18 GÖTEBORG. T: 031-7711398, F: 031-153815, E: Anders.Kullingsjo@teknik.skanska.se

Kurttila, P., M.Sc., Tjärhovsvägen 18, 2 tr, 116 21 STOCK-HOLM

Källström, R., M.Sc., J&W Samhällsbyggnad, Box 382, 401 26 GÖTEBORG. T: 031-614300, F: 031-614305, E: rolf. kallstrom@jw.se

Königsson, L., Eng., Vägverket, Box 1140, 631 80 ESKIL-STUNA. T: 016-157019, E: lars.konigsson@vv.se

Lakka, L., Gamla Rekekroken 32, 263 92 JONSTORP

Lanngren, A., M.Sc., Tyréns Infrakonsult AB, Box 8062, 700 08 ÖREBRO. T: 019-174435, F: 019-183710

Lantz, L., Eng., KM Bygg & Anläggning AB, Rullagergatan 6, 415 26 GÖTEBORG. T: 031-7272500, F: 031-7272501, E: lennart.lantz@km.se

Larsson, G., Geotechnical Eng., VBB VIAK AB, Box 385, 651 09 KARLSTAD. T: 054-141704, F: 054-158926, E: gunnar. larsson@sweco.se

Larsson, K., Essviksvägen 32, 862 34 KVISSLEBY. T: 060-560977

Larsson, M., Eng., J&W Samhällsbyggnad, Byggmästargatan 5, 803 24 GÄVLE. T: 026-663550, F: 026-663560

Larsson, M., Eng., Vägverket Konsult, 781 87 BORLÄNGE. T: 0243-94316, F: 0243-87927, E: mats.larsson@vv.se

Larsson, R., Researcher, Swedish Geotechnical Institute, 581 93 LINKÖPING. T: 013-201869, F: 013-201914, E: rolf. larsson@swedgeo.se

Larsson, S., Geotechnician, Tyréns Infrakonsult AB, 118 86 STOCKHOLM. T: 08-4290257, F: 08-4290020, E: stefan. larsson@tyrens.se

Larsson, Å., Eng., Swedish National Road Administration, Box 1023, 651 15 KARLSTAD. T: 054-140028, F: 054-140051, E: asa-k.larsson@vv.se

Laugesen, J., M.Sc., Det Norske Veritas AS, Veritasveien 1, NO-1322 HÖVIK. T: +47 67 579094, F: +47 67 577474, E: jens.laugesen@dnv.com

Lavemark, A., M.Sc., ABV Rock Group, P.O. Box 89426, RIYADH 11682. T: +966-1-4782239, F: +966-1-4773109, E: anders.lavemark@attglobal.net

Lehmann, A., M.Sc., Vägverket Konsult Öst, Box 4107, 171 04 SOLNA. T: 08-4041006, E: axel.lehmann@vv.se

Leking, B., Bohusläns Geoteknik AB, Box 15, 451 15 UDDEVALLA. T: 0522-39045, F: 0522-35978

Lemmeke, L., M.Sc., VBB VIAK AB, Geijersgatan 8, 216 18 MALMÖ. T: 040-167285, F: 040-154347, E: leif.lemmeke@sweco.se

Lennartsdotter, P.K., M.Sc., GF Konsult AB, Box 5056, 402 22 GÖTEBORG. T: 031-3355327, F: 031-3358955, E: katarina.parkkonen@gfkonsult.se

Leyon, B., Eng., Geo-gruppen AB, Marieholmsgatan 122, 415 02 GÖTEBORG. T: 031-438450, F: 031-489450, E: bo.leyon@geogruppen.se

Liedberg, S., D.Sc., Skanska Teknik AB, 405 18 GÖTEBORG. T: 031-7711474, F: 031-153815, E: sven.liedberg@teknik.skanska.se

Liljemark, A., M.Sc., VBB VIAK AB, Box 34044, 100 26 STOCKHOLM. T: 08-6956214, F: 08-6956240, E: anneli.liljemark@sweco.se

Liljestrand, B., Regional Manager, Lemminkäinen Construction, Torsgatan 12, 111 23 STOCKHOLM. T: 08-247861, F: 08-249866, E: bjarne.liljestrand@lemminkainen.se

Lindahl, A., M.Sc., J&W Mark och Anläggning, Slagthuset, 211 20 MALMÖ. T: 040-108288, F: 040-108281, E: annika.lindahl@jw.se

Lindberg, B., Hercules Grundläggning AB, Vikans Industriväg 14, 418 78 GÖTEBORG. T: 031-530110, F: 031-546085

Lindberg, G., D.Sc., Bergdalssvägen 4, 132 41 SALTSJÖ-BOO

Lindberg, L.-B., Eng., Ängsullsvägen 39, 162 46 VÄLLINGBY

Lindberg, M., Engineering Geologist, J&W Samhällsbyggnad, 121 88 STOCKHOLM-GLOBEN. T: 08-6886435, F: 08-6886914, E: michael.lindberg@jw.se

Lindblom, U., D.Sc., GECON AB, Viktor Rydbergsgatan 1A., 411 34 GÖTEBORG. T: 031-204570, E: gecon@surfzon.com

Lindbom, B., M.Sc., Statens oljelager, Box 16247, 103 24 STOCKHOLM. T: 08-54521506, F: 08-246814, E: bjorn.lindbom@nosa.se

Lindeborg, B., President, Consoil AB, Box 545, 175 26 JÄRFÄLLA. T: 08-58030295, F: 08-58026070, E: profile@consoil.se

Lindfred, M., M.Sc., Envipro Miljöteknik AB, Kungsgatan 28, 411 19 GÖTEBORG. T: 031-130380, F: 031-7110121, E: martin.lindfred@envipro.se

Lindgren, A., M.Sc., Massarsch Constructive IT, Björnnäsvägen 21, 113 47 STOCKHOLM. T: 08-6732239, F: 08-6732241, E: anders.lindgren@mcit.se

Lindgren, H., M.Sc., HL KONSULT, Ö Dalenvägen 7, 429 30 KULLAVIK. T: 031-932844, F: 031-932844

Lindh, B.G., Consulting Eng., BG Lindh AB, Gullbergs Strandgata 36, 411 04 GÖTEBORG. T: 031-150103, F: 031-150923, E: goran.lindh@mbox2.swipnet.se

Lindh, P., Head of Department, Swedish Geotechnical Institute, Hospitalsgatan 16 A, 211 33 MALMÖ. T: 040-356773, F: 040-938685, E: per.lindh@swedgeo.se

Lindkvist, P., M.Sc., VBB VIAK AB, Box 50120, 973 24 LULEÅ. T: 0920-33587, F: 0920-33545, E: peter.lindkvist@sweco.se

Lindmark, A., M.Sc., Solcon AB, Teknikringen 1, 583 30 LINKÖPING. T: 013-210024, F: 013-210424

Lindmark, P., Licentiate of Engineering, Norrköping Miljö & Energi, Box 193, 601 03 NORRKÖPING. T: 011-151517, F: 011-151545, E: per.lindmark@nme.se

Lindquist, H., Geologist, Geolog Hans Lindquist AB, Köpmansgatan 2, 652 26 KARLSTAD. T: 054-183091, F: 054-144402, E: geolindquist@telia.com

Lindqvist, H., M.Sc., Långrevsgatan 11, 133 43 SALTSJÖBADEN

Lindstrand, O., Sr. Hydrogeologist, VBB VIAK AB, Box 34044, 100 26 STOCKHOLM. T: 08-6956243, F: 08-6956240

Lindström, M., M.Sc., NCC AB, 170 80 SOLNA. T: 08-585 533 77, F: 08-585 533 60, E: marten.lindstrom@ncc.se

Ljung, S., M.Sc., District Court of Vänersborg, Box 1070, 462 28 VÄNERSBORG. T: 0521-270200, F: 0521-270340

Ljunggren, B., Eng., Björke, Igelfors, 612 95 FINSPÅNG

Ljungqvist, P., Geologist, Geomiljö Väst KB, Sönnerbovägen 6, 432 92 VARBERG. T: 0340-621461, F: 0340-621461, E: per.ljungqvist@geomiljovast.se

Lord, U., J&W Samhällsbyggnad, Box 8094, 700 08 ÖREBRO. T: 019-178950, F: 019-133200, E: urban.lord@jw.se

Ludvig, B., Managing Director, Petro Bloc AB, Stora Badhusgatan 6, 411 21 GÖTEBORG. T: 031-7113110, F: 031-7113113, E: bengt@petro-bloc.se

Lundahl, B., M.Sc., Lundahl Project Management, Sommarvägen 18, 177 61 JÄRFÄLLA. T: 08-58081215, F: 08-58036451, E: bjorn.lundahl@mbox301.swipnet.se

Lundgren, T., D.Sc., Envipro Miljöteknik AB, Repslagaregatan 19, 582 22 LINKÖPING. T: 013-357277, F: 013-357271, E: tom.lundgren@envipro.se

Lundin, J., M.Sc., Vattenfall Hydro Power AB, Box 527, 162 16 STOCKHOLM. T: 08-7397005, F: 08-7395332, E: jan.lundin@swedpower.vattenfall.se

Lundin, S.-E., Senior Consultant, Bjerking Ingenjörsbyrå AB, Box 2006, 750 02 UPPSALA. T: 018-651100, F: 018-710485, E: se.lundin@bjerking.se

Lundqvist, G., Grundförstärkningar AB, Box 11173, 404 24 GÖTEBORG. T: 031-152200, F: 031-159019, E: gunnar.lundqvist@fop.se

Lundström, H., M.Sc., Bohusläns Geoteknik AB, Box 15, 451 15 UDDEVALLA. T: 0522-39045, F: 0522-35978

Lundström, R., Professor, Rune Lundström Geotechnical Consultant, Yxstigen 5, 181 47 LIDINGÖ. T: 08-7672666, F: 08-7672666

Lundström, T., Project Manager, AVAkonsult i Klippan AB, Vallgatan 3, 264 32 KLIPPAN. T: 0435-14720, F: 0435-14720, E: ava@swipnet.se

Lyman, H., M.Sc., Sundsvalls kommun, 851 85 SUNDSVALL. T: 060-191189, F: 060-591889, E: hans.lyman@sundsvall.se

Löfling, P., M.Sc., Vägverket Konsult, 781 87 BORLÄNGE. T: 0243-94317, F: 0243-87927, E: per.lofling@vv.se

Löfroth, H., Research & Consulting Eng., Swedish Geotechnical Institute, 581 93 LINKÖPING. T: 013-201854, F: 013-201912, E: hjordis.lofroth@swedgeo.se

Löfvenberg, K., M.Sc., Ragn-Sells Miljökonsult AB, Box 744, 191 27 SOLLENTUNA. T: 08-6234275, F: 08-6234270, E: katarina.lofvenberg@ragnsells.se

Löwegren, B., Eng., VBB VIAK AB, Box 33, 721 03 VÄSTERÅS. T: 021-141100, F: 021-141121, E: bo.lowegren@sweco.se

Löwegren, N., Banverket, 781 85 BORLÄNGE. T: 0243-445638, F: 0243-445617, E: niklas.lowegren@banverket.se

Löwhagen, P., M.Sc., Skanska Teknik AB, Drottningtorget 14, 205 33 MALMÖ. T: 040-144736, F: 040-144556, E: per.lowhagen@teknik.skanska.se

Macsik, J., Lic. Eng., Luleå University of Technology, 971 87 LULEÅ. T: 0920-91330, F: 0920-72075, E: Josef.Macsik@ce.luth.se

Magnfors, N.-H., Soils Eng., Ängsullsvägen 93, 162 46 VÄLLINGBY

Magnusson, C., M.Sc., Sveriges geologiska undersökning, Box 16247, 103 24 STOCKHOLM. T: 08-54521508, F: 08-246814, E: caroline.magnusson@nosa.se

Magnusson E., Hammarby Sjöstad, Box 8311, 104 20 STOCKHOLM. T: 08-50826431, F: 08-50826066, E: erling.magnusson@hammarbysjostad.stockholm.se

Magnusson, G., Geologist, NCC AB, NCC Industri, Tagenevägen 25, 425 37 GÖTEBORG. T: 031-7715036, F: 031-152013, E: gustaf.magnusson@ncc.se

Magnusson, I., Eng., Pihl-Aarsleff Bygg-och Anläggning AB, Strandgatan 1, 216 12 LIMHAMN. T: 031-657550, F: 031-654979, E: inge.magnusson@pihl-aarsleff.se

Magnusson, O., Professor, Skanska, 169 83 SOLNA. T: 08-50435339, F: 08-7536048, E: ove.magnusson@teknik.skanska.se

Malmborg, B.S., Senior Lecturer, Lund Institute of Technology, Box 118, 221 00 LUND. T: 046-2227422, F: 070-6160721, E: Bo_Stigson.Malmborg@tg.lth.se

Malmros, L., M.Sc., J&W Samhällsbyggnad, Box 13052, 600 13 NORRKÖPING. T: 011-139551, F: 011-124552, E: lars.malmros@jw.se

Markström, I., Project Engineer, Sydkraft Konsult AB, 205 09 MALMÖ. T: 040-255280, F: 040-304696, E: ingemar.markstrom@sycon.se

Martinsson, S., Eng., Svedalavägen 15, 372 97 RONNEBY

Marxmeier, N., Research and Consulting Eng., Tyréns Infrakonsult AB, 118 86 STOCKHOLM. T: 08-4290351, F: 08-4290020, E: natascha.marxmeier@tyrens.se

Massarsch, A., M.Sc., Massarsch Constructive IT, Björnnäsvägen 21, 113 47 STOCKHOLM. T: 08-6732240, F: 08-6732241, E: andreas.massarsch@geo.se

Massarsch, R., Consultant, Geo Engineering AB, Ferievägen 25, 168 41 BROMMA. T: 08-871990, F: 08-878920, E: rainer.massarsch@geo.se

Mattsson, H., M.Sc., Luleå University of Technology, 971 87 LULEÅ. T: 0920-72147, F: 0920-72075, E: Hans.Mattsson@anl.luth.se

Mattsson, R., General Manager, De Neef Scandinavia AB, Långavallsgatan 15, 424 57 GUNNILSE. T: 031-3300590, F: 031-3308041, E: rolf.mattsson@deneef.se

Melander, K., Eng., Envi AB, Kungegårdsgatan 7, 441 57 ALINGSÅS. T: 0322-17101, F: 0322-636756, E: kenth@envi.se

Melhus, B., M.Sc., Räntmästargatan 92, 702 27 ÖREBRO

Minsér, A., Managing Director, Terra Mark AB, Box 356, 123 03 FARSTA. T: 08-6058120, F: 08-6058281, E: agne@engineer.com

Modin, C.-O., Geotechnical Consultant, Tyréns Infrakonsult AB, 851 71 SUNDSVALL. T: 060-169600, F: 060-169629, E: carl-olof.modin@tyrens.se

Mohlin, B., Eng., VBB VIAK AB, Box 259, 851 04 SUNDSVALL. T: 060-169910, F: 060-613007, E: benny.mohlin@sweco.se

Molin, J., Consulting Eng., VBB VIAK AB, Geijersgatan 8, 216 18 MALMÖ. T: 040-167000, F: 040-154347, E: jan.molin@sweco.se

Morin, H., M.Sc., Banverket, Box 43, 971 02 LULEÅ. T: 0660-294975, F: 0660-294970, E: helena.morin@banverket.se

Morin, M., M.Sc., Swedish National Rail Administration, Box 43, 971 02 LULEÅ. T: 0660-294977, F: 0660-294970, E: mattias.morin@banverket.se

Moritz, L., M.Sc., Vägverket Region Sydöst, VSÖte, Box 494, 581 06 LINKÖPING. T: 013-289034, F: 013-289010, E: lovisa.moritz@vv.se

Mosesson, J., M.Sc., Skogsvägen 13, 181 41 LIDINGÖ

Murén, P., NCC AB, 170 80 SOLNA. T: 08-6552000, F: 08-6243315, E: per.muren@ncc.se

Myhrberg, P., M.Sc., Tyréns Infrakonsult AB, Box 27, 291 21 KRISTIANSTAD. T: 044-186269, F: 044-211326, E: paul.myhrberg@tyrens.se

Myles, B., Development Manager, 24 Conifer Close, Reigate, Surrey RH2 9NN, UK

Möller, B., M.Sc., FmGeo AB, Södra Promenaden 9, 211 29 MALMÖ. T: 040-6616110, F: 040-6616120, E: bjorn.moller@fmgeo.se

Möller, H., M.Sc., Tyréns Infrakonsult AB, Drottninggatan 62, 252 21 HELSINGBORG. T: 042-289817, F: 042-289816, E: henrik.moller@tyrens.se

Möller, L., Technical Manager, Scandinavian Terra Tec AB, Ängen, Villa Ängviken, 663 92 HAMMARÖ. T: 054-522030, F: 054-522020

Möller, P., M.Sc., Terraling, Virkesgränd 20, 183 63 TÄBY. T: 08-7560920, F: 08-7560917, E: per_and_inga_moller@yahoo.com

Najder, T., D.Sc., Movägen 3, 133 36 SALTSJÖBADEN

Nehrfors, U., Stockholm Office of Housing and Real Estate, Box 8311, 104 20 STOCKHOLM. T: 08-50828079, F: 08-50827215, E: ulf.nehrfors@gfk.stockholm.se

Nelson, A., GEOTEC, Box 174, 243 00 HÖÖR. T: 0413-24460, E: anders@geotec.se

Nilsson, A., M.Sc., Det Norske Veritas Certification AB, Lilla Bommen 1, 411 04 GÖTEBORG. T: 031-7112661, F: 031-7212600, E: andrea.nilsson@dnv.com

Nilsson, B., M.Sc., ebab Byggadministration AB, Box 1045, 121 14 ENSKEDEDALEN. T: 08-390450, F: 08-392762, E: bernt.nilsson@ebab.se

Nilsson, G., Geotechnical Eng., AB Stockholm Konsult, Box 9611, 117 91 STOCKHOLM. T: 08-7858665, F: 08-7858504, E: gunnar.nilsson.geo@stockholmkonsult.se

Nilsson, G., Geologist, Envipro Miljöteknik AB, Rallarvägen 37, 184 40 ÅKERSBERGA. T: 08-54021742, F: 08-54061770, E: goran.nilsson@envipro.se

Nilsson, K.-E., M.Sc., VSL Norge A/S, 169 83 SOLNA. T: 08-50437200, F: 08-7534973, E: karl-erik.nilsson@spannarmering.se

Nilsson, L.Å., Tellstedt Geoteknik AB, Varbergsgatan 12, 412 65 GÖTEBORG. T: 031-405280, F: 031-404076, E: lars.nilsson@tellstedt.se

Nilsson, L., Eng., Scandiaconsult Sverige AB, Box 5343, 402 27 GÖTEBORG. T: 031-3353312, F: 031-403952, E: lnnmvg@scc.se

Nilsson, R., M.Sc., Valebergsvägen 49, 444 60 STORA HÖGA. E: karibo.jaderberg@stockholm.mail.telia.com

Nilsson, T., Eng., Swedish National Rail Administration, Box 43, 971 02 LULEÅ. T: 0920-35246, F: 0920-35205, E: torgny.nilsson@banverket.se

Nilsson, U., M.Sc., J&W Samhällsbyggnad, Robertsviksgatan 7, 972 41 LULEÅ. T: 0920-238303, F: 0920-238381, E: ulrika.nilsson@jw.se

Nilsson, Å., M.Sc., Swedpower AB, Box 527, 162 16 STOCKHOLM. T: 08-7397140, F: 08-7395332, E: ake. nilsson@swedpower.vattenfall.se

Nord, B., Consulting Engineer, COFEC Consulting Engineers, 100 Tue Tinh, HANOI. T: +84-4-8267748, F: +84-4-8260258, E: nord.cofec@fpt.vn

Nordal, S., Professor, Norwegian University of Science and Technology, Högskoleringen 7a, N-7491 TRONDHEIM. T: +47-73594594, F: +47-73594609

Nordén, H., Tyréns Infrakonsult AB, 118 86 STOCKHOLM. T: 08-4290294, E: henrik.norden@tyrens.se

Nordén, S., Akademiska Hus AB, Box 483, 401 27 GÖTEBORG. T: 031-632415, F: 031-632401, E: staffan.norden@akademiskahus.se

Norder, B., Research & Consulting Engineer, Blendas Gata 3, 422 51 HISINGS BACKA

Nordin, R., Eng., Tredim Geoteknik AB, Krukmakargatan 54, 117 41 STOCKHOLM. T: 08-6690102, F: 08-6690103, E: tredim@nitroz.se

Nordlander, H., Building Inspector, Östergraninge 5528, 882 95 GRANINGE

Nordlander, T., Civil Eng., VBB VIAK AB, Box 385, 651 09 KARLSTAD. T: 054-141717, F: 054-158926, E: tomas.nordlander@sweco.se

Nordqvist, F., B.Sc., SWECO – VBB VIAK, Box 34044, 100 26 STOCKHOLM. T: 08-6956435, F: 08-6956350, E: fredrik.nordqvist@sweco.se

Norén, C., Civil Eng., NCC Teknik AB, 405 14 GÖTEBORG. T: 031-7715000, E: christer.noren@ncc.se

Norman, S.-O., Eng., C H Swahn Anläggningar AB, Box 246, 781 23 BORLÄNGE. T: 0243-237085, F: 0243-63229, E: swahns@chswahns.se

Norstedt, U., Chief Eng., Swedish State Power Board, 162 87 VÄLLINGBY. T: 08-7396970, F: 08-176064, E: urban.norstedt@generation.vattenfall.se

Novak, J., M.Sc., Scandiaconsult Sverige AB, Box 4205, 102 65 STOCKHOLM. T: 08-6156192, F: 08-7021939, E: jnobos@scc.se

Nyberg, M., Lic. Eng., Swedish National Road Administration, 405 33 GÖTEBORG. T: 031-635105, F: 031-635212, E: marianne.nyberg@vv.se

Nygren, I., M.Sc., Högskolan Dalarna, 781 88 BORLÄNGE. T: 023-778832, F: 023-778050, E: iny@du.se

Nygren, J., M.Sc., Golder Grundteknik KB, Box 20127, 104 60 STOCKHOLM. T: 08-50630600, F: 08-50630601, E: jonas_nygren@golder.se

Nylén, F., VBB VIAK AB, Box 34044, 100 26 STOCK-HOLM. T: 08-6956444, F: 08-6956210, E: fredrik.nylen@sweco.se

Nyman, K.-E., Intergrund AB, Järngatan 7, 234 35 LOMMA. T: 040-412530, F: 040-414930

Nystrand, B.-Å., Assistant Prof., University of Uppsala, Villavägen 1B, 751 22 UPPSALA. T: 018-4712582, F: 018-555920, E: Bengt-Ake.Nystrand@natgeog.uu.se

Nyström, B., M.Sc., SwedPower AB, 971 77 LULEÅ. T: 0920-77365, F: 0920-77369, E: birgitta.nystrom@swedpower.vattenfall.se

Nyström, F., M.Sc., Swedish National Rail Administration, Box 1041, 405 21 GÖTEBORG. T: 031-104680, F: 031-103204, E: fredrik.nystrom@banverket.se

Näsman, L., Project Manager, J&W Samhällsbyggnad, Box 52, 851 02 SUNDSVALL. T: 060-671500, F: 060-116370, E: lennart.nasman@jw.se

Nätterdahl, K., Chalmers University of Technology, 412 96 GÖTEBORG. T: 031-7721000, F: 031-7722107, E: natte@vsect.chalmers.se

Odén, K., Research and Consulting Eng., Swedish Geotechnical Institute, Chalmers Vasa, hus 5, 412 96 GÖTEBORG. T: 031-7786569, F: 031-7786575, E: karin.oden@swedgeo.se

Odenstedt, S., Consulting Eng., Tallboängen 72, 436 44 ASKIM

Ohlsson, F., M.Sc., Gullregnsvägen 20, 435 37 MÖLNLYCKE. E: folke.ohlsson@harryda.mail.telia.com

Ohlsson, T., B.A., Landskrona kommun, Drottninggatan 7, 261 80 LANDSKRONA. T: 0418-470662, F: 0418-473576, E: tomas.ohlsson@tv.landskrona.se

Olandersson, L., Svenska Väg- och Vattenbyggares, Box 1334, 111 83 STOCKHOLM. T: 08-245450, F: 08-4119226, E: bomedia@algonet.se

Olofsson, E., M.Sc., Skanska Teknik AB, 405 18 GÖTEBORG. T: 031-7711354, F: 031-153815, E: Eleonor.Olofsson@teknik.skanska.se

Olofsson, K., VBB VIAK, Ringvägen 2, 831 34 ÖSTER-SUND. T: 063-150588, F: 063-133315, E: kjell.m.olofsson@sweco.se

Olofsson, S., Consulting Eng., Gavelåsvägen 3, 433 31 PARTILLE

Olovsson, C., Field Eng., Tyréns Infrakonsult AB, Box 27, 291 21 KRISTIANSTAD. T: 044-186297, F: 044-211326, E: christer.olovsson@tyrens.se

Olsson, A., M.Sc., Markervägen 8, 175 67 JÄRFÄLLA, E: arne.olsson@beta.telenordia.se

Olsson, B., Consulting Engineer, J&W Samhällsbyggnad, Södra Strandgatan 5, 1 tr, 503 35 BORÅS. T: 033-128200, F: 033-125845, E: bengt.olsson@jw.se

Olsson, C., Consulting Engineer, Vägverket, Box 543, 291 25 KRISTIANSTAD. T: 044-195154, F: 044-195155, E: connie.olsson@vv.se

Olsson, E., Vägverket Konsult, Södra Hamngatan 5, 411 14 GÖTEBORG. T: 031-3396120, F: 031-3396111, E: elisabeth.olsson@vv.se

Olsson, E.-L., Geotechnical Eng., Swedish National Rail Administration, 781 85 BORLÄNGE. T: 0243-445596, F: 0243-445617, E: eva-lotta.olsson@banverket.se

Olsson, G., Division Manager, Europile, Äsperedsgatan 9, 424 57 GUNNILSE. T: 031-943610, F: 031-943315, E: goran.olsson@europile.se

Olsson, L., Ph.D., GEOSTATISTIK Lars Olsson AB, Box 116, 147 22 TUMBA. T: 08-53038703, F: 08-53039206, E: lars.olsson@geostatistik.se

Olsson, M., M.Sc., Scandiaconsult Sverige AB, Stora Varvsgatan 11 N:3, 211 19 MALMÖ. T: 040-105432, F: 040-126650, E: monsm@scc.se

Olsson, M., Swedish National Road Administration, 405 33 GÖTEBORG. T: 031-635103, F: 031-635212, E: mats.olsson@vv.se

Olsson, M., Vägverket Produktion, Testvägen 7, 232 37 ARLÖV. T: 040-287968, F: 040-437440, E: mikael-a.olsson@vv.se

Olsson, S., Eng., Roddarvägen 13, 831 52 ÖSTERSUND

Olsson, T., Consulting Eng., J&W Samhällsbyggnad, Box 382, 401 26 GÖTEBORG. T: 031-614300, F: 031-614370, E: tord.olsson@jw.se

Orre, B., Eng., Bo Orre Markråd AB, Holmbodavägen 67, 192 77 SOLLENTUNA. T: 08-6226101, F: 08-6226116, E: orre.markrad@swipnet.se

Orrje, O., D.Sc., Brötvägen 39, 167 66 BROMMA. T: 08-254585, E: olle.orrje@swipnet.se

Oscarsson, M., M.Sc., Scandiaconsult Sverige AB, Box 4205, 102 65 STOCKHOLM. T: 08-6156331, F: 08-7021934, E: mats.oscarsson@scc.se

Oscarsson, R., M.Sc., Göteborgs Gatu AB, Box 1086, 405 23 GÖTEBORG. T: 031-628000, F: 031-628002, E: Roger.Oscarsson@gatubolaget.goteborg.se

Ottosson, E., Chief Engineer, Swedish Geotechnical Institute, 581 93 LINKÖPING. T: 013-201834, F: 013-201912, E: elvin.ottosson@swedgeo.se

Ouacha, M., M.Sc., Johan Helldén AB, Teknikringen 1E, 583 30 LINKÖPING. T: 013-210298, F: 013-213925, E: monica.ouacha@johanhelldenab.se

Palmberg, B., M.Sc., SwedPower AB, Box 527, 162 16 STOCKHOLM. T: 08-7396477, F: 08-7396983, E: bengt.palmberg@swedpower.vattenfall.se

Palmdahl, L., FB Engineering AB, Box 12076, 402 41 GÖTEBORG. T: 031-7751000, F: 031-122063

Palmgren, H., Design Eng., J&W Samhällsbyggnad, Box 382, 401 26 GÖTEBORG. T: 031-614327, F: 031-154668, E: helmer.palmgren@jw.se

Palmquist, E., M.Sc., SIGMA SÄVAB AB, Ystadvägen 17, 241 38 ESLÖV. T: 0413-60760, F: 0413-61490, E: erik.palmquist@sigmasavab.se

Palmqvist, G., Field Eng., Tyréns Infrakonsult AB, Drottninggatan 62, 252 21 HELSINGBORG. T: 042-289821, F: 042-289816, E: goran.palmqvist@tyrens.se

Pantzar, U., M.Sc., Geotekn Byggnadsbyrån Håpe AB, Fasanvägen 34, 131 44 NACKA. T: 08-7161501, F: 08-7161500

Parck, K., M.Sc., Skanska Teknik AB, 169 83 SOLNA. T: 08-50435346, F: 08-7536048, E: katarina.parck@skanska.se

Paulsson, C., Engineer, Hercules Grundläggning AB, Vikans Industriväg 14, 418 78 GÖTEBORG. T: 031-530110, F: 031-546085, E: christer.paulsson@hercules.se

Paus, K., M.Sc., K. Paus Konsulter, Milvägen 28, 181 47 LIDINGÖ. T: 08-7676705

Paus, O., B.A., J&W Energi och Miljö, 121 88 STOCKHOLM-GLOBEN. T: 08-6886402, F: 08-6886922, E: ole.paus@jw.se

Persson, B., Eng., Stamkullevägen 80, 461 42 TROLL-HÄTTAN

Persson, H., Geotechnical Engineer, J&W Samhällsbyggnad, Slagthuset, 211 20 MALMÖ. T: 040-108228, F: 040-108281, E: hans.persson@jw.se

Persson, K., Managing Director, SCANMIX AB, Industrihamnen 30, 234 34 LOMMA. T: 040-411033, F: 040-411088, E: kjell.persson@scanmix.com

Persson, L., M.Sc., Golder Grundteknik KB, Box 20127, 104 60 STOCKHOLM. T: 08-50630600, F: 08-50630601, E: lars_persson@golder.se

Persson, L., Head of Division, Geological Survey of Sweden, Box 670, 751 28 UPPSALA. T: 018-179148, F: 018-179210, E: lars.persson@sgu.se

Persson, L., Tyréns Infrakonsult AB, Drottninggatan 62, 252 21 HELSINGBORG. T: 042-289823, F: 042-289816, E: leif.persson@tyrens.se

Persson, M., J&W Samhällsbyggnad, Box 758, 851 22 SUNDSVALL. T: 060-149550, F: 060-159738, E: mikael.persson@jw.se

Persson, S.-E., Eng., Tyréns Infrakonsult AB, 118 86 STOCKHOLM. T: 08-4290000, F: 08-6448850

Persson, Ö., Eng., J&W Samhällsbyggnad, Box 52, 851 02 SUNDSVALL. T: 060-150140, F: 060-116370, E: orjan.persson@jw.se

Perzon, J., Consulting Eng., PEAB Sverige AB, 401 80 GÖTEBORG. T: 031-7008487, F: 031-7008440, E: jan.perzon@peab.se

Petersson, B., Engineer, Tyréns Infrakonsult AB, Box 27, 291 21 KRISTIANSTAD. T: 044-186297, F: 044-211326, E: bengt.petersson@tyrens.se
Petersson, E., M.Sc., J&W Samhällsbyggnad, Slagthuset, 211 20 MALMÖ. T: 040-108211, F: 040-108281, E: eva.petersson@jw.se
Petsonk, A., Sr Project Manager, J&W Energi och Miljö, 121 88 STOCKHOLM-GLOBEN. T: 08-6886416, F: 08-6886922, E: andrew.petsonk@jw.se
Pettersson, B., Eng., Swedish National Road Administration, Box 1062, 551 10 JÖNKÖPING. T: 036-151400, F: 036-151453, E: bjorn.pettersson@vv.se
Pettersson, J., Area Manager, Skanska Berg och Bro, 169 83 SOLNA. T: 08-50436834, F: 08-7534790, E: jan.r.pettersson@skanska.se
Pettersson, L., Head of Divsion, Swedish National Road Administration, 781 87 BORLÄNGE. T: 0243-75681, F: 0243-75442
Pettersson, S.-Å., Product Manager, Atlas Copco Craelius AB, 195 82 MÄRSTA. T: 08-58778548, F: 08-59118782, E: sten-ake.pettersson@atlascopco.com
Phung, D.L., Technical Manager, Skanska Berg och Bro, 169 83 SOLNA. T: 08-50436912, F: 08-7534790, E: long.phung@skanska.se
Pihl, B., B.A., Tyrens, 118 86 STOCKHOLM. T: 08-4290000, F: 08-6448850
Pletikos, C., M.Sc., VBB VIAK AB, Geijersgatan 8, 216 18 MALMÖ. T: 040-167107, F: 040-154347, E: carmen.pletikos@sweco.se
Ploman, K.-F., Blomstervägen 15, 790 15 SUNDBORN
Polla, P., M.Sc., Skanska AB, Box 1012, 581 10 LINKÖPING. T: 013-129850, E: peter.polla@skanska.se
Polugic, T., J&W Samhällsbyggnad, Box 758, 851 22 SUNDSVALL. T: 060-149562, F: 060-159738, E: tomislav.polugic@jw.se
Possfelt, U., AB J&W, Laholmsvägen 10, 302 48 HALMSTAD. T: 035-18 11 00, F: 035-18 11 01, E: ulf.possfelt@jw.se
Pousette, A., Angleterre, 186 97 BROTTBY
Pramborg, B., Consulting Eng., Karins Allé, 181 44 LIDINGÖ, E: bengt@applepaj.se
Pramsten, A., M.Sc., VBB VIAK AB, Box 34044, 100 26 STOCKHOLM. E: anna.pramsten@sweco.se
Pusch, R., Professor, Geodevelopment AB, IDEON Alfa, 223 70 LUND. T: 046-2862400, F: 046-2862405, E: pusch@geodevelopment.ideon.se
Puustinen, V., Field Engineer, Swedish Geotechnical Institute, 581 93 LINKÖPING. T: 013-201903, F: 013-201912, E: veijo.puustinen@swedgeo.se
Pyyny, G., M.Sc., Mariehemsvägen 37A, 906 52 UMEÅ
Rankka, W., D.Sc., VBB VIAK, Box 145, 551 13 JÖNKÖPING. T: 036-151818, F: 036-710965, E: wilhelm.rankka@sweco.se
Rausberger, B., BR Consulting AB, Vinterstigen 11, 144 40 RÖNNINGE
Reblin, T., VBB VIAK AB, Box 1902, 791 19 FALUN. T: 023-46400, F: 023-46401, E: thomas.reblin@sweco.se
Rehnman, S.-E., M.Sc., Royal Institute of Technology, 100 44 STOCKHOLM. T: 08-7908028, F: 08-7907928, E: svenerik@aom.kth.se
Ribberström, O., Contractor, Ove Ribberström Projektering, Norra Rydsbergsvägen 26, 443 50 LERUM. T: 0302-12025, F: 0302-16260, E: orp@swipnet.se
Riise, P., Civ. Eng., KM Geoteknik, Rullagergatan 6, 415 26 GÖTEBORG. T: 031-7272500, F: 031-7272501, E: per.riise@km.se
Rimbratt, J., Eng., Sjöfartsverket, Box 949, 461 29 TROLLHÄTTAN. T: 0520-472200
Risberg, M., Eng., J&W Samhällsbyggnad, Robertsviksgatan 7, 972 41 LULEÅ. T: 0920-238300, E: moje.risberg@jw.se
Rodebjer, M., M.Sc., Nokia Networks, Box 1113, 164 22 KISTA. T: 08-41008606, E: mats.rodebjer@nokia.com
Rogbeck, J., B.A., Envipro Miljöteknik AB, Repslagaregatan 19, 582 22 LINKÖPING. T: 013-357278, F: 013-357271, E: jan.rogbeck@envipro.se

Rogbeck, Y., M.Sc., Scandiaconsult Sverige AB, Teknikringen 1 C, 583 30 LINKÖPING. T: 013-212640, F: 013-210460, E: yvonne.rogbeck@scc.se
Roman, P., Eng., Sundsvalls kommun, 851 85 SUNDSVALL. T: 060-191315, F: 060-121019, E: per.roman@sundsvall.se
Romell, J., Specialist, Piling engng, Skanska Teknik AB, 405 18 GÖTEBORG. T: 031-7711000, F: 031-153815, E: jan.romell@teknik.skanska.se
Ronander, B., Managing Director, Bomo-Contract AB, Sutarebo, Mahult, 310 38 SIMLÅNGSDALEN. T: 035-78121, F: 035-78121, E: ronander.nordander@telia.com
Roos, O., Vatten och Samhällsteknik AB, Box 742, 391 27 KALMAR. T: 0480-61500, F: 0480-86739, E: olle.roos@vosteknik.se
Rosén, B., Hydrogeologist, Swedish Geotechnical Institute, 581 93 LINKÖPING. T: 013-201808, F: 013-201912, E: bengt.rosen@swedgeo.se
Rosén, H., M.Sc., J&W Samhällsbyggnad, Robertsviksgatan 7, 972 41 LULEÅ. T: 0920-238300, F: 0920-238381, E: hakan.rosen@jw.se
Rosén, R., M.Sc., Golder Grundteknik KB, Box 20127, 104 60 STOCKHOLM. T: 08-50630600, F: 08-50630601, E: rolf_rosen@golder.se
Rubensson, A., M.Sc., Skanska Teknik AB, 169 83 SOLNA. T: 08-50436950, F: 08-7536048, E: annika.rubensson@teknik.skanska.se
Ruin, M., M.Sc., Hercules Grundläggning AB, Box 1283, 164 29 KISTA. T: 08-7503319, F: 08-7506057, E: magnus.ruin@hercules.se
Ryberg, U., Vägverket Konsult, Box 1398, 801 38 GÄVLE. T: 026-149368, F: 026-129944, E: ulf.ryberg@vv.se
Rydell, B., Director, Swedish Geotechnical Institute, 581 93 LINKÖPING. T: 013-201824, F: 013-201909, E: bengt.rydell@swedgeo.se
Rydén, C.-G., Ph.D., Swedish National Road Administration, 781 87 BORLÄNGE. T: 0243-75670, E: clas-goran.ryden@vv.se
Rydström, H., Geologist, J&W Energi och Miljö, 121 88 STOCKHOLM-GLOBEN. T: 08-6886419, F: 08-6886922, E: hans.rydstrom@jw.se
Ryner, A., M.Sc., Golder Grundteknik KB, Box 20127, 104 60 STOCKHOLM. T: 08-50630600, F: 08-50630601, E: anders.ryner@golder.se
Röshoff, K., Managing Director, BergByggKonsult AB, Box 3439, 165 23 HÄSSELBY. T: 08-7595050, F: 08-7595065, E: kennert.roshoff@bergbyggkonsult.se
Sahlberg, O., M.Sc., Biblioteksgången 2, 183 70 TÄBY
Sahlström, B., Eng., Golder Grundteknik KB, Box 20127, 104 60 STOCKHOLM. T: 08-50630600, F: 08-50630601, E: birgitta_sahlstrom@golder.se
Sahlström, P., Affärsstigen 3, 182 74 STOCKSUND
Sahlström, P. Olof, M.Sc., Swedish National Road Administration, Box 4202, 171 04 SOLNA. T: 08-7576690, F: 08-7576390
Sandberg, E., M.Sc., Gatubolaget, Box 1086, 405 23 GÖTEBORG. T: 031-628000, F: 031-628002, E: emil.sandberg@gatubolaget.goteborg.se
Sandberg, U., Eng., Nordisk Kranuthyrning AB, Bastborregatan 11, 721 34 VÄSTERÅS. T: 021-127095, E: nordkran.u.sandberg@swipnet.se
Sandegren, E., Chief Engineer, Kräftvägen 11, 181 30 LIDINGÖ
Sandqvist, G., M.Sc., Repslagargatan 16, 118 46 STOCKHOLM
Sandros, C., M.Sc., Scandiaconsult Sverige AB, Box 5343, 402 27 GÖTEBORG. T: 031-3353308, F: 031-400571, E: cssmvg@scc.se
Sandström, T., Värdshusvägen 36, 145 50 NORSBORG
Schalin, L., M.Sc., Swedish National Road Administration, Box 1062, 551 10 JÖNKÖPING. T: lena.schalin@vv.se
Schulze, J.F., Senior Eng., & Consultant, Scand Pro AB, Korta gatan 7, 171 54 SOLNA. T: 08-56200316, F: 08-56200350, E: janfredrik.schulze@mailbox.swipnet.se
Schälin, J., M.Sc.
Secund, S., Bredängen 40, 427 00 BILLDAL

Sellberg, B., FORMAS, Box 1206, 111 82 STOCKHOLM. T: 08-7754028, F: 08-7754010, E: bjorn.sellberg@fomas.se

Sellgren, E., Ph.D., J&W Samhällsbyggnad, 121 88 STOCKHOLM-GLOBEN. T: 08-6886187, F: 08-6886916, E: eskil.sellgren@jw.se

Sigsäter, L., Geologist, VBB VIAK AB, Geijersgatan 8, 216 18 MALMÖ. T: 040-167000, F: 040-154347, E: lars.sigsater @sweco.se

Sigurdsson, T., Ph.D., Swedish National Road Administration, Södra Hamngatan 5, 411 14 GÖTEBORG. T: 031-3396110, F: 031-3396111, E: thrainn.sigurdsson@vv.se

Sjöberg, S.-G., Research & Consulting Eng., Swedish Geotechnical Institute, Box 4202, 171 04 SOLNA. T: 08-7576881, F: 08-288445, E: svengunnar.sjoberg@swedgeo.se

Sjödahl, T., Eng., Tyréns Infrakonsult AB, 118 86 STOCKHOLM. T: 08-4290251, F: 08-6448850, E: thord.sjodahl@tyrens.se

Sjödin, M., M.Sc., MS Civil, Avtagsvägen 1 A, 191 50 SOLLENTUNA. T: 08-962136, F: 08-962136, E: mats.sjodin@privat.utfors.se

Sjögren, M., Tyréns Infrakonsult AB, 851 71 SUNDSVALL. T: 060-169600, F: 060-169629, E: micael.sjogren@tyrens.se

Sjögren, S., Tyréns Infrakonsult AB, 851 71 SUNDSVALL. T: 060-169625, F: 060-169629, E: sune.sjogren@tyrens.se

Sjökvist, K., EurIng FEANI, CurtDok SA Ltd., Mölndalsbacken 62, Stureby, 122 66 ENSKEDE. T: 08-864510

Sjölin, F., M.Sc., BAT Geosystems AB, Box 1060, 186 26 VALLENTUNA. T: 08-51170600, F: 08-51173361, E: fredrik@batab.se

Sjölin, J., J&W, Rullagergatan 6, 415 26 GÖTEBORG. T: 031-7272500, F: 031-7272501, E: jan.sjolin@jw.se

Sjölund, G., M.Sc., EkoTec AB, Box 2092, 174 02 SUNDBYBERG. T: 08-6640545, F: 070-6196620, E: gustaf.sjolund@midroc.se

Sjöström, Ö.A., M.Sc., Orjan Sjostrom Ltd., T10-40A, South Horizons, Ap Lei Chau, HONG KONG. T: +852-2234-5156, F: +852-2234-5156, E: sjostrom@hknet.com

Skogsberg, J., M.Sc., J&W Samhällsbyggnad, Box 8094, 700 08 ÖREBRO. T: 019-178950, F: 019-133200

Smekal, A., M.Sc., Swedish National Rail Administration, 781 85 BORLÄNGE. T: 0243-445645, F: 0243-445617, E: alexander.smekal@banverket.se

Sommar, L., ASTRA, 151 85 SÖDERTÄLJE. T: 0755-26000

Spetz, T., Geotechnical Eng., FB Engineering AB, Box 12076, 402 41 GÖTEBORG. T: 031-7751174, F: 031-7751122, E: tsp@fbe.se

Spångberg, B., Eng., Bagarfruvägen 13, 1 tr, 128 67 SKÖNDAL

Stenersen, J., Skanska Berg och Bro, Industrivägen 6-8, 137 37 VÄSTERHANINGE. T: 08-50073811, F: 08-50073815, E: jorgen.stenersen@skanska.se

Stenqvist, F., VBB VIAK AB, Box 1848, 581 17 LINKÖPING. T: 013-252700

Stephansson, O., Professor, Royal Institute of Technology, 100 44 STOCKHOLM. T: 08-7907906, F: 08-7906810, E: ove@ce.kth.se

Stille, B., Civil Eng., Skanska Teknik AB, 169 83 SOLNA. T: 08-50435344, F: 08-7536048, E: bjorn.stille@teknik.skanska.se

Stille, H., Prof., Kungliga Tekniska Högskolan, 100 44 STOCKHOLM. T: 08-7906000, F: 08-7906500

Stjernkvist, J., Field Eng., Klabböle 82, 905 87 UMEÅ

Stjernkvist, S., Martin Jokus väg 47, 260 41 NYHAMN SLÄGE

Strindberg, P., FB Engineering AB, Box 12076, 402 41 GÖTEBORG. T: 031-7751000, F: 031-122063

Ström, B., HB Fundamenta Grundrådgivning, Örnsätrabacken 72, 127 36 SKÄRHOLMEN. T: 08-888208, E: carl.bertil@telia.com

Ström, C., B.Sc., Trollrunan 98, 423 46 TORSLANDA

Stål, F., VBB VIAK AB, Box 1902, 791 19 FALUN. T: 023-46436, F: 023-46401, E: fredrik.stal@sweco.se

Sundberg, J., Ph.D., Geo Innova AB, Teknikringen 1 C, 583 30 LINKÖPING. T: 013-363090, F: 013-363091, E: jan.sundberg@geoinnova.se

Sundquist, O., President, BORROS AB, Box 3063, 169 03 SOLNA. T: 08-272620, F: 08-836702, E: info@borros.se

Sundström, B., Managing Director, QTENT Konsult, Box 168, 921 23 LYCKSELE. T: 0950-26400, F: 0950-37700, E: borje@qtent.se

Sundström, L., Eng., VBB VIAK AB, Geijersgatan 8, 216 18 MALMÖ. T: 040-167000, F: 040-154347, E: larsake.sundstrom@sweco.se

Svedberg, B., M.Sc., Scandiaconsult Sverige AB, Box 4205, 102 65 STOCKHOLM. T: 08-6156416, F: 08-7021934, E: bo.svedberg@scc.se

Svenningsson, P., Vägverket Produktion, Nya Tingstadsgatan 12, 422 44 HISINGS BACKA. T: 031-656500, F: 031-656599, E: peter.svenningsson@vv.se

Svensk, I., M.Sc., Skanska, 405 18 GÖTEBORG. T: 031-7711000, F: 031-153815, E: ingmar.svensk@teknik.skan ska.se

Svensson, B., Brantingsgatan 52, 115 35 STOCKHOLM

Svensson, B.-E., Swedish National Road Administration, Box 601, 301 16 HALMSTAD. T: 035-151500, F: 035-128 197

Svensson, J., M.Sc., Grengatan 85, 589 55 LINKÖPING. E: jorgen. greng85@ebox.tninet.se

Svensson, L., Eng., GM Consult, Box 17071, 104 62 STOCKHOLM. T: 08-680935

Svensson, L.-G., Eng., S Djursgård, Lekaryd, 342 92 ALVESTA

Svensson, L., M.Sc., J&W Samhällsbyggnad, Box 382, 401 26 GÖTEBORG. T: 031-614450, F: 031-154668, E: lennart.svensson@jw.se

Svensson, M., Post Graduate Student, University of Lund, Box 118, 221 00 LUND. T: 046-2227426, F: 046-2229127, E: mats.svensson@tg.lth.se

Svensson, P., M.Sc., J&W Energi och Miljö, Slagthuset, 211 20 MALMÖ. T: 040-108393, F: 040-108281, E: paul.svensson@jw.se

Svensson, P.L., M.Sc., SUAB Spångarna Utveckling AB, 370 30 RÖDEBY. T: 0455-338476, F: 0455-338474, E: psnska@telia.com

Svensson, R., M.Sc., GeoExperten RS AB, Box 4155, 227 22 LUND. T: 046-307001, F: 046-184908, E: geo.experten@swipnet.se

Svensson, T., FB Engineering AB, Box 12076, 402 41 GÖTEBORG. T: 031-7751047, F: 031-7751166, E: ts@fbe.se

Swanson, O., Managing Director, Arkitekt- och Ingenjörs, Box 16105, 103 22 STOCKHOLM. T: 08-232300, F: 08-204913

Swedenborg, P.J., Margaretavägen 16 B, 187 74 TÄBY

Swedenborg, S., M.Sc., Höbacken 12, 432 52 VARBERG

Säfström, L., Swedish National Road Administration, Box 1140, 631 80 ESKILSTUNA. T: 016-157169, F: 016-157080, E: leif.safstrom@vv.se

Sällfors, G., Prof., Chalmers University of Technology, 412 96 GÖTEBORG. T: 031-7722100, F: 031-7722107, E: sallfors@vsect.chalmers.se

Säterkvist, A., M.Sc., Scandiaconsult Sverige AB, Box 1932, 791 19 FALUN. T: 023-84050, F: 023-19983, E: astbnf@scc.se

Sätterström, P.G., Eng., AB Jacobson & Widmark, Box 34, 371 21 KARLSKRONA. T: 0455-44750, F: 0455-44751, E: goran.satterstrom@jw.se

Söder, C.-O., Consulting Eng., SWECO, Box 34044, 100 26 STOCKHOLM. T: 08-6956038, F: 08-6956350, E: carl-olof.soder@sweco.se

Söderblom, R., Ph.D., Krokusvägen 15, 181 31 LIDINGÖ

Södergren, I., M.Sc., Swedish National Road Administration, 781 80 BORLÄNGE. T: 0243-75870, F: 0243-75565, E: ingrid.sodergren@vv.se

Sörgårn, L., Eng., Swedish National Road Administration, Box 6, 351 03 VÄXJÖ. T: 0470-755564, F: 0470-755555, E: leif.sorgarn@vv.se

Tengborg, P., Project Engineer, Sydkraft Konsult, 205 09 MALMÖ. T: 040-255804, F: 040-304696, E: per.tengborg@sycon.se

Tenne, M., Geologist, Geokonsult Tenne AB, Segeltorpsvägen 11, 125 53 ÄLVSJÖ. T: 08-861044, F: 08-998266, E: mats.tenne@mbox300.swipnet.se

Thorén, A., M.Sc., Atek mark och miljö, Dadelvägen 12, 741 30 KNIVSTA. T: 018-291533, E: at.atek@telia.com

Thorén, H., B.A., Swedish National Road Administration, 405 33 GÖTEBORG. T: 031-635311, F: 031-635212

Thun, N., Geotechnician, Banverket, Norra Regionen, 971 02 LULEÅ. T. 0920-35172, E: niklas.thun@banverket.se

Thunström, C., Eng., Scandiaconsult Sverige AB, Box 5343, 402 27 GÖTEBORG. T: 031-3357451, F: 031-400855

Thurner, H., Managing Director, Geodynamik AB, Box 7454, 103 92 STOCKHOLM. T: 08-206790, F: 08-206795, E: heinz.thurner@geodynamik.com

Tibblin, L.-G., VBB Viak, Box 34044, 100 26 STOCK-HOLM. T: 08-6956000, F: 08-6956010

Tilly, L., M.A., VAI VA-Projekt AB, Box 47005, 100 74 STOCKHOLM. T: 08-7758794, F: 08-7441624, E: lena. tilly@vaivaprojekt.se

Torkeli, C., M.Sc., Nya Tanneforsvägen 21A, 582 42 LINKÖPING

Torsteinsrud, K., M.Sc., Swedish National Rail Administration, Box 1346, 172 27 SUNDBYBERG. T: 08-7622020, F: 08-7622680, E: kjell.torsteinsrud@banverket.se

Torstensson, B.-A., D.Sc., BAT Geosystems AB, Box 1060, 186 26 VALLENTUNA. T: 08-51170600, F: 08-51173361

Torstensson, D., M.Sc., Kungliga tekniska högskolan, 100 44 STOCKHOLM. T: 08-7906649, F: 08-7908689, E: daniel@ce.kth.se

Toyrä, B., M.Sc., Swedish National Rail Administration, Box 43, 971 02 LULEÅ. T: 0920-35288, F: 0920-35205

Tremblay, M., Research & Consulting Eng., Tyréns Infrakonsult AB, Lilla Badhusgatan 4, 411 21 GÖTEBORG. T: 031-606318, F: 031-606301, E: marius.tremblay@tyrens.se

Tränk, R., Consulting Engineer, Swedish Geotechnical Institute, 581 93 LINKÖPING. T: 013-201841, 08-7576762, F: 013-201912, 08-983030, E: roland.trank@swedgeo.se

Törnell, J., JiTe Geo, Lilla Källstorp, 234 35 LOMMA. T: 040-411143, F: 040-531298, E: jonas.tornell@telia.com

Udén, D., M.Sc., Sycon Teknikkonsult AB, Box 9119, 961 19 BODEN. T: 0921-76838, F: 0921-76850, E: dag.uden@sycon.se

van, M.P., B.Sc., Stenolles backe 34, 134 62 INGARÖ

Vestgård, J., Eng., Golder Grundteknik KB, Box 20127, 104 60 STOCKHOLM. T: 08-50630600, F: 08-50630601

Viberg, L., Chief Engineer, Swedish Geotechnical Institute, 581 93 LINKÖPING. T: 013-201846, F: 013-201914, E: leif. viberg@swedgeo.se

Viking, K., Ph.D. Student, Royal Institute of Technology, 100 44 STOCKHOLM. T: 08-7908025, F: 08-7907928, E: viking@aom.kth.se

Viklander, P., Luleå University of Technology, 971 87 LULEÅ. T: 0920-91334, F: 0920-72075, E: Peter.Viklander@anl. luth.se

Vikström, L., Research & Development Eng., Luleå University of Technology, 971 87 LULEÅ. T: 0920-72537, F: 0920-72075, E: larsv@ce.luth.se

Vilkénas, A., M.Sc., Stockholm Konsult, Box 9611, 117 91 STOCKHOLM. T: 08-7858656, F: 08-7858504, E: algis. vilkenas@stockholmkonsult.se

Vinberg, P., Per Vinberg Projektledning AB, Gökbacksvägen 13, 135 68 TYRESÖ, E: per.vinberg@home.se

von Reedtz, P., M.Sc., Hercules Grundläggning AB, Box 623, 194 26 UPPLANDS-VÄSBY. T: 08-59078355, F: 08-59078358, E: peter.vonreedtz@hercules.se

Väst, G., B.Sc., Vägverket Konsult, Box 1008, 901 20 UMEÅ. T: 090-172793, F: 090-172610, E: goran.vast@vv.se

Wagenius, J., M.Sc., Golder Grundteknik KB, Box 20127, 104 60 STOCKHOLM. T: 08-50630600, F: 08-50630601, E: johan_wagenius@golder.se

Wahlstrom, L., M.Sc., Swedish National Rail Administration, Box 1346, 172 27 SUNDBYBERG. T: 08-7623990, F: 08-7623388, E: lennart.wahlstrom@banverket.se

Wallin, T., BESAB, Tagenevägen 7, 422 59 HISINGS BACKA. T: 031-7427000, F: 031-7427020

Wallmark, G., Eng., Swedish National Rail Administration, Box 1346, 172 27 SUNDBYBERG. T: 08-7623286, F: 08-7622175, E: goran.wallmark@banverket.se

Wallner, F., M.Sc., VBB VIAK AB, Box 259, 851 04 SUNDSVALL. T: 060-169910, F: 060-613007, E: fredrik. wallner@sweco.se

Wassenius, J., M.Sc., Scandiaconsult, Box 5343, 402 27 GÖTEBORG. T: 031-3353305, F: 031-400571, E: jwsmvg@scc.se

Weckman, H., Managing Director, VK-Infracom Oy, Kalevalavägen 5 C, FI-02130 ESBO. T: +358-9-4393150, F: +358-9-43931550, E: hans.weckman@vk-group.com

Wedel, P.O., Mossberget 10, 427 00 BILLDAL

Wennerstrand, J., M.Sc., KTH-Haninge, Marinens väg 30, 136 40 HANINGE. T: 08-7073149, F: 08-7073127, E: jawe@haninge.kth.se

Werkelin, D., AB PentaCon, Södra Kyrkogatan 3, 621 56 VISBY. T: 0498-279085, F: 0498-247415, E: daniel. werkelin@pentacon.se

Wernqvist, G., Engineer, Herkulesvägen 26, 135 47 TYRESÖ

Wesström, B., B.Sc., SWEDECONSULT B WESSTRÖM HB, Värtavägen 50, 183 63 TÄBY. T: 08-7568421, F: 08-7568421, E: swedeconsult.wesstrom@telia.com

Westerberg, B., Postgraduate Student, Luleå University of Technology, 971 87 LULEÅ. T: 0920-91358, F: 0920-72075, E: Bo.Westerberg@anl.luth.se

Westerlund, B., M.Sc., Vägverket, 405 33 GÖTEBORG. T: 031-635104, F: 031-635212, E: bo.westerlund@vv.se

Wetterholm, J., Engineer, Skansbergsvägen 6 A, 141 70 HUDDINGE

Wetterholt, L., M.Sc., Tyréns Infrakonsult AB, Box 27, 291 21 KRISTIANSTAD. T: 044-186266, E: lennart.wetterholt@tyrens.se

Wetterling, S., Managing Director, Soilex AB, Bjurholmsplan 26, 116 63 STOCKHOLM. T: 08-6418826, F: 08-6401821, E: soilex@mbox200.swipnet.se

Wiberg, N.-E., Prof., Chalmers University of Technology, 412 96 GÖTEBORG. T: 031-7721000

Widerström, J., Consulting Eng., Vargungevägen 21, 370 24 NÄTTRABY. F: 0455-49224, E: j.widerstrom@telia.com

Widing, S., M.Sc., Ringstedsgatan 63, 164 48 KISTA

Widmark, T., Eng., Swedish National Road Administration, Box 595, 721 10 VÄSTERÅS. T: 021-164045, F: 021-164056, E: torild.widmark@vv.se

Wiktorsson, P., M.Sc., Vägverket, Box 1140, 631 80 ESKILSTUNA. T: 016-157217, F: 016-157080, E: per.wiktorsson@vv.se

Wilén, P., Geohydrologist, Swedpower AB, Folkungagatan 20, 411 02 GÖTEBORG. T: 031-624900, F: 031-624950, E: peter.wilen@swedpower.vattenfall.se

Willer, F., M.Sc., AB Stockholm Konsult, Box 9611, 117 91 STOCKHOLM. E: frank.willer@stockholmkonsult.se

Willquist, K., Eng., J&W Samhällsbyggnad, Box 382, 401 26 GÖTEBORG. T: 031-614300, F: 031-614305, E: kristina. willquist@jw.se

Zackrisson, P., M.Sc., Vägverket Konsult, 781 87 BORLÄNGE. T: 0243-94318, F: 0243-87927, E: peter.zackrisson @vv.se

Zweifel, G., M.Sc., Vägverket Region Norr, Box 809, 971 25 LULEÅ. T: 0920-243927, F: 0920-243860, E: gunnar. zweifel@vv.se

Åhnberg, H., Research & Consulting Eng., Swedish Geotechnical Institute, 581 93 LINKÖPING. T: 013-201857, F: 013-201914, E: helen.ahnberg@swedgeo.se

Åkerblom, G., Geologist, Fiskarfjärdsstranden 11, 127 41 SKÄRHOLMEN. E: gustav.akerblom@swipnet.se

Änäs, M., Engineering Geologist, Swedish National Road Administration, Box 186, 871 24 HÄRNÖSAND. T: 0611-44181, F: 0611-44054, E: mikael.anas@vv.se

Åstedt, B., M.Sc., Swedish National Rail Administration, Box 366, 201 23 MALMÖ. T: 040-202775, F: 040-202019, E: bjorn.astedt@banverket.se

Åström, M., M.Sc., Tyréns Infrakonsult AB, 851 71 SUNDSVALL. T: 060-169060, F: 060-169629, E: maria. astrom@tyrens.se

Öhman, S.-Å., Eng., J&W Samhällsbyggnad, Laholmsvägen 10, 302 48 HALMSTAD. T: 035-181100, F: 035-181101, E: sven-ake.ohman@jw.se

Öhrner, P., M.Sc., VBB Anläggning AB, Box 34044, 100 26 STOCKHOLM. T: 08-6956229, F: 08-6956040, E: per. ohrner@sweco.se

Örtendahl, P., Ingenjörsfirman Geotech AB, Datavägen 53, 436 32 ASKIM. T: 031-289920, F: 031-681639

Östergren, T., M.Sc., Tellstedt Geoteknik AB, Varbergsgatan 12, 412 65 GÖTEBORG. T: 031-405280, F: 031-404076, E: thomas.ostergren@tellstedt.se

SWITZERLAND/SUISSE

Secretary: Dr. Markus Caprez, Swiss Federal Institute of Technology, Swiss Society for Soil and Rock Mechanics, CH-8093, Zürich. T: +(41) 1 371 66 56, F: +(41) 1 633 10 62, E: caprez@igt.baug.ethz.ch, sgbf@igt.baug.ethz.ch

Total number of members: 248

Abdulahi, S., Käferholzstr. 157, CH-8046, Zurich
Aeberli, U., Dipl. Geologe ETHZ, Im Lerchen, Heubachstr. 59, CH-8810, Horgen. T: 01 725 53 47
Amann, P., Prof. Dr., Institut fuer Geotechnik, ETH-Hoenggerberg, CH-8093, Zuerich. T: 01 633 26 00, F: 01 633 10 79
Amberg, G., Dipl. Masch.-Ing. ETH, Tuschgenweg 111, CH-8041, Zuerich. T: 01 482 47 33
Ammann, J., Dipl. Bauing. ETH, Fadenstrasse 18, CH-6300, Zug, F: 041 711 24 50
Anagnostou, G., Dr. sc. techn., Sonneggstr. 29, CH-8006, Zuerich. T: 01 261 83 17, F: 01 633 10 97
Andres, Geotechnik AG, Dipl.Ingenieure ETH/SIA, Kesselhaldenstr. 57a, CH-9016, St. Gallen. T: 071 288 27 88, F: 071 288 36 12
Bachmann, S., Dipl. Bau-Ing. ETH, Sonnhaldenstr. 64, CH-3210, Kerzers
Bachofen, R., Dipl. Bauing. ETH, Mueslistrasse 7, CH-8307, Effretikon. T: 052 343 03 82, F: 01 383 26 50
Baenziger, D.J., Baenziger+Bacchetta+Partner, Engmattstr. 11, CH-8027, Zuerich. T: 01 201 71 00, F: 01 202 29 63
Baer, H., Ing.HTL, Luetisaemetstrasse 100, CH-8706, Meilen. T: 01 739 18 19, F: 01 252 22 46
Balissat, M., Dipl. Bauing. ETH-L, Im Weiherhau 9, CH-5405, Baden-Daettwil. T: 056 493 17 30, F: 056 493 17 60
Bardill, H., Dipl. Bauing. ETH, Lerchenbergstr. 21, CH-8703, Erlenbach. T: 01 910 99 92, F: 01 211 22 16
Basler, M., Dipl. Bauing. SIA, Rebbergstr. 49, CH-8049, Zuerich. T: 01 342 18 31
Bazzi, G., Dr.sc.tech. Dipl.Ing.ETH, Eichstrasse 15, CH-8957, Spreitenbach. T: 056 401 47 03, F: 01 302 19 11
Bender, H., Dr., Dipl. Bauing. ETH, Wybergstr. 41, CH-8542, Wiesendangen. T: 052 337 33 47
Bergamin, S.F., Dipl. Bau-Ing. ETH/SIA, Kilchbergstr. 68, CH-8038, Zuerich. T: 01 482 89 93, F: 01 633 10 97
Bernardi, W., Dipl. Bauing. ETH, Preyenstrasse 1, CH-8623, Wetzikon, F: 061 467 67 01
Birchler, Pfyl+Partner AG, Herrn Max Birchler, Riedstr. 7, CH-6430, Schwyz. T: 041 811 20 44, F: 041 811 71 70
Bischof, R., Dipl. Bauing. ETH, Sonnhalde 319, CH-5425, Schneisingen. T: 056 241 19 73
Bischoff, N., Dipl. Bauing. ETH, Kirchackerstr. 14, CH-8608, Bubikon. T: 055 243 24 12
Bonnard, C.A., EPFL, CH-1015, Lausanne. T: 021 693 23 12, F: 021 693 41 53
Bonzanigo, L., Dr.sc.net, EURING Consulenze geologiche, Viale Stazione 16A, C.P. 1152, CH-650, Bellinzona. T: 091 825 94 50
Bourdeau, P.L., Purdue University, USA-47907, West Lafayette, INDIANA., F: 765-494-5031
Bourquin, M., Ing.dipl.EPFZ/SIA. Rue des Sablons 10, CH-2000, Neuchâtel. F: 032 721 18 92
Brenner, R.P., Dr., Dipl. Bauing. ETH, Rosenstr. 8, CH-8570, Weinfelden. T: 071 622 32 03, F: 071 622 32 03
Bretscher, E., Bauing. HTL, Mitglied, Sonnenbergstr. 9, CH-8542, Wiesendangen. T: 052 337 16 47
Brugger, M., Bauing. HTL, Hoehtalstr. 37, CH-5408, Ennetbaden. T: 056 222 18 25
Bucher, F., Prof. Dr., Institut fuer Geotechnik, ETH-Hoenggerberg, CH-8093, Zuerich. T: 01 633 25 36, F: 01 633 10 79
Buchs, T., Bonnard & Gardel, Ingénieurs-, Avenue de Cour 61, CH-1007, Lausanne. T: 021 618 11 11, F: 021 618 11 22
Buol, P., Dipl.Ing. ETH, In den Bueelen 17, CH-7260, Davos-Dorf. T: 081 416 20 82, F: 081 410 15 30
Buro, M., Losinger Sion SA, Rte. de Vissigen 110, C.P. 925, CH-1051, Sion. T: 027 203 43 61, F: 027 203 43 89

Buser, H.R., Dipl. Bauing. ETH, Buchenrain 4, CH-8634, Hombrechtikon. T: 055 244 21 28
Caprez, M., Dr., Institut fuer Geotechnik, ETH-Hoenggerberg, CH-8093, Zuerich. T: 01 633 25 32, F: 01 633 10 62
Conrad, R., Dipl. Bauing. ETH, Unterburg 76, CH-8158, Regensberg. T: 01 853 17 15
De Crenville, Geotechnique SA, 17 Chemin des Champs-Courbes, CH-1024, Ecublens. T: 021 691 24 91, F: 021 691 24 96
De Montmollin, M., Ing.civil.dipl. EPFZ/SIA, Quai Max-Petitpierre 42, CH-2000, Neuchâtel
De Rivaz, B., Géologue, Les Combes, CH-1971, Grimisuat. T: 027 398 26 27
Degen, W., Dr., Vibroflotation AG, Letzistr. 21, CH-8852, Altendorf. T: 055 451 90 20, F: 055 451 90 21
Delaloye, S., Ing.civ.dipl. EPFL, Rue du Scex 16, CH-1950, Sion
Demont, J.-B., C. von der Weid & Associés SA, Av. du Moléson 11, c.p. 977, CH-1701, Fribourg. T: 026 322 23 55, F: 026 323 13 05
Descoeudres, F., Prof., Lab. de mécanique des roches, EPFL-Ecublens, CH-1015, Lausanne. T: 021 693 23 21, F: 021 693 50 60
Dillo, M., FBF (Boden-und Felsmechanik), Technikumstrasse 21, CH-6048, Horw. T: 041 349 34 54, F: 041 349 39 61
Drack, E., Dipl. Bauing. ETH, Via Seghezzone 4a, CH-6512, Giubiasco. T: 091 857 70 05
Dubas, C., Ing.civil.dipl. EPFZ, Au Sichoz 39, CH-1814, La Tour-de-Peilz. T: 021 944 09 51
Dudli, H., Dipl. Bauing. ETH, Pischastrasse, CH-7205, Zizers. T: 081 322 71 44
Dudt, J.-P., Ecole Polytechnique Lausanne, LMR-ISRF-GC-EPFL, CH-1015, Lausanne. T: 021 693 23 25, F: 021 693 41 53
Duerst, R., Dipl. Bauing. ETH, Zuercherstr. 177, CH-8645, Jona/Rapperswil. T: 055 210 46 91
Duret, J.-J., Géologue, Rue du Lac 15, CH-1207, Genève. T: 022 735 10 84
Dysli, M., Ing. dipl. EPFL, Parc de la Rouvraire 22, CH-1018, Lausanne. T: 021 647 96 64, F: 021 693 41 53
Eberhard, M., Eberhard & Partner AG, Schachenallee 29, CH-5000, Aarau. T: 062 823 27 07, F: 062 823 27 06
Eder, P., Dipl. Bauing. ETH, Wabersackerstr. 55, CH-3097, Liebefeld. T: 031 971 21 55
Einstein, H.H., Prof. Dr., Environmental Engineering, Room 1-342, USA, Cambridge MA 02139
Energie, O.S., EOS, Case postale 570, CH-1001, Lausanne. T: 021 341 21 11, F: 021 341 20 49
Fechtig, R., Prof., Federal Institute of Technology, ETH-Zentrum/ vdf, CH-8092, Zuerich. T: 01 633 31 18, F: 01 422 85 21
Fellmann, W., Mengis+Lorenz AG, Schlossstrasse 3, CH-6005, Luzern. T: 041 310 51 02, F: 041 310 79 79
Figi, H., Dipl. Bauing. ETH, Teuchelweg 49, CH-7000, Chur. T: 081 252 53 29
Fleury, F., Fleury Ingenieur Conseil, Chemin de la Laiterie 15, CH-1066, Epalinges. T: 021 784 34 86, F: 021 784 34 89
Floerl, W., Maxit AG, Hauptstrasse 30, CH-4127, Birsfelden
Flury, F., MFR Géologie-Géotechnique SA, 9, rue de Chaux, CH-2800, Delémont. T: 032 422 61 14
Fontana, A., Ing.civ.dipl. EPF/SIA, 37, Crêts de Champel, CH-1206, Genève. T: 022 347 59 69
Fontana, O., Ing.civil.dipl.EPFL/SIA, Konsumstrasse 11, CH-3007, Bern. T: 031 382 52 02
Fonyo, B., Sytec Bausysteme AG, Meriedweg 11, Postfach, CH-3172, Niederwangen. T: 031 980 14 14
Frey, R., Dr., Ingenieurbuero, Mühlegasse 18, 6340 Baar. T: 041 7606 393, F: 041 7606 392

Friedli, P., Buero fuer Geotechnik, Faerberstr. 31, CH-8008, Zuerich. T: 01 251 85 93

Gautschi, M.A., Dipl.Bauing. ETH/SIA, Zuerichstr. 39, CH-8118, Pfaffhausen. T: 01 825 41 04, F: 01 825 41 04

Gencer, M., Ing. civil, Dr.ès.sc.techn., Av. de Cour 6A, CH-1007, Lausanne. T: 021 616 62 24, F: 021 617 41 24

Geotechnisches, Institut AG, Gartenstr. 13, Postfach 6258, CH-3001, Bern. T: 031 389 34 11, F: 031 381 31 15

Gicot, O., Gico. T: Géotechnique, Fort St. Jacques 25, C.P. 64, CH-1703, Fribourg. T: 026 424 26 41, F: 026 424 26 60

Gilly, M., Plaz 86, CH-7530, Zernez. T: 081 856 17 00, F: 081 856 18 22

Grob, H., Prof., Dipl. Bauing. ETH/SIA, Rychenbergstr. 106, CH-8400, Winterthur. T: 052 213 41 95

Grossmann, D., Dipl. Bauing. ETH/SIA, 18 Larkhall Place, GB, Bath BA1 6SF. T: 44(0)1225311160, F: 01 361 64 13

Gysi, W., Dipl. Bauing. ETH/SIA, Summerauweg 18, CH-8623, Wetzikon. T: 01 930 63 53, F: 01 970 22 64

Haeny & Cie. AG, MIT Division, Postfach, CH-8706, Meilen. T: 01 925 41 11, F: 01 923 62 45

Hafen, F., Dipl. Bau-Ing. ETH, Wangelenrain 53, CH-3400, Burgdorf. T: 034 422 26 59

Handke, A., Geologe ETH/SIA, Rohanstrasse, CH-7205, Zizers

Heitzmann, P., Bundesamt fuer Wasser und, CH-3003, Bern. T: 031 324 76 85, F: 031 324 76 81

Hermanns, S., Prof. Dr., Rita, Institut fuer Geotechnik, ETH-Hoenggerberg, CH-8093, Zuerich. T: 01 633 25 24, F: 01 633 10 79

Heubi, C., Bonnard Et:Gardel, Avenue de Cour 61, CH-1001, Lausanne. T: 021 618 11 11, F: 021 617 47 18

Hey, F., Ing.civil.dipl. EPFL, Ch. des Colombettes 52, CH-1740, Neyruz. T: 026 477 02 28, F: 026 460 74 79

Hoerler, J., Dipl. Bauing. ETH/SIA, Pilatusstrasse 16, CH-8203, Schaffhausen. T: 052 625 76 67

Hofacker, H., Stucki, Hofacker + Partner AG, Eng-weg 7, CH-8006, Zuerich. T: 01 361 03 14, F: 01 361 01 60

Hohberg, J.-M., IUB Ing.-Unternehmung AG, Thunstrasse 2, CH-3000, Bern 6. T: 031 357 11 11, F: 031 357 11 12

Hohl, J., Ingenieurbuero Th. Hohl, Untere Schmitten, CH-3036, Detligen. T: 031 825 61 53, F: 031 825 60 49

Hollenweger, R., Dipl. Bauing. ETH, Alte Landstrasse 187, CH-8800, Thalwil. T: 01 720 89 44

Honold, P., Dr., Dipl. Bauing. ETH, Langacherstr. 8, CH-8103, Unterengstringen. T: 01 750 57 10, F: 01 363 36 29

Horisberger, W., Horisberger + Nydegger AG, Klaraweg 1, CH-3006, Bern. T: 031 352 99 88, F: 031 352 12 17

Huber, M.P., smh Tunnelbau AG, Obere Bahnhofstr. 58, CH-8640, Rapperswil. T: 055 210 74 04, F: 055 210 68 28

Huder, J., Prof., Dr.sc.techn., Greifenseestr. 31, CH-8603, Schwerzenbach. T: 01 825 37 85, F: 01 633 10 79

Immer, W.F., SBB-CRM-QM, Parkterrasse 14, CH-3000, Bern 65. T: 0512 20 51 87, F: 0512 20 42 45

Iseli, B.-F., Bureau de Géologie, Hinterbueelstrasse 3, CH-8307, Effretikon. T: 052 343 60 40

Juhola, M.O., Ing.buero M O Juhola Oy, Lounaisvaeylae 2 a.4., SF-00200, Helsinki

Junker, J.P., Dr.sc.techn.dipl.Ing.ETH, Thun-strasse 93a, CH-3074, Muri. T: 031 951 45 30

Kaempfen, E., Dipl. Bauing. ETH, Wieggischmatta 27, CH-3911, Ried-Brig. T: 027 924 54 16

Kalt, L., Dipl. Bauing. ETH/SIA, Paradisstrasse 24a, CH-9402, Moerschwil. T: 071 866 17 67

Kapp, H., Grundbauberatung AG, Helvetiastrasse 41, CH-9000, St. Gallen. T: 071 244 88 44, F: 071 244 88 16

Kappeler, H., Dipl. Ing. ETH/SIA, Walchstr. 13, CH-3073, Guemligen., F: 031 951 10 12

Karakas, I., Ing.civ.dipl. M.S.C.E./SIA, Chemin du Bochet 7, CH-1025, St. Sulpice. T: 021 613 14 44, F: 021 617 41 34

Kiefer, P., Kiefer & Studer AG, Therwilerstr. 27, CH-4153, Reinach. T: 061 711 94 76, F: 061 711 96 34

Kipfer, H.-R., Dipl. Ing. ETH, Bubendoerferstr. 11, CH-4411, Seltisberg. T: 061 911 03 30, F: 061 921 55 42

Kobel, M., Dr. Phil., Dipl. Geologe, Breitenstrasse 50, CH-8832, Wilen/SZ. T: 01 784 04 97, F: 01 784 05 90

Kraehenbuehl, V., Dipl. Bauing. ETH, Kleinalbis 35, CH-8045, Zuerich. T: 01 461 22 63, F: 01 461 22 29

Kraft, H.-P., Frank + Kraft + Partner, Hofer-Str. 1, D-81737, Muenchen. T: 49-89-6700610, F: 49-89-67006133

Krumdieck, A., Dipl. Bauing. FIUC, Rebackerstr. 12, CH-8955, Oetwil a.d.Limmat. T: 01 748 03 50, F: 01 252 22 46

Laloui, L., Dr., EPFL, LMS-DGC, CH-1015, Lausanne. T: 021 693 23 14, F: 021 693 41 53

Lang, H.-J., Prof., Dipl. Bauing. ETH, Weidstr. 20, CH-8103, Unterengstringen. T: 01 750 26 24, F: 01 633 10 79

Lardelli, T., Buero fuer Technische Geologie, Steinbruchstrasse 12, CH-7002, Chur. T: 081 252 88 30, F: 081 253 55 13

Lefèbvre, H.L., Ing. civ. Reg.A, Rue de la Justice 5, CH-1096, Cully

Leoni, R., Leoni Gysi Sartori, Via alla Campagna 4, CH-6924, Sorengo. T: 091 967 62 67, F: 091 967 62 47

Lindemann, M., Dipl. Ing., Heideckstr. 10, D-80637, Muenchen

Locher, P., Ing. HTL, Studerstrasse 58, CH-3004, Bern. T: 031 301 21 54

Locher, T., Beratender Geologe, Waldstr. 4, CH-8118, Pfaffhausen. T: 01 383 87 66, F: 01 383 26 76

Luethi, F., Dipl. Bau-Ing. HTL, Nussbaumstrasse 48 A, CH-3006, Bern. T: 031 333 52 22, F: 034 422 39 60

Marbet, U., Theo Mueller & Partner AG, Bielstrasse 10, CH-4500, Solothurn. T: 032 625 85 25, F: 032 625 85 20

Marche, R., Dr., 18 Chemin de la Grande Piece, CH-1227, Carouge. T: 022 300 20 12

Margreth, S., Dipl. Bauing. ETH, Duchliweg 1B, CH-7260, Davos

Marmier, J.-P., Hunziker et Marmier SA, Fontenailles 21, CH-1007, Lausanne. T: 079 301 15 84, F: 021 616 88 80

Martinenghi, L., Dr., Studio d'ing. Martinenghi SA, Vicolo dei Saroli 1, CH-6944, Cureglia. T: 091 967 35 17, F: 091 967 35 19

Martinenghi, T., Istituto Consulenza Geotecnica, Via Besso 7, CH-6900, Lugano. T: 091 966 07 77, F: 091 967 22 24

Maurer, H., Dr.sc.nat., ETHZ, Institut für Geophysik, ETH-Hönggerberg, CH-8093, Zurich, T: 01 633 38 38, F: 01 633 10 65

Mayor, P.-A., Institut fuer Geotechnik, ETH-Hoenggerberg, CH-8093, Zuerich. T: 01 633 34 10

Meia, J., Dr., ès sc., géologue, 2, Villarets, CH-2036, Cormondrèche. T: 032 731 23 07, F: 032 731 23 07

Meier, P., Dipl. Bauing. ETH, Montjolweg 18, CH-9475, Sevelen. T: 081 785 28 55

Monigatti, F., Devaud Monigatti et Associés, Rte St-Nicolas-de-Fluee 16, CH-1709, Fribourg. T: 026 425 84 00, F: 026 424 19 41

Moore, D., Dipl. Bau-Ing., Branzistr. 21, 8303 Bassersdorf

Morelli, A., Dipl. Bau-Ing. ETH, Berneggweg 8, CH-8055, Zuerich. T: 01 450 78 37

Mueller, E.R., Dipl. Nat., Laubgasse 8, CH-8500, Frauenfeld. F: 052 720 14 10

Mueller, R.W., Bundesamt fuer Wasser und, Laendtestrasse 20, Postfach, CH-2501, Biel. T: 032 328 87 25, F: 032 328 87 12

Naterop, D., Solexperts AG, Ifangstrasse 12, Solexperts AG, CH-8603, Schwerzenbach. T: 01 825 29 29

Neuenschwander, M., Navarra & Neuenschwander Sagl, via Daro 6a, CH-6500, Bellinzona. T: 091 825 85 03, F: 091 826 12 34

Noher, H.-P., Geotechnisches Institut AG, Hochstr. 48, CH-4002, Basel. T: 061 365 28 00, F: 061 365 23 79

Oboni, F., Oboni & Associés SA, Les Biolettes, CH-1054, Morrens. T: 021 731 44 25, F: 021 731 44 25

Otta, L., Dr., Dipl. Bauing., Langacherstrasse 11a, CH-8127, Forch. T: 01 980 10 81

Pedrozzi, G., Studio di Geologia ing., Via Ligamo 20, 6963, Pregassona. T: 091 941 23 51, F: 091 942 80 16

Pellet, F., Université de Grenoble, BP 53, F-38041, Grenoble cedex 9. T: 33 04 76 825000, F: 33 04 76 827000

Perret, F.M., Ing.dipl.EPFZ, Falaises 140, CH-2000, Neuchâtel. T: 032 710 11 56

Perrin, F., Ingenieur ETS, 20 Av. des Amazones, CH-1224, Chêne-Bougeries. F: 022 786 18 30

Peyer, O., Dipl. Bauing. ETH, unt. Bubenholzstrasse 98, CH-8152, Opfikon. T: 01 811 40 47

Piguet, J.-M., Ing.civ.dipl. EPFL, Chale T: la Jaucheraie, CH-1972, Anzère. F: 021 693 41 53

Pradel, D., Dr. Ing. Senior Engineer, 340 S. Doheny Dr., CA 90211, Beverly Hills, CA 90211

Racine, C., Bergstr. 125, CH-8032, Zuerich. T: 01 252 41 11

Ramholt, T., Dipl. Bauing. ETH, Hegibachstrasse 114, CH-8032, Zuerich. T: 01 422 57 52

Reimbert, A.M., Ing. Conseil, 67, Bd. de Reuilly, F-75012, Paris. T: 0033/3430872

Rey, J.-P., Stump Sondages SA, En Viorens, CH-1037, Etagnières. T: 021 731 47 11, F: 021 731 15 13

Richardet, Y., Stucky Ingénieurs-Conseils SA, Rue du Lac 33, CH-1020, Renens 1. T: 021 637 15 54, F: 021 637 15 08

Rieder, U., Dipl. Bauing. ETH, M.S., Tulpenweg 3, CH-3324, Hindelbank. T: 034 411 10 35, F: 031 911 51 82

Roose, K., Dipl. Bauing. ETH/SIA, Wildenbuehlstrasse 64, CH-8135, Langnau a/Albis. T: 01 713 11 29

Rueck, P., Materialtechnik am Bau, Schlossgasse 2, CH-5600, Lenzburg. T: 062 892 11 31, F: 062 892 11 30

Rueegger, R., c/o Rueegger Systeme AG, Oberstrasse 200, CH-9000, St. Gallen. T: 071 277 53 65, F: 071 277 79 41

Rutishauser, W., ACS Partner, Gubelstrasse 28, CH-8050, Zuerich. T: 01 312 70 00, F: 01 311 60 97

Ryser, C., Dipl. Geologe, Rheintalweg 15, CH-4125, Riehen. T: 061 641 58 18, F: 061 921 52 76

Saegesser, R., Dipl. Bauing. ETH, MS, Im Baumgarten 28, CH-8606, Greifensee. T: 01 940 84 02, F: 01 311 67 66

Schanz, T., Dr., Professur Bodenmechanik, Marienstrasse 7, D-99421, Weimar. T: +49 364358 4558

Schindler, J., Dr., Schindler+Schindler, Albisstr. 103, CH-8038, Zuerich. T: 01 482 15 80, F: 01 482 15 84

Schlegel, H.-J., Dipl. Bauing. ETH/SIA, Liestalerstrasse 4, CH-4413, Bueren. T: 061 911 92 22

Schmid, O., Buero fuer beratende Geologie, Bahnhofstrasse 11, CH-3900, Brig. T: 027 923 09 00, F: 027 924 39 17

Schmidhalter, P., Dipl. Bauing. ETH/SIA, Neue Simplonstr. 38, CH-3900, Brig-Glis. T: 027 924 50 08, F: 027 924 50 09

Schmidiger, R., ELECTROWATT, Bellerivestr. 36, CH-8034, Zuerich. T: 01 385 22 11, F: 01 385 25 55

Schneider, H., Dr., Geo-Consulting, Bahnhofstr. 21, CH-6304, Zug. T: 041 729 60 50, F: 041 729 60 59

Schneider, J.F., Prof. Dr., Inst. fuer Angewandte Geologie, P. Jordan-Str. 70, A-1190, Wien

Schneider, T., Dr., Ruetihofstr. 53, CH-8713, Uerikon. T: 01 926 30 46, F: 01 926 37 79

Schneller, P., Dipl. Bauing. ETH, Termerweg 75, CH-3900, Brig. T: 027 923 36 14, F: 027 924 46 14

Scholer, C., Dipl. Bauing. ETH, Amselweg 5, CH-5070, Frick. T: 062 871 16 04

Schuerer, H., Dipl. Bauing. ETH/SIA, Spinnereistr. 12, 8135 Langnau A. T: 01 713 16 04

Schuler, E., WSP Wildberger Schuler, Muehletalstrasse 28, CH-8200, Schaffhausen. T: 052 630 13 00, F: 052 630 13 50

Schwegler, B., Dipl. Bauing. ETH, Tunnelweg 25, CH-6414, Oberarth. T: 041 855 54 40

Sieber, N., Sieber Cassina+Partner, Langstrasse 149, CH-8004, Zuerich. T: 01 291 07 00, F: 01 297 70 91

Sinniger, R., Prof., Dipl. Bauing. ETH/SIA, Sonnenbergstr. 36, CH-8708, Maennedorf. T: 01 920 08 96, F: 021 693 28 63

Solexperts, A.G., Postfach 230, CH-8603, Schwerzenbach. T: 01 825 29 29, F: 01 825 00 63

Spang, R.M., Dr.rer.nat. Dipl.-Geologe, Westfalenstrasse 5-9, D-58455, Witten. T: 2302 91 402-0, F: 230291402-20

Spiegel, U., Dipl. Bauing. ETH, Dorfstrasse 2, CH-8630, Rueti. T: 055 241 12 10

Springman, S.M., Prof. Dr., ETH Zuerich, ETH-Hoenggerberg, CH-8093, Zuerich. T: 01 633 38 05, F: 01 633 10 79

Staempfli, M., Bauing. HTL, Suttenweidli, 3536 Aeschau

Staerk, R., CES Bauingenieur AG, Seestr. 94, CH-6052, Hergiswil. T: 041 632 50 30, F: 041 632 50 32

Staeubli, R., Dipl. Bauing. ETH, Saegenstr. 145, CH-7000, Chur. T: 081 284 88 68

Steiger, A., Andreas Steiger & Partner, St. Karlistrasse 12, PF 7829, CH-6000, Luzern 7. T: 041 248 51 71, F: 041 248 51 72

Steiner, W., Dr.sc.; Dipl. Bauing. ETH, Buehlenstr. 8, CH-3132, Riggisberg. T: 031 809 17 41

Sterba, I., Institut fuer Geotechnik, ETH-Hoenggerberg, CH-8093, Zuerich. T: 01 633 34 26, F: 01 633 10 79

Stoessel, F., Dr.phil.nat., Wylerringstr. 43, 3014 Bern. T: 076 365 18 04

Stolz, M., Ingenieurbuero Stolz, Kranichweg 6, CH-3074, Muri. T: 031 951 82 12

Straubhaar, R., Bauing. Techn. HTL, Treppenweg 2, CH-8634, Hombrechtikon. T: 055 244 17 70

Stucki, D., Dipl. Bauing. ETH/SIA, Pfaffensteinstrasse 38, CH-8118, Pfaffhausen. T: 01 825 50 70, F: 01 825 50 70

Stucki, E., Dipl. Bauing. ETH/SIA, Pfaffensteinstrasse 38, CH-8118, Pfaffhausen. T: 01 825 50 70

Studer, J., Dr., Studer Engineering, Thujastrasse 4, CH-8038, Zuerich. T: 01 481 06 00, F: 01 481 06 02

Suter, Kurt E., Dipl.Ing. ETH, Im unteren Stieg, CH-7304, Maienfeld. T: 081 302 37 40, F: 081 330 73 08

Tappy, O., SGI-Piguet, Avenue du Temple 19, CP 325, CH-1000, Lausanne 12. T: 021 654 21 65, F: 021 312 58 09

Teysseire, J.-C., Teysseire & Candolfi Ag, Terbinerstrasse 18, CH-3930, Visp. T: 027 948 07 00, F: 027 948 07 01

Teysseire, J., Teysseire & Candolfi AG, Terbinerstrasse 18, CH-3930, Visp. T: 027 948 07 00, F: 027 948 07 01

Thut, A., Solexperts AG, Schulstrasse 5, Postfach 230, CH-8603, Schwerzenbach. T: 01 806 29 29, F: 01 806 29 30

Toscano, E., Edy Toscano AG, Nordstr. 114, CH-8037, Zuerich. T: 01 360 21 11, F: 01 360 21 12

Trommer, B., Dipl. Bauing. ETH/SIA, Restelbergstr. 79, CH-8044, Zuerich. T: 01 362 04 56

Tschirky, A., Dipl.Natw.ETH/SIA, Geologe, Sixer 6, 7320 Sargans

Uldry, J.-D., Bureau CER. T: SA, Av. Ritz 35, CH-1950, Sion 2 Nord. T: 027 322 75 45

Voborny, O., COLENCO Power Engineering AG, Mellingerstrasse 207, CH-5405, Baden. T: 056 483 15 51, F: 056 493 73 57

Vogt, N., Ingenieurbuero Norbert Vogt AG, Heiligkreuz 18, CH-9490, Vaduz. T: 075 232 22 02, F: 075 233 15 01

Vollenweider, U., Dr., VOLLENWEIDER AG, Hegarstr. 22, CH-8032, Zuerich. T: 01 387 90 90, F: 01 383 26 76

von Gunten, H., Geotechnisches Institut AG, Geotechn. Inst./Gartenstr. 13, CH-3007, Bern. T: 031 389 34 11, F: 031 381 31 15

Vuilleumier, F., BONNARD & GARDEL, Av. de Cour 61, c.p. 241, CH-1001, Lausanne. T: 021 618 11 11, F: 021 618 11 22

Vulliet, L., Prof. Dr., Lab. de Mécanique des sols, EPFL-Ecublens, CH-1015, Lausanne. T: 021 693 23 11, F: 021 693 41 53

Waldmeyer, J.-P., Dipl. Bauing. ETH/SIA, Rte du Grand Mont 59, CH-1052, Le Mont. s. Lausanne. T: 021 653 23 21, F: 021 653 23 21

Wanner, H., Dr. H. WANNER AG, Im Bahnhof, Postfach, CH-8607, Aathal-Seegraeben. T: 01 932 22 22, F: 01 932 27 81

Wanzenried, H., Dipl. Bauing. ETH/SIA, Schaelfelgrabenweg 54, CH-3033, Wohlen b. Bern. T: 031 829 26 64

Weber, R.A., Dipl. Bauing. ETH/SIA, Huerdliweg 4, CH-8155, Niederhasli. T: 01 850 05 16

Wei, Z.Q., Amberg Ingenieur Buero, Trockenloostrasse 21, CH-8105, Regensdorf-Watt. T: 01 870 91 11, F: 01 870 06 20

Werder, F., Emch + Berger AG, Gurtenstr. 1, 3001 Bern. T: 031 385 61 11

Wetterwald, V.M., Dipl. Bauing. ETH, Mooshaldenstr. 44, CH-8708, Maennedorf. T: 01 920 61 51

Wieser, S.A.G., Beratende Ingenieure ASIC, Rorschacher Strasse 21, CH-9000, St. Gallen. T: 071 244 78 12, F: 071 245 01 03

Winkler, M., Dipl. Bauing. ETH/MBA, Dorfstrasse 14, CH-8603, Schwerzenbach. T: 01 825 31 35

Woerlen, A., Dipl. Bauing. ETH, Naglerwiesenstr. 35, CH-8049, Zuerich. T: 01 341 82 44

Wuermli, P., Knecht & Wuermli, Badenerstrasse 18, CH-8004, Zuerich. T: 01 241 04 31

Wuhrmann, E., Terratest Técnicas Especiales, c/ Alcal 65, 4a planta, E-28014, Madrid. T: 3491 423 75 00, F: 3491 423 75 01

Zeller, J., Dipl. Bauing ETH, Im Hubaecker 10, CH-8967, Widen. T: 056 633 37 47

Zimmermann, T., ZACE SERVICES SA, Case postale 2, CH-1015, Lausanne. T: 021 802 46 05, F: 021 802 46 06

Zinebi, A., Geolab F. Perrin SA, Route de Divonne 48, CH-1260, Nyon

Corporate members

AIC-Ingénieurs conseils SA, Schaer Weibel et Meylan, Av. Tissot 2 bis, CH-1006, Lausanne. T: 021 320 20 22, F: 021 320 73 63

AJS SA, Bureau d'Ingénieurs civil, 4, Rue du Musée, CH-2001, Neuchâtel. T: 032 720 0100, F: 032 720 0101

ANDRES GEOTECHNIK AG, Dip. Ingenieure ETH/SIA, Kesselhaldenstr. 57a, CH-9016, St. Gallen. T: 071 288 27 88, F: 071 288 36 12

BALESTRA AG, Ingenieure und Planer, Sonnenplaetzli 7, CH-6430, Schwyz. T: 041 811 38 38, F: 041 811 20 22

BATIGROUP AG Spezialtiefbau, Binzmuehlestrasse 11, Postfach, CH-8050, Zuerich. T: 01 307 98 98, F: 01 307 98 99

BIRCHLER Pfyl and Partner AG, Herrn Max Birchler, Riedstr. 7, CH-6430, Schwyz. T: 041 811 20 44, F: 041 811 7170

BONNARD & GARDEL, Ingénieurs-Conseils SA, Av. de Cour 61, case postale, CH-1001, Lausanne. T: 021 618 11 11, F: 021 618 11 22

BUNDESAMT fuer Strassen, CH-3003, Bern. T: 031 322 94 11, F: 031 323 23 03

BUNDESAMT fuer Wasser und Geologie (BWG), Laendtestrasse 20, Postfach, CH-2501, Biel. T: 032 328 87 11, F: 032 328 87 12

CES Bauingenieur AG, Lehmann+Erni, Gueterstrasse 3, Postfach 523, CH-6060, Sarnen 2. T: 041 666 70 30, F: 041 666 70 31

CSD Ingénieurs Conseils SA, Ingénieurs, Géologues, Chemin de Montelly 78, CH-1007, Lausanne. T: 021 620 70 00, F: 021 620 70 01

COLENCO Power Engineering AG, Wasserkraftanlagen und, Mellingerstrasse 207, CH-5405, Baden. T: 056 483 17 17, F: 056 493 73 59

DE CÉRENVILLE GEOTECHNIQUE SA, 17 Chemin des Champs-Courbes, CH-1024, Ecublens. T: 021 691 24 91, F: 021 691 24 96

EICHENBERGER, HANS AG, Ingenieurbüro, Sumatrastr. 22, Postfach, CH-8023 Zuerich. T: 01 252 20 77, F: 01 252 20 25

FATZER AG/Geobrugg, Herrn Bernhard Eicher, Salmsacherstrasse 9, CH-8590, Romanshorn. T: 071 466 81 55, F: 071 466 81 50

FH-AARAU, Studiengang Bauingenieurwesen, CH-5210, Windisch. T: 056 462 44 11, F: 056 462 44 15

GEOTECHNISCHES INSTITUT AG, Gartenstr. 13, Postfach 6258, CH-3001, Bern, T: 031 389 34 11, F: 031 381 31 15

GEOTEST: AG, Baugrunduntersuchungen-, Birkenstr. 15, CH-3052, Zollikofen. T: 031 911 01 82, F: 031 911 51 82

GEOTECHNIQUE APPLIQUEE, Dériaz SA, 10, Rue Blavignac, CH-1227, Carouge. T: 022 342 30 00, F: 022 342 26 93

GRUNER AG, Ingenieure und Planer, Gellertstrasse 55/Postfach, CH-4020, Basel. T: 061 312 76 20, F: 061 312 40 09

HÄNY & CIE. AG, MIT Division, Postfach, CH-8706, Meilen. T: 01 925 41 11, F: 01 923 62 45

INGENIEURBUERO Galli+Partner, AG, Industriestr. 57, CH-8152, Glattbrugg. T: 01 809 99 99, F: 01 809 99 98

JET Injectobohr AG, Spezialtiefbau, Baechaustrasse 73, CH-8806, Baech. T: 01 786 55 48, F: 01 786 55 78

KISSLING & ZBINDEN AG, Ingenieure Planer USIC, Seftigenstrasse 22, CH-3000, Bern 17. T: 031 371 29 55, F: 031 372 16 63

LOCHER AG Zuerich, Bauingenieur u.Bauuntern., Pelikan-Platz 5, Postfach, CH-8022, Zuerich. T: 01 218 92 58, F: 01 218 92 61

LOMBARDI SA, Ingegneri consulenti, Via R. Simen 19, C.P. 1535, CH-6648, Minusio. T: 091 744 60 30, F: 091 743 97 37

MONIGATTI FRANCO, Devaud Mongatti et Associés SA, Rte St. Nicholas de Flüe 16, CH-1709, Fribourg. T: 026 425 84 00, F: 026 424 19 41

PERSS Ingénieurs SA, Bureau de planifications, Rte. du Levan. T: 8/cp 283, CH-1709, Fribourg. T: 026 424 50 19

SARNAFIL International AG, Business Area Tiefbau, Industriestrasse, CH-6060, Sarnen. T: 041 666 99 66, F: 041 666 97 00

STUCKY INGENIEURS-CONSEILS SA, Rue du Lac 33, Case postale, CH-1020, Renens VD 1. T: 021 637 15 13, F: 021 637 15 08

STUMP BOHR AG, Stationsstr. 57, CH-8606, Naenikon-Uster. T: 01 941 77 77, F: 01 941 78 00

TIEFBAUAMT, Graubuenden, Grabenstr. 30, CH-7001, Chur. T: 081 257 37 00, F: 081 257 21 57

TIEFBAUAMT der Stadt Zuerich, Werdmuehleplatz 3, Postfach, CH-8023, Zuerich. T: 01 216 51 11, F: 01 212 08 77

TIEFBAUAMT des Kantons Bern, Reiterstrasse 11, CH-3011, Bern

TIEFBAUAMT des Kantons Zuerich, Walchetor, CH-8090, Zuerich. T: 01 259 11 11, F: 01 259 31 29

WIESER STACHER AG, Beratende Ingenieure ASIC, Rorschacher Strasse 21. T: 026 244 78 12, F: 071 245 01 03

ZSCHOKKE LOCHER SA, Traveaux Spéciaux, Route de la Venoge 10, CP 81, 1025 Echandens. T: 703 6615, F: 703 6649

TUNISIA/TUNISIE

Secretary: Mounir Boussida, Association Tunisienne de Mécanique des Sols, E.N.I.T., B.P. 37 Le Belvedère, 1007 Tunis.
T: 2161 874 709, F: 2161 872 729, E: Mounir.bouassida@enit.rnu.tn

Total number of members: 12

Bouassida, M., 4, Passage Tejckestène, Riadh Ennasr 1 2037 Ariana. T: 2161 874700, 2161 871476, F: 2161 872729, E: mounir.bouassida@enit.rnu.tn

Jouini, Z., N°52, Rue Ibn Charef, Tunis. T: 2161 469146, F: 2161 469146, E: ziedjouini@voila.fr

Ounaies, W., Imb. Nasr 1, Appt.10, Cité Olympique, Tunis. T: 2161 770520

Ben Said, B., BP.129, 7000 Bizerte. T: 2161 422075, F: 2162 422075, E: BensaïdB@netcourrier.com

Jellali, B., Cité Eddir CNRPS, N°912, 7100 le Kef. T: 2168 203691, Mobile: 09292996, F: 2168 203691, E: Jellali_belgacem@excite.fr

Ben Zarrouk, H., 24, rue 6290, Omrane Supérieur. T: 2161 781088, F: 2161 780717

Kharrat, Y., 48, Avenue de Paris Megrine. T: 2161 425661, F: 2161 426672, 2161 427673, E: omc.soltech@gnet.tn

Boussetta, S., Imm 8, Appt.7, Rue Drissi, La Goulette. T: 2161 737527, F: 2161 872729

Sfar, A., Cité Hédi Nouira, FB03, Borj Louzir Ariana. T: 2161 735300, F: 2161 735812, E: OMMP@Email.Ati.tn

El Ouni, M.R., 4, Rue Erriadh Radès Meliane 2040. T: 2161 440709, F: 2161 440709, E: el_ouni@yahoo.fr

Guetif, Z., 18, Rue Echaref, Cité Echebi, Fouchena 2082. Mobile: 09217385, E: Guetif.mz@planet.tn

Kanoun, F., 12 Rue 7238, El Menzah 9, 1013. T: 2161 889043, Mobile: 09435583, F: 2161 889043

232

TURKEY/TURQUIE

Secretary: c/o Prof. Dr. A. Sağlamer, Turkish National Society for SMFE, İnşaat Fakültesi, İstanbul Teknik Üniversitesi, 80626 Ayazağa, İstanbul. T: +90 212 285 3744, F: +90 212 285 3582

Total number of members: 195

Acun, N., Prof. Dr., University Professor, Selvili Sok. 10/7, 80680 4.Levent, Istanbul. T: 212 264 1555
Adiloğlu İ., Director, Özışık İnşaat ve Taahhhüt A.Ş., 9 Gençlik Cad. Tandoğan, Ankara. T: 312 440 9906, F: 312 440 1844
Akçelik, N., Civil Engineer, Karayolları Genel Müdürlüğü Teknik Araştırma Dairesi F Blok, Yücetepe 06100, Ankara. T: 312 419 1430-398, F: 312 417 2851
Akdağ, E.N., Civil Engineer, Kasktaş Enderunlu İsmail Hakkı Bey Sok. No: 31 Balmumcu 80700 Beşiktaş, İstanbul
Akdeniz, G., 93 Yavuz Cad. Atatürk Mah. 35040 Bornova Izmir. T: 232 339 8619, E: goksinakdeniz@operamail.com
Akdoğan, M., Research Assistant, Orta Doğu Teknik Üniversitesi, Mühendislik Fakültesi, 06531 Ankara
Akman, S., Prof. Dr., University Professor, İstanbul Teknik Üniversitesi, İnşaat Fakültesi, 80626 Ayazağa, İstanbul. E: akmans@itu.edu.tr
Aköz, Y., Prof. Dr., University Professor, Istanbul Teknik Üniversitesi, İnşaat Fakültesi 80626 Ayazağa, İstanbul. T: 212 285 3701, E: akoz@itu.edu.tr
Aksoy, İ.H., Doç. Dr., University Dozent, Istanbul Teknik Üniversitesi, İnşaat Fakültesi, 80626 Ayazağa, İstanbul. T: 212 285 3741, E: ihaksoy@srv.ins.itu.edu.tr
Aksoy, S., Civil Engineer, Karayolları Genel Müdürlüğü Araştırma Dairesi, F BlokYücetepe 06100, Ankara. T: 312 417 8097
Aktan, M.H., Civil Engineer, Kasktaş Enderunlu İsmail Hakkı Bey Sok., No: 31 Balmumcu, 80700 Beşiktaş Istanbul
Alaçam, Ö., Contractor, İstiklal Cad. Baro Han 304, 80050 Beyoğlu, İstanbul. T: 212 249 4825, F: 212 244 0933
Alaner, U., Eng., Piri Mehmet Paşa Mah., Yoğurthane Sokağı, Otman Apt. 21, D.4, 34930 Silivri, İstanbul
Alkaya, D., Research Assistant, Pamukkale Üniversitesi Mühendislik Fakültesi, İstiklal Mah. İstiklal Cad. 145/9 Denizli
Alpagut, Y.E., Engineer, Kasktaş Enderunlu İsmail Hakkı Bey Sok. 31, 80700 Balmumcu, İstanbul. T: 212 274 5842
Alpaslan, A.H., Dr., Civil Engineer, DSİ 2. Bölge Müdürlüğü Bornova, Izmir. T: 232 486 1317
Altan, E., Civil Engineer, Büyükdere Cad. 42/5, 80290 Mecidiyeköy, İstanbul. T: 212 275 4860, F: 212 266 4755
Altan, O., Prof. Dr., University Professor, İstanbul Teknik Üniversitesi, İnşaat Fakültesi, 80626 Ayazağa, İstanbul. T: 212 285 3810
Alyanak, I., Prof. Dr., Eski Bağdat Cad. Asam Apt. 29/6 Küçükyalı, 81750 Maltepe, İstanbul. T: 252 343 1553 (home), Mobile: 542 584 9642
Alyanak, İ., Prof. Dr., University Professor, 1780/1 Sok. 14-8, 35530 Karşıyaka, Izmir. T: 232 381 3676, F: 258 266 2012
Anmaç, G., Civil Engineer Nurol İnşaat A.Ş. 7 Arjantin Cad., Gaziosmanpaşa, Ankara. T: 312 428 4250, F: 312 427 8501
Ansal, A., Prof. Dr., University Professor, İstanbul Teknik Üniversitesi, İnşaat Fakültesi, 80626 Ayazağa, İstanbul. T: 212 285 3702, F: 212 285 65 87, E: ansal@itu.edu.tr
Arıoğlu, E., Dr., Chairman, Yapı Merkezi Hacı Reşit Paşa Sok. 7, Çamlıca, Istanbul. T: 216 321 9000, F: 216 321 9013
Arslan, U., Prof. Dr., University Professor, Technische Hochschule Darmstadt Institut für Geotechnik, Fakültät für Bauingenieurweisen, 64287 Darmstadt, Petersenstr. 13 GERMANY. T: 615 116 3749, F: 615 116 6683, E: arslan@geotechnik.tu-darmstadt.de
Ataman, G., Civil Engineer, Karayolları Genel Müdürlüğü Teknik Araştırma Dairesi Başkanlığı, 06100 Yücetepe, Ankara. T: 312 418 5995
Atılgan, İ., Kasktaş Enderunlu İsmail Hakkı Bey Sok. No: 31 Balmumcu, 80700 Beşiktaş, Istanbul

Avcı, C., Doç. Dr., University Dozent, Boğaziçi Üniversitesi, Mühendislik Fakültesi, 80815 Bebek, Istanbul. T: 212 263 1500
Aydilek, A., Civil Engineer, University of Wisconsin, 765 W. Washington Av. 302, 53703, Madison, WI, USA
Aygit, R., Chairman, TEMAR Temel Mühendisliği Ltd., Şti. Karadut Sok. 23/8 Altıyol, 81310 Kadıköy, Istanbul. T: 216 338 4298, F: 216 348 9210, E: temar@temar.com
Bahar, M.A., Kasktaş, Enderunlu İsmail Hakkı Bey Sok. No: 31, Balmumcu, 80700, Beşiktaş, Istanbul
Bakioğlu, M., Prof. Dr., University Professor, İstanbul Teknik Üniversitesi, İnşaat Fakültesi, 80626 Ayazağa, Istanbul. T: 212 285 3799, E: bakioglu@itu.edu.tr
Bakır, S., Dr., Research Assistant, Orta Doğu Teknik Üniversitesi, Mühendislik Fakültesi, 06531 Ankara. T: 312 210 2406, F: 312 210 1262
Baykal, A., Dr., Civil Engineer, Boğaziçi Üniversitesi, Mühendislik Fakültesi, 80815 Bebek, Istanbul. T: 212 383 1500
Baykal, G., Prof. Dr., University Professor, Boğaziçi Üniversitesi, Mühendislik Fakültesi, 80815 Bebek, Istanbul. T/F: 212 263 2204, E: baykal@boun.edu.tr
Bektaş, F., Dr., Karadeniz Teknik Üniversitesi, Mühendislik Mimarlık Fakültesi, 61080 Trabzon. E: fozmen@risc01. ktu.edu.tr
Berilgen, M., Doç. Dr., Yıldız Teknik Üniversitesi, İnşaat Fakültesi, 80750 Yıldız, Istanbul. T: 212 259 7070-2759, F: 212 259 6762
Bilsel, H., Doç Akdeniz Üniversitesi, İnşaat Mühendisliği Bölümü Gazi Magosa, Kuzey Kıbrıs Türk Cumhuriyeti
Birand, A., Prof. Dr., University Professor, Orta Doğu Teknik Üniversitesi, Mühendislik Fakültesi, 06531 Ankara. T: 312 210 2412, F: 312 210 1105, E: abirand@metu.edu.tr
Birder, U., Dr., Chairman, Teknik İnşaat ve Tic.A.Ş. Hilmipaşa Cad. Gürman Apt. 32/3, 81090 Kozyatağı, Istanbul. T: 216 361 7187, F: 216 380 4863, E: usmet@superonline.com
Boduroğlu, H., Prof. Dr., University Professor, İstanbul Teknik Üniversitesi, İnşaat Fakültesi, 80626 Ayazağa, Istanbul. T: 212 285 3797, F: 212 285 6587, E: boduroglu@itu.edu.tr
Candoğan, A., Dr., Executive Advisor, Kasktaş Enderunlu İsmail Hakkı Bey Sok. 31, 80700 Istanbul. T: 212 274 5842, E: candogan@kasktas.com.tr
Cayıt, H.İ., Civil Engineer, Baytem Mühendislik Müşavirlik, Strasbourg Cad. Erler İş Hanı 30/22, 06430 Sıhhıye, Ankara. T: 312 229 1157, F: 312 229 2255
Çağatay, E., Civil Engineer, Akın Cad. 21 Burhaniye, Balıkesir. T: 232 425 1490
Çağlayan, M.H., Dr., MNG-Zemtaş Uğur Mumcu Cad. No. 88 Gaziosmanpaşa, Ankara
Çalışan, O., Engineer, Noktalı Sok. 16/15, 06700 Gaziosmanpaşa, Ankara
Çalışan, Ö.F., Engineer, Ahmet Rasim Sok. 10/6, Çankaya, Ankara
Çamlıbel, N., Prof. Dr., University Professor, Yıldız Teknik Üniversitesi, Mimarlık Fakültesi, 80750 Yıldız, Istanbul. T: 212 259 7070-2231, F: 212 261 0549
Çırak, M., Civil Engineer, STFA İnşaat A.Ş.19 Tophanelioğlu Cad., 81190 Altunizade, Istanbul. T: 216 326 4420, F: 216 325 5798
Çinicioğlu, F., Prof. Dr., University Professor, İstanbul Üniversitesi, Mühendislik Fakültesi, Avcılar Kampüsü, 34850 Avcılar, Istanbul. T: 212 591 1998-203, F: 591 1997
Çokça, E., Dr., Orta Doğu Teknik Üniversitesi, Mühendislik Fakültesi, 06531 Ankara
Çoruk, Ö., Engineer, Kocaeli Üniversitesi, Mühendislik Fakültesi, Jeoloji Mühendisliği Bölümü, 41300 İzmit, Kocaeli. T: 262 321 5968

Dadaşbilge, K., Dr., Geocon Zemin Uzmanları ve Mühendislik Ltd., Adnan Saygun Cad. Körkadı Sok. Güzel Konutlar Sitesi, F Blok D: 4, 2.Ulus, Istanbul. T: 212 257 6129, F: 212 257 8034

Dadaşbilge, O., Geocon Zemin Uzmanları ve Mühendislik Ltd., Adnan Saygun Cad. Körkadı Sok. Güzel Konutlar Sitesi, F Blok D: 4, 2.Ulus, Istanbul. T: 212 257 6129, F: 212 257 8034, E: ozan@nokta.net

Dalgıç, D., Research Resistant, Dokuz Eylül Üniversitesi, Mühendislik Fakültesi, Bornova, İzmir. T: 312 388 0110, F: 312 388 7864

Demirelli, K., Civil Engineer, Bağdat Cad. Çamlık apt. 176/14, 81580 İdealtepe, İstanbul

Dora, G., Civil Engineer, Göktepe Sok. Başak Apt. No: 4/10 Feneryolu, Kadıköy, Istanbul

Durgunoglu, T., Chairman, Geoteknik A.Ş. Mehmetçik Cad. 42/8, Mecidiyeköy, Istanbul. T: 212 212 7429, F: 212 212 7434, E: geoteknik@fornet.net.tr

Durgunoğlu, T., Prof. Dr., University Professor, Boğaziçi Üniversitesi, Mühendislik Fakültesi, 80815 Bebek, Istanbul. T: 212 274 3748, F: 212 266 1034, E: durgunt@boun.edu.tr

Durukan, Z., Engineer, Emin Ali Paşa Cad., Tülin Apt. 54/11, Bostancı, Istanbul

Düzceer, R., Civil Engineer, Kasktaş Enderunlu İsmail Hakkı Bey Sok. 31, 80700 Balmumcu, Istanbul. T: 212 274 5842, F: 212 266 3393, E: kasktas@kasktas.com.tr

Egeli, İ., Dr., Maunsell Geotechnical Services Ltd., World Finance Centre, Harbour City, 19 Canton Road Kowloon, HONG KONG. T: 852 2302 1013, F: 852 2730 7110, E: egelio@canada.com

Eğin, D., Dr., STFA 19 Tophanelioğlu Cad., 81190 Altunizade, Istanbul. T: 218 21 333 9225, F: 218 214 447 87

Engür, Ü., Civil Engineer, STFA Temel Araştırma A.Ş. STFA Grup Merkezi, 19 Tophanelioğlu Cad., 81190 Altunizade, Istanbul. T: 216 339 5409, F: 216 339 5579

Erdemgil, M., Prof. Dr., SMP 21/9 Çankaya Cad., 6700 Çankaya, Ankara. T: 312 446 8527, F: 312 446 8529

Ergun, U., Prof. Dr., University Professor, Orta Doğu Teknik Üniversitesi, Mühendislik Fakültesi, 06531 Ankara. T: 312 210 2415, F: 312 210 1262, E: eruf@metu.edu.tr

Erguvanlı, A., Doç. Dr., Civil Engineer, Tekyapim, Tevfik Erdönmez Sok. Birlik Apt. 20/7, 80280 Esentepe, Istanbul. T: 212 266 4347, F: 212 267 3178, E: Tekyapim@ixir.com

Ergül, B., Civil Engineer, Cengiz Topel Durağı Bebek Yolu, Kopuz Apt. 7/5, Etiler, Istanbul

Erken, A., Doç. Dr., University Dozent, İstanbul Teknik Üniversitesi, İnşaat Fakültesi, 80626 Ayazağa, İstanbul. T: 212 285 6567, E: erken@itu.edu.tr

Erol, O., Prof. Dr., University Professor, Orta Doğu Teknik Üniversitesi, Mühendislik Fakültesi, 06531 Ankara. T: 312 210 2408, E: orer@metu.edu.tr

Eroskay, O., Prof. Dr., University Professor, İstanbul Üniversitesi, Mühendislik Fakültesi, Avcılar Kampüsü, 34850 Avcılar, Istanbul. T: 212 279 8937, F: 212 279 9229, E: o.eroskay@iku.edu.tr

Ersöz, H., Partner, Ersöz İnş.San.Müş.Tic.Ltd.Şti., Prof. Nurettin Mazhar Öktel Sok. 16/2, Şişli, Istanbul. T: 212 248 4235, F: 212 248 4237

Etkesen, Z., Civil, Engineer, Karayolları Genel Müdürlüğü Teknik Araştırma Dairesi Başkanlığı F Blok, 06100 Yücetepe, Ankara. T: 312 419 1430

Gök, S., Research Assistant, İstanbul Teknik Üniversitesi, İnşaat Fakültesi, 80626 Ayazağa, Istanbul. T: 212 285 6867, E: sgok@srv.ins.itu.edu.tr

Gönen, S., Managing Director, STFA Zemin Grubu, 19 Tophanelioğlu Cad., 81190 Altunizade, Istanbul. T: 216 339 5408

Güler, E., Prof. Dr., University Professor, Mühendislik Fakültesi, 80815 Bebek, Istanbul. T: 212 263 1540-1452, E: eguler@boun.edu.tr

Gülçelik, A., Managing Director, Kasktaş, 31 Enderunlu İsmail Hakkı Bey Sok., 80700 Balmumcu, Istanbul. T: 212 274 5842, F: 212 266 3393

Güner, A., Prof. Dr., University Professor, Trakya Üniversitesi, Çorlu Mühendislik Fakültesi, 59860 Çorlu, Tekirdağ. T: 282 652 94 75

Güner, M., Civil Engineer, STFA Temel Kazıkları İnşaat A.Ş. 19, Tophanelioğlu Cad., 81190 Altunizade, İstanbul. T: 216 339 5408, F: 216 340 0695

Gürbüz, C., Doç. Dr., University Dozent, Kandilli Rasathanesi ve Deprem Araştirma Enstitüsü, 81120 Çengelköy, Istanbul

Gürpınar, O., Prof. Dr., University Professor, İstanbul Üniversitesi, Mühendislik Fakültesi, Avcılar Kampüsü, 34840 Avcılar, Istanbul

Güvendi, Y., Director, Koray Yapı End. ve Tic. A.Ş., Büyükdere Cad., Yapı Kredi Plaza C Blok Kat 19, 80620 Levent, Istanbul. T: 212 279 7082, F: 212 268 0097

Harbiyeli, S., Engineer, Gündüz Cad. 101/A Antakya Hatay

Hatipoğlu, B., Research Assistant, İstanbul Teknik Üniversitesi, İnşaat Fakültesi, 80626 Ayazağa, Istanbul.T: 212 285 3745, E: bulent@srv.ins.itu.edu.tr

Horoz, A., Director, Yüksel Proje, Ahmet Rasim Sok. No: 11, 06550 Çankaya, Ankara. T: 312 440 6020, F: 312 440 6677, E: ahoroz@yp.yukselproje.com.tr

Işıkara, A.M., Prof. Dr., University Professor, Kandilli Rasathanesi ve Deprem Araştırma Enstitüsü, 81120 Çengelköy, Istanbul

İbrahimiye, M., Managing Director, Kasktaş, Enderunlu İsmail Bey Sok., 3180700 Balmumcu, Istanbul. T: 212 274 5842, F: 212 266 3393

İhsan, M., Civil Engineer, Cihannuma Mh. Salih Efendi Sok. 3/2, Beşiktaş, İstanbul

İlbay, N., Managing Director, Ayson Sondaj ve Araştırma A.Ş., Büyükdere Cad. 42 Doğuş Han, 80290 Mecidiyeköy, Istanbul. T: 212 275 4860, F: 212 266 4755, E: nusreti@dogusins.com.tr

İnan, O., Civil Engineer, 1479 Sok. 8/204, 35220 Alsancak, İzmir

İnanır, M., Civil Engineer, Geotest İnşaat San. ve Tic. Ltd. Şti., Kıbrıs Şehitleri Cad. 1479 No: 8 TEV-1 İş Merkezi Kat:2 D:204, 35220 Alsancak, Izmir. T: 232 464 5701, F: 232 464 5541

İnanır, O.E., Civil Engineer, Geotest İnşaat San. ve Tic. Ltd. Şti., Kıbrıs Şehitleri Cad. 1479 No: 8 TEV-1 İş Merkezi Kat:2 D:204, 35220 Alsancak, Izmir. T: 232 464 5701, F: 232 464 5541, E: oinanir@member.asc.org

İncecik, M., Prof. Dr., University Professor, İstanbul Teknik Üniversitesi, İnşaat Fakültesi, 80626 Ayazağa, Istanbul. T: 212 285 3743, E: mincecik@srv.ins.itu.edu.tr

İsrafiloğlu, V.A. T: 39 14 28, 32 99 04, 32 00 69, F: 99412 32 9904

İyisan, R., Doç. Dr., University Dozent, İstanbul Teknik Üniversitesi, İnşaat Fakültesi, 80626 Ayazağa, Istanbul. T: 212 285 6580, E: iyisan@itu.edu.tr

Kandemir, A., Civil Engineer, Çetin Emeç Blv., Metiş Sitesi E Blok D.10, Balgat, Ankara

Karadede, E., Director, STFA Zemin Grubu, Tophanelioğlu Cad. No. 19, Altunizade, İstanbul. T: 216 339 5409, F: 216 339 3300

Karakimseli, Z., Tekke Sok. 14, 80870 Yeniköy, İstanbul. T: 212 262 4087, F: 212 262 9100

Karaşahin, M., Research Assistant, Gülistan Mah. 2821 Sok., 4732200 Isparta

Kaya, M.H., General Director, Erke Dış Tic. Ltd. Şti., Sıracevizler Cad.80/a, 80260 Şişli, Istanbul. T: 212 231 7250, F: 212 234 4707

Kaya, Z., Civil Engineer, Dikilitaş Sok., Setaltı Damla Apt. 46/1 D:4, Beşiktaş, Istanbul. T: 212 260 6293

Kayalar, A.Ş., Doç. Dr., University Dozent, Dokuz Eylül Üniversitesi, Mühendislik Fakültesi, 35100 Bornova, Izmir. T: 232 388 1047, F: 232 388 7864

Keskin, N., Doç. Dr., University Dozent, Süleyman Demirel Üniversitesi, Mühendislik Mimarlık Fakültesi. T: 246 237 0855, F: 246 237 0859, E: nilay@mmf.sdu.edu.tr

Kılıç, H., Civil Engineer, Garanti-Koza İnşaat, 28 Kısıklı Cad., 81180 Üsküdar, Istanbul

Kın, A.S., Dr., STFA Temel Kazıkları İnşaat A.Ş., 19 Tophanelioğlu Cad., 81190 Altunizade, Istanbul. T: 216 326 70 00

Kıran, F., STFA Temel Araştırma ve Sondaj Ltd. Şti., STFA Grup Merkezi, 19 Tophanelioğlu Cad., 81190 Altunizade, Istanbul. T: 216 326 7000

Kıymet, H., Temelkon Müh. İnş.San. ve Tic. Ltd. Şti., Muallim Naci Cad., Melek Han 41, K4, D63, Ortaköy, Istanbul. T: 212 258 9194, F: 212 258 9199 (116)

Kip, F., Civil Engineer, Istanbul Teknik Üniversitesi, İnşaat Fakültesi, 80626 Ayazağa, Istanbul. T: 212 285 3746

Konrat, A., Civil Engineer. Tekfen Mühendislik A.Ş., Moda Cad. 266/19, Kadıköy, Istanbul. T: 212 257 6130, F: 212 265 5796

Kulaç, F., Zetaş, Silahtarbahçe Soyak Sitesi, Selami Ali Mahallesi, B7 Blok D.1, 81150 Üsküdar, Istanbul

Kumbasar, F., Prof. Dr., Civil Engineer, Büyükdere Cad., Emlak Kredi Blokları B Blok, 80620 Levent, Istanbul. T: 212 264 7753

Kumbasar, V., University Professor, Istanbul Teknik Üniversitesi, İnşaat Fakültesi, 80626 Ayazağa, Istanbul. T: 212 285 3742

Kurtuluş, A., Research Assistant, Istanbul Teknik Üniversitesi, İnşaat Fakültesi, 80626 Ayazağa, Istanbul. T: 212 285 6867, E: akurtulus@srv.ins.itu.edu.tr

Kuruoğlu, M., Research Assistant, Dokuz Eylül Üniversitesi, Mühendislik Fakültesi, Bornova, Izmir. T: 232 388 0110, F: 232 388 7864, E: mehmet.kuruoglu@dev.edu.tr

Lav, A., Doç. Dr., University Dozent, Istanbul Teknik Üniversitesi, İnşaat Fakültesi, 80626 Ayazağa, Istanbul. T: 212 285 3700, F: 212 285 6587

Meral, V., Civil Engineer, Refik Belendir Sok., 91/9, 06540 Çankaya, Ankara. T: 312 440 2730

Mercan, M., Civil Engineer, 2. Sok., 5/6, Kısmetli Apt., AGE Blokları, Eryaman, Ankara

Mirata, T., Prof. Dr., Orta Doğu Teknik Üniversitesi, Mühendislik Fakültesi, 06531 Ankara

Mollamahmutoğlu, M., Doç. Dr., University Dozent, Gazi Üniversitesi, Mühendislik Mimarlık Fakültesi, Ankara. T: 312 231 7400, F: 312 230 8434

Mut, T., Geophysicist, STFA Temel Araştırma ve Sondaj Ltd., 19 Tophanelioğlu Cad., 81190 Altunizade, Istanbul. T: 216 339 5409

Müftüoğlu, H., Civil Engineer, I.Mediko Sitesi A17 D:8, Ortaköy, Istanbul. T: 212 275 6861, F: 212 212 9504

Okur, V., Research Assistant, Osmangazi Üniversitesi, Mühendislik Mimarlık Fakültesi, Bademlik, Eskişehir

Olgun, C.G., Civil Engineer, Mühendislik Fakültesi, Boğaziçi Üniversitesi, 80815 Bebek, Istanbul

Önal, O., Civil Engineer, Mustafa Kemal cad. 117/8, Bornova, İzmir

Önalp, A., Prof. Dr., University Professor, Sakarya Üniversitesi, Mühendislik Fakültesi, Adapazarı. T: 264 277 4003, E: aonalp@turk.net

Öntuna, A., Dr., Civil Engineer, Kalamış Cad., 103, D Blok3, 81030 Fenerbahçe, Istanbul

Ören, A.H., Civil Engineer, Dumlupınar cad., 148/7 Derya Sitesi A Blok, Bornova, İzmir (213)

Ören, F., Civil Engineer, Küplüce Yolu 17/3, Beylerbeyi, Istanbul. T: 216 321 7538

Örmeci, C., Prof. Dr., University Professor, Istanbul Teknik Üniversitesi, İnşaat Fakültesi, 80626 Ayazağa, Istanbul. T: 212 285 3807, E: cankut@itu.edu.tr

Özaydın, K., Prof. Dr., Yıldız Teknik Üniversitesi, Mühendislik Fakültesi, 80750 Yıldız, Istanbul. T: 212 259 7070-2711

Özbayoğlu, F., Doç. Dr., 60 Sokak, 111/9, Emek, Ankara. T: 312 222 5476, E: babaduygu@yahoo.com

Özçelik, H., Esat İnş.Dış Tic.A.Ş., Kuleli Sok. 1/5, 06700 Gaziosmanpaşa, Ankara

Özdoğan, A., Birlik Mah., 43 Sok. 13/2, Çankaya, Ankara

Özkan, T., Doç. Dr., University Dozent, Istanbul Teknik Üniversitesi, İnşaat Fakültesi, 80626 Ayazağa, Istanbul. T: 212 285 3748, E: tozkan@srv.ins.itu.edu.tr

Özkan, Y., Prof. Dr., Orta Doğu Teknik Üniversitesi, Mühendislik Fakültesi, 06531 Ankara. T: 312 210 2414

Özüdoğru, K., Prof. Dr., University Professor, Istanbul Teknik Üniversitesi, İnşaat Fakültesi, 80626 Ayazağa, Istanbul. T: 212 285 6628, F: 212 285 6587

Özüer, B., Doç. Dr., University Dozent, Istanbul Teknik Üniversitesi, İnşaat Fakültesi, 80626 Ayazağa, Istanbul

Özveren, H., Engineer, STFA Zemin Grubu, Tophanelioğlu Cad. No. 19, Altunizade, İstanbul

Pamuk, A., Civil Engineer, Department of Civil Engineering, Renssealer Polytechnic Institute, Troy, NY 12180, USA

Pamukçu, S., Prof. Dr., University Professor, Dept. of Civil and Environmental Engineering, Leigh University, 13 East Packer Avenue, Betlehem, PA 18015, USA

Sağlamer, A., Prof. Dr., University Professor, Istanbul Teknik Üniversitesi, İnşaat Fakültesi, 80626 Ayazağa, Istanbul. T: 212 285 3744, F: 212 285 6587, E: asaglam@itu.edu.tr

Saraç, H., Dr., Tekfen Mühendislik A.Ş., Tekfen Sitesi, Etiler, Istanbul. T: 212 257 6130, F: 212 265 5796

Sargın, İ., Research Assistant, Boğaziçi Üniversitesi, Mühendislik Fakültesi, 80815 Bebek, Istanbul. T: 212 263 1500 (1442), E: sargin@boun.edu.tr

Sayar, C., Prof. Dr., University Professor, Istanbul Teknik Üniversitesi, Maden Fakültesi, 80626 Ayazağa, Istanbul. T: 216 355 2587 (153)

Sezen, A., Research Assistant, Istanbul Teknik Üniversitesi, İnşaat Fakültesi, 80626 Ayazağa, Istanbul. T: 212 285 6581, F: 212 285 6587, E: asezen@srv.ins.itu. edu.tr

Sivrikaya, O., Reasearch Assistant, Istanbul Teknik Üniversitesi, İnşaat Fakültesi, 80626 Ayazağa, Istanbul. T: 285 3745, E: o1s2m3a4@yahoo.com

Siyahi, B., Doç., University Dozent, Osman Gazi Üniversitesi, İnşaat Mühendisliği Bölümü, 26030 Eskişehir

Soyaçıkgöz, E., Chairman, MENSOY İnşaat San. ve Tıc. Ltd. Şti., Cumhuriyet Bulvarı, Sevil İş Merkezi No: 302, 35220 Mecidiyeköy, Istanbul

Soygür, Ü., Senior Engineer, Paris Cad., Kızılay Birlik Apt. 31/7, Aşağıayrancı, Ankara. E: ahmutlu@mikasa.mmg. gazi.edu.tr

Sümerman, K., General Manager, İnsitu, 4/6, Kırkpınar Sok., 06690 Çankaya, Ankara. T: 312 440 2601, F: 312 440 2501

Şamandar, A., Engineer, Düzce Meslek Yüksek Okulu, Düzce, Bolu

Şatırlar, B., STFA Temel Kazıkları İnşaatı A.Ş., 19 Tophanelioğlu Cad., 81190 Altunizade, Istanbul. T: 216 326 7000

Şenol, A., Dr., Research Assistant, Istanbul Teknik Üniversitesi, İnşaat Fakültesi, 80626 Ayazağa, Istanbul. T: 212 285 3745, F: 212 285 6587, E: asenol@srv.ins.itu. edu.tr

Tan, O., Doç. Dr., University Dozent, Istanbul Teknik Üniversitesi, İnşaat Fakültesi, 80626 Ayazağa, Istanbul. T: 212 285 3739, F: 212 285 6587

Tan, Ö., University, Atatürk Üniversitesi, İnşaat Mühendisliği Bölümü, 25050 Erzurum

Tavlaşoğlu, A.D., Civil Engineer. Tavlaşoğlu İnş. A.Ş., İskele Caddesi 6/3, Caddebostan, Istanbul. T: 216 386 5661, F: 216 357 3386

Taymaz, T., Doç. Dr., University Dozent, Istanbul Teknik Üniversitesi, Maden Fakültesi, Jeofizik Mühendisliği Bölümü, 80626 Ayazağa, Istanbul. E: taymaz@itu.edu.tr

Tekinsoy, M.A., Prof. Dr., University Professor, Çukurova Üniversitesi, Mühendislik Mimarlık Fakültesi, 01330 Balcalı, Adana

Teymur, B., Research Assistant, Istanbul Teknik Üniversitesi, İnşaat Fakültesi, 80626 Ayazağa, Istanbul

Tezgören, Ö., General Coordinator, Baytur İnş. Taah. A.Ş., 19 Mayıs Cad. 34, 80220 Şişli, Istanbul. T: 212 212 8505, F: 212 212 6817

Togrol, E., Prof. Dr., University Professor, Istanbul Teknik Üniversitesi, İnşaat Fakültesi, 80626 Ayazağa, Istanbul. T: 212 285 3747, F: 212 285 3582, E: togrol@itu.edu.tr

Toker, M., Partner, Toker Sondaj ve İnş.Kol.Şti., Nenehatun Cad. No.60/7, Gaziosmanpaşa, Ankara. T: 312 136 3674

Trak, B., Dr., Managing Director, Freysaş, H. Reşit Paşa Sok., No. 15, 81180 Çamlıca, Istanbul. T: 33 1 475 53 756, F: 33 1 475 53 3752

Tuğlu, H., Civil Engineer, Karayolları Genel Müdürlüğü, Teknik Araştırma Dairesi Başkanlığı, 06100 Yücetepe, Ankara. T: 312 418 8016, F: 312 425 4738

Tuncan, A., Prof. Dr., University Professor, Anadolu Üniversitesi, Mühendislik Mimarlık Fakültesi, Yunus Emre Kampüsü, 26470 Eskişehir. T: 222 335 0580-2040, F: 222 335 3616, E: atuncan@anadolu.edu.tr

Tuncan, M., Prof. Dr., University Professor, Anadolu Üniversitesi, Mühendislik Mimarlık Fakültesi, Yunus Emre

Kampüsü, 26470 Eskişehir. T: 222 335 0580-2040, F: 222 335 3656

Turan, N., Engineer, Atatürk Cad., Vakıflar Bankası İş Hanı, No: 406, 16020 Bursa

Tümay, M., Prof. Dr., University Professor, Georgia Gulf Distinguished Professor, College of Engineering, CEBA 3304 V, Louisiana State University, Baton Rouge, LA 70803, USA

Tümer, H., Civil Engineer, Teksan Temel A.Ş., 27/9, Cinnah Cad., Çankaya, Ankara

Türeli, A., Engineer, Yeşilçimen Sok. 7, Levent, Istanbul

Uğurlu, A., Engineer, Istanbul Üniversitesi, Mühendislik Fakültesi, Jeoloji Müh. Bölümü, Avcılar, Istanbul

Ulukan, B., Civil Engineer, Karayolları Genel Müdürlüğü, Otoyolları Daire Başkanlığı, 06100 Yücetepe, Ankara. T: 312 419 1430, F: 312 229 4416

Ural, D., Doç. Dr., University Dozent, Istanbul Teknik Üniversitesi, İnşaat Fakültesi, 80626 Ayazağa, Istanbul

Uyanık, N., Sahra-i Cedid Durağı, İnönü Cad. 7/17, Erenköy, İstanbul

Uygurer, C., Mesa Koru Sitesi, Lale Blok D.13, Çayyolu, Ankara

Uysal, B., Civil Engineer, DSI 18. Bölge Müdürlüğü, Isparta

Uzal, M., Chairman, Yol ve Yapı End. Taah. Ve Tic. Ltd. Şti., 128/3 İstinye Cad., 80860 İstinye, İstanbul. T: 212 594 1744, F: 212 594 1788

Uzun, H.B., Civil Engineer, KASKTAŞ, Enderunlu İsmail Hakkı Bey Sok., No: 31, Balmumcu, 80700 Beşiktaş, İstanbul

Uzuner, B.A., Prof. Dr., University Professor, Karadeniz Teknik Üniversitesi, 61080 Trabzon. T: 462 325 3223, F: 462 325 3205, E: uzuner@ktu.edu.tr

Ülker, R., Prof. Dr., University Professor, Istanbul Teknik Üniversitesi, İnşaat Fakültesi, 80626 Ayazağa, Istanbul. Tel: 212 285 3419

Ünal, G., Dr., Anadolu Üniversitesi, Mühendislik Mimarlık Fakültesi, Eskişehir. T: 222 335 0580, F: 222 335 3616, E: gunal@anadolu.edu.tr

Ünver, M.A., Mng Zemtaş A.Ş., 88, Uğur Mumcu Cad., Gaziosmanpaşa, Ankara. E: zemtas@marketweb.net.tr

Üstay, A.H., Chairman, Üstay İnşaat, Abide-i Hürriyet Cad., Mecidiyeköy Yolu, 268, Boydaş Han K3, 80270 Şişli, Istanbul. T: 212 231 2181, F: 212 246 7964

Vardar, M., Prof. Dr., University Professor, Istanbul Teknik Üniversitesi, Maden Fakultesi, 80626 Ayazağa, Istanbul. T: 212 285 6147, E: vardar@itu.edu.tr

Wasti, Y., Prof. Dr., University Professor, Orta Doğu Teknik Üniversitesi, Mühendislik Fakültesi, 06531 Ankara. T: 312 210 2410

Yayla, N., Prof. Dr., University Professor, Istanbul Teknik Üniversitesi, İnşaat Fakültesi, 80626 Ayazağa, Istanbul. T: 212 285 3660, F: 212 285 6587, E: hyayla@srv.ins.itu. edu.tr

Yeşilçimen, Ö., Yapı Merkezi İnş.San.A.Ş., Yapı Merkezi, 81180 Çamlıca, Istanbul. T: 212 236 2400, E: oyesilcimen@ daristanbul.com

Yıldırım, H., Doç. Dr., University Dozent, Istanbul Teknik Üniversitesi, İnşaat Fakültesi, 80626 Ayazağa, Istanbul. T: 212 3703, F: 212 285 6587

Yıldırım, S., Prof. Dr., University Professor, Yıldız Teknik Universitesi, Muhendislik Fakültesi, Yıldız, Istanbul. T: 212 259 7070, F: 212 258 5140

Yıldız, İ.T., Civil Engineer, DSİ Lojmanları I.Kısım, E-3, Eskişehir Yolu, Ankara. T: 312 286 3882

Yılmaz, H.R., Doç. Dr., University Dozent, Dokuz Eylül Üniversitesi, Mühendislik Fakültesi, Bornova, Izmir

Yılmaz, E., Civil Engineer, ENAR Müh. Mim. Danışmanlık. Ltd. Şti., Ali Şir Nevai Sok., Baysal Apt. No: 30/2, Koca Mustafa Paşa, 34280 Istanbul. E: Eyilmaz@srv.ins.itu.edu.tr

Yöner, N., Civil Engineer, KASKTAŞ, Enderunlu İsmail Hakkı Bey Sok., No: 31, Balmumcu, 80700 Beşiktaş, Istanbul

Yurdumakan, S., KASKTAŞ, Milli Müdafaa Cad., D. Tokuz İş Merkezi, 12/13, Kat 6, Istanbul

Yüzer, E., Prof. Dr., University Professor, Istanbul Teknik Üniversitesi, Maden Fakültesi, 80626 Ayazağa, Istanbul. T: 212 285 6146, E: yuzer@itu.edu.tr

UNITED KINGDOM

Secretary: Mrs Dionne Dalgetty, British Geotechnical Association, The Institution of Civil Engineers, 1-7 Great George Street, London SW1P 3AA. T: 44 20 7665 2233, F: 44 20 7299 1325, E: admin@geo.org.uk

Total number of members: 1383

Abbiss, C.P., Dr, 51 Orchard Drive, Watford, WD1 3DX. T: 01923 226882, E: 101557.3035@compuserve.com

Abraham, R.N., Mr, 27 Huntsmead, Alton, Hampshire, GU34 2SF

Absolon, L., Mr, 7 Fallowfields, Crick, Northampton, NN6 7GA. T: 01788 523 415

Abu, T., Mr, University of Paisley, Dept of Civil, Struct. & Env. Eng., High Street, Paisley, Renfrewshire, PA1 2BE. T: 0141 8483277, F: 0141 8483275

Adam, C.H., Mr, 1 Queenslie Court, Summerlee Street, Glasgow, Lanarkshire, G33 4DB. T: 0141 7748828

Addenbrooke, T.I., Dr, 2 Balliol House, Manor Fields, London, SW15 3LL. T: 0207 594 6117 (W)

Airlie, D.C., Mr, 4 Easterfield Court, Kirkton, Livingston, West Lothian, EH54 7BZ. T: 1506413798

Ajzenkol, D., Mr, 84 Ulleries Road, Solihull, West Midlands, B92 8E. T: 0121 684 5260, E: gp92@dial.pipex.com

Ajzenkol, J., Mrs, 84 Ulleries Road, Solihull, West Midlands, B92 8EE. T: 0121 684 5260

Akagi, H., Prof., Department of Civil Engineering, Waseda University, 3-4-1 Ohkubo, Shinjuku, Tokyo 169, JAPAN. T: 81 3 52863405, F: 81 3 52720695

Akbar, S., Eur Ing, Mott MacDonald Group, St. Anne House, 20-26 Wellesley Road, Croydon, CR9 2UL

Akbar, A., Mr, Drummond Building, Newcastle University, Claremont Road, Newcastle Upon Tyne, NE1 7RU

Alderman, J.K., Dr, Sub Soil Surveys Ltd., Chaddock Lane, Astley, Tyldesley, Manchester, M29 7LD. T: 01942 883565, F: 01942 883566

Al-Dhahir, Z., Dr, 18 Highgrove Park, Gringer Hill, Maidenhead, SL6 7PQ

Aldridge, T.R., Mr, 9 Sebright Road, Boxmoor, Hemel Hempstead, HP1 1QY

Alexander, J.W., Mr, Blue Ball Cottage, Blue Ball Lane, Egham, Surrey, TW20 9EQ. T: 01784 433571

Allan, P.G., Dr, 4 Greystoke Gardens, Morpeth, Northumberland, NE61 1JB

Allen, R.G., Mr, 17 Albert Road, Ashford, Kent, TN24 8N. T: 01233 638279

Allenou, C.P.Y., Mr, 1 De Montfort Road, Streatham, London, SW16 1NF

Allison, D.P., Dr, 23 Frans Du Tiot Street, Phalaborwa, SOUTH AFRICA. T: 015 7810703, E: dallison@pmc.co.za

Allison, J.A., Dr, Bullen Consultantsltd, 3 Copthall House, Station Square, Coventry, CV1 2FZ. T: 01203 632299, F: 01203 632221

Allison, R.J., Mr, Snuggledown Cottage, The Street, Lawshall, Bury St. Edmunds, Suffolk, IP29 4QA. T: 01787 880218

Almeida, M.D.S.S., Prof., COPPE-U F R J, Caixa Postal 68506, Rio De Janeiro, 21945-970 BRAZIL. T: 55 21 2809993, F: 55 21 2809545

Alonso, E.E., Prof., Echegaray 8, 08950 Esplugas De Llobregat, Barcelona, SPAIN

Alsayed, M.I., Dr., 2R 1 Westbank Quadrant, Glasgow, Lanarkshire, G12 8NT. T: 0141 339 6004, F: 44 3396004, E: m.alsayed@lineone.net

Al-Shaikh-Ali, M.M.H., Dr, 4 Howeth Court, 2 Ribblesdale Avenue, London, N11 3GA. T/F: 0208 368 9990

Alston, C., Mr, 102 Senator Reesor's Drive, Markham, Ontario, L3P 3E5 CANADA. T: 905 2948508

Al-Tabbaa, A., Dr, Department of Engineering, University of Cambridge, Trumpington Street, Cambridge, CB2 1PZ

Amirsoleymani, T., Dr, 24a Russell Road, West Hendon, London, NW9 6AL. T: 020 8203 7474, F: 020 8203 7470

Amor, J., Mr, Maqueda 71-7b, 28024, Madrid, SPAIN

Amwell, The Lord, 11 Westfield Close, Hitchin, Hertfordshire, SG5 2HF. T: 01462 450852 (H)

Anagnostopoulos, A., Dr, 30 Philadelpheos Street, Kifissia, Athens 145 62, GREECE

Anderson, S.E., Miss, Flat 2 Finlay House, 13 Victoria Park Square, London, E2 9PB. T: 2089833726

Anderson, K., Mr, Struthers Farm, Edmundbyers, Consett, County Durham, DH8 9NL

Anderson, K.G., Mr, 1 Dale Court, Homesdale Road, Bromley, BR1 2QY

Anderson, R.J., Mr, 16 Myrtle Road, Hampton Hill, Hampton, Middlesex, TW12 1QE

Anderson, W.F., Prof., University of Sheffield, Sir Frederick Mappin Building, Mappin Street, Sheffield, S1 3JD

Andrei, A.F.J.M., Mr, 68 Rue Sainte Catherine, 59800 Lille, FRANCE. T: 00333 20 036891, E: antoine.andrei@waika9.com

Andreou, I.I., Mr, 28 Corri Avenue, Southgate, London, N14 7HL

Andrews, J.C., Mr, 10 Regal Drive, Rishworth, Sowerby Bridge, W Yorks, HX6 4RW. T: 01422 823647

Anketell-Jones, J.E., Mr, 198 Slag Lane, Lowton, Warrington, WA3 2EZ

Ansell, P., Dr, Conival, Sheriffhales, Shifnal, Shropshire, TF11 8RD. E: page@ansell3.freeserve.co.uk

Apted, J.P., Mr, 18 Bingley Grove, Woodley, Berks, RG5 4TT

Apted, R.W., Mr, 8 Belmont Terrace, Murrayfield, Edinburgh, Midlothian, EH12 6JF

Arber, N.R., Mr, 3 Cunliffe Drive, Sale, Cheshire, M33 3WS

Armstrong, K.M.C., Mr, Rear Workshop, 29 Chapel Street, West Auckland, Bishop Auckland, County Durham, DL14 9HP. T: 01388 8322684

Arnott, D.F., Mr, Flat 1, 46 Palatine Road, Withington, Manchester, M20 3JL. T: 077 489 63332

Arnould, R., Mr, Geotechnical Consulting Group, 1A Queensberry Place, London, SW7 2DL

Arthur, J.R.F., Dr, Fortune, Hill Road, Reydon, Southwold, Suffolk, IP18 6NL. T: 07502 723797

Arthur, J.C.R., Mr, Top-Hole Site Studies, The Gatehouse, 2 Holly Road, Twickenham, Middlesex, TW1 4EE

Ascari, P., Mr, Viale Lazio 8, 20135, Milan, ITALY

Atherton, D., Mr, Peter Brett Associates, 16 Westcote Road, Reading, Berkshire, RG30 2DE

Athorn, M.L., Dr, 2 Grimsthorpe Close, Grantham, Lincolnshire, NG31 8SU. T: 01476 569 789

Atkinson, J.H., Prof., 3 Battledean Road, London. N5 1UX. T: 020 7040 8153, F: 020 7040 8570

Attewell, P.B., Mr, 50 Archery Rise, Nevilles Cross, Durham, County Durham, DH1 4LA. T: 0191 3867298

Augarde, C.E., Dr, Manchester City Council, Technical Services Department, Town Hall, P.O. Box 488, Manchester, M60 2JT. T: 612363377

Auld, F.A., Dr, 113 Howden Close, West Bessacarr, Doncaster, DN4 7JN. E: alan.auld@alanauld.co.uk

Avalle, D.L., Mr, 35 Rushfield Road, Chester, CH4 7RE

Ayres, D.J., Mr, 46 Northumberland Place, Paddington, London, W2 5AS. T: 020 7727 4637

Babtie Geotechnical, 95 Bothwell Street, Glasgow, Lanarkshire, G2 7HX

Bachy Limited, Foundation Court, Godalming Business Centre, Cattershall Lane, Godalming, Surrey, GU7 1XW

Bailey, P.A., Mr, Merrydown Cottage, 19 Trodds Lane, Merrow, Surrey, GU1 2XY. T: 01483 455961

Bailey, R.C., Mr, Reinforced Earth Co. Ltd., 7 Hollinswood Court, Stafford Park 1, Telford, TF3 3DE. T: 01952 201901, F: 01952 2000047

Baker, D.A.B., Mr, Ramblers, Deepdene Avenue Road, Dorking, RH4 1ST

Baker, K.E., Mr, 20 Fontmell Park, Ashford, TW15 2NW

Baldwin, M.J., Mr, 21 Cheviot Way, Upper Hopton, Mirfield, West Yorkshire, WF14 8HW. T: 0532-711111

Ball, J.D., Mr, 13 Poplar Avenue, Altrincham, Cheshire, WA14 1LF. T: 1619285778, E: jon.ball@roger_bullivant. co.uk

Ball, P.J.H., Mr, 9 Dalehead Road, Leyland, Lancs., PR5 2AX

Balsubramaniam, A.S., Prof., G.E.Div., Asian Inst. of Tech., P.O. Box 2754, Bangkok, THAILAND

Baltzoglou, A., Mr, 10, Aimou str., Nea Smyrni, 171 23, Athens, GREECE. T: 020 8519 8864

Baluch, S.Q., Mr, Barnards, Harepath Hill, Seaton, Devon, EX12 2TF. T: 01223 235084

Bangs, K.W., Mr, 50 Gogh Road, Aylesbury, Buckinghamshire, HP19 3SH. T: 1296428745

Baple, J.P., Mr, 2 Ellis Farm Close, Woking, GU22 9QN

Barber, J.F., Eur Ing, 9 Lynton Road, London, N8 8SR. T: 834 10458

Bardanis, M., Mr, 29 Xanthippis Street, Athens, GREECE

Barker, D.H., Mr, Geostructures Consulting, Stables Block Trimworth, Model Farm, Crockham Hill, Edenbridge, Kent, TN8 6SR

Barker, J.E., Mr, Birmingham University, School of Civil Engineering, Edgbaston, Birmingham, B15 2RL. T: 0121 4145153 (W)

Barnes, G.E., Mr, 12 Southgate, Fixton, Urmston, Manchester, Lancashire, M41 9FS. T: 0161 0747 9560, F: 0161 2438, E: gbc@gebarnes.demon.co.uk

Barraclough, P.A., Mr, The Ballroom, Hawkswick, Skipton, North Yorkshire, BD23 5QA

Barry, A.J., Mr, 11-3-2 Menara Hartamas, Jalan 3/70A (Jln Sri Hartamas 3), Kuala Lumpur, 50480 MALAYSIA. E: tjb@pc.jaring.my

Barry, D.L., Mr, 18 Trelawne Drive, Cranleigh, Surrey, GU6 8BS. T: 01483 274465

Barton, M.E., Dr, Department of Civil & Environ. Engr., University of Southampton, University Road, Southampton, SO17 1BJ. T: 01703 592459, F: 01703 594986

Barton, D., Mr, 119-127 South Road, Haywards Heath, RH16 4LR. T: 01273 897000, F: 01273 897100, E: lewes@ owenw-dps.demon.co.uk

Barton, R.R., Mr, 42 Manor Road South, Hinchley Wood, Esher, KT10 0QL. T: 2083985414, E: ray.r.barton@talk21. com

Basile, F., Dr, 25 Upper Addison Gardens, London, W14 8AJ. E: BasileF1@halcrow.com

Bate, S.A., Miss, Brock Cottage, Prees Green, Whitchurch, Shropshire, SY13 2BL. T: 01948 840616, F: 840616, E: sam-bate@hotmail.com

Batt, A.G.H., Mr, High Lodge, The Cartway, Wedhampton, Devizes, Wiltshire, SN10 3QD. T: 1380840634

Batten, M., Dr, Tps Consult, The Lansdowne Building, Lansdowne Road, Croydon, CR0 2BX

Baudet, B.A., Miss, 23c Sumatra Road, London, NW6 1PS. T: 020 7435 1707

Bayes, I.S., Mr, 6 Kingsbridge Court, Peterborough, PE4 5BE

Beadman, D.R., Mr, 27 Orchard Way, Send, Woking, Surrey, GU23 7HS

Beasley, D.H., Dr, Lettons Way, Dinas Powys, South Glamorgan, CF64 4BY. T: 01793 812479 (W)

Beavan, G.C.G., Mr, 53 Cobton Drive, Hove, East Sussex, BN3 6WF. T: 01272 272450

Beeby, M.I., Mr, 1 The Paddock, 73 Mudeford, Christchurch, Dorset, BH23 3NJ. T: 01202 483766

Been, K., Dr, 11 Easthorpe View, Bottesford, Nottingham, NG13 0DL

Bell, A.G., Mr, Expanded Piling, Cheapside Works, Waltham, Grimsby, South Humberside, DN37 0J. T: 01472 822552, F: 01472 220675

Bell, F.G., Mr, Alwinton, Blyth Hall, Blyth, S81 8HL. E: AlexBell@netcomuk.co.uk

Belloni, L.G., Mr, Via V. Monti 41, 20123 Milano, ITALY. T: 02 4985213

Belokas, G., Mr, 94a Olympou Str., Chalandri, 15234, Athens, Hellas, GREECE. T: 0030 1 6854819, F: 0030 1 6810915

Benary, R., Mr, Meshek 6Y, Kfar Ben Nun, 99780 ISRAEL

Bennett, F., Mrs, Clonlea House, Field End Lane, Holmbridge, Huddersfield, HD7 1NH. T: 01484 689531, F: 01484 689932, E: goenviro@ashton-bennett.co.uk

Benson, N., Mr, 45 Cooloongatta Road, Camberwell, Melbourne, Victoria, 3124 AUSTRALIA. E: nbenson@ golder.com.au

Bentham, P., Mr, 9 Drayton Grove, Ealing, London, W13 0LA. T: 0044 020 8991 1492

Benton, L.J., Ms, Ove Arup & Partners, Admiral House, Rose Wharf, 78 East Street, Leeds, West Yorkshire, LS9 8EE

Berlogar, F., Mr, 5587 Sunol Blvd., Pleasanton, California, 94566 USA

Berry, A.J., Dr, 2 Spring Way, Alresford, Hampshire, SO24 9LN. T: 01962 732737, E: a.berry@kingston.ac.uk

Bertakis, G., Mr, 185 Doiranis Street, Kallithea 176 73, Athens, GREECE

Best, N.J., Miss, Peter Brett Associates, 66 Tilehurst Road, Reading, Berkshire, RG30 2JN

Biddle, A.R., Mr, Steel Construction Institute, Unit D, Silwood Park, Buckhurst Road, Ascot, Berkshire, SL5 7QN

Bienfait, C.M., Miss, 117 Westhall Road, Warlingham, Surrey, CR6 9HG

Billam, J., Dr, 23 Oldway Drive, Solihull, West Midlands, B91 3HP. T: 0121 705 1662

Binns, A., Mr, Flat 1, 8 Granby Road, Harrogate, North Yorkshire, HG1 4ST. T: 1423545921

Birch, G.P., Mr, 210 Chesterfield Drive, Sevenoaks, Kent, TN13 2EH

Birch, G.P., Mr, W S Atkins Consultants Ltd., 210 Chesterfield Drive, Sevenoaks, TN13 2EH

Birdsall, R.O., Mr, BC & T Consultants, Arundel House, Byland Close, Whitby, North Yorkshire, YO21 1JE

Birrell, J.W.C., Mr, Knight Piesold Pty. Ltd., Keck Seng Tower, 133 Cecil Street 315-02, 69535 SINGAPORE. T: +65 225 7006, E: kpsing@singnet.com.sg

Birtwhistle, J.S., Dr, 22 Richmond Road, Altrincham, Cheshire, WA14 4DU

Bitcheno, P.J., Mr, 43 Hillbrow Road, Ashford, TN23 4QH

Black, M.G., Mr, 18 Penns Road, Petersfield, Hampshire, GU32 2EN. T: 01730 302210

Blackford, S., Mr, 10 Ormsby Grange Road, Sutton, Surrey, SM2 6TH. T: 020 86429319

Blair-Fish, P.M., Dr, 6 Clarence Road, Harpending, Hertfordshire AL54AJ. T: 158 246 2114, E: hair-fish@clarence.demon. co.uk

Bland, S., Miss, 25 Addison Road, Guildford, Surrey, GU1 3QQ

Blight, I.C., Mr, 31 Caterham Road, London, SE13 5AP. T: 020 8852 6266

Blower, T., Mr, 18 Stockton Road, Wilmslow, Cheshire, SK9 6EU. T: 01625 537214

Boardman, D., Mr, 60 Springfield Road, Southwell, Nottinghamshire, NG25 0B

Boardman, M.F., Mr, Wayside, Quaker Brook Lane, Hoghton, Preston, Lancashire, PR5 0JA. T: 1254852393

Boden, D.G., Mr, T R L Ltd., Old Wokingham Road, Crowthorne, Berkshire, RG45 6AU

Boden, J.B., Mr, Woodlands, Millhouse Lane, Abbots Langley, WD5 0SF. T: 01923 269781

Boghosian, H.H.A., Dr, 209 Syon Lane, Osterley, Middlesex, TW7 5PU

Bolton, M.D., Mr, 232 Queen Ediths Way, Cambridge, Cambridgeshire, CB1 8NL

Bond, A.J., Dr, 54 Wilmot Way, Banstead, Surrey, SM7 2PY. T: 01737 354864

Boniface, T.A., Mr, 148 Montagu Street, Compstall, Stockport, Cheshire, SK6 5JE. E: edge@edgeuk.u-net.com

Bonney, J.E., Mr, 14 Neath Abbey Road, Neath, West Glamorgan, SA10 7BD. T: 01639 635766

Booth, A.R., Mr, Dunrobin, The Crescent, Dunblane, FK15 0DW

Borin, D.L., Dr, 69 Rodenhurst Road, London, SW4 8AE. T: 020 8674 7251, F: 020 8674 9685

Bourke, M., Mr, 14 Dunglass Avenue, Glasgow, G14 9DX. T: 0141 204 2130

Bowey, A.W., Dr, 19 Penny Lane, Barwell, Leicester, Leicestershire, LE9 8HU. T: 02476 511266 (W)

Bowman, E.T., Miss, Magdalene College, University of Cambridge, Cambridge, Cambridgeshire, CB3 0AG

Boyd, A.B.M., Mr, AB Geotechnics Limited, Centrix Park, Hardie Road, Livingston, EH54 8AR

Boyd, P.J.H., Mr, 20 Paddock Row, Elsworth, Cambridge, Cambridgeshire, CB3 8JG. T: 01954 267754, E: pjhboyd@elsworth20.freeserve.co.uk

Bracegirdle, D.R., Mr, Flat 15, The Gatehouse, The Moorings, Leamington Spa, Warwickshire, CV31 3QA. T: 02476 694664

Braddish, P., Mr, 11 Birkbeck Road, London, N12 8DZ

Bradshaw, S.M., Mr, Old Town Slack Farm, Wadsworth, Hebden Bridge, West Yorkshire, HX7 8TE

Branch, S., Mr, L B H Wembley, The Old School House, Bridge Road, Kings Langley, Herts, WD4 8RQ. T: 01923 263222, F: 01923 263888, E: ibhwembley@compuserve.com

Bransby, M.F., Dr, Dept of Civil Engineering, The University of Dundee, Perth Road, Dundee, DD1 4HN

Bransby, P.L., Eur Ing, Red Gables, Benslow Path, The Avenue, Hitchin, Hertfordshire, SG4 9RH. T: 01462 45779, E: peter.bronsby@which.net

Branston, W.A., Mr, 8 Andrews Grove, Ackworth, Pontefract, West Yorkshire, WF7 7NU. T: 01977 610590, E: webranston@leedsalumni.org.uk

Brenton, D.M., Mr, 7 Pitfield Close, Fenstanton, Huntingdon, Cambridgeshire, PE18 9EE. T: 01685 883140

Brice, M., Mr, 55 Cavendish Avenue, Cambridge, Cambridgeshire, CB1 4UR

Bridges, C.A., Mr, C/O SMEC Asia Ltd., 14/F, Hua Fu, 14/F, Hua Fu, Commercial Build., 111 Queens Road West, HONG KONG. T: 852 2517 1136, E: cabridges_2000@yahoo.com

Bridle, R.C., Mr, 111 High Street, Amersham, Buckinghamshire, HP7 0DZ. T: 44 1494 862082

Brightman, D., Mr, Hanburys, Shootersway, Berkhamstead, Herts, HP4 3NG

Bromhead, E.N., Mr, 9 Kevins Drive, Yateley, Hampshire, GU46 7TL

Bromhead, J.C., Mr, 36 Snailbeach, Mainsterley, Shrewsbury, Shropshire, SY5 0NX

Brook, D., Mr, Doe Minerals Division, 4/A3 Eland House, Bressenden Place, London, SW1E 5DU. T: 020 7890 3842

Brooks, R.G., Mr, 93 Halstead Road, Mountsorrel, Loughborough, Leicestershire, LE12 7HE

Brown, A.J., Mr, 12 Scillonian Road, Guildford, Surrey, GU2 5PS

Brown, D.F., Mr, 23K Claremont Street, Aberdeen, AB10 6QQ

Brown, D.P., Mr, Flat G, 1st Floor, Block 2, Belvedere Gardens, Phase 3, 625 Castle Peak Road, Tsuen Wan, HONG KONG. T: 85224021853

Brown, L.W., Mr, 11 Pine Grove, Maidstone, Kent, ME14 2AJ. T: 01622 683307

Brown, M.J., Mr, 20 Maple Close, Doveridge, Ashbourne, Derbyshire, DE6 5LU. T: 01189 569462, E: mbrown6610@aol.com

Brown, N.A., Mr, Tugela, Strawberry Gardens, Penally, Tenby, Dyfed., SA70 7QF

Brown, P.J., Mr, 53 Locksway Road, Milton, Southsea, Hampshire, PO4 8JW. T: 2392422644, E: p.j.brown@cwcom.net

Brown, P.R., Mr, 9 Falkland Mount, Leeds, West Yorkshire, LS17 6JG. T: 0113 2946320

Brown, S.F., Prof., University of Nottingham, Dept of Civil Engineering, University Park, Nottingham, Nottinghamshire, NG7 2RD

Brownlie, N.J., Mr, 17 Burdock Close, Burghfield Common, Reading, Berkshire, RG7 3YY. T: 01734 831371

Bruce, D.A., Dr, 161 Bittersweet Circle, Venetia, Pennsylvania, 15367 USA. T: 412 9423082

Bruggemann, D.A., Mr, 6 Cannon Close, Raynes Park, London, SW20 9HA. T: 01372 3591 (incomplete), E: bruggeman@cableinet.co.uk

Bryant, D.K., Mr, 18 Chapman Street, Llanelli, SA15 3EB. T: 1554756667, F: 01792 472019

Bryson, L.S., Mr, 1715 West Greenleaf Avenue, Chicago, Illinois, 60626-2411 USA

Building Research Establishment, Bucknalls Lane, Watford, Hertfordshire, WD25 9XX

Bull, J., Mr, 63 Bower Street, Maidstone, Kent, ME16 8BB. T: 1732771388

Burch, M.G., Mr, 23 Springfield Rd., Somersham, Nr. Ipswich, IP8 4PG

Burd, H.J., Mr, University of Oxford, Dept of Engineering Science, Parks Road, Oxford, Oxfordshire, OX1 3PJ

Burden, P.R., Mr, 6 Raglan Place, Bristol, BS7 8EQ

Burge, R.F., Mr, Parkman – Bua, P.O. Box 260, Chilumba, MALAWI

Burgess, K.W., Mr, 75 Orchid Crest, Upton, Pontefract, West Yorkshire, WF9 1NT. T: 7788114728

Burland, J.B., Mr, 13 Barton Road, Wheathampstead, St. Albans, Hertfordshire, AL4 8QG. T: 1582832366, F: 832366

Bursnoll, C.E., Miss, Graylaw, 51 Mold Road, Caergwrle, Wrexham, Clwyd, LL12 9LR

Burt, N.J., Dr, 59 Pendennis Road, Streatham, London, SW16 2SR. T: 2086771325, E: carolburt@nasuwt.net

Butcher, A.P., Mr, 3 Coniston Road, Kings Langley, WD4 8BT

Byrne, B.W., Mr, Holywell Manor, Manor Road, Oxford, Oxfordshire, OX1 3UH

Cabarkapa, Z., Mr, 1a Queensberry Place, London, SW7 2DL. T: 8673 1398

Cadden, A.J.R., Mr, Kingscote, Waltham Road, Maidenhead, Berkshire, SL6 3NH. T: 1628823027

Cale, S.A., Mr, Kanthack House, Station Road, Ashford, Kent, TN23 1PP

Calkin, D.W., Mr, 14 Manor Farm Close, Oakington, Cambridge, CB4 5AT

Callaghan, M.V., Mr, Knight Pie'sold Consulting, 34 Commerce Crescent, P.O. Box 10, North Bay, CANADA

Callington, H.W., Mr, 6 Damask Close, Weston, Nr Hitchin, SG4 7EL

Campton, A.L., Miss, 47 The Sanctuary, Culverthouse Cross, Cardiff, CF5 4RW

Candler, C.J., Mr, 14 Cliddesden Road, Basingstoke, Hampshire, RG21 3DU. T: 01256 474871, E: kim.candler@lineone.net

Cannon, G.G., Mr, 52 Bank Parade, Burnley, Lancashire, BB11 1TS

Cannon, W.C., Mr, 16 The Cuillins, Uddingston, Glasgow, Lanarkshire, G71 6EY. T: 0141 332 8754

Capps, C.T.F., Mr, T C E S, Roadstone House, 50 Waterloo Road, Wolverhampton, WV1 4RU. T: 01902 316130, F: 01902 316131

Card, G.B., Eur Ing, Card Geotechnics Ltd., Alexander House, 50 Station Road, Aldershot, Hampshire, GU11 1BG

Carr, P., Miss, 41 Beaconsfield Place, Epsom, Surrey, KT17 4BD

Carrier, W.D., Eur Ing, 76 Woodside Drive, Florida 33813-3557, Lakeland, USA. T: 18636468890

Carrington, R.W.M., Mr, P.O. Box 13886, Nairobi, KENYA. T: 254 2 583133, F: 254 582826, E: carrington@form net.com

Carrington, T.M., Mr, 5 Shepherd Close, Canford Place, Aylesbury, Buckinghamshire, HP20 1DR. T: 1296487435, F: 44 487435, E: timcarongton@lineone.net

Carson, A.M., Mr, 30 Coseley Avenue, Monkmoor, Shrewsbury, Shropshire, SY2 5UP. T: 1743356466

Carter, T.G., Dr, 396 Riverside Drive, Oakville, Ontario, CANADA

Carter, J.P., Prof., School of Civil Engineering, University of Sydney, N.S.W. 2006, AUSTRALIA. T: 61 2 3512109, F: 61 2 3513343

Casanovas-Rodriguez, J.S., Prof., Capitan Arenas 33, Barcelona 08034, SPAIN. T: 34 3 4016869, F: 34 3 4016504, E: casanova@gauguin.upc.es

Castello, R.R., Prof., Rua Romulo Leao Castello, 106, Ilha Do Boi, 29.052-740 Vitoria-E.Santo, BRAZIL. T: 027 2250622, F: 027 3971695, E: renocstl@npd1.lifes.br

Cavanagh, S., Mr, W S Atkins Consultants Ltd., Dunedin House, Riverside, Columbia Drive, Stockton-on-Tees, Cleveland, TS19 6BJ

Cavounidis, S., Dr, 4 Ang. Hadjimihali, Athens 10558, GREECE. T: 01 3222050, F: 01 3241607

Chadwick, N.C., Mr, 2 Verbania Way, Manor Glade, East Grinstead, West Sussex, RH19 3UP

Chalmers, A., Mr, 2 Hillbury Close, Chesham Bois, Amersham, Bucks, HP6 5LB

Chan, A., Dr, School of Civil Engineering, University of Birmingham, Edgbaston, Birmingham, West Midlands, B15 2TT

Chandler, R.J., Prof., Dept of Civil Engineering, Imperial College, London, SW7 2BU

Chang, J., Miss, Mott Macdonald Group, Underground & Foundation Div., St. Anne House, Wellesley Road, Croydon, Surrey, CR9 2UL

Chantler, C.A., Miss, 46 Frenches Farm Drive, Heathfield, East Sussex, TN21 8BZ. T: 01435 864274

Chaplin, T.K., Dr, 32 Allnutt Avenue, Basingstoke, Hampshire, RG21 4BW. T: 44 01256 322803

Chapman, D.N., Dr, 21 Badger Way, Blackwell, Bromsgrove, Worcestershire, B60 1EX. T: 1214456347, E: david. chapman@excite.co.uk

Chapman, P., Mr, 72 The Beeches, Beaminster, Dorset, DT8 3SN. T: 01308 862349

Chapman, T.J.P., Mr, 131 Wynchgate, Wichmore Hill, London, N21 1QT. T: 2088868176

Charles, J.A., Dr, 35 Greenways, Abbots Langley, Hertfordshire, WD5 0EU. T: 1923266949

Charman, J.H., Mr, Highfield, 45 Oxted Green, Godalming, Surrey, GU8 5DD. T: 01483 423020, E: johncharman@ compuserve.com

Chartres, F.R.D., Mr, The Smithy, Great Woodcote Drive, Purley, Surrey, CR8 3PL. T: 0044 20 86689859, F: 0044 87632511, E: sarahchartres@btconnect.com

Chaytor, J.M., Mr, 67 Cadwell Drive, Ockwells Park, Maidenhead, SL6 3YS. T: 01628 38476

Cheetham, J.A., Dr, 4 Moorfield Drive, Wilmslow, SK9 6DL

Cherrill, H.E., Mr, 106 Westbourne Road, Penarth, South Glamorgan, CF64 3HH. T: 207049863

Cheung, L.Y.G., Miss, Newham College, Cambridge, Cambridgeshire, CB3 9DF

Child, G.H., Mr, 105 Barkham Ride, Wokingham, Berkshire, RG40 4EP. T: 0118 9734620, F: 0118 9734620

Child, K.T., Mr, 1 Northgate Rise, Linton, Wetherby, LS22 4UB

Chipp, P.N., Mr, Mill Stones, Mill Lane, Walkeringham, Doncaster, South Yorkshire, DN10 4HY. T: 1427890612, E: peterchipp@hotmail.com

Chmoulian, A., Dr, 21 Sovereign Place, Peterborough, PE3 6DS

Chow, F., Dr, Flat 3, 12 Chiswick High Road, London, W4 1TH. T: 0207 581 8348 (W)

Chowdhury, R.N., Prof., Wollongong University, Wollongong, 2522 AUSTRALIA

Christian, J.T., Dr, 23 Fredana Road, Waban, M A 02468-1103, USA

Chung, S.G., Prof., Dong-A-University, Dept of Civil Engineering, 840 Hadan-Dong, Saha-Gu, SOUTH KOREA

Church, G.D., Mr, 20 Rusthall Road, Rusthall, Tunbridge Wells, Kent, TN4 8RD

Clapham, H.G., Mr, 42 Greenacres, Leverstock Green, Hemel Hempstead, Hertfordshire, HP2 4NA

Clark, R.G., Eur Ing, Marlowe, Broughton Hackett, Worcester, Worcestershire, WR7 4BE. T: 01905381898 (H) 44019 26431007 (W)

Clark, J., Miss, 11 The Middlings, Sevenoaks, Kent, TN13 2NW. T: 07930 581732, E: jc497@soton.ac.uk

Clark, A.R., Mr, 3 Belvedere Walk, Winnersh, Wokingham, Berkshire, RG41 5TQ

Clark, C.R., Mr, 7 Ridvan Grove, Ngaio, Wellington, NEW ZEALAND

Clark, T.A., Mr, Yew Tree Cottage, Snuff Lane, Shotteswell, Banbury, Oxfordshire, OX17 1JH. T: 01295 738143

Clarke, B.G., Prof., Eur Ing, Arncliffe, 3 Rectory Road, Newcastle upon Tyne, Tyne and Wear, NE3 1XR

Clarkeburn, J.W., Mr, Top Left, 21 Craigmillar Road, Glasgow, Lanarkshire, G42 9JZ. T: 0141 649 1468, E: h.clarkeburn@ phil.hull.ac.uk

Clarkson, J., Miss, 106-D London Road, Peterborough, PE2 9BY. T: 01733 334455

Clayton, C.R.I., Dr, 'Wildfield', Pond Hill, Wanborough, GU3 2JW

Clifton, A.J., Mr, 4 St. Vincents Hill, Redland, Bristol, BS6 6UP. E: CliftonA@rpspllc.co.uk

Clinton, Mr, D.B., 25 Thackeray Avenue, Clevedon, Avon, BS21 7JL

Clough, I.R., Mr, 106 Dluchouse Lane, Oxted, Surrey, RH8 0AR. T: 01883 715 854, F: 01883 716 038, E: ian@ riclough.globalnet.co.uk

Cobbe, M.I., Mr, 6 Teeton Road, Ravensthorpe, Northampton, NN6 8EJ. T: 01604 770636

Cockcroft, J.E.M., Mr, Chestnut House, Mitre Drive, Repton, Derby, Derbyshire, DE65 6FJ. T: 01283 701577 01283 701577 (F), F: 44 701577

Cocksedge, J.E., Mr, Sycamore Crescent, Church Crookham, Fleet, Hampshire, GU13 0NN

Coe, R.H., Mr, Dean Cottage, Miles Lane, Cobham, Surrey, KT11 2ED. T: 01932 863421, E: rhcoe@btinternet.com

Cogswell, M.V., Mr, 39 Inveralmond Drive, Cramond, Edinburgh, Midlothian, EH4 6JX. T: 1313128315, F: 3128315, E: michaelvcogswell@btinternet.com

Cole, K.W., Mr, Southern Down Nursing Home, Worcester Road, Chipping Norton, Oxfordshire, OX7 5YF

Cole-Baker, J.R., Mr, 19 Bannister Gardens, Storrington, Pulborough, West Sussex, RH20 4PU. T: 1993745557, F: 741898, E: jcolebaker@aol.com

Collins, R.S., Mr, 1 Charnwood Grove, West Bridgford, Nottingham, NG2 7NT. T: 0115 9061354

Colmenares, J.E., Mr, Calle 19, 4-56 Apt 809, Bogota, COLOMBIA

Comrie, R.J., Mr, 10 Underhill Drive, Guide Post, Choppington, Northumberland, NE62 5YT

Conn, G.M., Dr, 20 Wood Drive, Stevenage, Hertfordshire, SG2 8PA

Connolly, M.L., Mr, Ivy Cottage, 6 Pendicke Street, Southam, Leamington Spa, Warwickshire, CV33 0P. T: 01203 511 266 (W)

Constantine, P.P., Mr, 5 Natwoke Close, Beaconsfield, Bucks, HP9 2AR

Cook, A.W., Eur Ing, 3 Lyndhurst Close, The Ridgeway, Woking, Surrey, GU21 4RA. T: 01483 767673, F: 769673, E: jimcook@cwcom.net

Cook, A.F., Eur Ing, Quadrant Court, 44–45 Calthorpe Road, Edgbaston, Birmingham, B15 2TH. T: 0121 452 7411, F: 0121 452 1799

Cook, D.A., Mr, Ashton, Sion Road, Bath, Avon, BA1 5SG. T: 44 01225 311635, E: absdac@Bathac.uk

Cooke, R.W., Dr, The Old Stable, Cley, Holt, Norfolk, NR25 7SF. T: 01263 741122, E: r_w_cooke@hotmail.com

Cooke, J.P., Mr, 92 Oakfield Road, Croydon, Surrey, CR0 2UB

Cooling, C.M., Dr, Gartlet, 21 Talbot Avenue, Oxhey, Watford, Hertfordshire, WD1 4AX. T: 01923 231638

Coop, M.R., Dr, Flat C, 23 Sumatra Road, London, NW6 1PS

Cooper, M.R., Dr, 23 Fairview Drive, Romsey, Hampshire, SO51 7LQ. T: 01794-512780, E: mike@3rd-degree. freeserve.co.uk

Copping, S.D., Mr, 17 Queens Cottage, Reading, Berkshire, RG1 4BE. T: 01734 595498

Corbet, S.P., Mr, 72 Rowan Walk, Leigh-on-Sea, SS9 5PL. T: 01702 521461, E: spc@maunsell.co.uk

Corbett, B.O., Eur Ing, 42 Wykeham Way, Burgess Hill, West Sussex, RH15 0HF. T: 01444 236725

Corke, D.J., Mr, 31 Registry Close, Kingsmead, Northwich, Chesire, CW9 8UZ. T: 1606352284

Cormie, W.M., Mr, 13 Winton Drive, Glasgow, Lanarkshire, G12 0PZ. T: 1413342002

Cotecchia, F., Dr, Istituto Di Geologia, Applicata E, Geologia, Politecnico Di Bari, Via Orabona, 4, Bari, ITALY. T: 0039 80 5460338, F: 0039 80 5460675

Coulton, R.H., Dr, 5 Auden Close, Osbaston, Monmouth, Gwent, NP5 3NW

Coumoulos, D., Dr, 8 Dryadon Street, Gr-152 37 Philothei, Athens, GREECE. T: 3016717079

Coutts, J.S., Mr, 37 Bottrells Lane, Chalfont St. Giles, Buckinghamshire, HP8 4EY. T: 44 01494 875209, E: jscoutts@ cwcom.net

Covil, C.S., Mr, Ove Arup & Partners, Level 10, 201 Kent Street, Sydney NSW 2000, AUSTRALIA. T: W612 93209320/F93209321

Cowell, B., Mr, Weymouth House, 8 Victoria Street, Hetton-le-Hole, Houghton Le Spring, DH5 9DF. T: 0191 233 4261

Cowley, M.N., Mr, Slapton Hill Barn, Blakesley Road, Slapton, Towcester, Northamptonshire, NN12 8QD. T: 01327 860060, F: 01327 860060

Cowsill, P.A., Mr, 11a Leygate View, New Mills, High Peak, Derbyshire, SK22 3EF. T: 01663 744580, F: 01663 741309

Cox, D.W., Mr, 49 Basterfield House, Golden Lane Estate, London, EC1Y 0TR

Cox, P.B., Mr, 1/F, 15 Tai Yuen Village, Yung Shine Wan, Lamma Island, HONG KONG. T: 852 2982 2809, E: JUNINHO@HKSTAR.COM

Crabb, T.R., Mr, 175 Blind Lane, Flackwell Heath, High Wycombe, Buckinghamshire, HP10 9LE

Craggs, E.J., Miss, 8 Courthouse Road, Maidenhead, Berkshire, SL6 6JD

Craig, W.H., Mr, 8 Crescent Road, Hale, Altrincham, Cheshire, WA15 9NA. T: 1619281716

Crawford, R.J., Mr, 2 Stonefield Close, Shrivenham, Swindon, Wilts, SN6 8DY

Crilly, M.S., Mr, 183 Trafalgar Street, Walworth, London, SE17 2TP. T: 020 77031380, F: 01923 66408, E: mike. crilly@talk21.com

Crockford, R.M., Mr, 32 Grasmere Avenue, Wetherby, West Yorkshire, LS22 6YT. T: WK 01937 541118

Cross, M., Dr, 14 St. Martins Avenue, Otley, West Yorkshire, LS21 2AN. T: 01943 464516, E: mcrtrn@carlbro.dk

Crossland, R.H., Mr, 108 Gales Drive, Three Bridges, Crawley, West Sussex, RH10 1QE

Crowe, N.D., Mr, 14 Chawton Park Road, Alton, Hampshire, GU34 1RG. T: 0410 578 558

Cuccovillo, T., Miss, School of Construction, South Bank University, 202 Wandsworth Road, London, SW8 2JZ

Cullum-Kenyon, S., Mr, Thurber Engineering, 190, 550-71 Avenue SE, Calgary, AB T2H OS6 CANADA

Cummings, S.J., Dr, 3 Lord Wardens View, Bangor, County Down, BT19 1GN

Cunningham, M., Mr, Field Cottage, The Cross, Eastington, Glos, GL10 3AA

Curtis, D.C., Mr, 22 South Drive, Timperley, Altrincham, Cheshire, WA15 6QJ

Curtis, J.C., Mr, 4 Manchester Drive, Leegomery, Telford, TF1 4XY. T: 01952 641737, F: 01952 243757

Dabee, B., Mr, Michaell Barclay Partnership, 105-209 Strand, London, WC2R 0SS. T: 020 7240 1191, F: 020 7240 2241

Dailey, R.J., Mr, 54 Oakwood Avenue, Purley, Surrey, CR8 1AQ. T: 0208 6605797

Daley, P., Eur Ing, 2 Arundel Road, Tunbridge Wells, Kent, TN1 1TB

Daley, A.J., Mr, Tanah Tech Pte Ltd., P.O. Box 61, Robinson Road Post Office, 900111 SINGAPORE

Dalton, J.C.P., Mr, Rectory Farm, Little Eversden, Cambridge, CB3 7HE

Daniels, P.C., Mr, 77 Bashford Way, Pound Hill, Crawley, West Sussex, RH10 7YG

Danilewicz, C.J., Mr, Killara, 8 Grasmere Road, Huddersfield, West Yorkshire, HD1 4LH

Darton, C.E., Mr, 47 Roebuck Close, Horsham, RH13 5UL. T: 01372 862890, F: 01372 862616

Dauncey, P.C., Mr, 11 Etheldene Avenue, Muswell Hill, London, N10 3QG

Dauris, H.C., Mrs, Arup Moscow, Ove Arup & Partners, 13 Fitzroy Street, London, W1T 4BQ. T: 70 95 9562337

Davachi, M.M., Dr, 31 Brabourne Mews S W, Calgary, A B, T2w 2v9, CANADA

Davey-Wilson, I.E.G., Dr, 1 Spindlers, Church Street, Kidlington, Oxfordshire, OX5 2YP

Davies, C., Miss, Bullen Consultants, 3 Copthall House, Station Square, Coventry, CV1 2FZ

Davies, A.H.R., Mr, 3 Gwynfryn Road, Pontarddulais, Swansea, SA4 1LG

Davies, E., Mr, C/O Mitsubishi Electric, Leon House, High Street, Croydon, Surrey, CR0 2XT

Davies, J.A.G., Mr, Ove Arup & Partners, Level 5 Festival Walk, 80 Tat Chee Ave, Kowloon Tong, Kowloon, HONG KONG

Davies, J.P., Mr, 11 Tidenham Gardens, Park Hill, Croydon, Surrey, CR0 5UT. T: 020 8667 0673

Davies, R.V., Mr, 60 Murhill, Limpley Stoke, Bath, Avon, BA3 6HG

Davies, M.C.R., Prof., University of Dundee, Professor of Civil Engineering, Dept of Civil Engineering, Perth Road, Dundee, Angus, DD1 4HN

Davis, A.J., Mr, 2 Hartington Place, Reigate Hill, Reigate, Surrey, RH2 9NL

Davis, J.A., Mr, 44 Emmanuel Road, Balham, London, SW12 0HN

Davison, L.R., Mr, 23 Malmains Drive, Frenchay, Bristol, BS16 1PL

Davison, M.J., Mr, Woodpeckers, Poplar Grove, Woking, Surrey, GU22 7SD

Dawson, A.R., Eur Ing, Dept of Civil Engineering, University of Nottingham, University Park, Nottingham, NG7 2RD

Dawson, M.P., Mr, 23 Prospect Road, Sevenoaks, Kent, TN13 3UA. T: 0208 663 6565 (WK), Ext. 235

Day, R.A., Dr, 28 Denham Terrace, Tarragindi 4121, AUSTRALIA. T: 07 3878 6997, E: day@civil.uq.oz.au

Day, S., Mr, 9 Castledine Court, Woodfield Plantation, Balby, Doncaster, South Yorkshire, DN4 8SE. T: 01302 726730

Daynes, P.J., Mr, 36 Norborough Road, Wheatley, Doncaster, DN2 4AT

De Ambrosis, L.P., Dr, Longmac Associates Pty. Ltd., P.O. Box 940, Crows Nest, 2065 AUSTRALIA. T: 02 4394033, F: 02 4360606

De Freitas, M.H., Dr, Department of Civil Engineering, Imperial College of Science, Technology & Medicine, London, SW7 2BU. E: m.defreitas@ic.ac.uk

De Moor, E.K., Dr, 27 Denmark Avenue, Woodley, Reading, RG5 4RS

Dean, E.T.R., Dr, Soil Models Lts., 337 Northbrooks, Harlow, Essex, CM19 4DP. T: 01279 419934, F: 01279 425204, E: RDEAN51@HOTMAIL.COM

Deason, P.M., Mr, 27 The Terrace, Sunninghill, Ascot, Berkshire, SL5 9NH. T: 01344 28991

Dedic, M., Mr, 29 Whitehall Gardens, London, W3 9RD. T: 2089928092

Delahaye, N.A., Mr, 67A Horniman Drive, Forrest Hill, London, SE23 3BU

Dennis, J.A., Mr, 8 Branwell Drive, Haworth, West Yorkshire, BD22 8HG. T: 01535 644885

Department of Transport Highways Agency, Rm 3/28, St. Christopher House, 90-114 Southwark Street, London, SE1 0TE

Deshpande, S.C., Mr, Sewri Eng. Const. Co. Ltd., Apeeyay Chambers, Wallace Street, Fort, INDIA

D'Hollander, R.D., Mr, 7887 E.Ridge Pointe Dr., Fayetteville, 13066-9518, New York, USA

Dias, A., Mr, 2F, 47a Tsz Tong Tsuen, Kam Tin, New Territories, HONG KONG. T: (852) 9310 8271, E: alberto_dias@ hotmail.com

Diaz-Rodriques, A., Prof., Ciudad Universitaria, Apartado Postal 70-165, Mexico City, MEXICO 04510. T: 52 5 622 32 27, F: 52 5

Dibben, S.C., Dr, 22 Eaton Road, Sale, M33 7TZ

Dickin, E.A., Mr, Manor Croft, Mill Lane, Rainhill, Prescot, Merseyside, L35 6NE

Dieteren, M.J.M.M., Eur Ing, Flat 2, Candide Lodge, Edgeley Road, Clapham, London, SW4 6EP

Dimmock, P.S., Mr, 1 De Montfort Road, Streatham, London, SW16 1NF. T: 020 7581 8348

Dishington, A.N., Mr, 15 Cypress Road, Woodley, Reading, RG5 4BD. T: 0118 927 2594

Dixon, J.C., Mr, Glaramara, Claremont Avenue, Esher, Surrey, KT10 9JD

Dixon, N., Mr, Dept of Civl. & Bldg. Engineering, Loughborough University, Loughborough, Leicestershire, LE11 3TU

Djerbib, Y., Mr, 39 Brailsford Avenue, Sheffield, S5 9DL. F: 01 10 1996

Do Val, E.C., Mr, Rua Teodoro Sampaio 744/126, S P. São Paulo, 05406-000 BRAZIL. T: 55 11 2118626, F: 55 11 8700674

Doake, R.M., Mr, C/O R H Cuthbertson & Ptns., 13 Eglinton Crescent, Edinburgh, EH12 5DF

Dodd, C.P., Mr, 71 Dessancourt, Holmes Chapel, Cheshire, CW4 7NB

Doherty, G.K., Mr, Highways & Engineering Dept., Gwynedd Council, Council Offices, Shirehall St., Caernarfon, Gwynedd, LL55 1SH

Doran, I.G., Prof., 47 Derryvolgie Avenue, Belfast, County Antrim, BT9 6FP. T: 2890660521

Dounias, G.T., Dr, Edafos Ltd., 9 Yperidou, 10558 Athens, GREECE. T: 30 1 3222050, F: 30 1 3241607

Dowell, R.W.R., Mr, 125 Spring Pool, Warwick, CV34 4UP

Downing, M.C., Mr, 76 Collingworth Rise, Park Gate, Southampton, SO31 1DB

Doyle, P.M., Mr, 93 Skibo Court, Dunfermline, Fife, KY12 7EW. T: 01506 654375

Drake, J.J., Mr, 87 Grafton Street, Warrington, WA5 1QA. T: 01925 828 987, F: 01925 244 051, E: jdrake@wsatkins. co.uk

Driscoll, R.M.C., Mr, 6 Lyndhurst Avenue, London, NW7 2AB. T: 020 8959 2010, E: dvisvich@globalnet.co.uk

Dunbar, S.J., Mr, 50 Rounceval Street, Chipping Sodbury, Bristol, Avon, BS37 6AR

Dunelm Drilling Co., Somerville House, St. Johns Road, Meadowfield Industrial Estate, Durham, County Durham, DH7 8TZ

Dunnicliff, J., Mr, Little Leat, Whisselwell, Bovey Tracey, Newton Abbot, Devon, TQ13 9LA. T: 01626 836161, F: 01626 832919, E: johndunnicliff@ibm.net

Duriez, B.J., Mr, Pick Everard, Halford House, Charles Street, Leicester, LE1 1HA

Duthy, M.L., Dr, C/O Binnie Black & Veatch, Grosvenor House, 69 London Road, Redhill, RH1 1LQ

Dyer, M.R., Dr, University of Durham, School of Engineering, South Road, Durham, County Durham, DH1 3LE

Dyson, S., Mr, 17 High Street, Chrishall, Royston, Hertford-shire, SG8 8RN. T: 1763837302, F: 01763 837302

E.A.Robertson, E.A., Miss, 36 Norborough Road, Don-caster, South Yorkshire, DN2 4AT

Earl, M.J., Mr, 20 Grosvenor Road, Sketty, Swansea, West Glamorgan, SA2 0SP

Earle, D.A., Mr, 61 Straight Road, Old Windsor, Windsor, Berkshire, SL4 2RT

Eaton, P., Mr, Cotswold Geotech, Shab Hill, Birdlip, Glouces-ter, GL4 8JX. T: 01452 864550, F: 01452 864831

Ebeling, R., Dr, 115 Brookwood Drive, Vicksburg, Ms 39183-1622, USA

Eccles, C.S., Mr, 55 St. Catherine Drive, Hartford, Northwich, Cheshire, CW8 2FE

Economou, V., Miss, 11 Bizaniou Street, Papagos 15669, Athens, GREECE. T: 010301 6548371

Eddleston, M., Eur Ing, 32 Tintern Road, Cheadle Hulme, Cheadle, Cheshire, SK8 7QF

Edmonds, H.E., Miss, Geotechnical Consulting Group, 1a Queensberry Place, London, SW7 2DL

Edmonds, I.G., Mr, Halcrow U K, Arndale House, Otley Rd., Leeds, Headlingley, LS6 2UL

Edser, R.J., Mr, 171 Craigleith Road, Edinburgh, Midlothian, EH4 2EB

Edwards, D., Mr, Flat 99, Pond House, Pond Place, London, SW3 6QT

Egan, D., Mr, 7 Priory Close, Wetherby, Leeds, West Yorkshire, LS2 7TN

Egerton, R.H.L., Mr, 57 High Street, Godstone, RH9 8LT. T: 01883 742641, F: 01883 744804

Eldred, P.J.L., Mr, 12 Magnolia Way, Wokingham, Berkshire, RG41 4BN. T: 1189792350, E: peter@eldred.co.uk

Eldridge, J.J., Mr, Twr. 15 21/C Pacific Palsades, 1 Braemar Hill Road, North Point, HONG KONG

Elliott, G., Dr, 6 Agricola Way, Thatcham, Berkshire, RG19 4GB

Elliott, J.P.K., Mr, 28 Manor Road, Leamington Spa, CV32 7RJ. T: 01926 334229/07951495796

Ellis, E.A., Mr, 40B Windmill Road, Croydon, Surrey, CR0 2XP. T: 07713 293273

Elson, W.K., Dr, Orchard End, 7 Finches Lane, Lindfield, Haywards Heath, West Sussex, RH16 2PG. T: 01444 483250, F: 01444 483250

Emanuel, D., Mr, 39 Yr-Ysfa, Maesteg, Mid Glamorgan, CF34 9AG

England, M.G., Dr, 14 Scotts Avenue, Sunbury-on-Thames, Middlesex, TW16 7HZ. T: 01932 787300

Epps, R.J., Mr, Chy Fawwyth, 4 Goodyers, Ashdell Park, Alton, Hampshire, GU34 2SH. T: 01420 84616

Ergun, M.U., Prof., Civil Engineering Department, Middle East Technical University, Ankara 06531, TURKEY. T: 90 312 2102415, F: 90 312 2101262

Esper, P., Mr, Gibb Ltd., Gibb House, London Road, Reading, Berkshire, RG8 1BL. T: 020 7631 8363 (D)

Esrig, M.I., Dr, 43 Royden Road, Tanafly, New Jersey-07670, USA. T: 201 5698561, F: 201 5699716

Esslemont, N.A., Mr, 4 Park Farm Way, Lane End, High Wycombe, Buckinghamshire, HP14 3EG

Evans, C.W., Dr, Swiss Cottage, 7 Malthouse Yard, Reepham, Norwich, NR10 4NF. T: 01603 870351, E: evans@c+v-fs3. ac.uk

Ewen, C.A., Mr, 8 Harper Way, Scone, Perth, Perthshire, PH2 6PW. T: 01738 551731

Fair, P.I., Mr, 29 Algar Crescent, Sheffield, S2 2JH. T: 0114 264 6045

Fairlie, M.A.C., Mr, East View, Ossetts Hole Lane, Yarningale Common, Claverdon, Warwickshire, CV35 8HN

Farmer, I.W., Mr, Ian Farmer Associates Ltd., 11-12 Skinnerburn Road, Newcastle upon Tyne, Tyne and Wear, NE1 3RH

Farrar, D.M., Mr, 63 Headington Road, Maidenhead, Berk-shire, SL6 5JR. T: 01628 28694

Fawcett, A., Dr, 3 Avon Close, Barford, Warwick, CV35 8BX

Fellenius, B.H., Dr, 735 Ludgate Court, Ottawa, K1J 8K8 CANADA

Felton, P.J., Mr, 3 Crediton Way, Claygate, Esher, Surrey, KT10 0EB

Fernie, R., Mr, 18 Battlemead Close, Maidenhead, SL6 8LB

Fernley, J.C., Mrs, 85 Park Crescent, North Shields, Tyne and Wear, NE30 2HL

Fikiris, I., Mr, 48 Sojradous Street, 17672 Kalliphea, Athens, GREECE. T: 0030 1 9597 438, E: i.fikiris@ic.ac.uk

Finch, M., Mr, Coflexip Stena Offshore, Enterprise Drive, Westhill Industrial Estate, Westhill, Aberdeenshire, AB32 6TQ

Findlay, J.C., Mr, 18 Birch Drive, Attleborough, NR17 2HF

Findlay, J.D., Mr, Hill Place Farm, 132 Oak Hill, Wood Street Village, Guildford, Surrey, GU3 3ET. T: 01483 823234, F: 823234

Fingland, A.S., Mr, 2 Earlsdon Avenue, Acklam, Middles-brough, Cleveland, TS5 8JH

Finlan, J.S., Mr, 3 Pohutukawa Drive, Pukete, Hamilton, NEW ZEALAND. T: 07 834 1851, F: 07 838 9324, E: stuart. finlan@opus.co.nz

Finlay, S., Miss, 12 Newlands Close, Chandlers Ford, Eastleigh, Hampshire, SO53 4PD. T: 01703 263 400

Fioravante, V., Prof., Corso 22, Marzo 38, Milano 20135, Italy. T: 39 2 70001619

Fisher, A.R.J., Mr, 18b Willington Road, Cople, Bedford, MK44 3TH. T: 01234 838647, E: fisher@bre.co.uk

Fisher, S., Mr, 81 Selwyn Court, Long Meadow, Aylesbury, Buckinghamshire, HP21 7EQ

Fitch, T.R., Eur Ing, 8 Farm Close, Hoylake Crescent, Uxbridge, Middlesex, UB10 8JB. T: 1895678193

Fitzpatrick, W., Mr, 65 Chamberlain Street, St. Helens, Merseyside, WA10 4NJ

Fleming, D.G., Mr, Berryland Cottage, Holywood, Dumfries, Dumfriesshire, DG2 0JQ

Fleming, W.G.K., Mr, 9 Clements Road, Chorleywood, Rickmansworth, Hertfordshire, WD3 5JS. T: 1923282898, E: ken.fleming@btinternet.com

Flower, G.R., Mr, 39 Sydenham Road, Knowle, Bristol, BS4 3DG. T: 0117 9720843

Fookes, P.G., Prof., Lafonia, 11a Edgar Road, Winchester, Hampshire, SO23 9SJ. T: 1962863029, F: 842317

Forbes, P.J., Mr, Knight Piesold Zimbabwe, P.O. Box 383, Harare, ZIMBABWE. T: 263 4 882732, F: 263 4 882367

Forbes-King, C.J., Mr, 4 Woodland Close, Roffey, Horsham, West Sussex, RH13 6AN

Ford, C.J., Mr, 12 Whitehead Close, Basing, Basingstoke, Hampshire, RG24 8SG

Forde, M.C., Prof., Tintern House, 46 Cluny Gardens, Edinburgh, Midlothian, EH10 6BN. T: 1314474960, F: 444 4528596, E: m.forde@ed.ac.uk

Fort, D.S., Eur Ing, 19 Blythwood Road, Pinner, Middlesex, HA5 3QD. T: 2088669552

Forth, R.A., Mr, Dept of Civil Engineering, Drummond Building, University of Newcastle, Newcastle Upon Tyne, NE1 7RU. T: 0191 2227099, F: 0191 2226613

Foster, S.A., Mr, 51 Innis Road, Coventry, West Midlands, CV5 6AX. T: 02476 714103 (H)

Foundation & Exploration Services Ltd., Brunel House, Stephenson Road, Houndmills Estate, Basingstoke, Hampshire, RG21 6XR. T: 01256 340000, F: 01256 843101

Fowell, R.J., Dr, 12 Otters Holt, Durkar, Wakefield, West Yorkshire, WF4 3QE. T: 1924240320

Francescon, M., Mr, 14 Chalgrove Road, Thame, Oxfordshire, OX9 3TF. T: 01923 42 3346 (WK)

Frangoulides, A.C., Mr, P.O. Box 60180, Paphos 8101, Cyprus

Fraser, I.W., Mr, 42 Western Road, Stourbridge, West Midlands, DY8 3XU

Freake, R.G., Mr, 21 Allens Close, Boreham, Chelmsford, CM3 3DR

Free, M., Dr, 20b Shrwood Court Tower 1, 18 Kwai Sing Lane, Happy Valley, HONG KONG

French, D.J., Dr, 9 Parklawn Avenue, Epsom, Surrey, KT18 7SQ. T: 01372 742680

French, S.N., Mr, 43 Carr Hill Road, Calverley, Pudsey, LS28 5PZ

Friedman, M., Mr, 12 Laurel Hill Court, Summerville Avenue, Co. Limerick, REPUBLIC OF IRELAND. T: 01227 763405, E: mff@jle.lul.co.uk

Fryett, J.W.P., Mr, W H White Plc, Woodlands Manor Farm, Woodlands, Wimborne, Dorset, BH21 8ND. T: 01202 814 090

Fugro Ltd., 18 Frogmore Road Industrial Estate, Frogmore Road, Hemel Hempstead, Hertfordshire, HP3 9RT

Fyfe, J.F., Mr, 61 Grampian Road, Stirling, FK7 9JN

Gaba, A.R., Mr, 15 Brown Croft, Hook, Hampshire, RG27 9SY

Gabriel, K.R., Mr, Highfield House, Rolvenden Road, Benendon, Kent, TN17 4EH. T: 01580 240255

Gabryliszyn, J., Mr, Oak Lawn, Tredington, Shipston-on-Stour, Warwickshire, CV36 4NS. T: 01608 661516

Gallagher, E.M.G., Mr, 11 Danes Road, Rusholme, Manchester, Lancashire, M14 5LB

Gallipoli, D., Mr, 5 Striven Gardens, Kelvinbridge, Glasgow, Lanarkshire, G20 6DU

Gammage, P.J., Mr, 148 Duffield Road, Derby, Derbyshire, DE22 1BG

Gammon, J.R.A., Eur Ing, Atkins China Ltd., 16/F, The World Trade Centre, 280 Gloucester Road, Causeway Bay, HONG KONG

Ganendra, D., Dr, 14 Lorong 20/16a, Paramount Garden, 46300 Petaling Jaya, Selangor, MALAYSIA. T: 603 7765233, F: 603 7769676, E: minco@po.jaring.my

Ganendra, R., Miss, 14 Jalan 20/16a, Paramount Garden, 46300 Petaling Jaya, Selangor, MALAYSIA. T: 603 7765233, F: 603 7769676

Gannon, J.A., Mr, Babtie Group, Pearl House, 32 Queen Street, Wakefield, West Yorkshire, WF1 1LE

Garassino, A.L., Mr, Via Curtatone 25, Milano, ITALY

Garner, M.J., Mr, 6 Tyne Road, Oakham, Rutland, LE15 6SJ. T: 01222 497000

Garrad, G., Mr, 86 Sheaveshill Court, The Hyde, London, NW9 6SJ

Garvey, D.J., Mr, 239 Baldwins Lane, Croxley Green, Rickmansworth, Hertfordshire, WD3 3LH. T: 1923442431, E: david@dgarvey.freeserve.co.uk

Gazetas, G., Prof., 36 Asimakopoulou Str., A.Paraskevi 15342, GREECE

Geary, M., Mr, 6 Avon Court, 9 The Avenue, Pinner, HA5 4UU. T: 2084212092

Geddes, J.D., Prof., Stonefield, Church Hill, Easingwold, York, North Yorkshire, YO61 3JT. T: 01347-822907

Gee, P.A., Mr, 4 Lant Close, Kings Bromley, Nr Burton On Trent, DE13 7JW

Gens, A., Prof., Sant Ferran 3-H., Sant Just Desvern, Barcelona 08960, SPAIN. T: 34 3 4736391, E: gens@ etsecpb.upc.es

Ghosh, N., Mr, Nimrod, 24 Woodlands Avenue, New Malden, Surrey, KT3 3UN. T: 2089428048, E: nandaghosh@nimrod. freeserve.co.uk

Giannakogiorgos, A., Mr, 42 K.Konstantinidi Str., 56431, Ilioupoli, Thessaloniki, GREECE. T: 003031 662130, E: a.giannakogiorgioros@ic.ac.uk

Giannopoulou, D.M., Ms, 18 Miltiadou Street, 15561 Holargos, Attika, GREECE. T: 003 016528639

Gibney, D.N., Mr, 6 Homefield Road, Bilbrook, Codsall, Wolverhampton, West Midlands, WV8 1JN

Gibson, J.M., Mr, 4 High Street, West Cornforth, Ferryhill, County Durham, DL17 9HR. T: 01434 682 803

Gibson, R.E., Prof., 23 South Drive, Ferring, Worthing, West Sussex, BN12 5QU. T: 1903700386

Gilbert, P., Mr, 94 Middleton Hall Road, Birmingham, B30 1DG. E: pgilbert@wsatkins.co.uk

Gilbert, T.M.W., Mr, 15C Wallfield Crescent, Aberdeen, AB25 2LJ. T: 01224 621 996, E: ldguk111@aol.com

Gildea, P.A., Mr, Coffey Partners Int. Pty. Ltd., 142 Wicks Road, North Ryde, N S W, 2113 AUSTRALIA. T: 61 2 888, F: 61 2 8889977

Gill, A.J., Mr, 13A Northampton Road, Addiscombe, Croydon, CR0 7HB. T: 07775 565 471

Gill, S.A., Mr, 9107 Samoset Blvd, Illinois, Skokie, 60062 USA. T: 847 6748057

Gillon, R.J.B., Miss, 3 Clifton Place, North Hill, Plymouth, PL4 8HU. T: 01752 219018

Glazier, D., Mr, 2 BEECHWOOD CLOSE, Surbiton, Surrey, KT66PF. T: 020 8390 2779

Glendinning, S., Dr, University of Newcastle, Dept of Civil Engineering, Drummond Building, Newcastle Upon Tyne, NE1 7RU. T: 0191 2226888, F: 0191 222 6613

Glover, N.P., Mr, 36 Ian Road, Newchapel, Stoke-on-Trent, Staffordshire, ST7 4PW

Glynn, D.T., Mr, 12 Cooldarragh Park North, North Circular Road, Belfast, County Antrim, BT14 6TL

Gmerek, S., Mr, 3 Priors Walk, Evesham, Worcestershire, WR11 6GG. T: 01788 534 635 (W)

Godden, S.L., Mrs, 33 Anstey Road, London, SE15 4J

Godfrey, M., Mr, 26 Canada Road, Arundel, West Sussex, BN18 9HY

Good, M.W., Mr, Martingales, 69 Leas Road, Warlingham, CR6 9LP

Goodfellow, R.J.F., Mr, 1616 Forbes Street, Rockville, M.D. 20851-1411, USA. T: 1 301 6105473 (H) 6703330 (W), E: bob.goodfellow@mciworld.com

Goodings, D.J., Dr, Department of Civil Engineering, University of Maryland, College Park, Maryland 20742, USA. T: 301 4051960, F: 301 4052585

Goodwin, A.K., Dr, 3 Davids Drive, Wingerworth, Chesterfield, Derbyshire, S42 6TT. T: 0114 225 3177 (WK), E: andy@ goodwin_com.freeserve.co.uk

Gordon, M.A., Dr, Vedbaek Stationsvej 5, 2tv, 2950 Vedbaek, DENMARK. T: 45663534

Gordon, T., Mr, 8 Nantfawr Road, Cyncoed, Cardiff, South Glamorgan, CF23 6JR. T: 2920756271

Gorinas, J.L., Mr, Pear Tree Cottage, Church Road, Catsfield, TN339DP

Gowans, I.A.T., Mr, Robert H Cuthbertson & Partners, 13 Eglinton Crescent, Edinburgh, EH12 5DF

Grange, D.M., Mr, 20 Common Lane, East Ardsley, Wakefield, West Yorkshire, WF3 2EP

Grant, D.I., Dr, The Front Garden Flat, 26 Beaconsfield Road, Clifton, Bristol, Avon, BS8 2TS

Grant, R.J., Dr, Richard Davis Assoc. Cons., Geotechnical Engineers, The Old Brewery, Newtown, Bradford-on-Avon, Wiltshire, BA15 1NF

Grant, M.J., Mr, Evans Grant G & E, 2 Delta Business Park, Salterns Lane, Fareham, Hampshire, PO16 0QS. T: 01329 822021, F: 01329 825274

Green, G.E., Dr, 9835 41St Avenue N.E., Seattle, Washington, 98115 USA

Green, R.G.V., Mr, High Ash, Corland, Taunton, Somerset, TA3 5SA

Greenwood, D.A., Eur Ing, 49 St. Huberts Close, Gerrards Cross, Buckinghamshire, SL9 7EN. T: 1753882874, F: 88287 1753

Greenwood, J.R., Mr, Yew Tree House, 10 Nottingham Road, Cropwell Bishop, Nottingham, Nottinghamshire, NG12 3BQ. T: 01159899798/0115 8482045

Greenwood, W.G., Mr, Park Lodge, Ambridge Road, Coggeshall, Colchester, CO6 1QT

Gregory, B.J., Dr, 7 The Steadings, Drumbeg Road, Belfast, County Antrim, BT9 5NR

Gregory, A.W., Mr, 57 Northumberland Road, Coventry, West Midlands, CV1 3AP. T: 01788 534500 (W)

Griffiths, J.S., Dr, Dept of Geological Sciences, University of Plymouth, Drake Circus, Plymouth, Devon, PL4 8AA. T: 01752 233100, F: 01752 233117

Griffiths, T.C.,Eur Ing, 4 Dovedale Close, Langdon Hills, Basildon, Essex, SS16 6NS

Griffiths, A., Mr, 12 Coombe Drive, Pondtail, Fleet, Hampshire, GU13 9DY

Grimes, J.N., Dr, Bryn Hyfryd, Mount Pleasant, Venton, Plymouth, Devon, PL7 5DU. T: 01752 690533

Grose, W.J., Mr, 11 Sandhurst Road, Orpington, Kent, BR6 9HN

Gulati, S.K., Mr, High Views, Ballards Hill, Cranbrook, Kent., TN17 1JS

Gwede, D., Mr, 91 Severus Road, Newcastle Upon Tyne, NE4 9NP

Haddow, D.H., Mr, 1 Barrett Road, Birkdale, Southport, Merseyside, PR8 4PG. T: 01704 567391

Hake, S.S., Mr, Thynne Court, Thynne Street, West Bromwich, West Midlands, B70 6PH

Halcrow Group Ltd., 1 Deanway Technology Centre, Wilmslow Road, Handforth, Wilmslow, Cheshire, SK9 3FB

Hall, F.D., Mr, Saltersgate, Saltersgate Lane, Siddington, Macclesfield, Cheshire, SK11 9LH

Hall, P.A., Mr, 75 Cleveland Road, Sunderland, Tyne and Wear, SR4 7JP

Hallowes, G.R., Mr, 27 Westleigh Avenue, Putney, London, SW15 6RQ

Hamza, M.M., Prof., Hamza Associates, 5 Ibn Marawan St., Dokki, Cairo 12311, EGYPT. T: 202 3607238, F: 202 3607239

Handcock, A.R., Mr, 9a Glenavon Park, Stoke Bishop, Bristol, Avon, BS9 1RS. T: 1179684430, E: alec.handcock@talk21.com

Hanna, T.H., Prof., 288 Ecclesall Road South, Sheffield, South Yorkshire, S11 9PT. T: 0114 2365359, F: 2365359, E: thomas.h.hanna@talk21.com

Harbottle, M.J., Mr, St. Hughes College, Oxford, OX2 6LE

Hardie, T.N., Mr, 29 Spring Pool, Warwick, CV8 2TX

Hardingham, A.D., Mr, 41 Haldon Avenue, Teignmouth, TQ14 8JZ. T: 01626 772491

Hardy, S., Mr, 268 Bluebell Road, Norwich, NR4 7LW. E: stuart_hardy@ic.ac.uk

Harle, B.A., Mr, 7 Lavrelwood Close, Droitwich, WR9 7SF

Harris, A.J., Mr, 35 Grindstone Crescent, Knaphill, Woking, Surrey, GU21 2PL. T: 01483 832185

Harris, A.R., Mr, 306 Stoops Lane, Bessacarr, Doncaster, South Yorkshire, DN4 7JB. T: 40 01302 538764, F: 44 538764

Harris, D.I., Mr, 99 Kingsway, Chandlers Ford, Eastleigh, Hampshire, SO53 1FA. T: 02380 254813

Harris, J., Mr, 1 Violet Lane, Mount Tavy Road, Tavistock, Devon, PL19 9JD. T: 01822 613611

Harris, J.S., Mr, 5 Foxhill Road, Burton Joyce, Nottingham, Nottinghamshire, NG14 5DB

Harris, M.N., Mr, 109 Campion Drive, Bradley Stoke, Bristol, BS12 0EW. T: 01454 616415, F: 01454 620260

Harris, E., Mrs, The Whitehouse, Stanwardine In The Fields, Baschurch, Shropshire, SY4 2HD

Hartington, J.A., Mr, 6 Roughgrove Copse, Binfield, Bracknell, Berks, RG42 4E. T: 01344 300926

Hartwell, D.J., Mr, Warren Cottage, Dell Road, Finchampstead, Nr Wokingham, Berks, RG40 3T. T: 01252 875961

Harvey, S.J., Mr, 35 Dale End, Brancaster Staithe, King's Lynn, Norfolk, PE31 8DA

Harvey, J.A.F., Ms, Kampala, Rockley New Rd., Christ Church, BARBADOS

Harwood, N.K., Mr, 16 Park Street, Totterdown, Bristol, BS4 3BL. T: 01225 320600

Hassall, M.C., Mr, Wardell Armstrong, 2 The Avenue, Leigh, Lancashire, WN7 1ES. T: 01942 260101, F: 01942 261754

Hawkins, A.B., Mr, Charlotte House, 22 Charlotte Street, Bristol, BS1 5PN

Hawkins, K.F., Mr, 2 Ragman's Close, Marlowe Bottom, Bucks, SL7 3QW

Hayes, S., Mr, WSP Group plc, 54 Hagley Road, Edgbaston, Birmingham, B16 8PE. T: 0121 456 1177

Head, K.H., Mr, Sonnamarg, 12 Stoke Road, Cobham, Surrey, KT11 3AS. T: 01932 864209

Headling, M., Mr, 9 Front Road, Woodchurch, Near Ashford, Kent, TN26 3PA. T: 1233860035, F: 658299, E: mheadling@knightpiesold.co.uk

Healey, L., Mr, 10 Long Acre Place, Laytham, Lancashire, FY8 4PN

Heath, C.H., Eur Ing, Grove House, The Grove, Knapton, North Walsham, Norfolk, NR28 0RS. T: 01603 627281 (W)

Heffernan, J.B., Mr, University of Paisley, Department of Civil Engineering, High Street, Paisley, Renfrewshire, PA1 2BE

Hellings, J.E., Mr, 57 Mallard Place, Twickenham, Middlesex, TW1 4SW

Helmore, A., Mr, 6 Ovington View, Prudhoe, Nothumberland, NE42 6RG

Hencher, S.R., Dr, Halcrow China Ltd., Suite 301, 3/F Chinachem Golden Plaza, 77 Mody Road, Tsimshatsui East, Kowloon, HONG KONG

Henkel, D.J., Dr, 125 Cranley Gardens, London, N10 3AG

Herbert, S.M., Mrs, The Poplars, Croft Lane, Little Shrewley, Hatton, Warwickshire, CV35 7HL

Hickmott, J.M., Mr, Southern Testing Labs Ltd., Keeble House, Stuart Way, East Grinstead, West Sussex, RH19 4QA

Hicks, M.A., Dr, The University of Manchester, School of Engineering, Simon Building, Oxford Road, Manchester, Lancashire, M13 9PL

Higginbottom, I.E., Mr, 5 Templewood, Ealing, London, W13 8BA. T: 020 8998 1594, F: 020 8998 1594

Higgins, K.G., Mr, Geotechnical Consulting Group, 1a Queensberry Place, London, SW7 2DL. T: 020 7581 8348, F: 020 7584 0157

Higginson, R.B., Eur Ing, Soils Ltd., Newton House, Cross Road, Tadworth, Surrey, KT20 5SR

Higgs, A.J.F., Mr, 41 Heol y Banc, Bancffosfelen, Llanelli, SA15 5DH. T: 01372 361018, E: ahiggs@g_r.com

Hight, D.W., Mr, 7 The Moorings, Hindhead, Surrey, GU26 6SD. T: 01428 606949, F: 606949, E: d.w.hight@dial.pipex.com

Hill, S.J., Mr, Ove Arup And Partners, Level 5, Festival Walk, 80 Tat Chee Avenue, Kowloon Tong, HONG KONG

Hinsley, M.R., Mr, 74 Poplar Avenue, Sherwood, Nottingham, NG7 7EG

Hird, C.C., Mr, 5 Darwin Close, Ranmoor, Sheffield, South Yorkshire, S10 5RJ. T: 1142309455

Hird, G.S.E., Mr, Model Farm, Chapel Street, Rockland St. Peter, Attleborough, NR17 1UJ

Hislam, J.L., Mr, 2 Upper Ashlyns Road, Berkhamsted, Hertfordshire, HP4 3BW. T: 44 01442 871805, F: 44 873803, E: johnhislam@ageotery.co.uk

Hitchcock, A.R., Mr, Foxwood, Combpyne, Axminster, Devon, EX13 8SY. T: 01297 442768

Ho, K.F., Mr, 1911 Windsor House, 311 Gloucester Road, Causeway Bay, HONG KONG

Ho, S., Mr, Geotechnical Engineering Office, 12/F., Civil Engineering Building, 101 Princess Margaret Road, Ho Man Tin Kowloon, HONG KONG

Hobbs, R., Dr, Triptons, Hillwood Grove, Brentwood, CM13 2PD

Hocombe, T.G., Mr, Ove Arup & Partners, C/O Ove Arup & Partners, 13 Fitzroy Street, London, W1P 6BQ

Hodges, W.G.H., Eur Ing, 18 Abshot Close, Titchfield Common, Fareham, Hampshire, PO14 4LZ. T: 1489575671

Hodgon, R.M., Mr, 3 Benedict Close, Orpington, Kent, BR6 9TU. T: 1689826051

Hodgson, A.J., Mr, 1 Orchard Close, Longwick, Princes Risborough, Buckinghamshire, HP27 9SR. T: 0442 240781

Hodgson, R.L.P., Mr, Lower Beer, Uplowman, Tiverton, Devon, EX16 7PF. T: 1884821239, F: 44 821239, E: r.l.p.hodgson@exeter.ac.uk

Holden, J.M.W., Mr, Allen Hill House, 10 Woolley Road, Matlock, Derbyshire, DE4 3HS. T: 01629 55624

Holliday, J.K., Mr, 7 Newtown Road, Hove, East Sussex, BN3 6AA. T: 01273 203726, F: 01273 778071

Hollingsworth, J.R., Mr, Millen House, Graemesdyke Road, Berkhamsted, Hertfordshire, HP4 3LZ. T: 1442874116, F: 874116

Holloway-Strong, M.U., Dr, 35 Tillingbourne Road, Shalford, Guildford, Surrey, GU4 8EY

Holt, C.M., Mr, Grayswood, Platts Ave, Endon, Stoke On Trent, Staffs, ST9 9EG

Holt, D.A., Mr, 34 Halifax Road, Enfield, Middlesex, EN2 0PP. T: 0207 755 3788

Holt, D.N., Mr, 57 The Gardens, Cassiobury Park, Watford, WD17 3DN

Holt, J.R., Mr, 170 Langdale Road, Dunstable, LU6 3BT. T: 01582 601825, F: 01582 601825

Holt, S., Mr, Cleanaway Limited, 24 Norfolk Road, Maidenhead, SL6 7AT. T: 1189328888, F: 189328383

Holton, I.R., Mr, 67 Autumn Glades, Leverstck Green, Hemel Hempstead, Hertfordshire, HP3 8UB

Holtz, R.D., Prof., University of Washington, Dept of Civil Engineering, Box 352700, Seattle W A 98195, USA. T: 206 543 7614, F: 206 685 3836

Holtzer, F., Mr, Flat 15, St. Matthews Lodge, 50 Oakley Square, London, NW1 1NB

Home, R., Mr, 5418 Lake Margaret Drive 1008, Orlanda, Florida 32812, USA. E: rohome@hotmail.com

Hope, V.S., Dr, 3 Green Acre, Knaphill, Woking, Surrey, GU21 2JR

Hope, M.A., Mr, 9 Buckhurst Way, East Grinstead, West Sussex, RH19 2AQ. T: 01342 322560

Horgan, G.J., Mr, 28 White Rose Avenue, Garforth, Leeds, LS25 2EE

Horner, P.C., Mr, 10 Riverside Road, Luton, LU3 2LY. T: 0582-584298

Horswill, P., Mr, Montgomery Watson, Terriers House, 201 Amersham Road, High Wycombe, Buckinghamshire, HP13 5AJ. T: 01494 526240, F: 01494 522074

Hoskins, C.G., Mr, West Lodge, Nutfield Park, South Nutfield, Redhill, Surrey, RH1 5PA. T: 1737823326, E: chris-g.hoskins@virgin.net

Houlsby, G.T., Prof., 25 Purcell Road, New Marston, Oxford, Oxfordshire, OX3 0HB

Housden, R.T., Mr, 23 Wilton Grove, London, SW19 3QU. T: 020 8542 3873

Houston, C.R.M., Eur Ing, 60 Westwood Avenue, Harrow, Middlesex, HA2 8NS

Howard, T.J., Mr, 4 Vernon Road, Uckfield, TN22 5DY

Howarth, M., Mr, 16 Ashurst Road, Ash Vale, Aldershot, Hampshire, GU12 5AF

Howie, J.A., Dr, 952 West 22nd Avenue, Vancouver, B C V5Z 2A1, CANADA

Howland, A.F., Dr, 88 Newmarket Road, Norwich, Norfolk, NR2 2LB. T: 44 01603 458637

Hsiung, B., Mr, Dept of Civil Engineering, University of Bristol, 0.37a Queens Building, Bristol, BS8 1TR

Hsu, Y.S., Mr, Mott MacDonald Ltd., Foundations & Geotechnics, St. Anne House, 26 Wellesley Road, Croydon, Surrey, CR9 2UL

Hughes, S.E., Miss, 18 Townfield Close, Talke, Stoke-on-Trent, Staffordshire, ST7 1NF. T: 0044 01782 784709, E: sarah.hughes@surreyuni.ntl.com

Hughes, D.B., Mr, 6 The Pastures, Stocksfield, Northumberland, NE43 7NG. T: 01661 843848

Hull, D.A., Mr, 20 Westpark Gate, Saline, Dunfermline, KY12 9US

Humphrey, R.D., Mr, 102 Lincoln Road, Dunholme, Lincoln, Lincolnshire, LN2 3QY

Hunt, R.J., Mr, G/F, 32b Braga Circuit, Kadoorie Hill, Kowloon, HONG KONG. T: 85226247432 (F)

Hunter, S.D., Mr, Kvaerner Cementation Ltd., 112-114 Kelvin Road South, Granite Side, P.O. Box 1836, Harare, ZIMBABWE

Huston, W.S.J., Mr, 16 Viking Court, Beaver Close, Hampton, Middlesex, TW12 2BZ

Hutchinson, D.E., Mr, 1 Blakeney Close, Bearsted, Maidstone, ME14 4QF. T: 0208 779 2577

Hutchinson, J.N., Mr, Imperial College, Department of Civil Engineering, Imperial College Road, London, SW7 2BU

Hutchison, R., Mr, 8 Shirley Road, Luton, LU1 1NZ. T: 1582410401, F: 0208 784 5561

Hutchison, R.J., Mr, Keith Gillespe & Assoc., 753 Beach Road, Browns Bay, Auckland, NEW ZEALAND. T: 01926 431 00 (WK)

Hyde, A.F.L., Dr, University of Sheffield, Dept of Civil & Structural Eng., Sir Frederick Mappin Building, Sheffield, South Yorkshire, S1 3JD. T: 0114 222 5741, F: 0114 222 5700

Hytiris, N., Mr, 12 Glendowart Court, 100 Buccleuch Street, Glasgow, Lanarkshire, G3 6NS. T: 1413322723

Indrawan, Z., Dr, 25 Winterbell Court, Churchlands, W A 6018, AUSTRALIA

Ingold, T.S., Dr, Mulberry Lodge, St. Peters Close, St. Albans, AL1 3ES

Ingram, P.J., Mr, 19 Shaldon Drive, Ruislip, Middlesex, HA4 0UJ. T: 0208 426 4163

Ireland, R., Mr, Coach House, 6 Dun Park, Kirkintilloch, Glasgow, G66 2DU

Isnard, A., Mr, Sec.Gen.Du Comite Francais De, Mecanique Des Sols, 17 Bis Place, des Reflets, La Defence 2 92400, FRANCE

Iversen, N., Mr, Chestnut Cottage, Church Lane, Welford on Avon, Stratford-Upon-Avon, Warwickshire, CV37 8EL

Jackson, A.M.R., Mr, 9 Lambourn Gardens, Harpenden, Hertfordshire, AL5 4DQ. T: 01582 762 742

Jackson, N.W., Mr, Graigllwyd Fach, Bwlch, Brecon, LD3 7RX

Jackson, C.V., Mrs, Queensberry House, Ham Lane, Powick, Worcester, WR2 4RD

James, A.T., Mr, 25 Ouseburn Avenue, York, YO26 5NL

James, E.L., Mr, The Woodlands, 16 Bottrells Lane, Chalfornt St. Giles, Buckinghamshire, HP8 4CY

Jameson, R.N.J., Mr, Nicholson Construction Co., P.O. Box 98, Bridgeville, P.A. 15017, USA

Jardine, F.M., Mr, 20 Winding Wood Drive, Camberley, Surrey, GU15 1ER

Jardine, R.J., Prof., 24 Kingsway, London, SW14 7HS

Jarvis, S.T., Mr, 63 Kirkstone Way, Lakeside, Amblecote, W Midlands, DY5 3RZ. T: 0121 4546261 (W)

Javadi, A.A., Dr, 30 Stuart Court, Swarland Grove, Bradford, West Yorkshire, BD5 0SU. T: 01274 233 851, E: a.arabpourjavadi@bradford.ac.uk

Jayatilake, N.A.B., Mr, Boraluwe Niwasa, Nittala-Panala, Aranayake, SRI LANKA

Jefferis, S.A., Mr, Talbot Lodge, Ardley Road, Middleton Stoney, Bicester, Oxfordshire, OX6 8SE. T/F: 01869 343037

Jefferson, I., Dr, Dept of Civil & Structural Eng., Nottingham Trent University, Burton Street, Nottingham, NG1 4BU. T: 0115 9418418, Ext. 21, F: 0115 9486450

Jenkins, P., Dr, 115 Oxlease, Witney, Oxfordshire, OX8 6QY. T: 01993 778615, E: pljenkins@brookes.ac.uk

Jenner, C.G., Mr, 7 Russell Square, Chorley, Lancashire, PR6 0AS. T: 01257 278915

Jeremiah, K.B.C., Mr, 19 Birchwood Road, Petts Wood, Orpington, BR5 1NX

Jerram, C., Mr, Ground Risk Management, Bretby Business Park, Burton-on-Trent, Staffordshire, DE15 0QD. T: 01283 551249

Jessep, R.A., Mr, 14 Abbots Rise, Redhill, RH1 1LL. T: 1883712979, E: rob-jessep@jessep.freeserve.co.uk

Jewaskiewitz, S.M., Mr, 12 Joubert Street, Rynfield, Benoni, Gauteng, 1501 SOUTH AFRICA

Jewell, R.A., Dr, 14 Rue Tenbosch, Brussels, BELGIUM. T: 00 322 646 8969, F: 00 322 646 1238

Jewell, P.J., Mr, 56 The Park, Redbourn, St. Albans, Hertfordshire, AL3 7LS. T: 1582793029, F: 44 7930329, E: peter@p-a-jewell.freeserve.co.uk

John, M.B., Mr, Chaloner House, Station Road, Rossett, Wrexham, Clwyd, LL12 0HE

John, N.E., Mr, 11 Westfield Drive, Penarth, Vale of Glamorgan, CF64 3NT

Johnson, A., Mr, Dreyerstr. 10 A, Hannover, 30169 GERMANY

Johnson, J.G.A., Mr, Ove Arup & Partners, 30 Barker Pool, Sheffield, S1 2HB. T: 0114 272 8247, F: 0114 275 9553

Johnson, R., Mr, 3 Kennedie Park, Mid Calder, Livingston, West Lothian, EH53 0RG. T: 020 7774 6271, F: 020 7774 9280

Johnson, C., Mrs, 59A Dagnall Park, South Norwood, London, SE25 5EG

Johnston, J.D., Mr, 69 Castlemain Avenue, Bournemouth, BH6 5EW. T: 0208 539 1463 (W)

Johnston, T.P., Mr, 1 Milford Road, Leightonbridge, Co Carlow, REPUBLIC OF IRELAND. T: 0503 30060

Johnston, I.W., Prof., 32 Erasmus Street, Surrey Hills, Victoria 3127, AUSTRALIA. T: 61 3 98909421, F: 61 3 98909421, E: ian.johnston@vut.edu.au

Jones, C.J.F.P., Dr, 116 The Mount, York, YO2 2AS

Jones, T.N., Dr, 12 Birchway, Bramhall, Stockport, Cheshire, SK7 2AG. T: 0161 440 7303, F: 44 1339079, E: tnjones@lineone.net

Jones, B.D., Mr, 5 Hamilton Way, Wallington, Surrey, SM6 9NJ

Jones, D.E., Mr, 6 Bank Mews, Bank House Lane, Helsby, Cheshire, WA6 0PP

Jones, D.R.V., Mr, 39 Shelford Road, Radcliffe-on-Trent, Nottingham, Nottinghamshire, NG12 1AE. T: 0602 056544 (Work)

Jones, M.J., Mr, 7A Bangor Street, Roath Park, Cardiff, CF24 3LQ

Jones, W., Mr, U K A E A, Dounreay, Thurso, Caithness, KW14 7TZ. T: 01847 802679, F: 01847 802209

Jordan, D.P., Mr, Cedar Cottage, The Street, Crookham Village, Hants, GU13 0SJ

Jouni, M.Y., Dr, P.O. Box 11-1148, Riad El Solh Beiruit, 11072070 Lebanon

Jovicic, V., Mr, Ove Arup & Partners, Arup Geotechnics, 13 Fitzroy Street, London, W1P 6BQ. T: 020 7465 2075

Joyce, M.D., Mr, 87 Mountbatten Avenue, Wakefield, West Yorkshire, WF2 6HH. T: 1924250678

Juarez-Badillo, E., Prof., Tepanco 32, Coyoacan, 04030 Mexico D.F., Mexico. T: 5 6597054

Kageson-Loe, N., Dr, Norsk Hydro A S A, Forskningssenteret, Sandsliveien 90, N – 5020 Bergen, NORWAY. T: 0047 5599 6086, F: 0047 5599 6510

Kalasin, T., Mr, University of Bristol: Faculty of E, 0.37a, Queens Building, University Walk, Bristol, BS8 1TR. T: 0117 9288286

Kalaugher, P.G., Mr, 6 Merrion Avenue, Exmouth, Devon, EX8 2HX. T: 1395273344, E: kalaugher@talk21.com

Kambo, G.S., Mr, 8 Furness Road, West Harrow, Middlesex, HA2 0RL

Karkee, M.B., Dr, Dept of Architecture & Envir. Systems, Faculty of System Science & Tecnology, Akita Prefectural University, 84-4Tsuchiya, Honjo 015-2255 JAPAN. T: 81 184 27 2047, F: 81 184 27 2186

Karstunen, M.K., Miss, Department of Civil Engineering, University of Glasgow, Rankine Building, Glasgow, G12 8LT. T: 0141 330 5208, F: 0141 330 4557

Kathirgaman, K., Mr, 67, Davidson Rd., Colombo 4, SRI LANKA

Katsibokis, G., Mr, 1 Chrysanthemon Street, Halandri 152-33, Athens, GREECE. T: 0030-1-6826609, E: gkatspm@hotmail.com

Kaul, P.K., Mr, 34 Perth Avenue, London, NW9 7JT

Kavanagh, S.T., Mr, 82 Henzel Road, Green Point, NSW, 2251 AUSTRALIA. E: sianjude@hotmail.com

Kearsey, W.G., Mr, 1 Fairlea, Maidenhead, SL6 3AS

Keedwell, M.J., Dr, 8 Clive Road, Balsall Common, Coventry, CV7 7DW. T: 01676 532537

Keller Ground Engineering, 611 Thorp Arch, Trading Estate, Tadcaster, North Yorkshire, LS24 9LE

Kennedy, L.J., Eur Ing, Lt Col. St. Francis Cottage, Hatherden, Andover, Hampshire, SP11 0HT

Kenny, M.J., Mr, University of Strathclyde, Department of Civil Engineering, John Anderson Building, 107 Rottenrow East, Glasgow, Lanarkshire, G4 0NG

Khouri, K.N., Mr, National Services Contracting Est. (Nscc), P.O. Box 856, Abu Dhabi, UNITED ARAB EMIRATES

Kidd, A.L., Mr, 30 Claridge Drive, Midleton, Milton Keynes., MK10 9GB. T: 01234 796003

Kilby, A.D., Mr, 7 The Ridings, Emmer Green, Reading, Berks, RG4 8XL

Kilkenny, W.M., Mr, Hill Cottage, Sutton Place, Abinger Hammer, Dorking, Surrey, RH5 6RL

Kim, S.W., Mr, 69 Grand Avenue, Surbiton, Surrey, KT5 9HX

King, I.J., Mr, Stable Cottage, Sparrow, Hill Way, Axbridge, BS26 2LA

Kings, P.A.C., Mr, 77 Reforne, Easton, Portland, Dorset, DT5 2AW. T: 01305 860628

Kirk McClure Morton, Elmwood House, 74 Boucher Road, Belfast, County Antrim, BT12 6RZ

Kirkpatrick, P.J., Mr, Willow Brook, 28A Moorland Crescent, Menston, Ilkley, West Yorkshire, LS29 6AF. T: 01943 875640

Klotz, U., Mr, Furtwaengler Str. 48, 70195, Stuttgart, Germany. T: 4.97116E +11, F: 020 7040 8751, E: u.klotz@city.ac.uk

Knight, D.E.C., Mr, 6th Floor Yuntong Mansion, 168 Anxi Oad, Lanzhou, Gansu Province, 730050 CHINA. E: decknight@yahoo.com

Knight, D.J., Mr, 67 The Fairway, Leigh-on-Sea, Essex, SS9 4QW. T: 071202 525603

Knight, K.M., Mr, 45 Fulmar Walk, Whitburn, Tyne & Wesr, SR6 7BW

Knill, J.L., Sir, Highwood Farm, Long Lane, Shaw-Cum-Donnington, Newbury, Berkshire, RG14 2. T: 01635 552300, F: 01635 36826

Koe, A., Mr, Knight Piesold, Norman House, Heritage Gate, Derby, DE1 1NU. T: 01332 293 060, F: 01332 293 063

Kolovaris, E., Mr, Neg. Alexanderou 76, Maroussi, 15124 Athens, GREECE. E: e.koloraris@ic.ac.uk

Komiya, K., Dr, 2-17-1 Tsudanuma, Narashino, Chiba 275-8588, JAPAN

Kong, W.K., Mr, 18h, Lagoon Court, 18 Plover Cove Road, Tai Po, N.T. Hong Kong, HONG KONG. T: 852 26509611, E: kongalo@yahoo.com.hk

Koutsabeloulis, N.C., Mr, Elm Lodge, North Street, Winkfield, Windsor, Berkshire, SL4 4TE

Kovacevic, N., Dr, 53 Dukes Avenue, London, W4 2AG. T: 020 8400 1917, E: n.kovacevic@gcg.co.uk

Kuku, D.I., Mr, 16 Swete Street, Plaistow, London, E13 0BS

Kutter, H.K., Prof., Institut Fuer Geologie, Ruhr-Universitaet Bochum, D-44780 Bochum, GERMANY. T: 0234 7003296, F: 0049 234 709412

Ladd, C.C., Prof., 7 Thornton Lane, Concord, Ma 01742, USA. T: 508 3693886, E: ccladd@MIT.edu

Lade, P.V., Prof., Dept of Civil Engineering, The John Hopkins University, 3400 N. Charles Street, Baltimore, M D 21218, USA. T: 410 5164996, F: 410 5167473

Lagioia, R., Dr, Universita Di Brescia, Dip. Ing. Civile, Via Branze 38, Brescia, ITALY

Laing Technology Group, Maxted House, 13 Maxted Road, Hemel Hempstead Industrial Estate, Hemel Hempstead, Hertfordshire, HP2 7DX

Lake, L.M., Dr, Arden, Snows Ride, Windlesham, Surrey, GU20 6P. T: 01276-471672

Lamaris, H., Mr, 6 Konstantinoupoleos Str., Maroussi 15124, Athens, Greece. T: 01 8062581

Lamport, M.C., Mr, The Stone House, Church Street, West Chiltington, West Sussex, RH20 2JW. T: 01798 817209

Langdale-Smith, T.S., Mr, 38/39 Willingham Road, Market Rasen, Lincolnshire, LN8 3DX

Langley, G.E., Mr, 17 Casterbridge Lane, Weyhill, Andover, SP11 0SY. T: 01264 771034

Langridge, B.F., Mr, 9 Boscombe Avenue, Hornchurch, Essex, RM11 1JQ

Larsen, H., Dr, Danish Geotechnical Institute, Maglebjergvej 1, P.O. Box 119, D K-2800 Lyngby, DENMARK. T: 45 38 39 07 00, F: 45 38 77 28 29

Latimer, J.R.W., Mr, 13 Hitherwood, Cranleigh, GU6 8BW. T: 00 233 51 26456, F: 00 233 51 38741

Lau, C.K., Mr, 3 Forfar Road, 12/F, Block E, Kowloon, HONG KONG

Lau, F.S., Mr, L P Consultants Sdn Bhd, P.O. Box 10195, 50706 Kuala Lumpur, MALAYSIA

Lau, W., Mr, 996 Kings Road, 1st Floor, Qually Bay, HONG KONG

Le Masurier, J., Mr, 52 Yeomeads, Long Ashton, Bristol, BS18 9B. T: 01222 874000 (W)

Leach, B.A., Mr, 2 Tytherington Park Road, Macclesfield, Cheshire, SK10 2EL. T: 01625 422229, F: 422229

Lee, A.J.G., Mr, 61 Blake Dene Road, Lilliput, Poole, Dorset, BH14 8HF. T: 01202 732630, F: 737630

Lee, K.S., Mr, Rm 1609 Wang Wai House, Wang Tau Hom Estate, Kowloon, HONG KONG

Lee, R.W.F., Mr, Blk 2, 9/F, Room D, Sun Shing Centre, 141 Kowloon City Road, Tokwawan, Kowloon, HONG KONG

Lee, S.W., Mr, Churchill College, Cambridge, Cambridgeshire, CB3 0DS. E: siew.lee@ic.ac.uk

Leek, I.C., Mr, 6 Railway Terrace, Penarth, South Glamorgan, CF64 2TT

Lefroy Brooks, S.R., Mr, The Coach House, Ballinger Grove, Great Missenden, Buckinghamshire, HP16 9LQ. T: 01494 865083, E: lefroy.brooks@compuserve

Legge, J.D., Mr, Suite 301, 3/F Chinachem Golden Plaza, 77 Mody Road, T.S.T East Kowloon, HONG KONG

Lehane, B.M., Mr, Dept of Civil Engineering, Trinity College, Dublin 2, REPUBLIC OF IRELAND. T: 010 3531 702238

Leiper, Q.J., Eur Ing, Carillion PLC, Company Chief Engineer, Carillion PLC, 24 Birch Street, Wolverhampton, West Midlands, WV1 4HY

Lemmon, M.S.J., Mr, 2 Cole Court Lodge, London Road, Twickenham, TW1 1HB

Leonard, R.E., Mr, 11 Edgewood Close, Pine Ridge, Crowthorne, Berkshire, RG45 6TA. T: 01344 779420

Leshchinsky, D., Prof., Department of Civil Engineering, University of Delaware, Newark, D E 19716, USA. T: 302 8312446, F: 302 7311001

Leventis, G.E., Mr, 10 Kolokotroni Street, Filothei, Athens 15237, GREECE. T: 201 794 6900

Levy, S.M., Mr, Second Floor No 64, Tai Po Tsai Village, Yan Yee Road, Tai Mong Tsai, Saikung Nt, HONG KONG

Lewis, D.W., Mr, 1 Priory Grove, Stockwell, London, SW8 2PD

Lewis, J.C.H., Mr, 16 Wakefield Avenue, Billericay, Essex, CM12 9DN. T: 0207 930 4324 (W)

Lewis, W.H., Mr, 9 Long Furlong, Rugby, CV22 5QS. T: 021 456 1568

Liao, H.J., Mr, Dept of Construction Engineering, Nat.Taiwan Inst. of Technology, P.O. Box 90-130, Taipei, TAIWAN

Liddle, K.E., Miss, 68 Baslow Drive, Heald Green, Cheadle, SK8 3HP

Liew, C.K.P., Mr, Block 3, Flat 19, Ashcroft, Pritchatts Road, Edgbaston, Birmingham, B15 2QU

Lightbody, S.W., Mr, 23 Keep Hill Road, High Wycombe, Buckinghamshire, HP11 1. T: 01628 782225

Limbrick, A.J., Mr, 30 Strawberry Vale, Twickenham, TW1 4RU. T: 020 8892 3544

Lindsay, D.J., Mr, 9 Parkham Close, Cramlington, NE233TQ

Lindsay, R.M.D., Mr, 71 Marefield, Lower Earley, Reading, RG6 3DZ

Lines, J., Mrs, Allonby House, Dene Road, Rowlands Gill, Tyne and Wear, NE39 1DU. T: 01207 542216 (H)

Lings, M.L., Mr, University of Bristol, Dept of Civil Engineering, Queens Building, University Walk, Bristol, BS8 1TR

Littlejohn, G.S., Mr, Almsford House, Fulwith Mill Lane, Harrogate, HG28HJ

Littler-Jones, G.H., Mr, Coed Derw, Pwllglas, Ruthin, Clwyd, LL15 2P

Lloyd, P.J., Mr, Abbeydale Bec, 4 Rosedale Avenue, Wake-field, West Yorkshire, WF2 6EP. T: 0850 915372, F: 01924 250229

Lomax, C.D., Mr, Setech (Geographical Engineers) Ltd., Centre For Advanced Industries, Coble Dene, North Shields, Tyne and Wear, NE29 6DE

Long, M.M., Mr, 15 Castleside Drive, Rathfarnam, Dublin 14, REPUBLIC OF IRELAND. T: +353 1 4920142

Long, S.E., Ms, The Old Bakery, 111 Queens Road, Norwich, NR1 3PL. T: 01603 626327, F: 01603 667934, E: s.long@elsevier.co.uk

Lontzetidis, K., Mr, Lamprou Porfyra 2, Thessaloniki, 54646 GREECE

Lord, J.A., Dr, The Warren, Ivinghoe, Leighton Buzzard, Bedfordshire, LU7 9EJ

Love, J.P., Mr, Orchard Barn, The Green, Foxton, Cambridge, Cambridgeshire, CB2 6ST. T: 0207 5818 348 (W)

Lovegrove, G.W., Mr, 39D Seabird Lane, Discovery Bay, Lantau Island, HONG KONG. T: 852 2987 6902, F: 852 2987 6208

Lovenbury, H.T., Mr, 15 Brushwood Road, Horsham, West Sussex, RH12 4PE. T: 1403241803, E: htlovenbury@lineone.net

Loveridge, F.A., Miss, Cedar Lodge, Tivoli Copse, Woodside Avenue, Brighton, BN1 5NF. T: 020 7685 6003, E: FXLOVERI@ctrl.co.uk

Lowe, R.A., Mr, 17 Fox Covert, Colwick, Nottingham, NG4 2DD

Luck, R.C., Mr, 3 Alexandra Close, Crediton, Devon, EX17 2DY. T: 01363 773625

Lunardi, I.C., Mr, 13 Wallace Street, Kilmarnock, SCOTLAND, KA1 1SB

Maaith, O.N.K., Dr, Jarzeem Midcal Store, P.O. Box 1794, Amman 11941, JORDAN

Machon, A., Mr, 3 Walford Place, Hillmorton, Rugby, Warwickshire, CV22 5HA. T: 1788335077, E: andy@machona.freeserve.co.uk

Mackey, C.W., Mr, P.O. Box 90, Tieri, Queensland 4709, AUSTRALIA. T: 0061 07 4981 62

Macklin, S.R., Mr, 57 Wick Road, Teddington, Middlesex, TW11 9DN

Maconochie, A.J., Mr, 52 Springwood Avenue, Stirling, FK8 2PB. T: 01786 475949

Maguire, W.M., Dr, 4 Manor Road, Richmond Park, Gnosall, ST20 ORJ

Mair, R.J., Prof., Department of Engineering, University of Cambridge, Trumpington Street, Cambridge, CB2 1PZ. T: 01223 332631, F: 01223 339713, E: rjm50@eng.cam.ac.uk

Makarchian, M., Mr, Dept of Civil Eng.Faculty of Eng., Bu-Ali Sina University, P.O. Box 65178-4161, Hamadan, Iran. T: 9881 225046

Mallard, D.J., Mr, 26 Gloucester Street, Winchcombe, Cheltenham, Gloucestershire, GL54 5LX

Malone, A.W., Mr, Department of Earth Sciences, The University of Hong Kong, Pokfulam Road, HONG KONG. T: F. 00 852 2517 6912, F: W 852 2559 2555, E: awmalone@netvigator.com

Manby, C.N.D., Mr, 12 Ashley Court, Ashley Road, Epsom, Surrey, KT18 5AJ. T: 01372 722 170

Manby, A.D., Mrs, Inverness Lodge, Oakshade Road, Oxshott, Leatherhead, Surrey, KT22 0LE. T: 01372 842354

Manhas, I.S., Mr, 15 Church Mews, Church Road, Woodley, Reading, Berks, RG5 4RJ

Maniscalco, D., Mr, 1 Lower Park Road, Manchester, Lancashire, M14 5RS. T: 0161 2242582/ 07903 153123, E: d.maniscalco@usa.net

Manley, G.J., Mr, 190 Huddersfield Road, Stalybridge, Cheshire, SK15 3DL. T: 01904 62 0413

Mann, D.C., Mr, Brambles, Armscote, Stratford-upon-Avon, Warwickshire, CV37 8DE. T: 1608682302

Marcetteau, A.R.M., Mr, 69 Ramsbury Drive, Earley, Reading, Berkshire, RG6 7RS. T: 01734 262219

Marchi, G., Mr, V Baccarini 29, Faenza (Ra), 48018 ITALY
Margetson, N., Mr, 28 Pasture Lane, Malton, North Yorkshire, YO17 0BS. T: 01904 524214
Margetts, L., Mr, 43 Grosvenor Road, Manchester, M16 8JP
Marklew, N., Mr, Flat One, 31 West Street, Banbury, Oxon, OX16 3HA
Marsden, C., Mr, 8 Seville Court, Clifton Drive, Lytham St. Annes, Lancashire, FY8 5RG. T: H.01253 794192
Marsden, C.A., Mr, 131 Liverpool Old Road, Much Hoole, Preston, Lancashire, PR4 4GA. E: marson-family@hoole131.freserve.co.uk
Marshall, M.A., Dr, 46 Deerlands Road, Wingerworth, Chesterfield, Derbyshire, S42 6UW. F: 0114 2548436, E: mark@mamdoc.demon.co.uk
Marshall, A., Mr, C/O PHLC Consultants, P.O. Box 21, Tarbela Dam Proj, Tarbela, Dist. Haripur, PAKISTAN. T: 1604786113
Marsland, F.M., Miss, Flat 1, 39 Farnhall Road, Guildford, Surrey, GU2 5JN
Martin, A.B., Mr, The Green House, Wycombe Road, Stokenchurch, High Wycombe, Buckinghamshire, HP14 3RP. T: 1494 483924
Martin, C.J., Mr, 6 Hillcroft Crescent, Wembley, HA9 8EE. T: 020 8903 8340
Martin, J., Mr, 13 Wolviston Avenue, Osbaldwick, York, North Yorkshire, YO1 3D
Martin, P.L., Mr, 15 Park Avenue, West Wickham, Kent, BR4 9JU. T: 020 8777 2181
Martin, W.S., Mr, 322 Glasgow Road, Waterfoot, Eaglesham, Glasgow, Lanarkshire, G76 0EW. T: 1416442603, E: wstewartmartin@hotmail.com
Martins, I.S.M., Dr, Praia De Icarai 281 Ao.501, Icarai – Niteroi – Cep 24230-004, Rio De Janeiro, BRAZIL
Masters, J.R., Mr, 42 Victoria Avenue, Hillingdon, Middx., UB10 9AH
Mather, N.E.E., Mr, 52 Frensham Drive, London, SW15 3EA. T: 0208 789 0830
Mather, D.V., Mrs, 39 Seaton Road, Welling, Kent, DA16 1DT. T: 0410 603424
Mathieson, G., Mr, 83 Blackheath Park, London, SE3 0EU. T: 020 8852 6564, F: 82972819, E: gmath.assccs@dial.zipca.com
Mathieson, W.G., Mr, Flat 6, Lewes Crescent, Kemp Town, Brighton, BN2 1FH. T: 01273 36500
Matthews, M.C., Dr, 1 Islay Gardens, Cosham, Portsmouth, Hampshire, PO6 3UF. T: 01705-327267
Maugeri, M., Prof., Corso Umberto 208, Acireale, 95024 ITALY
Maurenbrecher, P.M., Mr, Groot Hertoginneln 15, Den Haag, 2517 EA NETHERLANDS
Mawditt, J.M., Mr, 14 Sprucedale Gardens, Wallington, Surrey, SM6 9LB. T: 020 8647 7073
Maxwell, Dr, A.S., 6A Island View, Discovery Bay, HONG KONG
May Gurney (Technical Services), Trowse, Norwich, Norfolk, NR14 8SZ
May, R.E., Dr, 17 Pavenham Close, Lower Earley, Reading, Berkshire, RG6 4DX. T: 1189610939
Mc Anally, C.J., Mr, 12 Rectory Green, West Boldon, Tyne and Wear, NE36 8QD
McAllister, J.R., Mr, Apartment 1008C, World Trade Centre Apartments, P.O. Box 9229, Dubai, UNITED ARAB EMIRATES. T/F: 9714 313800
Mccandless, G.K., Mr, 75 Station Road, Saintfield, Ballynahinch, County Down, BT24 7DZ
McCombie, P.F., Mr, 17 Fanshaw Way, Warminster, Wiltshire, BA12 9QX
McCrae, R.W., Mr, 23 Estridge Way, Tonbridge, TN10 4J
McCusker, A., Mr, 43 Hope Park Gardens, Bathgate, West Lothian, EH48 2Q. T: 01506 636189, E: EDI@DAMES.COM
McDermott, R.J.W., Dr, Plas Heulog, Mountain Lane, Penmaenmawr, Gwynedd, LL34 6YP
McDougall, J.R., Mr, 36 Shore Road, Aberdour, Burntisland, Fife, KY3 0TU. T: 01383 861114, E: jrmi@bolton.ac.uk
McEntee, J.M., Mr, Drumbrade, Spring Lane, Slough, Bucks., SL2 3EH

McGinnity, B.T., Mr, 7 Arnold Close, Stoke Mandeville, Aylesbury, Buckinghamshire, HP22 5XZ. T: 01296 614254
McGinty, K., Mr, Flat 28, St. Andrews Drive, Glasgow, G41 5SQ
McGough, D.J., Mr, Flat 3, East Court, 222 Portland Road, Hove, E Sussex, BN3 5QT. T: 01273 728 593
McGown, A., Prof., 39 Thrums Avenue, Dishopbriggs, Glasgow, G64 1DG. T: 0141 7725770
McKenzie, A., Mr, 8 Saxon Way, Woodbridge, Suffolk, IP12 1LG. T: 01394 384332
Mckenzie, I., Mr, Fondedile Foundations Ltd, Rigby Lane, Hayes, Middlesex, UB3 1ET
McKinlay, A.C.M., Mr, 10 Rowena Street, Kenmore, Brisbane, Queensland 4069, AUSTRALIA. T: 07 33782358
McKinley, J.D., Mr, Cardiff University: School of Engin, P.O. Box 925, Cardiff, South Glamorgan, CF24 0YF
McLaren, J.N., Mr, 11 Meadowcrft Close, East Grinstead, West Sussex, RH19 1NA. T: 01342 325785
McNamara, A.M., Mr, 10 Frankswood Avenue, Orpington, Kent, BR5 1BP. T: 01689 822522
Mcpherson, A., Mr, 11 Frankby Close, Wirral, Merseyside, CH49 3PT
Mehjoo, A.R., Mr, 18 Macleod Road, Winchmore Hill, London, N21 1SN. T: 020 8364 0939
Mellors, T.W., Dr, Brylls, West End, Waltham St. Lawrence, Reading, Berkshire, RG10 0NT. T: 1189343493
Menzies, B.K., Dr, E D S Instruments Ltd., Unit 12 Eversley Way, Egham, Surrey, TW20 8RB. T: 01784 439228, F: 01784 434644
Merrifield, C.M., Dr, The Manchester School of Eng., Simon Building, University of Manchester, Oxford Road, Manchester, Lancashire, M13 9PL
Merritt, A.S., Mr, St. John's College Cambridge, England, Cambridgeshire, CB2 1TP
Michael Anthony Paul, M.A.P., Prof., Dept Civil & Offshore Engineering, Heriot Watt University, Edinburgh, EH14 4AS. T: 0131 4513148, F: 0131 4515078
Middleton, J., Mr, 13 Norwood Court, Thornhill Road, Benton, Newcastle Upon Tyne, NE12 8AF. T: 0191 2663159
Mihalis, I., Mr, 14 Karaiskaki Street, 15562 Holargos, Athens, GREECE. T: 01 6542538
Millar, D.L., Mr, Camborne Scool of Mines, Trevenson Pool, Redruth, Cornwall, TR15 3S
Millar, M.J., Mr, Geniefa, Cuckfield Road, Hurstpierpoint, Hassocks, West Sussex, BN6 9LL. T: 01273 833258
Miller, A.B., Mr, Warrenden, Smallhythe Road, Tenterden, Kent, TN30 7NE. T: 0044 01580 763098, F: 44 763098, E: abm@warrenden.demon.co.uk
Miller, D., Mr, 6 Queens Drive, Cumbernauld, Glasgow, Lanarkshire, G68 0HN. T: 01698 504037
Miller, D.A., Mr, 11 Kirkton Road, Dumbarton, G82 4AS. T: 01389 731870, F: 01389 764445
Milligan, G.W.E., Mr, Eur Ing, 9 Lathbury Road, Oxford, Oxfordshire, OX2 7AT. T: 1865516407
Millmore, J.P., Mr, 7 High Standing, Chaldon, Caterham, Surrey, CR3 5DY. T: 01883 341066, E: millmore@callnetuk.com
Milne, A., Mr, Geotechnical Engineering Ltd., Rock House, Lower Tuffley Lane, Gloucester, GL2 6DT
Milne, C.A., Mr, 3 Ainsworth Street, Ulverston, Cumbria, LA12 7EU. T: 1229585716, F: 585716, E: c.milne@virgin.net
Milton, L.J., Mr, 115 Brentwood Road, Brighton, East Sussex, BN1 7ET. T: 01273 566271 (H)
Mitchell, A.R., Mr, Blk 49 Hume Avenue 06-02, Parc Palais, SINGAPORE. T: 65 764 4549, F: 65 764 4549, E: arm20@magix.com.sg
Mitchell, J.W.L.R., Mr, Chapel Garth, Bawtry Road, Hatfield Woodhouse, Doncaster, South Yorkshire, DN7 6PQ
Moffat, A.I.B., Mr, 43 Bishops Hill, Acomb, Hexham, Northumberland, NE46 4NH
Molenkamp, F., Prof., University of Manchester, Instutite of Science & Technology, Dept of Civil & Structural Eng., P.O. Box 88, Manchester, M60 1QD
Monahan, H.P., Miss, 3 St. Annes Close, Cheltenham, Gloucestershire, GL52 2JQ

Monteith, R.M.S., Mrs, 117 Heath Road, Beaconsfield, Buckinghamshire, HP9 1DJ

Moore, I.R., Mr, Crossways, Isfield, Uckfield, East Sussex, TN22 5UJ

Moore, R.J., Mr, 105 Kingsdown Road, Walmer, Deal, Kent, CT14 8BL. T: 01304 373663, E: Richard.Moore2@ Halliburton.com

Moorhead, M.S., Eur Ing, M. Moorhead Consulting, 2nd Floor, Holy Rood House, 11–13 Swan Street, West Malling, Kent, ME19 6JU

Moran, D.E., Mr, 150 South Prospect Avenue, Tustin, California 92780, USA. T: 714 5442215, F: 714 5447395

Mordhorst, C., Mr, P.O. Box 29129, Thunder Bay, Ontario, P7B 6P9 CANADA. T: 001 807 345 1531, F: 001 807 345 1531

Morgan, C.S., Mr, 18 Old Hay Close, Dore, Sheffield, South Yorkshire, S17 3GQ. T: 1147367713

Morgan, V.R., Mr, 3 Ings Close, Alton, GU34 1TB. T: 01483 427 311 (W)

Morgenstern, N.R., Prof., Dept of Civil Engineering, The University of Alberta, Edmonton, Alberta, CANADA

Morley, R.S., Mr, 36 Ormond Drive, Hampton, TW12 2NT

Morrell, S.A., Miss, 11 Fellview Close, Hathersage, Hope Valley, Derbyshire, S32 1DS. T: 1889563680

Morris, D.J., Mr, 21 Horseshoe Close, Colehill, Wimborne, Dorset, BH21 2UL. T: 44 01202 881517, E: john-morris@tusitala.freeserve.co.uk

Morrison, P.R., Dr, 3 Hogarth Hill, London, NW11 6AY

Mosforth, J.L., Eur Ing, 1104 Evesham Road, Astwood Bank, Redditch, Worcestershire, B96 6EA. T: 1527893157, E: jmosforth@hotmail.com

Moss, R.L., Mr, 57 Dalston Drive, Didsbury, Manchester, M20 5LQ

Mould, M.A., Mr, Flat 1, 46 Princes Gardens, London, SW7 1LU. T: 020 7594 9482, E: martin.mould@ic.ac.uk

Moynihan, T.P., Mr, Killaha West, Kenmare, Co. Kerry, IRELAND. T: 00353 64 42424

Muir Wood, D., Prof., Bristol University Department of Civil Engineering, Queens Building, University Walk, Bristol, BS8 1TR

Muir Wood, A.M., Sir, Franklands, Pangbourne, Reading, Berkshire, RG8 8JY. T: 0207 602 7282 (W)

Munachen, S.E., Mr, 38 Lincoln Way, Harlington, Dunstable, Bedfordshire, LU5 6N. T: 01525 872903, E: scott. munachen@eng.ox.ac.uk

Munro, C., Mr, D N 041, Dounreay, Nuclear Establishment, Thurso, Caithness, KW14 7TZ

Murphy, J.G., Mr, Oakwood House, 59 Howards Thicket, Gerrards Cross, Buckinghamshire, SL9 7NU

Murray, E.J., Mr, 188 Station Road, Earl Shilton, Leics., LE9 7GD

Mushet, S.E., Miss, 4 Stoneyhill Grove, Musselburgh, Midlothian, EH21 6SE. T: 0131 665 7230

Myles, B., Mr, 15 Greystones Drive, Reigate, Surrey, RH2 0HA

Nash, D.F.T., Mr, Dept of Civil Engineering, University of Bristol, Queens Building, University Walk, Bristol, BS8 1TR

Naylor, D.J., Dr, Copper Roof, 45 Pennard Road, Southgate, Swansea, SA3 2AA. T: 01792 233755

Needham, A.D., Mr, Thornbury, 75 Hazelwood Road, Duffield, Belper, Derbyshire, DE56 4AA. T: 01332 841131 (H). E: a.needham@virgin.net

Nettleton, I.M., Mr, Edge Consultants UK Ltd., Atlas House, Simonsway, Manchester, M22 5PP. T: 0161 4366767, F: 0161 4997987

Nevitt, S.C., Mrs, 4a School Road, Moseley, Birmingham, B13 9ET. T: 07947 539 192 (M)

Newman, R., Mr, 7 Hillside Road, Ashtead, Surrey, KT21 1R. T: 01372 276641

Newman, T.G., Mr, 185 Elm Road, New Malden, Surrey, KT3 3HX. T: 020 8949 7522

Newsom, T.A., Dr, Department of Civil Engineering, University of Dundee, Perth Road, Dundee, DD1 4HN

Newton, M.A., Mr, 58 Haigh Moor Road, Tingley, Wakefield, W Yorks, WF3 1EE

Ng, C.F., Dr, B.G Technology Ltd., Gas Research & Technology Centre, Ashby Road, Loughborough, Leicestershire, LE11 3GR

Ng, W.W., Dr, Dept Civil Eng., Hong Kong University (sci & Tech, Clear Water Bay, Kowloon, HONG KONG)

Ng, P.K.H., Mr, Rm 1004, 10/F, Peony House East Blk, 1-7 Pok Man Street, Tai Kok Tsui, Lowloon, HONG KONG

Ngambi, S., Dr, 41 Spray Close, Colwick, Nottingham, Nottinghamshire, NG4 2GT

Nichol, D., Dr, Geotechnical Section, Developement Services Dept Eng. Services, Wrexham County Borough Council, Guildhall, Wrexham, LL11 1AY. T: 01978 297104

Nicholls, R.A., Mr, 4 Priory Green, Highworth, Swindon, Wiltshire, SN6 7NU

Nicholls, R.M., Mrs, 15 The Dene, Abinger Hammer, Dorking, RH5 6PX. E: rachel.nicholls@ntlworld.com

Nicholson, D.P., Mr, 1 Byron Avenue, Colchester, Essex, CO3 4HG. T: 01206 544906, E: nicholson_duncan@msn.com

Nicholson, R., Mr, Ivy Farm Cottage, Swathwick Lane, Wingerworth, Chesterfield, Derbyshire, S42 6QP

Nielsen, H.L., Mr, Mollebakken 7, 8660 Skanderborg, DENMARK. T: 45 87 44 2290

Ninis, N., Mr, 23 Thasou Street, Athens, 112 57 GREECE

Nixon, I.K., Mr, Downsview, 28 Valley Close, Goring-On-Thames, Reading, Berkshire, RG8 0AN. T: 01491 872731

Noble Denton & Associates Ltd., Noble House, 131 Aldersgate Street, London, EC1A 4EB

Norbury, D.R., Eur Ing, Little Field Gate, Newton, Upper Basildon, Reading, Berkshire, RG8 8JG

Norman, C.L., Miss, 87 Thame Road, Warborough, Wallingford, Oxfordshire, OX10 7DU

Nuttall, F.S., Mr, 8 Bre, Garston, Watford, WD2 7JR

Nyirenda, Dr, Eur Ing, Z.M., 46 Swaledale, Wildridings, Bracknell, Berkshire, RG12 7ET. T: 0207 681 5141

O'Brien, A.S., Mr, 11 Meadowbrook, Oxted, Surrey, RH8 9LT. T: 020 8686 5041

O'Brien, M., Mr, 42 Aro Street, Aro Valley, Wellington, NEW ZEALAND

Ogura, H., Mr, Geotop Corporation, 1-16-3 Shinkawa Chuo-Ku, Tokyo, 104-0033, JAPAN

Older, J.L., Mr, 12 Bernard Avenue, London, W13 9TG

Olsen, H.W., Dr, P.O. Box 775730, Steamboat Springs, Colorado, 80477 USA. T: 970 879 7856, F: 970 879 7856

Omine, K., Dr, Department of Civil Engineering, Kyushu University, 6-10-1 Hankozaki, Higashi-ku, Fukuoka 801-81, JAPAN. T: 81 92 642 3285, F: 81 92 642 3322, E: OOMINE@CIVIL.KYUSHU-U.AC.JP

O'Neill, M.A., Mr, 30 Manorfield Drive, Horbury, Wakefield, West Yorkshire, WF4 6JZ

O'Neill, D., Ms, C/O Nefco, 77 Federal Avenue, Quincy, Mass, 2169 USA

Onions, K.R., Mr, 33 Sandon Avenue, Westlands, Newcastle-Under-Lyme, Staffordshire, ST5 3QB. T: 01782 636204

Oprandi, R.I.J., Mr, C/ Amaniel 23, 1 Ext. Izq, 28015 Madrid, SPAIN. T: 01273 476813 (H)

O'Reilly, M.P., Dr, 24 Marlborough Rise, Camberley, Surrey, GU15 2ED. T: 1276506896

O'Riordan, N.J., Mr, Barnfield, Brighton Road, Horsham, West Sussex, RH13 6ER. T: 044 01403 253020, E: noriordan @aol.com

Orr, T.L.L., Mr, Clova, Lordello Road, Shankill, Co Dublin, REPUBLIC OF IRELAND

Ottway, T.J., Mr, Halcrow Group Limited, Deanway Technology Centre, Wilmslow Road, Handforth, Wilmslow, Cheshire, SK9 3FB

Ouhadi, V.R., Mr, BU-Ali Sina University, Faculty of Engineering, P.O. Box 4161, Hamedan, IRAN. E: vahido@ Sadaf.basu.ac.ic

Overy, R.F., Mr, 12 Midmar View, Kingswells, Aberdeen, Aberdeenshire, AB15 8FE. T: 1224884196

Owen, M.W., Mr, 28 Heol Gerrig, Treboeth, Swansea, SA5 9BP

Owen, R.H., Mr, 13 Atwood Avenue, Kew, Richmond, Surrey, TW9 4HF

Pachakis, M.D., Mr, 15 Plastira Street, Melissia Attikis, 15127 GREECE. T: 301 8042688

Page, D.P., Mr, 31 Blatchington Road, Hove, BN3 3YL

Pagella, J.F., Mr, Elizabeth House, 2 Gregorys Court, Chagford, Devon, TQ13 8AP. T: 01647 433307

Pagett, S.F., Mr, 12 Dorchester Court, Mayfare, Croxley Green, Herts, WD3 3DQ

Palanque, C.M.H., Mr, 49 Taverner Close, Oakridge, Basingstoke, Hampshire, RG21 4JF

Palmeira, E.M., Dr, University of Brasilia, Dept of Civil Engineering, Brasilia, D F 70.910-900, BRAZIL. T: 55 61 2737313, F: 55 61 2721053

Palmer, J.S., Mr, Cal Dex Consultants Ltd., 46 Kimpton Road, Blackmore End, Hertfordshire, AL4 8LD. T: 01438 833089, F: 01438 833089

Palmer, T.M., Mr, Top Floor Flat, 485 Liverpool Road, London, N7 8PG

Pandis, K., Mr, Patriarchou Photiou 21, Maroussi, Athens, GREECE 15122. T: 01 8024234

Pang, P.L.R., Dr, Flat A, 10/F, 35–37 Macdonnell Road, Hk, HONG KONG. T: 286 83781, F: 286 84681

Pang, C.H., Mr, 626-B Taman Bukit Melaka, Bukit Beruang, 75450 Melaka, MALAYSIA. E: kazekumo@hotmail.com

Pang, L.S., Mr, 404 Hong Chung House, Mei Chung Court, Tai Wai, HONG KONG. T: 26067501, F: 26067501, E: ispgd@icable.com

Pantelidou, H., Miss, Ove Arup & Partners, Arup Geotechnics, 13 Fitzroy Street, London, W1P 6BQ. T: 020 7465 2075

Papacharalambous, G., Mr, 58 Plastira Str., 15451 Neo Psihiko, Athens, GREECE. T: 00 30 1 6719344

Papadopoulos, N., Mr, 4 Votsi Street, 66-100 Drama, GREECE

Papanastasiou, P., Dr, Schlumberger Cambridge Research, High Cross, Madingley Road, Cambridge, CB3 0EL

Pappin, J.W., Mr, 2/F, 19 Tat Chee Avenue, Yau Yat Chuen, Kowloon, HONG KONG. T: 852 25732404, F: 852 21427181, E: thepappins@hotmail.com

Parker, E.J., Dr, School of Science & environment, IT blk, Coventry University, Priory Street, Coventry, CV1 5FB

Parker, J.B., Mr, 19 Pyotts Copse, Old Basing, Basingstoke, Hampshire, RG24 8WE. T: 0118 950 0761 (W)

Parkes, D.B., Mr, Cedarwood, Wisborough Lane, Storrington, Pulborough, West Sussex, RH20 4ND. T: 44 1903 743106, F: 44 743106, E: douglas.parkes@which.net

Parkes, R.D., Mr, 5 St. Lukes Road, Maidstone, Kent, ME14 5AR. T: 1622674035

Parkhurst, S.A., Mr, 98 Blenheim Road, Caversham, Reading, Berkshire, RG4 7RP

Parnell, P.G., Mr, 27 Bower Mount Road, Maidstone, Kent, ME16 8AX

Parr, A.W., Eur Ing, 33 St. Michaels Road, Sandhurst, Camberley, Surrey, GU17 8HD

Parry, L.N., Mr, 18 Birch Avenue, Caterham, Surrey, CR3 5RW

Parry, N.V., Mr, Cherry Tree Cottage, Nunnery Lane, Woodmancote, Dursley, Gloucestershire, GL11 4AW. T: 01452 527743 (W), E: neil.parry@virgin.net

Parry, R.H.G., Dr, 5 Farm Rise, Whittlesford, Cambridge, Cambridgeshire, CB2 4LZ. T: 01223 832024, E: rhgp2@cam.ac.uk

Parry, S., Mr, Flat 3/F, Kennedy Court, 7A Shui Fai Terrace, Wanchai, HONG KONG. T: 2849 5592

Paskell, P.B., Mr, 13 Heron Close, Lower Halstow, ME9 7EF. T: 0771 8272056

Patel, D.C., Mr, 78 Alicia Gardens, Harrow, Middlesex, HA3 8JE. T: 0208-907 6508 WIFE

Pattinson, J.I., Mr, Devereux, 30 Rambling Way, Potten End, Berkhamsted, Herts, HP4 2SF. T: 01442 876584

Paul, J., Mr, Duck House Farm, Brow Top, Grindleton, Clitheroe, Lancashire, BB7 4QR. T: 01200 441 787 (H), E: jpaul@excite.co.uk

Paulson, R.M., Mr, 38 Westwood Avenue, Timperley, Altrincham, WA15 6QF

Peacock, W.S., Dr, Ground Risk Management, Stanhope, Bretby Business Park, Burton-on-Trent, Staffordshire, DE15 0QD

Peat, L.J., Miss, 16 Foxhill Road, Reading, RG1 5QS

Peck, D.A., Mr, 52 Chollerford Close, Gosforth, Newcastle-Upon-Tyne, NE3 4RN. T: 0191 213 0292, E: dpeck@uk.packardbell.org

Pedley, M.J., Dr, 104 Bennett Crescent, Cowley, Oxford, OX4 2UW

Pellegrini, P., Mr, 17 Rosemary Avenue, Finchley, London, N3 2QP

Pellew, A.L., Mr, 108 Tredegar Road, London, E3 2EP. T: 0780 103 3331, E: a.pellew@ic.ac.uk

Penman, A.D.M., Dr, Sladeleye, Chamberlaines, Harpenden, AL5 3PW

Penn, M., Dr, 12 Weybourne Close, Harpenden, Hertfordshire, AL5 5RE. T: 0208 774 2551 (W), E: mmp@mmcroy.mottmac.com

Pennington, D.S., Dr, Ove Arup & Partners, P.O. Box 629, West Perth, W A 6872, AUSTRALIA

Pennock, R.L., Mr, Flat 3, 4 West Cliffe Grove, Harrogate, North Yorkshire, HG2 0PL

Peronius, A.N., Mr, Apartado Postal 3052, Tegucigalpa, Honduras

Perry, J., Dr, Mott Macdonald, St. Anne House, 20-26 Wellesley Road, Croydon, CR9 2UL

Petley, D.J., Dr, Dept of Engineering, University of Warwick, Gibbet Hill Road, Coventry, CV4 7AL. T: 01203 523523, Ext. 31, F: 01203 418922

Petrasouits, G., Mr, Budapest Technical University, Geotechnical Department, 1521 Budapest, HUNGARY

Pettit, J.H., Mr, 15 Heol-Y-Ddol, Caerphilly, Mid-Glamorgan, CF83 3JF

Phear, A.G., Mr, 5 College Place, St. Albans, AL3 4PU

Phelan, J.A., Mr, G/F 23A Sha Po New Village, Yung Shue Wan, Lamma Island, HONG KONG

Phillips, A., Mr, 5 Ael-Y-Bryn, Pentyrch, Cardiff, South Glamorgan, CF15 9TD. T: 029 20473727 (W)

Phillips, K.A., Mr, 5 Lewesdon Court, Silver Street, Lyme Regis, Dorset, DT7 3HU

Phipps, P.J., Mr, 6 Lynne Court, 53 Cambridge Road, London, SW20 0PY

Pickering, N., Mr, 2 Beaumont Close, Weston-Super-Mare, North Somerset, BS23 4LL. T: 1934633443, E: nigelpickering@dolphinds.freeserve.co.uk

Pierpoint, N.D., Mr, 103 Lincoln Avenue, Twickenham, Middlesex, TW2 6NJ. T: 0208 686 5041 (H)

Pine, R.J., Dr, White Walls, Mongleath Road, Falmouth, TR11 4PN. T: 1326311788, E: rrpine@geomechanics.freeserve.co.uk

Pitts, J., Dr, 3 Fellows Yard, Plumtree, Nottingham, NG12 5NS. T: 0115 9376955

Plant, G.W., Dr, Londolozi, 3 Acacia Grove, Berkhamsted, Hertfordshire, HP4 3AJ. T: 1442865617, F: 879898, E: graham@grahamplant.com

Pollard, R.J., Mr, 38 Fisher Avenue, Rugby, CV22 5HW

Pollock, J.A.D., Mr, 11 Ireland Road, Pietermaritzburg 3201, SOUTH AFRICA. T: 27333471223, E: alwynpollock@futurenet.co

Pooley, A.J., Mr, 133 Rue Des Terrasses, Thoiry, 1710 France

Porter, A.B., Mr, 3 Brydeck Avenue, Penwortham, Preston, PR1 9QL. E: andyporter@teba.tsnet.co.uk

Poskitt, T.J., Prof., Tanglewood, Lunghurst Road, Woldingham, Caterham, Surrey, CR3 7EJ

Potter, L.J., Dr, 3 Church Street, Haslingfield, Cambridge, Cambridgeshire, CB3 7JE. T: 01223 504022

Potts, V.J., Miss, Russet Lodge, 52 Somerset Road, Basildon, Essex, SS15 6PE

Potts, D.M., Mr, Russett Lodge, 52 Somerset Road, Laindon, Basildon, Essex, SS15 6PE. T: 1268546262

Poulos, H.G., Prof., 142 Wicks Road, North Ryde, NSW, 2113 AUSTRALIA

Powderham, A.J., Mr, 12 Hawthorn Road, Wallington, Surrey, SM6 0SX

Powell, C., Mr, 3 Newby Road, Farnhill, Keighley, West Yorkshire, BD20 9AT. T: 01535 634319

Powell, J.J.M., Mr, 22 Leycroft Way, Harpenden, Hertfordshire, AL5 1JW. T: 44 01582 760671, E: john@powmail.freeserve.co.uk

Power, P.T., Mr, 14 Chestnut Avenue, Chesham, Buckinghamshire, HP5 3NA

Powrie, W., Mr, University of Southampton, C/O Dept of Civil & Env. Eng., University Road, Southampton, Hampshire, SO17 1BJ

Preene, M., Dr, Ove Arup & Partners, Admiral House, 78 East Street, Leeds, West Yorkshire, LS9 8EE

Price, C.Hd.G., Mr, Pakistan Hydro Consultants, P.O. Box 4, Tarbella Dam, Hazara Province, PAKISTAN. T: 92995660759

Price, G., Mr, 23 Reynards Way, Bricket Wood, St. Albans, AL2 3SG

Price, I., Mr, The Rosary, Longdon, Tewkesbury, Gloucestershire, GL20 6AS

Price, J.L., Mr, 67 Oakley Close, Thornbury Road, Isleworth, Middlesex, TW7 4HY. T: 2085687698, E: john.price@talk21.com

Price, M.S., Mr, The Paddock, 77 Rook Lane, Caterham, Surrey, CR3 5BN

Price, R.W.J., Mr, 75 Alford Road, West Bridgford, Nottingham, NG2 6HP

Pringle, D.R., Miss, 23 Blackthorn Way, Wakefield, West Yorkshire, WF2 0HN. T: 01924 314359

Privett, K.D., Dr, Samaras, Llandough, Cowbridge, South Glamorgan, CF71 7LR. T: 01446 775011, F: 01446 775011

Pryke, J.F.S., Mr, Swan House, 10 Swan Street, Ashwell, Baldock, Hertfordshire, SG7 5NX

Puller, D.J., Mr, Lantau, Seal Hollow Road, Sevenoaks, Kent, TN13 3SF

Puller, M.J., Mr, 12 Farnaby Drive, Sevenoaks, Kent, TN13 2LQ

Pycroft, A.S., Mr, 99 The Mall, Swindon, Wiltshire, SN1 4JE. T: 01793 812479 (W)

Pyle, S.M., Mr, Flat 15a, Wing Hing Court, 110–114 Tung Lo Wan Road, Tai Hang, Causeway Bay, HONG KONG

Pyrah, I.C., Prof., Napier University, Merchiston Campus, 10 Colinton Road, Edinburgh, EH10 5DT

Quarrell, S.E., Eur Ing, Soil Consultants Ltd., Inkerman Farm, Amersham Road, High Wycombe, Buckinghamshire, HP15 7JH

Racker, C., Mrs, Auf Dem Grend 12, D–53844, Troisdorf, GERMANY. E: raeker@hotmail.com

Radosevic, N., Mr, 34 Priory Avenue, Caversham, Reading, Bershire, RG4 7SE. T: 0118 954 6205

Rahim, A.A.A., Mr, 164 Cotton Avenue, Acton, London, W3 6YG. T: 2089936868

Rainey, T.P., Mr, 6 Fresham House, 12 Durham Road, Bromley, Kent, BR2 0SG. T: 020 8466 1747

Raison, C.A., Mr, 7 Riverford Croft, Coventry, West Midlands, CV4 7HB. T: 44 024 7647550, E: raison@globalnet.co.uk

Ramsbottom, C.M.J., Mr, 9 Parkfield Place, Highfield, Sheffield, S2 4TH

Ramsey, N.R., Mr, Fugro Limited, Frogmore Road Industrial Estate, 18 Frogmore Road, Hemel Hempstead, Hertfordshire, HP3 9RT

Ramshaw, C.L., Mrs, 5 Brancepeth Close, Durham, DH1 5XL. T: 0191 3862143 (H) 3747094 (W), E: martin. raushaw@which.net

Randolph, M.F., Prof., Geomechanics Group, University of Western Australia, Nedlands, WA 6907, AUSTRALIA. T: 61 9 3803075, F: 61 9 3801044

Rankilor, P.R., Prof., 9 Blairgowrie Drive, West Tytherington, Macclesfield, Cheshire, SK10 2UJ. T: 1625429003, F: 0044 619961, E: rankilor@compuserve.com

Rankin, W.J., Mr, Cherry Trees, 75 Buxton Lane, Caterham, Surrey, CR3 5HL. T: 01883 340291

Raven, K.P., Mr, 161 Marshall Lake Road, Shirley, Solihull, West Midlands, B90 4RB

Rawcliffe, J., Mr, Rawcliffe Associates, The Paddocks, Follifoot, Harrogate, N Yorks, HG3 1EA. T: 01423 879808, F: 01423 879808

Rawlings, C.G., Mr, Brown & Root, C/O Egnata Odos, A.E Km Thessaloniki – Thermi, P.O. Box 30, 57001 THERMI, Thessaloniki, GREECE

Raybould, M.J., Mr, Scott Wilson Kirkpatrick & Co, Rose Hill West, Chesterfield, Derbyshire, S40 1JF. T: 01246 210 325

Read, K.J., Mr, St Clair, Carey Hill Road, Stoney Stanton, Leicester, LE9 4LA. T: 01455 271 255

Reader, R.A., Mr, 115 Jalan Cengal, Serendah Golf Resort, Serendah 48200, Selangor Darul Ehsan, MALAYSIA. T: 36015725, F: 00 60 360641658

Reda, A.K., Dr, P.O. Box 114-5227, Beirut, LEBANON. T: 961 3 829 635

Redgers, J.D., Mr, 3 Post Cottages, Crawley, Winchester, SO21 2PT

Redhouse, I.A., Mr, 39 Sherborne Road, Peterborough, Cambridgeshire, PE1 4RJ. T: 4.40173E+12

Redmond, K., Mr, 91 Houstoun Gardens, Uphall, Broxburn, West Lothian, EH52 5SJ. T: 01506 855366, E: k.redmond@napier.ac.uk

Rees, S.W., Dr, Cardiff School of Engineering, Division of Civil Engineering, Queen's Buildings, P.O. Box 917, Cardiff, CF2 1XH. T: 01222 894000 et, F: 01222 874597

Rennison, M.E., Mr, 29 Station Road, Chessington, Surrey, KT9 1AX

Renton-Rose, D.G., Mr, The Lodge, Coopers Bridge, Tunbridge Lane, Liphook, Hampshire, GU30 7RF

Rhodes, S.J., Mr, 14 Calbourne Drive, Calcot, Reading, RG31 7DB. F: 08 08 1997

Rice, S.M.M., Dr, 47 Pentwyn, Radyr, Cardiff, CF4 8RE. T: 01222 842296

Richards, D.J., Dr, 29 Milton Grove, New Milton, Hampshire, BH25 6HB

Richards, L.R., Dr, Rock Engineering Consultant, The Old School House, Swamp Rd, Ellesmere, RD5 Christchurch, NEW ZEALAND. F: 64 3 329 5663

Richards, M.G., Mr, Guilden Arches, 98 Meldreth Road, Shepreth, Royston, Hertfordshire, SG8 6PX

Richardson, J.M., Mr, 50 Bricknell Avenue, Hull, HU5 4JS

Richardson, M., Mr, 11 Cedar Grove, Heaton Moor, Stockport, SK4 4RN

Rickeard, T.S., Mr, Pengelly, Peppard Road, Sonning Common, Reading, Berkshire, RG4 9NJ

Ridley, A.M., Dr, Soil Mechanics Section, Dept of Civil Eng, Imperial College, London, Imperial College Road, SW7 2BU

Riley, V.K., Miss, 6 Ravenglass Close, Lower Earley, Reading, Berkshire, RG6 3BX

Ritchies, Bowland House, Gadbrook Business Centre, Rudheath, Northwich, Cheshire, CW9 7TN

Ritsos, Mr, A., 17 Artemidos Str., Papagov, Athens, 15669 GREECE. T: 3016512132, F: 3016897581

Rix, D.W., Mr, 14 Alexandra Road, Heaton Moor, Stockport, SK4 2QE

Robb, M.F., Mr, Rocklands, 83 Front Road, Drumo, Lisburn, County Antrim, BT27 5JX. T: 2890826505, E: robbs@btclick.com

Roberts, E., Mr, 13 country court Thorntree Court, Talmers Villle, Newcastle Upon Tyne, NE12 9QX. T: 1912150722, E: tayl.roberts@WSPGROUP.COM

Roberts, K.J., Mr, 8 Thoresby Road, Bingham, Notts., NG13 8RE

Roberts, L.D., Mr, 6 Melrose Road, Merton Park, London, SW19 3HG. T: 7767797058

Roberts, P.A., Mr, 39 Long Acres, Durham, DH1 1JF

Roberts, R.B., Mr, 15 Kingsland Road, Cardiff, South Glamorgan, CF4 2EJ. T: 01222 610853

Roberts, T.O.L., Mr, 40 Twisden Road, London, NW5 1DN. T: 44 207 4853154

Robertson, C.W., Mr, 11 Rowan Drive, Bearsden, Glasgow, G61 3HQ

Robertson, I., Mr, 58 Meadowlands, West Clandon, Guildford, Surrey, GU4 7TB. T: 01483 535 000

Robinshaw, A.D., Mr, 45 Knoll Road, Fleet, Hampshire, GU13 8PT. T: 1252657701, E: alec.robinshaw@nt/world.com

Robinson, C.A.W., Mr, 41 Langdale Road, Wistaston, Crewe, Cheshire, CW2 8RS. T: 0161 4366767

Robinson, T.M., Mr, Tall Trees, Broadlayings, Woolton Hill, Newbury, Berkshire, RG20 9TS

Roche, D.P., Mr, 47 Rivermead Road, Exeter, Devon, EX2 4RH

Rochester, M.J., Mr, 56 Marshall Road, Sheffield, S8 0GP

Rodd, P.G., Mr, 18 Portway, Baughurst, Tadley, Hampshire, RG26 5PD. T: 01734 816617

Rodriguez Miranda, J.J., Mr, 8008 South Orange Avenue, Orlando, Florida, 32809 USA

Rodriguez, C.E., Mr, 7 Marston Avenue, Chessington, Surrey, KT9 2HF

Rogers, C.D.F., Dr, 11 Sinope A50, Nr.Coalville, Leicester, LE67 3AY

Rogers, I.T., Mr, Ary Schefferstraat 133, Den Haag, 2597 V R, NETHERLANDS

Rogers, R.J., Mr, 51 The Spinney, Moortown, Leeds, LS17 6SP

Rogers, S.H., Mr, Casanene, The Green, Badby, Daventry, Northamptonshire, NN11 3AF

Roll, C., Mr, 45 Pulham Avenue, Broxbourne, Hertfordshire, EN10 7TA

Rolo Gonzalez, R., Mr, 87 Nantes Close, London, SW18 1JL. T: 0788 748 1641, E: rrolo@rocketmail.com

Rooney, F., Ms, 1 The Orchard, Longfield Road, Tring, Hertfordshire, HP23 4D. T: 01442 828724

Roscoe, G.H., Eur Ing, 3 Park View Road, Berkhamsted, Hertfordshire, HP4 3EY

Rose, S.P., Miss, 39 Rokeby Road, London, SE4 1DE. T: 07967 029566 (M), E: sxrose@ctrl.co.uk

Rosenbaum, M.S., Prof., Dept of Civil & Structrul Eng., The Nottingham Trent University, Burton Street, Nottingham, Nottinghamshire, NG1 4BU

Rostron, B.J., Mr, Charnwood, Cromford Road, Wirksworth, Matlock, Derbyshire, DE4 4FH. T: 016929 823610

Rouainia, M., Dr, University of Glasgow, Department of Civil Engineering, Oakfield Avenue, Glasgow, G12 8LT

Rowell, M.P., Mr, 2 Laurel Cottages, Church Road, Cotton, Stowmarket, Suffolk, IP14 4RA

Ruddlesden, S., Mr, 22 Rosebery Road, Exeter, EX4 6LT. T: 07977 403273

Rudrum, D.M., Mr, 40 Dirdene Gardens, Epsom, Surrey, KT17 4AZ. T: 0207 465 3600

Russell, D., Mr, 26 Nelson Road, London, SW19 1HT

Russo, L., Mr, Via C.Correnti 35, 20035, Lossone (MI), ITALY. T: 0039 39 460979

Rust, M., Mr, Dept of Civil & Enviro. Eng., University of Southampton, University Road, Southampton, SO17 1BJ

Rutty, P.C., Mr, Beechwood House, Tierenane, Ballick-moyler, Co Laois, REPUBLIC OF IRELAND. E: prutty@bma.ie

Ryan, C., Mr, Ground Floor Flat, 114 Drayton Bridge Road, London, W7 1EP. T: 0118 9635000

Saada, A.S., Mr, Dept of Civil Engineering, Case Western Reserve University, Cleveland, Ohio, 44106 USA

Sabbagh, A., Mr, 52 Western Drive, Shepperton, Middlesex, TW17 8HW. T: 01932 882984

Saffari Shooshtari, N., Mr, 17 Monks Drive, London, W3 0EG

Sagar, K., Ms, 2nd Floor, 21, Wan Long Village, Yung Shue Wan, Lamma Island, HONG KONG. E: karen_sagar@hotmail.com

Sahasi, K.S., Mr, 9 Park Gate, Hitchin, SG4 9BP

Sakellariou, M., Mr, 25A Amfissis Street, Gr155 62, Holargos

Salgado, R., Prof., School of Civil Engineering, Purdue University, West Lafayette, I N, 47907-1284 USA

Sallis, D.K., Mr, Flat 35, Minerva Court, 124 Houldswurth Street, Finnieston, Glasgow, G3 8BY. T: 0141 204 4656

Salmon, D.E., Mr, 27 The Crescent, London, SW19 8AW. T: 0208-946-3210 (H)

Sammons, C.J., Mr, 34 Quentin Road, London, SE13 5DF. T: 020 8852 1147

Samuel, H.R., Mr, 11 Werter Road, London, SW15 2LL. T: 0208 246 6854 (W), F: 020 854, E: hsamuel@globalnet.co.uk

Samworth, I.D., Mr, 5 Goodwood Grove, York, YO24 1ER. F: 01 10 1996

Samy, K., Miss, Edgar Morton Lab. Rm 2.58, Simon Building, Manchester, Oxford Road, M13 9PL

Sanders, C.J., Mr, 25 Antonine Road, Bearsden, Dunbarton-shire., G61 4DP

Sanders, R.L., Mr, 30 Victoria Road, Woodbridge, IP12 1EJ

Sands, T.B., Mr, 54 Woodfield Crescent, Ealing, London, W5 1PB. T: 2089915453, E: e.b.sands@herts.ac.uk

Sarigiannis, D., Mr, Kerasountos 18, Kalamaria, 55131, Thessaloniki, GREECE

Sarma, S.K., Dr, Dept of Civil Engineering, Imperial College, Imperial College Road, London, SW7 2BU. T: 020 7594 6054, F: 020 7225 2716

Sartain, N.J., Mr, 3 Kirklington Road, Southwell, Nottinghamshire, NG25 0AR. T: 01636 814 264

Sawyer, J., Mr, 111 Winchester Road, Basingstoke, Hampshire, RG21 8XR

Saxton, R.H., Mr, 23 Amherst Road, Pennycomequick, Plymouth, Devon, PL3 4HH. T: 1712267967

Schina, S., Miss, 5 Demonogianni Street, Athens, GREECE

Schmidt, B., Dr, 1676 Yorktown Road, San Mateo, C A, 94402 USA

Schofield, A.N., Prof., 9 Little St. Marys Lane, Cambridge, Cambridgeshire, CB2 1RR

Scott, M.B., Mr, Southern Testing Laboratories, Stuart Way, Herontye, East Grinstead, West Sussex, RH19 4QA

Scott, M.J., Mr, 3 Westfield Court, Abenbury Fields, Wrexham, Clwyd, LL13 8JA

Scott, P.D., Mr, The Dormy House, Combe Hay, Bath, Avon, BA2 7EG

Seaman, J.W., Mr, 3 Whitewood Road, Berkhamsted, Hertfordshire, HP4 3LJ

Seddon, P.J., Mr, 19 Egmont Avenue, Surbiton, Surrey, KT6 7AU

Seddon, R.P., Mr, 717 Huddersfield Road, Lees, Oldham, Lancs, OL4 3PY

Selemetas, D., Mr, Room D, Flat G/2, 9 Kelvinhaugh Gate, Glasgow, Lanarkshire, G3 8PY

Sethi, P., Mr, 76 Rivets Close, Aylesbury, Buckinghamshire, HP21 8JR. T: 01895 422 485

Shah, S.A., Mr, Brown And Root Ltd., Unit 18, Mole Business Park, Randalls Road, Leatherhead, Surrey, KT22 7BA

Shanks, R., Mr, 100 Rothes Road, Rimbleton, Glenrothes, Fife, KY6 2AN

Sharma, R.S., Mr, Dept of Civil & Environ. Engr., University of Bradford, Richmond Road, Bradford, West Yorkshire, BD7 1DP. T: 01274 233856

Sharratt, M.J., Mr, Flat C, 7 Clifton Road, London, N8 8NY

Shaw, S.M., Mr, 9 Wentworth Road, Thame, Oxfordshire, OX9 3XG. T: 1844217205

Sheerman-Chase, A., Mr, Mulberry Cottage, 189 Faversham Road, Ashford, TN29 9A. T: 01233 623810, F: 01233 642862, E: schase@serveian.co.uk

Shein, A., Dr, Flat No 24, Elmwood Court, Pershore Road, Birmingham, B5 7PD

Shepherd, G., Mr, Geosol Limited, 14 Gateside Place, Kilbarchan, Johnstone, Renfrewshire, PA10 2LY

Sherrell, F.W., Mr, Redlands, Mount Tavy Road, Tavistock, Devon, PL19 9JL. T: 01822 612887 (H)

Sherwood, D.E., Mr, 45 Tekels Avenue, Camberley, Surrey, GU15 2LH. T: 0044 1276 61679, E: david.sherwood@soletande-bady.com

Shevelan, J., Mr, 17 Ibbotson Road, Sheffield, South Yorkshire, S6 5AD

Shi, Q., Dr, 8 Cavesson Court, Brownlow Road, Cambridge, CB4 3TB. T: 01223 369069, F: 01223 369069

Shibuya, S., Prof., 05/10/18, Shinohara-Nakamachi, Nada-Ku, Kobe 657, JAPAN. T: 81 78 8611826

Shields, J.G., Mr, 9 Buckstone Neuk, Edinburgh, Midlothian, EH10 6TU. T: 01236 822666

Shields, C.H., Mrs, 43 St. Lawrence Way, Bricket Wood, St. Albans, Hertfordshire, AL2 3XY. T: 0923 664182 WORK

Shilston, D., Mr, 3 Park Avenue, Staines, TW18 2E

Shires, R.H., Eur Ing, 12 Dimple Park, Egerton, Bolton, Lancashire, BL7 9QE

Shohet, D.C., Eur Ing, 9 Shiremeade, Elstree, Borehamwood, Hertfordshire, WD6 3JZ. T: 081 953 7539, F: 0044 3870341, E: d.shohet@virgin.net

Shrimpton, G.J., Mr, Trefewen, Pendarves Road, Camborne, Cornwall, TR14 7QF. T: 01209 714402

Shrubsall, A., Mr, Flat 1/B, Regency House, 22-26 Osnaburgh Street, London, NW1 3NB

Sidhu, H.K., Miss, Noble Denton & Associates, 131 Aldersgate Street, London, EC1A 4EB

Sills, G.C., Dr, Dept of Engineering Science, Oxford University, Parks Road, Oxford, OX1 3PJ

Simons, N.E., Prof., Eur Ing, Mullion House, 191 Sidney Road, Walton-on-Thames, Surrey, KT12 3SD. T: 01932 242415, F: 044 231179

Simpson, B., Eur Ing, 6 Coldfall Avenue, London, N10 1HS. T: 2088838227

Simpson, M.S., Miss, Holly House, 37 Suffield Close, Long Stratton, Norwich, NR15 2JL. T: 0798 9093149 (M)

Simpson, B.W., Mr, 38 Staffordshire Croft, Warfield, Bracknell, Berkshire, RG42 3HW

Simpson, D.P., Mr, 8 Hopgarth, Haxey, Doncaster, South Yorkshire, DN9 2QB

Sivakumar, V., Dr, Department of Civil Engineering, Queen's University of Belfast, College Park, Belfast, County Antrim, BT7 1NN. T: 01232 274009, F: 01232 663754

Sivasubramaniam, A., Mr, 23 Southbury, Lawn Road, Guildford, Surrey, GU2 4DD. T: 01483 571708

Skempton, A.W., Prof., Sir 16 The Boltons, London, SW10 9SU

Skinner, H.D., Miss, 7 Castano Court, Abbots Langley, Herts, WD5 0HP

Skinner, A.E., Mr, 56 Alfriston Road, Battersea, London, SW11 6NW. T: 2072284925

Skinner, P.J., Mr, Beech Tree Cottage, Cotswold Close, Staines, Middlesex, TW18 2DD

Skinner, R.W., Mr, 19 Trefoil Close, Hartley Wintney, Hook, Hampshire, RG27 8TS. T: 01252 843025

Skipp, B.O., Eur Ing Dr, 22 Amberley Crt, Angell Rd, London, SW9 7HL. T: 2072743028

Skulason, J., Mr, Almenna Verkfraedistofan, Fellsmula 26, 108 Reykjavik, ICELAND. T: 588 8100, F: 568 0284

Skuse, M.J., Mr, Ove Arup & Partners, 4 Pierhead Street, Cardiff, South Glamorgan, CF10 4QP

Sleight, A.F., Ms, Flat 10, Merrywood, Fortyfoot Road, Leatherhead, Surrey, KT22 8RN

Sloan, A., Mr, 30 Rhannan Road, Cathcart, Glasgow, G44 3BB. T: 0141 637 2767

Smalley, I.J., Dr, Tin Drum Bookstore, 68 Narborough Road, Leicester, LE3 0BR

Smart, P., Mr, Civil Engineering Department, Glasgow University, University Avenue, Glasgow, G12 8QQ

Smedley, M.I., Mr, 17 Seamons Road, Altrincham, Cheshire, WA14 4ND

Smethurst, J.A., Mr, Geo Group, Dept of Civil and Enviromental Engineering, University of Southampton, Southampton, SO17 1BJ

Smith, M.G., Dr, 122 New Road, Marlow Bottom, Buckinghamshire, SL7 3NW

Smith, P.R., Dr, Clematis, 23 Orchard Road, Barnack, Stamford, PE9 3DP. T: 01780 740 784, E: psmith@barnack.demon.co.uk

Smith, N.A., Eur Ing, 5 Ray Park Avenue, Maidenhead, Berkshire, SL6 8DP. T: 1628776665

Smith, A.J., Mr, 17 Parkland Court, Recreation Road, Colchester, CO1 2LB

Smith, A.K.C., Mr, 28 Boroughbridge Road, Knaresborough, HG5 0NJ. T: 01423 799033, F: 01423 799032

Smith, A.O., Mr, 17 Foxley Close, Lymm, Cheshire, WA13 0BS. T: 01925 752453, E: osmith@unsight-media.co.uk

Smith, C.C., Mr, Dept of Civil & Structural Eng., P.O. Box 600, Mappin Street, Sheffield, S1 4DT

Smith, D.M., Mr, 61 Foxhill Road, Reading, RG1 5QS. E: edge@edgeuk.u-netcom

Smith, M.V., Mr, 18 Nether Beacon, Lichfield, Staffs, WS13 7AT

Smith, P.G.C., Mr, 55 Chestnut Road, Raynes Park, London, SW20 8ED. T: 020 8542 6931

Smith, R.C., Mr, Southern Testing Laboratories, Keeble House, Stuart Way, East Grinstead, West Sussex, RH16 2SE

Smith, R.J.H., Mr, Elwood House, Cross Road, Albrighton, WV7 3RA

Smith, S.M., Mr, 22 Woodland Way, Crowborough, East Sussex, TN6 3BG

Smith, T.S., Mr, 10 Grenville Road, Pen-Y-Lan, Cardiff, CF2 5BP

Smith, H.L., Mrs, Southern Testing Labs Ltd., Keeble House, Stuart Way, East Grinstead, West Sussex, RH19 4QA

Smyth-Osbourne, K.R., Mr, 14 Sheridan Road, London, SW19 3HP. T: 0181 542 1639

Snedker, E.A., Mr, 19 Borrowdale Drive, Leamington Spa, Warwickshire, CV32 6NY. T: 4.40193E+12

Snelgrove, D.C., Mr, 11 Longview Close, Snettisham, Kings Lynn, PE31 7RD

So, E.K.M., Miss, 1 Vaughan Avenue, Hendon, London, NW4 4HT

Socarras, M., Dr, 27 Darlaston Road, London, SW19 4L. T: 020 8946 4992, F: 020 8946 5160, E: socarras@msn.com

Soga, K., Dr, Engineering Department, Cambridge University, Trumpington Street, Cambridge, CB2 1PZ. T: 01223 332 713, F: 01223 332 662

Soil Mechanics Ltd., Glossop House, Hogwood Lane, Finchampstead, Wokingham, RG40 4QW

Soils Ltd., Newton House, Cross Road, Tadworth, Surrey, KT20 5SR

Solera, S.A., Mr, 1 York Road, Colwyn Bay, Clwyd, LL29 7ED. T: 4.40149E+12

Southgate, G.J.T., Mr, Hedges, Park Road, Combs, Stowmarket, Suffolk, IP14 2JS. T: 01449 615507, E: greg@gjts.kene.co.uk

Spalton, C.D., Mr, 11 Springfield Close, Westham, Pevensey, East Sussex, BN24 5JF. T: 01825 75, F: 01825 761740, E: chriss@soil.co.uk

Sparks, A.D.W., Mr, 4 Orkney Street, Rondebosch 7700, SOUTH AFRICA

Sparks, R., Mr, Coed Bach, 10 Woodlands Way, Barton, Preston, Lancashire, PR3 5DU. T: 0802 755213 (M), F: 250163, E: robert.rsce@virgin.net

Spiesberger, G.A., Mr, 29 Crowlees Close, Cowlees Road, Mirfield, West Yorkshire, WF14 9JT

Spilman, H.D., Mr, 38 South Avenue, Stourbridge, West Midlands, DY8 3XY. T: 0384 392826

Spink, T.W., Mr, 25 St. Clair's Road, Parkhill, Croydon, Surrey, CR0 5NE

Spong, M., Mr, 99, Woodcote Road, Caversham, Reading, RG4 7EZ

Springman, S.M., Prof., Waisenhof, Bunghofstrasse 48, C.H.8157, Dielsdorf, SWITZERLAND. T: 4118540648

St John, H.D., Mr, Geotechnical Consultin Group, 1a Queensberry Place, London, SW7 2DL. T: 01923 449126 (H)

Stallebrass, S.E., Dr., Flat 15, 165 Cromwell Road, London, SW5 0SQ.

Standing, J.R., Mr, Downing College, Cambridge, Cambridgeshire, CB2 1DQ. T: 01223 321461 (H), 327551 (W)

Stansfield, L., Eur Ing, Hillsview, New Road, Forest Green, Dorking, Surrey, RH5 5SA. T: 01306 621371 (H)

Statham, I., Dr, 39 Heol-Y-Pentre, Pentyrch, Cardiff, South Glamorgan, CF4 8QD. T: 01222 890163

Stats Limited, Porters Wood House, Porters Wood, St. Albans, Hertfordshire, AL3 6PQ

Stead, D., Mr, Dept of Earth Sciences, Simon Fraser Univ., 8888 Univ., Drive, Burnaby, British Columbia, CANADA. T: 01209 714866, F: 01209 716977

Stearns, P.S., Mr, 13 Poles Hill, Chesham, Bucks, HP5 2QP. T: 01494 784345, E: sandyandpaul@hotmail.com

Steed, K.R.R., Mr, 19 Eastmead Lane, Stoke Bishop, Bristol, BS9 1HW

Steedman, R.S., Dr, 25 Eldon square, Reading, Berkshire, RG1 4DP. T: 0207 462 5336, F: 0207 462 5258, E: sstee@whitby-bird.co.uk

Steger, E.H., Mr, 16 Lingfield Road, Wimbledon Common, London., SW19

Steiner, W., Dr, Buhlenstrasse 8, Ch-3132 Riggisberg, SWITZERLAND. T: 41 31 8091741

Stenning, A.S., Dr, C/O W S Atkins China, 16th Floor, World Trade Centre, 280 Gloucester Road, Causeway Bay, HONG KONG

Stent Foundation Ltd., Osborn Way Industrial Estate, Osborn Way, Hook, Hampshire, RG27 9EX

Stephen, M., Mr, Dunvegan, Bagshot Road, Knaphill, Woking, Surrey, GU21 2SE. T: 1483475213

Stevenson, M.W., Mr, 17 Tainters Brook, Uckfield, East Sussex, TN22 1UQ. T: 01825 761476, E: monionstevenson@virgin.net

Stewart, D.I., Dr, 1-B Newlay Lane, Norsforth, Leeds, LS18 4LE

Stewart, W., Dr, Dept of Civil Engineering, Rankine Building, University of Glasgow, Oakfield Avenue, Glasgow, G12 8LT. T: 0141 330 5203, F: 0141 330 4557

Stidston, J.C., Miss, 39 Bloy Street, Bristol, BS5 6AX. T: 07773 700272 (M)

Stone, K.J.L., Dr, 119 Barrack Lane, Bognor Regis, PO21 4ED

Stork, R.F., Mr, 44 Crofton Road, Attenborough, Nottingham, Notts, NG9 5HW

Storr, T.R., Mr, 20 Milton Way, Ettiley Heath, Sandbach, CW11 3GJ

Straughton, M.G., Mr, 28 West Road, Midsomer Norton, Bath, BA3 2TR. T: 07879-478680

Stroud, M.A., Dr, 35 Queens Crescent, St. Albans, Herts., AL4 9QQ

Stuart, A.M., Mr, C/O 6/4 Inchgarvie Court, Ferry Road Drive, Edinburgh, Midlothian, EH4 4DA. T: 44 013 3322427, E: amstuart@globalnet.co.uk

Sture, S., Mr, Dept Civil Env. & Arch. Eng., Campus 428, Univ. of Colorado, Boulder, Colorado 80309

Suckling, T.P., Eur Ing, 2 Wear Road, Bicester, Oxfordshire, OX26 2FE. E: tony.suckling@talk21.com

Sun, H.W., Mr, Flat A, 19/F, Block 2, Scenecliff, 33 Conduit Road, Mid-Levels, HONG KONG. T: 00 852 25666878, F: 00 852 25666818

Surrey Geotechnical Consultants Ltd., Brunel University: Science Centre, Runnymede Campus, Englefield Green, Englefield Green, Egham, Surrey, TW20 0JZ

Surya, I., Mr, Jalan Oihapit No.11, Bandung, 40114 Indonesia

Sutherland, H.B., Prof., 22 Pendicle Road, Bearsden, Glasgow, Lanarkshire, G61 1DY. T: 1419425305, F: 9425305

Swain, A., Dr, 40 Parthia Close, Waterfield, Tadworth, Surrey, KT20 5LB

Swannell, N.G., Mr, 27 Home Close, Chiseldon, Swindon, Wiltshire, SN4 0ND. T: 44 1793 740105

Sweby, C.V., Mr, Sunyani, Send Marsh Road, send Marsh Ripley, Woking, Surrey, GU23 6JQ. T: 01483 225423

Sweeney, D.J., Mr, 1 Ridgeway, Epsom, Surrey, KT19 8LD. T: 1372720227, E: sweeney.snafu@virgin.net

Sweeney, M., Mr, Old Barn Cottage, 1 Rudds Lane, Haddenham, Aylesbury, Buckinghamshire, HP17 8JP. T: 1844292512

Sykes, P.R., Mr, 8 Woodlark Close, Wharton, Winsford, Cheshire, CW7 3HL. T: 01606 861548

Talbot, J.C.S., Eur Ing, Kilmuire, Oakdene Road, Godalming, Surrey, GU7 1QF

Tandy, S., Mr, 13 Arana, Murton Park, Arlecdon, Frizington, Cumbria, CA26 3U. T: 01946 861153

Tate, H.M., Mr, 25 St. Judes Road, Englefield Green, Egham, Surrey, TW20 0BY. T: 1784434894

Taunton, P.R., Dr, Rowan Cottage, Woolaston Common, Lydney, Gloucestershire, GL15 6NY. T: 01594 529455, E: paul.taunton@tesco.net

Taylor Woodrow Civil Engineering Ltd., 345 Ruislip Road, Southall, Middlesex, UB1 2QX

Taylor, I.K., Miss, Scott Wilson Kirkpatrick, 3 Pemberton House, Stafford Park, Telford, Shropshire, TF3 3BD

Taylor, A., Mr, Raeburn Drilling & Geotechnical Ltd., Raeburn Drilling & Geotechnical Ltd., Whistleberry Road, Hamilton, Lanarkshire, ML3 0HP

Taylor, D.I., Mr, 30 Saxon Road, Hoylake Wirral, Merseyside, L47 3AF

Taylor, E.H., Mr, Meadowside, Church Lane, Little Abington, CB1 6BQ

Taylor, G.R., Mr, Flat 22, Kingsmead Lodge, 50 Cedar Road, Sutton, Surrey, SM2 5AH. E: g.r.taylor@ic.ac.uk

Taylor, R.N., Prof., The City University, Northampton Square, London, EC1V 0HB. T: 020 7040 8157, F: 020 7040 8832

Tedd, P., Dr, West Leith Bungalow, West Leith, Tring, Hertfordshire, HP23 6JR

Teh, M.H., Mr, Wisma Ssp, Level 8-18, 1 Jln.Sr 8/3, Serdang Raya, Seksyen 8, 43300 Seri Kembangan, MALAYSIA. T: 39433366, F: 3.94327E+11

Terente, V.A., Mr, 10 Cherryburn Gardens, Newcastle Upon Tyne, NE4 9UQ

Terram Ltd., Mamhilad Park Estate, Pontypool, Gwent, NP4 0YR

Terrill, D.C., Mr, 50 Barrack Road, Christchurch, Dorset, BH23 1PF

Thames Water Utilities Ltd., Site Services, Ashford Common Works, Staines Road West, Ashford, TW15 1RU

Thomas, S.D., Dr, Eur Ing, Oxford Geotechnica International, Mountjoy Research Centre, University Science Park, Durham, DH1 3SW. T: 1913867319

Thomas, C.D., Mr, 2 Hereford Drive, Clitheroe, Lancashire, BB7 1JP. T: 44 1210 428184

Thomas, N.W.G., Mr, 110 Van Swietenstraat, SM 2518, Den Hagg, NETHERLANDS

Thomas, R.H., Mr, 23 Purfield Drive, Wargrave, Reading, RG10 8AP

Thomas, H.R., Prof., 4 Upper Cosmeston Farm, Penarth, South Glamorgan, CF64 5UB. T: 0222 874279 (W)

Thompson, P., Mr, Roseneath, Peat Common, Elstead, Godalming, Surrey, GU8 6DX

Thompson, P.A., Mr, 19/F Crystal Court Flat F, No. 4 Parkvale Village, Lantau Island, Discovery Bay, NT, HONG KONG

Thompson, R.P., Mr, 14 Kings Road, Wilmslow, Cheshire, SK9 5PZ

Thomson, F.M., Miss, Flat 10, Adams Court, 4 Luther King Close, London, E17 8RX. T: 01203 522 108

Thomson, G., Mr, 37 Lincoln Close, Woolston, Warrington, Cheshire, WA1 4LU

Thomson, G.H., Mr, 74 Eastwick Road, Walton-on-Thames, Surrey, KT12 5AR. T: 01932 848582, F: 01932 848582

Thorburn, J.Q., Mr, Strathdene, 62 Townhead Street, Strathaven, Lanarkshire, ML10 6DJ. T: 1357523409, F: 523409

Thorn, Mr, M.R.J., C/O Scott Wilson Kirkpatrick, 38th Floor, Metroplaza, 223 Hing Fong Road, Kwai Fong, HONG KONG

Thornton, C., Dr, Civil & Mechanical Engineering, School of Engr. & Applied Science, Aston Univ., Birmingham, West Midlands, B4 7ET

Threadgold, L., Mr, 57, Beverley Road, Leamington Spa, Warwickshire, CV32 6PW. T: 1926422177, F: 44 76694642, E: len@geotechnics.co.uk

Thurlow, P., Mr, Soldata Asia, 3/F Kowloon Centre, 29 Ashley Road, Tst, CHINA

To, W.J., Mr, 140 Wardrew Road, St Thomas, Exeter, EX4 1EY

Toll, D.G., Mr, Nanyang Technological University, University of Durham: School of Eng., South Road, Durham, DH1 3LE. T: (+44/0)191 374 2566, F: (+44/0)191 374 2550, E: d.g.toll@durham.ac.uk

Tomlin, L.J., Mr, 21 Rendells Meadow, Bovey Tracey, Newton Abbot, Devon, TQ13 9QW. T: 01626 836742

Tomlinson, M.J., Mr, Flat 17, Northumbria Court, 6 Sheen Road, Richmond, TW9 1AE. T: 0208 940 1432

Tonge, D., Mr, C/O The Library, Betchel Water Technology Limited, Chadwick House, Warrington, Risley, WA3 6AE. T: 01925 857333, F: 01925 857328

Tonks, D.M., Dr, 5 Hazelwood Road, Wilmslow, Cheshire, SK9 2QA. T: 1625533533, F: 0161 7987, E: edge@edgeuk.u-net.com

Toolan, F.E., Mr, Fugro Limited, 18 Frogmore Road, Hemel Hempstead, Hertfordshire, HP3 9RT. T: 01442 240781

Toye, M.G.E., Mr, Rosemead, 112 Junction Road, Norton, Stockton-on-Tees, TS20 1QB

Transport Research Laboratory, Crowthorne Enterprise Centre, Crowthorne Business Estate, Old Wokingham Road, Crowthorne, Berkshire, RG45 6AU

Treharne, G., Mr, Parc Castell-Y-Mynach, Creigiau, Cardiff, South Glamorgan, CF15 9NW

Trenter, N.A., Mr, 70 Park Road, Hampton Hill, Hampton, Middlesex, TW12 1HP

Trinder, S.K., Mrs, 26-A Warwards Lane, Birmingham, West Midlands, B29 7RB

Troughton, V.M., Mr, 70 Tudor Way, Uxbridge, Middlesex, UB10 9AB. T: 01895 235861

Tsatsanifos, C.P., Dr, 131 Xifissias Avenue, Athens, Gr-11524, GREECE. T: 30 1 6929484, F: 30 1 6928137

Tsui, K.L., Mr, 194 Tung Choi St., 1/f, Kowloon, HONG KONG. T: 23949388, F: 23818176

Turnbull, M.J., Mr, Kieowkoopseweg No 26, Nootdorp, 2631 PR NETHERLANDS

Turner, G.M., Mr, Millfield, Mill Lane, Boroughbridge, York, North Yorkshire, YO51 9LH. T: 01765 607 291 (WK)

Turner, M.J., Mr, 5 North End, Steeple Claydon, Buckingham., MK18 2NP

Twine, D., Mr, 149 Thorpedale Road, Stroud Green, London., N4 3BD

Tyler, A.J.H., Mr, Rose Cottage, Robin Hill Farm, Clay Lane, Chichester, Sussex, PO18 8AB

Tyrrell, A.P., Dr, 77 Lower Road, Fetcham, Leatherhead, Surrey, KT22 9HG. T: 01372 456400, F: 451123, E: andrew.tyrrell@wspgroup.com

Uglow, I.M., Mr, 2 Church Farm, Station Road, Congresbury, Bristol, BS19 5DX. F: 08 08 1997

Vaciago, G., Ing Studio Geotecnico Italiano, Via Ripamonti, 89, 20139 Milano, ITALY

Vadgama, N.J., Mr, 10 Strone Way, Off Boadmead Road, Hayes, UB4 9RU

Vandolas, V., Mr, 42 Vaou Street, Glyfada, Athens 16675, GREECE

Vanner, M.J., Mr, 'larch', 1 Blanford Road, Reigate, RH2 7DP

Vaughan, P.R., Prof., 101 Angel Street, Hadleigh, Ipswich, Suffolk, IP7 5DE. T: 1.47382E+11, F: 44 82492, E: prvaughan@care4free.net

Vavitsas, A., Mr, Flat 113, Churchill Court, Brydon Close, Salford, M6 5HB. T: 30944384501

Vaziri, M., Dr, 67 Beech Avenue, Ruislip, Middlesex, HA4 8UG

Vickery, K.W., Mr, Irving Cottage, 89 Main Road, High Wycombe, HP14 4RT. T: 01494 563977

Viggiani, C., Prof., Via Posillipo 281, 80123 Napoli, ITALY. T: 39 81 5750073, F: 39 81 5750073

Vincett, C., Mr, 35 Highcroft Crescent, Leamington Spa, CV32 6BN

Von Roon, S.O., Mr, Flat 9, Cardrew Court, Friern Park, London, N12 9LB. T: 01923 423100

Wade, C.M., Mr, Trolley Cottage, Venns Gate, Cheddar, Somerset, BS27 3LW. T: 0370 667 492 M

Wade, S.J., Mr, 117 Littleworth Road, Downley, High Wycombe, Buckinghamshire, HP13 5UZ. T: 01494 436339

Wagstaff, S.C., Mr, 46 Princess Diana Drive, St. Albans, AL4 0ED

Wakeling, T.R.M., Mr, 'springpark', 8 Devonshire Drive, Surbiton, KT6 5DR. T: 020 8398 4834

Walbancke, H.J., Dr, C/O Binnie Black & Veatch, Grosvenor House, 65 London Road, Redhill, Surrey, RH1 1LQ

Walker, A.D., Mr, 2323 Millgrove Road, Pittsburgh, PA 15241, USA. T: 412 8540286, F: 412 854 0238, E: yorkpud@ aol.com

Wallace, J.C., Mr, 7 Chiswick Court, Moss Lane, Pinner, Middlesex, HA5 3AP. T: 020 84293295 (H), 0207 6928606 (W)

Wallace, W.A., Mr, 6 Williamwood Park West, Netherlee, Glasgow, Lanarkshire, G44 3TE

Walton, G., Dr, Minster Cottage, Church Street, Charlbury, Chipping Norton, Oxfordshire, OX7 3PR. T: 01608 811096, E: gwpract@cix.compulink.co.uk

Warburton, C., Ms, 8 Bradbeers, Trull, Taunton, Somerset, TA3 7JQ. T: 01823 356026

Ward, H.A., Ms, Old Castletown South, Kildorrey, Co. Cork, REPUBLIC OF IRELAND. T: 021 4966 400 (W), E: hward@egpettit.ie

Wareham, B.F., Mr, 5 Oriel Hill, Camberley, Surrey, GU15 2JW. T: 127626072

Warland, P.F., Mr, Brookfield, Frogs Hall Lane, Woodgate, Swanton Morley, Dereham, Norfolk, NR20 4NU. T: 01362 638528 (T), F: 01362 638482

Warner, M.F., Mr, Amplus Ltd., Unit 1, Thistle Grove, St. Lawrence, Jersey, Channel Islands, JE3 1NN. T: 01534 863545, F: 01534 862474, E: amplus@psilink.co.je

Warren, C.D., Mr, 7 Willow Close, Woodham, Surrey, KT15 3S. T: 01932 344529

Warren, G., Mr, 38 South Western Road, St. Margarets, Middlesex, TW1 1LQ

Warwick, R.G., Mr, Healthwaite, Forest Moor Road, Knaresborough, North Yorkshire, HG5 8JP. T: 1423868770, E: warwicksmail@virgin.net

Watkins, S., Miss, 31 Hen Parc Avenue, Upper Killay, Swansea, SA2 7HA

Watson, P.C., Dr, 1 Paddock Villas, Gravesend Road, Higham, Rochester, Kent, ME3 7DP. T: 01474 824495, E: paul.watson@arup.com

Watson, P.D.J., Dr, 7 Church Close, Loughton, Essex, IG10 1LQ. E: p.d.watson@kingston.ac.uk

Watson, N.I., Mr, 177 Queens Road, Penkhull, Stoke-on-Trent, ST4 7LF

Watson, C., Ms, Flat 4, Houghton, Rookery Road, Staines, Middlesex, TW18 1BT

Watts, G.R.A., Mr, Ground Engineering, Transport Research Laboratory, Crowthorne, Berkshire, RG45 6AU. T: W44 01344 770367/F770648

Watts, K.S., Mr, 'bluepine', 19 Harthall Lane, Kings Langley, Hertfordshire, WD4 8JW. T: 44 01923 263515, E: watts_bluepine@connectfree.co.uk

Webb, S.A., Miss, 19 Old Farm Road, Guildford, GU1 1QN

Webber, I.P., Mr, 22 Tewit Well Road, Harrogate, North Yorkshire, HG2 8JE. T: 01423 561 543

Weeks Group plc, The Oasts, Newnham Court, Maidstone, Kent, ME14 5LH

Weeks, A.G., Dr, Compton Rise, 13 Faraday Road, Maidstone, Kent, ME14 2DB. T: 01622 758983

Weeks, R.C., Mr, Tall Pines, 27 High Street, Barford, Warwick, Warwickshire, CV35 8BU. T: 44 01926 624571, E: Robertcweeks@aol.com

Weltman, A.J., Mr, The Georgian House, Bellyard, Painswick, Gloucester., GL6 6XH

West, A.J., Miss, 22 The Ridgeway, Nettlebed, Henley-on-Thames, Oxfordshire, RG9 5AN

West, S.L., Mr, 15 Braeside Road, West Moors, Ferndown, Dorset, BH22 0JS

Weston, G.R., Mr, 37 Commercial Road, Eastbourne, East Sussex, BN21 3XF

Whapples, C.W., Mr, Hill Cannon Partnership, Royal Chambers, Station Parade, Harrogate, N Yorks, HG1 1EP

Wharmby, N.J., Mr, Bachy Soleanche Ltd., Henderson House, Langley Place, Higgin Lane, Burscough, Wigan, Lancashire, WN8 7DA

Wheeler, A.B.S., Mr, 79 Hargreave Terrace, Darlington, DL1 5LF

Wheeler, M., Mr, 5 Pinfold Court, Chester, CH4 7ES

Wheeler, M.S., Mr, 9 Bourne Close, Broxbourne, Hertfordshire, EN10 7NE

Wheeler, S.D., Mr, 70 Gainsborough Court, Walton-on-Thames, KT12 1NL

Wheeler, S.J., Prof., 51 Newark Drive, Glasgow, Lanarkshire, G41 4QA. T: 0141 330 5202 (D)

White, D., Mr, Churchill College, Cambridge, Cambridgeshire, CB3 0DS

White, J.A., Mr, 95 Crouch Street, Noak Bridge, Basildon, Essex, SS15 4AU. T: 01268 289309

White, J.K., Mr, 17 Ravenscroft Park, Barnet, Hertfordshire, EN5 4ND

White, T.P., Mr, 28 Harrington Hill, London, E5 9EY. T: 020 8806 9973

Whitesmith, W.J., Mr, 4 Glasshouse Lane, Kenilworth, Warwickshire, CV8 2AJ. T: 01926 859465, F: 01926 851113

Whitlam, R.I., Mr, Fairfax House, 38 The Grove, Ilkley, West Yorkshire, LS29 9EE. T: 01943 600009, F: 01943 816666

Whittington, K.A., Mr, 2 Church Road, Dover, Kent, CT17 9LW. T: 01304 240502, F: 01304 203898

Whittle, R.W., Mr, Harlyn, Drury Lane, Ridgewell, Halstead, Essex, CO9 4SL

Whitworth, L.J., Mr, 95 Geraldine Road, London, SW18 2N. T: 01284 725103

Whyley, P.J., Mr, 21 Kirkby Avenue, Sale, M33 3EP

Wickson, J.L., Mr, Goosey Cottage, Boot Lane, Dinton, Aylesbury, Buckinghamshire, HP17 8UJ. T: 01296 748816, E: john-w@dial.pipex.com

Wight, P., Mr, Coetmor, 24 Secontium Road South, Caernarfon, Gwynedd, LL55 2LL

Wightman, R.M., Miss, 10-A Mandalay Road, London, SW4 9ED

Wilcox, T., Mr, 8 Rosecomb Place, Shenley Brook End, Milton Keynes, Buckinghamshire, MK5 7HU. T: 01908 520992

Wilkinson, D.L., Mr, 5 The Willows, Bishop Auckland, County Durham, DL14 7HH. T: 1388661475

Wilkinson, G.D., Mr, 31 Park Green, Bookham, Leatherhead, Surrey, KT23 3NL. T: 1372454552

Wilkinson, P.D., Mr, 90 Mousehold Street, Norwich, NR3 1NX

Wilkinson, R.D., Mr, Les Nouradons, 13122 Ventabren, FRANCE

Wilks, J.A., Miss, Forest Holme, Broadwell, Coleford, Gloucestershire, GL16 7BN

Williams, S.G.O., Dr, 240 Clay Hill, Wiggington Bottom, Tring, Herts, HP23 6H. T: 01442 824390

Williams, H.M., Eur Ing, 8 Woodland Crescent, Abercynon, Mountain Ash, Mid Glamorgan, CF45 4UT. T: 1443740539, F: 740539, E: hmasters-williams@yahoo.co.uk

Williams, R.E., Mr, 182 Old Lodge Lane, Purley, Surrey, CR8 4AL. T: 020 8668 8892

Wilson, C.D.V., Dr, Dept of Geology, The University, Liverpool, L69 3BX

Wilson, J., Dr, Clover Cottage, Staplehurst Road, Carshalton Beeches, SM5 3JX. T: 081 642 6944

Wilson, R.L., Eur Ing, The Grove House, Little Bognor, Pulborough, West Sussex, RH20 1JT. T: 1798865569, F: 44 865672, E: rlwconsult@compuserve.com

Wilson, D.D., Mr, 32 Molyneux Street, London, W1H 5HW. T: 020 7723 3942, F: 020 7402 3969

Wilson, E.J., Mr, 36 Brunswick Road, Gloucester, Gloucestershire, GL1 1JJ. T: 01452 422843, F: 01452 304612

Wilson, G.C.G., Mr, Babtie Group, Sandling Block, Springfield, Maidstone, Kent, ME14 2TG. T: 01622 678808, F: 1622690547

Wilyman, M.F., Mr, 72 Riverside Park, Otley, LS21 2RW. T: 01942 824406

Winchester, A.J., Mr, Russell Way, Sutton, Surrey, SM1 2SP

Windle, J.C., Miss, 69 Dryden Court, Renfrew Road, London, SE11 4NH. T: 2077930731

Withers, A., Mr, 10 Middle Road, Barnet, Hertfordshire, EN4 8TE. T: 020 8441 4864, E: a.withers@virgin.net

Withiam, J.L., Dr, D'Appolonia, 275 Center Road, Monroeville P.A., 15146-1451 USA. T: 412 856 9440

Wood, C.E.J., Mr, The Bungalow, Hook Road, Epsom, Surrey, KT19 8TU. T: 01372 720152

Wood, D.C., Mr, Unit 17D-5-2, Casavista Condominiums, Jalan Kapas, Bangsar, Kuala Lumpur, 59100 MALAYSIA. T: 00603 2284 5714

Wood, H.J., Mr, 158 Kennington Road, London, SE11 6QR. T: 07747 795483, F: 0207 654 0591 (W)

Wood, J.D., Mr, Glanville & Associates, Porters Wood House, Porters Wood, St. Albans, AL3 6PQ

Wood, J.M., Mr, 11 Shoreham Avenue, Rotherham, South Yorkshire, S60 3DB. T: 1709363314

Wood, L.A., Prof., 33 Church Street, Isleworth, Middlesex, TW7 6BE. T: 2088470681

Woodland, A.R., Mr, Mayfield, Middleton Tyas, Richmond, North Yorkshire, DL10 6PG

Woodrow, L.K.R., Dr, C/O Civil Engineering Dept., 14Th Fl., Civil Engineering Bld., 101 Princess Margaret Road, Homantin

Woods, D., Mr, 30 Kenilworth Court, Coventry, West Midlands, CV3 6HZ. T: 2476501060, E: dickon@classicfin.net

Woods, E.S., Mr, 20 Downs Way, Great Bookham, Surrey, KT23 4LN. T: 0372 457406

Woods, K.G., Mr, Craig Cottage, Crown Terrace, Llanfairtalhaiarn, Abergele, Clwyd, LL22 8RY

Woods, R.I., Mr, University of Surrey, Department of Civil Engineering, Guildford, Surrey, GU2 5XH

Woodward, J.C., Mr, Redroofs, Upper Icknield Way, Whiteleaf, Princes Risborough, Buckinghamshire, HP27 0LY. T: 1844345562, F: 345362

Wooster, S.G., Mr, 17 Mill Road, Countess Wear, Exeter, EX2 6LH. T: 01392 271985

Worrall, P.K., Mr, Sunnyside, Wellington Road, Muxton, Telford, Shropshire, TF2 8NX

Wrightman, J., Mr, 20 Clare Road, Braintree, Essex, CM7 2PA

Wrigley, W., Mr, How Tallon, Barningham, Co. Durham, DL11 7DU. E: WRIGLEYSW@freeserve.co.uk

Wu, P.C.M., Mr, Flat B, 3/F1,Block D1,Tang Court, 168 Nga Tsin Wai Road, Kowloon, HONG KONG. T: 852 2493 7932, F: 852 2712 6357, E: paul_cm_wu@compuserve.com

Wynne, C.P., Eur Ing, 2 Hale Farm Cottage, Hale Farm, Bridgwater Road, Winscombe, Norrth Somerset, BS25 1NN. T: 01934 842420

Yaji, R., Dr, Karnataka Regional Engr. College, Surathkal, P O Srinivasnagar 574 157 (DK), Karnataka, INDIA. T: 91 0824 475984, F: 91 0824 476090

Yap, T.Y., Mr, Room 2.58, School of Engineering, University of Manchester, Oxford Road, Manchester, Lancashire, M13 9PL

Yeandle, N.W., Mr, 26 Belle Vue Road, Exmouth, Devon, EX8 3DP. T: 01395 276241

Yenn, R.J., Mr, 8 Wordsworth Rise, East Grinstead, West Sussex, RH19 1TW

Yilmaz, M., Mr, Hillside Lodge, Main Road, Holmesfield, Sheffield, South Yorkshire, S18 5WT. T: 0044 114 2899142

Yim, W.K., Mr, Flat C, 3/F, Fu Cheong House, 24-26 Tai Wai Road, Shatin, N.T., HONG KONG

Youdan, D.G., Eur Ing, Fugro Ltd., 18 Frogmore Road, Hemel Hempstead, Herts, HP3 9RT. T: 01442 240781, F: 01442 258961

Younan, N.A., Mr, 7 Adly Yeken, Pacha Street, Mazloom, Alexandria, EGYPT

Young, D.K., Mr, Kirkstyle, South Park, Hexham, Northumberland, NE46 1BT. T: 1434606278

Young, N.R., Mr, 10 Kingsmead, South Nutfield, Redhill, Surrey, RH1 5NN. T: 020 7681 5383

Young, R.A., Mr, C/O Ove Arup, New Oxford House, 30 Barkers Pool, Sheffield, South Yorkshire, S1 2HB

Yu, H.S., Prof., University of Nottingham, University Park, Nottingham, NG7 2RD

Zacharopoulos, A., Mr, Imperial College, Civil Engineering Dept, Exhibition Road, London, SW7 2BU. T: 0171 589 5111

Zampiras, E., Mr, Miltiadou 7-9, Holargos, Athens, 15562 GREECE

Zdravkovic, L., Dr, Imperial College, Dept of Civil Engineering, London, SW7 2BU. T: 0171 5946069, F: 0171 5946053

Zytynski, M., Mr, MZ Associates Ltd., 1 High Street, Little Eversden, Cambridge, Cambridgeshire, CB3 7HE

UNITED STATES OF AMERICA/ETATS-UNIS D'AMERIQUE

Secretary: Robert D. Holtz, Dept of Civil Engineering, University of Washington, Box 352700, Seattle, WA 98195-2700, USA. T: (206)543-7614, F: (206)685-3836, E: holtz@u.washington.edu

Total number of members: 2901

Abdelghaffar, M.E.M., Suez Canal University, 1 26 July St Apt 705, Cairo, 11111 EGYPT. T: 1.12E+11, F: 011-206-6-23186, E: magdy20@menanet.net

Abdel-Salam, S.S., Zagazig Univ, 47 Goal Gamal St, Agoza, Giza, EGYPT., F: 2023455904, E: s.salam@link.com.eg

Abdrabbo, F.M., Alexandria Univ, Civil Engr Dept Faculty of Eng, Alexandria, 21544 EGYPT. T: 2035 432480, F: 2035440386, E: frcu:iny%"fathi@alex.eun.eg"

Abdul Ghani, A.N.B., Univ Sains Malaysia, 1293 Permatang Ara, Sungai Bakap, Sps, 14200 MALAYSIA. T: 04-5824742, F: 04-6576523, E: anaser@usm.my

Abernathy, D.T., 7577 Quiet Cove Circle, Huntington Beach, CA 92648. T: 714-596-5871

Aboaziza, A.H., Woodward Clyde Cons, 12880 Sundance Ave, San Diego, CA 92129

Abramson, L.W., Hatch Mott Macdonald, 11661 Se First St Suite 205, Bellevue, WA 98005. T: 425-454-9608, F: 425-454-9613, E: labramson@hatchmott.com

Acree, R.G., Nodarse & Assoc, 123 North Orchard Street, Suite 1-B, Ormond Beach, FL 32174., F: 904-673-8357, E: ricka@nodarse.com

Adam, J.M., Soils Eng Serv Inc., 13 The Fairway, Montclair, NJ 7043. T: 973-744-6090, F: 9738089099, E: ja@sesi.org

Adams, Jr, W.R., Pennsylvania State of, 394 Burton Ave, Washington, PA 15301. T: 412-225-8305, F: 412-429-5069, E: bsgadams@telerama.lm.com

Adams, M.W., 606 Marshall, #21, Houston, TX 77006. T: 713-522-5248, F: 281-230-6739, E: madams@gseworld.com

Adams, M.A., Gse Lining Technology, 19103 Gundle Rd., Houston, TX 77073. T: 281-230-8690, F: 281-230-6704, E: meadams@gseworld.com

Addison, M.B., 3235 Kenilworth Dr., Arlington, TX 76001. T: 817-557-6464, F: 817-467-5535, E: mbaddison@home.com

Adelsohn, L., Frank H Lehr Assocs, 101 S Harrison St, East Orange, NJ 7018. T: 973-673-2520

Adkins, Jr, H.G., US Dept of Agriculture, 408 Winding Way, Columbia, SC 29212. T: 803-781-5249, F: 803-253-3670, E: grady.adkins@sc.usda.gov

Admay, T.V., 4700 B. Kerley Rd., Durham, NC 27705. T: 919-403-2865, F: 919-544-0810

Affi, L.A., Soil Structure Interface Cons, 1769 W Algonquin Rd., Mt Prospect, IL 60056. T: 847-427-0999, F: 847-690-0189, E: liibanaffi@aol.com

Aggour, M.S., Univ of Maryland, Dept of Civil Eng, College Park, MD 20742. T: 301-405-10942, F: 301-405-2585, E: msaggour@eng.umd.edu

Agrawal, G., BSK & Assocs, 1181 Quarry Ln Bldg 300, Pleasanton, CA 94566. T: 925-462-4000, F: 925-462-6283, E: girish_agrawal@hotmail.com

Aguirre, V.E., Aguirre Engrs Inc., P.O. Box 3814, Englewood, CO 80155. T: 303-799-8378

Ahmad, S., Progressive Cons Pvt Ltd., 25-D/1 Gulberg Iii, Lahore 54660, PAKISTAN

Ahmed, I., GEI Consultants, Inc., 1021 Main Street, Winchester, MA 1890. T: 781-721-4028, F: 781-721-4073, E: iahmed@geiconsultants.com

Akagi, T., Toyo Univ, Nishiki 1-22-9 Nerimaku, Tokyo, 179-0082 JAPAN. T: 81-3-3931-5931, F: 81-492-314482, E: akagi@eng.toyo.ac.jp

Akil, C.H., Edwards & Kelcey Inc., 299 Madison Ave, P.O. Box 936, Morristown, NJ 7962. T: 973-267-8830- 1243, F: 973-267-3555, E: cakil@ekmail.com

Akomolede, O.G., Ultratechnic Services Inc., 5803 Glenbrook, Mason, OH 45040. T: 513-755-0347, F: 513-755-8923, E: akomolede@aol.com

Al-Ali Al-Tearkawi, Z.H., Mushrif Trad Contracting Co, P.O. Box 32147, Rumeithiyah, 25552 KUWAIT. T: 965 5721798, F: -4743067

Alarcon, A., Facultad De Ingeniera, Calle 125 No 41-26, Bogota, COLOMBIA. T: 571-624-3053

Alber, G.C., Greg Alber Pe, 16009 Normandy Avenue, Cleveland, OH 44111. T/F: 216-251-5444, E: gregalber@hotmail.com

Albus, D.E., Albus-Keefe & Associates, 1403 N Batavia St, Suite 115, Orange, CA 92867. T: 714-744-9760, F: 714-744-9750, E: dalbus@earthlink.com

Aldinger, P.B., Paul B Aldinger & Assocs, 115 Laurel Ave, Providence, RI 2906. T: 401-521-7594, F: 401-435-5569, E: pba369@aol.com

Aldrich, C., Aldrich Geotechnical Inc., 906 E Palm Dr., Glendora, CA 91741. T: 818-335-3229

Alexander, J.L., Froehling & Robertson, 2806 N Graham St, Charlotte, NC 28206. T: 704-376-1596, F: 704-343-9026, E: frcharlott@aol.com

Alghazzawi, M.M., Bucknell University, Civil Engineering Department, Lewisburg, PA 17837. T: 570- 577-1683, F: 570-577-3415, E: mmurad@bucknell.edu

Al-Homoud, A.S., American Univ of Shj, School of Engineering, P.O. Box 26666, Sharjah, UNITED ARAB EMIRATES. T: 011-971-6-5055901, F: 011-971-6-50559, E: ahomoud@aus.ac.ae

Ali, R., Amec Earth & Environmental Inc., 14662 Nw Benny Drive, Portland, OR 97229. T: 503-466-4930, F: 503-620-7892

Allen, J.D., BP Amoco, 12222 Cabo Blanco, Houston, TX 77041. T: 832-467-4685

Allen, R.L., Ninyo & Moore, 953 N Capistrano Pl, Orange, CA 92869. T: 714-639-6559, F: 949-472-5445, E: lallen@ninyoandmoore.com

Allman, C.M., Geotechnical Specialities Inc., 804 Westminster Lane, Virginia Beach, VA 23454. T: 757-498-6304, F: 757-461-1436, E: maryce@aol.com

Alperstein, R., RA Consultants, 23 Langdale Rd., Wayne, NJ 7470. T: 973-942-3007, F: 973-942-0195, E: raconsultants@mindspring.com

Al-Shafei, K.A., Saudi Aramco, P.O. Box 10155, Dharan, 31311 SAUDI ARABIA. T: 011-966-3-874-6146, F: 011-966-3-873-1, E: shafeika@aramco.com.sa

Alshawabkeh, A.N., Northeastern Univ, Dept of Civil & Envir Eng., 420 Snell Eng Ctr, Boston, MA 2115. T: 617-373-3994, F: 617-373-4419, E: aalsha@neu.edu

Altschaeffl, A.G., Purdue Univ, Civil Eng Bldg, West Lafayette, IN 47907. T: 317-494-5028, E: altsch@ecn.purdue.edu

Alumbaugh, D.L., Univ of Wisconsin, 2258 Engineering Hall, 1415 Engineering Dr., Madison, WI 53705. T: 608-262-3835, F: 608-263-2453, E: alumbaugh@engr.wisc.edu

Alvarez, T.A., Horizon Consultants, 11 Nagle Ave., c/o Asc-3257, New York, NY 10040. T: 809-547-1152, F: 809-547-1052, E: talvarez@hotmail.com

Amadei, B., Univ of Colorado, 428 Ucb, Boulder, CO 80309., T: 303-492-7317, E: amadei@spot.colorado.edu

Amaning, K.O., Geotech Consultants Int'l Inc., 2290 North County Road 427, Suite 100, Longwood, FL 32750. T: 407-331-6332, F: 407-331-9066, E: owusua@aol.com

Ampollini, M., Spearin Preston & Burrows Inc., 1936 Tomlinson Ave, Bronx, NY 10461. T: 718-823-4318, F: 212-629-0582

Andersen, G.R., Burns Cooley Dennis, Inc., 109 Quail Run, Madison, MS 39110. T: 601-856-4911, F: 601-856-9774, E: glen_andersen@bcdgeo.com

Andersen, L.D., Ch2m Hill, P.O. Box 635, Chesterton, IN 46304. T: 219-926-8892, F: 972-385-0846, E: landerse@ch2m.com

Anderson, P.E., Richard O., Somat Engineering Inc., 50029 Joy Rd., Canton, MI 48187. T: 734-455-0029, F: 734-946-1147, E: roape1@aol.com

Anderson, B.K., 1053 Apollo Beach Blvd, Apt 2, Apollo Beach, FL 33572. T: 813-645-8197

Anderson, D.M., Electrical Density, 2701 Conestoga Dr., #127, Carson City, NV 89706. T/F: 775-883-1518, E: arai01@msu.com

Anderson, D.G., Ch2m Hill, 5204 137th Pl Se, Bellevue, WA 98006. T: 425-644-7852, F: 425-462-3100, E: danderso@ch2m.com

Anderson, E.K., Anderson Eng Consultants Inc., 10205 Rockwood Rd., Little Rock, AR 72204. T: 501-455-4545

Anderson, J.E., Ch2m Hill, 155 Grand Avenue, Suite 1000, Oakland, CA 94604. T: 510-251-2426, F: 510-622-9168, E: janders1@ch2m.com

Anderson, K., Fmsm Engineers, 1409 N Forbes Rd., Lexington, KY 40511. T: 606-422-3000 x3073, F: 606-422-3100, E: kanderson@fmsmengineers.com

Anderson, L.R., Utah State Univ, Utah State University, 1053 Fox Farm Rd., Logan, UT 84321. T: 801-753-5119, F: 435-797-1185, E: loren@lab.cee.usu.edu

Anderson, N.O., Neil Anderson & Assoc Inc., 22 N Houston Ln, Lodi, CA 95240. T: 209-367-3701, F: 209-333-8303

Anderson, S.A., URS Greiner Woodward Clyde, 4582 South Ulster St, Stanford Place 3 Ste 1000, Denver, CO 80237. T: 303-740-2666, F: 303-694-3946, E: scott_a_anderson@urscorp.com

Anderson, S.T., Geosystems Engineering Inc., 2404 Louisiana St, Lawrence, KS 66046. T: 785-842-6030, F: 913-962-0924, E: s_anderson@geosystemseng.com

Anderson, T.C., Schnabel Foundation Co, 818 Spring Park Loop, Celebration, FL 34747. T: 407-566-0803, F: 8476 392347, E: tom@schnabel.com

Andonyadi, M., ECS Ltd., 3104 Ashburton Avenue, Herndon, VA 20171. T: 703-264-3331

Andrews, D.C., Gannett Fleming Inc., Foster Plaza Iii – Suite 200, 601 Holiday Dr., Pittsburgh, PA 15220. T: 412-922-5575, F: 412-922-3717, E: dandrews@gfnet.com

Andromalos, K.B., Geocon Inc., 3919 Laurel Oak Cir, Murrysville, PA 15668. T: 724-325-1626, F: 412 373 3357, E: kandromalos@geocon.net

Andrus, R.D., Dept of Civil Engineering, Clemson University, 318 Lowry Hall Box 340911, Clemson, SC 29634. T: 864-656-0488, F: 864-656-2670, E: randrus@clemson.edu

Angulo, M., Harza Engineering Company, 233 S Wacker Dr., Chicago, IL 60606. T: 312-831-3862, F: 312-831-3999, E: mangulo@harza.com

Antes, D.R., Paulus Sokolowski & Sartor Inc., 5339 Lindley Ave #102, Tarzana, CA 91356. T: 818-345-7428, F: 818-881-4187, E: daveantes@erols.com

Apoldo, L.J., Seashore Associates, 116 Kingsland Circle, Monmouth Jct, NJ 8852. T: 908-329-8920, F: 732-280-3916, E: lapoldo@idt.net

Arango-G, I., Bechtel Corporation, 22 Bowling Dr., Oakland, CA 94618. T: 510-653-8948, F: 415-768-4955, E: iarango@bechtel.com

Arditti, H., Terra Insurance Co, 2 Fifer Ave, Corte Madera, CA 94925. T: 415-927-2901, F: 415-927-3204

Arellano, D., 1154 Pomona Drive, Champaign, IL 61821. T: 217-355-6023, E: darellan@uiuc.edu

Arman, A., GEC Inc., Gulf Engr & Consultant, 1148 Verdun Dr., Baton Rouge, LA 70810. T: 225-766-3750, F: 225-612-3022, E: aarman@gecinc.com

Armstrong, R.M., Geotechnical Engineer, P.O. Box 125, Savoy, IL 61874. T: 217-351-8712, F: 217-351-8700

Arndt, S.M., Sarndt Engineering, P.O. Box 1117, Palmer Lake, CO 80133. T: 719-668-4054, F: 719-668-3990, E: stevedonnajulia@earthlink.net

Arora, S.S., Arora & Assocs Pc, 3120 Princeton Pike, 3rd Floor, Lawrenceville, NJ 8648. T: 609-844-1111, F: 609-844-9799, E: sarora@arorapc.com

Arrigoni, E.L., Rock Soil, V Le Legioni Romane 44, Milano Mi, 20147 ITALY. T: 39 2 4079381, F: 39-2-6571, E: rocksoil@tin.it

Arroyo, J.R., Univ of Puerto Rico, Mayaguez Campus, General Eng Dept P.O. Box 9044, Mayaguez, PR 681. T: 787-832-4040 X3789, F: 787-265-3816, E: jose@ce.uprm.edu

Arulmoli, K., Earth Mechanics Inc., 17660 Newhope St Ste E, Fountain Valley, CA 92708. T: 714-751-3826, F: 714-751-3928, E: arulmoli@earthmech.com

Arulnathan, R., URS, 804 Hancock St, Hayward, CA 94544. T: 510-247-3349, F: 510-874-3268, E: rajendram_arulnathan@urscorp.com

Arze, Jr, E., Arze Recine Y Asociados Ings, Jose Pedro Alessandri 1495, Santiago, CHILE. T: 2388018

Arzi, A.A., 80 Krinizi Street, Suite 23, Ramat-Gan, ISRAEL. T: 011-972-3-5325059

Aschenbrenner, R.D., 2395 West Shore Drive, Lummi Island, WA 98262. T/F: 360-758-2371

Ascoli, R.G., AM Engineering and Testing Inc., 1586 Packwood Road, Juno Beach, FL 33408. T: 561-776-7546, F: 561-745-0981, E: amengineering@earthlink.net

Ashford, S.A., UC San Diego, Mail Code 0085, 9500 Gilman Di, La Jolla, CA 92093. T: 858-822-0431, F: 858-822-2260, E: sashford@ucsd.edu

Ashmawy, A.K., Univ of South Florida, Civil & Env Eng Enb118, 4202 E Fowler Ave, Tampa, FL 33620. T: 813-974-5598, F: 813-974-2957, E: ashmawy@eng.usf.edu

Asproudas, S.A., S Asproudas & Assocs, 82 Plastira St, Nea Smirni, Athens, 17121 GREECE. T: 01-932-2021, F: 01-935-9270, E: asper@tee.gr

Atalar, C., Near East University, Civil Engineering Dept, Dikmen Rd., Pk 670, Lefkosa Trnc, Mersin, 10 TURKEY. T: 90-392-2236464, F: 90-392-2236461, E: cavitatalar@hotmail.com

Atkins, P.V., Halmar Builders of Ny Inc., 160 W Lincoln Ave, Mount Vernon, NY 10550. T: 914-668-9500, F: 914-668-9414, E: paula@halmarbuilders.com

Aubeny, C.P., Texas A&M University, 2112 Pantera Dr., Bryan, TX 77807. T: 979-775-6071, F: 979-845-6554, E: caubeny@civilmail.tamu.edu

Aubry, K., Dunkelberger Engineering, 1175 Summerwood Circle, West Palm Bch, FL 33414. T: 407-790-0896, F: 561-689-5955

Audibert, J.M.E., Fugro Mc Clelland, 17407 Sugar Pine Dr., Houston, TX 77090. T: 281-444-7206, F: 713-778-5573, E: jaudibert@fugro.com

Augello, A.J., Golder Associates Inc., 10 Chrysler, Suite B, Irvine, CA 92618. T: 949-583-2700, F: 949-583-2770, E: aaugello@golder.com

Austin, J.S., 108 Haviland Rd., Chesapeake, VA 23320. T: 757- 421-0138, F: 757-432-9014, E: jaustin@members.asce.org

Axelrod, P.A., Axelrod Inc., 5732 E Marconi Ave, Scottsdale, AZ 85254. T: 602-494-8293, F: 602-494-8306, E: paulaxel@aol.com

Axten, G.W., American Geotech, 22725 Old Canal Road, Yorba Linda, CA 92887. T: 714-685-3900, F: 714-685-3909

Azzouz, A.S., 11 Colonial Rd., Apt 4, Milford, MA 1757. T: 508-478-5108, E: azzouz@thewayout.net

Babbitt, D.H., Genterra Cons, 3860 W Land Park Dr., Sacramento, CA 95822. T: 916-442-0990, F: 916-979-0786, E: donbabbitt@msn.com

Bachner, J.P., Asfe, 8811 Colesville Rd., Suite G106, Silver Spring, MD 20910. T: 301-565-2733, F: 301-589-2017, E: asfe@aol.com

Bachus, R.C., Geosyntec Consultants, 1100 Lake Hearn Drive, Suite 200, Atlanta, GA 30342. T: 404-705-9500, F: 404-705-9400, E: bobb@geosyntec.com

Badu-Tweneboah, K., Geosyntec Consultants, 4774 Willow View Court, Liberty Township, OH 45011. T: 513-894-6973, F: 513-648-3415, E: kwasib@geosyntec.com

Baez, J.I., Hayward Baker Inc., 2617 Military Ave, Los Angeles, CA 90054. T: 805-933-1331, F: 805 933 1338, E: jibaez@haywardbaker.com

Bagahi, K.H., 19 Portica, Newport Coast, CA 92657. T: 949-252-8292, F: 714-832-0825

Baglioni, V.P., Boseaux Inc., 7522 Summer Trail Dr., Sugar Land, TX 77479. T: 713-343-9953

Baha, A.S., Geotechnical Solutions Inc., 17935 Sky Park Circle Drive, Suite F., Irvine, CA 92614. T: 949-261-8328, F: 949-261-0449, E: asbaha@geotechnicalsolutions.com

Bahmanyar, G.H., Terra Technology & Instrumentation, P.O. Box 261473, Encino, CA 91426. T: 818-700-1122, F: 818-700-0500, E: bahmanyar@aol.com

Bailey, Jr, M.M., US Army Corps of Engrs, 3985 Cranbrook Ct, Lilburn, GA 30047. T: 770-921-1595, F: 770-919-5828

Bailey, B., Fugro South, Inc., 2880 Virgo Lane, Dallas, TX 75229. T: 972-484-8301, F: 972-620-7328, E: bbailey@ fugro.com

Bailey, P.F., New York State of, 14 Smith Ave, Colonie, NY 12205. T: 518-459-9386, F: 518-457-8080, E: paulb@ global2000.net

Bailey, R.J., Gibson Grouting Service, 1817 Chatham, Charlotte, NC 28205., F: 803-980-1060, E: gibsongrouting@ aol.com

Bajuniemi, R.L., Harza Engineering Co Inc., 425 Roland Way, Oakland, CA 94621. T: 510-568-4001, F: 510-636-2177, E: rbajuniemi@harza.com

Baker, Jr, C.N., STS Cons Ltd., 750 Corp Woods Pkwy, Vernon Hills, IL 60061. T: 847-279-2471, F: 847-279-2510, E: baker@stslto.com

Baker, J.A., Geotechnolgy Inc., 322 Meadow Pl Ct, Saint Charles, MO 63303. T: 314-926-3032, F: 314-241-3526, E: jabaker@i1.net

Baker, T., Amoco Fabrics & Fibers, 260 The Bluffs, Austell, GA 30168. T: 770-944-4718, F: 770-944- 4745, E: bakertl@ bp.com

Baker, V.A., Taber Consultants, 1077 N Gate Rd., Walnut Creek, CA 94598., F: 925-926-1058, E: vabaker@pacbel l.net

Bakhtar, K., Bakhtar Assocs, 3420 Via Oporto, Suite 201, Newport Beach, CA 92663. T: 949-675-2800, F: 949-675-2233, E: kbakhtar@aol.com

Baladi, G.Y., Dtra/Cpt, 11021 Bridgepoint Ne, Albuquerque, NM 87111. T: 505-822-0381, F: 505-846-2388, E: baladi@ dtra.mil

Baladi, G.Y., Michigan State Univ, College of Engineering, Dept of Civil Eng, E Lansing, MI 48824. T: 517-355-5147

Balconi, J.A., CTI & Associates Inc., 2436 Roselawn, Walled Lake, MI 48390. T: 248-624-9567, F: 248-486-5050, E: johnbalconi@cti.assoc.com

Baldwin, D.W., Susquehanna Electric Co, Conowingo Hydro Station, 2569 Shores Landing Rd., Darlington, MD 21034. T: 410-457-2513, F: 410-457-2473, E: don.baldwin@ exeloncorp.com

Ballard II, F.A., Ballard Construction Inc., 2500 Regency Parkway, Cary, NC 27511. T: 919-654-6814, F: 919-654-6779, E: fballard@ballardsports.com

Ballegeer, J.P., 6272 West Cross Dr., Littleton, CO 80123. T: 303-734-9682, F: 303 220 0442, E: jballegeer@ jacesare.com

Balsavage, E.L., Earth Eng & Science Inc., 492 Capitol Hill Rd., Dillsburg, PA 17019. T: 717-697-5701, F: 717-695-5702

Baltz, J.F., Acres Intl Corp, 1362 Underhill Rd., East Aurora, NY 14052. T: 716-741-9755, F: 716-689-3749

Bang, S., South Dakota School of Mines & Technology, Dept of Civil Eng, Rapid City, SD 57701. T: 605-394-2440, F: 605-394-5171, E: sbang@silver.sdsmt.edu

Banks, D.C., 302 Enchanted Dr., Vicksburg, MS 39180. T: 601-638-1975, E: donbanks@canufly.net

Banta, D.E., Donald E Banta & Assocs Inc., 415 Meridian Ave, San Jose, CA 95126., F: 408-275-9019

Barber, C.J., Armand Corporation, 181 W Patricia Rd., Holland, PA 18966. T: 215-355-0177, F: 856-489-8212, E: barbersitch@erols.com

Barbosa, R.E., Eng Solutions Inc., 8170 Sw 29th Ct, Davie, FL 33328. T: 954-476-0387, F: 954-370-0150, E: rbarbos@ aol.com

Barcelona, M.J., Univ of Michigan, Cee/1221 Ist Bldg, 2200 Bonisteel Blvd, Ann Arbor, MI 48109. T: 734-763-6512, F: 734-763-6513, E: mikebar@engin.umich.edu

Bardet, J.-P., University of Southern California, Dept of Civil Engineering, Los Angeles, CA 90089. T: 213-740-0608, F: 213-744-1426, E: bardet@usc.edu

Barkau, M.G., URS Greiner Woodward Clyde, 500 12th St, Ste 200, Oakland, CA 94607. T: 510-874-3039, F: 510-874-3268, E: mark_barkau@urscorp.com

Barker, D.H., Geostructures, Model Farm, Crockham Hill, Edenbrdge, Kent, TN8 6SR UK. T: 906-487-2454, F: 011-173-286-6858, E: geo-engineer@bigfoot.com

Barneich, J.A., Geopentech, 601 N Parkcenter Dr., Ste 110, Santa Ana, CA 92705. T: 714-796-9100, F: 714-796-9191, E: john_barneich@geopentech.com

Barnes, S.J., Cumberland Geotech Cons, 801 Belvedere St, Carlisle, PA 17013. T: 717-245-9100, F: 717-245-9656, E: jbarnes@cumberlandgeo.com

Barr, Jr, L.M., URS, 438 Railroad Ave, Souderton, PA 18964. T: 215-703-0342, F: 215-542-3888, E: lmb4@bellatlantic. net

Barreiro, D., K&B Consultants Inc., 4914 Sw 72nd Ave, Miami, FL 33155. T: 305-666-3563, F: 305-666-3069

Barrier, G.D., Barrier Engineering PC, 3901 Davis Lane, Charlotte, NC 28269. T: 704-599-0862

Barron, J.S., Buffalo Drilling Co, 10440 Main St, Clarence, NY 14031. T: 716-759-7821, F: 716-759-7823

Barton, G.W., European Comm Unit Scr/B/4, 2 Patrijzenstraat, Everberg, 3078 BELGIUM. T: 351-541-0000 FA0022, F: 32-2-299-5734, E: gerald.barton@cec.eu.int

Barton, R.E., Earthtec Testing & Eng, 1596 W 2650 S., Ste 108, Ogden, UT 84401. T: 801-399-9516, F: 801-399-9842, E: earthtecl@hotmail.com

Baryla, J.-M.P., Louis Berger Int'l Inc/Scao, 1 Place Du Sud, 1309 Eve, 92806 Puteaux 00101, FRANCE. T: 33-01-47170754, F: 33-01-472418, E: jm_baryla@yahoo.fr

Basnett, C.R., Ch2m Hill, 115 Perimeter Center Place NE, Suite 700, Atlanta, GA 30346. T: 770-604-9095, F: 770-604-9183, E: cbasnett@ch2m.com

Bassan, R., Raphael Bassan Cons Eng, 1560 Broadway, Room 710, New York, NY 10036. T: 212-764-5720

Bastani, S.A., Leighton & Associates, 65 Groveside Drive, Aliso Viejo, CA 92656. T: 949-425-9980, F: 949-250-1114, E: abastani@leightongeo.com

Bastick, M.J., Spie Batignolles, 54 Rue Charles Lorilleux, 92800 Puteaux, FRANCE. T: 331-47781421, F: 331-3424 3894, E: michel_bastick@spiebatignolles.fr

Batchelder-Adams, G.P., URS Corporation, 532 Clarkson St, Denver, CO 80218. T: 303-733-7943, F: 303-694-3946, E: gregg_batchelder_adams@urscorp.com

Bauer, R.A., Illinois State Geological Survey, 615 E Peabody Dr., Champaign, IL 61820. T: 217-244-2394, F: 217-244-0029, E: bauer@isgs.uiuc.edu

Baxter, C.D., Department of Ocean Engineering, University of Rhode Island, Narragansett, RI 2882

Bay, J.A., Utah State University, Dept of Civil & Environ Engr, Logan, UT 84322. T: 435-797-2947, F: 435-797-1185, E: jabay@lab.cee.usu.edu

Beakley, J.W., American Concrete Pipe Assoc, 222 W Las Colinas Blvd, Suite 641e, Irving, TX 75039. T: 972-506-7216, F: 972-506-7682, E: jbeakley@concrete-pipe.org

Bean, Jr, J.E., Univ of New Mexico, 10455 4th St Nw, Albuquerque, NM 87114. T: 505-897-4180, F: 505-296-3289, E: jebean@unm.edu

Beard, J.J., North Carlina State of, 5312 Pine Dr., Raleigh, NC 27606

Beard, R.M., Earth Systems So California, 223 Montebello Ave, Ventura, CA 93004. T: 805-647-9857, E: rbeard@ earthsys.com

Beaty, L.P., Beaty Construction Inc., 5292 W 100 North, Boggstown, IN 46110. T: 317-835-2254

Beck, B.F., PE La Moreaux & Assocs, 106 Administration Rd., Oak Ridge, TN 37830. T: 423-483-7483, F: 423-483-7639, E: pelaor@nsil.net

Bedian, M.P., Perini Corp, 200 Winston Dr., Apt 2311, Cliffside Pk, NJ 7010. T: 201-886-1194, F: 914-345-8211, E: papazian@papart.com

Bednarczyk, J.P., 1027 Walnut St, Utica, NY 13502. T: 315-735-9987, E: bednarcz@borg.com

Beech, J.F., Geosyntec Consultants, 1100 Lake Hearn Dr., Ne, Atlanta, GA 30342. T: 404-705-9500, F: 404-705-9400

Beenenga, C.R., Gannett Fleming Inc., P.O. Box 67100, Harrisburg, PA 17106. T: 717-763-7211, Ext. 2698, F: 717-763-8150, E: cbeenenga@gfnet.com

Beikae, M., Metropolitan Water District, 13 Son Morell, Laguna Niguel, CA 92677. T: 949-249-2696, F: 213-217-7456

Belaskas, D.P., GZA Geoenvironmental of NY, 8 Timway Ct, Fairport, NY 14450. T: 716-223-3881, F: 716-359-0162, E: dbelaskas@gza.com

Belfast, A.F., Geotechnics Inc., 9245 Activity Rd., Ste 103, San Diego, CA 92126. T: 619-536-1000, F: 619-536-8311, E: tbelfast@geotechnicsinc.com

Belgeri, J.J., S&Me Inc., P.O. Box 1118 Tcas, Blountville, TN 37617. T: 423-323-2101, F: 423-323-5272, E: jbelgeri@smeinc.com

Bell, R.A., 80 Oak Knoll Dr., San Anselmo, CA 94960. T: 415-454-0553, F: 415-454-1601, E: roybell@earthlink.net

Belsom, C.L., Southport Inc., 5208 Janice Ave, Kenner, LA 70065. T: 504-885-0548, F: 504-263-2444

Benak, J.V., Univ of Nebraska, 3417 S 79th St, Omaha, NE 68124. T: 402-393-4973

Benavides, R., Estodios Y Control, Lomas Alta 125b, Lomas De Rosales, Tampico Ta, 89100 MEXICO. T/F: 228-75-48

Benedict, C.M., Gannett Fleming Inc., 430 Elm St, Apt A 104, Pottstown, PA 19464. T: 610-705-8966, E: geo tech@usa.net

Bennert, T.A., Rutgers University, 153 Impatiens Court, Toms River, NJ 8753. T: 732-557-9502, F: 732-445-0577, E: bennert@eden.rutgers.edu

Bennett, P.E., Jonathan Kevin, Terratech Inc., 850 Tuscawilla Hills, Charles Town, WV 25414. T: 304-728-0438, F: 703-771-2662, E: jon@terratechva.com

Benoist, J.M., ADWR Dam Safety Section, 500 N Third St, Phoenix, AZ 85004. T: 602-417-2400 X7191, F: 602-417-2423

Benoit, J., Univ of New Hampshire, Dept of Civil Eng, Kingsbury Hall, Durham, NH 3824. T: 603-862-1419, F: 603-862-2364, E: jbemoit@maple.unh.edu

Benson, Jr, T.C., Lowney Associates, 251 East Imperial Highway #470, Fullerton, CA 92835. T: 714-441-3090, F: 714-441-3091, E: tbenson@lowney.com

Bentler, D.J., Univ of Kentucky, Dept of Civil Engineering, 161 Civil Engr/Trans Bldg, Lexington, KY 40506. T: 606-257-8037, F: 606-257-4404, E: dbentler@pop.engr.uky.edu

Bentley, Jr, A.L., 30003 Morningside Dr., Perrysburg, OH 43551. T: 419-666-6059, F: 419-666-6164

Bentley, K.J., Golder Associates Inc., 220 Santa Barbara, San Clemente, CA 92672. T: 949-940-9063, F: 949-583-2770, E: kbentley@golder.com

Benvie, D.A., Tectonic Eng Cons Pc, P.O. Box 37, Mountainville, NY 10953. T: 914-928-6531

Benzer, Jr, W.B., 117 Nw 83rd St, Seattle, WA 98117. T: 206-782-8577, E: wbenzer@yahoo.com

Berends, B.E., Kumar & Assocs Inc., 2540 Norwich Dr., Colorado Spgs, CO 80920. T: 719-590-7639, F: 719-632-1049

Berg, R.R., Ryan R Berg & Assocs Inc., 2192 Leyland Alcove, Woodbury, MN 55125. T: 651-735-7622, F: 651-735-7629, E: ryanberg@worldnet.att.net

Berggren, B.J., Rsv Engineering Inc., 31960 Sw Charbonneau Dr., Suite 101, Wilsonville, OR 97070. T: 503-694-6960, F: 503-694-6962, E: bjbrsv@aol.com

Bergman, K.H., Harding Ese, 9 Diego Dr., San Rafael, CA 94903. T: 415-479-2461, F: 415-884-3300, E: khbergman@mactec.com

Bergstrom, W.R., Nth Consultants Ltd., 517 West Meadowbrook, Midland, MI 48640. T: 517-832-7617, F: 517-835-5465, E: wbergstrom@nthconsultants.com

Berk III, M., Consultant, 39 Graeler Dr., Saint Louis, MO 63146. T: 314-432-3417

Berlogar III, F., Berlogar Geotech Cons, 5587 Sunol Blvd, Pleasanton, CA 94566. T/F: 925-484-0220, E: fberlogar@aol.com

Berry, R.M., Rembco Geotechnical Contractor, P.O. Box 23009, Knoxville, TN 37933. T: 865-690-6917, F: 865-690-9135, E: berry@rembco.com

Beverly, B.E., Haley & Aldrich Inc., 12 Lorna Dr., Auburn, MA 1501. T: 508-832-4817, F: 617-886-7690, E: beb@haleyaldrich.com

Bhowmik, S.K., Ardaman & Assoc, 7719 Dillsbury Court, Orlando, FL 32836. T: 407-226-9397, F: 407-859-8121

Bhushan, K., Group Delta Consultants, 92 Argonaut, Ste 120, Aliso Viejo, CA 92656. T: 949-609-1020, F: 949-609-1030, E: gdckul@aol.com

Bianchini, G.F., Superior Scaffold, 164 Applegate Dr., West Chester, PA 19382. T: 610-981-4714, F: 302-325-3605, E: vn/gwhite@veriomail.com

Biegel, C.K., P.O. Box 6380, Kahului, HI 96733. T: 808-879-5631

Bielak, J., Carnegie Mellon Univ, Dept of Civil & Env Eng, Schenley Park, Pittsburgh, PA 15213. T: 412-268-2958, F: 412-268-7813, E: jbielak@cmu.edu

Biesiadecki, G.L., Langan Engineering, 90 West Street Suite 1510, New York, NY 10006. T: 212-964-7888, F: 212-964-7885, E: gbiesiadecki@langan.com

Bigham, R.E., Buchanan Soil Mechanics Inc., 211 Tee Drive, Bryan, TX 77801. T: 979-822-0719, F: 979-822-7604, E: buchanansoil@aol.com

Bilgin, O., Camp Dresser Mckee, 300 Pond Street, Braintree, MA 2184. T: 781-843-5637, F: 617-452-8431, E: bilgino@cdm.com

Billings, H.J., 934 Evergreen Ave, San Leandro, CA 94577. T: 510-483-4274

Billington, E.D., Ag&E, 405-A Parkway Dr., Greensbobo, NC 27401. T: 336-274-9456, F: 336-274-9486, E: billingt@agrande.com

Bird, D.W., 14551 Autumn Wood Drive, Westfield, IN 46074. T: 317-573-9515, F: 317-844-2980, E: davidwbird@aol.com

Bird, G.R., 11898 Featherwood Drive, St Louis, MO 63146. T: 314-569-3455, F: 314-569-3383, E: r_bird70@post.harvard.edu

Bischoff, J.A., URS Corporation, 500 12th St, Oakland, CA 94607. T: 510-874-1701, F: 510-874-3268, E: john_bischoff@urscorp.com

Bischoff, W.A., Woodward Clyde Cons, 55 S Market St, Ste 1650, San Jose, CA 95113. T: 408-297-9585

Bishop, C.S., Bowser Morner Assoc Inc., 1190 Delaneys Ferry Rd., Versailles, KY 40383. T: 859-873-1189, F: 859-253-0183, E: cbishop@bowser-morner.com

Black, K.T., US Navy, 4500 Third Ave, Apt 104, San Diego, CA 92103. T: 619-296-7727, F: 619-437-2409, E: ktblack73@earthlink.net

Blackburn, Jr, L.D., Blackburn Eng, P.O. Box 100, Lowndesboro, AL 36752. T: 334-265-0206

Blackwood, T.W., Carlson Geotechnical, 19402 S Creek Rd., Oregon City, OR 97045. T: 503-631-7996, F: 503-670-9147, E: tblackwood@carlsontesting.com

Blake, K.M., Christian Testing Labs Inc., 1211 Newell Parkway, Montgomery, AL 36110. T: 334-260-9170, F: 334-260-9177, E: kevinblake@ctltesting.com

Blakita, P.M., Hayward Baker Inc., 1875 Mayfield Rd., Odenton, MD 21113. T: 410-551-1980, Ext. 204, F: 410-551-8206, E: pmblakita@haywardbaker.com

Blanchard, J.D., CFS Engineering Geology, 295 Cain Drive, Santa Maria, CA 93455. T: 805-937-4953, F: 805-937-0742

Blanchette, D.A., Boyle Eng, 1661 Garland St, Lakewood, CO 80215. T: 303-237-7038, F: 303-987-3908, E: dblanchette@boyleengineering.com

Blandin, G.T., Indeco Inc., 1101 Scenic Hill Ln, De Witt, IA 52742. T: 319-659-2284, F: 319-359-4318, E: gblandin@clinton.net

Blender, E.M., P.O. Box 851, Castle Rock, CO 80104., E: ladybugpro@netscape.net

Blessey, W.E., Walter E Blessey-Consulting En, 5546 Dayna Ct, New Orleans, LA 70124. T: 504-488-9357, F: 504-734-1195

Blickwedehl, R.R., Suny Canton, Cornell Drive Nn 105, Canton, NY 13617. T: 315-386-7220, E: blick-wer@canton.edu

Block, R.N., 89 Strawberry Hill Rd., Acton, MA 1720. T: 508-263-4402, F: 978-264-9182, E: rblock@alum.mit.edu

Blois, L., Tech Instit State for Bldg Surv, Via Luigi Catanelli No 26, Perugia, 6154 ITALY. T: 39-75-5990021, F: 39-75-32084/39755, E: lublois@tin.it

Bloom, H., Williams Atlas Assocs, 67-32 Harrow St, Forest Hills, NY 11375. T: 718-544-4599, F: 212-268-8960, E: blom.har@worldnet.att.net

Blystra, A.R., Thunder Bay Power Company, 15 West 6th Street, Holland, MI 49423. T: 616-394-0606, F: 616-394-5936, E: ablystra@thunderbaypower.com

Boakye, S.Y., Stone & Webster Engr Corp, Geotech Div, 245 Summers Street, Boston, MA 2210. T: 617-589-1194, F: 617-589-1008, E: samuel.boakye@stoneweb.com

Boatright, J.W., Tennessee State of, 460 Rose Rd., Somerville, TN 38068. T: 901-465-9078, F: 901-465- 8991, E: mrt117in@bellsouth.net

Boddie, P.J., URS Greiner Woodward Clyde, 55 S Market St, Ste 1650, San Jose, CA 95113. T: 408-297-9585, F: 408-297-6962

Boddy, J.R., Dames & Moore, 916 S 3rd Ave, Bozeman, MT 59715. T: 708-745-7465, F: 849-228-8372

Bodine, D.G., Geosyntec Consultants, 505 Lincoln St, Coal City, IL 60416. T: 815-634-4822, F: 312-658-0576, E: danb@geosyntec.com

Bodner, B.S., Law Eng & Envir Services Inc., 401 Rockport Lane, Birmingham, AL 35242. T: 205-980-8984

Bodocsi, A., 3246 Columbia Pky, Cincinnati, OH 45226. T: 513-871-0986, E: abodocsi@hcnutting.com

Boehm, D.W., Hayward Baker Inc., 2510 Decatur Ave, Fort Worth, TX 76106. T: 817-625-4241, F: 817-626-2749, E: dwboehm@haywardbaker.com

Boehm, R.G., STS Consultants, 702 Goldfinch Lane, Howards Grove, WI 53083. T: 414-565-2723, F: 920-468-3312, E: boehm@stsltd.com

Bogdan, L.I., Dywidag Systems Intl, 2154 South St, Long Beach, CA 90805. T: 562-531-6161, F: 562-531-2667

Boge, L.A., Stantec, 16 Rising Sun Court, Henderson, NV 89014. T: 702-263-9604, F: 702-361-0659, E: bogey@powernet.net

Boggs, E.W., 17 Broadmoor Dr., Mechanics-burg, PA 17055. T: 717-766-9438, E: edwardb715@aol.com

Bold, T.A., ND Dept of Transportation, 2028 N Bell St, Bismarck, ND 58501. T: 701-255-4764, F: 701-328-6913, E: tbold@state.nd.us

Bolinaga, F.J., Dragados Obras Y Proyectos Venezuela, Poba International #1-20092, P.O. Box 02-5255, Miami, FL 33102. T: 58-2-923111, F: 58-2-927724, E: fjb_dycvensa@attglobal.net

Bolton, J.M., Army Corps of Engineers, 711 Roundtree Ct, Sacramento, CA 95831. T: 916-422-3523, F: 916-557-6803, E: jbolton@spk.usace.army.mil

Bonaparte, R., Geosyntec Consultants, 1100 Lake Hearn Dr., Ne, Suite 200, Atlanta, GA 30342. T: 404-705-9500, F: 404-705-9400, E: rbonaparte@geosyntec.com

Bonaz, R., FC International, 17 Square Des Plantanes, Bailly 78870, FRANCE., F: 33146237117, E: rene.bonaz@fc international.com

Bonner, G.A., Dublin Corporation, 18 Charlemont, Griffith Ave, Dublin, 9 IRELAND (EIRE). T: 00-353-1-8376535, F: 353-1-6623290, E: gerald.bonner@dublincorp.ie

Bono, N.A., Batsman Stens Vag 70, Spanga, 163 41 SWEDEN. T: 46-8-760-4645, E: nabono@aol.com

Boodey, T.M., New Hampshire Dept of Transp, 29 Old Canaan Road, Barrington, NH 3825. T: 603-664-9757, E: t_boodey@hotmail.com

Booth, R.D., Kleinfelder Inc., 4905 Hawkins Ne, Albuquerque, NM 87109. T: 505-344-7373, F: 505-344-1711

Boratynec, D.J., Leighton & Assocs, 550 Paularino Ave, Apt D202, Costa Mesa, CA 92626. T: 714-966-1445, F: 949-250-1114, E: dboratynec@leightongeo.com

Borden, R.H., North Carolina State Univ, Civil Eng Dept, Box 7908, Raleigh, NC 27695. T: 919-515-7630, F: 919-515-7908, E: borden@eos.ncsu.edu

Borja, R.I., Stanford Univ, Dept of Civil Eng, Terman Eng Center, Stanford, CA 94305. T: 650-723-3664, F: 650-723-7514, E: borja@cive.stanford.edu

Boscardin, M.D., GEI Cons Inc., 1021 Main St, Winchester, MA 1890. T: 781-721-4000, F: 781-721-4073, E: mboscardin@geiconsultants.com

Bosscher, P.J., Univ of Wisconsin, 1415 Engineering Drive, Room 2218, Madison, WI 53706. T: 608-262-7245, F: 608-262-5199

Boswell, J.E.S., Bohlweki Environmental (Pty) Ltd., P.O. Box 11784, Vorna Valley, 1686 SOUTH AFRICA. T: 1656, F: 1680, E: bohlweki@pixie.co.za

Botz, J.J., Sts Consultants Ltd., 2133 Oakwood Dr., Green Bay, WI 54304., F: 414-468-3312, E: botz@stsltd.com

Bouazza, A., Monash Univ, Dept of Civil Eng, Clayton Melbourne, Vic 3168 Australia, AUSTRALIA. T: 011-61-3-99054956, F: 011-61-3-990549, E: malek.bouazza@eng.monash.edu.au

Boudra, L.H., Law Eng & Envir Services Inc., 316 Observatory Dr., Birmingham, AL 35206. T: 205-836-1905, F: 205-985-2951, E: lhboudra@aol.com

Bourdeau, P.L., Purdue Univ, School of Civil Eng, West Lafayette, IN 47907. T: 765-494-5031, F: 765-496-1364, E: bourdeau@ecn.purdue.edu

Bowders, Jr, J.J., Civil Engr, E 2509 Engr Bldg East, Univ of Missouri, Columbia, MO 65211. T: 573-882-8351, F: 573-882-4784, E: bowders@missouri.edu

Bowers, C.H., ASCE, Geo-Institute of ASCE, 1801 Alexander Bell Dr., 4th Floor, Reston, VA 20191. T: 703-295-6352, F: 703-295-6351, E: cbowers@asce.org

Bowers, M.T., Univ of Cincinnati, 515 St Thomas Ct, Fairfield, OH 45014. T: 513-858-3381, F: 513-556-2599, E: mbowers@uceng.uc.edu

Bowes, D.E., Donald E Bowes Pe Inc., 16225 Se 29th St, Bellevue, WA 98008. T: 425-562-6093, F: 425-641-3747

Bowles, J.E., Eng Computer Software, 1605 West Candletree Dr., Suite 106, Peoria, IL 61614. T: 309-692-9707, E: jbowles@fadse.com

Bowman, A.E., Kleinfelder, 263 E Robindale Rd., Las Vegas, NV 89123. T: 702-263-7259, F: 702-361-9094, E: abowman@kleinfelder.com

Bowman, D.R., BBC & M Eng Inc., 5732 Dunnwood Ct, Dublin, OH 43017. T: 614-792-3753, F: 614-793-2410, E: dbowman728@aol.com

Boyajian, R.T., BSK & Associates, 567 W Shaw, Suite B, Fresno, CA 93704. T: 559-497-2880, F: 559-497-2886

Boyce, A.M., SB Testing & Engineering, Mount Johnson, St Lucie 01, BARBADOS. T: 246-439-2215, F: 246-228-4781, E: sbtesting@seagrapes.com

Boyce, G.M., Parsons Brinckerhoff, 400 S W Sixth Ave, Suite 802, Portland, OR 97204. T: 503-417-9361, F: 503-274-1412, E: boyce@pbworld.com

Boyce, S.C., US Air Force, 2735 Roundtop Drive, Colorado Springs, CO 80918. T: 719-260-8634, E: scboyce@aol.com

Boyle, S.R., Shannon & Wilson Inc., 9317 42nd Avenue Ne, Seattle, WA 98115. T: 206-528-0878, F: 206-633-6777, E: srb@shanwil.com

Bozzo, A.M., Technomare SPA, San Marco 3584, Venice, 30124 ITALY. T: 011-39041796711, F: 390422306573 39, E: bognolo.e@tecnomare.it

Bradshaw, A.S., Hart Crowser, 1230 N 47 St, Seattle, WA 98103. T: 206-547-9556, F: 206-328-5581, E: asb@hart crowser.com

Bradshaw, S.M., Terrain Geotechnical Consultants Ltd., Old Town Slack Farm, Wadsworth Hebden Bridge, West Yorkshire, HX7 8TE UK. T: 011-0044-0-1422842682, F: 4.40E+12, E: terraingeotech@talk.com

Brand, A.H., Mueser Rutledge Cons Engrs, 80-67 Grenfell St, Kew Gardens, NY 11415. T: 718-846-8799

Brandes, H.G., Dept of Civil Engineering, Univ of Hawaii, 2540 Dole St, Honolulu, HI 96822. T: 808-956-8969, F: 808-596-5014, E: brandes@wiliki.eng.hawaii.edu

Brandl, E., Schnabel Foundation Co, 2388 Woodland Park, Houston, TX 77077., F: 281-531-7539

Brannen, R.E., Buena Engrs Inc., 4462 Amber Canyon Dr., Las Vegas, NV 89129. T: 702-656-9857, F: 702-837-1600, E: raymond_brannen@urscorp.com

Branson, R.U., CTL Thompson Inc., 7052 S Owens Way, Littleton, CO 80127. T: 303-979-7121, F: 719-528-5362, E: rbranson@ctlthompson.com

Branum, W.H., Professional Service Ind Inc., 30002 Nanton Dr., Spring, TX 77386. T: 281-362-9769, F: 630-691-8761, E: hbranum@psiusa.com

Bratton, J.L., Applied Research Assocs, 4300 San Mateo Blvd Ne, Suite A220, Albuquerque, NM 87110. T: 505-881-8074, F: 718-388-2602, E: jbratton@ara.com

Breitnauer, M.J., Sunbelt Laboratories, Inc., 1410 Gail Borden, Suite C-5, El Paso, TX 79935. T: 915-591-4647, F: 915-591-4805, E: sunbelt@dzn.com

Brengola, A.F., Hayward Baker, 1875 Mayfield Road, Odenton, MD 21113. T: 410-551-1980, E: afbrengola@haywardbaker.com

Brennan, J.J., Kansas Dept of Transp, 2950 Sw Wayne, #14, Topeka, KS 66611. T: 913-296-3008, F: 785-296-2526

Breslin, J.F., Harza Eng Co, 747 North Oriole, Park Ridge, IL 60068. T: 847-823-8214, F: 312-831-3999, E: jbreslin@harza.com

Brettmann, T.T., Berkel & Co Contractors Inc., 2330 Precinct Line Rd., Richmond, TX 77469. T: 281-344-1090, F: 281-344-1099, E: tbrettmann@berkelapg.com

Briaud, J.-L.C., Texas A&M Univ, Civil Eng Dept, College Station, TX 77843. T: 979-845-3795, F: 979-845-6554, E: briaud@tamu.edu

Bricknell, I., Tams Cons Inc., 6164 31st St Nw, Washington, DC 20015. T: 202-966-6680, F: 703-875-8360, E: ibricknell@tamsconsultants.com

Brigham, J.E., Amec Earth & Environmental, 115 South 8th St, Tacoma, WA 98402. T: 253-383-4940, F: 253-383-4923, E: jbrigham@agraus.com

Brinker, F.A., Site-Blauvelt Engineers, 291 Old New Road, Doylestown, PA 18901., T: 856-273-9244, E: frederickb@site-blauvelt.com

Brissette, K.M., Harrington Engineering & Const, 652 Roane Ln, Valparaiso, IN 46385. T: 219-926-8651, F: 219-926-8446, E: hec@niia.net

Brodbeck, D.C., Scherzinger Drilling, P.O. Box 202, Miamitown, OH 45041. T: 513-353-2400

Broderick, G.P., Univ of New Haven, 616 Norton Town Rd., Guilford, CT 6437. T: 203-458-7604, F: 203- 932-7158, E: broderic@charger.newhaven.edu

Brouillette, R.P., URS Greiner Woodward Clyde, 17521 E Fort Pickens, Baton Rouge, LA 70817. T: 504-753-8469, E: rickey_brouillette@urscorp.com

Brown, D.A., Auburn Univ, Harbert Eng Center, Dept of Civil Eng, Auburn, AL 36849. T: 334-844-6283, F: 334-844-6290, E: dbrown@eng.auburn.edu

Brown, M.J., SAIC, 945 E Maple St, Palmyra, PA 17078. T: 717-838-5995, E: mattb@paonline.com

Brown, R.P., University of Texas, 1713 B Burton Drive, Austin, TX 78741. T: 512-707-9135, F: 512-471-6548, E: rollins.brown@worldnet.att.net

Brown, T.D., Albuquerque Caisson, P.O. Box 91090, Albuquerque, NM 87199. T: 505-821-2166, F: 505-828-0970, E: tdbrown@aol.com

Brown, W.S., Fluoro-Seal Inc., 16360 Park Ten Place, Suite 325, Houston, TX 77804. T: 281-578-1440, F: 281-578-3159, E: bill@fsi.com

Bruce, D.A., Geosystems LP, 161 Bittersweet Cir, Venetia, PA 15367. T: 724-942-0570, F: 724-942-1911

Bruce, M.E.C., Geotechnica SA Inc., 161 Bittersweet Circle, Venetia, PA 15367. T: 724-942-4220, F: 724-942-1911, E: geotecnica@aol.com

Bruer, S.M., Ch2m Hill, 727 N First St, Suite 400, St Louis, MO 63102. T: 314-421-0900, F: 314-421-3927, E: sbruer@ch2m.com

Bruggers, D.E., Engeo, 2401 Crow Canyon Rd., Suite 200, San Ramon, CA 94583. T: 925-838-1600, F: 925-838-7425, E: dbrugger@engeo.com

Bruhn, R.W., GAI Cons Inc., 2365 Mount Vernon Ave, Export, PA 15632. T: 724-733-8111, F: 412-372-2161, E: rblexport@mindspring.com

Brumund, W.F., Golder Assocs Corporation, 3730 Chamblee Tucker Rd., Atlanta, GA 30341. T: 770-496-1893, F: 770-934-9560, E: bbrumund@golder.com

Brunette, B.E., Alaska State of, Dept of Transportation, 6860 Glacier Hwy, Juneau, AK 99801. T: 907-465-4198, F: 907-465-3506, E: brunette@alaska.net

Brungard, M.A., URS, 3676 Hartsfield Rd., Tallahassee, FL 32303. T: 850-574-3197, F: 850-576-3676, E: martin_brungard@urscorp.com

Brusey, W.G., 20 Ridge Rd., Apt 2, Ridgewood, NJ 7450. T: 201-447-2954, F: 212-435-6039

Brusso, K.R., STS Consultants Ltd., 11425 W Lake Park Dr., Milwaukee, WI 53224. T: 414-359-3030, F: 414-359-0822, E: brusso@stsltd.com

Bryan, D., Navfacengcom, 1421 Whittier Rd., Virginia Bch, VA 23454. T: 757-481-3418, F: 757-322-4354, E: dbryan1421@aol.com

Bryant, J.C., Day & Zimmermann Inc., 9210 Henry Harris Road, Fort Mill, SC 29715. T: 803-548-4129

Bryant, K.E., Southern Company Generation, 556 Russet Bend Dr., Birmingham, AL 35244. T: 205-424-0556, F: 205-257-1596, E: kebryant@southernco.com

Brylawski, Ed., Geonor Inc., P.O. Box 903, Milford, PA 18337. T: 570-296-4884, F: 570-296-4886, E: ry@geonor.com

Bucher, S.A., Terracon Inc., 2391 Oak Hill Dr., Lisle, IL 60532. T: 630-968-3028, E: sabucher@terracon.com

Budiman, J.S., Illinois Inst of Tech, Civil Eng Dept, Chicago, IL 60616. T: 312-567-3544, F: 312-567-3519, E: budiman@iit.edu

Buffington, D.L., Pincock Allen & Holt, 6170 Arapahoe Dr., Evergreen, CO 80439. T: 303-670-8053, F: 303-987-8907, E: dlb@hartcrowser.com

Buhr, C.A., Burns & Mc Donnell Eng Co, 9400 Ward Parkway, P.O. Box 419173, Kansas City, MO 64141. T: 816-822-3137, F: 816-822-3494, E: cbuhr@burnsmcd.com

Bullock, P.J., Univ of Florida, Dept of Civil Engineering, 345 Weil Hall, P.O. Box 116580, Gainesville, FL 32611. T: 352-372-9905, F: 352-392-3394, E: pjbullock@ce.ufl.edu

Bunnell, D.B., Bunnell-Lammons Eng Inc., 1200 Woodruff Rd., Ste B-7, Greenville, SC 29607. T: 864-288-1265, F: 864-288-4430, E: dan@blecorp.com

Buranek, D., Emcon Assocs, 1921 Ringwood Ave, San Jose, CA 95131. T: 408-453-7300

Burch, G., Raba Kistner Consultants Inc., 8434 Timberwilde Drive, San Antonio, TX 78250. T: 210-647-7852, F: 210-699-6426, E: gburch@rkci.com

Burgin, C.R., Bhate Eng Corp, 4335 Maplewood Dr., Trussville, AL 35173. T: 205-655-7926, F: 205-599-0229, E: cburgin@bhate.com

Burke, C.E., Law Eng & Envir Services Inc., 22455 Davis Dr., Ste 100, Sterling, VA 20164. T: 703-404-7000, F: 703-404-7070, E: eburke@lawco.com

Burke, G.K., Hayward Baker Inc., 1130 Annapolis Road, Odenton, MD 21113. T: 410-551-8200, F: 410-551-1900, E: gkburke@haywardbaker.com

Burke, S.D., Montgomery Watson Inc., 1610 Excelsior Ave, Oakland, CA 94602. T: 510-251-2530, E: scott.burke@mw.com

Burke, W.W., Paulus Sokolowski & Sartor Inc., 1 Locke Circle, Medford, NJ 8055. T: 609-953-1807, F: 732-560-9768, E: wburke@psands.com

Burns, S.E., University of Virginia, P.O. Box 400742, Thornton Hall, Charlottesville, VA 22904. T: 804-924-6370, F: 804-982-2951, E: sburns@virginia.edu

Burrell, C.D.A., Burrell – Burrell Associates, 7133 Bec Layout, P.O. Box 19, BELIZE. T: 501-2-24750, F: 501-2-24751, E: bba@btl.net

Burrous, C.M., NMG Geotechnical, 2 Boise, Irvine, CA 92604. T: 949-559-9273, F: 949-476-8322, E: cburrous@nmggeotech

Burtis III, J.J., Century Eng Inc., 32 West Road, Towson, MD 21204. T: 410-823-8070, F: 410-823-2184, E: jburtis@centuryeng.com

Bush, R.K., Dynegy Marketing & Trade, 1000 Louisiana St, Suite 5800, Houston, TX 77002. T: 713-767-8998, F: 713-767-8762, E: randy.bush@dynegy.com

Buss, K.G., Geoengineers Inc., 19229 303rd Place Ne, Duvall, WA 98019

Butail, A., Terra Assocs Inc., 12525 Willows Rd., Suite 101, Kirkland, WA 98034. T: 425-821-7777

Butikofer, M., Sieber Cassina & Partners, Schadaustr 35, Thun Ch 3604, SWITZERLAND. T: 033-336-9154, F: 031-382-3031, E: mbueti@csi.com

Butler, L.W., STS Cons Ltd., 1035 Kepler Dr., Green Bay, WI 54311. T: 920-468-1978, F: 920-468-3312, E: butler@stsltd.com

Butler, R.C., Golder Assocs Ltd., Suite 500, 4260 Still Creek Dr., Burnaby, V5C 6C6 CANADA BC. T: 604-298-6623, F: 604-298-5253, E: rbutler@golder.com

Butruille, F.R., Kleinfelder Inc., 12416 Sw 64th Ave, Portland, OR 97219. T: 503-977-2292, F: 503-643-1905

Butzke, M.R., Allender Butzke Engrs Inc., 3660 109th St, Urbandale, IA 50322. T: 515-252-1885, F: 515-252-1888, E: mbutzke@abengineers.com

Byerly, J.R., John R Byerly Inc., 2257 South Lilac Ave, Bloomington, CA 92316. T: 909-877-1324, F: 909-877-5210, E: byerlype@earthlink.net

Byerly, M.W., GEC, 1230 East Hillcrest St, Orlando, FL 32803. T: 407-898-1818, F: 407-898-1837, E: mwbyerly@g-e-c.com

Byle, M.J., Gannett Fleming Inc., 1411 Highland Rd., Downingtown, PA 19335. T: 610-873-8118, F: 610-650-8190, E: mbyle@gfnet.com

Caballero, D., Arcadis Geragaty & Miller, 308 Chestnut Rd., Linthicum Heights, MD 21090. T: 410-691-9878, F: 410-987-4392, E: dcaballero@arcadis-us.com

Cabanting, V.V., Royal Commission, P.O. Box 30031, Yanbu Al-Sinaiyah, SAUDI ARABIA. T: 011-966-4-321-6385, F: 011-966-4-321-6, E: vcabanting11@hotmail.com

Cadden, A., Schnabel Engr Assocs, 321 Highland Farm Rd., West Chester, PA 19382. T: 215-431-7237, F: 610-696-7771, E: acadden@schnabel-eng.com

Calabria, C.R., Geosystems Consultants Inc., 1500 Miller Pl, Yardley, PA 19067., F: 215-643-9440, E: calabrcr@geosystems.com

Calavera, J., Inst Tecnico De Materiales Y Construccio, Intemac, Monte Esquinza 30, Madrid, 28010 SPAIN. T: 1-3197202, F: 1-3083609, E: jcalavera@intemac.es

Caldwell, D.S., Michelucci & Assocs, 2455 Bennett Valley Rd., Suite 104b, Santa Rosa, CA 95404. T: 707-527-7434, F: 707-578-3195

Cameron, J.E., Ardaman & Associates, 137 W Christina Blvd, Lakeland, FL 33813. T: 863-644-5581

Camire, J.A., Architectural Skylight Co, Rr #1 Box 1020, Waterboro, ME 4087. T: 800-345-7899 X132, F: 207-247-6754

Cammack, C., Urs Corporation, 7911 Darnell Ln, Lenexa, KS 66215. T: 913-492-6520, F: 913-344-1011, E: charles_cammack@urscorp.com

Camp III, W.M., S&ME Inc., 840 Lowcountry Blvd, Mount Pleasant, SC 29464. T: 843-884-0005, F: 843-881-6149, E: bcamp@smeinc.com

Camp, Jr, G.N., Qore Property Sciences, 9115-E Old Statesville Road, Charlotte, NC 28269. T: 704-599-4548, F: 704-598-1050, E: ccamp@qore.net

Campagna, Jr, N.A., GZA Geoenvironmental Inc., 86 Countryside Lane, Norwood, MA 2062. T: 781-762-6866, F: 617-965-7769, E: ncampagna@gza.com

Campbell, Jr, W.R., Landtec, 316 Gloria, Keller, TX 76248. T: 817-431-1142, F: 817-572-3609, E: rayatlandtec@netzero.net

Campbell, B.D., D'appolonia Cons Engrs, 275 Center Rd., Monroeville, PA 15146. T: 412-856-9440, F: 412-856-9535

Campbell, D.J., Geoengineers Inc., 10229 167th Pl Ne, Redmond, WA 98052. T: 425-885-3180

Campbell, De W.A., Bureau of Reclamation, P.O. Box 280326, Lakewood, CO 80228. T: 303-979-2448, F: 303-445-6419, E: dacampbell@hotmail.com

Camper, K.E., TCDI A Division of Hayward Bak, 4976 Prairie Oak Rd., Gurnee, IL 60031. T: 847-625-7893, F: 847-634-8582, E: kecamper@haywardbaker.com

Campo, D.W., Parsons Brinckerhoff, One Penn Plaza, 3rd Floor, New York, NY 10119. T: 212-465-5773

Canady, T.B., Geotechnics Inc., 9245 Activity Rd., Ste 103, San Diego, CA 92126. T: 858-536-1000, F: 858-536-8311, E: tcanady@geotechnicsinc.com

Canino, M.C., Malcolm Pirnie, 104 Corporate Park Dr., Box 751, White Plains, NY 10602. T: 914-641-2559, F: 914-641-2455, E: mcanino@pirnie.com

Cannon, J.D., Tri State Testing, P.O. Box 5474, Fort Oglethorpe, GA 30742. T: 706-866-6911, F: 423-510-0237, E: dirtgun@aol.com

Caprio, D.W., Geotechnical Consultants Inc., 720 Greencrest Dr., Westerville, OH 43081. T: 614-895-1400, F: 614-895-1171, E: dcaprio@gci2000.com

Cardona-Yulfo, L.O., PCO Associates Caribe, P.O. Box 9066523, San Juan, PR 906. T: 787-723-0540, F: 787-721-7646

Carey, P.J., PJCarey & Associates, 5878 Valine Way, Sugar Hill, GA 30518. T: 678-482-5193, F: 678-482-5827, E: pjcarey@mindspring.com

Cargill, K.W., Geosyntec Consultants, 17112 Carrington Park Dr., Apt 913, Tampa, FL 33647. T: 813-632-0504, F: 813-558-9726, E: kcargill@geosyntec.com

Cargill, K.W., Schnabel Foundation Co, 2246 Nantucket Ct Ne, Marietta, GA 30066. T: 770-428-6089, F: 770-977-8530, E: kcargill@worldnet.att.net

Cargill, P.E., S&Me, 840 Lowcountry Blvd, Mt Pleasant, SC 29464. T: 843-884-0005, F: 843-881-6149

Carmichael, J.S., Carmichael Engineering Inc., 106 Creek Drive, Montgomery, AL 36117. T: 334-277-8051, F: 334-271-0763, E: steve@carmichaelengineering.com

Caronna, S., GCA, 941 Georgia St, Santa Rosa, CA 95404. T: 707-545-1807

Carpenter, G.C., Donald E Banta & Assocs Inc., 2481 Kilkare Rd., Sunol, CA 94586. T: 510-862-2978, F: 408-275-9019

Carpenter, J., ECS Ltd., 5433 Braddock Ridge Dr., Centreville, VA 20120. T: 703-830-4070, F: 703-834-5527, E: jcarpenter@ecslimited.com

Carpenter, M.D., STS Consultants Ltd., 3909 Concord Ave, Schofield, WI 54476. T: 715-355-4304, F: 715-355-4513, E: carpenter@stsltd.com

Carr, Z.F., NTH Consultants Ltd., 1898 Fleetwood, Grosse Pointe Woods, MI 48236. T: 313-884-0004

Carrier III, W.D., Argila Enterprises Inc., 76 Woodside Dr., Lakeland, FL 33813. T/F: 863-646-1842, E: dcarrier@tampabay.rr.com

Carter, Jr, C.G., Christian Wheeler Eng, 13835 Tierra Bonita Road, Poway, CA 92064. T: 858-486-2336, F: 858-496-9758, E: ccarter@christianwheeler.com

Carter, P.G., Mid Coast Geotechnical, 1030 Railroad St #103, P.O. Box 3125, Paso Robles, CA 93447. T: 805-237-1462, F: 805-237-1483, E: pgc@tcsn.net

Carter, R.L., NTNL Concrete Masonry Assoc, 7265 Mosby Drive, Warrenton, VA 20187. T: 540-428-2027, F: 703-713-1910, E: lancen3@hotmail.com

Casagrande, D.R., Casagrande Cons, 40 Massachusetts Ave, Arlington, MA 2474. T: 781-648-3630, F: 781-643-3850

Casco, A.A., Ingeneria de Pavimentos, Suelos Y Materiales De Constr, Apdo Postal U-9360 Tegucigalpa, HONDURAS., F: 504-486-6066

Cash, G.D., It Corp, 1425 S Victoria Ct, Suite A, San Bernardino, CA 92408. T: 909-478-1223, F: 909-799-7604, E: gcash@itcrp.com

Caskey, J.M., Geopier Foundation Company – Midsouth, 8978 Heath Cove, Cordova, TN 38018. T: 901-737-0275, F: 901-309-3373, E: geopier@att.net

Castelli, R.J., Parsons Brinckerhoff Quade &, 26 Astor Pl, Williston Pk, NY 11596., F: 212-465-5096

Castles, J.B., Pacific Soils Eng Inc., P.O. Box 2249, Cypress, CA 90630. T: 714-220-0770

Castro, A., 2208 Glenoak Dr., Champaign, IL 61821. T: 217-352-4483, F: 217-351-8700, E: acastro@prairienet.org

Castro, G.V., GEI Cons Inc., 1021 Main St, Winchester, MA 1890. T: 781-721-4036, F: 781-721-4073, E: gcastro@geiconsultants.com

Catalano, A., Treviicos Corp, 250 Summer St, 4th Floor, Boston, MA 2210., F: 617-357-5999

Cathie, D.N., Sage Engineering Sa/Nv, 41 Ave Du Chant D'oiseau, Brussels, 1150 BELGIUM. T: 32-27630469, F: 32-27760319, E: dcathie@sage-be.com

Cavanaugh, J., Schnabel Foundation Co., 1654 Lower Roswell Road, Marietta, GA 30068. T: 770-971-6455, F: 770-977-8530, E: cavanaugh.joseph@worldnet.att.net

Cavin, L.F., Foundation Cons Inc., 1840 Holmes St, Livermore, CA 94550. T: 925-443-7296, F: 925-455-1686, E: fredcavin@msn.com

Cavin, T.S., Andreyev Engineering Inc., 1170 W Minneola Ave, Clermont, FL 34711. T: 352-241-0508

Celeste, L.M., C&S Engineers Inc., 18 Mallard Ln, Oswego, NY 13126. T: 315-342-9440, F: 315-455-9667, E: lceleste @cscos.com

Cerny, M.R., Geotechnical Services Inc., 7050 S 110th Street, Omaha, NE 68128. T: 402-339-6104, F: 402-339-6297, E: mc@gsi.usa.com

Cerros, J.A., Geotechnical Exploration Inc., 7420 Trade St, San Diego, CA 92121. T: 858-549-7222, F: 619-549-1604, E: jcerros@san.rr.com

Chacko, M.J., Fugro West Inc., 4756 Sullivan Lane, Apt 205, Ventura, CA 93003. T: 805-658-0315, F: 805-650-7010, E: jchacko@fugro.com

Chadwick, K.R., Hayward Baker Inc., 129 Yawpo Ave, Oakland, NJ 7436. T: 201-651-1642, F: 914-966-0760, E: krchadwick @haywardbaker.com

Chainey, K.S., 6329 Bonita Ct, Richmond, CA 94806. T: 510-234-6361

Chamberlain, Jr, E.J., Edwin Chamberlain, 20 Wolfeboro Rd., Etna, NH 3750. T: 603-643-4150, F: 603-643-8317, E: edwin_chamberlain@valley.net

Champion, K.L., Champion Eng Inc., 2644 Nutwood Trace, Duluth, GA 30097. T: 770-622-0931, F: 770-622-0932, E: klccc@bellsouth.net

Chan, C.K., Univ of California, 636 4th Ave, San Francisco, CA 94118. T: 415-386-1005

Chan, F.K.H., BFC Civil, 301 Savage Road, Newmarket, L3X 1S4 CANADA ON. T: 905-898-3038, F: 416-754-8692, E: fchan@bfc.ca

Chan, H.C., Geotechnical Engrg office, 25 Man Cheong St, 10th Fl, Ferry Point, HONG KONG. T: 852-9041-5405, F: 852-2303-0212, E: hcmchan@alum.mit.edu

Chan, K.-C., Ove Arup & Partners, Level 5 Festival Walk, 80 Tat Chee Ave, Kowloon, HONG KONG. T: 852-2528-3031, F: 852-2865-, E: andrew.chan@arup.com

Chan, L.-Y., IT Corp, 2200 Cottontail Lane, Somerset, NJ 8873. T: 732-560-4243, F: 732-469-7275, E: lchan@ theitgroup.com

Chan, P.S.-C., PSC Assocs, 1185 Terra Bella Ave, P.O. Box 699, Mountain View, CA 94042. T: 650-969-1144, F: 650-969-5523

Chan, S.F., 33-22-3 Sri Penaga, Bukit Bandaraya, Kuala Lumpur, 59100 MALAYSIA., F: 60-3-4301567

Chan, W.L., Drainage Services Dept, Flat G Blk 4 18/F, Melody Garden, Tuen Mun NT, HONG KONG. T: 852-2404-6582, F: 852-2833-9162

Chandler, D.S., Allied Engineering Serv Inc., 2002 Knaab Drive, Bozeman, MT 59715. T: 406-585-8857, F: 406-582-5770, E: doug@alliedengineering.com

Chandwani, D.B., Fluor Daniel, P.O. Box 711265, Houston, TX 77271. T: 281-565-9249, F: 281-565-9251, E: dayalc@ yahoo.com

Chaney, R.C., Humboldt State Univ, Dept of Enviro Resources Engrg, Bldg 18, Arcata, CA 95521. T: 707-826-4992, F: 707-826-3616, E: rcc1@humboldt.edu

Chang, C.C.C., Sverdrup Corp, 260 Madison Avenue, Suite 1200, New York, NY 10016. T: 212-481-9460, F: 212-481-9484, E: changcc@sverdrup.com

Chang, H., Threetech, 9 403 Hyang Jeong Plaza, 185-7 Gumi-Dong, Budang-Gu, Seongnam, SOUTH KOREA. T: 31-713-7233, F: 31-713-3810, E: schang@unitel.co.kr

Chang, J.-D.L., Ceramic Technologies Pte Ltd., 1b Pine Grove #12-05, Singapore 591001, SINGAPORE. T: 650-463-2337, F: 65 863-8006, E: shsworks@singnet.com.sg

Chang, K.-H., 2120 N 120th St, Seattle, WA 98133. T: 206-361-8326, E: khc2@worldnet.att.net

Chang, M.-H., Natl Yunlin U of Sci & Tech, Dept of Constr Eng 123 Sec 3, Natl Yunlin U of Sci & Tech, Univ Rd, Touliu City, Yunlin, Taiwan, ROC. T: 886-5-5342601, 886-5-5344718, F: 886-5-5312049

Chang, T.-T.D., Chung Yuan Univ, Civil Eng Dept, Chung Li, 32023 Taiwan, ROC. T: 886-3-453-4464, F: 886-3-463-5758

Chang, T.T., 299 Chung Shan Rd., Hsinchu, 300 Taiwan, ROC. T: 03-522-2309

Chang, Y.C.E., Harza Eng Co, 21169 North 21st St, Barrington, IL 60010. T: 312-381-5835, F: 312-831- 3493, E: ychang@harza.com

Chao, K.-C., Shepherd Miller Inc., 1906 Bronson St, Fort Collins, CO 80526. T: 970-266-1166, F: 970-223-7171, E: kcchao@shepmill.com

Chapman, D.R., Lachel & Associates Inc., 17 Corn Hill Dr., Morristown, NJ 7960. T: 973-984-1808, F: 973-734-0055, E: lachelnj@aol.com

Chapman, D.R., Blakeslee Arpaia Chapman Inc., 19 Jenda Way, Madison, CT 6443. T: 203-421-4834, F: 203-488-3997, E: dchapman@bac-inc.com

Chapman, K.R., Schnabel Foundation Co, 3075 Citrus Circle, Suite 150, Walnut Creek, CA 94598. T: 925-947-1881, F: 925-997-0418, E: ron@schnabel.com

Chardaloupa, E.T., Otm Sa Engineering Cons Co, 84 Thoucididou Str, Alimos-Athens, 17455 GREECE., F: 003-01-8235288, E: ehar@hol.gr

Charles, R.D., Duffield Assocs Inc., 5400 Limestone Rd., Wilmington, DE 19808. T: 302-239-6634, F: 302-239-8485

Charlie, W.A., Colorado State Univ, Dept of Civil Eng, Erc-Csu, Fort Collins, CO 80523. T: 970-491-5048, E: wcharlie@engr.colostate.edu

Chatti, K., Michigan State Univer, Dept of Civ & Envir Engr 3546, Engr Bldg, East Lansing, MI 48824. T: 517-355-6534, F: 517-432-1827, E: chatti@egr.msu.edu

Chau, G.W.C., Lam Construction Co Ltd., Flat 7 2/F Wang King House, Tin Wang Court Ma Chai Hang, Kowloon, HONG KONG. T: 011-00852-23200590, F: 011-00852-28828, E: gchau207@hkstar.com

Chau, I.P.-W., Kowloon-Canton Railway Corpora, 35/Floor Flat C Block 4, Riviera Gardens, Tsuen Wan NT, HONG KONG. T: 011-852-24097374, F: 011-852-2409543

Chaudhry, V.K., 249 AGCR Enclave, Delhi, 110092 INDIA. T: 0091-11-2160458, F: 0091-11-8-45123, E: vijaycha@ bol.net.in

Che, E., Kleinfelder, Inc., 3459 Senasac Avenue, Long Beach, CA 90808. T: 562-420-4413, E: notickee@ hotmail.com

Chen, C.Y., US Dept of State, 11944 Riders Lane, Reston, VA 20191. T: 703-758-9405

Chen, J.-H., 26 Sunspree Pl, The Woodlands, TX 77382. T: 714-449-0665, F: 562-694-7746, E: jenc@chevron.com

Chen, J.-R., Cornell Univ, 1807 Hasbrouck Apts, Ithaca, NY 14850. T: 607-253-6472, E: jc67@cornell.edu

Chen, J.-W., National Cheng Kung Univ, No 4 Lane 50, Ku-Nan Street, Kaohsiung, Taiwan, ROC. E: geochen@ mail.ncku.edu.tw

Chen, L., Patel Chen Assocs Inc., P.O. Box 287, Grantville, PA 17028. T: 717-469-0937

Chen, T.-L., Amoeba Cons Engrs, 10 Fl No. 30 Lane 164, Fu-Lin St, Taipei, Taiwan, ROC. T: 886-227-583223, F: 886-227-583220

Chen, W.-H., Hoshin Engrg Consultants, 3f 63 Alley 4 Lane 36, Min-Sheng E Rd., Sec 5, Taipei, 105 Taiwan, ROC. T: 011-886-2-27627853, F: 011-886-2-27682-00

Cheney, J.A., Univ of Calif, 418 Anza Ave, Davis, CA 95616. T: 530-753-4928, F: 530-792-1180, E: jacheney@ ucdavis.edu

Cheng, H.H., Ardaman & Assoc, 1726 Sir John Court, Orlando, FL 32837. T: 407-857-7362, F: 407-859-8121, E: hhcheng2000@hotmail.com

Cheng, S., Naval Facilities Eng Command, 478 Kekauluohi St, Honolulu, HI 96825. T: 808-395-3105, F: 808-474-6306, E: chengkk@efdpac.navfac.navy.mil

Cheng, T.H., Falcon Piling Pte Ltd., No 47 Alley 25 Lane 446 An, Ping Rd., Tainan, Taiwan, ROC. T: 886-6-2289666, F: 65-8621350

Chernauskas, L.R., Geosciences Testing & Res Inc., 7 Gooseneck Lane, Westford, MA 1886. T: 978-692-5344, F: 978-251-9396, E: les@gtrinc.net

Cherrington, G.G., Cherrington Cons Engr Svcs, 859 Lurline Dr., Foster City, CA 94404. T: 650-574-7449, F: 650-574-4904, E: ggcherring@aol.com

Cheung, W.-T.W., California Dept of Transportation, 253 S. Ashton Ave, Millbrae, CA 94030. T: 650-652-5797, F: 415-356-6640, E: warwicksolar11@yahoo.com

Chew, R.Y., Robert Y Chew Geotech Inc., Suite 7, 26062 Eden Landing Rd., Hayward, CA 94545. T: 510-783-1881, F: 510-783-1912

Chin, C.-C., Geomatrix Consultants, 2101 Webster Street 12th Fl, Oakland, CA 94612. T: 510-663-4100, F: 510-663-4141, E: ccchin@geomatrix.com

Chin, C.-T., Moh & Assocs Inc., 7f 83 Chung-Hsiao E Rd Sec 1, Taipei, 100 Taiwan, ROC. T: 886-2-239-34167, F: 886-2-269-68688, E: ctchin@alum.mit.edu

Chin, J.T., JT Geodesign, 27 Jalan Bnrp 7/1b, Bukit Rahman Putra, Sg Buloh, 47000 MALAYSIA. T/F: 03-61567993, E: cjt@tm.net.my

Chitwood, D.E., US Army Corps of Engineers, 336 E 46th St, Long Beach, CA 90807. T: 213-836-2821, F: 213-452-4199, E: dchitwood@spl.usace.army.mil

Chiu, C.-W., Hong Kong Inst of Vocational Education, Flat 2 3/F Block C, 6 Dragon Terrace Dragon Court, Causeway Bay, HONG KONG. T: 011-852-2578-5383, F: 011-852-2578-53, E: justchiu@vtc.edu.hk

Chiu, S., 36 Kau Wah New Village, Lai Chi Kok, Kowloon, HONG KONG. T/F: 85223710119, E: drslchiu@hkstar.com

Cho, K.Y.-J., Gannett Fleming Engineers and, 136 Delmar Ave, Glen Rock, NJ 7452. T: 201-652-3583, F: 212-268-6684, E: kcho@gfnet.com

Chou, N.N., Genesis Group Taiwan, 11f-1 No 268, Kuang-Fu S Rd., Taipei, Taiwan, ROC. T: 886-2-2772-1031, F: 886-2-2772-2494, E: gengroup@ms4.hinet.net

Christensen, C., Kleinfelder Inc., 7490 Marylebone Rd., West Jordan, UT 84084. T: 801-566-7490, F: 801-466-6788, E: cchristensen@kleinfelder.com

Christensen, R., RW Christensen Inc., N 739 Old Hwy 26, Fort Atkinson, WI 53538. T: 920-568-0400, F: 920-568-0401, E: rchriste@compufort.com

Christensen, S.N., Soil and Materials Engineers, 550 Glendale Circle, Ann Arbor, MI 48103. F: 313-454-0629, E: christen@plymouth.soilmat.com

Christian, C.H., Christian Wheeler Engineering, 10854 Uvalde Ct, San Diego, CA 92124. T: 619-278-6534, F: 619-496-9758

Christian, J.T., 23 Fredana Rd., Waban, MA 2468. T: 617-969-2741, F: 617-244-0816, E: jtchrist@world.std.com

Christie, M.A., Geosyntec Consultants, 204 1/2 Miramar Ave, Long Beach, CA 90803. T: 562-621-0846, F: 714-969-0820

Christman, T.K., Ohio Epa, 3060 Rightmire Blvd, Columbus, OH 43221. T: 614-451-4338, F: 614-644-3146, E: timothy.christman@epa.state.oh.us

Christopher, B., Consultant, 210 Boxelder Lane, Roswell, GA 30076. T: 770-641-8696, F: 770-645-1383, E: barryc 325@aol.com

Christopherson, A.B., Peratrovich Nottingham & Drage, 1506 W 36th Ave, Anchorage, AK 99503. T: 907-561-1011, F: 907 563-4220, E: pnd@alaska.net

Christy, T.M., Geoprobe Systems, 601 N Broadway, Salina, KS 67401. T: 913-825-1842, F: 913-825-2097

Chrysovergis, S.P., SPC Geotechnical Inc., 1235 N Tustin Ave, Anaheim, CA 92807. T: 714-630-0321, F: 714-630-0326, E: stavros@spcgeo.com

Chryssafopoulos, H.S., HSCE Inc., 6642 Patio Ln, Boca Raton, FL 33433. T: 561-368-2354, F: 602-234-0699

Chu, D.B., Ninyo & Moore, 11 Washington, Irvine, CA 92606. T: 714-653-2463, F: 949-472-5445

Chu, W., Converse Consultants, 18809 Sherbourne Place, Rowland Hghts, CA 91748. T: 626-965-6258, F: 909-796-7675, E: wchu@converseconsultants.com

Chung, C.-K., Seoul Natnl Univ, Department of Civil Eng, Sinlim-Dong Gwanak-Gu, Seoul, 151-742 SOUTH KOREA. F: 822-887-0349, E: geolabs@gong.snu.ac.kr

Chung, K., Kaycie Geoconsultants Inc., 148-11 Ulchiro 2-Ga, Chung-Gu, Seoul, 100-192 SOUTH KOREA. F: 011-82-2-2266-0, E: koochung@kcgeo.co.kr

Chung, S.G., Dong A Univ, Dept of Civil Eng, Pusan, 604-714 SOUTH KOREA. T: 051-200-7625, F: 051-201-1419, E: sgchung@mail.donga.ac.kr

Ciance, M.A., Jaworski Geotech Inc., 150 Zachary Rd., Manchester, NH 3109. T: 603-647-9700

Ciancia, A.J., Langan Engineering & Environm, 165 Lake Drive West, Wayne, NJ 7470. T: 973-628-7810, F: 212-964-7885, E: aciancia@langan.com

Cibor, J.M., Fugro South Inc., 6100 Hillcroft, P.O. Box 740010, Houston, TX 77274. T: 713-778-5576, F: 713-773-5600, E: jcibor@fugro.com

Cikanek, E.M., Morrison Knudsen Corp, 1211 Town Center Drive, Las Vegas, NV 89134. T: 702-295-4439, F: 702-295-5198

Cioffi, A.L., NY City Techincal College, 267 Lafayette Ave, Cortlandt Manor, NY 10567. T: 914-739-5234, F: 718-260-5677, E: tcioffi@nyctc.cony.edu

Clark, D.M., Civil & Environmental Cons, 333 Baldwin Rd., Pittsburgh, PA 15205. T: 412-429-2324, F: 412-429-2114, E: dclark@cecinc.com

Clark, M.F., Sea Consultants Inc., 83 Western Ave, Delmar, NY 12054. T: 518-478-0472, F: 617-498-4775, E: kmclark1@msn.com

Clark, R.P., Tams Consultants Inc., 72 Mosadak St, Dokki, Cairo, 12311 EGYPT. T: 011-2064 332833, F: 2064. 332833, E: clark@egyptonline.com

Clarke, C.R., Oklahoma Dot/Material Division, 616 Coopers Hawk Dr., Norman, OK 73072. T: 405-872-8737, F: 405-522-0552, E: cclarke@odot.org

Cleary, T.F., New Hampshire State of, 29 Bartley Hill Rd., Londonderry, NH 3053. T: 603-434-4721, E: tfcleary@aol.com

Clemence, S.P., Syracuse Univ, 330 Berkeley Dr., Syracuse, NY 13210. T: 315-422-4596, F: 315-443-1324

Clemente, J.F., Tams Cons Inc., 655 Third Ave, New York, NY 10017. T: 212-867-1777, F: 212-697-6354

Clemente, J.L.M., 5275 Westview Dr., Bp2 3a10, Frederick, MD 20702. T: 301-228-7646, F: 301-682-6415, E: jlclemen @bechtel.com

Clements, K.A., 1718 Garden St, Santa Barbara, CA 93101. T: 805-569-1376, F: 805-564-1327, E: kclements@fugro.com

Cliano, D.F.Z., Alstom Power, 109 Poblacion, San Jose, Negros Oriental, 6202 PHILIPPINES. T: 011-63-35-4170358, E: dennis_cliano@hotmail.com

Clinton, S., Hayes Seay Mattern & Mattern, P.O. Box 13446, Roanoke, VA 24034. T: 703-857-3115, F: 540-857-3180, E: hsmm@roanoke.infi.net

Clinton, T.M., NAT Associates, 308 Fm 1830, Argyle, TX 76226. T: 940-455-2236, F: 940-455-2409, E: natassociates @worldnet.att.net

Coduto, D.L.J., Terra Insurance Co, 2 Fifer Ave, Suite 100, Corte Madera, CA 94925. T: 415-927-2901, F: 415-927-3204

Coduto, D.P., Ca Poly State Univ @Pomona, 1166 Eileen Ct, Upland, CA 91784. T: 909-982-1535

Coggins, R.T., Mississippi State Univ, 189 Fondren Dr., Columbus, MS 39702. T: 601-898-8262, E: msucog@yahoo.com

Colbaugh, E., David, Evans, Colbaugh & Assoc Inc., 2453 Impala Dr., Carlsbad, CA 92008. T: 760-438-4646, F: 760-438-4670, E: ecageo@aol.com

Coleman, Jr, J.E., Soiltech Engineering & Testing, 2502 Gravel Dr., Fort Worth, TX 76118. T: 817-595-0064, F: 817-595-0708

Coleman, D.M., Aquaterra Engineering Inc., 225 Woodrun Drive, Ridgeland, MS 39157. T: 601-853-2179, F: 601-956-9533, E: dcoleman@aquaterrainc.com

Coleman, J.R., Coleman Geotechnical, 22611 Marylhurst Court, El Toro, CA 92630. T: 949-770-4861, F: 949-461-5262, E: jcole21502@aol.com

Collett, D.L., Morrison Knudsen Corp, P.O. Box 73, Boise, ID 83729. T: 011-44-1925-85-4536, F: 011-44-1925-85, E: david.collett@wgint.com

Collins, B.M., Braun Intertec Corporation, 1820 39th St Sw #216, Fargo, ND 58103. T: 701-492-9379, F: 701-232-7817, E: bcollins@brauncorp.com

Collins, T.G., Huesker Inc., P.O. Box 411529, Charlotte, NC 28241. T: 704-588-5500, F: 704-588-5988

Collison, G.H., Golder Assocs Inc., 9100 Nesbit Lakes Dr., Alpharetta, GA 30022. T: 770-998-8156, F: 770-934-9476, E: gcollison@golder.com

Compton III, G., Deep Foundations Institute, 120 Charlotte Pl, 3rd Floor, Englewood Cliffs, NJ 7632. T: 201-567-4232, F: 201-567-4436, E: dfihg@dif.org

Comstock, Jr, D.C., Envirocare of Utah, 2550 Duportail Street, Apt K261, Richland, WA 99352. T: 509-943-5894, F: 801-537-7345, E: dccjmb@yahoo.com

Conlin, Jr, M.F., CSC Engineering & Environmental Consultn, 3008 Welsh Avenue, College Sta, TX 77845. T: 979-693-7401, F: 979-778-0820, E: rconlin@txcyber.com

Conlon, C.R., Schnabel Foundation Co, 210 Cleveland St, Cary, IL 60013. T: 847-639-8900, F: 847-639-2347

Conlon, R.J., Chi Energy Inc., 10 Dubeau Dr., Derry, NH 3038. T: 603-425-6326, F: 978-681-7727, E: rjconlon @aol.com

Connell, D.H., Harlan Tait Associates, 130 Produce Avenue Suite E, South San Francisco, CA 94080. T: 650-737-5440, F: 650-737-5441, E: dconnell@htageo.com

Conner, M.R., AG Wassenaar Inc., 2180 S Ivanhoe St, Suite 5, Denver, CO 80222. T: 303-759-8100, F: 303-756-2920, E: connerm@agwassenaar.com

Conner, S.E., Schnabel Eng Assoc, 101 Professional Park Dr., Se, Blacksburg, VA 24060. T: 540-953-1239, F: 540-953-3863, E: sconner@schnabel-eng.com

Conroy, J.M., 2011 Franklin Dr., Papillion, NE 68133. T: 402-593-9451, F: 402-221-4579, E: james.conroy@offutt.af.mil

Contreras, I.A., Barr Engineering, 4700 W 77th St, Minneapolis, MN 55435. T: 294-0329, F: 912-832-2601, E: icontreras@barr.com

Conway, P.J., P.O. Box 1782, Rohnert Park, CA 94927. T: 707-792-9221

Cook, B.E., Geoenvironmental Group, 216 Northwest Terrace, Silver Spring, MD 20901. T: 301-681-5623, F: 888-659-8068, E: becjck@erols.com

Cook, D.H., HNTB, 1402 Baylor Ln, Jacksonville, FL 32217. T: 904-737-2121, F: 904-448-9474, E: dcook@ hntb.com

Cook, R.F., 6738 Vanderbilt Place, Rancho Cucamonga, CA 91701. F: 818-568-6101

Cooley, L.A., Burns Cooley Dennis Inc., P.O. Box 12828, Jackson, MS 39236. T: 601-856-9911, F: 601-856-9774, E: bcd@netdoor.com

Cooper, J.D., Hillis Carnes Engineering Asso, 484 Hawk Ridge Lane, Sykesville, MD 21784. F: 301-631-2166, E: jcooper@hcea.com

Cope, T.M., Total Engineering Service Inc., 444 S 2nd St, La Salle, CO 80645. T: 970-284-9240, F: 970-284-9239, E: tmcope@concentric.net

Cording, E.J., Univ of Illinois, 2230 Newmark Ce Lab Mc 250, 205 N Mathews Ave, Urbana, IL 61801. T: 217-333-6938, F: 217-351-8700, E: ecording@uiuc.edu

Cordon, C.R., Cordon Y Merida Ings M-574, P.O. Box 02-5345, Miami, FL 33102. T: 502-334-3473, F: 502-334-3472, E: cymings@gua.net

Cornforth, D.H., Cornforth Cons Inc., The Lincoln Building, 10250 Sw Greenburg Rd., Ste 111, Portland, OR 97223. T: 503-452-1100, F: 503-452-1528, E: dcornforth@ cornforthconsultants.com

Corser, P.G., Montgomery Watson Inc., P.O. Box 774018, Steamboat Spgs, CO 80477. T: 970-879-6260, F: 970-879-9048, E: patrick.corser@mw.com

Cosgrove, K.M., Materials Testing Lab, 1529 Jericho Turnpike, New Hyde Park, NY 11040. T: 516-354-6600, F: 516-354-6690

Cosgrove, T.A., 1665 Brownstone Blvd, Apt 2, Toledo, OH 43614. T: 419-861-1198, E: tcosgrove@bowser-morner.com

Costantino, C.J., City College of New York, Four Rockingham Road, Spring Valley, NY 10977. T: 914-354-2602, F: 212-650-6965, E: ccostantino@earthlink.net

Costopoulos, S., Geoplan Ltd Cons Engrs, Scoufa Str 2, Athens, 15231 GREECE. T: 01-3646177, F: 01-3624182

Cottin, D.J., Geotest Inc., 53 Huntington Pkwy, Saint Charles, MO 63301. T: 314-946-4838

Cotton, B.E., Fugro South Inc., 6100 Hillcroft, Houston, TX 77081. T: 713-778-5596, F: 713-778-5544, E: bcotton@ fugro.com

Couch, Jr, F.B., Stanley D Lindsey & Assoc, 7843 Nolensville Road, Arrington, TN 37014. T: 615-395-4344, E: fcouch7574@aol.com

Courpon, J., Soletanche-Bachy, 102 Rue De Longchamp, 75116 Paris, FRANCE. T: 27117834381, F: 27114941604, E: jcourpon@soletanche-bachy.com

Couvrette, C.P., Cornerstone Geotechnical, 14709 226th Ave Ne, Woodinville, WA 98072. T: 425-788-5958, F: 425-844-1987, E: couvrette@aol.com/chuckc@cornerstone geotechnical.c

Cowell, M.J., Geostructures Inc., 107 Loudoun St Se, Leesburg, VA 20175. T: 703-771-9844, F: 703-771-9847, E: geostructures@erols.com

Cowen, J.A., Building & Earth Sciences Inc., 5545 Derby Drive, Birmingham, AL 35210. F: 205-836-9007, E: jcowen@wwisp.com

Craig, W.H., Univ of Manchester, School of Engineering, Manchester, M13 9PL UK. T: 0161-275-4321, F: 0161-275-4361, E: william.craig@man.ac.uk

Cramer, J.M., Nilex Corp, 15171 E Fremont Ave, Englewood, CO 80112. T: 303-766-2000, F: 303-766-1110, E: jcramer @nilex.com

Crampton, W., Group Delta Cons Inc., 4455 Murphy Canyon Rd., Suite 100, San Diego, CA 92123. T: 858-573-1777, F: 858-573-0069, E: walterc@groupdelta.com

Crandall, L. LeRoy, Crandall Consultants Inc., 330 Washington Blvd #601, Marina Del Rey, CA 90292. T: 310-578-7475, F: 310-823-7684, E: crancon@earthlink.net

Crapps, D.K., Schmertmann & Crapps Inc., 4509 NW 23rd Ave, Suite 19, Gainesville, FL 32606. T: 352-378-2792, F: 352-372-9808, E: sci@bellsouth.net

Crawforth, S.G., Kleinfelder Inc., 11347 W Hickory Rise Dr., Boise, ID 83713. T: 208-376-0490, F: 208-376-9703, E: sgcrawforth@kleinfelder.com

Creegan, P.J., Parsons Engineering Science, 44 Wild Rye Way, Napa, CA 94558. T: 707-251-9235, F: 707-251-9238, E: pcreegan@compuserve.com

Critz, W.W., Skamania County, P.O. Box 790, Courthouse Annex, Stevenson, WA 98648. T: 509-427-9448, F: 509-427-4839

Crosby, J., Jo Crosby & Assocs, P.O. Box 4220, Mountain View, CA 94040. T: 415-969-3268, F: 415-969-3345, E: jcassoc1@juno.com

Cross, G.R., Foundation Systems Engineering, 201 Colonial Hghts Rd., Suite A, P.O. Box 5267, Kingsport, TN 37663. T: 423-239-9226, F: 423 239 8677, E: fse@tricon.net

Cross, J.K., J Keith Cross Pe, 16332 Inglewood Lane Ne, Kenmore, WA 98028. T: 425-487-1860, F: 425-485-1621, E: jkcross@lightmail.com

Crotchett, D.R., Kelsey-Seybold Clinic Pa, P.O. Box 580649, Houston, TX 77258. T: 281-990-0645, F: 281-483-3395, E: denton.r.crotchett1@jsc.nasa.gov

Crowther, C.L., Whitaker Laboratory Inc., 18 Old Ferry Cove, Beaufort, SC 29902. T: 843-521-0134, F: 912-233-5061, E: ncrowther@islc.net

Crum, D.A., Corps of Engrs, 8768 Pheasant Run Road, Woodbury, MN 55125. T: 651-730-8820, F: 651-290-5805, E: douglas.a.crum@usace.army.mil

Crumley, A.R., Geoconsult Inc., #62 Caoba Street, TORRI-MAR, Guaynabo, PR 966. T: 787-793-6029, F: 809-793-0410, E: rcrumley@geoconsult-inc.com

Cunningham, P.E., James Colby, Team Services, 333 SW 9th St, Suite H., Des Moines, IA 50309. T: 515-282-8818, F: 515-282-8741, E: colby@teamsvcs.com

Curras, C.J., Dept Civil & Env Eng – UW Platteville, 1 University Plaza, Dept Civil & Env Eng, University of

Wisconsin, Platteville, WI 53818. T: 608-342-1541, F: 608-342-1566, E: currasc@uwplatt.edu

Curry, Jr, G.W., Malcolm Pirnie Inc., 11832 Rock Landing Dr., Newport News, VA 23606. T: 757-873-4355, F: 757-873-8723, E: gcurry@pirnie.com

Cushing, A.G., Dept of Civil & Environmental, 413 Cornell Street, Ithaca, NY 14850. T: 607-277-0113, E: agc5@cornell.edu

Cymanski, D.V., KC Engineering Co, 303 Stonyford Dr., Vacaville, CA 95687. T: 707-451-2096, F: 707-447-4143, E: kcengr@cwnet.com

Daemen, J.J.K., Univ of Nevada, 2620 Pioneer Dr., Reno, NV 89509. F: 702-784-1833

Dailey, J.H., JH Dailey Cons Geotech Engrs, P.O. Box 5345, Petaluma, CA 94955. T: 707-778-7978, F: 415-357-1205, E: jhdge@worldnet.att.net

Daly, J.J., Golder Assocs Inc., 3730 Chamblee Tucker Road, Atlanta, GA 30341. T: 706-496-1893

Damron, J.T., Padre Assocs Inc., 5450 Telegraph Rd., Suite 101, Ventura, CA 93003. T: 802-644-2200, F: 805-644-2050

Daniel, D.E., Univ of Illinois, Dept of Civil Engineering, 205 North Mathews Ave, Urbana, IL 61801. T: 217-333-1497, F: 217-333-9464, E: dedaniel@uiuc.edu

D'Antonio, R.G., LA County Public Works, 17434 Emelita Street, Encino, CA 91316. T: 818-774-9071, F: 626-458-4913, E: rdanton@dpw.co.la.ca.us

D'Appolonia, E., 1177 Mc Cully Dr., Pittsburgh, PA 15235. T: 412-373-3855

Dare, C., Harza Eng Co, 425 Roland Way, Oakland, CA 94621. T: 510-568-4001, F: 510-568-2205, E: cdare@harza.com

Darendeli, M.B., Univ of Texas, 11900 Hobby Horse Court, Apt 518, Austin, TX 78758. T: 512-719-3063, E: darendeli@mail.utexas.edu

Darling, J.W., URS Corp, 10975 El Monte St, Ste 100, Overland Park, KS 66211. T: 913-344-1000, F: 913-344-1011

Darnell, Jr, J.R., Reynolds-Schlattner-Chetter-Roll Inc., 14607 Hillside Ridge, San Antonio, TX 78233. T: 210-654-3863, F: 210-366-2324, E: rscr@flash.net

Darragh, Jr, R.D., 2721 Alida St, Oakland, CA 94602. T: 510-531-9746, F: 510-531-9702

Das, B.M., California State Univ, College of Eng & Computer Scie, 6000 J St, Sacramento, CA 95819. T: 916-278-6127, F: 916-278-5949, E: dasb@csus.edu

Dasenbrock, D.D., Minnesota Department of Trans, 414 7th Ave SE Apt B302, Minneapolis, MN 55414. T: 612-623-4061, F: 651-779-5510, E: derrickd@tc.umn.edu

Dasgupta, R., Planning & Design Bureau, M50 Market Block, Greater Kailash Ii, New Delhi, 110048 INDIA. T: 648-4406, F: 317-576-4070

Dash, U., Penn Dept of Transp, 2890 Sunset Dr., Camp Hill, PA 17011. T: 717-761-5508, F: 717-787-2882, E: udash@dashes.com

David, L.M., Malcolm Drilling Co, 515 Laurent Rd., Hillsborough, CA 94010. T: 650-348-8085, F: 650-952-5542

Davidian, Z.K., SEPTA, 302 Langford Rd., Broomall, PA 19008. T: 610-325-1019, F: 215-580-8177

Davidson, J.L., Univ of Florida, Dept of Civil Eng, P.O. Box 116580, Gainesville, FL 32611. T: 352-392-0957, E: jdavi@ce.ufl.edu

Davidson, L.K., Terracon Cons Inc., 5440 W 152nd Terrace, Leawood, KS 66224. T: 913-685-9709, F: 913-599-0574, E: lkd@terracon.com

Davidson, R.R., URS Australia Pty Ltd., 8 Kennedy Pl, St Ives NSW, 2075 AUSTRALIA. T/F: 1.16E+12, E: richard_davidson@urscorp.com

Davie, J.R., Bechtel Corp, 5275 Westview Drive, Frederick, MD 21703. T: 301-228-7647, F: 301-682-6415, E: jdavie@bechtel.com

Davis, C.W., Mississippi Dept of Transport, Box 1023, Raymond, MS 39154. T: 601-857-8655, F: 601-359-1797, E: cdavis@mdot.state.ms.us

Davis, C.A., Los Angeles Dept of Water and Power, 27017 Vista Encantada Dr., Valencia, CA 91354. T: 661-297-1215, F: 213-367-3792, E: craig.davis@water.ladwp.com

Davis, J.R., US Army Corps of Engineers, 13680 Mount Pleasant Road, Jacksonville, FL 32225. T: 904-220-8483, E: jacob.r.davis@saj02.usace.army.mil

Davis, S.L., URS Greiner Woodward Clyde, 106 York Drive, Princeton, NJ 8540. T: 609-688-3499, E: stuart_davis@urscorp.com

Davisson, M.T., MT Davisson Cons Eng, 15 Lake Park Rd., Champaign, IL 61822. T: 217-359-5206, F: 217-352-1495

Davit, A.J., 608 Plantation Ct, Panama City, FL 32404. T: 850-874-0005, E: anthony.davit@knology.net

Davit, G.C., Professional Consulting Corp, 19225 Gatlin Dr., Gaithersburg, MD 20879. T: 301-926-8569

Day, S.R., Geo-Solutions, 26 West Dry Creek Cir, Suite 600, Littleton, CO 80120. T: 720-283-0505, F: 720-283-8055, E: sday@geo-solutions.com

De Alba, P.A., Univ of New Hampshire, Dept of Civil Eng, Kingsbury Hall – Rm 236, Durham, NH 3824. T: 603-862-1428, F: 603-862-2364, E: padealba@hypatia.unh.edu

De Bondt, A.H., Ooms Avenhorn Holding Bv, P.O. Box 1, 1633 Z6, Auenhorn, NETHERLANDS ANTILLES. T: 011-31-229-547700, E: adebondt@ooms.nl

De Carvalho, L.H., CEC Eng Cons Sc Lida, Rua Joaquim Nabuco 1550, Apt 1001 Aldeota, Fortaleza – Ce, 60125-120 BRAZIL. T: 085-224-9837, F: 85-4581462, E: cec@secrel.com.br

De Donati, A., Via Damiani 67, Morbegno So, 23017 ITALY. T: ++39-342-612858, E: dedonati@libero.it

De Groff, W., Fugro Mc Clelland, 2619 Live Oak Dr., Rosenberg, TX 77471. T: 713-342-9763

De La ROCha, A.J., Western Technologies Inc., 16408 S 46th Way, Phoenix, AZ 85048. T: 480-753-0818, F: 602-470-1341

de Larios, J.C., Geomatrix Consultants, 116 Knoll Circle, S San Francisco, CA 94080. T: 415-952-9287, F: 510-663-4141, E: jdelarios@geomatrix.com

De Mascio, F.A., Froehling & Robertson Inc., 933 Nugent Dr., Chesapeake, VA 23322. T: 757-546-8581, F: 757-436-1674

De Natale, J.S., Ca State U @San Luis Obispo, Dept of Civil Eng, San Luis Obispo, CA 93407. T: 805-756-1370

De Puy, M.A., Panama Canal Commission, P.O. Box 5169, Zona 5, PANAMA. T: 507-260-7143, F: 507-272-2750, E: madepuya@pananet.com

De Rose, C.R., Test Lab Inc., 2901 N Dale Mabry Hwy, Apt 217, Tampa, FL 33607. T: 813-876-2932

De Rubertis, K.P., P.O. Box 506, 6318 Flowery Divide, Cashmere, WA 98815. T: 509-782-3434, F: 509-782-2247, E: derubertis@aol.com

De Santis, E., PSI Inc., 5329 Night Roost Ct, Columbia, MD 21045. T: 410-997-7154, F: 410-715-9319

de Verteuil, P.D., Dodson Stilson Inc., 6121 Huntley Rd., Columbus, OH 43229. T: 614-888-0576, E: pdeverteuil@dlzcorp.com

De, A., Geosyntec Consultants, 1500 Newell Ave, Suite 800, Walnut Creek, CA 94596. T: 925-943-3034, F: 925-943-2366, E: anirbande@geosyntec.com

Deal, C.E., Clifton E Deal, 735 Se Hale Place, Gresham, OR 97080. T/F: 503-661-4392, E: cdeal@compuserve.com

Dean, D.A., Carlton Engineering Inc., 8594 Black Kite Dr., Elk Grove, CA 95624. T: 916-682-8116, F: 530-677-6645, E: ddean@carlton-engineering.com

Deatherage, J.D., Copper State Engineering Inc., 7820 E Evans Rd., Ste 700, Scottsdale, AZ 85260. T: 480-368-1551, F: 480-368-1556, E: cse@copperstate-eng.com

Deaton III, H.J., Schnabel Foundation Co, 45240 Business Court, Suite 250, Sterling, VA 20166. T: 703-742-0020, F: 703-742-3319

Decker, M.D., Colchester Town of, 127 Norwich Ave, Colchester, CT 6415. T: 860-537-7288, F: 860-537-0547, E: publicworks@colchesterct.org

Dee, S.E., Atest Inc., P.O. Box 262121, Highlands Ranch, CO 80163. T: 303-841-6593, F: 303-841-0136

Deeker, M.R., Sinclair Oil Corporation, 9310 West 90 Street, Overland Park, KS 66212. T: 913-381-1758, F: 913-321-1750, E: mdeeker@sinclairoil.com

Deel II, C.C., SAIC, 2111 Eisenhower Ave, Ste 205, Alexandria, VA 22314. T: 703-683-6242, F: 703-683-6249, E: deelc@saic.com

Deere, D.U., 6834 Sw 35th Way, Gainesville, FL 32608. T: 352-378-3061

DeGroot, D.J., Univ of Massachusetts, Amherst- Dept of Civil Eng, 38 Marston Hall, Amherst, MA 1003. T: 413-545-0088, F: 413-545-2840, E: degroot@ecs. umass.edu

Del Val, J., Del Val Grouting Cons, 6299 Crampton Dr., N, Salem, OR 97303. T: 503-390-5159, F: 503-390-5178, E: dvconstl@org.org

Delea, D.J., 36 Shannon Rd., Hempstead, NH 3841. T: 603-993-1109, F: 603-889-3984, E: dan@geophysical.com

Demaree, M.M., Mc Phail Assocs, 30 Norfolk St, Cambridge, MA 2139. T: 617-868-1420

Demcsak, M.R., Lippincott & Jacobs Cons Engrs, 67 Harrowgate Dr., Cherry Hill, NJ 8003. T: 609-424-4170, F: 856-461-3166

Dennis, Jr, N.D., Univ of Arkansas, 4190 Bell Engineering Ctr, Dept of Civil Engr, Fayetteville, AR 72701. T: 501-575-2933, F: 501-575-7168, E: ndd@engr.uark.edu

Dennis, Jr, W.D., Burns Cooley Dennis Inc., P.O. Box 12828, Jackson, MS 39236. T: 601-856-9911, F: 601-856-9774, E: bcd@netdoor.com

Derick, R.K., Universal Eng Sciences, 3532 Maggie Blvd, Orlando, FL 32811. T: 407-423-0504, F: 407-423-3106, E: kderick@uesorl.com

Desai, C.S., Univ of Arizona, Dept of Civil Eng & Eng Mech, Tucson, AZ 85721. T: 520-621-6569, F: 520-621-6577, E: csdesai@engr.arizona.edu

Desai, M.B., PSS Cons Engrs, P.O. Box 4039, Warren, NJ 7059. T: 732-560-9700, E: mdesai@psands.com

Desai, N.D., Gammon India Ltd., A14 3rd Flr Divya Prakash Soc, Dadabhai Cross Rd1 Near Bhavan, College Andehri, Mumbai Mah, 400058 INDIA. T: 6237916/6233458, F: 9.10E+11, E: desain@bom5.vsnl.net.in

Deschamps, R.J., FMSM Engineers, 2702 Westmoreland Drive, Lexington, KY 40513. T: 859-381-9672, F: 859-422-3100

Desrosiers, R., Hart Crowser, 75 Montgomery Street, Fifth Floor, Jersey City, NJ 7302. T: 201-985-8100, F: 201-985-8182, E: rldesrosiers@netzero.net

Deterling, P.A., Earthwork Engineering Inc., 175 Ridge Road, Hollis, NH 3049. T: 603-465-3877, F: 603-465-9650, E: eworknh@aol.com

Dette, J.T., 22 Duer Pl, Weehawken, NJ 7087. T: 201-866-0692, E: jtdette@aol.com

Deutsch, Jr, W.L., 216 Woodland Dr., Downingtown, PA 19335. T: 610-269-9848, F: 610-269-5675

Devin, S.C., Steven C Devin Pe, P.O. Box 1782, Quincy, CA 95971. T/F: 530-283-2199, E: sdevin@psln.com

Dewey, C.S., Federal Highway Admin Group, Wflhd, 610 E Fifth St, Vancouver, WA 98661. T: 360-696-7702

Di Gioia, Jr, A.M., GAI Cons Inc., 11 Wisteria Dr., Pittsburgh, PA 15235. T: 412-793-9387, F: 412-856-4970, E: a.digioia@gaiconsultants.com

Di Pilato, M.A., Sanborn Head & Associates, 239 Littleton Rd., Suite 1c, Westford, MA 1886. T: 978-392-0900, F: 978-392-0987, E: mdipilato@sanborn-head.com

Diamond, J.A., Lockwood Greene Engrs, 400 Mall Blvd, Ste E, Savannah, GA 31406. T: 912-352-3000, F: 912-352-0492, E: jdiamond@lg.com

Diaz, G.M., Diaz Yourman & Assocs, 12622 Singing Wood, Santa Ana, CA 92705. T: 714-731-6352, F: 714-838-8741

Dickenson, S.E., Oregon State University, Dept of Civil Engineering, 307 Apperson Hall, Corvallis, OR 97331. T: 541-737-3111, F: 541-737-3052, E: sed@engr.orst.edu

Dief, H.M., GRL & Associates Inc., 8000 S Orange Ave, Orlando, FL 32809. T: 407-826-9539, F: 407-826-4747, E: hdief@lycos.com

Dietrich, R.J., 5240 Happy Pines Dr., Foresthill, CA 95631. T: 530-367-4490, E: rjdietrich@foothill.net

Dietz, J.T., Chicago Bridge & Iron, 30 West 110 Greenbrook, Warrenville, IL 60555. T: 630-393-2286, F: 815-439-3130, E: jdietz@chicagobridge.com

Digenova, J., Haley & Aldrich Inc., 82 Mulberry Lane, Chester, NH 3036. T: 603-483-5071, F: 603-624-8307, E: jgd@haleyaldrich.com

Dill, R.B., Haley & Aldrich Inc., 37 Sunset Rd., Winchester, MA 1890. T: 781-729-5086, F: 617-886-7600, E: rbd@haleyaldrich.com

Dilley, L.M., Shannon & Wilson Inc., 3151 E 64th, Anchorage, AK 99507. T: 907-562-5067, F: 907-561-4483, E: lmd@shanwil.com

Dirlam, J.K., Dept of The Interior, 1417 Key Blvd, P.O. Box 9822, Apt 107, Arlington, VA 22209. T: 703-522-7545

Divis, C.J., Treadwell & Rollo Inc., 109 Clement St, #3, San Francisco, CA 94118. T: 415-668-4628, F: 415-955-9041, E: cjdivis@treadwellrollo.com

Divito, R.C., Haley & Aldrich Inc., 950 East Ave, Rochester, NY 14607. T: 716-244-2946, F: 716-232-6768

Dobbs, W.B., Black & Veatch International-B, 8400 Ward Parkway W4, Kansas City, MO 64114. T: 001-591-278-4081, F: 001-591-278-408, E: bill.bv@puntoinformatico.com

Dobry, R., Rensselaer Polytech Inst, Dept of Civil Eng, Troy, NY 12180. T: 518-276-6934, E: dobryr@rpi.edu

Dodd, J.S., P.O. Box 240, Masonville, CO 80541. T: 970-481-5557, F: 303-721-9153

Dodsworth, P.R., Bace Geotechnical Inc., 6088 Bennett Valley Rd., Santa Rosa, CA 95404. T: 707-526-0205, F: 707-838-4420, E: victor@sonic.net

Dohms, P.H., Gallet & Assocs Inc., P.O. Box 30035, Pensacola, FL 32503. T: 850-477-0454, F: 850-477-0534, E: pdohms@aol.com

Dolcimascolo, A.R., Tams Cons Inc., 444 East 86th St, New York, NY 10028. T: 212-737-8434, F: 212-697-6354, E: ard@tamsconsultants.com

Dominguez, J., Unisersidad De Sevilla, Av Reina Mercedes, Sevilla, 41012 SPAIN. T: 011-34-5-455-6999, F: 011-34-5-455-69, E: pepoh@cica.es

Donald, V.R., P.O. Box 82160, Baton Rouge, LA 70884. T: 504-765-1800, F: 504-765-1810, E: vdonal@geoscience.eng.com

Dormier, R.A., Freeman-Millican, Inc., 504 Edgewood Ln, Ovilla, TX 75154. T: 972-217-1147, F: 214-503-1148, E: richard@fmi-dallas.com

Dornic, S.D., Exxonmobil Development Co, 119 W Hobbit Glen Dr., The Woodlands, TX 77384. T: 936-271-0495

Dotani, R.K., Govt of Balochistan, Commuication & Works Dept, Mt & Fc Laboratory Secr, Quetta, PAKISTAN. T: 844681

Dougherty, J.W., Geotechnical Consultants Inc., 26 Silverhorn Dr., San Antonio, TX 78216. T: 210-495-3202, F: 210-349-6151

Dove, J.E., Virginia Tech, 2901 Lancaster Drive, Blacksburg, VA 24060. T: 770-936-0371, F: 540-231-7532, E: jedove@vt.edu

Doven, A.G., Eastrn Medi Univ Dogu Akdeniz, Univ Insaat Muhendisligi Bolum, Gazimagosa Kktc, Mersin 10, TURKEY. T: -6302329, F: -3651012, E: gurhan.doven@emu.edu.tr

Dowding, C.H., Northwestern Univ, Dept of Civil Engr, Evanston, IL 60208. T: 847-491-4338, F: 847-491-4011, E: c-dowding@northwestern.edu

Doyle, E.H., 13811 Placid Woods Ct, Sugar Land, TX 77478. T: 281-494-1037, E: ehdoyle@flash.net

Doyle, J.V., Doyle Consultants, 10940 Wilshire Blvd, Suite 1600, Los Angeles, CA 90024. T: 310-443-4224

Doyle, L.C., Geotechnical Eng Testing Inc., 904 Butler Dr., Mobile, AL 36693. T: 334-666-7197

Dransfield, J.S., Amec Earth & Environmental Inc., 14623 189th Ave Ne, Woodinville, WA 98072. T: 425-881-0669, E: james.dransfield@amec.com

Drash, C.J., Drash Consulting Engineers Inc., P.O. Box 781208, San Antonio, TX 78278. T: 210-641-2112, F: 210 641-2124, E: cdrash@drashce.com

Dreessen, R.E., Thompson Dreessen & Dorner Inc., 10836 Old Mill Rd., Omaha, NE 68154

Drew, G.A., Ardaman & Assoc, 9970 Bavaria Rd., Ft Myers, FL 33913. T: 941-768-6600, F: 941-768-0409, E: grdaman@strato.net

Driscoll, D.D., Geotechnical Resources Inc., 9725 Sw Beaverton Hillsdale, Beaverton, OR 97005. T: 503-641-3478

Drnevich, V.P., Purdue Univ, 3309 Elkhart St, W Lafayette, IN 47906. T: 765-497-0511, F: 765-496-1364, E: drnevich@purdue.edu

Drooff, E.R., Hayward Baker Inc., 1875 Mayfield Rd., Odenton, MD 21113. T: 410-551-8200, F: 410-551-8206, E: erdrooff@haywardbaker.com

Druback, G.W., Malcolm Pirnie Inc., 128 Tick Tock Way, Stanfordville, NY 12581. F: 914-641-2645, E: gdruback@pirnie.com

Druebert, H.H., Shannon & Wilson Inc., 19329 47th Ave NE, Seattle, WA 98155. F: 206-633-6777, E: hhd@shanwil.com

Drumheller, J.C., Densification Inc., 40650 Hurley Ln, Paeonian Springs, VA 20129. T: 540-882-4404, F: 540-882-4190, E: filldoctor@aol.com

Drumm, E.C., Univ of Tennessee, Dept of Civil Eng, 223 Perkins Hall, Knoxville, TN 37996. T: 423-974-7715, F: 423-974-2608, E: edrumm@utk.edu

Druschel, S.J., Pathfinder Environmental, 64 Forest Ave, S Hamilton, MA 1982. T: 978-468-2181, E: druschel@mediaone.net

Druss, D.L., Parsons Brinckerhoff Inc., 185 Kneeland St, Boston, MA 2111. T: 617-951-6237, F: 617-346-7963, E: dldruss@home.com

Du, M., Tams Cons Inc., 180 Pearsall Drive, Apt 6e, Fleetwood, NY 10552. T: 914-663-4023, F: 212-697-6354, E: mdu@tamsconsultants.com

Dubbe, R.E., ECS Ltd., 814 Greenbrier Cir, Suite A, Chesapeake, VA 23320. T: 757-366-5100, F: 757-366-5203

Duevel, B.J., URS Greiner Woodward-Clyde, 111 Sw Columbia, Ste 990, Portland, OR 97201. T: 503-948-7250, F: 503-222-4292, E: bryan_duevel@urscorp.com

Dugan, Jr, J.P., Haley & Aldrich Inc., 93 Tall Timbers Rd., Glastonbury, CT 6033. F: 860-659-4003, E: jpd@haleyaldrich.com

Dunbar, R.A., Consulting Engineer, 2474 Buckley Rd., Columbus, OH 43220. T: 614-451-4144

Duncan, G.W., WK Engineering Inc., P.O. Box 87, Kuttawa, KY 42055. T: 502-388-0019, F: 502-388-7566, E: wallace@vci.net

Duncan, J.M., Virginia Tech, Dept of Civil Eng, 104 Patton Hall, Blacksburg, VA 24061. T: 540-231-5103, F: 540-231-7532, E: jmd@vt.edu

Duncan, L.C., Lance Duncan Pe Cpg, 10820 Chain of Rock St, Eagle River, AK 99577. T: 907-694-6731, F: 907-694-6730, E: lanceduncan@worldnet.att. net

Dunkelberger, C.E., Dunkelberger Eng & Testing Inc., 2309 Sw Webster Lane, Port St Lucie, FL 34953. T: 561-340-7036, F: 561-343-9404

Dunkelberger, D.S., Dunkelberger Eng & Testing, 6833 Vista Parkway North, W Palm Beach, FL 33411. T: 561-689-4299, F: 561-689-5955, E: dougd@detwpb.com

Dunlap, G.T., Burns Cooley Dennis Inc., P.O. Box 12828, Jackson, MS 39236. T: 601-856-9911, F: 601-856-9774, E: tdunlap@bcdgeo.com

Dunlap, S.L., Harris Cty Flood Control Dist, 16303 Mill Point, Houston, TX 77059. T: 281-480-1842, F: 713-684-4140, E: sld@hcfcd.co.harris.tx.us

Dunn, R.J., Geosyntec Consultants, 1500 Newell Ave, #800, Walnut Creek, CA 94596. T: 925-943-3034, F: 925-943-2366, E: jeffd@geosyntec.com

Dunn, T.C., Parsons Brinckerhoff Quade &, 11 Longfellow Rd., Shrewsbury, MA 1545. T: 508-842-5957, F: 617-951-0898

Dunne, T.N., Fugro West Inc., 5855 Olivas Park Drive, Ventura, CA 93003. T: 805-650-7000, F: 805-650-7010, E: tdunne@fugro.com

Durgunoglu, H.T., Bogazici Univ, Civil Eng Dept, PK 2 Bebek, Istanbul, 80815 TURKEY. T: 90-216-4923303, F: 90-216-4923306, E: durgunoglut@zetas.com

Durkee, D.B., Golder Associates, 2700 N Central Ave, Suite 300, Phoenix, AZ 85004. T: 602-728-0372, F: 602-728-0430, E: ddurkee@golder.com

Duryea, P.D., Knight Piesold And Co, 9441 W Walden Ave, Littleton, CO 80128. T: 303-904-1482, F: 303-629-8789, E: pete@kpco.com

Duryee, W.A., HNTB, 10301 Cody, Overland Park, KS 66214. T: 816-472-1201, F: 816-472-4060, E: wduryee@hntb.com

Dussault, M., Queformat Ltd., 591 Le Breton St, Longueuil, J4G 1R9 CANADA QC. T: 450-674-4901, F: 450-674-3370, E: info@queformat.com

Dvinoff, A.H., Geosystems Consultants Inc., 405 Burgess St, Philadelphia, PA 19116. T: 215-677-0838, F: 215-643-9440

Dwyre, E.M., Professional Service, 10918 Braewick Drive, Carmel, IN 46033. T: 317-843-0379, F: 317-216-7135, E: liza.dwyre@psiusa.com

Dyckman, D.F., 136 Barkentine St, Foster City, CA 94404. T: 650-349-3369

Dyregrov, A.O., Dyregrov Cons, 1666 Dublin Ave, Winnipeg, R3H 0H1 CANADA MB. T: 204-632-7252, F: 204-632-1442, E: dyregrov@autobahn.mb.ca

Easton, C., Freese and Nichols Inc., 4055 International Plaza, Suite 200, Ft Worth, TX 76109. T: 817-735-7335, F: 817-735-7491

Eastwood, D.A., Geotech Eng & Testing, 800 Victoria Drive, Houston, TX 77022. T: 713-699-4000, F: 713-699-9200, E: geo.tech@wt.net

Eckert, J.W., Eng Cons Svcs, 11446 Vale Spring Dr., Oakton, VA 22124. T: 703-620-2206, F: 703-834-5527, E: jeckert@ecslimited.com

Economides, J.A., San Diego County Water Auth, 3211 Fifth Ave, San Diego, CA 92103. T: 619-682-4100, F: 619-692-9356

Eddy, D.K., The Eddy Group, 6600 Kensal Ct, Springfield, VA 22152. T: 703-451-4094, F: 703-451-4962, E: soilengineer@members.asce.org

Eddy, G.L., Golder Associates Inc., 44 Union Blvd., Suite 300, Lakewood, CO 80228. T: 303-980-0540, F: 303-985-2080, E: greg_eddy@golder.com

Edelen, Jr, W.F., Froehling & Robertson Inc., 7138 Forrest Rader Drive, Charlotte, NC 28227. T: 704-573-4644, F: 704-596-3784, E: bedelen@fandr.com

Edfors, D.E., Lane Constr Corp, 14317 Blackmon Dr., Rockville, MD 20853. T: 301-460-3019, F: 703-893-1439

Edgar, T.V., Univ of Wyoming, 1860 North 23rd St, Laramie, WY 82072. T: 307-721-3830, F: 307-766-2221, E: tvedgar@uwyo.edu

Edgers, L., Tufts Univ, 26 Craftsland Rd., Chestnut Hill, MA 2467. T: 617-734-8324, F: 617-627-3994, E: ledgers@tufts.edu

Edil, T.B., Univ of Wisconsin, Madison Dept of Civil Eng, 1415 Engineering Dr., Madison, WI 53706. T: 608-262-3225, F: 608-263-2453, E: edil@engr.wisc.edu

Edinger, P.H., Mueser Rutledge Cons Engrs, 708 Third Ave, New York, NY 10017. T: 212-490-7110, F: 212-953-5626, E: pedinger@mrce.com

Edmunds, G.L., Westinghouse Savannah River Co, 4 Troon Way, Aiken, SC 29803. T: 803-649-1690

Edris, Jr, E.V., US Army Corps of Engrs, 416 Garden Grove St, Vicksburg, MS 39180. T: 601-638-5688, F: 601-634-3453, E: edrise@wes.army.mil

Egan, J.A., Geomatrix Consultants, 2101 Webster St, Suite 1200, Oakland, CA 94612. T: 510-663-4292, F: 510-663-4141, E: jegan@geomatrix.com

Egan, P.D., Tensar Earth Technologies Inc., 435 Laurel Chase Court, Atlanta, GA 30327. T: 404-256-5845, F: 404-250-9077

Ege, M., STFA Marine Construction Co, Bostanci Cami Sokak Hayat Apt, No 10/5, Bostanci, Istanbul, 81070 TURKEY. T: -550, F: +90-216-327-69-, E: muratzege@hotmail.com

Einstein, H.H., MIT, Room 1-342, Cambridge, MA 2139. T: 617-253-3598, F: 617-253-6044, E: einstein@mit.edu

Eith, P.E., Anthony W., Waste Management, 514 Parliament Rd., Marlton, NJ 8053. T: 856-596-5478, F: 215-428-3205, E: aeith@wm.com

Eksioglu, B.C., Langan Eng Services Inc., 90 West Street Suite 1210, New York, NY 10006. T: 212-964-7888, F: 212-964-7885, E: bersioglu@langan.com

El Abd, G.M., El Abd Eng, El-Abd Eng, 46 Al Sarwa St Dokki, Giza, EGYPT. F: 415-453-0343, E: galdeprom@in touch.com

El-Emam, M.M., Royal Military Collage CE Dept, P.O. Box 1700 Stan Forces, Kingston Ont, K7N 7B4 CANADA ON. T: 613-541-6000 Ext. 6347, F: 613-545-8336, E: elemanm@rmc.ca

Eliahu, S., SE Consulting Inc., 3315 Stage Coach Drive, Lafayette, CA 94549, T: 510-944-0250, F: 510-944-3378, E: secon@msn.com

Elioff, M.A., Parsons Brinckerhoff, 444 South Flower Street, Suite 1850, Los Angeles, CA 90071. T: 213-362-9470, F: 213-362-9480, E: elioff@pbworld.com

El-Kelesh, A.M., Osaka Univ Grad School of Eng, 2-1 Yamadaoka, Suita, Osaka, 565-0871 JAPAN. T: 810-6-6879-7626, F: 81-6-6879-7626, E: adel@civil.eng.osaka u.ac.jp

El-Khoury, F.C., Dar Al-Handasah Consultants, 15 Amr St, P.O. Box 895, Cairo, EGYPT. F: 3461170

El-Rousstom, A.K.H., Soil & Foundation Co Ltd., P.O. Box 8718, Jeddah, 21492 SAUDI ARABIA. T: 966-2 669-4987

Elton, D.J., Auburn Univ, Civil Eng Dept, 238 Harbert Eng Ctr, Auburn, AL 36849. T: 334-844-6285, F: 334-844-6290, E: elton@eng.auburn.edu

Engels, J.G., Earth Tech, 41 Madison Ave, Wakefield, MA 1880. T: 781-246-1418, F: 978-371-2468, E: jengels@earthtech.com

Ennis, B.P., 18 Tennyson Ave, North Haver, CT 6473. T: 203-239-1524, E: bennis8039@aol.com

Enrique, E.C., Toltest Inc., 8622 Misty Ridge Circle, Sylvania, OH 43560. T: 419-824-3432, F: 419-321-6259, E: eenrique@aol.com

Erel, B., Engineering Technologies Inc., 208 Milford Haven Cove, Longwood, FL 32779. T: 407-788-0452, F: 407-648-0092, E: bilgin@et-eng.com

Ergun, M.U., Middle East Tech Univ Odtu, Civil Engr Dept, Ankara, 6531 TURKEY. T: 0090-312-2102415, F: 0090-312-210126, E: eruf@metu.edu.tr

Erickson, A.E., Ch2m Hill, 135 S84th #325, Milwaukee, WI 53214. T: 414-272-2426, F: 414-272-4408, E: aerickso@ch2m.com

Ericson, W.A., BCI Engineers & Scientists Inc., 5150 Ewing Rd., Bartow, FL 33830. T: 941-537-5544, F: 941 667 2662, E: wericson@bcieng.com

Erikson, C.M., Mcphail Associates, 58 Grove St, Winchester, MA 1890. T: 781-729-7249, E: cerikson@mcphailgeo.com

Ernst, P.J., Pacific Soils Eng Inc., 6111 Killarney Ave, Garden Grove, CA 92845. T: 714-893-6578, F: 714-220-9589

Erol, O., Insaat Muh Bol Odtu, A., Ankara, 6531 TURKEY. T: 903122408, F: 9.03E+11, E: orer@rorqual.metu.edu.tr

Eruchalu, Jr, B.C., Milwaukee County, P.O. Box 250632, Milwaukee, WI 53225. T: 414-536-3449, F: 414-223-1850

Eskandari, A.Y., BSK & Assocs, 1181 Quarry Lane Building 300, Pleasanton, CA 94566. T: 925-462-4000, F: 925-462-6283

Esmiol, E.E., 6142 Dudley Ct, Arvada, CO 80004. T: 303-424-2610, F: 519-888-6197

Esrig, M.I., Melvin I Esrig Consulting Eng, 43 Royden Rd., Tenafly, NJ 7670. T: 201-569-8561, F: 201-569-9716, E: esrigs@aol.com

Esser, A.J., EDP Consultants Inc., 9375 Chillicothe Rd., Kirtland, OH 44094. T: 440-256-6500, F: 440-256-6507, E: ajesser@edpconsultants.com

Esser, J.S., 1167 Stratford Lane, Lake Zurich, IL 60047. T: 847-540-7203

Estalrich-Lopez, J., DPT Geologia, Universidad Autonoma, Bellaterra, 8193 SPAIN. T: 034-93-5811270, F: 034-93-5811263, E: joan.estalrich@uab.es

Esterhuizen, J.J., Chzm Hill, 3687 Twinberry Place, Corvallis, OR 97330. T: 703-951-2963, F: 541-752-0276, E: jesterhu@ch2m.com

Eto, H., P.O. Box 468, Sacramento, CA 95812. T: 916-557-6610, F: 916-557-6803, E: heto@spk.usace.army.mil

Eustis, C.L., Louis J Capozzoli & Assocs, 4747 South Park Street, Apt 203, Baton Rouge, LA 70816. T: 225-293-8446, F: 225-293-2463, E: ljca@mindspring.com

Evans III, C.H., URS Corp, 5106 Central Ave, Tampa, FL 33603. T: 813-237-8222, F: 813-286-6587, E: charlie_evans@urscorp.com

Evans, Jr, L.T., Willdan Assocs, 2813 Magna Vista St, Pasadena, CA 91107. T: 626-793-4341, F: 626-337-2103, E: tevans@willdan.com

Evans, A.D., Kleinfelder Inc., P.O. Box 813, Redlands, CA 92373. T: 909-307-9741, F: 909-792-1704, E: aevans@kleinfelder.com

Evans, D.A., Evans, Colbaugh & Assocs Inc., 2453 Impala Drive, Carlsbad, CA 92008. T: 619-438-4646, F: 619-438-4670

Evans, E.B., S&Me Inc., 1981 Oak Branch Way, Stone Mtn, GA 30087. T: 770-985-1960, F: 770-209-9370, E: eevans@smeinc.com

Evans, J.C., Bucknell University, Dept of Civil Eng, Lewisburg, PA 17837. T: 570-577-1371, F: 570-577-1822, E: evans@bucknell.edu

Ezzeldine, O.Y., Cairo Univ, 23 Amer St 7, Dokki Giza, Cairo, EGYPT. T: 011-202-3367087, F: 011-202-5729124

Failmezger, R.A., 2762 White Chapel Rd., Lancaster, VA 22503. T/F: 804-462-6189, E: insitusoil@prodigy.net

Fairhurst, C., Itasca Consulting Group Inc., 417 5th Ave N, South St Paul, MN 55075. T: 651-451-1234, F: 651-552-9624, E: fairh001@tc.umn.edu

Faris, J.R., Naval Facilities Eng Command, 2131 Avy Ave, Menlo Park, CA 94025. T: 650-233-9014, F: 650-244-2553, E: farisjr@efawest.navfac.navy.mil

Fark, R., Hdr Eng Inc., 541 Clemson Dr., Pittsburgh, PA 15243. T: 412-278-3735, F: 412-497-6080, E: rfark@hdrinc.com

Farmer II, M.L., Enviro Geotech Cons Inc., 154 Jester Pkwy, Rainbow City, AL 35906. T: 256-442-1221, F: 256-442-1242, E: mfarmer@tds.net

Farouz, E.E., Ch2m Hill, 13921 Park Center Rd., Herndon, VA 20171. T: 201-798-0871, E: farouz@msn.com

Farrar, J.A., US Bureau of Reclamation, 12253 West Arlington Ave, Littleton, CO 80127. T: 303-973-0359, F: 303-445-6341, E: jfarrar@do.usbr.gov

Farrell, T.F., 806 Lexington Dr., Aliquippa, PA 15001. T: 412-378-3260

Farrell, T.M., Farrell Design-Build Co Inc., P.O. Box 2374, Placerville, CA 95667. T: 916-622-8885, F: 530-621-4837, E: 4tom@farrellinc.com

Faucett, B., Universal Eng Sciences, P.O. Box 561281, Rockledge, FL 32956. T: 321-632-0179, F: 321-638-0978

Fay, S.M., Joseph B Fay Co, 1374 Freeport Rd., Pittsburgh, PA 15238. T: 412-963-1870, F: 412-963-7968, E: sfjbfay@aol.com

Fayad, P.H., EDRAFOR, Jdeidet El Metn, Metn, 12022140 LEBANON. T: 00-961-1-878313, F: 00-961-1-888707, E: edrafor@dm.net.lb

Feferbaum-Zyto, S., Geodesign Ltd., Dept of Public Works, P.O. Box 2321, Natanya, 42122 ISRAEL. T: 972-98616724, E: feferbaum_sa@yahoo.com

Feist, P., Nova Engr & Environmen, 5425 Mill Valley Dr., Douglasville, GA 30135. T: 770-942-3573, F: 770-425-1113

Feldsher, T.B., URS Corporation, 500 12th St, Suite 200, Oakland, CA 94607. T: 510-874-3245, F: 510-874-3268, E: theodore_feldsher@urscorp.com

Felice, C.W., AMEC Earth & Environmental, 14150 227th Ave Ne, Woodinville, WA 98072. T: 206-869-1690, F: 425-883-3684, E: cwf5156@aol.com

Feng, T.-W., P.O. Box 12-102, Chung-Li City, Taiwan, ROC. F: 886-3-4563160, E: twfeng@cycu.edu.tw

Feng, W.-L., LFR Levine Fricke, 515 Cottonwood Ln, Schaumburg, IL 60193. T: 847-891-2493, F: 847-695-7799, E: wlfeng@aol.com

Fennick, T.J., Mcphail Associates Inc., 51 Pudding Brook Dr., Pembroke, MA 2359. T: 617-826-9455, F: 617-868-1423

Ferguson, G., Geosystems Engr Inc., 7802 Barton St, Lenexa, KS 66214. T: 913-962-0909

Ferguson, K.A., GEI Cons Inc., 10135 Summit View Pointe, Highlands Ranch, CO 80126. T: 303-471-4094, F: 303-662-8757, E: kferguson@geiconsultants.com

Fernandez, G., G Fenandez Geotechnical Cons, 207 E. Sherwin Cir, Urbana, IL 61802. T: 217-351-8704, F: 217-351-8700

Feroz, S.M., JE Sverdrup, 227 Ridgemont Lane, Walnut, CA 91789. F: 714 549 5800, E: ferozms@sverdrup.com

Ferris, W.R., 106 Paseo Way, Greenbrae, CA 94904. T: 415-461-5927

Fiegel, G.L., Civil & Enviro Eng Dept, Ca Polytechnic State Unvsty, San Luis Obispo, CA 93407. T: 805-756-1307, F: 805-756-6330, E: gfiegel@calpoly.edu

Filz, G.M., Virginia Tech, Civil Eng Dept, Blacksburg, VA 24061. T: 540-231-7151, F: 540-231-7532, E: filz@vt.edu

Findlay, R.C., Findlay Engineering Inc., 450 Saco Street, Westbrook, ME 4092. T: 207-854-1970, F: 207-854-4964, E: cfindlay@findlayengineering.com

Fink, H.I., New York Dept of Transp, 75-40 Bell Blvd, Apt 50, Flushing, NY 11364. E: hifl11@aol.com

Finno, R.J., Northwestern Univ, Technological Inst, Dept of Civil Eng, Evanston, IL 60208. T: 847-491-5885, F: 847-491-4011, E: r-finnoenorthwestern.edu

Fiorillo, M.A., Van Der Horst, 33 Dundee Dr., Rochester, NY 14626. T: 716-225-1877, F: 716-359-9668, E: mf@rochester.rr.com

Fischer, G.R., Shannon & Wilson Inc., 5383 S Lamar Street, Littleton, CO 80123. T: 303-707-1693, F: 303-825-3801, E: grf@shanwil.com

Fischer, J.A., Geoscience Services, 129 Longview Terr, Gillette, NJ 7933. T: 908-647-5710, F: 908-221-0406, E: geoserv@hotmail.com

Fishman, K.L., Mcmahon & Mann Consulting Eng, 104 Regency Drive, Grand Island, NY 14072. T: 716-773-8047, F: 716-834-8934, E: klf@buffnet.net

Fitchett, Jr, W.W., Whitman Requardt & Assocs, 2315 Saint Paul St, Baltimore, MD 21218. T: 410-235-3450, F: 410-243-5716

Fiteni, Jr, J.J., TAMS Cons Inc., 3 Bryants Brook Rd., Wilton, CT 6897. T: 212-867-1777, F: 212-697-6354

Fitzwater, R.R., Eng Constr Support Inc., 8023 Dulins Ford Road, Marshall, VA 22115. T: 703-347-0646

Flaate, K.S., Road Technology Department, Bernh Herresv 6, Oslo, 376 NORWAY. T: 47-22491087, F: 47-22499687, E: kflaate@online.no

Flanagan, R.F., Parsons Brinckerhoff Inc., One Penn Plaza #2, New York, NY 10119. T: 212-465-5000

Fleming, P., Geo-Technologies Inc., 4978 E Aire Libre Ave, Scottsdale, AZ 85254. T: 602-493-6876, F: 602-404-2375, E: pfgti@home.com

Fletcher, C.S., Qore Property Science, 1210 Taramore Dr., Suwanee, GA 30024. T: 770-623-0712, F: 770-476-0213

Flissar, S.G., City of Sierra Vista, 1011 N Coronado Drive, Sierra Vista, AZ 85635. T: 520-458-3315, F: 520-452-7099, E: sflissar@ci.sierra-vista.az.us

Floess, C.H., Earth Tech., 12 Metro Park Rd., Albany, NY 12205. T: 518-437-8372, F: 518-458-2472, E: carstenfloess@earthtech.com

Flood, P.J., Earth Tech, 119 Chenoweth Dr., Simpsonville, SC 29681. T: 864-967-7648, F: 864-234-3069, E: pat_flood @earthtech.com

Flores, J.A., F Consultant International, P.O. Box 183, Sugar Land, TX 77478. T: 281-450-7817

Floyd, D., Spencer White & Prentis, 6 Colletti Ln, Swansra, MA 2777. T: 508-675-6511

Focht III, J.A., Focht Consultants Inc., P.O. Box 461047, San Antonio, TX 78246. T: 210-495-3624, F: 210-495-2727, E: focht@wireweb.net

Focht, Jr, J.A., Focht Consultants Inc., 12226 Perthshire, Houston, TX 77024. T: 713-464-2477, F: 713-464-8783, E: jafocht@flash.net

Fok, T.D.-Y., Thomas Fok & Assocs Ltd., 325 South Canfield-Niles Rd., Youngstown, OH 44515. F: 330-799-2519

Foley, P.M., Edwars and Kelcey Inc., 604 3rd St, Towanda, PA 18848. T: 717-265-7211, F: 610-696-3550, E: pfoley@ekmail.com

Fong, C.K.-T., Fong & Associates Inc., P.O. Box 62305, 1004 W 8th Ave Suite-C, King of Prussia, PA 19406. T: 610-768-0877, F: 610-768-0879

Fong, F., 3253 Falcon Ridge Rd., Diamond Bar, CA 91765. T: 909-861-1943, E: ffongge@aol.com

Foody, K.J., St Lawrence Cement, 3 Columbia Circle, Albany, NY 12203. T: 518-452-3563 3051, E: kfoody@stlawrence cement.com

Foose, G.J., Dept Civil/Environ Eng, Univ of Cincinati, P.O. Box 210071, Cincinati, OH 45221. T: 850-410-6453, F: 850-410-6142, E: foose@eng.fsu.edu

Fornek, J.T., US Army Corps of Engrs, 8241 West Agatite, Norridge, IL 60706. F: 312-353-2156, E: john.t.fornek@ usace.army.mil

Forrest, C.L., URS Corporation, 1615 Murray Canyon Rd., Suite 1000, San Diego, CA 92108. T: 619-294-9400, F: 619 293-7920, E: carol_forrest@urscorp.com

Forsythe III, R.G., 118 Westmoreland St, Summerville, SC 29483. T: 803-821-0357, F: 843-723-3648, E: rgf3@ bellsouth.net

Forte, E.P., Underpinning & Foundation, 107 Vanderbilt Ave, Manhasset, NY 11030. F: 718-786-6981, E: ed.forte@ underpinning.com

Foshee, F.W., S&ME Inc., 840 Low Country Blvd, Mt Pleasant, SC 29464. T: 803-884-0005, Ext. 225, F: 803-881-6149

Foster, O.C., Fugro South Inc., 117 South Main Street, Mcgregor, TX 76657. T: 254-840-2252, F: 254-840-3730, E: fugrowaco@aol.com

Foti, Jr, V.C., O&G Industries, 112 Wall Street, Torrington, CT 6790. T: 860-485-6612, F: 860-485-6602

Fotoohi, K., Exponent, 5401 Mcconnell Ave, Los Angeles, CA 90066. T: 310-302-7207, F: 310-823-7045, E: kfotoohi@ exponent.com

Fowler, M.E., Parsons Brinckerhoff Quade &, 77 Ford St, San Francisco, CA 94114. T: 415-225-8852, F: 415-243-9501, E: fowlerm@pbworld.com

Fox, P.J., UCLA, Dept of Civil & Env Engrg, 5731 Boelter Hall, Los Angeles, CA 90095. T: 310-794-4805, F: 310-206-2222, E: pfox@seas.ucla.edu

Fox, T.J., Fox Geotechnical Engr & Cons, 34071 Glouster Cir, Farmingtn Hls, MI 48331. T: 248-661-2582, F: 248-788-2753, E: foxenginee@aol.com

Fragaszy, R.J., National Science Foundation, 4201 Wilson Blvd, Room 545, Arlington, VA 22230. T: 703-306-1361, F: 703-306-0291, E: rfragasz@nsf.gov

Fraine, K.G., The Tech Group, Inc., 8332 Glastonbury Court, Annandale, VA 22003. T: 703-698-0120, F: 703-995-4680, E: kenfraine@netzero.net

France, J.W., URS Corporation, 6553 East Costilla Pl, Engle-wood, CO 80112. T: 303-770-0558, F: 303-694-3946, E: john_france@urscorp.com

Francis, M.J., URS, 615 Piikoi Street, Suite 900, Honolulu, HI 96814. T: 808-593-1116, F: 805-933-1198, E: mathew_ francis@urscorp.com

Francke, C.T., Rocksol, 921 ROCky Mountain Pl, Longmont, CO 80501. T: 303-485-5478, F: 303-485-5479

Franco B., O.A., Orlando Franco & Assocs, P.O. Box 786, Santiago, DOMINICAN REPUBLIC. T: 580 1962, Ext. 357, F: 809-971-8354, E: o.franco@codetel.net.do

Franco, J.C., Cte Construction Testing Eng, 4640 Santa Monica Ave, San Diego, CA 92107. T: 619-223-9249, F: 619-746-9806

Frankidakis, D.J., Hydrosystems Cons Engs, 89 Kallistratou St, Athens, 15771 GREECE. T: 775 6510, F: 775 9977

Frano, A.J., RSV Engineering, Inc., 2 North Dee Road, #107, Park Ridge, IL 60068. T: 847-685-2024, F: 847-843-3047, E: ajfrsvil@rsv-engineering.com

Franz, B.J., Civil & Environ Consult, Inc., 3600 Park 42 Dr., Ste 130b, Cincinnati, OH 45241. T: 513-985-0226, F: 513-985-0228, E: bfranz@cecinc.com

Franz, R.J., ECS Ltd., 229 Bingham Cir, Mundelein, IL 60060. T: 609-273-1110, F: 847-279-0369, E: rfranz@ ecslimited.com

Fraser, R.A., Golder Associates, P.O. Box 1734, Milton Bc Qld, 4064 AUSTRALIA. T: 07 32176444, F: 07 32176700, E: rfraser@golder.com.au

Freed, D.L., Gibble Norden Champion Cons En, 82 Edgewood Lane, Glastonbury, CT 6033. T: 860-659-8435, F: 860-388-4613, E: freed@gncengineers.com

Freeman, W.S., 623 Ne Goldie Dr., Hillsboro, OR 97124. T: 503-640-4334, F: 503-823-5433, E: freemanb@ci.portland.or.us

Freer, Jr, R.H., 725 Capt'n Kate Ct, Naples, FL 34110. T: 941-597-3511, F: 941-597-3541

Freiman, T.J., Amec Earth & Environmental Inc., 3232 W Virginia Ave, Phoenix, AZ 85009. T: 602-272-6848, F: 602-272-7239, E: tony.freiman@amec.com

Freire, E.M., Craig Test Boring Co Inc., 460 Howellville Rd., Berwyn, PA 19312. T: 610-644-1921, F: 609-625-4306, E: emfreire@hotmail.com

Freitas, M.J., Geomatrix Consultants, 2101 Webster St, 12th Floor, Oakland, CA 94612. T: 510-663-4238, F: 510-663-4141, E: mfreitas@geomatrix.com

French, J.B., Harding Ese, 1697 Visalia Ave, Berkeley, CA 94707. T: 510-528-3072, E: jbfrench@mactec.com

French, S.L., Wallace Kuhl & Assocs, P.O. Box 1137, W Sacramento, CA 95691. T: 916-372-1434, F: 916-372-2565, E: sfrench@wallace.kuhl.com

Fries, H.F., 6362 S Wilbur Wright Rd., Straughn, IN 47387. T: 317-332-2302, E: bohica.fries@bigfoot.com

Frizzi, R.P., Langan Eng Assocs Inc., River Dr., Center 1, Elmwood Park, NJ 7407. T: 201-794-6900, F: 201-794-0366, E: rfrizzi@langan.com

Frueh, R., Shannon & Wilson Inc., 2043 Westport Center Drive, St Louis, MO 63146. T: 314-392-0050, F: 314-392-0051, E: rhf@shanwil.com

Fry, C.J., 7255 Wilrose Ct, North Tonawanda, NY 14120. T: 716-692-9862, E: cjjfry@aol.com

Fuentes III, G., Fuentes Concrete Pile, P.O. Box 373, Puerto Real, PR 740. E: kp4bjd@barf80.nshore.org

Fuentes, J.J., P.O. Box 363825, San Juan, PR 936. T: 787-785-9065, F: 787-740-4366, E: fuentes@tld. net

Fuerstenberg, G.J., BL Companies, 76 Applewood Drive, Meriden, CT 6450. T: 203-440-3092, F: 203- 630-2615, E: gfuerst@home.com

Fuhriman, M.D., Fugro West, 7700 Edgewater Drive, Suite 848, Oakland, CA 94621. T: 510-633-5100, F: 510-633-5101, E: mfuhriman@fugro.com

Fujitani, K.F., Fujitani Hilts & Assocs, 2255 Sw Canyon Rd., Portland, OR 97201. T: 503-223-6147, F: 503-223-6140, E: fujitani@fhainc.com

Fung, K.K., Earthmax Consultants Inc., 40 Schooner Hill, Oakland, CA 94618. T: 510-540-6188, F: 510-540-6187, E: kkfgeo@aol.com

Futrell, Jr, A.L., Qore Property Sciences, 515 Oakley Dr., Nashville, TN 37220. T: 615-833-5305, F: 615-244-6023

Gabriele, P.E., John M., 6352 S Starlight Dr., Morrison, CO 80465. T: 303-697-9507, F: 303-697-3072, E: gabrieleco@aol.com

Gaikwad, P.N., Caltrans – Environmental, 3337 Michelson Dr., Ste 380, Irvine, CA 92612. T: 949-724-2279, E: prakashgaikwad@hotmail.com

Galavis-M., L.E., CA Oficina De Suelos, Suites Profesionales 5-A, Av Buenos Aires Los Caobos, Caracas, VENEZUELA. T: 02-7811622, F: 02-7811828, E: galavisl@telcel.net.ve

Gallet, A.J., Gallet & Assocs Inc., 320 Beacon Parkway West, Birmingham, AL 35209. T: 205-942-1289

Gallup, M.L., Environmental Decision Group, 5665 Flatiron Pkwy, Boulder, CO 80301. T: 303-938-5500, Ext. 5565, F: 303-938-5520, E: mgollup@lescorp.com

Gambin, M.P., Apageo-Segelm, 21 Quai D'anjou, 75004 Paris, FRANCE. T: 011-331-43548646, F: 1.13E+12, E: mgambin@magic.fr

Gamboa, J., Constructora Titan, SACV, San Lorenzo 153-205, Col Del Valle, Mexico Df, 3100 MEXICO. T: 407-700-0000-0000-057, F: 52-5/5-750691, E: isa_99@infdsel.net.mx

Ganeshwara, V., Leighton & Associates, 22014 Craggyview St, Chatsworth, CA 91311. T: 818-701-6723, F: 818-707-7280, E: ganeshwara@hotmail.com

Gangopadhyay, K., EBA Engineering Inc., 9423 Seven Cts Dr., Baltimore, MD 21236. T: 410-529-1327, F: 410-358-7213, E: kunal@ebaengineering.com

Garcia, L.O., Geo Cim, Inc-Luis O Garcia & Associates, P.O. Box 10872, San Juan, PR 922. T: 787-792-2626, F: 787-782-5990, E: logarcia@geocim.com

Garcia-Echevarria, C., Geo-Cim Inc., Palos Grandes S-9, Garden Hills, Guaynabo, PR 966. T: 787-783-4695, F: 787-782-5990

Garner, C.W., Garner Engineering Inc., 9300 Professor Dr., Little Rock, AR 72227. T: 501-225-8581, F: 501-225-8181

Garner, L.S., Usda-Nrcs, 8752 Independence Way, Arvada, CO 80005. T: 303-940-3835, F: 318-473-7771

Gary, T.M., CIS Engineering, Inc., 20371 Middlebury Street, Ashburn, VA 20147. T: 703-589-1066, F: 703-729-5740, E: cis@flash.net

Gassman, S.L., USC CEE Dept, 300 Main St, C226, Columbia, SC 29208. T: 803-777-8160, F: 803-777-0670, E: gassman@engr.sc.edu

Gausseres, R.F., 735 Owens Street, Rockville, MD 20850. T: 301-424-6727

Gautreau, G.P., Gautreau & Gonzalez Inc., 2320 Drusilla Lane, Suite D, Baton Rouge, LA 70809. T: 225-926-3582, F: 225-925-1339, E: gautreaugeotec@eatel.net

Gazioglu, S.M., Dames & Moore, 5151 Beltline Road, Suite 700, Dallas, TX 75240. T: 972-980-4961, F: 972-991-7665, E: dalsmg@dames.com

Gebhardt, F.C.E., Leighton & Associates, 6 Pamlico, Irvine, CA 92620. F: 949-250-1114, E: fgebhardt@gtginc.com

Geisser, M.F., Beta Group Inc., 39 Major Potter Rd., Warwick, RI 2886. T: 401-884-8183, F: 401-333-9225, E: mgeisser@beta-eng.com

Gemperline, M.C., 4817 South Zang Way, Morrison, CO 80465. T: 303-973-2660, E: mgemperline@do.usbr.gov

Gendron, Jr, G.N., Miller Engineering Inc., 100 Sheffield Rd., Manchester, NH 3108. T: 603-668-6016, E: ggendron@millerengandtesting.com

Gentner, M.C., STS Cons Ltd., 1502 Randolph St Suite 100, Detroit, MI 48226. T: 313-963-2990, F: 313-963-2890, E: gentner@stsltd.com

Gentry, B.W., Geotechnical Resources Inc., 9725 Sw Beaverton Hillsdale, Hwy, Beaverton, OR 97005. T: 503-641-3478, F: 503-644-8034, E: bgentry@gri.com

Gentsch, M.L., King Tool, 103 Briar Meadow, Longview, TX 75604. T: 903-297-3673, F: 903-759-6781, E: mgentsch@swb.net

Gerath, R.F., 4603 Hoskins Rd., North Vancouver Bc, V7K 2R2 CANADA BC. T: 604-985-8983, F: 604-684-5124, E: bgerath@istar.ca

Gerhart, P.C., Gerhart Consultants Inc., 24 Altawood Lane, Sandy, UT 84092. T: 801-942-7639, F: 801-733-6732, E: pgerhart@members.asce.org

Gerlach, J.A., US Army Engineer Dist Japan, Usaed-J, Box 88, Apo, AP 96338. E: john.a.gerlach@poj02.usace.army.mil

Germaine, J.T., MIT, 110 Parker Street, Wilmington, MA 1887. T: 978-988-0178, F: 617-253-6044, E: jgermain@mit.edu

Gery, J.R., Allwest Geoscience Inc., 5976 Calle Cuervo, Yorba Linda, CA 92887. T: 714-970-8262, F: 714-238-1105, E: jgery-aw@pacbell.net

Ghadiali, B.M., Dmjm Acet, 2605 Armacost Ave, Apt 4, Los Angeles, CA 90064. T: 310-473-3574, F: 310-816-0464, E: ghadiali@trenchteam.com

Ghahraman, V.G., Northeastern University, 32 Warwick Road, Watertown, MA 2472. T: 617-923-1679, F: 617-373-4419, E: vahe@neu.edu

Ghazinoor, A., Petra Geotechnical, 3185-A Airway Ave, Costa Mesa, CA 92626. T: 714-549-8921, F: 714-549-1438, E: cdmg@ix.netcom.com

Ghobadi, R., New York City, 4489 Broadway, Apt 4c, New York, NY 10040. T: 212-304-1180, F: 718-595-5193

Ghofory-Ashtiany, M., Intl Instit of Eq Engr & Seism, P.O. Box 19395/3913, Tehran Islamic Rep, IRAN

Ghosh, P.E., Partha, Professional Service Ind Inc., 470 Executive Center Dr., 2j, West Palm Bch, FL 33401. T: 561-686-8751, F: 561-844-2474, E: parthavg@aol.com

Giampaolo, S.C., McMahon Assocs Inc., 425 Commerce Dr., Ste 200, Ft Washington, PA 19034. E: steve.giampaolo@mcmtrans.com

Gibbons, C., General Services Admin, 1159 Tennyson Pl, Atlanta, GA 30319. T/F: 404-250-9829, E: cgibbons@mindspring.com

Giblin, J.A., Giblin Assocs, P.O. Box 517, Fulton, CA 95439. T: 707-545-8327, E: gacivil@aol.com

Gibson, D.A., 217 East 23rd St, Big Stone Gap, VA 24219. T: 540-523-3013, F: 540-679-1843, E: davidg@mounet.com

Gibson, G.W., Consolidated Engr Laboratories, 649 Catalina Dr., Livermore, CA 94550. T: 925-443-6430, F: 925-485-5018, E: GIB@celhq.com

Giddings, A.M., State of Alaska Dept of, Meadow Lakes Rd., P.O. Box 872024, Wasilla, AK 99687. T: 907-373-0270

Giesler, D.P., Willicher Str 15, GERMANY. T: 49-211-594776, F: 49-211-635360

Gifford, A.B., Geoengineers-Gifford, 523 East Second Avenue, Spokane, WA 99202. T: 509-363-3125, F: 509-534-2925

Giger, M.W., ESDF, c/o Swiss Embassy of Cairo, Eda-Kuri-ersektion, 3003 Bern, Egpt, EGYPT. T: 202-341-0342, F: 202 338-1277, E: esdf@gega.net

Gilbert, J.W., Amec Earth & Environmental Inc., 4137 S 500 W, Salt Lake, UT 84123. T: 801-266-0720, F: 801-266-0727, E: wade.gilbert@amec.com

Gilbert, M.M., Anderson Consulting Group, 8510 Nob Hill Ln, Granite Bay, CA 95746. T: 916-660-0925, F: 916-786-7891, E: mmg@acgconsulting.com

Gilbert, R.B., Univ of Texas at Austin, Dept of Civil Engi-neering, 9.227 Cockrell Hall, Austin, TX 78712. T: 512-471-4929, F: 512-471-6548, E: bob_gilbert@mail.utexas.edu

Gildea, H.M., 751 Millwood Ln, Keswick, VA 22947

Giles, T.L., Giles Eng Assocs Inc., N8w22350 Johnson Dr., Waukesha, WI 53186. T: 262-544-0118, F: 262-549-5868, E: tlgiles@gilesengr.com

Gill, S.A., Ground Engineering Cnslt Inc., 9107 Samoset Blvd, Skokie, IL 60076. T: 847-674-8057, F: 847-559-0181

Gillespie, R.W., RW Gillespie & Assocs Inc., 80 Rochester Ave Suite 101, Portsmouth, NH 3801. T: 603-427-0244, F: 603-430-2041

Gilley, C.W., Terrain Engineering Inc., 25740 Washington Ave, Murrieta, CA 92562. T: 909-698-7890 1, F: 909-698-7898, E: cwgilley@earthlink.net

Gilman, T.A., Geoengineers, Inc., 8607 N Sally Ct, Spokane, WA 99208. T: 509-466-9222, F: 509-363-3126, E: tdugger @geoengineers.com

Ginsbach, T.S., Northwest Geotech Inc., 9120 Sw Pioneer Ct, Suite B, Wilsonville, OR 97070. T: 503-682-1880, F: 503-682-2753, E: ngipdx@aol.com

Girault, P., Pagri SA, Av Volcan 120, Mexico DF 11000, MEXICO. T: 011-525-5404329, F: 525-5204408

Giroud, J.-P., Geosyntec Consultants, 621 Nw 53rd Street, Suite 650, Boca Raton, FL 33487. T: 561-995-0900, F: 561-995-0925

Gizzi, R.G., 4 Holly Hill Lane, Katonah, NY 10536. T: 914-232-4429, F: 914-476-8705, E: eccopjn@aol.com

Gladstone, R.A., AMSE, 11406 Tanbark Dr., Reston, VA 20191. T: 703-620-4521, F: 703-749-3034, E: bobgladstone @compuserve.com

Glaser, M.B., Levine Fricke Inc., 12011 Us Hwy 50 West, Gunnison, CO 81230. T: 970-209-6718

Glaser, S.D., Dept of Civil and Envirn Eng, 440 Davis Hall, Univ of Ca, Berkeley, CA 94720. T: 510-642-1264, F: 510-642-7476, E: glaser@ce.berkeley.edu

Glos, G.H., Gray Wolf Engineering, 1081 W Graythorn Rd., Tucson, AZ 85737. T: 520-229-6151, E: nomad@azstar net.com

Glueck, M.G., E2SI, 3401 Carlins Park Drive, Baltimore, MD 21215. T: 410-466-1400, F: 410-466-7371, E: e2si@erols.com

Glynn, E.F., Villanova Univ, Cee Dept, Villanova, PA 19085. T: 610-519-7398, F: 610-519-6754, E: eglynn@email.vill.edu

Gnanapragasam, N., Seattle University, Dept of Civil & Env Eng, 900 Broadway, Seattle, WA 98122. T: 206-296-5522, F: 206-296-2179, E: nirmalag@seattleu.edu

Goble, G.G., George G Goble Consulting Engi, 3775 Moffit Court, Boulder, CO 80304. T: 303-449-6512, F: 303-492-7317, E: goble@bridgetest.com

Goble, J.A., Bowser-Morner Associates, 105 Hawthorne Ct, Georgetown, KY 40324. T: 502-863-7466, F: 859-253-0183, E: jgoble@bowser-morner.com

Godlewski, P.M., Shannon & Wilson Inc., 400 North 34th St Suite 100, P.O. Box 300303, Seattle, WA 98103. T: 206-633-6853, F: 206-633-6777, E: pmg@shanwil.com

Godshall, J.A., Foremost Industries, 14201 Se Mc Gillivray Blvd, Vancouver, WA 98683. T: 360-607-2501, E: jardenrigs@aol.com

Goette III, R.J., Richard Goettle Inc., 12071 Hamilton Ave, Cincinnati, OH 45231. T: 513-825-8100

Goin III, W.H., Professional Service Ind Inc., 540 Snowdon Ct, Highland Vill, TX 75077. T: 972-317-6744, F: 972-317-0043, E: walter.goin@psiusa.com

Gol, A., Nadir Apt 26/1, P.O. Box 81070, Caddebostan Istan-bul, TURKEY. E: gol@turk.net

Goldstein, A.E., 4 Barkly Rd., Parktown W 2193, P.O. Box 91904, Auckland Park, 2006 SOUTH AFRICA. T: 011 482 1090, F: 2711 7261862, E: aegolds@mweb.co.za

Golesorkhi, R., Treadwell & Rollo, 2947 Gibbons Dr., Alameda, CA 94501. T: 510-769-0141, F: 415-955-9041, E: rgolesorkhi@treadwellrollo.com

Golisch, R.G., Frederic R Harris, 115 Maple St, Toms River, NJ 8753. T: 732-505-9437, F: 201-866-3661

Gomez-Achecar, M.O., Epsa-Labco Centro De Los Heroe, Horacio Vicioso Esq JB Perez, Santo Domingo, DOMINI-CAN REPUBLIC. T: 809-535-8989, F: 809-535-8991, E: c.epsa@codetel.net.do

Gomm, R.T., Agra Earth & Environmental Inc., 116 Weath-erwood Ct, Henderson, NV 89014. T: 702-263-8872, F: 702-252-4494

Gonzales, R.P., Fugro South Inc., 11009 Osgood, San Antonio, TX 78233. T: 210-655-9516, F: 210-655-9519, E: rgonzales@fugro.com

Gonzalez, C.M., Langan Engineering & Envir Ser, 7900 Miami Lakes Drive West, Suite 102, Miami Lakes, FL 33016. T: 305-362-1166, F: 305-362-5212, E: cgonzalez@langan.com

Gonzalez-Munoz, D.A., Cordon & Merida Engrs M-574, P.O. Box 02-5345, Miami, FL 33102. T: 502-331-8631, F: 502-334-3472

Goode, Jr, J.C., Entech Eng Inc., 505 Elkton Dr., Colorado Springs, CO 80907. T: 719-531-5599, F: 719-531-5238

Goodings, D.J., Univ of Maryland, Dept of Civil Eng, College Park, MD 20742. T: 301-405-1960, F: 301-405-2585, E: goodings@eng.umd.edu

Gordon, B.B., L Robert Kimball & Associates, R D 1 Box 198, Huntingdon, PA 16652. T: 814-643-5447, F: 814-472-7712, E: gordobb@hotmail.com

Gordon, J.R., Geoengineers Inc., 2015 Evening Star Lane, Bellingham, WA 98226. T: 360-734-3792, F: 360-647-5044, E: jgordon@geoengineers.com

Gosain, J.N., Everest Eng Co, 915 W Liberty Dr., Wheaton, IL 60187. T: 630-462-9797, F: 630-462-9941

Goss, G.C., Glenn C Goss Ph.D., Pe, 6002 Twinhill Dr., Arlington, TX 76016. F: 817-516-8076, E: ggoss@fastlane.net

Goughnour, R.R., 705 Duff Rd NE, Leesburg, VA 20176. T: 703-771-3297, F: 703-771-0975, E: bgough@erols.com

Govil, S., Assoc Soils Engr, 4 Tularosa Ct, Aliso Viejo, CA 92656. T: 949-448-0739

Goyal, B.L., Foster Wheeler Usa Corp, 6804e Highway 6 S, Houston, TX 77083. T: 713-561-5988, F: 713-597-3050

Graham, D.V., Froehling & Robertson Inc., 7559 Studley Rd., Mechanicsville, VA 23116. T: 804-746-3079, F: 804-264-7862

Graham, L.D., Shafe Kline & Warren Inc., 11100 W 91st St, Overland Park, KS 66214. T: 913-888-7800, F: 913-888-7868, E: ldg@skw_inc.com

Grant, B.R., Maxim Technologies, 303 Irene, P.O. Box 4699, Helena, MT 59604. T: 406-443-5210, F: 406-449-3729

Grant, M.J., Nysdot, 63 Dorchester, Selkirk, NY 12158. T: 518-439-8142

Grant, W.P., Pangeo Inc., 4302 43rd Ave Ne, Seattle, WA 98105. T: 206-525-1639, F: 206-262-0374, E: paulg@pangeoinc.com

Graterol, M.J., Geotecnica De Venezuela Ca, Apdo P.O. Box 80912, Prados Del Espe, Caracas, 1080-A VENEZUELA. T: 011-058-02-9455880, F: 011-058-02-9414, E: geotec@cantv.net

Gray, B.P., SW Cole Eng Inc., Six Liberty Dr., Bangor, ME 4401. T: 207-848-5714, F: 817- 540-2209

Gray, C.W., Seeco Cons Inc., 7350 Duvan Dr., Tinley Park, IL 60477. T: 708-429-1666, F: 708-429-1689, E: www.seeco.com

Gray, D.H., Univ of Michigan, Dept of Civil Eng, Ann Arbor, MI 48109. T: 734-764-4354, F: 734-764-4292, E: dhgray@engin.umich.edu

Gray, R.E., GAI Cons Inc., 570 Beatty Rd., Monroeville, PA 15146. T: 412-856-6400, F: 412-856-4970

Grazel, J.A., John Grazel Inc., P.O. Box 6187, Santurce, PR 914. T: 787-761-2250, F: 787-761-1665

Green, G.E., Gordon Green Ph.D., Pe, 9835 41st Ave NE, Seattle, WA 98115. T: 206-527-2742

Green, M.L., Waterways Experiment Station, 102 Sunset Ave, Vicksburg, MS 39180. T: 601-638-2114

Green, R.K., URS Corporation, 85 Shuey Dr., Moraga, CA 94556. T: 925-376-4381, F: 510-874-3268, E: robert_green@urscorp.com

Green, R.K., ENSR Corp, 2507 Willowby, Houston, TX 77008. T: 713-974-0423, F: 713-520-6802, E: rgreen@ensr.com

Greene, B.H., US Army Corps of Engrs, 1207 Colonial Place, Sewickley, PA 15143. T: 412-741-1928

Green-Heffern, J.M., Ch2m Hill, 5930 East Caley Dr., Englewood, CO 80111. T: 303-771-2885, F: 303-754-0194, E: jgreenhe@ch2m.com

Greenman, J., Haner Ross & Sporsen Inc., 15 Se 82nd Dr., Gladstone, OR 97027. T: 503-657-1384, F: 503-657-1387, E: hrs@northwest.com

Greer, D.J., Kentucky Transp Cabinet, 209 Gray Hawk Ct, Versailles, KY 40383. T: 606-873-0912, F: 502-564-2865, E: dgreer@mail.kytc.state.ky.us

Grefe, R.P., Wisconsin Dept of Natural Resources, 4602 Turner Ave, Madison, WI 53716. T: 608-222-5273, F: 608-267-2768, E: greferp@execpc.com

Gregory, G.H., Gregory Geotechnical, P.O. Box 121128, Fort Worth, TX 76121. T: 817-244-5569, F: 817-244-5579, E: gre-geo-ghg@att.net

Gregory, P.O., CAL Engineering & Geology, 1870 Olympic Blvd #100, Walnut Creek, CA 94596. T: 925-935-9771, F: 925-935-9773, E: pgregory@caleng.com

Greguras, F.R., URS Corporation, 72 Shaw Place, San Ramon, CA 94583. T: 925-829-6689, F: 415-882-9261, E: frank_greguras

Gribaldo, A.C., Earth Systems Cons Inc., 2275 E Bayshore Rd., Suite 100, Palo Alto, CA 94303. T: 650-856-6750, F: 650-858-2783

Grieco, Jr, D., Eng Mechanics Inc., 4636 Campbells Run, Pittsburgh, PA 15205. T: 412-923-1950, F: 412 787-5891

Grieder, J.K., Agra Foundations Inc., 10108 32 Ave West Bldg C-3, Suite A-2, Everett, WA 98204. T: 435-353-5506, F: 435-353-4151, E: jgrieder@agrafoundations.net

Griffin, G.D., UMA Eng Ltd., 151 Riverside Way Se, Calgary, T2C 3V8 CANADA Ab. T: 403-279-0231, F: 403-270-0399, E: ggriffin@umagroup.com

Griffin, J.G., George Harms Constr Co Inc., 280 Randolph Rd., Freehold, NJ 7728. T: 732-431-0254, F: 732-938-2782, E: jgriffin@ghcci.com

Griffin, P.M., Bechtel, 8180 Greensboro Dr., Ste 900, Mclean, VA 22102. T: 703-748-9461, F: 703-748-9443, E: pmgriffi@bechtel.com

Griffith, A.H., Saudi Aramco, P.O. Box 5450, Dhahran, 31311 SAUDI ARABIA. F: 966-3-873-1183, E: griffiah@aramco.com.sa

Griffiths, D.V., Colorado School of Mines, Dept of Engineering, Illinois Street, Golden, CO 80401. T: 303-273-3669, F: 303-273-3602, E: d.v.griffiths@mines.edu

Grillasca, A.G., Geotechnical Eng Services Psc, M.Sc., 343 Winston Churchill 138, San Juan, PR 926. T: 787-648-2224, F: 787-755-6643, E: ges@caribe.net

Grishaber, P.A., PG Soils Inc., 901 Rose Ct, Burlingame, CA 94010. F: 650-344-6772

Gronseth, M.P., Montgomery Watson, 8017 S Showcase Ln, Sandy, UT 84094. T: 801-569-2256, F: 801-272-0430, E: michael.gronseth@mw.com

Gross, B.A., Geosyntec Consultants, 1004 E 43rd St, Austin, TX 78751. T: 512-451-7836, F: 512-322-3953, E: bethg@geosyntec.com

Grosser, A.T., Barr Engineering Co, 4700 W 77th St, Minneapolis, MN 55435. T: 612-832-2600, F: 612-832-2601, E: agrosser@barr.com

Grove, Jr, B.D., IT Corporation, 4100 Council Rock Road, Marietta, GA 30068. T: 770-977-5821, F: 770-442-7399, E: bgrove@theitgroup.com

Groves, C.B., Shannon & Wilson Inc., 12470 Sparrowwood Dr., Saint Louis, MO 63146. T: 314-576-4136

Gruen, Jr, H.A., Earth Mechanics, 963 Heathergreen Ct, Concord, CA 94521. T: 925-609-8509, F: 510-839-0765

Grumberg, R.A., US Army, Cmr 467, Box 3847, Germany, Apo, AE 9096. T/F: 1.15E+13, E: grumbergra@hq.7arcom.army.mil

Gruner, L.B., CPR/Geoconstruction Div., 505 Page Ave, Allenhurst, NJ 7711. T: 732-531-3205, F: 410-298-4086, E: grunegrout@aol.com

Gucunski, N., Rutgers Univ, Dept of Civil & Env Eng, 623 Bowser Rd., Piscataway, NJ 8854. T: 732-445-4413, F: 732-445-0577, E: gucunski@rci.rutgers.edu

Guertin, Jr, J., Gza Geoenvironmental Inc., P.O. Box 2094, Acton, MA 1720. T: 978-635-1699, F: 617-965-7769, E: jguertin@gza.com

Guglielmetti, J.L., Dupont Engineering, 818 Waverly Rd., Kennet Sq, PA 19348. F: 302-695-0734, E: john.l.gugliel metti@usa.dupont.com

Guha, S., Petra Geotechnical Inc., 406 Massachusetts Lane, Placentia, CA 92870. T: 714-996-3236, F: 714-549-1438, E: soumitra_guha@hotmail.com

Guido, V.A., Cooper Union, 51 Astor Pl, New York, NY 10003. T: 212-353-4304, F: 212-353-4341, E: guido@cooper.edu

Gularte, F.B., Hayward Baker Inc., 6005 Cobblestone Dr., Ventura, CA 93003. F: 805-933-1338

Gunalan, K.N., Parsons Brinckerhoff, 488 E Winchester St, Ste 400, Murray, UT 84107. T: 801-262-3735, F: 801-262-4303, E: gunalan@jobworld.com

Gunn, T.P., Turner Constr Co, 212 Greenwood Drive, Manchester, CT 6040. T: 860-646-1504, F: 203-783-8899, E: tgunn@tcco.com

Gupta, L.P., Urban Engineers Inc., 5938 Cobblestone Dr., Erie, PA 16509. T: 814-868-2151, F: 814-453-2020

Gupta, S.N., Earth Eng & Sciences Inc., 3401 Carlins Park Dr., Baltimore, MD 21215. T: 410-337-7260, F: 410-466-7371, E: e2si@erols.com

Guros, F.B., Washington Group International, 2615 Kiowa Ct, Walnut Creek, CA 94598. T: 925-934-9029, E: fbguros@pacbell.net

Gurtowski, T.M., Shannon & Wilson Inc., 400 N 34th St, Ste 100, P.O. Box 300303, Seattle, WA 98103. T: 206-789-4779, F: 206-633-6777, E: tmg@shanwil.com

Gurtowski, T.M., Shannon & Wilson Inc., 400 North 34th St, Suite 100, P.O. Box 300303, Seattle, WA 98103. T: 206-695-6801, F: 206-633-6777, E: tmg@shanwil.com

Gutierrez, M.S.M., Virginia Tech, Civil Engineering, 200 Patton Hall, Blacksburg, VA 24061. T: 540-231-6357, F: 540-231-7532, E: magutier@vt.edu

Gwaltney, C.A., Univ of Evansville, 1800 Lincoln Ave, Civil Engineering Dept, Evansville, IN 47722. T:812-479-2691, F: 812-479-2780, E: cg2@evansville.edu

Gygax, A.U., P.O. Box 579, Baileys Harbor, WI 54202. T/F: 920-839-2398

Ha, K.-H., Ssangyong Engineering Co, 12-1002 Keuk Dong Apt, Kwang Jang Dong 218-1, Kwang Jin-Gu Seoul, South Korea, SOUTH AFRICA. T: 02-454-5237, F: 031-750-6693, E: hkh1213@ssyeng.co.kr

Haas, C., Burstwiesenstr 62, Zurich 8055, SWITZERLAND. T: 41-1-450-4482, E: haas@alum.mit.edu

Hackman, R.E., Eng Cons Svcs Ltd., 1340-P Charwood Rd., Hanover, MD 21076. T: 410-859-4300, F: 410-859-4324, E: rhackman@ecslimited.com

Hadala, P.F., Louisiana Tech University, P.O. Box 821684, Vicksburg, MS 39182. T: 601-636-5862, F: 318-257-2652, E: paulh@engr.latech.edu

Hadj-Hamou, T., Geosyntec, 2100 Main St, Ste 150, Huntington Beach, CA 92648. T: 714-969-0800, F: 714-969-0820, E: tarik@geosyntec.com

Haertle, J.E., Giles Eng Assocs Inc., 2300 Timberwood, Irvine, CA 92620. T: 714-731-6292, F: 714-779-0068, E: dj_dmarco@msn.com

Haestad, R.J., Roald Haestad Inc., 37 Brookside Rd., Waterbury, CT 6708. T: 203-753-9800

Hagerty, D.J., Univ of Louisville, Speed Scientific School, Dept of Civil & Environ Eng, Louisville, KY 40292. T: 502-852-4565, E: hagerty@louisville.edu

Hale, H.C., 4317 Alta Vista Lane, Dallas, TX 75229

Halim, R.A., ACRES Intl Corp, 6th Floor 500 Portage Ave, Winnipeg, R3C 3Y8 CANADA MB. T: 204-786-8751, F: 204-786-2242, E: rhalim@wpg.acres.com

Hall, K.M., Hall Blake & Assocs Inc., P.O. Box 271099, Memphis, TN 38167. T: 901-353-1981, F: 901-357 8126, E: hba01@msn.com

Hall, L.O., FB Engineering Ab, Haga Ostergata 21, Goteborg, 413 01 SWEDEN. T: 011-46-31-13-99-12, F: 011-31-775-11-2, E: lhl@fbe.se

Hall, R.E., Underground Construction Co In, 139 Hillcrest Rd., Berkeley, CA 94705. T: 510-652-4854, F: 510-652-3703 (H), E: hallre@msn.com

Hamel, J., Hamel Geotech Cons, 1992 Butler Dr., Monroeville, PA 15146. T: 412-824-5943

Hamidieh, E., Bay Area Geotech Group, 950 Industrial Ave, Palo Alto, CA 94303. T: 415-852-9133, F: 415-852-9138, E: ebbih@ibm.net

Hamilton, P.J., Mountain Pacific Inc., P.O. Box 5007, Laguna Beach, CA 92652. T: 949-497-3487, F: 949-376-8957, E: info@mountainpacificusa.com

Hamje, J.T., Delaware Valley Construction C, 784 Moccasin Dr., Harleysville, PA 19438. T: 215-513-0896, F: 215-513-9402, E: fxpertservices.net

Hammam, M.H., Bauer-Egypt, 197-26 July St Agouza, Giza, 12411 EGYPT. T: 202-302-6083, F: 202-302-3805

Hammer, D.P., Xcorps Llc, 72 Sentry Dr., Wilder, KY 41076. T: 859-781-4949, F: 859-781-8876, E: dphxcorps@aol.com

Hammer, G.G., Colorado State of, 4111 W. 20th St Rd., Greeley, CO 80634. T: 970-352-0259, E: greg.hammer@state.co.us

Hammond, D.J., HNTB Companies, 1201 Walnut Suite 700, Kansas City, MO 64106. T: 816-527-2574, F: 816-221-9016, E: dhammond@hntb.com

Hampton, P.E., Delon, Delon Hampton & Assocs, 800 K Street Nw, Suite 720/North Lobby, Washington, DC 20001. T: 202-898-1999, F: 202-371-2073, E: drhampton@delonhampton.com

Hampton, D.R., Geosciences Dept, Western Michigan University, 1903 West Michigan Ave, Kalamazoo, MI 49008. T: 616-387-5496, F: 616-387-5513, E: duane.hampton@wmich.edu

Han, J., Tensar Earth Technologies Inc., 3860 Angora Pl, Duluth, GA 30096. T: 770-248-7399, F: 404-250-9185, E: jhan@tensarcorp.com

Handfelt, L.D., URS Corporation, 1615 Murray Canyon Rd., Suite 1000, San Diego, CA 92108. T: 619-294-9400, F: 619-293-7920, E: leo-handfelt@urscorp.com

Handy, R.L., 1502 270th Street, Madrid, IA 50156. T: 515-795-3355, F: 515-795-3998, E: rlhandy@pionet.net

Hanford, Jr, R.W., 4455 Boardwalk Dr., Huntington Beach, CA 92649. T: 714-625-0457, F: 714-625-0467, E: richanford @msn.com

Hanks, G.A., Rhon Ernest Jones Cons Eng Inc., 11060 Sw 1st St, Coral Springs, FL 33071. T: 954-255-9785, F: 954-341-5961

Hanna III, E.R., GZA Geoenvironmental Inc., 78 S Ellicott St, Williamsville, NY 14221. T: 716-632-0924, F: 716-685-3629, E: ehanna@gza.com

Hanna III, R.E., Sithe Energies Inc., 622 East End Ave, Pittsburgh, PA 15221. T: 412-371-3233, F: 721-763-9068

Hanna, B.J., Marvin E. Davis & Associates, 1890 Marlette Avenue, Reno, NV 89503. T: 775-787-9316, F: 775-853-9199, E: bjh1211@aol.com

Hannalla, A.E.K., M., Arab Foundations Co, 67 Tarik El Horria, Alexandria, EGYPT

Hannan, R.W., US Army Corps of Engrs, 29146 South Cramer Rd., Molalla, OR 97038. T: 503-651-2973, F: 503-808-4845

Hannen, W.R., Wiss Janney Elstner & Assocs, 330 Pfingsten Rd., Northbrook, IL 60062. T: 847-272-7400, F: 847-291-9599, E: whannen@wje.com

Hansen, L.A., AMEC Earth & Environmental Inc., 3232 West Virginia Ave, Phoenix, AZ 85009. T: 602-272-6848, F: 602-272-7239, E: lawrence.hansen@amec.com

Hansmire, P.E., William Henry, Jacobs Associates, 500 Sansome Street, Suite 700, San Francisco, CA 94111. T: 415-434-1822, F: 415-956-8502, E: hansmire@jacobssf.com

Hanson, Jr, K.E., URS Corporation, 10327 Lightner Bridge Dr., Tampa, FL 33626. T: 813-920-1514, F: 813-874-7424, E: kenneth_hanson@urscopr.com

Hanson, D., Hanson Engineering PC, 9366 Lilley Road, Plymouth, MI 48170. T: 313-454-6560

Hanson, G.J., USDA-ARS, 1301 N Western St, Stillwater, OK 74075. T: 405-624-4135, F: 405-624-4136

Hanson, J.L., Lawrence Technological Univ, 21000 West 10 Mile Road, Southfield, MI 48075. T: 248-204-2538, F: 248-204-2509, E: hanson@ltu.edu

Hanson, W.E., Hanson Engrs, 1525 S 6th St, Springfield, IL 62703. T: 217-788-2450, E: hanson@family-net.net

Haq, S.E., Parsons Brinckerhoff, 35 Phelps Ave, Apt B, New Brunswick, NJ 8901. T: 732-745-2082, F: 212-465-5592

Harada, T., Miyazaki Univ-Dept Civil Eng, 1-1 Gakuen Kibanadai, Nishi, 889-2192 JAPAN. T: 0985-58-7325, F: 0985-58-7344, E: harada@civil.miyazaki-u.ac.jp

Harakas, J.B., Geoengineers Inc., S 1817 Rockwood Blvd, Spokane, WA 99203. T: 509-535-2707, F: 509-363-3126, E: jharakas@geoengineers.com

Harb, J.N., EMC, Fanar Hai El Moteur, Harb Bldg 3rd Flr, Beirut, LEBANON. T: 011-961-3-652565, F: 961-1-895454, E: jharb@lynx.neu.edu

Hardcastle, J.H., Univ of Idaho, Dept of Civil Eng, Moscow, ID 83844. T: 208-885-6302, F: 208-885-6608, E: jimhard @uidaho.edu

Harder, L.F., California State of, 3510 Creekwood Dr., Rocklin, CA 95677. F: 916-657-2467, E: harder@water.ca.gov

Hardin, C.D., Geo-Environmental Consultants, 13414 Cocheco Ct, Huntersville, NC 28078. T: 704-947-0395, F: 704-596-8770

Hardin, D.E., Hepworth Pawlak Geotech Inc., 5020 County Road 154, Glenwood Springs, CO 81601. T: 970-945-7988

Hardman, S.L., Geopacific Engineering Inc., 23891 Sw Warbler Place, Sherwood, OR 97140. T: 503-925-9474, F: 503-598-8705, E: shardman@geopacificeng.com

Hargraves, J.J., H&F Consultants, 8130 Tumberknoll Lane, Cordova, TN 38018. T: 901-754-9323, E: Timber525@aol.com

Harlan, R.C., Consulting Engineer, 150 Lombard St #505, San Francisco, CA 94111. T: 415-982-3522, F: 415-982-5403, E: rharlan@attglobal.net

Harmel, R.M., 8001 W 10th Ave, Apt 5, Lakewood, CO 80215. T: 303-234-0916

Harney, M.D., University of Washington, 19835 32nd Ave Ne, Lake Forest Park, WA 98155. T: 206-363-9227, F: 206-543-1543, E: harney@u.washing ton.edu

Harnly, J.R., Foundation Design PC, 2 Chipping Ridge, Fairport, NY 14450. T: 716-223-5252, F: 716-458-3323

Haro, J.A., Haro Kasunich & Assocs, 116 E Lake Ave, Watsonville, CA 95076. T: 831-722-4175, F: 831-722-3202

Harrington, T.J., Harrington Engineering Inc., 455 Rigg Road, Valparaiso, IN 46383. F: 219-926-8446, E: hec@niia.net

Harris IV, W.A., Clough Harbour & Assoc, 3 Jay St Apt B, Cohoes, NY 12047. T: 518-899-4787, F: 518-453-4773, E: wharris4@aol.com

Harris, C.W., ECI, 6741 E Eagle Place, Hghlnds Ranch, CO 80130. T: 303-470-0223, F: 303-740-8671, E: charris@frharris.com

Harris, M.J., Cornforth Cons Inc., The Lincoln Bldg – Suite 111, 10250 Sw Greenburg Rd., Portland, OR 97223. T: 503-452-1100

Harrison, P.J., Maxim Technologies Inc., 10078 Rolling Glen Dr., Cedarburg, WI 53012. T: 414-675-6224, F: 414-675-6238, E: patrickharrison@csi.com

Hart, L.J., Geoscience Testing & Research, 28 Kenwood Road, Everett, MA 2149. T: 617-387-7817, F: 978-251-9396, E: hartinc@mailcity.com

Hartfeil, K.M., Economic & Engineering Services, 111 SW 5th Ave, Suite 1670, Portland, OR 97204. T: 503-223-3033, F: 503-274-6248

Hartzell, P.E., E. Paul, 112 St Andrews Drive, Harbor Pines, Egg Harbor Township, NJ 8234. T: 609-601-8426, F: 609-601-8427, E: geotekdoctor@aol.com

Harwood, D.D., Pacific Soils Engineering Inc., 4172 Center Park Drive, Colorado Spring, CO 80196. T: 714-220-0770, F: 714-220-9589, E: gintguruguy@cs.com

Hasen, M., Hvj Associates Inc., 2611 Hodges Bend Circle, Sugar Land, TX 77479. T: 281-980-2454, F: 281-933-7293, E: mhasen@hvj.com

Haser, L.M., MFG Inc., 13319 Oddom Court, Cypress, TX 77429. T: 281-251-0769, E: lmhaser@msn.com

Hashash, Y.M.A., University of Illinois @Urban, Room 2230c Newmark Ce Lab, 205 N Matthews Ave, Urbana, IL 61801. T: 217-333-6986, F: 217-265-8041, E: hashash@uiuc.edu

Hatch, G.E., 8321 Fort Hunt Rd., Alexandria, VA 22308. T: 703-780-4228, F: 516-427-5376

Hatem, D.J., Burns & Levinson, 125 Summer St, Boston, MA 2110. T: 617-345-3368, F: 617-345-3299

Hatheway, A.W., Professor Emeritus, 10256 Stoltz Dr., Rolla, MO 65401. T: 573-364-0818, F: 573-341-2071, E: allen@hatheway.net

Hatton, C.N., URS Greiner Woodward Clyde, 201 Durham Ct, Castle Rock, CO 80104. T: 303-688-1540, F: 303-694-3946, E: christopher_hatton@urscorp.com

Haubert, A.E., STS Consultants Ltd., 1440 Eagle Ridge Drive, Antioch, IL 60002. T: 847-838-1440, F: 847-279-2579, E: haubert@stsltd.com

Hauser, M.A., Richard Goettle Inc., 12071 Hamilton Ave, Cincinnati, OH 45231. T: 513-825-8100, F: 513-825-8107

Hawk, T.S., Pick Corporation, 125 Green Meadow Ln, Port Matilda, PA 16870. F: 814-861-7113

Hawkes, J.W., City of Taylorsville, 2520 W 4700 St, Ste A2, Taylorsville, UT 84118. T: 801-963-5400, F: 801-963-7891, E: jhawkes@ci.taylorsville.ut.us

Hawkins, Jr, R.J., Geosystems Engineering Inc., P.O. Box 14544, Lenexa, KS 66285. T: 913-962-0975, F: 913-962-9024, E: jhawkins@geosystemseng.com

Hayashi, K.K., Pacnavfacengcom, 95-1093 Milia St, Mililani, HI 96789. T: 808-626-1468, F: 808-474-1104, E: hayashikk@efdpac.navfac.navy.mil

Hayat, T.M., Nespak, 16 Bridge Colony, Abid Majeed Rd., Lahore Cantonement, PAKISTAN. T: 42-6672990, F: 042-5160509, E: tmhayat@hotmail.com

Hayden, J.M., EGS, 6725 Bucklake Rd., Tallahassee, FL 32311. T: 850-942-7260, F: 850-385-8050

Hayden, M., EGS, 6725 Buck Lake Rd., Tallahassee, FL 32311. T: 850-942-7260, F: 850-385-8050, E: 110342.1457@compuserve.com

Hayes, C.K., Kleinfelder Inc., 315 3rd St, Ste A, Huntington Beach, CA 92648. T: 714-688-2513

Hayes, G., 6122 Wood Haven, Carmichael, CA 95608. T: 916-967-1193

Haynes, D.J., Chevron Usa Production Co, 8900 Oak Hills Ave, Bakersfield, CA 93312. T: 661-387-1447, F: 661-392-3715, E: dojh@chevron.com

Hazenberg, C.A., Northstar Vinyl Products, 240 Farm Ct, Roswell, GA 30075. T: 404-640-0187, F: 770-794-1105, E: hazenberg@northstarvinyl.com

Head, C.L., Sanborn Head & Assocs Inc., Suite 1, 6 Garvins Falls Rd., Concord, NH 3301. T: 603-229-1900, F: 603-229-1919

Healy, J.M., Hanson Engrs, 1525 South Sixth St, Springfield, IL 62703. T: 217-788-2450, F: 217-788-2503, E: dhealy@hansonengineers.com

Heckel, R.D., Qore Propert Scien-Atlanta, Testing, 820 Fesslers Pkwy, Ste 240, Nashville, TN 37210. T: 615-244-6020, F: 615-244-6023, E: rheckel@qore.net

Hedges, C.S., Soil Testing Engineers, 316 Highlandia, Baton Rouge, LA 70810. T: 504-293-6402

Hedien, J.E., Harza Eng Co, Sears Tower, 233 South Wacker Drive, Chicago, IL 60606. T: 312-831-3095, F: 312-831-3999, E: jhedien@harza.com

Hegazy, Y.A., D'appolonia, 1135 Fox Hill Dr., Apt 109, Monroeville, PA 15146. T: 412-374-7132, F: 412-856-9535, E: yahegazy@dappolonia.com

Heiberg, S., Statoil, N., Stavanger, 4035 NORWAY. T: 47-51990000, F: 47-51990050, E: shei@statoil.com

Heinert, K.D., Williams Form Eng, 280 Ann St NW, Grand Springs, MI 49510. T: 616-365-9220, F: 616-365-2668, E: williams@williamsform.cm

Held, Jr, L.A., 3011 28th St, Metairie, LA 70002. T: 504-454-1501

Helfrich, S.C., 267 Candy Lane, Redlands, CA 92373. T: 909-792-7366, F: 909-792-3566, E: schfun@aol.com

Hendron, Jr, A., P.O. Box 125, Savoy, IL 61874. T: 217-351-8701, F: 217-351-8700

Hendron, M.A., P.O. Box 40, Fithian, IL 61844. T: 217-548-2072, F: 217-351-8700, E: mahe@earthlink.net

Hengen, D.L., Pete Lien & Sons, 4909 Innsbruck Ct, Rapid City, SD 57702. T: 605-341-7783

Henke, R., Dynamic In Situ Geotechnical, 7 Wyndam Ct, Lutherville, MD 21093. T: 410-252-4474, F: 410-252-4474, E: whenke@home.com

Henry, M.P., US Army Corps of Engineers, Cmr 470 Box 7085, Apo, AE 9165

Heo, N.Y., Kyungpook National Univ, Dept of Civil Eng, 1370 Sankyuk-Dong, Puk-Gu, Taegu, 702-701 SOUTH KOREA. T: 011-81-53-950-7556, F: 011-81-53-950-6, E: tg2169@hanmail.net

Hepworth, R.C., Hepworth-Pawlak Geotechnical, 6859 N Village Rd., Parker, CO 80134. T: 303-841-4164, F: 303-841-7556, E: hpgeo2@hpgeotech.com

Herbert, M.V., Kleinfelder, 20350 Empire Ave, Suite A-1, Bend, OR 97701. T: 541-382-4707, F: 541-383-8118

Herlache, W.A., 388-44th St, Oakland, CA 94609. T: 510-601-1508, F: 510-601-1534, E: aherlache@hotmail.com

Herrmann III, H.G., 9120 Lyon Park Ct, Burke, VA 22015. T: 703-978-8365, F: 202-433-2280, E: herrmannhg@nfesc.navy.mil

Hertlein, B.H., 3478 Glen Flora Ave, Gurnee, IL 60031. F: 847-267-8040, E: hertlein@stsltd.com

Hetherington, M.D., Hetherington Engr Inc., 5205 Avenida Encinas, Ste A, Carlsbad, CA 92008. T: 760-931-1917, F: 760-931-0545, E: hengineer@aol.com

Heung, W., Parsons Brinckerhoff, 10849 Ravel Ct, Boca Raton, FL 33498. T: 561-883-0681, F: 954-583-6570

Heup, T.H., Black & Veatch, P.O. Box 8405, Kansas City, MO 64114. T: 913-339-2000

Heydinger, A.G., Univ of Toledo, Dept of Civil Eng, Mail Stop 307, Toledo, OH 43606. T: 419-530-8133, F: 419 530-8116, E: aheyding@eng.utoledo.edu

Heynen, P.M., Heynen Engineers, 11 Peck Street, North Haven, CT 6473. T: 203-985-8133, F: 203-985-8101

Hick, B.A., Earth Systems Cons Inc., 1024 West Ave M-4, Palmdale, CA 93551. T: 661-948-7538, F: 661-948-7538, E: bhick@ptw.com

Hickey, E.W., HJ Foundation Inc., 8510 Nw 68th St, Miami, FL 33166. T: 305-592-8181, F: 305-592-7881, E: ewh@jhfoundation.com

Hill, C.S., Earth Systems Southwest, 79-811 B Country Club Drive, Bermuda Dunes, CA 92201. T: 760-345-1588, F: 760-345-7315

Hill, D.E., United Consulting, 625 Holcomb Bridge Rd., Norcross, GA 30071. T: 770-582-2863, F: 770-582-2912, E: donhill@unitedconsulting.com

Hill, R.J., Cornforth Cons Inc., 21725 Sw Imperial Ct, Beaverton, OR 97006. T: 503-642-4038, F: 503-452-1528

Hillebrandt, D.H., Don Hillebrandt Assocs, 6219 Clive Ave, Oakland, CA 94611. T: 510-531-2008, F: 510-531-2795, E: dhillassoc@aol.com

Hillis, R.M., Hillis-Carnes Engr Assoc Inc., 12011 Guilford Rd., Annapolis Jct, MD 20701. T: 410-880-4788, F: 410-880-4098

Hiltunen, D.R., Penn State Univ, Dept of Civil Eng, 212 Sackett Bldg, University Park, PA 16802. T: 814-863-2936, F: 814-863-7304, E: drh5@psu.edu

Hinderliter, M.H., Terracon, 832 Nw 67th St, Suite 1, Oklahoma City, OK 73116. T: 405-848-1607, F: 405-840-2713, E: mhhinderliter@terracon.com

Hine, R.C., 914 Rambling Dr., Baltimore, MD 21228. T: 410-747-7596, F: 410-243-0041, E: djhine@aol.com

Hipps, K.H., CTE Engineers Inc., 303 E Wacker Dr., Suite 600, Chicago, IL 60601. T: 312-861-4199, Ext. 4384, F: 312-861-4152, E: kirk.hipps@cte-eng.com

Hirani, J.S., P.O. Box 5940, Miami, FL 33283. T: 305-386-1381

Hlobil, Z., Les Biolettes, 1054 Morrens S Lausanne, SWITZERLAND. F: 41217314426, E: zina.h@bluewin.ch

Ho, S.-Z.J., LFR Levine Fricke, 3150 Bristol St, Suite 250, Costa Mesa, CA 92626. T: 714-444-0111, F: 714-444-0117, E: john.ho@lfr.com

Ho, Y.-M., Maa Eng Cons Hk Ltd., 93c Broadway 12/F, Mei Foo Sun Shuen, Kowloon, HONG KONG. T: 852-274-49322, F: 852-286-12081, E: maahk@netvigator.com

Hoar, R.J., TRW, 30508 Los Altos Dr., Redlands, CA 92373. T: 909-794-5463, F: 909-382-6178

Hodek, R.J., Michigan Tech Univ, Dept of Civil Eng, Houghton, MI 49931. T: 906-487-2797, F: 906-487-2943

Hodnett, D.L., Tri-State Testing & Drilling, 2132 North Fork Dr., Soddy Daisy, TN 37379. T: 423-842-8397, F: 423-510-0237, E: dchod2@aol.com

Hoey III, F.J., Tighe & Bond, 53 Southampton Rd., Westfield, MA 1085. T: 413-562-1600, F: 413-562-5317, E: fjhoey@tighebond.com

Hoffmann, Jr, A.G., Gannett Fleming Inc., 752 N Meadowcroft Ave, Pittsburgh, PA 15216. T: 412-343-7898, F: 412-922-3717, E: ahoffmann@gfnet.com

Hoffmann, Jr, W.C., CTL Thompson Inc., 5240 Mark Dabling Blvd, Colorado Springs, CO 80918. T: 719-528-8300, F: 719-528-5362, E: ctlthmpsn@aol.com

Holbrook, R.M., Geo/Environmental Assoc, P.O. Box 70343, Knoxville, TN 37938. T: 865-922-5732, F: 303-845-7497, E: rmichael@geoe.com

Holland, K.L., Haley & Aldrich, 110 16th St, Suite 900, Denver, CO 80202. T: 303-534-5789, Ext. 220, F: 303-534-1777

Holliday, F.J., CTL Thompson Inc., 3500 Swanstone Drive, #10, Fort Collins, CO 80525. T: 970-207-9673

Hollingsworth, R.A., Grover-Hollingsworth, 31129 Via Colinas, Suite 707, Westlake Village, CA 91362. T: 818-889-0844, F: 818-889-4170, E: grover15@ix.netcom.com

Holm, L.A., Ch2m Hill, 136 Bay Path Dr., Oak Ridge, TN 37830. T: 865-482-7380, F: 678-579-8066, E: lholm@ch2m.com

Holman, T.P., Site-Blauvelt Engineers Inc., 16000 Commerce Parkway, Suite B, Mt Laurel, NJ 8054. T: 856-273-1224, F: 856-273-9244, E: terryh@site-blauvelt.com

Holmberg, B.O., Testing Engineers Inc., P.O. Box 548, Dixon, IL 61021. T: 815-288-1489, F: 815-288-6279, E: tei@essex1.com

Holmes, A.L., Arkansas State of, Hwy & Transp Dept, 10324 Interstate 30 P.O. Box 2261, Little Rock, AR 72203. T: 501-569-2251, F: 501-569-2119

Holmquist, D.V., CTL Thompson Inc., 1971 West 12th Ave, Denver, CO 80204. T: 303-825-0777, F: 303-825-4252, E: dvhpims@aol.com

Holtz, R.D., Univ of Washington, Dept of Civil Engr, Box 352700, Seattle, WA 98195. T: 206-543-7614, F: 206-685-3836, E: holtz@u.washington.edu

Holzbach, J.F., Monroe County of, 50 Westminster Rd., Rochester, NY 14607. T: 716-442-5341, F: 716-324-1228

Hom, J.C., John C Hom & Assocs, 1618 2nd Street, San Rafael, CA 94901. T: 415-258-9027, F: 415-258-9309

Honegger, E., Pipl Bauing Eth/Sia, Zur Brunnenstube, Ch 8914 Aeugst A. Albis, SWITZERLAND. F: 41628877270, E: schulung@tfb.ch

Hong, S.-W., Korean Inst of Constr Tech, 2311 Daewha-Dong, Ilsan Gu Goyang-Si, Kyunggi-Do, 411-712 SOUTH KOREA. T: 82-031-910-0485, F: 82-031-910-0211, E: jhbyun@kict.re.kr

Hont, N.S., Public Works, Engineering Division, P.O. Box 7000, Kingman, AZ 86402. T: 520-757-0910, F: 520-757-0912

Hoppe, K.B., NTH Consultants Ltd., 38955 Hills Tech Drive, Farmington Hills, MI 48331. T: 248-324-5290, F: 248-324-5178, E: khoppe@nthconsultants.com

Hoppenjans, J.R., Bowser Morner Assoc Inc., P.O. Box 838, Toledo, OH 43696. T: 419-691-4800, F: 419-691-4805

Horiuchi, S., Shimizu Corp, 3-4-17 Etchujima, Koto-Ku, Tokyo, JAPAN. T: 03-3820-5519, F: 03-3643-7260, E: horiuchi@sit.shimz.co.jp

Hornbeck, D.E., Southern Polytechnic St Univ, Civil Engr Tech Dept, 1100 S. Marietta Parkway, Marietta, GA 30060. T: 770-528-7252, F: 770-528-5455, E: dhornbec@spsu.edu

Horninger, G.M., Schnabel Eng Assoc, 510 East Gay Street, West Chester, PA 19380. T: 610-696-6066, F: 610-696-7771, E: ghorninger@schnabel-eng.com

Horta, E.N., Ardaman & Assoc, 790 Sw 174 Th Terrace, Pmbk Pines, FL 33029. T: 305-438-6012, F: 305-825-2686

Horvath, J.S., 148 Johnson Rd., Scarsdale, NY 10583

Hoshiya, M., Musashi Inst of Tech, 1-28 Tamazutsumi, Setagaya-Ku, Tokyo, 158-8557 JAPAN. T: 03-3703-3111, F: 03-5707-2187

Hossain, D., Soil Probe Ltd., 24 Beacon Rd., Scarborough, M1P 1G7 CANADA ON. T: 416-750-8483, F: 416-754-1259, E: delhoss@pathcom.com

Hotchkiss, A.W., Colorado State of, Dept. of Transp, 4201 E Arksnsas Ave, Denver, CO 80222. T: 303-512-4043, F: 303-757-9242, E: awhotchkiss@qwest.net

Houghton, R., Spencer White & Prentis, 15 Carriage Lane, Bedford, NH 3110. T: 603-472-8507

Housley, S.A., TTL, Inc., 3516 Greensboro Ave, Tuscaloosa, AL 35401. T: 205-345-0816, F: 205-343-0619, E: shousley@ttlinc.com

Housner, G.W., California Inst of Tech, 1201 E California Blvd, Mc 104-44, Pasadena, CA 91125

Houston, S.L., Arizona State Univ, Dept of Civil Eng, Tempe, AZ 85287. T: 602-965-2790, F: 602-965-0557, E: sandra.houston@asu.edu

Houston, T.W., Structural Dynamics Engr, 119 Davis Rd., Ste 6a, Augusta, GA 30907. T: 706-860-5355, F: 706-860-5596

Houston, W.N., Arizona State Univ, Dept of Civil Eng, Ms-5306, Tempe, AZ 85282. T: 602-965-2891, F: 602-965-0557, E: bill.houston@asu.edu

Hover, W.H., GZA Geoenvironmental Inc., 51 Sportsmens Trail, Whitman, MA 2382. T: 617-857-1796, F: 617-965-7769, E: whover@gza.com

Hovland, H.J., 781 Alvarado Rd., Berkeley, CA 94705. T: 510-549-1672

Howard, T.R., 1011 Lyon Rd., Moscow, ID 83843. T: 208-882-1006, F: 852-2581-1033, E: trhoward@uidaho.edu

Howell, R.B., California State of, 4101 Giselle Ct, Sacramento, CA 95821. T: 916-489-2602, E: howelltalk@aol.com

Hoyler, R.C., Mc Phail Assocs Inc., 22 Whitney Woods Lane, Cohasset, MA 2025. T: 617-383-6212, F: 617-868-1423, E: rhoyler@mcphailgeo.com

Hoyos, L.R., Univ of Texas, Dept of Civil Eng, Box 19308, Arlington, TX 76019. T: 817-272-3879, F: 817-272-2630, E: lhoyas@ce.uta.edu

Hoyt, R.M., Atlas Systems Inc., 10400 Walnut Dr., Centralia, MO 65240. T: 573-682-5938, F: 816-796-0919

Hradilek, P.J., US Bureau of Reclamation, Unit 3500/Box 16, Apo Miami, AA 34030. T: 055-61-226-4536, F: 5561-225-9564, E: burec@brnet.com.br

Hryciw, R.D., Univ of Michigan, 1118 Ferdon, Ann Arbor, MI 48104. T: 734-662-9414, F: 734-764-4292, E: romanh@umich.edu

Hsu, C.-C., Tam Kang Univ, Dept of Hydraulic Eng, Tam-Sue, Taipei, Taiwan, ROC. T: 02-26220476, F: 02-26209651, E. 069184@mail.tku.edu.tw

Hsu, D.C.-Y., IT Corp, 2790 Mosside Blvd, Monroeville, PA 15146. T: 412-858-3928, F: 412-858-3979, E: dhsu@itcrp.com

Hsuan, Y.G., 700 Ardmore Ave, #422, Ardmore, PA 19003. T: 610-896-4782, F: 215-895-1363, E: ghsuan@coe.drexel.edu

Hu, R.E., Hu Associates Inc., 11955 Rivera Rd., Santa Fe Springs, CA 90670. T: 562-696-6062, F: 562-698-5771, E: richhu888@aol.com

Huang, A.-B., National Chiao-Tung Univ, Dept of Civil Engineering, Hsinchu, Taiwan, ROC. T: 88635722803, F: 88635734116, E: abhuang@cc.nctu.edu.tw

Huang, F.-C., American Geotech, 150 S Majorca Pl, Placentia, CA 92870. T: 714-577-8618, F: 714-685-3909, E: huangj@amgt.com

Huang, S.-W.S., URS Greiner Woodward Clyde, 173 Ortega Avenue, Mountain View, CA 94040. T: 650-965-1680, F: 408-297-6962, E: stephen_huang@urscorp.com

Hubbard, Jr, D.A., Virginia State of, 40 Woodlake Dr., Charlottesvle, VA 22901. T: 804-978-7998, F: 804-951-6366, E: dhubbard@geology.state.va.us

Huber, K.A., Langan Eng Assocs Inc., 2850 Cowpath Road, Hatfield, PA 19440. T: 215-412-2227, F: 215-864-0640, E: khuber@langan.com

Huber, T.R., P.O. Box 230, Pebble Beach, CA 93953. T: 831-624-4326, E: huber2k@aol.com

Hudock, G.W., Golder Associates, 3730 Chamblee Tucker Rd., Atlanta, GA 30341. T: 770-496-1893, F: 404-894-2281

Hueckel, T.A., Duke Univ, Dept of Civil & Envir Eng, Durham, NC 27708. T: 919-660-5205, F: 919-660-5219

Hughes, M.L., Oklahoma State University, 5923 Woodlike Drive, Stillwater, OK 74074. T: 405-743-8178

Hui, T.W., Dr., WH Ting Cons S/B, 18 Jln Ss20/10 Damansara Kim, Petaling Jaya Selangor, 47400 MALAYSIA. T: 9.01E+14, F: 9.01E+14, E: tingwh@pop.jaring.my

Huie, M.R., Huie Construction Services, Inc., 117 Water St, Suite 202, Milford, MA 1757. T: 508-478-2840, F: 508-478-3524

Hull, J.A., Golder Associates Ltd., 2671 Panorama Dr., North Vancouver, V7G 1V7 CANADA BC. T: 604-929-7244, F: 604 298 5253

Hull, J.H., 3401 Glendale Ave, Ste 300, Toledo, OH 43614. T: 419-385-2918, F: 419-385-5487, E: jhull@hullinc.com

Hulley, M.E., XCG Consultants Limited, 33 Earl St, Kingston, K7L 2G4 CANADA ON. T: 613-542-5888, F: 613-542-0844, E: mikeh@xcg.com

Humphrey, D.N., Univ of Maine, Dept of Civil Eng Rm 103, 5711 Boardman Hall, Orono, ME 4469. T: 207-581-2170, F: 207-581-3888, E: dana.humphrey@umit.maine.edu

Hung, J.C.-J., Parsons Brinckerhoff Quade &, 110 West 90th St, Apt 3h, New York, NY 10024. T: 212-877-6189, F: 212-465-5592, E: hung@pbworld.com

Hunt, C.E., Geosyntec Consultants, 1500 Newell Ave, Suite 800, Walnut Creek, CA 94596. T: 925-943-3034, F: 925-943-2366, E: cehunt@geosyntec.com

Hunt, S.W., Harza Engineering Co, S 74 W12971 Courtland Lane, Muskego, WI 53150. T: 414-425-5243, F: 414-773-8999, E: shunt@harza.com

Hunter, Sr., K.C., Hunter Agri-Sales Inc., P.O. Box 2, Coatesville, IN 46121. T: 317-539-4400, F: 317-539-4131, E: khunter@hunter-inc.com

Hunter, M.F., Taylor Hunter Associates Inc., 818 Civic Center Dr., Oceanside, CA 92054. T: 760-721-9990, F: 760-721-9991, E: mfhunter@taylor-hunter.com

Huntsman, S.R., Ninyo & Moore, 675 Hegenberger Road, Suite 220, Oakland, CA 94621. T: 510-633-5640, F: 510-633-5646, E: srhuntsman@aol.com

Hurley, C.H., Claude H Hurley Co, 175 West First St, Elmhurst, IL 60126. T: 630-279-7762, F: 630 279 7795

Hurlocker, A.L., US Army Corps of Engrs, 4489 Southwood Place, Dumfries, VA 22026. T: 703-878-0704, F: 202-761-0476, E: aehurlocker@earthlink.net

Hussein, M.H., GRL & Assocs Inc., 8000 S Orange Ave, Suite 108, Orlando, FL 32809. F: 407-826-4747, E: mhgrlfl@aol.com

Hussin, J.D., Hayward Baker Inc., 6850 Benjamin Rd., Tampa, FL 33634. T: 813-884-3441, F: 813-884-3820, E: jdhussin@haywardbaker.com

Huzjak, R.J., 1079 Yarnell Dr., Palmer Lake, CO 80133. T: 719-481-9189, F: 303-662-8757, E: rhuzjak@geiconsultants.com

Hwu, B.-L., National Defense University, Office of Academic Affairs, Long-Tang, Tao-Yuan, 325 Taiwan, ROC. T: 011-8863-4890540, F: 011-8863-489053, E: blhwu@ndu.edu.tw

Hynes, C.S., 74 Honey Hollow Road, Earlton, NY 12058. T: 518-634-7116, E: chynes@nicholson-rodio.com

Hynes, M.E., US Army Corps of Engrs, Waterways Experiment Sta, 3909 Halls Ferry Rd., Vicksburg, MS 39180. T: 601-634-2280, F: 601-634-3453, E: hynesm@wes.army.mil

Hyslip, J.P., Ernest T Selig Inc., P.O. Box 201, Williamsburg, MA 1096. T: 413-268-3437, F: 413-585-9392, E: hyslip@etselig.com

Ibarra, H.V., Louis Berger Int'l Inc., c/o Horacio V Ibarra Pe, 1819 H Street Nw, Ste 900, Washington, DC 20006. T: 1.15E+11, F: 1.15E+11, E: hibarra@ibw.com.ni

Ibrahim, I., 210b West Coast Way, Hong Leong Garden Shopping Ctr, SINGAPORE, 127102 SINGAPORE. T: 011-65-94560954, F: 011-65-774-2794, E: icham@hotmail.com

Ichikawa, Y., Nagoya Univ, Dept of Geotech & Env Engrg, Chikusa, Nagoya, 464-8603 JAPAN. T: 52-789-3829, F: 52-789-3837, E: a40346a@nucc.cc.nagoya-u.ac.jp

Idriss, I.M., Univ of California, P.O. Box 330, Davis, CA 95617. T: 916-758-5739, F: 530-758-1104, E: imidriss@aol.com

Iglesia, G.R., 11580 Kirby Place, San Diego, CA 92126. T/F: 858-635-9583, E: grigles@juno.com

Ilmudeen, S.M., STS Consultants Ltd., P.O. Box 6051, Evanston, IL 60204. T: 847-424-8849, F: 847-279-2510, E: ilmudeen@stsltd.com

Inanir, O.E., Geotest Soil Mech & Geotec Eng, Kibris Sehitleri Cad 1479 Sk, Tev1 Is Merkezi #8 D 204, Alsancak, Izmir, 35220 TURKEY. T: 011-90-232-4645701, F: 9.02E+11, E: oinanir@members.asce.org

Inci, G., 5200 Anthony Wayne Dr., Apt 716, Detroit, MI 48202. T: 734-647-1780, E: gokhan@ce.eng.wayne.edu

Ingersoll, R.W., Exxon Mobile Develepoment Co, 22511 Trailwood Ln, Tomball, TX 77375. T: 281-374-7891, F: 281-423-4801, E: roger.w.ingersoll@exxonmoblie.com

Inouye, K.S., US Forest Service, 1323 Club Dr., Vallejo, CA 94592. T: 707-562-8876, E: kinouye@fs.fed.us

Ip, V.P.-K., 331 Amberwick Lane, Brea, CA 92821. T: 714-585-1954, F: 714-529-1087, E: ipvincent@aol.com

Iravani, S., Geo Syntec, 14025 Riveredge Dr., Ste 280, Tampa, FL 33637. T: 813-558-0990, F: 813-558-9726, E: said@geosyntec.com

Isa, R., Apartado 61970, Caracas, 1060-A VENEZUELA. T: 58166300570, F: 582-2655016, E: richisa@cantv.net

Isenberg, R.H., SCS Engrs, 13439 Lake Shore Dr., Herndon, VA 20171. T: 703-793-0321, E: risenberg@scseng.com

Isenhower, W.M., 10604 Fountainbleu Circle, Austin, TX 78750. T: 512-257-1218, F: 512-467-1384, E: firm-ensoft@ensoftinc.com

Ishibashi, I., Old Dominion Univ, Dept of Civil & Environ Eng, Norfolk, VA 23529. T: 757-683-4641, F: 757-683-5354, E: iishibas@odu.edu

Italiano, S.P., SPI Consulting, 3332 Mildred Ln, Lafayette, CA 94549. T: 510-283-2465, F: 510-536-3320, E: spicon@ix.netcom.com

Iwasaki, Y.T., Geo-Rsch Inst, 1-88-417, Koyoen-Sannoh-Cho, Nishinomiya, Hyogo-Pre, 662-0018 JAPAN. T: 81-798-70-0428, F: 81-6-6578-6255, E: iwasaki@geor.or.jp

Jacobi, C.L., Jacobi Geotechnical Engineering Inc., 425 Round Tower Drive, Saint Charles, MO 63304. F: 636-978-7113, E: jacobiengineer@aol.com

Jacobs, L.M., Larry M Jacobs & Assocs, 328 East Gadsden Street, Pensacola, FL 32501. T: 850-434-0846, F: 850-433-7027, E: operations.lmja@worldnet.att.net

Jacobs, W.S., Kellogg Brown & Root Co, P.O. Box 3, Houston, TX 77001. T: 713-753-4641, F: 713-753-3877, E: wes.jacobs@halliburton.com

Jaffe, A.H., David V Lewin Corp, Caxton Bldg/Ste 540, 812 Huron Rd., Cleveland, OH 44115

Jafroudi, S., Petra Geotechnical Inc., 3185-A Airway Ave, Costa Mesa, CA 92626. T: 714-549-8921, F: 714-549-1438, E: sjafroudi@petra-inc.com

James, J.R., Gannett Fleming, 1007 Lindendale Drive, Pittsburgh, PA 15243. T: 412-571-2217, F: 412-922-3717, E: jjames@gfnet.com

James, M., Michael James & Associates, 1400 Carpentier St, Apt 307, San Leandro, CA 94577. T: 510-663-5764, F: 510-251-1599, E: engineer@sirius.com

Jamiolkowski, M.B., Technical Univ, 24 Duca Degli Abruzzi, Torino, 10136 ITALY. T: 39-011-5644840, F: 39-011-5644893, E: sgi-jamiolkowski@studio-geotecnica.it

Jansen, R.B., 509 Briar Rd., Bellingham, WA 98225. T/F: 360-647-0983

Jao, M., P.O. Box 11184, Lamar Univ, Beaumont, TX 77710. T: 703-849-0595, F: 703-849-0675

Jaramillo, C.A., Harza Eng Co, 299 N Dunton Ave, Apt 523, Arlington Heights, IL 60004. T: 847-797-9008, F: 312-831-3999, E: cjaramillo@harza.com

Javete, D.F., DF Javete & Assocs, 5398 Belgrave Pl, Oakland, CA 94618. T: 510-655-3964, F: 510-655-0600, E: djavete@jps.net

Jaworski, Jr, W.E., GZA Geoenvironmental Inc., P.O. Box 25, Norfolk, MA 2056. F: 617-965-7769

Jaworski, G.W., 150 Zachary Rd., Unit 1, Manchester, NH 3109. T: 603-647-9700, F: 603-647-4432

Jayawickrama, P.W., Texas Tech Univ, P.O. Box 41023 Dept of Ce, Lubbock, TX 79409. T: 806-742-3471, Ext. 245, F: 806-742-3488, E: priyantha.jayawickrama@coe.ttu.edu

Jemtrud, E.M., 191 Fairmount St, San Francisco, CA 94131

Jenevein, D.R., Golder Assocs Inc., 368 Goldco Cir, Golden, CO 80403. T: 303-277-0815, F: 303-745-0887, E: dougsfc17@aol.com

Jennings, A.T., Schnabel Engineering Associate, 20682 Pomeroy Ct, Ashburn, VA 20147. T: 703-858-0647, F: 703-589-1246

Jensen, C.N., 1840 Alcatraz Ave, Suite C, Berkeley, CA 94703. T: 510-658-9111, F: 510-658-8918

Jeong, S., Yonsei Univ College of Eng, Dept of Civil Eng, Seoul, 120-749 SOUTH KOREA. T: 82-2-2123-2807, F: 822-364-5300, E: soj9081@yonsei.ac.kr

Jermstad, D.B., Carlton Engineering Inc., 3932 Ponderosa Rd., Suite 200, Shingle Springs, CA 95682. T: 530-677-5515, F: 530-677-6645, E: djermstad@carlton-engineering.com

Jernigan, R.L., URS Greiner/Woodward Clyde, 2490 Prairie Lane, Castle Rock, CO 80104. T: 303-663-5623, E: russel_jernigan@urscorp.com

Jewell, A.C., SHE-ESA, 845 Jefferson Ave, Loveland, CO 80537. T: 303-225-9652, F: 970-484-4118, E: acj@o-day.com

Jiang, C.T.-W., Morhol Inc., 1646 Clementine St, Anaheim, CA 92802. T: 714-815-8785, F: 714-991-1326, E: morhol@earthlink.net

Jimenez Quinones, P., Univ of Puerto Rico, Civil Eng Dept, Box 3263, Mayaguez, PR 681. T: 787-832-4040, F: 787-833-4309, E: pluis@coqui.net

Jin, J.S., US Dept of Labor, P.O. Box 10457, Rockville, MD 20849. T: 301-294-3056, F: 301-294-3057

Joachim, C.E., US Army Corps of Engrs, P.O. Box 820061, Vicksburg, MS 39182. T: 601-636-0623, E: joachim.charles.e@wes.army.mil

Jobe, L.O., Gamworks Agency, 17 Cape Point Rd., Bakau, P.O. Box 2642 S/K, West Africa, GAMBIA. T: 011-220-496-355, F: 011-220-375-344, E: lamin_jobe@hotmail.com

Johansen, N.I., Univ of Southern Indiana, 8600 University Blvd, Evansville, IN 47712. T: 812-465-1606, F: 812-421-9880, E: johansen@usi.edu

Johnsen, L.F., Heller and Johnsen, Foot of Broad Street, Stratford, CT 6615. T: 203-380-8188, F: 230-380-8198

Johnson, Jr, J.M., Golder Assocs Inc., 7664 W Plymouth Place, Littleton, CO 80128. T: 303-933-0523

Johnson, Jr, M.L., URS Corporation, 8181 E Tufts Ave, Denver, CO 80237. T: 303-740-3822, F: 303-694-3946, E: max_johnson@urscorp.com

Johnson, A.I., AIJ Consulting, 7474 Upham Court, Arvada, CO 80003. T: 303-425-5610, F: 303-425-5655

Johnson, D.L., URS, 6121 S Marion Way, Littleton, CO 80121. T: 303-730-8737, E: dan_johnson@urscorp.com

Johnson, D.L., EI Du Pont De Nemours, 11 Rising Road, Newark, DE 19711. T: 302-738-3115, F: 302-695-0734, E: donald.l.johnson@usa.dupont.com

Johnson, G., Santa Rosa Group, 76945 Sheffield Ct, Palm Desert, CA 92211. T: 760-345-4448, F: 619-778-5144, E: gwjohnson30@hotmail.com

Johnson, J.B., Seminole County of, 520 Lake Mary Blvd, Suite 200, Sanford, FL 32773. T: 407-665-5653, F: 407-665-5789, E: bjohnson@co.seminole.fl.us

Johnson, M.P., Hillis-Carnes Engineering Asso, 2361 Putnam Lane, Crofton, MD 21114. T: 301-261-3234, F: 410-880-4098, E: mikej@hcea.com

Johnson, P.D., Missouri Dept of Natural Res, 1750 Bluebird Lane, Columbia, MO 65201. T: 573-657-4095

Johnson, S.L., Swinerton & Walberg Co, 1295 W Washington Suite 201, Tempe, AZ 85281. T: 602-629-1295, F: 602-629-1296, E: sjohnson@swinerton.com

Johnston, J.K., PMK Group, 25 Ireland Brook Dr., Rd 4, N Brunswick, NJ 8902. T: 732-940-1906, F: 908-686-0035, E: jjohnston@pmkgroup.com

Johnston, J.W., Henley Johnston & Assocs Inc., 133 W Greenbriar Lane, Dallas, TX 75208. T: 214-946-6707, F: 214-943-7645, E: johnston_pwjw@msn.com

Jonasson, A.B., Linuhonnun Hf, Sudurlandsbraut 4-A, Reykjavik, ICELAND

Jones, A., GZA Geo Environmental, 108 Londonderry Way, Uxbridge, MA 1569. T: 508-278-6793, E: ajones@gza.com

Jones, C.P., Robert P Jones Co, P.O. Box 3368, Boise, ID 83703. T: 208-385-0151

Jones, J., Harza Eng Co, 248 Arlington, Elmhurst, IL 60126. T: 630-832-8217

Jones, R., GEC Inc., 500 Wilson Pike Cir, Ste 122, Brentwood, TN 37027. T: 615-373-4040, F: 615-377-0187

Jong, H., EGL, 881 Monte Verde, Arcadia, CA 91007. T: 626-574-1066, F: 562-945-0364, E: kellyjong@aol.com

Jordan, K.M., Jordan Engineers Inc., 600 South Andreasen Dr., Suite E, Escondido, CA 92029. T: 760-233 2626, F: 760-233-2628, E: jeiengrs@aol.com

Jorenby, B.N., Alfred Benesch & Co, 1908 E Racine Street, Apt #4, Janesville, WI 53545. T: 608-752-9404

Journeaux, N.L., Journeaux Bedard & Assocs Inc., 429 Beaconsfield Blvd, Beaconsfield, H9W 4B8 CANADA QC. T: 514-695-7135, F: 514-636-8447, E: jba@journeauxbedard.com

Joyal, N.A., DCM/Joyal Engineering, 484 N Wiget Ln, Walnut Creek, CA 94598. T: 925-945-0677, F: 925-945-1294, E: dcmjoyal@dcmjoyal.com

Joyet, R.A., ECI, 5350 Yellowstone St, Littleton, CO 80123. T: 303-347-8705, F: 303-740-8671, E: bjoyet@frharris.com

Juang, C.H., Clemson Univ, Dept of Civil Engineering, 214 Lowry Hall, Clemson, SC 29634. T: 803-656-3322, F: 803-656-2670, E: hsein.juang@ces.clemson.edu

Juel, E.A., Thomas Dean & Hoskins Inc., 1200 25th Street South, Great Falls, MT 59405. T: 406-761-3010, F: 406-727-2872, E: erling.juel@tdandh.com

Jun Shan, S., 3512 Lancaster Ave, Apt 1w, Philadelphia, PA 19104. T: 215-382-0269, E: jshan@usa.net

Kabir, M.G., PS&S Inc., 29 Harrison Road, Parsippany, NJ 7054. T: 973-781-0528, F: 732-560-9768, E: gkabir@psands.com

Kaelin, J.J., Maienberg 1, Ch-8852 Altendorf, SWITZER-LAND. E: jjk@jkaelin.ch

Kaffezakis, G.J., Hdr/Wljorden, 490 Lakeshore Dr., Duluth, GA 30096. T: 770-447-6419, F: 678-775-4848

Kagawa, T., Wayne State Univ, 554 Rolling Rock, Bloomfield, MI 48304 T: 248-335-7104, F: 313-577-3881, E: tkagawa @ce.eng.wayne.edu

Kahle, J.G., Ground Engineering Solutions, 3534 Rutherford Road, Taylors, SC 29687. T: 864-292-2901, F: 864 292-6361, E: ges@innova.net

Kaiser, P.K., Laurentian Univ, 411 Kaireen St, Sudbury, P3E 5T3 CANADA ON. T: 705-522-4652, F: 705-675-4838, E: pkaiser@nickel.laurentian.ca

Kalavar, S.R., Technology & Mgmt Applctns Inc., 8935 Shady Grove Court, Gaithersburg, MD 20877. T: 301-258-5053

Kald, L., TRC Engineers Inc., 72-16 32nd Ave, East Elmhurst, NY 11370. F: 212-563-5561, E: lkald@trcsolutions.com

Kaldveer, P., 75 Cumberland Ct, Danville, CA 94526. T: 925-837-7597, F: 925-937-0750, E: pkaldveer1@cs.com

Kallio, B.F., Thermoretec Corporation, 2376 Western Ave, Roseville, MN 55113. T: 651-484-3951, F: 651-222-8914, E: bka1066198@aol.com

Kallio, T.A., Schneider Eng Corp, 14167 Williamsburg Dr., Carmel, IN 46033. T: 317-574-0585, F: 317-899-8010

Kamisky, T., 2852 Grinnel Dr., Davis, CA 95616. T: 916-759-8475, F: 916-372-2565, E: tkamisky@wallace-kuhl.com

Kancharla, V.R., Testwell Laboratories Inc., 212 Ridgecrest Rd., Briarcliff, NY 10510. T: 914-944-8115, F: 914-762-9638, E: kancharla@aol.com

Kane, W.F., Kane Geotech Inc., P.O. Box 7526, Stockton, CA 95267. T: 209-472-1822, F: 209-472-0802, E: wkane@kanegeotech.com

Kaniarz, D., Mc Dowell & Assocs, 7438 Steadman, Dearborn, MI 48126

Kanipe, J.F., Ph.D., Pe, P.O. Box 321267, Cocoa Beach, FL 32932. T: 321-784-0648

Kannon, W.B., 4460 Glencannon, Suisun, CA 94585. T: 707-864-6132, F: 415-642-5387

Kanu, J., Shun Tat Construction Engrg Ltd., Flat C 6/F Block 10, Phase 2 Tai Hing Garden, Tuen Mun Nt, HONG KONG. T: 852-2535-4784, F: 852-2435-9232, E: kanu20@hkabc.net

Kaplan, P., Amec Earth & Environmental Inc., 780 Vista Blvd, Suite 100, Sparks, NV 89434. T: 775-331-2375, F: 775-331-4153, E: paul.kaplan@amec.com

Karakouzian, M., Univ of Nevada Las Vegas, 1751 East Reno Ave, Apt 125, Las Vegas, NV 89119. T: 702-736-6357, F: 702-895-0959, E: mkar@nevada.edu

Karem, W., Qore Inc., 482 Clear Creek Road, Shelbyville, KY 40065. T: 502-633-9589, F: 859-299-2481, E: waynekarem@mindspring.com

Karim, U.F., Univ Twente, Civil Eng Dept, P.O. Box 217 7500 Ae Enschede, NETHERLANDS. T: 33-53-48941518, F: 33-53-4892511, E: u.f.a.karim@sms.utwente.nl

Karimi, S., NMG Geotechnical, 17991 Fitch, Irvine, CA 92614. T: 949-442-2442, F: 949-476-8322

Karna, U.L., Arora & Assocs Pc, 52 Fieldcrest Dr., Westampton, NJ 8060. T: 609-877-4298, F: 609-844-9799, E: ukarna@arorapc.com

Karnik, B.A., Iowa State University, 28 Schilletter Village, #C, Ames, IA 50010. T: 515-572-4202, E: karnikb@usa.net

Karns, D.R., US Army Corps of Engrs, 5040 Reeds Rd., Shawnee Msn, KS 66202. T: 913-262-8189, F: 816-426-5509, E: dennis.r.karn@nwk01.usace.army.mil

Karp, L.B., 100 Tres Mesas, Orinda, CA 94563. T: 925-254-1222, F: 925-254-2825, E: lbk@geoplex.org

Kartofilis, C., Nicholson Construction Co, 792 Killarney Dr., Pittsburgh, PA 15234. T: 412-343-2616, F: 412-221-3127, E: dkartofilis@nicholson-rodio.com

Karunaratne, G.P., Natl Univ of Singapore, Dept of Civil Engrg, Eia #07-03, 1 Engineering Drive 2, 117576 SINGAPORE. T: 65-8742170, F: 65-7791635, E: cvegpk@nus.edu.sg

Karzulovic, A.L., A Karzulovic & Assoc, Alfredo Rio Seco 0238, Providencia Santiago, 6441356, CHILE. T: 562-2229011, F: 562-2227890, E: akl99@vtr.net

Kasali, G., Rutherford & Chekene, 427 Thirteenth Street, Oakland, CA 94612. T: 510-740-3200, F: 510-740-3340, E: gkasali@ruthchek.com

Kassouf, C.J., Triad Eng Inc., 7575 Northfield Rd., Walton Hills, OH 44146. T: 440-786-1000, F: 440-786-1133, E: triad@modex.com

Katti, D.R., North Dakota State University, Civil Engineering Department, Cie Building Rm 201c, Fargo, ND 58105. T: 701-231-7245, F: 701-231-6185, E: dkatti@badlands.nodah.edu

Katzir, M., Foundation & Soil Lt, P.O. Box 10107, Zahala, Tel Aviv, 61100 ISRAEL. T: 972-36472356, F: 972-36495170

Kaufman, L.D., County Sanitation Dist, P.O. Box 4166, San Dimas, CA 91773. T: 909-592-8618, F: 562-692-2941, E: lkaufman@lacsd.org

Kavazanjian, Jr, Edward, Geosyntec Consultants, 2100 Main St, Suite 150, Huntington Beach, CA 92648. T: 714-969-0800, F: 714-969-0820, E: edkavy@geosyntec.com

Kawakami, M.Y., IT Corporation, 46-305 Ikiiki St,Kaneohe, HI 96744. T: 808-247-2121, F: 808-839-0339, E: mkawakami @theitgroup.com

Kawamura, N., Geoconsult Geotech Eng, P.O. Box 362040, San Juan, PR 936. T: 787-782-3554, F: 787-793-0410, E: kawamura@geoconsult-inc.com

Kayes, T.J., Tonkin & Taylor Ltd., P.O. Box 5271, Auckland 1, NEW ZEALAND. E: hmurphy@tonkin.co.nz

Kazaniwsky, P.W., Site-Blauvelt Engineers Inc., 1091 Pederson Blvd, Atco, NJ 8004. T: 856-768-4680, F: 856-273-9244, E: petrok@site-blauvelt.com

Kazmi, Q.A., Haley & Aldrich, 8819 Sundale Drive, Silver Spring, MD 20910. T: 301-585-9419, F: 301-622-7822, E: qaok@haleyaldrich.com

Kazuya, Y., Ibaraki University, Dept of Urban & Civil Eng, Hitachi, Ibaraki, 316 JAPAN. T: 81-294-38-5166, F: 81-294-38-8268, E: yasuhara@civil.ibaraki.ac.jp

Keane, J.M., Tigh Na Gile, Ballyoughtera, Skibbereen, Co Cork, IRELAND (Eire). E: andmearse@cswebmail.com

Keane, J.J., Inc., Village of Garden City, 351 Stewart Ave, Garden City, NY 11530. T: 516-465-4005, F: 516-377-2383, E: johnkeanepe@yahoo.com

Keating, J.P., Gregg Drilling & Testing Inc., 10161 Suntan Cir, Huntington Bh, CA 92646. T: 714-968-9527, F: 562 427 3314, E: pkeating@greggdrilling.com

Keener, Q.R., 64191 Tumalo Rim Dr., Bend, OR 97701. T: 541-382-1193, E: qkeener@bendnet.com

Keller, G.R., US Forest Service, P.O. Box 37, 5506 Genesee Road, Taylorsville, CA 95983. T: 530-284-6441, F: 530 283-4156, E: gkeller@fs.fed.us

Kelley III, W., Memphis Stone & Gravel Co, P.O. Box 1683, 1111 Wilson St, Memphis, TN 38101. T: 901-774-7874, F: 901-774-4028, E: wkelley1@midsouth.rr.com

Kelley, G.P., Langan Eng Assocs Inc., River Dr., Center 1, Elmwood Park, NJ 7407. T: 201-794-6900

Kelley, M., Patton Burke & Thompson Llc, 2209 Rita Court, Irving, TX 75060. T: 972-790-9494, F: 972-831-0800, E: monica@nspemail.com

Kelley, S.P., University of Massachusetts, P.O. Box 3373, Amherst, MA 1004. T: 413-323-1057, F: 413-545-2840, E: skelley@ecs.umass.edu

Kelly, P.B., Patrick B Kelly Consultng Engr, 127 NW 22, Camas, WA 98607. T: 503-232-2787, F: 503-235-2885, E: pkellygeo@aol.com

Kelly, W.E., Catholic Univ of America, Dean of Engineering, 102 Pangborn Hall, Washington, DC 20064. T: 202-319-5160, F: 202-319-4499, E: kellyw@cua.edu

Kelsay, H.S., Geotechnical Resources Inc., 640 South West Viewmont Dr., Portland, OR 97225. T: 503-292-8380, F: 503-644-8034, E: skelsay@gri.com

Kennedy III, C.M., Qore Inc., 345 Forest Valley Ct Ne, Atlanta, GA 30342. T: 404-252-4322, F: 770-476-0213, E: atlanta@gore.net

Kenter, R.J., Civil & Env Cons Inc., 9912 Carver Rd., Cincinnati, OH 45242. T: 513-677-2550, F: 513-985-0228, E: rjkenter@eos.net

Keong, S.M., Singapore Polytechnic, Block 285d #11-62, Toh Guan Rd., Singapore 604285, SINGAPORE. T: 655 638807, F: 657721973, E: sammk@sp.edu.sg

Kerr, M.L., AMEC, 3800 Ezell Road, Ste 100, Nashville, TN 37211. T: 615-333-0630, F: 615-781-0655, E: michael. kerr@amec.com/mlkerr1@concentric.net

Kersich, A.T., 2744 N Gregory Dr., Billings, MT 59102. T: 406-652-5758, F: 406-655-7926

Kesavan, S.K., DW Kozera Inc., 4507 Wilkens Ave, Baltimore, MD 21229. T: 410-536-0336, F: 410-823-1062

Keshian, Jr, B., Roy F Weston Inc., 13 Eagle Nest Drive Ne, Albuquerque, NM 87122. T: 505-797-4578, F: 505-837-6595, E: keshianb@mail.rfweston.com

Kessler, P.E., Richard S., Cons Geotech Engr, 59 Jacobus Ave, Little Falls, NJ 7424. T: 973-890-1039, F: 973-812-0789, E: dickkpe@bellatlantic.net

Khambhati, N.N., Arora & Assocs Pc, 40 Ginnie Lane, West Windsor, NJ 8550. T: 609-275-4740, F: 609-844-9799, E: nkhambhati@arorapc.com

Khan, S.D., Delaware Apartments, 211 E Delaware Place, Chicago, IL 60611

Khattak, M.J., Univ of La At Lafayette, P.O. Box 42291, Department of Civil Engr, Layfayette, LA 70504. T: 337-482-5356, F: 337-482-6688, E: khattakm99@hotmail.com

Khiabani, R., Leighton & Associates, 28772 Walnut Grove, Mission Viejo, CA 92692. T: 949-455-0408, F: 949-250-1114, E: rkhiabani@leightongeo.com

Khilnani, K., Advanced Earth Sciences Inc., 9261 Irivine Blvd, Irvine, CA 92618. T: 949-458-3832, F: 949-458-1046, E: kkhilnani@aesciences.com

Khosla, V.K., PSI, 6530 Duneden Ave, Solon, OH 44139. T: 440-248-1215, F: 216-642-7010, E: vijay. khosla@psiusa.com

Khoury, M.A., URS Corporation, 201 Willowbrook Blvd, Wayne, NJ 7470. T: 973-785-0700, F: 973-785-0023, E: majed_khoury@urscorp.com

Kidd, J.W., GZA Geoenvironmental Inc., 3 Huntington Rd., Arlington, MA 2474. T: 781-646-2376, F: 617-965-7769, E: jkidd@gza.com

Kiefer, F.W., 2035 N 1350 E, North Logan, UT 84341. T: 435-752-8747, E: fjkiefer@cache.net

Kiefer, T.A., STS Consultants Ltd., 1415 Indigo Dr., Mt Prospect, IL 60056. T: 847-296-3218, F: 847-279-2510, E: kiefer@stsltd.com

Kim, B.T., Kyungpook Natl Univ, Dept of Civil Engrg., Inkyuk-Dong, Pook-Gu, Taegu, 702-701 SOUTH KOREA. T: 011-8253-950-5609, F: 011-8253-950-65, E: btkim@members.asce.org

Kim, N.K., Sungkyunkwan Univ Civil Eng, Janangu Chunchun-Dong 300, Kyungkeedo, Suwon, 440-746 SOUTH KOREA. T: 82-331-290-7521, F: 82-331-290-7549, E: nkkim@yurim.skku.ac.kr

Kim, W.C., Seo Yeong Eng Co Ltd., Song Pa Gu Song Pa Dong, 27-Nov, Seoul, SOUTH KOREA, 138170 SOUTH AFRICA. T: 002-412-0737, F: 002-2298-3270, E: logankim@yahoo.co.kr

Kimura, S.A., Control Point Surveying Inc., 930 Kaheka St, Apt 1401, Honolulu, HI 96814. T: 808-946-4738, E: skimura@controlpointsurveying.com

Kinard, D.T., Haley & Aldrich Inc., 34 Linwold Dr., W Hartford, CT 6107. T: 860-570-0235, F: 860-659-4003, E: dtk@haleyaldrich.com

King, B.R., Rembco Geotech Contractors, P.O. Box 23009, Knoxville, TN 37933. T: 865-690-6917, F: 865-690-9135, E: bkingdth@ntown.net

King, D.G., KC Engineering Company, 8798 Airport Rd., Redding, CA 96002. T: 530-222-0832, F: 530-222-1611

Kinner, E.B., Haley & Aldrich Inc., 20 Robbins Rd., Lexington, MA 2421. T: 617-886-7600, E: bos@haleyaldrich.com

Kirkland, T.E., Shannon & Wilson Inc., 400 North 34th Street, P.O. Box 300303, Seattle, WA 98103. T: 206-632-8020, F: 206-695-6777

Kirmani, M.A., Weidlinger Assocs, One Broadway, 11th Fl, Cambridge, MA 2142. T: 617-374-0000, F: 617-374-0010, E: mkirmani@ma.wai.com

Kirschner, A.R., CGK Environmental Inc., 5 Whetstone Dr., Middleborough, MA 2346. T: 508-947-5697, F: 508-923-0894

Kismetian, A.A., Caltrans – District 6, 2623 E Solar Ave, Fresno, CA 93720. T: 559-299-6475

Kitamura, D.M., Ernest K Hirata & Assocs Inc., 44-132 Nanamoana St, Kaneohe, HI 96744. T: 808-247-0940

Kitch, W.A., 2579 Mardella Drive, Beavercreek, OH 45434. T: 937-426-5396, F: 937-656-2949, E: william.kitch@mcguire.af.mil

Kitsonas, T.M., A Zacharopoulos Sa, Dodekanissou 20, Vrilissia, Athens, 15235 GREECE. T: 0030-1-6827346, F: 3018822688

Kjartanson, B.H., Iowa State Univ, Civ & Constr Engr, 486 Town Engr Bldg, Ames, IA 50011. T: 515-294-3925, F: 515-294-8216, E: bkjartan@iastate. edu

Klein III, J.P., Law Eng Inc., 22455 Davis Dr., Ste 100, Sterling, VA 20164. T: 703-404-7000, F: 703-404-7070, E: jklein@lawco.com

Klein, E.M., KCI Technology Inc., 4017 Star Wreath Way, Ellicott City, MD 21042. T: 410-825-2951, F: 301-621-4873, E: bbqkfg@erols.com

Klein, S.J., Jacobs Associates, 500 12th Street, Suite 200, Oakland, CA 94607. T: 510-874-3281, F: 510-874-3268, E: stephen_klein@urscorp.com

Kleinberg, S., Samuel N Kleinberg Cons Engr, 61 Beechwood Ln, Berkeley Hts, NJ 7922. T: 908-322-9678, F: 908-322-5808, E: snkce@aol.com

Kleiner, D.E., Harza Eng Co, 2120 Hickory La, Schaumburg, IL 60195. T: 847-358-4552, F: 312-831-3889, E: dkleiner@harza.com

Klevberg, P., Maxim Technologies Inc., 1601 Second Ave N, Suite 116, Great Falls, MT 59401. T: 406-453-1641, F: 406-771-0743, E: maximgf@mocc.com

Klinedinst, G.L., Federal Highway Admin Group, 37359 Koerner Ln, Purcellville, VA 20132. T: 540-668-6730, E: gklinedinst@starpower.net

Kling, H.F., Zeiser Kling Consultants Inc., 1221 East Dyer Road, Suite 105, Santa Ana, CA 92705. T: 714-755-1355, F: 714-755-1366, E: hkling@zkci.com

Klinzing, D.W., City Public Service, P.O. Box 1771, San Antonio, TX 78296. T: 210-353-2554, F: 210-353-4449

Klosky, J.L., US Military Academy, Dept of Civil and Mech Engr, West Point, NY 10996. T: 914-938-2471, F: 202-319-4499, E: il7354@usma.edu

Knadler, D.J., Traylor Bros Inc., 8244 Lincoln Ave, Apt #C, Evansville, IN 47715. T: 812-473-9031, F: 812-474-3251, E: djkengrd@netscape.net

Knight, Jr, S.C., Knight Cons Engrs Inc., 1586 Old Stage Rd., Williston, VT 5495. T: 802-878-3226, F: 802-879-6376

Knott, R.A., Law Eng & Envir Services Inc., 1302 Oak Bluff Ct, Canton, GA 30114. T: 770-924-9599, F: 770-499-6601, E: rknott@lawco.com

Knutson, L.C., Lowney Associates, 936 Kiely Blvd, Unit D, Santa Clara, CA 95051. T: 408-241-0277, F: 650-967-2785, E: lknutson@lowney.com

Ko, J.-M., Faculty of Construction and La, The Hong Kong Polytechnic Uni, Hunghom Kowloon, HONG KONG. T: 852-2766-5037, F: 852-2362-2574, E: cejmko@polyu. edu.hk

Koerner, R.M., Geosynthetic Institute, 130 Wood Rd., Springfield, PA 19064. T: 610-543-3213, F: 610-522-8441, E: robert.koerner@coe.dresel.edu

Koester, S.R., Testing Service Corp, 3 S 016 Williams Rd., Warrenville, IL 60555. T: 630-393-6147, F: 630-633-2726

Kokosa, V.R., Sanborn Head & Associates Inc., 239 Littleton Rd., Ste 1c, Westford, MA 1886. T: 978-392-0900, F: 978-392-0987, E:vkokosa@sanborn-head.com

Kolbasuk, G.M., Serrot Intl, 125 Cassia Way, Henderson, NV 89014. T: 702-566-8600, Ext. 291, F: 702-567-6755, E: kolbasukg@serrot.com

Komisarek, G.D., Geotechnical Services Inc., 2947 East 41st Court, Davenport, IA 52807. T: 319-441-9572, F: 319-285-8545

Kon, C.-M., Ensr Corporation, 18 Jalan 9/17, Shah Alam 40100, Selangor, MALAYSIA. T: 603-55100027, F: 603-77250381, E: cheemkon@tm.net.my

Koragappa, N., IT Corporation, 10867 Northridge Sq, Cupertino, CA 95014. T: 408-252-6172, E: nkoragappa @theitgroup.com

Kornegay, N.E., Law Eng Inc., 3049 Alan Shepard Dr., Hueytown, AL 35023. T: 205-491-4753, F: 205-985-2951, E: bkornega@kennesaw.lawco.com

Kort, I.K., Valenzuela Castillo 1597, Santiago-9, Providencia, 6640638 CHILE. T: 056-2-2360495, F: 56-2-2358407, E: ikort@rdc.cl

Kothawala, M.A., CTE Engineers Inc., 213 West Harding Road, Lombard, IL 60148. T: 630-627-1040

Koutsabeloulis, N.C., VIPS Limited, Elm Lodge, N St Winkfield, Windsor, Berkshire, SL 44TE UK. T: 44134489174, F: 5.44E+12, E: vips@vips.co.uk

Koutsoftas, D.C., URS Corporation, 60 Joost Ave, San Francisco, CA 94131. T: 415-584-1360, F: 415-882-9261

Kovacs, J.W., Foster Plaza Iii, Foster Plaza 111, Suite 200, 601 Holiday Dr., Pittsburgh, PA 15220. T: 412-922-5575, F: 412-922-3717, E: jkovacs@gfnet.com

Kovacs, W.D., Univ of Rhode Island, 25 Oak Hill Rd., Peace Dale, RI 2879. T: 401-783-0603, F: 401-874-2786, E: kovacsw@egr.uri.edu

Kozal, J.P., Universal Forest Products, 9356 Laubach Nw, Sparta, MI 49345. T: 616-887-8137, F: 616-554-1215, E: jkozalair@aol.com

Kozera, D.W., DW Kozera Inc., 1408 Bare Hills Rd., Baltimore, MD 21209. T: 410-823-1060, F: 410-823-1062, E: dwkozera @bellatlantic.net

Kporku, G.K., Embassy of Ghana, 1-5-21 Nishi-Azabu, Minato-Ku, Tokyo, 106-0031 JAPAN. T: 81-3-54108631, F: 81-3-54108635, E: h.e@ghanaembassy.or.jp

Kraft, Jr, L.M., Advance Geotechnical Services, 2497 Parkway Dr., Camarillo, CA 93010. T: 805-987-1689, F: 805-388-6167

Kramer, P.E., Richard W., Kramer Cons, 5926 Urban St, Arvada, CO 80004. T/F: 303-424-2509, E: rwkcedams@ aol.com

Kramer, S.L., Univ of Washington, Dept of Civil Eng, 265 Wilcox Hall Box 352700, Seattle, WA 98195. T: 206-685-2642, F: 206-685-3836, E: kramer@u.washington.edu

Kreipke, M.V., 12191 Clipper Dr., Apt 403, Woodbridge, VA 22192. T: 703-491-2575, F: 703-643-9812, E: mukreipke@ prodigy.net

Krivanec, C.E., Geomatrix Cons, 3730 Sundale Rd., Lafayette, CA 94549. T: 925-284-8136, E: ckrivanec@ geomatrix

Krizek, R.J., Northwestern Univ, Technological Inst, Dept of Civil Eng, Evanston, IL 60208. T: 847-491-4040, F: 847-491-4011, E: rjkrizek@nwu.edu

Kroeger, E.B., Southern Illinois Univ, Mail Code 6603, Carbondale, IL 62901. T: 907-474-5150, F: 907-474-6635, E: ffebk1@uaf.edu

Kroll, S.N., Pacific Soils Eng Inc., 3002 Dow Ave, Ste 514, Tustin, CA 92780. T: 714-730-2122, F: 714- 730-5191

Kropp, A.L., Alan Kropp & Assocs Inc., 2140 Shattuck Ave, Suite 810, Berkeley, CA 94704. T: 510-841-5095, F: 510-841-8357, E: akropp@akropp.com

Krusinga, J.M., Soil & Materials Engrs, 3629 Frederick Drive, Ann Arbor, MI 48105. T: 734-669-0544, F: 734-454-0629, E: krusinga@plmouth.soilmat.com

Krutz, N.C., City of Sparks – Public Works, 1325 Majestic Dr., Reno, NV 89503. T: 702-746-3945, F: 775-353-7874, E: nkrutz@ci.sparks.nv.us

Krzewinski, T.G., American Eng Testing, P.O. Box 16008, 4431 W. Michigan Street, Duluth, MN 55816. T: 218-628-1518, F: 218-628-1580, E: aetdul@cp.duluth.mn.us

Ksouri, I., Colorado State of, Dept of Transportation, 4340 E Louisiana, Denver, CO 80222. T: 303-757-9747, F: 303-757-9242, E: ilyess.kosuri@dot.state.co.us

Ku, C.-C., 8340 Tally Ho Rd., Lutherville, MD 21093

Kubena, M.E., SEA Inc., 15423 Vantage Pkwy E, Ste 114, Houston, TX 77032. T: 281-442-3473, F: 281-590-9675

Kuhn, M.R., Univ of Portland, School of Eng, 5000 N Willamette Blvd, Portland, OR 97203. T: 503-943-7361, F: 503-283-7316, E: kuhu@up.edu

Kuhns, G.L., Geotechnical & Enviro Cons, 1230 East Hillcrest St, Orlando, FL 32803. T: 407-420-1010, F: 407-420-1060

Kulchin, L., Kulchin & Assoc Inc., 1402 Oxen Run, Truckee, CA 96161. T: 530-562-1874

Kulchin, S.A., 7507 245th Way NE, Redmond, WA 98053. T: 425-868-9833, F: 425-898-8631, E: kulchin@msn.com

Kulesza, R.L., Bechtel Corp, 313 Catamaran St, Foster City, CA 94404. T: 650-341-0344, F: 415-768-4955, E: r741@ aol.com

Kulhawy, F.H., Cornell Univ, 113 Orchard St, Ithaca, NY 14850. T: 607-257-2891, F: 607-255-9004, E: fhk1@ cornell.edu

Kulikowski, J.J., Genterra Consultants Inc., 4 Precipice, Laguna Niguel, CA 92677. T: 949-495-9459, F: 949-753-8887, E: joekul@genterra.com

Kulkarni, S.S., US Navy, Cfay Code 1000e, Psc 473 Box 1, Fpo, AP 96349. T: 81-45-661-4922, F: 8.13E+11, E: c1000e @cfay.navy.mil

Kuo, C.L., Professional Service Ind Inc., 2713 Falling Leaves Dr., Valrico, FL 33594. T: 813-653-2104

Kupper, A.G., Amec Earth & Enviromental Inc., 4810 93rd St, Edmonton, T6E 5M4 CANADA Ab. T: 780-436-2152, F: 780-435-8425, E: angela.kupper@amec.com

Kurth, N.J., New York State of, 5419 Pittsford Palmyra Rd., Pittsford, NY 14534. T: 716-385-9597, F: 716-214-5206, E: nkurth@frontiernet.net

Kurup, P.U., Univ of Massachusetts Lowell, Civil & Environmental Engr, 1 University Avenue, Lowell, MA 1854. T: 978-934-2278, F: 978-934-3052, E: pradeep_kurup@ uml.edu

Kushner, J.J., Seismic Technologies, P.O. Box 83143, Portland, OR 97283. T: 503-283-4798, F: 503 242 1017, E: builder@ teleport.com

Kusky, P.J., URS Greiner Woodward-Clyde, 30775 Bainbridge Rd., Ste 220, Solon, OH 44139. T: 440-349-2708, F: 4403 491514, E: peter_kusky@urscorp.com

Kwiatkowski, T.M., Gza Geoenvironmental Inc., One Edgewater Dr., Norwood, MA 2062. T: 781-278-3817, F: 781-278-5701, E: tkwiatkowski@gza.com

Kwok, C.K., Fugro (Hong Kong) Limited, 22 Floor Hopewell Centre, 183 Queen's Road East, Wanchai Hong Kong, HONG KONG. T: 852 25779023, F: 852 28952379

Kwon, O., Keimyung Univ, Dept of Civil Eng, 1000 Sindang-Dong, Dalseo-Gu, Taegu, 704-701 SOUTH KOREA. T: 82-53-580-5280, F: 82-53-580-5165, E: ohkyun@kmucc. keimyung.ac.kr

Kyfor, Z.G., New York State of, 16 Payne Ct, Clifton Park, NY 12065. T: 518-371-3540

La Penta, B.A., Roy F Weston Inc., 243 Rose St, Metuchen, NJ 8840. T: 732-548-5748, F: 732-417-5801

La Vassar, J.M., DOE Dam Safety Sec, 517 Bulldog St Se, Lacey, WA 98503. T: 360-456-1370, F: 360-407-7162, E: jlsd461@ecy.wa.gov

Labban, M.J., University of Virginia, 3958 Persimmon Drive, Apt #202, Fairfax, VA 22031. T/F: 703-425-5187, E: mjlabban@hotmail.com

Lacasse, S.M., Norwegian Geotechnical Inst, P.O. Box 3930 Ullevaal Stadion, Oslo, 806 NORWAY. T: 47 22 02 31 03, F: 47 22 23 04 48, E: sl@ngi.no

Lacy, H.S., Mueser Rutledge Cons Engrs, 708 Third Ave, New York, NY 10017. T: 212-490-7110, F: 212-490-6654

Lacz, S.W., Moretrench Geocisa Llc, 51 Longview Rd., Rockaway, NJ 7866. T: 973-983-1149, F: 973-586-7265, E: slacz@mtac.com

Ladd, B.E., Langan Engrg Inc., 146 A Donor Ave, Elmwood Park, NJ 7407. T: 201-794-8350, F: 212-964-7885, E: bladd @langan.com

Ladd, C.C., MIT, 7 Thornton Ln, Concord, MA 1742. T: 978-369-3886, F: 617-253-6044, E: ccladd@mit.edu

Ladd, D., Landau Assocs, 130 2nd Ave South, Edmonds, WA 98020. T: 425-778-0907, E: dladd@landauinc.com

Ladd, R.S., 1930 Lime Tree Dr., Edgewater, FL 32141. T: 904-424-0332, F: 904-424-0590

Laforte, M.-A., Neil A Levac Associates, 366 Charron St, Rockland, K4K 1B2 CANADA ON. T: 613-446-4512, F: 613-446-1427, E: mlaf101658@aol.com

Lafronz, N.J., AMEC Earth & Environmental Inc., 3232 W Virginia Ave, Phoenix, AZ 85009. T: 602-272-6848, F: 602-272-7239, E: nick.lafronz@amec.com

LaGatta, M.D., Langan Engineering, One Penn Center, 1617 Jfk Blvd, Ste 960, Philidelphia, PA 19103. T: 215-864-0640, Ext. 297, E: mlagatta@langan.com

Lai, S.-W.P., Berlogar Geotech Cons, 355 Calle La Montana, Moraga, CA 94556. T: 925-299-1627, F: 925-846-9645

Laier, J., Southern Earth Sciences Inc., 762 Downtowner Loop West, Mobile, AL 36609. T: 334-344-7711, F: 334-341-9488

Lam, P.C.K., 2304-6 World Trade Ctr, 280 Gloucester Road, Causeway Bay, HONG KONG

Lam, S.-H., 2013 St Julien Ct, Mountain View, CA 94043. T: 650-965-8958, F: 408-945-1012, E: alshlam@aol.com

Lam, Y.-S., Hong Kong Govt of, Flat B 22nd Fl, Block 24 Baguilo Villa, Victoria Rd., HONG KONG

Lamb, J.H., Mc Dowell & Assocs, 21355 Hatcher, Ferndale, MI 48220. T: 248-399-2066, F: 248-399-2157

Lamb, Jr, J.H., Mc Dowell & Assocs, 1326 Rock Valley Dr., Rochester, MI 48307. T: 248-608-8892, F: 248-399-2157, E: jlamb2@flash.net

Lamb, R.O., Austin City of, 2708 Sherwood Ln, Austin, TX 78704. T: 512-326-3034, F: 512-499-7203, E: robert.lamb@ci.austin.tx.us

Lamb, R.C., Stoney Miller Cons Inc., 14 Hughes, Ste B101, Irvine, CA 92618. T: 949-380-4886

Lamb, T.J., Stone & Webster Eng Corp, 17 Martins Cove Road, Hingham, MA 2043

Lambe, P.C., 1805 Lodestar Drive, Raleigh, NC 27615. T: 919-847-0509, F: 919-954-1428, E: plambe@geotechpa.com

Lambert, D.D., Lambert & Assocs, P.O. Box 0045, Montrose, CO 81402. T: 970-249-2154, F: 970-249-3262

Lambert, M.T., Shannon & Wilson Inc., 2043 Westport Center Dr., Suite 276, St Louis, MO 63146. T: 314-392-0050, F: 314-392-0051, E: mtl@shanwil.com

Lambrechts, J.R., Haley & Aldrich Inc., 145 Florence Ave, Arlington, MA 2476. T: 781-646-6210, F: 617-886-7688, E: jrl@haleyaldrich.com

Lamont, J., URS Corporation, 4233 Ne 75th St, Seattle, WA 98115. T: 206-524-2052, F: 206-727-3350, E: joseph_lamont@urscorp.com

Lamoureux, C., Soprema Inc., 1675 Rue Haggerty, Drummondville Que, J2C 5P7 CANADA QC. T: 819-478-8163, F: 819-478-0163, E: clamoureux@sopremacanada.com

Landau, H.G., Geosphere Engineering Llc, 23829 115th Pl W, Edmonds, WA 98020. T: 206-546-7939, F: 206-546-3959, E: hglandau@aol.com

Landau, R.E, Richard E Landau PEPC, 768 Springfield Ave B6, Summit, NJ 7901. T/F: 908-522-1185, E: relandau@juno.com

Landers, Jr, P.D., ATC Assoc, 1300 Williams Dr., Marietta, GA 30066. T: 770-427-9456, F: 770-427-1907, E: paul.landers@usa.net

Lane, J.D., Brigham Young University, 230 Snlb, Provo, UT 84602. T: 801-378-9320, F: 801-378-7519, E: jdlane@et.byu.edu

Lane, S.B., URS Corporation, 36 E 7th St, Suite 2300, Cincinnati, OH 45202. T: 513-651-3440, F: 513-651-3452

Lang, G.W., PSI, 850 Poplar Street, Pittsburgh, PA 15220. T: 412-922-4010, F: 412-922-4043

Langer, J.A., Gannett Fleming Inc., P.O. Box 67100, Harrisburg, PA 17106. T: 717-763-7211, F: 717-763-1808, E: jlanger@gfnet.com

Langpap, T.V., Geocon Inc., 6960 Flanders, San Diego, CA 92121. T: 858-558-6900

Langston, R.E., Harris County Flood Control District, 746 Langwood, Houston, TX 77079. T: 713-531-9780, F: 713-692-8502, E: langston@hal-pc.org

Laplante, J.P., Hart Crowser Inc., 4409 45th Avenue Sw, Seattle, WA 98116. T: 206-322-8285, F: 206-328-5581, E: jpl@hartcrowser.com

Larrague, J.A., Larrague Galladini & Assoc Ei, Montaneses 2641 10b, Capital Federal Cp, 1428 ARGENTINA. T: 5411-47884457, F: 5411-47876406, E: jparrague@lgingenieria.com.ar

Larsen, P.E., Paul C., Professional Service Ind Inc., 5089 Briarstone Trace, Carmel, IN 46033. T: 317-581-0043, F: 317-216-7135, E: larfam@aol.com

Larsen, D.J., 5311 Springmeadow Dr., Dallas, TX 75229. T: 214-987-9141, F: 214-987-4301, E: larsenintl@cwix.mail

Larson, M.L., URS Corporation, 1131 Thorntree Court, San Jose, CA 95120. T: 408-268-3712, F: 408-297-6962

Larson, R.E., Kleinfelder Inc., 9405 Oviedo St, San Diego, CA 92129. T: 858-484-0534, F: 858-320-2001, E: rlarson@kleinfelder.com

Larsson, K.J., TAMS Cons Inc., 59 Elmwood Rd., Verona, NJ 7044. T: 973-239-2103, F: 973-338-1052

LaRue, B.K., Spencer White & Prentis, 117 Morris Town Rd., Bernardsville, NJ 7924. T: 908-696-9200, F: 908-696-9212, E: swpfc-keith@worldnet.att.net

Lau, G.Y.F., Pacific Geotech Engrs Inc., 99-871 Aumakiki Loop, Aiea, HI 96701. T: 808-484-2237, F: 808-848-5102, E: pge@pacificgeotechnical.com

Lau, K.W.K., Hong Kong Govt, 101 Princess Rd., Homantin, Kowloon, HONG KONG. T: 852-27625288, F: 852-27140214, E: ken_wk_lau@hotmail.com

Lau, P.K., Unicon Concrte Product Hk Ltd., Flat 305 G/F 139 Caine Rd., Hong Kong, CHINA. T: 852-25474629, F: 852-24786002, E: ppklau@attglobal.net

Lau, W.-S., Univ of Canterbury, 1161 Falcon Drive, Coquitlam, V3E 2C2 CANADA BC

Lauckner, M.S., 11911 66th Street N, #830, Largo, FL 33773. T: 727-524-3727, F: 813-289-4405, E: lauckner@pbworld.com

Laughlin, R.D., Midwest Testing Inc., 3377 Hollenberg Dr., Bridgeton, MO 63044. T: 314-739-2727, F: 314-739-5429, E: rlaughlin@mwtesting.com

Lavergne-Ramirez, H., GMTS, P.O. Box 195374, San Juan, PR 919. T: 787-792-8904, F: 787-782-7455, E: hlavergne@caribe.net

Lawrence, C.A., Ch2m Hill, 1890 Art School Rd., Chester Sprgs, PA 19425. T: 215-569-4735, E: clawrenc@ch2m.com

Lawrence, J.D., ATC Associates Inc., 1300 Williams Dr., Marietta, GA 30066. T: 770-427-9456, F: 770-427-1907, E: lawrence66@atc-enviro.com

Lawrence, T.S., Professional Engineers Inc., 1108 Cliff Court, Maryville, TN 37803. T: 423-681-2084, E: tlaw@icX.com

Lawrence, V.M., Jentech Consultants Ltd., 14a Hope Road, Kingston 10, JAMAICA. T: 011-876-926-2201, F: 011-876-929-251, E: jets@cwjamaica.com

Lawson, P.E., Rebecca Sue, Washinton State of, 4318 Meridian Rd Ne, Olympia, WA 98516. T: 360-493-2270, F: 360-407-6305, E: rlaw461@ecy.wa.gov

Lawson, C.R., Royal Ten Cate 11th Flr, Menara Glomac Kelana Bus Ctre, 97 Jalan Ss 7/2, Petaling Jaya, 47301 MALAYSIA. T: -74928226, F: -74928228, E: royal@po.jaring.my

Lawton, E.C., Univ of Utah, Dept of Civil Eng, 122 S Central Campus Dr., Rm 104, Salt Lake City, UT 84112. T: 801-585-3947, F: 801-585-5477, E: lawton@civil.utah.edu

Lay, C.-H.G., Dames & Moore, 27383 Provident Rd., Agoura Hills, CA 91301. T: 818-706-0337, F: 213-996-2458, E: snacgl@dames.com

Lay, M.E., Shn Consulting Engrs & Geol, 1227 Parkside Dr., Mckinleyville, CA 95519. T: 707-839-2389, F: 707-441-8877

Lazarte, C.A., URS Greiner Woodward Clyde, 820 Kains Ave, Apt 302, Albany, CA 94706. T: 510-527-0473, F: 510-874-3268, E: carlos_lazarte@urscorp.com

Lazcano-Diaz, S., Suelo-Estructura, Cauda 964-1 J Del Bosque, Guadalajara Jal, 44520 MEXICO. T: 523-647-7981, F: 523-122-3557, E: slazcano@micronet.com.mx

Leaf, C.L., Golder Associates Inc., 14 Eustace Road, Marlton, NJ 8053. T: 856-797-1217, F: 609 273 9132, E: chris_leaf @golder.com

Leary, D.J., Langan Eng Assocs Inc., 83 Old Cow Pasture Ln, Kinnelon, NJ 7405. F: 201-794-0366

Leary, R.M., US Fish & Wildlife Service, 247 Magoon Road, Orange, MA 1364. T: 978-544-8597, F: 413-253-8451, E: robert_leary@fws.gov

Leathers, F.D., GEI Cons Inc., 1021 Main Street, Winchester, MA 1890. T: 781-721-4000, F: 781-721-4073, E: fleathers@ geiconsultants.com

Leavens, R.W., Enterprise Eng Inc., 220 Westwood Dr., Southlake, TX 76092. T: 817-488-3007, F: 817-410-9951, E: russrascal@aol.com

Lecaros De Cossio, J., Landell Mills Ltd., 120 Arthur, Wimbledon, London, SW1 98AA UK. T: 84-4-856-1269, E: javierlecaros@yahoo.com

Leckband, S., The Circuit Inc., 8751 E Mulberry St, Scottsdale, AZ 85251. T: 602-947-1640, F: 602-970-4329, E: susanne.leckband@worldnet.att.net

Leckrone, J.E., STS Cons Ltd., 7402 Westshire Dr., Ste 100, Lansing, MI 48917. T: 517-321-4964, F: 517-321-2132, E: leckrone@stsltd.com

Ledbetter, Jr, J.F., North Carolina State of, 4504 Eliot Pl, Raleigh, NC 27609. T: 919-872-6946, F: 919-250-4128, E: jledbetter@dot.state.nc.us

Ledbetter, R., Leobetter & Associates, 1662 Oak Ridge, Vicksburg, MS 39183. T: 601-636-8975, F: 601-661-8020, E: richard@ledbetter.com

Lee, B.-S., Namwon Keonseol Engineering Co, Banpo Apt 66-302, Banpo-Dong, Seocho-Gu, Seoul, 137-040 SOUTH KOREA. T: 02-593-0423FA0008, F: 02-537-6379

Lee, D.H., David H Lee & Associates Inc., 21311 Cupar Ln, Huntington Bh, CA 92646. F: 949-461-7901, E: dhlee@ pacbell.net

Lee, H.J., US Geological Survey, Box 370041, Montara, CA 94037. T: 650-728-7999, F: 650 3295411, E: homa@octopus. wr.usgs.gov

Lee, I.K., Unisearch Ltd., 76 Coral Tree Dr., Carlingford, Sydney Nsw, 2118 AUSTRALIA. T/F: 02-9871-3614

Lee, L.J., 2116 New Bedford Drive, Sun City Ctr, FL 33573. T: 813-633-2309

Lee, R., 29 Richard Lee Rd., Phoenixville, PA 19460. T: 610-917-0109, F: 610-917-9108, E: quantgeo@aco.com

Lee, S.R., Korea Advanced Inst of Science & Tech, 373-1 Kusong-Dong Yusong-Ku, Taejon, 305-701 SOUTH KOREA. T: 8242-869-3617, F: 8242-869-3610, E: srlee@ cais.kaist.ac.kr

Lee, S.-M., 2f #19 Alley 1 Lane 5, Section 3 Jen-Ai Rd., Taipei, 106 Taiwan, ROC. T/F: 1.19E+13, E: lsm59192@ ceci.org.tw

Lee, W.F., Taiwan Constr Research Inst, 11 Fl No 190 Sec 2 Chung-Hsing, Hsientien, Taipei, 231 Taiwan, ROC. T: 011-886-2-2912-1323, Ext. F: 011-886-2-8665, E: weilee@ tcri.org.tw

Lee, W.B., Hong Kong Univ, 72-1 Sangsu-Dong, Mapo-Gu, Seoul, 121-791 SOUTH KOREA. T: 02-320-1667, F: 02-3141-0774

Legaspi, Jr, D.E., Geomatrix Consultants, 2101 Webster Street, Suite 1200, Oakland, CA 94612. T: 510-663-4100, F: 510-663-4141, E: dlegaspi@geomatrix.com

Legatski, L.A., Elastizell Corp of America, P.O. Box 1462, 267 Collingwood, Ann Arbor, MI 48106. T: 734-761-6900, F: 734-761-8016, E: legatski@umich.edu

Lehan, P.E., George R., Dmjm & Harris, 605 Third Avenue, New York, NY 10158. T/F: 212-973-2900, E: george. lehan@dmjmharris.com

Leiendecker, H., 5108 Stearns Hill Dr., Waltham, MA 2451. T: 781-899-1109, F: 617-452-8000, E: harald_leiendecker @yahoo.com

Lemke, J.F., GEZ Inc., 4400 G St, Sacramento, CA 95819. T: 916-451-7325

Lens, J.E., Geodesign Inc., P.O. Box 699, 54 Main St, Windsor, VT 5089. T: 802-674-2033, F: 802-674-5943, E: lens@ geodesign.net

Lentz, E.R., Wayne County, 9149 Harrison Street, Livonia, MI 48150. T: 734-525-3656, E: el1141@aol.com

Lentz, R.W., Univ of Missouri, Dept of Civil Engr, Rolla, MO 65409. T: 573-341-4488, E: lentz@novell.cicil. umr.edu

Leo, E., Schnabel Engineering Assoc, 2073-J Lake Park Dr., Smyrna, GA 30080. T: 770-438-2560, F: 770-619-5601, E: eleo@schnabel-eng.com

Leonard, B.D., Strata Consultants, 772 Southhampton Court, Farmington, UT 84025. T: 801-451-8866, F: 801-530-3150, E: bleonard@vbfa.com

Leonard, D.M., Pinnacle Engineering Inc., P.O. Box 605, Winchester, OR 97495. T: 541-440-4871, F: 541-672-0677, E: pinnacle@rosenet.net

Leonard, J.W., 412 Royale Pk Dr., San Jose, CA 95136. T: 408-629-3089, F: 408-629-9529, E: jwlge@earthlink. net

Lepore, S.M., URS Corp, 714 North Chippewa Circle, Boynton Beach, FL 33436. T: 317-849-4990 1712, F: 561-994-6524, E: susan_lepore@urscorp.com

Leps, T.M., Thomas M Leps Inc., P.O. Box 217, Dinuba, CA 93618. T: 209-595-9001, F: 209-595-1232

Leshchinsky, D., Univ of Delaware, Dept of Civil Engr, Newark, DE 19716. T: 302-831-2446, F: 302-831-3640, E: dov@ce.udel.edu

Lester, M., Patton Burke & Thompson, 829 Cambridge Pl, Grand Prairie, TX 75051. T: 972-647-1744, F: 972-831-0800

Leventis, G.E., Langan Engrg & Environ Svc, 90 West Street, Suite 1510, New York, NY 10006. T: 212-964-7888, F: 212-964-7885, E: gleventis@langan.com

Levitt, L.J., Levitt Engrs Inc., 4523 Vance, Fort Worth, TX 76180. T/F: 817-485-0307, E: levitt1@flash.net

Levy, W., Mass Water Resources Authority, Charlestown Navy Yard Bldg 39, 100 First Ave, Boston, MA 2129. T: 617-242-6000, F: 617-788-4889

Lew, M., Law/Crandall Inc., 200 Citadel Drive, Los Angeles, CA 90040. T: 323-889-5300, F: 323-889-5398, E: mlew@ lawco.com

Lewis, Jr, N.F., JA Cesare and Associates Inc., 6486 South Kipling Court, Littleton, CO 80127. T: 303-973-0419, F: 303-220-0442, E: nflewis@deseretonline.com

Lewis, C.L., Gorian and Associates, 3851 Elkwood Ave, Newbury Park, CA 91320. T: 805-376-9403, F: 805 373 6938, E: ctlewis@gte.net

Lewis, K.H., Univ of Pittsburgh, Dept of Civil Eng, 949 Benedum Eng Hall, Pittsburgh, PA 15261. T: 412-624-9889, F: 412-624-0135, E: lewis@engrng.pitt.edu

Lewis, L.E., Kleinfelder Inc., 715 Glenhill Ct, Novato, CA 94947. T: 415-892-9913, F: 415 472 6773

Lewis, M.R., Bechtel Corp, 4469 Dogwood Way, Evans, GA 30809. T: 706-863-0035, F: 803-952-6849, E: mike.lewis@ srs.gov

Lewis, S.P., Lewis Engr, 98 Ramona Ave, Piedmont, CA 94611. T/F: 510-601-7206, E: lewisengineering@mindspring.com

Leznicki, J.K., URS Corp, 74 Pease Ave, Verona, NJ 7044. T: 973-857-4602, F: 973-785-0023, E: jacek_leznicki@ urscorp.com

Li, P.Y.C., Morrison Knudsen Corp, 355 Calle La Montana, Moraga, CA 94556. T: 510-299-1627, F: 415-442-7673, E: pam-li@mk.com

Li, X.-S., Dept of Civil Eng, Hong Kong Univ of Sci & Tech, Clear Water Bay Kowloon, HONG KONG. T: 852-235-87177, F: 852-235-81534, E: xsli@ust.hk

Li, Y., 1265 E Univeristy, Bldg H, Apt 2015, Tempe, AZ 85281. T: 225-336-4195, F: 602-255-8138, E: liyong@ unix1.sncc.lsu.edu

Liang, R.Y., Univ of Akron, Dept of Civil Eng, Akron, OH 44325. T: 330-972-7190, F: 330-972-6020, E: rliang@ uakron.edu

Liao, S.S.C., Parsons Brinckerhoff Quade &, 12 Turning Mill Lane, Sharon, MA 2067. T: 781-784-0421, F: 617-482-8487, E: liao@pbworld.com

Liberatore, M.G., Terra Firma Geotechnical, 26910 Fern Ridge Rd., Sweet Home, OR 97386. T: 541-367-7768, F: 541-367-8178

Lightwood, G.T., Shannon & Wilson, P.O. Box 300303, Seattle, WA 98103. T: 206-695-6825, F: 206-695-6777, E: gtl@shanwil.com

Likins, J.E., Geocon Inc., 14121 Recuerdo Dr., Del Mar, CA 92014. T: 619-755-0679, F: 858-558-6159, E: likins@geoconinc.com

Lim, R.M.-T., Geolabs-Hawaii, 2006 Kalihi St, Honolulu, HI 96819. T: 808-841-5064, F: 808-847-1749, E: robin@geolabs.net

Lima, D.C., Univ Federal De Viscosa, Dept De Engenharia Civil, Vicosa – Mg, 36570-000 BRAZIL. T: 055-31-8992765, F: 055-31-8992830, E: declima@mail.ufv.br

Limbach, F.W., JA Cesare & Assocs, 16637 W 73rd Dr., Arvada, CO 80007. T: 303-421-4871, F: 303-220-0442, E: imbach@uswest.net

Lin, B.M., Cyme Inc., 2401 Wexford Ave, South San Francisco, CA 94080. T: 650-616-9696, F: 650-616-9698, E: cyme@aol.com

Lin, G., S&ME Inc., 905 East 69th Street, Savannah, GA 31405. T: 912-353-8885, F: 912-353-8878, E: glin@smeinc.com

Lin, K.-C., Geotest Eng Inc., 16923 Meadowleigh Ct, Sugar Land, TX 77479. T: 281-265-1298, F: 713-266-2977

Lin, S.-S., National Taiwan Ocean Univ, Dept of Harbor & River Eng, Keelung, 20224 Taiwan, ROC. T: 1.19E+17, F: 886-2-24623679, E: sslin@mail.ntou.edu.tw

Lin, W.Y., MK Engineers & Constructors G, 944 Seville Pl, Fremont, CA 94539. T: 510-657-4899, F: 415-442-7673, E: wei_lin@mk.com

Lincoln, H.S., 5613 Tilia Ct, Burke, VA 20152. T: 703-426-2926, E: howard.lincoln@erols.com

Lindquist, D.D., 1910 Fairview Ave E, Seattle, WA 98102. T: 206-324-9530, F: 206-328-5581, E: ddl@hartcrowser.com

Lindquist, E.S., Eqe International, 5953 Thornhill Dr., Oakland, CA 94611. T: 510-339-7963, F: 510-663-1046, E: esl@eqe.com

Ling, H.I.P., Columbia Univ, 18 Burns Place, Cresskill, NJ 7026. T: 302-738-1617, F: 212-854-6267, E: ling@civil.columbia.edu

Linnan, B.D., Woodward-Clyde Cons, 10975 El Monte, Suite 100, Overland Park, KS 66211. T: 913-681-5043, F: 913-344-1011, E: bdlinna@wcc.com

Lithman, M.L., Ellis & Assocs Inc., 9919 Chelsea Lake Rd., Jacksonville, FL 32256. F: 904-880-0970, E: lithman@mediaone.net

Littell, R., Sage Cons Inc., 1978 Ventura Blvd, Camarillo, CA 93010. T: 805-482-6088, F: 805-389-9815, E: robert.littell@sagecon.com

Littlechild, B.D., Ove Arup & Partners, 5 Festival Walk 80 tatchee Ave, Kowloon Tong, HONG KONG. T: 852-2268-3436, F: 852-2268-3970, E: brian.littlechild@arup.com

Liu, C.-T.T., Caltrans, 720-D N Golden Springs Dr., Diamond Bar, CA 91765. T: 909-396-0876, F: 559-244-2852, E: ctedliu@msn.com

Liu, D.C., 340 E Oakdale Ave, Lake Forest, IL 60045. T: 847-234-3780, E: dcliu3@worldnet.att.net

Liu, T.K., 16 Audubon Rd., Lexington, MA 2421. T: 781-862-8667

Livingston, J.E., Ch2m Hill, 2525 Airpark Dr., Redding, CA 96001. T: 530-243-5831, F: 530-243-1654, E: jlivings@ch2m.com

Llewellyn, J.F., Foundation Syst Eng, 3025 Hodges Landing Dr., Knoxville, TN 37920. T: 423-579-9633, F: 423-573-6031, E: fseknx@aol.com

Lo, C.S., Room 507-515, Trade Square, No 681 Cheung Sha Wan Road, Kowloon Hong Kong, HONG KONG. T: 28511816, F: 28052902, E: simonlo@cigwh.com.hk

Lo, D.O.K., Flat E 10/F Pak Hee Court, Bedford Gardens, 159 Tin Hau Temple Rd., Northpoi, HONG KONG. T: 011-852-2882-2842, F: 011-852-2882-84

Lo, K.-H., Maa Engrg Consults Ltd., Flat C 13th Fl Block 1, Woodview Ct 75 Kung Lok Rd., Kwun Tong Kowloon, HONG KONG. T: 852-2342-8507, F: 852-2861-2081, E: maahk@netvigator.com

Locke, G.E., IT Corp, 990 Snyder Ln, Walnut Creek, CA 94598. T: 925-935-8035, F: 925-827-2029, E: glocke@theitgroup.com

Locke, L.J., URS Greiner Woodward Clyde, 500 12th St, Suite 200, Oakland, CA 94607. T: 510-874-3161, F: 510-874-3268, E: linda_locke@urscorp.com

Lockhart, C.W., Golder Assocs Inc., 18300 NE Union Hill Road, Suite 200, Redmond, WA 98052. T: 425-883-0777, F: 425-882-5498, E: clockhart@golder.com

Loehr, J.E., Univ of Missouri at Columbia, Dept of Civil Engr, E2509 Engr Bldg East, Columbia, MO 65211. T: 573-882-6380, F: 573-882-4784, E: eloehr@missouri.edu

Logan, T.J., Dames & Moore, 7800 Congress Ave, Suite 200, Boca Raton, FL 33487. T: 561-994-6500, F: 561-994-6524, E: tom_logan@urscorp.com

Loh-Doyle, M.K., Doyle Consultants, 10940 Wilshire Blvd, Suite 1600, Los Angeles, CA 90024. T: 310-443-4224

Lohman, J.S., Kleinfelder Inc., 1370 S Valley Vista Dr., Suite 150, Diamond Bar, CA 91765. T: 909-396-0335, F: 909-396-1324, E: jlohman@kleinfelder.com

Loigman, H., Hal Loigman Pe, 150 Lantern Ln, King of Prussia, PA 19406. T: 610-525-7390, E: halloigman@aol.com

Lommler, J.C., AMEC Earth & Environmental Inc., P.O. Box 1229, Sandia Park, NM 87047. T: 505-281-8185, F: 505-821-7371, E: john.lommler@amec.com

Long, E.L., Arctic Foundations Inc., 5621 Arctic Blvd, Anchorage, AK 99518. T: 907-562-2741, F: 907-562-0153, E: elong@arcticfoundations.com

Long, J.H., Univ of Illinois, Dept of Civil Eng, 205 North Mathews, Urbana, IL 61801. T: 217-333-2543, F: 217-333-9464, E: j-long@uiuc.edu

Long, J.D., Law Engineering & Env Serv, 535 Pimlico Dr., Seymour, TN 37865. T: 423-579-3194, F: 423-588-8026, E: jlong@lawco.com

Long, R.A., Environmental Strategies Corp, 3624 Heather Glen Dr., Colo Springs, CO 80922. T: 719-638-4342, F: 303-850-9214, E: rlong@escden.com

Lopez-Saiz, J.M., FCC Construccion SA Informacion Y, Documentaco C/Acanto 22-3, Madrid, 28045 SPAIN. T: 34-1-5390797, F: 34-1-5272431, E: jasantos@fcc.es

Lorenzen, T.A., Harding-Ese A Mactec Company, Box 65, Jefferson Cty, MT 59638. T: 406-933-8671, F: 406-443-0452, E: tlorenzen@harding.com

Losada, M.A., Universidad De Granada, Cuesta Escoriaza 1, Granada, 18009 SPAIN. T: 34-958-220198, F: 34-958-248374, E: mlosada@platon.ugr.es

Lou, C.N., Civil Konsult Snd Bhd, P.O. Box 10588, 88806 Kota Kinabalu, Sabah, MALAYSIA. T: 60-88-434085, F: 60-89-271391

Loucks, D.G., Daniel G Loucks Pe, P.O. Box 163, Ballston Spa, NY 12020. T: 518-371-7622, F: 518-383-2059, E: el007salem@aol.com

Loughney, R.W., Loughney Associates Inc., 25 Arrowpoint Road, New Preston, CT 6777. T: 860-868-9995, F: 860-868-1025

Lourie, P.E., David Eugene, Lourie Consultants, 3924 Haddon St, Metairie, LA 70002. T: 504-887-5531, F: 504-888-1994, E: lcon1@aol.com

Loush, K.C., Eastern Exterior Wall Systems Inc., 1324 Prospect Avenue, Bethlehem, PA 18018. T: 610-866-7603, F: 610-866-4672, E: kenclpe@aol.com

Lowe III, J., 26 Grandview Blvd, Yonkers, NY 10710. T: 914-779-3720, F: 914-779-0928, E: jloweiii@juno.com

Lowery, P.S., Facility Engineering Assoc, 125 E Clifford Ave, Alexandria, VA 22305. T: 703-739-9838, F: 703-591-4857, E: plowery@erols.com

Lowney, J.V., Lowney Assocs, 405 Clyde Avenue, Mountain View, CA 94043. T: 650-967-2365, F: 650-967-2785, E: jlowney@lowney.com

Luccioni, L.X., 2100 Main Street, Suite 150, Huntington Beach, CA 92647. T: 714-969-0800, Ext. 258, E: luccioni@geosyntec.com

Ludlow, S.J., Earth Exploration Inc., 7770 West New York Street, Indianapolis, IN 46214. T: 317-273-1690, F: 317-273-2250, E: sjl@earthengr.com

Ludwig, C.J., Schnabel Foundation Co, 3075 Citrus Circle, Suite 150, Walnut Creek, CA 94598. T: 925-947-1881, F: 925-947-0418

Ludwig, H.P., Schnabel Foundation Co, 5210 River Rd., Bethesda, MD 20816. T: 301-657-3060, F: 301-657-4080, E: sfc05@erols.com

Luebbers, M., URS Corporation, 14201 Cherrywood Ln, Tustin, CA 92780. F: 714-973-4086, E: michael_luebbers@urscorp.com

Lukas, R.G., Ground Engineering Consultants, 4555 Forestview Dr., Northbrook, IL 60062. T: 847-537-2355, F: 847-559-0181

Luke, B.A., Univ of Nevada Las Vegas, Unlv Dept of Civil & Env Eng, 4505 Maryland Pkwy, Box 4015, Las Vegas, NV 89154. T: 702-895-1568, F: 702-895-3936, E: bluke@ce.unlv.edu

Lum, W.B., 1712 S King St, #202, Honolulu, HI 96826. T/F: 808-944-9429

Luna, R., Dept of Civil Engr, Univ of Missouri – Rolla, 306 Butler Carlton Hall, Rolla, MO 65409. T: 573-341-4484, F: 573-341-2749, E: rluna@umr.edu

Lundberg, R.P., Forge Engineering Inc., 6200 Shirley Street, Suite 204, Naples, FL 34109. T: 941-514-4100, F: 941-514-4161, E: forge@gate.net

Lupo, J.F., Golder Associates, 10372 Beas Ln, Conifer, CO 80433. T: 303-838-4868, F: 303-290-9904

Luque, C., Casilla 3275, Guayaquil, South America, ECUADOR

Luscher, U., 312 Camino Sobrante, Orinda, CA 94563. T: 925-254-5281, E: jlusc@aol.com

Lustig, M.T., Geotechnical Services Inc., 2501 33rd St, Des Moines, IA 50310. T: 515-255-8648, F: 515-270-1911

Lutenegger, A., Univ of Massachusetts, Dept of Civil Eng, Amherst, MA 1003. T: 413-545-2872, F: 413-545-4525, E: lutenegg@ecs.umass.edu

Lytton, R.L., Texas A&M Univ, 503a Ce/Tti Bldg, College Station, TX 77843. T: 979-845-9964, F: 979-845-0278, E: r-lytton@tamu.edu

Mabry, P.N., Georizon Inc., 9006 Tennga Lane, Chattanooga, TN 37421. T: 423-899-1922, F: 423-296-0260, E: pmabry@georizon.com

Mabry, R.E., 1635 Ludwell Dr., Maple Glen, PA 19002. T: 215-643-4390

Mac Kenzie, I.M., Ove Arup & Partners, P.O. Box 76, Millers Point Nsw, 2000 AUSTRALIA. F: 02-9320 9321

Mac Pherson, H.H., Harza Eng Co, 750 N Dearborn St, Apt 1907, Chicago, IL 60610. T: 312-573-1481

MacConochie, A.F., Froehling & Robertson Inc., 212 Oak Grove Road, Norfolk, VA 23505. T: 757-533-9546, F: 757-436-1674, E: fmacconochie@fandr.com

Mack, W.R., Hurt & Proffitt, 2524 Langhorne Rd., Lynchburg, VA 24501. T: 804-847-7796, F: 804-847-0047, E: wrm@handp.com

Mackiewicz, S.M., S&Me Inc., 840 Low Country Blvd, Mt Pleasant, SC 29464. T: 843-884-0005, F: 843-881-6149, E: smackiewicz@smeinc.com

Macklin, Jr, P.R., Yenter Companies Inc., 20300 West Higway 72, Arvada, CO 80007. T: 303-279-4458, F: 303-279-0908, E: pmacklin@yenter.com

Magginas, V.K., 6 Turnham Lane, Gaithersburg, MD 20878. T: 301-279-7810, F: 301-279-0357, E: vmagginas@aol.com

Maharaj, R.Jr, Sopac Secretariat, Private Mail Bag, Gpo, Suva, FIJI ISLANDS. T: 679-381-377, F: 679-370-040, E: rossi@sopac.org.fj

Mahdavian, M., HNTB-CDMG, 24761 Glenwood Dr., Lake Forest, CA 92630. T: 949-454-8888, F: 949-789-0819, E: mmahdavian@aol.com

Mahiquez, L.F., Tierra Inc., 5909b Breckinridge Pkwy, Tampa, FL 33610. T: 813-626-7775, F: 813-621-1496, E: tierrafl@mindspring.com

Maitland, J.K., Foundation Eng Co, 820 NW Cornell, Corvallis, OR 97330. T: 541-757-7645, F: 541-757-7650

Majchrzak, M.F., Kleinfelder Inc., 7133 Koll Center Pkwy, Suite 100, Pleasanton, CA 94566. T: 925-484-1700 221, F: 925-484-5838, E: mmajchrzak@kleinfelder.com

Majeski, P.J., Earth Tech Corp, 26 Baker Rd., Arlington, MA 2474. T: 781-646-5540, F: 978-371-2468

Makarechi, H., MCE Group, 28761 Via Pasatiempo, Laguna Niguel, CA 92677. F: 714-560-9018, E: mcehm@flash.net

Makdisi, F.I., Geomatrix Cons Inc., 262 Birchwood Dr., Moraga, CA 94556. T: 925-376-2748, F: 570-663-4141, E: fmakdisi@geomatrix.com

Makdissy, T.S., Terrasearch Inc., 6840 Via Del Oro, Ste 110, San Jose, CA 95119. T: 408-362-4920, F: 408-362-4926

Malak, J.T., URS Corporation, 41 B Mathews Ave, Riverdale, NJ 7457. T: 908-560-5709, F: 973-785-0023

Malivuk, J., Summit Testing & Insp Co, P.O. Box 2231, Akron, OH 44309. T: 330-869-6606, F: 330-869-6437, E: jmal910@aol.com

Man, F.-T., 50 Chau Tau Village, Yuen Long, NT, HONG KONG

Mana, A.I., King Fahd Univ of Petroleum, P.O. Box 306, Dhahran, 31261 SAUDI ARABIA. T: 9661-405-1177, F: 9661-405-9974, E: almana@kacst.edu.sa

Manassero, M.E., Studio Geotecnico Italiano, Via Ripamonti 89, Milano, ITALY. T: 39-2-5220141, F: 39-2-5691845, E: sgi_manassero@studio-geotecnico.it

Manegold, W.J., Pacific Gas & Electric Co, 41885 Corte Santa Ines, Fremont, CA 94539. T: 510-651-2004

Mangla, A.K., Atlanta Testing & Eng, 11420 Johns Creek Pkwy, Duluth, GA 30097. T: 770-476-3555, F: 770-476-0213, E: ateatlanta@mindspring.com

Mank, P.R., Dunkelberger Engineering & Testing, Inc., 8260 Vico Court, Unit B, Sarasota, FL 34240. T: 941-379-0621, F: 941-379-5061, E: prmank@aol.com

Mann, D.K., Harlan Tait Associates, 1366 La Playa St, San Francisco, CA 94122. T: 415-731-4572, F: 650-737-5441, E: dona_72@yahoo.com

Mann, G., Creative Eng Options, 5418 159th Place Ne, Redmond, WA 98052. F: 425-867-9664

Mann, M.J., Mcmahon & Mann Cons Engrs Pc, 2495 Main St, Suite 432, Buffalo, NY 14214. T: 716-834-8932, F: 716-834-8934

Marasa, M.J., Geosciences Design Group Llc, 804 Overhills Dr., Old Hickory, TN 37138. T: 615-754-0123, F: 615-883-3767, E: gdgllc@bellsouth.net

Marble, D.K., Div of Water Rights/Dam Safety, Box 146300, Salt Lake City, UT 84114. T: 801-538-7376, F: 801-538-7467, E: nrwrt.dmarble@state.ut.us

Marche, R., Geos Ingenieurs Conseils SA, 7 Route De Drize, Ch-1227 Carouge Geneve, SWITZERLAND. T: 41-22-309-30-60, F: 41-22-309-3070, E: geos@geos.ch

Marciano, E., Malcolm Pirnie Inc., 104 Corporate Park Drive, Box 751, White Plains, NY 10602. T: 914-641-2796, F: 914-641-2455, E: emarciano@pirnie.com

Marcik, L.J., Criscuolo Shepard Assoc Pc, 420 E Main St, Branford, CT 6405. T: 203-481-8749

Marcin, B.G., 136 Chinook Trail, P.O. Box 1316, Eastsound, WA 98245. T: 360-376-4690

Marcus, Jr, G.R., Tams Consultants Inc., 137-27 68th Dr., Apt B, Flushing, NY 11367. T: 718-544-5842, F: 212-697-6354, E: gmarcus@tamsconsultants.com

Marcuson III, W.F., Wf Marcuson Iii & Asscos, 95 Plantation Dr., Vicksburg, MS 39183. T: 601-638-1704, F: 601-634-4656, E: marcusw@mail.wes.army.mil

Mare, A.D., Pacrim Geotechnical Inc., 506 SW Sixth Ave, Suite 1006, Portland, OR 97204. T: 503-771-7342, F: 503-223-2306, E: amare@pacrimgeo.com

Marinho, F., Rua Cel Oscar Parto 167, Apt 63, Sao Paulo Sp 04003, BRAZIL. T: 55-11-38185703, E: fmarinho@usp.br

Marinucci, A., Schnabel Foundation Co, 117 Green St, Landsdale, PA 19446. T: 215-361-0921, F: 610-277-7932, E: a_marinucci@hotmail.com

Markouizos, K.G., NMG Geotechnical, 17991 Fitch, Irvine, CA 92614. T: 949-442-2442, F: 949-476-8322, E: kmarkouizos@nmggeotech.com

Marley, W.M., Earthtech, Unit 1 51 Secam St, Mansfield Qld, 4122 AUSTRALIA. T: 617-334-33166, F: 617-334-94705, E: mmarley@earthtech.com.au

Marquardt, J.R., HAS, 639 Apalalchee Cir NE, St Petersburg, FL 33702. T: 813-576-2444, F: 813-971- 1862, E: engrx@msn.com

Marquez, J.L., Hillsborough Cnty Public Works, Eng Division, P.O. Box 1110, Tampa, FL 33601. T: 813-272-5912-3631, E: marquezj@hillsboroughcounty.org

Marr, Jr, W.A., Geocomp Corp, 1145 Massachusetts Ave, Boxboro, MA 1719. T: 978-635-0012, E: wam@geocomp.com

Marr, L.S., Dannenbaum Eng Corp, 2915 Blue Glen Ln, Houston, TX 77073. T: 281-443-6814, F: 713-527-6338, E: lsm@insync.net

Marra, M.R., 136 Glasswycke Dr., Glassboro, NJ 8028. T: 856-863-8815, F: 363-6271

Marsh, E.T., American Geotech, 1725 Crescent Knolls Gln, Escondido, CA 92029. T: 760-839-2787, F: 858-457-0814

Marshall, J.P., Marshall Engineering Inc., 2010 Industrial Dr., Annapolis, MD 21401. T: 410-573-0939, F: 410-224-8630

Marshall, P.C., Univ of Memphis, 4925 Brentdale, Memphis, TN 38118. T: 901-368-0375, F: 901-373-7927

Marti, J., Principia Ing Consultores SA, Velazquez 94, Madrid, 28006 SPAIN. T: 341-2091482, F: 341-5751026, E: marti@principia.es

Martin, Jr, A.C., 1293 SW Timberline Dr., Lake Oswego, OR 97034. T: 503-635-1472, F: 503-635-6484, E: aarmart@aol.com

Martin, G.R., Univ of Southern California, Dept of Civil Engr, University Park, Los Angeles, CA 90089. T: 213-740-9124, F: 213-744-1426

Martin, J.W., Amec Earth and Environmental, 5546 Shady Trail, Old Hickory, TN 37138. T: 615-754-2722, E: jmartin@oees.com

Martin, J.P., Drexel Univ, 252 Sagamore Rd., Havertown, PA 19083. T: 610-446-6477, F: 215-895-1363, E: martinjp@droxol.edu

Martin, R.E., Schnabel Eng Assoc, P.O. Box 1360, Ashland, VA 23005. T: 804-798-0081, F: 804-798-0048, E: rmartin@schnabel-eng.com

Martin, T.L., Golder Associates Inc., 1904 Byrd Ave, Ste 100, Richmond, VA 23230. T: 804-288-8749, F: 804-282-4249, E: tmartin@golder.com

Martinez, D.T., Pacific Soils Eng Inc., 655 Ave B, Redondo Beach, CA 90277. T: 310-540-0613, F: 714-220-9589

Martinez, R.E., URS Consultants Inc., 9702 Kings Canyon Place, Tampa, FL 33634. T: 813-881-0580, F: 813-874-7424, E: tparem@dames.com

Martinez-Ramirez, G., San Martin De Porres #3716, Sanernesto 3755 Jardsnignacio, Guadalajara, Jalisco, 45000 MEXICO. T: 523-121-6471, E: gabrichma@yahoo.com

Marulanda, A., Ingetec SA, Apartado Aereo 5099, Bogota, COLOMBIA. T: 571-288-4498, F: 571-288-4531, E: amarula@ingetec.com.co

Masatsugu, G., 99 020 Kaamilo St, Aiea, HI 96701. T: 808-488-8710

Mashhour, M.M.A., 13 26 July St, Lebanon Square, Mohandessin, Cairo, EGYPT. T: 011-202-3035404, F: 00202-3469604, E: mmashhour@purenet.com.eg

Massoudi, N., URS Corp, 321 Market St W, Apt 201, Gaithersburg, MD 20878. T: 215-669-5146, F: 301-869-8728

Mast, K.C., URS Corporation, 800 West St. Clair Ave, Cleveland, OH 44113. T: 216-622-2400, F: 216-622-2428, E: keith_mast@urscorp.com

Maswoswe, J.J., Parsons Brinckerhoff Quade &, 31 Lorraine Dr., Ashland, MA 1721. F: 617-346-7963, E: jjmaswos@bigdig.com

Matasovic, N., Geosyntec Consultants, 2100 Main St, Suite #150, Huntington Beach, CA 92648. T: 714-969-0800, F: 714-969-0820, E: nevenm@geosyntec.com

Matey, J.G., L Robert Kimball & Associates Inc., 1225 Old Concord Rd., Monroeville, PA 15146. T: 412-373-3626, F: 814-472-7712

Matheny, S.C., Civil Solutions Inc., 6008 Petros Drive, W Bloomfield, MI 48324. T: 248-363-0234, F: 248-366-9855, E: smath17@aol.com

Matheson, G.M., Schnabel Eng Assoc, 16201 Orchard View Ct, Gaithersburg, MD 20878. F: 601-530-63 76

Mathews, D.L., US Army Corps of Engrs, 7812 Chadwick, Prairie Vlg, KS 66208. T: 913-341-3894, F: 816-426-5462, E: david.l.mathews@usace.army.mil

Mathews, W.A., Bunnell-Lammons Engineering, 1200 Woodruff Rd., Suite B-7, Greenville, SC 29607. T: 864-288-1265, F: 864-288-4430

Mathis, H.A., Mathis Geotechnical Consulting, Inc., 561 Marblerock Way, Lexington, KY 40503. T: 859-296-5664, F: 859-296-9135, E: hmathis@iglou. com

Mathur, J.N., Geologic Associates, 6587 Calle Del Norte, Anaheim, CA 92807. F: 949-260-1830, E: jmathur@geo-logic.com

Mathy, D.C., Dcm/Joyal Engineering, 484 N Wiget Ln, Walnut Creek, CA 94598. T: 925-945-0677, F: 925-945-1294, E: dcmjoyal@dcmjoyal.com

Matthews, D.M., CDE Resources Inc., 36153 Escena Dr., Yucaipa, CA 92399. T: 909-795-9146, F: 909-474-1097

Matthews, G.P., Malcolm Pirnie Inc., 104 Corporate Park Dr., White Plains, NY 10602. T: 914-694-2100

Maxwell, B., Maxwell Eng, P.O. Box 9527, Jackson, MS 39286. T: 601-982-5641, F: 601-982-5641, E: maxeng@msn.com

May, I.P., 2610 St Paul St, Baltimore, MD 21218. T: 410-889-7842, F: 410-436-6836, E: ira.may@jhu.edu

Mayer, J.H., 2 Crabapple Ln, Plainfield, NJ 7060. T: 908-755-0436, F: 908-755-5636

Maynard, J.K., Maguire Group Inc., 225 Foxborough Blvd, Foxborough, MA 2035. T: 508-543-1700, F: 508-543-5157, E: jmaynard@maguiregroup.com

Maynard, R.D., Camp Dresser & Mc Kee Inc., P.O. Box 3885, Bellevue, WA 98009. T: 425-453-8383, F: 425-646-9523, E: maynardrd@cdm.com

Maynard, T.R., 6261 N Oriole Ave, Chicago, IL 60631. T/F: 773-763-4059, E: trmagm@earthlink.net

Mayne, P.W., Georgia Inst of Tech, School of Civil & Env Engrg, Atlanta, GA 30332. T: 404-894-6226, F: 404-894-2281

Mayu, P.H., GEOS Ingenieurs Conseils Sa, 7 Route De Drize, Ch-1227 Carouge Geneve, SWITZERLAND. T: 22-309-3060, F: 22-309-3070, E: geos@geos.ch

Maza, J.A., INCYH, Centro Reg Andino Belgrand, Oste 210 Mendoza 5500, ARGENTINA. T/F: 54-261-4288251, E: mazaja@impsat1.com.ar

Mazhar, F.M., Hazra Engineering Company, 233 S Wacker Dr., Chicago, IL 60606. T: 312-831-3800, F: 312-831-3999, F: fmazhar@harza.com

Mc Cabe, W.M., URS Corporation, 505 N 73rd St, Seattle, WA 98103. T: 206-781-1067, F: 206-727-3350, E: seawmm@dames.com

Mc Carthy, D.F., Mohawk Valley Comm College, 16 Toggletown Rd., Clinton, NY 13323. T: 315-853-6641, E: dfmpe@dellnet.com

Mc Carthy, R.J., Stoney Miller Cons Inc., 14 Hughes, B-101, Irvine, CA 92618. T: 949-380-4886, F: 949-455-9371

Mc Caskie, S.L., Sverdrup Civil Inc., 13723 Riverport Dr., Maryland Heights, MO 63043. T: 314-770-4554

Mc Clelland, D.E., US Forest Service, Engineering Section, P.O. Box 7669, Missoula, MT 59807. T: 406-329-3351, F: 406-329-3198, E: demcclelland@fs.fed.us

Mc Clure, J.G., 4957 Northdale Dr., Fremont, CA 94536. T: 510-792-8061, F: 510-894-1245, E: jamesmcclure@home.com

Mc Clymont, A.A., Mc Clymont & Rak Engrs, 1059 Thomas Busch Hwy, Pennsauken, NJ 8110. T: 609-488-1700, F: 609-488-4501

Mc Comb, C.L., STS Cons Ltd., 519 S Prairie Ave, Barrington, IL 60010. T: 847-381-6809, F: 847-279-2550, E: mccomb@stsltd.com

Mc Cormack, T.C., Saint Martins College, 2104 Se Sherman, Portland, OR 97214. T: 503-234-4580, F: 360-438-4548, E: tmccormack@stmartin.edu

Mc Curdy, K.L., Thelen Associates, P.O. Box 232, Batesville, IN 47006. T: 812-933-1741, F: 859-746-9408

Mc Daniel, G.T., Law Eng & Envir Services Inc., 5845 Nw 158 St, Miami Lakes, FL 33014. T: 305-826-5588, F: 305-826-1799, E: tmcdanie@lawco.com

Mc Dermott, M.P., Mc Kinney Drilling Co, 3150 Walnut St, Colmar, PA 18915. T: 215-643-0238, F: 215-822-3329

Mc Donald, D.J., University of Toledo, 2465 Manchester, Toledo, OH 43606. T: 419-535-9943, F: 419-530-8116, E: bmcdonal@eng.utoledo.edu

Mc Donnell, W.J., Camp Dresser & Mc Kee Inc., 60 Johnson Circle, North Andover, MA 1845. T: 603-362-9260

Mc Dowell, R., Mc Dowell & Assocs, 21355 Hatcher, Ferndale, MI 48220. T: 313-399-2066

Mc Elroy, Jr, J.J., Langan Eng & Env Srvcs Inc., 433 Maple Ave, Haddonfield, NJ 8033. T: 856-857-9255, F: 215-864-0671, E: jmcelroy@langan.com

Mc Gillivray, R.T., Ardaman & Assoc, 10904 Kewanee Dr., Tampa, FL 33617. T: 813-988-7859, F: 813-628-4008, E: rg13@gte.net

Mc Grath, F.S., S&Me, 3109 Spring Forest Rd., Raleigh, NC 27616. T: 919-872-2660, F: 919-876-3958

Mc Grath, T.J., Simpson Gumpertz & Heger Inc., 297 Broadway, Arlington, MA 2474. T: 781-641-7240, F: 781-643-7560, E: tjmcgrath@sgh.com

Mc Guffey, V.C., 22 Lombard St, Schenectady, NY 12304. T: 518-393-3565, F: 518-393-1005, E: geo96pop@aol.com

Mc Keen, D.W., Mueser Rutledge Cons Engrs, 1 Wood Ln, Valley Stream, NY 11581. T: 516-791-8041, F: 212-953-5626

Mc Keen, R.G., Univ of New Mexico, 12808 Hugh Graham Road, Albuquerque, NM 87111. T: 505-294-3183, F: 505-246-6001, E: gmckeen@unm.edu

Mc Kown, A.F., Haley & Aldrich Inc., 465 Medford St, Boston, MA 2129. T: 617-886-7410, F: 617-886-7710, E: afm@haleyaldrich.com

Mc Lain, M.E., Zannino Engineering, 1650 A Mountain Road, Glen Allen, VA 23060. T: 804-262-0299, F: 804-262-8479, E: mrkml@earthlink.net

Mc Mahon, D.R., Mcmahon & Mann Consulting, 27 Irving Terrace, Buffalo, NY 14223. T: 716-874-0082, F: 716 834-8934, E: dmcmahon@mmce.net

Mc Manis, K.L., Univ of New Orleans, 5809 Flower Drive, Metairie, LA 70003. T: 504-455-3386, F: 504-280-5586, E: klmce@uno.edu

Mc Millan, J.C., ICF Kaiser Engrs Inc., 630 Countryside Ct, Brentwood, CA 94513. T: 415-634-1555, F: 510-419-5355, E: jmcmillan@icfkaiser.com

Mc Namara, E.J., Underpinning & Foundation, 46 Butterfield Dr., Greenlawn, NY 11740. T: 516-757-1675, F: 718-786-6981

Mc Neilan, T.W., Fugro Mc Clelland, 1243 Sunnycrest, Ventura, CA 93003. T: 805-644-6271, F: 805-650-7010, E: tmcneilan@fugro-usa.com

Mc Neill, P.B., Geosolutions, 220 High St, Sanluisobispo, CA 93401. T: 805-543-8539, F: 805-543-2171, E: tass@thegrid.net

Mc Neill, R.L., AN Engineering Consultancy, 38 Nichols Hill Dr., Asheville, NC 28804. T: 828-658-8886, F: 828-645-8886, E: robertmcneill@msn.com

Mc Neilly, P.E., Mark F., Golder Associates Inc., 5303 Boulevard East, West New York, NJ 7093. T: 201-583-0540, F: 856-616-1874, E: mark_mcneilly@golder.com

Mc Vey, G.S., Morrison Knudsen Corp, 13331 Walke Pointe Way, Chesterfield, VA 23832. T: 804-639-3927

Mc Wee, J.M., 28121 Alava, Mission Viejo, CA 92692. T/F: 949-859-1993, E: geocon@fea.net

McElroy, J.A., CDM Jessberger, 4414 Ne 77th Street, Seattle, WA 98115. T: 206-524-4192, F: 425-646-9523, E: mcelroyja@cdm.com

McFerron, D.J., J Ryan Engineering, 1045 Hanover Ave, Norfolk, VA 23508. T: 757-423-4599, F: 757-423-7077, E: ewmcf@aol.com

McGinnis, J.T., Law Eng & Envir Services, 22455 Davis Drive, Suite 100, Sterling, VA 20164. T: 703-404-7014, F: 703-404-7074, E: jmcginni@lawco.com

McKelvey, J.G., Whitney Bailey Cox & Magnanillp, 57 Windersal Ln, Baltimore, MD 21234. T: 410-661-9239, F: 410-324-4100, E: tmckelvey@wbcm.com

McKittrick, D.P., Reinforced Earth Co, 11526 Hemingway Dr., Reston, VA 20194. T: 703-318-9637, F: 703-318-9632, E: dpmckitt@aol.com

McNutt, J.E., Terracon Cons Inc., 7604 Capstick Ave, Las Vegas, NV 89129. T: 702-645-7437, F: 702-597-9009

McRae, M.T., Jacobs Associates, 500 Sansome St, 7th Floor, San Francisco, CA 94111. T: 415-434-1822, F: 415-956-8502

Meade, R., USAF Academy, 2830 Boxwood Pl, Colorado Springs, CO 80920. T: 719-264-9596, F: 719-333-2234, E: ron.meade@usafa.af.mil

Mech, G.J., US Army Corps of Engrs, 1538 West 49th St, Davenport, IA 52806. T: 319-388-9907, F: 309-794-5659, E: george.j.mech@usace.army.mil

Meek, J.W., Fr Holst Hoch U Tiefbau, Brahms Allee 33 Xi, Hamburg, D-20144 GERMANY. F: 040-756-16-199

Mehta, H.C., USS-Posco Ind, 178 Carriage Ln, Pacheco, CA 94553. T: 510-674-1139, F: 510-439-6354, E: kalpana hemantmehta@worldnet.att.net

Meisenheimer, J.K., Stone & Webster Eng Corp, 6033 S Lima St, Englewood, CO 80111. T: 303-741-7631, F: 303-741-7670, E:james.meisenheimer@stoneweb.com

Meissner, H.E., Univ of Kaiserslautern, Elfen Weg 24, Karlsruhe, D-76199 GERMANY. T: 721-883-376, F: 631-205-3806

Mejia, L.H., URS Greiner Woodward Clyde, 72 Camino Sobrante, Orinda, CA 94563. T: 925-258-1900, F: 510-874-3268, E: lelio-mejia@urscorp.com

Melick, C.T., Melick Tully & Assocs Inc., 117 Canal Rd., S Bound Brook, NJ 8880. T: 908-356-3400, F: 908-356-9054

Mercado, C., Geoexplor, Box 928, 138 Winston Churchill Ave, San Juan, PR 926. T: 787-753-6621, F: 787 763-4386, E: geoexplor@msn.com

Merino, M.P., North American Coal, 2412 Lawnmeadow Dr., Richardson, TX 75080. T: 214-235-2116

Merjan, S., Underpinning Foundation &, Constr Inc., 46-36 54th Road, Maspeth, NY 11378. T: 718-786-6557

Merkle, D.H., 269 Hugh Thomas Dr., Panama City, FL 32404. T: 850-871-3132, E: dhmerkle@aol.com

Merl, J.H., STS Consultants, 750 Corporate Woods Parkway, Vernon Hills, IL 60061. T: 847-279-2534, F: 847-279-2510, E: merl@stsltd.com

Merritt, Jr, J.L., Merritt Cases Inc., 13086 Oak Crest Dr., Yucaipa, CA 92399. T: 909-797-9769, F: 909-797-9511, E: jmer13086@aol.com

Merry, S.M., University of Arizona, Civil Engineering Bldg Rm 324e, Tucson, AZ 85721. T: 520-626-6892, F: 520-621-2550, E: merry@engr.arizona.edu

Merten, P.M., TSI Engineering Inc., 1300 Greentree Ln, Saint Louis, MO 63122. T: 314-962-9144, F: 314-644-3135, E: pmerten@tsi-engineering.com

Mesenbrink, P.E., 35585 Industrial Rd., Livonia, MI 48150. T: 734-453-6006, F: 734-453-6201, E: pmesenb01@worldnet.att.net

Mesri, G., Univ of Illinois, 2230 NCEL, 205 North Mathews Ave, Urbana, IL 61801. T: 217-333-6934, F: 217-265-8041, E: g-mesri@uiuc.edu

Metcalf, Jr, T.E., Kleinfelder Inc., 1522 Charles Dr., Redding, CA 96003. T: 916-244-7203, F: 916-244-3031

Metcalf, M.R., 2211 Poland Avenue, Youngstown, OH 44502. T: 330-747-0152, F: 330-549-3639, E: michael_metcalf@praxair.com

Meuris, J.C., Engeo Inc., 64 Colorado Ave, Berkeley, CA 94707. T: 510-838-7425, E: engstaff@engeo.com

Meyer, C.P., GPA Inc., 1920 Hamilton Lane, Orlando, FL 32806. T: 407-648-9979, F: 407-671-5540, E: gpameyer@aol.com

Meyer, J.E., Meyer Borgman & Johnson, 12 South 6th Street, Ste 810, Minneapolis, MN 55402. T: 612-338-0713, F: 612-337-5325, E: meyer123@tc.umn.edu

Meyer, K.T., Mlaw, 4807 Buckskin Pass, Austin, TX 78745. T: 512-442-3171, F: 512-651-0098, E: ktmeyer@mlaw-eng.com

Meyer, M.E., Langan Engrg & Enviro Svcs Inc., 7900 Miami Lakes Dr., W, Ste 102, Miami Lakes, FL 33016. T: 305-362-1166, F: 305-362-5212, E: mmeyer@langan.com

Meyers, M.S., Univ of Wi Platteville, 590 Pyrite Road, Platteville, WI 53818. E: meyersm@awplatt.edu

Meyersohn, W.D., Dames & Moore, 221 Main St, Ste 600, San Francisco, CA 94105. T: 415-243-3738, F: 415-882-9261, E: daniel_meyersohn@urscorp.com

Mian, B.R., Wilbur Smith Associates, P.O. Box 92, Columbia, SC 29202. T: 803-758-4545, F: 803-458-4561, E: bmian@wilbursmith.com

Michael, M.J., Kitsap County of, E308 Leffler Lp, Grapeview, WA 98546. T: 360-275-3078, F: 360-337-5790, E: mike5767@aol.com

Michalowski, R.L., Univ of Michigan, Dept of Civil & Environ Eng, 2340 G.G. Brown Lab, Ann Arbor, MI 48109. T: 734-763-2146, F: 734-764-4292, E: rlmich@umich.edu

Micklus, P.G., Sor Testing Labs Inc., 98 Sandpark Rd., Cedar Grove, NJ 7009. T: 973-239-6001, F: 973-239-8380

Middleton, T.J., Ontario Ministry Nat. Resource, Hwy 101 E, Ontario Gov. Complex, P.O. Bag 3020, S Porcupine, P0N-1H0 CANADA ON. T: 705-235-1161, F: 705-235-1246, E: tim.middleton@mnr.gov.on.ca

Mikkelsen, P.E., 16483 Se 57th Pl, Bellevue, WA 98006. T: 425-653-9315, F: 425-746-9577, E: mikkel@teleport.com

Mikulas, P.M., Los Angeles Co Sanitation Dist, Solid Waste Mgt Dept, P.O. Box 4998, Whittier, CA 90607. T: 562-699-7411 x2451, F: 562-692-2941, E: pmikulas@lacsd.org

Miles, S.B., US Geological Survey, 4506 Fremont Ave N, Apt B, Seattle, WA 98103. T: 206-633-1678, F: 206-543-3313, E: smyles@u.washington.edu

Millan, A.A., Ingenieria Y Estudios Ltd., Calle 12 No 13-53, Pereira, COLOMBIA. T: 576-335-6296, F: 576-335-6679, E: amillan@interco.net.co

Miller, C.M., Geo-Test Inc., 5203 Camino Sandia, Albuquerque, NM 87111. T: 505-299-9302, F: 505-857-0803, E: geotest@geo-test.com

Miller, D.J., Gannett Fleming, 2323 Ridgecrest Rd., Fort Collins, CO 80524. T: 970-493-3483, E: djmiller@gfnet.com

Miller, D.G., Colorado State of, 8849 Weld County Rd., 70, Windsor, CO 80550. T: 970-674-0240, F: 970-392-1816, E: dennis.miller@state.co.us

Miller, D.L., Duane Miller & Associates, 9720 Hillside Dr., Anchorage, AK 99516. T: 907-346-1021, E: duane_miller@compuserve.com

Miller, E.A., Miller Pacific Eng Group, 165 N Redwood Dr., Suite 120, San Rafael, CA 94903. T: 415-491-1338, F: 415-491-1831

Miller, G.H., Ralston Purina Co, 1548 Blue Roan Ct, Chesterfield, MO 63005. T: 636-537-2558, F: 314-982-4199, E: ghmiller@ralston.com

Miller, G.A., University of Oklahoma, School of Civil Eng & Env Sci, 202 West Boyd Street – Rm 334, Norman, OK 73019. T: 405-325-5911, F: 405-325-4217, E: gamiller@ou.edu

Miller, G.E., Iowa Dept of Transp, 1111 Plum, Atlantic, IA 50022. T: 712-243-1405, F: 712-243-6788, E: glen.miller@dot.state.us

Miller, H., Univ of Mass, 20 Harbor Road, Mattatoisett Neck, Mattatoisett, MA 2739. T: 508-758-2951

Miller, J.S., Giles Eng Assocs Inc., 227 Taliesin Rd., Wales, WI 53183. T: 414-968-4625, F: 414-549-5868, E: jsmiller@mail.execpc.com

Miller, J.P., Southern Co Services, 5030 English Turn, Birmingham, AL 35242. T: 205-981-1755, F: 205-992-0356

Miller, M.J., URS Corporation, 8181 E Tufts Ave, Denver, CO 80237. T: 303-694-2770

Miller, R.P., 3235 Tenth Ave West, Seattle, WA 98119. T: 206-284-1025, F: 206-284-0869, E: rpm3235@aol.com

Millet, R.A., URS Corp, 8181 E Tufts Ave, Denver, CO 80237. T: 303-694-2770, F: 303-694-3946, E: richard_millet@urscorp.com

Mills, S.M., USDA-NRCS, 53 Mill Road, Jefferson, GA 30549. T: 706-367-8943, F: 706-546-2145, E: stacey@dlmenterprises.com

Milne, Jr, G., 3219 Alden Dr., Lansing, MI 48910. T: 517-882-0573

Milstone, B.S., Milstone Geotechnical, 17020 Melody Ln, Los Gatos, CA 95030. T: 408-353-5528, F: 408-353-9690, E: asce@milstonegeo.com

Mimura, C.S., Geolabs Inc., 98-1832 Apelekoka St, Aiea, HI 96701. F: 808-847-1749, E: clayton@geolabs.net

Mindess, M., Sts Consultants Ltd., 7522 Vinewood Ct, Maple Grove, MN 55311. T: 612-420-3550, F: 612-315-1836, E: mindess@stsltd.com

Miner, R.F., Robert Miner Dynamic, Testing Inc., Box 340, Manchester, WA 98353. T: 360-871-5480, F: 360-871-5483

Mintiens, J.G., Rummel Klepper & Kahl, 20 Mariners Walk Way, Baltimore, MD 21220. T: 410-238-1990, F: 410-427-6171, E: jmintiens@rkkengineers.com

Misfeldt, G.A., Clifton Assocs Ltd., 2513 Taylor St E, Saskatoon, S7H 1W9 CANADA Sk. T: 306-373-7383, E: greg_misfeldt@clifton.ca

Mishu, J.R., Geotek Eng Co, 2909 Elizabeth St, Nashville, TN 37211. T: 615-833-3800, F: 615-833-4097, E: rmishu@home.com

Misra, A., Univ of Missouri, Dept of Civil Engineering, 5100 Rockhill Rd., Kansas City, MO 64110. T: 816-235-1285, F: 816-235-1260, E: misraa@umkc.edu

Mitchell, B.A., Citizens Utilities Co, 2419 Lillie Ave, Kingman, AZ 86401. T: 520-753-5622, F: 520-753-6072, E: monam@ctaz.com

Mitchell, J.K., Virginia Tech, 209 Mateer Circle, Blacksburg, VA 24060. T: 540-552-3992, F: 540-231-7532, E: jkm@vt.edu

Mitchell, T.J., Terratech Engineering, P.O. Box 686, Shutesbury, MA 1072. T: 413-221-6802, E: tjm@terratech engineering.com

Mlynarek, J., Sageos Geosynthetics Tech, 3000 Boulle Rd., Saint Hyacinthe, CANADA QC. T: 514-369-0182, F: 450-778-3901, E: jacek@sageos.ca

Moakler, R.J., Fuller Mossbarger Scott & May, 62 Wildwood Rd., Jeffersonvlle, IN 47130. T: 812-282-5508, F: 502-625-7314, E: rjmoakler@netscape.net

Mockridge, R., Qore Inc., 391 Johannah Pl, Lilburn, GA 30247. T: 770-923-5785, F: 770-476-0213

Moeller, D.T., Parsons Brinckerhoff, 50 Lake Front Blvd, Suite 111, Buffalo, NY 14202. T: 716-853-1220, F: 716-853-1322, E: moeller@pbworld.com

Mohamed, F.K., Misr Raymond Foundations, P.O. Box 309, Rabie El-Gizee, 12515 EGYPT. T: 011-002-011-357525, F: 011-002-02-3491, E: fkm01@hotmail.com

Mohammad, A.M., Frederic Harris Inc., 245 E 44th Street, Apt 16-F, New York, NY 10017. T: 212-883-0837, F: 212-973-2935, E: amohammad@frharris.com

Mokwa, R.L., Terracon, 11849 W Executive Dr., Boise, ID 83713. T: 208-323-9520, F: 208-323-9592

Momeni, M.H., Group Alpha Inc., 3011 Rigel Ave, Las Vegas, NV 89102. T: 702-248-8477, F: 702-248-8495

Mooney, M.A., School of Civil & Enviro Sci, University Oklahoma, 202 W Boyd St Rm334, Norman, OK 73019. T: 405-325-3550, F: 405-325-4217, E: mooney@ou.edu

Moore, J.F., Exxon Mobil Corp, 16111 Sir William Dr., Spring, TX 77379. T: 281-379-1153

Moore, T.L., San Diego Gas & Electric, 8316 Century Park Court Cp52g, San Diego, CA 92123. T: 619-654-1656, F: 619-654-8635, E: tmoore@sdge.com

Moossazadeh, J.K., 5015 Shoreham Place, San Diego, CA 92122. T: 858-320-2000 x2204, F: 858-320-2001, E: jmoossazadeh@kleinfelder.com

Morales, E.M., Em2a Partners & Co, 17 Scoutde Guia Corner, Scout Reyes, Quezon City, PHILIPPINES. F: 632-3744338, E: em2apart@pgatech.com.ph

Moran, M.S., Gallet & Assocs Inc., 320 Beacon Pkwy West, Birmingham, AL 35209. T: 205-942-1289, F: 205-942-1266, E: alabamastaff@gallett.com

Morgan, R.J., GZA Geoenvironmental Inc., 139 Little Rest Rd., Kingston, RI 2881. T: 401-783-9619

Morin, W.J., 6805 Lemon Rd., Mc Lean, VA 22101. T: 703-790-8814, F: 703-790-8283, E: wjmorin@worldnet.att.net

Morioka, B.T., URS Corporation, 1335 Akiahala St, Kailua, HI 96734. T: 808-263-0194, F: 808-593-1198, E: bmorioka@gte.net

Morisoli, M.P., 165 N Redwood Drive St 120, San Rafael, CA 94903. T: 415-491-1338, F: 415-491-1831, E: millerpac@aol.com

Moriwaki, Y., Geopentech, 23105 Sonoita, Mission Viejo, CA 92691. T: 949-951-7292, F: 714-796-9191, E: yoshimoriwaki @home.com

Morman, S.C., Ricker Atkinson Mcbee & Assoc, 2105 South Hardy Dr., Suite 13, Tempe, AZ 85282. T: 480-921-8100, F: 480-921-4081, E: smorman@r-a-m.net

Morris, P.E., Edward Mark, Grand Junction Lincoln Devore, 1441 Motor St, Grand Junction, CO 81505. T: 970-242-8968, F: 970-242-1561, E: gjldem@gj.net

Morris, D., Cuthbertson Maunsell Ltd., 35 Shandon Crescent, Edinburgh, EH11 1QF SCOTLAND. T: 44-131-346-0965, F: 44-131-311-4090, E: david.morris@maunsell.com

Morris, D.V., 4393 Westgrove Drive, Addison, TX 75001. T: 972-713-9109, F: 972-713-9171

Morris, M.B., Gannett Fleming Inc., Foster Plaza 3 – Suite 200, 601 Holiday Dr., Pittsburgh, PA 15220. T: 412-922-5575, F: 412-922-3717, E: mmorris@gfnet.com

Morrison II, J.A., Kiewit Engineering Co, 1000 Kiewit Plaza, Suite 800, Omaha, NE 68131. T: 402-342-2052, F: 402-271-2996, E: jim.morrison@kiewit.com

Morrison, C.S., NCDOT, 6808 Jean Drive, Raleigh, NC 27612. T: 919-870-7229, F: 919-250-4098

Morrison, G.M., State of California Rwqcb R1, 655 Arlington Circle, Novato, CA 94947. T: 415-883-7326

Morrison, P.A., Gage Group, 7499 Parklane Rd., Ste 112, Columbia, SC 29223. T: 803-741-9000, F: 803-741-9900

Morrow, K.P., Fugro-Mcclelland Marine Geosciences Inc., 9940 Richmond Ave, #2122, Houston, TX 77042. T: 713-278-7238, F: 713-778-5573, E: kpmorrow@worldnet.att.com

Mortazavi, M.H.S., Peto Mac Callum Ltd., 53 Kilkenny Drive, Toronto, M1W 1J9 CANADA ON. T: 416-492-5050

Moser, K.R., Geostructures Inc., 12975 Hampton Forest Ct, Fairfax, VA 22030. T: 703-631-5885, F: 703-771-9847, E: geomo@erols.com

Moskowitz, J., Mueser Rutledge Cons Engrs, 708 Third Ave, New York, NY 10017. T: 212-490-7110, F: 212-953-5626, E: jmoskowitz@mrce.com

Moss, A.L., Mega Inc., 1238 Island Drive, Logan, UT 84321. T/F: 435-755-3294, E: mega@cache.net

Motamed, F., URS Corp, 911 Wilshire Blvd, 8th Floor, Los Angeles, CA 90017. T: 213-996-2284, F: 213-996-2374, E: farid_motamed@urscorp.com

Moura, L., Geotecnia, 1624 Armstrong Ct, Concord, CA 94521. T: 925-686-9181, F: 925-686-6556, E: luis@geotecnia.com

Mouradian, A.G., Gannett Fleming, P.O. Box 80794, Valley Forge, PA 19484. T: 610-650-8101, F: 610-650-8190, E: amouradian@gfnet.com

Moya, L.A., Bel Ingenieria SA, Residencial Tulin, Casa 8f, San Jose, COSTA RICA. T: 011-506-283-1525, F: 506-283-2723/50

Mueller, C.G., URS Greiner Woodward Clyde, 682 Dougherty Terrace Dr., Ballwin, MO 63021. T: 314-227-9603, E: cgm@urscorp.com

Mueller, J.L., Conoco Inc., Ns 3044, P.O. Box 2197, Houston, TX 77252. T: 281-293-4093, F: 281-293-2158, E: jeffrey.l.mueller@usa.conoco.com

Mulhem, P.E., Aziz M., Dow Chemical, 4897 Knoll Crest Dr., Antioch, CA 94509. T: 925-757-1546, F: 925-432-5722, E: amulh@cs.com

Mullin, T.F., Dames & Moore, 7800 Congress Ave, Suite 200, Boca Raton, FL 33487. T: 561-994-6500, F: 561-994-6524

Mundell, J.A., Mundell & Associates, 429 E Vermont St, Ste 200, Indianapolis, IN 46202. T: 317-630-9060, F: 317-630-9065, E: mundell@indy.net

Mundy, P.K., Peters & Ross, 114 Hopeco Rd., Pleasant Hill, CA 94523. T: 925-942-3649, F: 603-308-6649, E: PetersRoss@aol.com

Munfakh, G.A., Parsons Brinckerhoff Quade &, Douglas Inc., One Penn Plaza, New York, NY 10119. T: 212-465-5205

Muraleetharan, K.K., University of Oklahoma, 313 Baker Street, Norman, OK 73072. T: 405-360-5490, F: 405-325-4217, E: muralee@ou.edu

Murdock, R.F., GEI Consultants Inc., 1021 Main St, Winchester, MA 1890. T: 781-721-4038, F: 781-720-4073, E: rmurdock@geiconsultants.com

Murff, J.D., Texas A&M University, 12 Sparrowglen Lane, Austin, TX 78738. T: 512-261-4617, E: dmurff@isp channel.com

Muriel, J.G., Geo-Engineering, F-12 Roble Blanco St, Santa Clara, Guaynabo, PR 969. T: 787-790-9276, F: 787-764-6325, E: murisan@prw.net

Murphy, P.E., William J., Schnabel Engineering Associate, 656 Quince Orchard Road, Suite 700, Gaithersburg, MD 20878. E: bmurphy@schnabeleng.com

Murphy, D.J., Law Engrg & Eviron Svc Inc., 5845 Nw 158th St, Miami Lakes, FL 33014. T: 305-826-5588, F: 305-826-1799, E: dmurphy@lawco.com

Murphy, M.C., 73 Regis Rd., Braintree, MA 2184. T: 781-848-0642

Mursch, R.D., 222 Ridge Top Drive, Connelly Spgs, NC 28612. T: 828-879-1186, F: 828-879-8407, E: davidmursch @hci.net

Murtagh, K.M., Camp Dresser & Mckee Inc., 1 Cambridge Place, 50 Hampshire St, Cambridge, MA 2139. T: 617-452-6000, F: 617-452-8000, E: murtaghkm@cdm.com

Murthy, R.T., 2700 North Central Avenue, Suite 300, Phoenix, AZ 85004. T: 602-728-0400, F: 602-728-0430, E: ivar@primenet.com

Murvosh, H., Branagan & Assocs, P.O. Box 96177, Las Vegas, NV 89193. T: 702-454-3394, F: 702-434-7594, E: banda@intermind.net

Musial, M.R., Golder Assocs Inc., 1750 Abbott Rd., Ste 200, Anchorage, AK 99507. T: 907-344-6001, F: 907-344-6011, E: mmusial@golder.com

Muzzy, M.W., Sevee & Maher Eng Inc., 4 Blanchard Rd., P.O. Box 85-A, Cumberland Ctr, ME 4021. T: 207-829-5016, F: 207-829-5692, E: mwm@smemaine.com

Mydlinski, J.L., Schnabel Engineering Assoc Inc., 533 Clove Terrace, Leesburg, VA 20176. T: 703-737-0117, F: 703-589-1246, E: jmydlinski@schnabel-eng.com

Myers, J.N., Myers Biodynamics Inc., Rolling Bay Mercantile Bldg, 11254 Sunrise Drive, Bainbridge Isl, WA 98110. T: 206-842-6073

Myers, J.T., Davey Kent Inc., 200 West Williams St, P.O. Box 400, Kent, OH 44240. T: 330-673-5400, F: 330-673-9178, E: tmyers@daveykent.com

Myers, K.W., Piedmont Geotechnical Consult, 1104 Luther Palmer Rd., Cleveland, GA 30528. T: 706-865-7703, F: 770-752-0890, E: kmyers@pgci.com

Myers, K.L., Myers Design Engineering Inc., P.O. Box 3287, Incline Vlg, NV 89450. T: 702-831-6835, F: 702-833-2413, E: kmyers4978@aol.com

Myers, M.W., Cal Engineering & Geology Inc., 128 Lees Place, Martinez, CA 94553. T: 925-229-5308, F: 925-935-9773, E: mmyers@caleng.com

Myers, T.J., Layne Christensen Company, 3611 Noble Ave, Richmond, VA 23222. T: 804-329-1147, F: 804-448-1771, E: tjmlayne@aol.com

Myint, U.T., Ayeyeikmon Co Ltd., No 28 E Inyarmyaing Rd., Bahan Tsp, Yangoon, MYANMAR. T: 011-95-1-514723, F: 011-95-1-254859

Mylonakis, G., City College of New York, 360 Central Park West, Apt 10l, New York, NY 10025. T: 212-222-3561, F: 212-650-6965, E: mylonakis@ccny.cuny.edu

Nacamuli, R.L., Nacamuli Associates Llc, 5 Marigold Ln, Califon, NJ 7830. T: 908-832-4218, F: 908-289-4112, E: rlnac@nacamuli.com

Nacci, D.M., The Maguire Group Inc., 30 Heritage Rd., N Kingstown, RI 2852. T: 401-884-5998, F: 508-543-5157, E: dnacci@maguiregroup.com

Nagle, G.S., URS Corporation, 904 Beech Dr., Walnut Creek, CA 94596. T: 925-935-9697, F: 510-874-3268, E: galen_nagle@urscorp.com

Nagle, M.A., Rieth-Riley Constr Co Inc., 28342 Stanton Rd., Walkerton, IN 46574. T: 219-656-8775, F: 219-233-3464, E: managle@rieth-riley.com

Nagle, S.P., Braun Intertec Eng Inc., P.O. Box 9296, Fargo, ND 58106. T: 701-232-8701, F: 701-232-7817

Nakai, S., Faculty of Eng Chiba Univ, 1-33 Yayoi-Cho, Inage-Ku, Chiba, 263-8522 JAPAN. T: -3382, F: -3663, E: nakai@tu.chiba-u.ac.jp

Nakashima, T., 1240 West Carmen Ave, Chicago, IL 60640. T: 773-334-1844

Nallainathan, C., Municipal Engineering Sv Co Pa, 9613 Leslieshire Drive, Raleigh, NC 27615. T: 919-848-9744, F: 919-772-1176, E: canags@aol.com

Nam, C.C., Juruter Perunding Maju, 7b Jalan 20/14 Paramount Gdn, Petaling Jaya, Selangor, 46300 MALAYSIA. T: 603-7874-8741, F: 603-7876-0734

Nardi, C.R., Lfr Levine Fricke, 125 Woodglen Ln, Martinez, CA 94553. T: 925-372-3720, F: 510-652-2246, E: chris.nardi@lfr.com

Narduzzo, L., Toronto Transit Commission, 170 Calvington Dr., Toronto, M3M 2M9 CANADA ON. T: 416-242-7969, F: 416-338-0129, E: luigi.narduzzo@ttc.ca

Narin van court, W.A., URS Corporation, 7 University Drive, Augusta, ME 4330. T: 207-623-9188, F: 207-622-6085, E: augwan@dames.com

Nava, R.C., 2-2-1 Fuchinobe, Sagamihara-Shi, Kanagawa-Ken, 229-0006 JAPAN. T: 81-427-59-5202, F: 81-427-59-5398, E: Rnava@slb.com

Nazarian, S., Univ of Texas At El Paso, Civ Engr Dept, El Paso, TX 79968. T: 915-747-6911, F: 915-747-8037, E: nazarian@utep.edu

Ndonga, C.M., Gado Consulting Engineers Ltd., 1252 Olde Fram Rd., Apt 303, Schaumburg, IL 60173. T: 847-397-7742, E: ndonga@yahoo.co.uk

Nehybka, J.D., Mueser Rutledge C E., 23 - 60 29th St, Apt 4-I, Astoria, NY 11105. T: 718-545-6441 & 518-2634, F: 212-490-6654, E: jnehybka@mrce.com

Nelson, D.L., BT Squared Inc., 2 Shea Court, Madison, WI 53717. T: 608-836-1243, F: 608-224-2839

Nelson, J.D., Colorado State Univ, 51361 Wcr 17, Wellington, CO 80549. T: 970-897-2444, F: 970-491-7727, E: john@engr.colostate.edu

Nelson, P.P., National Science Foundation, 4201 Wilson Blvd, Room 545.17, Arlington, VA 22230. T: 703-292-7018, F: 703-292-9053, E: pnelson@nsf.gov

Nelson, W.S., Waldemar S Nelson & Co Inc., 1200 St Charles Ave, New Orleans, LA 70130. T: 504-523-5281, F: 504-593-5261, E: waldemar.nelson@wsnelson.com

NeSmith, W.M., 1503 Milner Crescent, Birmingham, AL 35205. T: 205-933-7957, F: 205-933-8979, E: wnesmith@bellsouth.net

Netherton, D.E., Streamflow Power Canada & Stre, P.O. Box 1151, North Bay, P1B 8K4 CANADA ON. T: 705-475-1994, F: 705-475-0774, E: sfpcanada@sympatico.ca

Nevels, Jr, J.B., Oklahoma Dept of Transp, 605 Mimosa, Norman, OK 73069. T: 405-364-7004, F: 405-521-2453

Nevin, T.J., Carlson Testing, Inc., P.O. Box 23814, Tigard, OR 97281. T: 503-684-3460, F: 904-396-5703, E: tnevin@carlsontesting.com

Newlin, Jr, E.C., Grubbs Hoskyn Barton & Wyatt, 3000 Windrift Cove, Fort Smith, AR 72903. T: 501-452-5269, F: 501-646-8127, E: edgailnewlin@aol.com

Newlin, C.W., 7223 North 15th Pl, Phoenix, AZ 85020. T: 602-861-1071, E: newlin@home.com

Neyer, J.C., Nth Cons Ltd., P.O. Box 9173, Farmington Hills, MI 48333. T: 248-324-5178, E: jneyer@nthconsul nthconsultants.com

Ng Kwok Hei, P., Rm 407 4/F Lee On Estate, Lee Wah House Ma On Shan, N T Kowloon, HONG KONG. T: 011-852-263-33805, F: 011-852-239-194

Niber, R.J., Whitlock Dalrymple Poston Ass, 8832 Rixlew Ln, Manassas, VA 20109. T: 703-257-9280, F: 703-257-7589, E: rniber@wdpa.com

Nickel, S.H., Nickel Engineering Inc., 7941 Portsche Lane, Lincoln, NE 68516. T/F: 402-423-0559, E: nickel@inebraska.com

Nickerson, J.F., VA Geotechnical Services Pc, 8211 Hermitage Rd., Richmond, VA 23228. T: 804-266-2199, F: 804-261-5569, E: jfnickerson@vgspc.com

Nicklin, M.S., 10 St Elmo St, Clifton Gardens, Sydney Nsw, 2088 AUSTRALIA. T: 612-9969-6558, F: 612-9969-3379

Niedzielski, J.C., Somat Engineering Inc., 16557 Blue Skies Dr., Livonia, MI 48154. T: 734-464-9306, F: 734-946-1147, E: jniedzielski@somateng.com

Niehoff, J.W., PSI Inc., 4765 Independence St, Wheat Ridge, CO 80033. T: 303-424-5578, F: 303-423-5625, E: jniehoff@aol.com

Niehus, D.P., Lillard & Clark Construction, 3775 South Knox Court, Denver, CO 80236. T: 303-761-3170, F: 303-762-1710, E: dniehus@lillardclark.com

Nigbor, R.L., University of So Calif, 1763 Braeburn Rd., Altadena, CA 91001. T: 626-398-2047, F: 213-744-1426, E: nigbor@usc.edu

Nilsson, R.K., RMT Inc., 1 Doverdale Road, Greenville, SC 29615. T: 864-676-9709, F: 864-281-0288, E: kent.nilsson@rmtinc.com

Ninyo, H., NMG Geotechnical Inc., 17791 Fitch, Irvine, CA 92614. T: 949-442-2442

Noel, E.W., Kleinfelder Inc., 1370 Valley Vista Drive, Suite 150, Diamond Bar, CA 91765. T: 909-396-0335, F: 909-396-1324, E: enoel@kleinfelder.com

Nolen-Hoeksema, R.C., 5406 Red Fox Run, Ann Arbor, MI 48105. F: 734-764-4292, E: rcnh@umich.edu

Noorany, I., San Diego State Univ, Dept of Civil Eng, San Diego, CA 92182. T: 619-594-5932, F: 619-594-6005, E: noorany@mail.sdsu.edu

Nordmark, T.S., 4160 Pleasant Valley Rd., Chantilly, VA 20151. T: 703-631-5322, F: 703-631-6041, E: snordmark@burnip.com

Nordquist, J.E., AGEC, 600 W Sandy Pkwy, Sandy, UT 84070. T: 801-566-6399

Nordstrom, T.D., Bsumek MU, 4329 Rose Garden Lane, Salt Lake City, UT 84124. T: 801-269-8439, E: timandellen@earthlink.net

North, J.S., Kleinfelder Inc., 4905 Hawkins Ne, Albuquerque, NM 87109. T: 505-344-7373

North, R.B., Waste Management Inc., 5200 Sw Macadam Ave, Ste 270, Portland, OR 97201. T: 503-243-9493, F: 503-242-9123, E: rbnorth@bigfoot.com

Nothdurft, D.B., Dept of Construction Mgt, East Carolina Univ, P.O. Box 2681, Greenville, NC 27836. T: 252-328-4140, F: 252-328-1165, E: nothdurftd@mail.ecu.edu

Nottingham, L.C., Triad Engineering Inc., P.O. Box 1435, St Albans, WV 25177. T: 304-755-0721, F: 304-755-1880, E: lcn@triadeng.com

Nowatzki, E.A., 5012 North Amapola Dr., Tucson, AZ 85745. T: 520-743-8123, F: 520-621-2550, E: nowatzki@u.arizona.edu

Nuyens, J.E.G., OREX, Ave Des Noisetiers 5, Brussels, 1170 BELGIUM. T: 322-660-2193, F: 322-537-3940, E: jean.nuyens@orex.be

O'Brien, B.S., Universal Engineering Sciences, 3532 Maggie Blvd, Orlando, FL 32811. T: 407-423-0504, F: 407-423-3106

Obrien, J.P., Haley & Aldrich Inc., 110 National Drive, Glastonbury, CT 6033. T: 860-659-4248, Ext. 3151, F: 860-659-4003, E: jpo@haleyaldrich.com

Ochieng, L.A. Nyandondo, Gath Consulting Engineers, P.O. Box 5637, Eldoret, KENYA. T: 011-254-035-43631, F: 011-254-02-4438, E: ochienglan@swiftkisumu.com

O'Connell, D.P., North East Engineers & Consult, 42 Valley Rd., Middletown, RI 2842. T: 401-849-0810, F: 401-846-4169, E: dano@northeastengineers.com

O'Connor, M.J., Arias & Kezar Inc., 3542 Rock Creek Run St, San Antonio, TX 78230. T: 210-349-1441, F: 210-308-5886

O'Connor, R.W., URS Corporation, 1736 Shanwick Court, Forest Hill, MD 21050. T: 410-803-9937, F: 410-785-6818, E: robert_o'connor@urscorp.com

O'Connor, S.E., Engineered Soil Repairs Inc., 1267 Springbrook Road, Walnut Creek, CA 94596. T: 510-210-2150, F: 510-210-2158

O'Donnell, C.L., Federal Hwy Admin, 55 Broadway 10th Floor, Cambridge, MA 2142. T: 617-494-3281, F: 617-494-3355, E: chuck.o'donnell@fhwa.dot.gov

Ogren, P.J., Hayes Eng Inc., 603 Salem St, Wakefield, MA 1880. T: 617-246-2800, F: 617-246-7596

Okcuoglu, C.A., Geotech Inc., 125 N Rt 73, P.O. Box 266, Maple Shade, NJ 8052. T: 609-667-7892, F: 609-677-7892, E: cao@geotechincnj.com

Okolo, O.O., Fcda Abuja Nigeria, Federal Capital Deu Auth, Engrg Svcs Dept Garki Abuja, Area Eleven Abuja Fct, Nigeria

Okonkwo, P.E., Ignatius O., Earth Design Associates Inc., P.O. Box 1455, Glastonbury, CT 6033. T: 860-951-3895, F: 860-236-9126, E: edainc@aol.com

Okoye, C.N., Pile Mechanics Inc., P.O. Box 249, New Burnswick, NJ 8901. T: 732-393-1756, F: 732-846-0565

Olsen, H.W., Colorado School of Mines, P.O. Box 775730, Steamboat Spr, CO 80477. T: 970-879-7856, F: 303-273-3602, E: holsen@mines.edu

Olsen, J.M., Univ of South Alabama, 1321 Westbury Dr., Mobile, AL 36609. T: 334-343-3208, F: 334-461-1400, E: jolsen@jaguarl.usouthal.edu

Olson, J.P., Univ of Vermont, Dept of Civil & Environ Engrg, 213 Votey Hall, Burlington, VT 5405. T: 802-656-1927, F: 802-656-8446, E: olson@emba.uvm.edu

Olson, J.C., Geo Co, Geo Co, 4767 S Ichabod Pl, Salt Lake City, UT 84117. T: 801-272-2064, F: 801-278-2612, E: jconly@webgny-prod.com

Olson, L.D., Olson Eng Inc., 5191 Ward Rd., #1, Wheat Ridge, CO 80033. T: 303-423-1212, F: 303 423 6071, E: ldolson@olsonengineering.com

Olson, R.D., Earth Exploration Inc., 7770 W New York St, Indianapolis, IN 46214. T: 317-273-1690, F: 317-273-2250, E: mail@earthengr.com

Olson, R.E., Univ of Texas @Austin, Dept of Civ Engr, Ecj9.227, Austin, TX 78712. T: 512-471-4149

O'Malley, E.S., Facility Engineering Assoc Pc, 2017 Golf Course Drive, Reston, VA 20191. F: 703-591-4857, E: omalley@feapc.com

O'Neill, B.E., United Mission To Nepal, P.O. Box 126, Kathmandu, NEPAL. T: 977-1-228118, F: 977-1-255559, E: bruce-sue@hydroconsult.com.np

O'Neill, D.A., NEFCO, 77 Federal Ave, Quincy, MA 2169. T: 617-689-0550, F: 617-689-0551, E: daoneill@nefco.com

O'Neill, M.W., Univ of Houston, Dept of Civil Eng, Houston, TX 77204. T: 713-743-4252, F: 713-743-4260, E: oneill@uh.edu

Ooi, J.Y., University of Edinburgh, Dept of Civil And Env Eng, The King Building, Edinburgh, EH9 3JN SCOTLAND. T: 011-44-131-6505725, F: 011-44-131-650-, E: jin@ed.ac.uk

Orlando, R., Terra Dynamic, 2727 E Aster Drive, Phoenix, AZ 85032. T: 602-404-3780, F: 602-788-5855, E: orlandosicily@worldnet.att.com

Orman, M.E., Sierra Geotechnical, 15572 Sunnyglade Rd., Grass Valley, CA 95949. T: 530-273-5580, F: 530-273-9402

Orozco, L.F., Luis F Orozco & Co, Apartado Aereo 102191, Bogota 10, COLOMBIA. T: 571-544-9444, F: 571-310-4759, E: lorozco@colomsat.net.co

Orr, W.S., 9 Sunset Dr., Northborough, MA 1532. T: 508-393-9614, F: 508-393-6694

Ortiz, A., 1684 Locust Street, #441, Walnut Creek, CA 94596. T: 925-743-8615

Ortiz, C.H., Corporacion Geotec, P.O. Box 10412 Caparra Sta, Caparra Heights, San Juan, PR 922. T: 787-751-4994, F: 787-763-4386, E: geotec@caribe.net

Ortiz, E.M., Arroyo Engineering, 6200 Utsa Boulevard, San Antonio, TX 78244. T: 210-558-3101, E: aarroyo@onr.com

Ortiz, F. De Paula, Virginia Department, 2812 Bywater Dr., Apt 112, Richmond, VA 23233. T: 804-360-1513, F: 804-786-2988, E: ortizland_f@vdot.state.va.us

Ortiz-Suarez, C.A., Corporacion Geotec, P.O. Box 10412, Caparra Station, Caparra Heights, PR 922. T: 787-763-4753, F: 787-763-4386, E: geotec@caribe.net

Osaimi, A.E., Osaimi Engineering office, P.O. Box 41876, Jeddah, 21531 SAUDI ARABIA. T: 966-3-857-3664, F: 966-3-857-3144, E: osaimi@zajil.net

Osborne, T.C., T-Oz Construction Inc., P.O. Box 729, Wauna, WA 98395. T: 253-884-1114, F: 253-884-1179, E: tozcon@aol.com

Osinubi, K.J., Ahmadu Bello Univ, Dept of Civil Engr, Zaria, Kaduna State, NIGERIA. E: osinubik@abu.edu.ng

Oskvig, C.C., Geoconsult, P.O. Box 362040, San Juan, PR 936. T: 787-782-3554, F: 787-793-0410, E: coskvig@geoconsult-inc.com

Osman, E.A.M., 119 King Faisal St, Giza, EGYPT. T: 02-3820813

Osterberg, J.O., 16416 East Powers Pl, Aurora, CO 80015. T: 303-699-1816, F: 303-680-0659, E: jooltd@aol.com

Otani, J., Kumamoto Univ, Kurokami 2-39-1, Kumamoto, 860-8555 JAPAN. T/F: 81-96-342-3535, E: junotani@gpo.kumamoto-u.ac.jp

Otani, S., Univ of Tokyo, Dept of Arch Eng 7-3-1 Hongo, Bunkyo-Ku, Tokyo, 113-8656 JAPAN. T: +81-3-3812-2111, Ext. 6158, F: -10284, E: otani@rcs.arch.t.u-tokyo.ac.jp

Othman, M.A., Geosyntec Consultants, 1100 Lake Hearn Drive, Suite 200, Atlanta, GA 30342. T: 404-705-9500, F: 404-705-9400

Ott, K.R., Parsons Brinckerhoff Quade &, 52 Crane Cir, New Providnce, NJ 7974. T: 908-665-1790, F: 212-465-5592

Otten, M.T., Foster Wheeler Env, 12100 Ne 195th St, #200, Bothell, WA 98011. T: 425-688-3797, F: 206-365-9973

Overton, D.D., 3801 Automation Way, #100, Fort Collins, CO 80525. T: 970-223-9600, F: 970-223-7171, E: doverton@shepmill.com

Ovesen, N.K., Danish Geotechnical Inst, P.O. Box 119, Lyngby, Denmark 2800dk, DENMARK. T: 454-5884444, F: 454-5881240, E: nko@geotekrisk.dk

Oviedo, J.S., 934 S Oakland Ct, Aurora, CO 80012. T: 303-360-7206, F: 303-841-0136

Oweis, N.Y., Ellis & Associates Inc., 9010 Cumberland Forest Way, Jacksonville, FL 32257. T: 904-737-4799, F: 904-880-0970, E: n.oweis@ellisassoc.com

Ozarowski, P.P., Haley & Aldrich Inc., 72 Raymond Rd., Concord, MA 1742. T: 978-369-5267, F: 617-886-7647, E: ppo@haleyaldrich.com

Ozier, J.M., Omi Inc., 5151 Research Dr., Nw, Huntsville, AL 35805. T: 256-837-7664, F: 256-837-7677, E: jozier@omi-eng.com

Ozkan, M.Y., Middle East Tech Univ, Civil Eng Dept, Insaat Muh Bol, Ankara, 6531 TURKEY. T: 90-312-2102414, F: 90-312-2101262, E: myozkan@metu.edu.tr

Ozols, V., 135 Eighth Ave, San Francisco, CA 94118

Pace, Jr, F.P., Synthetic Industries, 6025 Lee Hwy, Suite 435, Chattanooga, TN 37421. T: 800-621-0444, F: 423-485-9068, E: frank_pace@sind.com

Pace, A.V., Golder Assocs, 14011-G Mango Dr., Delmar, CA 92014. T: 858-259-0145, F: 858-571-3943, E: apace@golder.com

Pack, J.S., D&B Drilling, 6753 Zinnia St, Arvada, CO 80004. T: 303-424-4343, F: 303-423-9155, E: imrpack@aol.com

Padgett, Jr, J. (Jay) Arthur, Geoservices Corp, 8001 Cryden Way, Forestville, MD 20747. T: 301-967-1942, F: 301-967-0358, E: japjr@geoservicescorp.com

Pagone, R.J., Richard Goettle Inc., 8105 Perry Hwy, 2nd Fl, Pittsburgh, PA 15237. T: 412-635-7155, F: 412-635-7156, E: goettlerjp@aol.com

Pahler, P.E., Thomas A.H., Thomas A H Pahler Pe, 7081 State Hwy 56, Norwood, NY 13668. T: 315-265-8746, E: thpahler@northnet.org

Paikowsky, S.G., Univ of Massachusetts, Dept of Civil & Env Eng, 1 University Ave, Lowell, MA 1854. T: 978-934-2277, E: samuel_paikowsky@uml.edu

Palermo, R.J., Gzageoenvironmental Inc., 115 Derby Rd., Revere, MA 2151. T: 781-289-2648, F: 781-278-5701, E: rpalermo@gza.com

Palmason, P.R., VST Consulting Engrs, Armuli 4, Reykjavik, 108 ICELAND. T: +354 5695000, F: +354 6595010, E: prp@vst.is

Palmer, B.G., Malcolm Pirnie, 1900 Polaris Parkway, Suite 200, Columbus, OH 43240. T: 614-888-4953, F: 614-888-5638, E: bpalmer@pirnie.com

Palmer, J.A., 1886 Jitney Dr., Sparks, NV 89434. T/F: 702-356-1618, E: jpal376@aol.com

Palmer, L., Ch2m Hill, 9418 White Dove Ct, Charlotte, NC 28277. T: 334-286-0088, F: 704-329-0141, E: lpalmer@ch2m.com

Palmerton, F.E., Palmerton & Parrish Inc., 4166 W Kearney St, Springfield, MO 65803. T: 417-864-6000, F: 417-864-6004, E: fred.palmerton@ppimo.com

Pamukcu, S., Lehigh Univ, Dept of Civil & Envir Eng, 13 E Packer Ave, Bethlehem, PA 18015. T: 610-758-3220, F: 610-758-6405

Pan, K.-L., Water Resources Bureau, 2 F No 4 Allen St, Lane 265 Hsin Yi Rd Sec 4, Taipei, Taiwan, ROC. T: 886-2-27057028, F: 886-2-27552412, E: klpan@wrb.gov.tw

Pan, Y.-W., National Chiao-Tung Univ, Dept of Civil Eng, 1001 Ta Hsueh Rd/Hsinchu, Taiwan, Taiwan, ROC. T: 011-886-3-5731931, F: 011-886-3-57162, E: ywpan@cc.nctu.edu.tw

Pandis, K., Pangaea Consulting Engineers, Patriarchou Photiou 21, Maroussi, Athens, 151 22 GREECE. T: 011-301-8024234, F: 011-301-692-813

Paniagua, J.G., Terratech Inc., 14918 Sw 104 St Unit 37, Miami, FL 33196. T: 305-380-9246, F: 305-383-2422, E: federicop@adelphia.net

Pankovs, I., IP Perforaciones Geotecnia Ca, P.O. Box 75908 El Marques, Caracas, 1070A VENEZUELA. T: 58-2-7811483, F: 58-2-2396095, E: pankovs@etheron.net

Panozzo, G.L., Ch2m Hill, 613 Nw Loop 410, Suite 200 Sprectrum Bldg, San Antonio, TX 78216. T: 210-377-3085, Ext. 265, F: 210-349-8944, E: gpanozzo@ch2m.com

Pantazidou, M., Carnegie Mellon Univ, Dept of Civil Engineering, 5000 Forbes Avenue, Pittsburgh, PA 15213. T: 412-268-7813, F: 412-268-2943, E: marina@cmu.edu

Panuccio, C.M., 454 Parker Ave, Buffalo, NY 14216. T: 716-837-5004

Papaloizou, Z., J & A Philippou, Vrasida1, Strovolos, Nicosia, 2038 CYPRUS. T: 357-2-313212, F: 357-2-662097, E: japhil@logos.cy.net

Pappas, J.L., Froehling & Robertson Inc., 1224 White Hill Road, Stuarts Draft, VA 24477. T: 540-337-0470, F: 804-823-4764, E: frroanoke@aol.com

Parenteau, D.M., Wenck Assocs Inc., 1800 Pioneer Creek Ctr, P.O. Box 249, Maple Plain, MN 55359. T: 612-479-4243, F: 612-479-4242

Parikh, G.G., Parikh Consultants Inc., 45796 Vinehill Terr, Fremont, CA 94539. T: 510-770-8770, F: 408-945-1012, E: parikh@pacbell.net

Parisi, P.A., 245 Standish Ave, Hackensack, NJ 7601. T: 201-489-4492, F: 201-489-9042, E: pappe28@aol.com

Park, I., Hanseo Univ, 20-1002 Woosung Apt 3cha, Secho Dong Secho Gu, Seoul, 137-773 SOUTH KOREA. T: 82-2-3471-8533, E: geotech@hanseo.ac.kr

Parker, E.J., D'appolonia Spa, Via Croce Rosa 4/15b, Genova, 16159 ITALY. T: -2962, E: eric.parker@dappolonia.it

Parker, H.W., Harvey Parker & Associates Inc., 10900 Ne 8th Street, Suite 900, Bellevue, WA 98004. T: 425-451-8516, F: 425-454-7891, E: harveyparker@compuserve.com

Parker, R.C., Dick Pacific Constr Co Ltd., 536 Hao St, Honolulu, HI 96821. T: 808-373-9291, F: 504-343-3804

Parkington, T.S., URS Corporation, 2025 First Avenue, Suite 500, Seattle, WA 98121. T: 206-728-0744, F: 206-727-3350, E: todd_parkington@urscorp.com

Parks, J.N., Dames & Moore, 7101 Wisconsin Ave, Suite 700, Bethesda, MD 20814. T: 301-652-2215, F: 301-656-8059, E: wasjnp@dames.com

Parmantier, D.M., Hayward Baker Inc., 3617 Northeast 65th Street, Vancouver, WA 98661. T: 360-695-6343, E: dmparmantier@haywardbaker.com

Parola, J.F., Case Foundation Co, 441 Harvey St, Grayslake, IL 60030. T: 847-223-6107, F: 630-529-2995, E: jfparola@casefoundation.com

Parrello, A.T., French & Parrello Assocs, 670 N Beers St, Bldg 3, Holmdel, NJ 7733. T: 732-888-7700, F: 732-888-7623, E: argo@fpawww.com

Parsons, W.H., F&Me Consultants, 517 9th Ave N, Myrtle Beach, SC 29577. T: 843-626-9253, F: 843-448-0681, E: fmebch@sccoast.net

Pasha, S.A., Environmental Risk Ltd., 1373 Broad Street, Clifton, NJ 7013. T: 973-773-8322, F: 973-773-9422, E: spasha@erl.com

Pass, D.G., Geosyntec Consultants, 1100 Lk Hearn Dr., Ne Ste 200, Atlanta, GA 30342. T: 404-705-9500, F: 404-705-9400, E: danp@geosyntec.com

Pasternack, S.C., BBC & M Eng, 6190 Enterprise Ct, Dublin, OH 43017. T: 614-793-2226

Patel, K.R., Kiran & Musonda Associate, P.O. Box 31406, Lusaka, ZAMBIA. T: 260-1-227665, F: 260-1-226364, E: sujata@zamnet.zm

Patel, R.B., Patel & Associates, P.O. Box 1121, Victorville, CA 92393. T: 619-243-1436, F: 619-243-1471

Patterson-Kane, K.J., Murray-North (Sea) Pte Ltd., 80 Marine Parade Rd., 19-20 Pkwy Parade 449269, Singapore, 449269 SINGAPORE. T: 3453055, F: 3448441, E: kenpk@mnsea.com.sg

Patton, Jr, R.S., Patton Burke & Thompson, Llc, 10575 Newkirk St, Suite 780, Dallas, TX 75220. T: 972-831-1111, F: 972-831-0800, E: rpatton@pbt.cc

Patton, B.W., Kleinfelder Inc., 16 Technology Dr., Suite 150, Irvine, CA 92618. T: 949-727-4466, F: 949-727-9242, E: bpatton@kleinfelder.com

Patton, F.D., Westbay Instruments Inc., 115 949 W 3rd St, N Vancouver, V7P 3P7 CANADA BC. T: 604-984-4215, F: 604-984-3538, E: fpatton@slb.com

Paulson, J.N., Anchor Wall Systems, 5959 Baker Road, Minnetonka, MN 55345. T: 952-933-8855, F: 952-938-4114, E: jpaulson@anchorwall.com

Pavlakis, M., M Pavlakis & Assocs, 7 Danya Rd., Victory Park 2195, Johannesburg, SOUTH AFRICA. F: 11888 7428, E: mpavlaki@iafrica.com

Pawlak, S.L., Hepworth Pawlak, Geotechnical Inc., 5020 Road 154, Glenwood Springs, CO 81601. T: 970-945-7988, F: 970-945-8454

Paxton, J.F., URS Corporation, 500 12th St Ste 200, Oakland, CA 94607. T: 510-874-3231, F: 510-874-3268, E: john_paxton@urscorp.com

Payiatakis, S., B/Pb, 191–193 Mesogion Ave, Athens, 115 25 GREECE. E: sxpayiat@bechtel.com

Pearlman, S.L., Nicholson Construction Co, P.O. Box 98, Bridgeville, PA 15017. T: 412-221-4500, F: 412-221-3127, E: spearlman@nicholson-rodio.com

Pearson, C.L., Geotech & Enviro Cons Inc., 1900 West Broadway Road, Tempe, AZ 85282. T: 480-966-8631, E: cpearson@gecaz.com

Pease, J.W., URS Corporation, 221 Main St Suite 600, San Francisco, CA 94105. T: 415-243-3731, F: 415-882-9261

Pease, K.A., Haley & Aldrich, 110 16th Street, #900, Denver, CO 80202. T: 303-534-1100, F: 303-534-1777

Peck, R., 1101 Warm Sands Dr., Se, Albuquerque, NM 87123. T: 505-293-2484, F: 505 323 7760

Pedersen, R.C., 3995 Fm535, Bastrop, TX 78602. T: 713-861-0661

Pegnam, M.L., URS Greiner Woodward Clyde, 4742 North Oracle Road, Suite 310, Tucson, AZ 85705. T: 520-887-1800, F: 520-887-8438, E: mpegna@aol.com

Pegues, Jr, J.C., PSI Inc., 3237 Valley Park Pl, Birmingham, AL 35243. T: 205-967-5244, F: 205-951-5537, E: jim.pegues@psiusa.com

Peiris, M.N., Ove Arup & Partners, 69 Shirley Park Rd., Croydon, Surrey, CR0 7EW UK. T: 4.42E+11, F: 4.42E+11, E: navin.peiris@arup.com

Penney, M.C., Geoinsight Inc., Bldg 2 Ste 210, 75 Gilcreast Rd., Londonderry, NH 3053. T: 603-434-3116, F: 603-432-2445

Penumadu, D., Box 5710, Clarkson University, Potsdam, NY 13699. T: 315-268-6506, F: 315-268-7985, E: penumadu@clarkson.edu

Perdomo, D., Koch Materials Co, 2219 N Stoneybrook Ct, Wichita, KS 67226. T: 316-631-3849, F: 316-828-7385, E: perdomod@kochind.com

Perez, A., Geosyntec Cons, 2100 Main St, Ste 150, Huntington Beach, CA 92648. T: 714-969-0800 x212, F: 714-969-0820, E: adamp@geosyntec.com

Perez, J.-Y., URS Corporation, 8181 E Tufts Avenue, Denver, CO 80237. T: 303-740-3898, F: 303-930-6085, E: jean-yves-perez@urscorp.com

Perez-Guerra, J.B., Perez Guerra Y Assoc Ca, Campo Alegre Res Campo Alegre, Apt 5a Primera Avenida, Caracas, VENEZUELA. T: 582-951-4879, F: 582-272-7035, E: perezguerra@cantv.net

Perin, R.J., Consol Inc Consolidation Coal Company, 4000 Brownsville Road, South Park, PA 15129. T: 412-854-6784, F: 412-854-6753, E: richperin@consolenergy.com

Perisho, R.J., Geo Focus, 49 Quail Court, Suite 200, Walnut Creek, CA 94596. T: 925-945-1995, F: 925-945-4989

Perkins, S.F., Wilson & Co, 7821 Rc Gorman Ne, Albuquerque, NM 87122. T: 505-821-9119, F: 505-348-4055, E: sfperkins@wilsonco.com

Perricone, G.J., PMK Ferris & Perricone, 68 1st Avenue, Secaucus, NJ 7094. T: 201-863-3386

Perrone, V.J., URS Corporation, 11220 Fieldstone Lane Ne, Bainbridge Island, WA 98110. T: 206-855-1588, F: 206-727-3350, E: vperrone@aol.com

Peters, R.VL., 29 The Fairway, Northwood, Middlesex, HA6 3DZ UK. T: 01923 823149

Petersen, M.M., Black & Veatch, 13102 Hemlock, Overland Park, KS 66213. T: 913-685-1145, F: 913-458-2934, E: petersenmm@bv.com

Peterson, J.L., AESI., 4210 Lytle Rd Ne, Bainbridge Is, WA 98110. T: 206-842-7192, F: 206-780-9438, E: jpeterson@aesibi.com

Peterson, M.S., Peterson-Rabasca Geoengineers, 12 Merrill Road, Freeport, ME 4032. T: 207-865-4348, F: 207-8464348, E: prg@nlis.net

Peterson, R.L., 704 Edgar Ave, Saint Paul, MN 55117. T: 651-488-1452, E: pete0838@tc.umn.edu

Petraborg, G.R., USA Waste Services Inc., 833 San Luis Road, Berkeley, CA 94707. T: 510-524-3033, F: 415-479-3737

Petrovsky, M.B., Mbp Consulting, 10 Woodfield Rd., Portland, ME 4102. T: 207-773-5425, F: 207-773-9590, E: mbpcnslt@javanet.com

Petry, T.M., Univ of Missouri, Dept of Civil Engr, 1870 Miner Cir, Rolla, MO 65409. T: 573-341-4472, F: 573 341-4729, E: petryt@umr.edu

Pezza, D.A., 54 Batcheller Avenue, Cranston, RI 2920. T: 978-269-3628, F: 978-318-8663, E: david.a.pezza@usace.army.mil

Pfeiffer, T.J., Foundation Engineering Inc., 7187 Sw 181st Pl, Aloha, OR 97007. T: 503-642-3271, F: 503-598-9343

Pfingsten, C.W., STS Cons Ltd., 29176 Gilmer Rd., Mundelein, IL 60060. F: 847-279-2510, E: pfingsten@stsltd.com

Pham, L.N., Peter & Assocs Inc., 14645 Seron Ave, Irvine, CA 92606. T: 949-551-9153, F: 949-492-1891, E: lanpham999 @hotmail.com

Phillips, J.S., 925 East Navajo, Farmington, NM 87401. T: 505-325-9581, F: 505-325-5367, E: jphillips@cyberport.com

Phillips, J.T., Ghd Engineers, P.O. Box Y3106, Perth Wa, 6832 AUSTRALIA. T: 61 9 4296666, F: 61 9 4296555

Picco, Jan, 154 E Boston Post Rd., Mamatoneck, NY 10543. T: 914-698-7735, F: 914-698-4215

Pickering, D.J., 4955 River Rd Ste 111, Delta, V4K 4V9 CANADA BC. T: 604-940-1230

Pilath, H.M., 919 5th Street, Golden, CO 80403. T: 303-986-3948, F: 303-742-2328

Pilling, P.E., Patrick Andrew, Black Eagle Consulting Inc., 1345 Capital Blvd, Ste A., Reno, NV 89502. T: 702-359-6600, F: 702-359-7766

Pina, R.M., Wingerter Labs Inc., 150 S.E. 25th Road, Apt 7k, Miami, FL 33129. T: 305-250-5696, F: 305 949 8698, E: chilopina@aol.com

Pinciotti, R.D., Haley & Aldrich Inc., 8909 Bradford Way, Frederick, MD 21701. T: 301-631-0638, F: 703-356-9445, E: rdp@haleyaldrich.com

Pircher, W.J.K., Sonnenbichlweg 5, Goetzens, A-6091 AUSTRIA. T: 43-5234-33856, F: 0043-5234-34115

Pires, J.A., Applied Research Associates, 7216 Lew Wallace Dr., Ne, Albuquerque, NM 87109. T: 505-797-7513, F: 505-872-0794

Pirog, R.J., URS Greiner, 62 Melody Lane, Cheektowaga, NY 14225. T: 716-681-9727

Pischer, D.A., Landau Assocs, Sound View Plaza, 130 2nd Avenue South, Edmonds, WA 98020. T: 425-778-0907, F: 425-778-6409, E: dpischer@landauinc.com

Pitts, E.C., Georgia Dept of Transp, 1225 Manley Road, Griffin, GA 30223. T: 770-229-2472, F: 404-362-6555, E: eric.pitts@dot.state.ga.us

Plaskett, M.E., Hayward Baker Inc., 6850 Benjamin Road, Tampa, FL 33634. T: 813-884-3441, F: 813-884-3820, E: meplaskett@haywardbaker.com

Plauson, S.P., Kleinfelder Inc., 1410 F St, Fresno, CA 93706. T: 559-486-0750, F: 559-442-5081, E: splauson@kleinfelder.com

Player, R.S.V., Ch2m Hill, Inc., 1260 W Broadway, Polk City, IA 50226. T: 515-984-9147, F: 515-289-7598, E: rplayer@ch2m.com

Plehn, D., Geosystems Eng, 7802 Barton, Lenexa, KS 66214. T: 913-962-0909, F: 913-962-0924, E: dplehn@geosystemseng.com

Pniewski, M.D., Soil & Materials Engrs, 414 N Wilson Ave, Royal Oak, MI 48067. T: 248-541-7905, F: 734-454-0629, E: imiak@concentric.net

Poepsel, P.H., HDR Eng, 8404 Indian Hills Drive, Omaha, NE 68114. T: 402-399-1000

Poland, C.D., Degenkolb Engineers, 225 Bush St, Suite1000, San Francisco, CA 94104. T: 415-392-6952, F: 415-981-3157, E: cpoland@degenkobl.com

Pond, E.C., Kleinfelder Inc., 6805 La Lucena Ave Ne, Albuquerque, NM 87113. T: 505-823-6575, F: 505-344-1711, E: pond@flash.net

Pool, R.G., Federal Energy Reg Comm, 1032 Oakdale Road Ne, Atlanta, GA 30307. T: 770-979-7451, F: 770-452-3810, E: rp62736@aol.com

Poon, C.-Y., China Light & Powder Co Ltd., Flat C 6th Fl Twr 1 Parc Oasis, Yau Yat Chuen, Kowloon, HONG KONG. T: 011-852-2571-4500, F: 011-852-2571-47, E: cyp88@yahoo.com

Poormand, I., Leighton & Associates, 17781 Cowan, Irvine, CA 92614. T: 949-250-1421, F: 949-250-1114, E: lpoormand @peightongeo.com

Popescu, M., Illinois Inst of Tech, 425 Elm St, Unit D., Deerfield, IL 60015. T: 847-236-1804, F: 312-567-3519, E: mepopescu@usa.net

Porciello, D.M., Langan Engineering, 1114 Lake Shore Drive, Massapequa Park, NY 11762. T: 516-541-6948, E: dporciello@langan.com

Porrazzo, V.F., Porrazzo & Assocs, P.O. Box 7444, Domestic Airport Post Office, Pasay City, PHILIPPINES. T: 632-304-3137, F: 632-527-0167, E: vporrazzo@aol.com

Porter, D.S., Douglas S Porter Jr, P.E., 1605 Old Settlement Rd., Round Rock, TX 78664. T: 512-218-8340, F: 512-288-4244, E: porter814@aol.com

Post, G.R., Ingenieur-Conseil, 5 Rue Henri Regnault, St Cloud 92210, FRANCE. T: 331-47717958, F: 331-460-20180

Poulos, H.G., Coffey Geosciences, 142 Wicks Rd., North Ryde Nsw, 2113 AUSTRALIA. T: 011-61-2-9888-7444, F: 612 9888-9977, E: harry_poulos@coffey.com.au

Power, M.S., Geomatrix Cons Inc., 2101 Webster St, Ste 1200, Oakland, CA 94612. T: 510-663-4100, F: 510-663-4141, E: cmusacchia@geomatrix.com

Powers III, W.F., Hayes Drilling Inc., 12705 El Monte St, Leawood, KS 66209. T: 913-851-4873, F: 816-363-3060, E: wpowers@hayesdrillinginc.com

Powlesland, A.G., Mott Macdonald, St Anne House, 20–26 Wellesley, Croydon, CR9 2UL UK. T: 44-181-6865041, F: 44-181-6815706, E: tp1@mm-croy.mottmac.com

Pradel, D.E., Praad Geotechnical, 5465 South Centinela Ave, Los Angeles, CA 90066. T: 310-313-3111, F: 310-313-4441, E: daniel_pradel@praad.com

Pradilla, R., Damon G Douglas Company, 245 Birchwood Ave, Cranford, NJ 7016. T: 908-272-0100, F: 908-272-1211

Prager, R.D., Intuition & Logic, 31 Sylvester Ave, Webster Groves, MO 63119. T: 314-963-9581, F: 314-963-0270, E: IandL@ix.netcom.com

Prakash, S., Univ of Missouri, @Rolla, 308 Civil Eng Dept, Rolla, MO 65409. T: 573-341-4489, F: 573-341-4729, E: prakash@umr.edu

Prat, P.C., Dept of Geotechnical Eng, Etseccpb-Upc Jordi Girona 1–3, Edif D2, Barcelona, 8034 SPAIN. T: +34- 934016511, F: +34-934017251, E: pere.prat@upc.es

Pratt, W.E., URS Corporation, 500 12th Street, Suite 200, Oakland, CA 94607. T: 510-874-3068, F: 510-874-3268, E: william_pratt@urscorp.com

Preber, T., South Dakota School of, Sdsm&T Dept of Cee, 501 E St Joe, Rapid City, SD 57701. T: 605-394-2446, F: 605-394-5171, E: tpreber@taz.sdsmt.edu

Press, M.J., Curtin Univ of Tech, 1 Bunning Pl, Kardinya Wa, 6163 AUSTRALIA. T: 893141697, F: 089 266 2818

Pressey, E.D., County of Santa Barbara, 1170 Mariano Dr., Ojai, CA 93023. T: 805-646-1967, F: 805-646-9765

Preston, K.M., AMEC Earth & Environmental Inc., 329 C R 2900, Aztec, NM 87410. T: 505-334-2041, F: 505-326-5721

Pretti, C.L., 9320 East Summer Trail, Tucson, AZ 85749. F: 520-573-0528

Prevost, J.H., Princeton Univ, Dept of Civil Eng, Princeton, NJ 8544. T: 609-258-5424, F: 609-258-1270, E: prevost@princeton.edu

Price, B.E., Rb&G Engineering Inc., 4551 Killarney Dr., Highland, UT 84003. T: 801-756-1732

Price, L.L., Applied Cons Tech Inc., 210 Hayes Dr., Suite C., Cleveland, OH 44131. T: 216-459-8378, F: 216-459-8954

Prieto-Portar, L.A., Florida Intl Univ, 4025 Irvington Ave, Coconut Grove, FL 33133. T: 305-442-2109

Prochaska, B.R., Soils & Fndtn Engrs Inc., 9935 Hillyard Ave, Baton Rouge, LA 70809. F: 504-752-1467

Prommer, P.J., Geosyntec Consultants, 55 West Wacker Drive, Suite 1100, Chicago, IL 60601. T: 312-658-0500, F: 312-658-0576, E: petep@geosyntec.com

Proskovec, G.E., HWS Consulting Inc., P.O. Box 80358, 825 J St, Lincoln, NE 68501. T: 402-479-2278, F: 402-479-2276, E: gproskov@hws-con.com

Pruett, J.S., Universal Engineering, 6950 Phillips Highway, Ste 1, Jacksonville, FL 32216. T: 904-296-0757, F: 904-296-0748, E: erikapruet@aol.com

Prymus, R.E., Raymond E Prymus Pc, 212 Edgewater Ave, Bayport, NY 11705. T: 516-472-4964

Pujol-Rius, A., Gei Consultants, Inc., 2201 Broadway, Suite 321, Oakland, CA 94612. T: 510-835-9838 105, F: 510-835-9842, E: apujol@geiconsultants.com

Puri, V.K., Southern Illinois Univ, Dept of Civil Eng & Mechanics, Carbondale, IL 62901. T: 618-453-7818, F: 618-453-7455

Pyke, R.M., 1076 Carol Lane, 136, Lafayette, CA 94549. E: bobpyke@attglobal.net

Pyles, M.R., Oregon State Univ, Dept of Forest Eng, Peavy Hall 213, Corvallis, OR 97331. T: 541-737-4571, F: 541-737-4316, E: marvin.pyles@orst.edu

Pyles, R.L., Specialized Engineering, 13006 Hiney Rd., Keymar, MD 21757. T: 301-898-7073, F: 301-607-4331, E: specengr@aol.com

Qiu, Y., Southwest Jiaotong University, School of Civil Engineering, Southwest Jiaotong University, Chengdu, 610031 CHINA. T: 86-28-7603091, F: 86-28-7609007, E: giuyanjun @members.asce.org

Quazza, K.J., Sesi Consulting Engs, 56 Skyview Rd., West Milford, NJ 7480. T: 201-728-5846, F: 201-808-9099

Qubain, B.S., Valley Forge Labs, 6 Berkeley Rd., Devon, PA 19333. T: 610-688-8517, F: 610-688-8143, E: bqubain@aol.com

Queen, F.A., Foundation Technologies Inc., 3300 Montreal Industrial Way, Suite 8, Tucker, GA 30084. T: 770-723-9887, F: 770-723-0884, E: ftinc@mindspring.com

Queiroz, L.A., Alameda Sarutaia 381, Apt O 162 Jardim Paulista, San Paulo – Sp, 01403-010 BRAZIL. T/F: 11-2514546

Quigley, D.W., Harding Ese, 28 2nd Street, Suite #700, San Francisco, CA 94105. T: 415-278-2104, F: 415-777-9706, E: dwquigley@mactec.com

Quinn, G.A., Ntl Engineering & Geoscience, 1300 24th Ave S.W, Great Falls, MT 59404. F: 406-761-6655, E: ntl1@valcom.net

Quiros, G., Marsco Inc., 1717 Country Club Blvd, Sugar Land, TX 77478. T: 713-980-8423, F: 713-465-2787

Quiroz, J.D., West Virginia University, 313 Eleanor Drive, Morgantown, WV 26508. T: 304-599-8408, F: 304-293-7109, E: jdquiroz@mail.wvu.edu

Raas, S.M., Steven Raas & Assocs, 444 Airport Blvd, Suite 106, Watsonville, CA 95076. T: 831-722-9446, F: 011 605 313 414

Raba, Jr, C.F., Raba Kistner Cons Inc., P.O. Box 690287, 12821 W Golden Ln, San Antonio, TX 78269. T: 210-699-9090, F: 210-699-6426, E: craba@rkci.com

Rabasca, S.J., Peterson-Rabasca Geoengineers, 3 Birch Knolls, Cape Elizabeth, ME 4107. T: 207-767-4514, F: 207-8464348, E: prg@nlis.net

Rabus, R.W., Genterra Consultants Inc., 15375 Barranca Pkwy, Ste A-107, Irvine, CA 92618. T: 949-753-8766, F: 949-753-8887, E: central@genterra.com

Raghu, D., New Jersey Inst of Tech, 38 East Fairchild Pl, Whippany, NJ 7981. T: 973-887-9206, F: 973-596-5790, E: raghu@adm.njit.edu

Rahardjo, P.P., Parahyangan Catholic Univ, Dept of Ce & Arch/Jalan, Ciumbuleuit 94 Bandung 40142, INDONESIA

Rahe, J.H., MFG Inc., 1070 N Youngfield Street, Golden, CO 80401. T: 303-238-1690, F: 303-447-1832, E: john.rahe@mfgenv.com

Ramage, J., Ch2m Hill, 121 Kennedy Ave, San Antonio, TX 78209. T: 210-805-8015, F: 49-721-354-4580, E: ramages@msn.com

Ramakrishna, A., Larsen & Toubro Limited, Ecc Constr Gp–P.O. Box 979, Mt Poonamalle Rd Manapakkam, Chennai TN, 600089 INDIA. T: 011-91-44-2343564, E: anmol@giasmd01.vsnl.net.in

Randall, O.L., Ft Collins City of, Utilities, P.O. Box 580, Ft Collins, CO 80522. T: 970-221-6809, F: 970-221-6619, E: orandall@ci.fort-collins.co.us

Randolph, B.W., Univ of Toledo, Dept of Civil Eng, 2801 W Bancroft St, Toledo, OH 43606. T: 419-530-8115, F: 419-530-8116, E: brandolp@eng.utoledo.edu

Ranken, R.E., Pace Civil Inc., 2239 Crestview Ave, Redding, CA 96001. T: 530-243-1178, F: 530-244-1978, E: rranken@pacecivil.com

Rao, P.V., Tata Cons Engrs, 73/1 St Marks Rd., Bangalore Krn, 560001 INDIA. T: 91-80-2274721, F: 91-802274873, E: tatbws@bln.vsnl.net.in

Rao, VVS, Nagadi Cons (P) Ltd., 106-1-D Kishengarh, Vasantkunj, New Delhi, 110 070 INDIA. T: 91-116891980, F: 91-11-6897403, E: nagadi@vsnl.com

Rashid, A.K., Bowser-Morner Associates Inc., 4994 S Main St, #14, Sylvania, OH 43560. T: 419-824-0015, F: 419-691-4805, E: arashid@bowser-morner.com

Rashidi, H.K., Earthspectives, 3 Savannah St, Irvine, CA 92620. T: 714-505-3160, F: 949-654-2759, E: earthspect@aol.com

Rathje, E.M., University of Texas, Ecj 9 227b Mc1792, Austin, TX 78712. T: 512-232-3683, F: 512-471-6548, E: e.rathje@mail.utexas.edu

Rattanawangcharoen, N., Chiang Mai University, Faculty of Engr, Chiang Mai, 50200 THAILAND. T: 665-3944159, F: 665-3217287, E: nipon_r@ds90.intanon.nectec.or.th

Rau, G.C., Rau & Assocs, 100 N Pine St, P.O. Box M., Ukiah, CA 95482. T: 707-462-6536, F: 707-463-2729, E: rau@saber.net

Rau, G.A., Exponent Failure Analysis Asso, 149 Commonwealth Drive, Menlo Park, CA 94025. T: 650-688-7381, F: 650-328-3094, E: grau@exponent.com

Rausche, F., GRL & Assocs Inc., 4535 Renaissance Pkwy, Cleveland, OH 44128. T: 216-831-6131, F: 216-831-0916, E: ralsche@pile.com

Reader, M.D., LKR Group, 4830 Sharynne Lane, Torrance, CA 90505. T: 310-540-4656, E: miker@groupdelta.com

Reades, D.W., c/o Golder Assocs Ltd., 2180 Meadowvale Blvd, Mississauga, L5N 5S3 CANADA ON. T: 905-567-4444, F: 905-567-6561, E: dreades@golder.com

Reagan, L.K., 7817 Wooddale Way, Citrus Hts, CA 95610. T: 916-721-8593, F: 916-366-7013, E: lreagan@kleinfelder.com

Reavis, G.T., Foundation Technology Inc., P.O. Box 570522, Houston, TX 77257. T: 281-414-0774, F: 281-754-4435, E: ftinc@mail.com

Rebull, P.M., Rebull & Assocs Ltd., P.O. Box 7727, Arlington, VA 22207. T: 703-522-7770, F: 703-522-2562, E: molefex@aol.com

Rechenmacher, A.L., Johns Hopkins University, Johns Hopkins University, Dept of Civil Engrg, 3400 N Charles St, Baltimore, MD 21218. T: 410-516-6107, F: 410-516-7473, E: alr@jhu.edu

Recker, K.L., Haley & Aldrich Inc., 81 Johnson Rd., Falmouth, ME 4105. T: 207-781-5902, F: 207-871-5999, E: klr@haleyaldrich.com

Reddi, L.N., Kansas State Univ, 2118 Fiedler Hall, Manhattan, KS 66506. T: 785-532-1586, F: 785-532-7717, E: reddi@ksu.edu

Redmond, P.L., Piedmont Engineering, 23 Snow Crest Drive, Belgrade, MT 59714. T: 406-388-4062, F: 406-388-9653, E: piedmont@in-tch.com

Reed, P.M., Earth Sciences (S) Pte Ltd., Tanglin P.O. Box 357, SINGAPORE. T: 65-97800475, F: 65-338-0816, E: ptesi@hotmail.com

Reed, R.A., Pacific Soils Eng Inc., 10653 Progress Way Box 2249, Cypress, CA 90630. T: 714-220-0770, F: 714-220-9589, E: psecyp@aol.com

Reed, R., Reed Eng Group, 6257 Lupton Dr., Dallas, TX 75225. T: 214-691-4445, F: 214-350-0019

Reeder, T.Z., Svedala Reedrill, P.O. Box 998, Sherman, TX 75091. T: 903-786-2981, F: 903-786-6405

Reese, J.D., Independent Consul & Engrs, 8214 Grainfield Rd., Severn, MD 21144. T: 410-969-7630, F: 301-218-1170, E: drdirt@erols.com

Reese, L.C., Lymon C Reese & Associates, 8805 Point West Drive, Austin, TX 78759. T: 512-346-5833, F: 512-467-1384, E: icreese@ensoftinc.com

Reeves, J.K., TC Baycor, 2170 Satellite Blvd #350, Duluth, GA 30097. T: 770-448-1518770446-06, F: jkreeve@ibm.net

Reif, M.A., Ch2m Hill, 11265 Center Harbor Rd., Reston, VA 20194. T: 703-787-0650, F: 703-471-1508, E: mreif@ch2m.com

Reineke, R.T., K Singh & Assocs Inc., 2028 N 114th St, Wauwatosa, WI 53226. T: 414-774-9275, F: 262-821-1174, E: reineke@execpc.com

Reinfurt, J.E., Sverdrup Civil Inc., 16719 Heather Moor Dr., Florissant, MO 63034. T: 314-921-9264, F: 314-770-5130, E: john.reinfurt@jacobs.com

Reith, C.M., Geo Technology Associates, 18 Boulden Cir, Ste 34, Newcastle, DE 19720. T: 302-326-2100, F: 302-326-2399, E: creith@mragta.com

Rempe, D.M., Geotechnical Cons, 8 Bel-Aire Court, Champaign, IL 61820. T: 217-352-1395, F: 217-352-1394, E: dmrempe@net66.com

Renfrey, G.J., Urs Australia, Apo Box 302, Brisbane Qld, 4001 AUSTRALIA. T: 07-32432111, F: 07-32432199, E: gavan_renfrey@urscorp.com

Rennie, D.C., Ca Department of Water Resources, 400 Munroe St, #36, Sacramento, CA 95825. T: 916-486-6138, F: 916-653-0698, E: dave_ren@yahoo.com

Rentenbach II, T.M., Rentenbach Eng Co, 2400 Sutherland Ave, Knoxville, TN 37919. T: 865-546-2400, F: 865-546-3414

Repetto, P.C., URS Corporation, Stanford Place 3 Ste 1000, 4582 S Ulster St, Denver, CO 80237. T: 303-740-2667, F: 303-694-3946, E: pedro-repetto@urscorp.com

Resnick, G.S., P.O. Box 1451, Ventura, CA 93002. T: 805-648-4888, E: gsresnick@compuserve.com

Restrepo, C.J., Espinosa & Restrepo, Trans 1era No. 83-83 Apt 403, Bogota, Cundinamarca, COLOMBIA. T: 57-1-6227267, F: 57-1-6227301

Reuss, R.F., Professional Service Ind Inc., 679 Harbor Island, Clearwater, FL 33767. T: 727-443-2880, F: 727-461-4680, E: rfreuss@aol.com

Reuter, G.R., GME Consultants Inc., 14677 92nd Place North, Maple Grove, MN 55369. T: 763-420-5725, F: 763-559-0720, E: reute005@tc.umn.edu

Reutter, J.C., Iowa Dept of Transp, 2051 220th St, Boone, IA 50036. T: 515-432-6789, F: 515-961-6352

Reynolds, R.T., Richard T Reynolds, Consulting Geotechnical Engr, 77 A Revere St Suite 3, Boston, MA 2114. T: 617-523-0873, F: 617-367-9882, E: reynoldsr@prodigy.net

Riad, A.H., Ardaman & Assoc, 8008 S Orange Ave, P.O. Box 593003, Orlando, FL 32859. T: 407-855-3860, F: 407-859-8121, E: ariad@ardaman.com

Rice, A.C., Seattle City of, 9001 45th Avenue Ne, Seattle, WA 98115. T: 206-517-2939, F: 206-386-1168

Rice, P.M., Bechtel/Parsons Brinckerhoff, 185 Kneeland St, Mailstop 05-7x-01, Boston, MA 2111. T: 617-482-2933, F: 617-482-2878, E: rice@pbworld.com

Richards, Jr, T.D., Nicholson Construction Co, 85 Gensler Road, Pittsburgh, PA 15236. T: 412-882-7547, F: 412-221-3127, E: trichards@nicholson-rodio.com

Richards, D.P., Tejon Engineering, Inc., 551 Everett Ln, Clarkdale, AZ 86324. T: 520-649-1547, F: 520-634-4431, E: donprichards1@earthlink.net

Richardson, E.V., Ayres Associates, 824 Gregory Rd., Fort Collins, CO 80524. T: 970-484-3158, F: 970-223-5578, E: richardsone@ayresassociates.com

Richter, T.J., URS., 111 Sw Columbia, Ste 900, Portland, OR 97201. T: 503-948-7213, F: 503-222-4292, E: trichter@alumni.princeton.edu

Richters, G., JE Sverdrup Civil Inc., 2 Center Plz, Boston, MA 2108. T: 617-482-7880, F: 617-742-8830, E: richteg@sverdrup.com

Ridlen, P.W., Smith & Co, 1417 E Lakeland Lane, Carbondale, IL 62901. T: 618-457-0105, F: 573-785-2651, E: pwridlen@juno.com

Ridley, C.A., Treadwell And Rollo Inc., 519 Belvedere St, San Francisco, CA 94117. T: 415-566-4545, F: 415-955-9041, E: cridley@pacbell.net

Riegel, M.D., Hatch Mott Macdonald, 27 Bleeker St, P.O. Box 1028, Millburn, NJ 7041. T: 973-912-7527, E: mrie gel@hatchmott.com

Riemer, M.F., Dept of Civil Engineering, 440 Davis Hall, U C Berkeley, Berkeley, CA 94720. T: 510-642-7457, E: riemer@ce.berkeley.edu

Riemer, W.H., Risch-Haff, Ehner/Redingen, L 8529 LUXEMBOURG. T: 352-630518, F: 352-639612, E: wriemer@pt.lu

Ries, C., 2175 Sky View Court, Moraga, CA 94556. T: 510-376-6971

Ries, E.R., Fbo/Istanbul, Amcongen Istanbul, Psc 97 Box 0002, Apo, AE 9827. T: 90-212-251-3602, F: 90-212-251-3632, E: rieserist@state.gov

Rigby, D.B., Hong Kong Univ of Sci & Tech, 7400 E Golf Links Rd Apt 273, Tucson, AZ 85730. T: 520-747-7055, F: 852-2358-1534, E: rigbydb@home.com

Rigby, E.H., Forbes Rigby Pty Ltd., 278 Keira St, Wollongong NSW, 2500 AUSTRALIA. T: 242284133, F: 242286811, E: ted.rigby@forbesrigby.com.au

Rimoldi, P., Corso Garibaldi 125, Milano 20121, ITALY. T: 39-039-9219307, F: 39-921-9200, E: pietro.rimoldi@temax.net

Ringen, A.R., Hayward Baker Inc., 265 N Brookshire Ave, Ventura, CA 93003. T: 805-654-8976, F: 805-933-1338, E: arringen@haywardbaker.com

Ringholz, R.P., Fugro South Inc., 23002 Grand Rapids, Spring, TX 77373. T: 281-288-6647, F: 713-778-5544, E: bringholz@fugro.com

Rinker, B.C., Harza Engineering Co, 132 Monterey Ave, Pacific Grove, CA 93950. T: 831-375-1834, F: 831-455-8181

Rinne, P.E., Edward E., Kleinfelder Inc., 6490 Ascot Dr., Oakland, CA 94611. T: 510-482-4382, F: 510-531-7802, E: rinnee@sprynet.com

Riordan, P.J., 5903 Elmbrook Rd., P.O. Box 2847, North Conway, NH 3860. T: 603-356-0881, E: riordanpj@prodigy.net

Rippe, P.E., Arlan H., Squier Associates, 4260 Galewood St, Lake Oswego, OR 97035. T: 503-635-4419, F: 503-635-1430, E: arlanr@squier.com

Risitano, J.R., Metcalf & Eddy Inc., P.O. Box 4071, Wakefield, MA 1880. T: 781-224-6313, F: 781-224-6665, E: john_risitano@metcalfeddy.com

Ritchie, Jr, L.R., Ritchie & Montgomery, 2329 Chester Rd., Birmingham, AL 35223. T: 205-879-8151, F: 205 879-8525

Rivard-Lentz, P.E., David John, GEI Consultants Inc., 138 Hurd Park Road, East Hampton, CT 6424. T: 860-365-0928, F: 860-537-6347, E: drientz@geiconsultants.com

Rivera Roldan, V.E., Victor E Rivera Assocs, P.O. Box 32198, Ponce, PR 732. T: 787-259-1410, F: 787-259-1604

Rixner, J.J., Haley & Aldrich of New York, 75 Brentwood Lane, Fairport, NY 14450. T: 716-377-1242, F: 716-232-6768, E: jjr@haleyaldrich.com

Rizzo, P.C., PC Rizzo Assocs, Expo Mart Ste 270e, 105 Mall Blvd, Monroeville, PA 15146. T: 412-856-9700, F: 412-856-9749, E: ed.zullo@rizzoassoc.com

Roarty, Jr, C.J., NTH Consultants Ltd., 277 Gratiot Suite 600, Detroit, MI 48226. T: 313-237-3936, F: 313-237-3900, E: croarty@nthconsultants.com

Robblee, G.C., GEI Cons Inc., 2141 Palomar Airport Road, Suite 160, Carlsbad, CA 92009. T: 760-929-9136, F: 760-929-0836, E: jrobblee@geiconsultants.com

Roberds, W.J., Golder Assocs Inc., 547 11th Ave West, Kirkland, WA 98033. T: 425-822-8344, F: 425-882-5498, E: broberds@golder.com

Roberto, G., University of Cincinnati, 168 Asbury Rd., Cincinnati, OH 45255. T: 513-231-3147, F: 5135562522, E: groberto@uceng.uc.edu

Roberts, J.E., Ca State Univ @Northridge, 319 E Sycamore Ave, El Segundo, CA 90245. T: 310-640-7233

Robertson, A.M.G., 580 Hornby St, Ste 330, Vancouver, V6C 3B6 CANADA BC. T: 604-684-8072, F: 604-681-4166, E: arobertson@info-mine.com

Robertson, D.T., Applied Foundation Testing, 1060 Roland Ave, Green Cove Springs, FL 32043. T: 904-284-1337, F: 904-284-1339, E: drobertson@testpile.com

Robertson, R.J., Ch2m Hill, 2300 Nw Walnut Blvd, Corvallis, OR 97330. T: 541-752-4271, F: 541-752-0276, E: rroberts@ch2m.com

Robinson, B.A., Brian A Robinson & Assoc, 4932 Casa Dr., Tarzana, CA 91356. T: 818-345-8292, F: 818-782-8282, E: geobar@prodigy.net

Robinson, K.E., Jacques Whitford & Assocs Ltd., 3711 N Fraser Way Unit 1, Burnaby Bc, V5J 5G5 CANADA BC. T: 304-436-3014, E: krobinson@jacqueswhitford.com

Robinson, R.A., Brian A Robinson & Assocs, 4932 Casa Dr., Tarzana, CA 91356. T: 818-345-8292, F: 818-782-8282, E: geobar@prodigy.net

Robison, M.J., Jordan Jones & Goulding, 5195 Blue Yarrow Run, Norcross, GA 30092. T: 770-582-0189, F: 770-455-7391, E: mrobison@jjg.com

Roblee, C.J., California State of, 1303 Beech Lane, Davis, CA 95616. T: 530-753-5038, F: 916-227-6974, E: cliff.roblee@dot.ca.gov

Rochford, W.A., US Army Corps of Engrs, Celrc-Ed-G., 111 N Canal St #600, Chicago, IL 60606. T: 312-353-6400 Ext. 3030, F: 312 353 2156, E: william.a.rochford@usace.army.mil

Rodgers, R.D., Treadwell & Rollo Inc., 136 Arapho Circle, San Ramon, CA 94583. T: 415-829-1736, F: 415-955-9041, E: rdrodgers@treadwellrollo.com

Rodrigo, C.S., Louis Berger & Assocs, 15514 Turtle Oak Ct, Houston, TX 77059. T: 281-990-6585, F: 281-412-4623, E: crodrigo@louisberger.com

Rodrigo, M., Rodrigo & Rjapakse, 159 Jacoby St., Maplewood, NJ 7040. T: 281-990-6585, F: 281-412-4623, E: mrodrigo@louisberger.com

Rodriguez Amaya, C., Hidroconsulta Ltd., Apartadeo Aereo 76319, Santafe De Bogota, COLOMBIA. T: 618-563861 85557236120, F: 6164024

Rodriguez Zelaya, L., Asoc Consultores En Ingenieria, 9 Ave No N 57 Blvd Cal Univer, San Pedro Sula, HONDURAS. T/F: 566-1656, E: aci@sigmanet.hn

Rodriguez, J.P., Central Industrial Services, Los Flamboyanes 162 Calle Pino, Gurabo, PR 778. T: 787-737-8214, F: 787-846-5425, E: jprodz@prtc.net

Rodriguez-Marek, A., Washington State Univ, P.O. Box 642910, Pullman, WA 99164. T: 509-335-7088, F: 509-335-7621, E: adrian@wsu.edu

Rodriguez-Perez, C., Geo-Engineering Inc., 10-3 Calle Tulip Parq, Monteverde I., San Juan, PR 926. F: 787-764-6325

Roesler, A.C., 619 W Mifflin St, Apt F., Madison, WI 53703. E: roesler@cae.wisc.edu

Rogers, J.D., Geolith Consultants, 396 Civic Dr., Pleasant Hill, CA 94523. T: 925-682-7601, F: 925-682-7605, E: jdrogers@geolith.com

Rogers, R.B., Civil Geotechnical Eng, 1614 Hillcrest Ave, Antioch, CA 94509. T: 925-754-0106, F: 925-779-9890

Rohe, F.P., Environmental Protection Inc., P.O. Box 333, Mancelona, MI 49659. T: 616-587-9108, F: 616-587-8020, E: pvcliner@geomembrane.com

Rohlf, R.A., Ky Dept For Surface Mining, 748 Wellington Way, Lexington, KY 40503. T: 606-223-8989, F: 502-564-5698, E: rarohlf@prodigy.net

Rokoff, M.D., URS Corporation, 4765 Mayfield Rd., Cleveland, OH 44124. T: 216-381-8216, F: 440-349-1514, E: mark_rokoff@urscorp.com

Rollings, M.P., Cold Regions Research And Eng Lab, P.O. Box 313, Lyme, NH 3768. T: 603-795-9301, F: 603-646-4640, E: mrollings@crrel.vsace.army.mil

Rollins, K.M., Brigham Young Univ, Dept of Civil Eng, 368 Clyde Bldg, Provo, UT 84602. T: 801-378-6334, F: 801-378-4449, E: rollinsk@byu.edu

Rolston, J.W., Earth Systems Consultants, 7949 Woodley Ave, 7949 Woodley Ave, CA 91406. F: 818-345-5283, E: geotek@earthlink.net

Rom, D.S., US Nuclear Regulatory Comm, 14735 Wrights Lane, Waterford, VA 20197. T: 540-882-3143, F: 540-882-3629, E: dsr@nrc.gov

Roman, W., Roman & Lougee, 311 Hawthorne Ave, Los Altos, CA 94022. T: 650-941-2781, F: 650-941-0916

Romero, S., School of Civil Engineering, 4371 Winters Chapel Rd., Apt 223, Atlanta, GA 30360. T: 770-798-9017, F: 404-894-2281, E: gt6829b@prism.gatech.edu

Romero-Requena, I.M., Geodec Sa, Av Ppal Urb San Luis Res, Atlanta 7a, Caracas, VENEZUELA. T: 58-2-9876796, F: 58-2-2656632, E: geodec@telcel.net.ve

Romhild, C.J., Bredebovej 35 St Tv, Bredebovej 35 St Tv, Dk 2800, Lyngby, DENMARK. T: 45-45-870-650, F: 817-481-0369

Romig, G.A., Romig Engineers Inc., 611 Veterans Blvd Ste 110, Redwood City, CA 94063. T: 415-364-2511, E: glenn@romigengineers.com

Rooney, J.W., R & M Cons Inc., 9101 Vanguard, Anchorage, AK 99507. T: 907-522-1707, F: 907-522-3403, E: jrooney@rmconsult.com

Roos, C.J., 4711 South Himes Ave, Apt 114, Tampa, FL 33611. T: 813-832-5110, E: cjroos@bigfoot.com

Root, M., Root Eng, 305 Summer St, Springfield, VT 5156. T: 802-885-5335, F: 802-885-8419, E: rooteng@compuserve.com

Roote, D.R., AG Environmental Services, 87 Vendome Dr., S., Rochester, NY 14606. T: 716-429-6986, F: 716-742-2249, E: dcroot@rpa.net

Rose, A.T., Univ of Pittsburgh-Johnstown, 225 Engineering & Science Bldg, Johnstown, PA 15904. T: 814-269-7249, F: 814-269-7245

Rosen, D.S., Israel Natnl Inst For, 21 Golomb St, Haifa, 33391 ISRAEL. T: 972-483-60179, F: 972-485-11911, E: rosen@ocean.org.il

Rosen, W.J., Tcdi, 10732 Dutchtown Rd., Knoxville, TN 37932. T: 865-966-0294, F: 865-966-0073

Rosenblatt, L., L Rosenblatt & Assocs, P.O. Box 57-6246, Modesto, CA 95357. T: 209-523-0289

Ross, A.J.B., Parsons Brinckerhoff, 1129 Country Shadows Way, Las Vegas, NV 89123. T: 702-270-6074, F: 310-522-9945, E: rossan@pbworld.com

Ross, T.E., Hayward Baker Inc., 5765 Millstone Drive, Cumming, GA 30040. T: 770-887-4916, F: 770-442-8344, E: teross@haywardbaker.com

Rossie, M., Rembco Geotech Contrs Inc., P.O. Box 23009, Knoxville, TN 37933. T: 423-690-6917, F: 423-690-9135, E: mrossie@rembco.com

Roth, B.C., Municon Consultants, 775 Congo Street, San Francisco, CA 94131. T: 415-584-2430, F: 415-584-4097, E: municon@municon.net

Roth, L.H., ASCE., 1801 Alexander Bell Dr., Reston, VA 20191. T: 703-295-6102, F: 703-295-6125, E: lroth@asce.org

Roth, M.J.S., Lafayette College, Dept of Civil Eng, Easton, PA 18042. T: 610-330-5427, F: 610-330-5059, E: rothm@lafayette.edu

Rotundo, M.A., Marcus A Rotundo PC, P.O. Box 405, Central Square, NY 13036. T: 315-668-3263, F: 315-668-0256, E: marpe59582@aol.com

Rourke, J.G., Fay Spofford & Thorndike Inc., 5 Burlington Woods, Burlington, MA 1805. T: 781-221-1123, E: jrourke@fstinc.com

Roussel, Jr, H.J., Roussel Eng Inc., 2700 Lake Villa Dr., Rm 220, Metairie, LA 70002. T: 504-887-8870, F: 504-887-8852, E: roueng@bellsouth.net

Rowe, E.A., EC Rowe & Associates, Monjitas 454 9th Floor, Santiago, CHILE. T: 011-562-638-1603, F: 011-562-638-185, E: ecrowe@entelchile.net

Rowe, L.W., 5200 River Avenue, Newport Beach, CA 92663. T: 949-548-3935, F: 949-631-8108

Rowland, M.G., BBC & M Engineering Inc., 1353 Meadow Road, Columbus, OH 43212. T: 614-486-2469, F: 614-793-2410, E: bbcm@infinet.com

Rowley, B.J., California Department of Transportation, 708 L St, Davis, CA 95616. T: 530-756-2938, F: 916-227-7075, E: bjrowley@hotmail.com

Roycroft, G.A., Waste Management Inc., 155 N Redwood Drive, Suite 250, San Rafael, CA 94903. T: 415-479-3700, F: 415-479-3737, E: groycroft@wm.com

Rubak, C.G., Waste Management, 860 Warwick Lane, Lake Zurich, IL 60047. T: 708-550-8203, F: 847-223-3188, E: crubak@wm.com

Rubright, R., Hayward Baker Inc., 10036 Inkpen Pl, Ellicott City, MD 21042. T: 410-461-7733, F: 852-2549-5668, E: rmrubright@haywardbaker.com

Rucker, M.L., Agra Earth & Environmental Inc., 3232 West Virginia, Phoenix, AZ 85009. T: 602-272-6848, F: 602-272-7239, E: mrucker@agraus.com

Ruckman, P.E., Albert Charles, 11488 Nucla Street, Commerce City, CO 80022. T: 303-287-1864, F: 303-287-1879

Rudianto, S., Geo-Optima Inc., 139 Heather Ave, Hercules, CA 94547. T: 510-245 2139, F: 062-21-5814523

Rudig, D.A., Hntb Corporation, W153 N 7745 Meadowlark Ln, Menomonee Fal, WI 53051. T: 414-253-4627, F: 414-359-2315, E: drudig@hntb.com

Rudolph, R.W., Subsurface Cons Inc., 3736 Mount Diablo Blvd, # 200, Lafayette, CA 94549. T: 925-268-0461

Ruggeri, E.A., Power Engineers, 3940 Glenbrook Dr., P.O. Box 1066, Hailey, ID 83333. T: 208-788-0595, F: 208-788-0525, E: eruggeri@powereng.com

Ruggiero, V.J., Apgar Associates, 102 South Appletree Rd., Howell, NJ 7731. T: 201-938-6630, F: 908- 234-1086

Rumpelt, T.K., Smoltczyk & Partner, Stossaecker Str 120, Stuttgart, D-70563 GERMANY. T: 49-711-7351824, F: 011-49-71113164, E: smoltczykpartner@csi.com

Ruse, R.S., City of Findlay, 1110 Hurd Avenue, Findlay, OH 45840. T: 419-424-9282, F: 419-424-7245, E: rruse@ci.findlay.oh.us

Russo, R.J., 86 Barlow Street, Clayfield, Brisbane Qld, 4011 AUSTRALIA. F: 301-6472267

Ruth, B.E., Univ of Florida, 2221 N W 97th St, Gainesville, FL 32606. T: 352-332-1775, F: 352-392-3394, E: bruth@ce.ufl.edu

Rwebyogo, M.F., Geo Engineering & Testing, 5119 Nw 34th Terrace, Gainesville, FL 32605. T: 352-375-2419, F: 352-336-7630, E: agc@gdn.net

Ryan, W.F., 2715 West Prospect St, Hood River, OR 97031. T: 541-386-2822, F: 503-823-2342, E: billr@bes.ci.portland.or.us

Rydin, S.M., Meridian Management Inc., 3940 Arctic Blvd, Suite 102, Anchorage, AK 99503. T: 907-677-2601, F: 907-677-2605, E: srydin@meridianak.com

Saada, A.S., Case Western Reserve Univ, Dept of Civil Eng, University Circle, Cleveland, OH 44106. T: 216-368-2427, F: 216-368-5229, E: axs31@po.cwru.edu

Saeb, S., Rocksol Consulting Group Inc., 848 Yellow Pine Ave, Boulder, CO 80304. T: 303-440-0117, F: 303-998-0698, E: saeb@rocksol.com

Sagaseta, C.M., ETS Ing Caminos, La Mina 52-A., Puente Arce, Cantabria, 39478 SPAIN. F: 34-942-201821, E: sagasetac@besaya.unican.es

Sahle, D., LCU Min of Works, P.O. Box 11574, Maseru, 100 LESOTHO. T: 266-316301, F: 266-310242, E: djsahle@lesoff.co.za

Saidin, F., Univ of Washington, Dept of Ce Sgem Program, Uw Box 352700, Seattle, WA 98195. T: 206-543-7363, E: fadzilah@u.washington.edu

Sak, A., Golder Associates Inc., 1484 Ida Lane, Lawrenceville, GA 30043. T: 770-237-0764, F: 770-934-9476, E: tony_sak@golder.com

Salazar Espinosa, J.H., Promocion Alter Sa De Cv, De La Concordia 11, Hermosillo, Sonora, 83200 MEXICO. T: 062-16-5130, F: 62-12-0117, E: jsalazar@mail.cybermex.net

Salazar, A., Delta No 10, Pallares Y Portillo 156-E-401, Col Parque San Andres, Del Coyoacan Cp, 4040 MEXICO. T: 011-525 5440515, F: 011-525-6632559

Salcido, R.F., 5794 Vista Linda, El Paso, TX 79932. T: 915-584-1484

Salembier, M., I S L., 75 Bd Mac Donald, 75019 Paris, FRANCE. T: 33-1-5526-9999, F: 33-1-4034-6336, E: michel.salembier@isl-ingenierie.fr

Salencon, J.C., Laboratoire De Mecanique Des, Solides Ecole Polytechnique, 91128 Palaiseau Cedex, FRANCE. T: 33169333303, F: 33169333026, E: salencon@lms.polytechnique.fr

Salgado, R., School of Civil Engineering, Purdue University, West Lafayette, IN 47907. T: 765-494-5030, F: 765-496-1364, E: rodrigo@ecn.purdue.edu

Salley, J.R., 921 Tramway Lane Ne, Albuquerque, NM 87122. T: 505-798-9904, F: 505-798-9673, E: jrs@shanwil.com

Salzman, G.S., 26019 Glen Eagle Drive, Leesburg, FL 34748. T: 352-323-6138, E: garels@lcia.com

Samford, A.M., Virginia Geotechnical Svces Pc, 8211 Hermitage Rd., Richmond, VA 23228. T: 804-266-2199, F: 804-261-5569, E: amsamford@vgspc.com

Samford, W.J., Virginia Geotechnical Svcs, 8211 Hermitage Rd., Richmond, VA 23228. T: 804-266-2199, F: 804-261-5569, E: wjsamford@vgspc.com

Sampaco, C.L., Ch2m Hill, 777 108th Avenue Ne, P.O. Box 91500, Bellevue, WA 98009. T: 425-453-5000, Ext. 5096, F: 425-462-5957, E: ksampaco@ch2m.com

Sams, C.E., Law Eng & Envir Services Inc., 4215 Windwood Cir, Charlotte, NC 28226. F: 704 357 1622

Samuels, R., Parsons Brinckerhoff Quade &, 195 Harwood Pl, Paramus, NJ 7652. T: 201-262-6587, F: 212-465-5592

Sanabria, A.G., Ing Inmobiliaria 491 Ca, Avenida Principal Lomas De, Chuao Quinta San Onofre, Urb Lomas De Chuao, Caracas, VENEZUELA. T: 9918872, F: 2619168, E: davos1@telcel.net.ve

Sanchez, Jr, F.O., FT Sanchez Construction, Maxilom Ext. Cor Osmena St, Cebu City, PHILIPPINES. T: 633-223-27881, F: 633-223-22153, E: socor@skyinet.net

Sanchez-Salinero, I., Matinsa, Rosalia De Castro 84, Madrid, 28035 SPAIN. T: 34-91-3738459, E: isanchez@fcc.es

Sander, E.J., Pennoni Associates Inc., 1559 Promise Lane, Wescosville, PA 18106. T: 610-481-9484, F: 610-262-0886, E: esander@pennoni.com

Sanders, S.H., Sanders & Associates, Geostructural Engineering, 7649 Sunrise Blvd Ste G., Citrus Heights, CA 95610. T: 916-729-8050, F: 916-729-7706, E: sanders@jps.net

Sandford, T.C., Univ of Maine, Civil Eng Dept, 5711 Boardman Hall, Orono, ME 4469. T: 207-581-2183, F: 207-581-3888, E: sandford@umit.maine.edu

Santamarina, J.C., School of Civil And Envir. Eng, Georgia Instit. of Tech., Atlanta, GA 30332. T: 404-894-7605, F: 404-894-2281, E: carlos@ce.gatech.edu

Santoni, J.P., Tocno America Cca, Box 450, Santo Domingo, DOMINICAN REPUBLIC. T: 809-689-2959

Sapp, D.R., Donald R Sapp & Son Inc., 7291 N W 43rd St, Miami, FL 33166. T: 305-593-0035, F: 305-593-0058

Saran, D., DLZ, 37891 Windwood Dr., Farmingtn Hls, MI 48335. T: 248-474-5821

Sargand, S.M., Ohio Univ, Dept of Civil Eng, Room 121, Athens, OH 45701. T: 614-593-1467, F: 614-593-0625, E: ssargand@bobcat.ent.ohiou.edu

Sarkar, S., Parsons Brinckerhoff, 3 Jason Court, Princeton Jct, NJ 8550. T: 609-799-4096, F: 212-465-5592, E: sarkars@pbworld.com

Sassila, A.E., CCA, 24313 Hemlock Drive, Plainfield, IL 60544. T: 815-254-0883, F: 312-782-5145, E: alasas@aol.com

Sathialingam, N., Law/Crandall Inc., 246 Monroe St, Irvine, CA 92620. T: 949-733-0853, F: 323-889-5398, E: nsathi@earthlink.net

Sauceda, J.B., US Army Corps of Engineers, 3307 Seppala Dr., A., Anchorage, AK 99517. T: 907-243-4773, F: 907-753-2688, E: james.b.sauceda@usace.army.mil

Saucier, C.L., Eustis Engr Co, 3011 28th Street, Metairie, LA 70002. T: 504-834-0157, F: 504-834-0354, E: csaucier@eustiseng.com

Sauer, E.K., Univ of Saskatchewan, 22-315 Bayview Cres, Saskatoon, S7V 1B5 CANADA SK. T: 306-242-4360, F: 306-934-7528, E: eksauer@innovationploce.com

Sauls, D.P., Louis J. Capozzoli & Assoc Inc., 10555 Airline Highway, Baton Rouge, LA 70816. T: 225-293-2460, F: 225-293-2463, E: sauls@mindspring.com

Sawyers, P.E., Jonathan Michael, ECS Ltd., 1900 Mountain Rd., Haymarket, VA 20169. T: 703-753-0020, F: 703-834-5529, E: jsawyers@ecslimited.com

Saxena, A., ASC Geosciences Inc., 3055 Drane Field Rd., Lakeland, FL 33811. T: 863-644-8300, F: 863-644-8203, E: anu@ascworld.net

Saye, S.R., Geotechnical Services, Inc., 7050 South 110th Street, La Vista, NE 68128. T: 402-339-6104, F: 402-339-6297, E: srs@gsi.usa.com

Sayed, S.M., GCI Incorporated, 2290 N Cr-427, Suite 100, Longwood, FL 32750. T: 407-331-6332, F: 407-331-9066, E: smsgci@aol

Scapuzzi, D.W., Parsons Brinckerhoff, 1451 Crestview Dr., San Carlos, CA 94070. T: 650-363-8111, F: 650 689 8317

Scarlett, M.J., Esp Associates Pa, 1107 Barrow Nook Ct, Apex, NC 27502. T: 919-363-0179, F: 919-858-9300, E: mscarlett@espassoc.com

Schaefer, V.R., South Dakota State Univ, Box 2219, Crothers Eng Hall/Room 112, Brookings, SD 57007. T: 605-688-6307, F: 605-688-5878, E: vernon_mschaefer@sdstate.edu

Schaeffer, J.A., Ch2m Hill, P.O. Box 91500, Bellevue, WA 98009. T: 425-453-5000, F: 425-468-3100, E: jschaeff@ch2m.com

Schaff, M.C., MC Schaff Assocs, 818 S Beltline Hwy E., Scottsbluff, NE 69361. T: 308-635-1926, F: 308 635-7807, E: mcschaff@mailbox-1.com

Schatz, R.H.W., Intecap Inc., 39 Wildwood Ave, Arlington, MA 2476. T: 781-648-2578, F: 617-423-0389, E: hschatz@intecap.com

Scheibel, L.L., Geomatrix Cons Inc., 2101 Webster St, Suite 1200, Oakland, CA 94612. T: 415-434-9400, F: 415-434-1365, E: lschelbel@geomatrix.com

Scheller, P.R., Geo-Science Eng Co, 1252 Mid Valley Dr., Jessup, PA 18434. T: 717-489-8717, F: 717-489-8714

Schellhorn, V.L., Aerial Industrial Inc., 32987 Hwy One South, Gualala, CA 95445. T: 707-884-4807, F: 707-884-9133, E: www.geojet.com geojet@mcn.org

Scherer, S.D., TCDI, 1062 Ash St, Winnetka, IL 60093. T: 847-441-7528, F: 847-634-8582, E: sdscherer@haywardbaker.com

Scheurenberg, R.J., Knight Piesold Consultores, Ave San Borja Sur 143, San Borja Lima 41, PERU. T: 511-226-0044, F: 511-226-0062, E: rscheurenberg@knightpiesold.com

Schmall, P.C., Moretrench American Corp, 11 Peachtree Rd., Basking Ridge, NJ 7920. T: 908-000-0000, F: 973-627-6078, E: pschmall@mtac.com

Schmehr, D.D., Exxon Mobil Res & Engineering, 13 Pepper Ln, Succasunna, NJ 7876.

Schmertmann, J.H., Johnhschmertmann Inc., Cons Geotech Engrs, 4509 N W 23rd Ave Ste 19, Gainesville, FL 32606. T: 904-378-2792, E: sci@bellsouth.net

Schmidt, B., Parsons Brinckerhoff, 1676 Yorktown Rd., San Mateo, CA 94402. T: 650-574-4550, F: 415-243-9501, E: schmidtbirger@msn.com

Schmidt, D.A., CMD Associates Llc, 259 E Nolley, Collierville, TN 38017. T: 901-853-3885, F: 901-259-2364

Schmidt, G.H., Ardaman & Assoc, 2500 Bee Ridge Rd., Sarasota, FL 34239. F: 313-522-6427

Schmitt, F.C., Temple Univ, 165 William Penn Blvd, West Chester, PA 19382. T: 610-399-3689, F: 215-204-6936

Schmitt, K.E., Geosyntec Consultants, 200 E Del Mar Blvd, Suite 250, Pasadena, CA 91105. T: 626-449-0664, F: 626-449-0411, E: kschmitt@geosyntec.com

Schnabel, Jr, H.V., Schnabel Foundation Co, 45240 Business Ct, Ste 250, Sterling, VA 20166. T: 703-742-0020, F: 703-742-3319

Schnabel, H.W., 200 Turnpike Road, Second Floor, Southborough, MA 1772. F: 508-303-3646, E: hank@schnab el.com

Schneeberger, C.E., Geos Ingenieurs Conseils SA, Route De Drize 7, 1227 Carouge/Geneve, SWITZERLAND. T: 4122-309-3060, F: 4122-309-3070, E: schneeberger@geos.ch

Schneider, H.R., Hochschule Rapperswil Hsr, Schutzengelstrasse 34d, Ch-6340 Baar, SWITZERLAND. T: 414-1760-1488, E: hansruedi.schneider@hsr.ch

Schneider, J.R., Ch2m Hill, 6686 E Jamison Ave, Englewood, CO 80112. T: 303-721-7983, F: 303-846-5243, E: jschneider@mho.net

Schnorr, R.J., Ralph J Schnorr-Farms, P.O. Box 11910, Saint Paul, MN 55111. T: 605-226-3879, F: 612-961-2910, E: schno003@tc.umn.edu

Schock, J.E., 2085 Water St, Lebanon, PA 17046. T: 717-273-3141, F: 717-766-5516

Schoenwolf, D.A., Haley & Aldrich Inc., One Plantation Ct, Rockville, MD 20852. T: 301-816-0318, F: 703-356-9445, E: das@haleyaldrich.com

Schonewald, I.V., GZA Geoenvironmental Inc., 4 Free St, Portland, ME 4101. T: 207-879-9190, F: 207-879-0099, E: schonewald@gza.com

Schreiber, M.O., Geo/Hydro Engrs Inc., 1000 Cobb Pl Blvd, Suite 290, Kennesaw, GA 30144. T: 770-426-7100, F: 770-426-5209, E: milt@geohydro.com

Schrieber, M.F., West Coast Geotech Inc., 6133 Atkinson St, West Linn, OR 97068. T: 503-656-6013

Schrier, D.T., Cotton Shires & Associates, 330 Village Ln, Los Gatos, CA 95030. T: 408-354-5542, F: 408 354-1852, E: wcageo@aol.com

Schroeder, J.A., HC Nutting Co, 5801 Lengwood Dr., Cincinnati, OH 45244. T: 513-624-0083, F: 513-321-4540

Schroeder, W.L., Oregon State University, 529 Nw 34th Street, Corvallis, OR 97330. T: 541-752-6612

Schubert, G.P., Giles Eng Assocs Inc., 7904 E Longlake Drive, Wind Lake, WI 53185. T: 262-895-6174, F: 262-549-5868, E: gpschub@aol.com

Schuler, M.L., Hayes Drilling Inc., 8845 Prospect, Kansas City, MO 64132. T: 913-381-1228, F: 816-756-5701

Schultz, J.M., 11130 Poplar St Apt C., Loma Linda, CA 92354. T: 909-796-0180

Schultz, M.S., Cdm Jessberger, One Cambridge Place, 50 Hampshire Street, Cambridge, MA 2139. T: 617-452-6399, F: 617-452-8399, E: schultzms@cdm.com

Schulz, C.B., TVGA Engineering Surveying, 113 Sunset Court Apt 6, Hamburg, NY 14075. T: 716-649-1153, F: 716-655-0937, E: cschulz@tvga.com

Schuster, R.L., 1941 Golden Vue Dr., Golden, CO 80401. T: 303-279-9849, F: 303-279-0281, E: rschuster@usgs.gov
Schuyler, J.N., Applied Geomechanics, 1336 Brommer St, Santa Cruz, CA 95062. T: 831-462-2801
Schwantes, Jr, E.D., Cons Geotech Eng, 19402 Woodlands Dr., Huntington Beach, CA 92648. T: 714-969-8935, F: 714-969-8945
Schwarm, D.R., Geoengineers, 839 Ellis Ave, Lake Oswego, OR 97034. T: 503-636-7007, F: 503-620-5940, E: dschwarm @geoengineers.com
Schwarz, L.G., 5308 S 44th Place, Rogers, AR 72758. T: 501-986-0909, E: schwarz@engr.uark.edu
Schweppe, M.G., URS Corporation, 3508 Marlborough Ave, Las Vegas, NV 89110. T: 702-452-5789, F: 702-295-3554, E: merrill_s_1999@yahoo.com
Sciandrone, J.C., 8234 Mediterranean Way, Sacramento, CA 95826. T: 916-383-2478
Scott, H.F., 4123 Rte 68, Rimersburg, PA 16248
Scott, J.C., URS Greiner Woodward Clyde, 6 Blue Sage, Littleton, CO 80127. T: 303-932-9219, F: 303-694-3946
Screwvala, F.N., Farrokh N Screwvala Inc., 27016 Knickerbocker Road, Bay Village, OH 44140. T/F: 440-892-9797, E: ellens@mciworld.com
Seals, R.K., Louisiana State Univ, Dept of Civil & Enviro Engrg, College of Engrg, Baton Rouge, LA 70803. T: 225-388-6503, F: 225-388-4945, E: ceseal@lsu.edu
Sealy, C.O., Unetco Minerals Corp, P.O. Box 60265, Grand Junction, CO 81506. T: 970-256-8836, F: 970-245-7543, E: sealyco@ucarb.com
Seaton, K., Soil & Material Consultants In, 4629 E 115th Ct, Thornton, CO 80233. T: 303-254-6709, F: 303-4312594
Secor, K.E., Hawley Mills Secor Consultants, 5901 Poso Ct, Bakersfield, CA 93309. T: 805-325-3329, F: 661-328-9615, E: ksecor@lightspeed.net
Sehn, A.L., Univ of Akron, Civil Engrg Dept, Akron, OH 44325. T: 330-972-7292, F: 330-972-6020, E: sehn@ uakron.edu
Seki, G.Y., Geolabs Inc., 6710 Hawaii Kai Dr., Apt 1608, Honolulu, HI 96825. T: 808-396-9731
Seli, J.J., 1 West Cary St, Richmond, VA 23220. T: 804-649-7035, F: 804-783-8023, E: jseli@schnabel-eng.com
Seminara, P.E., John Adrian, Southern California Geotech-nic, 1067 N El Dorado Dr., Fullerton, CA 92835. T: 714-773-4903, F: 714-777-0398, E: jaseminara@socalgeo.com
Semple, R.M., Mueser Rutledge Consulting, P.O. Box 4005, New York, NY 10163. T: 201-653-6255, F: 212-490-6654, E: rsemple@mrce.com
Sen, K.K., Nicholson Construction Co, 225 Friend St, Boston, MA 2114. T: 617-589-0787, Ext. 230, F: 617-589-0788, E: ksen@nicholson-rodio.com
Sener, J.C., Boise State University College, 4745 Wildrye Dr., Boise, ID 83703. T: 208-345-4387, F: 208-426-4800, E: jsener@boisestate.edu
Senter, J.K., HBC Engineering Inc., 7811 Bankside, Houston, TX 77071. F: 713-690-8787, E: ksenter@hbc engineering.com
Serbis, G., Ydro-Seka Oe, Jean Soutsou 8, Athens, 11474 GREECE
Serpico, Jr, J.J., Maser Consulting Pa, 11 Ireton Key, Colts Neck, NJ 7722. T: 732-845-1931, F: 732-583-6619
Sert, S., SERT Consulting, P.O. Box 4140, El Dorado Hills, CA 95762. T: 530-676-9162, F: 530-676-9163, E: sert consulting@yahoo.com
Sevilla-Larrea, J.F., Fondo De Solidaridad, Calle Alcabalas (300) Oe4-200, Quito, ECUADOR. T: 593-2-454-989, F: 593-2-503-100, E: jsevilla@uio.satnet.net
Seymour, J., URS Corporation, 24414 Farmington Rd., Farmington Hls, MI 48336. T: 248-478-8154, F: 313-961-3480
Shackelford, C.D., Colorado State Univ, Dept of Civil Eng, Fort Collins, CO 80523. T: 970-491-5051, F: 970-491-3584, E: shackel@engr.colostate.edu
Shah, H.R., 12 Redwood Mt, Reigate, Surrey, RH2 9NB UK. F: 01737-772767, E: shahh@bv.com
Shallenberger, D.C., Earth Systems Pacific, 288 El Dorado Way, Shell Beach, CA 93449. T: 805-773-0186, F: 805-544-1786

Shannon, W.L., 1630 43rd Ave E., Apt 825, Seattle, WA 98112. T: 206-720-8328
Shao, L., Hayward Baker Inc., 1780 Lemonwood Dr., Santa Paula, CA 93060. T: 805-933-1331, F: 805-933-1338, E: ishao@haywardgaker.com
Sharafkhani, B., Louisiana State of, Deq, P.O. Box 82178, Baton Rouge, LA 70884. T: 225-765-0355, F: 225-765-0617
Sharma, S., Univ of Idaho, Box 441022, Moscow, ID 83844. T: 208-885-6403, F: 208-885-6608, E: ssharma@ uidaho.edu
Sharp, K.D., Ch2m Hill, 777 108th Ave Ne, P.O. Box 91500, Bellevue, WA 98007. T: 206-453-5005, F: 425-468-3100, E: ksharp@ch2m.com
Sharp, M.R., URS Greiner Inc., 12211 Count Pl, Thonotosassa, FL 33592. T: 813-986-5934, F: 813-874-7424, E: tpamrs@dames.com
Shbib, R., Okayama University, 2-1-11tsushima-Higashi, Okayama-Shi, 700-0081 JAPAN. T: 0081-86-255-8454, F: 0081-86-2518580, E: randa@d3.dion.ne.jp
Sheedy, Jr, G.J., 2551 S Seymour Pl, Milwaukee, WI 53227. T: 414-425-4944, F: 414-769-3135
Sheffield, W.J., ATC Associates Inc., 4620 Sims Court, Tucker, GA 30084. T: 770-934-7825, F: 770-427-1907, E: sheffield66@atc-enviro.com
Sheldon, D.R., Rhode Island State of, 2888 Kingstown Road, Kingston, RI 2881. T: 401-783-3492, F: 401-222-2599, E: debdav25@aol.com
Sheleheda, M.J., Almes & Assocs Inc., Four Triangle Lane, Ste 200, Export, PA 15632. T: 724-327-5200, F: 724-327-5280
Shellhammer, J.H., 169 Urick Ln, Monroeville, PA 15146
Shepardson, D.E., 10035 Prospect Ave., Santee, CA 92071. T: 619-449-9830, E: des@shepardson.com
Sherman, Jr, W.C., 4705 Green Acres Ct, Metairie, LA 70003. T: 504-887-5547
Sherman, I., Cons Civil & Geotech Engr, 24030 Killion St, Woodland Hls, CA 91367. T: 818-348-5522, F: 818-348-5522, E: engrirv@earthlink.net
Sherring, J.E., URS Corporation, 4 North Park Dr., Suite 300, Hunt Valley, MD 21030. T: 410-785-7220, F: 410-785-6818, E: james_sherring@urscorp.com
Shervington, R.A., Roger A Shervington Pe Inc., 332 East Radcliffe Dr., Claremont, CA 91711. T: 909-621-1527, F: 909-621-7869, E: kc6lue@hotmail.com
Sherwani, F.K., Overseas Pakistanis Foundation, House No 182-A., Street No 68, Sector F-, Islamabad, SE103 PAKISTAN. T: 051-299684,-280837, F: 051-293768, E: sherwani@isb.comsats.net.pk
Shewbridge, S.E., Scott Shewbridge Assoc Inc., 49 Danville Oak Place, Danville, CA 94526. T: 925-314-3054, F: 925-314-3058, E: scott-shewbridge@ssha.com
Shibata, H., Kokushikan Univ, Dept of Civil Engr 4-28-1, Setagaya-Ku, Tokyo, 154-8515 JAPAN. T: 03-5481-3277, F: 03-5481-3277, E: hshibata@kokushikan.ac.jp
Shibata, T.D., Secor Intl Inc., 1046 Viewpointe Ln, Corona, CA 91719. T: 909-735-5724, F: 909-335-6120, E: tshibata @secor.com
Shie, C.-F., China Jr, College of Ind & Comm Mn, P.O. Box 75-17, Taipei, Taiwan, Taiwan, ROC. T: 011-886-2-2930-4586, E: robert@msl2.hinet.net
Shields, D.R., GEI Cons Inc., 1021 Main St, Winchester, MA 1890. T: 781-721-4032, F: 781-721-4073, E: dshields @geiconsultants.com
Shiflett, M.M., Trinity Engineering, 6103 Waterview Dr., Arlington, TX 76016. T: 817-451-3576, F: 817-429-7869, E: mmshiflett@tetco.net
Shillaber, R.S., Sanborn Head & Associates Inc., 25 Meetinghouse Hill Rd., Deerfield, NH 3037. T: 603-463-7676, F: 603 229 1919
Shimel, C., Charles Shimel Consulting Engs, 118 Main St, Tappan, NY 10983. T: 845-398-2714, F: 845-365-2607
Shimel, S.A., Univ of Washington, 7510 31st Ave Nw, Seattle, WA 98117. T: 206-784-0907, F: 206-633-6777, E: shimel@u.washington.edu
Shin, E.C., Univ of Inchon, College of Eng Civil Eng Dept, 177 Dowha Dong Nam Gu, Incheon, 402 749 NORTH

KOREA. T: 011-8232-770-8466, F: 011-8232-770-84, E: ecshin@lion.inchon.ac.kr

Shipton, B.C., G H Cook Assocs Inc., P.O. Box 240, Dawson Creek, V1G 4G3 CANADA BC. T: 250-782-9275, F: 250-782-9229, E: bshipton@neonet.bc.ca

Shiraishi, S., Tashi Fudosan Co.Ltd., 8 Minami-Motomachi, Shinjuku-Ku, 160-0012 JAPAN. F: 03-3341-6455, E: xmb00215@nifty.ne.jp

Shoop, S.A., USA Crrel, 72 Lyme Rd., Hanover, NH 3755. T: 603-646-4100, F: 603-646-4640, E: sshoop@crrel. usace.army.mil

Shuttle, D.A., Golder Associates Inc., 13859 Ne 65th St, Apartment 580, Redmond, WA 98052. T: 425-881-8071, F: 425-882-5498, E: dshuttle@golder.com

Siddiqi, F.H., Geo-Environmental Inc., 126 Lakepines, Irvine, CA 92620. T: 714-832-3366, F: 714-634-2513

Siedschlag, M.D., Geotechnical Services Inc., 9305 Drexel, Omaha, NE 68127. T: 402-331-7545, F: 520 299 4477, E: ms@gsi.usa.com

Siegel, R.A., Ronald A Siegel Pe, 7109 Porlamar Court Nw, Albuquerque, NM 87120. T: 505-831-3869, F: 505-831-3869, E: raslpe@swcp.com

Siegel, T.C., S & Me Inc., 840 Low Country Blvd, Mt Pleasant, SC 29464. T: 843-884-0005, F: 843-881-6149, E: tsiegel@ smeinc.com

Siller, G.A., Syracuse University, 27 Colonial Ct, Staten Island, NY 10310. T: 718-273-3609, E: gasiller15@ yahoo.com

Silva, F.J., Kleinfelder, 2404 Robles Dr., Antioch, CA 94509. T: 925-754-0953, F: 925-427-6478, E: fsilva@kleinfelder. com

Silva-Pena, J., Ruitoque Condominio, La Bahia Casa 28, Santander, COLOMBIA. T: 6786382, F: 6389410, E: fasilpen@col1.telecom.com.co

Silva-Tulla, F., Arthur D Little Inc., 12 Baskin Rd., Lexington, MA 2421. T: 781-862-1515, F: 617-498-7019, E: silvatulla. f@adlittle.com

Silver, M.L., 3 Ngo Quyen St, Hanoi, VIETNAM. T: 84-4-825-3029, F: 84-4-825-3704, E: msilver@hn.vnn.vn

Silvestri, V., Ecole Polytech Montreal, 380 Outremont, Outremont, H2V 3M2 CANADA QC. T: 279-3080, F: 340-5841

Simac, M.R., Earth Improvement Technologies, P.O. Box 423, Cramerton, NC 28032. T: 704-824-0137, F: 704-824-0151, E: simac.eit@loclnet.com

Simmonds, A.J., Geokon Inc., 48 Spencer St, Lebanon, NH 3766. T: 603-448-1562, F: 603-448-3216, E: tony@ geokon.com

Simon P.E., Richard M., VA Geotech Svcs Pc, 8211 Hermitage Road, Richmond, VA 23228. T: 804-261-5569, F: 804-752-7548, E: rmsimon@vgspc.com

Simon, S.D., Intl Constr Equipment Inc., 301 Warehouse Dr., Matthews, NC 28104. T: 704-821-8200, Ext. 116, F: 704-821-8201, E: ssimon@iceusa.com

Simone, E.M., Paulus Sokolowski & Sartor Inc., 28 Land St, Gillette, NJ 7933. T: 908-647-4122, F: 732-560-9768, E: esimone@psands.com

Simpson, B.E., TRC., 10024 S Gwendelyn Ln, Hghlnds Ranch, CO 80126. T: 303-471-1105, F: 303-792-0122, E: bsimpson@trcsolutions.com

Simpson, J.M., Simpson Bridge Co Inc., 3159 Chatata Valley Rd Ne, Charleston, TN 37310. T: 423-336-2108, F: 423-559-1862, E: jmsimpson@simpsonbridge.com

Simpson, P.T., Pickett T Simpson Pe, 4335 Buckingham Dr., Schenectady, NY 12304. T: 518-393-5150, E: simpsp2 @rpi.edu

Sims, J.M., Earth Systems Pacific, 4378 Santa Fe Rd., San Luis Obispo, CA 93401. T: 805-544-3276

Sinats, J.A., US Navy, 807 Rose Dr., Benicia, CA 94510. T: 707-745-1082, F: 650-244-2553, E: sinatsja@efawest. navfac.navy.mil

Singh, A., Lockwood Singh & Assocs, 5401 Mcconnell Ave, Los Angeles, CA 90066

Singh, S., Santa Clara Univ, Dept of Civil Eng, Santa Clara, CA 95053. T: 408-554-6869, F: 408-554-5474, E: ssingh@ mailer.scu.edu

Singhanet, V., Geosoils Inc., 610 Union St, Encinitas, CA 92024. T: 619-753-0652, F: 714-647-0745, E: geosoilsa@ bigplanet.com

Sirak, M.J., Michael J Sirak, 70 Oakwood Dr., Ste 6, Glastonbury, CT 6033. T: 860-633-5793, E: mjsirak@ aol.com

Sirota, E.B., MFG Inc., 1071 Castlegate Lane, Santa Ana, CA 92705. T: 714-544-4714, F: 949-253-2954, E: ed.sirota@ mfgenv.com

Sitar, N., Univ of Ca-Berkely, Dept of Civil Eng, 440 Davis Hall, Berkeley, CA 94720. T: 510-643-8623, F: 510-642-7476, E: nsitar@ce.berkeley.edu

Sittenfeld, M., Bel Ingenieria S A, Apdo 1034, San Jose, 1000 COSTA RICA. T: 011-506-253-0565, F:0115062245 244/5, E: beling@sol.racsa.co.cr

Sivey, P.M., Sivey Enterprises, 4790 Hayden Crest Dr., Hilliard, OH 43026. T: 614-876-8826, F: 614-850-0232, E: SiveyPM @aol.com

Skaggs, R.L., Condor Earth Technologies Inc., 188 Frank West Circle, Suite 1, Stockton, CA 95206. T: 209-234-0518, F: 209-234-0538, E: rskaggs@condorstockton.com

Skelly, P.E., Michael J., Lawler Matusky & Skelly, One Blue Hill Plaza, Pearl River, NY 10965. T: 845-735-8300, F: 845-735-7466, E: mskelly@lmseng.com

Skouby, M.C., 17150 State Rt B., St James, MO 65559. T: 573-265-0509

Sloan, S.W., Univ of Newcastle, Dept of Civil, Surv, & Env Engr, NSW, 2308 AUSTRALIA. T: 011-61-0249-216040, F: 011-61-0249-216, E: scott.sloan@newcastle.edu.au

Slocombe, B.C., Keller Ground Engineering, Oxford Rd., Ryton-On-Dunsmore, Coventry, CV8 3EG UK. T: 1-2476-511266, F: 1-2476-305230

Slowey, A.D.L., Elsevier/Geo Abstracts, The Old Bakery, 111 Queens Rd., Norwich, NR1 3PL UK. T: 44-1603-626327, F: 44-1603-667934, E: p.tarrant@elsevier.co.uk

Smeallie, P.H., The Geo Council, 600 Woodland Terrace, Alexandria, VA 22302. T: 703-683-1808, F: 703-683-1815, E: geoetmn.com

Smirnoff, T.P., Parsons Brinckerhoff, 7300 Elmsbury Lane, West Hills, CA 91307. T: 818-999-6411, F: 213-362-9480, E: smirnoff@pbworld.com

Smith, Jr, R.E., Law Eng Inc., 1900 Starbrook Dr., Charlotte, NC 28210. T: 704-552-2828, F: 704-357-8624, E: rsmith@ lawco.com

Smith, P.E., Elizabeth M., Kleinfelder Inc., 5015 Shoreham Place, San Diego, CA 92122. T: 858-320-2000, F: 858-320-2001, E: lsmith@kleinfelder.com

Smith, A.C., Stone & Webster Eng Corp, 196 Summer St, Arlington, MA 2474. E: smith196@aol. com

Smith, B.A., Stearns & Wheler, 1161 Skyhigh Rd., Tully, NY 13159. T: 315-696-8608, F: 315-655-4180, E: beth.ann. smith@stearnswheler.com

Smith, B.L., Stearns & Wheler, One Remington Park Dr., Cazenovia, NY 13035. T: 315-655-8161, F: 315-655-4180, E: brad.smith@stearnswheler.com

Smith, B.E., Froehling And Robertson, 16948 Frederick Road, Mount Airy, MD 21771. T: 301-854-5141, F: 410-947-6503, E: bsmith@fanor.com

Smith, D.L., Trumbull Corporation, 121 Audbert Drive, Pittsburgh, PA 15236. T: 412-882-1374

Smith, D.H., Smith Geotechnical, 1225 Red Cedar Cir Ste H., Fort Collins, CO 80524. T: 970-490-2620, F: 970-490-2851, E: dsmith@smithgeotech.com

Smith, G.T., 61 Cunningham Dr., Lutterworth, Leics, LE17 4YR UK. F: 01455 559828, E: terencesmith2@compu serve.com

Smith, G.M., Mactec-Ers, 2597b 3/4ths Rd., Grand Junction, CO 81503. T: 970-248-6394, F: 970-248-6040, E: gsmith@ doegjpo.com

Smith, J.G., Wisconsin Testing Lab, 1426 Mayfield Rd., Hubertus, WI 53033. T: 262-628-1167, F: 262-252-5373, E: jekama@nconnect.net

Smith, M.W., Geo Engineers, 31727 Ne 114th St, Carnation, WA 98014. T: 206-882-8266, F: 425-861-6050, E: msmith@ geoengineers.com

Smith, M.G., URS Greiner Woodward-Clyde, 1038 S Cascade Lane, Anaheim, CA 92808

Smith, R.W., CDM Jessberger, 11811 Ne First St, Suite 201, Bellevue, WA 98005. T: 425-453-8383, F: 425-646-9523, E: smithrw@cdm.com

Smith, R.E., Smith Geoconsultants, 1605 Benchley, Henderson, NV 89052. T: 702-228-2698, E: resmith113@aol.com

Smith, S.L., 165 N 400 W., Lindon, UT 84042. T: 801-225-5711, F: 801-225-3363, E: slsmith@earthtectesting.com

Smith, T.V., Qore Property Sciences, Inc., 3922 Cedar Cove Lane, Jacksonville, FL 32257. T: 904-268-7726, F: 904-262-9996

Smith, T.D., Portland State Univ, Dept of Civil Engineering, P.O. Box 751, Portland, OR 97207. T: 503-725-3225, F: 503-725-5950, E: trevor@eas.pdx.edu

Smith, W.D., URS Greiner Woodward Clyde, 12321 West 101st Street, Lenexa, KS 66215. T: 913-888-3162, F: 913-344-1011, E: wayne_smith@urscorp.com

Snedegar, L.W., Le Gregg Associates Inc., 2909 Middlesex Ct, Lexington, KY 40503. T: 859-277-9601, F: 859-255-0940, E: legregg@lex.infi.net

Snethen, D.R., Oklahoma State Univ, School of Civil Eng, Es 207, Stillwater, OK 74078. T: 405-744-6328, F: 405-744-7554

Snow, P.E., Christopher L., Gza Geo Environmental, 4 Free St, Portland, ME 4101. T: 207-879-9190

Snow, M.S., Golder Associates, 24892 Avenida Avalon, Laguna Hills, CA 92653. T: 949-458-6503, F: 949-583-2770, E: msnow@golder.com

Snyder, D.B., RBS Inc., P.O. Box 490, White Sulpher Spring, WV 24986. T: 304-645-2277, F: 304-645-2597, E: dbs@inetone.net

Snyder, G.E., TW Consultants Inc., Box 262, Timblin, PA 15778. T: 814-256-3545

Snyder, K.P., Kumar & Associates Inc., 2390 S Lipan Street, Denver, CO 80223. T: 303-742-9700, F: 303-742-9666, E: eksnyder@worldnet.att.net

Sohn, K.C., Soil Foundation Systems Inc., 749 Edge Ln, Los Altos, CA 94024

Solanki, H.T., Kimley Horn & Assocs Inc., 3012 Bucida Dr., Sarasota, FL 34232. T: 941-371-6191, F: 941-922-2351, E: hsolanki@kimley-horn.com

Soliman, N.N., Parsons Brinckerhoff Inc., 25 Whitman Ave, Metuchen, NJ 8840. T: 732-494-3320, F: 609-734-6949

Soltis, P.C., Geosyntec Consultants, 8707 Walker St Unit 10, Cypress, CA 90630. T: 714-226-9105, F: 714-969-0820, E: psoltis@geosyntec.com

Sommerfeld, G.D., Black & Veatch, 301 Ne Chelmsford Court, Lees Summit, MO 64064. T: 816-478-9648, F: 913-458-2934, E: sommerfeldgd@bv.com

Sordo, R., Univ Autonoma Metropolitana, Av Parque Lira 75, Col San Miguel Chapultepec, Mexico Df, 11850 MEXICO. T: 525-318-9082, F: 525-382-3998, E: esz@correo.azc.uam.mx

Sosnowski, R.S., Malcolm Pirnie Inc., 3860 Marcy St, Mohegan Lake, NY 10547. T: 914-526-8544, F: 914-641-2645, E: rsosnowski@pirnie.com

Soto, A.A., Black & Veatch, Mco 140 P.O. Box 25233, Miami, FL 33102. T: 913-458-4202, E: alfonsoto@yahoo.com

Souther, J.T., HKM Engineering Inc., 222 N. 32nd Street, Suite 700, P.O. Box 31318, Billings, MT 59107. T: 406-656-6399, F: 406-656-6398, E: jsouther@hkminc.com

Souto, F.L.G., Forte Inc., 29720 Medbury Ave, Farmington Hills, MI 48336. T: 810-473-2527, F: 810-473-6972, E: forte79295@aol.com

Spang, A.W., 1635 Nw 138th Ave, Portland, OR 97229. T: 503-626-9889, F: 503-626-8611, E: wspangi@home.com

Spears, J.F., Spears Engineering, P.O. Box 1007, Auburn, WA 98071. T: 206-833-7967, F: 206-735-2867, E: spearseng@earthlink.net

Speedie, J.A., Speedie & Assocs, 3331 E Wood St, Phoenix, AZ 85040. T: 602-997-6391, F: 602-943-5508, E: jspeedie@speedie.net

Spikula, D.R., Energy Developments Inc., 7007 Cherry Hills Rd., Houston, TX 77069. T: 281-397-6739, E: daniel.spikula@energydi.com

Sprague, C.J., Sprague & Sprague Consulting, Engineers, P.O. Box 9192, Greenville, SC 29604. T: 864-242-3106, F: 864-242-3107, E: cjoelsprague@cs.com

Squier, L.R., 3943 Tempest Dr., Lake Oswego, OR 97035. T: 503-636-0028, E: rads@connw.net

Sraders, G.A., Great Lakes Dredge & Dock Co, 2122 York Rd., Oakbrook, IL 60523. T: 630-574-2969, F: 630-574-2980, E: gasraders@gldd.com

Stadler, A.T., Univ of Nc Charlotte, 9014 Coleshire Ct, Charlotte, NC 28269. T: 704-598-3040, F: 704-687-6953, E: stadler@uncc.edu

Staggs, D.G., 106 Interpromontory Rd., Great Falls, VA 22066. T: 703-759-9244, F: 703-759-6646, E: davanst@aol.com

Stanley, M.H., Subsurface Consultants Inc., 155 Kearney Ave, Sonoma, CA 95476. T: 707-935-7799, F: 510-299-7970, E: markstanley@sprynet.com

Starich, M.J., Paradigm Consultants Inc., 4125 Southwestern, Houston, TX 77005. T: 713-668-0619, F: 713-686-6795, E: marie@thelabtexas.com

Stauffer, P.A., Michael W West & Assoc Inc., 5988 S Eudora Ct, Littleton, CO 80121. T: 303-804-0609, F: 720-529-5335, E: petestauffer@yahoo.com

Steele, J.A., Steele Foundations Inc., 3299 "K" St Nw, Suite 601, Washington, DC 20007. T: 202-342-1194, E: mjw steele@aol.com

Steele, R.W., Steele Foundations Inc., 3299 K St Nw, Suite 601, Washington, DC 20007. T: 202-342-1194

Stegman, B.G., Stegman Engineering, 24 Coupler Dr., Stewartstown, PA 17363. T: 717-993-6559, F: 717-993-0087, E: stegman@nfdc.net

Steiger, A.K., Andreas Steiger & Partner, St Karlistr 12 Pf 7829, Ch-6000 Luzern 7, SWITZERLAND. T: 011-041-248-51-71, F: 011-041-248-51-, E: andreas.steiger.partner@swissonline.ch

Stein, J.E., John Wiley & Sons Inc., 605 Third Avenue, New York, NY 10158. T: 212-850-6541, F: 212-850-6103, E: jstein@wiley.com

Steinberg, P.E., Malcolm L., Steinberg & Associates, 1201 Prospect St, El Paso, TX 79902. T: 915-544-8659

Stella, C.G., Interbox Sa Marinas Division, Via Mazza 35, Gavirate, Varese, 21026 ITALY. T: 39 0332 746747, F: -9942310, E: martini@energy.it

Stemple, K.D., Triad Eng Inc., 2200 Universe Drive, Martinsburg, WV 25401. T: 540-678-0232, F: 540-667-2260, E: kevins@triad-winc.com

Stephens, M.H., Engineering Consulting Service, 6909 International Dr., Ste 103, Greensboro, NC 27409. T: 336-856-7150 3009, F: 336-856-7160, E: mstephens@ecslimited.com

Stephens, R.S., Stephens Associates, 147 County Road, Tewksbury, MA 1876. T: 978-988-2115, F: 978-988-2117, E: rsstephens@earthlink.net

Stephenson, R.W., Univ of Missouri, Civil Engineering, Rolla, MO 65401. T: 314-341-4458, F: 314-341-4729

Stephenson, R.J., 399 Seminole Dr., NE, Marietta, GA 30060. T: 770-428-5568

Stettler, D.R., Landau Assocs, 7917 Cyrus Place, Edmonds, WA 98026. F: 425-778-6409, E: dstettler@landauinc.com

Stevens, J., Norwich University, 122 Smith Hill Rd., Northfield, VT 5663. F: 802-485-2260, E: stevens@norwich.edu

Stevens, P.D., Stevens Ferrone & Bailey, 1470 Enea Cir, Ste 1551, Concord, CA 94520. T: 925-688-1001

Stevens, R.F., Fugro Mc Clelland, 8843 Chelsworth Dr., Houston, TX 77083. T: 281-879-4342, F: 7137785573, E: bstevens@fugro.com

Stevens, W.R., Consolidated Engineering Labs, 4289 Colgate Way, Livermore, CA 94550. T: 925-454-0177, F: 925-485-5018, E: ste@celhq.com

Stewart, H.E., Cornell Univ, School of Civil & Envir Eng, Hollister Hall, Ithaca, NY 14853. T: 607-255-4734, F: 607-255-9004, E: hes1@cornell.edu

Stewart, J.P., John Stopen Engrs Pc, 206 Scott Ave, Syracuse, NY 13224. T: 315-446-4638, F: 315-472-8430, E: jstewar3 @twcny.rr.com

Stewart, J.P., Univ of Ca Los Angeles, 5731-H Boelter Hall, Los Angeles, CA 90095. T: 310-206-2990, F: 310-206-2222, E: jstewart@seas.ucla.edu

Stewart, W.L., URS Greiner Woodward Clyde, 230 Lafayette Ave, Cortlandt Manor, NY 10567. T: 914-739-9232, F: 973-785-0023, E: winstonstewart@urscorp.com

Stieben, G.P., Fugro South, 26014 Apache Creek, San Antonio, TX 78258. T: 210-497-7362, F: 210-655-9519

Stier, R.R., Affholder Inc., 17988 Edison Ave, Chesterfield, MO 63005. T: 636-532-2622, F: 636-537-2533, E: rstier@ insituform.com

Stilson, M.K., Rb&G Engineering Inc., 210 N Mall Dr., #48, St George, UT 84790. T: 435-656-2783, F: 801-374-5773

Stimac, T.J., Stimac & Associates, 912 Tulare Ave, Albany, CA 94707. T: 510-528-2230, F: 510-527-930, E: tstimacgeo@aol.com

Stobbe, M.W., 706 Fairmount Ave, Santa Cruz, CA 95062. T: 831-426-0446

Stoffel, M.A., Grubbs Hoskyn Barton & Wyatt, Inc., P.O. Box 55105, Little Rock, AR 72215. T: 501-455-2536, F: 501-455-4137, E: rreed1152@aol.com

Stone Jr, R.C., Western Technologies Inc., 3611 W Tompkins Ave, Las Vegas, NV 89103. T: 702-798-8050, F: 702-798-7664, E: wt-lv@primenet.com

Stone Sheridan, A.M., Houston City of, 17703 Quiet Dawn, Houston, TX 77095. T: 281-861-7666, F: 713-837-0464, E: annmarie.sheridan@cityofhouston.net

Stone, J.E., Geodesign Inc., 14945 Sw Sequoia Pkwy, Suite 170, Portland, OR 97224. T: 503-968-8787, F: 503-968-3068, E: stonzrwe@teleport.com

Stormont, J.C., Univ of New Mexico, Dept of Civil Engrg, Albuquerque, NM 87131. T: 505-277-6063, F: 505-277-1988

Strachan, C.L., Shepherd Miller Inc., 3801 Automation Way Ste 100, Fort Collins, CO 80525. T: 970-223-9600, F: 970-223-7171, E: cstrachan@shep mill.com

Strachan, G.C., Squier Assocs Inc., 7981 168th Ave Ne, Redmond, WA 98052. T: 425-702-6616, F: 425-881-6616, E: squihgi1@gte.net

Straley, C.F., GAI Cons Inc., 2243 Second St, Culloden, WV 25510. T: 304-743-5340, F: 304 926 8180, E: c_straley@ vakron.alumlink.com

Strange, A.C., DLZ Corporation, 266 Greenglade Ave, Worthington, OH 43085. T: 614-885-0886, F: 614-431-3864, E: astrange@columbus.rr.com

Strauss, T.V., Terracon, 14730 Weir St, Omaha, NE 68137. T: 402-641-5967, F: 402-330-7606, E: ts64436@alltel.net

Strazer, R.J., Applied Geotech Inc., 541 Ne 20th Ave-Suite 103, Portland, OR 97232. T: 503-232-1800, F: 503-232-9272, E: strazerrj@cdm.com

Stringfellow, Jr, H.W., P.O. Box 20334, Jackson, MS 39289. T: 601-922-6673

Sture, S., Univ of Colorado, Dept of Civ Envir & Archt Eng, Campus Box 428, Boulder, CO 80309. T: 303-492-7651, F: 303-492-7317, E: sture@bechtel.colorado.edu

Styler, A.N., A G E S Inc., 108 Cedar Brook Court, Mcmurray, PA 15317. T: 724-942-0654, F: 412-221-6695

Styron, J.P., Fugro-Mcclelland (Southwest), 3902 Perrin Central Blvd, Apt 1010, San Antonio, TX 78217. T: 210-646-0323, F: 210-655-9519

Sud, V.K., B2 Thapar, Inst of Eng & Tech, Patalia Pu, 147001 INDIA

Sudduth, C.G., 1851 Boca Ave, Los Angeles, CA 90032. F: 323-223-9348, E: csudduth@earthlink.net

Suen, W.Y., Mass Transit Railway Corp, Flat A 35/F Twr 1 Tierra Verde, 33 Tsing King Road, Tsing Yi, HONG KONG. T: 852-25672121, E: wys@mtrcorp.com.hk

Suhan, C.J., Testing Engrs & Cons Inc., 1333 Rochester Rd., Troy, MI 48083. T: 248-588-6200, F: 248-588-6232, E: csuhan@tectest.com

Suits, L.D., 5 Willoughby Dr., Albany, NY 12205. T: 518-869-9436, E: ldsuits@aol.com

Sullivan, P.E., Thomas P., Raamot Assocs, 154 Pine Bush Road, Stone Ridge, NY 12484. T: 914-687-7741, F: 212 279 8180

Sullivan, J.P., The Geotechnical Group Inc., 100 Crescent Road, Needham, MA 2494. T: 781-449-6450, F: 781-449-1283, E: engineers@tgge.com

Sullivan, W.J., Geoscience Group Inc., 500 Clanton Rd., Ste K., Charlotte, NC 28217. T: 704-525-2003, F: 704-525-2051, E: wsullivan@geosciencegroup.com

Sullivan, W.R., Golder Assocs Inc., 225 Lochan Cove, Alpharetta, GA 30022. F: 770-934-9476, E: rsullivan@ golder.com

Sum, E., EPC Consultants Inc., P.O. Box 149, San Mateo, CA 94401. T: 650-571-9393, F: 415-675-7586, E: edsum@ cheerful.com

Sun, M.C., Facility Engineering Assoc Pc, 11001 Lee Hwy, Suite D., Fairfax, VA 22030. T: 703-591-4855, F: 703-591-4857, E: sun@feapc.com

Sundaram, A.V., Harza Eng Co, 136 E Russet Way, Palatine, IL 60067. T: 847-359-5857, F: 312-831-3999, E: asundaram @harza.com

Sung, D.Y., MSI Inc., Cpo Box 3699, Seoul, 100-636 SOUTH KOREA. F: 02-237-0081

Surendra, M., Earth Engineering & Sciences, 7481 Hickory Log Cir, Columbia, MD 21045. T: 410-381-3757, F: 410-466-7371, E: suriss@hotmail.com

Suroor, A.H., Earth Engineering, P.O. Box 924112, Houston TX 77292. T: 713-460-2155, E: suroor@work mail.com

Surti, N.M., NC Dept of Transp, 100 East Spring Hollow Ln, Cary, NC 27511. T: 919-859-2433, F: 919-250-4119, E: nsurti@dot.state.nc.us

Surya, I., Pt Nasuma Putra, Jalan Cihapit 11, Bandung 40114, INDONESIA. T: 22-4238719, F: 22-4238719, E: isurya@maranatha.edu

Sussmann, Jr, T.R., US Dot Volpe Center, Dts 76, 55 Broadway Kendall Sq, Cambridge, MA 2142. T: 617-494-3663, F: 617-494-3616, E: sussmannt@volpe.dot.gov

Suter, K.E., PSI Inc., 705 Pleasant Way, Chesapeake, VA 23322. T: 757-546-1581, F: 757-546-1581, E: ksuterpsi@ aol.com

Sutherland, E.D., Soil & Materials Engrs, 2663 Eaton Rapids Rd., Lansing, MI 48911. T: 517-887-9181, F: 517-887-2666, E: sutherla@lansing.soilmat.com

Sutterer, K.G., Rose Hulman Inst of Tech, 38 Ferndale Dr., Terre Haute, IN 47803. T: 812-877-4042, F: 812-877-8440, E: kevin.g.sutterer@rose-hulman.edu

Sutton, J.R., The Sutton Group, 51 Shuey Dr., Moraga, CA 94556. T: 510-376-5255, F: 510-631-1371, E: jrsgroup@ ix.netcom.com

Suzuki, M., Shimizu Corp Izumi Resrch Inst, Fukoku-Seimei Bldg 27f, 222 Uchisaiwai-Cho Chiyoda-Ku, Tokyo, 100-0011 JAPAN. T: 03-3508-8101, F: 03-3508-2196, E: suzuki@ori.shimz.co.jp

Svinkin, M.R., 13821 Cedar Rd., Apt 205, Cleveland, OH 44118. T: 216-397-9625, F: 216-397-1175

Swaffar, K., Nth Cons Ltd., 277 Gratiot Ave, Suite 600, Detroit, MI 48226. T: 313-965-0036, F: 313-965-0683, E: kswaffar@nthconsultants.com

Swaisgood, J.R., Swaisgood Consulting, P.O. Box 1083, Conifer, CO 80433. T: 303-838-8774, E: j.swaisgood @att.net

Swallow, M.A., Golder Assocs Inc., 8933 Western Way Ste 12, Jacksonville, FL 32256. T: 904-363-3430, F: 904-363-3445, E: mswallow@golder.com

Swan, C.W., Tufts University, Dept of Civil and Envir Eng, Anderson Hall, Medford, MA 2155. T: 617-627-2212, F: 617-627-3994, E: cswan@tufts.edu

Swanson, E.A., Marquess & Assocs, P.O. Box 490, Medford, OR 97501. T: 541-772-7115, E: rs@marquess.com

Swanson, W.R., CDM, 84 Trenton St, Melrose, MA 2176. T: 617-665-2106, F: 617-452-8274, E: swansonwr@ cdm.com

Swantko, T.D., Group Delta Consultants Inc., 2341 W 205th Street Suite 103, Torrance, CA 90501. T: 310-320-5100, F: 310-320-2118, E: toms@groupdelt.com

Sweeney, S.E., New York State Thruway Auth, 88 Gipp Rd., Albany, NY 12203. T: 518-456-7385, F: 518-471-5034, E: steve_sweeney@thruway.state.ny.us

Swekosky, F.J., 717 A Old Barn Rd., Lake Barrington, IL 60010. T: 847-381-9815

Swift, P.E., John Raymond, Golder Assocs Inc., P.O. Box 142, Carnation, WA 98014. T: 425-333-4731, F: 425-882-5498, E: hswift@golder.com

Sydlik, M., Earth Inc., 8150 Perry Hwy, Suite 102, Pittsburgh, PA 15237. T: 412-366-8005, F: 412-366-7320

Sykora, D.W., Bing Yen & Assoc Inc., 17701 Mitchell N., Irvine, CA 92614. T: 949-757-1941, F: 949-757-1943, E: sykora@bingyen.com

Szymoniak, T.S., Oregon Department of Transportation, 60594 Springtree Court, Bend, OR 97702. T: 541-312-8365, F: 541-312-2231, E: tszymoniak@bendnet.com

Tabba, M.M., TTI-CEO-President, P.O. Box 746, Riyadh, 11323 SAUDI ARABIA. T: 96614609514, F: 96614602135, E: mmtabba@awalnet.net.sa

Taber, H.R., Taber Consultants, 3911 W Capitol Ave, West Sacramento, CA 95691. T: 916-371-1690

Tai, T.L., Tai & Assocs, P.O. Box 52073, Raleigh, NC 27612. T: 919-782-9525, F: 919-782-9540, E: dtai168@aol.com

Talbot, J.R., GEI Consultants Inc., 1301 Wagner Street, Saint Leonard, MD 20685. T: 410-586-8772, F: 301-494-3824, E: jtalbot@mindspring.com

Tallard, G.R., Liquid Earth Support Inc., 128 Corlies Ave, Pelham, NY 10803. T: 914-738-4880, F: 914-738-4804, E: impermix@aol.com

Tam, K.T., 3571 Kilby Ct, Richmond, V6X 3M9 CANADA BC. T: 604-303-0963, F: 604-303-0960

Tamaro Jr, G., Mueser Rutledge Cons Engrs, 708 Third Ave, New York, NY 10017. T: 212-490-7110, F: 212-490-6654, E: tamaro@mrce.com

Tamr, R.A., Levine Fricke Inc., 1007 Bayside Dr., Palatine, IL 60067. T: 847-506-1619, F: 847-695-7799, E: radwan. tamr@lfr.com

Tan, C.-K., Langan Eng & Env Servs Inc., 15 Shawnee Trail, Danville, NJ 7834. T: 973-575-6557, F: 201-794-7501, E: ctan@langan.com

Tan, C.G., Land Transport Authority, Blk 109 Pasir Ris St 11, #04-595 Singapore 510109, SINGAPORE. T: 065-584-7509, F: 065-3342092, E: chin_gee_tan@lta.gov.sg

Tan, K.P., Ascendas Land International Pte Ltd., 3 Rivervale Link 04-26, The Rivervale, 545119 SINGAPORE. T: 65-3121560, F: 65-5647458, E: kingpheow@ascendas.com

Tan, S.A., National Univ of Singapore, Dept of Civil Eng, Kent Ridge 0511, SINGAPORE. T: 65-772-2278, F: 65-779-1635, E: cvetansa@nus.edu.sg

Tan, T.-L., Kedah Kilang Papan, 15 Taman Purnama, Jalan Purnama, Alor Setar, Malaysia, 5050 MALAYSIA. T: 04-771-0411, F: 04-772-8388, E: tloong@pc.jaring.my

Tanaka, S., The Constr Engr Rsrch Inst, 11-42 Kamitsutsui-Dori, Chuoh-Ku, Kobe, JAPAN. T: 011-81-78-221-5722, F: 011-78-851-5454, E: kensetsu@pearl.ocn.ne.jp

Tanal, V., Parsons Brinckerhoff Quade &, Douglas, One Penn Plaza, New York, NY 10119. T: 212-465-5208, F: 212-465-5753

Tanenbaum, R.J., Amec Earth & Enviromental Inc., 16760 W Bernardo Dr., San Diego, CA 92127. T: 858-487-2113, F: 858-487-2357, E: ronald.tanenbaum@amec. com

Tang, T.-S., Eng & Envir Cons Sdn Bhd, 5 Jalan 17/62, 46400 Petaling Jaya, Selangor, MALAYSIA. F: 603-717 3164, E: tinseng@pop4.jaring.my

Tang, W.H.C., Hong Kong Uni of Sci & Tech, Ce Dept, Clear Water Bay Kowloon, HONG KONG. T: 011852-23587152, F: 011852-23355493, E: wtang@ust.hk

Tantala, A.M., Tantala Associates, 4903 Frankford Ave, Philadelphia, PA 19124. T: 215-289-4600, F: 215-288-1885

Tape, R.T., Esse Nova Engineering, 8147 Comox Rd., Blaine, WA 98230. T: 360-371-0557, F: 360-371-2673

Tapia-Illanes, J.C., Geosonda Ltda, Av General Bustamante 32, 6640755 Providencia Santiago, CHILE. T: 56-2-2094830/2510264, F: 2510264

Tappendorf, D.E., Midwest Engineering, 501 Mercury Ave, Champaign, IL 61822. T: 217-359-2128, F: 217-359-8446, E: dtappendorf@midwesteng.com

Tart, Jr, R.G., Golder Assocs Inc., 18030 Cambridge Drive, Arlington, WA 98223. T: 360-403-8743, F: 907-344-6011, E: btart@golder.com

Tarvin, P.A , STS Cons Ltd., 818 Century Court, Slinger, WI 53086. T: 262-644-7084, F: 414-359-0822, E: tarvin@stsltd.com

Tate, D.E., Ogden Environmental, 3735 Hillbrook Ct, Nashville, TN 37211. T: 615-781-8322, F: 615-331-4715, E: detate@oees.com

Taubert, D.H., Capitol Cement, P.O. Box 33240, San Antonio, TX 78265. T: 210-655-3010, F: 210-599-1709

Tavassoli, M., P.S.I., 724 Central Ave, Jefferson, LA 70121. T: 504-733-9411, F: 504-733-9415, E: mohammad.tavassoli @psiusa.com

Tay, K.-A., Terracon, 1436 Glynview Cir, Lawrenceville, GA 30043. T: 770-513-2324, F: 770-623-9628, E: ektay@terracon.com

Taylor, Jr, D.L., AG Wassenaar Inc., 2180 S Ivanhoe, Suite 5, Denver, CO 80222. T: 303-759-8100

Taylor, G.E., Hayward Baker Inc., 1780 Lemonwood Dr., Santa Paula, CA 93060. T: 805-933-1331, F: 805-933-1338, E: getaylor@haywardbaker.com

Taylor, J.E., Westpoint Stevens Inc., 1500 Pinehaven Ct, Opelika, AL 36801. T: 205-749-5650

Taylor, L.R., Taylor-Hunter Associates Inc., 818 Civic Center Drive, Oceanside, CA 92054. T: 760-721-9990, F: 760-721-9991

Taylor, M.A., The Geotechnical Group, 100 Crescent Rd., Needham, MA 2494. T: 781-449-6450, F: 781-449-1283, E: mtaylor@tggonline.com

Taylor, T.H., Rocksol Cons Group Inc., 2525 Arapahoe Ave, Apt E4, Boulder, CO 80302. T: 303-786-1413, E: taylor@rocksol.com

Taylor, T.R.B., 151 D Raymond Bldg, University of Kentucky, Lexington, KY 40506. F: 859-257-4404, E: tbtaylo@pop.uky.edu

Teale, C.E., Hte Northeast Inc., 50 Grafton Drive, Bedford, NH 3110. T: 603-668-1654, F: 608-668-0608, E: htenh@xtdl.com

Teixeira, A.H., Ah Teixeira & Partners, R Managua 82 Cid Ade Jardim, Sao Paulo – Sp, 5601050 BRAZIL. F: 11-3813-3984, E: aht@originet.com.br

Tejidor, F.J., 10440 Leslie Dr., Raleigh, NC 27615. T: 919-848-2014, F: 919-876-6840, E: fjtejidor@pbsj.com

Tepper, T.J., Dunkelberger Eng & Testing Inc., 4818 Sea Oats Circle #201, West Palm Bch, FL 33417. T: 561-478-3041

Terashi, M., Nikken Sekkei, 4-11-1 Minami-Kase, Saiwai-Ku, Kawasaki, 212-0055 JAPAN. T: 011-81-44-599-1151, F: 011-81-44-599-9, E: terashi@nikken.co.jp

Terrill, L.J., Geotechnika Inc., 562 Christine Ct, Loveland, CO 80537. T: 970-667-6748, E: terrill@earthlink.net

Terry, M.W., Hayward Baker, 5178 Shotwell St, Woodstock, GA 30188. F: 813 884 3820

Terry, T., GEI Consultants Inc., P.O. Box 101416, Denver, CO 80250. F: 303-662-8757, E: tterry@geiconsultants. com

Ter-Stepanian, G., 4 Sanford Dr., Shelton, CT 6484. T: 203-922-8165, F: 203-922-8142, E: gterstepanian@hotmail.com

Thiele, D.J., Thiele Geotech Inc., 13478 Chandler Rd., Omaha, NE 68138. T: 402-556-2171, F: 4025567831, E: dthiele@thielegeotech.com

Thielen, D.L., Landau Associates, 5285 Sw Meadows Rd., Suite 315, Lake Oswego, OR 97035. T: 503-443-6010, E: d.thielen@landauinc.com

Thien, P., Public Works Dept, Structural Section, Kota Kinabalu Sabah, 88582 MALAYSIA. F: 60 3 7179637, E: pthien@pc.jaring.my

Thom, J.H., HKM Engineering, P.O. Box 31318, Billings, MT 59107. T: 406-656-6399, F: 406-656-6398, E: jthom@hkminc.com

Thomann, T.G., URS Corporation, 522 Watchung Ave, Bloomfield, NJ 7003. T: 201-893-0633, F: 973-785-0023, E: thomas-thomann@urscorp.com

Thomas, D.B., CTL Thompson Inc., 12916 W Arlington Pl, Littleton, CO 80127. T: 303-979-3148, F: 303-825-4252

Thomas, D.L., Terracon Inc., 6614 W 175th St, Tinley Park, IL 60477. T: 708-802-9808, F: 708-478-8867, E: dlthomas @terracon.com

Thomas, G.E., Geotesting Services Inc., 45 J Commerce Way, Totowa, NJ 7512. T: 973-812-1818, F: 973-812-8640, E: greg-thomas@urscorp.com

Thomas, H.P., Geoengineers Inc., 2611 Brittany Dr., Anchorage, AK 99504. T: 907-333-9101, F: 907-561-5123, E: hthomas @geoengineers.com

Thomas, M., Eee Consulting Inc., P.O. Box 354, Montpelier, VA 23192. T: 804-883-0016, F: 804-883-0018, E: eeethomas@aol.com

Thomas, P.C., Chicago City of, Dept of The Environment, 30 N Lasalle St 25th Flr, Chicago, IL 60602. T: 312-744-4018, E: pthomas@ci.chi.il.us

Thomas, R.R., Malcolm Drilling Co Inc., 317 Estrella Way, San Mateo, CA 94403. T: 650-349-3409, F: 650-952-5542, E: Constengineer@aol.com

Thome, D.A., 117 S Wisner, Jackson, MI 49203. T: 517-788-7332, F: 412-221-3127, E: dthome@nicholson. rodio.com

Thompson III, James B., Zipper Zeman Assoc, 15211 152nd Ave Ne, Woodinville, WA 98072. T: 425-485-4856

Thompson III, William Robert, TTL., Inc., 4154 Lomac St, Montgomery, AL 36106. T: 334-244-0766, F: 334-244-6668, E: rthompson@ttlinc.com

Thompson, P.E., Robert W., CTL Thompson Inc., 1971 W 12th Ave, Denver, CO 80204. T: 303-825-0777, F: 303-825-4252, E: denver@ctlt.com

Thompson, C.N., Bernalillo County Public Works, 7540 Deerfield Rd Nw, Albuquerque, NM 87120. T: 505-899-0417, F: 505-848-1510, E: chuckthompson@ juno.com

Thomson, Jr, T.A., Duffield Associates Inc., 4728 Weatherhill Drive, Wilmington, DE 19808. T: 302-235-0430, F: 302-239-8485, E: tthomson@duffnet.com

Thunberg, C.W., Nobis Engineering, Inc., 9 Hampshire Avenue, Rochester, NH 3867. T: 603-335-3495, F: 603-224-2507

Tice, J.A., Law Eng Inc., 11821 Edgewater Court, Raleigh, NC 27614. T: 919-847-0022, F: 919-831-8137, E: atice@ lawco.com

Tijmann, W.B., WB Tijmann Cons, WB Tijmann Cons, P.O. Box 2367, Woburn, MA 1888. T: 617-438-3889, F: 617-438-9414, E: tijmann_70@yahoo.com

Tillis, R.K., Hultgren-Tillis Engineers, 2520 Stanwell Dr., Ste 100, Concord, CA 94520. T: 925-685-6300, E: tillis@ hultgrentillis.com

Timmerman, D.H., Timmerman Geotechnical Group, 473 Spaulding Dr., Kent, OH 44240. F: 330-434-3496, E: dtimgeo@aol.com

Tinjum, J.M., RMT inc., 744 Heartland Trail, Madison, WI 53717. T: 608-831-4444, F: 608-831-3334, E: jim.tinjum@ rmtinc.com

Tinucci, J.P., Pantechnica Corp, 9187 Victoria Dr., Eden Prairie, MN 55347. T: 952-937-5879, F: 612-632-4824, E: jtinucci@pantechnica.com

Tirolo, Jr, V., Slattery Skanska Inc., 435 Fifth St, Brooklyn, NY 11215. T: 718-768-3683, F: 718-767-2668, E: vincetun@ aol.com

Titus, Jr, L.J., ECS Ltd., 44240 Ginghamsburg Pl, Ashburn, VA 20147. T: 703-724-1798, F: 703-834-5527, E: ltitus@ ecslimited.com

Tjandra, S., Pt Engitama Nusa Geotestindo, Jl Tebet Barat Iv No 33, Jakarta 12810, INDONESIA. T: 62-21-8301646, F: 62-21-8290163, E: pteng@indo.net.id

Tobin, R.F., GEI Cons Inc., 1021 Main St, Winchester, MA 1890. T: 781-721-4000, F: 781-721-4073, E: rtobin@ geiconsultants.com

Tobin, T.A., Geoengineers Inc., 1803 203rd Ave Se, Sammamish, WA 98075. T: 425-392-3027, F: 425-861-6050

Tokashiki, R.S., La Dept of Water & Power, 2009 Manzanita Ln, Manhattan Bch, CA 90266. T: 310-796-1168, F: 213-367-6792

Tokimatsu, K., Tokyo Inst of Tech, O-Okayama Meguro-Ku, Tokyo, 152-8552 JAPAN. T: 03-5734-3160, F: 03-5734-2925, E: kohji@o.cc.titech.ac.jp

Tolliver, P.L., Morris & Ritchie Assoc Inc., P.O. Box 1661, Bel Air, MD 21014. T: 410-569-4084, F: 410-879-1820

Tom, E.C.-K., Washington Group International, 44152 Glendora Drive, Fremont, CA 94539. T: 510-656-5743, F: 415-442-7673, E: edward.tom@wgint.com

Tong, M.M.S., Engeo Inc., 2401 Crow Canyon Road, Suite 200, San Ramon, CA 94583. T: 925-838-1600, F: 925-838-7425, E: mmt@engeo.com

Torres, R.A., California Department of, 1844 Merced Way, W Sacramento, CA 95691. T: 916-372-2617, E: rtorres@ water.ca.gov

Torrey, V.H., 410 Confederate Lane, Windsor, NC 27983. T: 252-794-5855, E: torreyv@coastalnet.com

Toth, P., Ontario Power Generation, 700 University Ave, Taranto, H8F 6 CANADA ON. T: 416-592-7348, F: 416-592-4446

Totten, L.J., Johnson Western Gunite, 940 Doolittle Dr., San Leandro, CA 94577. T: 510-568-8112

Townsend, F.C., Univ of Florida, Civil Eng Dept Box 116580, Gainesville, FL 32611. T: 352-392-0926, F: 352-392-3394, E: ftown@ce.ufl.edu

Trast, J.M., STS Consultants Ltd., 1035 Kepler Dr., Green Bay, WI 54311. T: 920-406-3114, F: 920-468-3312, E: trast@ stsltd.com

Traughber, E.B., EB Traughber & Assocs Inc., 3320 N Hullen St, Ste B., Metairie, LA 70002. T: 504-887-4482, F: 504-887-4487

Trautwein, S., Trautwein Soil Testing Equi, P.O. Box 31429, Houston, TX 77231. T: 713-721-1866

Travis, C.B., Apex Environmental Inc., 1306 Pine Drive, Aiken, SC 29801. T: 803-649-0396, F: 803-652-1418, E: ctravis@mindspring.com

Triano, S.J., Malcolm Pirnie Inc., 98 Cross Road, Holmes, NY 12531. T: 914-878-7635, F: 914-641-2645

Triggs, Jr, J.F., Triggs Technologies Inc., 33977 Chardon Rd., Willoughby, OH 44094. T: 440-585-9995

Trimbath, W.D., Michael Baker Jr, Inc., 142 Colson Dr., Pittsburgh, PA 15236. T: 412-655-3688, F: 724-495-4112, E: btrimbath@mbakercorp.com

Troxell, D.E., IT Corp, 4975 Longview Ct, Murrysville, PA 15668. T: 724-325-3051, F: 412-858-3418

Trudeau, P.J., Stone & Webster, Inc., 12 Anawon Rd., Plymouth, MA 2360. F: 617-589-2959, E: paul.trudeau @stoneweb.com

Trumbull, C.D., Harza Engineering Co, 301 Brookhaven Ct, Hayward, CA 94544. T: 510-429-1932, F: 510-636-2177, E: ctrumbull@harza.com

Tsai, C.-F., Ca State Univ @Long Beach, 512 S Old Ranch Rd., Arcadia, CA 91007. F: 626-445-9683, E: stasai@ csuib.edu

Tsai, C.-N., Soil Testing Eng Inc., 3837 Lake Latania, Baton Rouge, LA 70816. T: 225-291-3328, F: 225-752-4878

Tsai, K.-W., College of Engineering, San Jose State University, San Jose, CA 95192. T: 408-924-3902, F: 408-924-3818, E: kutsai@email.sjsu.edu

Tsau, P.E., Wen Shiung, USFK US Army, Unit 15707, Box 3, Apo, AP 96258. T: 001-82-31-870-6696, F: 001-82-31-870-8, E: tsauw@usfk.korea.army.mil, wenstsau@hotmail. com

Tse, W.M., Architectural Service Dept, Flat C 24/Fl Hong Shing Ct, Healthy Village 668 Kings Rd., North Point, HONG KONG. T: 852-2570-8778, F: 2561-3685, E: waibo @netvigator.com

Tso, H.C.M., China State Engr Corp, 11 Gh King Fook Court, 173 Tin Hau Temple Rd., North Point, HONG KONG. T: 011-852-28652757, F: 011-852-2866933, E: t57214@ctimail.com

Tsuzuki, M., Fugro Japan Co Ltd., 1-21-2 Jingumae Shibuya-Ku, Tokyo, 150 JAPAN. T: 81-3-3497-0298, F: 81-3-3423-9662, E: tsuzuki@fugro.co.jp

Tucker, F.C., Dewberry & Davis, 115 Nautical Cove, Stafford, VA 22554. T: 540-720-5444, F: 703-849-0103, E: ftucker @dewberry.com

Tufenkjian, M.R., Calif State Univ Los Angeles, 1630 San Pasqual, Pasadena, CA 91106. T: 626-793-6276, F: 323-343-4555, E: mtufenk@calstatela.edu

Tumay, M.T., Louisiana State University, College of Engineering, 3304 V Ceba, Baton Rouge, LA 70803. T: 225-334-5937, F: 225-578-4845, E: mtumay@eng.lsu.edu

Turnbull, C.B., Dames & Moore, 500 Market Place Tower, 2025 First Ave, Seattle, WA 98121. T: 206-728-0744, F: 206-727-3350, E: seaclot@dames.com

Turner, W.G., Terracon Cons Inc., 6973 S Lexington Dr., West Jordan, UT 84084. T: 801-562-1436, F: 801-545-8600, E: wgturner@terracon.com

Tweedie, J.J., Maine Dept of Trans, 30 Leighton Park Dr., Winthrop, ME 4364. T: 207-377-5520, F: 207-287-6737, E: jeff.tweedie@state.me.us

Tweedie, R.W., Thurbur Engineering Ltd., 9636 51st Ave, Ste 200, Edmonton, T6E 6A5 CANADA Ab. T: 403-438-1460, F: 780-437-7125, E: rtweedie@edm.thurbergroup.com

Udaka, T., Earthquake Engineering, 107 Woodualley Ct, Danville, CA 94506. T: 925-648-4522, F: 925-964-9245, E: udaka@flush.co.jp

Ueng, T.S., National Taiwan Univ, Dept of Civil Engineering, Taipei 106, Taiwan Roc, Taiwan, ROC. T: 886-2-3621734, F: 886-2-3631558, E: ueng@ce.ntu.edu.tw

Ullrich, C.R., Univ of Louisville, Dept of Civil And Env Eng, 101 W S Speed Hall, Louisville, KY 40292. T: 502-852-4616, F: 502-852-8851, E: crullrich@louisville.edu

Underbrink Sr., D.M., Naismith Engineering Inc., P.O. Box 3099, Corpus Chrsti, TX 78463. T: 361-814-9900, F: 361-814-4401, E: dunderbrink@naismith-engineering.com

Urban, G.K., Goffman Mc Cormick & Urban Inc., 26977 Moro Azul, Mission Viejo, CA 92691. T: 949-582-8103, F: 949-888-1380, E: info@gmugeo.com

Urton, S.D., GTG-PSI., 6820 Monaco St, Commerce City, CO 80022. T: 303-289-8658

Urzua, A., Prototype Eng Inc., P.O. Box 868, Winchester, MA 1890. T: 781-729-2363, F: 781-729-2369, E: urzua@bc.edu

Usnick, S.C., Alpha-Omega Geotech, P.O. Box 2670, Kansas City, KS 66110. T: 913-371-0000, F: 913-371-6710

Usseglio-Polatera, J.-M., Sogreah, 6 Rue De Lorraine, Echirolles 38130, FRANCE. T: 011-3376-334208, F: 1.13E+11

Vachris, C.F., Vachris Engineering, 370 Old Country Rd., Garden City, NY 11530. T: 516-747-5096, F: 516-747-1933

Vacula, S.H., FA Dente Engineering, 65 Oliver St, Cohoes, NY 12047. T: 518-238-2915, F: 518-238-2614

Vakili-Mirzamani, J., Ninyo & Moore, 28791 Via Pasatiempo, Laguna Niguel, CA 92677. T: 949-831-8279, F: 949-472-5445, E: jvakili@ninyoandmore.com

Valceschini, R.B., Amec Earth And Environmental, 10250 Silver Knolls Blvd, Reno, NV 89506. T: 775-972-3839, F: 775-331-4153, E: robval@quixnet.net

Valera, J.E., Knight Piesold Llc, 1050 Seventeenth St, Suite 500, Denver, CO 80265. T: 303-629-8788, F: 303 629-8789, E: julio@kpco.com

Valleie, J., 180 North Napoleon Ave, Agoura, CA 91377. T: 818-707-2465

Vallejo, L.E., Univ of Pittsburgh, 767 College Avenue, Apt 104, Pittsburgh, PA 15232. T: 412-362-2761, F: 412-624-0135, E: vallejo@engrng.pitt.edu

Van Aller, H., Maryland Dept of Environment, 140 Overlook Dr., Queenstown, MD 21658. T: 410-827-6097, F: 410-221-6317, E: hvanaller@yahoo.com

Van Beveren, J.L., Van Beveren & Butelo Inc., 706 W Broadway Suite 201, Glendale, CA 91206. T: 818-543-4560, F: 818-543-4565, E: jvanbeve@lawco.com

Van Dillen, D.E., Macneal-Schwendler Corp, 2975 Redhill Ave, Costa Mesa, CA 92626. T: 714-444-5057, F: 714-545-9434, E: david.vandillen@mscsoftware.com

Van Orden, R.J., Melick-Tully & Assoc Pc, 3 Tunis Cox Rd., White Hse Sta, NJ 8889. F: 732-356-9054, E: rjvan orden@aol.com

Van Ravenstein, H.J., Unitech Consultants Pty Ltd., P.O. Box 390, Mentone Vic, 3132 AUSTRALIA. T: 011-613-977-61137, F: 61397760646/613, E: raven@vds.net.au

Van Riessen, G.J., Geotechnical Services Inc., 917 Southeast Fifth St, Lees Summit, MO 64063. T: 816-524-0219, F: 816-461-4888

Van Roosendaal, D.J., Haley & Aldrich, 465 Medford Street, Ste 2200, Boston, MA 2129. T: 617-886-7470, F: 617-886-7770, E: djvr@haleyaldrich.com

Van Zyl, D.I.A., Mining Life-Cycle Center, 150 University Terrace, Reno, NV 89503. T/F: 775-337-0971, E: dirkja@attglobal.net

Vander Linde, D.L., 7185 Farm House Lane, Cumming, GA 30040. T: 770-781-5050, E: fullgrip@mindspring.com

VanMarcke, E.H., Princeton Univ, 578 Province Line Rd., Hopewell, NJ 8525. T: 609-333-0256, F: 609-333-0257, E: evm@princeton.edu

Vasquez-Herrera, A.R., VVA Consultores Ca, 1408 Nw 82nd Ave, (C-516), Miami, FL 33126. T: 011-58-2-985-0011, F: 011-58-2-985-11, E: arvasquezh@yahoo.com

Vaughn, K.C., PSI Inc., 3355 Martinique Ave, Boulder, CO 80301. T: 303-449-4463, F: 303-423-5625, E: craig.vaughn2@gte.net

Vedula, R.V., Langan Eng & Env Services, 11949 Sw 14 Street, Pembroke Pines, FL 33025. T: 954-964-7556, F: 305-362-5212, E: kvedula@langan.com

Vejvoda, M., Suncoast Post-Tension, 1528 East Cedar St, Ontario, CA 91761. T: 909-673-0490, F: 909-947-9420, E: mvejroda@hotmail.com

Verduin III, J.R., Anchor Environmental Llc, 1411 4th Ave, Suite 1210, Seattle, WA 98101. T: 206-287- 9130, F: 206-287-9131, E: jverduin@anchorenv.com

Verigin, W.M., 487 Pimental Way, Sacramento, CA 95831. T: 916-421-4833

Veyera, G.E., Univ of Rhode Island, 58 Wildwood Trail, E Greenwich, RI 2818. T: 401-885-3851, F: 401-874-2786, E: veyerag@egr.uri.edu

Vicente, E.E., Vicente Geotechnical Services, 17582 Brent Ln, Tustin, CA 92780. T: 714-573-1142, F: 714-731-0526, E: evicente@mediaone.net

Vick, S.G., 42 Holmes Gulch Way, Bailey, CO 80421. T: 303-697-5433

Vidal, S.F.F., Selwyn Vidal & Assocs Ltd., 20 El Carmen St, Arima, TRINIDAD-TOBAGO. T: 809-667-3139, F: 304210

Viessman, M.W., US Army Corps of Engrs, 10610 Kennedy Ln, Jacksonville, FL 32223. T: 904-292-0025, F: 904-232-3665

Vijayvergiya, V.N., Geotest Eng Inc., 207 Sanderling Ln, Sugar Land, TX 77478. T: 281-494-7195, F: 713-266-2977

Vilkenas, A.A., Stockholm Konsult, Sondagsuageu 30, Farsta, 123 60 SWEDEN. T: 08-785-86-56, F: 08-785-85-04, E: algis.vilkenas@stockholmkonsult.se

Vipulanandan, C., Univ of Houston, Dept of Civil Eng/Cigmat, 4800 Calhoun Rd., Houston, TX 77204. T: 713-743-4278, F: 713-743-4260, E: cvipulanandan@uh.edu

Virely, D., Hydro Quebec, 1257 Avenue Sauve, Sainte Foy, G1W 3C9 CANADA QC. T: 418-652-0175, F: 3017722853, E: didier.virely@wanadoo.fr

Vito, D., EBS Eng & Constr, 350 Woolwich St S., P.O. Box 188, Breslau, N0B 1M0 CANADA ON. T: 519-648-3613, F: 519-648-3505, E: dnvito@golden.ent

Vitton, S.J., Michigan Tech Univ, Dept of Civil Eng, 1400 Townsend Dr., Houghton, MI 49931. T: 906-487-2527, F: 906-487-2943, E: vitton@mtu.edu

Voegele, D.M., Arcadis Geraghty & Miller, 611 Chestnut St, Ste 200, Chattanooga, TN 37450. T: 423-756-7193 X206, F: 423-756-7197, E: dvoegele@arcadis-us.com

Vogt, W.L., Paradigm Consultants Inc., 2501 Central Parkway, Ste A3, Houston, TX 77092. T: 713- 686-6771, F: 713-686-6795, E: woody@thelabtexas.com

Volk, J.C., URS Corp, 1400 Union Meeting Rd., Ste 202, Blue Bell, PA 19422. T: 215-542-3800, F: 215-542-3888, E: john_volk@urscorp.com

Volpe, R.L., RL Volpe & Associates, 110 Atwood Ct, Los Gatos, CA 95032. T: 408-356-9363, F: 408-356-4155, E: dickvolpe@msn.com

Von Fange, R.M., Patton Burke & Thompson Llc, 403 Ranger Dr., Buda, TX 78610. T: 512-295-2471, F: 512-832-8886, E: rmvonf@hotmail.com

von Rosenvinge IV, T., Geodesign Inc., 984 Southford Rd., Middlebury, CT 6762. T: 203-758-8836, Ext. 1, F: 203-758-8842, E: t_vonrose@geodesign.net

Voor III, B., Ogden Environmental, 3800 Ezell Rd Suite 100, Nashville, TN 37211. T: 615-333-0630

Voskuilen, M.R., Sverdrup, 21912 8th Pl W., Bothell, WA 98021. T: 425-486-8619, F: 425-452-1212, E: voskuimr@sverdrup.com

Vote, B.E., P.O. Box 2012, Sparks, NV 89432.

Voyen, J.K., American Eng Testing, 856 Tanglewood Drive, Shoreview, MN 55126. T: 651-484-4126, F: 651-659-1379, E: jvoyen@amengtest.com

Vucetic, M., Univ of California-Los Angeles, Civil Eng Dept, 5732-D Boelter Hall, Los Angeles, CA 90095. T: 310-206-6260, F: 310-267-0171, E: vucetic@seas. ucla.edu

Wachholz, M.J., Wilcox Associates Inc., 143 Arbutus Ave, Cadillac, MI 49601. T: 231-779-8916, F: 231-775-3135, E: mattw@wilcoxassociates.com

Wagner, J.E., 3229 First Place North, Arlington, VA 22201. T: 703-276-8287, F: 703 243-2925, E: wagnerj6@erols.com

Wahls, H.E., North Carolina State Univ, Dept of Civil Eng, P.O. Box 7908, Raleigh, NC 27695. T: 919-515-7244, F: 919-515-7908, E: wahls@eos.ncsu.edu

Walberg, F.C., US Army Corps of Engrs, 3604 Ne 77th St, Kansas City, MO 64119. T: 816-436-8618, F: 816-426-5462, E: francke.c.walberg@usace.army.mil

Waldman, D.P., Forester Communications, P.O. Box 3100, Santa Barbara, CA 93130. T: 805-681-1312, E: dw@forester.net

Walker III, L.A., 28 Davis St, Phillipsburg, NJ 8865. T: 908-213-0214, E: lawalker@eden.rutgers.edu

Walker, W.K., Call & Nicholas Inc., 2919 E Beverly Dr., Tucson, AZ 85716. T: 520-881-6193, F: 520-670-9251, E: cni@goodnet.com

Wallace, J.F., IGES Inc., 182 S 600 E., Suite 206, Salt Lake City, UT 84102. T: 801-521-1800, F: 801-521-2800, E: john@igesinc.com

Wallace, T.S., Wallace Kuhl & Assocs, P.O. Box 1137, West Sacramento, CA 95691. T: 916-372-1434, F: 916-372-2565

Waller, A.D., Aspera Assocs Inc., 65 E India Row, Apt 32 G., Boston, MA 2110. T: 617-522-3670

Waller, F., Woodward Clyde Cons, 6 Lexington Dr., Bluffton, SC 29910. T: 843-757-1475, F: 843-757-1476, E: fswaller@aol.com

Walsh, K.D., Del E Webb School of Constr, 14644 S 20th Pl, Phoenix, AZ 85048. T: 602-704-1345, F: 480-965-1769, E: ken.walsh@asu.edu

Walter, S.J., Geo Concepts Inc., 14401 Gilmore Street, Suite 200, Van Nuys, CA 91401. T: 818-994-8895

Walton, J.C., Site-Blauvelt Engineers Inc., 16000 Commerce Parkway, Mount Laurel, NJ 8054. T: 856-273-1224, F: 856-273-9244, E:johnw@siteblauvelt.com

Wambold, W.S., Short Elliott Hendrickson Inc., 176 8th Ave Ne, Forest Lake, MN 55025. T: 651-982-0791, F: 651-490-2150, E: wwambold@sehinc.com

Wang, J.-N.N., Parsons Brinckerhoff Quade &, 456 Elwood St, Piscataway, NJ 8854. F: 212-465-5592

Wang, M.C., Penn State Univ, Dept of Civil Eng, 212 Sackett Bldg, University Park, PA 16802. T: 814-863-0026, F: 814 863-7304, E: mcw@psu.edu

Wangombe, C.K., Uniconsult(Kenya) Ltd., P.O. Box 2228, Nyeri, KENYA

Wantland, G.M., URS Corp, 7650 W Courtney Cambell Cauway, Tampa, FL 33607. T: 813-286-1711, F: 813-286-6587

Warder, D.L., ATC Associates Inc., 11344 Fieldstone Ct, Carmel, IN 46033. T: 317-844-2121, F: 317-849-4278, E: warder86@atc-enviro.com

Wardlaw, E.G., Geoscience Engineers Llc, P.O. Box 9144, Jackson, MS 39286. T: 601-956-0851, F: 601-956-1116

Wardwell, R.E., RE Wardwell Inc., P.O. Box 426, Orono, ME 4473. T: 207-866-3482

Warner, J., Warner Eng Services, P.O. Box 1208, Mariposa, CA 95338

Warner, R.M., Paul C Rizzo Assoc., 4334 Gladstone St., Pittsburgh, PA 15207. T: 412-422-2952, F: 412-856-9749, E: robbiewarner@yahoo.com

Warrington, D.C., Church of God Lay Ministries, 2403 Lennox Ct, Chattanooga, TN 37421. T: 423-892-2159, F: 423-892-6478, E: sales@vulcanhammer.com

Watkins, R.M., Willmer Engineering Inc., P.O. Box 941009, Atlanta, GA 31141. F: 770-939-4299, E: wei06@willmerengineering.com

Watry, S.M., 820 Idylberry Road, San Rafael, CA 94903. T: 415-492-1850, F: 415-472-0948, E: swatgeo@ix. netcom.com

Watson, C.H., Cheryl Watson Consulting, 9246 Pamunkey Crest Dr., Mechanicsvlle, VA 23111. T: 804-779-3537, E: chwatson403@mindspring.com

Watson, R.P., Delaware Solid Waste Auth, P.O. Box 455, 1128 South Bradford St, Dover, DE 19903. T: 302-739-5361, F: 302 739 4287

Way, P.E., Grover C., Consulting Soils Engineer, 504 S 11th St, Tacoma, WA 98402

Wayne, M.H., Tensar Earth Technologies Inc., 1958 Rotherham Way, Atlanta, GA 30338. T: 404-667-3885, F: 404-250-9056, E: drwayne@aol.com

Weatherby, D.E., Schnabel Foundation Co, 45240 Business Ct, Suite 250, Sterling, VA 20166. T: 703-742-0020, F: 703-742-3319

Weatherer, D.J., Berkel & Co Contractors Inc., 3910 Knowles Ave, Kensington, MD 20895. T: 301-946-0020, F: 301-946-0260, E: weatherer@mindspring.com

Weaver II, J.W., Weaver Boos Consultants, Inc., 200 S Michigan Avenue, Suite 900, Chicago, IL 60604. T: 312-922-1030, F: 312-922-0201, E: weavboos@concentric. net

Weaver, J.J., Geomatrix Cons Inc., 3580 Griffith Park Blvd, Los Angeles, CA 90027. F: 949-642-4474, E: jweaver@geo matrix.com

Weaver, K.D., Ken Weaver Ceg, 40442 Valencia Ct, Fremont, CA 94539. T: 510-657-5127, F: 510-657-5127

Weaver, R.J., Power Engineers Inc., 3940 Glenbrook Dr., P.O. Box 1066, Hailey, ID 83333. T: 208-788-3456, F: 208-788-0525, E: rweaver@powereng.com

Webb II, G., H C Nutting Co, 611 Lunken Park Dr., P.O. Box C., Cincinnati, OH 45226. T: 513-321-5816, F: 513-321-4540

Webb, Jr, G.F., Webb Eng & Surveying Inc., P.O. Box 42, Lake Charles, LA 70602. T: 337-439-1463, F: 337-439-1464

Webb, R.E., CTL., 680 Becket Place, Colorado Springs, CO 80906. T: 719-538-4589, F: 719-528-5362, E: rwebb58@aol.com

Weber, J.R., Geosciences Inc., 810 W 1st Ave, Apt A., Albany, GA 31701. T: 912-878-6306, F: 912-432-7018, E: jim. weber@bigfoot.com

Wedding, C.N., Frontier Kemper Constructors Inc., 6601 Highway 500, Owensboro, KY 42301. T: 270-229-4905, F: 812-425-3186, E: nealwedding@iname.com

Wegrzyn, M., Box 5407, Mayaguez, PR 681. T: 787-831-1867 EX 000

Weidig, P.C., Weidig Geoanalysts, 46-035 Konohiki Street, Apt 3864, Kaneohe, HI 96744. T: 808-234-5656, F: 808-524-5658, E: weidig@lava.net

Weinstein, G.M., NAWC, Rt 547 Bldg 596, 2nd Fl, Lakehurst, NJ 7762. T: 732-323-2230, F: 732-323-2882

Welch, B.F., ESP Assocs PA., 372 Crompton St, Charlotte, NC 28273. T: 704-504-1015, F: 704-504-0917, E: bwelch@espassoc.com

Welfer, C.T., GTS Technologies Inc., 100 S 17th Street, Camp Hill, PA 17011. T: 717-975-8809, F: 717-233-0994, E: cwelfer@gtstech.com

Welsh, J.P., Joseph P Welsh & Assoc Inc., 4404 Island View Road, Snow Hill, MD 21863. T/F: 410-632-4625, E: welshjd @yahoo.com

Welti, C.W., Clarence Welti Assocs, 257 Timrod Rd., Manchester, CT 6040. T: 203-649-9502, F: 860-657-2514

Wendland, S.A., Geosystems Engineering, 7802 Barton, Lenexa, KS 66214. T: 913-962-0909, F: 913-962-0924, E: swendland@geosystemseng.com

Wentz, Jr, F.J., Carlton Engineering Inc., 3932 Ponderosa Rd., Shingle Springs, CA 95682. T: 530-677-5515

Werner, P., Montana State of, 17200 Rocky Mountain Rd., Belgrade, MT 59714. T: 406-388-5197, F: 520-425-0468, E: pwerner@mt.gov

Weyandt, S.S., SEH Inc., Rt 2, Box 43 C., Iron River, WI 54847. T: 715-372-5005, F: 218-722-1514

Whalen, J.P., Stanford University, 1787 7th St, Los Osos, CA 93402. T: 805-528-0475, F: 805-788-2332, E: jwhalen@co.slo.ca.us

Whalley, W.B., 137 Coggins Point Rd., Port Royal Plantation, Hilton Head, SC 29928. T/F: 843-681-5707, E: wmwhalley@hargray.com

Wheeler, J.R., Hayward Baker Inc., 151 Randall Rd., Stow, MA 1775. T: 978-562-6585, F: 410-551-8206, E: jrwheeler@haywardbaker.com

White, D.M., Amec, 5119 136th St Sw, Edmonds, WA 98026. T: 425-742-5891, F: 425-821-3914, E: dean.white@amec.com

Whiting, N.M., Minnesota State of, Dept of Transp, 1400 Gervais Ave, Maplewood, MN 55109. T: 651-779-5603, F: 651-779-5616, E: nancy.whiting@dot.state.mn.us

Whittle, A.J., MIT, 30 Woodward Lane, Boxborough, MA 1719. T: 978-264-0614, F: 617-253-6044

Whitty, J.E., URS Corp., Barley Mill Plaza, Building 27, 4417 Lancaster Pike, Willmington, DE 19805. T: 302-992-6846, F: 302-892-7637, E: juhitty@aecinfo.com

Wiberg, E.J., Summit Environmental Consultants, 640 Main St, Lewiston, ME 4240. T/F: 207-795-6009, E: ewiberg@summitenv.com

Wikar, K.C., Potomac Crossing Consultants, 1768 Baltimore-Annapolis Blvd, Annapolis, MD 21401. T: 410-349-8957, F: 301-686-0000, E: wikark@wwbgec.com

Wikstrom, S.A., Subsurface Consultants, Inc., 2231 Coloma St, Oakland, CA 94602. T: 510-531-4245, F: 510-239-9010, E: swikstrom@golder.com

Wilbur, B.C., Wilbur Engineering Inc., 2916 E 5th Ave, Durango, CO 81301. T: 970-247-4844, E: wilbeng@rmi.net

Wild, P.A., 38 Peach St, Walpole, MA 2081. T: 508-668-1634

Wilk, P.A., HNTB Corporation, 17106 California Ave, Hazel Crest, IL 60429. T: 708-335-0323

Willar, P.J., Tonkin & Taylor, P.O. Box 5271, Wellesley St, Auckland, NEW ZEALAND. T: 00649-3556058, F: 00649-3070265

Williams, G.S., GSW Enterprises Inc., 17310 Bending Cypress Rd., Cypress, TX 77429. T: 281-855-3422, F: 281-859-3351

Williams, J.W., 1021 Crestview Dr., San Carlos, CA 94070. T: 650-592-4930, F: 408-924-5053, E: williams@geosun1.sjsu.edu

Williams, N.D., Globex Engineering & Develpmnt, 1239 E Newport Center Dr., Ste 117, Deerfield Beach, FL 33442. T: 954-571-9200, F: 954-418-9800, E: globex@gate.net

Williamson Jr, A.R., Gts Technologies Inc., 919 Old Silver Spring Road, Mechanicsburg, PA 17055. T: 717-697-7748, E: awilliamson@earthlink.net

Willmer, J.L., Willmer Engineering Inc., 3772 Pleasantdale Rd., Suite 165, Atlanta, GA 30340. T: 404-633-3588

Wilson Jr, H.P., Russo Corp, 1421 Mims Ave SW, Birmingham, AL 35211. T: 205-923-4434, F: 205-925-0665

Wilson Jr, J.H., Corrpro, 01181 North Hampton, Westminster, CA 92683. T: 714-775-6807, F: 562-942-2834, E: jwilson@corrpro.com

Wilson, D.W., UC Davis, 1659 Colusa Ave, Davis, CA 95616. T: 530-757-2299, F: 530-752-7872, E: dxwilson@ucdavis.edu

Wilson, D.D., Trigon Engineering Consultants, 110 Royal Oak Ct, Greer, SC 29650. T: 803-268-7368, F: 864 297 4225, E: dwilson@trigoneng.com

Wilson, M.S., Ardaman & Assoc, 3175 West Tharpe Street, Tallahassee, FL 32303. T: 850-576-6131, F: 850-574-0735, E: ardaman@aol.com

Wilson, W.E., WE Wilson Engrs, 4102 Eden Ave, Nashville, TN 37215. T: 615-292-8700

Wilton, J.L., Jacobs Associates, 500 Sansome St Ste 700, San Francisco, CA 94111. T: 415-434-1822, F: 415-956-8502

Wiltshire, R.L., US Bureau of Reclamation, 8053 S Zephyr St, Littleton, CO 80128. T: 303-978-1994, F: 303-445-6472, E: dgwiltshire@prolynx.com

Wimberly III, P.M., Wimberly Associates Inc., 2807 Lincoln Way #3, White Oak, PA 15131. T: 412-751-8049, F: 412-751-1186

Windom, F.S., Porter W Yett Co, 5949 Ne Cully Blvd, Portland, OR 97218. T: 360-891-9115, F: 503-460-2923, E: flindt@cnnw.net

Wineland, J.D., 6102 N Wheeling Ave, Tulsa, OK 74130. T: 918-371-4328, F: 918-610-9519, E: john.wineland@cpc-ltd.com

Winge, G.A., Directorate of Public Roads, Road Technology Department, P.O. Box 8142 Dep, Oslo, 33 NORWAY. F: 011-47-22073444, E: grethe.winge@vegvesen.no

Winters, W.J., US Geological Survey, 384 Woods Hole Rd., Woods Hole, MA 2543. T: 508-457-2358, F: 508 457-2310, E: bwinters@usgs.gov

Wirt, J.R., Virginia Geotechnical Services, 8211 Hermitage Rd., Richmond, VA 23228. T: 804-266-2199, F: 804 261-556, E: jrwirt@vgspc.com

Wirth, G.E., Schnabel Eng Assoc, 2522 Lawnside Road, Timonium, MD 21093. T: 410-252-0577, F: 410-625-9764, E: gwirth@schnabel-eng.com

Wirth, J.L., Geotechnology Assocs Inc., 2504 Rochelle Dr., Fallston, MD 21047. T: 410-877-3031, F: 410-893-3437

Wisniewski II, R.T., Mueser Rutledge Cons Engrs, 1170 Cambridge Ln, Bridgewater, NJ 8807. T: 908-658-9211, F: 212-953-5626, E: rwisniewski@mrce.com

Wissmann, K.J., Geopier Foundation Co, 515 Sunrise Dr., Blacksburg, VA 24060. T: 540-951-4561, F: 540-951-8078, E: kordw@earthlink.net

Withiam, J.L., D'appolonia Cons Engrs, 275 Center Road, Monroeville, PA 15146. T: 412-856-9440, F: 412-856-9535, E: jlwithiam@dappolonia.com

Witsman, G.R., ECS Ltd., 814 Greenbrier Cir, Ste A., Chesapeake, VA 23320. T: 757-366-5100, F: 757-366-5203, E: gwitsman@ecslimited.com

Wittreich, C.D., Ch2m Hill, 5102 Chukar Dr., West Richland, WA 99353. T: 509-967-3540, F: 509-372-9292

Wold, R.J., Blackhawk Geometrics, 301 Commercial Road, Ste B., Golden, CO 80401. T: 303-278-8700, F: 303-278-0789, E: rwold@blackhawkgeo.com

Wolfe, J.T., 3016 Brook St, Oakland, CA 94611. T: 510-268-8968, F: 925-299-7970, E: wolfe@subsurfaceconsultants.com

Wolff, T.F., Michigan State Univ, 4326 Manitou Dr., Okemos, MI 48864. T: 517-349-1617, F: 517-432-1356, E: wolff@egr.msu.edu

Wolfskill, L.A., 4 Waterwood, Huntsville, TX 77320. F: 936-891-5103

Woloshin, B.H., Universal Eng Sciences, 1258 Climbing Rose Dr., Orlando, FL 32818. T: 407-292-6653, F: 407-423-3106, E: bwoloshin@vesorl.com

Wong, A.W., Innovative Technical Solutions Inc (Itsi, 1389 Stanton Way, San Jose, CA 95131. T: 408-441-1219, F: 925-256-8996, E: www.itsi.com

Wong, F.C.-h., The Chinese Univ of Hong Kong, Flat A 22/F Block 2 Avon Park, 15 Yat Ming Rd Fanling, New Territories, HONG KONG. T: 011-852-2683-2407, F: 011-852-2683-24, E: wonchf@cuhk.edu.hk

Wong, I.H., 95 Cashew Road, #03-03, Cashew Heights, 679666 SINGAPORE. T: 766-4307, F: 762-6924, E: singnet.com.sg

Wong, Y.F., 43-39 168th Street, Flushing, NY 11358. T: 718-461-5450, F: 718-461-3710, E: jjy41@aol.com

Woo, E.P., Subsurface Cons Inc., 636 Church St, San Francisco, CA 94114. T: 415-626-3304, F: 510-268-0137, E: ewoo@subsurfaceconsultants.com

Wood, A.M., Dominion Generation, Hcr 01 Box 280, Warm Springs, VA 24484. T: 540-279-3204, F: 540-279-3299, E: mike_wood@dom.com

Wood, J.R., Portland State Univ, 2336 Se Pine, Portland, OR 97214. T: 503-238-6872, F: 503-823-2342, E: jwoodeit@aol.com

Woods, C.B., Langan Engineering, 5 Phoenix Pky, Oakhurst, NJ 7755. T: 732-728-0485, E: dchipurdue@hotmail.com

Woods, R.D., Univ of Michigan, 2340 Gg Brown Lab, 2350 Hayward St, Ann Arbor, MI 48109. T: 313-764-8495, F: 313-764-4292, E: rdw@engin.umich.edu

Woodward, M.B., Law Eng & Envir Services Inc., 11920 Remsen Rd., Jacksonville, FL 32223. T: 904-262-1474, F: 904-396-5703, E: mwoodwar@kennesaw.lawco.com

Wool, J.M., 650 Serrano Dr., Sn Luis Obisp, CA 93405. T: 805-546-9887, F: 496-202-18465

Wooten, R.L., Gei Consultants Inc., 1021 Main Street, Winchester, MA 1890. T: 781-721-4034

Wordes, D.R., Postner & Rubin, 17 Battery Pl, New York, NY 10004. T: 212-269-2510, F: 212-425-0968, E: dwordes@postner.com

Wortley, C.A., Univ of Wisconsin, 206 Everglade Dr., Madison, WI 53717. T: 608-833-1290, F: 608-263-3160, E: wortley@engr.wisc.edu

Wray, W.K., Michigan Tech Univ, Academic And Student Affairs, 1400 Townsend Dr., Houghton, MI 49931. T: 906-487-2440, F: 906-487-2935, E: wkwray@mtu.edu

Wright, R.S., Cons Engr, 1913 Victoria Dr., Santa Ana, CA 92706. T: 714-836-0834, E: rswright@pacbell.net

Wright, S.G., Univ of Texas, 3406 Shinoak, Austin, TX 78731. T: 512-452-2095, F: 512-471-6548, E: swright@mail.utexas.edu

Wright, W.B., Wright Padgett Christopher, 1017 Chuck Dawley Blvd, Mt. Pleasant, SC 29464. T: 843-884-1234, F: 843-884-9234, E: bwright@wpceng.com

Wu, A., Nfesc Ec Det, 10805 Willow Run Ct., Potomac, MD 20854. T: 301-983-9104, F: 301-983-1710

Wu, J.Y.-H., Chung Hua University, 2f #21 Lane 52, Jinchen I Road Hsin Chu, Taiwan 30077, Taiwan, Taiwan, ROC. T: 886-35-728854, F: 886-35-372188, E: jasonwu@chu.edu.tw

Wu, M.-J., Shannon & Wilson Inc., 400 N 34th St Suite 100, P.O. Box 300303, Seattle, WA 98103. T: 206-632-8020, F: 206-695-6777, E: jw@shanwil.com

Wu, T.H., Ohio State Univ, 2070 Neil Ave, Columbus, OH 43210. T: 614-292-1071, F: 614-292-3780, E: tienwu@magnus.acs.ohio-state.edu

Wurster, D.W., Wurster Engineering & Construction Inc., 102 Ponders Rd., Greenville, SC 29615. T/F: 864-292-1993, E: wurster@geotechnicalresources.com

Wyatt, P.E., Mark E., GHBW Engineers Inc., P.O. Box 55105, Little Rock, AZ 72215. T: 501-455-2536, E: mwyatt@grubbsengineers.com

Yacko, D.G., Site Environmental Services, 15774 S La Grange Rd., Pmb 990, Orland Park, IL 60462. T: 815-464-6560, F: 815-464-6481

Yaghoubian, N.F., P.O. Box 55684, Sherman Oaks, CA 91413. T: 818-907-9095, F: 818-907-1305

Yaldo, H.S., NTH Consultants Ltd., 38955 Hills Tech Drive, Farmington Hills, MI 48331. T: 248-553-6300, F: 248-324-5390, E: hyaldo@nthconsultants.com

Yamada, M., Iwakura Chuzaiji Cho No 156, Sakyo Ku, Kyoto, 606-0021 JAPAN. T: 075-724-5821, F: 075-724-8043

Yamamoto, E., 7-19 6 Chome Chuo-Cho, Higashi Kurume City, Tokyo, 203-0054 JAPAN. T: 0424-74-7315, F: 03-3563-5034, E: dogi@mx2.alpha-web.ne.jp

Yamamuro, J.A., Univ of Delaware, Dept of Civil & Env Engr, Newark, DE 19716. T: 302-831-6074, F: 302-831-3640, E: yamamuro@ce.udel.edu

Yamanouchi, T., The Yamanouchi Research Lab, 5-25-25 Morooka, Fukuoka, 816-0094 JAPAN. T/F: 092-571-7521

Yamasaki, K., Cornforth Cons Inc., 11835 Sw Tuckerwood Ct, Beaverton, OR 97008. T: 503-643-8126, F: 503-452-1528, E: kyamasaki@cornforthconsultants.com

Yang, D.S.-L., Raito Inc., 1660 Factor Ave, San Leandro, CA 94577. T: 510-346-9840, F: 510-346-9841, E: dsyang@jaccess.com

Yang, M.Z., Michael Baker **Jr,** Inc., 555 Business Center Dr., Suite 100, Horsham, PA 19044. T: 215-444-0888, F: 215-444-0889, E: yang@engr.utk.edu

Yarmak Jr, E., Arctic Foundations Inc., 5621 Arctic Blvd, Anchorage, AK 99518. T: 907-562-2741, F: 907/562-0153, E: eyarmak@arcticfoundations.com

Yegian, M.K., Northeastern Univ, 8 King St, Lexington, MA 2421. T: 781-862-6680, F: 617-373-4419, E: myegian@neu.edu

Yeosock, M.M., Norwalk City of, 5 Country Club Dr., White Plains, NY 10607. T: 914-345-7157, F: 203-857-0143

Yesiller, N., Wayne State Univ, Dept of Civil Engineering, 5050 Anthony Wayne Dr., Detroit, MI 48202. T: 313-577-3766, F: 313-577-3881, E: yesiller@eng.wayne.edu

Yeung, M.R., Univ of Hong Kong, Dept of Civil Engineering, Pokfulam Rd., Hong Kong, HONG KONG. T: 28593498, F: 25596189, E: mryeung@hku.hk

Yilmaz, E., Enar Muh. Mim Danismanlik Ltd Sti, Ali Sir Nevai Sk Baysal Ap, No 30/2 Kmp Fatih, Istanbul, 34280 TURKEY. T/F: 9.02E+11, E: eyilmaz@srv.ins.itu.edu.tr

Yilmaz, R., Fugro Geosciences Inc., 6105 Rookin St, Houston, TX 77074. T: 713-778-5580, F: 713-778-5501, E: ryilmaz@fugro.com

Yoakum, D.D., Geo Soils Consultants Inc., 6634 Valjean Ave, Van Nuys, CA 91406. T: 818-785-2158

Yokel, F.Y., Felix Y. Yokel Ph.D., Pe Cnsltng Eng, 8208 Fenway Rd., Bethesda, MD 20817. T: 301-365-2931, F: 301-365-2847, E: fyokel@juno.com

Yoo, C., Sungkyunkwan Univ, Dept of Civil Eng, 300 Chun-Chun Dong Jangan-Gu, Kyonggi-Do, Suwon, 440-746 SOUTH KOREA. T: 011-82-331-290-7518, F: 011-82-331-290-, E: csyoo@yurim.skku.ac.kr

Yoon, G.L., Korea Ocean Research & Dvmt, Sadong 1270 Kordi Eng, Amsan, 425744 SOUTH KOREA. T: 011-82-502-140-5316, F: 1.18E+12, E: glyoon@kordi.re.kr

Youd, T.L., Brigham Young Univ, Dept of Civil Eng, 368 Clyde Building, Provo, UT 84602. T: 801-378-6327, F: 801-378-4449, E: tyoud@byu.edu

Young, A.G., Marsco Inc., 2619 Hodges Bend Cir, Sugar Land, TX 77479. T: 281-980-1166, F: 713-465-2787, E: marsco@marsco.com

Young, D.A., Hepworth-Pawlak Geotechnical, 263 Lupine Dr., New Castle, CO 81647. T: 970-984-3756, F: 970-945-8454, E: npgeo@hggeotech.com

Young, L.W., Bechtel Corp, 10404 Peakview Ct, Damascus, MD 20872. T: 301-253-6266, F: 301-682-6415, E: lwyoung@bechtel.com

Youngs, R.R., Geomatrix Cons Inc., 1147 High Court, Berkeley, CA 94708. F: 510-663-4141, E: byoungs@geomatrix.com

Yourman, A.M., Diaz Yourman & Associates, 17421 Irvine Blvd, ******, Tustin, CA 92780. T: 714-838-8565, F: 714-838-8741

Yow, Jr, J.L., Lawrence Livermore Natl Lab, University of California, 7000 East Ave, Livermore, CA 94550. T: 925-422-3521, F: 925-422-5514, E: yow1@llnl.gov

Yow, M.G., Chelan County Pud, 1412 Orchard St, Wenatchee, WA 98801. T: 509-665-3659, F: 509-664-2881, E: gene@chelanpud.org

Yu, K.Y.P., Yu & Associates Inc., 619 River Drive 3/F., Elmwood Park, NJ 7407. T: 201-791-0075, F: 201-791-4533, E: pyu@yu-associates.com

Zaccheo, P.F., Parsons Brinckerhoff, 8 E Ridge Ln, Mount Kisco, NY 10549. T: 914-666-7736, F: 212-465-5592, E: zaccheo@pbworld.com

Zaman, M.M., Univ of Oklahoma, School of Civil Eng, 202 West Boyd, Rm 334, Norman, OK 73019. T: 405-325-5911

Zaman, P., Black & Veatch, 5900 W 86th St, Shawnee Mission, KS 66207. T: 9136481547, F: 9134583730, E: zamanpr@bv.com

Zamojski, L.D., Acres Intl Corp, 50 Constance Lane, Cheektowaga, NY 14227. T: 716-895-9638, F: 716-689-3749, E: lzamojski@acres.com

Zdinak, A.L., Virginia Geotechnical Services, 8211 Hermitage Road, Richmond, VA 23228. T: 804-266-2199, F: 261-5569, E: alzdinak@vgspc.com

Zeevaert, L., Isabel La Catolica 67, Delegacion Cuauhtemoc, Mexico 06080 D.F., MEXICO. T: 709-4208, F: 5709-42-08

Zeghal, M., Rensselaer Polytechnic Inst, Dept of Civil Engineering, Jec 4049, Troy, NY 12180. T: 518-276-2836, F: 518-276-4833, E: zeghal@rpi.edu

Zehrbach, B.E., Earth Systems Cons Norcal, 6170 De Palma Ct, San Jose, CA 95120. T: 408-268-1536, F: 510-353-0344, E: bzehrbach@earthsys.com

Zeman, A.R., Zipper Zeman Assoc Inc., 814 Eagle Cliff Rd Ne, Bainbridge Is, WA 98110. T: 206-855-9003, E: zem an@zipperzeman.com

Zeman, J.R., Kleinfelder, 10221 West Emerald St, Ste 180, Boise, ID 83704. T: 208-376-5083, F: 208-376-9703

Zeng, X., Case Western Reserve Univ, Dept of Civil Eng, 10900 Euclid Ave, Cleveland, OH 44106. T: 216-368-2923, F: 216-368-5229, E: xxz16@po.cwru.edu

Zepeda-Garrido, A., Univ of Queretaro, J Guadalupe Velaquez 10, Queretaro Queretaro Cp 76020, MEXICO. T: 5242-230853, E: zepeda@sunserver.uaq.mx

Zerr, E.G., Zerr Engineering, 1006 Court Terr, Colby, KS 67701. F: 785-462-6992

Zimmie, T.F., Rensselaer Polytech Inst, Civil Eng Dept, Troy, NY 12180. T: 518-276-6939, F: 518-276-4833, E: zimmit@rpi.edu

Zipper, J.E., John E Zipper, 9111 Cascade Dr., ***, Edmonds, WA 98026. T: 425-778-6096, E: zipperjk@aol.com

Zisman, P.E., Edward D., BTL Engineering Services Inc., 5802 N Occident St, Tampa, FL 33614. T: 813-884-0755, F: 813-886-5377, E: btleng@gte.net

Zmijewski, D., Schnabel Engrg Assoc, 101 Bayard St, New Brunswick, NJ 8901. T: 732-435-1212, F: 732-435-1203, E: davez@schnabel-eng.com

Znidarcic, D., Univ of Colorado, Campus Box 428, Boulder, CO 80309. T: 303-492-7577, F: 303-492-7317, E: znidarci @spot.colorado.edu

Zoino, W.S., Consulting Geotechnical Engr, 93 Riverside Drive, P.O. Box 308, West Harwich, MA 2671. T: 508-432-1844, F: 508-828-1060, E: wszengineer@aol.com

Zornberg, J.G., U of Colorado At Boulder, Dept of Ceae, Campus Box 0428, Boulder, CO 80309. T: 303-492-4699, F: 303-492-7317, E: zornberg@spot.colorado.edu

Zreik, D.A., Lebanese American University, c/o Gebran Karam/ Shcl of Engr, 475 Riverside Drive Rm-1846, New York, NY 10115. T: 961-1-203648, F: 961-1-336399, E: dazk@cyberia.net.lb

Zuckerman, B.R., Harding Ese Inc., 28 2nd St, Ste 700, San Francisco, CA 94105. T: 415-278-2117, F: 415-777-9706, E: brzuckerman@mactec.com

Zvanut, P., Zag Slovenian Natl Bldg & Civ Eng Inst, Dimiceva 12, Ljubljana, 1000 SLOVANIA. T: 388-1-588-84-91, F: 386-1*588-82-64, E: pavel.zvant@zag.si

Zweigler, R.I., The J Byer Group Inc., 1461 E Chevy Chase Dr., #200, Glendale, CA 91206. T: 818-549-9959, F: 818-543-3747, E: zweigler@byergroup.com

VENEZUELA

Secretary: José Ignacio Amundaray Pietri, Sociedad Venezolana de Geotecnia, Colegio de Ingenieros de Venezuela, Av. Ppal. Parque Los Caobos, Caracas. T/F: +58 1 571.3824, E: svdg@telcel.net.ve

Total number of members: 23

Bajetti, B., Sergio, Apto.Postal 70857, Los Ruices 1071-A, Caracas

Barrios, D., M. Rosalba, Urb. Macaracuay, Av. Principal., Qta. La Pelusita. Caracas 1071

Carrillo, P., Pedro, Av. Don Bosco. Qta, Urupagua, N°. 19. La Florida, Caracas

Centeno, P., Francisco, Urb. El Rosal. Av. Sojo, Qta. Acacias B. Caracas

Centeno, W., Roberto, Urb. El Rosal. Av. Sojo, Qta. Acacias B. Caracas

De Marco, Z., Pietro, Av. Francisco de Miranda. La California, Edif. Sara. Ofic. 11-10 Caracas

Falcon-Ascanio, Marco, Urb. El Rosal. Av. Sojo, Qta. Acacias B. Caracas

Ferrer, F., Diego, Apto.Postal 67012, Plaza Las Americas, Caracas

Graterol, M., Jaime, Urb. Sorocaima.La Trinidad-1031, Av. Cristobal Colón. Qta. Los Muchachitos, Caracas

Hiedra, C., Juan C., Chacao. Apto.60417-Caracas

Lupini, B., Juan F., C.C. Los Chaguaramos, Calle Edison, Piso 18, Caracas

Martin, F., Antonio, Edif. Lex. Piso 4. Av. Libertador, Chacao. Caracas

Martinez, F., Jorge, C.C. Los Chaguaramos, Calle Edison, Piso 18, Caracas

Pankovs, K., Igor, Av. Fco.de Miranda.Edif.Irene, Piso 19. Ofic.196.Cortijos de Lourdes, Caracas

Perri, A., Gianfranco, Calle Beethoven.Torre Financiera.Piso 10, Ofic.10 E.Colinas de Bello Monte, Caracas

Pesti, J., Andrés, Av. Caurimare.Qta.Morroca.Colinas de Bello Monte, Caracas

Rodriguez, A., Carlos, Urb. El Rosal. Av. Sojo, Qta. Acacias B. Caracas

Rodriguez, D., Nelson, Calle Beethoven.Torre Financiera. Piso 10, Ofic.10 E.Colinas de Bello Monte, Caracas

Rodriguez, D.C., Juan, C.C. Los Chaguaramos, Calle Edison, Piso 18, Caracas

Salcedo, R., Daniel, Av. Caurimare.Qta.Morroca.Colinas de Bello Monte, Caracas

Sanchez, N., Freddy, Av. Buchivacoa. Coro, Estado Falcón

Sancio, T., Rodolfo, Av. Qta.Autana. Urb. Caurimare, Apto. Postal 75987 Caracas

Tapia, G., Manuel, Av. Principal Los Ruices.Centro Empresarial, Los Ruices.Piso 2.Ofic. 204, Caracas

VIETNAM

Secretary: Prof. Nguyen Truong Tien, Vietnamese Society of Soil Mechanics and Geotechnics, Centre for Transfer of Technology of Construction, 243A La Tahnh Str. Dong Da, Hanoi. T: (84)4 7660477, F: (84)4 8341863, E: tien.cofec@fmail.vnn.vn

Total number of members: 23

Bui Anh Dinh, Prof. Dr. Hanoi University of Transportation and Communication, Cau Giay, Hanoi. T: (84)4 7661371 (home)

Cao Van Chi, Prof. Dr. Hanoi Water Resource University, 297 Tay Son, Dong Da, Hanoi

Doan The Tuong, Dr. Institute for Building Science and Technology, Add, Nghia Tan, Cau Giay, Hanoi. T: (84)4 7544014 (work)

Le Duc Thang, Prof. Dr. 669 H7 Tan Mai, Hanoi. T: (84)4 8642028 (home)

Le Quy An, Prof. Dr. President 130 Bui Thi Xuan Str., Hanoi

Nghiem Huu Hanh, Prof. Dr. Hanoi Water Resource University, 171 Tay Son, Dong Da, Hanoi. T: (84)4 5633339 (work), (84)4 8523055 (home)

Nguyen Anh Dung, Dr. Company for Foundation Engineering and Construction, 21A Phan Chu Trinh, Hanoi. T: (84)4 8267748

Nguyen Ba Ke, Prof. Dr. Vice President 264 Hue Str., Hanoi. T: (84)4 8525021 (home)

Nguyen Bao Huan, Prof. Dr. A2 Khuong Thuong, Hanoi. T: (84)4 8692975 (home)

Nguyen Cong Chinh, Dr. Hanoi Construction Department, Lien Co Zone, Van Ho, Hanoi

Nguyen Cong Man, Prof. 75 Khuong Thuong, Dong Da, Hanoi. T: (84)4 8534198 (work), (84)4 8528512 (home), E: ncman@ffitovn

Nguyen Huu Dau, Dr. Standing Committee Institute of Transportation Science and Technology Science, 1252 Lang Road, Hanoi

Nguyen Huy Phuong, Prof. Dr. University of Mine and Geology, Dong Ngac, Tu Liem, Hanoi. T: (84)4 8364810 (home)

Nguyen Sy Ngoc, Dr. Hanoi University of Transportation and Communication, Cau Giay, Hanoi

Nguyen Truong Tien, Prof. Dr. Vice President and Secretary, Centre for Transfer of Technology of Construction. 243A La Tahnh Str., Dong Da, Hanoi. T: (84)4 7661897 (work), 090405769 (home), F:(84)4 8341863, (84)4 8240286, E: tien.cofec@fmail.vnn.vn

Nguyen Van Quang, Prof. Dr. A12 Nguyen Hong Road, Hanoi. T: (84)4 8351959 (home)

Nguyen Van Quang, Prof. Dr. 1252 Lang Road, Hanoi. T: (84)4 8342822 (work)

Pham, T. Quang Hao, Eng. 37 Chau Long Str., Ba Dinh, Hanoi. T: (84)4 7660477 (work)

Pham Van Ty, Prof. Dr. 24 Le Quy Don Str., Quan Hai Ba Trung, Hanoi. T: (84)4 8211899 (home)

Tran Dinh Ngo, Eng. Standing Committee, B9/2 Hostel 30/4 Dien Bien Phu, Ward 25, Binh Thanh Dist., Ho Chi Minh city

Trinh Van Cuong, Dr. Standing Committee, Hanoi Water Resource University, 297 Tay Son, Dong Da, Hanoi. T: (84)4 5651376

Vu Cong Ngu, Prof. Dr. Hanoi Civil Engineering University. T: (84)4 8699679 (home)

Vuong Van Thanh, Dr. Hanoi Architecture University, Km No. 9 Nguyen Trai Road, Hanoi

YUGOSLAVIA/YUGOSLAVIE

President: Prof. Dr Maksimovic Milan, Yugoslav Society of SMGE, Faculty of Civ. Engr. University of Belgrade, Bul. Kralja Aleksandra 73, YU-11000 Belgrade. T: 011/3238-569, E: max@EUnet.yu

Total number of members: 50

Anagnosti, P., Prof. Dr, Faculty of Civ. Engr. Uni. of Belgrade, Bul. Kralja Aleksandra 73, YU-11000 Belgrade. T: 011/156-704

Coric, Sla., Prof. Dr, Faculty of Mining and Geology, Uni. of Belgrade, Djusina 7, YU-11000 Belgrade. T: 011/482-171

Coric, Slo., Prof. Dr, Faculty of Mining and Geology, Uni. of Belgrade, Djusina 7, YU-11000 Belgrade. T: 011/3243-335

Cabarkapa, Z., M.Sc., Faculty of Civ. Engr. Uni. of Belgrade, Bul. Kralja Aleksandra 73, YU-11000 Belgrade. T: 011/430-098

Caki, L., Dr, Faculty of Mining and Geology, Uni. of Belgrade, Djusina 7, YU-11000 Belgrade. T: 011/3243-337

Cekerevac, C., B.Sc., Faculty of Civ. Engr. Uni. of Belgrade, Bul. Kralja Aleksandra 73, YU-11000 Belgrade

Divac, D., Dr, Institute Jaroslav Cerni, Bul. Vojvode Misica 43, YU-11000 Belgrade. T: 011/645-335

Dragas, J., Faculty of Mining and Geology, Uni. of Belgrade, Djusina 7, YU-11000 Belgrade. T: 011/3243-334

Folic, R., Prof. Dr, Faculty of Civil Eng., Novi Sad, Trg Dos. Obradovica 26, YU-21000 Novi sad. T: 021/59-798

Gojgic, D., Dr, Energoprojekt, Bulevar Lenjina 12, YU-11000 Belgrade. T: 011/131-516

Gojkovic, S., The Highway Institute, Kumodraska 257, YU-11000 Belgrade. T: 466-522

Grujic, M., The Highway Institute, Kumodraska 257, YU-11000 Belgrade. T: 011/493-402

Hadzi-Nikovic, G., Faculty of Mining and Geology, Uni. of Belgrade, Djusina 7, YU-11000 Belgrade. T: 011/3243-334

Jokanovic, I., M.Sc., Civ. Eng., Faculty of Civ. Engr. Uni. of Belgrade, Bul. Kralja Aleksandra 73, YU-11000 Belgrade. T: 011/604-955

Jotic, S., The highway Institute, Kumodraska 257, YU-11000 Belgrade

Jovanovic, V., M.Sc., Faculty of Civ. Engr. Uni. of Belgrade, Bul. Kralja Aleksandra 73, YU-11000 Belgrade. T: 011/643-846

Kordic-Dikovic, N., M.Sc., Institute for Testing Materials RS, Bul. Vojv. Misica 43, YU-11000 Belgrade. T: 011/650-322

Kovacevic, V., Faculty of Civ. Engr. Uni. of Podgorica, Cetinjski put bb, YU-81000 Podgorica. T: 081/12-082

Lazarevic, V., Jugofund, Bul. AVNOJ-a 181, YU-11000 Belgrade. T: 011/604-855

Lazovic, M., Dr, Prof., Faculty of Civ. Engr. Uni. of Belgrade, Bul. Kralja Aleksandra 73, YU-11000 Belgrade. T: 011/541-207

Lelovic, S., M.Sc., Faculty of Civ. Engr. Uni. of Belgrade, Bul. Kralja Aleksandra 73, YU-11000 Belgrade. T: 011/508-290

Lukic, D., Dr. Skupstina grada Beograda, 27 marta 43-45

Maksimovic Milan, Prof. Dr, President Yugoslav Soc. SMFE, Faculty of Civ. Engr. Uni. of Belgrade, Bul. Kralja Aleksandra 73, YU-11000 Belgrade. T: 011/3238-569

Maksimovic Mihajlo, Geosonda, 27 marta 97-99, YU-11000 Belgrade. T: 3228-922

Maletic, V., Geosonda, 27 marta 97-99, YU-11000 Belgrade. T: 3228-922

Manojlovic, M., Dr, Institute for Testing Materials RS, Bul. Vojvode Misica 43, YU-11000 Belgrade. T: 011/650-322

Maras-Dragojevic, S., M.Sc., Civ. Eng., Faculty of Civ. Engr. Uni. of Belgrade, Bul. Kralja Aleksandra 73, Belgrade

Markovic, G., Prof. Dr, Faculty of Mining and Geology, Uni. of Belgrade, Djusina 7, YU-11000 Belgrade. T: 011/3243-337

Markovic, V., Geosonda, 27 marta 97-99, YU-11000 Belgrade. T: 3228-922

Melentijevic, S., Faculty of Civ. Engr. Uni. of Belgrade, Bul. Kralja Aleksandra 73, YU-11000 Belgrade.

Milovic, D., Prof. Dr, Faculty of Civil Eng., Novi Sad, Str. Ustanicka 152, YU-11000 Belgrade. T: 011/4882-352

Mitrovic, P., Dr, The Highway Institute, Kumodraska 257, YU-11000 Belgrade. T: 011/493-402

Nagulj, V., B.Sc., Mostogradnja, Vlajkoviceva 19, YU-11000 Belgrade. T: 011/3232-310

Nikolovski, T., Vojvode Dobrnca 22, 11000 Belgrade. T:011/759-642

Petrovic, R., Dr, The Highway Institute, Kumodraska 257, YU-11000 Belgrade. T: 011/493-402

Radic, Z., M.Sc., Faculty of Civ. Engr. Uni. of Belgrade, Bul. Kralja Aleksandra 73, YU-11000 Belgrade

Rakic, D., M.Sc., Faculty of Mining and Geology, Uni. of Belgrade, Djusina 7, YU-11000 Belgrade

Samardakovic, M., Faculty of Civ. Eng. Uni. of Nis, Branka Krsmanovica 29/4, YU-18000 Nis. T: 018/332-120

Santic, Lj., The Highway Institute, Kumodraska 257, YU-11000 Belgrade. T: 011/493-402

Santrac, P., Dr, Doc. Faculty of Civ. Engr. Uni. of Subotica, Kozaracka bb, YU-24000 Subotica. T: 024/27-723

Skoric, S., Geosonda, 27 marta 97-99, YU-11000 Belgrade. T: 3228-922

Spasojevic, A., Dr, Civ. Eng. Dept., Cambridge, UK

Stevanovic, S., Prof. Dr, Faculty of Civ. Engr. Uni. of Belgrade, Bul. Kralja Aleksandra 73, YU-11000 Belgrade. T: 011/4889-166

Susic, N., Dr, Institute for Testing Materials RS, Bul. Vojvode Misica 43, YU-11000 Belgrade

Sutic, J., Prof., Faculty of Mining and Geology, Uni. of Belgrade, Dalmatinska 8, YU-11000 Belgrade. T: 011/3247-801

Tomanovic, Z., M.Sc., Faculty of Civ. Engr. Uni. of Podgorica, Cetinjski put bb, YU-81000 Podgorica. T: 081/13-982

Todorovic, B., Geosonda, 27 marta 97-99, YU-11000 Belgrade. T: 011/650-322

Vujanic, V., M.Sc., The highway Institute, Kumodraska 257, YU-11000 Belgrade. T: 011/440-784

Vukicevic, M., M.Sc., Faculty of Civ. Engr. Uni. of Belgrade, Bul. Kralja Aleksandra 73, YU-11000 Belgrade. T: 011/4446-417

Zdravkovic, L., Dr, Lecturer, Imperial College, Civ. Eng. Dept. London, UK